Handbook of

International Bridge Engineering

Handbook of
International
Bridge Engineering

Edited by
Wai-Fah Chen
Lian Duan

CRC Press
Taylor & Francis Group
Boca Raton London New York

CRC Press is an imprint of the
Taylor & Francis Group, an **informa** business

Cover Photo: Qingdao Bay Bridge provided by *China Highway Magazine*

CRC Press
Taylor & Francis Group
6000 Broken Sound Parkway NW, Suite 300
Boca Raton, FL 33487-2742

© 2014 by Taylor & Francis Group, LLC
CRC Press is an imprint of Taylor & Francis Group, an Informa business

No claim to original U.S. Government works

Printed on acid-free paper
Version Date: 20130808

International Standard Book Number-13: 978-1-4398-1029-3 (Hardback)

Library of Congress Cataloging-in-Publication Data

Handbook of international bridge engineering / editors, Wai-Fah Chen, Lian Duan.
        pages cm
    Includes bibliographical references and index.
    ISBN 978-1-4398-1029-3 (hardcover : alk. paper)
    1. Bridges--Design and construction. I. Chen, Wai-Fah, 1936- II. Duan, Lian.

TG145.H28 2014
624.2--dc23   2013028504

Visit the Taylor & Francis Web site at
http://www.taylorandfrancis.com

and the CRC Press Web site at
http://www.crcpress.com

Contents

Foreword

Much more than a travel guide to the world's biggest and most beautiful bridges, this *Handbook of International Bridge Engineering* takes us on a world tour of the art of bridge conception and construction. Professor Wai-Fah Chen and Dr. Lian Duan, the editors of this fine handbook, had the remarkable idea of compiling contributions from 64 top bridge experts, each presenting an in-depth study of his or her native country and the wonderful bridges there. Bridges, as we know, carry a multitude of symbols, their beauty and the spectacular technical achievements employed in their conception awe the greater public, and as shown in this book, they inspire researchers and engineers to make materials more efficient and to fight natural hazards.

Each country is the subject of a chapter, all of which clearly present the principal geographical aspects of the country, the fundamental stages of bridge development there, communication means through the ages, the evolution of rules and regulations, various types of road construction and road construction materials, railway transportation and pedestrian passages, and, last but not least, such modern preoccupations as durability, monitoring, maintenance, professional training, economic considerations, financing, and future challenges. The reader of this marvelous handbook will not fail to appreciate the variety of factors that determine any given country's bridge history, to evaluate the cultural, economic, and industrial impact on the evolution of infrastructures, and to consider the immense progress made in bridge design worldwide. A formidable reference guide, this work will be without a doubt a welcome acquisition to fine research libraries everywhere.

Jacques Combault
2007–2010 President of the International Association for
Bridges and Structural Engineering
Professor at the École des Ponts, ParisTech
Paris, France

This volume on international bridge engineering provides contributions from authors in 26 countries. It provides a resource covering engineering practice for major bridges in the authors' countries. For each country there is a chapter that provides a historical summary of design specifications, philosophy, and loads applied to all types of bridge systems for highway, railway, and pedestrian applications. Although the primary building materials of steel and concrete are highlighted, other types of materials such as timber, stone, and advanced composites are also dealt with.

During ancient times, wood, bricks, and stones were used to build bridges. Since the industrial revolution, bridges have evolved primarily with the invention of cast iron, wrought iron, steel, and reinforced and prestressed concrete. Great Britain from the early nineteenth century, the United States from the late nineteenth century through the mid-twentieth century, and Germany, France, Switzerland, Scandinavia, Japan, and Russia in subsequent decades built the most significant bridges. During the last two decades, China has built the largest number of major bridges in the world. They include many of the longest spans of nearly all types of bridges.

Time-dependent deterioration of both steel and concrete elements remains a challenge even today as corrosion and fatigue must be considered. Many of our theoretical assumptions for materials, such as being homogeneous, isotropic, and ductile, are often not true at the joints and connections. As a result, we have to rely on large-scale experimentation to develop simple design rules especially for connections. Failures of structures or their components have provided knowledge that has contributed to our understanding and ability to provide more durable and safer structures. In the United States, the 1940 Tacoma Narrows Bridge collapse from aero-elastic flutter from wind had a significant impact on the science and engineering of large bridge structures and elements. The collapse of the Ohio River Silver Bridge in 1967 as a result of a small stress corrosion crack in a steel eye bar had a lasting influence on inspection and maintenance as well as the quality and toughness of the steel materials used in bridges. Since the extensive use of welded steel components and connections after World War II, fatigue crack growth from truck and train loads has resulted in the need to detect and repair or retrofit many bridge structures and to improve their fatigue resistance. A notable case was the Hoan Bridge in Milwaukee, which experienced major fractures in all three girders at a cross section in 2001. All of these examples provide knowledge and enhance our ability to provide better details and connections in bridges that will be built in the future and improve our ability to maintain older structures that remain in service around the world.

For example, as I write this brief note in July 2011 alone, the Internet revealed the collapse of a major multiple-span arch bridge in China's Fujian Province that was built in 1999, as well as the collapse of three other bridges, two of which were attributed to overloaded vehicles. These events remind us of the role that connection design, quality, materials, loads, and time-dependent performance factors have on bridge durability and service. This handbook will provide readers with valuable information and enhance the performance of future bridges. In addition to the chapters for each of the 26 countries, a chapter is devoted to highway bridges of composite steel–concrete girder designs in 10 countries, a chapter features the highest bridges, and a chapter lists the longest bridges and bridge spans in the world. Many photographs and drawings are provided in all the chapters.

This volume provides a rich resource for practicing bridge engineers as well as for those interested in unique bridge structures that have been built to provide passage over major rivers and other crossings in the world. I have known Dr. Wai-Fah Chen since our student days together at Lehigh University in the 1960s, and as a fellow faculty member during his decade at Lehigh. We have had the opportunity to serve together on many professional committees throughout our careers. I am pleased to provide these introductory comments to his latest handbook.

John W. Fisher
Professor Emeritus of Civil Engineering, Lehigh University
Bethlehem, Pennsylvania

The purpose of science is to discover the truth of nature, while the purpose of engineering is to enhance and improve nature to make our living environment more enjoyable. The foundation of science is facts; the foundation of engineering is an accumulation of experiences. The truth discovered by science, such as mathematics, physics, and chemistry, so far does provide practical tools for engineers in the design of bridges, but it still falls far short of being able to explain all engineering phenomena. The design itself is a succession of decisions based on the direct or indirect applied experience of the engineer. Making decisions is art, not a science. Therefore, engineering is an art, not a science! In the design of a bridge, it is more important to understand what has been done successfully than to try to be scientifically accurate. This handbook is a perfect tool that provides engineers with information on past successful bridge designs. This important work helps bridge engineers to make intelligent decisions throughout their design processes.

It is a great pleasure to recommend to my fellow bridge engineers this remarkably comprehensive resource book encompassing state-of-the-art and best practices for bridge engineering worldwide. You will find an informative historical sketch and geographical characteristics for each country, along with a careful development of each country's design considerations, specifications, philosophy, and loads; landmark and recorded bridges, including girder, arch, cable-stayed, and suspension; bridge management and maintenance; and trends addressing new developments, milestone bridges under planning and construction, and the latest construction materials.

Perhaps you have only a marginal interest in bridge design and construction, but wish to learn more about the development of benchmark bridges around the world. You will find 10 benchmark comparisons for highway composite girder design along with a listing of world's highest bridges, longest bridges, and longest bridge spans from around the world. In addition, more than 1650 beautiful bridge photos and drawings are collected to illustrate the great achievements of modern bridge engineers.

I wish to congratulate both editors, whom I have known for many years, for having convinced and gathered a great team of internationally recognized bridge engineers to contribute their various chapters. I wish also to congratulate the 64 contributors for devoting their time, energy, experience, and talent to complete this landmark project. For the first time, this handbook has brought together, in a unified manner, so much of what until now was known only to a few experts in the field in their respective countries. Students, researchers, practitioners, novices, and experts alike will profit greatly from having a copy of this handbook, as it will help them to keep abreast of modern bridge engineering developments and state-of-the-art practices around the world.

Man-Chung Tang
Chairman of the Board and Technical Director
T. Y. Lin International
San Francisco, California

Preface

An international team of bridge experts and internationally known authors from 26 countries has joined forces to produce the *Handbook of International Bridge Engineering*, with the theme "bridge the world in the twenty-first century." The handbook is a unique, comprehensive, and up-to-date reference work and resource book covering bridge engineering practice and landmark and recorded bridges around the world, including 26 major countries and areas such as: Canada and the United States in North America; Argentina and Brazil in South America; Bosnia, Bulgaria, Croatia, Czech, Denmark, Finland, France, Greece, Macedonia, Poland, Russia, Serbia, Slovakia, and Ukraine in the European continent; China, Indonesia, Japan, Chinese Taipei, and Thailand in Asia; and Egypt, Iran, and Turkey in the Middle East.

Each country's chapter presents a historical sketch and geographical characteristics; design considerations, specifications, philosophy, and loads; various types of bridges including girder, truss, arch, cable-stayed, suspension, and so on, in various types of materials (stone, timber, concrete, steel, advanced composite) and in various purposes (highway, railway, and pedestrian); bridge management, maintenance, and monitoring including repair, professional education, cost analysis, and funding; and future trends addressing special topics and new developments, new milestone bridges under planning and construction, and new materials. Ten benchmark comparisons for highway composite girder design from different countries are also presented in Chapter 27. The highest bridges around the world are highlighted in Chapter 28. The top 100 longest bridges, and the top 20 longest bridge spans for various bridge types including suspension bridges, cable-stayed bridges, extradosed bridges, truss bridges, arch bridges, steel girder bridges, concrete girder bridges, movable bridges (vertical lift, swing, and bascule), floating bridges, stress ribbon bridges, and timber bridges, are listed in Chapter 29. More than 1650 beautiful bridge photos and drawings illustrate the great achievements of engineering professions.

This handbook is aimed squarely at practicing bridge engineers around the world. The ideal reader will be a structural and bridge engineer, researcher, or student with a need for a single reference source to keep abreast of bridge development and the state of the practice around the world. The editors acknowledge with thanks the comments, suggestions, and recommendations made during the development of the handbook by Professor Ivan Baláž, Slovak University of Technology, Slovak Republic; Dr. Reidar Bjorhovde, The Bjorhovde Group, United States; Mr. John Bors, ChemCo Systems, United States; Professor Jean Armand Calgaro, France; Professor Estevam Las Casas, University Federal De Minas Gerais, Brazil; Professor Ernani Diaz, Federal University of Rio de Janeiro, Brazil; Professor Radomir Folić, Faculty of Technical Sciences, Serbia; Dr. Jianping Jiang, MMM Group, Canada; Mr. Charles King, Buckland & Taylor, Ltd., Canada; Professor Xila Liu, Shanghai Jiaotong University, China; Professor David A Nethercot, Imperial College, United Kingdom; Professor Doncho Partov, Higher School of Civil Engineering, Lyuben Karavelov, Bulgaria; Professor Michelle Pfeil, University Federal do Rio de Janeiro, Brazil; Ms. Helena Russell, editor, *Bridge Design and Engineering*, United Kingdom; Mr. Eric Sakowski, Technicolor, United States; Mr. Yusuf Saleh, transportation engineer, California Department of Transportation, United States; Professor Jiri Strasky, Brno University of

Technology, Czech Republic; Mr. Juhani Virola, consulting engineer, Finland; Mrs. Analia Wlazlo, editor, *Vial* magazine, Argentina; Mr. Roman Wolchuk, consulting engineer, United States. We wish to thank all the authors for their contributions and also to acknowledge at CRC Press/Taylor and Francis Group, Joseph Clements, acquiring editor, Kari Budyk, senior project coordinator, and Glen Butler, project editor.

Wai-Fah Chen and Lian Duan
May 2013

Editors

Dr. Wai-Fah Chen is a research professor of civil engineering at the University of Hawaii. He was dean of the College of Engineering at the University of Hawaii from 1999 to 2007, and a George E. Goodwin Distinguished Professor of Civil Engineering and head of the Department of Structural Engineering at Purdue University from 1976 to 1999. He received his BS in civil engineering from the National Cheng-Kung University, Taiwan, in 1959, MS in structural engineering from Lehigh University, Pennsylvania, in 1963, and PhD in solid mechanics from Brown University, Rhode Island, in 1966. He received the Distinguished Alumnus Award from the National Cheng-Kung University in 1988 and the Distinguished Engineering Alumnus Medal from Brown University in 1999.

Dr. Chen's research interests cover several areas, including constitutive modeling of engineering materials, soil and concrete plasticity, structural connections, and structural stability. He is the recipient of several national engineering awards, including the Raymond Reese Research Prize and the Shortridge Hardesty Award, both from the American Society of Civil Engineers, and the T. R. Higgins Lectureship Award in 1985 and the Lifetime Achievement Award, both from the American Institute of Steel Construction. In 1995, he was elected to the U.S. National Academy of Engineering. In 1997, he was awarded honorary membership by the American Society of Civil Engineers. In 1998, he was elected to the Academia Sinica (National Academy of Science) in Taiwan.

A widely respected author, Dr. Chen has authored and coauthored more than 20 engineering books and 500 technical papers. His books include several classical works such as *Limit Analysis and Soil Plasticity* (Elsevier, 1975), the two-volume *Theory of Beam-Columns* (McGraw-Hill, 1976 and 1967), *Plasticity in Reinforced Concrete* (McGraw-Hill, 1982), and the two-volume *Constitutive Equations for Engineering Materials* (Elsevier, 1994). He currently serves on the editorial boards of more than 15 technical journals. Dr. Chen was the editor-in-chief for the popular 1995 *Civil Engineering Handbook* (CRC Press, 1995 and 2003), the *Handbook of Structural Engineering* (CRC Press, 1997 and 2005), the *Bridge Engineering Handbook* (CRC Press, 2000 and 2013) and the *Earthquake Engineering Handbook* (CRC Press, 2003) and most recently the *Semi-Rigid Connections Handbook* (J. Ross Publishing, 2011). He currently serves as the consulting editor for McGraw-Hill *Encyclopedia of Science and Technology*.

Dr. Chen was a longtime member of the executive committee of the Structural Stability Research Council and the specification committee of the American Institute of Steel Construction. He was a consultant for Exxon Production Research on offshore structures; for Skidmore, Owings, and Merrill in Chicago on tall steel buildings; and for the World Bank on the Chinese University Development Projects, among many others. Dr. Chen has taught at Lehigh University, Purdue University, and the University of Hawaii.

Dr. Lian Duan is a senior bridge engineer and Structural Steel Committee chair with the California Department of Transportation (Caltrans), United States. He worked at the North China Power Design Institute from 1975 to 1978 and taught at Taiyuan University of Technology from 1981 to 1985. He received his diploma in civil engineering in 1975, MS in structural engineering in 1981 from Taiyuan University of Technology, and PhD in structural engineering from Purdue University in 1990.

Dr. Duan's research interests cover areas including inelastic behavior of reinforced concrete and steel structures, structural stability, seismic bridge analysis, and design. With more than 70 authored and coauthored papers, chapters, and reports, his research focuses on the development of unified interaction equations for steel beam-columns, flexural stiffness of reinforced concrete members, effective length factors of compression members, and design of bridge structures.

Dr. Duan has over 30 years experience in structural and bridge engineering. He was lead engineer for the development of *Caltrans Guide Specifications for Seismic Design of Steel Bridges*. He is a registered professional engineer in California. He served as a member of several National Cooperative Highway Research Program (NCHRP) panels and was a Transportation Research Board (TRB) steel committee member from 2000 to 2006.

He was the coeditor of the *Bridge Engineering Handbook* (CRC Press, 2000 and 2013) and the winner of *Choice* magazine's Outstanding Academic Title Award for 2000. In 2001, he received the prestigious Arthur M. Wellington Prize from the American Society of Civil Engineers (ASCE) for the paper "Section Properties for Latticed Members of San Francisco-Oakland Bay Bridge" in the *Journal of Bridge Engineering*, May 2000. He received the Professional Achievement Award from Professional Engineers in California Government in 2007 and the Distinguished Engineering Achievement Award from the Engineers' Council in 2010.

Contributors

Mourad M. Bakhoum
Cairo University
Cairo, Egypt

Ivan Baláž
Slovak University of
Technology
Bratislava, Slovak Republic

Wojciech Barcik
Mosty-Wrocław Research &
Design Office
Wrocław, Poland

Jan Bień
Wrocław University of
Technology
Wrocław, Poland

Jan Biliszczuk
Wrocław University of
Technology
Wrocław, Poland

Simon A. Blank
Consulting Engineer
Castro Valley, CA

Michael Britt
Modjeski and Masters, Inc.
Harrisburg, Pennsylvania, USA

Jean-Armand Calgaro
Ministry of Ecology, Energy,
Sustainable Development,
and Town and County
Planning
La Defense, France

Alp Caner
Middle East Technical
University
Ankara, Turkey

Dyi-Wei Chang
CECI Engineering
Consultants, Inc.
Taipei, Taiwan

Tomás A. del Carril
del Carril – Fazio Civil
Engineers
Buenos Aires, Argentina

**Augusto Carlos de
Vasconcelos**
Consulting and Eng. SS Ltda
São Paulo, Brazil

Wiryanto Dewobroto
Universitas Pelita Harapan
Tangerang, Indonesia

Dobromir Dinev
University of Architecture
Sofia, Bulgaria

Lian Duan
California Department of
Transportation
Sacramento, California, USA

Dzong-Chwang Dzeng
CECI Engineering
Consultants, Inc.
Taipei, Taiwan

Radomir Folić
Faculty of Technical Sciences
Obradovića, Serbia

Lubin Gao
Federal Highway
Administration
Washington, D.C., USA

Niels Jørgen Gimsing
Technical University of
Denmark
Lyngby, Denmark

Paweł Hawryszków
Wrocław University of
Technology
Wrocław, Poland

Lanny Hidayat
Department of Public Works
Cibinong, Indonesia

Maciej Hildebrand
Wrocław University of
Technology
Wrocław, Poland

Reggie Holt
Federal Highway
Administration
Washington, D.C., USA

Bisera Karalić-Hromić
Faculty of Civil Engineering
Sarajevo
Road Directorate of Federation
Sarajevo, Bosnia and Herzegovina

Ping-Hsun Huang
CECI Engineering
    Consultants, Inc.
Taipei, Taiwan

Dragan Ivanov
University Sts Cyril &
    Methodius, Skopje,
    Macedonia

Esko Järvenpää
WSP Finland Ltd
Oulu, Finland

Yutaka Kawai
Nihon University
Chiba, Japan

Boris Koboević
Faculty of Civil Engineering
    Sarajevo
Sarajevo, Bosnia and
    Herzegovina

Mykhailo Korniev
JSC "Mostobud"
Kyiv, Ukraine

Brian Kozy
Federal Highway
    Administration
Washington, D.C., USA

John M. Kulicki
Modjeski and Masters, Inc.
Harrisburg, Pennsylvania, USA

Paul Liles
Georgia Department of
    Transportation
Atlanta, Georgia, USA

Ekasit Limsuwan
Chulalongkorn University
Bangkok, Thailand

M. Myint Lwin
Federal Highway
    Administration
Washington, D.C., USA

Shervin Maleki
Sharif University of Technology
Tehran, Iran

Gilson L. Marchesini
Sobrenco Engenharia E
    Comercio Ltda
Sao Paulo, Brazil

Gang Mei
China Highway Planning &
    Design Institute
    Consultants, Inc.
Beijing, China

Joost Meyboom
MMM Group
Vancouver, BC, Canada

Masatsugu Nagai
Nagaoka University of
    Technology
Niigata, Japan

Tihomir Nikolovski
FAKOM AD-Skopje
Skopje, Macedonia

Yoshiaki Okui
Saitama University
Saitama, Japan

Doncho Partov
Higher School of Civil
    Engineering "Lyuben
    Karavelov"
Sofia, Bulgaria

Amorn Pimanmas
Thammasat University
Pathum Thani, Thailand

Goran Puž
Institute IGH
Zagreb, Croatia

Quan Qin
Tsinghua University
Beijing, China

Jure Radić
University of Zagreb
Zagreb, Croatia

Betsy Reiner
Modjeski and Masters, Inc.
Harrisburg, Pennsylvania, USA

Kimio Saito
Kajima Corporation
Tokyo, Japan

Eric Sakowski
Technicolor
Hollywood, CA
USA

Vadim A. Seliverstov
Joint Stock Company
    Giprotransmost
Moscow, Russian Federation

Stamatios Stathopoulos
DOMI S.A.
Athens, Greece

Jiri Strasky
Brno University of Technology
Brno, Czech Republic

Robert A.P. Sweeney
Modjeski and Masters, Inc.
Harrisburg, Pennsylvania, USA

Júlio Timerman
Engeti Consulting and Eng.
    SS Ltda.
São Paulo, Brazil

Shouji Toma
Kokkai Gakuen University
Sapporo, Japan

Ahmet Turer
Middle East Technical
    University
Ankara, Turkey

Herry Vaza
Department of Public Works
Bekasi, Indonesia

Kenneth J. Wright
HDR Engineering, Inc.
Pittsburgh, Pennsylvania, USA

Gongyi Xu
China Railway Major Bridge
    Reconnaissance and Design
    Institute, Co., Ltd.
Wuhan, China

Masaaki Yamamoto
Kajima Corporation
Tokyo, Japan

Yeong-Bin Yang
National Taiwan University
Taipei, Taiwan

Cetin Yilmaz
Middle East Technical
    University
Ankara, Turkey

Damir Zenunović
University of Tuzla
Tuzla, Bosnia and Herzegovina

1

Bridge Engineering in Canada

Joost Meyboom
MMM Group

1.1 Historical Development

1.1.1 Canada's Geography

Canada is the world's second-largest country by area and extends in the east–west direction from the Atlantic Ocean to the Pacific Ocean and in the north–south direction from the Arctic Ocean to the border with the United States. The population density of Canada is 3.3 inhabitants per square kilometer, which is among the lowest in the world. The most densely populated part of the country is the Québec City in Windsor Corridor, situated in Southern Québec and Southern Ontario along the Great Lakes and the Saint Lawrence River. Major population centers are typically found in a southern band near the border of the United States.

From a bridge design and construction perspective, Canada has a large variety of design challenges given the diversity of its geography and climate. For example, there are major rivers to be crossed, such as the Fraser, Mackenzie, and St. Laurence—all of which are in many places more than 2 km wide. In addition, there are numerous significant seismic zones, such as along the West Coast, in the Yukon, and in the Ottawa-Montréal region. Canadian bridge engineers also have to contend with vastly differing geotechnical conditions, including the endless soft soil conditions of the Fraser River Delta, swelling clays in the prairies, permafrost in the northern regions and variable and unpredictable bedrock profiles in the Canadian Shield. Other important considerations in building bridges in Canada are seasonal restrictions with regard to remote construction sites, fish migrations, bird nesting, extreme winter conditions, heavily trafficked urban areas, hurricane force winds, and ice loading.

1.1.2 Canada's First Bridges

Canada's population centers have developed based on available modes of transportation. First Nations, or indigenous, peoples typically traveled on foot or by canoe and as such their villages and towns were located near water. Permanent European settlement was initially restricted to areas that could be

1

accessed by sea-going boats such at Québec City, Montréal, Kingston, Toronto, and Churchill. Smaller trading posts were established based on Canada's extensive natural waterways, which were traversed by teams of traders in canoes. Relatively simple timber bridge structures would have been required at this time in Canada's history, but no significant bridge building took place.

In the 1830s, railway technology was introduced to Canada and various short line railways were constructed in Eastern Canada, particularly between Montréal and Ontario. The introduction of railways required sophisticated engineering and the corresponding construction of bridges to span waterways. By 1880 the Canadian Pacific Railway had been extended to the Pacific Ocean and at the time was the world's longest railway. This railway was an engineering marvel that passed through complex geotechnical conditions, crossed massive rivers such as the North Saskatchewan and the Fraser, and winded through tunnels in the Rocky Mountains. Railway expansion continued in Canada until the early 1900s, with the construction of many impressive structures to carry trains over wild and rugged terrain.

Bridges were initially of timber construction given the abundance of available timber. Possibly the oldest timber bridge in Canada is the Percy Covered Bridge in Powerscourt, Québec, which was built in 1861 to carry people and horse-drawn wagons over the Châteauguay River (see Figure 1.1). The bridge is an example of an inflexible arched truss, a design pioneered by Daniel McCallum, the general superintendent of the New York and Erie Railroad until he founded the McCallum Bridge Company in 1858. This design was used throughout Canada and the United States for timber railway bridges, but was made obsolete with the advent of steel bridges.

Another notable bridge from the mid-1800s is Montréal's Victoria Bridge (see Figure 1.2). The Victoria Bridge was the first crossing of the St. Laurence River and was completed in 1859. Although completed before the Percy Bridge, it is not considered Canada's oldest bridge because it has been significantly modified over its life. It was constructed to provide a fixed railway link between the Island of Montréal and Montréal's South Shore. When it was opened it was the longest bridge in the world, with a length of 3 kilometers.

The project was initially considered to be too ambitious and not feasible. With the success of the Britannia Bridge in Wales, however, where a ductile steel tubular superstructure was successfully used, the project gained credibility and was started in the early 1950s. The crossing alignment was established by one of Canada's preeminent engineers of the time, Thomas Keefer. The chief engineer for the project was James Hodges and the steel superstructure was designed in England by Robert Stephenson. The project was constructed by Peto, Brassey and Betts for a cost of $6.6 million. The project required ductile steel to be shipped from England to site in "tube" modules on a just-in-time schedule, the development

FIGURE 1.1   (**See color insert.**) Canada's oldest bridge, the Percy Covered Bridge, Powerscourt, Québec.

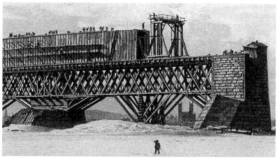 

FIGURE 1.2 Victoria Bridge, Montréal.

of special lifting derricks, the use of "floating cofferdams," and the coordination of a labor force of more than 3000 workers. The bridge was built in 5 years.

The original superstructure of the Victoria Bridge was a closed steel tube, similar to that used for the Britannia Bridge. This tube was replaced in the 1890s with open trusses to relieve the noise and smoke created in the tube by locomotives. In addition, the tube construction limited the crossing to a single track. The superstructure replacement was carried out by constructing the new trusses around the original tube construction without disruption to the railway operations.

1.2 Design Practice

Bridges in Canada are typically designed to the Canadian Highway Bridge Design Code (CHBDC) or, for railway bridges, the American Railway Engineering and Maintenance-of-Way Association (AREMA) Manual of Railway Engineering. These codes provide the design requirements for relatively standard bridge types. Where unusual bridge configurations or long-span structures are required, additional codes are often also considered and sophisticated analysis and testing methods are used to inform the design. The CHBDC is a Limit States–based code and closely corresponds to similar codes in the United States and Europe as well as other, more general structural design codes in Canada. It provides the requirements for loading in terms of magnitude, configuration, and combinations, and factors for live loads, dead loads, seismic loads, wind loads, soil pressure, ice and snow loads, restraint loads, thermal loads, secondary prestress loads, hydraulic induced loads, and loads caused by settlement. In addition the CHBDC provides the requirements for material resistance in terms of strengths and corresponding factors. Many provinces in Canada have developed a supplement to the CHBDC to account for local conditions.

Most bridge design offices use some type of finite element software for design including MIDAS, SAP, S-FRAME, LUSAS, and others. Drawings are typically prepared using either AUTOCAD or Microstation. Design is often undertaken in 3 stages—conceptual, preliminary, and detailed—with cost estimates refined as the design is progressed. An owner will often develop a preliminary design to deliver a project using a design-build model, whereas detailed design corresponds to drawings that can be issued for construction.

Design is carried out by engineering companies for a range of clients, including government agencies for traditional design-bid-build delivery and for contractors if the project is delivered as a design-build or public-private partnership (P3) project. Design-bid-build is typically used where owners need to keep design control. Design-build procurement is used where there is scope for innovation from a contractor–designer team. Many of Canada's recent big bridges have been designed and built as part of a P3 project.

1.3 Major Canadian Bridges

A selection of major Canadian bridges is given in Table 1.1:

TABLE 1.1   Major Canadian Bridges

Bridge	Year	Length, Bridge Type
	Arch Bridges	
Whirlpool Rapids Bridge, Ontario/New York	1897	329 m, thrust arch.
Alexandra Bridge, Ontario/Québec	1901	563 m, five-span through arch.
University Bridge, Saskatchewan	1916	335 m, spandrel arch.
Center Street Bridge, Alberta	1916	178 m, continuous multiple arches.
Bloor Street Viaduct, Toronto	1918	494 m deck arch.
Peace Bridge, Ontario/New York	1927	1768 m, deck arches with a through-arch main span.
Broadway Bridge, Saskatchewan	1932	355 m continuous deck arches.
Pattullo Bridge, British Columbia	1937	1227 m, steel trussed arch main span.
Rainbow Bridge, Ontario/New York	1941	290 m, arch main span.
Thousand Islands Bridge, Ontario/New York	1937	13.7 km, two suspension bridges and a thrust arch.
Viau Bridge, Québec	1962	Multiple arches.
Queenston Lewiston Bridge, Ontario/New York	1962	488 m, thrust arch.
Sault St. Marie International Bridge, Ontario/Michigan	1962	4500 m, trussed arch main span.
Old Burlington Bay Skyway, Ontario	1958	2200 m, trussed arch main span.
Sea Island Bridge, Nova Scotia	1962	744 m, 152 m trussed arch main span.
Centennial Bridge, New Brunswick	1967	1000 m, trussed arch main span.
Port Mann Bridge, British Columbia	1964	2000 m, 603 m tied arch, 365 m main span. Longest tied arch bridge at completion of construction.
Laviolette Bridge, Québec	1967	2707 m, 335 m trussed arch main span.
Burton Bridge, New Brunswick	1973	765 m, trussed arch main span.
New Blue Water Bridge, Ontario/Michigan	1997	1862 m, 285 m tied arch main span.
	Truss Bridges	
Hartland Bridge, New Brunswick	1901	391 m, timber-covered bridge.
South Saskatchewan River, CPR River	1908	341 m, steel truss.
Lethbridge Viaduct, Alberta	1909	1624 m, railway bridge.
Dawson Bridge, Alberta	1912	236 m, continuous deck trusses.
Sky Trail Bridge, Saskatchewan	1912	910 m, multiple steel-truss spans.
High Level Bridge, Alberta	1915	777 m, 80 m deck trusses.
Québec Bridge, Québec	1919	987 m, cantilever truss with a 549 m main span.
Jacques Cartier Bridge, Québec	1930	2687 m, cantilever truss.
Burrard Bridge, British Columbia	1932	950 m, deck and through trusses.
Mercier Bridge, Québec	1934	1326 m, deck truss with a through-truss main span.
Old Blue Water Bridge, Ontario/Michigan	1938	1883 m, cantilever truss span with a main span of 265 m
Princess Margaret Bridge, New Brunswick	1959	1075 m, continuous truss.
Iron Workers' Memorial Bridge, British Columbia	1960	1292 m, continuous deck trusses.
J.C. van Horne Bridge, New Brunswick	1961	805 m, cantilever through truss.
Peace Bridge, Alberta	2011	131 m, pedestrian bridge.
	Suspension Bridges	
Ambassador Bridge, Ontario	1927	2300 m, 560 m main span.
Île d'Orléans Bridge, Québec	1935	4430 m, 677 m main span.

TABLE 1.1   Major Canadian Bridges (*Continued*)

Bridge	Year	Length, Bridge Type
Lion's Gate Bridge, British Columbia	1937	1823 m, 472 m main span.
Thousand Islands Bridge, Ontario/New York	1937	13.7 km, two suspension bridges and a thrust arch.
Angus L. Macdonald Bridge, Nova Scotia	1955	1300 m, 441 m main span.
Ogdensburg-Prescott Bridge, Ontario/New York	1960	351 m main span.
Pierre Laporte Bridge, Québec	1970	1041 m, 667 m main span.
Murray MacKay Bridge, Nova Scotia	1970	1200 m, 426 m main span.
	Cable-stayed Bridges	
Papineau-Leblanc Bridge, Québec	1969	241 m main span. Longest cable-stayed bridge at construction.
Alex Fraser Bridge, British Columbia	1983	932 m, 465 m main span. Longest cable-stayed bridge at construction.
SkyTrain Bridge, British Columbia	1990	616 m, 340 m main span.
Provencher Bridge, Winnipeg, Manitoba	2004	192 m, 106 m main span, pedestrian bridge.
Pitt River Bridge, British Columbia	2009	500 m, 190 m main span.
A25 Bridge, Québec	2013	1200 m, 280 m main span.
New Port Mann Bridge, British Columbia	2013	2000 m, 470 m main span.
Deh Cho Bridge, Northwest Territories	2014	1100 m, 190 m main span.
	Extradosed Bridges	
Golden Ears Bridge, British Columbia	2009	968 m, three 242 m main spans.
North Arm Bridge, British Columbia	2010	562 m, 180 m main span.
	Girder Bridges	
Champlain Bridge, Ontario/Québec	1928	1100 m, cantilever truss.
Albert Memorial Bridge, Saskatchewan	1930	256 m reinforced concrete girders.
Île aux Tourtes Bridge, Québec	1965	2000 m, continuous girder bridge.
Macdonald Cartier Bridge, Ontario/Québec	1965	614 m, continuous haunched girder bridge.
New Burlington Bay Skyway, Ontario	1985	2200 m, balanced cantilever main span.
Cambie Street Bridge, British Columbia	1985	1100 m, continuous post-tensioned concrete girder.
Dudley Menzies Bridge, Alberta	1993	530 m, four-span light rail transit bridge.
Confederation Bridge, PEI/New Brunswick	1997	12.91 km, 250 m main span, precast segmental.
Tsable River Bridge, British Columbia	1999	400 m, 118 m main span, cast-in-place.
Jemseg River Bridge, New Brunswick	2001	950 m, continuous plate girder bridge.
St. John River Bridge, New Brunswick	2002	1000 m, continuous plate girder bridge with a 120 m main span.
New Park Bridge, British Columbia	2007	405 m, five-span curved plate girder.
	Floating Bridges	
William Bennett Bridge, British Columbia	2008	650 m.

1.4 Recent Bridge Projects

The following project descriptions provide an overview of some recent bridge projects undertaken in Canada. Unfortunately, not all recent Canadian bridge projects are described. Notable exceptions include the new William Bennett Bridge, the new Bluewater Bridge, the new A30 Bridge, the Lion's Gate Bridge Deck Replacement, the Provencher Bridge, and the Deh Cho Bridge, as well as other important projects.

1.4.1 Cable-Stayed Bridges

1.4.1.1 Pitt River Bridge

As part of British Columbia's Gateway Program, the need for a more reliable crossing of the Pitt River was identified. The existing crossing was provided by two aging swing bridges with a history of mechanical and electrical malfunction. The crossing is used by more than 100,000 vehicles per day and severe congestion was experienced at the crossing with the existing swing bridges.

As part of the project development, bridge capacity requirements were established using sophisticated traffic modeling. It was found that a seven-lane structure would be required to meet 2031 traffic demands. Other key issues affecting the design and construction of the new Pitt River Bridge included:

- Liquefiable sands and deep compressible clays with soft soils to a depth of more than 100 m
- The need for preloading and lightweight fills to construct embankments
- Close proximity to two existing swing bridges with a history of settlement-induced problems
- Maintenance of the existing swing bridges during construction
- Staging of construction to minimize traffic disruption on the existing bridges
- Marine works subject to environmental schedule constraints
- A 100 m × 15 m/16 m navigation channel
- Significant vessel impact loads
- Lifeline seismic performance
- Provision for a future Light Rapid Transit facility
- Liaison with numerous stakeholder groups, including two municipalities, marine users' groups, cyclists, private developers, and environmental regulatory agencies

A number of bridge configurations were considered by the owner [1]. Options included arrangements for balanced cantilever concrete construction, steel plate girders, and various cable-stayed arrangements (see Figure 1.3).

Because the project was procured and delivered using a design-build model the bridge type was left to the design-build team. The winning team found that a cable-stayed bridge (see Figure 1.4) allowed for a significant reduction in the vessel impact loads given that piers could be located further back from the navigation channel than with a conventional plate girder bridge. As such, a cable-stayed solution was found to be more economical than a plate girder bridge with shorter spans.

The New Pitt River Bridge consists of the main, cable-stayed portion and girder approach structures. The cable-stayed superstructure width varies from 40.5 m to 48 m to accommodate a flare in the road alignment required for the interchange located on the north approach. A 190 m cable-stayed main span with 80 m side spans was used. Because of the very wide deck, three planes of cables were used to support the steel–concrete composite. Cables were arranged in a harp configuration. The bridge is supported on 1.8 m diameter piles driven into the till located approximately 100 m below the mud line of the channel. Pile caps are located at water level. Other elements of the Pit River Bridge design are described next.

- Use of a lightweight structure such as a cable-stayed bridge meant fewer piles, fewer piers, and simplified construction for the constrained site. Only a short trestle bridge was required for construction as the majority of the work could be done by beam and winch and small barges. The large middle span greatly exceeded navigational clearance requirements and was able to completely avoid the operation of the existing swing bridges.
- Three planes of cables, rather than the conventional two, allowed for a reduction of structural steel quantities in the deck.
- The west end of the bridge was flared to accommodate the necessary highway geometry; out-of-plane forces were eliminated by rotating one of the towers in plan view. This eliminated the need to widen the entire bridge and greatly reduced construction costs.

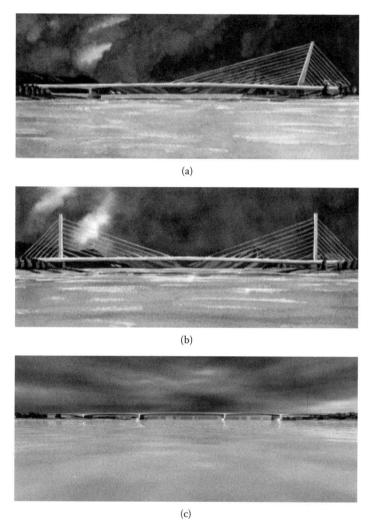

(a)

(b)

(c)

FIGURE 1.3 Options considered for the Pitt River Bridge: (a) single tower cable-stayed bridge, (b) double-tower cable-stayed bridge, and (c) girder bridges [1].

- Slender solid concrete towers were utilized with cable anchorages crossing inside the towers, rather than conventional larger hollow towers with tension elements between anchorages. At each level of cables the crossover was alternated to avoid inducing torsion in the towers. The towers were also designed to eliminate crossbeams between the towers above the deck. These strategies greatly simplified construction and reduced materials quantities.

A detailed three-dimensional structural analysis of the bridge was undertaken for the design. This model included live-load influence line generation with multiple lane combinations for optimal load cases, seismic analyses using response spectra analysis and nonlinear time-history, vessel impact loading on river piers, loss-of-stay cable condition with accumulation of different static stress states, and staged construction of the bridge deck in balanced cantilever. The contractor chose to lift each of the steel frames in separate halves, which required the sequence be modeled explicitly.

The bridge foundations were a major part of the design effort, particularly given the tight allowable settlement requirements specified in the contract. The project site is underlain by soft compressible soils with significant liquefiable zones. Satellite data indicates the region is undergoing subsidence at a rate of

5 mm per year. Survey data from the existing bridge structures indicates settlements of up to 900 mm. These geotechnical challenges were compounded by the owner's contract requirement to support the main river bridge piers on piles terminating in dense, glaciated deposits at nominally 100 m depth. To accommodate these potentially large differential settlements and avoid the creation of a significant bump at the end of the bridge, the Pitt River Bridge design included two 30 to 33 m approach spans at each end with abutments on spread footings and the intermediate piers on friction piles.

To support the use of higher design pile capacities and minimize costs, a top-down, static pile load test was carried out using river bridge pier production piles for both the test and reaction piles. The piles comprised a 1.8 m diameter, open-toe steel pipe driven to approximately 100 m depth. The load test was completed successfully to a load of 45,000 kN, which is one of the largest top-down tests ever conducted. This historic test has become a new standard in metro Vancouver, and has subsequently been applied on similar large-scale projects of this nature.

The bridge became operational in 2009 (refer to Figures 1.4 and 1.5). It is owned by the British Columbia Ministry of Transportation and the design-build team was led by Peter Kiewit & Sons. MMM Group was the lead designer for the project as supported International Bridge Technologies for the cable-stayed design and by Associated Engineering.

(a)   (b)

(c)

FIGURE 1.4   Pitt River Bridge: (a) tower construction, (b) cable anchorage, (c) deck construction. (Photo courtesy of MMM Group.)

FIGURE 1.5   Pitt River Bridge near completion. (Photo courtesy of MMM Group.)

FIGURE 1.6   Construction of A25 Bridge, Montréal [2].

1.4.1.2 A25 Bridge

The completion of Autoroute 25 in Montréal is a $400 million project and is the first private-public partnership transportation project for the province of Québec. The project involves 7 km of new expressway and 13 new bridges, including a cable-stayed bridge over the Riviére des Prairies, between the islands of Laval and Montréal in eastern Montréal.

The main bridge is about 1200 m long and comprised of nine approach spans varying in length from 24 m to 96 m, and a three-span cable-stayed bridge with a 280 m main span and 115 m side spans [2]. The bridge spans the environmentally-sensitive Sturgeon Pool of the Riviére des Prairies. The bridge is somewhat unique in this way in that the cable-stayed structure is used to span an environmentally sensitive area rather than a navigation channel. As such the vertical clearance under the bridge could be quite low. The designers were challenged with large ice loads from the river, vertical height restrictions, wind loads, and seismic requirements for a lifeline structure.

The concrete towers for the cable-stayed bridge are 70 m tall and support two planes of 20 cables in a semi-harp pattern, which carry the six-lane composite steel and concrete deck and a pedestrian walkway. Transverse connection between the towers was provided below the deck to provide a clean and elegant design. Box girders were used for the edge girders. These were framed around the towers by locally providing web stiffeners for the inner web, diaphragms, and dropping the outer web and flanges. Stay-cable anchorages were bolted to the edge girders (see Figure 1.6). The bridge is being constructed by Peter Kiewit & Sons and designed by Parsons Transportation.

1.4.1.3 New Port Mann Bridge

As part of the Province of British Columbia's Gateway Program a major upgrade to the Trans-Canada Highway segment through the Greater Vancouver area has been undertaken. This project included a $2.4 billion on-land segment as well as a new crossing of the Fraser River directly adjacent to the existing Port Mann Bridge—a tied arch, which when constructed in 1964, was the world's longest tied arch bridge.

Key design issues identified for the New Port Mann Bridge included [3]:

- Liquefiable sands, deep compressible clays, and glacial deposits at depths greater than 60 m below ground
- Risk of settlement of the existing adjacent bridge when installing new foundations
- Lifeline performance for 1:475, 1:1000, and 1:2475 seismic events
- Provision for future Light Rail Transit crossing
- Bridge aesthetics

1.4.1.3.1 Owner's Concept

In developing the bridge for a public-private partnership procurement, a reference design was developed for the owner that, as directed by the owner, twinned the existing bridge. As part of the procurement process, however, proponents were free to replace the existing bridge with a new wider crossing. Development of the owner's reference design included several studies related to ship impact loads, river hydraulics and bathymetry, and geotechnical conditions. Numerous span and bridge configurations were explored and considered including multispan cable-stayed spans, an arch span, and an extradosed bridge (see Figure 1.7).

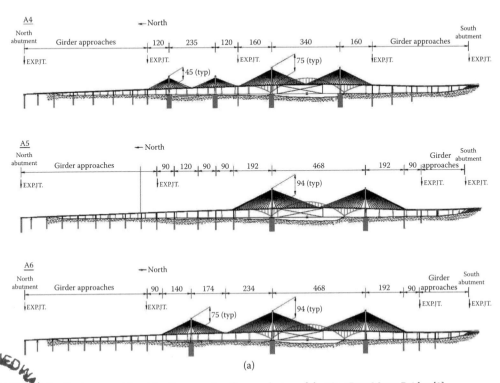

(a)

FIGURE 1.7 Options considered for the owner's reference design of the New Port Mann Bridge [3].

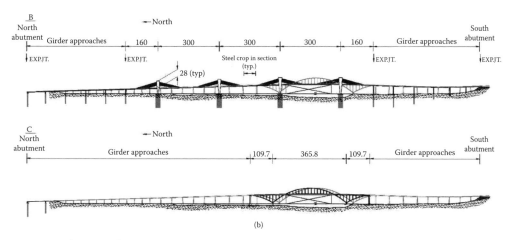

FIGURE 1.7 (*Continued*)

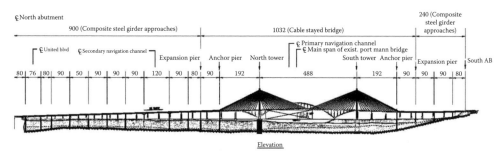

FIGURE 1.8 Owner's reference design for New Port Mann Bridge [3].

The owner's reference design consisted of an approximately 2 km long crossing with a 848 m long cable-stayed main bridge (see Figure 1.8). The cable-stayed bridge had a 468 m main span with 192 m side spans. Although the required navigation channel width is 200 m, a main span considerably longer than the 200 m wide was considered necessary to avoid unstable slope conditions at the south bank of the river and to span the Canadian National Railway (CNR) yard. Span arrangements for the approaches were restricted by secondary navigation channels.

A diamond-shaped tower was developed with a total height of approximately 150 m above the pile cap and 94 m above the road surface (see Figure 1.9). The diamond-shaped tower was selected to concentrate the tower cable anchorages and facilitate cable installation. It was also selected to provide torsional stiffness to the plate girder deck system and for aesthetics. Float-in, precast pile caps were envisaged to be installed at the water level. Ship impact loads of 40 mN required the base of the towers to be solid. The towers were designed to be hollow above the deck and equipped with elevators and platforms to allow maintenance of the cable anchorages. Cables were provided with dead anchorages at deck level and jacking points at the top of the tower.

Steel edge girders with transverse steel floor beams made composite with a concrete deck were supported by two inclined cable planes (see Figures 1.10 and 1.11). Cables were provided with a fixed anchorage at deck level and locations for jacking in the tower head. A deck width suitable for five lanes of traffic was used. The edge girders were dimensioned to be consistent with a 12 m cable spacing along the deck. Three floor beams were located between cable anchorages.

The cable-stayed bridge was made continuous with the first approach span to allow separation of the expansion joints and back stay anchorage. This was done to minimize congestion. Sixty meter long deep foundations were designed with precast pile caps to allow for ease of construction.

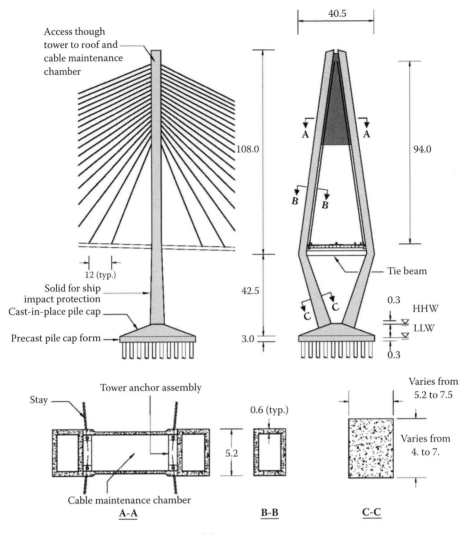

FIGURE 1.9   Owner's reference design for tower [3].

1.4.1.3.2 Design-Build Solution

The winning design-build proposal provided for a new 10-lane bridge and called for the demolition of the existing Port Mann Bridge. The winning design comprised an 850 m long cable-stayed bridge with 1223 m long precast concrete segmental box girder approaches (see Figure 1.12). The cable-stayed structure has a main span of 470 m and side spans of 190 m.

The bridge is characterized by a very wide superstructure that has a width of 65 m and consists of two five-lane decks, separated by a 10 m median (see Figure 1.13a). The median is required to locate the central pylons. Each deck consists of a composite structure with steel edge girders, floor beams, and precast concrete deck panels. The towers are about 160 m above the water. All cable anchorages are concentrated in the upper 40 m of the tower, which comprises a composite steel concrete anchor housing. Concentration of the cable anchorages greatly increased the efficiency of cable installation and will increase the efficiency of future maintenance and inspection operations. Concentrating the cable anchorages in a single central tower also provides considerable torsional stiffness to the bridge. See Figure 1.13b through e for tower construction photographs.

The cable-stayed bridge was conceived to be continuous with expansion joints provided at the end piers and longitudinal fixity provided at the on-land pylons. As such the on-land pylon design is governed by seismic loads. Lateral restraint is provided to the cable-stayed spans with wind keys and bearings at each pier (see Figure 1.14). The approach spans consist of three parallel precast segmental box girders with cantilever construction above the water and span-by-span construction on land. The concrete segmental approach span nominal depths vary with span length from a typical 3 m to 5 m in the longer cantilever spans over the North Channel (see Figure 1.15).

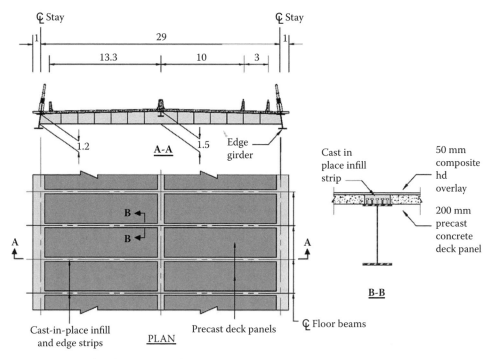

FIGURE 1.10 Owner's reference design for deck [3].

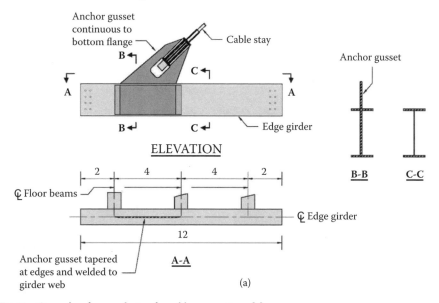

FIGURE 1.11 Owner's reference design for cable connections [3].

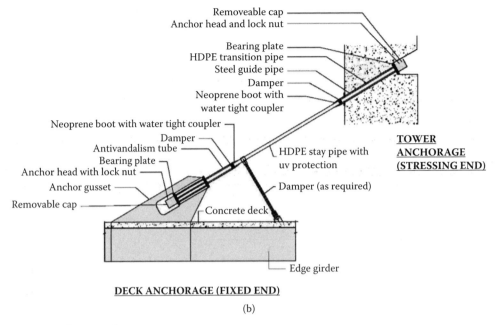

DECK ANCHORAGE (FIXED END)

(b)

FIGURE 1.11 (*Continued*)

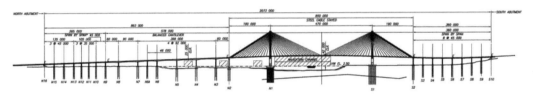

FIGURE 1.12 New Port Mann Bridge [4].

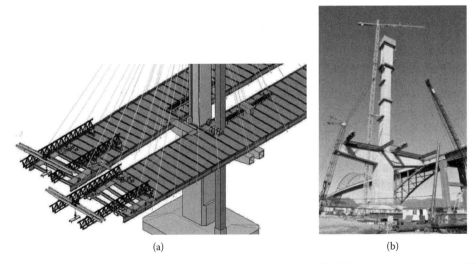

(a)   (b)

FIGURE 1.13 New Port Mann Bridge: (a) tower and deck arrangement [4], (b)–(e) tower construction. (Photo courtesy of Scott Marshall.)

(c)

(d)

(e)

FIGURE 1.13 (*Continued*)

FIGURE 1.14 Construction of New Port Mann Bridge. (Photo courtesy of Scott Marshall.)

FIGURE 1.15   New Port Mann Bridge [4].

Foundations for the New Port Mann Bridge are generally Φ 1.8 m steel piles or drilled shafts, supported on a firm ground till layer under the loose sand deposits at a depth below the river.

The bridge is owned by the Province of British Columbia, is being constructed by a Kiewit-Flatiron Joint Venture, and was designed by T.Y. Lin and International Bridge Technologies. Geotechnical design was carried out by Shannon & Wilson.

1.4.2 Extradosed Bridges

1.4.2.1 North Arm Bridge

Canada Line is the newest component in Vancouver's SkyTrain System and was designed and built between 2005 and 2009 using a public-private partnership. It provides a 19 km link between the Vancouver International Airport and downtown Vancouver. Given the required trip time, a fully grade-separated system was selected consisting of 9 km of bored and cut-and-cover tunnel and 10 km of elevated guideway. The elevated guideway crosses the Fraser River at two locations where significant dedicated transit bridge structures were required. The larger of these bridges, the North Arm Bridge, is Canada's first extradosed bridge (see Figure 1.16).

An extradosed bridge was selected to cross the Fraser River to meet a number of conflicting project requirements. These requirements included [5]:

- Navigation clearance
- Aviation clearance
- Rail grades
- Significant seismic loads
- Rapid construction process

A haunched girder would have pushed the LRT profile higher to accommodate the required shipping clearance, whereas the tower height of a cable-stayed bridge would have interfered with the flight path of Vancouver International Airport. As such, neither of these two options was considered desirable. An extradosed bridge with constant cross section, however, allowed the LRT profile to be kept as low as possible without infringing on marine clearance and, given the short towers required for an extradosed design, flight path interference could be avoided. In addition, because the extradosed option allowed for

FIGURE 1.16 **(See color insert.)** North Arm Bridge. (Photo courtesy of MMM Group.)

a constant cross section that was similar to that used for the elevated guideway approaches, the extradosed solution provided an economical solution that minimized the prefabrication schedule. Another advantage of the extradosed solution was the reduced effects of fatigue on the cables as compared to a cable-stayed solution [5].

Defining characteristics of the North Arm Bridge include:

- Total length of 562 m long, with a180 m main span, 139 m side spans and 52 m haunched transition spans.
- Two rail tracks and a pedestrian/bikeway.
- A single post-tensioned box cross section used for the main and side spans.
- Two 22 m tall pylons support the main and side spans with a single plane of cables with 24 cables. The superstructure's box section is designed to carry torsional loads and is assisted to some extent by the vertical component of the cable forces as well as the benefits of the post-tensioning.
- The main piers were split to provide sufficient flexibility under construction loads as well as under the significant seismic loads experienced in British Columbia's Lower Mainland.
- A constant cross section was used for the main bridge to minimize the elevation of the superstructure while respecting the navigation clearance.
- Spans were arranged to limit the project to a single marine foundation.

Φ 0.915 m and Φ 2 m steel pipe piles were driven to depths of 20 to 45 m below the ground surface into till-like material. The piles were cleaned out and filled with concrete. Split piers were used to facilitate balanced cantilever construction while providing sufficient longitudinal flexibility for seismic loads. The pylons consist of twin post-tensioned concrete columns linked by structural steel tension ties and supported by a single precast concrete base.

Precast segment weights were limited to 70 tonnes to conform to the contractor's construction methods (see Figure 1.17a). This meant that in the haunched transition spans a maximum segment length of 2.8 m was used. For the main bridge, segment length was dictated by this weight limit as well as the physical space required to anchor the extradosed cables. As a result the main span segments were dimensioned to be 3.4 m deep and 3.6 m long [5].

Prestressing was provided in the superstructure through cables embedded in the webs of the section as well as through the extradosed cables. The vertical component of the extradosed cables was transferred to the box section webs using "V"-shaped steel struts anchored into the concrete (see Figure 1.17b). A concrete anchor block at deck level transfers the longitudinal component cable force into the deck.

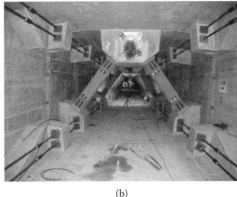

(a)   (b)

FIGURE 1.17 North Arm Bridge: (a) typical precast deck segment, (b) detail at central cable anchorage in deck. (Photos courtesy of MMM Group.)

Steel struts were used for this detail rather than more conventional concrete tension ties to minimize weight and reduce congestion within the cross section. The steel struts were installed after casting of the cross section to simplify the construction method. Although a haunched girder near the pylons would be more efficient from a pure design perspective and thus result in less material, overall economy was improved with the constant section by eliminating costs associated with the adjustment of forms required to fabricate haunches.

Fifty-eight 15.7 mm individually sheathed, galvanized, and waxed strands were used for the 24 extra-dosed cables (see Figure 1.18). Cables were sheathed in a Φ 225 mm high-density polyethylene pipe. The cables were designed to be installed with light monostrand jacking equipment and to be replaceable. Anchorages were provided in the pylon and the deck. High-damping internal rubber dampers were installed at the end of the deck anchorage guide pipe to manage cable vibrations. Double helix ribs were provided on the exterior surface of the cable sleeves to further limit rain- and wind-induced vibration. The project was constructed by SNC Laval with RSL Joint Venture (Rizzani de Eccher/SNC-Lavalin). Buckland & Taylor designed the bridge and provided construction engineering services. MMM Group was the owner's engineer.

1.4.2.2 Golden Ears Bridge

With significant population growth in the Fraser Valley within the Greater Vancouver area, it was decided that an existing ferry service should be replaced with a fixed link. A design-build-finance-operate (DBFO) project was developed to accomplish this objective. The project includes a significant new crossing of the Fraser River as well as considerable new highway infrastructure on either approach. Important site constraints that needed to be addressed in the design of the new bridge included:

- Two widely separated navigation channels with a vertical clearance of 45 m for the main channel
- An adjacent airport
- High river flows
- Sensitive environmental areas
- Poor foundation conditions
- The potential for significant seismic and ship impact loads

As part of the design development by the winning DBFO team, a "hybrid" multispan composite, cable-stayed bridge was developed (see Figure 1.19) and found to be extremely advantageous from a constructability and economic perspective. Key characteristics of the bridge include [6]:

FIGURE 1.18 Construction of the North Arm Bridge. (Photos courtesy of MMM Group.)

FIGURE 1.19 Golden Ears Bridge under construction.

- A five-span continuous 968 m long hybrid cable-stayed/extradosed bridge with two navigation spans. Three equal main spans of 242 m with side spans of 121 m were used with piers up to 80 m tall (see Figure 1.19).
- A composite steel and concrete deck was used with precast concrete deck panels for the main bridge.
- The deck of the river crossing bridge is supported with two parallel vertical cable planes from pylons. The cable-stays are arranged at a relatively flat angle in a harped configuration, resulting in 40 m high pylons above deck level.
- The main bridge pylons are braced with shallow transverse crossbeams hidden within the depth of the deck section to provide transverse frame action. This arrangement permits the use of a single access/maintenance gantry for the full length of the main crossing.
- Approach roads and viaducts of 1.4 km were required. Structures consisted of concrete precast, prestressed girders with concrete decks.

In developing the design, several bridge arrangements were considered, including cable-stayed, concrete segmental box girder, and the "hybrid" cable-stayed/extradosed. Selection criteria was established based on weight, seismic performance, constructability, and cost-effectiveness. Based on close cooperation between designers and construction experts, the continuous, "hybrid" solution was selected [6].

The geometric constraints imposed by the two navigation channels led to three main spans: an over-the-river span of 242 m and two 121 m end spans. Although a true cable-stayed arrangement would have provided a very good solution for these span requirements, the tower height for a cable-stayed bridge would have encroached on the flight path for the nearby Pitt Meadows Airport. A fully extradosed bridge would have resulted in acceptably low towers but the required single 242 m span was considered to be uneconomical for a true extradosed bridge.

Consequently, a hybrid bridge form was developed that combined the lightweight superstructure advantages of a cable-stayed bridge with the compact towers of an extradosed bridge. The combination of the parallel harped cable-stays and the low profile towers that do not need crossbeams above the deck provides a clean, aesthetically pleasing view.

A flexible pier concept was used to address the project's seismic requirements given the site's weak soils and the consequent potential for liquefaction at the site. Pier flexibility was provided by splitting the pier into two walls separated in the longitudinal direction [6]. This pier arrangement provides a ductile substructure that for seismic loads helps isolate the superstructure from the foundations. The piers are designed to perform elastically with plastic hinges at their top and bottom sections. They are designed for the 1:475 earthquake and to perform inelastically for the 1:2500 earthquake.

An asymmetrical vertical curve was required for the road design to accommodate the main navigation channel. This resulted in a significant difference in the height of the main bridge piers, with a resultant difference in stiffness. The shorter piers, because they are much stiffer than those directly adjacent to the navigation channel, would be disproportionately at risk during seismic events. To solve this problem a permanent steel "hinge" detail that effectively acts as a pin at the bottom of the pier was developed [6].

The underlying deep soft silt and clay deposits posed a significant settlement risk [6]. To maintain an efficient design, it was found that differential settlements could not be allowed to exceed 250 mm. To control differential settlements, a system was developed to allow the bridge towers to be raised using hydraulic jacks. This system consists of 3 m deep reinforced concrete settlement slabs installed under the towers. These slabs are post-tensioned to the pile caps and if excessive differential movement occurs, the post-tensioning can be released and the settlement jacks jacked to compensate for the settlement. The gap between the pile caps and the settlement slabs would then be grouted and the settlement slab re-post-tensioned to the pile caps.

Geotechnical investigations indicated that the site is underlain by weak soils and, notwithstanding installation of very deep boreholes, neither till nor bedrock was found. To deal with these challenging geotechnical conditions, the foundation design solution required the installation of Φ 2.3 m to Φ 2.5 m bored cast-in-place concrete piles to depths of between 40 m and 89 m. The upper portions of these piles were cased with steel pipes while grab drilling was carried out. The lower, noncased lengths of the piles were supported using a polymer suspension while they were cleaned out. Prefabricated reinforcement cages with weights of up to 82 tonnes were installed in the excavated piles and connected to previously installed parts of the cage prior to backfilling with tremie concrete. Sonic measurements were taken on each pile to ensure the absence of voids [7].

The bridge is owned by TransLink. The design-builder consisted of a joint venture between Bilfinger Berger and CH2MHill. Buckland & Taylor provided the design services for the bridge and its approaches.

1.4.3 Concrete Segmental Construction

1.4.3.1 Calgary West LRT Extension

This project consists of a $700 million, 8 km extension to the Calgary LRT Transit System between Calgary's downtown area, to 73rd Street SW in the western part of the city. A key element of the new system is the 1.5 km long elevated guideway, including a balanced cantilever bridge as well as various bridges required to separate the new LRT system from existing facilities. In general, a high architectural content was required by the city for all bridge design.

The elevated guideway section extends 1.5 km from downtown Calgary to Bow Trail. The construction of the elevated guideway included working in close proximity to major roadways and alongside and over the Canadian Pacific Rail; this involved detailed coordination between numerous stakeholders and companies. The elevated section is 9.6 m wide, the span-by-span section is 1100 m long, and the balanced cantilever section is 212 m long.

The elevated guideway was constructed using span-by-span segmental construction and consists of 36 piers, one bent structure, and 444 precast segments. Segment production began in April 2010 with seven precast, short-line forms. The precast segments were erected in place using a launching truss. Segment erection began in July 2010 and was completed by March 2011. Both the span-by-span and the balanced cantilever precast segments were being erected with the 110 m long truss. Certain sections of the cantilever structure were erected with a conventional crane. Segment weights varied from 25 to 54 tonnes. Segments were loaded on low-boy trailers and shipped 15 km from the precast yard (see Figure 1.20).

FIGURE 1.20 Span-by-span construction of the precast segmental elevated guideway structure. (Photos courtesy of MMM Group.)

FIGURE 1.21 Balanced cantilever construction for the Calgary West LRT Project. (Photo courtesy of MMM Group.)

Both the design and the construction of the structures were finished ahead of schedule. Design challenges included special loads introduced by continuous direct-connected rail, drilled shaft deep foundations, aesthetic requirements from the city of Calgary, and the need to construct significant parts of the structures in winter. Due to Calgary's cold winter temperatures, materials suitable for this climate were used. In particular, the epoxy used in the segment joints and certain launching truss components were selected to operate at temperatures of up to −15°C (see Figure 1.21).

The project was awarded to SNC-Lavalin in late 2009 as a design-build project. Construction started in March 2010 and is scheduled to be complete in early 2013. The elevated guideway was constructed by SLG—a joint venture formed between SNC-Lavalin Constructors (Western) Inc. and Graham Construction. Design management and design for the elevated structures was provided by MMM Group.

1.4.3.2 Tsable River Bridge

The Tsable River Upstream Bridge is a 400 m long, 54 m tall highway bridge crossing the Tsable River Valley on Vancouver Island near Courtenay, British Columbia. It was built as part of the new 128 km four-lane Inland Highway (Highway 19) (see Figure 1.22). Initially it was thought that a steel bridge would provide the most economical solution given the bridge's somewhat remote location and its height above the valley. The decision to undertake a dual design was consequently made reluctantly. Notwithstanding, the British Columbia Ministry of Transportation commissioned and tendered two complete bridge designs, one in concrete and one in steel, for this major bridge crossing. The lowest-priced design was the concrete option, based on six independent bids.

In addition to dealing with seismic conditions and economic considerations, construction impact on the pristine local environment had to be minimized. For example, an old-growth forest of Douglas firs fills the valley, some of the trees reaching above the roadway level of the bridge. Moreover, the river flowing along the valley bottom is a vital salmon resource. Compounding these constraints, the valley sides are very steep and the terrain is potentially unstable. In addition the area is home to Roosevelt elk, black bears, cougars, cutthroat trout, steelhead, and several species of salmon. Hummingbirds, warblers, bald eagles, and blue herons live along the banks.

The bridge's span arrangement was affected by restrictions on the location of the bridge foundations as specified by the environmental regulatory agency. These limitations, together with the steep terrain

FIGURE 1.22 Tsable River Bridge during construction. (Photos courtesy of MMM Group.)

and potential instability of the ground, made selecting the optimum pier locations and span arrangements very difficult and consequently the chosen design has long spans to minimize the number of foundations. A four-span arrangement of 82 m–118 m–118 m–82 m was found to provide an optimal solution given the site constraints.

Various superstructure types were considered. These included cable-stayed box girders, cellular arch configurations, and trapezoidal box girders. The last option was found to be the most economical. A cast-in-place segmental, balanced cantilever method was selected to allow the superstructure construction to proceed above the sensitive valley slopes and river, and have little or no impact on the valley throughout most of its length.

Given that this tall concrete structure is in an area of seismic activity (peak ground acceleration 0.33 g) the superstructure's weight was minimized and it was found that a single-cell configuration provided the most economical solution. It is one of the largest single-cell box structures in North America. Measures taken to reduce the superstructure weight included:

- Elimination of a central web
- Incorporation of stiffening ribs into the undersides of the deck and deck cantilevers

The use of post-tensioned concrete in a single-cell box girder bridge results in a structure which is inherently very durable and will require less maintenance than steel. Moreover, using a single cell as opposed to the more typical double cell construction simplified the construction.

Designing and building this major crossing with the associated environmental, geotechnical, seismic, and site topographical constraints was a major challenge. The design of the superstructure had to account for secondary effects, including complex creep and shrinkage movements associated with the segmental, time-dependent nature of the construction. During the building phase it was important to carefully monitor the deflections of the cantilevers and to calculate compensating formwork settings in order to ensure that the cantilevers met at the centers of the spans.

Since the project was completed, any vegetation that was affected has begun to grow again and the bridge harmonizes well with its surroundings. Ironically, the environmental success of the structure prevents its aesthetic success from being fully seen. An observer cannot appreciate the scale and slender beauty of the crossing in its entirety because too many of the tall trees remain and prevent an overall panoramic view. The bridge is owned by the Province of British Columbia and was designed by MMM Group (formerly ND Lea) and TY Lin.

1.4.3.3 Confederation Bridge

Confederation Bridge is 12.91 km long and provides a two-lane highway link for the Trans-Canada Highway across the Northumberland Strait between the provinces of Prince Edward Island and New Brunswick (see Figure 1.23). A number of important project requirements had to be addressed in the design and construction of the project, including the following [8]:

- A fixed opening date was specified in the contract that required an aggressive design and construction schedule.
- A 100-year design life was specified.
- Progressive collapse of the adjacent spans had to be avoided in the event of the loss of a single span.
- Given the climate at the site, a maximum construction season of 8 months is available.
- Environmental impacts had to be minimized.
- Ice bridging between the piers was not permitted as this would affect ice flows in the St. Laurence River.
- Ice loads in the order of 25 mN had to be resisted.
- A 49 m × 200 m navigation channel had to be accommodated.
- Significant winds occur at the site, potentially posing a safety issue for bridge users. This concern needed to be addressed in the design and operation of the bridge.

A multispan post-tensioned concrete box girder structure was developed to meet these criteria. The bridge consists of:

- Seven approach spans on the Prince Edward Island side with a total length of 555 m
- Forty-five marine spans with a total length of 11,080 m
- Fourteen approach spans on the New Brunswick side with a total length of 1275 m

The design and construction of the bridge had to account for severe weather conditions, and the bridge is the longest structure across ice-covered waters. Ice loading was in fact one of the major design considerations and solutions developed to accommodate these very significant loads led to a number of the innovations for which this bridge is known [8].

1.4.3.3.1 Approaches

The approaches to the main bridge span over relatively shallow water that limited the size of marine equipment that could be used for construction. As such, approach span lengths were limited to 93 m. A haunched, trapezoidal box girder was used for the approaches with a structural depth of between 3.0 m and 5.06 m. Spans were supported on hollow rectangular piers. Conical pile caps were developed

FIGURE 1.23   Confederation Bridge. (Photo courtesy of MMM Group.)

to mitigate ice loads and piles were designed to withstand lateral ice loads. A temporary causeway was constructed to allow the approach foundations and substructures to be constructed within cofferdams. Launching gantries were used to construct the superstructure [8].

1.4.3.3.2 Main Bridge

Given the length of the crossing and the cost associated with marine construction, the span length of the main bridge was maximized. Spans of 250 m were found to be feasible, efficient, durable, and economical. Transition spans of 165 m were used between the main bridge and the approach spans. In addition to minimizing the scope of marine works, the long spans provide considerable mass on top of the piers to help resist lateral ice loads.

The main bridge spans are haunched, trapezoidal box girders which vary in depth from 4.5 m at midspan to 14 m over the piers. The bottom flange of the box is 5 m wide and the clear distance between the box webs at the top of the box is 7 m. Φ 8 m hollow hexagonal piers were used with a maximum height of almost 46 m. Foundations consist of footings on bedrock [8].

The design concept for the bridge was developed to allow rapid construction. This was achieved by prefabrication of almost all elements, minimization of the number of different precast elements required, and maximization of the size of the precast elements such that on-site time was minimized. This strategy included the following:

- Only four field sections were required, namely the pier base, pier shaft/ice shield, the main span cantilever segment, and the drop-in segment. The cantilever segments were 192.5 m long and connected to each other using drop-in spans.
- Field segments were connected using field cast joints and continuity post-tensioning.
- Standardized precast elements were fabricated to assemble the field sections in the casting yards.
- Large marine cranes were used to allow very heavy elements to be placed (see Figure 1.24).

The bridge is a series of portal frames given that the superstructure is rigidly connected to the piers. Provision for thermal movement was provided at midspan of certain spans by using a modified drop-in section.

FIGURE 1.24 Heavy lift vessel Svanen. (Photo courtesy of Ballast Nedam.)

All bridge components were constructed on land, in purpose-built staging yards. High-performance, 55 MPa concrete with fly ash, silica fume, and other admixtures was used to meet the required 100 year design life. Considerable effort was expended on the development of the concrete mix design and quality control/assurance. Conventional reinforcement was used rather than epoxy coated, galvanized, or other specialized reinforcing steel.

The bridge was designed with several safety features, including strictly enforced speed limits, a road surface made of long-lasting bituminous mixture that minimizes vehicle spray during wet weather, and more than 7000 drain ports that allow rainwater and slush to run off the bridge. Twenty-two closed-circuit television cameras and a crew of dedicated personnel provide full surveillance of the bridge 24 hours a day, and traffic signals, emergency alarms, and call boxes are supported by an uninterruptable power supply.

The bridge was built using a public-private partnership procurement model. The design and construction was carried out by a joint venture of Ballast Nedam, GTMI (Canada), Northern Construction, and Straight Crossing Inc. Bridge components were fabricated between 1994 and mid-1996, and placement of components began in fall 1994 until late 1996. Approach roads, toll plazas, and final work on the structure continued until the spring of 1997. The project was constructed for approximately $1 billion. Current stakeholders in the bridge concession are OMERS, VINCI Concessions Canada Inc., BPC Maritime Corporation, Strait Crossing Inc. (Calgary), and Ballast Nedam Canada Limited.

1.4.4 Steel Girders

1.4.4.1 New Park Bridge

The Province of British Columbia and the federal government of Canada have undertaken a multiyear program to upgrade the section of Trans-Canada Highway between Cache Creek in British Columbia and the Alberta border. This section of highway runs through the Rocky Mountains and was originally constructed in the 1950s. It was identified as having a number of substandard alignments and higher than average accident rates. The highway provides a vital link for Canada's Prairie Provinces to the western sea board and is as such of strategic importance to Canada's economy. About 10,000 vehicles per day use the road, almost 25% of which are trucks.

As part of the highway upgrading program, a new high-level bridge, the New Park Bridge, was required to replace the existing low-level bridge over the Kicking Horse River. The New Park Bridge is a 404 m long bridge that is about 95 m above the Kicking Horse River. The bridge is on a curved alignment and on a steady 5.921% grade descending from east to west. The bridge deck is 23 m wide and is super elevated at 5.5%. The superstructure consists of multispan constant depth continuous steel plate girders with a composite concrete deck. A span arrangement from west to east of 50 m–70 m–70 m–80 m–80 m–54 m was used to meet site constraints and balance the structure. Site constraints included geotechnical conditions, limits to in-stream works, and an active railway line (see Figure 1.25).

An important element in the success of the project was an optimization process to find the best balance between bridge length, east embankment height, and the depth of rock cut on the west approach. It was found that a 405 m long bridge with a 30 m high west embankment and a 90 m deep rock cut on the west approach provided the most cost-effective solution.

The bridge cross section was arranged to allow an efficient construction operation. In this regard, three main girder lines were provided with intermediate girder lines (see Figure 1.26a). This arrangement minimized steel quantities and allowed the use of transverse precast stay-in-place formwork spanning between the main and intermediate girder lines. The bridge foundations typically required up to 16 Φ 900 piles per pier. Some of those piles had to be bored and steel cased for up to 45 m through overburden before being socketed 8 m into the bedrock.

The Park Bridge was incrementally launched through a curve (see Figure 1.26b and c). This approach significantly reduced the construction time and cost. The girders were fabricated in at different plants

FIGURE 1.25 Construction of the New Park Bridge. (Photos courtesy of Flatiron Constructors.)

(a) (b)

(c)

FIGURE 1.26 Construction of New Park Bridge: (a) steel superstructure, (b)–(c) launching.

across Canada, assembled behind the down slope abutment and, with the use of a launching nose, pushed out over the Kicking Horse Canyon. The launching hardware included hydraulic jacks, rollers, and sliding plates that allowed the girders to be moved at a rate of 20 m/h. The launching operation was completed in 115 days and was followed by the installation of partial depth, precast concrete planks that were made composite with the girders and provided with a cast-in-place topping.

The project was completed 21 months ahead of schedule and opened in August 2007 for a total cost of $143 million. The project was done as a public-private partnership between the Trans Park Highway Group (TPHG) and the British Columbia Ministry of Transportation. On behalf of the TPHG Group, Belfinger Berger provided concessionaire services. Flatiron Constructors was the design-builder. Parsons Transportation with Delcan, Stantec, and Golder Associates provided engineering services.

1.4.5 Movable Bridges

1.4.5.1 Johnson Street Bridge

In 2009 the city of Victoria decided to replace two existing Strauss bascule bridges located on the edge of the city's downtown core. Before making this decision, investigations and preliminary designs were prepared to compare the cost of rehabilitating and strengthening the existing bridges with the cost of building a new architecturally significant bridge. It was found that either option would have similar costs and that new construction would have fewer unknowns than a rehabilitation project. The decision to replace the existing bridges was endorsed by a referendum. Important requirements for the new bridge included:

- Realignment of the approach roads to eliminate an "S" curve and eliminate pedestrian/cyclist/vehicular conflicts at the east bridge head
- Construction of a new bascule bridge to carry a 5 m multiuse trail, three 3 m lanes of traffic, two 1.8 m wide on-street cycle lanes, and a 2.5 m wide pedestrian sidewalk
- Lifeline seismic performance
- Increase of the navigation channel width to meet the requirements of the Navigable Waters Act
- Integration with the existing and proposed adjacent path and trail systems
- Design in accordance with current accessibility requirements
- Provision for a future 5 m wide rail corridor
- Decommissioning of the existing bridges

The replacement bridge design (see Figure 1.27) was developed in the context of the city's Old Town Design Guidelines as well as with respect to the historic nature of the site. The new bridge design was developed to provide view corriors of the old town, the Upper Harbor, and the Inner Harbor that are currently blocked by the existing bridge superstructure and counterweights. The design of the new bridge reflects the truss and heavy construction of traditional railway bridges and as such provides a memory of this important historical element of the site while providing for a new, modern design.

In developing the replacement bridge concept, it was decided that a single leaf bascule bridge would be appropriate from an urban and architectural perspective. Three concepts were developed and presented to the public. Concepts included a cable-stayed option, an option with an overhead counterweight, and a truss (see Figure 1.28). The truss option was selected for design and implementation.

The replacement bridge will have three spans: a west approach, the bascule span, and the east approach. An inclined rest pier is proposed, while the east pier provides the counterweight pit and houses the bearings and motors required to drive the bridge. A corbel on the east pier provides support for the east approach span. The bascule superstructure comprises a tapered truss connected to a Φ 12 m wheel that provides for the required rotating movement.

FIGURE 1.27 Proposed Johnson Street Bridge. (Image courtesy of Wilkinson Eyre Architects.)

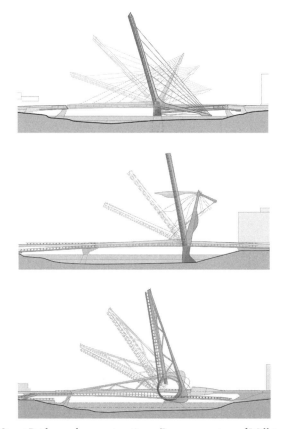

FIGURE 1.28 Johnson Street Bridge replacement options. (Images courtesy of Wilkinson Eyre Architects.)

The bascule spans needs to provide for a 41 m wide navigation channel. Thus, a clear distance of 45 m was established between substructures. To accommodate these dimensions, the bascule span measures 51 m between its tip and the center of rotation and about 65 m between the tip and the end of the counterweight. The counterweight is located below the deck and attached to the wheel. Traffic drives directly over the counterweight. The wheels are connected under the deck to provide lateral stiffness. Bearings mounted in the counterweight pit support the wheel and allow it to rotate. The motors and drives are attached to the tail of the counterweight and "walk" down a rack that is mounted in the counterweight pier to open the bridge (see Figure 1.29). In addition to being architecturally important, the "lobe" on the wheel is necessary to adjust the bridge's center of gravity.

The bridge deck is separated into three distinct decks: the road deck, the multiuse path deck, and the sidewalk (see Figure 1.30). The road deck on the bascule span will be constructed using a steel orthotropic plate supported by transverse floor beams while on the approaches a concrete deck will be used. The orthotropic deck will be provided with an epoxy asphalt wearing surface. Aluminum planks will be used for the path and sidewalk decks. In addition to the orthotropic deck, a number of deck options were considered for the bascule span. These included fiber-reinforced plastic (FRP), exodermic, concrete-filled grillages, and open grating decks. An evaluation based on weight, initial cost, and life cycle cost was made and the orthotropic solution was identified as most appropriate.

The approach spans were designed to be similar to the bascule span from an architectural perspective. This was achieved by using the same transverse floor beam arrangement as used for the bascule span. Steel edge beams are used to bring floor beam loads to the abutments and piers.

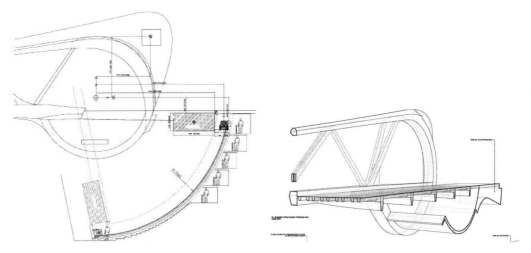

FIGURE 1.29 Johnson Street Bridge mechanical and counterweight arrangements. (Images courtesy of Wilkinson Eyre Architects.)

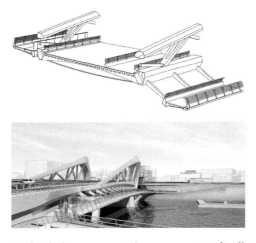

FIGURE 1.30 Johnson Street Bridge deck arrangement. (Images courtesy of Wilkinson Eyre Architects.)

The site is underlain by silts and clays that were deposited on bedrock. The bedrock elevation varies from being at-grade at the bridge's west abutment to being about 20 m below water at the middle of the navigation channel. As such, spread footings are required for the west abutment and pier whereas drilled, large-diameter shafts are required for the east pier and east abutment. The bridge is currently being designed for the city of Victoria by MMM Group, together with Wilkinson Eyre Architects and Stafford Bandlow. Construction is scheduled to be complete in 2016.

1.4.6 Pedestrian Bridges

1.4.6.1 Humber River Pedestrian Bridge

The Humber River Pedestrian Bridge was conceived to be a gateway between the city of Toronto and the neighboring city of Etobicoke. The bridge was also designed to be an important marker for the place on the Humber River where the Ojibwa First Nation used to spend winters and begin their summer hunting expeditions. To reflect this element of the bridge, the infill plates on the upper lateral arch bracing were designed to resemble the thunderbird, one of the important symbols of the Ojibawa people. From a technical perspective, the bridge has a number of unique and innovative characteristics.

The Humber River Pedestrian Bridge is over 130 m in length, with a span of 100 m between thrust blocks. It is a tied arch structure, with architectural abutments, lighting, stainless steel hangers, and other unique structural features. It incorporates a number of historic and cultural features by means of unique iconography. The arches themselves are slender tubular ribs, inclined inwards.

A post-tensioned concrete deck provides the tension tie for the arch. The horizontal thrust from the arch ribs is transferred through the abutments up to the deck, where it is equilibrated with the deck post-tensioning. A thrust arch was not possible at the site given that the site is overlain by about 40 m of very weak soils. Forty m long, Φ 1.22 m driven pipe piles were used with rock sockets. The steel fabrication of the Φ 1.22 m diameter arch ribs was affected using a carefully fabricated saddle and heat treatment of the steel.

Erection of the bridge was also noteworthy. The entire arch was erected on land, and then using a track, guided towards the water with two large cranes, where one end of the arch was placed on a barge. The barge was then winched across the river until both arch ends could be hoisted onto the preprepared anchors. During this operation a temporary tension tie was used. This temporary arrangement was released as the deck post-tensioning was installed. A silica fume concrete deck was cast under very strict quality-control procedures to ensure the long-term durability of the deck. Stainless steel hangers and architectural lighting add to the beauty of this bridge (see Figure 1.31). The bridge was designed by an integrated team consisting of Delcan Corporation, Montgomery Sisam Architects, Ferris and Quinn Landscape Architects, and Brad Golden (public artist).

1.4.6.2 Mimico Creek Bridge

This bridge is based on a similar structure constructed in Spain and designed by Dr. Santiago Calatrava (see Figure 1.32). The bridge fits gently in the landscape of Mimico Creek, which is characterized by wetlands, water birds, and riparian vegetation. The landscape is intimate and is a gem of green space in the city of Toronto.

The bridge is a link in Toronto's Waterfront Trail system and is used by pedestrians, cyclists, and roller bladers, and is fully accessible for people who have difficulty with mobility. As part of the project, an urban artist was engaged to help with the integration of the bridge into the landscape. Technically, the bridge is a finely tuned structural sculpture that is held in equilibrium by the torsion tube that connects the arch rib and the deck's steel framing. Torsion is further equilibrated at the abutments. Out-of-plane forces are introduced into the deck by the inclined arch hangers and are equilibrated by

FIGURE 1.31   Humber River Pedestrian Bridge. (Photos courtesy of Delcan Corporation.)

FIGURE 1.32   Mimico Creek Bridge. (Photos courtesy of Delcan Corporation.)

an in-plane truss located in the deck framing system. Design of the bridge required consideration of second-order effects to accurately model the structural behavior.

The bridge deck is finished with a stainless steel handrail and a timber deck that was fitted with non-slip strips. The bridge foundations include Φ 600 mm diameter steel pipe piles socketed into the underlying bedrock. The superstructure is a structural steel space frame in the form of a single inclined arch. The main structural elements of this space frame include a Φ 762 mm torsion tube, a Φ 273 mm edge tube, and a Φ 273 mm arch tube. These main structural steel elements are connected by means of a number of floor beams, bracing members, and struts using a combination of welded and bolted connections.

The bridge was designed so that it can be erected in pieces or assembled in one piece and subsequently lifted into its final position by means of cranes. The contractor elected to fabricate the bridge in four pieces, to move these four pieces to the site, to carry out field assembly of the entire bridge into a single unit on shore, and finally, to erect the bridge in a single piece using a crane with a 600 tonne capacity. The entire steel structure weighed 50 tonnes. The bridge is owned by the city of Toronto and was designed by Delcan Corporation together with Calatrava SA.

1.4.6.3 Recent Footbridge Design Competitions

Recently there have been a number of very interesting design competitions in Canada for footbridges. Of particular note were those for a new crossing of the Bow River in Calgary and for a new bridge across the railway tracks at the Sky Dome in downtown Toronto. The following describes one submission from the Toronto competition.

The bridge structure for the span over the rail yards (see Figure 1.33) is derived from the rhythm of the rail tracks and uses a multiplicity of lighter structural elements acting together in a complex pattern to create its support. As with a box truss, the depth of the structure is used for support and inhabited, however in this case the structure is more fragmented and shaped into a more contemporary form. The structure responds to span lengths by adding depth at midspan where the moment of the span is highest. It is completely fabricated from plate steel, which has an easy translation into built form using current fabrication technology; plate steel can be cut into any form required by the computer-aided design (CAD) files provided from the architectural modeling software.

Approaches to the bridge have been placed on fill. This is an economic as well as a design decision. The fill may necessitate a small amount of retaining wall, however it is generally much more economical in bridge building than structure. The approaches have further been shaped to allow an easy transition on bicycle from the line of travel of the approaches to the direction of the bridge. A shorter stair entrance is also found on either end of the structure for those users who may choose a shorter approach to the structure. The structure extends out over the traveled pedestrian way and in this way the structure becomes a land feature and defines a space for those who are passing the bridge (see Figure 1.34).

Enclosure has been designed with meshes, one denser than the other, to provide patterning of the structure. The meshes allow air passage, keeping the enclosed space from becoming stagnant. They

FIGURE 1.33 Toronto Pedestrian Bridge-Bird View. (Image courtesy of Infrastructure Studio.)

FIGURE 1.34   Toronto Pedestrian Bridge Passing View. (Images courtesy of Infrastructure Studio.)

also allow rain to penetrate and wash the deck and in this way require less maintenance than a glazed enclosure. The mesh is also vandal resistant and easy to clean.

Erection will take advantage of the staging area near the site. The structure will be primarily shop-fabricated steel shipped to site in segments and then field welded into larger parts, which can then be erected by crane. The final craned pieces could be bolted into place with overlapping members, if necessary, rather than field welded. The deck could be cast after the steel structure has been erected, saving lifting capacity on the cranes.

Architectural lighting was included. The lighting accents the structure, provides a night time presence, and provides additional safety. This design was developed by Infrastructure Studio, a Vancouver-based engineering and architectural practice.

1.4.6.4 Winston Street Overpass

This $5 million bridge is part of the Greater Vancouver area's greenway and spans a railroad corridor and urban arterial road. It is also envisaged that, as the area is developed, the bridge will form the heart of an urban village. The Greater Vancouver Greenway is being developed as part of Canada's efforts to reduce greenhouse gas emissions in accordance with the Kyoto Accord and is aimed at encouraging alternate, personally powered forms of transportation. Thus, the bridge was constructed using dimensions suitable for bicycling, roller blading, and walking [9].

A rigorous analysis of site conditions, establishment of design criteria, and research into sustainable construction materials was undertaken at the commencement of the project in order to develop a bridge configuration that reflects the intent of the project and the character of the site. Important considerations in this regard included:

- Soft, compressible soils
- Grades appropriate for wheelchair access
- Integration with a highly urban environment on one side of the project and with a very natural environment on the other
- Seismic and dynamic loads on the bridge
- Construction over railway tracks and a major urban arterial road
- A clearly defined budget
- Use of environmentally sustainable materials
- Environmental restrictions posed by adjacent fish habitat
- Limited opportunities for pier locations

Numerous concepts and structural systems were developed for the project, including a very interesting woven "cocoon" concept (see Figure 1.35).

The final bridge arrangement is characterized by relatively simple precast concrete approach spans and a more dramatic main truss span of 75 m (see Figure 1.36). Precast approach spans were developed to be either straight or with a constant radius to make the design more cost-effective. The truss is unique in that the panel points on either side of the deck are off-set to provide an interesting experience for people on the bridge. The bridge is located in very poor soil and in a very highly seismic zone. Foundation and superstructure design was integrated to ensure life-safety under the design earthquake. The truss piers are steel pipes arranged like trees [9].

The superstructure was developed to maximize repetition in truss member dimensions and incorporate standardized joints between members. Painted steel was used for architectural and durability reasons. The truss panels are offset from one another on either side of the bridge to give the impression of a "woven" structure. Seven piers were used. Each pier consisted of a single Φ 910 mm driven pile. Closed-end piles were used. Piles were filled with concrete and have an approximately 600 mm pipe embedded in their tops to provide very simple and flexible piers. Each pier has a cross head that is fitted with simple neoprene bearings to receive the superstructure (see Figure 1.37).

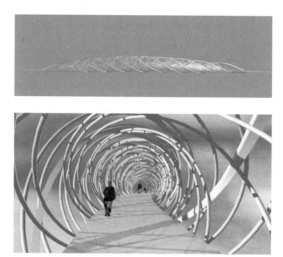

FIGURE 1.35 Early concept for the Winston Pedestrian Bridge. (Images courtesy of Patkau Architects.)

FIGURE 1.36 Winston Street Overpass. (Images courtesy of Patkau Architects.)

FIGURE 1.37   Construction of Winston Street Overpass.

At the south end the bridge is terminated at a 30 m long, inverted U-shaped abutment. This arrangement ensures minimum weight and therefore minimizes settlement of the approach. At the north end, the truss structure is terminated with about 2.4 m clearance between existing grade and the truss soffit. At this end the existing grade was used together with a retaining wall and backfill to provide the required approach and the bridge end can be set on a pile-supported strip footing. The abutment structure was blended with the existing grade and a curb let-down was provided to facilitate access to the adjacent LRT station. The bridge is owned by the city of Burnaby and design was carried out by Delcan Corporation and Patkau Architects.

1.4.7 Heritage Conservation

1.4.7.1 Kinsol Trestle

The Kinsol Trestle, which crosses the Koksilah River near Shawnigan Lake on Vancouver Island, was built by the CNR in 1920 and renewed in stages until 1958. The bridge was constructed using heavy Douglas fir timbers and carried both passenger and freight traffic, primarily logs and forest products. CNR discontinued rail service in 1979. The trestle has been unused since then, and was in a very deteriorated condition.

The crossing of the Koksilah River is the only missing link in the Cowichan Valley Trail between Shawnigan Lake and the town of Lake Cowichan. This recreational pedestrian, equestrian, and bicycle path, much of which follows the CNR right of way, forms a part of the Trans-Canada Trail. The Cowichan Valley Regional District (CVRD) determined that the trail will cross the Koksilah River at the site of the Kinsol Trestle, whether by restoring the historic trestle or by removing it and building a new crossing. The CVRD decided to rehabilitate and preserve the existing timber structure using best heritage conservation practices and by installing a new superstructure that will be supported by new timber to provide a pedestrian, equestrian, and bicycle crossing of the Koksilah River (see Figure 1.38).

FIGURE 1.38 Reconstruction of the Kinsol Trestle. (Images Courtesy of MMM Group.)

The existing trestle is a multispan, timber trestle structure built in plan to a 7° curve (R = 248.48 m). The structure has a total length of 187.15 m and consists of 46 bents, including the two abutment bents. Bents consist of 30 ft (9.14 m) high stories constructed using braced 12 in × 12 in posts and a 12 in × 12 in sill timber between stories. Longitudinal and diagonal bracing was originally provided between bents. Bent heights vary between 0 m at the abutments to about 41 m at the river. The river crossing consists of a Howe truss with a clear span of 29.06 m and a structural depth of approximately 9 m. There are six truss lines in the Howe truss that support nine bents across the river.

A new superstructure was provided supported by two new abutments and by "active bents" consisting of new timbers. The new superstructure was a deck truss consisting of atmospheric steel with a timber deck. Bents that are not used to carry loads from the new superstructure are labeled as inactive bents. These as well as the Howe trusses were restored but not to carry load. The project was carried out using an alliance contract to provide the owner with a guaranteed maximum price and cost certainty for what would normally have been a very risky undertaking. The refurbished trestle was opened to the public on July 26, 2011 (see Figure 1.39).

The bridge is owned by the province of British Columbia and operated by the Cowichan Valley Regional District. The design was carried out by a team consisting of MMM Group, Commonwealth Historic Resource Management, Jonathon Yardley Architects, Macdonald & Laurence, Cascade Engineering, Stantec, and Rysck Geotechnical. Construction was implemented by Knappet Construction.

FIGURE 1.39   Reconstructed Kinsol Trestle. (Image Courtesy of MMM Group.)

1.4.8 Canadian Bridge Engineers Overseas

Canadian bridge engineers have been active overseas for many years. For example, Buckland & Taylor have provided services for a number of complex bridges around the world, including the Rama 8 Bridge in Thailand and the Sheikh Zayed Bridge in the United Arab Emirates. MMM Group also has a long history of overseas work and has recently completed a major bridge project in Nepal and are representing owner's engineer for a major cable-stayed bridge in Slovania.

Another example of Canadian bridge engineering overseas is the recent series of projects of iconic bridges in China, which were delivered by a Dutch-Canadian team in partnership with a local Chinese design institute [11]. These projects were developed and won as part of an international design competition and have included bridge trusses, arches, and suspension bridges. In particular these bridges were

- Tongnan Bridge (Tianjin): A spine-like truss with a cantilevered orthotropic steel deck
- Liulin Bridge (Tianjin): A two-span cable-supported bridge consisting of a deck suspended from three half arches
- Phoenix Bridges (Guanzhou): Three unique arch structures with spans ranging from 100 m to 400 m and with helix-like and inclined arch arrangements
- Tuanbo Bridge (Tianjin): A suspension bridge with a 500 m main span and inclined, centrally arranged towers

These bridge designs were developed by Verburg Hoogendijk Architects, and Delcan Corporation.

1.4.8.1 Tongnan Bridge

The bridge consists of a spine-like central truss that supports two parallel steel box girders separated by a 3 m wide opening along the length of the bridge. The boxes are 8.6 m wide with a maximum depth of 1.35 m. The boxes are provided with internal webs and the top plate will consist of an orthotropic plate [10] (see Figure 1.40).

Transverse ribs spaced at 4 m connect the two boxes and extend to the edge of the sidewalk. A central, 13 m deep truss will be engaged by the ribs to limit main span deflections. The truss consists of tree-shaped vertical posts. Hangers are provided to connect the edge of the deck with the top of the truss posts. Truss panels were 16 m long.

FIGURE 1.40 Tongnan Bridge.

FIGURE 1.41 Luilin Bridge. (Image courtesy of Verburg Hoogendijk Architects.)

1.4.8.2 Luilin Bridge

The Luilin Bridge has two 87 m spans. The form of the bridge was developed to mimic a dragonfly and consists of an arch that has been cut in half and separated. The ends of the arch are supported by props and the arch ribs are referred to as "flying girders" [11] (see Figure 1.41). Load on the deck is carried to transverse ribs that are suspended from cables that carry the load into the flying girders. Cables are spaced at 5 m and discontinued 20 m from the centerline of the central pier. The flying girders are propped at one end and fixed at the central pier to give them a span of about 77 m. The flying girders extend past the prop by about 10 m and cables are provided in the extension.

1.4.8.3 Phoenix Bridges

The Phoenix Bridges were developed as a series of three arch structures over the Jiaomen and Lower Hengli waterways in Guongzhou. In Chinese mythology the phoenix symbolizes the feminine, and thus elegant, flowing arches were chosen for these bridges. The bridges are the Oscillation Bridge over the Jiaomen Waterway, the Kinetic Bridge over the Upper Hengli Waterway, and the Inseparable Bridge over the Lower Hengli Waterway [11] (see Figure 1.42).

FIGURE 1.42   Phoenix Bridges. (Images courtesy of Verburg Hoogendijk Architects.)

One goal of the design was to ensure that the separate functions and characters of the bridges would be clearly identifiable yet harmonious with one another. The main structural element of each bridge is an elegant and "light" arch. To make sure that this character is conveyed it was proposed to use steel coated with white paint. The deck is prefabricated concrete to enhance quality and smoothness and give the bridges a solid and sturdy appearance. The bridges will carry a light rail system, highway traffic, and bicycle/pedestrian users.

1.4.8.4 Tuanbo Bridge

The Tuanbo Bridge was designed to carry vehicular traffic, bicycle and pedestrian traffic, and LRT loads. The main bridge consists of two towers, a three-span continuous precast box girder, two main cables, two side stabilizer cables, saddles, ground anchors, and cable hangers [11] (see Figure 1.43). The towers are built up using arch structures and the cross section of each tower leg is a nominal 5.0 m × 2.5 m hollow rectangular section of reinforced concrete.

The two main bridge cables are 500 mm diameter each and are located at each side of the LRT line, which is located along the centerline of the bridge. The cables are thus approximately 11 m apart. The deck is suspended from the main cable with double cable hangers. These hangers are vertical and have a diameter of 80 mm. The hangers are spaced at 25 m in the longitudinal direction.

FIGURE 1.43 Tuanbo Bridge. (Image courtesy of Verburg Hoogendijk Architects.)

References

1. British Columbia Ministry of Transportation, "Concept Advice Report for Pitt River Bridge," 2006, Victoria, BC, Canada.
2. Spoth, T. and J. Viola, "Autoroute A-25 Project—Main Bridge Innivations," IABSE Symposium, Large Structures and Infrastructure for Environmnetally Constrained and Urban Areas, 2010, Venice, Italy, pp. 77–84.
3. British Columbia Ministry of Transportation, "Concept Advice Report for the New Port Mann Bridge," 2008, Victoria, BC, Canada.
4. Port Mann Highway Improvement Project, *Construction Update*, Summer 2010, Coquitlam, BC, Canada, 4 pp.
5. Griezic, A., C. R. Scollard, and D. W. Bergman, "Design of the Canada Line Extradosed Transit Bridge," 7th International Conference on Short- and Medium-Span Bridges, Montréal, 2006.
6. Welch, R., "Golden Ears Bridge," *Canadian Consulting Engineer*, June 6, 2010 51(5):20–25.
7. Belfinger Berger Foundations, "Golden Ears Bridge Mono Piles," Corporate Project Description, 2009, Wiesbaden, Germany.
8. Combault, J., "Fixed-Link and Long-Span Bridges," 1st International Symposium on Bridges and Large Structures, May 2-8, 2008, São Paulo, Brazil, 20 pp.
9. Meyboom, J., A. -L. Meyboom, and M. Sutton, "The Greenway Bridge," Proceedings of the Canadian Society of Civil Engineers, International Conference on Short- and Medium-Span Bridges, Montréal, Canada, 2006.
10. Meyboom, J. and W. Victor Anderson, "The Tongnan Bridge," Proceedings of the Canadian Society of Civil Engineers, International Conference on Short- and Medium-Span Bridges, Montréal, Canada, 2006.
11. Meyboom, J., T. Verburg, and H. Hawk, "Six Signature Bridges in the People's Republic of China," Institute of Civil Engineering, International Conference on Bridge Engineering, Beijing, 2007.

2

Bridge Engineering in the United States

M. Myint Lwin
Federal Highway
Administration

John M. Kulicki
Modjeski and Masters, Inc.

2.1 Introduction

John Kulicki and Betsy Reiner

2.1.1 Geographical Characteristics

The United States of America is situated in central North America, between the Pacific and Atlantic Oceans, bordered by Canada to the north and Mexico to the south. The state of Alaska is in the northwest of the continent, with Canada to the east and Russia to the west across the Bering Strait. The state of Hawaii is an archipelago in the mid-Pacific.

At 3.79 million square miles (9.83 million km^2) and with over 308 million people, the United States is the third largest country both by total area and population. It is one of the world's most ethnically diverse and multicultural nations, the product of large-scale immigration from many countries. The U.S. economy is the world's largest national economy, with an estimated 2010 GDP of $14.799 trillion.

2.1.2 Historical Development

The evolution of bridges in the United States is probably not much different from anywhere else in the world. Civilizations have borrowed their bridging ideas from each other for centuries. Primitive bridges constructed of fallen logs, rope, or stones have led to the modern-day girder bridges, cable-supported bridges, and arch bridges.

2.1.2.1 Early American Bridges

The earliest roads were simply dirt, stone paths, or trails, where horse-drawn wagons or pack trains of mules transported people and goods. Transport by these means was generally slow and the likelihood of the dirt roads to be dusty in dry weather and muddy in wet weather made transport difficult.

As trade became more economically important, a need arose to transport goods further distances. The first U.S. freight hauler, known as the Conestoga wagon, was developed by German immigrants in Pennsylvania in the late 1700s. This wagon was capable of hauling six tons of freight and was specially designed to withstand the rough terrain of the dirt roads (Donley 2005). Soon after the introduction of the Conestoga wagon, the Philadelphia and Lancaster Turnpike was constructed as the nation's first macadamized road (HSP 1908).

The earliest bridges in America took the form of logs placed over an obstacle. In roads through swampy or low-lying areas, logs were placed side-by-side, perpendicular to the direction of travel. This type of construction, known as corduroy construction, consumed a substantial amount of timber and resulted in an uneven roadway surface. Narrow waterways or valleys were crossed by placing large stones or fallen logs parallel to the direction of travel. The arch form, as borrowed from the ancient Romans, was also prevalent in early American bridge construction. Stone arch bridges were common, with the Frankford Avenue Bridge constructed in 1697 being the oldest stone arch bridge in the country (Figure 2.1) (ASCE 1977).

Beginning in the early 1800s, truss bridges were introduced as a means capable of spanning longer distances. These truss bridges were mostly constructed using timber, since the material was plentiful and workers did not have to be highly skilled. Despite the advances made in land transportation during this time, it was easier and more economical to transport certain goods by rivers and seas. Since

FIGURE 2.1 Frankford Avenue Bridge in Philadelphia, Pennsylvania. (Courtesy of Joseph Elliott, Historic American Engineering Record.)

transportation by water was limited to cities along the coast or by a navigable river, many settlements were concentrated in these areas.

2.1.2.2 Canal Era

By the late 1700s, American canal builders were taking notes from their European counterparts. Successful canal systems constructed by the Dutch, French, and English inspired a new vision of a water-connected society in America (Shaw 1990). This new network of canals enabled the United States to begin expanding inland from the eastern seaboard.

Canal builders in the late 1700s and early 1800s were the first to construct American bridges of any consequence. Bridges were not only required for transportation over the newly built canals, but also as a means of conveying the canal across another waterway or road. These bridges, known as aqueducts, were commonly constructed using the familiar stone arch borrowed from Roman times. Figure 2.2 shows the Schoharie Creek Aqueduct, the Erie Canal over Schoharie Creek at Fort Hunter, New York.

2.1.2.3 Railroad Era

With the introduction of the first steam locomotive in the United States around 1830, the railroad industry began to thrive. Less than 40 years later, the First Transcontinental Railroad connected the Pacific Coast with the already established railroad network in the eastern United States. Railroads soon became the dominant mode of transportation for both passengers and freight. Smaller towpath canals were all but abandoned and wagon roads went into a 50-year period of neglect. Bridge building flourished during this time of great expansion. Not only were bridges required on the thousands of miles of newly laid track, but the inability of locomotives to climb grades required the construction of bridges on even the slightest of hills.

2.1.2.3.1 Trusses

While the truss form was used prior to this time period, truss bridges became very popular due to their ability to provide strength with considerable savings in materials and weight. Wrought iron became the material of choice, which soon gave way to the use of mild steel. Many engineers took the opportunity to develop their own patented configurations of diagonals and verticals. Several of the truss forms developed during the Railroad Era are still in use today, such as the Warren, Pratt, and Howe, to name a few.

FIGURE 2.2 Schoharie Creek Aqueduct at Fort Hunter, New York. (Courtesy of Jack Boucher, Historic American Engineering Record.)

Some bridge manufacturers, such as the American Bridge Company, Phoenix Iron Works, and Wrought Iron Bridge Company, began producing catalog bridges. These bridges were comprised of pre-fabricated wrought iron or steel components that were designed based on commonly used span lengths. The bridge would be selected from a catalog, delivered to the site, and erected on abutments placed to match the span length. Figure 2.3 shows the Laughery Creek Bridge near Aurora, Indiana, which was built by the Wrought Iron Bridge Company of Canton, Ohio, in 1878.

2.1.2.3.2 Trestle Bridges

The rapid construction of trestle bridges made them the ideal structure type for the vast expansion of the railroad during this time period. Trestles were mostly built of timber and were considered tempo-rary until more permanent structures could be built. Many trestles were covered with earth fill, forming an embankment that would remain long after the timber had rotted away. Figure 2.4 shows the Secret Town Trestle in the California Sierras, built in 1865, being buried by earth fill.

2.1.2.4 Motor Car Era

Around the turn of the twentieth century, mass production of automobiles ushered in the era of the motor car. Massive road improvements began with the construction of the Lincoln Highway (Figure 2.5) in 1913, which was the first road for the automobile across the United States (Fridell 2010). In 1919, a lieutenant colonel in the Army named Dwight D. Eisenhower accompanied the First Transcontinental Motor Convoy across the country. For a young Eisenhower, the unsafe bridges and muddy, dusty, nar-row, and uneven roads left a lasting impression and drew attention to the need for improvements of the nation's highway system (ASCE 2002).

Around the same time as the beginning of the Motor Car Era, the concept of reinforced concrete was introduced. The benefits of reinforced concrete as a building material became apparent after the San Francisco earthquake of 1906, where the few reinforced concrete buildings were the only structures to survive. Since then, reinforced concrete has been widely used in the construction of bridges and buildings.

FIGURE 2.3 Laughery Creek Bridge. (Courtesy of Jack Boucher, Historic American Engineering Record.)

FIGURE 2.4 Secret Town Trestle in the California Sierras. (Courtesy of California State Library.)

FIGURE 2.5 Lincoln Highway route in 1916.

2.1.2.4.1 Steel Truss Bridges

Steel truss bridges retained their popularity, but the need for longer spans necessitated the use of continuous span trusses. Continuous span truss bridges required a shallower truss depth than their simple span equivalents, and, therefore, were more economical. The Huey P. Long Bridge just west of New Orleans, Louisiana, was opened in 1935. Figure 2.6 illustrates the proposed structure after completion of widening of this bridge, which is currently underway.

2.1.2.4.2 Concrete Arches

Reinforced concrete was the modern material and arches were a well-understood structural form dating back to Roman times. The compressive strength of concrete made it the ideal material for the arch shape and was easier to use than stone or masonry. The Columbia-Wrightsville Bridge, shown in Figure 2.7, opened in 1930 and spans the Susquehanna River between Columbia and Wrightsville, Pennsylvania. The Columbia-Wrightsville Bridge is on the original Lincoln Highway route.

2.1.2.4.3 Suspension Bridges

While suspension bridges are based upon one of the oldest concepts in the world, they continued to be a favored type into this era. Suspension bridges can be constructed without falsework, so they are the practical bridge type to cross gorges or wide waterways where access is restricted, difficult, or impossible. The west span of the San Francisco-Oakland Bay Bridge (see Section 2.8.3) is actually two suspension bridges end to end with a central anchorage between the two. When opened in 1936, it was the only double-suspension bridge in the world.

2.1.2.5 Interstate Highway Era

Eisenhower's experiences during the First Transcontinental Motor Convoy and his admiration of Germany's autobahn during his service in World War II inspired his interest in a network of interstate highways in the United States. This support eventually led to the 1956 Federal Aid Highway Act and the beginning of the interstate system as we know it today (Figure 2.8).

During this time, bridge engineers found themselves building bridges over dry land, at ridges, and over the highways themselves. Engineers began to utilize the tensile strength of steel and the compressive strength of concrete by pairing these materials in the form of prestressed concrete and composite steel bridges.

FIGURE 2.6   Proposed Huey P. Long Bridge after completion of widening. (Courtesy of Modjeski and Masters, Inc.)

FIGURE 2.7 Columbia-Wrightsville Bridge. (Courtesy of Joseph Elliott, Historic American Engineering Record.)

NATIONAL SYSTEM OF INTERSTATE AND DEFENSE HIGHWAYS

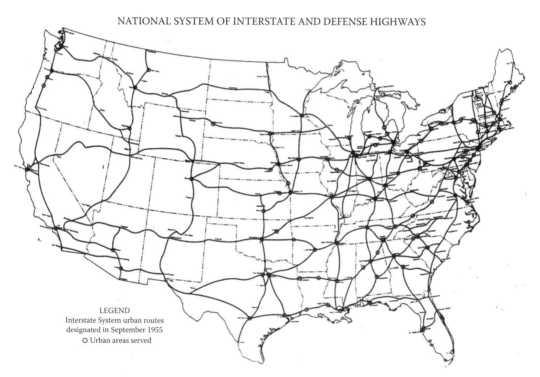

LEGEND
Interstate System urban routes
designated in September 1955
O Urban areas served

FIGURE 2.8 Proposed National System of Interstate and Defense Highways Plan developed in September 1955.

2.1.2.5.1 Prestressed Concrete

Prestressed concrete makes the best use of the compressive qualities of concrete and the tensile properties of steel. Prestressing allows shallower structure depth and a tremendous savings in approach roadway earthwork for interstate separations. Prestressed concrete can either be pretensioned or posttensioned, precast or cast-in-place. Box-shaped concrete beams, which were often prestressed, became favorable due to their increased torsional resistance and excellent wheel load distribution.

FIGURE 2.9 A precast concrete segmental box for the Palmetto Section 5 - SR-82/6SR-836 Interchange in Miami, Florida. (From Finely Engineering Group. With permission.)

2.1.2.5.2 Composite Steel

Composite steel girders, where a concrete deck is attached to the top flange of a steel girder through mechanical connectors, utilize the best advantages of the compressive properties of concrete and the tensile properties of steel.

2.1.2.6 The New Millennium

As we transition into the new millennium, the focus of bridge engineers and bridge owners is shifting from building new structures to maintaining the existing structures in service. Existing bridges are often retrofitted to meet the demands of today's truck loads. Accelerated bridge construction, using prefabricated elements, has become popular due to its improved level of safety and minimized impacts on traffic and the environment. Precast concrete beams, precast segmental concrete box girders, and precast concrete decks are commonly used for bridge construction today.

2.1.2.6.1 Segmental Bridges

Segmental concrete bridges are constructed using segments of concrete box girders that are tied together with post-tensioning tendons. The segments of concrete can be either precast or cast-in-place, although the benefits of rapid construction with a precast segmental bridge often make it the favored type of construction (Figure 2.9). Section 2.3.4 discusses the Pine Valley Creek concrete segmental bridge in California.

2.2 Design Practice

John Kulicki and Robert Sweeney

2.2.1 Introduction

Bridge engineering is almost as old as humankind. The need to cross obstacles resulted in a progression from logs and trees to stone and masonry to iron and steel to reinforced and prestressed concrete. Structural forms progressed from beams to arches to trusses to cable-supported bridges. Spans and loads increased, necessitating engineering analysis and design that became codified in design specifications. By the mid- to late-1800s, there were a multitude of design specifications for bridges in the United States. Most major municipalities, railroad companies, major consultants, and manufacturers of predesigned bridges subscribed to formalized loadings and design methods.

In the early 1900s, leaders of the profession called for more formalized codification of design requirements. The Cooper loading, adopted by the American Railway Engineering Association (AREA, later AREMA) for railroad bridges, served as a model for standardizing on a design live load. Proposals for design highway bridge loadings were debated and by the early 1920s there was consensus on what became known as the H20 truck and lane loading. By 1931 a printed version of the first edition of the *Standard Specifications for Highway Bridges* was released by the American Society of State Highway Officials (AASHO, later AASHTO). The *Standard Specifications* continued to evolve after release of the first edition, eventually leading to 17 editions, and the need to increase the design loading resulted in adoption of the HS20 truck in 1944; however, the lane load was retained at the H20 level. By the late 1970s some states were increasing the truck and lane load by 25%, resulting in the HS25 loading.

Design methods for highway bridges evolved as well. In the early 1970s, load factor design joined the traditional allowable stress design methods for some types of bridges. In 1986, following the lead of the Ontario Ministry of Transport and Communication and others, AASHTO began to study the possibility of a major change in its bridge design specification and in 1988 started development of a probability-based limit states design specification. These new specifications were adopted as an alternative to the *Standard Specifications* in 1993 and became the required specifications for federally funded bridges in 2007. Once again the live load model was changed to be more reflective of actual truck traffic by combining a truck configuration with a uniformly distributed load, which became known as the HL93 loading.

The railroad engineering community continues to maintain and evolve its own bridge design specification, known as the *Manual of Recommended Practice*, through AREMA. That specification retains the allowable stress design procedures, although concrete design uses a strength-based resistance model.

2.2.2 Highway Bridges: AASHTO Design Specifications Development

2.2.2.1 Introduction

Several bridge design specifications will be referred to repeatedly herein. In order to simplify the references, the "Standard Specifications" means the AASHTO *Standard Specifications for Highway Bridges* (AASHTO 2002), and the Seventeenth Edition will be referenced unless otherwise stated. The "Load and Resistance Factor Design (LRFD) Specifications" means the AASHTO *LRFD Bridge Design Specifications* (AASHTO 2010a), and the Fifth Edition will be referenced, unless otherwise stated. This latter document was developed in the period 1988–1993 when statistically based probability methods were available, which became the basis of quantifying safety. Because this is a more modern philosophy than either the load factor design method or the allowable stress design method, both of which are available in the Standard Specifications and neither of which have a mathematical basis for establishing safety, much of this section will deal primarily with the LRFD Specifications.

There are many issues that comprise a design philosophy. For example, the expected service life of a structure, the degree to which future maintenance should be assumed to preserve the original resistance of the structure or should be assumed to be relatively nonexistent, how brittle behavior can be avoided, how much redundancy and ductility are needed, the degree to which analysis is expected to accurately represent the force effects actually experienced by the structure, the extent to which loads are thought to be understood and predictable, the degree to which the designers' intent will be upheld by vigorous material testing requirements and thorough inspection during construction, the balance between the need for high precision during construction in terms of alignment and positioning compared to allowing for misalignment and compensating for it in the design, and, perhaps most fundamentally, the basis for establishing safety in the design specifications. It is this last issue, the way that specifications seek to establish safety, with which this section deals.

2.2.2.2 Limit States

All comprehensive design specifications are written to establish an acceptable level of safety. There are many methods of attempting to provide safety and the method inherent in many modern bridge design specifications, including the LRFD Specifications, the Ontario Highway Bridge Design Code (OMTC 1994), and the Canadian Highway Bridge Design Code (CAS 1998), is probability-based reliability analysis. The method for treating safety issues in modern specifications is the establishment of "limit states" to define groups of events or circumstances which could cause a structure to be unserviceable for its original intent.

The LRFD Specifications are written in a probability-based limit state format requiring examination of some, or all, of the four limit states defined below for each design component of a bridge:

1. The service limit state deals with restrictions on stress, deformation, and crack width under regular service conditions. These provisions are intended to ensure the bridge performs acceptably during its design life.
2. The fatigue and fracture limit state deals with restrictions on stress range under regular service conditions, reflecting the number of expected stress range excursions. These provisions are intended to limit crack growth under repetitive loads to prevent fracture during the design life of the bridge.
3. The strength limit state is intended to ensure that strength and stability, both local and global, are provided to resist the statistically significant load combinations that a bridge will experience in its design life. Extensive distress and structural damage may occur under strength limit state conditions, but overall structural integrity is expected to be maintained.
4. The extreme event limit state is intended to ensure the structural survival of a bridge during a major earthquake; when a vessel, vehicle, or ice flow collides with it; or where the foundation is subject to the scour that would accompany a flood of extreme recurrence, usually considered to be 500 years. These provisions deal with circumstances considered to be unique occurrences whose return period is significantly greater than the design life of the bridge. The joint probability of these events is extremely low, and, therefore, they are specified to be applied separately. Under these extreme conditions, the structure is expected to undergo considerable inelastic deformation by which locked-in force effects due to temperature effects, creep, shrinkage, and settlement will be relieved.

2.2.2.3 Design Objectives

2.2.2.3.1 Safety

Public safety is the primary responsibility of the design engineer. All other aspects of design, including serviceability, maintainability, economics, and aesthetics, are secondary to the requirement for safety. This does not mean that other objectives are not important, but safety is paramount. In design specifications the issue of safety is usually codified by an application of the general statement the design resistances must be greater than or equal to the design load effects. In Allowable Stress Design (ASD), this requirement can be formulated as:

$$\Sigma Q_i \leq \frac{R_E}{FS} \qquad\qquad (2.1)$$

where

Q_i = a load
R_E = elastic resistance
FS = factor of safety

In Load Factor Design (LFD), the formulation is

$$\Sigma \gamma_i Q_i \leq \varphi R \qquad\qquad (2.2)$$

where

γ_i = a load factor
Q_i = a load
R = resistance
φ = a strength reduction factor

In Load and Resistance Factor Design (LRFD), the formulation is

$$\Sigma\eta_i\gamma_iQ_i \le \varphi R_n = R_r \qquad (2.3)$$

where

$\eta_i = \eta_D\,\eta_R\,\eta_I$, limited such that $\eta = \eta_D\,\eta_R\,\eta_I \ge 0.95$ for loads for which a maximum value of γ_i is appropriate, and, $\eta_i = \dfrac{1}{\eta_I\eta_D\eta_R} = 1.0$ for loads for which a minimum value of γ_i is appropriate

η_D = a factor relating to ductility
η_R = a factor relating to redundancy
η_I = a factor relating to operational importance
γ_i = load factor, a statistically based multiplier on force effects
φ = resistance factor, a statistically based multiplier applied to nominal resistance
η_i = load modifier
Q_i = nominal force effect, a deformation, stress or stress resultant
R_n = nominal resistance, based on the dimensions as shown on the plans and on permissible stresses, deformations, or specified strength of materials
R_r = factored resistance, φR_n

Equation 2.3 is applied to each designed component and connection as appropriate for each limit state under consideration.

2.2.2.3.2 Special Requirements of the LRFD Specifications

Comparison of the equation of sufficiency as it was written above for ASD, LFD, and LRFD shows that as the design philosophy evolves through these three stages, more aspects of the component under design and its relation to its environment and its function to society must be expressly considered. This is not to say that a designer using ASD necessarily considers less than a designer using LFD or LRFD. The specification provisions are the minimum requirements and prudent designers often consider additional aspects. However, as specifications mature and become more reflective of the real world, additional criteria are often needed to assure adequate safety that may have been provided, albeit nonuniformly, by simpler provisions. Therefore, it is not surprising to find that the LRFD Specifications require explicit consideration of ductility, redundancy, and operational importance (Equation 2.3), while the Standard Specifications do not.

Ductility, redundancy, and operational importance are significant aspects affecting the margin of safety of bridges. While the first two directly relate to the physical behavior, the last concerns the consequences of the bridge being out of service. The grouping of these aspects is, therefore, arbitrary, however, it constitutes a first effort of codification. In the absence of more precise information, each effect, except fatigue and fracture, is estimated as ±5%, accumulated geometrically, a clearly subjective approach. With time, improved quantification of ductility, redundancy, and operational importance, and their interaction, may be attained.

2.2.2.3.3 Ductility

The response of structural components or connections beyond the elastic limit can be characterized by either brittle or ductile behavior. Brittle behavior is undesirable because it implies the sudden loss of load-carrying capacity immediately when the elastic limit is exceeded. Ductile behavior is characterized by significant inelastic deformations before any loss of load-carrying capacity occurs. Ductile behavior provides warning of structural failure by large inelastic deformations. Under cyclic loading, large reversed cycles of inelastic deformation dissipate energy and have a beneficial effect on structure response.

If, by means of confinement or other measures, a structural component or connection made of brittle materials can sustain inelastic deformations without significant loss of load-carrying capacity, this component can be considered ductile. Such ductile performance should be verified by experimental testing. Behavior that is ductile in a static context, but that is not ductile during dynamic response, should also be avoided. Examples of this behavior are shear and bond failures in concrete members, and loss of composite action in flexural members. The ductility capacity of structural components or connections may either be established by full- or large-scale experimental testing or with analytical models that are based on realistic material behavior. The ductility capacity for a structural system may be determined by integrating local deformations over the entire structural system.

Given proper controls on the innate ductility of basic materials, proper proportioning and detailing of a structural system are the key considerations in ensuring the development of significant, visible, inelastic deformations, prior to failure, at the strength and extreme event limit states. For the fatigue and fracture limit state for fracture-critical members and for the strength limit state for all members:

$\eta_D \geq 1.05$ for nonductile components and connections
$= 1.00$ for conventional designs and details complying with these specifications
≥ 0.95 for components and connections for which additional ductility-enhancing measures have been specified beyond those required by these specifications

For all other limit states:

$\eta_D = 1.00$

2.2.2.3.4 Redundancy

Redundancy is usually defined by stating the opposite, for example, a nonredundant structure is one in which the loss of a component results in collapse, or, a nonredundant component is one whose loss results in complete or partial collapse. Multiple load path structures should be used, unless there are compelling reasons to the contrary. The LRFD Specifications require additional resistance in order to reduce probability of loss of a nonredundant component, and to provide additional resistance to accommodate load redistribution. For the strength limit state:

$\eta_R \geq 1.05$ for nonredundant members
$= 1.00$ for conventional levels of redundancy
≥ 0.95 for exceptional levels of redundancy

For all other limit states:

$\eta_R = 1.00$

The factors currently specified were based solely on judgment and were included to require more explicit consideration of redundancy. Research is underway by Ghosn and Moses (2001) to provide a more rational requirement based on reliability indices thought to be acceptable in damaged bridges that must remain in service for a period of about two years. The "reverse engineering" concept is being applied to develop values similar in intent to η_R.

2.2.2.3.5 Operational Importance

The concept of operational importance is applied to the strength and extreme-event limit states. The owner may declare a bridge or any structural component or connection thereof to be of operational importance. Such classification should be based on social/survival and/or security/defense requirements. If a bridge is deemed of operational importance, η_I is taken as ≥ 1.05. Otherwise, η_I is taken as 1.0 for typical bridges and may be reduced to 0.95 for relatively less important bridges.

2.2.2.4 Design Load Combinations in the LRFD Specifications

The permanent and transient loads and forces listed in Table 2.1 are considered in the AASHTO LRFD Specifications.

The vehicular live load HL93 consists of either one or, for force effects at interior supports of continuous beams, two of the truck loads with three axles or the tandem load, combined with the uniform lane load as shown in Figure 2.10.

The load factors for various loads, comprising a design load combination, are indicated in Tables 2.2 through 2.4 for LRFD, all of the load combinations are related to the appropriate limit state. Any, or all, of the four limit states may be required in the design of any particular component, and those which are the minimum necessary for consideration are indicated in the specifications where appropriate. Thus, a design might involve any load combination in Table 2.2.

All relevant subsets of the load combinations in Table 2.2 should be investigated. The factors should be selected to produce the total factored extreme force effect. For each load combination, both positive and negative extremes should be investigated. In load combinations where one force effect decreases the effect of another, the minimum value should be applied to load-reducing the force effect. For each load

TABLE 2.1   Load Designations

Name of Load	LRFD Designation
Downdrag	DD
Dead Load of Structural Components Attachments	DC
Dead Load of Wearing Surfaces and Utilities	DW
Dead Load of Earth Fill	EF
Horizontal Earth Pressure	EH
Locked-In Force Effects from Construction	EL
Earth Surcharge Load	ES
Vertical Earth Pressure	EV
Vehicular Braking Force	BR
Vehicular Centrifugal Force	CE
Creep	CR
Vehicular Collision Force	CT
Vessel Collision Force	CV
Earthquake	EQ
Friction	FR
Ice Load	IC
Vehicular Dynamic Load Allowance	IM
Vehicular Live Load	LL
Live Load Surcharge	LS
Pedestrian Live Load	PL
Secondary Forces from Post-tensioning	PS
Settlement	SE
Shrinkage	SH
Temperature Gradient	TG
Uniform Temperature	TU
Water Load and Stream Pressure	WA
Wind on Live Load	WL
Wind Load on Structure	WS

Source: AASHTO LRFD Bridge Design Specifications, 5th Edition with Interims through 2010, American Association of State Highway and Transportation Officials, Washington, DC. With permission.

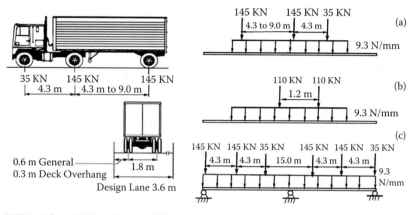

35 KN = 8.0 kip;   145 KN = 32.0 kip;   110 KN = 24 kip;   0.3 m = 1.0 ft.;   0.6 m = 2.0 ft.;
1.2 m = 4.0 ft.;   1.8 m = 6.0 ft.;   3.6 m = 12.0 ft.;   15.0 m = 50.0 ft.;   9.3 N/mm = 0.64 kip/ft.

FIGURE 2.10   AASHTO-LRFD live load: HL93. (From *AASHTO LRFD Bridge Design Specifications*, 5th Edition with Interims through 2010, American Association of State Highway and Transportation Officials, Washington, DC. With permission.)

TABLE 2.2   Load Combinations and Load Factors in AASHTO LRFD

Load Combination Limit State	DC DD DW EH EV ES EL PS CR SH	LL IM CE BR PL LS	WA	WS	WL	FR	TU	TG	SE	Use One of These at a Time			
										EQ	IC	CT	CV
Strength I (unless noted)	γ_p	1.75	1.00	—	—	1.00	0.50/1.20	γ_{TG}	γ_{SE}	—	—	—	—
Strength II	γ_p	1.35	1.00	—	—	1.00	0.50/1.20	γ_{TG}	γ_{SE}	—	—	—	—
Strength III	γ_p	—	1.00	1.40	—	1.00	0.50/1.20	γ_{TG}	γ_{SE}	—	—	—	—
Strength IV	γ_p	—	1.00	—	—	1.00	0.50/1.20	—	—	—	—	—	—
Strength V	γ_p	1.35	1.00	0.40	1.0	1.00	0.50/1.20	γ_{TG}	γ_{SE}	—	—	—	—
Extreme Event I	γ_p	γ_{EQ}	1.00	—	—	1.00	—	—	—	1.00	—	—	—
Extreme Event II	γ_p	0.50	1.00	—	—	1.00	—	—	—	—	1.00	1.00	1.00
Service I	1.00	1.00	1.00	0.30	1.0	1.00	1.00/1.20	γ_{TG}	γ_{SE}	—	—	—	—
Service II	1.00	1.30	1.00	—	—	1.00	1.00/1.20	—	—	—	—	—	—
Service III	1.00	0.80	1.00	—	—	1.00	1.00/1.20	γ_{TG}	γ_{SE}	—	—	—	—
Service IV	1.00	—	1.00	0.70	—	1.00	1.00/1.20	—	1.0	—	—	—	—
Fatigue I— *LL, IM & CE* only	—	1.50	—	—	—	—	—	—	—	—	—	—	—
Fatigue I II— *LL, IM & CE* only	—	0.75	—	—	—	—	—	—	—	—	—	—	—

Source: *AASHTO LRFD Bridge Design Specifications*, 5th Edition with Interims through 2010, American Association of State Highway and Transportation Officials, Washington, DC. With permission.

TABLE 2.3 Load Factors for Permanent Loads, γ_p, in AASHTO LRFD

Type of Load, Foundation Type, and Method Used to Calculate Downdrag	Load Factor	
	Maximum	Minimum
DC: Component and Attachments	1.25	0.90
DC: Strength IV Only	1.50	0.90
DD: Downdrag Piles, α Tomlinson Method	1.4	0.25
Piles, λ Method	1.05	0.30
Drilled Shafts, O'Neill and Reese Method	1.25	0.35
DW: Wearing Surfaces and Utilities	1.50	0.65
EH: Horizontal Earth Pressure	1.50	0.90
• Active	1.35	0.90
• At Rest	1.35	N/A
• AEP for Anchored Walls		
EL: Locked-in Construction Stresses	1.00	1.00
• *EV*: Vertical Earth Pressure		
• Overall Stability	1.00	N/A
• Retaining Walls and Abutments	1.35	1.00
• Rigid Buried Structure	1.30	0.90
• Rigid Frames	1.35	0.90
• Flexible Buried Structures	1.5	0.9
• Metal Box Culverts and Structural Plate Culverts with Deep Corrugations	1.3	0.9
• Thermoplastic Culverts	1.95	0.9
• All Others		
ES: Earth Surcharge	1.50	0.75

Source: AASHTO LRFD Bridge Design Specifications, 5th Edition with Interims through 2010, American Association of State Highway and Transportation Officials, Washington, DC. With permission.

TABLE 2.4 Load Factors for Permanent Loads Due to Superimposed Deformations, γ_p, in AASHTO LRFD

Bridge Component	*PS*	*CR, SH*
Superstructures—Segmental Concrete Substructures supporting Segmental Superstructures (see 3.12.4, 3.12.5)	1.0	See γ_p for *DC*
Concrete Superstructures— nonsegmental	1.0	1.0
Substructures supporting nonsegmental Superstructures	0.5	0.5
• using I_g	1.0	1.0
• using $I_{effective}$		
Steel Substructures	1.0	1.0

Source: AASHTO LRFD Bridge Design Specifications, 5th Edition with Interims through 2010, American Association of State Highway and Transportation Officials, Washington, DC. With permission.

combination, every load that is indicated, including all significant effects due to distortion, should be multiplied by the appropriate load factor.

Table 2.2 shows that some of the load combinations have a choice of two load factors. The larger of the two values for load factors shown for TU, TG, CR, SH, and SE are to be used when calculating deformations; the smaller value shall be used when calculating all other force effects. Where movements are calculated for the sizing of expansion dams, the design of bearing, or similar situations where consideration of unexpectedly large movements is advisable, the larger factor should be used. When considering the effect of these loads on forces which are compatibility-generated, the lower factor may be used. This latter use requires structural insight.

Consideration of the variability of loads in nature indicates that loads may be either larger or smaller than the nominal load used in the design specifications. The LRFD Specifications recognize the variability of permanent loads by providing both maximum and minimum load factors for the permanent loads, as indicated in Table 2.3. For permanent force effects, the load factor which produces the more critical combination shall be selected from Table 2.2. In the application of permanent loads, force effects for each of the specified six load types should be computed separately. Assuming variation of one type of load by span, length, or component within a bridge is not necessary. For each force effect, both extreme combinations may need to be investigated by applying either the high or low load factor as appropriate. The algebraic sums of these products are the total force effects for which the bridge and its components should be designed. This reinforces the traditional method of selecting load combinations to obtain realistic extreme effects.

When the permanent load increases the stability or load-carrying capacity of a component or bridge, the minimum value of the load factor for that permanent load shall also be investigated. Uplift, which is treated as a separate load case in past editions of the AASHTO *Standard Specifications for Highway Bridges*, becomes a Strength I load combination. For example, when the dead load reaction is positive and live load can cause a negative reaction, the load combination for maximum uplift force would be $0.9DC + 0.65DW + 1.75(LL+IM)$. If both reactions were negative, the load combination would be $1.25DC + 1.50DW + 1.75(LL+IM)$. The load combinations for various limit states shown in Table 2.2 are described below.

- Strength I: Basic load combination relating to the normal vehicular use of the bridge without wind.
- Strength II: Load combination relating to the use of the bridge by permit vehicles without wind. If a permit vehicle is traveling unescorted, or if control is not provided by the escorts, the other lanes may be assumed to be occupied by the vehicular live load herein specified. For bridges longer than the permit vehicle, addition of the lane load, preceding and following the permit load in its lane, should be considered.
- Strength III: Load combination relating to the bridge exposed to maximum wind velocity, which prevents the presence of significant live load on the bridge.
- Strength IV: Load combination relating to very high dead load to live load force effect ratios. This calibration process had been carried out for a large number of bridges with spans not exceeding 60 m. Spot checks had also been made on a few bridges up to 180 m spans. For the primary components of large bridges, the ratio of dead and live load force effects is rather high, and could result in a set of resistance factors different from those found acceptable for small- and medium-span bridges. It is believed to be more practical to investigate one more load case, rather than requiring the use of two sets of resistance factors with the load factors provided in Strength I, depending on other permanent loads present. Strength IV is expected to govern when the dead load to live load force effect ratio exceeds about 7.0.
- Strength V: Load combination relating to normal vehicular use of the bridge with wind of 90 km/h velocity.
- Extreme Event I: Load combination relating to earthquake. The designer-supplied live load factor signifies a low probability of the presence of maximum vehicular live load at the time when the earthquake occurs. In ASD and LFD the live load is ignored when designing for earthquake.
- Extreme Event II: Load combination relating to reduced live load in combination with a major ice event, vessel collision, or a vehicular impact.
- Service I: Load combination relating to the normal operational use of the bridge with 90 km/h wind. All loads are taken at their nominal values and extreme load conditions are excluded. This combination is also used for checking deflection of certain buried structures, investigation of slope stability, and investigation of transverse bending stresses in segmental concrete girders.
- Service II: Load combination whose objective is to prevent yielding of steel structures due to vehicular live load, approximately halfway between that used for Service I and Strength I limit state, for which the effect of wind is of no significance. This load combination corresponds to the

overload provision for steel structures in past editions of the AASHTO *Standard Specifications for the Design of Highway Bridges.*

- Service III: Load combination relating only to prestressed concrete structures with the primary objective of crack control. The addition of this load combination followed a series of trial designs done by 14 states and several industry groups during 1991 and early 1992. Trial designs for pre-stressed concrete elements indicated significantly more prestressing would be needed to support the loads specified in the proposed specifications. There is no nationwide physical evidence that the vehicles used to develop the notional live loads have caused detrimental cracking in existing prestressed concrete components. The statistical significance of the 0.80 factor on live load is that the event is expected to occur about once a year for bridges with two design lanes, less often for bridges with more than two design lanes, and about once a day for the bridges with a single design lane.
- Fatigue I: Infinite life fatigue and fracture load combination relating to gravitational vehicular live load and dynamic response.
- Fatigue II: Finite life fatigue load case for which the load factor reflects a load level that has been found to be representative of the truck population.

2.2.2.5 Serviceability

The LRFD Specification treats serviceability from the viewpoints of durability, inspectibility, maintainability, rideability, deformation control, and future widening. Contract documents should call for high-quality materials and require that those materials that are subject to deterioration from moisture content and/or salt attack be protected. Inspectibility is to be assured through adequate means for permitting inspectors to view all parts of the structure that have structural or maintenance significance. The provisions related to inspectibility are relatively short, but as all departments of transportation have begun to realize, bridge inspection can be very expensive and is a recurring cost due to the need for biennial inspections. Therefore, the cost of providing walkways and other access means and adequate room for people and inspection equipment to be moved about on the structure is usually a good investment. Maintainability is treated in the specification in a similar manner to durability; there is a list of desirable attributes to be considered.

The subject of live load deflections and other deformations remains a very difficult issue. On the one hand, there is very little direct correlation between live load deflection and premature deterioration of bridges. There is much speculation that "excessive" live load deflection contributes to premature deck deterioration, but, no causative relationship has been statistically established.

Rider comfort is often advanced as a basis for deflection control. Studies in human response to motion have shown that it is not the magnitude of the motion, but rather the acceleration that most people perceive, especially in moving vehicles. Many people have experienced the sensation of being on a bridge and feeling a definite movement, especially when traffic is stopped. This movement is often related to the movement of floor systems, which are really quite small in magnitude, but noticeable nonetheless. There being no direct correlation between magnitude (not acceleration) of movement and discomfort has not prevented the design profession from finding comfort in controlling the gross stiffness of bridges through a deflection limit. As a compromise between the need for establishing comfort levels and the lack of compelling evidence that deflection is cause of structural distress, the deflection criteria, other than those pertaining to relative deflections of ribs of orthotropic decks and components of some wood decks, were written as voluntary provisions to be activated by those states that so chose. Deflection limits, stated as span divided by some number, were established for most cases, and additional provisions of absolute relative displacement between planks and panels of wooden decks and ribs of orthotropic decks were also added. Similarly, optional criteria were established for a span-to-depth ratio for guidance primarily in starting preliminary designs, but also as a mechanism for checking when a given design deviated significantly from past successful practice.

2.2.2.6 Constructability

Several new provisions were included in the LRFD Specification related to:

- The need to design bridges so that they can be fabricated and built without undue difficulty and with control over locked-in construction force effects
- The need to document one feasible method of construction in the contract documents, unless the type of construction is self-evident
- A clear indication of the need to provide strengthening and/or temporary bracing or support during erection, but not requiring the complete design thereof

2.2.3 Railroad Bridges: The American Railway Engineering and Maintenance-of-Way Association (AREMA) Specification Development

2.2.3.1 Introduction

The U.S. railroad network consists predominantly of privately owned freight railroad systems classified according to operating revenue, the government-owned National Railroad Passenger Corporation (Amtrak), and numerous transit systems owned by local agencies and municipalities. The Federal Register in 2010 listed 693 railroad entries, some of them subsidiaries of larger railroads. Based on data from the Association of American Railroads (AAR), there are 7 class 1 (major) railroads, 23 regional railroads, and 533 local railroads operating over approximately 140,000 mi (225,308 km) of track. Amtrak operates approximately 23,000 mi of railroad. The seven Class 1 railroads comprise only 1.25% of the number of railroads in the United States but account for 67.5% of the trackage and 93.3% of the freight revenue.

The AREMA *Manual of Recommended Practice* has been developed to meet the needs of the freight railroads in North America, and conventional passenger trains up to 90 mph (144.8 km/h). Bridge design philosophy is controlled by the predominant limit states that are relevant:

- Fatigue
- Robustness
- Stability
- Minimum movement compatible with track alignment and tolerable deviation for it

The *Manual* continues to use a working stress formulation for steel, concrete, and timber with an optional limited load factor type of analysis for some concrete and prestressed concrete members.

Railroad routes are well established and the construction of new railroad routes is not common; thus, the majority of railroad bridges built or rehabilitated are on existing routes and on existing right of way. Simply stated, the railroad industry first extends the life of existing bridges as long as economically justified. It is not uncommon for a railroad to evaluate an 80- or 90-year-old bridge, estimate its remaining life, and then rehabilitate it sufficiently to extend its life for some economical period of time.

Bridge replacement generally is determined as a result of a lack of load-carrying capacity, restrictive clearance, or deteriorated physical condition. If bridge replacement is necessary, then simplicity, cost, future maintenance, and ease of construction without significant rail traffic disruptions typically governs the design. The type of bridge chosen is most often based on the capability of the railroad to do its own construction work. Low-maintenance structures, such as ballasted deck prestressed concrete box girder spans with concrete caps and piles, are preferred by some railroads. Others may prefer weathering steel elements.

In review of the existing railroad industry bridge inventory, by far the majority of bridges are simple span structures over streams and roadways. Complex bridges are generally associated with crossing major waterways or other significant topographical features. Signature bridges are rarely constructed by railroads. The enormity of train live loads generally preclude the use of double leaf bascule bridges and suspension and cable stay bridges due to bridge deflection and shear load transfer, respectively.

Railroads, where possible, avoid designing skewed or curved bridges, which also have inherent deflection problems. When planning the replacement of smaller bridges, railroads first determine if the bridge can be eliminated using culverts. A hydrographic review of the site will determine if the bridge opening needs to be either increased or can be decreased.

The *Manual* provides complete details for common timber structures and for concrete box girder spans. Many of the larger railroads develop common standards, which provide complete detailed plans for the construction of bridges. These plans include piling, pile bents, abutments and wingwalls, spans (timber, concrete, and steel), and other elements in sufficient detail for construction by in-house forces or by contract. Only site-specific details such as permits, survey data, and soil conditions are needed to augment these plans.

Timber trestles are most often replaced by other materials rather than in-kind. However, it is often necessary to renew portions of timber structures to extend the life of a bridge for budgetary reasons. Replacing pile bents with framed bents to eliminate the need to drive piles or the adding of a timber stringer and recentering a chord to increase capacity is common. The replacement of timber trestles is commonly done by driving either concrete or steel piling through the existing trestle, at twice the present timber span length and offset from the existing bents. This is done between train movements. Either precast or cast-in-place caps are installed atop the piling beneath the existing timber deck. During a track outage period, the existing track and timber deck is removed and new spans (concrete box girders or rolled steel beams) are placed. In this type of bridge renewal, key factors are use of prefabricated bridge elements light enough to be lifted by railroad track–mounted equipment (piles, caps, and spans), speed of installation of bridge elements between train movements, bridge elements that can be installed in remote site locations without outside support, and overall simplicity in performing the work.

The railroad industry has a large number of 150 to 200 ft (45.7 to 61.0 m) span pin-connected steel trusses, many with worn joints, restrictive clearances, and low carrying capacity, for which rehabilitation cannot be economically justified. Depending on site specifics, a common replacement scenario may be to install an intermediate pier or bent and replace the span with two girder spans. Railroad forces have perfected the technique of laterally rolling out old spans and rolling in new prefabricated spans between train movements.

Railroads frequently will relocate existing bridge spans to other sites in lieu of constructing new spans, if economically feasible. This primarily applies to beam spans and plate girder spans up to 100 ft (30.5 m) in length. For this reason the requirement for heavy and densely loaded lines are generally applied to all new bridges as the bridge span could end up anywhere at some time in the future. Furthermore, a new industry may locate on a line and completely change the traffic volume.

In general, railroads prefer to construct new bridges on-line rather than relocating or doglegging to an adjacent alignment. Where site conditions do not allow ready access for direct span replacement, a site bypass, or runaround, called a "shoo-fly" is constructed, which provides a temporary bridge while the permanent bridge is constructed. The design and construction of larger and complex bridges is done on an individual basis.

2.2.3.2 Basic Differences between Railroad and Highway Bridges

A number of differences exist between railroad and highway bridges:

- The ratio of live load to dead load is much higher for a railroad bridge than for a similarly sized highway structure. This can lead to serviceability issues such as fatigue and deflection control governing designs rather than strength. Robustness is essential to deal with the harsh railway environment.
- The design impact load on railroad bridges is higher than on highway structures.
- Simple span structures are typically preferred over continuous structures for railroad bridges. Many of the factors that make continuous spans attractive for highway structures are not as advantageous for railroad use. Continuous spans are also more difficult to replace in emergencies than simple spans.

- Interruptions in service are typically much more critical for railroads than for highway agencies. Therefore, constructibility and maintainability without interruption to traffic are crucial for railroad bridges.
- Since the bridge supports the track structure, the combination of track and bridge movement cannot exceed the tolerances in track standards. Interaction between the track and bridge should be considered in design and detailing.
- Seismic performance of highway and railroad bridges can vary significantly. Railroad bridges have performed well during seismic events.
- Railroad bridge owners typically expect a longer service life from their structures than highway bridge owners expect from theirs.

2.2.3.3 Typical Railroad Bridge Types

Railroad bridges are nearly always simple span structures. Listed below in groupings by span length are the more common types of bridges and materials used by the railroad industry for those span lengths.

- Short spans to 16 ft (4.9 m)
  Timber stringers
  Concrete slabs
  Rolled steel beams
- Short spans to 32 ft (9.8 m)
  Conventional and prestressed concrete box girders and beams
  Rolled steel beams
- Short spans to 50 ft (15.2 m)
  Prestressed concrete box girders and beams
  Rolled steel beams, deck and through girders
- Medium spans, 80–125 ft (24.4–38.1 m)
  Prestressed concrete beams
  Deck and through plate girders
- Long spans
  Deck and through trusses (simple, cantilever, and arches)

Suspension bridges are not used by freight railroads due to excessive deflection.

2.2.3.4 Live Load

Historically, freight railroads have used the Cooper E load configuration as a live load model. Cooper E80 is currently the minimum design live load recommended by AREMA for new structures. The E80 load model is shown in Figure 2.11. The 80 in E80 refers to the 80 kip weight of the locomotive drive axles. An E60 load has the same axle locations, but the loads are factored by 60/80. The designated steel bridge design live load also includes an "alternate E80" load, consisting of four 100-kip axles. This is shown in Figure 2.12. This load controls on shorter spans and is factored for other E loads in a similar manner. Most of the larger railroads are designing new structures to carry E90 or E100 loads.

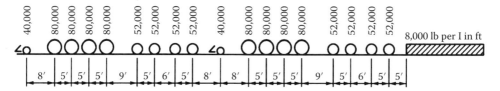

FIGURE 2.11   AREMA Cooper E80 live load. (Courtesy of AREMA.)

FIGURE 2.12   AREMA alternative live load. (Courtesy of AREMA.)

An E90 design would be based on a scaled loading of 112.5% (i.e., 90/80) of both the E80 and the alternative loads.

The Cooper live load model does not match the axle loads and wheel spacing of locomotives currently in service. It did not even reflect all locomotives at the turn of the twentieth century, when it was introduced by Theodore Cooper, an early railroad bridge engineer. Nevertheless, it has remained in use throughout the past century. One of the reasons for its longevity is the wide variety of rail rolling stock that has been and is currently in service. The load effects of this equipment on given spans is compared to the Cooper load. The Cooper live load model gives a universal system with which all other load configurations can be compared. Engineering personnel of each railroad calculate how the load effects of each piece of equipment compare to the Cooper loading. A table of maximum load effects over various span lengths is included in Chapter 15, Part 1 of the AREMA *Manual*.

2.2.3.5 Other Live Loads

Provisions are also made for loads due to impact, longitudinal force, and centrifugal force, together with an allowance for lateral loads from equipment. The effect of wind on railway equipment is also required.

2.2.3.6 Other Loads

Other loads that effect most bridges are included; the principle ones are wind loads and the effect of seismic loading.

2.2.3.7 High-Speed Railway Considerations

The AREMA *Manual* (2011) does not as yet have criteria for high-speed rail. These are under development and, due to the need to meet higher crash worthiness standards in North America, are likely to be different than those developed in Europe or Asia.

2.3 Concrete Girder Bridges

Reggie Holt and Lian Duan

2.3.1 Introduction

Concrete is the most-used construction material for bridges in the United States. The number of bridges built over the past 100-plus years is shown in the Figure 2.13. The figure divides the bridges built each year into five bridge type categories: concrete, steel, prestressed concrete, timber, and other. As can be seen from the data, the number of bridges utilizing prestressing has grown steadily. Prestressed concrete bridges have gone from being almost nonexistent in 1940 to clearly being the predominant bridge type currently being built in the United States. When combining the reinforced concrete and prestressed concrete data from Figure 2.13, one can see that concrete is by far the most common material type used for bridges in the United States.

One of the advancements that has been made in recent years is the use of high-performance concrete (HPC). HPC is an engineered concrete that has been designed to be more durable and, if necessary, provide increased compressive strengths. The increased compressive strengths will allow longer spans and

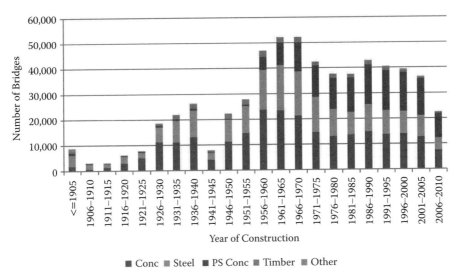

FIGURE 2.13   Bridge types constructed in the United States.

smaller or fewer structural components along with a more impermeable concrete that is more resilient to deleterious environmental loadings.

2.3.2 Precast Prestressed Concrete Girders

Prestressing has also played an important role in extending the span capability of concrete bridges. Single-piece precast prestressed concrete girders can accommodate span lengths up to approximately 160 ft (48.8 m). Figure 2.14 shows a few of the more common prestressed concrete beam shapes used in the United States. The butted beam sections are typically used for shorter spans and are popular because they do not require extensive formwork to cast a deck or topping.

In an effort to extend the acceptable span range for precast prestressed concrete girders, spliced-girder spans are gaining popularity. By splicing precast girder segments together, spans in excess of 300 ft (91.4 m) can be attained. A spliced girder (Figure 2.15) can be defined as a precast prestressed concrete member fabricated in multiple pieces called girder segments that are assembled into a single girder. Post-tensioning is used in the field to make the multiple piece assembly a continuous girder unit. Typically "I" and "U" girder sections are used to form these spliced-girder assemblies.

Another recent development with spliced-girder spans are the use of horizontally curved precast concrete "U" girders (Figure 2.16). The torsional rigidity of this cross section allows this bridge type to accommodate the tight horizontal radiuses that are typically seen on interchange alignments and their corresponding flyover bridge structures. In many cases these curved precast bridges use spliced-girder segments, which can accommodate spans in excess of 200 ft (61 m).

2.3.3 Segmental Concrete Bridges

Construction of segmental concrete bridges began in the United States in the early 1970s and currently there are close to 200 segmental bridges built in the United States. Segmental concrete bridges become advantageous for projects with constrained sites, multiple repetitive bridge spans, and bridges requiring significant span lengths. National Bridge Inspection (NBI) data has shown that segmental bridges are very durable. One reason for this is that most segmental bridges have post-tensioned bridge decks that limit the service stresses and accompanying cracking.

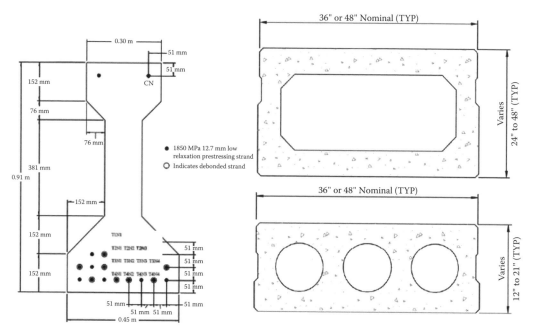

FIGURE 2.14 Standard precast prestressed concrete girder shapes: (a) "I" girder section and (b) butted prestressed concrete beam section.

FIGURE 2.15 Spliced precast girder bridges.

2.3.4 Pine Valley Creek Bridge

The Pine Valley Creek Bridge is the first cast-in-place prestressed concrete bridge in the United States built by the segmental balanced cantilever method (Figure 2.17). It is located in the state of California and carries I-8 highway traffic (Figure 2.18). The bridge was designed and is owned by the California Department of Transportation (DOT). The bridge was constructed by S. J. Groves & Sons Company of Liverpool, New York, and opened to traffic in 1974.

FIGURE 2.16   Example of U-girder bridge. (Courtesy of the Federal Highway Administration [FHWA].)

FIGURE 2.17   Pine Valley Creek Bridge under construction. (Courtesy of the California DOT.)

The bridge rises 450 ft (137.2 m) above the valley floor, and is 1700 ft (518.2 m) long, including five spans of 270 ft (82.3 m), 340 ft (103.6 m), 450 ft (137.2 m), 380 ft (115.8 m), and 270 ft (82.3 m.) The bridge has a center span of 450 ft (137.2 m), probably the longest concrete box girder span in the United States at the time of its completion, according to the *Engineering News Record*, July 1, 1971. The superstructure consists of two parallel boxes each carrying a 50 ft (12.8 m) roadway separated by a 38 ft (11.6 m) median.

FIGURE 2.18 Overview of Pine Valley Creek Bridge. (Courtesy of the California DOT.)

Wide deck overhangs were selected to minimize the apparent depth of the box of 19 ft (5.8 m.) The deck slab was prestressed in the transverse direction to reduce the dead load. The piers are cellular columns. The columns for the tallest pier of 380 ft (115.8 m) are 14 ft (4.3 m) × 32 ft (8 m) at the bottom, decrease to 14 ft (4.3 m) × 16 ft (4.9 m) at about 5/8 of the way up, and increase again to arrive at 23 ft (7.0 m) × 24 ft (7.3 m) at the soffit of the box girders. The pier footings of 7.9 ft (2.4 m) depth were anchored by prestressing tendons into rock 25 ft (7.62 m) deep.

2.3.5 Kanawha River Bridge

The innovative Kanawha River Bridge (Figures 2.19 and 2.20) is a record-setting, low-cost, and aesthetically pleasing segmental box girder structure built as part of the I-64 widening project in Kanawha County, West Virginia. The bridge carries I-64 eastbound traffic consisting of three through lanes, one auxiliary lane, and shoulders for a total width of 64 ft (19.5 m). The overall bridge length is 2975 ft (906.8 m), including a record 760 ft (231.7 m) main span. The structure crosses over railroad tracks, the Kanawha River, and three active roadways.

The eight-span structure has span lengths of 144 ft (43.9 m), 247 ft (75.3 m), 295 ft (89 m), 295 ft (89 m), 460 ft (140 m), 760 ft (231.6 m), 540 ft (164.6 m), and 209 ft (63.7 m). The bridge has an S-curve alignment including two circular curves, spiral transitions, and a tangent alignment at the river span. The 760 ft (231.6 m) long main span—the longest box girder span in the United States—resulted from the need to locate the main piers outside the main channel of the Kanawha River in order to avoid interference with barge traffic. The bridge section is an optimized single-cell box girder with 16 ft long (4.9 m) overhangs and structural depth varying between 38 ft (11.6 m) and 16 ft (4.9 m).

A continuous concrete box girder superstructure was chosen for the full length of the bridge. This 2975 ft (906.8 m) continuous structure has expansion joints at the abutments only. The advantages of this design are to reduce maintenance, improve serviceability, and simplify construction, as intermediate hinges are not needed. In order to address the potentially large transverse displacements due to creep, shrinkage, and temperature changes on a curved alignment, the bridge bearings are restrained in the radial direction. The approach piers and abutments have multirotational replaceable disc bearings with service capacities up to 6900 kips (30.7 millinewton) that restrain the transverse displacements while allowing longitudinal displacements. The bearings and the substructures were designed for the resulting radial forces.

FIGURE 2.19 Cantilever construction of Kanawha River Bridge. (Courtesy of Santiago Rodriguez, T. Y. Lin International.)

FIGURE 2.20 Kanawha River Bridge. (Courtesy of Santiago Rodriguez, T. Y. Lin International.)

Materials and design details were selected to achieve a 100-year service life. High-performance, low-permeability 6500 psi (41.4 megapascal) concrete was used throughout the superstructure. The concrete box section is post-tensioned longitudinally, transversely, and vertically in order to minimize cracking. The post-tensioning system consists of high-density polypropylene corrugated ducts and permanent fiber-reinforced caps with air- and water-tight connections. Prebagged grouts with thixotropic, non-shrinking, and nonbleeding characteristics were used. Mass concrete specifications were followed to prevent thermal cracking during construction. The bridge was designed by T. Y. Lin International and is owned by the West Virginia DOT.

2.3.6 Washington Bypass: US17 over the Pamlico-Tar River

The Washington Bypass is a North Carolina DOT design-build project consisting of a 6.8 mi (10.9 km) bypass route around the city of Washington, North Carolina, on US-17. The project, located on the North Carolina's coastal plain in Beaufort County, features a 2.8 mi (4.5 km) structure over the Tar River and the adjoining environmentally sensitive wetlands. To minimize the construction footprint in these environmentally sensitive areas, a new and innovative top-down construction approach using a unique overhead gantry prototype specially designed and built was developed for this project. This approach resulted in a minimal impact to the wetlands and an accelerated construction schedule when compared to conventional construction techniques.

The process consists of two self-contained gantries capable of performing all the tasks associated with the bridge construction, including driving 124 ft long (37.8 m) precast piles, erecting 50 ton bent caps, erecting 121 ft long (36.9 m) precast girders, and supporting deck pouring operations. The two self-launching 594 ft long (181.1 m) gantries weighing about 750 tons each were used, one at each end of the bridge, and working towards the middle of the structure. The world's first application of the pile driving operation from an erection gantry is the most unique feature of the system and is the essential element that truly eliminates the need for equipment and temporary access trestles and groundwork in the fragile wetlands (Figures 2.21 and 2.22).

FIGURE 2.21   Pile driving from an erection gantry. (Courtesy of Flatiron Construction Corporation.)

FIGURE 2.22   Gantry launched ahead of pile driving. (Courtesy of Flatiron Construction Corporation.)

The bridge was designed and built by Flatiron Construction Corporation, located in Longmont, Colorado. Construction activities were ongoing simultaneously across three spans (typically 121 ft [36.9 m] in length) in an assembly line progression. As a span is completed and the deck is cured, the gantry is launched ahead to begin the pile driving on the next span. The dramatic reduction in wetland disturbance offered by this "true top-down" construction operation was well received by the U.S. Army Corps of Engineers, North Carolina Division of Water Quality, North Carolina Department of Natural Resources, U.S. Coast Guard, and other environmental agencies during the permitting process. The bridge is owned by the North Carolina DOT.

2.4 Steel Girder Bridges

Brian Kozy and Lian Duan

2.4.1 Introduction

The steel and concrete girder bridges are clearly the most common of all bridge types in the United States, since the vast majority of bridges have spans less than 300 ft (91.4 m), where it is the most economical option. Steel girder bridges are typically made from either rolled I-beams, I-shaped plate girders, or so-called "tub" girders, which are U-shaped steel sections enclosed with a composite concrete deck. Steel girders with orthotropic deck are usually used for longer spans. There are subtle differences in design details for these bridge types found across the country, depending on the preferences of the state and the experience of builders in the region. However, the general features of the design have proven successful and remained largely unchanged for decades.

Some of the advancements that have been made in recent years include the use of high-performance steel (HPS) with improved material properties, integral abutments and piers to minimize deck joints and bearings, and application of simple spans made continuous for live load to ease construction. Uncoated weathering steel is used whenever the environment and site conditions allow for it, to minimize first cost and lifecycle maintenance. There is current motivation in industry to move more toward standardized and modular design to provide for accelerated fabrication and construction and economy from mass production of components.

New designs continue to push the limits with more severe support skews and horizontal curvature, which has presented new challenges for design, fabrication, and construction. Studies have shown that wider girder spacing of 11–14 ft (3.35–4.27 m) is more economical, however designs often use smaller girder spacing of 7–10 ft (2.13–3.05 m) with additional girders to improve redundancy and ability to perform future redecking in side-by-side staged construction. HPS has demonstrated ability to provide much higher fracture toughness than conventional steels, and research is underway that is expected to provide analytical techniques to engineer structural safety in low-redundancy options such as two-girder and/or single-girder cross sections, which are often the most economical.

2.4.2 Marquette Interchange Ramps

Milwaukee Transportation Partners (a joint venture between HNTB and CH2M HILL), provided preliminary and final design services for the $810 million reconstruction and reconfiguration of the Marquette Interchange and adjacent freeways in downtown Milwaukee, Wisconsin. The interchange as shown in Figure 2.23 includes three interstate highways: I-94, I-794 and I-43. It is the cornerstone of the southeastern Wisconsin freeway system. Figure 2.24 shows the underside of ramps showing the twin box superstructure analytically demonstrated to provide redundancy. The project includes a 5-level system interchange, 5 mi of interstate highway, 28 two-lane ramps, more than 60 bridges totaling 2.1 million square feet of bridge deck, and 5 mi of retaining wall. The construction of the interchange started in 2005. It was opened to traffic on August 19, 2008.

FIGURE 2.23 Aerial view of Marquette Interchange: 5 levels and 28 ramps. (Courtesy of HNTB.)

FIGURE 2.24 Marquette Interchange underside of ramps: twin box superstructure. (Courtesy of HNTB.)

The preliminary engineering phase identified two feasible bridge types for the system ramps: trapezoidal steel box girders and concrete segmental box girders. After 14 months of engineering and cost analysis the decision was made to move forward with the steel option. The centerpiece of the project consists of eight high-level system ramps, all curved multispan twin steel composite box girder bridges up to 2400 ft (731.5 m) long. Individual bents range up to 1600 ft (487.7 m) long between movement joints. The maximum spans of the steel box girders are in excess of 200 ft (61 m). Over 200 steel box girders were used in the construction.

Two-girder systems have historically been considered to be nonredundant by many owners, and hence fracture critical. Design and construction of nonredundant and/or fracture critical bridge structures has been avoided out of fear that failure of one main structural element would lead to total structural collapse. The redundancy of the proposed structures was addressed and proven by performing detailed three-dimensional (3D) collapse analysis after total fracture of one of the two main members of the structure.

HPS is a superior product, with higher yield strength, improved weldability, and greater levels of toughness, ductility, and corrosion resistance. It also has improved weathering resistance, all of which can lead to more economical bridges than conventional 50W designs. The design uses HPS70W and HPS50W, respectively, for adjacent field section elements and provided both the function and cost-benefit needed on a mega project like the Marquette Interchange.

2.4.3 Woodrow Wilson Bridge

The Woodrow Wilson Bridge (Figures 2.25 and 2.26), opened to traffic in 2007, carries the Outer and Inner Loop of I-95 and I-495 over the Potomac River in the Washington, DC metropolitan area, connecting the bridge's bascule spans to the Maryland shoreline. The bridge features 12 lanes, 6 lanes in

FIGURE 2.25 Aerial view of Woodrow Wilson Bridge, Potomac River crossing and movable bascule span. (Courtesy of URS Corporation.)

FIGURE 2.26 Woodrow Wilson Bridge, haunched girders with irregular spacing to accommodate pier tension struts. (Courtesy of the Virginia DOT. Photo by Will Torres.)

each direction plus shoulders and provision for future rail transit. After an initial contract bid for the Woodrow Wilson Bridge came in above budget in 2001, the bridge's original haunched parabolic tub girder design was modified to a more economical haunched plate girder design, allowing the designer (Parsons Transportation Group) to keep the aesthetics of the original design while lowering fabrication and shipping costs. Attention to steel detailing, fabrication, and constructability attributed to lowering fabrication and construction costs.

The bridge has a total length of 6736 ft (2,053 m). The unique steel framing of pairs of girders spaces at 5 ft (2.9 m) and intermediate floorbeam supported stringers in between was dictated by the need to clear the concrete tie beam of the V pier providing 19 ft (5.8 m) clearance, thereby requiring an intermediate stringer to support the concrete slab. The use of HPS material, while slightly more expensive and requiring a longer lead time, was well offset by providing better weldability and improved toughness and lighter weight of the superstructure. The concrete slab made use of epoxy coated reinforcement for additional corrosion protection consistent with the 75-year design life.

The bridge project was also split into three contracts. Some of the innovative features are the V-shaped piers and the movable bridge span. The Woodrow Wilson Bridge is one of only nine bridges on the U.S. Interstate Highway System that contains a movable span.

2.4.4 San Mateo-Hayward Bridge

The San Mateo-Hayward Bridge (Figure 2.27) is a part of California State Route 92 and crosses San Francisco Bay, linking the San Francisco Peninsula with East Bay. With a total length of 7 mi (11.3 km), it is the longest bridge in the San Francisco Bay Area. The 1.9 mi (3.1 km) high rise section, forming the western end of the bridge, is composed of steel orthotropic box girder spans. The eastern trestle portion accounts for the remaining 5.1 mi (8.2 km) of the overall length. The shipping channel beneath the high rise is 750 ft (229 m) wide with a vertical clearance of 135 ft (41 m).

The superstructure of the high rise section includes two main rectangular orthotropic box girders with open rib stiffeners of 3.43 m wide × 4.567 m deep at midspan and 135 m deep at supports. The orthotropic box deck with open rib stiffeners varies in thickness with deck stresses. The bridge was designed and is owned by the California DOT. The steel structure was constructed by fabricated by Murphy Pacific. The bridge opened to traffic in 1967.

FIGURE 2.27 High rise section of the San Mateo-Hayward Bridge. (Courtesy of Lian Duan.)

FIGURE 2.28   San Diego–Coronado Bridge. (Courtesy of the California DOT.)

2.4.5 San Diego–Coronado Bridge

The San Diego–Coronado Bridge (Figure 2.28) is a part of California State Route 75 and crosses San Diego Bay, linking San Diego with Coronado. The 2.2 mi long (3407 m) bridge ascends from Coronado at a 4.67% grade before curving 80 degrees toward San Diego. It is a prestressed concrete/steel girder bridge. The steel girder portion has a length of 2263 m. The main span reaches a maximum height of 200 ft (61 m), allowing warship navigation, and consists of three continuous spans (201 m, 201 m, and 171 m) of constant-depth single-cell orthotropic box girder 25 ft (7.62 m) high. The roadway has 65 ft (18 m) wide and carries five traffic lanes. Steel plate girders with composite concrete deck are used on the remaining length of 1690 m. It is notable among the world's great bridges for the number and size of its concrete towers. There are 30 towers designed with a curved cap to echo the mission arch shape, associated historically with regional architecture. The towers rest on 487 prestressed concrete piles of 54 in diameter and with walls 5 in thick.

The bridge featured the longest steel box girders in the world until 2008. The bridge was designed and is owned by the California DOT. The steel structure was constructed by fabricated by Murphy Pacific. The bridge opened to traffic in 1969.

2.5 Arch Bridges

Reggie Holt, Brian Kozy, Myint Lwin, Kenneth Wright and Lian Duan

2.5.1 Introduction

Arches have been admired and treasured by people throughout the world from ancient times to the present. They have inherent beauty and are technically efficient and competitive in building medium- and long-span bridges across rivers and deep valleys. Architects, engineers, and communities can work together to build beautiful and functional arches with lasting value to be a source of pride for the people. Several types of arches are illustrated in this section.

2.5.2 Steel Arch Bridges

2.5.2.1 Blennerhassett Island Bridge

The Blennerhassett Island Bridge (Figure 2.29), which spans the Ohio River between West Virginia and Ohio, was identified as the critical remaining "missing link" of the final segment of Appalachian Highway Corridor D, a major economic development highway that traverses approximately 240 mi (386 km) along US-50 from Cincinnati, Ohio, to Clarksburg, West Virginia.

The total length of the signature bridge is 4008 ft (1221.7 m). The structure includes an 878 ft (267.6 m), tied arch main span with a rise of 175 ft (53.5 m) and approach spans that consist of variably spaced steel plate girders with spans up to 401 ft (122.2 m) in length. The tied arch bridge includes a 100 ft (30.5 m), 6 inch wide deck that carries four 12 ft (3.7 m) wide lanes of traffic, with 18 ft (5.5 m) shoulders and a 14 ft (4.3 m) median.

To minimize the size and weight of the approach span superstructure, the design utilizes hybrid girders and high-strength steel. Post-tensioned pier caps support the main tied arch span and contribute to the structure's cost efficiency. The bridge's tied arch ranks as the longest networked tied arch structure in the United States and is among the longest in the world.

The central challenge of this project involved the need to develop a design solution that would satisfy the significant engineering constraints imposed by the bridge's massive proportions and maximize structural stability and safety, while minimizing costs. By pursuing a hybrid design for the Blennerhassett Island Bridge, engineers achieved their goal. The arch span is suspended by post-tensioned, seven-wire-strand steel cables configured in a unique X-shaped network, which enhances stiffness and redundancy in the superstructure. The post-tensioned, networked cables allow the structure to redistribute some of the arch rib horizontal load, so that the members function similarly to those in a truss structure.

To evaluate stress distribution within the structure under normal conditions, as well as during catastrophic events such as cable loss, a 3D finite element model of the bridge was created. The 3D model was used to refine the construction sequence. Each time the survey points on the arch were measured, the 3D model was updated to obtain data on the actual stresses to the members. The networked cables were carefully adjusted to optimize deck elevations and stress distribution for the structure, based on the results of the 3D model.

FIGURE 2.29 Blennerhassett Island Bridge. (Courtesy of Michael Baker Corp.)

The arch tie is a box-shaped tension tie that was specially designed to withstand cracking and not collapse. The tension tie was mechanically fastened together with bolts for redundancy, rather than welded together, which enables it to withstand loads, even if it is one of the four plates that comprise the box fractures. Mill-to-bear connections were established on the arch ribs to reduce the number of bolts required at the connection and redirect the load to load-bearing members.

In preparation for construction of the tie girders and arch, the contractor constructed eight temporary drilled caissons in the river as shown in Figure 2.30. The tie girders and arch were constructed in segments, from each main river pier, halfway across the channel. The most efficient method was to construct a significant portion of the tie girders and use this as a base for building out the arch ribs until the cantilevers reached the center of the span. To avoid disruption of river traffic on the Ohio River, a major commercial shipping channel, the tied arch was constructed while maintaining a reduced waterway opening. Temporary adjustable stays were used to brace the arch segments during erection, prior to installation of the cable hangers. As each cable hanger was installed, the supporting temporary stay was removed.

The Ohio side of the arch was constructed six inches out of position longitudinally, and then jacked into place during installation of the arch's keystone section. The ends of the arch were temporarily post-tensioned to the pier caps to ensure the stability of the cantilevered sections during jacking. Sand jacks with steel shims and polytetrafluoroethylene sliders were mounted on top of the river caissons and served as temporary supports. The jacks and sliders also could be quickly and easily removed after the arch was constructed. Large, barge-mounted cranes were necessary to install the heavy steel segments (which weighed up to 60 tons) for the arch and the West Virginia approach. The parabolic arch rib segments had to be precisely balanced during lifting to enable in situ connection to the erected segments.

The contractor designed a temporary bridge and used a "barge bridge" to cross the back channel of the Ohio River to access the island from the West Virginia shore. On the island, the contractor erected 70 ft high (21.3 m) falsework towers, designed to withstand a 75 mph (122.3 km/h) wind load, to support the girder segments. The towers were anchored by guy wires connected to concrete deadmen embedded in the island soil.

FIGURE 2.30 Tied arch construction while maintaining river traffic. (Courtesy of Michael Baker Corp.)

2.5.2.2 Fremont Bridge

The Fremont Bridge (Figure 2.31) is a steel tied arch bridge over the Willamette River located in Portland, Oregon. This double-deck bridge carries a total of eight lanes of I-405 and US-30 traffic between downtown and North Portland, where it intersects with I-5. It has the longest main span of any bridge in Oregon, and was the second-longest steel tied arch bridge in the world at the time of its completion. The bridge was opened to traffic on November 15, 1973.

The main span between supports is 1255 ft (382.5 m), and the two side spans are 448 ft (136.5 m) each. The roadway is 170 ft (51.8 m) above the river. Arch ribs are tied together transversely with K-bracing. The 902 ft (274.9 m) length of the main span where the arch is above the girder was constructed off site, floated on the river, and lifted into final position.

This is a two-hinged parabolic arch system design, with the ends of the side arches tied to the continuous 18 ft deep steel plate box girder at road level. The tie girder provides flexural stiffening as well as tension resistance for the arch thrust. The main arch and side-span arches are continuous at the supports. The final form of the arch was the result of extensive collaboration between the Oregon DOT; Parsons, Brinckerhoff, Quade & Douglas; the Portland Art Commission; and the FHWA.

The contract imposed strict restrictions on the interruption of river navigation. In order to have the least impact on navigation and lowest erection cost of the main span, the contractor built 902 ft (274.9 m) of the main span in California, and assembled it at Swan Island, about 1.7 mi (2.7 km) downstream from the bridge site. The preassembled section weighed 6000 tons (Figure 2.32).

After completion, the 6000 ton section of the main span was floated into place on a barge. On March 16, 1973 the 6000 ton steel arch span was lifted 170 ft (52 m) to the final position using 32 hydraulic jacks. At the time, it was the heaviest and highest lift ever completed. The contractor was credited for innovation in the use of technology and techniques to successfully accomplish a project of this scale.

2.5.2.3 New River Gorge Bridge

The US-19 New River Gorge Bridge (Figure 2.33) was completed near Fayetteville, West Virginia, on October 22, 1977, and this completed the last link in Appalachian Corridor "L." It was designed by the Michael Baker Company and was constructed by the American Bridge Division of United States Steel.

FIGURE 2.31 Fremont Bridge. (Courtesy of Bob Heims, U.S. Army Corps of Engineers.)

FIGURE 2.32   Preassembled main span, weighing 6000 tons. (Courtesy of City of Portland Archives.)

FIGURE 2.33   New River Gorge Bridge. (Courtesy of Wikipedia.)

The bridge, with a total length of 3030 ft (924 m), currently has the third-longest steel arch span in the world at 1700 ft (518.2 m) in length. The rise of the arch is 360 ft (109.7 m) and the overall height of the deck above the gorge is 876 ft (267 m). The total weight of steel in the bridge is 44,000 kips (195.7 millinewton) of which 21,066 kips (93.7 millinewton) is in the arch rib itself. The construction cost was $37 million. This bridge provides a spectacular view of the gorge, and yet does not detract from the appearance of the area due to the open structure and the use of unpainted weathering steel, which fits into the pristine surroundings.

This bridge is included in a discussion of truss bridges because the arch rib is a trussed arch, and the superstructure is supported by constant-depth deck trusses supported by the spandrel columns. The trussed arch was important in achieving the long span in a cost-effective manner. This was also one of the first major bridges that was fabricated using unpainted weathering steel.

2.5.3 Concrete Arch Bridges

2.5.3.1 Hoover Dam Bypass Bridge

The new 1900 ft long (571 m) Hoover Dam Bypass Bridge (Figure 2.34) spans the Black Canyon at about 1500 ft (457.2 m) south of the Hoover Dam, connecting the Arizona and Nevada Approach highways nearly 900 ft (274.3 m) above the Colorado River. The bridge carries four lanes of traffic and a sidewalk and is a part of US-93. It is a composite steel–concrete deck arch bridge with a main span of 1060 ft (323 m). The arch rib is made of reinforced concrete. It is the second-longest concrete arch bridge span in the world. The bridge is owned by Arizona DOT and the Nevada DOT.

Prior to the completion of the new bridge, the existing US-93 used the top of the Hoover Dam to cross the Colorado River. US-93 is the major commercial corridor between the states of Arizona, Nevada, and Utah. The traffic volumes, combined with the sharp curves on US-93 in the vicinity of the Hoover Dam, created a potentially dangerous situation. A major catastrophe could occur, involving innocent bystanders, millions of dollars in property damage to the dam and its facilities, contamination of the waters of Lake Mead or the Colorado River, and interruption of the power and water supply for people in the Southwest. By developing an alternate crossing of the river near the Hoover Dam, through-vehicle and truck traffic are removed from the top of the dam. This new route eliminates the problems with the former highway: sharp turns, narrow roadways, inadequate shoulders, poor sight distance, and low travel speeds.

The type study for the Hoover Dam Bypass Bridge was developed by the design team, comprised of Central Federal Lands Highway Division (CFLHD) and the design consultant, and was guided by direction from the FHWA, the Arizona DOT, the Nevada DOT, the Bureau of Reclamation, the National Park Service, and the Western Area Power Administration. The public had the opportunity to comment through the project website and by casting ballots at the visitor center at the Hoover Dam. A composite concrete–steel deck arch bridge was selected to address the specific design issues inherent to the Hoover Dam site. It was selected on the merits of cost, schedule, aesthetics, and technical excellence.

FIGURE 2.34 Hoover Dam Bypass Bridge. (Courtesy of FHWA.)

The bridge construction contract was awarded in October of 2004 for $114 million to Obayashi Corporation/PSM Construction USA, Inc. (Joint Venture). Construction began in early 2005 and was completed in October 14, 2010, meeting construction requirements and challenges within budget and on schedule. The bridge was opened to pedestrians on October 16, 2010, with a dedication ceremony for the communities. The bridge was opened to traffic on October 19, 2010.

The United States Congress officially named the new Hoover Dam Bypass Bridge the Mike O'Callaghan–Pat Tillman Memorial Bridge after two prominent local citizens who dedicated themselves to public service and the greater good. Mike O'Callaghan was a longtime Nevadan, former governor, community leader, and businessman. He died in March 2004 at the age of 74. Pat Tillman graduated with honors from Arizona State University and played professional football for the Arizona Cardinals before joining the Army. He was killed in Afghanistan in 2004 at the age of 27.

2.5.3.2 Bixby Creek Bridge

Bixby Creek Bridge (Figure 2.35), a reinforced concrete open-spandrel arch bridge, is a part of California Highway Route 1, carrying two traffic lanes in Big Sur, California. The bridge is located 120 mi (190 km) south of San Francisco along the west coast. It is 714 ft (218 m) long, 24 ft (7.3 m) wide and has a main span of 320 ft (98 m), with two heavy abutments unnecessary to support the structure. Bixby Bridge has 10 evenly spaced column supports from the arch ribs to its roadway, and supports outside the arch of longer, even spacing than that of the supports between the abutments. The bridge is about 280 ft high (85 m) above the canyon, and construction of the bridge required 26 stories of falsework.

Bixby Creek Bridge is aesthetically pleasing and one of the most photographed bridges in the world. It was one of the largest single-arch concrete bridges in the world when it was completed in 1932. The bridge was seismically retrofitted in 2000. The bridge was designed and is owned by the California DOT.

FIGURE 2.35 **(See color insert.)** Bixby Creek Bridge. (Courtesy of the California DOT.)

2.6 Truss Bridges

Kenneth Wright

2.6.1 Introduction

Trusses have been a staple type of steel bridges for more than a century. From a structural perspective, trusses use material very efficiently, resulting in lower material costs than many other bridge types. At the same time, the labor required to fabricate and erect a truss bridge is much higher than for many other bridge types. At the most basic level, trusses carry load similarly to a girder. The top and bottom chords of the truss primarily carry the bending moments in the truss, much as the flanges do in a girder. The diagonals and/or verticals primarily carry the shear forces that are developed, much as the web does in a girder.

Trusses were very common in the first half of the twentieth century when the ratio of material cost to labor cost was high. The efficiency of trusses rendered them the bridge of choice in many locations over a wide range of span lengths. Many trusses from this era were composed of eyebar tension members and built-up compression members that were connected by large pins at the joint locations. A majority of the other trusses were composed of riveted built-up members connected at the joints by riveted connections. The members were most commonly I-sections or box sections comprised of angles or channels tied together by lacing bars or batten plates. These built-up members were relatively light and efficient.

Lacing bars are narrow steel plates, typically 2–3 in (51–76 mm) wide (Figure 2.36) that were typically used in a lattice pattern to brace the light members that provided the structural strength of the members. Batten plates are much wider than lacing bars, typically 4–12 in (102–305 mm) wide (Figure 2.37) that were installed perpendicular to the member axis to brace the primary member components. Batten plates were typically used for built-up sections using heavier rolled shapes as basic components that did not require bracing at the close spacing required for laced members.

As the twentieth century progressed and labor began to surpass material as the largest portion of bridge cost, it became more common for trusses to be constructed using rolled W-shapes or welded I-shapes and box girders. While heavier, the reduction in labor cost associated with these member types still allowed trusses to be a viable and cost-effective bridge type. Eventually, simpler bridge types such as steel plate girders and prestressed or segmental concrete became more cost-effective in the most

FIGURE 2.36   Lateral bracing members using lacing bars. (Courtesy of HDR Engineering, Inc.)

FIGURE 2.37   Member using batten plates. (Courtesy of HDR Engineering, Inc.)

common bridge spans. Trusses remain cost-effective for longer span structures, particularly in the span range of 500–700 ft. Trusses can also be cost-effective for shorter span lengths, particularly if other constraints are present, such as limitations on structure depth.

Many of the longer span trusses that were built in the United States used suspended spans to achieve these long span lengths. When using this approach to achieve a continuous truss, the anchor spans and a portion of the main channel span were constructed first, leaving an opening in the center span. The closure section of the center span was then erected on barges and floated into place under the cantilevered portions of the bridge. The closure section was then raised into place using pulley systems supported by the cantilevers and connections made to complete the bridge. This method of construction allowed the use of very long main spans without requiring a significant amount of falsework in the navigation channel during construction.

Analysis of trusses was based on the presumption that they could be analyzed assuming the connections are pinned. Pinned trusses are easily reconciled with this analysis method. Trusses with riveted or bolted end connections to gusset plates have historically been analyzed assuming that they have pinned connections also. While this is not strictly correct, the end connections are compact enough relative to the length and stiffness of the truss members that the pinned end connection approximation is reasonable. The validity of this assumption has been confirmed through computer analyses of trusses.

Floor systems generally consist of transverse floor beams with longitudinal stringers that can be either framed into the floor beams or a stacked system in which the stringers can be run continuously over multiple floor beams. Framed-in floor systems (Figure 2.38) are ideal when there are significant depth restrictions for the superstructure. The stringer sections are short and usually connected to the floor beam webs via simple shear connections. Stacked floor systems (Figure 2.39) provide several advantages. Most notably, the stringer design can be optimized by using a continuous design as opposed to the simple span design that is used for a framed-in system, and the stringer-to-floor beam connections are much simpler and require less labor to fabricate and erect.

Bracing is an important component of truss bridges because they have very limited inherent lateral stiffness. In most trusses, particularly those with longer spans, both top chord and bottom chord lateral bracing are provided. The primary functions of top and bottom lateral bracing are to provide

FIGURE 2.38   Framed-in floor system. (Courtesy of HDR Engineering, Inc.)

FIGURE 2.39   Stacked floor system. (Courtesy of HDR Engineering, Inc.)

lateral stiffness for the truss system and to provide a load path for lateral wind loads to be carried back to the support locations. This lateral bracing is typically in some variation of X- or K-bracing. The other bracing component typically used in trusses is sway bracing. Sway bracing is typically installed along vertical or diagonal members of the truss. The primary purposes of sway bracing are to distribute loads between trusses to maintain the verticality and relative position of the trusses. The sway braces at the end posts of the truss, called portal braces, tend to be relatively heavy in order to transfer wind loads from the top chord down to the bearing locations.

Deck trusses and through trusses are the most common types. For deck trusses, the transverse floor beams typically rest on the top chord. The trusses are often located inside the edges of the bridge deck, reducing the cost of the sway bracing and lateral bracing systems for the bridge. The floor beams overhang the trusses by several feet, thus providing some continuity that leads to an efficient floor beam design. Through trusses are more typically used in locations where depth restrictions prohibit having the truss below the deck, such as over a navigable waterway. The floor system typically is supported at the bottom chord level. The trusses must be spaced much farther apart since the deck is located between the trusses. This leads to more costly sway bracing and lateral bracing systems because of the additional member lengths. Additionally, the floor beams are less efficient as they are generally designed as simple spans between trusses.

Trusses typically are fracture critical structures, that is, they contain members wherein the failure of one truss member could lead directly to the collapse of the bridge. Yet failure of truss bridges has been rare due to the generally favorable member details and the relatively low stress ranges in the truss members. While riveted connections would only be considered a Category D fatigue detail, the longer spans generally associated with truss bridges lead to stress ranges that are acceptable even for a Category D detail.

Trusses also provide great aesthetic flexibility. The top and bottom chords can take a variety of shapes— both parallel chord and variable depth trusses are common. Additionally, the member types and sizes also can impact the appearance of the truss. Trusses prior to the 1960s primarily consisted of riveted or bolted built-up members. Since the mid-1960s, many trusses have been constructed primarily using welded box shapes for many of the members and rolled shapes or welded I-shapes for the other members. The welded construction used in modern trusses provides a cleaner, more contemporary appearance to the truss when compared with the built-up member fabrication prevalent prior to the 1960s.

The truss bridge form has proven to be adaptable to other types of structures. Arch bridges have been constructed using trussed arch ribs as opposed to solid rib arches. This has allowed the arch bridge form to be extended to much longer spans than would be possible using solid rib arches. The New River Gorge Bridge (Figure 2.33), highlighted in greater depth in the arch section of this handbook, is an example of a truss-ribbed arch that is currently the third-longest arch span in the world. Additionally, the floor system of this bridge is a truss system, allowing the weight to be minimized.

Another application of trusses in highway bridges is in suspension bridges. It is important for bridges as long and slender as suspension bridges to have some lateral and torsional stiffness at the deck level to minimize dynamic movements and excitation under wind and/or seismic loads. A classic example of this is the Golden Gate Bridge (see Section 2.8.2) in San Francisco, California. As with the New River Gorge Bridge, the stiffening truss also helps to minimize the overall weight of the floor system.

2.6.2 Commodore Barry Bridge

The Commodore Barry Bridge (Figure 2.40), named after the Revolutionary War hero, spans the Delaware River near Chester, Pennsylvania. Designed by the engineering firm E. Lionel Pavlo, Inc., the bridge has a main span of 1644 ft (501.1 m), which ranks as the third-longest main cantilever span in the world. The three main spans are 822 ft, 1644 ft, and 822 ft (250.5 m, 501.1 m, and 205.5 m), with more than 10,000 ft (3048 m) of approach spans. The vertical clearance over the channel is 192 ft (58.5 m) and the height to the top of the towers is 418 ft (127.4 m) above mean high water.

The overall bridge length including approaches is 2.6 mi (4.18 km), and it carries five lanes of traffic. The bridge was completed on February 1, 1974, at a cost of $115 million. There is a total of 49,000 tons

FIGURE 2.40   Commodore Barry Bridge. (Courtesy of Jim Dietrich, Wikipedia.)

of steel in the bridge. The bridge was retrofitted with a network of cables shortly after it opened to alleviate vibration issues in the main cantilever span that were leading to premature cracking of certain members. Additionally, tuned mass dampers were installed on many of the I-shaped members. The pin hanger assemblies in the main suspended span hangers were retroffitted to add redundancy.

2.6.3 Richmond–San Rafael Bridge

The Richmond–San Rafael Bridge (Figure 2.41) crosses San Francisco Bay north of the San Francisco-Oakland area. Opened in September of 1956, it crosses between Marin and Contra Costa Counties. The total bridge length is approximately 5.5 mi (8.9 km) and was constructed at a cost of $66 million. The bridge is a double-deck structure, carrying westbound traffic on the upper deck and eastbound traffic on the lower deck.

The bridge crosses two primary shipping channels, so the two three-span trusses were designed to be identical through trusses in order to save money. This resulted in a dip in the bridge profile between the main channel bridges, and the bridge has been nicknamed the "roller coaster span." The main spans of the cantilever trusses are 1070 ft (326.1 m) long, with a vertical clearance of 185 ft (56.5 m) and a tower height of 325 ft (99.1 m). The approach spans are primarily deck trusses.

2.6.4 Corpus Christi Harbor Bridge

The Corpus Christi Harbor Bridge (Figure 2.42) carries six lanes of US-181 over the Port of Corpus Christi ship channel entrance. The structure, built in 1959, consists of 5 truss spans, 15 welded plate girder spans, and 37 prestressed concrete girder spans. The total bridge length measures 5819 ft (1773.6 m). The

FIGURE 2.41 Richmond–San Rafael Bridge. (Courtesy of the California DOT.)

FIGURE 2.42　Corpus Christi Harbor Bridge. (Courtesy of HDR Engineering, Inc.)

Ztruss spans consist of two 272 ft (82.9 m) simple span deck truss units (spans 27 and 31) and a three-span, 1244 ft (379.2 m) continuous truss unit (spans 28, 29, and 30, with spans of 312 ft, 620 ft, and 312 ft [95.1 m, 189.0 m, and 95.1 m]). The 1244 ft (379.2 m) truss unit consists of a 388 ft (118.3 m) suspended tied arch segment supported by two 116 ft (35.4 m) cantilever truss segments and two 312 ft (95.1 m) anchor spans. The trusses are comprised almost entirely of built-up members with riveted connections.

This bridge is the only significant highway link across the ship channel between the city of Corpus Christi and recreational, residential, and industrial areas to the north. It can never be completely closed to highway traffic because there is no viable detour. In addition, maritime access must be maintained continuously because the Corpus Christi ship channel is one of the busiest in the United States.

2.7　Cable-Stayed Bridges

Lubin Gao

2.7.1　Introduction

The John O'Connell Memorial Bridge in Sitka, Alaska, is usually considered the first cable-stayed bridge in the United States carrying highway vehicles. It opened in 1972 with a main span of 450 ft (137 m). Its steel pylons support the steel deck with three stays at each side. The first concrete cable-stayed highway bridge in the United States is the Ed Hendler Bridge over the Columbia River, connecting Pasco and Kennewick, Washington, with a main span of 752 ft (222 m). Its 80 ft wide (24.4 m) concrete deck was built by cast-in-place segmental construction. It opened in 1978.

In 1983, the first cable-stayed bridge on the interstate highway system, the Luling Bridge, opened to traffic. The bridge, also known as the Hale Boggs Memorial Bridge, is in St. Charles Parish, Louisiana, and carries four lanes of I-310 across the Mississippi River. It is the only cable-stayed bridge with a steel orthotropic deck in the United States. Its main span reaches 1222 ft (372.5 m) with a steel deck 76 ft (23.2 m) wide. The pylons were built with weathering steel. Since then, more than 30 cable-stayed highway bridges (Figure 2.43) have been constructed in the country. The maximum span length ranges from 300 ft (90 m) to 1600 ft (480 m). Engineering practice has demonstrated that a cable-stayed bridge is the most competitive solution for spans longer than 1000 ft (300 m) both in aesthetics and economy.

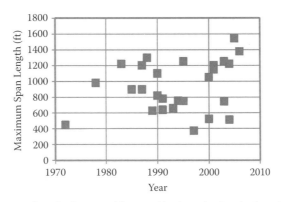

FIGURE 2.43   Maximum span length of major cable-stayed highway bridges built in the United States.

A cable-stayed bridge consists of a deck, stay cables, piers, pylons, foundations, and other structural components. The deck supports the roadway while providing the passageway for traffic. Stay cables typically anchored at the deck girder at the bottom and the pylon at the top are stressed to transfer the loads from the deck to the pylons. These loads, including the weight of the superstructure and superimposed dead load, are ultimately carried by the piers and pylons and further transferred to the foundations primarily through compression. Active stressing of the stay cables during the construction stages and fine tuning of the cable forces before completing the adjustment of load effects, especially the bending moments in the deck girder and pylons, are necessary to arrive at an optimized state under dead loads. These actions yield the effective use of materials and generate cost savings as a result.

Three types of superstructures are generally used as the deck of cable-stayed bridges in the United States:

1. Steel edge girders framed with steel floor beams composite with a concrete deck slab (Cooper River Bridge)
2. Concrete edge girders with concrete floor beams and a cast-in-place concrete deck slab (Sidney Lanier Bridge)
3. Concrete box girders (Sunshine Skyway Bridge)

Steel plate girders and rectangular concrete beams are two widely utilized types of edge girders. The concrete edge beams are normally cast in place with form travelers. The steel edge girders are mostly preassembled with the floor beams to form steel frame segments determined by lifting capacity of hoisting equipment. The segments are then lifted into place and splice connected to the previously erected segment.

In the United States, concrete box girders used in cable-stayed bridges are typically precast. The specific strength of the concrete of concrete edge beams and box girders ranges from 5,000 psi (35 MPa) to 10,000 psi (69 MPa). The structural steel conforming to ASTM A709 Grade 50 (AASHTO M270 Grade 50) with minimum yield strength of 50 ksi (345 MPa) is generally utilized for the steel edge girder and floor beams. When the ratio of side span to main span is small (e.g., the Charles River Bridge), the optimized solution is a hybrid construction with a steel main span and concrete side spans.

High-strength, low-relaxation prestressing strands are used for stay cables as industry standard practice in the United States. The 0.6 in (15 mm) diameter strands conformed to ASTM A722 (AASHTO M275) Grade 270 with a minimum tensile strength of 270 ksi (1862 MPa). To meet the design life requirements for durability and safety, multilayer corrosion-protective measures are normally implemented in stay cables. The first barrier for corrosion protection is the outer pipe, typically made of HDPE (high-density polyethylene) or stainless steel. The last barrier is the individual sheathing of each strand. Under dead load conditions, the pylons are primarily in compression. Therefore, they are mostly built with concrete and cast in place with formworks. Hollow sections are often used for better bending properties and ease to facilitate the anchorages of stay cables. Large-diameter drilled shafts and steel pipe piles are two types of deep foundations generally utilized under the pylons of cable-stayed bridges in the United States, required by the large reactions resulting from the long main span.

Designers of long-span cable-stayed bridges always face challenges. The first challenge is the aero-dynamic stability and wind responses during both construction stages and in service. Due to its greater flexibility as a result of shallower deck and longer spans, a cable-stayed bridge typically has lower bending and torsional natural frequencies. The limiting unlimited amplitude vibration from wind, such as flutter, should be carefully studied through structure modifications based on analysis and wind tunnel testing. Measures should also be investigated during the design phase and implemented in the construction or service stages as deemed necessary to control the limited amplitude vibration due to wind. Vibration due to vehicle loading and other environmental factors exerts another challenge on serviceability. The variation of stresses due to these vibrations and vehicle traffic affects the design, detailing, and fabrication of stay cables and anchorages that meet the fatigue requirements.

In the United States, the design and construction of cable-stayed bridges are governed by the AASHTO *Standard Specifications for Highway Bridges* (AASHTO 2002) or the *LRFD Bridge Design Specifications* (AASHTO 2010a) and *Construction Specifications* (AASHTO 2010b). Because of the special features of cable-stayed bridges, project-specific design criteria and special provisions are always developed for each project to supplement the AASHTO Specifications. The "Recommendations for Stay Cable Design, Testing and Installation" by the Post-Tensioning Institute (PTI 2008) and the AASHTO *Guide Specifications for Design and Construction of Segmental Concrete Bridges* (AASHTO 2003) are two important documents for the design of cable-stayed bridges.

2.7.2 Sunshine Skyway Bridge

The Bob Graham Sunshine Skyway Bridge (Figure 2.44) is in St. Petersburg, Florida, across Tampa Bay, with a 1200 ft (366 m) main span. When the bridge opened in 1987, it was considered a milestone in the construction of cable-stayed bridges in the United States. It was the world's longest concrete cable-stayed bridge at the time of its completion.

The skyway bridge (PCI 1986) was designed to replace the old steel truss spans constructed in 1954 that were partly destroyed in 1980, when the more than 1200 ft long (366 m) steel structure was struck by a ship during a storm. The bridge was required to provide a 1000 ft × 175 ft (305 m × 53 m) navigational channel. The main structural unit consists of 11 spans of 140 ft, 3 spans of 240 ft, 3 spans of 540 ft, 1200 ft and 540 ft, 3 spans of 240 ft and 1 span of 140 ft (3 spans of 73.2 m, 3 spans of 164.6 m, 366.7 m and 164.6 m, 3 spans of 3 73.2 m and a span of 42.7 m). The skyway carries two lanes of highway traffic and a full-width shoulder in each direction. In addition, a median of 12.5 ft (3.81 m), inclusive of parapet widths, is needed to anchor the stay cables. This requires a total deck width of 95.3 ft (29 m).

The superstructure is a single-cell concrete box girder with a depth of 14.67 ft (4.47 m) and a bottom slab width of 32.83 ft (10 m). The top slab of the box girder was transversely post-tensioned with tendons consisting of 0.6 in (15 mm) diameter strands, after the concrete strength reached 2500 psi (17 MPa). The precast box girder segments were match-cast using the short-line method. Internal struts were provided to support the wide top slab of the box girder. These struts were precast separately and post-tensioned at the cable anchor segments to distribute the cable forces. A self-advancing beam-winch system was utilized in erecting the precast segments.

FIGURE 2.44   Bob Graham Sunshine Skyway Bridge. (Courtesy of FIGG Bridge Engineers.)

The two pylons are reinforced concrete, 431.36 ft (131.48 m) above water and 242.38 ft (73.88 m) above the roadway surface. Each pylon carries 21 stay cables. The high-level approach span piers consist of match-cast box segments, manufactured in the casting yard, barged to the site, lifted into place, and post-tensioned vertically. Piers of the 4000 ft (1,219.2 m) main span structural unit were cast in place. Piles were used, except for the pylon foundations where shafts 5 ft (1.52 m) in diameter were utilized. Stay cables consist of strands 0.6 in. (15 mm) in diameter, encased in 6 5/8 in or 8 5/8 in (168 mm or 219 mm) steel pipes and grouted after final stressing. The size of the stay cables ranges from 38 to 82 strands.

The high-level approach and the cable-stayed main spans were designed by Figg & Muller Engineers. Paschen Contractors, Inc., American Bridge, and Morrison-Knudsen were the contractors for the precast segmental high-level approach and main spans. Construction inspection was performed by Skyway Construction Engineering and Inspection Consultants, a consortium consisting of Parsons, Brinckerhoff, Quade & Douglas; DRC Consultants (later merged with T. Y. Lin International); Kissinger, Campo & Associates; and H. W. Lochner, Inc. LoBuono Armstrong & Associates provided construction engineering services to the contractors for erecting the segmental and cable-stayed spans. The Florida DOT owns the bridge.

2.7.3 Sidney Lanier Bridge

The Sidney Lanier Bridge (Figure 2.45) in Brunswick, Georgia, carries four lanes of US-17 across the South Brunswick River. It replaced the original bridge built in 1956 and partially destroyed by ship collisions in 1972 and 1987. The new bridge provides a vertical clearance of 185 ft (56.4 m) at the centerline of the navigational channel, allowing large ships to access the Port of Brunswick. The bridge opened in 2003.

The navigational clearance requires a long-span bridge. The cable-stayed bridge consists of a main span of 1250 ft (381 m) and two side spans of 625 ft (190.5 m) each. To accommodate four 12 ft (3.66 m) traffic lanes, two 8 ft (2.44 m) shoulders, and spaces to anchor the stay cables, the deck was designed to have a total width of 79.5 ft (24.2 m). It consists of two concrete edge beams (4.5 ft [1.37 m] wide by 5 ft [1.52 m] deep) and an 11 in thick (0.279 m) concrete deck. At each cable anchorage location, a transverse concrete floor beam is provided. The deck was cast in place by segmental construction with form travelers. The length of a typical segment is approximately 27.7 ft (8.4 m).

The two H-shaped concrete pylons were cast in place 486 ft tall (148 m) measured from the top of footing to the top of pylon. The legs of the pylons are rectangular hollow sections providing access for

FIGURE 2.45   Sidney Lanier Bridge. (Courtesy of Tim Ross, Wikipedia.)

inspecting the cable anchorages inside the legs. Each of the pylons supports the deck with 44 stay cables anchored to the concrete edge beams on either side of the deck. There are 176 stay cables in total supporting the concrete superstructure. Each stay cable consists of 0.6 in (15 mm) diameter low-relaxation strands, encased in HDPE pipes. The longest cable measures approximately 687.5 ft (209.6 m). This bridge is the last of the major bridges that used cement-grouted stay cables in the United States. There have been corrosion issues attributed to grouting itself.

The cable-stayed bridge was designed by DRC Consultants (later merged with T. Y. Lin International). Finley McNary provided the construction engineering services to the contractor, a joint venture of Recchi America and GLF Construction Corporation. Construction inspection was provided by Figg Engineering Group. The Georgia DOT owns the bridge.

2.7.4 Charles River Bridge

The Charles River Bridge (Figure 2.46) is a cable-stayed bridge in Boston, Massachusetts, carrying I-93 and US-1 traffic across the Charles River. Its formal name is the Leonard P. Zakim Bunker Hill Memorial Bridge (Kumarasena, et al. 2003) to commemorate both Boston's civil rights activist Leonard P. Zakim and the Battle of Bunker Hill during the Revolutionary War. It is the widest and the first steel–concrete hybrid cable-stay bridge in the United States. The bridge replaced the Charlestown High Bridge, a steel truss bridge constructed in 1954.

The cable-stayed structure consists of a main span and two side spans: 267 ft (81.4 m) at the downtown side and 745 ft (227.1 m) and 420 ft (128 m) at Charlestown side. The bridge carries 10 lanes of traffic: 8 lanes passing through the legs of the pylons and 2 lanes cantilevered on the east side, exterior of the stay cables. The cantilevered exterior lanes, which accommodate northbound traffic from the Sumner Tunnel to the North End, provide the unique and asymmetrical design.

The deck is 183 ft (56 m) wide in the main span and 126 ft (38.4 m) wide in the side spans. The superstructure of the main span is a single steel box girder 18 ft (5.49 m) deep at the piers and 9 ft (2.74 m) deep at center span, composited with precast concrete deck panels. The steel box girder consists of two steel box edge girders, a longitudinal steel fascia girder, floor beams, cantilever floor beam extensions, and precast concrete deck panels. The superstructure of the side spans is a 10 ft deep (3.05 m) cast-in-place concrete box girder, partially filled with heavyweight concrete to balance the weight in the main span. To accommodate the longitudinal grade, as much as 5%, the two inverted Y-shaped pylons are different heights of 322 ft (98.1 m) for the north pylon and 295 ft (89.9 m) for the south pylon, measured from the

FIGURE 2.46 Charles River Bridge. (Courtesy of VidTheKid, Wikipedia.)

water. However, the heights of both pylons measured from the top of the roadway to the top of pylons are the same. The pylons were built by cast-in-place construction. The Grade 70 HPS anchor box was cast in the hollow pylon legs to anchor the cables. Each pylon is supported by drilled shafts 8 ft (2.4 m) in diameter.

There are 116 stay cables supporting the roadway from two pylons. On the main span, there are four sets of 17 cables from each side of the roadway to the top of the pylons. On each side span, there are 24 stay cables in single-plane from the median to the top of the pylons. Each stay cable consists of low-relaxation seven-wire strands 0.6 in (15 mm) in diameter, encased in a HDPE pipe treated with a double helical fillet to mitigate potential wind-induced cable vibrations. Each strand is individually sheathed and protected from corrosion. The longest cable is close to 500 ft (152.4 m) in length. The size of cables ranges from 14 strands to 73 strands.

The concept of the bridge was initially developed by Christian Menn. The final design was engineered by HNTB and Figg Bridge Engineers. Bechtel and Parsons Brinkerhoff served as construction management consultants. The general contractor was Atkinson-Kiewit Joint Venture. T. Y. Lin International provided construction engineering to the contractor. The Massachusetts Turnpike Authority owns the bridge.

2.7.5 Cooper River Bridge

The new Cooper River Bridge (Figure 2.47) was opened to traffic and named the Arthur Ravenel Jr. Bridge on July 16, 2005. It connects downtown Charleston to Mt. Pleasant, South Carolina, carrying eight traffic lanes of US-17 across the Cooper River. The new bridge (Abrahams, et al. 2003) replaced the two deficient steel truss bridges built in 1929 and 1966, respectively. The high seismicity, hurricanes, and potential ship collisions created critical challenges to the design of the bridge.

The navigational clearances of 1000 ft (304.8 m) horizontal and 186 ft (56.7 m) vertical required a long-span bridge. The cable-stayed bridge consists of a main span of 1546 ft (471.2 m), two side spans of 650 ft (198.1 m) each, and two end spans of 225 ft (68.6 m) each, that is, 225 ft (68.6 m), 650 ft (198.1 m), 1546 ft (471.2 m), 650 ft (198.1 m), and 225 ft (68.6 m), resulting in a total length of 3296 ft (1004.6 m). Its main span is the longest in North America.

FIGURE 2.47   Arthur Ravenel Jr. Bridge. (Courtesy of Xinghua Lu.)

To accommodate four 12 ft (3.66 m) traffic lanes in each direction, a 12 ft (3.66 m) bicycle/pedestrian lane at the south side of the deck, and spaces to anchor the stay cables, the deck was designed to have a total width of 126 ft (38.4 m). It consists of two 6.5 ft (1.98 m) deep steel plate edge girders and steel floor beams spaced at 15.7 ft (4.78 m) composite with a 9 1/2 in. (241 mm) concrete deck slab. The deck slab is composed of 8000 psi (55 MPa) precast panels, with cast-in-place closure strips over the girders and floor beams, and a 2 in thick (50.8 mm) latex-modified concrete wearing surface. The two diamond-shaped reinforced concrete pylons are each 570 ft (174 m) tall, built by cast-in-place construction. Each pylon is supported on 11 drilled shafts 10 ft (3.05 m) in diameter, each measured 230 ft (70.1 m) in length. The two main foundations are protected from vessel impact with rock islands.

There are 128 stay cables in total. Each cable consists of 31–90 low-relaxation seven-wire strands 0.6 in (15 mm) in diameter, encased in a HDPE pipe of 8 in (203 mm) to 12 in (305 mm) in diameter. Each strand is individually sheathed and protected from corrosion. The exterior surface of the HDPE pipes was treated for ultraviolet protection and to reduce rain- and wind-induced vibration.

The new Cooper River Bridge was constructed with the design-build method by Palmetto Bridge Constructors, a joint venture between Tidewater Skanska and Flatiron Constructors, Inc. The cable-stayed bridge was designed by Parsons Brinckerhoff, in association with Buckland & Taylor Ltd. Design review and construction inspection were provided by the joint venture of HDR, Inc. and T. Y. Lin International. The South Carolina DOT owns the bridge.

2.8 Suspension Bridges

Lian Duan and Myint Lwin

2.8.1 Introduction

Suspension bridges have the longest spans of any type of bridges. This type of bridge has cables suspended between towers and suspender cables that carry the weight of the deck below, upon which traffic crosses. Table 2.5 lists major suspension bridges built in the United States.

TABLE 2.5 List of Major Suspension Bridges in the United States

Bridge Name	Location	Owner	Total Length ft (m)	Main Span ft (m)	Year Opened to Traffic
Wheeling	West Virginia	West Virginia DOT	1307 (398.4)	949.2 (289.3)	1849
John A. Roebling	Kentucky-Ohio	Kentucky Transportation Cabinet	2162 (659.0)	1057 (322.2)	1866
Brooklyn	New York	New York City DOT	5989 (1825) anchored at each end of the bridge, plus vertical	1595 (486.2)	1883
Williamsburg	New York	New York City DOT	7308 (2227.5)	1600 (487.7)	1903
Manhattan	New York	New York City DOT	6855 (2089.4)	1470 (448.1)	1909
Bear Mountain	New York	New York State Bridge Authority	2255 (678.2)	1632 (497.4)	1924
Benjamin Franklin	New Jersey/ Pennsylvania	Delaware River Port Authority	9573 (2917.9)	1750 (533.4)	1926
Mid Hudson	New York	New York State Bridge Authority	3000 (914.4)	1500 (457.2)	1930
St. Johns	Oregon	Oregon State DOT	2067 (630.0)	1207 (367.9)	1931
George Washington	New York/New Jersey	Port Authority of New York and New Jersey	4760 (1450.9)	3500 (1066.8)	1931
San Francisco- Oakland Bay, West Spans	California	California DOT	10,122 (3085.2)	2310 (704.1 m)	1936
Golden Gate	California	Golden Gate Bridge Highway and Transportation District	8981 (2737.4)	4200 (1280.2)	1937
Bronx-Whitestone	New York	MTA Bridges and Tunnels	3770 (1149.1)	2300 (701.0)	1939
Tacoma Narrows	Washington	Washington State DOT	5939 (1810) 5979 (1822) 5900 (1736)	2800 (853) 2800 (853) 2800 (853)	1940 1950 2007
Delaware Memorial I and II	Delaware	Delaware River and Bay Authority	10,786 (3287.6)	2150 (655.3)	1951
Mackinac Bridge	Michigan	Michigan DOT	26,372 (8038)	3800 (1158)	1957
Verrazano Narrows	New York	MTA Bridges and Tunnels	13,700 (4175.8)	4260 (1298.5)	1964
Carquinez	California	California DOT	3465 (1056.1)	2388 (727.9)	2003
San Francisco- Oakland Bay Self-Anchored Suspension Span	California	California DOT	2049 (624.4)	1263 (385)	2013
Three Sisters in Pittsburgh, Self-Anchored	Pennsylvania	Allegheny County	884 (269.4)	430 (131.1)	1926–1928

2.8.2 Golden Gate Bridge

The Golden Gate Bridge (Figure 2.48) is a part of US-101 and California State Route 1 connecting the San Francisco Peninsula to Marin County on the other side of the strait. It consists of six structures: San Francisco (south) approach viaduct, San Francisco (south) anchorage housing and pylons S1 and S2, Fort Point arch, suspension bridge, Marin (north) anchorage housing and pylons N1 and N2, and Marin (north) approach viaduct, with a total length of 9150 ft (2788 m). The landmark structure spanning the strait at the entrance to San Francisco Bay onto the Pacific Ocean is a three-span suspension bridge with a center span of 4200 ft (1280 m) and two side spans of 1125 ft (343 m). The Golden Gate Bridge was the longest suspension bridge in the world and held that distinction until the 1964 completion of the Verrazano-Narrows Bridge (main span 1298 m) in New York City. It is one of the best-known engineering structures in the world and an internationally recognized icon of San Francisco, California, and the United States. Even today, it is still the second-longest suspension bridge main span in the United States.

The Golden Gate Bridge is 90 ft (27 m) wide with six traffic lanes and pedestrian/bicycle lanes at both sides. The main cable diameter is 36.376 in (0.92 m). The 746 ft tall (227 m) towers consist of multicellular steel shafts braced with struts and were the tallest in the world for over 60 years until completion of the Akashi-Kaikyo Bridge (with towers 298 m tall) in Japan in 1998. The suspended structure consists of two parallel 25 ft deep (7.6 m) stiffening trusses, spaced at 90 ft (27 m).

The bridge was designed by Charles Alton Ellis (Griggs and Francis 2010; Meiners 2001; Van Der Zee 1986). The suspension bridge was constructed by Pacific Bridge Company (main towers), Bethlehem Steel Company (structural steel of suspension span), John A. Roebling Sons Co. (cables of the suspension span), and Barrett & Hilp (anchorages). The bridge opened on May 28, 1937. The bridge is owned by the Golden Gate Bridge Highway and Transportation District. The original reinforced concrete deck was replaced by the orthotropic steel plate deck in 1986. The Golden Gate Bridge has been seismically retrofitted since 1997 and the last phase of the retrofit project is expected to be completed in 2015.

2.8.3 San Francisco–Oakland Bay Bridge West Span

The San Francisco–Oakland Bay Bridge (SFOBB) West Span (Figure 2.49) connects the city of San Francisco to Yerba Buena Island in the state of California. The West Span together with the tunnel and the East Span of truss bridges provides the only direct I-80 highway link between the city of San

FIGURE 2.48   (See color insert.) Golden Gate Bridge. (Courtesy of Lian Duan.)

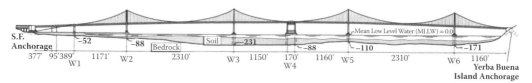

FIGURE 2.49 Overview of the San Francisco–Oakland Bay Bridge West Span. (Courtesy of the California DOT.)

FIGURE 2.50 San Francisco–Oakland Bay Bridge West Span. (Courtesy of the California DOT.)

Francisco and the East Bay communities. The 10,122 ft (3085.2 m) West Span includes three continuous truss spans of 389 ft, 95 ft, and 377 ft (118.6 m, 29.0 m, and 114.9 m) and the twin suspension bridges arranged back to back around a center anchorage. The twin bridges have main spans of 2310 ft (704.1 m) and back spans of 1160 ft (353.6 m) and are virtually identical.

The superstructure is made up of the main cables, the suspender cables, the suspended-span stiffening trusses, and the three-span continuous trusses. Each of the main cables was spun in place and consists of 37 strands of 472 wires for a total of 17,464 wires compacted into a circular cross section 28.75 in (730 mm) in diameter. Each panel point of the stiffening truss is hung from a group of four suspender cables, or wire ropes. Each suspender cable is 2.25 in (57.15 mm) in diameter and consists of six 19-wire strands wrapped around a center rope of seven 7-wire strands. The bridge has upper and lower concrete decks carrying five lanes in each direction and provides service to over 280,000 vehicles daily. The double-deck stiffening truss is made up of built-up members, laced members, and some rolled sections (Figure 2.50).

All the towers are similar, except that Towers W2 and W6 are about 420 ft (128.0 m) tall, while Towers W3 and W5 are about 470 ft (143.3 m). The tower legs are cellular in cross section, made up of 1 in thick (25.4 mm) vertical web plates connected along their edges with angles. Tower W2 is supported by a gravity concrete pier which was constructed in a sheet pile cofferdam 90 ft (27.4 m) below water. Towers W3, W5, and W6 are supported by cellular, hollow, reinforced concrete caissons which extend from 110 ft (33.5 m) to 230 ft (70.1 m) below water level. Pier W4 is actually a central anchorage for the twin, end-to-end suspension bridges and supported by a hollow cross section caisson of 92 ft × 197 ft (28.0 m × 60.0 m) which is formed using 55 steel cylinders 15 ft (4.6 m) in diameter by 5/16 in (8 mm) thick. It extends 220 ft (67.1 m) below water and 280 ft (85.3 m) above water. It was the largest pier in the world at the time of its construction. All the caissons were socketed into the underlying bedrock.

The West Span was seismically retrofitted in 2004 to improve operational and safety standards to the greatest extent possible and to achieve the project-specific performance-based seismic design criteria (Caltrans 1997, Reno and Pohll 1998). The SFOBB was designed and is owned by the California DOT. It was constructed by American Bridge Company and opened to traffic in 1936.

2.8.4 Tacoma Narrows Bridges

The Tacoma Narrows, in Tacoma, Washington, separates the beautiful Olympic Peninsula to the north and the city of Tacoma to the south with swift tidal currents and deep waters. The winds, tides, and water depth posed challenges to the designers and builders. The original 1940 Tacoma Narrows Bridge was built and opened to traffic on July 1, 1940. Unfortunately, on November 7, 1940, the bridge floor system and the stiffening girders failed under high and steady winds. After 10 years of studies, investigations, and planning, the 1950 Tacoma Narrows Bridge was rebuilt using the original main piers and anchorages. It was opened to traffic on October 14, 1950. Due to increased traffic demand, a new Tacoma Narrows Bridge was built alongside the 1950 Tacoma Narrows Bridge and opened to traffic on July 16, 2007.

2.8.4.1 1940 Tacoma Narrows Bridge

In 1938, the Washington State Highways Department (WHSD), on behalf of the Toll Bridge Authority, designed the suspension bridge with a main span of 2600 ft (792.5 m) and 1300 ft (392.2 m) side spans. The superstructure consisted of a 39 ft wide (11.9 m) floor and a 22 ft deep (6.7 m) stiffening truss. Under a Public Works Administration (PWA) grant, the Toll Bridge Authority was required to have an independent review of the WSHD's design. The reviewers recommended major changes to the design: lengthening the main span to 2800 ft (853.4 m), with shorter side spans of 1100 ft (225.3 m) each, and changing the stiffening truss to solid plate girders only 8 ft (2.4 m) deep. The Toll Bridge Authority accepted the recommendations and made changes accordingly. The main span ranked third in length behind the Golden Gate and George Washington Bridges at that time. The design of the 1940 Tacoma Narrows Bridge was influenced by the deflection theory, which made it feasible to design long-span suspension bridges with shallow plate girders. By the mid-1930s several of the world's longest suspension bridges were built of plate girders.

The design of the Tacoma Narrows Bridge set records in depth-to-span ratio at an unprecedented 1:350. The Golden Gate Bridge, with a stiffening truss system, has less than half the depth-to-span ratio, at 1:168. The width-to-span ratio of the Tacoma Narrows Bridge was a record-breaking 1:72. The Golden Gate Bridge has a width-to-span ratio of 1:47. The Bronx-Whitestone Bridge has a ratio of 1:31.

Construction bids were opened in September 1938. The contract for building the Tacoma Narrows Bridge was award to the Pacific Bridge Co. at a low bid of $5.6 million. Associate contractors were Bethlehem Steel Co. and John A. Roebling Sons Co. Construction on the bridge officially started in November 1938. The bridge was completed in 19 months, and opened to traffic on July 16, 1940 (Figure 2.51).

From the time the deck was built, the bridge began to move vertically in windy conditions. It was reported that the bridge bounced up and down even in mild to moderate winds. The construction workers gave the bridge the nickname "Galloping Gertie." Shock absorbers and temporary "tie-downs" were installed to control the vertical motions, but they were not effective. The vertical oscillation continued to occur after the bridge was opened to traffic. The bridge would bounce vertically even in light wind as low as 4 to 5 mph (6.44 to 8.05 km/h). It was common to see up and down movements of 1 to 5 ft (0.31 to 1.53 m), giving a total of rise and fall of 2 to 10 ft (0.61 to 3.05 m). Motorists had reported getting "seasick" driving over the bridge. Several measures aimed at stopping the motion were ineffective. The bridge

FIGURE 2.51 The original Tacoma Narrows Bridge opened to traffic on July 16, 1940. (Courtesy of Washington State DOT.)

FIGURE 2.52 Collapse of the 1940 Tacoma Narrows Bridge. (Courtesy of Barney Elliott, Wikipedia.)

collapsed on November 7, 1940 (Figure 2.52). The collapse of the bridge left a lasting impact in the understanding of the performance of suspension bridges under winds, in subsequent research in aerodynamics and harmonics, and in the improvement in the design and wind-tunnel testing of suspension bridges.

2.8.4.2 1950 Tacoma Narrows Bridge

The design of the replacement bridge was completed in 1944, subject to review and approval by the consulting board. However, due to materials and labor shortages as a result of the involvement of the United States in World War II, it wasn't until 10 years after the collapse of the original bridge that the replacement bridge was opened to traffic, on October 14, 1950 (Figure 2.53).

FIGURE 2.53   1950 Tacoma Narrows Bridge. (Courtesy of Washington State DOT.)

The 1950 Tacoma Narrows Bridge carries four lanes of traffic, two more than the 1940 bridge. The roadway width is 50 ft (15.24 m) as compared to 26 ft (7.93 m) of the 1940 bridge. The design incorporated engineering knowledge gained from the catastrophe of 1940, and the subsequent research carried out at the University of Washington and the California Institute of Technology. By this time, aerodynamic testing had become standard practice in the design of suspension bridges. The depth of the stiffening truss is 33 ft (10.01 m), according to the approved drawing signed by members of the consulting board in 1945. This bridge has open grating in the deck that allows the wind to pass through. The 1950 Tacoma Narrows Bridge has a depth-to-span ratio of 1:85, much lower than the 1:350 of the 1940 bridge.

The design engineer of the 1950 Tacoma Narrows Bridge was Dexter R. Smith and the principal engineer was Charles E. Andrew. They collaborated on the design with Professor Ray Fletcher Farquharson's research group at the University of Washington, and used the work of Theodore von Karman in wind tunnel analysis at the California Institute of Technology. The team did elaborate tests on a model bridge using a wind tunnel.

Construction bids were opened in August 1947. Contracts for building the replacement bridge were awarded to the primary contractors, Bethlehem Pacific Coast Steel Corporation and John A. Roebling Sons Co., for a total low bid of $11.2 million. Construction began in April 1948. Bethlehem Pacific Coast Steel fabricated and erected the steel truss part of the bridge. John A. Roebling Sons Co. spun the cable. The piers used for the 1940 bridge were reused for the new towers of the new bridge. The pedestals for the old bridge had to be removed and enlarged to accommodate the wider base and heavier towers of the new bridge. The height of the pedestals was also raised to reduce the salt spray on the towers. The bridge was completed in 29 months, and opened to traffic on October 14, 1950. During the months of January through March 1951, several wind storms with sustained wind speeds up to 75 mph swept through the Tacoma Narrows. The new bridge stood strong and firm against the winds, showing no sign of vertical or torsional movement, except for a slight lateral deflection.

2.8.4.3 2007 Tacoma Narrows Bridge

The 1950 bridge was designed to carry 60,000 vehicles per day. By 2004, traffic had increased to 90,000 vehicles per day. In 1998 voters in several Washington counties approved an advisory measure to create a second Narrows span and the Washington State DOT began developing financial, environmental, and technical plans for the new bridge. The communities wanted a new bridge to relieve congestion and improve safety. The newest 2007 bridge (Figure 2.54) is located about 185 ft (56.4 m) apart from and parallel to the south of the 1950 bridge.

FIGURE 2.54 2007 Tacoma Narrows Bridge in the foreground. (Courtesy of Washington State DOT.)

The new Narrows Bridge carries four 11 ft wide (3.35 m) lanes of eastbound traffic toward Tacoma. In addition, the bridge has a 10 ft right shoulder for disabled vehicles and a 10 ft (3.05 m) barrier-separated bicycle/pedestrian lane. The new bridge is designed to have the capability of adding a lower deck for more traffic lanes or for light rail/transit bus in the future.

The bridge is designed for a 150-year design service life. Performance under winds and physical wind tunnel testing are a part of the design and construction criteria. Major earthquakes in Washington state occur in the Puget Sound, where the bridge is located. The design criteria included a performance-based, two-level seismic design approach: a Safety Evaluation Earthquake (SEE) and a lower-level Functional Evaluation Earthquake (FEE). The SEE considers ground motions from an earthquake with a mean return period of 2500 years. During the SEE, there should be no collapse, primary structural members may incur minimal or repairable damage, and some significant damage to secondary structural components may occur. The FEE considers a lower-level earthquake event with a mean return period of 100 years. During FEE, there should be no damage to the structural components. The design criteria also require that the seismic performance be evaluated at one-half the estimated scour depth at the lower foundations.

Construction began in October 2002 with the award of a design-build contract to Tacoma Narrows Constructors (TNC), a joint venture of Bechtel Infrastructure and Peter Kiewit & Sons at a cost of $849 million. TNC engaged the design team of Parsons and HNTB, a joint venture, to provide design and construction support throughout the project, and subcontracted the superstructure steel fabrication and erection engineering to the joint venture of Nippon Steel and Kawada Bridge (NSKB) of Japan. NSKB further subcontracted to Samsung Heavy Industries (SHI) in South Korea to fabricate the stiffening trusses and the orthotropic deck.

The orthotropic deck and stiffening truss system were fabricated by Samsung Heavy Industries (SHI) on Koje Island, South Korea. The mile-long superstructure was fabricated in 46 sections, each measuring 120 ft (36.6 m) long × 78 ft (23.8 m) wide × 30 ft (9.1 m) high, and weighing 400 tons. The sections were shipped to the jobsite in three shipments. Each section is fully outfitted with the orthotropic deck, maintenance traveler rails, access platforms, and utility supports and painted before shipping across the Pacific Ocean. Prefabrication of the sections significantly accelerated erection at the jobsite (Figure 2.55).

2.8.4.4 Key Facts

Table 2.6 lists key facts of the three Tacoma Narrows Bridges.

FIGURE 2.55 2007 Tacoma Narrows Bridge during erection of the prefabricated deck sections. (Courtesy of Washington State DOT.)

TABLE 2.6 Key Facts of Tacoma Narrows Bridges

	1940 Bridge	1950 Bridge	2007 Bridge
Stiffening System	Plate Girder	Trusses	Trusses
Total Structure Length, ft (m)	5939 (1810.2)	5979 (1822.4)	5700 (1737.4)
Suspension Bridge Section, ft (m)	5000 (1524.0)	5000 (1524.0)	5400 (1645.9)
Center Span Length, ft (m)	2800 (853.4)	2800 (853.4)	2800 (853.4)
Span Depth, ft (m)	8 (2.4)	33 (10.0)	23.5 (7.2)
Depth-to-Span Ratio	1:350	1:85	1:119
Cable Diameter, in (cm)	17.5 (44.5)	20.1 (51.1)	20.5 (52.1)
Number of No. 6 Wires	6308	8702	8816

2.8.5 George Washington Bridge

The George Washington Bridge (Figure 2.56) is a suspension bridge with a total length of 4760 ft (1451 m) spanning the Hudson River, connecting Manhattan in New York City to Fort Lee in New Jersey. It carries I-95, US-1, and US-9. The suspension bridge is supported by four main cables, each 36 in (0.91 m) in diameter and composed of 26,474 wires. The main span, which is 3500 ft (1067 m) long between two 604 ft tall (184 m) steel towers, is twice as long as any previous suspension bridge. The bridge has an upper level with four lanes in each direction and a lower level with three lanes in each direction, for a total of 14 lanes of travel. A path on each side of the bridge's upper level carries pedestrian and bicycle traffic.

The bridge was first opened to traffic with six traffic lanes on October 25, 1931. Two additional lanes were created in the in 1946. The lower level of six lanes opened on August 29, 1962. This made the George Washington Bridge one of the world's busiest bridges and the world's only 14-lane suspension bridge. The bridge was designed by Othmar H. Ammann, and is constructed and owned by the Port Authority of New York and New Jersey.

FIGURE 2.56   The George Washington Bridge. (Courtesy of Historic American Engineering Record.)

2.8.6 San Francisco–Oakland Bay Bridge Self-Anchored Suspension Span

The San Francisco–Oakland Bay Bridge (SFOBB) self-anchored suspension (SAS) span (Figure 2.57) is a signature span for the SFOBB east span seismic replacement project. It connects the Yerba Buena Island tunnel with the newly constructed east skyway.

The 624.4 m long (2055.1 ft) SAS span includes a main span of 385 m (1263.1 ft) and a side span of 180 m (590.6 ft). The single tower is 160 m (524.9 ft) high with four legs connected by seismic shear link beams. The superstructure is 77.8 m (255.2 ft) wide and carries 10 traffic lanes and a bike path and sidewalk. The superstructure is made of two orthotropic boxes connected by crossbeams. The orthotropic box is 27 m (88.6 ft) wide and 5.55 m (18.2 ft) high.

The SFOBB SAS span will be the world's largest SAS bridge and the world's only SAS bridge with only one tower when completed. It was designed by a joint venture of T. Y. Lin International and Moffatt & Nichol and is owned by the California DOT. It is under construction by a joint venture of American Bridge and FLUOR. The SAS span is expected to open to traffic in 2013.

2.8.7 The Three Sisters Suspension Bridges

The Three Sisters are three very similar self-anchored suspension bridges spanning the Allegheny River in Pittsburgh, Pennsylvania at 6th, 7th, and 9th Streets. They are generally known as the Sixth Street, Seventh Street, and Ninth Street bridges (Figures 2.58 through 2.62), and are located in the home of the International Bridge Conference held annually in Pittsburgh since 1984. Recently they have been named after important citizens of Pittsburgh: the Roberto Clemente Bridge, the Andy Warhol Bridge, and the Rachel Carson Bridge, respectively.

FIGURE 2.57   Rendering of San Francisco–Oakland Bay Bridge self-anchored suspension span. (Courtesy of the California DOT.)

FIGURE 2.58   View of the Three Sisters. (Courtesy of Juliet Lwin.)

The Three Sisters set records in the United States as the only trio of nearly identical bridges, the first use of self-anchored suspension bridge design, and the only few remaining bridges with large and multiple steel eyebar chains for the suspension system. The steel bridges were designed by the Allegheny Department of Public Works, the superstructures were built by the American Bridge Company, and the substructures were built by the Foundation Company. They were constructed in 1924 to 1928, replacing older bridges. The bridges are owned by Allegheny County, Pennsylvania.

Key facts of the Three Sisters Bridges are shown in Table 2.7.

FIGURE 2.59 Sixth Street (Roberto Clemente) Bridge. (Courtesy of Juliet Lwin.)

FIGURE 2.60 Seventh Street (Andy Warhol) Bridge. (Courtesy of Juliet Lwin.)

FIGURE 2.61 Eyebars of the Seventh Street Bridge. (Courtesy of Juliet Lwin.)

FIGURE 2.62   Ninth Street (Rachel Carson) Bridge. (Courtesy of Juliet Lwin.)

TABLE 2.7   Key Facts of the Three Sisters

	Sixth Street Bridge	Seventh Street Bridge	Ninth Street Bridge
Official Name	Roberto Clemente	Andy Warhol	Rachel Carson
Total Length	884 ft (269.4 m)	884 ft (269.4 m)	840 ft (256.0 m)
Main Span	430 ft (131.1 m)	442 ft (134.7 m)	410 ft (125.0 m)
Side Spans	215 ft (65.5 m) each	221 ft (67.4 m) each	215 ft (65.5 m) each
Roadway Width	38 ft (11.6 m)	38 ft (11.6 m)	38 ft (11.6 m)
Vehicular Lanes	2	2	2
Sidewalks	2	2	2
Navigation Clearance	40 ft (12.2 m)	40 ft (12.2 m)	40 ft (12.2 m)
Opened to Traffic	October 19, 1928	June 17, 1926	November 26, 1926

2.9 Movable Bridges

Paul Liles

2.9.1 Introduction

Movable bridges are bridges that allow marine traffic to go under or through the bridge opening by altering the position of the superstructure during passage of traffic beneath the bridge. The most common use for movable bridges involves allowing for passage of boat or marine traffic through the bridge opening. The principal advantage of this type of structure is that a very flat grade can be maintained on the roadway as vertical clearance for boats or ships is provided by moving or opening the bridge to marine traffic. The main disadvantages to this type of structure result from the disruption of highway or rail traffic when the bridge is open, the impediment to marine traffic when the bridge is closed, and the expense of having personnel on duty to maintain bridge operations.

There are many types of movable bridges and the diversity of these structures is a tribute to engineers and their various designs. Movement is usually provided by a mechanical system that is powered by electric motors using operating winches, gearing, or hydraulic pistons.

Most of the movable bridges can be grouped into three main types: vertical lift bridges, swing span bridges, and bascule or draw bridges. Specific examples of these types of bridges with a brief description of the movements are discussed in the following sections.

2.9.2 Sacramento Tower Bridge

The Tower Bridge (Figure 2.63) in Sacramento, California is an example of a vertical lift bridge. This is a type of movable bridge where the span rises vertically while remaining parallel with the bridge deck. The span is raised on towers by cables and counterweights located at the ends of the movable span. The bridge carries State Route 275 over the Sacramento River and was one of the first vertical lift bridges on the California Highway System. The bridge carries four lanes of vehicular traffic over the river. The overall bridge length is 737 ft (225 m) while the lift span portion of the bridge is 209 ft (63.7 m) and provides 100 ft (30.5 m) of vertical clearance above high water when raised. Two towers are supported by two concrete piers at a depth of 50 ft (15.2 m) below water.

The bridge was designed and is owned by the California DOT. The bridge was constructed by Sir William Arrol and Company and George Pollock and Company. The bridge was opened to traffic in December 1935 at a cost of approximately $666,000.

2.9.3 George P. Coleman Memorial Bridge

The George P. Coleman Memorial Bridge (Figure 2.64) in coastal Virginia is an example of a swing span bridge. This is a type of movable bridge where the superstructure turns or pivots horizontally about a support pier to allow marine traffic to pass through the bridge opening. The bridge is a double swing span bridge and carries US-17 over the York River between Yorktown and Gloucester Point, Virginia. The bridge carries four lanes of vehicular traffic over the river. The overall bridge length is 3750 ft (1140 m) while the turn span portion of the bridge is 450 ft (150 m) and provides 60 ft (18 m) vertical clearance above high water in the closed position.

The bridge was originally built in 1952 but was reconstructed and widened in 1995. For the 1995 reconstruction, the new swing spans were floated into position in a finished state to minimize disruption to traffic on the bridge. The 1995 reconstruction was designed by Parsons Brinkerhoff and the construction was performed by Tidewater Construction Corporation. The reconstructed bridge was reopened to traffic in 1995 at a cost of approximately $76.8 million.

2.9.4 Woodrow Wilson Memorial Bridge

The Woodrow Wilson Memorial Bridge (Figure 2.65) located near Washington, DC, is an example of a bascule or draw bridge. This is a type of movable bridge where the bridge opens by rotating about a horizontal axis parallel to the waterway. The Woodrow Wilson Memorial Bridge crosses the Potomac

FIGURE 2.63 Sacramento Tower Bridge. (Courtesy of California DOT.)

FIGURE 2.64   George P. Coleman Memorial Bridge. (Courtesy of Parsons Brinkerhoff, Inc.)

FIGURE 2.65   Woodrow Wilson Memorial Bridge. (Courtesy of Hardesty & Hanover, LLP.)

River between Alexandria, Virginia, and Oxon Hill, Maryland. The bridge carries I-95 and I-495 over the Potomac River and is one of only a handful of movable bridges on the U.S. interstate system.

The Woodrow Wilson Memorial Bridge consists of two parallel double leaf bascules that carry twelve lanes of vehicular traffic over the Potomac River. The overall bridge length is 6736 ft (2053 m) while the Bascule portion of the bridge spans 269 ft (82 m) and provides 70 ft (21 m) of vertical clearance in the closed position. At the present time, the bridge opens approximately 65 times a year to allow ship traffic to pass the bridge.

The movable portions of the bridge were designed by Hardesty & Hanover, LLP with Finley McNary Engineers designing the unique precast, post-tensioned V-piers. The bridge was constructed by the American Bridge Company and Edward Kraemer & Sons, with PDM Bridge, LLC doing the steel construction for the bascule span. The bridge was completed in May 2008 at a cost of approximately $826 million.

2.10 Floating Bridges

Myint Lwin

2.10.1 Introduction

A floating bridge may be constructed of wood, concrete, steel, or a combination of these and other materials, depending on the design requirements. A 406 ft long (123.8 m) movable wooden pontoon railroad bridge was built in 1874 across the Mississippi River in Wisconsin. It was rebuilt several times before it

TABLE 2.8   Major Floating Bridges in the United States

Bridge Name	Location	Owner	Length of Floating Bridge ft (m)	Drawspan ft (m)	Year Opened to Traffic
First Lake Washington	Lake Washington	Washington State DOT	6561 (1999.8)	200 (61.0)	1940
Hood Canal	Hood Canal	Washington State DOT	6530 (1990.3)	600 (182.9)	1962
Albert D. Rosselini	Lake Washington	Washington State DOT	7578 (2309.8)	200 (61.0)	1963
Homer M. Hadley	Lake Washington	Washington State DOT	5736 (1748.3)	None	1989
New Lacey V. Murrow	Lake Washington	Washington State DOT	6561 (1999.8)	None	1993
Eastbank Esplanade	Portland, Oregon		1200 (365.8)	None	2001
Ford Island Bridge	Pearl Harbor, Honolulu Hawaii	U.S. Navy	930 (283.5)	930 (283.5)	1998

was abandoned. A 320 ft long (97.5 m) wood floating bridge is still in service in Brookfield, Vermont. The present Brookfield Floating Bridge is the seventh replacement structure, and was built by the Vermont Agency of Transportation in 1978. A 720 ft long (219.5 m) concrete floating drawspan was built as part of the Ford Island Bridge in Honolulu, Hawaii. A floating drawspan is needed to provide unlimited vertical clearance for the naval vessels going into Pearl Harbor.

The most significant floating bridges in the United States are located in Washington state (Gloyd 1988; Lwin 1993a). Currently, there are three concrete floating bridges on Lake Washington in Seattle, and one concrete floating bridge on Hood Canal in the Olympic Peninsula. These floating bridges form major transportation links in the state and interstate highway systems. Hundreds of thousands of people use these bridges daily to get to destinations of work, commerce, and leisure. These bridges are listed in Table 2.8, and will be briefly introduced in the following paragraphs. More detailed information on these floating bridges is covered in the references.

Lake Washington is a fresh water lake. It is about 1 to 3 mi (1.6 to 4.8 km) wide and 20 mi (32.3 km) long. The water in most parts is 100 to 200 ft (30.5 to 60.9 m) deep. The bottom of the lake consists of soft clay and peat extending another 100 to 200 ft (30.5 to 60.9 m) in thickness.

Hood Canal is a westerly arm of the tidal waters of Puget Sound in Washington state. It is not a man-made canal. It is about 55 mi (88.5 km) long, 1 to 2 mi (1.6 to 4.8 km) wide and more than 300 ft (91.4 m) deep. There is a tidal variation of 16 ft (4.9 m) and a maximum current of 3.5 mph (5.6 km/h).

2.10.2 First Lake Washington Floating Bridge

The First Lake Washington Floating Bridge (Figure 2.66) was opened to traffic on July 2, 1940, one day after the opening of the famous and fateful Tacoma Narrows Suspension Bridge (Andrew 1939; Murrow 1938). The First Lake Washington Floating Bridge (Andrew 1939) was constructed of 25 reinforced concrete pontoons connected rigidly to form a continuous floating structure. It carried four lanes of traffic, two sidewalks, and a drawspan for the passage of large vessels. The drawspan was replaced with a straight and fixed span in 1981 to improve safety. Since then large vessels have to bypass the floating bridge and go under a high-level bridge constructed over the East Channel of Lake Washington. A typical concrete pontoon measured 350 ft (106.7 m) long, 59 ft (18.0 m) wide, and 14.5 ft (4.4 m) deep. The interior of the pontoon was divided into compartments with watertight bulkheads to control flooding and progressive failures.

Lacey V. Murrow was the director of highways and the chief engineer of the Toll Bridge Authority of Washington state when the First Lake Washington Floating Bridge was planned, financed, and built in 1940. He was instrumental in promoting and adopting the innovative idea of a concrete floating bridge. In 1967, the First Lake Washington Floating Bridge was dedicated to the memory of the late Lacey V. Murrow for his wisdom in recognizing the structural feasibility and cost-effectiveness of a floating bridge, and for his courage to build one in face of skepticism.

FIGURE 2.66   First Lake Washington Floating Bridge. (Courtesy of Washington State DOT.)

2.10.3 Evergreen Point Bridge

The Evergreen Point Bridge (Figure 2.67) is the second floating bridge built on Lake Washington. The population and traffic on Mercer Island and the eastern shore of Lake Washington were growing very fast. By the beginning of the 1950s the need for another bridge across Lake Washington was clear. This time there was no question about what type of bridge to build, but there was heated debate over the location of the new bridge. The final choice was a second floating bridge located about 3 mi (4.8 km) north of the First Lake Washington Floating Bridge. The Second Lake Washington Floating Bridge, commonly known as the Evergreen Point Bridge, was opened to traffic on August 8, 1963.

A typical concrete pontoon measures 360 ft (109.7 m) long, 60 ft (18.3 m) wide, and 14.75 ft (4.5 m) deep. Each pontoon is prestressed longitudinally before connecting rigidly end-to-end to form a continuous floating structure. The interior of each pontoon is divided into compartments with watertight bulkheads to control flooding and progressive failures. This bridge is formally named after the former governor of Washington state: the Albert D. Rosselini Bridge. However, it is more popularly known as the Evergreen Point Bridge.

2.10.4 Hood Canal Bridge

The Hood Canal Floating Bridge, shown in Figure 2.68 (Nichols 1962), was built to span the northern end of Hood Canal and was opened to traffic in August 1961 to replace a ferry system. It was constructed of 23 reinforced and longitudinally prestressed concrete pontoons. The pontoons were connected rigidly to form a continuous floating structure with a navigation opening at mid-channel where the water depth is over 300 ft (91.4 m). It carried two lanes of traffic. The roadway was elevated 20 ft above the water surface to minimize spray from winds and waves. The reinforced concrete roadway was supported on reinforced concrete columns rising from the pontoon deck.

The western half of the original Hood Canal Floating Bridge was destroyed by an unusually strong and long-duration storm in February 1979. The loss of the bridge caused great economic and social impact in the region. The western half of the bridge was replaced quickly and the bridge was reopened to traffic in October 1982, only 3 years and 8 months after the sinking of the west half.

The new West Half (Lwin and Gloyd 1984) was designed and constructed with better criteria and more durable materials. It is three times stronger than the remaining East Half. Meanwhile, the East Half continued to deteriorate in the harsh marine environment. The deterioration of the East Half was at such an advanced state that the Washington State DOT replaced the East Half with design similar to the West Half. The new East Half (Figure 2.69) was opened to traffic in June 2009.

FIGURE 2.67 Second Lake Washington Bridge. (Courtesy of Washington State DOT.)

FIGURE 2.68 The original Hood Canal Floating Bridge. (Courtesy of Washington State DOT.)

2.10.5 Homer M. Hadley Memorial Bridge

By 1965, more capacity was needed to carry traffic across the First Lake Washington Floating Bridge, which had become a major part of I-90. I-90 links Seattle to eastern Washington state and the rest of the I-90 corridor to Boston, Massachusetts. The Washington State Department of Highways decided to build a third floating bridge (Lwin 1989) just 60 ft (18.3 m) north of the First Lake Washington Floating Bridge as part of a 6.9 mi (11.1 km) and $1.2 billion public works project to improve mobility on I-90 between Seattle and Bellevue. The new bridge was completed and opened to traffic on June 4, 1989.

The bridge was named the Homer M. Hadley Memorial Bridge (Figure 2.70) in recognition of engineer Hadley's innovations and influence in the design and construction of concrete bridges in Washington state. Engineer Hadley was the first to conceive the idea of a concrete floating bridge across

FIGURE 2.69   The new East Half of the Hood Canal Bridge. (Courtesy of Washington State DOT.)

FIGURE 2.70   Homer M. Hadley Bridge on the left and new Lacey V. Murrow Bridge on the right. (Courtesy of Washington State DOT.)

Lake Washington in 1920. He formally presented his concept at a meeting of the American Society of Civil Engineers in 1921. His proposal generated debate and skepticism in the community. The bankers labeled his concept "Hadley's Folly." However, Hadley's big dream came true when Lacey V. Murrow, director of the State Department of Highways, confirmed the technical feasibility of his proposal, and advanced the design and construction of the first concrete floating bridge across Lake Washington in the middle of 1930s. Hadley's floating bridge design paved the way for floating bridges in Washington state.

The bridge carries five lanes of traffic (three westbound and two reversible) and one sidewalk for pedestrians and bicycles. The bridge carries in excess of 100,000 vehicles a day. There is no drawspan. Large vessels go under a new high-level bridge on the east of Mercer Island. A typical pontoon measures 354 ft (107.9 m) long, 75 ft (22.9 m) wide, including cantilevered roadway slabs, 16 ft (4.9 m) deep, and has a water draft of 9 ft (2.7 m). The interior is divided into compartments with watertight bulkheads to control flooding and progressive failures.

2.10.6 New Lacey V. Murrow Bridge

Soon after the Homer M. Hadley Memorial Bridge (Figure 2.70) was opened to traffic in June 1989, the First Lake Washington Floating Bridge (Lacey V. Murrow Bridge) was closed for renovation. The renovation would upgrade the bridge to modern standards for a three-lane one-way eastbound traffic with shoulders. Unfortunately, while undergoing renovation some pontoons took in excessive water during a long rainstorm over the Thanksgiving weekend in November 1990. As a result, eight pontoons sank and the remaining pontoons suffered major damages and were considered structurally unsuitable for reuse in a major highway system. Because of the traffic demand, it was necessary to rebuild the bridge as quickly as possible. Construction of the replacement bridge (Lwin 1993b; Lwin, Bruesch, and Evans 1995) started in January 1992 and the bridge was opened to traffic in September 1993. The contract had an incentive clause to pay the contractor $18,500 a day for early completion. The contractor completed the project one year early, earning a $6.7 million bonus.

One important feature of the new bridge was the use of HPC to assure low permeability and shrinkage. The HPC contained fly ash and silica fume. The concrete had an average 28-day compressive strength of over 10,000 psi (69 MPa). The permeability was less than 1000 Coulombs as tested in accordance with the AASHTO T-277 Rapid Chloride Permeability Test. The shrinkage was less than 400 microstrains. The workability of the concrete was good.

The new Lacey V. Murrow Bridge is constructed of 20 prestressed concrete pontoons rigidly connected together to form a continuous floating structure. A typical pontoon measures 360 ft (109.7 m) long, 60 ft (18.3 m) wide, and 16.75 ft (6.1 m) deep, and has a draft of 9.75 ft (3.0 m). The interior is divided into compartments with watertight bulkheads to control flooding and progressive failures. Water sensors are installed in each compartment for early detection and early warning of water entry. When water is detected, an alarm system is activated to alert emergency response personnel. A bilge piping system is installed in the compartments for pumping out water when necessary. These are special precautionary features incorporated into the new bridge to provide early warning of water entry and to safeguard against progressive failures.

2.11 Pedestrian Bridges

Lubin Gao

2.11.1 Introduction

Pedestrian bridges are those designed for carrying pedestrians and bicyclists rather than vehicular traffic. In some circumstances, a pedestrian bridge may carry animals, horse riders, or other nonvehicular live loads. If a bridge carries both vehicular traffic and pedestrians, it is normally not classified as a pedestrian bridge. However, a pedestrian bridge may be required to be designed for a maintenance vehicle if vehicular access to the bridge is not prevented by permanent physical methods.

A pedestrian bridge may span a small creek or a large river. All structural types of bridges can be found in pedestrian bridges. In general, beams or girders are used in short-span pedestrian bridges; trusses and arches are used in the medium-span range; and cable-supported structures are used for long spans. In the United States, a structural type called stress ribbon has been utilized in building a number of pedestrian bridges in the last two decades. The Lake Hodges Pedestrian Bridge, completed in 2009 in California, set a world record for this type of bridge with a total length of 990 ft (302 m).

Wood has been the traditional material for building pedestrian bridges. Steel and concrete, the two materials typically used in highway bridges, now dominate pedestrian bridges. In recent years, the use of fiber-reinforced polymer (FRP) composites in pedestrian bridges has been gradually and constantly growing. With its low-maintenance, lightweight, modular construction and ease of installation, FRP composite material has been used in prefabricated modular trusses and deck panels (Tang and Podolny 1998).

Special considerations to aesthetics and environment generally weigh more in designing pedestrian bridges than vehicular bridges. In urban areas, aesthetics are usually the controlling factor in selecting the bridge type. Occasionally, the aesthetic consideration ultimately leads to the design of architectural masterpieces, such as the Chicago's BP Pedestrian Bridge (opened in 2004 and designed by Pritzker Prize–winning architect Frank Gehry). In some situations, a pedestrian bridge may be required to overwhelm its surroundings to become a new landmark, like the Turtle Bay Sundial Bridge; in other situations, a bridge may be required to blend into the environment with minimal visual impact, like the Lake Hodges Pedestrian Bridge.

Pedestrian bridges are typically narrower, with a greater span-to-width ratio than highway bridges. The depth of the superstructure is normally small. The flexibility resulting from a shallow superstructure makes pedestrian bridges more susceptible to vibrations due to environmental factors, such as wind and the pedestrians themselves. Designers of long-span pedestrian bridges are challenged to control the magnitude of acceleration at deck level to assure pedestrian comfort.

In the United States, the AASHTO *Standard Specifications for Highway Bridges* (AASHTO 2002) are also used for the repair and rehabilitation of older, existing pedestrian structures. For new pedestrian bridge design, the AASHTO *LRFD Bridge Design Specifications* (AASHTO 2010a) are the governing specifications for designing pedestrian bridges. The AASHTO *Guide Specifications for Design of FRP Pedestrian Bridges* (AASHTO 2008) and the AASHTO *LRFD Guide Specifications for Design of Pedestrian Bridges* (AASHTO 2009) provide additional guidance on the issues requiring additional or different treatment due to the nature of pedestrian bridges and their loadings.

2.11.2 Peter DeFazio Pedestrian Bridge

The Peter DeFazio Pedestrian Bridge (Figure 2.71) in Eugene, Oregon is a pedestrian suspension bridge with a precast concrete deck. The bridge is one of the five pedestrian bridges crossing the Willamette River in the Ruth Bascom Riverbank Path System (Strasky 2002). The bridge consists of two 75 ft long

FIGURE 2.71   Peter DeFazio Pedestrian Bridge, Eugene, Oregon. (Courtesy of Robert Cortright, Bridge Ink.)

(23 m) cable-stayed side spans, a 338 ft long (103 m) suspended main span, and cast-in-place approach spans. The concrete deck is 21 ft (6.5 m) in width and 1.33 ft (0.42 m) in thickness maximum.

The suspended spans of the deck were composed of 44 precast segments longitudinally post-tensioned after erection with 19 0.6 in (15 mm) multistrand tendons. Each main suspension cable consists of 43 0.6 in (15 mm) strands with multiple layers of corrosion protection. The strands were internally greased and sheathed in HDPE, installed in a galvanized carrier pipe, and cement grouted upon completion. The cables were anchored at both ends with the bond anchorage system. The two pylons were fabricated with structural steel. The bridge was designed by OBEC Consulting and Jiri Strasky. Kiewit Pacific Co. was the general contractor of this project. The owner is the city of Eugene, Oregon. It cost approximately $2.4 million.

2.11.3 Turtle Bay Sundial Bridge

The Sundial Bridge at Turtle Bay (Figure 2.72) is a cantilevered cable-stayed pedestrian bridge across the Sacramento River in Redding, California, connecting the north and south sections of Turtle Bay Exploration Park and serving as a gateway to the Sacramento River Trail system (Melnick 2004). It opened on July 4, 2004.

The bridge is located in an environmentally sensitive area that requires no piers in the water of the Sacramento River. This leads to a bridge of 722 ft (220.1 m) in length with a main span of 413 ft (126 m). The superstructure consists of a steel triangular pipe truss and a 23 ft wide (7.01 m) translucent deck. The truss has a bottom chord 14 in (356 mm) in diameter and two top chords 11 in (279 mm) in diameter. The deck is made up of nonskid structural glass panels with granite accents.

There are 14 locked coil stay cables in total supporting the deck from the pylon. The cables are connected to the transverse bulkheads in the deck truss and to the plate brackets on the inclined pylon. The stay cables are not anchored along the centerline of the deck but instead divide the deck into two walkways. The 217 ft tall (66.1 m) triangular pylon is inclined at 42 degrees and runs in a true north–south direction. It also serves as a working sundial, casting its shadow on the north side of the bridge. The pylon is made of steel plates varying from 1 in (25 mm) at the base to 5/8 in (16 mm) at the top. The tapered pylon is composed of double walls stiffened with horizontal, vertical, and skewed stiffeners.

The bridge was designed by the Spanish architect Santiago Calatrava. Kiewit Pacific Co. was the general contractor of this project. Buckland & Taylor Ltd. provided erection engineering services to the

FIGURE 2.72 Sundial Bridge, Redding, California. (Courtesy of Chad Kearney.)

contractor for both the pylon and the deck. The owner of this project is the city of Redding, California. It cost approximately $23.5 million.

2.11.4 Bob Kerrey Pedestrian Bridge

The Bob Kerrey Pedestrian Bridge (Figure 2.73) is a landmark cable-stayed footbridge across the Missouri River connecting Council Bluffs, Iowa to Omaha, Nebraska (Brown 2008). The bridge opened on September 28, 2008, and was named after the former Nebraska senator who secured funding for the project. It is one of the longest pedestrian bridges in the United States.

The bridge is more than 2224 ft (678 m) in length with a main span of 506 ft (154.2 m), providing a navigational channel of 467 ft (142.3 m) with a minimum clearance of 53 ft (16.2 m). The two side spans of the cable-stayed bridge are 253 ft (77.1 m) each in length. Its S-curved alignment symbolizes the flowing waters of the Missouri River below. The bridge bends from one side of the first pylon to the opposite side of the second pylon. Although the bridge alignment is horizontally curved, the superstructure segments and precast deck panels are straight-edged and identical in size and shape, and arranged to create the "S" curve.

The deck is typically 16.33 ft (4.98 m) wide, but is widened at the midspan of the main span and two pylon locations to provide observation areas. The depth of the precast concrete deck panels is 12 in (305 mm) at the curbs and 3 in (76 mm) elsewhere. The steel framing consists of two W21 × 182 rolled beams as edge girders spaced at 24 ft (7.32 m) from center to center, W21 × 62 floor beams, and W7 × 26.5 diagonal braces. The Grade 50 steel superstructure reduces the dead load and wind load that lead to the reduction in the size of the foundation and the cables, and the amount of falsework as a result.

There are 80 stay cables in total in this bridge. Forty cables per pylon are arranged in two planes. The cables range in diameter from 1 1/4 in (32 mm) to 2 1/8 in (54 mm) and are spaced approximately 23 ft (7.01 m) apart at the deck. Each pylon is three-sided, 203 ft (61.9 m) tall, and supported by a single drilled shaft of 13 ft (3.96 m) in diameter that extends approximately 85 ft (26 m) into the riverbed. The bridge was built by the design-build method. HNTB designed the bridge. APAC Kansas, Inc. was the general contractor of this project. The owner of this project is the city of Omaha, Nebraska. It cost approximately $22 million.

2.11.5 Lake Hodges Pedestrian Bridge

The Lake Hodges Pedestrian Bridge (Figure 2.74) is in the city of Escondido in San Diego County, California. Its official name is the David Kreitzer Lake Hodges Bicycle Pedestrian Bridge. It was the world's longest stress-ribbon bridge when it opened in 2009. The bridge provides a crucial link between

FIGURE 2.73 Bob Kerrey Pedestrian Bridge, Omaha, Nebraska. (Courtesy of Wikipedia.)

FIGURE 2.74   Lake Hodges Bicycle Pedestrian Bridge, Escondido, California. (Courtesy of Susan Hunt Williams.)

trails in Escondido and Rancho Bernardo, eliminates a 9 mi (15 km) detour, and furnishes a safe route for hikers and bicyclists. The bridge is located at an environmentally sensitive area. The unusual stress-ribbon design with long spans and a thin deck created a solution with minimal ecological and visual impact.

The 990 ft long (302 m) bridge consists of three 330 ft (100.6 m) spans. It crosses Lake Hodges with a 15 ft (4.57 m) clearance. The deck is 12 ft (3.66 m) wide. It consists of 87 precast concrete panels, with 29 panels in each span. A typical panel is 14 ft × 10 ft (4.27 m × 3.05 m). The deck panels were post-tensioned to ensure continuity in the spans, close the transverse joints, and give the bridge its required stiffness for live loads. The use of precast panels eliminated the need for falsework within the lake and reduced the construction footprint. The bridge utilized 114 0.6 in. (15 mm) diameter seven-wire low-relaxation strands in two cables as the stress ribbons.

The north abutment is supported by 15 rock anchors, 75 ft (22.9 m) deep in 11 in (279 mm) diameter holes. The south abutment uses four drilled shafts 8 ft (2.44 m) in diameter and 90 ft (27.4 m) deep. Each abutment measures 30 ft × 30 ft (9.14 m × 9.14 m). The bridge was designed by T. Y. Lin International. Flatiron Construction Corporation was the general contractor of this project. The project architect was Safdie Rabines Architects, San Diego. San Dieguito River Park Joint Powers Authority is the owner of this project. It cost approximately $10.3 million.

2.11.6 Aurora Arch Pedestrian Bridge

The Aurora Arch Pedestrian Bridge (Figure 2.75) is in Aurora, Nebraska across Lincoln Creek, approximately 65 mi (105 km) west of Lincoln. It opened in 2004 and features the innovative use of steel and confined concrete for the arch ribs and tension ties, and the use of FRP composite for the bridge deck (Tuan 2004).

The bridge consists of a single span of 100 ft (30 m). The circular arch ribs have a radius of 72.5 ft (22 m) and a rise of 20 ft (6.1 m). The arch measures 18 ft (5.5 m) from the walkway to the crown. The inclined ribs and the ties (or bottom chords) are all 8 in (203 mm) steel pipes filled with concrete. For the ties (bottom chords), a high-strength prestressing tendon with a polyethylene sheath is inserted inside each steel pipe, and positioned at the center by spacers along the length of the tie (bottom chord). Expansive concrete is pumped into the steel pipe. After the concrete is hardened, the tendon is post-tensioned to 90 kips (400 kN) to exert precompression on the concrete inside the pipe. At same time, the steel pipe provides lateral confinement on the concrete. As a result, the ties can carry much higher tensile forces without causing tensile stresses in the concrete. The steel pipe is used solely to provide lateral confinement for the concrete, and not intended to carry external loading.

FIGURE 2.75   Aurora Arch Pedestrian Bridge, Aurora, Nebraska. (Courtesy of Christopher Tuan.)

The 10 ft wide (3 m) deck was constructed with 10 ft × 10 ft (3 m × 3 m) FRP composite honeycomb panels for ease of installation, low maintenance, and improved durability. Two different hanger configurations were used, epoxy coated strands and 1 in steel rod. Maximum loads and tension in the hangers were monitored.

The research and development of the arch bridge design were sponsored by the Mid-America Transportation Center (MATC), the Center for Infrastructure Research (CIR) at the University of Nebraska-Lincoln, and the city of Aurora. The final design of the superstructure was performed by InfraStructure, LLC. It cost approximately $100,600.

2.12 Accelerated Bridge Construction

Myint Lwin

2.12.1 Introduction

There are over 600,000 highway bridges in the National Bridge Inventory in the United States. These bridges are over 20 ft long and on public roads. The average age of these bridges is about 43 years. Many of these bridges were designed with a service life of 50 years. The United States is confronting an aging infrastructure and the need to inspect, repair, rehabilitate, or replace bridges. In the last two decades the traffic demand has grown tremendously, while highway capacity has increased little, resulting in congestion on the highways. Construction activities on roads and bridges compound the traffic problems. Innovative techniques, strategies, and technologies in construction are needed to improve quality in construction, reduce traffic congestion, improve work zone safety, and achieve economy. The FHWA (2006, 2007, 2009 and 2011) and AASHTO are working with the state transportation agencies, industry, and academia to accelerate the adoption of new technologies and innovative practices in the renewal of the aging infrastructure.

2.12.2 Accelerated Bridge Construction Concepts

Accelerated bridge construction (ABC) is an innovative technology to reduce construction time and cost on highway projects, improve construction quality and work zone safety, and reduce adverse impacts on the traveling public. ABC uses prefabricated elements and systems extensively to ensure quality in the constructed projects, minimize on-site disruption to traffic, and improve worker safety in the work zone. Prefabricated elements for substructure and superstructure and complete bridge systems for rapid bridge removal and replacement have been used for many years. With the availability of self-propelled modular transports and high-capacity cranes, large and heavy bridge components and systems can be moved steadily and safely. Prefabricated systems allow bridges to be opened to traffic in days or weeks rather than months or years.

FHWA recently launched the Every Day Counts (EDC) initiative to shorten project delivery and accelerate the deployment of innovative technologies. One of the innovative technologies included in the initial phase of EDC is prefabricated bridge elements and systems (PBES). PBES is defined as bridge structural elements and systems that are built off the bridge alignment to accelerate on-site construction time relative to conventional practice. With PBES, many time-consuming construction tasks no longer need to be accomplished sequentially in the work zone. Instead, PBES are constructed concurrently, off-site and/or off-alignment, and brought to the project location ready to erect. Because PBES are usually fabricated under controlled climate conditions, weather has a smaller impact on the quality, safety, and duration of a project. Through the use of standardized bridge elements, PBES offers cost savings in both small and large projects. The use of rapid on-site installation of PBES can reduce the environmental impact in sensitive areas. EDC Innovation Summits have been held throughout the United States to promote implementation of EDC initiatives.

Another effort to promote ABC is the Accelerated Construction Technology Transfer (ACTT). FHWA, in collaboration with AASHTO, has conducted workshops on the use of ACTT to reduce construction time, dramatically save money, and improve safety and quality by minimizing delays and hazards associated with work zones. The ACTT process begins with a 2- to 3-day workshop in which a multidisciplinary team of transportation experts works with their local counterparts to evaluate all aspects of a project and develop recommendations to the sponsoring agency for reducing construction time and cost and enhancing safety and quality. The sponsoring agency is responsible for evaluating and adopting the recommendations to the extent feasible to the agency. The process has benefitted many states.

2.12.3 States Implementing ABC

ABC is gaining popularity across the United States. Many states are using various methods for implementing ABC. At present, the primary methods are using prefabricated components that are built off-site and can be quickly put in place once at the job site, or building the entire structure off-site and moving it into place using a high-capacity self-propelled modular transporter (SPMT) or crane. ABC can be extended to other time-consuming activities during planning, permitting, and contracting to reduce project delivery time. Some examples of activities are right-of-way acquisition, utility relocation, material procurement, permits, and others.

The Utah State DOT (UDOT) is one of the forerunners in embracing ABC techniques. ABC is now standard practice in Utah. Since 1997 Utah has used ABC in over 200 settings. By accelerating project delivery and minimizing traffic interruptions, UDOT has gained trust from political representatives and praise from the community. For example, on I-80 at Mountain Dell and Lambs Canyon near Salt Lake City, UDOT replaced four bridge superstructures in 37 hours over two weekends (Figure 2.76). The bridges were built adjacent to the existing structures in the median of I-80 over a four-month period. They were then transported to their final position by SPMTs. Using off-site construction and SPMTs, UDOT estimated that motorist delay was decreased by 180,000 hours, equating to a cost savings of over $2.5 million.

The Florida State DOT used SPMTs to remove an old bridge and install a new one in a matter of minutes (Figure 2.77). The old Graves Avenue Bridge was moved from its current position across I-4 to the side of the road for demolition in 22 minutes in 2006. This was the first use of SPMTs in the United States to replace a bridge across an interstate. Then, SPMTs were used to move the new spans from their fabrication site along I-4 to the bridge location, limiting the impact on motorists to only two weekend nights of detours and closures along the corridor.

The Connecticut State DOT used the largest mobile, land-based high-capacity crane to lift a 320 ft long truss weighing over 850 tons into position to form the main segment of the 1280 ft long bridge that

FIGURE 2.76   Utah uses SPMTs to replace bridges on weekends. (Courtesy of Utah DOT.)

FIGURE 2.77   Using SPMTs to remove the I-4 West Graves Avenue Bridge at night. (Courtesy of Florida DOT.)

carries Church Street South Extension over the New Haven Rail Yard (Figure 2.78). The Connecticut State DOT specified that this portion of the bridge be completed in a single night operation over a weekend.

The Massachusetts State DOT (MassDOT) has established an accelerated bridge program (ABP) to significantly reduce the number of structurally deficient highway bridges in the state system. MassDOT is relying on the use of ABC and SPMTs in construction to accelerate project development and delivery. Since 2008, the ABP has completed 28 bridge projects, with another 61 bridge projects currently in construction, and an additional 69 bridge projects scheduled to start construction in 2011. By the end of the eight-year ABP, more than 200 bridges are planned to be repaired or replaced. A method of accelerated bridge replacement is shown in Figure 2.79 for the replacement of the Cedar Street Bridge.

The New York State DOT used prefabricated components to replace the Belt Parkway Bridge in Brooklyn without impacting traffic during rush hour. Using traditional construction techniques, this replacement was scheduled to take 3–4 years to complete. Using ABC techniques, the entire project was completed in 14 months at a final cost of 8% less than the original estimate.

The Virginia State DOT replaced the Coleman Bridge along Highway 17 over a period of nine days. The truss and swing spans were constructed off-site at a nearby manufacturing facility down the river

FIGURE 2.78 High-capacity crane lifting 850 tons of bridge. (Courtesy of Connecticut DOT.)

FIGURE 2.79 Accelerated replacement of the Cedar Street Bridge. (Courtesy of Massachusetts DOT.)

from the bridge, then floated to the construction site on barges (Figure 2.80). Originally, the contractor estimated the entire process would take 12 days, but finished three days earlier than anticipated.

The Washington State DOT reduced the construction time for the replacement of the deck of the Lewis and Clark Bridge across the Columbia River from four years to four months of nighttime closures and three weekend closures by using SPMTs to bring in prefabricated elements for the deck replacement

FIGURE 2.80 Prefabricated bridge being transported to the job site. (Courtesy of Virginia DOT.)

FIGURE 2.81 Using SPMT to replace deck panels. (Courtesy of Washington DOT)

project. The same SPMTs were used to remove the old deck panels off the bridge (Figure 2.81). More than 3900 ft of concrete deck paneling was installed using the SPMTs.

These are just a few examples of states implementing ABC to meet budget constraints, structural needs in replacement or rehabilitation of structures, and the expectations of motorists for high-quality, longer-lasting, and less-disruptive highways. Accelerated bridge construction technologies have demonstrated benefits in building bridges safer, faster, and better. There is a need to balance speed, quality, and economy to achieve safe, durable, efficient, and sustainable bridges.

2.13 Future Bridge Design

John Kulicki and Michael Britt

2.13.1 Introduction

The importance of maintaining and enhancing the U.S. transportation infrastructure with limited funding resources is an ever-growing concern. At the same time, owners are demanding that bridge designs remain functional, yet meet the aesthetic desires of affected stakeholders. They are also demanding that today's new and rehabilitated bridges be constructed more quickly, require less maintenance, and last longer.

The bridge community is rising up to meet these demands. Several trends are emerging that will affect future bridges. To allocate resources more efficiently to an aging facility, asset management plans are being developed. More collaborative bridge type selection processes are evolving. To deliver projects more quickly, developers of major public and private transportation projects are using a variety of project delivery methods to reduce design and construction time. These emerging initiatives are briefly explained in the following text, along with a practical example of their application in the transportation community.

2.13.2 Asset Management

2.13.2.1 Overview

Faced with shrinking budgets, aging systems, increased traffic demands, and increased loads, many U.S. bridge owners are developing new strategies to extend the useful service life of their facilities and get the most out of their transportation dollars. One solution is an asset management plan that essentially follows the business processes that embody the principles of performance-based planning, programming, and management as promoted by FHWA and AASHTO.

Essentially, asset management involves taking what is already there, caring for it, upgrading and improving it where practical, and making it last as long as possible. As applied to bridges, this concept means examining an existing bridge and performing all the necessary maintenance and preventive treatments to make it last as long as possible, or until it costs more to keep up than building a new one.

2.13.2.2 Definition of Asset Management

Transportation asset management is a set of guiding principles and best practice methods for making informed transportation resource allocation decisions and improving accountability for these decisions. The term "resource allocation" covers not only allocation of money to program areas, projects, and activities but also covers deployment of other resources that add value (staff, equipment, materials, information, real estate, etc.). While several of these principles and practices were initially developed and applied within the domain of infrastructure preservation, most established definitions of asset management are considerably broader. *AASHTO Transportation Asset Management Guide* (AASHTO 2011), defines asset management as

> a strategic approach to managing transportation infrastructure. It focuses on ... business processes for resource allocation and utilization with the objective of better decision-making based upon quality information and well-defined objectives.

Asset management is concerned with the entire life cycle of transportation decisions, including planning, programming, construction, maintenance, and operations. It emphasizes integration across these functions, reinforcing the fact that actions taken across this life cycle are interrelated. It also recognizes that investments in transportation assets must be made considering a broad set of objectives, including physical preservation, congestion relief, safety, security, economic productivity, and environmental stewardship.

2.13.2.3 Summary of Process

Figure 2.82 illustrates the strategic resource allocation process that embodies the following elements:

- *Goals and Objectives* are established and supported by performance measures through the system planning process and used to guide the overall resource allocation process.
- *Analysis of Options and Tradeoffs* includes examination of options within each investment area, as well as tradeoffs across different investment areas.
- *Resource Allocation Decisions* are based on the results of tradeoff analyses. These decisions involve allocations of financial, staff, equipment, and other resources to the different investment areas and/or to different strategies, programs, projects, or asset classes within an individual investment area.

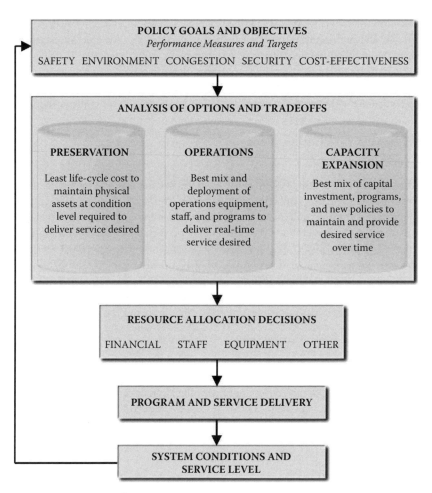

FIGURE 2.82   Strategic resource allocation process.

- *Program and Service Delivery* is accomplished in the most cost-effective manner, which again involves consideration of different delivery options (e.g., use of contractors, consultants, in-house forces), as well as a delivery tracking process involving recording of actions taken, costs, effectiveness, and lessons learned to guide future activity.
- *System Conditions and Service Levels* are tracked to see the extent to which established performance objectives are being addressed. This information is used to refine goals and priorities (e.g., put more emphasis on safety in response to an increase in crash rates).

2.13.2.4 Transportation Investment Categories

In Figure 2.82, the box labeled "Analysis of Options and Tradeoffs" shows three types of investment categories: preservation, operations, and capacity expansion. These are defined as follows:

1. *Preservation* encompasses work to extend the life of existing facilities (and associated hardware and equipment), or to repair damage that impedes mobility or safety. The purpose of system preservation is to retain the existing value of an asset and its ability to perform as designed. System preservation counters the wear and tear of physical infrastructure that occurs over time due to traffic loading, climate, crashes, and aging. It is accomplished through both capital projects and maintenance actions.

2. *Operations* focus on the real-time service and operational efficiency provided by the transportation system for both people and freight movement on a day-to-day basis. Examples of operations actions include real-time traffic surveillance, monitoring, control, and response; intelligent transportation systems (ITS); HOV lane monitoring and control; ramp metering; weigh-in motion; road weather management; and traveler information systems. Operations will not be discussed in this section.

3. *Capacity expansion* focuses on the actions needed to expand the service provided by the existing system for both people and freight. Capacity expansion can be achieved either by adding physical capacity to an existing asset or acquiring/constructing a new facility.

These categories are defined in order to show that:

- Asset management is not just about preservation of highway network assets; it is about making investment decisions that address a wide range of policy goals.
- The three categories provide a simple, useful way for decision-makers to align program investment categories and priorities with key policy objectives. For example, many owners have established "preservation first" goals or favor maximizing efficiency of operations prior to investing in new capacity. The categories may present alternative ways of meeting a policy goal. For example, it may be appropriate to consider operational improvements to address a congestion problem as an alternative to adding a new lane.
- Decisions about the resources allocated to each category cannot be made independently. Meeting many goals (e.g., safety) may require a mix of investments across these categories. Similarly, an increase in capacity expansion investments may require increased operations and preservation expenditures at some point in time.

Tradeoff analysis may be done across investment categories as well as within them. An owner might wish to define investment areas coincident with the categories discussed above, or they may define a different set of categories.

2.13.2.5 Representative Project: Asset Management Plan for a Suspension Bridge Owner

Many suspension bridges are approaching 100 years of age. Their load capacity and ability to handle increasing traffic capacity are being challenged. Some will see an increasing need for significant rehabilitation, while others must provide additional lanes to handle the increasing traffic demand. The implementation of asset management plans are seen as a means of examining the existing bridges and performing all the necessary maintenance and preventive treatments to make them last as long as possible, or until it costs more to keep them up than to build a new one. What follows is the framework for an asset management plan that considers preservation as well as the possibility of capacity expansion.

This asset management plan has been developed with the entire life cycle of the bridge in mind. It focuses on the existing bridge but takes into consideration that its remaining useful life, until a major rehabilitation is undertaken, may be limited. The plan includes planning, programming, engineering, construction, maintenance, and operations. It recognizes that continued and sometimes significant investment must be made to considering a broad set of objectives, including physical preservation, congestion relief, safety, security, economic productivity, and environmental stewardship.

The key stakeholders in the development and implementation of the plan are the management of the owner and the consultant team developing the plan. The plan is usually supported and reviewed by the affected stakeholders and agencies. The plan is usually undertaken to meet the needs of the transportation facility itself and also the needs of the transportation system in and around the facility. The asset management plan presented herein focuses on preservation and the possibility of crossing enhancement. Crossing enhancement could simply mean constructing a new facility. In a more aggressive interpretation, crossing enhancement could mean adding physical capacity to an existing asset. Both interpretations are accounted for in the plan.

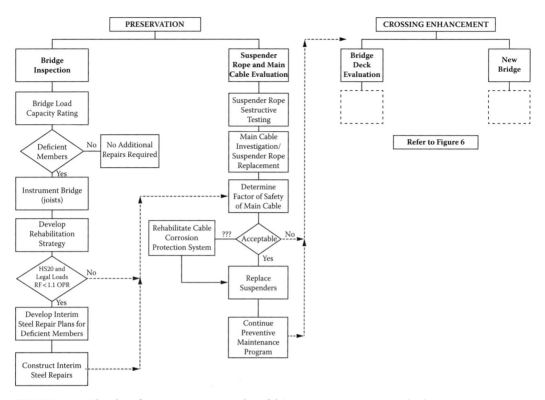

FIGURE 2.83   Flowchart for asset management plan of the representative suspension bridge.

The plan has four legs: two dealing with preservation and two dealing with enhancing the crossing. The initial preservation leg, bridge inspection, is usually completed with the construction of some type of interim repairs. The second preservation leg, suspender rope and main cable evaluation, follows. The two legs of the crossing enhancement may be undertaken simultaneously with the first two or immediately after the results of the first two legs are evaluated. This usually includes the completion of the new bridge study, the conclusion of the permitting process, and conceptual design of the new crossing. A conceptual flowchart for the asset management plan for this suspension bridge owner is presented in Figure 2.83.

2.13.3 Collaborative Bridge Type Selection Process

2.13.3.1 Overview

Tasked with developing new bridges, owners have had to incorporate the desires of affected stakeholders along with the traditional constraints of site-specific engineering principles. Successful teams have developed processes to balance these two often divergent forms of input. The most successful approaches have the following benefits:

- They have a track record of success
- They are unbiased
- They provide a cost-conscious, technical, and innovative approach with a goal of addressing the interests of all participants
- They provide a mechanism to keep the public informed of project progress
- They afford opportunities for public input as deemed necessary by the client

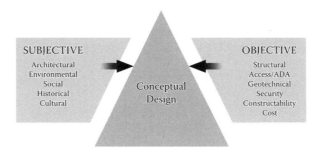

FIGURE 2.84 Inputs to the collaborative process.

Process leaders understand the importance of making facilitated stakeholder opinion the foundation of the process, while maintaining a balance of state-of-the-art engineering, sensitivity to the landscape, and overall cost.

There are two very different types of input (Figure 2.84) that drive these processes from concept to reality. These are subjective and objective and actually balance the conceptual design process to ensure the final design is functional, within budget, constructible, and aesthetically, culturally, and environmentally appealing. Subjective criteria may include aesthetic, architectural, and historical issues specific to the project area. Objective criteria may include technical issues such as structural analysis, geotechnical studies, security assessments, constructability, and cost.

To ensure eventual stakeholder acceptance of both the process of defining the visual quality of the bridge and the result, it is vital to start with no preconceived notions. To fairly and impartially explore options in a logical and methodical way, a structured, team-oriented approach is used as the framework of analysis. These processes encompass distinct steps, and they do allow for procedural modification as they evolve. A debriefing is usually held after the completion of each step, as well as meetings, workshops, and open houses. This offers the owners and/or stakeholders an opportunity to "tweak" the remaining steps based on real-time progress, findings, and budget and schedule constraints. The common steps of the most successful processes are explained in the following text.

2.13.3.2 Project Definition

Prior to taking a project to the public for dialogue and development, the project itself must be defined. The process team establishes the technical criteria for the bridge, such as span lengths, pier locations, vertical clearances, superstructure types, and budget. Constraints on structural design, foundation conditions, grade criteria, and economic feasibility must be established as a basis for future discussions with stakeholders and as a platform from which to begin the development of a bridge concept study report. At this time the goals, objectives, tasks, and deliverables are reaffirmed. The work plan and schedule will be mutually established, and administration procedures will also be established along with proper communication procedures. At this point, interaction is primarily between the process team and the owner's personnel.

2.13.3.3 Education and Information Exchange

This is an information-sharing period that begins with the listening sessions and culminates with an interactive workshop. The process team works with the owner to refine the process proposed here and develop the preferred conceptual design. However, prior to the workshop, the process team will work with the owner to frame out the workshop. The first workshop is critical—it is the best opportunity to develop a plan to connect positively with the stakeholders. It is extremely important to the process team that they establish a cooperative, trusting relationship with all parties involved. This is where the process is refined, if necessary, and where the needs, desires, and constraints of project are drawn out.

At the start of the workshop, specific bridge concepts will not yet have been developed. Rather, broad structural, architectural, landscape, and other concepts will be exhibited and explained. This should initiate open-minded information gathering. During this step, possible activities include:

- Site visits to both sides of the waterway to identify view sheds and cultural and environmental issues
- An architectural tour of local bridges and landmarks in the region
- Guest speakers such as architectural historians, planners, and bridge maintenance engineers

During the first workshop, the owner is usually asked what features or aspects they want to see in their bridge. These responses are often thematic in nature, such as "iconic," "vintage," or "gateway." At the end of this step, the process team and the owner will have a better understanding of each other's needs, have developed a collective vision for the bridge, understand the process for designing a bridge, and have refined the required stakeholder and/or public interaction process. The stage will be set for the first interaction with the general public. The development of baseline guidelines and evaluation criteria usually begins to evolve. At this point, interaction expands to include stakeholders and affected communities. The process team usually briefs other governmental and public agencies and public representatives as appropriate and necessary.

2.13.3.4 Concept Development

During the concept development stage, all ideas are encouraged, keeping in mind the goal of arriving at a manageable number of concepts for the public open house. The process team will develop a variety of bridge concepts to spark interest and discussion. These concepts are based on structure types befitting the project site and based on the constraints established during the education and information exchange. Once the team has compiled the options that appear to meet the vision of the project, fulfill the goals and objectives, and meet any other mutually agreed upon criteria, they are ready to prepare draft concept development manuals and have the first public open house.

At the end of the public open houses, concepts will be evaluated using a matrix approach and assigning weight to each criterion. The concepts will be compared against subjective criteria such as aesthetics, architecture, and historic preservation, as well as objective criteria such as durability, constructability, inspection, maintenance, relative qualitative first costs, and lifecycle costs. This approach aids in evaluating concepts to separate the less desirable options from the more favored concepts. At the end of this step, a consensus is reached on approximately three to five alternatives.

2.13.3.5 Concept Refinement

The goal of this stage is to improve performance and refine the favored alternatives. The process requires more rigorous technical input such as preliminary loading, seismic and aerodynamic analysis, refinement of vessel collision forces, and sufficient structural analysis to develop major component sizes. At this stage the process team can also begin the preliminary quantitative cost estimates.

By now the alternatives have passed the global architectural, environmental, and cultural trials in the previous step. The architectural character will begin to take form and the appropriate amenities or incorporation of any public art can be visualized through more detailed renderings. Typically the process team may prepare one or more of the following for each of the remaining alternatives:

- Photo montage renderings
- Physical models
- Computer-generated videos
- Large presentation boards of the options
- Design quality plan and elevation drawings

2.13.3.6 Preferred Alternative Selection

At the end of this stage, with the final public open house, the project stakeholders will select one river bridge type to move into final design and ultimately construction by using the same evaluation matrix from the public workshops.

2.13.3.7 Summary

As you can see, the proposed process (Figure 2.85) requires two very different types of input: subjective input such as aesthetic, architectural, and historical issues specific to the project area; and objective (technical) input to support the implementation of subjective concepts. The conceptual facilitated process is ideally suited for this project for several reasons:

- Collaboration and public "ownership" of the design
- Starts with no preconceptions
- Remains open to all possibilities
- Provides a forum so that all parties are heard

2.13.3.8 Representative Project: Monongahela River Bridge (Design Section 53D)

Motorists depend on Pennsylvania's highways and bridges to stay connected—to both their communities and industries. That's why the Pennsylvania Turnpike Commission established an ambitious, 13-section design plan as part of the Mon-Fayette PA Route 51 to I-376 network to reduce congestion, improve safety, and stimulate social and economic development in the Monongahela River Valley. A critical part of the plan, Section 53D, called for the design of a long-span bridge over the Monongahela River. The commission turned to Modjeski and Masters to design Section 53D, the crown jewel of the 24 mi Mon-Fayette Expressway.

The first step involved a rigorous bridge type selection study to identify the most suitable bridge crossing. Working with the commission and various stakeholder groups, our technical experts led a team that established evaluation criteria for comparing the alternatives, including cost, constructability, lifecycle maintenance, and aesthetics. Next, the team used this criteria to develop multiple-truss, arch and cable-stayed bridge alternatives. In the end, the dual-composite cable-stayed bridge alternative proved to be the most cost-competitive, durable, and aesthetic choice. Preliminary design of the cable-stayed bridge was completed (Figure 2.86) including a study to optimize the tower geometry for both performance

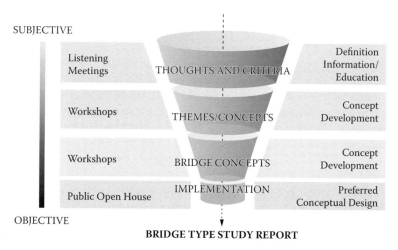

FIGURE 2.85 Summary of collaborative process.

FIGURE 2.86 Consensus on the bridge type and span arrangement for proposed Monongahela River Bridge. (Courtesy of Modjeski and Masters, Inc.)

and appearance. From this study, our engineers identified a hexagonal tower base cross section—an innovative solution that reduced the crossing's overall footprint and impact on the waterway.

2.13.4 Project Delivery Methods

2.13.4.1 Procurement Approaches

Developers of major public and private transportation projects in the United States and elsewhere are using a variety of project delivery methods to complete these projects. In the United States, transportation projects have been traditionally procured through a design-bid-build process. There is considerable interest on the part of transportation agencies in alternative forms of procurement and their benefits.

2.13.4.1.1 Design-Bid-Build

Design-bid-build is the traditional procurement approach for a project. The owner provides the completed plans and specifications and procures the construction services based on the lowest bid. The primary intent of a design-bid-build project is to build the project exactly as the owner specifies. The ability of the contractor to innovate is limited because of this.

2.13.4.1.2 Design-Build

Design-build is a procurement approach where the contractor provides both design and construction through a single contract between the owner and the design-build contractor. The owner will prepare a portion of the design, usually 15 to 35%, before bid. The major benefit of design-build is that it allows the contractor to be innovative during the design phase because the designer and the contractor are on the same team and constructability related issues can be addressed during design. When used with performance/end result specifications, design-build allows the contractor greater innovation.

2.13.4.1.3 Privatization

Privatization is a procurement approach in which a private entity finances or invests in a transportation project and develops, design, builds, and maintains a roadway or bridge for a specified duration in return for toll revenue, the cost of long-term financing, or development rights. The finance can take the form of a public-private partnership or a wholly private venture. Privatization also applies to contracting with a private entity for maintenance or other services traditionally performed in-house by the public agency. The benefit of this type of procurement is that it encourages the private entity to innovate or implement research findings to reduce maintenance costs, to ensure long performance, and to allow for recovery costs.

2.13.4.1.4 Construction Manager/General Contractor

Construction management is a contracting approach in which an owner contracts with an independent construction manager to provide program or project management or administrative services. The construction manager acts on behalf of the owner in overseeing and coordinating design and construction. With an owner-construction manager scheme, the construction manager does not contract directly with the construction contractors. This method of procurement can bring innovative or state-of-the-art techniques to the management of large or complex projects, such as improvement of cost control, identification and resolution of potential changes earlier in construction, or the reduction of construction time through fast-tracking design and construction activities and closely coordinating and tracking multiple design and construction contracts in a master schedule.

2.13.4.2 Contracting Approaches

The difference between a procurement approach and a contracting approach is one of scope. A procurement approach is a general scheme for purchasing services. A contracting approach is a specific technique used under the larger umbrella of a procurement approach to provide techniques for bidding, managing and specifying a project. There is a growing list of contracting approaches that are being incorporated into today's transportation projects. The top four are explained next, in order of their perceived acceptance in the transportation industry.

2.13.4.2.1 Partnering

Partnering is a method for establishing a relationship that focuses on achieving mutually beneficial goals through a formal procedure for better communication, shared risks, and resolving disputes at the lowest level. Partnering in the broadest sense can be performed between the owner, contractor, and other parties at the project level, among disciplines within the contracting agency, or between the owner and industry organizations on a broader scale. There are several benefits to partnering:

- Develops trust and communication
- Helps reduce perceived risk by building confidence in the project participants
- Helps the owner and the contractor reach a mutually acceptable decision on the proposed innovation
- Provides open communication and tools for working through avoidable problems and the owner's concerns

2.13.4.2.2 Value Engineering

Value engineering (VE) in design is the analysis of a design or process to identify individual components or functions that can be provided with improved quality or the same quality at a reduced lifecycle cost. VE in construction takes the form of proposals by the contractor that could produce a savings to the owner without impairing the essential functions and characteristics of the facility, including service life, economy of operation, ease of maintenance, desired appearance, and safety. VE in design encourages innovations from an in-house or process VE team that may result in a reduction of project cost or in the functionality of the design. VE in construction provides a means for the contractor to suggest innovative changes to the design or construction of the project with a value engineering change proposal (VECP), resulting in a savings in cost or time.

2.13.4.2.3 *Incentives/Disincentives*

Incentives/disincentives (I/Ds) are contract provisions intended to motivate a designer/contractor to complete the work on or ahead of schedule or to provide a product at a higher level of quality, safety, or overall performance. The benefits of using I/Ds include:

- Provides opportunities to try an innovative technique to meet I/D requirements that have an associated financial bonus
- Promotes innovations to reduce construction time in areas such as construction phasing or maintenance of traffic or the use of innovative materials
- Encourages innovation to improve quality or safety through new methods or materials in other ways

2.13.4.2.4 *Performance/End Result Specifications*

Performance/end result specifications define the required results of construction using measurable criteria or properties of the finished product. The measureable criteria or properties are used to verify compliance with the specifications when the work is completed. Performance/end result specifications differ from material and method specifications in that they specify the performance requirements of the end product and let the contractor determine how these performance criteria are to be met. These types of specifications provide flexibility in the design or construction of a project by allowing the contractor to be creative in its approach to construction to meet the specifications.

2.13.4.2.5 *Other Contracting Approaches*

Other contracting approaches that are gaining ground include multiparameter bidding, construction warranties, lane rental, constructability reviews, and bid alternate/design alternate.

2.13.4.3 Representative Project: I-15 Reconstruction

This project (Figure 2.87) consisted of the demolition, redesign, and reconstruction of 17 miles of I-15 in Salt Lake City, Utah. The existing highway was deteriorating and it carried a traffic volume higher than the design volume. Based on these conditions, the Utah DOT (UDOT) decided to completely rebuild the highway instead of making repairs. This work included widening I-15 to 10 lanes and building 137 bridges. The work, which had a construction cost of just over $1.5 billion, was completed in 4 1/2 years in order to be ready for the winter Olympics held in Salt Lake City in 2002.

FIGURE 2.87　Typical view of a portion of the completed 17 mile reconstruction of I-15. (Courtesy of the FHWA.)

A design-bid-build approach would have required 10 years to complete. UDOT decided to perform the project on a design-build basis to reduce this time. In addition to the time constraints, UDOT was interested in receiving a high-quality highway. Therefore, UDOT used performance/end-result specifications, construction warranties, and a preventative maintenance program. UDOT developed the performance/end-result specifications using task forces composed of consultants, professional associations, UDOT, and the FHWA. This partnering approach ensured that the performance/end-result specifications addressed the needs of the contractors, designers, and UDOT. UDOT also provided a $950,000 stipend for each unsuccessful bidder on a short list of qualified bidders.

The contractor introduced a number of innovations to the project through the design-build process and the performance/end-result specifications. For example, on the design-build side, the contractor was able to address traffic maintenance concerns as the design was developed. The contractor planned to install portions of the advanced traffic management system (ATMS) early in the project so that the ATMS could be used to reduce construction impacts on the public.

On the performance/end-result specifications side, the contractor used the flexibility of the specifications to introduce innovations and improve the quality of the project. For example, the geotechnical specification stipulated maximum allowable soil settlement values for the embankments and substructures, but did not specify the means for mitigating settlement. The contractor chose to take an aggressive approach that combined the use of wick drains, surcharge loads, and lime cement stabilization with the use of Styrofoam fill in the embankments to meet the settlement requirements. The contractor also introduced innovations to address the maintenance requirements of the projects. For example, the contractor used pipes embedded in the bridge abutments to allow grout to be injected into the abutments, to stabilize them if settlement occurs.

References

AASHTO. 2002. *Standard Specifications for Highway Bridges*, 17th Edition, American Association of State Highway and Transportation Officials, Washington, DC.

AASHTO. 2003. *Guide Specifications for Design and Construction of Segmental Concrete Bridges*, 2nd Edition, with 2003 Interim Revisions, American Association of State Highway and Transportation Officials, Washington, DC.

AASHTO. 2008. *AASHTO Guide Specifications for Design of FRP Pedestrian Bridges*, 1st Edition, American Association of State Highway and Transportation Officials, Washington, DC.

AASHTO. 2009. *LRFD Guide Specifications for Design of Pedestrian Bridges*, 2nd Edition, American Association of State Highway and Transportation Officials, Washington, DC.

AASHTO. 2010a. *AASHTO LRFD Bridge Design Specifications*, 5th Edition with Interims through 2010, American Association of State Highway and Transportation Officials, Washington, DC.

AASHTO. 2010b. *AASHTO LRFD Bridge Construction Specifications*, 3rd Edition, American Association of State Highway and Transportation Officials, Washington, DC.

AASHTO. 2011. *AASHTO Transportation Asset Management Guide: A Focus on Implementation*, 1st Edition, American Association of State Highway and Transportation Officials, Washington, DC.

Abrahams, M., Bryson, J., Wahl, P. and Stoyanoff, S. 2003. "Replacement of the Cooper River Bridges, Charleston, South Carolina," *Roads and Bridges Magazine*, November.

Andrew, C.E. 1939. "The Lake Washington Pontoon Bridge," *ASCE Civil Engineering*, 9(12), New York.

AREMA. 2011. *Manual for Railway Engineering*, American Railway Engineering and Maintenance of Way Association, Lanham, MD.

ASCE. 1977. *Civil Engineering*, 47(10), October. American Society of Civil Engineers.

ASCE. 2002. *Civil Engineering*, 72(11/12), November/December 2002. American Society of Civil Engineers.

Brown, C. 2008. "Meandering Across the Missouri," *Modern Steel Construction*, March. American Institute of Steel Construction, Chicago, IL.

Caltrans. 1997. *San Francisco-Oakland Bay Bridge West Spans Seismic Retrofit Design Criteria*, L. Duan, ed., California DOT, Sacramento, CA.

CAS. 1998. *Canadian Highway Bridge Design Code*, Canadian Standards Association, Rexdale, Ontario, Canada.

Donley, S. 2005. *Pennsylvania, Our Home*, Gibbs Smith, Layton, UT.

FHWA. 2006 *Prefabricated Bridge Elements and Systems Cost Study: Accelerated Bridge Construction Success Stories*, Federal Highway Administration, Washington, DC.

FHWA. 2007. *Manual on Use of Self-Propelled Modular Transporters to Remove and Replace Bridges*, FHWA-HIF-07-22, Federal Highway Administration, Washington, DC.

FHWA. 2007. *Connection Details for prefabricated Bridge Elements and System*, FHWA-00-010, Federal Highway Administration, Washington, DC.

FHWA. 2011. *Framework for Decision-Making*, FHWA-IF-06-30, Federal Highway Administration, Washington, DC.

Fridell, R. 2010. *Seven Wonders of Transportation*, Twenty-First Century Books, Minneapolis, MN.

Ghosn, M. and Moses, F. 2001. *Redundancy in Highway Bridge Superstructures*, NCHRP Report 458, Transportation Research Board, National Research Council, Washington, DC.

Gloyd, C.S. 1988. "Concrete Floating Bridges," *ACI Concrete International*, 10(5): 17–24.

Griggs, F.E. Jr. 2010. "Joseph B. Strauss, Charles A. Ellis, and the Golden Gate Bridge: Justice at Last," *Journal of Professional Issues in Engineering Education and Practice*, 136(2): 71–83.

HSP. 1908. *The Pennsylvania Magazine of History and Biography*, 32(1), Historical Society of Pennsylvania, Philadelphia, PA.

Kumarasena, S., McCabe, R.J., Zoli, T. and Pate, W.D. 2003. "Zakim Bunker Hill Bridge, Boston, Massachusetts," *Structural Engineering International*, 13(2): 90–94, May.

Lwin, M.M. 1989. "Design of the Third Lake Washington Floating Bridge," *ACI Concrete International*, 11(2): 50–53.

Lwin, M.M. 1993a. "Floating Bridges—Solutions to A Difficult Terrain," *Proceedings of the Conference on Transportation Facilities through Difficult Terrain*, Wu, J.T.H. and Barrett, R.K., eds., Aspen-Snowmass, Colorado, August.

Lwin, M.M. 1993b. "The Lacey V. Murrow Floating Bridge, USA," *Structural Engineering International*, 3(3): 145–148.

Lwin, M.M., Bruesch, A.W. and Evans, C.F. 1995. "High-Performance Concrete for a Floating Bridge," *Proceedings of the Fourth International Bridge Engineering Conference*, Vol. 1, Federal Highway Administration, Washington DC.

Lwin, M.M. and Gloyd, C.S. 1984. "Rebuilding the Hood Canal Floating Bridge," *ACI Concrete International*, 6(6): 30–35.

Meiners, W. 2001. "Credit Where Credit Was Due: Charles Ellis, a Purdue Civil Engineering Professor from 1934–1946, is Finally Recognized as the True Designer of the Golden Gate Bridge," *Purdue Engineering Extrapolations*, Purdue University Schools of Engineering, Summer, 6–11.

Melnick, S. 2004. "Sun Sculpture," *Modern Steel Construction*, October, American Institute of Steel Construction, Chicago, IL.

Murrow, L.V. 1938. "A Concrete Pontoon Bridge to Solve Washington Highway Location Problem," *Western Construction News*.

Nichols, C.C. 1962. "Construction and Performance of Hood Canal Floating Bridge," *Proceedings of Symposium on Concrete Construction in Aqueous Environment*, ACI Publication SP-8, Detroit, MI.

OMTC. 1994. *Ontario Highway Bridge Design Code*, Ontario Ministry of Transportation and Communications, Toronto, Ontario, Canada.

PCI. 1986. "Sunshine Skyway Bridge Closes the Gap," *PCI Journal*, November/December. pp. 168–176.

PTI. 2008. *Recommendations for Stay Cable Design*, Testing and Installation, 5th Edition, Post-Tensioning Institute, Farmington Hills, MI.

Reno, M. and M. Pohll. 1998. "Seismic Retrofit of San Francisco-Oakland Bay Bridge West Span," *Transportation Research Record: Journal of the Transportation Research Board*, 1624: 73–81.

Shaw, R.E. 1990. *Canals for a Nation: The Canal Era in the United States 1790–1860*, The University Press of Kentucky, Lexington, KY.

Strasky, J. 2002. "Long-Span, Slender Pedestrian Bridges," *ACI Concrete International*, 24(2): 42–48.

Tang, B. and Podolny, W. Jr. 1998. "A Successful Beginning for Fiber Reinforced Polymer (FRP) Composite Materials in Bridge Applications," *FHWA Proceedings, International Conference on Corrosion and Rehabilitation of Reinforced Concrete Structures*, December 7–11, Orlando, FL.

Tuan, C. 2004. "Aurora Arch Bridge," *ACI Concrete International*, 26(4): 64–67.

Van Der Zee, J. 1986. *The Gate: The True Story of the Design and Construction of the Golden Gate Bridge*, Simon and Schuster, New York, NY.

Relevant Web Sites

http://goldengatebridge.org
http://www.dot.ca.gov/hq/esc/tollbridge/SFOBB/Sfobb.html
http://www.panynj.gov/bridges-tunnels/george-washington-bridge.html
http://en.wikipedia.org/wiki/Three_Sisters_(Pittsburgh)
http://www.dot.ca.gov/hq/esc/tollbridge/SM-Hay/SMfacts.html
http://www.dot.ca.gov/hq/esc/tollbridge/Coronado/Corofacts.html
http://en.wikipedia.org/wiki/New_River_Gorge_Bridge
http://www.phillyroads.com/crossings/commodore-barry
http://bata.mtc.ca.gov/bridges/richmond-sr.htm
http://en.wikipedia.org/wiki/Richmond_%E2%80%93_San_Rafael_Bridge
http://www.fhwa.dot.gov/bridge/abc/index.cfm
http://www.fhwa.dot.gov/bridge/prefab/projects.cfm

3
Bridge Engineering in Argentina

Tomás A. del Carril
Buenos Aires

3.1 Introduction

Argentina is a large country, with a continental area of approximately 2.8 million km² and has a population of about 40 million people. It is located at the extreme south of the American continent, between Chile on the west and the Atlantic Ocean on the east. It is about 3900 km long from north to south and 1400 km wide from east to west (Figure 3.1).

3.1.1 Geographic Characteristics

Argentina's terrain includes the Andes Mountain chain all along the west border with Chile, a large plain in the center of the country interrupted by only a few hills, and a large fertile extension of land known as the Pampas region to the east. There are only two rivers in the country that are navigable by freights: the Uruguay River and the Paraná River. The enclosed area between these two rivers is known as the Mesopotamian region. There are several other important rivers that cross the country from west to east, flowing into the Atlantic Ocean. Said rivers are the result of the confluence of many small tributaries that descend from the mountains, forming low-speed plain rivers.

FIGURE 3.1   Physical map of Argentina, Buenos Aires City, and their main rivers.

3.1.2 Historical Development

The transportation infrastructure of the country is derived of its history and geography. During the Spanish colonization (sixteenth and seventeenth centuries), Argentina was part of the Viceroyalty of Peru, and the road transportation of goods came from all over to the capital Lima, across the high plateau of the northern Andes, and from there to Spain via Panamá. Only small bridges were required by these very scarce transportation demands. In 1776 the viceroyalty of Rio de la Plata was created by Spain in order to simplify the trade that was redirected to its capital, Buenos Aires, which grew as unique port of the southern colonies.

Around 1810, Spain began to lose its power and all the South American colonies became independent countries. Trade had to go through the port of Buenos Aires, which became large and rich thanks to taxes imposed on trade. The road network also was improved and changed accompanying these transformations, becoming a radial system in which all the roads went towards the port of Buenos Aires. Trade was essentially composed of products of the land, agricultural products, and meat. The country soon became one of the most important food and leather suppliers in the world.

Although there was a lot of immigration in the nineteenth century, the population was still very small. At present, two-thirds of the 40 million people that form the total population of the country live and work near Buenos Aires and its surroundings, in the principal cities that grew in the littoral of the country as main ports. The average density for the total area is only 14.3 people per square kilometer, leaving the country ranked 200th among the nations in the world. Also, due to the independence war, the Paraná River and all of Mesopotamia were isolated, leaving them disconnected from the whole area of the country, as a natural barrier against eventual attacks from the neighbors. There was no interest in building bridges over the main rivers (Paraná and Uruguay) until well into the twentieth century.

Many civil wars took place in the first half of the nineteenth century, preventing the economic and technical development of the country. The first engineering school was founded in 1877, more than 60 years after independence. Between 1870 and 1914, nearly all of the country's railway network was built by British and French companies. Local engineers did not play any role in designing and building the most important bridges of the network. To allow goods to be exported through the port of Buenos Aires, the network was developed, according the interests of the foreign companies, as a radial one. The main river (Paraná) was crossed by ferry. The first railway crossing over the Paraná River was not built until the 1970s (see Section 3.7.2).

In 1932 the Dirección Nacional de Vialidad (DNV; National Highway Administration) was created and the construction of roads and bridges received a major boost. Previous bridges were built by local communities or by railway companies to allow the freight to reach the port. Until late 1930s the industrial development of the country was almost nonexistent and local construction companies were not able to build large bridges. Most of the important bridges were designed and built by foreign engineers and contractors. They usually came from Europe, mainly from Germany. Germany had a strong influence on Argentine engineering and, as a matter of fact, the Argentine standards for concrete, steel, and bridge structural design are influenced or even translated directly from the DIN code (Deutsches Institut für Normung e. V.). Since almost the year 2000, there has been a strong tendency in the engineering organizations to switch towards the American Standards (ACI, AISC, and AASHTO) but, due to the great inertia of Argentina's institutions, this change might take too long.

3.1.3 Historical Bridges

3.1.3.1 Old Bridge of Areco

One of the very few bridges that survived from the days of the independence wars is the "Puente Viejo de Areco" (Areco Old Bridge), shown in Figure 3.2, built in San Antonio de Areco between 1854 and 1856 and designed by a French engineer whose last name was Mollard. It was the first toll bridge in the country.

FIGURE 3.2 Old bridge of Areco River, before its restoration (2004).

The arch bridge is 18 m long. It is composed by two parallel masonry walls that are filled with soil. Two additional circular holes at both sides of the arch increase its hydraulic capacity. Its rise-to-span ratio is 1:5. The arch is 60 cm deep at the center. It rests on spread foundations. The bridge was restored in 2002.

3.1.3.2 La Boca Aerial Transporter Bridge

This old bridge spans the Riachuelo River connecting Buenos Aires City Federal District to Maciel Island (Buenos Aires Province), and is an important tourist attraction of the La Boca neighborhood at present. The British South Railway Company was authorized to build the bridge in 1908 and donated it in return for various tax exemptions granted by the national government. When completed in 1914, it was named Nicolás Avellaneda, after the Argentine president who played an important role in the federalization of Buenos Aires in 1880. The original purpose was to serve pedestrians, automobiles, street cars, wagons, and trams on its 8 m × 12 m platform, across a length of 65 m (Figure 3.3).

The structure leaves a 50 m × 40 m navigational clearance. The iron parts of the bridge were built in England and then assembled in Buenos Aires and rest on the foundations built by National Port Authority, eight 4 m diameter 24 m high concrete piles. The bridge was opened to traffic on May 30, 1914. It has been in disuse since 1939 and was closed in 1960. It was declared a Cultural and Historical National Monument in 1993 to prevent it from being dismantled. These types of bridges are of a valuable typology of engineering, developed during the industrial revolution. Little by little, technological advances in construction made them disappear, leaving, at present, only eight of this kind worldwide.

3.1.4 Bridge Infrastructure

3.1.4.1 Roadway Bridges

The Argentine Republic has approximately 500,000 km (310,000 mi) of roads of different types or categories. Around 40,000 km (25,000) of them constitute the primary or national network and 180,000 km are the secondary or provincial network. The remaining 280,000 km (175,000 mi) forms the tertiary network, dependent on the municipalities. Only 61,000 km (38,000 mi) from the 220,000 (138,000 mi) that form the primary and secondary networks are paved, whereas 37,000 km (23,000 mi) have had some type of improvement (with gravel or some type of stabilization). The remaining 120,000 km (75,000 mi) are rural roads, without any stabilization.

FIGURE 3.3 The old Nicolás Avellaneda Transporter Bridge in the foreground. The homonymous lift built in 1937 can be seen behind it (See Section 3.9.3).

Recalling that Argentina possesses a continental surface of 2.8 million km^2 (1.1 million mi^2), it is seen that the country has an underdeveloped road structure, which owes itself partly to its scanty population, a problem that is aggravated by the high demographic concentration in certain zones. The primary network of the country, which includes 3100 bridges, is under the administration of the Dirección Nacional de Vialidad (DNV), an entity of the National Secretary of Public Works, whereas the secondary network depends on the DNV of each province, under the domain of each one of the provincial governments. There are around 5000 bridges in the secondary network.

The primary and secondary networks were formed in the 1940s, and, under internal financing, underwent an epoch of relative improvement in the 1960s until the middle of the 1970s. As the financial situation of the State was deteriorating throughout the years, the successive governments were turning the funds from the DNV towards other destinations, diminishing the road construction and maintenance. In the 1990s, near 10,000 km (6200 mi) of routes were granted in concession by toll to private companies. There are around 1800 km (1100 mi) of expressways, all of them granted in concession by toll. The concession system shares 75% of the whole traffic of the country.

3.1.4.2 Railway Bridges

The construction of the railway network in Argentina began in the second half of the nineteenth century. Most of the railway system was constructed between 1870 and 1914 by the English and French, leading to the country to occupy the tenth place in the world, with approximately 47,000 km (29,000 mi) of railroad. Rail transport was an engine of the development of the country. After the first world war, there were no increases in the railway network and in 1946, the extensive network went to the hands of the State, suffering a yearly deterioration because of the erroneous political decisions of the authorities. At present, the railway Argentine network has an extension of 31,000 km (19,000 mi), with three track gauges, which is an inconvenient situation for its operation.

In 1930, Argentina's railway network of 40,000 km (25,000 mi) represented 45% of the total of South America and it was the third overall, after the United States (400,000 km) and Canada (65,000 km). At this time, trains were carrying more than 50 million tons/year, whereas now, only the 36% of that value is moved by railways. The decline also reached to the inter-city traffic of passengers, which nowadays is 2.5 millions of persons per year, a tenth of those who were mobilizing by train 25 years ago.

The development of railway bridges has been according to the development of the rail network. Almost no railway bridges have been constructed in the last 70 years, with the exception of the cable-stayed Zárate–Brazo Largo mixed bridges (see Section 3.7.2). On the contrary, there are many cases of railway bridges that have been modified to adapt them for road traffic (see Section 3.6.1). There is no inventory of the railway bridges in the country but an estimation of around 2500 bridges serves as a good figure.

3.2 Design Practice

The evolution of bridge design practices in Argentina has followed a path directly related to the evolution of the country. During colonial times, the few bridges we know about were built in wood and haven't survived the test of time. However, since the eighteenth century, some documents and design plans about the construction of important bridges have survived. Until 1810, bridge construction was executed in the state-of-the-art knowledge of the colonial government, the Kingdom of Spain.

After independence, the country intensified commerce with other nations of Europe, mainly with England, France, and Germany, which strongly invested in the country. Said investments were accompanied by new technological advances in design and construction that were absorbed by the locals. These countries also had great academic influence on the universities of Argentina and particularly on the training of engineers. After the creation of the DNV in 1932 and with the schools of engineering well consolidated, many excellent bridge designers had the opportunity to apply their knowledge and creativity in building bridges all around the country.

3.2.1 Design Specifications for Highway Bridges

In 1952, the DNV published and put into effect the "Basis for Calculation of Reinforced Concrete Bridges," which contains specifications for loads and their combinations and specifications for the design of reinforced concrete sections based on the allowable stresses design method. This regulation was an almost literal translation of the German Standards (DIN) of the 1930s.

In reference to the design of reinforced concrete sections, this regulation became rapidly obsolete and the design of reinforced concrete for bridges was carried out following the requirements of the standards for the design of building concrete structures, also an adaptation of German DIN standards. By contrast, specifications of loads are now still used with minimal change to the distributed load, decreasing as a function of the length of the influence line-loaded length to get maximum stresses.

The inconsistency involved in using loads from one standard and dimensioning according to another standard has not been an obstacle to the design of bridges of moderate span so far, and these have produced a great degree of safety. However, for large projects, DNV has accepted the use of foreign standards, mainly those of AASHTO and Eurocodes, in its different editions at the time of design. At present, the government administration office CIRSOC (Research Center for Safety Regulations of Civil Works) has created a commission to update the regulations for bridge design based on an adaptation of AASHTO standards.

3.2.2 Design Specifications for Railway Bridges

Since the beginning of railway construction in the country, British companies built and operated them, as well as designed and manufactured in England a large number of bridges that were then assembled in Argentina, according to their own specifications. In 1909, a National Administration of Railways was created to manage the rail network. In 1949, the property of the railways was transferred to the government. The new Administration of Railways (Ferrocarriles Argentinos) issued standards for the structural design of steel and concrete bridges that are still in use.

Because the railways haven't had further development in Argentina and have been losing share in the transport of cargo and passengers, there are virtually no modern railroad bridges. Only a few structures for passage of streets or roads under railway lines in operation are now constructed, but these small structures don't require special technologies or designs.

3.3 Concrete Girder Bridge

3.3.1 Agustín P. Justo–Getulio Vargas Bridge

This bridge, shown in Figure 3.4, was built over the Uruguay River, a natural boundary between the east of Argentina and southwest of Brazil, linking Paso de los Libres city (Argentina) with Uruguayana city (Brazil). It is a parallel and separate roadway–railroad dual-purpose bridge.

When completed in 1945, it became the first physical and a very important connection between the two largest countries in South America and the first long bridge over the Uruguay River. Its importance meant both countries made huge efforts to build it during World War II.

The bridge was designed by a joint committee that consisted of three members of each country. Among them was the famous Argentinean military engineer Manuel N. Savio. Each country worked independently on its construction and met halfway. On the Argentinean side, the contractors were Administración de Vialidad and Ferrocarriles del Estado. The work began in December 1942 and ended in February 1945. It was opened to traffic in December 1945.

The bridge's total length is 1419 m (35 m + 40 × 38 m + 35 m + 40 × 38 m + 35 m), symmetrically composed by a simply supported 35 m side span, 10 continuous four-span frames of 38 m and a simply supported center span of 35 m. The superstructure comprises two independent decks with an overall width of 12.9 m. The roadway deck carries two 3.15 m wide lanes and 1.2 m sidewalks on each side. The remaining 3.85 m are occupied by the railway track. Both decks are supported by two haunched box beams 2.5 m deep at midspan and 4 m at the abutments. The whole structure is cast-in-place reinforced concrete. The piers are wall type and integrated to the superstructure. They are founded on spread foundations on a sound rock stratum. Due to the fact that this section of the river is not navigable, the vertical clearance over water level is only 3.7 m.

FIGURE 3.4 The Agustín P. Justo–Getulio Vargas Bridge in a stamp from 1947, as a testimony of its importance for the country.

3.3.2 Manuel Elordi Bridge

Built on National Route 34 to span the Bermejo River in Salta Province, the Manuel Elordi Bridge was the country's first prestressed concrete bridge. Designed by engineer Helmuth Cabjolsky (a German resident in Argentina), the bridge (Figure 3.5) has a total length of 336 m composed by two continuous beams of three 56 m spans. The cross section is two cast-in-place T-beams. The roadway deck is 7.2 m wide. The piers are concrete rectangular solid walls. The owner is the DNV. The bridge built was an alternative to the original design provided by DNV on reinforced concrete. The contractor was Zarazaga y De Gregorio. It was constructed between the years 1958 and 1961. After this bridge was built, the prestressed concrete technique was expanded all around the country for bridges with spans over 20 m.

3.3.3 Colastiné River, Old Bridge

In 1967, the Colastiné River was the only remaining obstacle for the completion of the land transportation between the country's Mediterranean region and the Mesopotamian region on National Route 168. In December 1967 the completion of the tunnel below the Paraná River was expected to establish an important road connection between the cities, so it was imperative to solve the crossing of the Setubal Lake (see Sections 3.3.4 and 3.3.13) and the Colastiné River.

The bridge (Figure 3.6) is 522 m long. It comprises 10 simply supported equal spans. The superstructure is composed of four 50 m long and 3 m deep precast prestressed T-beams, with an overall weight of 120 tons. The deck slab was cast-in-place. It carries two lanes 4.15 m wide and two sidewalks of 2 m each. The piers are of hollow concrete and supported by spread footings.

There were many technical difficulties with the material of the river bed that affected the construction schedule. The works began in 1957 by the owner DNV, and the bridge was opened to traffic on March 1967. It was designed by contractor Compañía General de Construcciones. At the time of its construction, this bridge established a country record in terms of the length of the precast prestressed beams.

FIGURE 3.5 The Manuel Elordi Bridge, over Bermejo River in Salta Province, is the country's first prestressed concrete bridge.

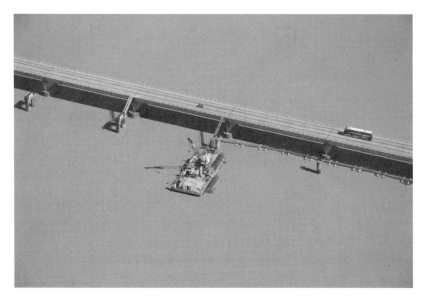

FIGURE 3.6   The Colastiné River old bridge. The new bridge (see Section 3.3.13) is under construction upstream. (Courtesy of Supercemento – Chediack, UTE, Empresa Constructora.)

3.3.4 Nicasio Oroño Bridge (New Setubal Lake Bridge)

One transportation obstacle between the Mesopotamian and Mediterranean regions was a transit restriction on National Route N° 168 due to an old suspension bridge over Setubal Lake (see Section 3.9.1), linking the cities of Santa Fe and Paraná. The overall bridge length is 298 m (84 m + 130 m + 84 m) composed of three continuous spans (Figure 3.7). The superstructure was built using the balanced cantilever cast-in-place method and its cross section comprises two 0.32 thick box beams of variable height between 2.2 m and 7 m. The top slab is 10 m wide and the bottom slab is 5.7 m wide. The deck accommodates four lanes 3.75 m wide, two sidewalks of 2.6 m, and a median of 1.5 m which rests on top of a simply supported slab that closes the gap between the two box sections. Two large aqueducts that provide fresh water to Santa Fe are suspended under the deck, between the two box beams.

Both main piers supported on Franki piles are composed of twin thin vertical walls monolithically connected to the superstructure in order to provide the required flexibility for long-term displacements and the necessary stiffness to withstand the high flexural moments transmitted by live load. A curved approach viaduct is on the Santa Fe side. The bridge was designed by Europe Etudes, Societe de Techniques pour l'Utilization du Precontraint (STUP) Argentine branch, and Pilotes Franki.

Construction works began in 1967 by the owner DNV. Contracting was a joint venture between the companies Cristiani Nielsen SA and Pilotes Franki SAIC. The bridge was opened to traffic in March 1970. The large flood of 1983 produced the collapse of the north tower of the old suspended bridge, located only 60 m upstream of the new bridge and an important scouring on the north pier of this bridge. After the preventive closure to traffic, new bored piles were built to retrofit the north pier to a safer situation.

3.3.5 New Pueyrredón Bridge

The New Pueyrredón Bridge (Figure 3.8) was built over the Riachuelo River and links Buenos Aires City Federal District with Avellaneda in Buenos Aires Province, two highly populated metropolitan areas. It constitutes the main southwest access to the city. The overall length of the bridge is around 1300 m. The

FIGURE 3.7 View of the New Setubal Lake Bridge from the Santa Fe side. The retrofitted north main pier is visible in the background.

FIGURE 3.8 View from downstream of the New Pueyrredón Bridge over the navigable section of the Riachuelo River.

main bridge is 187.3 m long (24.5 m + 43.1 m + 46.6 m + 43.1 m + 30.2 m), with five simply supported spans. Approach viaducts are 1026 m long, covering a span range from 8.9 m to 33.4 m. The girders are eight 2.35 m deep, precast prestressed trapezoidal box beams, with weight varying from 100 tons to 200 tons. It was erected using floating cranes. Two independent cast-in-place deck slabs are linked together by a simply supported median 1 m long concrete slab. The deck accommodates four lanes 3.75 m wide and two sidewalks 2.50 m wide, on both sides.

Two bents with two lines of eight circular columns of 1 m diameter are rigidly connected to pile caps. Each pier is supported by 161 m diameter precast prestressed concrete piles composed of 10 segments of 3 m long precast hollow cylinders, brought together by prestressed tendons and reaching a dense sand stratum called Puelchense formation. The northern approach viaduct is composed of simply supported 17.85 m long, 33.40 m wide spans, with 14 precast prestressed concrete T-beams. The southern approach viaduct is a continuous cast-in-place curved trapezoidal box beam, with variable spans ranging between 16.5 m and 37.8 m. Piers are circular columns 1.6 m in diameter and are founded on the Puelchense stratum.

Construction began in 1969 by owner DNV. It was designed by the contractor Empresa Argentina de Cemento Armado (EACA) and was developed by Pretensac. The bridge was opened in 1971 and became a great traffic relief during rush hours. At the time of its construction, this bridge established a country record due to the fact that precast beams of 200 tons were lifted up into place. The precast segments were erected by a new floating crane that was incorporated to the Port Authority of Buenos Aires fleet, originally conceived for extracting old shipwrecks.

3.3.6 Libertador General José Gervasio Artigas Bridge

This bridge (Figure 3.9) was built over the Uruguay River, a natural boundary between the east of Argentina and western Uruguay. The bridge is located near the cities of Paysandú (Uruguay) and Colón (Argentina). The governments appointed a new committee, called COTEPAYCO (Technical Committee for the Paysandú–Colón Bridge), in 1966, which became the owner of this bridge.

The bridge total length reaches 2350.44 m. The main structure is 335 m long (97.5 m + 140 m + 97.5 m), composed by a three-span continuous beam built by the segmental balanced cantilever cast-in-place method. The Uruguayan side approach viaduct is 460 m long with ten simply supported spans of 46 m each. The Argentine approach viaduct is 1555.44 m long, composed by 22 pairs of cast-in-place concrete continuous spans with π sections. The overall deck width is 11.6 m. Roadway lanes are 8 m wide and pedestrian sidewalks are 1.8 m wide per side. Central span piers are composed by two concrete thin-walls. The abutments are open end, spill-through. The vertical clearance for navigation under the center span is 34 m.

The bridge was designed by the engineering firm Cabjolsky–Heckhausen of Argentina. Construction began by a joint venture of EACA, Ing. Odemar H. Soler SA, and Zarazaga y De Gregorio SAIC in September 1970. It was financed by the Ministries of Public Works of both countries. The bridge was inaugurated in December 1975.

3.3.7 Guachipas River Bridge

This bridge (Figure 3.10) was built on National Route 22 over the Guachipas River, in Salta Province, and was part of the Cabra Corral Hydroelectric Project. The construction of a dam required that the bridge would have to be built very high, taking into account that an artificial lake would be formed by flooding most of the existing route and the existing bridge over the Guachipas River. The selection of the type of bridge was the result of a combination of economic and safety criteria as well as construction speed.

FIGURE 3.9 Overview of General Jose Gervasio Artigas Bridge.

(a)

(b)

FIGURE 3.10 The Guachipas River Bridge under construction, before (a) and after (b) the filling of the Cabra Corral Dam.

Safety was extremely important due to the fact that the bridge was to be located in a highly seismic area and any reparation would be impossible after the lake was formed.

The overall length of the bridge is 378.6 m. It comprises eight simply supported spans (46.1 m + 6 × 48.9 m + 39.1 m). The superstructure is constituted by three precast prestressed T-beams 2.5 m deep. The deck has a total width of 10.7 m. It accommodates two lanes 4.15 m wide and two sidewalks of 1.2 m per side. Piers are composed of large hollow cylinders 6 m in diameter, 300 mm thick, and nearly 60 m tall, which rest on a square (6.2 m × 6.2 m) cross-sectional base, 6 m tall. Due to its great height, the piers are among the most outstanding feature of this bridge. The piers are topped with square slabs (7 m × 7 m × 1 m) to support the T-beams. Triangular counterforts assure a very stiff connection in sections of high flexural moments. Some piers are supported by a pile cap (15 m × 15 m × 1 m) and 48 Franki piles of 0.53 m diameter and approximately 8 m long. Some of the piers are founded with spread foundations on a highly resisting sandstone stratum.

Construction works began in 1971 by owner Agua y Energía Eléctrica de la Nación, and the bridge was designed by contractors EACA and Panedile Argentina SA. The bridge was opened to traffic in 1974, well before the hydroelectric project schedule. At the time of its construction, this bridge established a country record in terms of the height of the piers. Unfortunately, they can't be seen after the lake has reached its final water level.

3.3.8 Libertador General San Martín Bridge

This bridge (Figure 3.11) is located over the Uruguay River, a natural boundary between the east of Argentina and the west of Uruguay, linking Fray Bentos (Uruguay) and Puerto Unzué (Argentina). In order to carry the roadway networks of both countries, the governments signed an agreement to build two bridges over the Uruguay River in 1960: the one linking the cities or Fray Bentos (Uruguay) and Puerto Unzué (Argentina) and the other between the cities Paysandú (Uruguay) and Colón (Argentina). On its completion, it became the first physical connection between the two countries and it is still part of the shortest route between the capital cities of Buenos Aires and Montevideo.

Construction began on August 1971 by the owner, COMPAU (Mixed Technical Commission for Bridges between Argentina and Uruguay), with funds submitted by IADB (Inter-American Development Bank). Joint venture COPUI (International Bridge over the Uruguay River Consortium) were the contractors, composed of Entrecanales y Tavora from Spain and Hochthieff from Germany. It was designed by the Uruguayan engineer Alberto Ponce Delgado, owner of the firm INVIAL of Urugay, and the Italian professor Riccardo Morandi acted as consulting engineer. Technical project management was carried out by a joint venture between designer INVIAL (Uruguay), and SAE (Sociedad Argentina de Estudios). It was opened to traffic in September 1976. Due to its main span of 220 m, it was world's longest segmental bridge at the time its construction began.

The total bridge length is 3408 m. The main structure is 510 m long (145 m + 220 m + 145 m), with three spans structured as a gerber beam type, with two double cast-in-place arms, built by the segmental

FIGURE 3.11   (See color insert.) Overview of General San Martín Internacional Bridge, between Fray Bentos (Uruguay) and Puerto Unzué (Argentina).

balanced cantilever method, and three 40 m isostatic precast prestressed beams. The two approach viaducts, also gerber-type structures, are composed of 30 m precast prestressed double cantilever arms placed over the piers, which support 40 m isostatic precast prestressed beams spans, thus forming secondary spans of 70 m. Seven secondary spans are placed on the Uruguayan side and the other 17 are on the Argentinean side. A long low-level transition viaduct (1×55 m $+ 27 \times 41$ m) completes the structure on the Argentinean side. The overall deck is width is 11.3 m, with two lanes of 3.65 m, narrow emergency shoulders of 0.50 m, and pedestrian walkways of 1.5 m width.

The two piers of the main span are composed by two concrete U-shaped walls facing each other. They are 36 m tall, built with vertical sliding forms in order to achieve a navigation clearance of 45 m at the main span. Each of these piers where founded on four, 10 m diameter, cylindrical caissons, driven with the use of compressed air 24 m below water level and reaching a sound sandstone stratum. The abutments are spill-through.

The piers of the main span, over the navigation channel, are protected against vessel collision by an independent protection system. The fences are placed upstream and are designed to absorb the energy of impact by plastic deformation of its reinforced concrete components: piles and plates with heavy reinforcement. There were no approved international standards for the design of that type of protection system during that time. Thus, it was considered an important innovation and it was the first time that a great bridge was protected against vessel collisions of such characteristics in the country.

3.3.9 25 de Mayo Urban Highway Viaduct

This highway viaduct (Figure 3.12) is 10 km long from the junction between Lafuente Ave. and Del Trabajo Ave. to Ing. Huergo Ave. The viaduct, with four 3.5 m wide lanes and shoulders 2.5 m wide in each direction, passes over a highly populated area, and it is the main access to the Buenos Aires international airport. The structure is comprised of two separate viaducts parallel to each other. Each of them consists of continuous beams of six spans of variable lengths from 25 to 30 m depending on the streets and railways to overpass. The cross section is a trapezoidal box lightened by the use of plastic hollow tubes and was cast in place with a sliding formwork adapted to small changes in the length and curvature of the alignment.

FIGURE 3.12 A sector of the 10 km 25 de Mayo Highway (AU-1). (Courtesy of ATEC Ingenieros Consultores.)

The owner is the city of Buenos Aires. Feasibility studies and original design were carried out by ATEC Ingenieros Consultores, who also acted as project managers. The construction and detail design were assigned to Concessionaire AUSA (Autopistas Urbanas SA), with the 20 year usufruct of the tolls. This was the country's first toll-operated urban highway. Construction works began in 1978 and it was opened to traffic in December 1981. The traffic was 70,000 average daily traffic (ADT) at the time it was inaugurated. At present 236,000 vehicles use the highway in peak hours, every day.

3.3.10 Presidente Tancredo Neves Bridge

This bridge (Figure 3.13) is placed over the Iguazú River, a natural boundary between northeast Argentina and southwest Brazil, very near Iguazú Falls (see Section 3.10.4), which was declared a World Heritage Site by UNESCO in 1984. Although the agreement to build this bridge was signed by the presidents of both countries in 1972, construction began in January 1983 by COMIX (a mixed committee between countries).

The bridge was built using the segmental, cast-in-place, balanced cantilever method. The superstructure comprises a three-span continuous box beam of 480 m total length (130 m + 220 m + 130 m), and a deck which is 16.5 m wide. The column bents are 48 m tall. The piers are rectangular at their base and then subdivide into two thin-wall columns, monolithically connected to the superstructure on the top to provide the required flexibility to the continuous beam. The piers are founded on 20 1.8 m diameter bored piles, clamped 4 m into bed rock (basalt). The pile caps are prestressed. The abutments are founded on reinforced concrete spread footings.

The bridge was designed by consulting firm Consulbaries (Argentina) and Figueredo Ferraz (Brazil). Construction was carried out by joint venture between contractors Supercemento (Argentina) and Sobrenco (Brazil) and a joint venture between the designers and the firm ETEL (Brazil) was responsible for the technical project management. It was opened to traffic in December 1985, only one month after schedule, which was a real achievement taking into account a huge flood in 1983 that flooded the area for several months.

FIGURE 3.13 Overview of Presidente Tancredo Neves Bridge almost completed. (Courtesy of Supercemento SAIC.)

3.3.11 9 de Julio Urban Highway

9 de Julio Ave. is an important traffic route of Buenos Aires. It crosses downtown from north to south, with a total length of 18 km. Only two parts of this gigantic project were built, which constitute the present southwest and northern access to downtown. The structures of these two branches are elevated viaducts (Figure 3.14); the south branch is 2.2 km long and links up with the new Pueyrredón Bridge (see Section 3.3.5) and the north branch is 2.7 km long and crosses over the whole railway gridiron while approaching the main station of Retiro and the railway system of the port of Buenos Aires. Two separate viaducts are parallel to each other, composed of simply supported spans with precast prestressed V-shaped cross sections of variable lengths ranging between 24 m and 42 m. The piers consist of octagonal columns founded on bored piles 1.2 m in diameter, driven to dense sand stratum at 28 m depth.

The owner is the Buenos Aires city government. The feasibility studies and original design were carried out by ATEC Ingenieros Consultores, who also acted as project managers. The construction works and detail design were performed by COVIMET (a joint venture between Argentinean and Spanish construction and structural design firms), who began the works as concessionaires responsible for building and managing the toll project for 20 years, until the project was transferred to the government of the city. At present 130,000 vehicles use the highway every day.

3.3.12 La Plata–Buenos Aires Highway over Riachuelo Bridge

The 54 km of the La Plata–Buenos Aires highway connects the capital of the Buenos Aires Province (La Plata) with the city of Buenos Aires (Federal District). It needs to cross the Riachuelo near its flow into the Río de la Plata, where its width increases very much, in a harbor zone. It was necessary to design two identical bridges with severe impositions: a free span not less than 75 m, a vertical clearance for navigation of 27.5 m, an adjusted schedule for construction, and consideration that low construction cost was a decisive variable for alternative selection.

The whole bridge has five spans, with three continuous central spans and two simply supported end spans. Total length is 236.5 m; the central span length is 76.5 m. The superstructure is made of prestressed concrete precast segments of 24 m, and 59 m for the central span. Erected as a gerber beam, the

FIGURE 3.14   A sector of the 9 de Julio South branch of the highway.

FIGURE 3.15 Bridge over the Riachuelo for the La Plata–Buenos Aires highway.

central and side spans were transformed in a continuous beam with the inclusion of prestressed tendons across the support sections to obtain continuity. The slabs were cast-in-place. Two bridges were built with a 16.9 m width to accommodate traffic in each direction. Piers are reinforced concrete columns with constant sections, cast with traveling forms, joined with a lintel to support the superstructure. The foundation was built with bored, cast-in-place piles 36 m long, to reach a dense sand stratum.

The bridge (Figure 3.15) was designed by the engineers Luis J. Lima, Edgardo L. Lima, R. González Saleme, and C. González Lima (La Plata, Argentina). The design was a simple and economic, easy to build in the imposed time with the disposable technology, and it fulfilled perfectly all the initial imposed conditions. It is adequate for the whole highway project and the surrounding landscape, without lack of measure, pomposity, uselessness, or useless cost enhancement. The owner is DNV and the contractor is a joint venture called COVIARES (Consorcio Vial Argentino Español). Construction works began in 1989 and it was opened to traffic in 1991. Piles were done by TREVI Argentina.

3.3.13 New Colastiné Bridge

The greatly increased traffic between the capital cities of the provinces of Entre Ríos and Santa Fe required that the National Route N° 168 be upgraded into a two-lane highway in each direction. It was necessary to build a new bridge over the Colastiné River beside the existing bridge (see Section 3.3.3) to accommodate the new traffic lanes (Figure 3.16).

The new bridge has the same configuration as the old one: 10 spans of 52.3 m. It is composed of a 523 m long, 3 m deep prestressed box beam. The box section has a 12.73 m wide top slab and a 5.2 m wide bottom slab. The substructure is composed of two 2 diameter column-shaft bents, 44 m long. Abutments are of the closed type and founded on five 1.4 m diameter piles. The new design reflects bridge construction technology advances in the country over the last 40 years. At present, the new bridge is in its final construction stage using the incremental launching method.

The bridge was designed by In-Group and engineer Carlos Gerbaudo (Córdoba, Argentina) was the technical project manager. The owner is DNV and the contractor is a joint venture between Supercemento SAIC and Jose J. Chediack SA. Construction work began in 2008 and it was opened to traffic in 2011.

FIGURE 3.16 The New Colastiné River Bridge (left) is being built beside the old one (right). (Courtesy of Supercemento–Chediack, UTE, Empresa Constructora.)

3.4 Steel Girder Bridges

3.4.1 La Polvorilla Viaduct

Considered to be one of the most difficult railroads routes in the world, the Huaytiquina Railway, today called Tren a las Nubes (Train to the Clouds), is a train service in Salta Province that links the Argentine northwest with the Chilean border in the Andes mountains. The train track line is 571 km long and at its highest point reaches 4220 m above mean sea level. It is the world's third highest railway line. Originally built for economic and social reasons, it is now only exploited for tourism due to its heritage value. It includes 11 viaducts, La Polvorilla being the most astonishing of them.

The construction of the railway started in 1921. Its purpose was to serve the borax mines of the area. La Polvorilla viaduct, the highest of the line, was finished on November 7, 1932. The complete railway was opened on February 20, 1948, but it was not until the late 1970s that it became a tourist attraction. The route was laid out by American engineer Richard Fontaine Maury. Located 4220 m (13,845 ft) above sea level, the curved viaduct is 224 m long (7 × 20 m + 6 × 14 m) and 70 m high (Figure 3.17).

The steel needed for the construction came from Cosulich steel mills in Trieste, Italy. All elements of the structure were shipped to Buenos Aires Harbor and transported by land (1500 km), to the construction site. León Gubbioni was site manager. The owner was Ferrocarriles del Estado but nowadays it is managed by a private concessionaire for tourism exploitation. During the construction, three workmen lost their lives.

3.5 Arch Bridges

3.5.1 Quequén Salado Bridge

This roadway bridge (Figure 3.18) was built over the Quequén Salado River. It is part of the route that links the villages of Oriente and Copetonas, located at the south of Buenos Aires Province. Taking into consideration the depth and width of the river and water flow conditions at the site, the most suitable

FIGURE 3.17 La Polvorilla Viaduct, "Train to the Clouds," Salta. (Courtesy of Gobierno de la Provincia de Salta. Photographer Antonio Tita.)

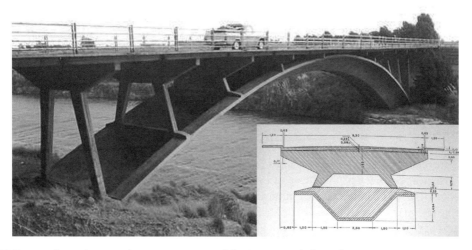

FIGURE 3.18 General view and transverse section of the Quequén Salado Bridge in Buenos Aires Province.

design was a bridge without intermediate piers, which led to the construction of an arch-type bridge. It has the unusual characteristic that the arch rib cross section is a reinforced concrete folded plate 0.25 m thick, 1.75 m deep at supports, 1.5 m at the crown, and 7 m wide. The bridge's span is 60 m, with a rise of 9 m. Spandrel spans are spaced every 5 m. The deck is a 0.22 m thick folded shell. The road width is 8.3 m with two sidewalks 1.2 m wide on both sides.

It was designed by the Bridge Division of Dirección Provincial de Vialidad de Buenos Aires, who are also the owners. Construction work began in 1961 by the contractor CODI SA. The bridge was opened to traffic in 1964.

3.5.2 San Francisco Bridge

This bridge (Figure 3.19) was built on Route 5 over the San Francisco River, linking the cities of Las Lumbreras and Pichanal in the province of Salta. The total length of the bridge is 720 m, composed of 12 equal spans of 60 m each. Each span is a tied arch with the deck acting as a tension chord. The use of prestress on the deck's girder beams and the prefabrication of the prestressed hanger resulted

FIGURE 3.19   General view of the San Francisco Bridge in Salta Province.

in a high-stiffness structure with well-distributed flexural moments in the arch under the action of partial live loads. It was a great innovation at the time it was built. After many years of service, there was a collapse in one of the spans. After that, all arches were reinforced with tensors, as seen in Figure 3.19.

The superstructure is pairs of hollow frustums cones linked at natural terrain level by beams, which provide great stability to the system against overturning induced by earthquake loads. The piers are cylinder shaped and founded 18 m deep. A cylindrical cofferdam with concrete sheet piles was used, which allowed the drainage of the excavation and remained as lost formwork. Construction work began in 1964 by owner Dirección Provincial de Vialidad de Salta. Contractors were Roffo, Irisiso y Cía. The bridge was opened to traffic in 1972.

3.6 Truss Bridges

3.6.1 Dulce River Bridge

The first studies for the construction of this railroad and roadway bridge began in 1880, but it was not until 1924 that works began. It was considered an important and very costly construction work for the country during that period. The structure was designed for dual purposes: railroad and roadway traffic. The bridge (Figure 3.20) is 840 m (12 × 70 m) total length. It spans the Dulce River in Santiago del Estero Province and links the cities of La Banda and Santiago del Estero. It is located near the latter, which is the oldest city in the country, founded in 1550 by Spanish colonizers.

It was designed by engineer Pedro Mendiondo and contractor was Baglioto, Binda y Cía. Steel trusses (6400 tons) were built by Gutehoffnuugshüte (Ruhr, Germany). In October 1927, after load testing by utilizing a cargo train, it was opened to the public. Each 70 m span has two truss beams built 11 m apart. The original design comprised 6.7 m width for the roadway and 3 m width for the tracks and sidewalks. In 1950 a bicycle lane of 1.5 m was added.

The piers are composed of three cylindrical caissons linked by a lintel to support the spans.

FIGURE 3.20   Dulce River Bridge between the cities of La Banda and Santiago del Estero, built in 1927.

There are two curious legends about the construction of this bridge. One is that all the steel needed for the bridge was given by Germany to Argentina free of charge as a compensation for the sinking of three Argentinean cargo vessels, Toro, Monte Protegido, and Oriana, by German submarines in the Atlantic Ocean during World War I in 1917. The other myth tells us that there is a solid gold rivet placed somewhere among the thousands of the structure. Yet nowadays, one can seldom see people searching for said rivet.

3.6.2 Superí and Zapiola Streets Bridges

The existing superstructures (Figure 3.21a) were six continuous 18 m long prestressed concrete spans over an important expressway in Buenos Aires. The cross section was transversely prestressed. The piers are solid wall type. Both bridges were built in 1969. To relieve traffic congestion of more than 250,000 ADT, the expressway was widened in 2001. Three exiting central columns of two continuous prestressed concrete overpasses bridges, of six spans each, were removed. At least four lanes were kept open to traffic in each direction, at all stages of the construction work. Different structural solutions were analyzed, but finally, one named "the hanger" was adopted. Rectangular steel tubes were chosen for the truss elements cross sections. An orthotropic steel deck supports the roadway surface and acts at the same time as the tension chord of the main truss from which the old structure was hanged.

The Superí Street Bridge (Figure 3.21b), the largest one, is composed by two truss beams 41.11 m long and 6.8 m high. Two transversal beams link both trusses at the upper vertex to avoid lateral buckling. The completed structure can be considered as two superimposed systems. The first is the prestressed concrete continuous beam of six spans, with its three central spans elastically supported by the new structure. The second is the new structure, the simply supported steel truss from which the existing concrete deck is suspended. The old deck was suspended by high-strength steel bars.

The two bridges were constructed between January 2001 and April 2002, leaving four lanes of the expressway opened to traffic permanently during the works. The bridge was designed by Del Carril–Fontán Balestra y Asoc. (Buenos Aires, Argentina). The contractor was AUSOL SA (Autopistas del Sol). The project was awarded the national structural engineering prize Ing. José Luis Delpini.

(a)

(b)

FIGURE 3.21 (a) Existing bridge before the widening works. (b) The bridge after removing three central supports.

3.7 Cable-Stayed Bridges

3.7.1 General Belgrano Bridge

This was the first bridge built over the Paraná River and the second physical connection between the Mesopotamian region and the rest of the country. This roadway bridge links Resistencia, capital of the Chaco Province, with the city of Corrientes, capital of the province of the same name. Due to the fact that in the 1960s the only existing link to the Mesopotamian region was a tunnel that wouldn't allow transit of cattle or fuel across the Paraná River, the bridge was very important for the three provinces of the region, which are enclosed by the Paraná and Uruguay rivers and isolated from the rest of the country. A new bridge would also serve international communications with two neighboring countries, Paraguay and Brazil.

FIGURE 3.22 The General Belgrano Bridge, viewed from the Corrientes shore of the Paraná River.

The project includes a cable-stayed bridge and two approach viaducts: one on the Chaco side on National Route 16 and another one on the Corrientes side on National Route 12 (Figure 3.22). Those structures are 2000 m long (1666 m over water) and 8.3 m wide, with one lane in each direction. The vertical navigation clearance is 35 m. The required horizontal clearance is 200 m. The cable-stayed bridge comprises three spans (163 m + 245 m + 163 m). It is composed of two 225 m long suspended structures, placed symmetrically along the main pylons axes, which are 245 m apart. A simply supported span of 20 m links these two structures to form the main span of 245 m. This suspended span reduces the effect of deflections on the two main structures. Each main pylon is a W-shaped frame that rests on a pile cap for its foundation. A set of eight stays completes each main 83 m high pylon.

The cross section is a two-box girder. The deck is completed with precast transverse slabs, 6.9 m long and 2 m wide, supported on the main beams. When the deck was finished, the transverse beams were prestressed. The original stays were of the locked coil type, but they had to be replaced after 25 years of service. The longitudinal beams of 3.5 m deep were prestressed. The approach viaducts over the river are composed of nine spans of 82.6 m each on the Chaco side, and three spans of 82.6 m each on the Corrientes side. They were built by the balanced cantilever method with precast segments 4.1 m long and 2 m deep at center span and 4.5 m deep over the supports. The overall deck width is 11.3 m, carrying two 3.65 m wide lanes, narrow emergency shoulders of 0.50 m and pedestrian sidewalks 1.5 m wide. The piers of the approach viaducts are constant deep box sections and were built by the sliding formwork method. The foundations were made using 1.8 m diameter bored piles with variable lengths between 38 m and 60 m, and with preloading cells, they are clamped by penetration into the hard clay stratum.

At around 20 years of service, the replacement of all the stays was required. Construction works were carried out by Freyssinet SA (Spain) in 1995. Parallel strands stays were installed. The link between both cities, with only two lanes, has proven to be insufficient for the traffic, which has been ever-growing since its construction. The construction of a new bridge is now under consideration.

Construction work began in August of 1968 by the initiative of the governments of the provinces involved: Corrientes, Chaco, and Formosa. The owner was DNV, who called for international bids in 1965. It was designed by Professor Jean Courbon from the Sociètè D´Etudes et D´Equipements D´Entreprises de París, France. The contractor was a joint venture between Ferrocemento SPA (Rome, Italy), Umberto Girola (Milan, Italy), Impresit (Salerno, Italy), and Sideco SACIC (Buenos Aires, Argentina). The bridge was opened to traffic in May 1973.

3.7.2 Zárate–Brazo Largo Bridges

The Zárate–Brazo Largo Bridges link the north of Buenos Aires Province and the south of the Mesopotamian region. Placed nearly 80 km away from Buenos Aires, they are of paramount importance for the transportation system of the country. The bridges comprise a 13 km highway that crosses over two branches of the Paraná River: Paraná de las Palmas and Paraná Guazú. They link the city of Zárate and a place called Brazo Largo. Between the branches, which are navigable by Panamax vessels, there is an island called Talavera.

Two bridges were built over the respective branches. Both structures serve a dual purpose: roadway and railway traffic. The latter was very challenging for structural designers due to the asymmetrically loaded deck, because there had never been a cable-stayed bridge with asymmetric loads. The original design was created by the famous Italian professor Ricardo Morandi in 1966, but the owner, DNV, finally awarded the detail engineering contract to joint venture TECHINT–Albano SA in December of 1970. The original design of the two main cable-stayed bridges was reformulated by the contractors under the supervision of Leonhardt, Andra & Partners (Stuttgart, Germany) and Fabrizio de Miranda (Milan, Italy). Only minor adaptations were introduced to the railway and roadway approach viaducts, but the two cable-stayed main structures were redesigned. The original design comprised two separate decks for railway below and for roadway in the upper part. The new design modified this criteria and placed a unique deck with the rails at the side of the railroad, thus introducing strong asymmetry in the load path.

The main bridges (Figure 3.23) are cable stayed with an overall length of 550 m (110 m + 330 m + 110 m). Navigational clearance is 50 m high, allowing the navigation of Panamax vessels. Two external high-strength welded steel orthotropic box beams are linked together by truss girder beams in the transverse direction. The asymmetrical 22.6 m wide structure has the following configuration: a four-lane roadway with 1.3 m external sidewalks, a median of 0.50 m that separates two lanes in each direction, and the remaining 4.5 m to accommodate the railway track. This asymmetry is also reflected in the stays configuration: two stays for every position on the railway side and only one for the roadway side.

Pylons are H-shaped frames with hollow concrete columns 130 m tall. A transverse beam at middle height is used to support the main steel beams and a steel X links both columns at the top.

The structure rests on bored piles of 2 m diameter with variable lengths between 30 and 70 m. Some of the piles were driven 30 m deep into the river. All of them have preloaded cells.

The original stays were parallel 7 mm diameter bars, but had to be replaced after 20 years of service. The new stays, which are at present being installed by Freyssinet SA (Spain), consist of parallel strands.

FIGURE 3.23   One of the two equal main cable-stayed bridges of the Zárate–Brazo Largo project.

Approach viaducts for roadway and railway traffic are quite similar in their structural configuration, but differ in dimensions and alignment.

Construction work began in November 1971 and the first bridge, over Paraná de las Palmas branch, was opened to traffic in February 1977, while the second, over the Paraná Guazú branch, was opened to traffic in November 1977. This project broke many records when completed. The bridge had the world's longest cable-stayed span for dual-purpose traffic and the first with an asymmetrically loaded deck. Its design and construction were a real challenge for engineers. It is still the country's longest cable-stayed span.

3.7.3 San Roque González de la Cruz Bridge

This bridge (Figure 3.24) connects the city of Posadas (Misiones, Argentina) with Encarnación (Paraguay), across the Paraná River, a natural boundary between northern Argentina and southern Paraguay. Its structure comprises a cable-stayed bridge with two approach viaducts. Its function is to serve international railway and roadway traffic.

The main structure comprises a three-span cable-stayed bridge (115 m + 300 m + 115 m), and a super-structure of a three-cell prestressed concrete box girder 19 m wide and 3 m high, providing high torsional stiffness to resist the highly asymmetric railway loads. The vertical clearance, once the Yacyretá Dam is completed, will be 18 m, which is considered to be sufficient for navigation purposes on that section of the river. The method of balanced cantilever with precast segments 10 m long was used. The foundations of the pylons are cylindrical caissons driven to the sound basaltic rock.

Four other structures in the project include a continuous slab for the roadway approach viaduct on the Argentinean shore, of 112 m (20 m + 3 × 24 m + 20 m), built over land; a 520 m (26 × 20 m) railway approach viaduct, also built over land; an Argentinean approach viaduct of 1505 m (29 × 55 m), built utilizing sliding formwork; and the Paraguayan approach viaduct of 385 m (7 × 55 m), built by the launching method for the first time in the country.

The bridge was structurally designed by joint venture COPPEN integrated by Cabiholsky-Hechausen, CONSULAR, CADIA, and Beccrra Ferrer-Lange (all from Argentina), and Leonhardt

FIGURE 3.24 San Roque González de la Cruz Bridge. (Courtesy of IECSA, Argentina.)

(Stuttgart, Germany) was consulting engineer. The owner is DNV. The bridge was built by a joint venture between the Argentinean construction firms Sideco Americana, EACA, and Girola between December 1980 and 1988. Construction work suffered many interruptions due to several reasons, among them the extraordinary floods of the Paraná River in 1982 and 1983, the Malvinas War, and Argentina's economic crisis with hyperinflation, which produced restrictions on the importation of the necessary elements for the bridge. The bridge won the International Prize Puente de Alcántara awarded by Fundación San Benito de Alcántara (Spain).

3.7.4 Nuestra Señora del Rosario Bridge

This cable-stayed bridge (Figure 3.25) is the main connection between the cities of Rosario (Santa Fe Province) and Victoria (Entre Ríos Province) crossing the Paraná River and its floodable plains. The whole project covers an extension of 59.4 km and is composed of various bridges over the flood areas, with an accumulated length of 8.1 km, and a main structure that crosses the Paraná River and is 3.49 km long. This structure is composed of a cable-stayed bridge 608 m long (124 m + 350 m + 124 m), which spans the navigation channel with a vertical clearance of 50 m, a 1130 m approach viaduct on the Rosario side, and a 2380 m approach viaduct on the Victoria side.

Pylons are H-shaped frames. The prestressed concrete superstructure is pi-shaped. It was built by the balanced cantilever cast-in-place method. Its segments are 10.4 m long. The deck is 11.3 m wide, with two lanes of 3.65 m, narrow emergency shoulders of 0.50 m, and pedestrian sidewalks 1.5 m wide.

Construction began in 1998, but was repeatedly interrupted due to lack of funding. It was opened to traffic in May 2003. Construction works were carried out by a joint venture Puentes del Litoral SA integrated by IMPREGILO SPA (Italy), YGLIS SA (Italy), HOCHTIEF AG (Germany), Benito Rogio e Hijos, SIDECO-IECSA (Argentina), and TECHINT (Argentina). The main bridge was designed by Leonhardt & Andrä und Partner GmbH. This is the lastest bridge built over the Paraná River.

FIGURE 3.25   **(See color insert.)** The Nuestra Señora del Rosario Bridge, viewed from the Rosario side. Protections against vessel collisions can be seen.

3.8 Suspension Bridges

3.8.1 Marcial Candioti Bridge

This bridge (Figure 3.26), located north of Santa Fe Province on National Route 168, spans Setubal Lake. It is an important cultural heritage symbol. It was built in the early 1920s and was opened to traffic in 1928. The main purpose of the bridge was to support an aqueduct, which used to be supported by two bridges that collapsed due to various reasons in 1904 and 1920.

The owner, Obras Sanitarias de la Nación, assigned the design to Casa Wattinne Bossut et Fils, who represented the Societe des Chantiers et Ateliers de la Gironde (Paris, France), who built all the parts of the bridge with the collaboration of the prestigious engineer M. G. Leinekuge. The design follows the system created by a lieutenant colonel of the French Army, Albert Gisclard, which consists in a configuration of nondeformable cable triangles and results in a very stiff structure. Additionally, the designer has added an Ordish suspension system to avoid the nonlinear deformation of the main cables. By applying this system, compression on the stiffness beam is not introduced.

The bridge has a total length of 295.26 m (73.88 m + 147.7 m + 73.88 m), and support an aqueduct and a 6 m wide deck for roadway transit. It rests on large piers founded on piles of nearly 12 m diameter. At the time of its construction it became an emblem of the city of Santa Fe. In 1973, 100 m downstream, the new Nicasio Oroño Bridge (see Section 3.3.4) was opened to traffic, so only pedestrian traffic was allowed on the old suspension bridge, which still remained as an icon.

The bridge almost collapsed due to the scouring of the foundation of one of the main towers in a significant flood. The bridge remained semi-sunk between the years 1983 and 2000. City authorities called for bids for its reconstruction in 1988. The contractor Concesiones y Construcciones de Infraestructura (CCI) was assigned the job. EEPP (Estudio Estructural Polimeni Perez y Asoc.) were the structural designers. It was imperative to maintain the original appearance of the bridge. The replacement tower was manufactured in the Ferma Shop (Esperanza, Santa Fe) and had to be placed 6 m from the original location. Thus, the bridge was forced to increase its main span, reaching 153.7 m long, and decrease the east span to 67.65 m. The retrofitted bridge was opened to traffic in 2002.

3.8.2 Hipólito Irigoyen Bridge

This self-anchored suspension bridge (Figure 3.27), one of the two only bridges of this type built in the country, spans Quequén River and links Necochea city to the Quequén beaches on the Atlantic Ocean,

FIGURE 3.26 The suspension bridge of Santa Fe, after its reconstruction.

FIGURE 3.27   The Hipólito Irigoyen Bridge, one of two suspension bridges in Argentina.

south of Buenos Aires Province. The total length is 270 m. The central span is 150 m with two side spans of 60 m. Pylons are 26 m tall. The bridge is self anchored. This technique was not very common around the world, due to the fact that it doesn't apply to long-span bridges. The stiffening beams are 8.8 m apart from each other and the deck has a 6 m roadway width and two 1 m wide sidewalks per side. This bridge has a better-known twin brother, the Chelsea Bridge in London. The structure elements were provided by the Chantiers et Ateliers de la Gironde, France, who also provided the materials for the other suspended bridge (see Section 3.8.1)

The bridge is a milestone for a self-anchored suspension bridge. When it was inaugurated only a few bridges of this type existed worldwide: one over the Rhine, opposite Cologne and three others over the Alleghany River in Pittsburgh, Pennsylvania in the United States. It was built by contractor Weyss & Freytagg (Germany), by commission of the owner, the Bridge and Roads Department of Buenos Aires Province, in 1928.

3.9 Movable Bridges

3.9.1 Old Pueyrredón Bridge

The Old Pueyrredón double-leaf bascule bridge (Figure 3.28) crosses the Riachuelo River. The total length of the bridge is 88.8 m. The center span is a double-leaf bascule of 38 m (two cantilever spans of 19 m), and the fixed spans are 25.4 m long. The original overall width was 12 m with two lanes of 3 m and two 3 m sidewalks. In 1931 the mechanical system failed and the bridge remained fixed in its closed position. It was widened in 1958, but in 1970 it was closed to traffic due to safety reasons. It was saved from dismantlement when city government designated it a Historic Landmark. After that, it was reopened for light vehicles. The bridge location was previously occupied by six other bridges since the

FIGURE 3.28 The Old Pueyrredón Bridge reopened in December 2009 after its last retrofitting. (Courtesy of EEPP, Estudio Estructural Polimeni Perez, Buenos Aires, Argentina.)

first wooden bridge over the Riachuelo River was built in 1791. Some of them collapsed and others were replaced when new technology rendered them obsolete.

The bridge was built between 1925 and 1931. The onsite contractor was Dickeroff & Widmann SA. Steel members were manufactured in Oberhaussen (Germany) by Gutehofnunghütte, who also designed and provided the mechanical parts for the movable span. The bridge was completely retrofitted in 2008. New mechanisms were installed and a new strengthened metallic structure was designed by EEPP (Estudio Estructural Polimeni Perez, Buenos Aires, Argentina), who were also site supervisors. The main contractor for the retrofitting was DALCO SA–EMA SA UTE. The retrofitted bridge was inaugurated on January 2010.

3.9.2 Alsina Bridge

This 72-year-old bridge (Figure 3.29) spans the Riachuelo River, connecting Nueva Pompeya neighborhood of Buenos Aires City Federal District with Valentín Alsina, a neighborhood of Lanús (Buenos Aires Province). It is a movable bascule bridge, which allows navigation traffic through the Riachuelo River, in La Boca neighborhood, one of the city's main tourist attractions at present.

The total length of the bridge is 173.16 m, composed by two fixed spans of 65.33 m and a bascule span of 42.5 m. The total width is 24 m, accommodating a roadway 18 m wide and two sidewalks 3 m meters wide on each side. The bridge crosses the navigation channel with a skew angle of 64 degrees. On both ends of the bridge, great portals were built, in colonial style, where policemen had guard precints.

The bridge was designed by engineer José María Páez and construction took six years, between 1932 and 1938. Foundations, piers, and abutments were designed by Parodi and Figini. The metallic trusses and the bascule mechanism were manufactured in Hannover (Germany) by Louis Eilers. The owner is the Buenos Aires city government.

3.9.3 Nicolás Avellaneda Bridge

This magnificent movable lifting bridge (Figure 3.30), built in 1937, spans the Riachuelo River and connects Maciel Island with Buenos Aires City Federal District. It allows navigation of vessels up to 21 m high when closed, and up to 43 m when open. The lifting operation required 3 minutes and served the vehicle traffic flow very well in those days. Vessels demanded its lifting three times a day. At present, the

FIGURE 3.29   The Alsina movable bridge over the Riachuelo River, with its colonial style portals, built in the 1930s.

FIGURE 3.30   Nicolás Avellaneda Bridge over the Riachuelo River (See also Figure 3.3).

high volume of traffic of the main south access to Buenos Aires uses a new bridge over the Riachuelo (see Section 3.3.12), close to the Avellaneda bridge.

The main bridge is a high-strength steel structure composed by two towers and a lifting span of 65 m. The roadway width is 12 m, accommodating two 3 m wide lanes in each direction. The project includes two approach viaducts with 1633 m total length. The lifting of the main span requires two 230 ton counterweights and four chains powered by two 55 HP electric engines. Foundations required the use of pneumatic caissons 20 m deep, resting on dense sand stratum.

The owner is DNV. Construction work began in 1937 and the bridge was opened to traffic in 1940. Engineer Juan Agustín Valle and architect Eduardo Rodríguez Videla were the architectonic designers.

3.10 Pedestrian Bridges

3.10.1 Puente del Sesquicentenario Bridge

This pedestrian bridge was built over Figueroa Alcorta Ave. for the sesquicentennial commemoration of the first revolt against the Spanish Crown (May 25, 1810), as part of the great exposition held in the parks north of Buenos Aires. It was demolished in 1974 to leave room for a large monument called Altar de la Patria, which was never completed. Finally, a concrete bridge was rebuilt almost 100 m away from its original position, using the original drawings. This elegant arch was designed by engineer Atilio Gallo and architect César Jannello. The owner is the Buenos Aires city government.

Its cross section is composed of a parabola on the underside and a horizontal surface on the top, extruded on a 50 m long circular arch path. Its gradient is 15% to allow for easy pedestrian transit. The arch depth is 50 cm at the center. The deck is 10 m wide. The cross section is composed by three hollow cells with the exception of the central 10 m, where the deck is solid. Two tendons are anchored at ends of the arch. The vertical reactions are transmitted to a spread footing through neoprene bearing pads.

The first construction was carried out by Bava, Seery, and Litmayer (Bs. As., Argentina) and the second by Hindoust–Klein (Bs. As., Argentina) (Figure 3.31). The footbridge is a city monument and provides service to many students who attend the law school of Buenos Aires University.

3.10.2 Kuñataí Bridge

This pedestrian bridge (Figure 3.32) crosses above National Route 12 in the city of Posadas, Province of Misiones. The superstructure, a prestressed precast trapezoidal box beam 25 m long, is supported by two helical staircases, which act at the same time as piers. It is actually a unique curved beam with a total length of 60 m, but due to its shape, the maximum flexural moments are almost identical to a simply supported beam of 25 m. After all main beams were erected, the whole structure was prestressed together. The width of the staircase and walkway is 2.4 m. The cross-sectional depth is variable between 0.60 m, at the center, and 1.1 m at the supports. The foundations required the installation of prestressed anchorages into the rock to guarantee stability.

The bridge was built by SETA Hidrovial SRL (Rosario, Argentina) in 1988. Professor Dante Seta was its designer and construction supervisor. Although it isn't a large structure, construction had some complexities and the finished structure is really impressive. Locals call it the "unsupported bridge."

FIGURE 3.31 The Puente del Sesquicentenario Pedestrian Bridge, which has been built twice.

FIGURE 3.32 The Kuñataí Pedestrian Bridge.

3.10.3 Puente de la Mujer Bridge

The old port of Buenos Aires has recently been urbanized with high-quality residential buildings, hotels, and pedestrian walkways. The real estate development is managed by Corporación Antiguo Puerto Madero. Within this large area, a subzone named Dock 3 was assigned to real estate developer Emprendimientos Inmobiliarios Arenales SA who, in turn, hired architect and engineer Santiago Calatrava (Spain) to design an iconic pedestrian bridge (Figure 3.33). It was Calatrava's first assignment in Latin America. He designed a movable steel cable-stayed bridge, which allows navigation traffic by rotating around its pier. The bridge has an overall length of 160 m. Its span arrangement comprises a 90 m central pivoting span, of which 70 m are stayed and the remaining 20 m act as a counterweight on the opposite side of the pylon, and two fixed spans. The deck width is variable, reaching 6.2 m. The superstructure is composed of a steel box beam. It is hollow at the central span of the bridge, but filled using high-density concrete and steel shots at the counterweight span. The pylon is inclined and 67 m high. Stays are parallel strands.

The creator said that he conceived this bridge-sculpture inspired by a couple dancing the tango, a popular music genre originated in the Río de La Plata region and which has become an emblem of Buenos Aires. The bridge was built by URSSA, who manufactured the steel structure at a mill in Bilbao (Spain). Contractor Trevi was in charge of foundations, using bored piles 1.2 m diameter and 26 m deep to reach a dense sand stratum. Manessman-Dematic-Demag (Germany) provided the mechanical parts. The site contractor was Brambles (Lastra, Spain).

3.10.4 Balconies Iguazú Footbridges

Iguazú Falls was declared a World Heritage Site by UNESCO in 1984. These bridges consist of the Upper, Lower, and Garganta del Diablo footbridges. They are located on the Iguazú River, which is a natural boundary between Argentina and Brazil, and are visited by thousands of tourists every year. They are within the Iguazú National Park. The Iguazú River flows through a smooth topography until it reaches a series of faults and rapidly descends into an 80 m high canyon in La Garganta del Diablo, where the water produces a thundering sound and an astonishing spectacle. There are more than 270 falls in the area, where cliffs and islets are scattered in a half-moon shape. There are three basic circuits for visiting the falls: an upper path, a lower path, and Garganta del Diablo path (Figure 3.34), all with several balconies to appreciate the beautiful landscape. The total length of the three circuits is approximately 3 km.

The owner is the National Parks Administration, and the manager is the Concessionaire Iguazu Argentina SA (Carlos E. Enríquez and Partners, Posadas, Argentina). Pedestrian walkways of 3 km

FIGURE 3.33   Puente de la Mujer Pedestrian Bridge.

FIGURE 3.34   Balcony at Garganta del Diablo, Iguazú Falls.

total length were designed by Del Carril-Fontan Balestra Engineers (Buenos Aires, Argentina) and construction works were carried out by Albano Construcciones (Buenos Aires, Argentina). The footbridges were built on piles embedded and anchored in the rock (basalt), the deck is 50 cm above ground level and built with grating to allow free movement of animals and the access of sunlight to the flora growing under them. The footbridges are 1.2 m wide and are designed to be circulated in one way, so that thousands of tourists may walk the circuits without interfering with each other. They have no stairs or steep slopes in order to allow easy movement of handicapped people.

The structures consist of a succession of 6 m, 9 m, and 12 m long metal spans, with two rolled steel beams and transverse grating plates. The deck of the walkways are placed 60 cm high with respect to the natural soil, so that small animals, snakes, and insects can live without disruption of their environment. Also, the grating used in the floor allows natural sunlight to reach the soils for the benefit of the flora. For the comfort and safety of children and handicapped people, the railings have two handrails at different heights.

Railings can be folded over the deck in case of a large flood, offering low resistance to the water passing over the bridges. In case of a very large flow, the bearings are designed to act as fuses, letting the spans be carried away by the water and leaving the foundations undamaged. This design was based on experience with the two previous walkways, which were destroyed by floods since parts of the foundations remained scattered across the landscape, causing a high environmental impact before and during their demolition. Since they have been in service, several spans of the walkways have twice been washed away by floodwaters, demonstrating the correct performance of the above-mentioned fuses, and the infrastructure remained in perfect condition. As the concessionaire has spans stored in his warehouses, the lost ones may be replaced in a few hours after the flood ends. As it is a highly ecologically driven project, flora and fauna were taken into account in the plan. Several pillars had to be moved a few feet away from the original plan in order to not disturb an alligator den or avoid destroying an important plant.

3.10.5 Dorrego Ave. Pedestrian Bridge

In 2008, the Buenos Aires city government called for bids a pedestrian bridge. The Dorrego Ave. Pedestrian Bridge is located in a high-density forested area of the city, where it is common to see people jogging or practicing various sports during the day (Figure 3.35). The recently inaugurated bridge still had no official name at the publication of this book. It is said that this bridge will be named Puente de Miguel (Miguel's Bridge) after a handicapped person who is very much admired by the locals for his striving to achieve, day by day, personal improvement.

The bridge is cable stayed. Its total length is 165.76 m, composed by a main bridge with a central span 49.14 m long, two side spans 36.47 m long, and completed with two simply supported approach spans

FIGURE 3.35   Dorrego Ave. Pedestrian Bridge.

of 21.84 m. The pedestrian walkway width is 3 m. The composite superstructure comprises a steel trapezoidal box beam 0.50 m deep and a concrete slab 0.12 m thick. Its two main towers are 17 m high and the foundations are 0.50 m diameter, 15 m deep bored piles. The maximum grade is 7% and the vehicle clearance is 4.95 m in the center and 4.43 m at the ends.

Structural designers were EEPP (Estudio Estructural Polimeni Perez) and the main contractor was Andersch Ingeniería SRL.

3.11 Special Bridges

3.11.1 Futaleufú Pipe Bridge

The Futaleufú Pipe Bridge (Figure 3.36) is part of the adduction system for the Futaleufú Hydroelectric Plant. The purpose of this special bridge is to carry a 7.7 m diameter pressurized pipe over very low-support-value soil over a deep valley. As a result of the analysis of many alternatives, a solution was adopted taking into consideration the short schedule for the project and unexpected risks from the construction process.

The total length of the bridge is 260 m (60 m + 7.8 m + 62.2 m + 62.2 m + 7.8 m + 60 m) composed of a central continuous beam and two simply supported spans 60 m long. The main structure is composed by a centrally located pylon and two pairs of rigid stays separated longitudinally at 124.4 m, leaving a 7.8 m long cantilever beam in each side that supports the simply supported side spans. The central span is composed of two pairs of reinforced concrete 7 m deep I-beams placed at the sides. This pairs of beams are separated 12.5 m and are transversally stiffened with diaphragms in correspondence with the steel pipe supports. Side spans are composed of two reinforced concrete beams 7 m deep, 7.2 m wide, box sections.

The main pylon is 44.3 m high and founded on a 16 m diameter, 1.2 m thick, and 31 m deep cylindrical caisson.

An exigent deflection limitation of the steel pipe had to be taken into account for the design of this special bridge. This rendered the project very expensive due to the fact it had to be done with high-strength

FIGURE 3.36 General view of the Futaleufú Pipe Bridge. (Courtesy of Hidroeléctrica Futaleufú.)

steel that was not manufactured in Argentina. Another unusual condition was the high percentage of live loads, reaching up to 50% of the total load. This condition lead to the design of rigid stays instead of cables.

The owner, Agua y Energía Eléctrica, built this project between 1971 and 1978. The bridge was designed by engineer Ricardo Wagner of Tecnoproyectos Consultora (Buenos Aires, Argentina), who was also technical project manager. The contractors were a joint venture between EACA and Panedile (both from Buenos Aires, Argentina). The firm Prentesac SA (Buenos Aires, Argentina) was in charge of detail engineering design and all of the prestressed operations were made by said firm with Carlos Heckhausen (Argentina) as a consulting engineer.

3.11.2 Cangrejillo Creek Pipeline Bridge

This bridge (Figure 3.37) was built over the Cangrejillo Creek in Catamarca Province in northwest Argentina. Its main purpose is to carry a copper concentrate pipe 0.20 m in diameter and a 1 m footpath for maintenance of the facility. The bridge will serve the transit of gold and copper through the pipe, which will connect a mine located to the north of Andalgalá with a rail facility in San Miguel de Tucumán. It spans 337 m over a 90 m deep valley. It is a cable suspension structure that supports an open steel deck frame. Catenary's cables sag 7.85 m at midspan.

The main structure is composed of two pairs of cables anchored in reinforced concrete abutments and will allow the future transit of a 2 ton truck, and it has a 3 m wide deck. The abutments provide a secure anchorage of the cables to the rock by means of a concrete block that is fixed to the side of the mountain with a system of 24 and 28 post-tensioned rock anchors. Each anchor is a 15 mm strand, 30 m to 60 m long, installed with a 45-degree slope.

As the bridge is placed in a highly seismic region, it was designed according to AASHTO standards and IMPRES-CIRSOC (Argentine standards for earthquake-resistant structures). This type of structure was considered the best alternative to a buried pipe and road, which would have produced a high environmental impact on the waterfalls and rainforest valley. The owner is Minera Alumbrera, and it was designed by OPAC Consulting Engineers Inc. (United States) and built by Albano Construcciones. It was opened to service in October 1998.

FIGURE 3.37 Overview of the Cangrejillo Pipe Bridge constructed over a 90 m deep valley. (Courtesy of Horacio O. Albano Ingeniería y Construcciones.)

3.12 Future Bridges

3.12.1 Reconquista–Goya Bridge

The road between the cities of Reconquista (Santa Fe Province) and Goya (Corrientes Province) is an important 42 km long project that crosses the Paraná River and its flood areas. This project required 16 bridges on the plateau and a main structure 4 km long to cross the Paraná River. This structure consists of a main cable-stayed bridge (Figure 3.38) over the navigation channel and two approach viaducts. The main bridge has a total length of 930 m with a five-span continuous prestressed concrete beam (90 m + 180 m + 390 m + 180 m + 90 m), trapezoidal box-type section 4.5 m deep. Its three central spans are supported by a set of stays arranged in a single plane along the longitudinal axis of the bridge.

A striking feature of this design is that it comprises five continuous spans instead of the conventional three spans common in such structures. This scheme reduces the tension force in the first pile next to the pylon. Furthermore, it decreases rotations at the junction with the viaducts. The deck has an overall width of 21.7 m, accommodating two lanes 8.3 m wide in each direction, a 2 m median, New Jersey type transit barriers, and two sidewalks of 1.5 m per side.

There are two different types of approach viaducts to the east and west, with six spans of 90 m, and low viaducts. Among the latter, the one on the west bank is a 1500 m structure composed by 60 m continuous spans. Its trapezoidal box-type cross section is similar to the one used on the main bridge. The east side low viaduct is a conventional structure of 35 m long precast prestressed beams to complete a structure of 245 m.

The project was developed by a joint venture between some of the most important consulting firms of the country: Consular, IATASA, ATEC, Oscar Grimaux y Asoc., and INCOCIV. The main bridge was designed by ATEC, COPIGA, and Del Carril–Fontán Balestra y Asoc. with the advising of Leonhardt, Andrä, und Partners (Germany) as consulting engineers and responsible for the preliminary designs and evaluation of alternatives for the structure of the Paraná River crossing. The approach viaducts and the bridges in the plateau were designed by these firms.

FIGURE 3.38 A rendered image of the future Reconquista–Goya main bridge.

The tenders of the bridge are the country's largest and are now complete. Its construction will require approximately 4 years. It is of vital importance for communication between the Mesopotamian region and the rest of the country and an important route of the Mercosur.

3.12.2 Buenos Aires–Colonia Bridge

The Buenos Aires–Colonia Bridge project was designed to provide a permanent link between Argentina and Uruguay, over the Río de la Plata River, a wide estuary formed by the junction of the Paraná and Uruguay Rivers. The governments of the two countries signed an agreement in 1996 that created the Binational Bridge Commission of Buenos Aires–Colonia to manage the project.

The bridge will join the city of Punta Lara, 40 km to the south of Buenos Aires, to a point 8 km away from Colonia in Uruguay. This location was considered the best alternative after the economic evaluation and environmental impact studies made by Louis Berger (leader of the consortium and engineering); Bear, Stearns & Co. Inc. (investment bank); and Latham & Watkins (law firm), who were given the ACEC (American Consulting Engineers Council) excellence award in Washington in 1988 for this study. The results also indicated that the construction of the bridge would result in a significant economic benefit for both countries. It was planned to involve a private partner for design, construction, operation, and maintenance, and the governments of Argentina and Uruguay would provide supporting roles.

Once the bridge is finished, with a total length of 42 km, it will be the longest bridge in the world. It includes access viaducts with spans of 40 m and 8 m over the highest water level, approach viaducts with spans of 100 m, three secondary bridges with main spans of 200 m, and vertical clearance of 32 m and a main cable-stayed bridge over the navigation dredged channel with a main span of 500 m and 65 m of vertical clearance. The decks will include a total of four lanes, shoulders, pedestrian corridors, and a New Jersey barrier, which add up to a total width of 20 m (Figure 3.39).

The project has a significant geopolitical importance for the integration of Uruguay and Argentina and will boost trade between Brazil, Uruguay, Argentina, and Chile. For economic and political reasons the construction of the project has been delayed. The parliament of Uruguay has approved the project but Argentina has not, perhaps because the country has other priorities. The internal rate of return still

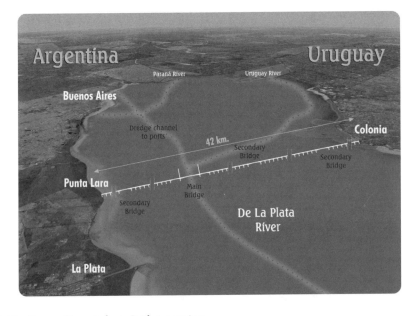

FIGURE 3.39 Buenos Aires–Colonia Bridge overview.

FIGURE 3.40 The future bridge over the Paraná River, joining Santa Fe (foreground) and Paraná (background). The entrance to the existing subfluvial tunnel can be seen at lower right. The picture was taken during one of the frequent floods of the Paraná River.

low, and there are no much interest for private companies to build and recovery the investment by toll revenues. The project can be completed in 10 years.

3.12.3 Santa Fe–Paraná Bridge

Since 1969, the vital connection between the capital cities of the provinces of Santa Fe (Santa Fe) and Entre Ríos (Paraná) has been made by the Hernandarias subfluvial tunnel. It goes under the Paraná River, with a length of 3.6 km. It was great engineering accomplishment in its time but today is obsolete due to the fact that it has only one lane in each direction. Furthermore, fuel and livestock truck traffic is not allowed.

Due to the permanent growth of the surrounding areas, the deceased civil engineer Alejandro Vega proposed, through the Colegio de Profesionales de la Ingeneiría Civil de la Provincia de Entre Ríos, the construction of a new bridge over the Paraná River. At present, feasibility studies have been completed, and the decision to build a new link between cities has been made by the government authorities. The new structure (Figure 3.40) will have a total length of 3.3 km, including a main 700 m cable-stayed bridge over the navigation channel and 2.6 km of approach viaducts. In 2011 tender engineering began for this fundamental new bridge for the integration of the Mesopotamian region and Mercosur countries, but has since been postponed.

3.13 Natural Bridges

3.13.1 Puente del Inca

The Puente del Inca Bridge is located on the eastern slopes of the High Andes, over the Las Cuevas Valley. Horizontal layers of carbonate ferruginous deposits from springs between rock formations were carved by the Las Cuevas River, forming a large natural arch. The arch (Figure 3.41) is 40 m long, 15 m wide, and has a rise of 20 m. It has variable thickness of between 5 and 8 m. It is located 2720 m above sea level. This type of natural arch can generally be found in mountainous or sea coast regions. Several of them are known, both in Argentina and elsewhere. However, the Puente del Inca Bridge stands out among others due to the fact that the smoothness of its shape seems to have been made by man, and the

FIGURE 3.41   Puente del Inca natural arch bridge.

horizontal road over it that was used by habitants of the area until recent years. It is a unique geological wonder, declared a Natural Monument. To take advantage of the hot springs, in 1925 a luxury hotel was built that was unfortunately destroyed by a devastating flood in 1965. At present, the site is still an important tourist attraction.

3.14 Bridge Management Systems

During the last 6 years, the DNV has been developing a Bridge Management System (BMS) called SIGMA Puentes. At present, there is a limited inventory of bridges in the primary network. Only a few of them were qualified according to their conditions, no regular inspections are carried out, and registers are scarce for all bridges. The implementation of a BMS in Argentina grows slowly because of lack of resources for bridge rehabilitation. Nevertheless, there is a BMS developed by Argentinean Consulting Engineers that is especially applicable to bridge networks and conditions prevailing in developing countries.

The system, called SIGEP (Spanish for BMS: Sistema de Gestión de Puentes) by the authors, was implemented under different local names in Guatemala, El Salvador, and Bolivia, and also in a privatized highway network for the accesses to Buenos Aires in Argentina. In its first presentation SIGEP was awarded the prize Ing. Luis De Carli by the Transit and Road Engineering Argentinean Congress in 1998. The implementation of SIGEP in Guatemala was awarded by the International Road Federation (IRF) in the Australia meeting of 2002. SIGEP implements an algorithm based on so-called fuzzy logic, considering the uncertainty per se of evaluations, the great amount of variables determining the condition state and vulnerability of a bridge, and the broader variables, which are usually not measurable.

Bibliography

AASHTO. 2012. *AASHO LRFD Bridge Design Specifications*, Customary U.S. Units, 2012, American Association of State Highway Officials, Washington, DC.

ACI. 2011. *Building Code Requirements for Structural Concrete (ACI 318-11)*. American Concrete Institute, Farmington Hills, MI.

AISC. 2010. *Steel Construction Manual*, 14th edition. American Iron and Steel Construction, Chicago, IL.

National Department of Transportation of Argentina 1952. *Reinforced Concrete Bridges Design Standards*, Comisión Nacional de Regulación del Transporte (CNRT), Buenos Aires, Argentina. (In Spanish)

National Department of Transportation of Argentina. 1952. *Argentina Standards for Concrete Railway Bridges Design*, Comisión Nacional de Regulación del Transporte (CNRT), Buenos Aires, Argentina. (In Spanish)

National Department of Transportation of Argentina. 1952. *Complementary Specifications for Railway Brides Design*, Comisión Nacional de Regulación del Transporte (CNRT), Buenos Aires, Argentina. (In Spanish)

Del Carril, T., Fazio, J., and Bignoli, A. J. 1997. *Management System for Conservation of Bridges*, XII, Argentine Congress on Roads and Transit. Oct., Highway Association of Argentina, Buenos Aires, Argentina. (In Spanish).

Del Carril, T., Fazio J., Berditchevsky, G. and Cronenbold, J. F. 2001. *Bridge Management System in Guatemala, Guatemala*, XIII Argentine Congress of Roads and Transit, Highway Association of Argentina, Buenos Aires, Argentina. (In Spanish).

Del Carril, T. and J. Fontán Balestra, J. F. 2002. Creative Solutions for Two Bridges, in *Structural Engineer*, May. pp. 21–30. Institution of Structural Engineers, UK.

DIN 1071. 1935. *Roadway Bridges Dimensions*. Interim 2018. 1935, Steel Construction Association, Berlin, Germany. (In German).

DIN 1072. 1935. *Roadway Bridges Live Loads*. Interim 2019. 1935, Steel Construction Association, Berlin, Germany. (In German).

DIN 1072. 1972. *Roadway and pedestrian bridges*, Beuth-Vertrieb Company, Berlin, Germany. (In German)

DIN 1073. 1935. *Specifications for Steel Bridge Design*, Interim 2020. 1935, Steel Construction Association, Berlin, Germany. (In German).

DIN 1075. 1936. *German Standard*, Department of Highways, Department of Transportation and Work, Publicas, tr. Eng. J. Augusto Junqueira, S. Paulo, Brazil. (In Portuguese).

ICPA 1964. Reinforced Concrete Arch Bridge over the Quequen Salado River, in *Cemento Portland*. Journal of the Portland Cement Institute of Argentina (ICPA), No. 57, July, pp. 20–22. (In Spanish)

Lima, L. and Lima, E. 2000. The Bridge Over "Riachuelo" River in La Plata-Buenos Aires Highway, in *Proceedings of the Bridge Engineering Conference* 2000, Sharm El-Sheikh, Sinai, Egypt.

Llarrul, M. and Del Carril, T. 2004. The Railway and Roadway Bridge over the Rio Dulce in Santiago del Estero City. In *Ingenieria Estructural*, Structural Engineers Association (AIE), Argentina Year 12, No. 28, April, pp. 19–21. (In Spanish)

MPW. 1952. *Basis for Calculating Reinforced Concrete Bridges*, General Administration of National Highways, Ministry of Public Works, Buenos Aires, Argentina. (In Spanish)

Polimeni, F. and Polimeni, M. 2002 Retrofitting of the Marcial Candiotti Suspended Bridge, in *XVII Argentine Congress of Structural Engineering*, Rosario, Argentina. (In Spanish).

Polimeni, F. and Polimeni, M. 2009. *Parks Connection, in Summa magazine*, Donn SA, No. 103, Sept., pp. 8–14. Buenos Aires, Argentina. (In Spanish)

Ponce Delgado, A. 1992. *Concept, Design and Construction of the Libertador General San Martin Bridge between Fray Bentos and Puerto Unzue*, ed. AMESUR, Montevideo, Uruguay. (In Spanish).

Ruiz, M., Castelli, E. and Prato, T. 2008. A New Bridge Management System for the National Department of Transportation of Argentina, *Proceedings of the Fourth International Conference on Bridge Maintenance, Safety and Management*, July, Seoul, South Korea.

4

Bridge Engineering in Brazil

Augusto Carlos
de Vasconcelos
Consulting and Eng. SS Ltda

Gilson L. Marchesini
*Sobrenco Engenharia E
Comercio Ltda*

Júlio Timerman
*Engeti Consulting
and Eng. SS Ltda*

4.1 Introduction

4.1.1 Geographical Characteristics

Brazil is the largest South American country, with a vast seashore area and bordered by the Atlantic Ocean. Its neighboring countries are Venezuela, Guyana, Suriname, French Guyana, Colombia, Peru, Bolivia, Paraguay, Argentina, and Uruguay. The name "Brazil" has a very controversial origin, although most explanations refer to the brazilwood extracted here. For some philologists, it derives from the Tupi word *ibira-ciri*, "prickly wood." When the Europeans heard this name the natives gave to the red-colored wood, they called it "Brazil." The total population in 2010 was 190 million and the overall GDP as of 2009 was about $2.194 trillion (CIA 2013).

From its earliest colonial history, transportation has always been a challenge for Brazil because of its size and topography. In the last 40 years this challenge has finally been met: a systematic approach has been adopted to plan and implement a national system of integrated surface transport—road, rail, air, and water. Although road or highway transportation is often more expensive than other modes, it is virtually unmatched as a fast means of moving comparatively small amounts of cargo and passengers over short distances, and it constitutes the main transport system in Brazil: 56% of freight in the country moves by road, as opposed to 21% by rail and 18% by water. In 2000, the total road network of Brazil was 1,071,821 mi (1,724,929 km), of which 9.6% was paved.

As the most appropriate method of moving nonperishable cargo over long distances, railways are the second most important transportation system in Brazil. The country has an 18,196 mi (29,283 km) railway network. Almost half of this network is concentrated in three states: São Paulo, Minas Gerais, and Rio Grande do Sul. They are largely concerned with freight transport of iron ore, petroleum derivatives, grain, and steel. Long-distance passenger railway services are practically nonexistent in Brazil. Passenger lines are limited to the suburbs of the great urban centers. The government sold off its controlling shares of the railways in 1997, although many states and cities retain control of local lines.

Brazil's long coastline and vast waterways in most of the hinterland have not been fully exploited for waterborne transport. The government invested significantly in the 1990s to promote the integration of road, rail, and water transportation systems in order to reinforce the trend toward intermodality. Brazil's major port facilities were significantly improved in the late 1990s, mainly through privatization. Brazil has 46 organized ports, 24 of which are ocean ports. Among the busiest are Santos, Rio de Janeiro, and Porto Alegre.

Brazil's physical characteristics and the requirements of fast economic growth led (starting in the 1930s) to the establishment of a vast network of air services. Although Brazilian air transportation still faces some challenges, the last 30 years have witnessed extraordinary progress. In 1994, the total passenger movement was calculated at 43 million passengers per year. In 2001, approximately 75 million passengers embarked or disembarked in Brazil. The demand for construction has been enormous. Major projects include the 5000 km Trans-Amazonian Highway, running from Recife and Cabedelo to the Peruvian border; the 4138 km north-south Cuiabá-Santarém Highway; and the 3555 km Trans-Brasiliana Project, which will link Marabá, on the Trans-Amazonian Highway, with Aceguá, on the Uruguayan border.

The period between 1995 and 2001 saw the inauguration of 14 newly constructed or renovated major airports. In 2002–2003, this sector grew at the rate of 8% a year, double the world average. Numerous airlines have flourished in Brazil at one time or another, but they have been consolidated into two major companies that compete nationwide: Tam and Gol. The busiest gateway in the country is the Congonhas domestic airport (São Paulo), which manages an average of 22,000 embarkations and disembarkations a month.

4.1.2 Historical Development

The great variability of types and structural forms of bridges requires, by the professional who conceives them, great experience, deep technical knowledge, and a high aesthetic sensibility. Brazil is known to have some of the most notable structural engineers in the world, designing memorable bridges despite the frequent difficulties encountered. Examples include the set of railway bridges designed for the railway Mayrink–Santos in 1940s; the first presstressed bridge in South America (Galeão Bridge in the harbor of Rio de Janeiro), built in 1949; and President Costa e Silva Bridge (also called Rio–Niterói Bridge), inaugurated in 1974, the longest span in the world with a continuous straight beam.

4.2 Design Practice

4.2.1 Live Loads

Brazilian standards used in the structural projects are described below:

- 1943–1960: NB1/43, NB2/43, NB6/43
- 1960–1978: NB1/60, NB2/61, NB6/60

- 1978–1984: NB1/78, NB2/61, NB6/60
- 1984–present: NBR6118/04, NBR7187/87, NBR7188/84, NBR8681/04

4.2.1.1 Live Loads 1943–1960 (NB6/43)

In the period between 1943 and 1960 live loads on road bridges specified in Norm NB6/43 are as follows.

4.2.1.1.1 Classes of Bridges

Road bridges were grouped in three classes:

Class I: Bridges located on federal and state trunk roads or on the main connection roads between such trunks

Class II: Bridges located on secondary connection road but on which, meeting the special circumstances of the place, there is convenience to foresee the passage of heavy vehicles

Class III: Bridges located on secondary connection roads not included in class II

4.2.1.1.2 Live Load

Live load was specified as a set of mobile loads to be applied to the structure in its most unfavorable position. Live loads are composed of vehicles as described in Figure 4.1.

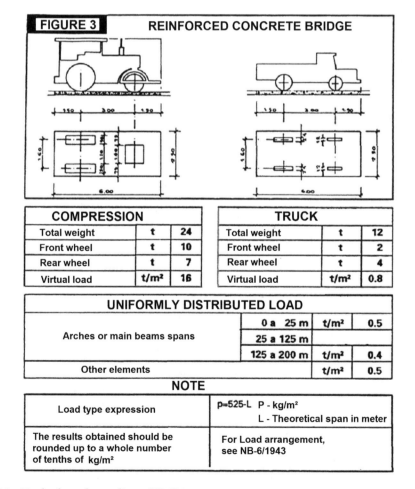

FIGURE 4.1 Live loads used according to NB6/43.

4.2.1.1.2.1 Live Loads for Highways Class III The live loads for class III bridges include a distributed load g_o = 400 kgf/m² (4.0 kN/m²) of a compressor type A. The restrictions were that one vehicle shall not be placed on each traffic lane, nor in a position that grants a distance of less than 2.5 m between the longitudinal axles of two vehicles.

4.2.1.1.2.2 Live Loads for Class II The live loads for class II bridges is the same as in Section 4.2.1.1.2.1, but one must verify the resistance of the structure for a type B (Figure 4.1), placed isolated on the bridge, in the most unfavorable position for the studied element but always guided in the traffic direction.

4.2.1.1.2.3 Live Loads for Class I The live loads for class I bridges include a distributed load g_o = 450 kgf/m² (4.5 kN/m²), of a type B compressor as described in Section 4.2.1.1.2.1. The structure resistance must be verified for a type C compressor (Figure 4.1), placed as in Section 4.2.1.1.2.1.

4.2.1.2 Live Loads 1960–1984 (NB6/60)

In the 1960–1984 period, live loads on road bridges specified in the Standard NB6/60 (ABNT 1960) were as follows (Figure 4.2).

4.2.1.2.1 Classes

- Class 36: Vehicle weight = 360 kN
- Class 24: Vehicle weight = 240 kN
- Class 12: Vehicle weight = 120 kN

4.2.1.2.2 Live Load Positioning

The vehicles were always placed in the traffic direction and must be placed in the most unfavorable position without considering axle load or the wheel that produces reduction of demanded efforts. The uniform load P must be applied to the longitudinal lane that the vehicle does not occupy. The P load must be applied on the remaining part of the road and on the sidewalks.

BRIDGE CLASS	VEHICLE Type	VEHICLE Total weight (T)	UNIFORMLY DISTRIBUTED LOAD p (kg/m²)	UNIFORMLY DISTRIBUTED LOAD p' (kg.m²)	UNIFORMLY DISTRIBUTED LOAD Load Distribution	HIGHWAY CLASS
36	36	36	500	300	p load - front and back of the vehicle p' load - outside vehicle area	Class I
24	24	24	400	300		Class II
12	12	12	300	300		Class III

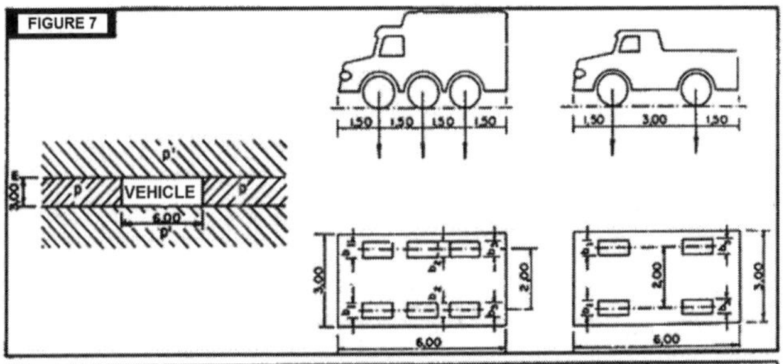

	AXIS	TYPE 36 t	TYPE 24 t	TYPE 12 t
Number of axes		3	3	2
Total vehicle weight	t	36	24	12
Weight of front wheel	t	6	4	2
Weight of rear wheel	t	6	4	4

FIGURE 4.2 Live loads used according to NB6/60.

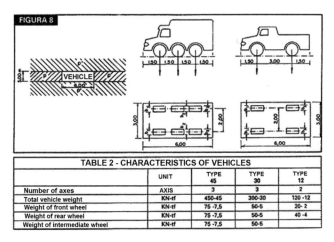

TABLE 1 - VEHICLE LOAD							
BRIDGE CLASS	VEHICLE			UNIFORMLY DISTRIBUTED LOAD			
	TYPE	TOTAL WEIGHT		P		P'	Load distribution
		KN	tf	KN/m²	kgf/m²	KN/m²	kgf/m²	
45	45	450	45	5	500	3	300	p load - front and back of vehicle
30	30	300	30	5	500	3	300	p' load - outside vehicle area
12	12	120	12	4	400	3	300	

TABLE 2 - CHARACTERISTICS OF VEHICLES				
	UNIT	TYPE 45	TYPE 30	TYPE 12
Number of axes	AXIS	3	3	2
Total vehicle weight	KN-tf	450-45	300-30	120 -12
Weight of front wheel	KN-tf	75 -7,5	50-5	20- 2
Weight of rear wheel	KN-tf	75 -7,5	50-5	40 -4
Weight of intermediate wheel	KN-tf	75 -7,5	50-5	

FIGURE 4.3 Live loads used according to ABNT NBR 7188/84.

4.2.1.3 From 1984–Present (NBR 7188/84)

From 1984 until the present, live loads on road bridges specified in the Norm NBR 7188/84 (ABNT 1984) are as follows (Figure 4.3).

4.2.1.3.1 Classes

- Class 45: Vehicle weight = 450 kN
- Class 30: Vehicle weight = 300 kN
- Class 12: Vehicle weight = 120 kN

4.2.2 Designing Criteria

Until 1960, the allowable stress design method was used for concrete bridge structures on stadium II. In 1960 the revision of the NB-1 "Calculus and Execution of Reinforced Concrete Structures" was published. The revision of the NB-2 "Bridges Standards—Design and Execution of Reinforced Concrete Bridges" was published in 1961. The safety criteria established by NB-2/61 are described next,

Bridges could be designed using either the criterion of ultimate load (stadium III) or the criterion of the allowable stress (stadium II). Such safety criteria were also established for the structures of reinforced concrete by the NB-1/60. With the revision of the NB-1, the NBR-6118, from 1978, the former safety criteria for the structures of reinforced concrete were modified, introducing the semi-probabilistic safety criterion, with coefficients γ_f of actions increase: $Y_g = 1.4$ for permanent loads and $Y_q = 1.4$ for accidental loads and coefficients Y_m of decrease of materials resistance; $Y_c = 1.4$ for concrete and $Y_s = 1.15$ for steel. Even with the introduction of this new safety concept, it was still permissible to use NB-2/61 for bridge design. The bridge standards NBR-7187/1982 and NBR-7187/1987 were published.

NBR-7197/1987 modified the former safety criteria, also introducing the semiprobalistic method of NBR-6118/1978 with load factors $Y_g = 1.35$ for permanent loads and $Y_q = 1.5$ for live loads, and decreased resistance factors $Y_c = 1.5$ for concrete and $Y_s = 1.15$ for steel. In 2003, the new concrete standards NBR-6118/2003 and the bridge standard NBR-7187/2003 were published.

The safety criteria adopted by the current standard for bridges, NBR-7187/2003, its load factors and combinations, as well as its respective resistance factor follow NBR-6118/2003, which in turn reports to the standards NBR-8681/2003, "Actions and Safety on Structures." NBR-8681/2003 establishes for normal combination a load factor of big bridges whose own weight exceeds 75% of the totality of the permanent load $Y_g = 1.3$ and, for bridges in general, $Y_g = 1.35$ and for the live loads $Y_q = 1.5$."

4.3 Concrete Girder Bridges

4.3.1 Galeão Bridge (Eng. Eugene Freyssinet)

Galeão Bridge (Figure 4.4), in the harbor of Rio de Janeiro, is Brazil's contribution to the science of prestressed concrete, and opened to traffic in 1949.

The crossing ties Governor's island, site of International Airport, to a smaller island.

The roadway has six traffic lanes with two sidewalks, and its total width is 20.60 m.

The bridge (Figure 4.5) includes 15 simply supported girders spans (43.4 m + 2 × 37.2 m + 2 × 28.3 m + 10 × 19.4 m). The bridge's total length is 369.4 m and the longest span is 43.40 m. Each span contains 19 prestressed concrete I-girders spaced at 1.1 m.

The height of the I-girder varies from 0.95 m to 1.9 m at the longest span. Figures 4.6 and 4.7 show the bridge under construction.

The prestressing cables, made up of 12 strands of wire 5 mm in diameter, are placed in each girder in a parabolic curve. In each center-span girder, for instance, there are 20 cables, which are placed in the bottom flange at midspan, but which curve upward from there, one by one, to be anchored at successive points along the top of the girder. At the supports, only eight cables remain in the section, and these are anchored at the girders end.

Working stresses used were 13 MPa for concrete and 1500 MPa for steel. Effective prestress in each cable after losses is 250 kN. Hydraulic jacks and patented anchorages were used in accordance with the Freyssinet method of prestressing (ENR 1948, 1945), by agreement with the Societé Technique pour l'Ùtilisation de la Precontraite, of Paris.

FIGURE 4.4   General view of the Galeão Bridge.

FIGURE 4.5 Side view of the Galeão Bridge.

FIGURE 4.6 Detail of the placement of a precast girder.

The superstructure is supported by three reinforced concrete pillars as shown in Figure 4.8, which sit in three caissons 2 m in diameter. Concrete pillars are capped with a prestressed concrete crossbeam. No heavy floor slabs were required to support the roadway, since the girder flanges are wide enough to nearly touch. A layer of concrete was cast to protect the girders, then asphalt surfacing was placed on top.

Construction of the bridge was authorized by the Aeronautics Ministry. The contractor was Companhia Nacional de Construções Civis e Hidráulicas, of Rio de Janeiro. Despite the lack of adherent prestressed cables, the performance of the bridge remains perfect today.

FIGURE 4.7   View of the assembly of the Galeão Bridge deck.

FIGURE 4.8   Bottom view of the Galeão Bridge.

4.4 Steel Box Bridges

4.4.1 President Costa e Silva Bridge (Engs. Antônio A. Noronha, Benjamin Ernani Diaz, and Bruno Contarini)

The President Costa e Silva Bridge, better known as the Rio-Niterói Bridge (Figures 4.9 and 4.10), was inaugurated on March 4, 1974 and is a national engineering landmark because of its size, signature design, and the creativity of the processes for executing the work. This mega construction is the biggest

FIGURE 4.9   Side view of the Rio-Niterói Bridge.

FIGURE 4.10   Side view of the Rio-Niterói Bridge.

bridge in the southern hemisphere; the longest span in the world with a continuous straight steel beam (the central span is 300 m long); and the biggest prestressed structure in the South America (with over 2000 km of cables in its inner structure). It is the seventh longest steel girder bridge in the world, but experts insist that it is still the biggest, in terms of spatial volume, because of its gigantic pillars and the caissons sunk into rock, over 60 m below the water level. Above all, it highlights the setting of Guanabara Bay.

The Rio-Niterói Bridge is 13 km long and has 10 km approach spans. When it was opened to traffic, the traffic volume was expected to be 50,000 vehicles per day, which was soon surpassed due to the region's growth. Currently, there are over 150,000 vehicles per day and 170,000 on the eve of holidays and during mid-summer. The main structure (Figures 4.10 and 4.11) is composed of three continuous spans (200 m + 300 m + 200 m). The cross section consists two steel box girders, each with a width of 6.86 m and variable height (13 m on the supports and 7.5 m in the midspan). The deck has a full width of 26.6 m and the freeboard is 60 m. The total weight of the steel structure is 13,000 tons.

FIGURE 4.11 (**See color insert.**) Side view of the Rio-Niterói Bridge main span.

The prestressed concrete superstructure of the 7884 m long and 26.6 m wide bridge over the Guanabara Bay, connecting the cities of Rio de Janeiro and Niteroi, was constructed by the cantilever method with the help of precast segments that were glued together with epoxy resin. An experimental investigation of the structural behavior of the glued joints was made in a model of the bridge. Important conclusions resulted from the experiment. It was shown that the glued joints do not affect the safety of the structure. Several investigations were made during the construction for determination of various important features of the glued joints for erection procedures. The behavior of the glued joints was studied with the help of a specially developed concrete test piece.

4.5 Arch Bridges

4.5.1 International Bridge Brazil–Paraguay (Eng. José Rodrigues Leite de Almeida)

This bridge (Figure 4.12) is located over Paraná River, linking the city of Foz do Iguaçu (Brazil) with Ciudad del Leste (Paraguay). The bridge is made of reinforced concrete and has a main arch span of 303 m and a total length of 552.4 m. To meet the clearance requirements for shipping at the maximum floods, the grade, at the closure, was set at the absolute elevation of 162 m. The bridge is composed of a central arch with two access viaducts. The deck on the arch area has a stretch of 79.4 m in middle span and fiver spans to each side, successively of 20, 21, 22, 22, and 22 m up to the pillars. Each of the two access viaducts is a continuous straight beam over five spans of 22, 22 22, 20, and 18 m, supported by circular columns.

Considering the large height of the deck, the consequent action of strong winds, and a maximum flood, around 40 m of arches, measured along their axles, are immersed in the stream, suffering great dynamic pressures. It was thought to give a great cross rigidity to the structure horizontally, not only at the widening of the arch at the end, out also at the deck, which has unmovable supports transversally over the pylons and joints. For this, the designers projected the construction of two big pylons with hollow sections leaning on the foundation blocks of the arches. The pylons and joints would acquire great transversal rigidity and the intermediary columns of the viaducts and the arches could be very slender.

FIGURE 4.12   Side view of the International Bridge.

FIGURE 4.13   Detail of the construction of the International Bridge.

The width of the deck is 9.5 m; the sidewalks, which are separated from the road by a fence of galvanized screen in order to allow cattle to pass, are 1.6 m wide on each side. The total deck width is 13.59 m, widened over the pylons, where it reaches 22.6 m. The viaducts were constructed by wood falsework, and reinforced concrete was used for towers up to the height of the average floods on the Brazilian bank. The Volta Redonda Group built metallic falsework with 1200 tons of steel that was 157.3 m long to provide the connection of the two stretches of the concrete falsework already built.

The steel falsework structure shown in Figure 4.13 is composed of four interconnected polygonal lanes by means of systems of horizontal and vertical wind bracing, disposed orthogonally and formed with crossed diagonals. Each lane is composed of two semi-arches, articulated at the joint and at the support extremities. Twelve interconnected panels, disposed successively according to inflection angles that vary from a maximum of 2°2' to a minimum of 56', make each semi-arch. Such stretches weigh on average 8500 kg, and the falsework is a result of the association of 96 of these panels. The steel cables are (1.125 in) in diameter for revestment and anchorage of the falsework during assembling, and are anchored on the concrete columns and suffer elevatory inflection by means of two steel towers of around 200 tons, assembled on the extremities of the concrete stretches. There are 256 such cables that are 12779 m long.

All devices of support, articulations, and anchorage, as well as two mobile derrick-type cranes with 20 ton capacities were manufactured to erect and position through a progressive advance over the metallic arch—the panels that constitute the polygonal lanes. Such devices gave the steel part a total weight of 1275 tons.

4.6 Progressive Cantilever Bridges

4.6.1 Bridge over the Peixe River (Eng. Emílio H. Baumgart)

4.6.1.1 Introduction

The bridge over the Peixe River, initially called the Herval Bridge and later renamed the Emílio Henrique Baumgart Bridge (Figure 4.14) in honor of its creator, was built in 1930. The casting of its final stretch took place on October 29. Its 68.3 m of central span represented, at that time, a record span of reinforced concrete beam which, together with the construction method of progressive cantilever, brought ample recognition all over the world, becoming the most well-known Brazilian work abroad. The bridge was destroyed by a flood in June 1983 after 53 years of service in perfect conditions of use. It was located at the connection between the towns of Joaçaba and Herval D'Oeste, in the central region of the state of Santa Catarina.

"It was the most remarkable bridge in Brazil until 1983," said Professor Augusto Carlos de Vasconcelos. Emílio Baumgart became known, after that bridge and other works of no lesser importance, as "the father of reinforced concrete in Brazil."

4.6.1.2 Description of the Bridge

The bridge featured a continuous superstructure of three spans of 26.8 m + 68.3 m + 26.8 m, cross section with track width of 7.5 m and sidewalks of 1 m each. Facing the sidewalks to the outside there were two longitudinal beams with variable height and width (0.30 m × 1.7 m in the middle of the span and 1 m × 4 m in the support). In the central area there were two minor longitudinal beams located under the track spaced at 4.15 m. Between the two main beams there were crossbeams spaced at 3.09 m along the whole bridge.

FIGURE 4.14   Side view of the Emílio Henrique Baumgart Bridge.

4.6.1.3 Construction Method

The cantilever method (Figure 4.15) was used to construct the bridge because the bridge site is a near a mountain frequent flood zone, and the level of the river water may rise 10 to 11 m in a very short period of time. In addition to that, there was the difficulty of taking materials to the site and the low cost of local labor, all of which were decisive factors for the choice of the executive process.

The external spans were built over conventional falsework, and the first 9.3 m of the central span, on both sides, was constructed over falsework in a fan shape. The rest of the central span (50.3 m) was built by concreting 30 voussoirs of 1.55 m each, simultaneously executed from the supports to the center of the river the central closing stretch, linking both sides with 3.20 m was concreted leaning on both side parts.

4.6.1.4 Reinforcement

The main reinforcements (Figure 4.16) were made of steel bars 38 mm in diameter in CA25 steel. On the supports, 46 bars were placed on the upper part of the section and 19 bars on the lower part. On the span, six bars were placed on the lower part and two bars on the upper part. On the stretch to be executed in cantilever, the bars had 4.65 m (length of two voussoirs), placed in an alternated way in order to not have all joints in the same section. The joints were constructed using threaded sleeves on the edges of the bars. On the edges (the extremities or concrete joints) short and thin bars were used. They had a diameter of 6.35 mm, disposed at each 5 cm, and were called a "beard", in order to avoid fissuring.

4.6.1.5 Deformation Control

To correct the differences of displacements on the edge in cantilever advancing into the middle of the river, and to promote the perfect liaison between the parts upon the work closure, Emílio Baumgart developed an extraordinarily creative and innovative process. Inspired by the executive processes until then applied to the construction of metallic bridges in cantilever and gerber beams, he implanted steel pins 76 mm in diameter transversely in the interior of the main beams, on the axle of the intermediary supports. Such pins crossed the main beams with 1 m of width at this point and leaned on bearings with 40 cm of width for each side, which were supported in the pillar in a Y shape, becoming the supports of such bearings.

On the edges (the final stretch of the outer spans), cylindrical foundations positioned on the edge of the main beams became an anchorage system to balance the cantilever of the main span as the advance

FIGURE 4.15 Aerial view of the Emílio Henrique Baumgart Bridge under construction.

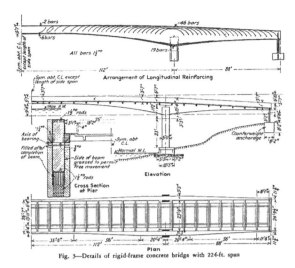

Fig. 3—Details of rigid-frame concrete bridge with 224-ft. span

FIGURE 4.16   Detail of original Emílio Henrique Baumgart Bridge design.

of the voussoirs proceeded for the closure of the span. This way, the system made up of the outer span "tilted" with the spin around the pin on the internal support, and with a system of additional counterweight on the edges of the beams, as well as the use of a system of nuts and screws on such edges, the following was produced: control of displacements on the cantilever edge caused by either immediate arrows or differed by aiming at, upon the closure of the work, no differences on the levels of both parts.

4.6.1.6 Occurrences During Construction and the Collapse of the Bridge

Regarding construction, two meaningful facts must be cited. The first is the occurrence of a sudden flood that washed away the falsework of the stretch concreted "in loco." Fortunately, the already-launched concrete had enough strength to resist its own dead load effects. Another occurrence was the stoppage of the work for a week due to combat at the site related to the 1930 revolution. Although floods were frequent and raised the water level a lot, the positioning of the work allowed it to resist to such floods for many years.

The collapse of the bridge in 1983 was due to the construction of a four-story building nearby which led to the termination of the section of the canal and, as a consequence, the increase of the outflow conditions as well as the collapse of the bridge with the undermining of the foundations of the structure in 1983, after 53 years of service.

4.6.2 Estreito Bridge over the Tocantins River (Eng. Sérgio Marques de Souza)

4.6.2.1 Introduction

The new capital of Brazil, Brasília, in the Brazilian central plains, demanded, for the full development of its role in the national integration, a land connection with all regions of the national territory. To achieve this objective, the roads Brasília-Belo Horizonte, Brasília-São Paulo, Brasília-Acre, and Brasília-Belém were built. This last road, 2200 km long, crossed unexplored stretches of the Brazilian territory and presented problems of all kinds for its realization, being the biggest obstacle to the intended connection between Belém and Brasília the bridging of Tocantins River.

Prior aerial studies made during the design definition showed that, at the Carolina downstream, the Tocantins River featured a major narrowing in its width, thus providing ideal conditions for the implementation of the work of art necessary to its bridging. The site chosen for the bridging is 100 km away from any source of supplies. The nearest town is Belém, situated 700 km from that point.

4.6.2.2 Bridging Problems

The canal characteristics, with a width of 130 m and minimum depth of 45 m, led to the solution of crossing the river with an only span of 140 m and showed the impossibility of using conventional direct cross-bracing. In order to make the conclusion of the work coincide with the one of the highway, it became necessary to seek a solution that demanded less construction time, as the total available time was 9 months. Making use of the solution imagined and adopted by the eminent Brazilian engineer Emílio Baumgart for the construction of the bridge over the Peixe River in 1928, the progressive canti-lever system, by means of indirect cross-bracing leaned on the stretches already concreted of the work, was used again. The bridge was then designed with a central span of 140 m in straight prestressed con-crete beams. Such a central span was a world record at that time.

4.6.2.3 Project

The bridge has a total length of 532.7 m, with two viaducts of access in reinforced concrete, with 20 m spans and central spans in prestressed concrete with a central span of 140 m, two side spans of 53 m, and two cantilevers of 5 m. To allow river sailing, the deck elevation enables a minimum free height of 10 m at flood times. A solution using a gerber beam (Figure 4.17) with two cantilevers from the margins as a support on its edges, versus a beam simply supported, was studied.

The constructive system adopted enabled the executions of cantilevers in voussoirs of 6.66 m each. Thus, the simply supported beam would have a length of 14 m to be executed with the same indirect cross-bracing adopted in the balance construction.

The small span of this beam would lead to two support balances of 63 m of free span each, featuring a major vertical displacement to the mobile loads. The unevenness between the two extremities of the central beam was 20 cm for only one of the loaded cantilevers, which led to the abandonment of such a design. Finally, they came to the ideal solution: sectioning the beam on its central stretch, however establishing a vertical connection by means of a pendulum which, besides making the arrows equal from the two cantilevers, also allowed horizontal displacements due to temperature and retraction.

The access viaducts were built in rigid squares of reinforced concrete by conventional methods, with-out presenting major technical problems in construction. The launched box prestressed concrete girder has a depth of 3.4 m at the edge of the cantilevers and 8 m at the supports, with the thickness of the girder web varying from 55 cm to 25 cm. The soffit slab is 70 cm at the support and 15 cm at the balance edges. Both top and bottom slabs are in reinforced concrete.

The Freyssinet system, a widely used system in Brazil, was adopted because it offered the possibility to allow progressive anchorages, following the cantilever growth. On the whole, each cantilever was set with 276 tendons of 12 Ø 7 mm to resist the bending moment of 65,200 t.m over the support. The shear force at the support is 2100 tons. The piers over the river have a size of 1.4 m × 7 m, supporting the load of 3800 tons from the superstructure reaction. Its connection with the beam is made by Freyssinet finger to allow the rotation of the support as the balance is being executed.

The column of the counterbalance span works as an anchor and has its larger foundation working as a counterweight. In order to provide room for the displacement of the structure, an intercalary pendulum is placed between the column and the beam, working over lead plates of 30 mm × 670 mm × 1.5 mm. The pendulum is placed in a concrete box facing the column in order to protect it and preserve the

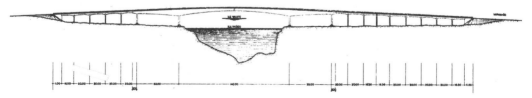

FIGURE 4.17 Original draft of the Estreito Bridge.

aesthetic of the structure. 480 tons of prestressing were applied through 12 tendons of 12 Ø 7 mm, which, starting from the top of the deck anchor on the concrete block of the foundation, is able to resist the maximum tension on this pendulum of 310 tons.

4.6.2.4 Construction of the Prestressed Spans

Two spans of 53 m of the counterbalance were built over wood falsework. To make the falsework safe during a possible sudden river flood, two concrete supports were constructed for each span, and the beams on this stretch had reinforcing CA-37 steel to meet support efforts on these two spots. The central span of 140 m was constructed by means of two sets of mobile trusses which, leaning on the already-concreted spans, supported the bottom mold of the cantilever element to be concreted.

Each truss set consisted of three units that moved over the main beams. The bottom mold of the new element was suspended on the stretch in balance, and underneath there was a scaffolding to allow the placement of the seizing beams and the anchors that supported the bottom mold during concreting. The platform used as the bottom mold of the new element to be concreted hung from the truss through steel cables that allowed its correct adjustment. After leveling and alignment, 12 steel anchors of Ø 32 mm were placed, connecting the truss to the platform in order to support the weight from the concrete of the new element. Each cantilever of the bridge on the main span was built in eight elements of 6.66 m each. The prestressing cables were placed such that they enable the anchorage of part of the cables at each advance of the cantilever. Some were anchored on the top deck slab and others in the web.

In order to facilitate the construction of each element over the river, spaces were left over the beams, place the major extension cables that ended in the most advanced stretches in balance. On the other extremity, on the stretch of 53 m on the edges, such cables were fixed to previously concreted anchorages. The shorter cables were left floating in the river until the continuation of the work enabled their anchorage.

The construction was executed 24 hours a day and the prestressing was performed after only 24 hours of concrete casting. This was possible due to the careful study of the concrete mixing and a well-equipped field lab. The factor water/cement ratio varied from 0.37 to 0.39, and liquid Plastiment was used in the proportion of 0.5% over the cement. The concrete used had a consumption of 450 kg/m³ of cement. The prestressing steel was composed by tendons of 12 Ø 7 mm with a tensile strength of 1,600 kg/cm². The initial prestressing tension applied was of 1,100 kg/cm².

The first element of the balance was built in 10 days, but they were able to gain speed, obtaining an amazing average of 5 days per element. The central span of 140 m was so constructed in only 40 consecutive days. This bridge (Figure 4.18) was designed, calculated, and executed by Prof. Dr. Sérgio Marques

FIGURE 4.18   General view of the Estreito Bridge.

de Souza SA, Engenharia e Comércio. Around 500 workers were recruited all over the country, due to the lack of skilled labor in the region.

4.6.3 Colombo Salles Bridge in Florianópolis, Santa Catarina (Eng. José Carlos de Figueiredo Ferraz)

This bridge is the second that connects Florianópolis, located in Santa Catarina Island, to the continent. It is part of the BR-282 that goes to Lajes. The first bridge was built in 1926 and became the landmark of Florianópolis—the cable-suspended bridge Hercílio Luz. There is already a third bridge, next to the second one, designed as a mixed structure in 1983. Colombo Salles Bridge (Figure 4.19) was, at the date of its completion in February 1975, the longest prestressed concrete beam in the world, with a total length of 315 m, built in the cantilever construction process.

The bridge is 1227 m long, with 15 simply supported spans and 3 continuous spans. The simply supported spans have eight spans with lengths of 75 m, two spans with lengths of 50 m, and five spans of 42.5 m. The bridge was designed by Figueiredo Ferraz Consultoria e Engenharia de Projetos SA and its construction, after a fierce competition, was won by Construtora Norberto Odebrecht.

The construction of such a huge bridge could have caused endless discussions that would eventually postpone decisions. However, the corrosion problems that occurred at the Hercílio Luz Bridge were similar to two other American suspension bridges that collapsed for the same reason, and the decision for a new project was accelerated and efforts were made to obtain resources. The challenge of designing and constructing this bridge was quickly accepted. It has consumed 28,200 m³ of concrete and 61% of it was made of prestressed concrete. The underwater foundations used 4700 m³ of concrete in the connection with the steel piles to the caissons.

4.6.3.1 Continuous Beam

The continuous beam (Figure 4.20) has three girder spans, symmetrical, with spans of 77.5 m and 160 m, and a total width of 17 m. The central span, the largest one, was initially built in two separate halves: two double cantilevers 77.5 m long were built simultaneously above each main column and finally the central 5 m connection established continuity.

For the self-weight load, the structure works in the first phase as two independent cantilevers, rigidly connected to the column. This link was achieved using two temporary concrete walls spaced at 16 m, built with the initial portion of the superstructure using normal scaffolding. The 18 successive segments

FIGURE 4.19 General view of the Colombo Salles Bridge.

FIGURE 4.20   Side view of the Colombo Salles Bridge.

4 m long were cast simultaneously on both sides, supported by formworks on special vehicles. The segments have two cells and vary in height from 3.66 m at midspan up to 9.5 m at the supports.

During the second phase new cables were driven in the central part, the final 5 m of closure was concreted, and then finally the structure worked as a continuous one. Once cables were prestressed for continuity, it was necessary to give the bridge the possibility of rotation on the central supports. This was the most delicate part of the work: the demolition of the temporary walls without violent impacts. The change of support system had to be slow and gradual. To prevent spolling and cracking of the concrete walls during the process of the prestressing cables for continuity, with minimum rotations on the support section, it was necessary to make the prestressing of the central cables as soon as possible, before significant creep deformations occurred. High-strength concrete was used to permit prestressing the first cable after just a few hours.

The values adopted were 1.2 for creep and 0.14 mm/m for shrinkage, applied in the design phase. With these parameters the theoretical calculation of the redistribution of self-weight load has been made, coming to the conclusion that 35% of that load is transferred to the continuous beam structure, the remaining 65% staying in the simply supported structure. It is therefore necessary to verify the service situation with two alternatives: 0% and 35% of redistribution.

The total 17 m of width was divided in 14 m for the two traffic lanes (7 m each), separated by a central reservation of 1 m and side refuges of 1 m each. Pedestrians use the bridge in catwalks hanged on the girders, with heights varying from 2.62 m to 2.98 m, the minimal value resulting from maximum utilization of available space in regions with lower segment height. The cells are not perfectly rectangular. The central wall is slightly higher than the others, facilitating the design of hanged catwalks on beams of variable height. The total width of the coffin is 11.2 m; therefore, part of the roadway is a cantilever, covering the sidewalks of the lower level.

The architectural design, developed with advice from Croce, Aflalo & Gasperini, was very good, separating pedestrians from vehicles and allowing passage protected from the sun, rain, and wind. A monumental staircase leads pedestrians to the suspended platform.

4.6.3.2 Simply Supported Spans

There are eight simply supported 75 m spans, consisting of two girders as in the continuous part, but interconnected by five transverse beams, three in the span and two over the columns. These transverse beams, with the same height as the girders, are seen from the underside between the girders. The height of the girders is 4.5 m and the catwalk beams are tied to the walls by 7 Ø 12.7 mm tendons. These beams were precast and assembled after the erection of the girders, holding on to them at the bottom. In the midspan there are nine tendons that were all lifted, prestressed, and anchored in the upper slab. For this reason, the slab was thickened over a length of 14 m at each end to 65 cm, to allow a good transfer of prestressing force to the girder section.

4.6.3.3 Foundations

The major construction difficulties were the foundations that, in some cases, reached a depth of 57 m. For a depth up to 25 m, it was possible to execute the caissons with compressed air equipment, drilling to the rock and then doing the shaft belling. For greater depths, steel forms were placed and driven in rock until a possible depth. Three strong steel beams (200 kg/m) were placed inside the caissons, driven until practically zero refusal. After that, underwater concrete was poured with the steel beams inside. The solution was created with the advice of Sigmund Golombek, who supervised the construction and defined the bases and depths.

The foundations of the Colombo Salles Bridge can be divided into three distinct groups: foundations using compressed air, foundations using steel piles, and foundations consisting of steel piles in the bottom part and caissons in lower depths. The foundations using compressed air could not be implemented beyond a depth of 33 m. Twelve steel forms of 1.5 m diameter and thickness of 9.5 mm were driven with Delmag equipment. The force for the driving was transmitted by a structure specially designed for this purpose, which received the designation "Apollo."

After placing the steel case at the desired depth, with the follow up of the related surveys, a pneumatic steel chamber with two compressors was mounted. The great pressure required to dewater the case and the little external friction between the case and the ground made the case work like an end closed tube with a tendency to float. To avoid floatation, counterweights up to 16 tones in the form of pieces of concrete were used. The work was nicknamed "crackers and toothpicks." For this reason the process only became advantageous to a depth of 25 m. After the level was reached, it was necessary to enlarge the base inside the rock to limit the soil bearing capacity to 50 kgf/cm², a value considered high. Only 24 caissons were erected with this system, for loads up to 1000 tf. Human sacrifice for such executions was extremely painful and slow. Men worked at pressures above 3 kgf/cm².

The other type of foundation used steel piles with welded TR32 rails. These piles had a capacity of 120 tf for four rails. Drilling was done using D-12 equipment until zero refusal was reached. The depths varied widely. This type of foundation was mainly used in the island area with hydraulic earthfill. The piles were encapsulated in an extension of 4 m in the tide zone. The bearing capacity was verified by load test using a reaction beam.

The last type of foundation was the most important and the most difficult. It was applied on 38 caissons where the survey showed the existence of rock at depths exceeding 25 m. The piles, however, were driven from inside the caissons from 25 m deep, staying therefore 13 m inside the caisson, which was considered satisfactory embedment. To construct these foundations special equipment was manufactured, consisting of four legs that could reach the bottom and a mobile platform on the four legs. The set, named "Straddle," was towed up to the approximate foundation location, accepting a tolerance of 50 to 100 cm.

At the local arrival, the equipment was attached to buoys to be fixed. The legs were then lowered by winches until they stopped at the bottom. With the same winches the platform was raised until 2 m above the water level, beyond the reach of the waves. For the straight placement of the steel cases, fork-shaped equipments were used, running on rails. The cases were placed vertically by a crane installed on a floating base and driven using Delmag equipment. The length of these cases did not vary much from around 45 m. The maximum depth of 57 m occurred at the supports of the 160 m span; the caisson was drilled until 40 m, and inside them, the piles had up to 57 m to achieve zero refusal.

After the piles had been driven, the caissons were cleaned and gauges removed. The underwater concrete pouring required the use of two 5 m³ mixers and a conveyor belt on floats.

Each caisson pile was designed for a 600 tf load, 200 tf for each pile. Two piles were subjected to a load test inside the caisson, which was also very complicated. One can therefore say that these foundations were a great milestone of engineering equipment. Before the construction of the Hercílio Luz Bridge a feat of this magnitude into the sea was totally impossible.

4.6.4 International Bridge over the Iguazú River, Linking Brazil to Argentina (Eng. José Carlos de Figueiredo Ferraz)

4.6.4.1 Introduction

The Tancredo Neves Bridge (Figure 4.21) is the latest of five international bridges in Brazil, completed in November 1985. It is located at the Iguazú River, quite near the Parana River, where the three countries, Brazil, Argentina, and Paraguay, have a triple frontier. The bridge is downstream from the famous Iguazú Falls. The city of Puerto Iguazú is located on the left bank of the Rio Iguazú (Argentina side) and, on the Brazilian side, not far from the bridge, there is the city of Foz do Iguaçu. The bridge, linking the south of the territory of Misiones in Argentina to the Parana State in Brazil, reduces the road distance between the most important economic centers of the two countries—São Paulo and Buenos Aires—by 300 km.

The construction of this bridge brought great development to the regions surrounding the three countries and increased tourism. A committee for construction called Comix began operating in December 1982. The complete design of the bridge was under the responsibility of Figueiredo Ferraz Consultoria e Engenharia de Projetos SA. The construction was done by a consortium of firms in Brazil and Argentina, Sobrenco and Supercemento Supervision, and monitoring and management were under the responsibility of another consortium of companies from the two countries, Consulbaires, Etel, and Figueiredo Ferraz. The authorization to begin work was given in January 1983 and the total duration of the construction was 35 months.

The bridge has a total length of 480 m with a main span of 220 m and two side spans of 130 m and is one of the longest in the world. The superstructure cross section (Figure 4.22) of 16.5 m consists of a single cell 8 m wide and height varying from 3.8 to 12.3 m. The top slab of the cell is the roadway, which has, besides the 8 m of the cell, two 4.25 m cantilevers.

The bridge was built by balanced cantilever construction (Figure 4.23) with segments 4 m long. The construction was initialized with two double symmetrical cantilevers of 110 m each, and continuity was

FIGURE 4.21   (See color insert.) Aerial view of the Tancredo Neves Bridge.

FIGURE 4.22 Internal view from box section.

FIGURE 4.23 Detail of progressive cantilever system.

provided with the addition of positive tendons. The remaining 20 m of lateral spans were concreted over scaffolding, and after completion, the continuity in the central span was held.

The substructure consists of two piers and two abutments. The piers have a square section of 8 m × 8 m, with wall thickness of 1.5 m in the first 26 m (Figure 4.24). Two parallel walls 8.0 m × 1.5 m extend upwards for more 22 m. This configuration allows a degree of mobility that supports the longitudinal displacement due to thermal effects, creep, and shrinkage of concrete. Each cap has 20 concrete piles 1.8 m in diameter carried out with a steel jack, 1 cm thick, nailed to the rock. The Brazilian side has water depth of 20 m, and therefore a drilling rig mounted on a floating island was used. After spiking the steel jack and rock drilling 4 m, the piles were braced and filled with concrete. On the Argentine side a landfill was made and then the steel jacks were located. Then, excavation and pile driving occurred simultaneously until rock was reached. Explosives were used in the final preparation of the bases, 4 m below the rock surface.

The girder prestressing was made using a Freyssinet system with cables of 22 Ø 12.7 mm strands. In the region of negative moments over the piers, 120 tendons were used. 42 tendons of continuity were used in the central span and 26 tendons in the side spans.

FIGURE 4.24   Detail of the main piers.

Each cable was post-tensioned with the force of 320 tf, providing an initial loss of 9% and a progressive loss of 18%, on average.

4.6.4.2 Deflection Control

During the construction of the segments, prestress was applied to the concrete at an early age (3 days), when it did not have the strength provided in the project to resist the prestressing. The modulus of elasticity was too small, resulting in deflections difficult to control analytically. The correction of the final elevation was done by parameter variations due to site measurements.

4.7 Cable-Stayed Bridges

4.7.1 Royal Park Complex, São Paolo (Eng. Catão Francisco Ribeiro)

This bridge created a new alternative entrance between Marginal Pinheiros and Jornalista Roberto Marinho Avenue, reducing the traffic of the crossing of this avenue and Luiz Carlos Berrini, and also the Morumbi Bridge. In the future, it should minimize the traffic of Bandeirantes Avenue, when Jornalista Roberto Marinho Avenue arrives to Imigrantes Road. There are two curves linking Jornalista Roberto Marinho Avenue and the express road of Marginal Pinheiros (passing over the Pinheiros River by stayed decks). These curves extend and overpass Luis Carlos Berrini Avenue. For the purpose of obtaining greater efficiency in the road system project, both curves were overlapped, so that the geometry of the main pole had to be in an "X" format.

The bridge has a total length of 2910 m with a main cable-stayed span of 580 m. The bridge carries four traffic lanes and has a width of 16 m. The stayed decks have a curvature with a span of 140 m over the United Nation Avenue and 150 m over the Pinheiros River. The prestressed concrete deck slab is 16 m wide, 48 mm thick, and is supported by two prestressed concrete main beams 1.5 m high.

Each one of the four stayed decks is supported by 18 pairs of stays, totaling 144 stays. The number of strands in each stay varies from 10 to 25. Besides the individual protection that each strand had, each stay has a high-density polyethylene yellow tube from which all of the strands evolve. The bridge consumed 375,000 m of stays (462,000 kg of steel).

The main pylon has a height of 132 m (Figure 4.25). Above the first clamping two rectangular columns with cellular cross sections depart, and at 11.4 m high, there is a new clamping with a prestressed concrete platform formed by two beams and one slab (from this platform the second deck departs).

FIGURE 4.25 Detail of the main pylon.

From the level of 23.4 m (the reference level is the upper surface of the foundation block), two leaning towers that are connected at the level 81 m depart, creating the joint of the "X." In this patch 42 m high are the anchorage devices of the stays.

The main pylon foundation is composed of 4 blocks supported by 28 drilled piles (diameter 41 cm), inserted in rock. From each of these blocks depart 4 pillars with 12 m height and rectangular cross sections, which are connected transversely by rizzed slabs and longitudinally by a prestressed concrete platform composed by two beams and one slab (from this platform departs the first deck).

The extreme supports are formed by pillars with rectangular cross sections and rounded corners, which support a beam that connects the pillar with the stayed deck and also supports the precast concrete beams that form the access of the bridge. The extreme supports from the stayed decks have 11 shafts 110 cm in diameter on one riverbank and on the other side, have 50 root piles 41 cm in diameter.

4.7.2 Bridge over the Guamá River, Pará (Eng. Catão Francisco Ribeiro and Eng. Fabrizio de Miranda)

This bridge (Figure 4.26) is over the Guamá River in Pará state, near the river mouth and 40 km from Belém, capital of the state. It is the third largest cable-stayed bridge in the country, after Hercílio Luz Chain Bridge (Florianópolis, 340 m span), and the cable-stayed bridge over the Paranaíba River (Carneirinho, 350 m of span).

The total length of this bridge is 1934 m. It consists of two approach sections of 719 m and 619 m, and a central cable-stayed section of 588 m. The main cable-stayed section consists of three continuous spans of 134 m + 320 m + 134 m. The superstructure in the main spans includes two main girders with rectangular crossbeams (1.45 m high) and a concrete deck slab. The approach sections include typical simply supported spans of 45 m with three precast prestressed I-girders. The precast slabs were positioned on the top of main girders and main concrete slabs were cast. The bridge deck is 12 m wide and 14.2 m wide for the approach and cable-stayed sections, respectively.

FIGURE 4.26   View of the Guamá River Bridge main span.

The bridge is part of the new highway connecting the main port of Vila do Conde state with the state capital, Belém. This highway cuts part of the Amazon forest and is very important to state development, because the ports of this region are not deep enough and are mainly used by small passengers and cargo ships.

The foundations are precast prestressed concrete piles with circular transversal sections measuring 80 cm in diameter and 15 cm of wall. The majority of these piles were battered with a ratio of 1:5. Pile caps were fabricated without joints (50 m) or within a special joint (metallic ring, when the length exceeded 50 m). The pylons (two each) were constructed on 2 concrete pile caps with 50 piles each and have a rectangular concrete section with rounded corners and a cellular cross section (115 m high) with 2 bracing beams, one of them under the deck and the other 50 cm above the deck. Two planes of stays were designed to support the deck.

The bridge over the Guamá River was constructed by the Novo Guamá joint venture, including Construbase, Paulitec, Pró-Base, and Cidade contractors. The project was managed by Consorcio Studio de Miranda and Enescil.

References

ABNT (Brazilian Association of Technical Standards) 1960 and 1984. NB6/43, NB6/60, NBR7188/84 e NBR 6118/03.

CIA. 2013. https://www.cia.gov/library/publications/ the-world-factbook/geos/br.html. Accessed May 25, 2013.

Enescil. Bridge over the Guamá River and the Park Royal Complex Bridge. Internal Files.

Engineering News Record Magazine, editions of September 16, 1948, p. 104, and April 05, 1945, p. 455.

Office Technician J. Carlos de Figueiredo Ferraz. Colombo Salles Bridge and Bridge over the Iguaçu River. Internal Files.

Sobrenco. Bridge over the Tocantins River. Internal Files.

Vasconcelos, Carlos, A. 1993. Brazilian Bridges: Notable Viaducts and Walkways. São Paulo, Brazil: Pini Publishing.

5

Bridge Engineering in Bosnia and Herzegovina

Boris Koboević
Faculty of Civil Engineering

Bisera Karalić–Hromić
Faculty of Civil Engineering Road Directorate of Federation

Damir Zenunović
University of Tuzla

5.1 Introduction

5.1.1 Geographical Characteristics

Bosnia and Herzegovina is a country in southeast Europe, on the Balkan Peninsula. Bordered by Croatia to the north, west, and south, Serbia to the east, and Montenegro to the southeast, Bosnia and Herzegovina is almost landlocked, except for 26 km of Adriatic Sea coastline, centered on the town of Neum (Figure 5.1). The country is mostly mountainous, encompassing the central Dinaric Alps. The northeastern parts reach into the Pannonian basin, while in the south borders the Adriatic.

There are seven major rivers in Bosnia and Herzegovina:

1. Sava is the largest river of the country, but it only forms its northern natural border with Croatia. It drains 76% of the country's territory into the Danube and the Black Sea.
2. Una, Sana, and Vrbas are right tributaries of Sava River. They are located in the northwestern region of Bosanska Krajina.

201

FIGURE 5.1   Geographical location of Bosnia and Herzegovina in Europe.

3. Bosnia River gave its name to the country, and is the longest river fully contained within it. It stretches through central Bosnia, from its source near Sarajevo to Sava in the north.
4. Drina flows through the eastern part of Bosnia, and for the most part it forms a natural border with Serbia.
5. Neretva is the major river of Herzegovina and the only major river that flows south, into the Adriatic Sea.

5.1.2 Road and Rail Networks

Bosnia and Herzegovina has about 22,500 km of roads categorized. The most important roads through Bosnia and Herzegovina are (Figure 5.2):

- Croatia-(Bos. Šamac/Brod/Vukovar/Orasje)-Doboj/Tuzla-Zenica-Sarajevo-
- Mostar-Ploce (Croatia)
- Croatia-Bihac-Banja Luka-Doboj-Tuzla-Bijeljina-Bos.Raca/Zvornik-Serbia
- Banja Luka-Jajce-Travnik-Zenica-Sarajevo-Gorazde-Visegrád-Serbia

The total length of railways is 1031 km, with industrial tracks to every significant manufacturing capacity. The most important railway lines are (Figure 5.3):

- Ploce (ports in Croatia on the Adriatic Sea)-Mostar-Sarajevo-Zenica-Doboj-Bosanski Šamac-Vinkovci (Croatia junction on the main railway Zagreb-Belgrade), and further to Central and Eastern Europe
- Bosanski Novi (with connection to Croatia)-Bihac-Prijedor-Banja Luka-Doboj-Tuzla (with a branch of the Brčko District/port of Brčko on river Sava)-Zvornik-Serbia.

Through Bosnia and Herzegovina stretch part of European route E73 or Corridor 5C, a branch of the fifth Pan-European Corridor, 340 km in length (Figure 5.4). E73 connects three countries, stretching from Budapest and Hungary via eastern Croatia, bisecting Bosnia and Herzegovina, and ending in the Croatian port of Ploče. Corridor 5C through Bosnia and Herzegovina includes the road Šamac-Doboj-Sarajevo-Mostar-Čapljina-Doljani, with an exit to the Adriatic Sea in Ploče, the railway Šamac-Doboj-Sarajevo-Mostar-Čapljina-Metković, waterways and quays on the Sava, and the Bosnia and Neretva Rivers.

5.1.3 Historical Development

Although the first organized road network in Bosnia and Herzegovina was recorded during the presence of Roman civilization from 229 BC to the 9th century AD, and although significant Roman roads passed through Bosnia and Herzegovina, almost none of the old Roman Bridge is preserved. The only visible parts of the structure are the remains of the bridge over the river Buna in Kosor (Figure 5.5). The bridge

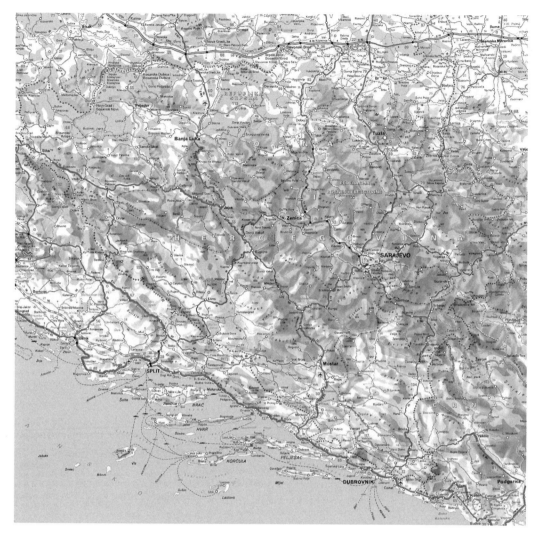

FIGURE 5.2 Road network of Bosnia and Herzegovina.

is an arch made of limestone and travertine, with seven spans of 2.6 m to 6.5 m, with a total length of 57 m and a width of 4 m. The bridge was destroyed in World War II in 1945 and today the only visible remains are a coastal arch structure and some piers.

The development of building bridges in Bosnia and Herzegovina began with the arrival of the Turks. The Ottoman Empire declared state interest in building bridges, which led to expansion of construction, particularly in the 16th century, when they built the largest and most important bridges. In this period all bridges were built of stone and wood. Wooden bridges were not preserved, because of their lack of durability, and they are based on stories that are transmitted for generations.

The most significant bridges of this period were the stone arch bridges. The characteristic elements of the the stone bridges are presented in Figure 5.6. One of the most famous is Old Bridge in Mostar, which is located in the old part of Mostar, listed on UNESCO's World Heritage List (Figure 5.7). The Mehmed Pasa Sokolovic Bridge in Visegrad was built by Koca Mimar Sinanin the period from the 1571 to 1577 (Figure 5.8). The bridge was built of tufa. The supporting structure consists of 8 river piers width 3.5 m to 4 m and length 11.5 m, which support 10 arches with spans of 10.7 m to 14.8 m. The total bridge length is 179.44 m. The width of the bridge with the boundary walls is 7.2 m (6 m + 2 × 0.60 m). The thickness

FIGURE 5.3 Rail network of Bosnia and Herzegovina.

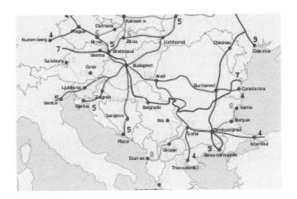

FIGURE 5.4 Pan-European transport corridors.

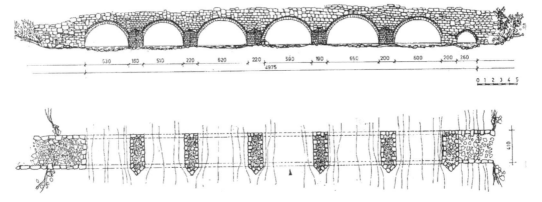

FIGURE 5.5 Graphic reconstruction of the bridge over the Buna River in Kosor.

FIGURE 5.6 Schematic axonometric view of characteristic elements of the stone Old Bridge.

FIGURE 5.7 **(See color insert.)** Old Bridge in Mostar.

FIGURE 5.8   Mehmed Pasa Sokolovic Bridge in Visegrad.

of the vaults (arches) in the crown is up to 85 cm. Figure 5.9 shows several bridges, Goat Bridge, near Sarajevo, the bridge at the mouth of Žepa, Arslanagića Bridge in Trebinje, the bridge in Plandiste, Šeher-ćehajin Bridge in Sarajevo, the Latin Bridge in Sarajevo, Old Bridge in Konjic, and Podgradska Bridge in Stolac, all built in this period.

Today there are about 3000 bridges in Bosnia and Herzegovina, of which 1000 are on the federal roads and the rest are regional and local roads. The approximate value of all bridges is 1 billion EUR. The bridges are of different ages, shapes, and structural systems, mostly built of concrete, stone, and steel. About 70% of the bridges were built in the period from 1955 to 1985, and 90% of bridges were constructed mainly of reinforced concrete and prestressed reinforced concrete. Other bridges are mostly composite steel–concrete. Thoses bridges destroyed during the war from 1991 to 1995 have been mostly reconstructed. Since bridges are generally made according to old Yugoslavia regulation (Temporary Technical Regulation), bridge reconstruction and their compliance with new regulations is implied.

5.2 Design Practice

Bridges are designed in accordance with the regulations of the former state of Yugoslavia (Serbia, Croatia, Bosnia and Herzegovina, Montenegro, Macedonia, and Slovenia), which are still valid regulations together with European Union regulations. The chapter on Serbia (Chapter 16) discusses calculation procedure settings used to build bridges in the former state of Yugoslavia.

5.3 Girder Bridges

The largest number of bridges in Bosnia and Herzegovina are beam bridges and reinforced concrete and prestressed concrete girder bridges. The superstructure of the bridges consists of beam girders with an "I" or "T" cross section with a recessed deck or deck over a beam girder or box girder. Experience building bridges in Bosnia and Herzegovina has shown that the rational ranges of prestressed concrete girder bridges ranges from 25 m to 40 m. Thus, the superstructure of bridges built in situ or semiprefabricated

FIGURE 5.9 (a) Goat Bridge in Sarajevo, (b) bridge at the mouth of Žepa, (c) Šeher-ćehajin Bridge in Sarajevo, (d) Latin Bridge in Sarajevo, (e) Arslanagić Bridge near Trebinje, (f) Old Bridge in Konjic.

is prefabricated beam girders. Most concrete decks are built in situ. The substructure of the bridge consists of abutments and river piers, which are reinforced concrete piers with various forms, depending on aesthetic requirements.

It should be pointed out that the geomorphology and geological structure of the location of bridges in Bosnia and Herzegovina, which are mostly unfavorable, with substrates at depths greater than 5 m, requires frequent use of deep foundations on piles. Selected representative examples of concrete girder bridges in Bosnia and Herzegovina are discussed next, which include a range of input parameters and compliance of the designed solution with the geological parameters of sites, specific local conditions (this especially refers to bridges in urban areas), and existing regulations. Reconstruction of war-damaged bridges is also discussed.

5.3.1 Prestressed Concrete Girder Bridges

5.3.1.1 Aleksin Han Bridge

Aleksin Han Bridge is a continuous prestressed frame structure with span 35 m + 70 m + 35 m (Figure 5.10a). The bridge was completed in 1969 (Figure 5.10b) and the midspan was destroyed (Figure 5.10e) during the war in 1992. The river piers are 34 and 23 m high wall panels of constant width and variable thickness. The superstructure is presetressed two-cell box girder of varying heights and parabolic intrados. The end spans of the bridge were constructed with a fixed scaffold, while the midspan was built by the cantilever construction method (Figure 5.10d).

The bridge experienced a change of environmental conditions. Specifically, the construction of the Grabovica hydroelectric plant, 3 km downstream from the bridge, formed the accumulation in the canyon of the Neretva River, depth 35–40 m (Figure 5.10c). Reconstruction after the war was conditioned by the morphology of the site. The Implementation Force (IFOR) troops have set the steel superstructure (using the system of Mabey & Jonson) with a span of 70 m, and have thus established limited-speed traffic on the bridge. The midspan was reconstructed using the modified cantilever construction method with a working platform above, supported by using steel scaffolding supported on river piers.

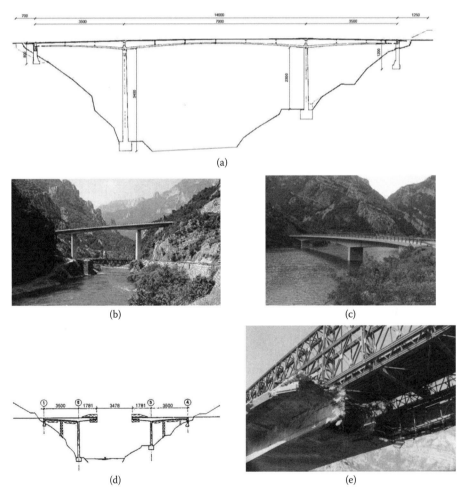

(a)

(b)   (c)

(d)   (e)

FIGURE 5.10   Aleksin Han bridge. (a) Elevation, (b) bridge in low water level, (c) bridge in high water level, (d) bridge under construction, and (e) bridge damage.

5.3.1.2 Ostrožac Viaduct

Ostrožac Viaduct (Figure 5.11a) is located in the center of the settlement Ostrožac and very nicely fits into the landscape (Figure 5.11b). The bridge is horizontally curved and crosses the valley with a continuous prestressed concrete box girder (Figure 5.11c) of five spans of 22 m + 3 × 28 m + 22 m, for a total length with parallel wings of 137.6 m.

5.3.1.3 Krečane Bridge

Krečane Bridge is a prestressed continuous box girder bridge with expansion and fixed bearings on the abutments (Figure 5.12a). It crosses the river Miljacka near Sarajevo, on the new route Sarajevo-Belgrade. The box girder is 1.7 m high and rigidly connected to the intermediate piers (Figure 5.12b). The bridge was built in 1979. The vertical alignment of the new road is significantly elevated in comparison to the existing old road, which resulted in several new bridges and tunnels. Part of the abandoned old road was turned into a promenade of the city of Sarajevo. Given that the bridge was located in the town and surroundings of the canyon Miljacka, a special request of the Urban Institute of Sarajevo was the formation of intermediate piers and their integration into the environment. A solution with a special fork shape was adopted (Figure 5.12c). The columns are supported on the foundations wells.

5.3.1.4 Korija Bridge

Korija Bridge is the largest and most interesting bridge of 12 bridges in the loop Korija connecting the bypass with regional road Sarajevo-Visegrad. The loop is situated in a rocky and picturesque canyon of the river Miljacka, about 0.5 km downstream of the Goat Bridge and along the route of the abandoned railroad Belgrade-Sarajevo. It is a special tourist route allowing quick access to the Old Town of Sarajevo (Bascarsija), as well as sporting and service facilities built during the XIV Olympic Winter Games in

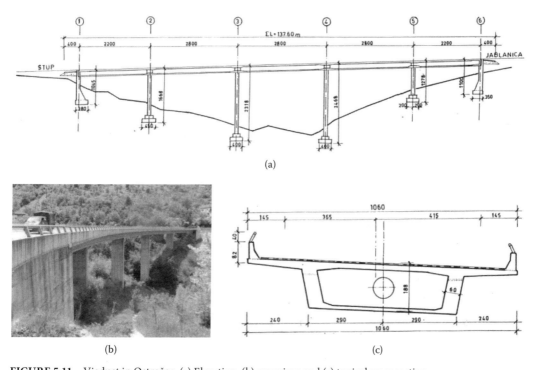

FIGURE 5.11   Viaduct in Ostrožac. (a) Elevation, (b) overview, and (c) typical cross section.

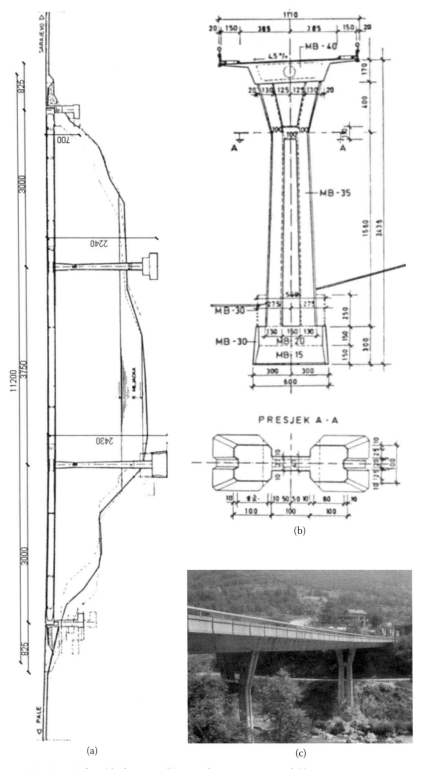

FIGURE 5.12   Krečane Bridge. (a) Elevation, (b) typical cross section, and (c) overview.

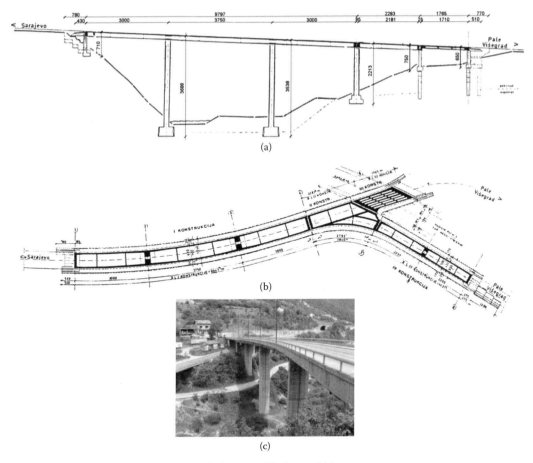

FIGURE 5.13 Bridge on the loop Korija. (a) Elevation, (b) plan, and (c) overview.

Sarajevo in 1980. The bypass part emerges from the tunnel Korija at the steep bank of the river Miljacka and crosses the river at a height of 40 m (Figure 5.13a and c). The Korija Bridge superstructure consists of four structural units (Figure 5.13b). The first and second units are the prestressed box girders of height 1.8 m, the third is a composite steel plate girder, and the fourth is a reinforced continuous beam. The second unit is particularly complex because it is in a forked part of the loop, and because very complex geometry was performed as a separate free-supported reinforced concrete structure with spans of 21.8 m and 20.3 m. The bridge has a total of three abutments, of which there are two shallow based and one based on bored piles Ø 1200 mm. Piers 2 and 3 are based on the foundation wells.

5.3.1.5 Koševo Viaduct

Koševo Viaduct was constructed at the beginning of 1980. It is located on the urban highway through Sarajevo, with a total length of 15 km. The highway route passes mainly through the central part of the city and links the roads from the northeast with those in the southwest. There are 5.5 km of bridge structures and 2.5 km of tunnels on the highway route due to geomorphological and urban conditions. Koševo Viaduct is a cast-in-place prestressed concrete structure of 12 spans ranging from 30 to 35 m (Figure 5.14a) with a total length of 381 m. The viaduct consists of five structure segments (Figure 5.14b). The superstructure of the bridge consists of multicellular box cross sections of constant height 1.52 m, which were adopted as the optimal constructive, rational, and aesthetic solution in urban

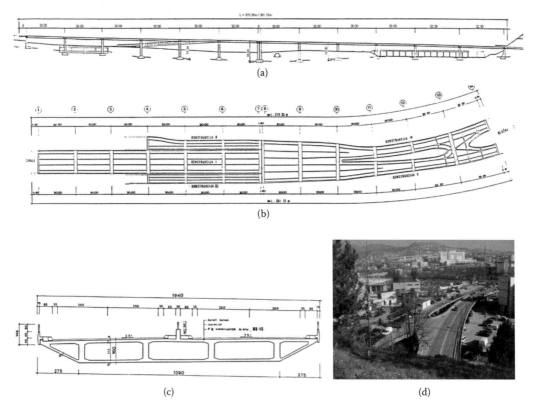

FIGURE 5.14   Koševo Viaduct. (a) Elevation, (b) plan, (c) typical section, and (d) overview.

areas (Figure 5.14c). Modern scaffolding and formwork (Figure 5.14d) were used. Because of the large number of piers, special attention was paid to their design. In all structures a cross section in the form of an elongated six corner section with contour dimensions 1.4 m × 2.2 m was adopted. At each location the number of piers depends on the superstructure width and moving from one to four columns.

5.3.1.6 Velešići Viaduct

Velešići Viaduct is a continuous prestressed concrete box girder bridge with four to six fields spanning from 24 m to 37 m (Figure 5.15a). It includes 11 structural units (Figure 5.15b). The first six structures carry mainly highway traffic, four structures are connecting ramps, and the eleventh structure carries street traffic. The box girder is 1.7 m high, with a variable number of webs from two on highway sections to three or four webs on connecting ramps (Figure 5.15c).

The deck is prestressed with transverse cables 6 Ø 7 mm. The total length of bridges on the Velešići Loop is 1752 m, while the total area is 1.8 ha. The piers are rectangular cross sections with fan form, extended to the upper part at the point of the support superstructure (Figure 5.15c) and supported on bored piles (Figure 5.15d). The Velešići Loop is an elongated rhombus, which requires a minimum of valuable urban space. It bridges the future city transversal and future railway station with 15 tracks. The bridge has a total of 7 abutments and 91 piers. The abutments are massive, with parallel wings, and are based on bored piles Ø 1200 mm at the beginning of the loop, or the shallow at the end of the loop.

5.3.1.7 Nedžarići Viaduct

Nedžarići Viaduct is a continuous prestressed box girder bridge with five spans of 25 m + 30 m + 35 m + 30 m + 25 m = 145 m (Figure 5.16a and b). The first phase, Stupska Loop in Sarajevo, crosses the tram

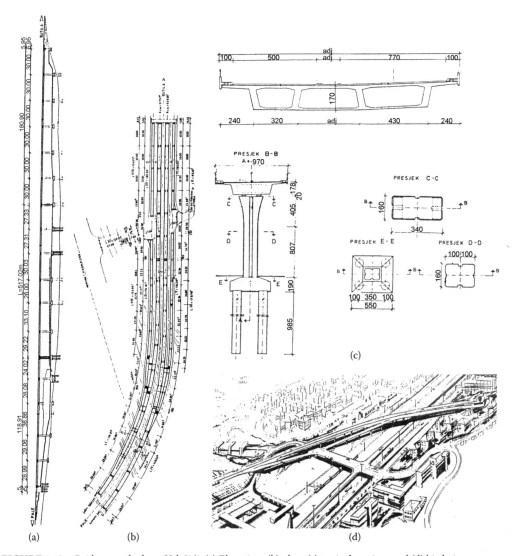

FIGURE 5.15 Bridges on the loop Velešići. (a) Elevation, (b) plan, (c) typical section, and (d) bird view.

line with two separate pavements (Figure 5.16c). The second loop, with another viaduct, which is parallel with the existing loop, and new branch loops have already been designed and are expected to start construction soon.

5.3.1.8 Bridge over the River Vrbas at Banja Luka

This bridge has a specific structural system of a two-pinned frame with ties spanning 78.8 m and a total length of 108.2 m with wings (Figure 5.17a). The superstructure is a box girder with three web parabolic intrados, slanted V-shaped forked twin pillars, and slanted ties partially hidden in the embankment (Figure 5.17b and c). The bridge was built in the form of duplicate structures connected by an elevated dividing lane. Foundations for the bridge were made on the spread footing on unfavorable soil layers of clay with medium to high plasticity. Given the low level of fixity in the slanting supports of the span structure, the bridge is a practical two-pinned frame with a joint formed at the intersection of column and tie. The system is insensitive to significant vertical settlement of footings.

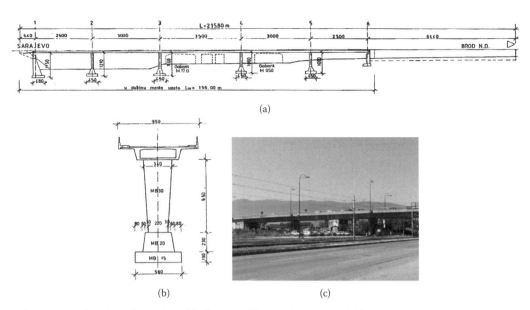

FIGURE 5.16   Stupskapetlja Viaduct. (a) Elevation, (b) typical section, and (c) overview.

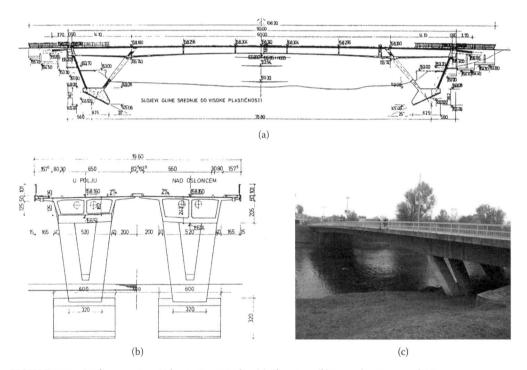

FIGURE 5.17   Bridge over river Vrbas in Banja Luka. (a) Elevation, (b) typical section, and (c) overview.

5.3.1.9 Bridge in the Canyon of the River Vrbas

The bridge is located in a picturesque environment immediately downstream of the beautiful Pliva waterfalls. Construction of the bridge began in May 1967 and opened to traffic in November 1968. The bridge consists of two parts aesthetically fitting the bridge silhouettes into the natural environment (Figure 5.18b). The first part is a continuous beam with spans 39 m+ 39 m and 34 m + 28 m + 22 m,

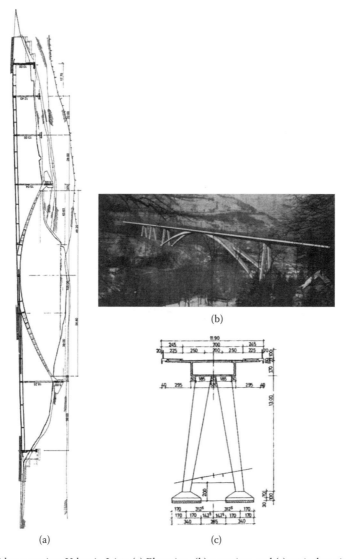

(a) (b) (c)

FIGURE 5.18 Bridge over river Vrbas in Jajce. (a) Elevation, (b) overview, and (c) typical section.

rigidly connected with the asymmetric two-pinned structures with span 105 m. The second part is a simple beam span of 17.7 m, which connects the first part with abutment on the right bank of the river Vrbas (Figure 5.18a). The continuous beams are box girder 1.7 m high and 5.2 m wide with 3 m cantilivers. The cross section of the arch is box, variable cross-section, height 1–2.50 m, width 1.2–5.2 m (Figure 5.18c). The total length of the bridge structure is 215 m. The bridge piers are from 12.4–18.3 m high with hollow trapezoidal cross sections, which narrow down.

5.3.1.10 Hasan Brkić Bridge over the Neretva River in Mostar

This bridge has spans of 11.25 m + 90 m + 11.25 m, but two short side spans are hidden behind the abutments. Both side spans are designed of the beam-tie by an angle of approximately 45° and anchor in the massive foundations of abutments, which were designed under the main span of 90 m (Figure 5.19a). The total length of the bridge is 112.5 m. The superstructure is a box with three chambers of variable height. The width of the box section is 13.74 m with cantilevers 3.75 m. The total width of the bridge is

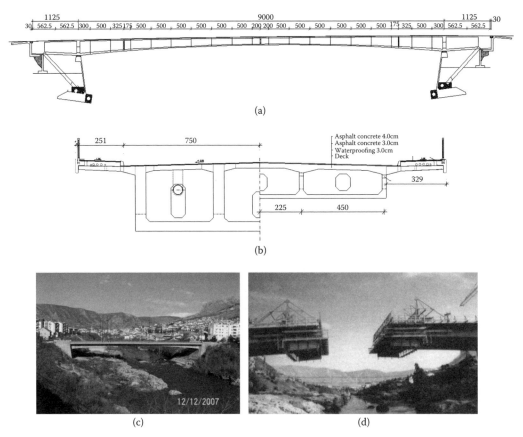

FIGURE 5.19 Hasan Brkić Bridge over river Neretva in Mostar. (a) Elevation, (b) typical section, (c) overview, and (d) under construction.

20.32 m. The bridge was built in 1980. This slim and elegant bridge was designed as a complex frame structure, as a result of urban conditions and requirements for bridging the river Neretva without piers (Figure 5.19c).

The main superstructure was constructed by the free cantiliver building method (Figure 5.19d). It was necessary to bridge the 90 m span with a variable-height structure. The height in the middle of the span is 1875 mm (1/25), and supports are 3625 mm (1/48), also caused by urban conditions (Figure 5.19b). The bridge superstructure including ties was destroyed on November 11, 1992. The bridge was rebuilt in 1996 using the remaining foundations and piers. In order to anchor the prestressing ties into the foundations of the existing abutments, concrete tie rods (two in the pair) were prefabricated and anchored into the existing foundation.

The remaining parts of the structure were in built on site: eight segment cantilivers 5 m long and the final piece 4 m long, which conected the left and right cantiliver structures, were built on both coasts. The span structure was prestressed with a DYWIDAG cable system with ropes of quality 1570/1770, with 1784 kN force in one cable. At each break, the end of the cantiliver segment is anchored with eight cables. Construction of the restored bridge took less than 11 months.

5.3.1.11 Bridge over the Neretva River in Capljina

This bridge consists of three parts (Figure 5.20a). The first part, on the east coast, is a continuous girder bridge with four spans of 4 m × 40 m. The cross section of the bridge consists of four precast T-beams. The deck is 7 m wide with 2.25 m footways on both sides (Figure 5.20b). The second part, crossing

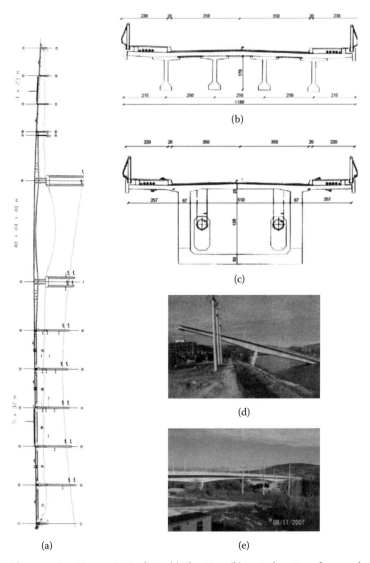

FIGURE 5.20 Bridge over river Neretva in Capljina. (a) Elevation, (b) typical section of approach spans, (c) typical section of main spans, (d) damage, and (e) reconstruction.

the Neretva River, is a continuous prestressed box girder of variable height with spans 40 m + 80 m + 40 m = 160 m (Figure 5.20c). The third part, on the right bank, is an I-beam structure with three spans 23 m long = 69 m.

The superstructure consists of precast T-beams which are continuous over the intermediate piers of subsequent prestressing. All of the span structures are made from prestressed concrete. Piers 1, 5, and 6 are based on wells and the others on foundation footings. The bridge was built in 1971 and was destroyed during the war in 1992 (Figure 5.20d). The present view of the reconstructed bridge is shown in Figure 5.20e.

5.3.1.12 Bridge over River Una in Kostajnica

This bridge is a simply supported system with five spans of 32.775 m + 3 × 32 m, 95 m + 32.775 m = 164.4 m (Figure 5.21a). It was the first prestressed concrete bridge, designed in 1964. The bridge is a replacement

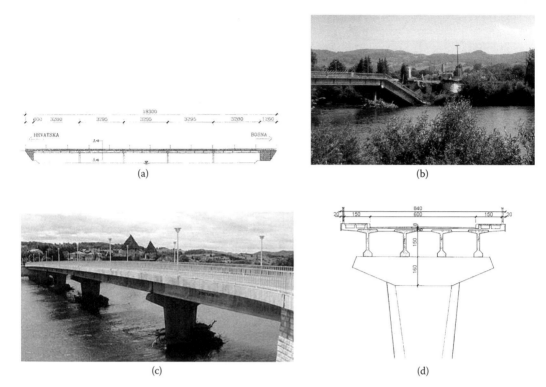

FIGURE 5.21 Bridge over river Una in Kostajnica. (a) Elevation, (b) damage, (c) overview, and (d) typical section.

of the old wooden bridge (Figure 5.21c). Superstructure consists of four main precast prestressed beams which together with partly recessed deck formedgrillagestructure (Figure 5.21d). The width of the bridge is 9.4 m. The abutments are massive reinforced concrete lined with stone. Piers are tapered from top to bottom in the transverse direction. The river piers are supported on caissons. During the war in 1991, the bridge was destroyed (Figure 5.21b). The bridge was rebuilt, creating a new main beam in the end fields associated with abutments and new equipment.

5.3.1.13 Bridge "Zli Brijeg"

Bridge "Zli Brijeg" near Kakanj crosses the river Bosnia, the Doboj-Sarajevo railroad, and local roads (Figure 5.22d). The superstructure with eight simply supported spans of 30.4 m + 6 × 30.8 m + 35.4 m = 250.6 m was built in 1974 (Figure 5.22a). In the early seventies in the former Yugoslavia, many precast prestressed concrete girder bridges were built. Evil Hill Bridge is one of the bridges built on the Highway Zenica-Sarajevo, the 65 km route crossing the river Bosnia 18 times, mostly under the oblique angle and with a relatively low-grade line. It is characterized by a high passing grade line. Six 30 m main girder spans in the curvature R = 700 m bridge the railway.

The superstructure for the bridging of the Bosnia River consists of four I-section beams with a concrete deck built on the site over a prefabricated plate that is the formwork and also an integral part of the section (Figure 5.22b). The medium-sized poles of the body are circular Ø 2 m beam headsets with a console at the top. The chosen form of piers is hydraulically advantageous in terms of obliquely intersecting rivers. The total width of bridge is 10.7 m.

For the purposes of European Corridor 5C, the second phase of bridges on Highway Zenica-Sarajevo was designed. The design also includes reconstruction of existing bridges in terms of their compliance

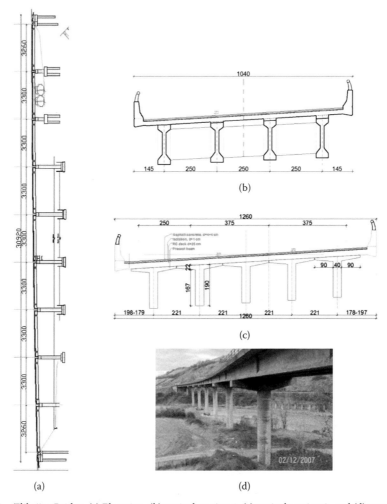

FIGURE 5.22 Zlibrijeg Bridge. (a) Elevation, (b) typical section 1, (c) typical section 2, and (d) overview.

with new regulations. Thus, new decks on the existing bridges established static continuity over the intermediate piers, and structures were strengthened with carbon strips and fibers. Individual intermediate piers and their foundations were also strengthened. The total width of the bridges of the second phase is 12.6 m (Figure 5.22c). The axis of the second-phase structure is parallel with the first-phase structure. The second phase of the Evil Hill Bridge has nine spans of 32.6 m + 7 × 33 m + 32.6 m = 296.2 m.

5.3.1.14 Bridge No. 6 over the Bosnia River

This bridge is horizontally curved with an "S" shape and radius R_r = 2400 m and R_l = 2000 m. The bridge consists of four structures. The first structure along the left bank of the Bosnia River has a span of 30.4 m + 3 × 30.8 m = 122.8 m. The second structure crosses the Bosnia River with a span of 5 × 30.8 m = 154 m. The third structure on the right bank of the Bosnia River spans 5 × 30.8 m = 154 m. The fourth structure recrosses the river with a span of 4 × 30.8 m + 30.4 m = 153.6 m. The total length of the bridge is 598.18 m (Figure 5.23a).

All of the span structures consist of four main precast prestressed girders with cast-in-place concrete decks above the main girders (Figure 5.23c). The total width of the bridge is 10.7 m. The abutments are

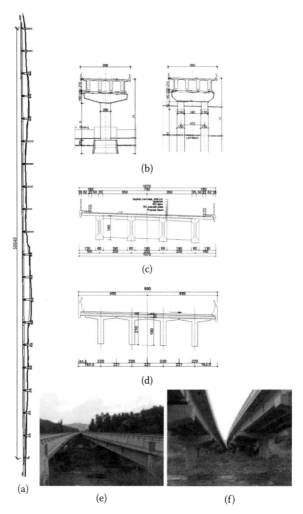

FIGURE 5.23 Bridge No. 6. (a) Elevation, (b) typical section 1, (c) typical girder section 1, (d) typical girder section 2, (e) overview, and (f) substructure.

with parallel wings. Intermediate columns are circular Ø 2 m, with a cantiliver headset beam at the top, based on wells in the river or out on the footings (Figure 5.23b). The third structure is based on piles 2 × Ø 1500 mm (Figure 5.23b and f).

The first phase of the bridge, on the right bank of the river, and the Zenica-Sarajevo railway were built in 1974. (Figure 5.23e). Design and construction of the second phase of the bridges on this section followed 30 years later and are aligned parallel to the first phase with the same span of 30 m. The cross section of the superstructure is made from four T-beams with a deck above (Figure 5.23d).

5.3.1.15 Bridge in Jošanica near Sarajevo

Disadvantages of previous precast prestressed concrete girder bridges are overcome in the reconstruction of the bridge in Jošanica near Sarajevo, which crosses the river Bosnia, railway, and highway. The original bridge was built in 1969 (Figure 5.24a and c). Reconstruction of the bridge was performed in 2010. After 40 years the main problem is degradation of the joint connections in the deck above

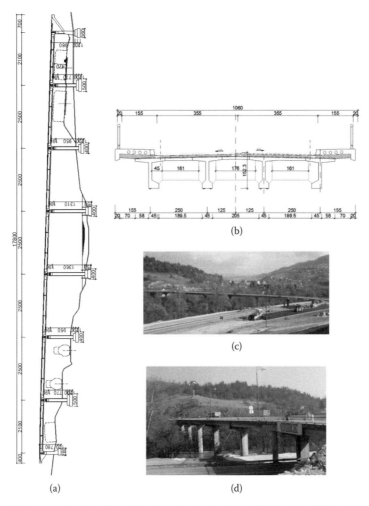

FIGURE 5.24 Bridge over river Bosnia in Jošanica, close to Sarajevo. (a) Elevation, (b) typical section, (c) overview, and (d) substructure.

the intermediate piers, and the passage of aggressive water to the pier cap beams and intermediate piers, or the passage of water through permeable dilatation devices, and deterioration of abutments (Figure 5.24d). Also, the bridge equipment must be replaced. Continuity of the deck was established by adding a new 10 cm thick plate, coupled with the existing plate, with additional reinforcement to take over the tensile stress from the intermediate piers (Figure 5.24b). On that occasion the cantilever pier cap beams and piers were also repaired with the addition of reinforcement and prestressing beams.

5.3.1.16 Gradište Bridge

Gradište Bridge near Ostrožac, on the road Konjic-Jablanica, which crosses a deep valley (Figure 5.25c, d), is a continuous girder over intermediate piers for additional dead load (layers above the bearing structure and equipment of the bridge) and live load (Figure 5.25a and b). Due to the poor geotechnical conditions of the soil, the foundations of all columns, except abutment 1 and intermediate pier 2, are supported with drilled piles Ø 1200 mm (Figure 5.25a).

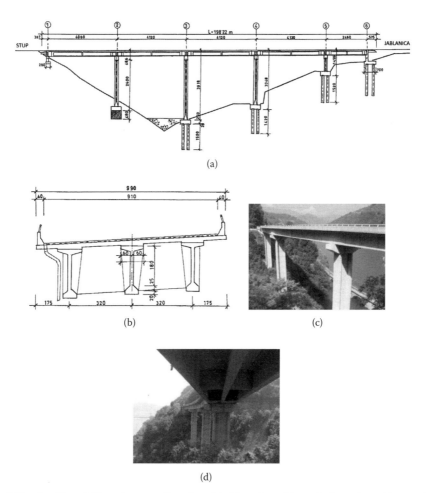

FIGURE 5.25 Gradište Bridge close to Ostrožac. (a) Elevation, (b) typical section, (c) overview, and (d) substructure.

5.3.1.17 Bridge over River Usora in Karuše near Doboj

This bridge is a precast prestressed concrete girder bridge with seven spans of 30.6 m + 5 × 31.2 m + 30.6 m = 217.2 m (Figure 5.26a), with a width of 12.67 m. The superstructure is in counterflow circular curves of radius R = 1450 m and consists of three prefabricated double-T prestressed girders with 2.10 m high and a deck thickness of 22 cm (Figure 5.26b). The main girders are connected with a continuous deck 22 cm thick over intermediate piers, which can take over the necessary actions arising from the permanent load of the second phase and the live load. Abutments are massive with parallel wings. The intermediate piers of the bridge are circular cross section Ø 2 m. Piers 2, 3, and 7 are directly based on the footings, and piers 3, 4, 5, and 6 on the wells (Figure 5.26c). The bridge was built in 1998 at the site of the existing reinforced concrete bridge, which had a sharp curve and caused accidents.

5.3.2 Composite Girder Bridges

5.3.2.1 Bridge over the Bosnia River in Zenica

The original bridge over the Bosnia River in Zenica was built on the main roads linking the city center with the new settlement Crkvice and had nine spans of approximate 17 m, with massive piers, which

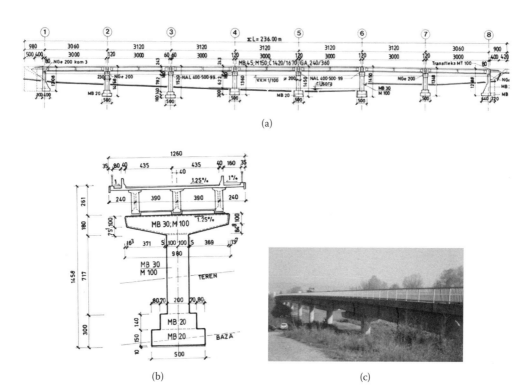

(a)

(b)   (c)

FIGURE 5.26 Bridge over river Usora in Karuše, close to Doboj. (a) Elevation, (b) typical section, and (c) overview.

significantly narrowed the profile of the water flow. The superstructure was made of rolled steel sections and timber. In the town of Zenica steel industry is located and new bridge is designed as a continuous prestressed composite steel structure with three spans of 45.00 + 60.00 + 45.00 = 150.00 m (Figure 5.27a and c). The total width of the bridge is 12.00 m. Abutments are covered with massive stone (Figure 5.27b). Intermediate piers are based on the wells, and were resolved as a separate (Figure 5.27d).

5.3.2.2 Bridge over the Bosnia River in Maglaj

This bridge is a steel truss structure of 2×77.85 m and width of 4.7 m. The bridge was built in 1912 and closed due to deterioration of traffic. An attempt to remedy the problem with Bailey bridge proved to be very expensive, due to the large lease. Therefore it was decided to build a new bridge at the same location (Figure 5.28b and e). The new bridge is a prestressed continuous composite structure with three spans 45 m + 60 m + 45 m = 150 m (Figure 5.28a). The total width of the new bridge is 12 m (Figure 5.28b). The axis of the bridge is straight and the finished level is a vertical convex curvature R = 4690 m. The abutments are directly based, and the intermediate piers use wells. The middle span of the bridge was destroyed during the war (Figure 5.28d). The bridge was successfully rehabilitated and opened to traffic.

5.3.2.3 Bridge over the Bosnia River in Zepce

The Zepce Bridge (Figure 5.29d) is a continuous composite prestressed steel girder bridge with three spans 30.8 m + 38.5 m + 30.8 m = 100.1 m (Figure 5.29a and b). The steel structure was assembled wholly in the river embankment behind the abutment using rollers and winches pulled on the piers of the bridge. The main girders are of constant height (Figure 5.29b). In the midst of all spans are yokes for

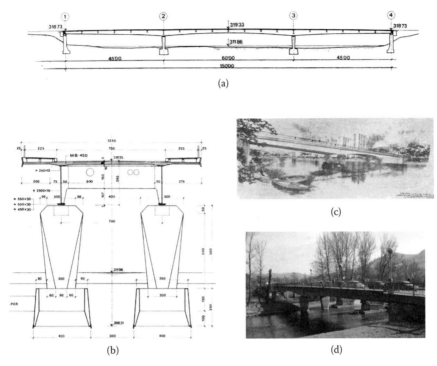

FIGURE 5.27   Bridge over river Bosnia in Zenica. (a) Elevation, (b) typical section, (c) rendering, and (d) overview.

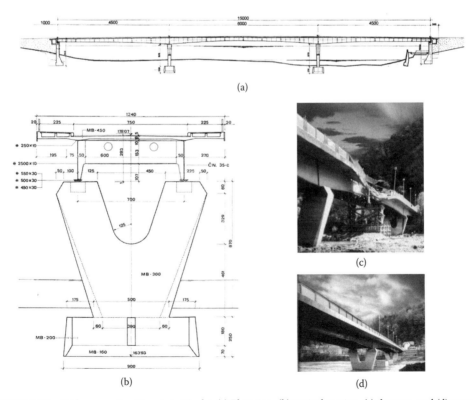

FIGURE 5.28   Bridge over river Bosnia in Maglaj. (a) Elevation, (b) typical section, (c) damage, and (d) overview.

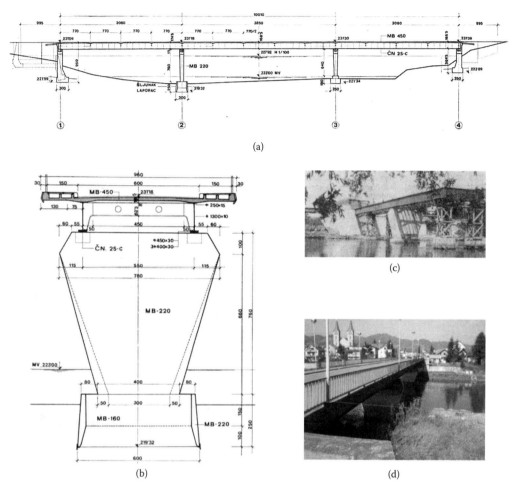

FIGURE 5.29 Bridge over river Bosnia in Zepce. (a) Elevation, (b) typical section, (c) under construction, and (d) overview.

launching the steel structure and the coupling of the bridge's own weight. The construction of the yoke is in the form of two independent towers below the main girder (Figure 5.29c).

5.3.2.4 White Bridge

The White Bridge crosses the backwater accumulation of the Salakovac hydro plant (Figure 5.30d). The original bridge was a simply supported precast prestressed concrete girder bridge with five spans 20 m + 3 × 36.2 m + 35.6 m = 164.2 m (Figure 5.30a). Bridging of backwater of accumulation hydro plant "Salakovac" is done by the construction of the bridge "White" (Figure 5.30b). The intermediate piers are the double-T cross section with cantilever cap beams. The piers are supported on spread except that the pier 4 is is based on a well. The fifth span, to Mostar, was demolished in the 1992 war (Figure 5.30c). The three middle spans and two intermediate piers were subsequently destroyed in the 1993 war. The bridge was reconstructed from 1995 to 1996. The new superstructure consists of two main steel girders and cross girders that make a grillage structure, which is coupled with reinforced deck. The total width of the bridge is 9.6 m.

5.3.2.5 Celebici-Lisicici Bridge

This bridge is a continuous composite girder bridge with 11 symmetrically arranged spans as follows: 32 m + 39.5 m + 7 × 55 m + 39.5 m + 32 m = 528 m (Figure 5.31a and c). It crosses Jablanica's

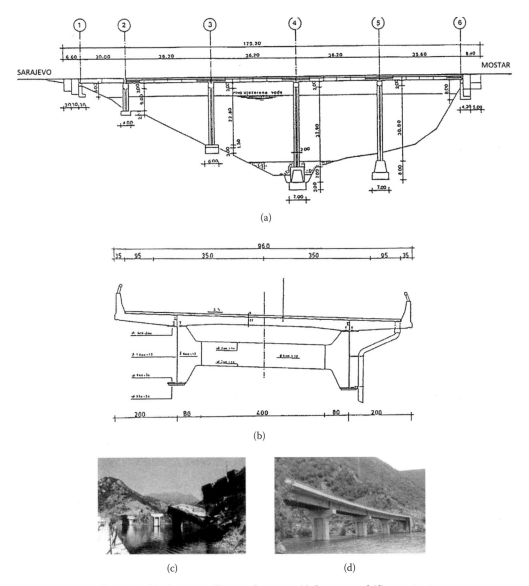

FIGURE 5.30 Bijela Bridge. (a) Elevation, (b) typical section, (c) damage, and (d) overview.

hydroelectric dam lake, with a water depth of 10–20 m, linking the densely populated right bank with the main road Sarajevo-Mostar (Figure 5.31d). All piers of the bridge are supported indirectly on bored piles Ø 1500 mm.

In the design phase two variants were proposed: a continuous composite structure (Figure 5.31a) and a suspended structure (Figure 5.31b). In conditions where the length of the bridging is approximately 500 m, with a water depth of 10–20 m, and the supporting soil at a depth of 5–25 m, this has has been adopted as a technically and economically favorable solution.

5.3.2.6 Bridge over the Sava and Una Rivers at Jasenovac

The original bridge consisted of seven independent continuous prestressed beams (Figure 5.32a). The total length of the bridge structures was 676.25 m + 129.1 m and the width was 2 × 3.5 m = 7 m (Figure 5.32b). The piers are made of reinforced concrete and founded on bored piles Ø 1200 and

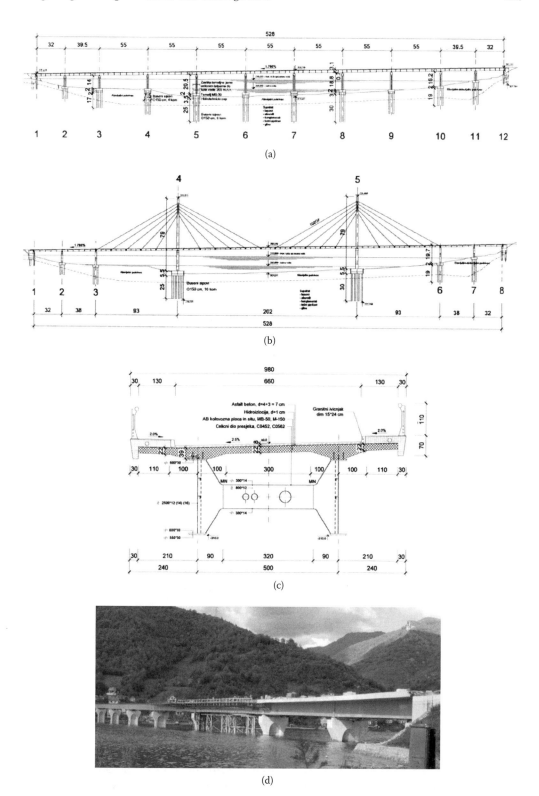

FIGURE 5.31 Bridge over Lake Jablanica. (a) Approach span elevation, (b) main span elevation, (c) typical section, and (d) under construction.

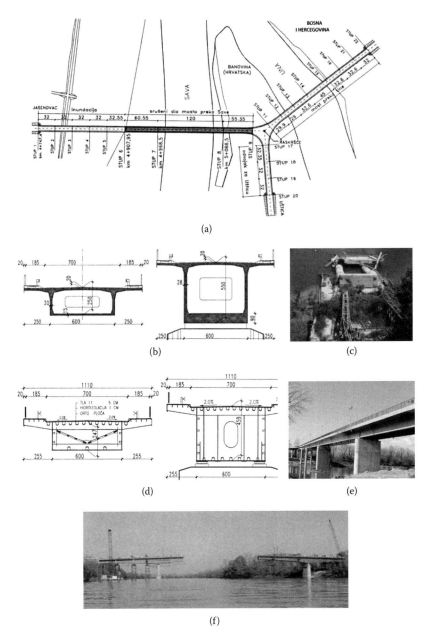

FIGURE 5.32 Bridge over rivers Una and Sava in Jasenovac. (a) Plan, (b) typical section 1, (c) damage, (d) typical section 2, (e) overview, and (f) under construction.

Ø 1500 mm. The bridge linking Bosnia and Herzegovina and Croatia was built from 1971 to 1973 and was destroyed twice during the war. Part of the structure on the left bank of the Una River and across the Sava River (Figure 5.32c) was first demolished. The bridge over the Una River was reconstructed using much lighter welded steel composite structures (Figure 5.32e). The main span structure and associated piers over the Sava River were reconstructed after the reconstruction of the bridge over the Una River. The superstructure is a steel orthotropic plate (Figure 5.32d). The side spans were made over the temporary piers and central span of 120 m with the cantilever method (Figure 5.32f). The bridge was completely renovated and opened for traffic in 2005.

5.3.2.7 Bridge over the Sava River in Orašje

This bridge (Figure 5.33c) consists of three parts with a total length of 79,175 m. The main spans are a continuous steel frame with three orthotropic deck spans of 85 m + 134 m + 85 m = 304 m. The superstructure over the floodplain is continuous composite steel girder and reinforced concrete deck (Figure 5.33d). The span on the left bank is 32.8 m + 9 × 33.6 m = 335.2 m, and on the right is 3 × 33.6 m + 32.8 m = 133.6 m (Figure 5.33a, b). The total width of the bridge is 10.3 m. The axis of the bridge is straight and the finished level has a vertical convex curvature. Intermediate piers 12 and 13 in the river Sava are based on the caissons. All other intermediate piers are based on bored piles Ø 1200 mm. All piers are made from reinforced concrete. The main structure of the bridge over the Sava River was completely destroyed by war activities in 1992. The bridge was renovated in 1998–1999 with the same structure as the original design.

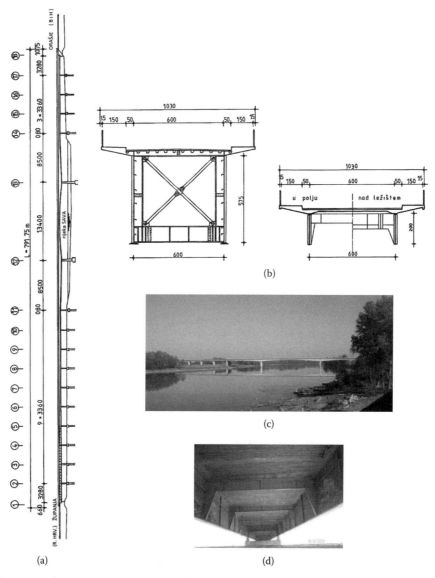

(a)   (d)

FIGURE 5.33 Bridge over river Sava in Orašje. (a) Elevation, (b) typical section, and (c) overview.

5.3.2.8 Gazelle Bridge over the Neretva River in Jablanica

The railway Sarajevo-Ploce was opened to traffic in 1965. The total track length is 195 km, with 108 tunnels with a total length of 36 km, 64 concrete and reinforced concrete bridges with a total length of 4.6 km, and 10 steel bridges with a total length of 0.9 km. The total length of the railway bridges and tunnels is 41.5 km, or 21.2% of the total length of the railway. These data indicate the complexity of construction works on the Sarajevo-Ploce railway line and very difficult geomorphologic conditions.

Among these bridges, the Gazelle Bridge over Neretva River in Jablanica in particular stands out (Figure 5.34c), with a total length of 300 m consisting of six structural units (Figure 5.34a). The approach structure in the direction of Sarajevo has four simply supported steel composite box girder spans of 36 m. The steel girders' height is 3.32 m and the reinforced concrete deck width is 5 m (Figure 5.34b). The railway sits in a road bed with wooden sleepers. The main span of the bridge is 121 m and crosses the river Neretva, and it is supported with angled piers with spans of 100 m, which are jointly supported on a special foundation (Figure 5.34d).

The superstructure of the main span is a steel box with a trapezoidal cross section with a height of 3.6 m and width of 2 m to 3.57 m (Figure 5.34b). The angled piers are also steel box sections, which are linearly tapered down, while the top has a circular haunch. The height of the piers is 40 m, and the height of the structures above the water of the Neretva River is about 55 m. In this part of the bridge there is no road bed, and the track was laid directly on the sleepers.

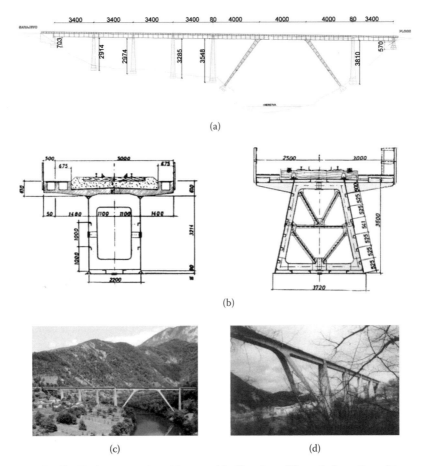

FIGURE 5.34 Gazelle Bridge over river Neretva. (a) Elevation, (b) typical section, (c) overview, and (d) substructure.

5.3.3 Steel Girder Bridges

5.3.3.1 Island Bridge

The Island Bridge crosses the broad and deep Jablanica Lake and connects the village on right bank of the Jablanica Lake with the main road Sarajevo-Mostar-Jablanica (Figure 5.35a). The superstructure of the bridge is a continuous steel box girder span 20 m + 140 m + 20 m = 180 m (Figure 5.35a). The short endspan represents a specific method of fixity. Negative reactions are transferred to the rocky ground with prestressed anchors. The total width of the bridge is 7.8 m. The height of the box girder in the middle section is 2.75 m, and over the abutments 5.5 m (Figure 5.35b). Abutments with coffer sections are made of reinforced concrete over the maximum possible water level. Assembling of the span structure was done using the free cantilever construction method with the use of boats on the lake (Figure 5.35c).

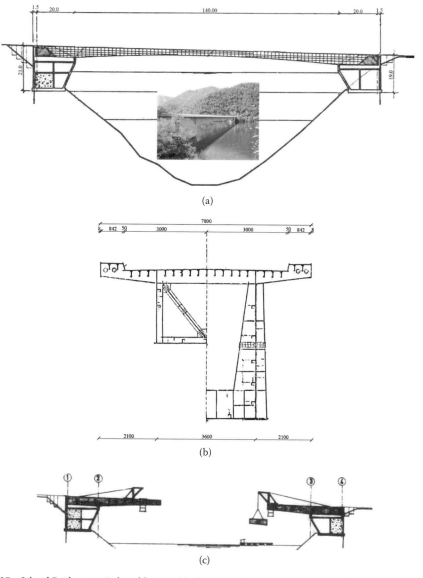

FIGURE 5.35 Island Bridge over Lake Jablanica. (a) Elevation, (b) typical section, and (c) under construction.

5.3.3.2 Bridge over Neretva River in Raštani

The main superstructure above the Neretva River is two continuous spans of 2×66 m and box girder of variable height of 3.2 m at the ends of up to 5 m over the intermediate support (Figure 5.36a) and carries the railway track Sarajevo-Mostar-Ploce (Figure 5.36c). The bridge was open with no road bed and direct support sleepers on a steel structure. The bridge was destroyed during the war in 1992 and was reconstructed from February 1995 to May 1996.

The new superstructure cross section of the bridge is rectangular instead of trapezoidal (Figure 5.36b). The orthotropic deck plate allows laying of wooden sleepers in the road bed. A temporary bridge made of heavy steel scaffold was built on the left bank of the river to assemble the Raštani Bridge. The segments were transported below the temporary bridge, lifted with a crane, and connected, and then racked over adjacent spans with a cantilever console with a span up to 55 m (Figure 5.36d). The last segment was erected on the right bank.

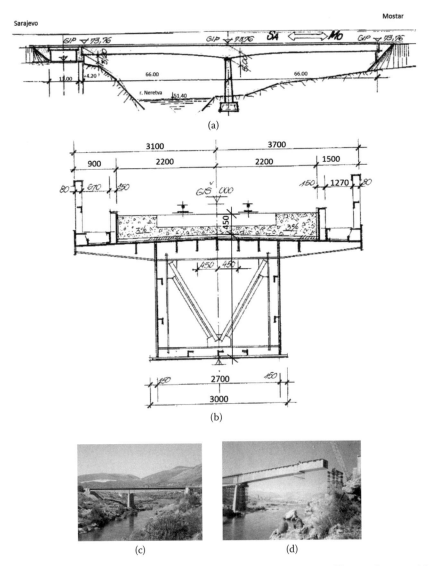

FIGURE 5.36 Bridge over river Neretva in Raštani, close to Mostar. (a) Elevation, (b) typical section, (c) overview, and (d) under construction.

5.4 Truss Bridges

5.4.1 Railroad Bridge across the Sava River between Bosnian and Slavonian Šamac

This bridge was built after World War II as an integral part of the railway Šamac-Sarajevo (Figure 5.37c). It is located on an important international railway corridor, Budapest-Sarajevo-Ploce. One the road Budapest-Sarajevo-Ploce, trans-European Corridor 5C passes over the bridge. Therefore, the bridge over the Sava River near Šamac is extremely important for the international relations of Bosnia and Herzegovina to the north–south and northwest–southeast (Corridor X).

The bridge was damaged during the recent war and was completely destroyed in 1993 (Figure 5.37d). The bridge was rebuilt in the period 2000–2002. Of the total length of the restored bridge 200 m is steel truss structure, which crosses the Sava River with three spans 55 m + 88 m + 55 m (Figure 5.37a). The remaining 400 m over the floodplain of the Sava River on the Bosnian bank is prestressed concrete

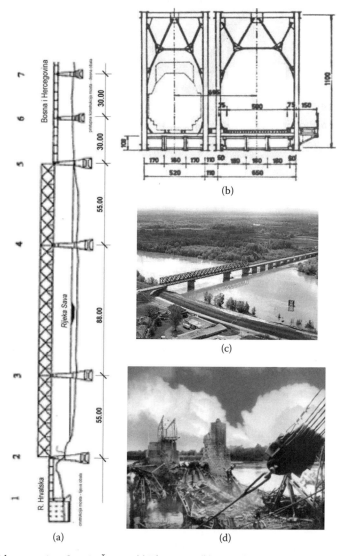

FIGURE 5.37 Bridge over river Sava in Šamac. (a) Elevation, (b) typical section, (c) overview, and (d) damage.

girder structure. The original bridge consisted of two parallel span structures for rail and road transport, which were based on common piers based on the caissons. The total number of piers is 16. The main structure over the Sava River was the first in Bosnia designed with three truss girders (Figure 5.37b).

5.4.2 Road Bridge over the Sava River in Bosanska Gradiška

The main bridge structure over the Sava River is continuous steel truss with a constant height of 6.5 m, with horizontal upper and lower chords, diagonals and bracing against the wind (Figure 5.38a). The piers are made of concrete. The central piers are based on caissons. The reinforced concrete deck is supported on the transverse and longitudinal secondary beams (Figure 5.38b). Side floodplain structures on both banks are made of reinforced concrete. It was built in late 1939 or the beginning of 1940. It was completely destroyed in World War II.

The new bridge was built in 1956 with a total length of 283.85 m (9.8 m + 93.795 m + 72.9 m + 72.9 m + 9.8 m) (Figure 5.38d). The steel truss of the first span, about 44 m in length, was destroyed in the 1992 war. Pier 1 was destroyed to 1 to 2 m above ground level with cracks in the remaining part. In addition, a floodplain reinforced concrete structure on the left bank and the associated abutment was also destroyed. Reconstruction of destroyed parts of the bridge was done in the eight phases, with one temporary pier. Reconstruction started on May 1, 1999 and finished on May 4, 2000.

5.4.3 Road Bridge over the Sava River in Brčko

This bridge was built in the late 19th century and reconstructed in 1986 (Figure 5.39a). The superstructure of the bridge is steel truss and consists of several types of structures of different spans (Figure 5.39b and c). On the left bank (Croatia) the bridge has 20 spans 20.34 m = 406.8 m. The superstructure of this

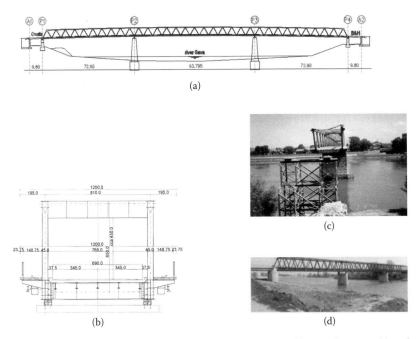

FIGURE 5.38 Bridge over river Sava in Bosanska Gradiška. (a) Elevation, (b) typical section, (c) under construction, and (d) overview.

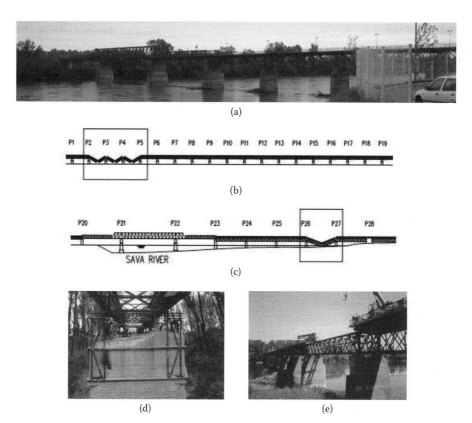

FIGURE 5.39 Bridge over river Sava in Brčko. (a) Reconstruction, (b) damaged spans 1, (c) damaged spans 2, (d) damaged piers, and (e) damaged trusses.

bridge is simply longitudinal steel girders with steel cross girders. The total width of the bridge over the left bank is 7.9 m. The middle part of the structure is a gerber steel truss structure. The central section is 64 m long with two 12 m overhangs on each side, on which the 35 m steel truss on each side length is supported, which makes a total of 47 m + 64 m + 47 m = 158 m. The total Width of the central part of the bridge (part over Sava River) is 9.2 m and width of the part of the bridge over left bank is 7.9 m.

On the right bank of the Sava River (Bosnia and Herzegovina) the bridge has five spans of 35 m and a span of 33 m = 208.3 m. The superstructure is composed of a simply supported steel truss system with transverse steel girders. Three spans of 20.34 m on the left bank (Croatia) and the third span on the right bank were destroyed during the war (Figure 5.39d). Mabey and Johnson temporary structures with spans of 70 m and 40 m were placed over the damaged areas (Figure 5.39e). Destroyed parts of the bridge were reconstructed in 2000. Complete reconstruction lasted 26 weeks. The superstructure remained the same.

5.5 Arch Bridges

5.5.1 Lučki Bridge over the Neretva River in Mostar

This bridge, also known as the Bridge of Mujaga Komadina, is located 200–300 m downstream from the famous Old Bridge and has a special aesthetic value harmoniously incorporated in the volatile profile of the River Neretva. It had a reinforced concrete fixed-arch span of 71.8 m. Dimensions of heel of the arch were 9.5 m × 1.6 m, and the crown of the arch was 8.2 m × 1.1 m. The width of the bridge was 12.1 m

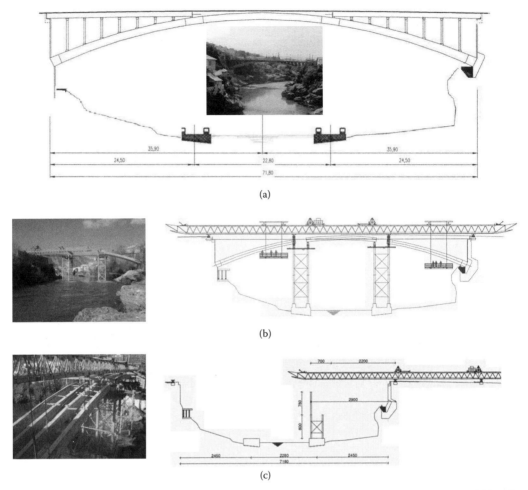

FIGURE 5.40 Lučki Bridge over river Neretva in Mostar. (a) Elevation, (b) under construction, and (c) under construction.

(Figure 5.40a). It was built in 1913 during the reign of the Austro-Hungarian monarchy and was the first reinforced concrete bridge in Bosnia and Herzegovina.

Unfortunately the bridge in Mostar was destroyed during the war in 1992 and was rebuilt in 2004 with strict adherence to the original layout and dimensions. The new arch is composed of four arched ribs made of three prefabricated segments in the longitudinal direction (Figure 5.40c). The arch rib is 0.5 m wide spaced from 2 m to 2.36 m and length 24.5 m + 22.8 m + 24.5 m. Two central yokes from prefabricated reinforced concrete columns with steel stiffeners supported on an existing foundation and another on a new foundation in the riverbed (Figure 5.40b) were built. Erection of the individual elements of yoke was performed by overthrust truss steel structures. Built yokes were used to support the overthrust structure (Figure 5.40b).

5.5.2 Concrete Jelovac Arch Bridge

Jelovac Bridge is one of five bridges with a total length 1050 m, on the slope route of railway line Sarajevo-Ploce in fromt of the Aries tunnel with length of 2850 m. It consists of three parts (Figure 5.41a) and fits in the natural environment (Figure 5.41b). The bridge is horizontally curved with R = 300 m. The finished level of the bridge is inclined 12.5%. The left approach structure to Sarajevo is a continuous

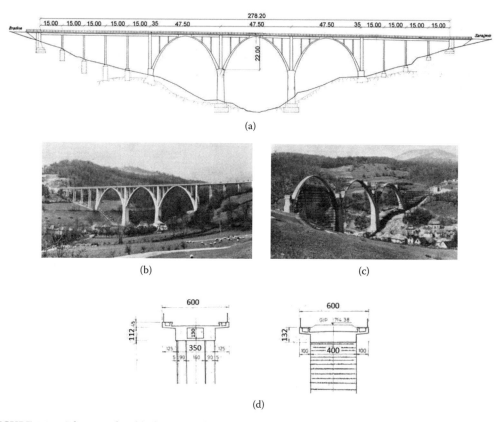

FIGURE 5.41 Jelovac Bridge. (a) Elevation, (b) overview, (c) under construction, and (d) typical sections.

reinforced concrete box girder with four spans of 15 m = 60 m. It has a height of 1.3 m and width of 4 m (Figure 5.41d). The total width of bridge is 6 m. The columns are dual with diameter 90 cm. The main structure was designed as continuous reinforced concrete vaults with axial distance of piers 3 × 47.5 m = 142.5 m (Figure 5.41d). The total length of the bridge is 308 m. The scaffold for the vaults and approach structure is piping systems (Figure 5.41c).

5.5.3 Concrete Arch Bridge over Neretva River in Jablanica

This bridge consists of three parts: approach structure on the right side, arched structure with a span of 52 m and continuous superstructure, and an access structure on the left bank, for a total length of 121.05 m (Figure 5.42a). Two fixed arch span of 52 m are stiffened with cross girders.

It was built in 1955 and was used for highway traffic and trains on Sarajevo-Ploce railway. With construction of the Sarajevo-Ploce railway track, the bridge was windened to the overall width of 9.99 m, in order to meet traffic conditions. The cantilivers of the existing bridge were removed and built a new with span of 3.3 m on both sides, which have over the new deck composed with the existing structure (Figure 5.42b). The bridge was reconstructed in 1968.

5.5.4 Customs Bridge

The bridge over Neretva River in Mostar has two arch spans, a larger span of 54 m over the Neretva River and a smaller span of 35 m over the floodplain on the left side (Figure 5.43a).

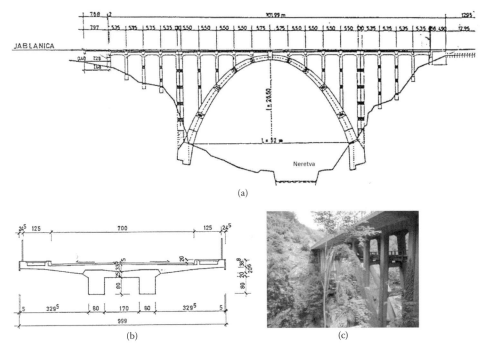

FIGURE 5.42   Bridge over river Neretva in Jablanica. (a) Elevation, (b) typical sections, and (c) overview.

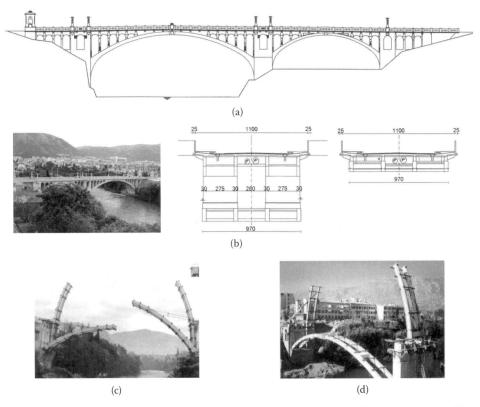

FIGURE 5.43   Customs Bridge over river Neretva in Jablanica. (a) Elevation, (b) typical sections, (c) under construction stage 1, and (d) under construction stage 2.

The reinforced concrete arch bridge was built in 1917 and the larger arch span was destroyed during the war in 1992. The entire bridge was reconstructed in its original form with an increased width of 11.5 m (Figure 5.43b) in 1996. Since the bridge was rebuilt in wartime, it was decided to use as much prefabricated construction technology as possible. All major precast components, such as half arches, piers, and plates, were prefabricated on a special erecting polygon away from the site, while other smaller components were prefabricated in a factory and shipped to the site. Special difficulties were that a drawing-out destroyed large parts of the structure from the bed of the river Neretva, as well as the rehabilitation of the abutment of the new arch. Half of the larger arch was transported to the site, connected by two and lowered using a hydraulic press at a definitive position (Figure 5.43c and d). Customs Bridge was the first rebuilt bridge in Mostar after the war and is the only link between the western and eastern parts of the city.

5.6 Cable-Supported Bridges

5.6.1 Pedestrian Cable-Stayed Bridge over the Bosnia River

This bridge consists of stiffening beams, pylon, suspenders, and abutments (Figure 5.44a). The stiffening beam, with total length 4×25 m $= 100$ m, is a box trapezoidal cross section (Figure 5.44b). The pylon

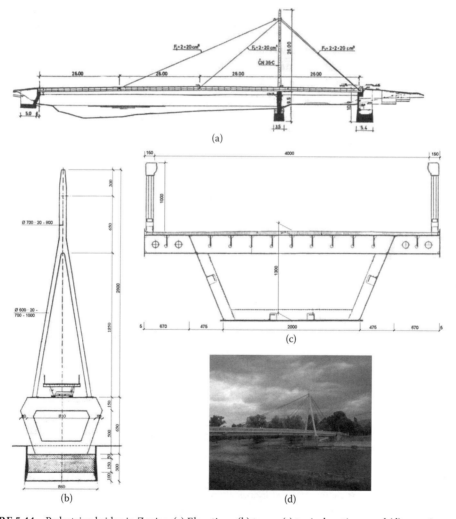

FIGURE 5.44 Pedestrian bridge in Zenica. (a) Elevation, (b) tower, (c) typical sections, and (d) overview.

(a)   (b)   (c)

FIGURE 5.45 Suspension bridge over Lake Jablanica in Ostrožac. (a) Overview 1, (b) overview 2, and (c) superstructure.

has an A-shaped form based on the well (Figure 5.44c). The suspenders are high-class parallel wires 52 Ø 7 mm, in a fan-shaped form. The abutments are reinforced concrete directly based on the footings. The two such bridges connecting the city center with Kamberovic Field, the future cultural and sports center, were built in 1985 in Zenica.

5.6.2 Suspension Bridge in Ostrožac

The construction of the Jablanica hydroelectric power plant, the first hydropower plant in Bosnia and Herzegovina after World War II, formed Jablanica Lake, which is the accumulation of water from the hydro power plant. In order to connect lake settlements on the right bank of the lake with the main road M-17, the Sarajevo-Mostar suspension bridge in Ostrožac was built, which has long been the only link between the inhabitants of the right bank with the world (Figure 5.45a).

The bridge in Ostrožac is a suspension bridge with a span of 207 m. The pylons are located on the bank of the lake and are made of reinforced concrete. The main cables are composed of five strands that are anchored in reinforced concrete blocks behind the pylon (Figure 5.45b). The bearing structure of pavement is formed from two-wall single-story Bailey elements with steel plates as the pavement surface, placed on a grill of steel beams (Figure 5.45c). The bridge is used for pedestrian and motor vehicle traffic. The total width of the bridge is 6.1 m.

5.7 Future Bridges

The highest activity in terms of building bridges in Bosnia and Herzegovina is related to the construction of European Corridor 5C. The basic design of the highway on Corridor 5C through Bosnia and Herzegovina was completed in 2006. Two innovations in relation to the current construction of bridges in Bosnia and Herzegovina came about due to the construction of Corridor 5C. The first is the application of thrust construction technology, as will be used for bridging the river Bosnia on route Johovac-Rudanka (Rudanka loop) (Figure 5.46). It is designed as a continuous prestressed superstructure with box girder spans of 30 m + 7 × 38 m + 30 m = 326 m (Figure 5.46). Prestressing of the span structure is carried out in two phases with cables 19 Ø 15.7 (St 1570/1770). Abutments and piers are supported on piles Ø 1500 (Figure 5.46). The superstructure consists of 17 segments with the station behind pier 10 and thrust from its left bank of the Bosnia River (pier 10) to the right bank (pier 1).

Another innovation is the semi-integral bridge structure with a continuous prestressed composite superstructure with plate cross section, running field by field on the scaffold (Figure 5.47). Two example bridges are the Bosnia Bridge in the loop Rudanka and bridges on the loop Butila (Sarajevo bypass). Bosnia Bridge is located at the junction of loop Rudanka and regional road Bosanski Brod-Doboj, and bridges the natural riverbed of Bosnia and the floodplain area along the right bank. The piers of the

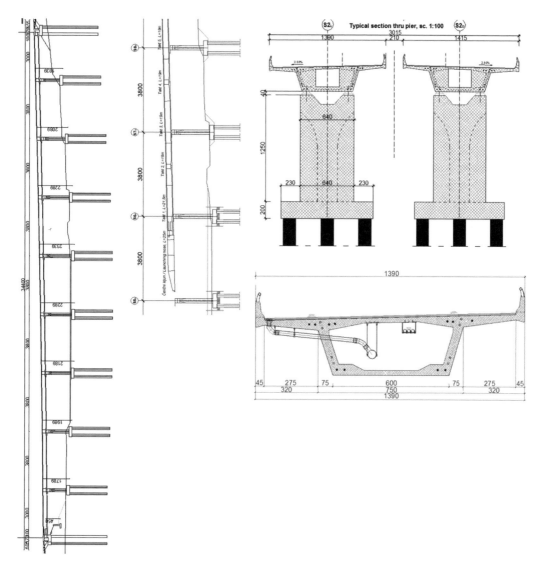

FIGURE 5.46 Bridge on the loop Rudanka.

bridge are based using bored piles Ø 1200 mm, with burial in geological substrate 6 Ø = 6 × 120 cm = 720 cm (Figure 5.47). Prestressing of the superstructure is made of 15 cable strands Ø 15.2 mm, which protects water through pipes Ø 90/97 mm quality (St 1570/1770). The structure is divided into eight segments to be performed in nine phases (Figure 5.47). An integrated monolithic superstructure was used to improve the performance of the joints in the superstructure, which was the weak point of the bridge in terms of durability.

Loop Butila on the Bosnia River is a significant and complex structure, Loop Butila was designed with traffic lines in three levels and four ramps to come on three levels, consisting of ramps A, B, C, and D (Figure 5.48a through e). The loop will take traffic on the Sarajevo bypass, route Zenica-Mostar. All four bridges were designed with integral semicontinuous composite prestressed structures with plate cross sections (Figure 5.48f through j). All the bridges are based on bored piles Ø 1200 mm. The total length of the loop Butila superstructures is 1991.96 m.

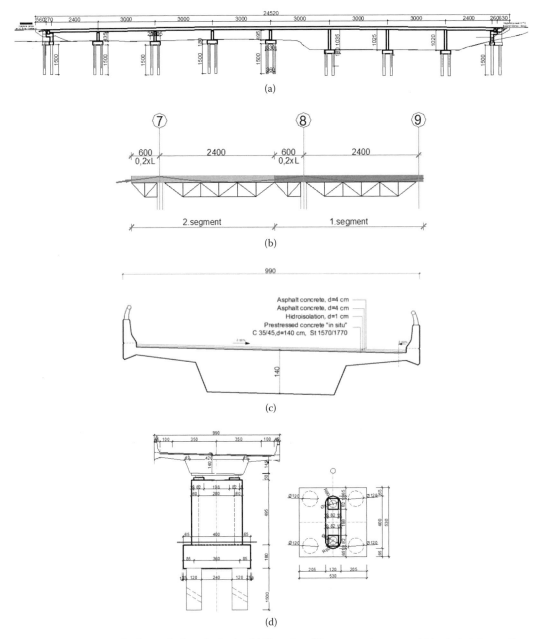

FIGURE 5.47 Bosnia Bridge on the loop Rudanka. (a) Elevation, (b) construction stages, (c) typical girder section, and (d) typical section.

5.8 Monumental Bridges

Old Bridge on the Neretva River in Mostar was built in the period from 1557 to 1566, and to this day is a symbol of the city of Mostar (Figure 5.49). The bridge was built by Mimar Hajrudin Aga, a student of the great Ottoman architect Koca Mimar Sinan, and by order of Sultan Suleyman the Magnificent. The bridge was built of limestone with the abutments in rock. The abutments are 6.53 m from the summer water level of the Neretva River, and at that point the span is 28.7 m with an arch arrow of 12.02 m

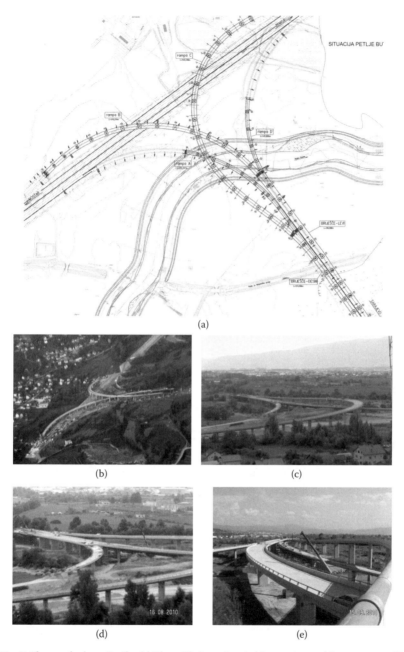

FIGURE 5.48 Bridge on the loop Butila. (a) Plan, (b) Overview 1, (c) overview 2, (d) overview 3, (e) overview 4, (f) elevation 1, (g) elevation 2, (i) typical girder section, and (j) typical section.

(Figure 5.49a). The thickness of the arch at the crown is 77 cm, with a width of 397 cm. With its harmonious shape of the white stone arch over the green river and its unique atmosphere, Old Bridge has impressed observers from around the world for centuries.

All of the bridges over the Neretva River in Mostar were destroyed in the recent war. The Old Bridge was demolished in September 1993 (Figure 5.49c). The city of Mostar signed an agreement on financing the reconstruction of the Old Bridge with the World Bank in 1997. The project was included by UNESCO. Special attention was given to creating designs and selecting appropriate scaffolds for the

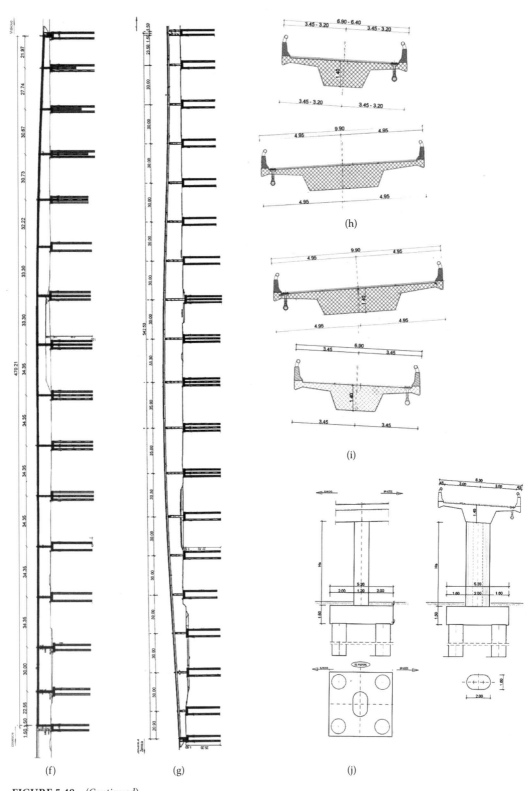

FIGURE 5.48   *(Continued)*

FIGURE 5.49 Old Bridge in Mostar. (a) Elevation, (b) overview, (c) destroyed bridge, (d) under construction, and (e) reconstructed bridge.

reconstruction of the bridge. Heavy metal scaffolding was selected for the massive structure, which was supported on a temporary cantilever of concrete columns that were later removed (Figure 5.49d). The opening ceremony of the reconstructed Old Bridge and Tower on the banks of the river Neretva was on July 23, 2004. After expending great effort and knowledge in the reconstruction, the new Old Bridge has excelled in all its glory, to the delight of Mostar's citizens and many visitors from around the world (Figure 5.49e).

6

Bridge Engineering in Bulgaria

Doncho Partov
Higher School of Civil Engineering "Lyuben Karavelov"

Dobromir Dinev
University of Architecture Sofia

6.1 Introduction

6.1.1 Geographical Characteristics

The Republic of Bulgaria is located on the Balkan Peninsula in southern Europe. The country borders five other countries: Romania to the north (mostly along the Danube River), Serbia and the former Yugoslav Republic of Macedonia to the west, and Greece and Turkey to the south. Black Sea coastline covers the east border of the country for 354 km.

Bulgaria has a territory of 110,879 km² and a population of about 7.2 million. The country's river system is characterized by relatively small rivers, with exception of the Danube. The major rivers are the Maritsa (480 km long), Iskar (360 km), Struma (415 km), Arda (241 km), Tundzha (390 km), and Yantra (285 km). Overall more than half of the runoff drains to the Aegean Sea and the rest flows to the Danube and the Black Sea.

The land transportation system of Bulgaria includes 4,294 km of railways (Figure 6.1) and 37,300 km of roadways (including 459 km of highways) (Figure 6.2). Nowadays, the owner of the railroad infrastructure

FIGURE 6.1   Railroad map of Bulgaria.

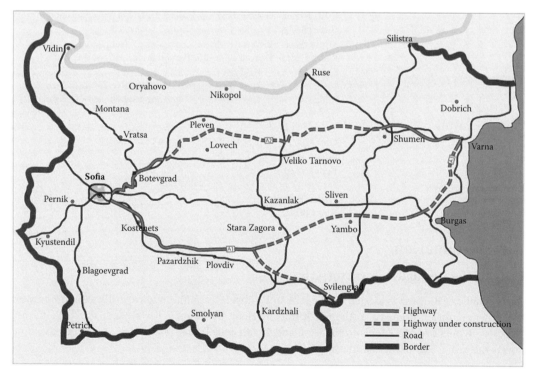

FIGURE 6.2   Road and highway map of Bulgaria.

is the Railway Infrastructure State Company and the roadway owner is the Road Infrastructure State Agency. According to the Road Infrastructure State Agency the total number of the road bridges in Bulgaria is approximately 5000. The Railroad Infrastructure State Company maintains about 623 concrete bridges and 327 steel bridges.

6.1.2 Historical Development

The modern development of the road and railroad infrastructure began in 1866 with the official opening of the 223 km long Rouse-Varna railway. A brief historical development is presented in Table 6.1.

The oldest stone bridges in Bulgaria date from the age of the Ottoman Empire (1386–1876). The arch bridges built during the fifteenth and sixteenth centuries are notable for their perfect proportions of the span-to-arch height ratio and the high quality of the stone masonry. The remarkable examples of this era are Kadin Most (see Section 6.4.1) over the Struma River with spans of 14.8 m + 20 m + 13.5 m, and

TABLE 6.1 Bridge Development in Bulgaria

Bridge Name	Year	Location	Designer	Main Features
Kadin Most	1470	Nevestino	Unknown	Three-span stone arch bridge; main span of 20 m; total length of 100 m
Dyavolski Most	1518	Ardino	Usta Dimitar	Three-span stone arch bridge; main span of 18.2 m; total length of 65.7 m
Old Bridge	1529	Svilengrad	Unknown	18-span stone arch bridge; main span of 18 m; total length of 286 m
Belinski Most	1867	Byala	Kolyo Ficheto	14-span stone arch bridge; main span of 12.6 m; total length of 276 m
Lom Bridge	1890	Lom	Unknown	Five-span truss bridge; main span of 23.5; total length of 117.5 m
Stambolov Bridge	1892	Veliko Tarnovo	Waagner-Biro Company	Truss arch bridge; span of 81.6 m; first modern bridge in Bulgaria
Tarnovo arch bridge	1956	Veliko Tarnovo	Patproject Company	Three-span concrete arch bridge; main span of 54 m; total length of 139.75 m
Balaban Dere arch bridge	1959	Studen Kladenec	L. Raykov, M. Minev	Three-span deck-stiffened arch bridge; spans of 26 m; first Maillart system bridge in Bugaria
Lakatnik arch bridge	1962	Lakatnik	M. Minev	Deck-stiffened concrete arch bridge; main span of 38 m; total length of 62 m
Opletnya arch bridge	1962	Opletnya	L. Raykov, H. Popov	Deck-stiffened arch bridge; main span of 80 m; total length of 212 m
Cherepish arch bridge	1964	Cherepish	M. Minev, I. Ivanchev	Two-span concrete arch bridge; spans of 40 m; total length of 105 m
Zverino arch bridge	1966	Zverino	D. Gocheva	Deck-stiffened arch bridge; main span of 50 m; total length of 141 m
Svilengrad steel bridge	1968	Svilengrad	B. Bankov, K. Kostov et al.	Seven-span steel girder bridge; main span of 36.5 m; total length of 212.5 m; first welded bridge in Bulgaria
Boyana overpass	1974	Boyana	Injproject Company	Three-span prestressed concrete box girder bridge; main span of 38.5 m; total length of 102.5 m
Varna Lake movable bridge	1975	Varna Lake	B. Bankov	Three-span truss bridge; main movable span of 31.62 m; total length of 80.34 m
Camerton viaducts	1976	Pirdop-Karlovo	L. Raykov, K. Tzekov et al.	Multispan concrete girder bridges; main span of 35 m; total length of 243 m

(Continued)

TABLE 6.1 *(Continued)* Bridge Development in Bulgaria

Bridge Name	Year	Location	Designer	Main Features
Asparuhov Bridge	1976	Varna	Patproject Company, M. Braynov	Multispan concrete girder bridge with a central steel box girder part; main span of 160 m; total length of 2050 m
Hemus viaduct	1977	Hemus Highway	Patproject Company	Multispan concrete girder bridge; main span of 22.5 m; total length of 425 m
Stamboliyski overpass	1980	Stamboliyski	Patproject Company	Multispan concrete girder bridge; main span of 25.6 m; total length of 449 m
Elin Pelin Bridge	1981	Elin Pelin	Gosha Company	Multispan steel orthotropic plate girder bridge; main span of 30 m; total length of 300 m
Gincy Bridge	1982	Gincy	I. Angelov	One-span girder bridge; span of 12 m; first in the world fiber-reinforced polymer (FRP) roadway bridge
Brussels Blvd. overpass	1983	Sofia	L. Raykov, M. Minev et al.	Multispan concrete box-frame bridge; main span of 49 m; total length of 2114 m; the longest bridge in Bulgaria
Brussels Blvd. steel bridge	1983	Sofia	B. Bankov	Two-span steel orthotropic cell box girder bridge; main span of 45 m; total length of 90 m
Trakia viaducts	1983	Trakia Highway	Patproject Company	Multispan concrete girder bridge; main span of 20 m; total length of 140 m
Saedinenie Bridge	1984	Plovdiv	M. Minev, I. Yakimov	Three-span concrete box girder bridge; main span of 86 m; total length of 174 m
Bebresh viaduct	1984	Hemus Highway	Patproject Company	Multispan concrete girder bridge; main span of 60 m; total length of 720 m
Hemus Highway viaduct 1	1986	Hemus Highway	D. Dimitrov, P. Staykov	Six-span continuous steel orthotropic plate girder bridge; main span of 72.5 m; total length of 410 m
Hemus Highway viaduct 2	1986	Hemus Highway, km 27.5	D. Dimitrov, P. Staykov	Three-span continuous steel orthotropic box girder bridge; main span of 162 m; total length of 362 m; the bridge with a longest span in BG
Hemus Highway viaduct, km 48.5	1986	Hemus Highway	Patproject Company	Multispan prestressed concrete girder bridge; main span of 40.5 m; total length of 402 m
Composite girder bridge in Varna	1986	Varna	B. Bankov	Two-span steel–concrete composite box girder bridge; main span of 44 m; total length of 88 m
Rogletz Bridge	1987	Rogletz	B. Bankov	Three-span steel truss structure; main span of 32 m; total length of 95.4 m
Rowing Canal Bridge	1989	Plovdiv	A. Georgiev, P. Petkov et al.	Three-span prestressed-ribbon footbridge; main span of 150 m; total length of 246 m; longest prestressed ribbon span bridge in the world
Rodopi overpass	1999	Plovdiv	Patproject Company	Multispan prestressed concrete girder bridge; main span of 46.59 m; total length of 129.53 m
Vartopa Bridge	1999	Sofia	P. Staykov and R. Mladjov	Five-span continuous steel orthotropic box girder bridge; main span of 62 m; total length of 272 m
Hemus Highway viaduct, km 48	1999	Hemus Highway, km 48	M. Minev, I. Yakimov et al.	Six-span prestressed concrete box girder bridge; main span of 140 m; total length of 656.2 m

TABLE 6.1 (*Continued*) Bridge Development in Bulgaria

Bridge Name	Year	Location	Designer	Main Features
Sofia Airport access bridge	2005	Sofia Airport	I. Lalov, M. Stakev et al.	Five-span concrete girder bridge; main span of 18 m; total length of 90 m
Makaza viaduct	2006	Rodopi Mountain	D. Kisov, M. Reshkov	Three-span box girder bridge; main span of 90 m; total length of 200 m
Varbica footbridge	2007	Momchilgrad	E. Dulevski, K. Topurov	Steel suspension footbridge; main span of 169 m
Chaya Bridge	2009	Katunica	E. Dulevski, K. Topurov et al.	Three-span continuous steel-concrete composite bridge; main span of 37 m; total length of 98 m
Lulin Highway viaduct, 3.5 km	2009	Malo Buchino	Mostconsult Company	Multispan prestressed concrete girder bridge; main span of 26.5 m; total length of 263.70 m

the Old Bridge (see Section 6.4.3) over the Maritsa River near Svilengrad, which has 18 arches with the longest span of 18 m. Some well-preserved arch bridges at the eastern part of the Rodopi Mountains are valuable historical heritage of the Middle Ages. Detailed research of the 15 bridges shows that most of the arches have an approximately semicircular shape and rise up to 8.5 m above the river. The arch spans vary from 4 m to 18 m and their widths vary from 2.2 m to 3 m. An example of these bridges is Dyavolski Most over the River Arda (see Section 6.4.2).

The Bulgarian Renaissance age (XVIII-XIX-th century) was remarkable, with the masterpieces of the self-educated builder Master Kolyo Ficheto. The Belenski Most (see Section 6.4.4) over the river Yantra, with 14 spans of 12.6 m, built in 1867; the seven-span stone bridge near Sevlievo with a total length of 110 m, built in 1857; and the wooden bridge over the river Osam in Lovech with six spans of 11 m are examples of the talent and experience of the local master.

During railway construction in Bulgaria after 1866, bridges were mostly steel trusses with riveted joints and spans between 20 m and 50 m. The bridge structures were imported from Belgium, Germany, France, Austria, and Hungary. The railway bridges of these years were completed with arch stone viaducts built by Italian and Belgium companies. Examples of these viaducts are the viaduct over the river Badechka near Zmeyovo, built in 1911 with a total length of 78 m and that includes six arches that rise 21 m above the river; the viaduct over the river Gabra near Vakarel, built in 1910 with a total length of 156.4 m and that includes eight 45 m high arches with spans of 15 m; and the viaduct near Klisura, built in 1952 with a total length of 120 m and that includes seven 36 m high arches with spans of 17 m (Figure 6.3).

The second half of the twentieth century can be described as a progress of bridge construction. The arch bridges built in these years were big challenges for Bulgarian engineers: the five-span bridge over the river Iskar near Gigen with the largest span of 50 m (Figure 6.4); the bridge over the river Roussenski Lom near Rousse with eight spans of 44 m built in 1957 (Figure 6.5); the bridge over the river Yantra near Veliko Tarnovo with spans of 44 m + 54 m + 44 m (see Section 6.1.2), built in 1956; and the arch bridge near the Koprivshtitsa railway station with a span of 67 m, built in 1952, are the representative viaducts of this age.

During the construction of the railway near the Studen Kladenec reservoir and the reconstruction of the railway Sofia-Mezdra, many deck-stiffened arch bridges (the Maillart bridge system) were built. The Lakatnik Bridge with a span of 80 m, built in 1962, best represents these railway bridges (see Section 6.5.3).

The construction of the double-deck steel bridge over the river Danube between Rousse and Giurgiu was completed in 1954. The total length of the bridge is 2360 m and includes 37 spans. The main part of the bridge over the river includes spans of 4 × 80 m + 2 × 160 m + 86 m + 2 × 160 m + 4 × 80 m. The middle span of the bridge is movable and can be lifted 20 m high. The bridge is a truss structure with bolted joints. The trusses were fabricated in Romania, Hungary, Czechoslovakia, Germany, and Poland. The bridge was built in 2 years (Figure 6.6).

FIGURE 6.3   Klisura viaduct.

FIGURE 6.4   Gigen Bridge. (Courtesy of Il. Ivanchev.)

The first precast concrete road bridge was built near Dragoman in 1968. The superstructure consists of precast prestressed concrete multicell box girders with cross-sectional dimensions of 0.98 m × 0.75 m, which are placed close to each other to form the bridge deck. The total length of the bridge is 18 m and the width is 12 m. During the 1970s the precast prestressed concrete girder was a common solution for road and the highway viaducts. The viaducts of the Hemus Highway (18 bridges) and Trakia Highway (5 bridges), the Asparuhov Bridge (Section 6.3.2), and the viaducts on the Pirdop-Rozino first-class road represent bridges of these years (Section 6.2.2).

FIGURE 6.5 Roussenski Lom Bridge.

FIGURE 6.6 Danube Bridge.

The precast prestressed concrete girders of these bridges have spans of 27 m and 39 m and weight of 1200 kN. The assembling of the superstructure was implemented by using the Italian launching truss girder. The girders of the Bebresh viaducts (Section 6.2.7) and Korenishki Dol at the Hemus Highway have a span of 58 m and weight of 2200 kN. The main girders were launched using the Bulgarian launching truss girder. The Hemus Highway has 27 viaducts with total lengths varying from 125 m to 714 m. The Trakia Highway has 24 viaducts with lengths varying from 126 m to 526.5 m. All highway bridges built between 1976 and 1990 were constructed by the Moststroy Company. In the late 1990s the typical Trakia Highway overpasses were precast concrete bridges. The main span of the overpass structure is 16 m with a total length of 62 m. The bridge is 9 m wide (Figure 6.7). Precast concrete bridge structures are used also as urban overpasses: the Stamboliyski overpass with spans of 20 m and a total length of 450 m (Section 6.2.4) and the Rodopi overpass (Section 6.2.8) with spans of 42.5 m and a total length of 139 m represent these urban structures.

Innovative construction methods were used to build bridges during the 1980s. The Saedinenie Bridge with spans of 44 m + 86 m + 44 m was constructed by "cantilever launching" (Section 6.7.1). The viaduct at the Hemus Highway with spans of 68 m + 130 m + 140 m + 130 m + 120 m + 68 m was built by cantilever concreting (Section 6.2.10). Several city overpasses in Sofia were completed by using the patented construction method called span-by-span concreting with lifting. Examples of these bridges are the Brussels Blvd. overpass with spans of 49 m and total length of 2200 m (Section 6.2.5); the pedestrian bridge in front of the National Palace of Culture (Figure 6.8); the bridges on the Tsarigradsko Shose city highway (Figure 6.9); and the curved bridge located at the Boyana–Sofia ring-highway interchange (Figure 6.10).

FIGURE 6.7   Trakia Highway overpasses.

FIGURE 6.8   National Palace of Culture Bridge.

FIGURE 6.9 Tsarigradsko Shose Bridge.

FIGURE 6.10 Boyana Curved Bridge.

The construction of steel railway bridges began in the late 1960s. The first welded steel plate girder railway bridge was built at Svilengrad with a total length of 395 m and opened to traffic in 1968 (Section 6.3.1). The bridge consists of two continuous girders with 10 spans of 36.5 m. The first steel continuous box girder railway bridge with spans of 42 m + 54 m + 42 m was built over the Iskar River. The box cross section has dimensions of 2 m × 3 m. The Varna Lake movable bridge, destroyed by a ship accident, was reconstructed in 1975 (Section 6.1.3). The bridge is a three-span structure with a movable central truss part. A new steel–concrete composite girder bridge used by Sofia city metropolitan railway with a total length 8 × 25.5 m = 204 m was opened to traffic in 2009 (Figure 6.11). The railway steel-concrete composite bridge over the Chaya River with a total length of 98 m was also opened to traffic in 2009 (Section 6.3.9). The composite overpass in Parvomay over the Plovdiv-Sofia railway with a total length of 94 m was opened for traffic in 2010 (Figure 6.12).

Steel orthotropic box girder bridges have been built since the 1970s. Examples of these bridges are the Asparuhov Bridge in Varna with a total length of 321 m, built in 1976 (Section 6.3.2); the Elin Pelin overpass with a total length of 300 m (Section 6.3.3); the central part of the Brussels Blvd. overpass with two spans of 45 m, built in 1983 (Section 6.3.4); Viaduct 1 (410 m long) and Viaduct 2 (362 m long) of the Hemus Highway (Section 6.3.5 and 6.3.6), built in 1986; and the Vartopa Bridge in Sofia with length of 267 m, built in 1999 (Section 6.3.8).

FIGURE 6.11 Iztok metropolitan bridge. (Courtesy of P. Staykov.)

FIGURE 6.12 Parvomay overpass. (Courtesy of E. Dulevski.)

A pedestrian Rowing Canal bridge was built in 1989 near Plovdiv (Section 6.7.2). The bridge is a prestressed reinforced concrete stress-ribbon structure with main span of 150 m, which is a world record for this type of structure. 17 bridges were constructed on Liulin Highway during 2007–2011 (Figure 6.13). The bridges are multispan, simply supported prestressed structures with maximum spans of 39 m and the total lengths varying from 240 m to 720 m. The new Calafat-Vidin Bridge over the Danube River is under construction (2009–2013). The bridge is an extradosed cable-stayed structure with a total length of 1790 m and main span of 180 m. The segmental construction method is used (Figure 6.14).

At the end of the 1970s, Bulgarian engineers built a lot of bridges and viaducts in Syria. Twelve steel–concrete composite bridges with spans of 34.5 m and piers 70 m tall, and 30 reinforced concrete bridges with maximum spans of 16.5 m were built at the Orintes-Latakia railway. In the 1990s in

FIGURE 6.13 Liulin Highway Bridge. (Courtesy of D. Kisov.)

FIGURE 6.14 Calafat-Vidin Bridge.

Zimbabwe Bulgarian builders constructed a composite steel–concrete bridge over the river Limpopo (Figure 6.15). The bridge is 16 spans of 33.8 m which consist of steel girders 1125 m long, produced in Bulgaria, and which are assembled to each other by high-strength bolts (Daalov and Daalov 2006).

6.1.3 Design Practice

In Bulgaria, there are no specific code regulations regarding bridge design. Prestressed concrete bridges are designed in accordance with German code (DIN 4227: 1988); reinforced concrete bridges in accordance with Russian code (SNiP 1974); and steel bridges in accordance with German codes (DIN 4114: 1961, DIN 1072: 1967, DIN 4101: 1970, DIN 1073: 1973, DIN 1079: 1979 and DIN 1045: 1988).

FIGURE 6.15   Limpopo Bridge. (Courtesy of T. Daalov. With permission.)

The Eurocode standard system (BDS EN 1990 Eurocode 0, BDS EN 1991 Eurocode 1, BDS EN 1992 Eurocode 2, BDS EN 1993 Eurocode 3, BDS EN 1994 Eurocode 4, BDS EN 1997 Eurocode 7 and BDS EN 1998 Eurocode 8) began to be implemented as the national standard in early 1990s. The Eurocode system was developed by the European Committee for Standardization (CEN) where Bulgaria is a member and has a position as an observer. All parts of the Eurocodes have been translated and harmonized with the Bulgarian standards and will be adopted as national standards in 2014.

6.2 Concrete Girder Bridges

6.2.1 Boyana Overpass

The Boyana overpass is a three-span (32 m + 38.50 m + 32 m) prestressed concrete continuous box girder bridge (Figure 6.16), which carries the Sofia ring-highway traffic. It is located at the Boyana–Sofia ring-highway interchange. The bridge was designed by Injproject with head designer L. Raykov. The bridge builder was Injstroy. The bridge owner is the Municipality of Sofia. It opened to traffic in 1974.

The bridge is 102.5 m long and 19.25 m wide. The superstructure is a five-cell box cross section with a depth of 1.3 m and two 3.6 m long overhangs. The box cells are 2.2 m wide. The walls of the box section in the longitudinal direction are each prestressed by seven tendons. The support zones of the superstructure are solid concrete structure which is additionally prestressed into the transversal direction by 10 tendons.

The bridge piers are single-column hammerhead bents. The dimensions of the column are 7.2 m × 1.25 m at the top, 4 m × 0.70 m at the bottom, and 9.2 m tall. The pier's footings have dimensions of 11.4 m × 2.2 m. The bridge abutments are solid concrete structures 4.65 m high. The abutment bearings are steel rollers and pier bearings are neoprene pads (Georgiev 2007).

6.2.2 Camerton Viaduct

The Camerton viaduct is located at the Pirdop-Karlovo first-class road (Figure 6.17). The designers were L. Raykov, K. Tzekov, G. Ganchev, K. Kotsev, and E. Milchev. The viaducts were built by Road Construction Company. The owner is the Road Infrastructure State Agency. The viaducts opened to traffic in 1976.

FIGURE 6.16   Boyana overpass.

FIGURE 6.17   Camerton viaduct.

The superstructure and the piers are of a standardized system, which is applied at km 219.641 and km 220.373 with spans of 33.93 m + 5 × 35 m + 33.93 m, and at km 221.059 with spans of 33.93 m + 7 × 35 m + 33.93 m. The main girders of the superstructure are precast prestressed concrete I-beams, 1.8 m deep. The top flange of the beams is 2.7 m wide and 0.15 m thick. The bottom flange is 1 m wide and 0.40 m thick. The web thickness is 0.25 m. The prestressing was performed by a Freyssinet system. Each of the spans has four main girders and three diaphragm beams (at the span's ends and at the middle of the span). The main girders are spaced at 3.1 m. The total width of the bridge structure is 12 m.

The bridge bearings are steel hinge supports. The expansion deck joints are placed on each of the piers. The bridge piers are two-column bents whose height varies from 19 m to 70 m. The columns have a box cross section with dimensions of 3 m × 3 m. The box walls are 0.25 m thick. The upper 12 m ends of the columns with the cap beams are split by 50 mm gaps. The shape of the pier's columns is similar to the musical tuning fork called a Camerton. This allows the upper parts of the columns to bend independently and provides longitudinal movement of the bridge deck without roller supports. The bridge pier footings have dimensions of 12 m × 10 m × 2.5 m. The abutments and wing-walls are monolithic reinforced concrete elements (Raykov 1970).

The concrete used for the superstructure is class C45/55 and for the substructure class C30/37.

In 2005 the viaducts were partially retrofitted and rehabilitated. The rehabilitation included replacement of the expansion joints, waterproofing of the superstructure, and cleaning and lubrication of the hinge bearings. Most XXI was the design company for the rehabilitation (Georgiev and Georgiev 2005).

6.2.3 Hemus Highway Viaduct, km 1.1

This viaduct is a precast prestressed concrete multispan bridge structure (Figure 6.18). It is located at km 1.1 of Hemus Highway. The design company was Patproject with designers B. Stoyanov, G. Todorov, D. Dragoev, P. Minchev, and N. Nenov. The bridge builder was Moststroy company. The owner of the viaduct is the Road Infrastructure State Agency. The bridge opened to traffic in 1977.

The viaduct consists of two parallel structures which rise up to 20 m above the gorge. The total length of the structure is 425 m and includes 18 spans of 22.5 m. Each of the spans has eight I-shaped girders spaced at 1.5 m. The girders are 1.2 m deep with flanges of 0.4 m wide and a web 0.18 m thick. The beams are made by concrete class C30/37 for the inner and C 40/45 for the outer girders (Topurov 2005). The viaduct is divided into four parts with expansion deck joints placed on the abutments and piers 5, 9, and 13. The bearings in the deck joints are steel roller supports. The remaining pier bearings are elastomeric bearing pads with dimensions of 200 mm × 300 mm × 47 mm. The total width of the deck is 13.7 m with a slab thickness of 0.14 m.

The main girders are placed on one-column hammerhead bents. The cap beams have dimensions of 0.90 m × 2 m × 10.8 m. The tallest pier, 15 m high, is a circular column with a diameter of 2 m. The pier footings of 4 m depth are divided into two steps each with diameters of 6.8 m and 5 m. Part of the piers has a pile foundation which consists of four cast-in-place piles with diameters of 1.2 m (Nikolov 2009).

In 2005 the viaduct structures were retrofitted and strengthened. The rehabilitation process included adding new piles to the foundation, replacement of the eight main girders with six new prestressed concrete beams, and replacement of the deck slab. The rehabilitation was designed by Mostconsult with designers D. Kisov and M. Reshkov and constructed by the Freyssinet-Moststroy consortium (Godiniachki and Reshkov 2009).

FIGURE 6.18 Hemus Highway viaduct, km 1.1. (Courtesy of P. Nikolov.)

6.2.4 Stamboliyski Overpass

The Stamboliyski overpass is a multispan, prestressed concrete girder bridge located in Stamboliyski (Figure 6.19). The bridge carries city traffic and serves as a road interchange over the complex railway junctions of the Stamboliyski railway station. The design company was Patproject with designers A. Georgiev, P. Petkov, V. Barzakova, A. Ormanjieva, T. Angelov, and C. Malinova. The bridge was built by Moststroy Company. The owner is Bulgarian State Railways. The overpass opened to traffic in 1980.

The superstructure is a 20-span structure that is 448.7 m long and includes spans of 20.75 m + 8 × 22.5 m + 25.6 m + 9 × 22.5 m + 20.75 m. The middle span consists of simply supported prestressed concrete girders. The remaining parts of the bridge are two continuous beam structures. The superstructure consists of 11 I-shaped girders 1.1 m deep spaced at 1.6 m, and the total width is 17.9 m. The bridge slab is 0.14 m thick with overhangs of 0.82 m. The superstructure is divided with expansion joints at the two middle piers and at the abutments. The expansion joints used for the bridge structure were produced by the Maurer Company.

The continuous superstructure is supported on the central piers and on the abutments by steel rollers. The remaining pier supports are lead bearings. The central span is supported by elastomeric bearing pads with dimensions of 300 mm × 200 mm × 48 mm (Georgiev and Draganov 1981).

The bridge piers are two-column bents. The columns are placed at distance of 9.4 m from each other and have a diameter of 1.2 m. They are 6.8 m tall above the ground. The pier cap beams have a rectangular cross section with dimensions of 1.24 m × 3.04 m and overhangs 3.5 m long. The cap beams ensure the continuity of the main girders over the intermediate piers. Each of the pier columns has a pile foundation which consists of one cast-in-place Benoto pile with a diameter of 1.2 m and a foundation depth of 13 m (Georgiev 1990). The concrete used for the bridge structure is class C20/25.

During 20 years of overpass service the bridge showed some serious defects in the structural components. Most of them are due to the insufficiency of the piers' foundation and beams' lead bearings. In 2002 the bridge structure was strengthened by two additional space frames. The frames consist of four rectangular columns with dimensions of 1.4 m x1 m, connected to each other by square beams with dimensions of 0.70 m × 0.70 m. The distances between columns are 4.8 m in the longitudinal direction and 4.92 m in the transversal direction. The frames are placed on mat foundations. The designers of the bridge strengthening were I. Ivanchev, D. Kisov, and M. Reshkov (Ivanchev, Kisov, and Reshkov 2004).

FIGURE 6.19 Stamboliyski overpass. (Courtesy of D. Kisov.)

6.2.5 Brussels Blvd. Overpass

The Brussels Blvd. overpass is longest bridge structure in Bulgaria (Figure 6.20). The bridge serves as a connection between the Sofia Airport and the entrance highway of the city. The designers were L. Raykov, M. Minev, H. Konstantinov, M. Genov, V. Yochev, and A. Georgiev. The bridge was built by Moststroy and opened to traffic in 1983. The owner is the Municipality of Sofia.

The overpass consists of two parallel structures separated with a gap of 1 m. Part of the structure is placed in a horizontal curve with a radius of 1000 m. The superstructure is a seven-span prestressed concrete frame structure with a total length of 2114 m. The longest span is 49 m, measured at the column's toes. The bridge width is 21.75 m. The bridge superstructure has a nine-cell box cross section 1.2 m deep and 5.64 m wide. The box cells are 0.56 m wide with overhangs 2.55 m long. The deck section was prestressed by Stobet-300 hydraulic jacks. The prestressing tendons are 18 Ø 7 of steel (St150/170). The expansion joints are placed every 98 m.

The bridge piers are V-shaped bents with inclined columns. The columns have a box cross section with dimensions varying from 5.65 m × 1.347 m to 4.428 m × 0.6 m. The column heights vary from 7.5 m to 10 m. The central part of the bridge at the railway overpass is steel orthotropic girders. The concrete used for the bridge structure is class C45/55 (Konstantinov 1984).

6.2.6 Trakia Highway Viaducts, km 37.4 and km 38.2

These viaducts are located at km 37.4 and km 38.2 of Trakia Highway near Ihtiman (Figure 6.21). The bridge was designed by Patproject with head designer B. Stoyanov. The owner of the viaduct is the Road Infrastructure State Agency. The bridge opened to traffic in 1983.

The viaduct consists of two parallel structures with a gap between them of 1 m. The total length of the viaduct is 140 m and includes seven spans of 20 m. The main girders of the superstructure are precast pre-stressed concrete beams, 1.1 m deep, 0.18 m thick for the web, and with 0.40 m wide flanges. Each of the spans has eight girders spaced at 1.6 m. The roadway is 11.5 m wide. The total width of the deck is 13.5 m with slab thickness of 0.14 m. The expansion joints are placed on each of the piers. The bridge bearings are lead pads at the pier supports and steel rollers at the abutments (Godiniachki and Reshkov 2009).

The main girders are placed on three-column rigid frame piers. The columns are precast with a square cross section with dimensions of 0.80 m × 0.80 m. They are 10.6 m tall. The distance between the columns in the transversal direction is 4.8 m. The cap beam has a cross section with dimensions of 0.80 m × 1.1 m × 13.5 m. The abutments consist of two columns with rectangular cross sections. The foundation of the piers consists

FIGURE 6.20   Brussels Blvd. overpass.

FIGURE 6.21 Trakia Highway viaducts, km 37.4. (Courtesy of D. Kisov. With permission.)

of three cast-in-place piles with diameter of 1.2 m and distances between them of 4.8 m. The pile caps have dimensions of 1.5 m × 1.9 m × 11.6 m. The abutments footings have dimensions 7.2 m × 13.5 m × 2.5 m.

In 2005 the viaducts were retrofitted and strengthened. The rehabilitation process included adding new piles to the foundation, replacement of the main girders with new prestressed concrete beams, and replacement of the deck slab. The rehabilitation was designed by Mostconsult with designers D. Kisov and M. Reshkov and constructed by the Freyssinet-Moststroy consortium (Kisov and Reshkov 2005).

6.2.7 Bebresh Viaduct

The Bebresh viaduct is a multispan precast prestressed concrete I-beam structure located at km 31.9 of Hemus Highway near Vitinya pass (Figure 6.22). The design company was Patproject with designers G. Todorov, D. Dragoev, and P. Minchev. The bridge builder was Moststroy. The owner of the viaduct is the Road Infrastructure State Agency. The bridge opened to traffic in 1984.

The viaduct consists of two parallel structures which rise up to 140 m above the valley. The total length of the bridge is 720 m and includes 12 spans of 60 m. The gap between two structures is 0.80 m. The main girders of the superstructure are 3.3 m deep (Lalov and Christov 1999). The top flange of the beams is 2.2 m wide and the bottom flange is 0.80 m wide. The thickness of the top flange varies from 0.20 m to 0.45 m and the thickness of the bottom flange is from 0.25 to 0.55 m. The web thickness is 0.20 m (Ivanchev and Topurov 2004). The bottom flange of the beams is prestressed by 18 tendons of 7 Ø 5 mm. The tendon steel is St 145/160. Each of the spans has three girders spaced at 4.5 m. The superstructure has four cross diaphragms per span. Two of them are placed at the beam's ends and at others are placed at 1/3 of the beam's length.

The bridge is divided into two parts with expansion joints placed on the abutments and on the central pier. The bearings at the deck joints are steel rollers. The remaining bearings are elastomeric bearing pads. The total width of the deck is 13.7 m with slab thickness of 0.22 m. The slab has cantilever parts which are 2.35 m long (Todorov, Dragoev, Minchev et al. 1990).

The bridge piers are concrete single-column bents with a box cross section of 10.5 m × 5.28 m. The box wall thickness is 0.30 m. The highest pier is 125 m tall. The bridge abutments are solid concrete walls 14 m long. Two of the intermediate piers have shaft foundations due to the existing landslides. The concrete used for the bridge structure is class C30/37 and class C45/55. The main girders are placed on the piers using a launching truss girder designed by B. Bankov (Figure 6.23) (Bankov and Partov 2004a).

FIGURE 6.22   Bebresh viaduct. (Courtesy of Moststroy.)

FIGURE 6.23   Bebresh viaduct during the girder launching.

6.2.8 Rodopi Overpass

The Rodopi overpass is a multispan precast prestressed concrete girder bridge structure carrying the traffic of the first-class road Sofia-Bourgas. It is located in Plovdiv and serves as a road interchange over the complex railway junctions of the Plovdiv goods railway station and the city highway (Figure 6.24). The design companies were Patproject and Mostconsult Ltd. with designers P. Petkov, V. Barzakova, D. Kisov, M. Reshkov, and K. Jordanov. The bridge was built by Moststroy. The bridge owner is the Road Infrastructure State Agency. The overpass opened to traffic in 1999.

FIGURE 6.24   Rodopi overpass. (Courtesy of D. Kisov.)

The Rodopi overpass structure was designed and built in two stages. The first stage included the four spans of 42.5 m outside of the railway junctions. The main girders are 1.35 m high and the distance between them is 1.5 m. In 1990 the construction process was stopped due to lack of financial support. The second stage of the design and construction began in 1997. The central part of the bridge structure over the railways has a new design: over the existing piers and by adding a new pier in the middle, the bridge structure includes five-spans of 46.59 m + 19.045 m + 30 m + 16 m + 18 m. The total length of the overpass is 129.53 m. The main girders are precast prestressed concrete beams 1.15 m deep. The steel of the tendons is St 145/160. Each of the spans has 20 beams spaced at 1.53 m. The total width of the bridge is 32 m and the deck slab is 0.20 m thick. The pier bearings are elastomeric pads with dimensions of 233 mm × 300 mm × 86 mm and 900 mm × 900 mm × 189 mm.

The overpass piers are three-column bents. The columns have a circular cross section with a diameter of 2 m and height of 7 m. The distance between them is 12 m. The cap beams with overhangs were post-tensioned by Stobet-200 hydraulic jacks. The columns have a shallow foundation. The concrete used for the bridge is class C40/45 (Ivanchev, Kissov, Reshkov et al. 2000).

6.2.9 Hemus Highway Viaduct, km 48.5

This viaduct is a precast prestressed concrete multi-span simply supported I-beam structure (Figure 6.25). It is located at km 48.5 of Hemus Highway near Botevgrad. The designing company was Patproject with head designer B. Stoyanov. The owner of the viaduct is the Road Infrastructure State Agency. The bridge opened to traffic in 1986.

The viaduct consists of two parallel structures which rise up to 50 m above the valley. The total length of the bridge is 402 m and includes 10 spans of 39.1 m + 8 × 40.5 m + 39.1 m. The precast prestressed concrete beams are 2.25 m deep with 0.84 m wide flanges and 0.3 m thick web. Each of the spans has three girders spaced at 4 m. The beams are made by concrete class C40/45.

The viaduct is divided into two parts with expansion joints placed on the abutments and the middle of the deck. The bearings in the deck joints are steel rollers. The remaining pier bearings are neoprene bearing pads with dimensions of 400 mm × 600 mm × 89 mm. The total width of the deck is 11.3 m with slab thickness of 0.22 m. The deck slab has two cantilever parts that are 1.75 m and 1.55 m long. The viaduct structure is situated in a horizontal curve with a radius of 1100 m.

The main girders are placed on one-column hammerhead piers. The columns have a rectangular box cross section with dimensions of 5 m × 3.1 m. The thickness of the box walls is 0.30 m. The piers are made by concrete class C30/37. The bridge abutments are solid concrete walls (Nikolov 2009).

FIGURE 6.25   Hemus Highway viaduct, km 48.5.

6.2.10 Hemus Highway Viaduct, km 48

This viaduct is located at km 48 of Hemus Highway near Pravetz (Figure 6.26). The designers were M. Minev, I. Yakimov, I. Ivanchev, K. Topurov, D. Kisov, M. Reshkov, R. Yovchev, and D. Dimitrov. The construction company was Moststroy. The viaduct owner is the Road Infrastructure State Agency. The bridge opened to traffic in 1999.

The bridge is 656.2 m long including a six-span prestressed concrete frame structure (68.1 m + 130 m + 140 m + 130 m + 120 m + 68.1 m). The bridge crosses the valley up to 80 m above the ground. The superstructure has a double cell box cross section with total width of 21.7 m, which includes a top slab 6.1 m wide and two 3.5 m long overhangs. The bottom slab of the box is 7.4 m wide. The box section has depths varying from 3 m (at center of span) to 8.5 m (at support zones). The middle wall of the box is 0.40 m thick and the side inclined walls are 0.50 m thick. The thickness of the bottom of the box section varies from 0.25 m at the middle of the span to 1.4 m at supports. The top slab is 0.25 m thick. The bottom of the box was prestressed by 12 tendons of 7 Ø 15 mm of St 145/160. The prestressing was performed by Stobet 200 hydraulic jacks, which are patented in the United Kingdom and France. The expansion joints are placed on the abutments, which have Teflon sliding bearings.

The intermediate bridge piers 2, 3, and 4 have cross sections that are two separate boxes with dimensions of 7.4 m × 3 m and wall thickness of 0.30 m. The center distance between two cells is 8 m. The end piers 1 and 5 are concrete columns with dimensions of 7.4 m × 0.80 m. The pier heights are 24.38 m, 63.22 m, 77.66 m, 73.28 m, and 39.07 m. The bridge abutments are solid concrete structures with shallow foundations 6.8 m wide. The concrete used for the bridge deck and the end piers is class C40/45. The intermediate piers are concrete class C30/37. Most of the pier's footings have dimensions of 16 m × 12.5 m × 3.5 m. Pier 5 has a shaft foundation because of an existing landslide. The double cell box bridge girder was constructed using an incremental launching method (Yovchev, Kisov and Reshkov 2000).

6.2.11 Sofia Airport Access Bridge

The Sofia Airport access bridge serves as a connection of the new terminal of the Sofia Airport and the city highway (Figure 6.27). The bridge designers were I. Lalov, M. Stakev, D. Hubchev, and P. Krastev. The bridge builder was Mix PS. The owner of the access bridge is the Municipality of Sofia. The bridge opened to traffic in 2005.

FIGURE 6.26 Hemus Highway viaduct, km 48. (Courtesy of D. Kisov. With permission.)

FIGURE 6.27 Sofia Airport access bridge. (Courtesy of I. Lalov.)

The bridge is a five-span precast prestressed concrete structure. Two are placed in a horizontal curve with a radius of 60 m and transversal slope of 2.5%. The superstructure consists of precast prestressed concrete main beams 18 m long. Each of the spans has five main girders spaced at 2.8 m and three diaphragm beams (at the span's ends and at the middle of the span). The deck slab is partially prefabricated with the top surface cast in place. The expansion joints are placed on the abutments. The total width of the bridge is 13.25 m.

The bridge bearings are elastomeric bearing pads produced by Maurer GmbH with dimensions of 200 mm × 300 mm × 74 mm. The bridge piers are two-column bents. The precast columns have a triple-box cross section with dimensions of 1.2 m × 1.6 m × 4.2 m. The end cells of the box section are additionally reinforced and filled with a concrete class C45/55. The pier cap beams are precast elements with widths varying from 1.2 m to 1.5 m. The piers have a pile foundation that includes 8 piles with diameters of 0.80 m and length of 13 m. The abutments' foundations includes 12 piles with the same size. The concrete used for the superstructure is class C35/45. The high ductility steel St 500S(B) is used for the reinforcement (Lalov, Stakev, Hubchev et al. 2009).

6.2.12 Makaza Viaduct

This viaduct is located in and named after the Makaza pass, which is situated at the eastern part of the Rodopi Mountains near the Bulgaria–Greece border (Figure 6.28). The bridge was designed by Mostconsult with designers D. Kisov and M. Reshkov. The viaduct builder was Kiska-Turan. The owner is the Road Infrastructure State Agency. It opened to traffic in 2006.

The bridge is a 200 m long concrete frame structure including three spans of 55 m + 90 m + 55 m. The superstructure has a prestressed box cross section 5.9 m wide with two cantilever parts 1.85 m long. The box section has a variable height from 2.5 m to 6 m. The box walls are 0.35 m thick, the top slab thickness varies from 0.25 to 0.30 m, and the bottom slab thickness varies from 0.25 to 0.5 m. The total width of the deck is 10.5 m. The bottom slab of the box section is prestressed.

The bridge piers have a box cross section with dimensions of 3 m × 6.8 m and a wall thickness of 0.35 m. The pier heights are 14 m and 34 m. The pier footings have dimensions of 8.5 m × 8.5 m × 2.5 m. The concrete used for the bridge deck is class C45/55, for the piers is class C35/45, and for the abutments and footings is class C25/30. The box bridge girder was constructed using an incremental launching method (Reshkov and Kisov 2005).

6.2.13 Lulin Highway Viaduct, km 3.5

This viaduct is a prestressed concrete multispan beam structure (Figure 6.29). It is located at km 3.5 of Lulin Highway near Malo Buchino. The bridge was designed by Mostconsult with head designer D. Kisov. The bridge builder was Mapa Insaat and the owner is the Road Infrastructure State Agency. The viaduct opened to traffic in 2009.

The viaduct consists of two parallel structures which rise up to 10 m above the valley. The total length of the bridge is 263.7 m and includes 10 spans of 26.5 m. The gap between the parallel structures is 0.10 m. The superstructure is precast prestressed concrete I-girders. The beams are 1.15 m deep with top flanges 1.25 m wide and bottom flanges 0.60 m wide. The web thickness is 0.18 m. The thickness of the top flange varies from 0.08 m to 0.13 m; the thickness of the bottom flange varies from 0.15 m to 0.25 m. The bottom flange was prestressed by 18 tendons with diameters of 7 Ø 5 mm made of steel class

FIGURE 6.28   Makaza viaduct. (Courtesy of D. Kisov.)

FIGURE 6.29 Lulin Highway viaduct, km 3.5. (Courtesy of D. Kisov.)

St 150/170. The Freyssinet system was used as a pretensioning method. The total bridge width is 15.2 m with a slab 0.20 m thick. The bridge structure is made by concrete class C30/37 and class C45/55. The viaduct is divided into two parts with expansion joints placed on the abutments and on the middle pier. The bridge bearings are elastomeric pads with dimensions of 200 mm × 300 mm × 85 mm. The bridge bearings are neoprene pads.

The bridge piers are single-column hammerhead bents. The columns have an oval cross section with dimensions of 1.6 m × 5 m. The column height varies from 22 m to 8.8 m. The pier footings have dimensions of 9.2 m × 9.2 m × 2 m. The footings have a pile foundation with the diameter of the piles equal to 1.2 m and 22 m deep. The bridge abutments are solid concrete structures which are 4.5 m and 6 m high (Kisov and Naydenov 2009).

6.3 Steel Girder Bridges

6.3.1 Svilengrad Steel Bridge

The Svilengrad steel railway bridge was the first welded plate girder bridge structure in Bulgaria, located at Svilengrad over the Maritsa River (Figure 6.30). The structure carries the international railway Sofia-Plovdiv-Istanbul. The bridge designers were B. Bankov, K. Kostov, and Ch. Grozdanov. The bridge builder was Stroymontaj. The owner is the Bulgarian State Railways company. It opened to traffic in 1968.

The bridge consists of two five-span continuous I-beams spaced at 2 m with spans of 36.5 m and two end simply supported beam structures with a span of 15 m. The bridge is 5 m wide. The middle part of the I-beam is 2.5 m deep and the bottom flange of the girder is 500 mm wide with varying thickness from 20 to 80 mm. The web thickness is 12 mm. The top flange of the girder is 750 mm wide and 30 mm thick. The flanges are stiffened by plates with dimensions of 500 mm × 20 mm and 2 mm × 245 mm × 20 mm. The I-beams of the end spans are 1.25 m deep. The section flanges have dimensions of 500 mm × 20 mm. The bottom flange is strengthened by an additional steel plate with dimensions of 500 mm × 14 mm and the top flange with a steel plate with dimensions of 350 mm × 20 mm. The web has transversal stiffeners placed at 1.825 m and longitudinal stiffeners placed at the inside part of the girders.

FIGURE 6.30 Svilengrad steel bridge.

The floor beams of the deck are placed at the ends and at the quarters of the span. The lateral bracings at the top and bottom flanges of the beam ensure against stability problems of the entire structure. The sidewalk cantilever beams are attached to the girder web stiffeners. The sidewalks are covered by prefabricated reinforced concrete slabs. The steel used for producing the superstructure is class M16S. The piers and the abutments supports are steel hinged and roller supports were made by the Creutz company. The assembling parts of the structure are connected each other by rivets (Bankov 1971).

6.3.2 Asparuhov Bridge

The Asparuhov Bridge is located at the entrance of the Black Sea Highway to the city of Varna (Figure 6.31). The bridge caries highway traffic over the ship canal that connects Varna Lake with the Black Sea. The bridge structure is composed of two parts: the reinforced concrete part was designed by Patprojectwith head designer B. Stoyanov; the steel part was designed by M. Braynov, D. Dimitrov, P. Staykov, S. Stoynov, E. Dulevski, D. Dakov, E. Pampulov, L. Raykov, and K. Tsekov. The bridge owner is the Road Infrastructure State Agency. The Asparuhov Bridge opened to traffic in 1976 (Stoyanov 1990).

The bridge consists of two parallel parts that rise up to 50 m above the canal. The total length of the bridge is 2050 m and includes 39 spans of 40.2 m of precast prestressed concrete T-beams 2.4 m deep and 3 spans of 80.5 m + 160 m + 80.5 m steel structure. The 18 tendons of high-strength steel (St 150/170) with diameters of 24 Ø 5 mm are used for prestressing. The top flange of the beams is 2.2 m wide; the web thickness is 0.20 m. Each of the spans has three main beams spaced at 3.5 m. The total width of the bridge deck is 21 m. The expansion joints are placed every five spans. The concrete used for the main beams is C40/45 (Ivanchev and Topurov 2004).

The steel part of the bridge passes over the ship canal. It is a three-span continuous welded orthotropic box beam. The box section is 5.5 m wide and has a varying depth from 2.8 m to 6.6 m. The deck plate has a varying thickness from 12 mm to 20 mm. The longitudinal ribs have cross-section dimensions of 12 mm × 200 mm and a distance between them of 300 mm; the transversal beams are T-shaped fabricated sections with web dimensions of 500 mm × 8 mm and flange dimensions of 160 mm × 10 mm. The stability problems of the box section are solved by adding of V-shaped braces placed every 4 m. The connections between the assembly units are made by high-strength bolts (Kasianov, Batanov, and Georgiev 1976; Stoyanov and Mihailov 1977).

FIGURE 6.31 The Asparuhov Bridge.

The pier bearings are steel rollers produced by Creutz. The steel used for fabrication of the superstructure is St52 (fy = 330 MPa). The bridge piers are two-column hammerhead bents. The pier foundation is set on very complex soil conditions. The columns are supported on pile foundation which consists of six cast-in-place concrete piles with a diameters of 1.2 m and up to 53 m depth. The bridge superstructure was constructed using a semicantilever launching method (Braynov, Dimitrov, and Staykov 1977).

6.3.3 Elin Pelin Bridge

The Elin Pelin Bridge is an overpass structure located in Elin Pelin (Figure 6.32). It carries the traffic of a second-class road 165 from Yordankino to Novi Han and crosses over the railway of the local railway station. The bridge was designed and built by Gosha and supervised by B. Bankov. The bridge owner is the Municipality of Elin Pelin. The overpass opened to traffic in 1981.

The overpass is a steel orthotropic plate girder structure, gerber system, with a total length of 300 m, and include 15 spans of different lengths. The largest span has a length of 30 m; the rest of the spans vary from 18.766 to 18.793 m. The superstructure is 7 m wide and consists of four main girders. The deck plate is 12 mm thick and stiffened with longitudinal ribs and transversal beams. The longitudinal stiffeners are cold-formed trapezoidal sections fabricated by a 6 mm thick metal sheet. The main girders of the central span are 1.1 m deep. The bottom flange has dimensions of 28 mm × 350 mm and is reinforced at the middle of the span by an additional 28 mm × 320 mm cover plate. The girders web is 10 mm thick. The rest of the spans have girders with same dimensions as above, except of the bottom flange whose dimensions are 20 mm × 350 mm. At 400 mm away from the bridge piers there are 100 mm diameter openings for the hinges of the gerber system. The transversal beams with overhangs are 2.85 m long, 600 mm deep, and are spaced at 3 m. The flanges have dimensions of 10 mm × 100 mm. The overhangs have varying depths.

The road part of the deck in each span consists of two symmetrical assembly units each 3 m wide. They are connected to each other by 24 mm diameter high-strength bolts. Two pedestrian lanes (2.35 and 1.5 m wide) are attached to the road parts. Each pedestrian lane is designed as an orthotropic slab with a 10 mm thick top plate, and stiffeners with cross sections of 6 mm × 150 mm, spaced at 435 mm. The pedestrian units are connected to the road units by bolts.

The piers are steel single-column hammerhead bents. The columns for the part of bridge with spans of 18.766 m were designed as steel I-sections. The columns for the largest span were designed as steel box sections. The steel base plates of the columns have dimensions of 20 mm × 1600 mm × 2600 mm for the box-section columns and 20 mm × 740 mm × 2500 mm for the I-section columns. Eighteen BM 36 anchor bolts were used to connect the base plates to the footings. The cap beam is a double-armed

FIGURE 6.32 The Elin Pelin Bridge.

cantilever with a varying depth, connected to the column by bolts. The transversal beams and the stiffeners of the orthotropic deck are welded to the steel plate. The orthotropic deck units and secondary beams are made from steel St37-2 and the main girders and piers are made from steel St52-3 (Partov and Ivanov 2008).

6.3.4 Brussels Blvd. Steel Bridge

This steel bridge is a part of the Brussels Blvd. overpass and serves as a connection between the Sofia Airport and the entrance highway of the city (Figure 6.33). The bridge is placed over the Sofia-Istanbul railway. The designer is B. Bankov, and the owner is the Municipality of Sofia. The overpass opened to traffic in 1983.

The overpass consists of two parallel structures separated with a gap of 1.56 m. The superstructure is a 90 m long steel orthotropic three-cell box girder and includes two spans of 45 m. The bridge is placed in a horizontal curve with a radius of 1000 m. The box girder is 1.3 m deep and has a total width of 21.75 m. These dimensions were required and limited by aesthetic considerations, because the steel part should fully follow the silhouette of the concrete part with the same depth.

In the longitudinal direction the box walls are connected to each other by upper and lower transversal steel beams at intervals of 1.493 m. The upper transversal steel beams are connected at the top by a 12 mm thick steel plate. The lower transversal beams are connected at the bottom by a 10 mm thick steel plate. The steel orthotropic deck plate is stiffened with longitudinal flat ribs 12 mm × 150 mm with a distance between them of 313 mm. The upper transversal beams have "T" cross sections with web dimensions of 8 mm × 300 mm and flange dimensions of 10 mm × 120 mm. The lower beams have "T" cross sections with web dimensions of 8 mm × 200 mm and flange dimensions of 10 mm × 100 mm. The box walls are 12 mm thick and stiffened by vertical ribs of 10 mm × 100 mm spaced at 1.493 m. The lateral bracings are placed at intervals of 7.465 m and at supports regions (Bankov, Partov, and Christov 1999).

The bridge end piers are V-shaped bents with inclined columns. The columns are triple-box cross sections with wall thicknesses of 12 mm. The columns are rigidly connected to the deck and have hinged connections to the footings. The central pier is a concrete hammerhead bent (Partov and Dinev 2007).

FIGURE 6.33 Brussels Blvd. steel bridge.

6.3.5 Hemus Highway Viaduct 1, 26.884 km

Viaduct 1 of the Hemus Highway is a steel continuous orthotropic plate girder structure (Figure 6.34). The bridge designers were D. Dimitrov, P. Staykov, S. Stoynov, E. Dulevsky, E. Pampulov, L. Raykov, Hr. Genchev, and K. Tsekov. It was produced by Kremikovtsy and constructed by Steel Structures using an incremental launching method by roller chains. It opened to traffic in 1986.

The bridge is 410 m long including six spans of 60 m + 4 × 72.5 m + 60 m. The superstructure consists of two parallel structures with a gap between them of 0.80 m. Each of the structures carries a 13.7 m roadway and consists of two steel main girders 3.6 m high spaced at 9.2 m. The web is stiffened by hot-rolled channel sections 160 mm deep.

The main girders are joined together by braces and transversal beams with overhangs, placed at distances of 3.15 m from each other. An additional longitudinal beam placed at the center of the span connects the transversal beams. The bottom flanges of the main girders are connected by lateral X-braces. The 12 mm orthotropic deck is stiffened by longitudinal trapezoidal ribs.

The viaduct is placed at a horizontal curve with a radius of 850 m and a longitudinal slope of 4.61%. The transversal slopes of each of the roadways are 4.5%.

The intermediate supports of the structure are reinforced concrete piers with a maximum height of 28 m. The piers have a rectangular cross section and individual footings. The steel used for the producing of the bridge superstructure is St09 G2B according to Bulgarian standard (BDS), with yielding strength equal to 330 MPa. The total weight of the superstructure is 21850 kN (Mladjov, Bonchev, and Grechenliev 1985).

Segments of orthotropic plate girders 12.85 m long were welded, prefabricated, and assembled using high-strength bolts. The bridge structure is supported on the abutments by Creutz steel roller supports, and the pier supports by elastomeric bearing pads produced by Gumba GmbH (Staykov 1985, 1986).

FIGURE 6.34   Hemus Highway viaduct 1, 26.884 km. (Courtesy of M. Minev and P. Staykov.)

6.3.6 Hemus Highway Viaduct 2, 27.5 km

Viaduct 2 of the Hemus Highway is a steel continuous orthotropic box girder structure, crossing the valley 60 m above the ground (Figure 6.35). The bridge designers are D. Dimitrov, P. Staykov, S. Stoynov, E. Dulevsky, E. Pampulov, and K. Tsekov. It was constructed by Steel Structures using an incremental launching method with hydraulic jacks and opened to traffic in 1986.

The bridge is 362 m long including spans of 100 m + 162 m + 100 m. The main span of 162 m is the largest span in Bulgaria. The superstructure consists of two parallel steel boxes with a height of 5 m and width of 6.5 m. The gap between the structures is 0.8 m. The viaduct is placed at a horizontal curve with a radius of 1407.25 m and has 4.61% of a longitudinal slope. The total width each of the roadway is 13.7 m.

The top and the bottom plates of the boxes are orthotropic structures stiffened by flat ribs. The thicknesses of the top plate vary from 12 mm to 16 mm; the webs vary from 10 mm to 14 mm and the bottom plate from 10 mm to 30 mm. The distances between the transversal beams of the orthotropic decks are 1.85 m and 2 m (Mladjov, Bonchev, and Grechenliev 1985).

The intermediate supports of the structure are two piers with a maximum height of 42.50 m. The reinforced concrete piers have a rectangular cross section and individual footings. The steel used for producing the bridge superstructure is St52.3 according to Deutsches Institut für Normung (DIN). The total weight of the superstructure is 22,000 kN. The steel consumption for the superstructure was 4.20 kN/m².

Segments of the orthotropic box girders were prefabricated by welding. The top and bottom plates of the segments were assembled by welding. The remaining joints were high-strength bolt connections. The bridge structure is supported on the abutments by Creutz steel roller supports, and the pier supports by neoprene bearing pads produced by Gumba GmbH (Staykov 1985).

6.3.7 Composite Girder Bridge in Varna

The steel–concrete composite box girder bridge in Varna is a road overpass carrying traffic over the Sofia-Varna railway and the canal of the central heating of Varna (Figure 6.36). The bridge was designed by B. Bankov and built by Transstroy. The bridge owner is the Municipality of Varna. The bridge opened to traffic in 1986.

FIGURE 6.35   Hemus Highway viaduct 2, 27.5 km. (Courtesy of M. Minev and P. Staykov.)

FIGURE 6.36   Composite bridge in Varna.

The total length of the bridge is 88 m and includes two spans of 44 m. The main girders of the superstructure are I-shaped steel beams 1.5 m deep. The bottom flange is 400 mm wide and 20 mm thick; the top flange is 300 mm wide and 14 mm thick. The web thickness is 10 mm. The distances between girders are 2.3 m, 2.5 m, and 2.3 m. Steel plates 12 mm thick connect the bottom flanges of the each pair of opposite beams in a longitudinal direction. The deck slab consists of precast concrete strips with dimensions of 9.5 m × 2 m × 0.20 m. The total bridge width is 9.5 m. The composite effect between steel girders and the concrete slab is provided by 120 mm high-shear stud connectors placed at intervals of 0.50 m. The studs have a diameter of 19 mm. The bridge was the first composite bridge in Bulgaria and used Nelson stud connectors (Bankov, Kuneva, Partov et al. 1990).

The pier is a single-column hammerhead bent. The bridge abutments are solid concrete walls. The superstructure was assembled by prestressing the steel beams. The steel box structure was pushed up every 1/5 of the span by hydraulic jacks to get a hogging deflection of the steel beams. Then the precast concrete slab strips were placed on the girders. Each of the strips has a gap at places of the shear studs. These gaps were filled with concrete into the next step. In the final step the hydraulic jacks were removed after the minimum strength of the concrete was reached. The steel used for the main girders is 09G2B (fy = 330 MPa); the concrete used for the slab is class C30 (fc = 30 MPa) and the reinforcement is AIII (fy = 375 MPa).

6.3.8 Vartopa Bridge

The Vartopa Bridge is a five-span continuous steel orthotropic trapezoidal box girder structure connecting Mladost, a large Sofia district, with the downtown (Figure 6.37). The bridge passes over the Vartopa River and carries the city metropolitan railway and highway traffic. The structure was designed by P. Staykov and R. Mladjov. The bridge was built by Metal Structures Company and opened to traffic in 1999. The owner is the Road Infrastructure State Agency.

The bridge has a total length of 276 m and includes spans of 48 m + 60 m + 60 m + 60 m + 48 m. It is placed in a horizontal curve with a radius of 1000 m and the longitudinal slope of 2.578%. The superstructure consists of two parallel steel boxes 2.8 m high and spaced at 13.6 m. The top plates of the boxes are 6.4 m wide and the bottom plates are 4 m wide. The orthotropic deck is formed by longitudinal and transversal stiffening beams. The transversal beams are 645 mm deep and spaced at 3 m. The longitudinal beams are also 645 mm deep and are placed at 1.5 m. The deck is additionally strengthened by longitudinal stiffeners with a trapezoidal cross section 250 mm deep. The thicknesses of the deck slabs vary from 12 mm to 16 mm; the box walls vary from 10 mm to 14 mm; the bottom of the box varies from 10 mm to 20 mm. The total width of the deck is 27.5 m. The expansion joints were placed on the bridge abutments. The box girders were spliced by welding.

The bridge piers are two concrete-filled steel tubular column bents with a maximum height of 15.2 m and a diameter of 1.2 m. The distance between columns is 13.6 m. The cap beams have an I-shaped cross section. The abutment supports are steel rollers produced by the Creutz company; the pier supports are elastomeric bearing pads produced by Gumba GmbH. The bridge piers have concrete footings. The welded structure of the bridge is manufactured by steel grade St 52.3 according to DIN standard. The superstructure was constructed by a cantilever launching.

FIGURE 6.37   Vartopa Bridge. (Courtesy of P. Staykov.)

FIGURE 6.38 The Chaya Bridge. (Courtesy of E. Dulevsky.)

6.3.9 Chaya Bridge

The Chaya Bridge is steel railway bridge over the Chaya River, located at km 167.847 of the Sofia-Istanbul railway near Katunica (Figure 6.38). The bridge was designed by E. Dulevski, K. Topurov, P. Nikolov, L. Georgiev, S. Ivanov, A. Marinov, and Z. Dueva. The builder was Terna-Bulgaria and the owner is the Bulgarian State Railways. The bridge opened to traffic in 2009.

The superstructure is a continuous steel–concrete composite I-girder structure with a total length of 98 m and includes spans of 30.5 m + 37 m + 30.5 m. Two main girders are 2.475 m deep and the concrete deck slab is 0.22 m thick. The distance between girders at the level of the bottom flanges is 4.8 m; at the top of the flanges it is 5.26 m. The top flanges of the beams have dimensions of 35 mm × 700 mm; bottom flanges are 40 mm × 700 mm. The floor beams have a span of 4.8 m with a distance between them of 2 m. The expansion joints are placed on the abutments.

The bridge piers are single-column hammerhead piers. The columns have rectangular cross sections of 7.56 m × 1.4 m. The abutments are solid concrete structures with dimensions of 5.9 m × 6.58 m × 0.40 m for the west abutment and 6.85 m × 6.2 m × 0.50 m for the east abutment. The pier footings have dimensions of 9.6 m × 6.7 m × 2 m; the west abutment footing is 7 m × 6 m × 1.5 m; the east abutment footing is 9.85 m × 6.2 m × 2 m. The piers have pile foundations that consist of five 26 m long piles with diameters of 1.2 m. The abutments each have two piles with the same dimensions. The bridge supports are steel rollers supplied by the Maurer company. The bridge consists of seven assembly units, connected to each other by high-strength bolts. The total weight of the superstructure is 2200 kN. The steel used for producing the structural members has a grade S235J2; the concrete for the composite slab is C30/37; the concrete for the piers and the abutments is C35/45; the concrete for the piles is C40/50 (Topurov and Dulevski 2009b).

6.4 Stone Arch Bridges

6.4.1 Kadin Most

Kadin Most is an arch bridge (Figure 6.39) over the Struma River near Nevestino in the Kiustendil region. The bridge was built in 1470 on the order of Vizier Isaac Pasha during the reign of Ottoman sultan Mehmed II. Today the bridge is still in use with a limited load-carrying capacity of 200 kN vehicles. The bridge was proclaimed a monument of culture in 1968.

The structure is a closed spandrel deck arch stone bridge with a total length of 100 m and includes three main spans of 14.8 m + 20 m + 13.5 m (Ivanchev and Topurov 2004). An additional two spans expand the capacity for the high level of the river water in the spring. They have spans of 4.8 m and 3 m. All arches are approximately semicircular shaped and have a total width of 6 m. The bridge deck is 5.6 m wide and has symmetrical longitudinal slopes of 10%. The stone parapet is 0.40 m wide and the height varies from 0.40 to 1 m.

FIGURE 6.39   Kadin Most.

The main piers have dimensions of 5 m × 6 m. The material used for the bridge construction is high-quality granite. The piers are supported by timber piles. The bridge structure survived one of the strongest earthquakes in Bulgaria in 1904 with a magnitude of 7.8 of the Richter scale and an epicenter in the Struma River region (Penev 2004).

6.4.2 Dyavolski Most

Dyavolski Most is an arch bridge over the Arda River (Figure 6.40). It is situated in a narrow gorge near Ardino in the Rodopi Mountains and is part of the ancient Roman road connecting the lowlands of Thrace with the north Aegean Sea trough the Makaza Pass. The bridge was constructed in 1515–1518 by the by the famous local builder Usta Dimitar from Nedelino. The bridge was proclaimed a monument of culture in 1984.

The structure is a closed spandrel deck arch stone bridge with a total length of 65.7 m and includes three main spans of 9 m + 18.2 m + 8.5 m. An additional four spans expand the capacity for the high level of the river water in the spring. They have varying spans from 1.85 to 3.52 m. The middle arch rises 8.2 m above the ground. All arches are approximately semicircular shaped and have a total width of 3.55 m. The bridge deck is 3.4 m wide and has symmetrical longitudinal slopes of 18%. The stone parapet has dimensions of 200 mm × 120 mm. The main arch piers have dimensions of 3.3 m × 3.35 m and 3.66 m × 3.35 m. The pier and the abutment are bedded on the solid gneiss rocks. The bridge structure survived one of the strongest earthquakes in Bulgaria in 1909 with a magnitude of 5.9 on the Richter scale and an epicenter in Yambol (Minev 2002a).

6.4.3 The Old Bridge

The Old Bridge or Mustafa Pasha Bridge is a stone arch bridge (Figure 6.41) over the Maritsa River near Svilengrad (old name Mustafa Pasha). The bridge was built in 1529 on the order of Vizier Mustafa Pasha. The bridge was proclaimed a monument of culture and is owned by the Road Infrastructure State Agency.

The structure is a closed spandrel deck arch bridge with a total length of 286 m and includes 18 main spans with the widest span of 18 m. An additional two spans expand the capacity for the high level of the river water in the spring. The bridge arches are shaped according to the typical Arabian style with tapered crowns of arches. The total width of the bridge is 7 m. The deck has a stone parapet 0.33 m wide and the height varies from 0.40 m to 0.80 m.

FIGURE 6.40 Dyavolski Most.

FIGURE 6.41 The Old Bridge.

The three middle piers have dimensions of 9.5 m × 6.8 m; the remaining piers have dimensions of 3.5 m × 6.8 m. The piers stone footings are timber pile foundations. The bridge structure survived one of the strongest earthquakes in Bulgaria in 1928 with a magnitude of 7.0 on the Richter scale and an epicenter near Popovica.

6.4.4 Belenski Most

Belenski Most is a stone arch bridge over the Yantra River near Byala in the Rousse region (Figure 6.42). The bridge was built in 1867 on the order of Mithad Pasha, statesman of the Danube Province in the Ottoman Empire. The bridge builder was the famous local master Kolyo Ficheto. The bridge owner is the Road Infrastructure State Agency and it was proclaimed a monument of culture in 1980.

FIGURE 6.42 Belenski Most.

The structure is a closed spandrel deck arch bridge with a total length of 276 m and includes 14 spans of 12.6 m. All arches rise 4.335 m above the ground. All arches are approximately semicircular shaped and the roadway is 9 m wide. The arches are built of two rows of stones with a total height of 0.8 m. The bridge deck has a stone parapet and with dimensions of 0.2 m × 0.9 m. The material used for the bridge construction is a high-quality limestone (Stoykov 1976).

The arch piers have dimensions of 5.4 m × 9 m. Part of the pier footings have a pile foundation consisting of timber piles 6 m long with a diameter of 0.20 m and distance between them of 0.30 m. The remaining part of the piers footings are bedded on a solid gneiss rock. The bridge structure survived one of the strongest earthquakes in Bulgaria in 1901 with a magnitude of 7.2 on the Richter scale and an epicenter near Shabla (Minev 2002b).

6.5 Concrete Arch Bridges

6.5.1 Tarnovo Arch Bridge

Tarnovo arch bridge is a complex overpass structure in Tarnovo (Figure 6.43). The main span of the structure crosses the Yantra River; other spans cross the railway junctions and the city highways. The bridge was designed by Patproject with head designer B. Stoyanov. The bridge builder was Moststroy and the owner is the Road Infrastructure State Agency. The overpass opened to traffic in 1956.

The bridge is an open spandrel deck arch structure with a total length of 139.75 m and includes three spans of 43.25 m + 54 m + 42.5 m. The main arch rises 19 m above the ground. The arch structure consists of two arches with a distance between them of 6.7 m. The arch has a rectangular cross section with a constant width of 1.1 m and varying depths of 0.80 m to 1.2 m. The arch bridge has a Vierendeel type of bracing with a distance between braces of 6.9 m. The deck-supporting columns are coupled by horizontal beams. The deck structure consists of three concrete girders with dimensions of 1.2 m × 0.40 m. The distance between them is 3.35 m. The girders are supported by the transversal cap beams with dimensions of 1.2 m × 0.90 m, placed on the columns. The columns have a square cross section with dimensions of 0.90 m × 0.90 m. The total width of the bridge is 10.9 m.

FIGURE 6.43 Tarnovo arch bridge.

The abutment's roller bearings are steel plates with dimensions of 400 mm × 100 mm × 20 mm. The hinged supports are realized by the lead plate with dimensions of 400 mm × 100 mm × 20 mm. The arch footings have dimensions of 5 m × 5 m. The concrete used for bridge structure is class C16/20. The arch structure was constructed by a wood truss scaffolding (Kissov, Reshkov, and Yordanov 2001).

6.5.2 Balaban Dere Arch Bridge

The Balaban Dere arch bridge crosses one of the feeders of the Studen Kladenec reservoir and carries the Kardjali-Perperek railway at km 81.2 (Figure 6.44). The bridge was designed by L. Raykov and M. Minev. The builder and owner is Bulgarian State Railways. It opened to traffic in 1959.

The bridge is a deck-stiffened arch structure (the Maillart bridge system) with a total length of 106.66 m. The arches rise 7.6 m above the ground. This was the first bridge in Bulgaria built using the Maillart bridge system. The bridge deck is a continuous beam structure with spans of 2 × 5 m + 26 m + 4.33 m + 26 m + 4.33 m + 26 m + 2 × 5 m (Raykov 1957). It consists of two rectangular concrete girders with dimensions of 1.3 × 0.50 m and a distance between them of 3.5 m. The deck slab is 0.25 m thick with a total width of 5.79 m. The bridge arch has a rectangular cross section with dimensions of 4 m × 0.30 m. The connection walls between the deck and the arch have dimensions of 4 m × 0.20 m. The distance between them is 4.333 m. The walls are rigidly connected to the bridge deck and arch. The expansion joints are placed over the abutments. The bearings are steel rollers.

The bridge piers outside of the arches are two column bents. The columns have a rectangular cross section with dimensions of 0.30 m × 0.70 m; the cap beam is 0.65 m deep. The columns have additional braces with dimensions of 0.50 m × 0.30 m. The tallest pier rises 8.3 m above the ground. The bridge abutments are frame structures. The arches have footings with dimensions of 4.5 × 6.8 m and are 1.7 m high bedded on solid rocks. The footings of the bridge piers have dimensions of 1.5 m × 2.2 m and are 1.5 m high. The concrete used for the bridge structure is class C20/25; the reinforcing steel has yield strengths of 235 MPa and 390 MPa (Kolchakov 1957, 1958).

FIGURE 6.44 Balaban Dere arch bridge. (Courtesy of M. Minev.)

6.5.3 Lakatnik Arch Bridge

The Lakatnik arch bridge crosses the Iskar River near the Lakatnik railway station and carries the Sofia-Mezdra railway at km 51.1 (Figure 6.45). The bridge was designed by M. Minev. The builder and owner is Bulgarian State Railways. It opened to traffic in 1962.

The bridge is a continuous superstructure combined with a reinforced concrete arch (the Maillart bridge system) with a total length of 62 m and a main span of 38 m. The arch has a polygonal shape whose vertices rise 4.746 m, 8.064 m, 10.223 m, 11.463 m, and 11.905 m above the ground. The continuous superstructure has spans of 6.5 m + 5.5 m + 38 m + 5.5 m + 6.5 m and consists of two rectangular concrete girders with dimensions of 1.2 m × 0.60 m with a distance between them of 2.9 m. The deck slab is 0.20 m thick at the span and 0.35 m at the overhangs. The bridge deck has a total width of 5.25 m. The arch has a rectangular cross section with dimensions of 4.3 m × 0.30 m. The connection walls between the deck and the arch have dimensions of 4.3 m × 0.25 m. The distance between them is 3.8 m. The walls are rigidly connected to the bridge deck and arch.

The bridge has expansion joints over the abutments. The abutment bearings are steel rollers. The bridge piers have dimensions of 4.3 m × 0.30 m and are 10.5 m tall. The concrete used for the bridge structure is class C35/45; the reinforcing steel has yield strengths of 235 MPa and 390 MPa.

6.5.4 Opletnya Arch Bridge

The Opletnya arch bridge (Figure 6.46), crosses the gorges of the Iskar River near the Opletnya railway station and carries the Sofia-Mezdra railway at km 54.4. The bridge was designed by L. Raykov, H. Popov, K. Tzekov, P. Chavov, and A. Ganev. The bridge builder and owner is Bulgarian State Railways and it opened to traffic in 1962.

The bridge is a deck-stiffened arch structure (the Maillart bridge system) with a total length of 212 m. The bridge is a continuous reinforced concrete box girder structure stiffened by a polygonal reinforced concrete arch with a span of 80 m. The arch crown rises 14.4 m above the gorge. The bridge is placed in a horizontal curve with a radius of 275 m. The magnitude of the centrifugal forces lead to the adoption of a nontraditional solution for the bridge structure. The arch structure is split into two parts at the arch base, like a swallow tail. The arch is 4.5 m wide and thickness varies from 0.75 m to 0.80 m. The superstructure box cross section has dimensions of 1.65 m × 4.5 m. The box walls have thicknesses of 0.60 m; the box slabs are 0.25 m thick. The total width of the deck is 6 m. The connection walls between the deck

FIGURE 6.45 Lakatnik arch bridge. (Courtesy of M. Minev.)

FIGURE 6.46 Opletnya arch bridge. (Courtesy of D. Gocheva.)

and the arch have dimensions of 4.5 m × 0.20 m. The distance between the walls is 8 m and they also are split at the lower ends. Outside of the arch the box deck has an additional four spans 12 m long. The remaining parts of the bridge (three and four spans 12 m long) are continuous beams with a rectangular cross section. The expansion joints are placed on the abutments and at the ends of the box deck. The bridge piers are two-column bents. The tallest pier rises 16 m above the ground.

The concrete used for the bridge structure is class C30/37; the reinforcing steel has yield strengths of 235 MPa and 390 MPa. During the construction of the bridge arch the scaffolding collapsed. The reason for the accident was that the load-carrying capacity of the scaffolding elements was exceeded due to unsymmetrical concrete laying.

6.5.5 Cherepish Arch Bridge

The Cherepish arch bridge crosses the Iskar River near the Cherepish railway station and carries the Sofia-Mezdra railway at km 74.1 (Figure 6.47). The bridge was designed by M. Minev and I. Ivanchev. The bridge builder was Moststroy; the owner is Bulgaria State Railways. It opened to traffic in 1964.

FIGURE 6.47   Cherepish arch bridge. (Courtesy of M. Minev and I. Ivanchev.)

The bridge is an open spandrel deck arch structure and consists of two reinforced concrete arches with equal spans of 40 m. The arches rise 11.2 m above the ground. The bridge deck is a continuous beam structure with a total length of 105 m and includes spans of 2 × 5 m + 40 m + 5 m + 40 m + 2 × 5 m. The superstructure consists of two rectangular girders. The cross section of the arch structure has dimensions of 4 m × 0.40 m. The deck slab is 0.25 m thick and 6 m wide. The deck-supporting walls have dimensions of 4 m × 0.20 m. The distance between them is 4 m. The walls are rigidly connected to the bridge deck and the arch. The expansion joints are placed on the bridge abutments.

The bridge bearings are steel rollers. The arch and the pier footings are bedded on solid rocks. The concrete used for the bridge structure is class C25/30; the reinforcing steel has yield strengths of 235 MPa and 390 MPa.

6.5.6 Zverino Arch Bridge

The Zverino arch bridge crosses the Iskar River near the Zverino railway station and carries the Sofia-Mezdra railway (Figure 6.48). The bridge designers were D. Gocheva and G. Kocev. The bridge contractor and owner is Bulgarian State Railways. It opened to traffic in 1966.

The bridge is a deck-stiffened arch structure (the Maillart bridge system) with a total length of 141 m. The bridge consists of a continuous multicell box girder with a main span of 50 m and combined with a reinforced concrete arch. The arch crown rises 10 m above the river. It has a polynomial shape and the cross section is 4.5 m wide; the thickness varies from 0.50 to 0.45 m. The bridge deck has a cell-box cross section with dimensions of 1.65 m × 3.5 m. The upper slab has a thickness of 0.25 m and overhangs 1.25 m long. The total width of the deck is 6 m. The lower slab is 0.20 m thick and the wall thickness is 0.60 m.

The superstructure box outside of the arch is prestressed by the Freyssinet system with 15 cables at the box walls and 24 cables at the box bottom slab. The cables have diameters of 12 Ø 7 mm, steel class St 150/170. Ten walls are rigidly connected to the deck with the arch structure and have dimensions of 3.5 m × 0.20 m. The distance between the walls is 5 m. The bridge has expansion joints over the abutments and the end piers. The arch structure was constructed by wood truss scaffolding. The pier bearings outside of the arch are steel rollers. The arch and the piers have reinforced concrete footings bedded on solid rocks (Gocheva and Kocev 1967).

FIGURE 6.48   Zverino arch bridge. (Courtesy of D. Gocheva.)

6.6 Truss Bridges

6.6.1 Lom Bridge

The Lom Bridge is a steel trough-truss structure (Figure 6.49) over the Lom River near the Danube riverside town of Lom. The bridge designer and contractor are unknown. The bridge owner is the Municipality of Lom and it opened to traffic in 1890. Today the bridge is still in use with a limited load-carrying capacity of 200 kN vehicles.

The bridge is a five-span structure with a total length of 117.5 m. Each of the spans consists of two parallel Pratt-type trusses with a distance between them of 5.7 m. The trusses have 10 panels with a total length of 23.5 m and height of 2.2 m. The bridge deck is 8.7 m wide. The truss chords are parallel and have T-shaped cross sections fabricated by two angle sections, back to back with a steel plate between them. The angle sections are additionally strengthened by steel plates. The truss diagonal members have I-shaped cross sections fabricated by four angle sections, back to back with a web of steel plate. The diagonal members are riveted directly to the chord members without gusset plates. All fabricated cross sections are fastened by a riveting.

The floor beams are placed at the lower joints of the trusses and are hot-rolled I-sections, 500 mm deep. The bridge deck has five stringers, placed at a distance of 1.14 m. The stringers also have an I-shaped cross-section, 270 mm deep. The concrete deck slab is 120 mm thick. The bridge superstructure has a lower lateral bracing made of hot-rolled angle sections. The truss supports are steel hinged on one pier and steel rollers on the other pier. The bridge abutments and piers are made of stone masonry. The piers are supported by timber piles (Bankov, Partov, and Momchilov 2002).

6.6.2 Stambolov Bridge

The Stambolov Bridge is a steel deck-truss arch structure over the Yantra River at Veliko Tarnovo (Figure 6.50). The bridge is named after the eminent Bulgarian prime minister Stefan Stambolov, who gave the idea to build a new modern steel bridge in the city. The bridge was designed and built by Waagner-Biro. The bridge owner is the Municipality of Veliko Tarnovo. The bridge opened to traffic in 1892.

The structure consists of two parallel Pratt-type truss arches with a distance between them of 5.6 m. The truss arches are 8.6 m high at the abutments and 1.2 m at the middle. They have 26 panels with a

FIGURE 6.49   Lom Bridge.

FIGURE 6.50   Stambolov Bridge.

total length of 81.6 m. The bridge deck is 7.3 m wide. The upper horizontal chord and lower parabolic arch chord of the trusses are U-sections, 500 mm deep and fabricated by four angle sections with dimensions of 80 mm × 80 mm × 10 mm, back to back with a steel plate between them of 20 mm thick. The outside flanges are additionally strengthened by steel plates with dimensions of 50 mm × 240 mm. The vertical and the diagonal members of the trusses have I-shaped cross sections fabricated by four angle sections with dimensions of 120 mm × 70 mm × 7 mm, back to back with a web of 10 mm thick. The diagonal members are riveted directly to the chord members without gusset plates.

The floor beams of the bridge are I-shaped, 600 mm high, fabricated by angle sections. The stringers also have an I-shaped cross section. All fabricated cross sections are fastened by riveting. Lateral "X" bracings are placed at the upper and lower chords and at the panel ends. The bridge is supported on the abutments by steel hinged bearings. The abutment's foundations are bedded on solid rocks. The bridge structure survived one of strongest earthquakes in Bulgaria in 1913, with a magnitude of 7.0 on the Richter scale (Penev 2001).

6.6.3 Varna Lake Movable Bridge

The Varna Lake movable bridge is a steel structure near Varna (Figure 6.51). The bridge carries railway and road traffic of the industrial part of Varna over the ship canal which connects Varna Lake with the Black Sea. The bridge was designed and built by the MAN Company. The bridge owner is the Municipality of Varna. It opened to traffic in 1939.

The bridge is a three-span riveted steel structure with a total length of 80.34 m and consists of two stationary parts with spans of 24.36 m and a movable central span of 31.62 m. In 1975 the movable part of the bridge was completely destroyed by a ship accident and the stationary parts were seriously damaged. The new Varna Lake movable bridge was designed by B. Bankov. The Varna side stationary part was designed as a simply supported bridge with an orthotropic deck. The superstructure consists of two main girders 2.8 m deep with a distance between them of 8 m. The deck slab is 12 mm thick and 13.86 m wide. The girder has bottom flanges with dimensions of 12 mm × 400 mm and a web 12 mm thick.

The floor beams are 1.55 m deep with a distance between them of 1.975 m and have bottom flanges with dimensions of 20 mm × 300 mm and webs 8 mm thick. The centerline of the railroad is offset from the centerline of the bridge by 1.55 m, and is supported by two longitudinal girders spaced at 1.5 m, with a span of 1.975 m and depth equal to the depth of the secondary beams. Their web has dimensions of 8 mm × 540 mm. The longitudinal girders also serve as deck stiffeners, and are designed as continuous beams supported by the secondary beams (at 1.975 m centers). The lateral bracings of the deck are K-braces placed at the bottom flanges of the main girders. The central movable part is a trough-truss structure which consists of two identical Warren trusses with spans of 31.62 m, including six panels 3.162 m long. The members of the upper and lower chords and compression diagonals have box cross sections; the tension diagonals have I-shaped cross sections (Bankov and Partov 2004b).

The truss deck is also an orthotropic plate structure and consists of bottom chord members, a 12 mm thick steel plate. The deck is stiffened by longitudinal ribs of 10 mm × 160 mm spaced at 300 mm. The floor beams are placed at the lower joints of the trusses and have a span of 8.4 m. The distances between the stringers are 1.757 m. The portal bracing of the trusses is formed by box section members placed

FIGURE 6.51 Varna Lake movable bridge.

at the upper joints of the trusses. The Asparuhovo side stationary part remains as the original riveted structure. The old bridge piers were used for the new movable part. The bearings are elastomeric bearing pads. Two types of steel grade were used for the steel bridge: steel grade M16C and steel grade 10G2CF (Bankov, Partov and Ivanov 2006).

6.6.4 Rogletz Bridge

The Rogletz Bridge is a steel deck-truss structure (Figure 6.52) over the Lom River near Rogletz. It was designed by B. Bankov, built by Remontstroy, and opened to traffic in 1987. The bridge owner is the Municipality of Vidin.

The bridge is a three-span simply supported truss structure (31.7 m + 32 m + 31.7 m) with a total length of 95.4 m. The structure consists of two "second-hand" parallel Pratt-type trusses with a distance between them of 4 m. These trusses originally were used for a bridge over the Jebel River, which was built in 1914. The trusses have parallel chords divided into 10 panels. They are 3.17 m deep and all truss members are connected by rivets. The truss chords are T-shaped compound sections which consist of steel plates with dimensions of 6 mm × 310 mm, 9 mm × 310 mm, 10 mm × 310 mm and 12 mm × 310 mm, two angle sections 90 mm × 90 mm × 9 mm and web with dimensions of 10 mm × 360 mm. The diagonals consist of two channel sections, back to back, 200 mm and 140 mm deep. The vertical members are four angles with dimensions of 90 mm × 90 mm × 9 mm, 75 mm × 75 mm × 8 mm, and 70 mm × 70 mm × 7 mm.

The trusses was carefully inspected and rehabilitated by replacement of the corroded members, gusset plates, and rivets. Additionally they were prestressed by uplifting the center of the trusses before the concrete deck slab was cast. The deck slab is a steel–concrete composite structure, 0.20 m thick. The floor beams are placed on the upper chord truss joints. They have I-shaped cross sections 0.55 m deep. The flanges consist of two angle sections, back to back with dimensions of 80 mm × 80 mm × 8 mm. The web is 6 mm thick.

The stringers are Pratt trusses 0.55 m deep. Chord members consist of two angled sections with dimensions of 70 mm × 70 mm × 7 mm and diagonals of steel plates with dimensions of 60 mm × 6 mm. The lateral braces are placed at each of the truss panels. The bridge piers have dimensions of 8 m × 8 m × 2 m. The piers footings are solid reinforced concrete with dimensions of 4 m × 10 m × 3 m. The lateral "X" bracings are placed at the top and the bottom of truss chords (Partov, Donchev, and Momchilov 2004).

FIGURE 6.52   Rogletz Bridge.

6.7 Pedestrian Bridges

6.7.1 Saedinenie Bridge

The Saedinenie Bridge is a pedestrian bridge structure located at Plovdiv over Maritsa River (Figure 6.53). This was the first prestressed concrete box girder bridge in Bulgaria built by the cantilever launching method. The bridge designers were M. Minev, I. Yakimov, I. Ivanchev, T. Etimov, and A. Chakarova. The bridge was built by Moststroy and opened to traffic in 1984. The owner is the Municipality of Plovdiv.

The bridge deck is a continuous beam structure that is 174 m long and includes spans of 44 m + 86 m + 44 m. The bridge consists of two parallel structures with a distance between them of 3.4 m. The superstructure has a box cross section with height varying from 2.8 m to 4 m. Box section widths vary from 2.94 m to 3.44 m. The box walls have a variable thickness from 0.28 m to 0.53 m. The bottom slab of the box section varies from 0.20 m to 0.90 m. The top slab thickness varies from 0.25 m to 0.50 m (Ivanchev and Topurov 2004). The cantilever parts of the top slab are 1.35 m long and 0.80 m thick. The total width of the bridge is 16 m. The deck structure is divided into segments 2.4 m long. Each segment was prestressed by 12 tendons of 4 ⌀ 5 mm made of steel St 145/160. Prestressing was performed by Stobet-200 hydraulic jacks (Yakimov 1990).

The superstructure has expansion joints placed on the abutments. The abutment's bearings are steel rollers. The bridge piers consist of two concrete walls placed at distance of 1.90 m. The walls are 5.0 m wide, 0.45 m thick and 7 m tall. Each pier has a pile foundation which consists of 14 piles with a diameter of 1.20 m. The concrete used for a bridge structure is class C45/55 (Minev and Yakimov 1987).

6.7.2 Rowing Canal Bridge

The Rowing Canal Bridge is a stressed-ribbon footbridge located at Plovdiv over the Rowing Canal (Figure 6.54). The bridge was designed by A. Georgiev, P. Petkov, V. Barzakova, R. Yovchev, T. Angelov, V. Kukushinkov, and P. Petrov and constructed by Moststroy. The owner is the Bulgarian Rowing Federation. The bridge opened to usage in 1989.

The superstructure is a prestressed concrete structure with a total length of 246 m and includes three spans of 48 m + 150 m + 48 m. The deck is 4 m wide and has maximum sag of 3.2 m. The deck slab has a thickness of 0.32 m. The bridge deck has 36 prestressing tendons with dimensions of 3 × 12 ⌀ 15P7 mm and made of high-strength steel St 150/170. The maximum prestressing force was 2000 kN. Prestressing was performed by Stobet-200 hydraulic jacks.

FIGURE 6.53 Saedinenie Bridge. (Courtesy of I. Yakimov.)

FIGURE 6.54 **(See color insert.)** Rowing Canal Bridge. (Courtesy of A. Georgiev.)

FIGURE 6.55 Varbica Footbridge. (Courtesy of E. Dulevski.)

The bridge piers are 6.2 m wide, their thicknesses vary from 0.85 m to 1.15 m, and they are 6.4 m tall. They serve as load-carrying structures for the bridge's staircases. The pier footings are caisson structures. The abutments are 7 m wide, 7 m long, and 9.5 m high. They have caisson foundations with dimensions of 16 m × 18 m and are 12 m deep. The concrete used for the bridge structure is class C50/60. The bridge slab is casted on a wooden formwork with steel scaffolding.

6.7.3 Varbica Footbridge

The Varbica Footbridge is a suspension structure over the Varbica River near Momchilgrad (Figure 6.55). The bridge was designed by E. Dulevski and K. Topurov. The builders were Savarona Ltd. and Boyadjiev Ltd. The owner is the Municipality of Momchilgrad, and it opened to traffic in 2007.

The bridge has a main span of 169 m and a total width of 2.5 m. The superstructure has two trough Warren trusses. The truss members are cold-formed box sections: the upper chord is a 160 mm × 160 mm × 5 mm box and the lower chord is a 150 mm × 100 mm × 5 mm box. The trusses are 1.5 m high and have panels 2.54 m wide. The floor beams are box sections with dimensions of 120 mm × 120 mm × 5 mm spaced at 1.27 m. The bridge slab is 50 mm thick, cast in place on a trapezoidal steel sheet with dimensions of 40 mm × 0.60 mm. The superstructure is suspended on four main cables with diameters of 32 mm. The vertical suspenders have diameters of 11 mm. The assembly parts of the bridge are connected to each other by high-strength bolts.

The main towers of the bridge are A-shaped concrete structures and consist of four columns with a rectangular cross section with dimensions of 460 mm × 460 mm and height of 16.81 m. The columns are connected to each other by horizontal beams. The superstructure is produced of steel grade S235J2; the concrete is class C25/30. The total weight of the steel is 320 kN (Topurov and Dulevski 2009a).

6.8 Other Bridges

6.8.1 Gincy Bridge

The Gincy Bridge is the first fiber-reinforced polymer (FRP) roadway bridge in the world (Kelly and Zweben 2003). It is located in Gincy over the river Nishava and serves as a connection between Gincy and the first class road Sofia-Vidin (Figure 6.56). The bridge designer was I. Angelov. The builder was GUSV and the owner is the Municipality of Gincy. The bridge opened to traffic in 1982.

The superstructure is a one-span fully glass fiber reinforced polymer structure with a length of 12 m and width of 7.6 m, including six channel-shaped girders with a trapezoidal cross section 1.2 m high, 0.60 wide at the top, and 0.40 m wide at the bottom. The distance between the girders is 1.2 m. The channels walls are 20 mm thick. The lateral stability of the girders is provided by transversal ribs placed at the abutments and every 2.4 m. The rib thickness is 20 mm. The deck slab is also an FRP structure with a thickness of 30 mm. An asphalt wearing surface of 100 mm thick covers the deck slab. The bridge parapets are also made by FRP.

All parts of the superstructure are precast and glued to each other in the workshop. The composite material used for the manufacturing of the superstructure consists of fabrics of randomly oriented glass fibers with a polymer matrix of epoxy resin. The components of the superstructure are produced by wet lay-up. The bridge bearings are steel plates. The bridge has reinforced concrete abutments. The bridge was load tested by a military tank after its completion. Today the bridge is in perfect condition without any rehabilitation (Dinev 2006).

FIGURE 6.56 Gincy FRP bridge.

References

Bankov, B. 1971. The first welded steel railway bridge in Bulgaria. *J. of Industrial Structures*, 1(1): 5–16.

Bankov, B., Kuneva, N., Partov, D. et al. 1990. Composite steel-concrete prefabricated bridge deck. *Proc. IABSE Symp. of Mixed Structures Including New Materials*, 1: 249–250.

Bankov, B. and Partov, D. 2004a. Design, static analysis and testing of frame-crane for installing of bridge beams on the viaducts along Hemus motorway. *J. Roads*, 44 (2): 20–26.

Bankov, B. and Partov, D. 2004b. Project of rehabilitation of the steel structures of the Small Asparuhov bridge. *J. Roads*, 44(1): 13–23.

Bankov, B., Partov, D. and Christov, Ch. 1999. Structure, design and construction of a reinforced concrete–steel scaffold bridge in Sofia. *Proc. of 11th International Scientific Conference*, October 18–20, pp. 25–28, Brno University of Technology, Czech Republic.

Bankov, B., Partov, D. and Ivanov, R. 2006. Structural rehabilitation of the Small Asparuhov Bridge. *Proc. of the 21st International Symposium on Steel Construction and Bridges*, Bratislava, September 20–23, pp. 77–84, Slovak Republic.

Bankov, B., Partov, D. and Momchilov, M. 2002. Rehabilitation of steel structures and their application for construction of road bridge over Lom River. *J. Roads*, 42(6): 20–25.

BDS EN. 1990. Eurocode 0—Basis of structural design.

BDS EN. 1991. Eurocode 1—Actions on structures; Part 1-1: General actions: Densities, self-weight, imposed loads for buildings; Part 1-2: General actions: Actions on structures exposed to fire; Part 1-3: General actions: Snow loads; Part 1-4: General actions: Wind actions; Part 1-5: General actions: Thermal actions; Part 1-6: General actions: Actions during execution; Part 1-7: General actions: Accidental actions; Part 2: Traffic loads on bridges; Part 3: Actions induced by cranes and machinery; Part 4: Actions on silos and tanks.

BDS EN. 1992. Eurocode 2—Design of concrete structures; Part 1-1: General rules and rules for buildings; Part 1-2: General rules: Structural fire design; Part 2: Concrete bridges: Design and detailing rules; Part 3: Liquid retaining and containment structures.

BDS EN. 1993. Eurocode 3—Design of steel structures; Part 1-1: General rules and rules for buildings; Part 1-2: General rules: Structural fire design; Part 1-3: General rules: Supplementary rules for cold formed thin gauge members and sheeting; Part 1-4: General rules: Supplementary rules for stainless steels; Part 1-5: General rules: Supplementary rules for planar plated structures without transverse loading; Part 1-6: General rules: Supplementary rules for the shell structures; Part 1-7: General rules: Supplementary rules for planar plated structural elements with out of plane loading; Part 1-8: Design of joints; Part 1-9: Fatigue; Part 1-10: Material toughness and through-thickness properties; Part 1-11: Design of structures with tension components; Part 1-12: Supplementary rules for high strength steels; Part 2: Steel bridges; Part 3-1: Towers, masts and chimneys: Towers and masts; Part 3-2: Towers, masts and chimneys: Chimneys; Part 4-1: Silos, tanks and pipelines: Silos; Part 4-2: Silos, tanks and pipelines: Tanks; Part 4-3: Silos, tanks and pipelines: Pipelines; Part 5: Piling; Part 6: Crane supporting structures.

BDS EN. 1994. Eurocode 4—Design of composite steel and concrete structures; Part 1-1: General rules and rules for buildings; Part 1-2: General rules: Structural fire design; Part 2: Composite bridges.

BDS EN. 1997. Eurocode 7—Geotechnical design; Part 1: General rules; Part 2: Ground investigation and testing.

BDS EN. 1998. Eurocode 8—Design of structures for earthquake resistance; Part 1: General rules, seismic actions and rules for buildings; Part 2: Bridges; Part 3: Assessment and retrofitting of buildings; Part 4: Silos, tanks and pipelines; Part 5: Foundations, retaining structures and geotechnical aspects; Part 6: Towers, masts and chimneys.

Braynov, M., Dimitrov, D. and Staykov, P. 1977. Steel structure of the new Asparuhov Bridge-Varna. *J. Roads*, 16(2): 4–8.

Daalov, T. and Daalov, B. 2006. Upper construction of the bridge on the river Limpopo, Zimbabwe. *Proc. of the 6th International Symposium on Steel Bridges*, May 31–June 2, Prague, Czech Republic.

DIN 1045. 1988. Concrete and reinforced concrete: design and construction, Publisher Beuth, Berlin.

DIN 1072. 1967. Streets and road bridges (Loading), Publisher Beuth, Berlin.

DIN 1073. 1973. Steel road bridges (basics of structural design), Publisher Beuth, Berlin.

DIN 1079. 1979. Steel road bridges (basics for construction), Publisher Beuth, Berlin.

DIN 4101. 1970. Welded steel road bridges (analysis, construction and erection), Publisher Beuth, Berlin.

DIN 4114. 1961. Stability analysis, Publisher Beuth, Berlin.

DIN 4227. 1988. Pre-stressed concrete: Structural elements from reinforced concrete with partial and complete pre-stressing, Publisher Beuth, Berlin.

Dinev, D. 2006. Structural analysis of elements strengthened with composite materials. Ph.D. thesis, UACEG, Sofia, BG.

Georgiev, A. 1990. Continuous bridge structures made by prefabricated elements. *J. Roads*, 30(4): 13–16.

Georgiev, CH. 2007. Static and dynamic investigation of standard bridge structures with a view of practical application of systems for reducing seismic effects. *Proc. of the 7th International Scientific Conference*, 2: 130–135.

Georgiev, A. and Draganov, D. 1981. A new structural system in our bridge construction. *J. Roads*, 20(4): 1–4.

Georgiev, A. and Georgiev, S. 2005. Rehabilitation of large bridges in Lot 5, Pirdop-Karlovo. *Proc. of First Symposium of Bridges*, pp. 91–97, UACEG, Sofia, Bulgaria.

Gocheva, D. and Kocev, G. 1967. Design, static analysis and assembly of an arch bridge near railway station Zverino on double railway Sofia-Mezdra. *J. Construction*, 14(12): 1–6.

Godiniachki, G. and Reshkov, M. 2009. Reinforced concrete bridges-problems with the deformations. *Proc. of the 9th International Scientific Conference*, 1: 212–218.

Ivanchev, I., Kisov, D. and Reshkov, M. 2004. Uncontrolled movement of the overpass superstructure and its securing against horizontal forces. *J. Roads*, 43(2): 8–15.

Ivanchev, I., Kissov, D., Reshkov, M. et al. 2000. Rodopi overpass crossing in Plovdiv: Design and construction of main structure. *J. Construction*, 57(5): 2–7.

Ivanchev, I. and Topurov, K. 2004. *Reinforced concrete bridges*. Sofia: Avrio OOD.

Kasianov, I., Batanov, T. and Georgiev, G. 1976. Producing and assembling of 120 t weight girders for Asparuhov bridge structure in Varna. *J. Roads*, 15(9): 14–19.

Kelly, A. and Zweben, C. 2003. *Comprehensive composite materials*. Elsevier, Amsterdam, UK.

Kisov, D. and Naydenov, M. 2009. Earthquake investigation of bridges. *Proc. of the 9th International Scientific Conference*, 1: 199–205.

Kisov, D. and Reshkov, M. 2005. Strengthening of the Trakia motorway viaducts from km 37.4 to 38.2. *Proc. of the First Symposium of Bridges*, pp. 75–89, UACEG, Sofia, Bulgaria.

Kissov, D., Reshkov, M. and Yordanov, K. 2001. Southern overpass in Veliko Tarnovo: Reconstruction and strengthening of the arch bridge. *J. Construction*, 58(2): 7–12.

Kolchakov, M. 1957. A rational method for solution of arch bridge. *J. Construction*, 6(8): 12–19.

Kolchakov, M. 1958. Continuous beam, stiffened at middle span by rod's chain. *J. Construction*, 6(5): 17–23.

Konstantinov, H. 1984. Specific features of the design and construction of the trestle bridge Lyudmila Zhivkova in Sofia. *J. Roads*, 23(7): 13–14.

Lalov, I. and Christov, Ch. 1999. Reinforced concrete bridges. Sofia: T. Kableshkov University of Transport.

Lalov, I., Stakev, M., Hubchev, D. et al. 2009. Design and construction of the access bridge at the Airport of Sofia. *Proc. of the 3rd Scientific Symposium*, UACEG, May 29, pp. 62–68, Sofia, Bulgaria.

Minev, M. 2002a. Devil's Bridge over the Arda River: Investigation related to the construction of the water power system Gorna Arda. *J. Roads*, 42(5): 15–18.

Minev, M. 2002b. Investigation about the capacity and protection of the Kolyo Ficheto Bridge over Yantra river near Byala. *J. Roads*, 42(6): 12–17.

Minev, M. and Yakimov, I. 1987. In-situ test of bridge Saedinenie over river Maritza in Plovdiv. *J. Roads*, 26(4): 1–6.

Mladjov, R., Bonchev, M. and Grechenliev A. 1985. The steel bridge structure on the Hemus motorway. *J. Construction*, 32(12): 11–18.

Nikolov, P. 2009. Admissible damages in elastomeric bearings due to seismic actions. *Proc. of the 3rd Scientific Symposium*, UACEG, May 29, pp.166–177, Sofia, Bulgaria.

Partov, D. and Dinev, D. 2007. Structure, design and construction of a steel orthotropic bridge in Sofia. *Int. J. Advanced Steel Construction*, 3(4): 752–764.

Partov, D., Donchev, V. and Momchilov, M. 2004. Reconstruction of the 80-year-old steel bridge structures and their implementation in a road bridge over the Lom River. *Proc. of the 5th Int. Conf. on Bridges across the Danube*, pp. 317–324, Novi Sad, Serbia.

Partov, D. and Ivanov, R. 2008. Design and construction of the orthotropic steel bridge. *International Orthotropic Bridge Conference*, Sacramento, August 25–30, pp. 116–124, California.

Penev. V. 2001. The Stambolov bridge. *J. Roads*, 41(2): 8–10.

Penev, V. 2004. Balkan arch bridges masterpieces. *J. Roads*, 44(1): 24–25.

Raykov, L. 1957. Elastic continuous beam connected with a rigidly connected strut chain for railway transport. *J. Construction*, 5(6): 1–4.

Raykov, L. 1970. Bridges No 7,8 and 9 on the road "Pirdop-Rozino". *J. Roads*, 9(4): 1–7.

Reshkov, M. and Kisov, D. 2005. Makaza viaduct: Particularities of the support stresses and deformations during cantilever concreting. *J. Roads*, 45(6): 9–12.

SNiP. 1974. Building norms and regulations. Technical requirements for bridge design, Moscow.

Staykov, P. 1985. Steel bridge structures along the Hemus motorway: Viaduct 1 and Viaduct 2. *J. Construction*, 32(12): 19–22.

Staykov, P. 1986. Repair of the left carriageway superstructure of the Pionerski Lager viaduct on the Hemus motorway. *J. Construction*, 32(1): 29–31.

Stoyanov, B. 1990. Asparuhov bridge. *J. Roads*, 29(4): 4–5.

Stoyanov, L. and Mihailov, Ch. 1977. Specific features in organization and technology of assemblage of steel structure of Asparuhov Bridge. *J. Construction*, 24(10): 1–7.

Stoykov, G. 1976. Master Alexis and master Kolyu Ficheto. Septemvri, Sofia, Bulgaria.

Todorov, G., Dragoev, D., Minchev, P. et al. 1990. The viaducts on the Hemus motorway. *J. Roads*, 29(4): 5–10.

Topurov, K. 2005. Some specific features of seismic retrofit of existing bridges. *Proc. of the 1st Symposium of Bridges*, UACEG, pp. 19–30, Sofia, Bulgaria.

Topurov, K. and Dulevski, E. 2009a. Bridge over Varbica River at Momchilgrad. *Proc. of the 3rd Scientific Symposium*, UACEG, May 29, pp.69–78, Sofia, Bulgaria.

Topurov, K. and Dulevski, E. 2009b. Bridges on Plovdiv-Svilengrad railway. *Proc. of the 3rd Scientific Symposium*, UACEG, May 29, pp. 5–26, Sofia, Bulgaria.

Yakimov, I. 1990. The bridge Saedinenie in Plovdiv. *J. Roads*, 30(4): 21–23.

Yovchev, R., Kisov, D. and Reshkov, M. 2000. Viaduct at the 48th km Hemus Highway and the technology cantilevered bridge cast in-situ in Bulgaria. *J. Construction*, 57(4): 13–18.

Bridge Engineering in Croatia

Jure Radić
University of Zagreb

Goran Puž
Institute IGH

7.1 Introduction

7.1.1 Geographic Characteristics

The Republic of Croatia is a medium-sized Central European and Mediterranean, that is, Pannonian–Adriatic, country. It has a surface area of 56,500 km², with a population of 4.7 million. The capital city is Zagreb.

One of the most significant values of the Croatian state territory is its position within the Central European, Mediterranean, and Danube regions, at the crossroads of ancient and modern paths from the east to the west and from the north to the south. The Adriatic Sea is part of the Mediterranean, deeply cut into the European continent at its northernmost part, giving the routes connecting the Adriatic ports and northern destinations international significance. The importance of transportation routes

FIGURE 7.1   Overview of Maslenica Gorge Bridges: state road steel bridge (front) and motorway bridge (back). (Courtesy of the Croatian Motorways Limited Company–HAC d.o.o., Croatia.)

and the demanding relief characteristics required the construction of significant bridges on Croatian territories from ancient times to the present (Radić 2004).

The relief characteristics strongly influence the diversity of structural types being designed and constructed. The northeast region, consisting of the Danube Plain or the Pannonian Plain, is mostly bound by bordering, navigable rivers: the Danube, Sava, and Drava Rivers. Long bridges have been built over these lowland rivers, on mostly gravel soil, occasionally sandy or clayish. The bridges have larger central spans, the sizes of which have been adapted to the navigational waterways.

The central part of the state is hilly, mountainous terrain and calcareous massifs spreading in the direction perpendicular to the main transport routes connecting the plain and the coast. The mountains are not very high but are marked by karst phenomena—canyons and river valleys that needed to be bridged by medium- and large-span viaducts, frequently on very high piers. Some of these karst phenomena are found at locations that fall under the natural monuments category and are protected as such, representing an additional challenge for bridge engineers.

The most significant Croatian bridges have been built in the coastal region, over the straits (Figure 7.1). There are karst slopes steeply immerging into the sea, and some of the 1300 islands in the Adriatic are connected or can be connected with the mainland by large-span bridges. It must also be emphasized that most Croatian territory is in a seismically active zone, classifying the design accelerations for new bridges mostly between 0.20 g and 0.40 g.

7.1.2 Historical Development

7.1.2.1 General

Considering the large diversity of obstacles, Croatia has developed a school of bridge design and construction whose accomplishments have worldwide significance. Keeping in mind the long historical period during which Croatia had not achieved full political independence, the first significant structures were designed and constructed by experts educated in the neighboring West European countries.

Technical university education started to develop at the beginning of the twentieth century with the establishment of the High Technical School in Zagreb. Since then, bridge design and construction systematically developed into a technical branch whose activities and achievements in Croatia do not lag behind those of developed world countries; on the contrary, some achievements are recognized worldwide as renowned original achievements.

The question arises: How do we recognize original achievements in bridge construction? The answer is through a pronounced approach to functional and design values, through rational utilization of different materials, attention toward durability, and interconnection of the design and construction processes. These principles are hard to attain during actual work, where compromise is unavoidable and original design ideas are difficult to realize. A wider overview of bridges designed by Croatian engineers is directed toward structures with special characteristics. It should be mentioned, however, that such structures appear after numerous typical structures. From a large group of designers, those stand out who are willing to abandon the everyday routine, and design a larger, aesthetically different type of bridge. This is why this overview also presents the usual types of structures with their specific characteristics resulting from Croatian experiences.

Today, one can say that the foundations of bridge design lay in a high level of education and theoretical based knowledge, as well as knowledge acquired during construction works. The end of the twentieth and the beginning of the twenty-first century were marked by extreme growth of bridge construction in Croatia, because of numerous post-war reconstructions and intensive development of road network (Radić 2006). There are several construction companies active in Croatia that have the know-how and technologies to design and build large, prestressed concrete bridges, as well as an industrial plant able to design and manufacture large steel bridge superstructures. The development of bridge construction has been especially enhanced by bridge rehabilitation and remediation after the Croatian War of Independence from 1991 to 1996, and the construction of a system of motorways initiated in 1999 as a high-priority state development project. Croatian builders today successfully trade their knowledge on the markets in their immediate neighborhood in Europe, but also on the Russian, Asian, and African markets.

7.1.2.2 Concrete Girder Bridges

Reinforced or prestressed concrete superstructures are most utilized in Croatia. For small spans, cast-in-place solid slabs are constructed. Only a few prestressed slabs were built, due to cost-effectiveness. The use of cored (voided) slabs is not advised, at least not for roads with high traffic volume, due to some bad experiences encountered in practice. Reinforced slab bridges up to 18 m in span are deemed acceptable. If longer spans are required and construction on falsework is acceptable reinforced concrete beam superstructures are to be used.

Viaducts of various lengths are typical for hilly and mountainous terrain. The most frequently used typical structures on roads are grillages composed of precast prestressed concrete girders composited with cast-in-place deck slabs and transversal girders. Such cross sections successfully combine the advantages of precasting and monolithic construction. Two girder shapes, bulb tee and single tee, are usually designed. Both pretensioning and post-tensioning are utilized, leaving the final choice to the contractors. The span lengths of such structures vary mostly between 30 and 40 m; only exceptionally, girders 50 m long are designed. Certain doubts about the durability of such structures arose from bad experiences with the deterioration of old structures at the connections between longitudinal spans over piers. For this reason, the older types of continuous slab were replaced by structures with full continuity achieved over solid cross girders. The designers' intention to extend continuous superstructure segments in order to reduce the number of expansion joints is also evident. A small number of overpasses and a few viaducts were constructed as monolithic structures by concreting on scaffold.

For span length over 50 m or more complex span lengths, bridges are typically built by the longitudinal launching and free cantilevering method. Monolithic prestressed box structures built by these technologies became common superstructures on Croatian motorways at locations high above the ground. Furthermore, contractors have mastered the construction of such structures in horizontal and vertical curves.

The simplest and safest method for forming the superstructure is by placing precast girders in such a way that their upper flanges make contact. Usually, a 20 cm thick deck slab (25 cm for highway viaducts) is cast in situ over the precast girders, acting compositely with them for an additional dead load and live load. Cross girders are cast at supports only.

7.1.2.3 Steel and Composite Girder Bridges

Today, composite-beam bridges are not built as frequently, although at the beginning of the twenty-first century several structures of this type were constructed. However, as far as achievements go, one has to mention the world's first bridge where efficiency of composite steel and concrete was scientifically confirmed. The bridge over the Sava River in Zagreb is a pioneering steel–concrete composite girder bridge. Girder is a completely welded structure, and this bridge was built at a time when welding of steel structures was at its beginnings. The bridge was completed in 1937 as a grillage structure made of longitudinal and cross girders over four spans, with a composite concrete carriageway deck and longitudinal steel beams. Load testing was performed on the completed structure to verify composite action and role of the carriageway deck in the main load transfer process. The bridge designers were Jure Erega and Milivoj Frković.

At the beginning of the twentieth century steel bridges designed in Croatia were very similar to the typical truss structures of the time, mostly railway bridges. However, diverse structures started to appear and steel bridge design in the second half of the twentieth century stagnated due to a strong pervasion of concrete and prestressed structures. A new uplift in steel construction followed after the Croatian War of Independence, during the rehabilitation of bridges destroyed in the war, when the heavy demolished structures had to be replaced by lighter ones, along with upgrading the design traffic load and road width.

7.1.2.4 Arch Bridges

Croatia has a long tradition in arch bridge construction, from ancient times until the present. Some arch bridges constructed in Croatia are acknowledged worldwide for their innovative technology and also for their span length (Figure 7.2). These structures in their immediate environment improve the

FIGURE 7.2   (See color insert.) Overview of Krk Bridge. (Courtesy of the Institute IGH, Croatia.)

overall line of sight of the landscape, adding a new dimension to the engineering profession in the field of environmental protection and sustainable development. New, more cost-effective structural systems replaced arch bridges in the span range, a field in which they dominated. Nevertheless, arch bridges can be competitive under certain circumstances and this is visible in examples of some new bridges built in Croatia.

7.1.2.5 Cable-Supported Bridges

Cable-stayed and suspension bridges are rare in Croatia, but the achievements witnessed in this field show the readiness of builders to tackle such challenges.

7.1.2.6 Bridge Aesthetics

It should be noted that little attention has been paid to the visual appearance of bridges outside towns, where architects did not participate in the design stage. The situation in towns is somewhat different, due to the occasional international tenders for bridge design taking place. Teams participating in these tenders as a rule have an employed architect, making bridge aesthetics a more regarded and studied subject.

An excellent example of unison between architects and engineers is the construction of the somewhat smaller-size Memorial Bridge in Rijeka (Figure 7.3) in 2001. It is a pedestrian bridge, at the same time a monument dedicated to the Croatian defenders—Croatian soldiers who fought in the War for Independence. The bridge had to be nested into the existing road network on one side, with minimum intrusion into the canal space under the structure. An extremely slender, steel, box-type fixed superstructure was chosen. With a 36.6 m span, the structure is only 65 cm high, giving the bridge a span-to-height ratio of 1:55.5. The bridge was designed by the 3LHD Architectural studio from Zagreb. One curiosity is that the span structure was made in a nearby shipyard (the Third May Shipyard in Rijeka).

FIGURE 7.3 The Memorial Bridge in Rijeka. (Courtesy of Institute IGH, Croatia.)

7.2 Design Practice

7.2.1 Loads on Road and Pedestrian Bridges

Loads on road bridges in Croatia are to be determined according to series of standards, HRN ENV, which are based on European prestandards (ENV) and adjusted to specific climate, environment, and traffic circumstances given in the Croatian National Application Document (NAD). Road traffic actions are represented by a series of load models that represent different traffic situations and different components of traffic action. Load model 1 is the main model for general and local verifications of the structure, which covers most of the effects of heavy traffic (Figure 7.4a). Model 1 prescribes a uniform load of 9 kN/m² in main traffic lanes and a uniform load of 2.5 kN/m² on the rest of the traffic surface. The lanes are 3 m wide. The concentrated load, representing a heavy vehicle, is simulated by 4 point loads of 300 kN each in the main traffic lane, with 4 point loads of 200 kN each in the secondary traffic lane and with 4 point loads of 100 kN each in the third (the last) traffic lane. The distance between loads is 1.2 m longitudinally and 2 m transversally. Dynamic actions are included in the models, but for fatigue check loads are to be multiplied by an additional dynamic factor.

Pedestrian bridges are to be calculated using uniformly distributed load of 5 kN/m², which should be lowered for structural elements longer than 10 m to a minimum of 2.5 kN/m².

Almost all of the Croatian territory lies in seismically active zones, so that seismic actions often govern the design of bridge parts, material consumption, detailing, and the general structural resistance and stability of bridges. Seismic design procedures are based on the ductile nonlinear behavior of bridges, assuming that the bridge structure may dissipate seismic energy through formation of plastic hinges, normally in piers. This allows the application of linear seismic analysis based on the design response spectrum. Response spectra are defined by four expressions, for the four ranges of vibration period, and

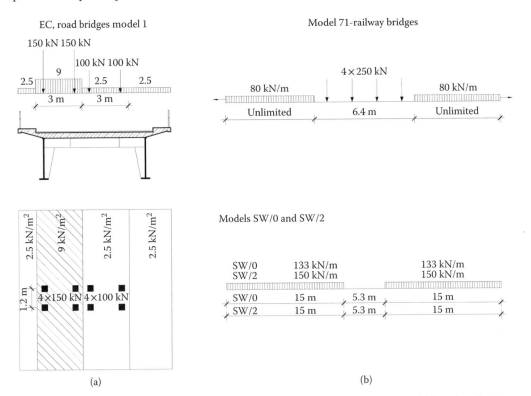

FIGURE 7.4 (a) Load model for road traffic based on European prestandards. (b) Load model for railway bridges—Model 71 for normal railway transport along principal railway lines, and models SW/0 I SW/2.

are modified according to the type of soil. Soil acceleration on soil type A (solid rock) is given in NAD for four different areas of seismic intensity based on MKS-64, from 6 to 9. Design soil acceleration varies from 0.05 g (seismic intensity 6) to 0.30 g (seismic intensity 9) and has a 500-year return period.

Parts of the bridge deck outside the roadway are to be designed for snow load. The Croatian territory is divided into four snow-related climatic zones with corresponding characteristic snow load values at various altitudes. Snow loads vary from 0.35 kN/m^2 to 13.2 kN/m^2. Wind forces are to be determined using an expression that takes into account wind velocity, reference area, and correction factors. According to the map of reference wind velocity in Croatia is divided into five wind-related zones and ten different regions. Reference wind velocities for bridges without traffic loading vary between 22 m/s and 50 m/s. For bridges under traffic a maximum wind velocity of 30.5 m/s is recommended. For bridges in specific areas of strong winds, wind-tunnel testing is recommended. The highest wind speed in Croatia was measured at the location of the Maslenica Highway Bridge in December 2003, with a gust speed of 85 m/s. Thermal actions on bridges are to be determined using temperature-based zoning of Croatia, given in NAD following new design codes (Radić et al. 2008c).

7.2.2 Loads on Railway Bridges

European standards dealing with railway transport through four loading schemes are applied in the design of railway bridges. Model 71 is used for normal railway transport along principal railway lines (Figure 7.4b). Model SW/0 simulates normal railway transport on continuous beams, while model SW/2 is used for railway lines characterized by heavy traffic. The empty-train model contains vertical load amounting to 12.5 kN/m, and is used in the verification of overturning stability, with full horizontal wind load applied in a 4 m in high surface, and for unlimited lengths. Dynamic influences are also taken into account during determination of vertical load. This is accomplished by multiplication with the dynamic factor, which normally varies between 1 and 2 and is dependent on the length of individual structural elements.

7.2.3 Bridge Detailing Manual

Bridge furnishings, such as pathways, handrails, safety barriers, waterproofing, drainage, lighting, bearings, and expansion joints are given special attention, because although they are functionally essential and aesthetically important, they are often considered belatedly in the design. A compendium of drawings depicting recommended bridge furnishing details is provided as a help to the designer. Only details proven reliable in current practice in terms of durability, maintenance, and repair are utilized, enhanced by explanatory notes and instructions.

7.3 Concrete Girder Bridges

7.3.1 Drežnik Viaduct

The longest viaduct in Croatia is the Drežnik viaduct (Figure 7.5). It is 2485 m long and was constructed utilizing precast prestressed concrete girder superstructure, taking the Rijeka–Zagreb Motorway over the Karlovac boundary areas. In order to enable free expansion of the town of Karlovac, the motorway was elevated from the ground level using this viaduct, crossing over several local roads, a railway line, and the Kupa River. Its design was developed by the IGH Institute, and it is presently managed by the Concession Company of the Rijeka–Zagreb Motorway. The viaduct was constructed in a very short period of time. The bridge was built by four leading Croatian bridge construction companies: Hidroelektra–Niskogradnja, Industrogradnja, Konstruktor, and Viadukt. A very interesting fact is that each of these companies constructed a separate bridge section, from the foundations to the road equipment.

FIGURE 7.5 Overview of the Drežnik viaduct. (Courtesy of the Croatian Motorways Limited Company–HAC d.o.o., Croatia.)

The viaduct is divided into seven segments, lengths 225 m + 270 m + 4 × 408 m + 358 m, with spans of 35 m on average. Part six has two-times-larger spans to cross the Kupa River. This extension was constructed with precast girder elements over the bank piers, acting as cantilevers and reaching 17 m on each side. After standard girders were erected between the cantilevers, continuity cables were placed. These cables were tensioned after in situ joining the precast cantilevers and girders. The piers are standard hollow, hammerhead piers with hollow box-type cross sections, with piled foundations (Friedl and Runjić 2001).

7.3.2 Dabar Viaduct

The largest span-precast prestressed concrete girders, almost 50 m, were used in the Dabar viaduct (Figure 7.6), which is the ultimate achievement in this kind of construction. The viaduct is on the A1 Motorway, which clearly states its importance: the route connecting Zagreb, the capital, with Split, the center of south Croatia. The viaduct was designed by the Faculty of Civil Engineering in Split, and is managed by the state-owned company for motorway operation, construction, and maintenance, Croatian Motorways Ltd.

The motorway centerline of the viaduct is in a horizontal curve with a 2400 m radius and a gradient of 1.857%. The superstructure consists of girders with an approximate length of 50 m (with differences in length due to the horizontal curve) weighing nearly 150 tons. The standard practice of joining the precast girders at the middle supports had been to utilize the continuous separated deck slab, thus eliminating the deck joints, but the durability aspect of this solution was questionable. During service the separated deck slab cracks and seeping water endangered the precast girder ends with tendon anchors, bearings, and piers. Full flexural continuity is therefore utilized on Dabar and other newer structures (Figure 7.7).

The superstructure is designed as a series of simple beams under self-weight and deck slab weight, and continuous beams under live traffic load. The post-tensioning is partial, that is, 40% of the traffic load is included in the boundary decompression moment. Sliding pot bearings are mounted on abutments and end piers, while the remaining piers are rigidly connected with the superstructure, ensuring the structure against seismic action forces and the movement of vehicles (Vlašić et al. 2006).

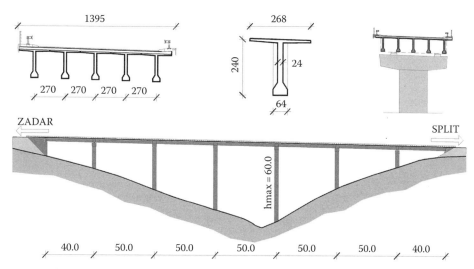

FIGURE 7.6 Sketches of the Dabar viaduct. (Courtesy of the Croatian Motorways Limited Company–HAC d.o.o., Croatia.)

FIGURE 7.7 Overview of the Dabar viaduct during construction. (Courtesy of the Croatian Motorways Limited Company–HAC d.o.o., Croatia.)

7.3.3 Jezerane and Mokro Polje Viaducts

These viaducts are located on the Tunnel Mala Kapela–Žuta Lokva section of the A1 (Zagreb–Split–Dubrovnik) Motorway. The given post-tensioned grillage system bridges are made with beams of a similar shape, however, different types of beams are used as well (Figure 7.8). The superstructures on a large number of overpasses are made of hollow plate beams connected by a shear key. Some larger viaducts have beams with a U-shaped cross section (Crnjak et al. 2008).

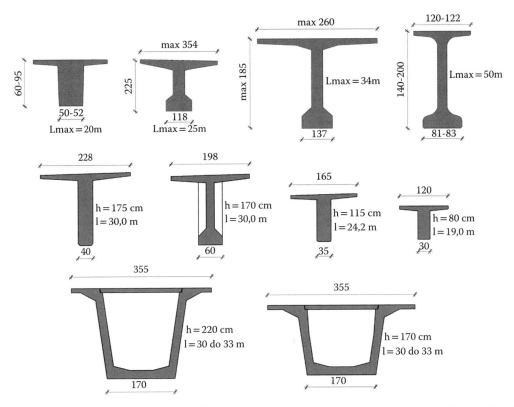

FIGURE 7.8 Cross sections of prestressed beam bridge girders utilized in Croatia. (Courtesy of the Faculty of Civil Engineering, University of Zagreb, Croatia.)

For the first time in Croatia, U-beams were used in the construction of superstructures on the A1 Motorway (Zagreb–Split), on sections passing through the Lika region. This type of beam was imposed by the foreign contractor, but it has proven efficient and has shown some static and construction benefits. The Jezerane and Mokro Polje viaducts (Figure 7.9) were designed by the IGH Institute.

7.3.4 Dobra Bridge

Grillages composed of precast prestressed concrete girders composited with cast-in-place deck slab and transversal girders, as discussed in Section 7.1.2.2, are considered to be a lot more economical than the ones used in other technologies adjusted for larger spans. Therefore, on some structures the main span, which is vaulted by prefabricated girders, has been extended by using piers with hammerhead beams. The basic principle of construction of such structures is the combination of free cantilevering, which is applied when concreting the segments above piers, with the subsequent assembling of girders to close the span. The overhangs above the piers have been subsequently made monolithic by girders. Dobra Bridge (Figure 7.10) on the A1 Motorway (Zagreb–Split) is an example of such a structure: its main span of 70 m is achieved using 40 m girders that were fixed to pier extensions (Crnjak et al. 2008).

7.3.5 Zečeve Drage Viaduct

Monolithic prestressed box-type superstructures constructed by incremental launching or free cantilevering erection procedures are becoming standard in Croatia. The Zečeve Drage viaduct is the longest viaduct built using incremental launching, with very complex geometry (Figure 7.11). The Zečeve Drage

FIGURE 7.9 Overview of the Jezerane and Mokro Polje viaducts. (Courtesy of the Croatian Motorways Limited Company–HAC d.o.o., Croatia.)

FIGURE 7.10 Overview of Dobra Bridge. (Courtesy of the Croatian Motorways Limited Company–HAC d.o.o., Croatia.)

FIGURE 7.11 Overview of the Zečeve Drage viaduct. (Courtesy of the Croatian Motorways Limited Company–HAC d.o.o., Croatia.)

viaduct carries the motorway connecting the Croatian capital and its largest port, Rijeka, over a deep valley. The motorway is managed by the Concession Company of the Rijeka–Zagreb Motorway.

The viaduct comprises two parallel structures, each providing a one-way multilane motorway carriageway. The first viaduct was completed in 2004, and the second four years later. Both structures were designed and constructed by Viadukt Construction Company from Zagreb. The total length of the structure is 920 m. The superstructure is a continuous prestressed concrete trapezoidal box girder with a constant depth of 4.2 m. A typical span is 50 m long.

The motorway is horizontally curved with radius of 2500 m and vertically concaved in the elevation R = 26.500 m, mostly overpassing its obstacles at approximately 50 m above ground level. The spatial curve of the bridge centerline and its length were the aggravating factors for implementation of the chosen construction technique, more so because the excavation was in a gradient ranging between 0.78% and 4.38%. The difficulty regarding the spatial curve was solved by choosing a spiral for the launching axis, but the cylinder over which the spiral was being lifted was wrapped around the axis passing through the points of interception of abutment-bearing axes and the superstructure axis. The length was overcome by a downhill boost, and arrestment was needed due to the longitudinal gradient, for which a special device was designed. The synchronized push and arrest actions were necessary to place a 22,000 ton box into the required position.

The bridge superstructure was prestressed in two stages. Centric prestressing was adopted for the first stage, with tendons placed in the deck slab and the bottom slab. After the launching had been completed the superstructure was prestressed by the eccentric tendons of the second stage, designed for the additional dead load and the traffic load. These tendons were placed within the box girder. The advantages are in thinner webs and reduction of the deadweight of the superstructure; also, tendons are accessible for visual inspection and replacing them is easier. Piers have a box-like cross section with constant dimensions. Foundations are shallow spread footings (Radić et al. 2006).

The incremental launching construction method enabled very quick erecting and construction: 37 segments were precast and placed in exactly 37 weeks, and each of the two viaducts was built in 2 years. In order to reduce cantilever bending moments during launching, a steel nose 32 m long was added to the front part of the first segment. During launching the superstructure was placed on auxiliary sliding bearings which were replaced with the permanent ones after the completion of the erection (Figure 7.12).

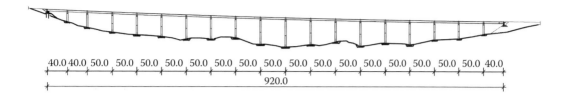

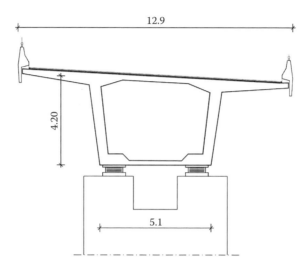

FIGURE 7.12 Sketch of the Zečeve Drage viaduct. (Courtesy of the Croatian Motorways Limited Company–HAC d.o.o., Croatia.)

7.3.6 Kamačnik Bridge

Kamačnik Bridge is on the same road as the previously described Zečeve Drage viaduct, and is managed by the same company. The bridge was designed by a design team from the Civil Engineering Faculty in Zagreb lead by Zlatko Šavor, and built by Konstruktor, from Split. The bridge design was influenced strongly by environmental protection of the canyon. This restraint resulted in a span of 125 m at a height of approximately 60 m above the river bed. Two parallel bridges were constructed, one for each motorway carriageway, approximately 240 m long. The main 125 m bridge span is the largest bridge of this type in Croatia, while the side spans are 70 m and 24 m respectively (Radić et al. 2006).

The bridge grade line is in a 5.7% vertical slope. The superstructure is a horizontally curved (R = 750 m) continuous prestressed concrete box girder, with variable heights ranging from 7.2 m above the piers to 3.2 m in the middle center, and a constant width (Figure 7.13). Location restraints necessitated an extremely nonsymmetrical layout with the short northern span supported by heavy counterweight abutment.

The superstructure was built using the cantilever method with segments cast in place, on the form carrier. Auxiliary piers in the left span were built because of nonsymmetrical disposition. Part of the left side span was cast on scaffolding. The right cantilever was cast after the counterweight abutment was finished, together with short span structure on the right side, cast on scaffolding (Figure 7.14). The key segment in the midspan was constructed after the removal of the form carriers, in special scaffolding.

Significant difficulties arose during the shallow foundation work on a rock base. Not only were the higher pier foundations on a fault, but there was a cave, investigated and filled with concrete,

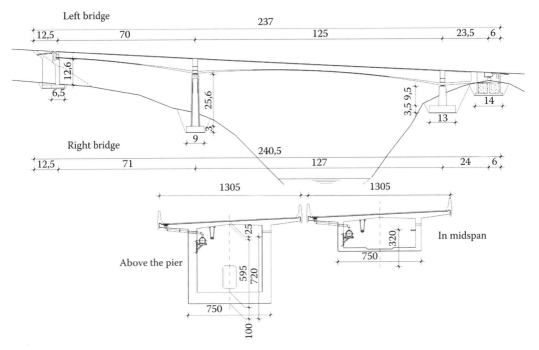

FIGURE 7.13   Sketch of Kamačnik Bridge (Courtesy of the Croatian Motorways Limited Company–HAC d.o.o., Croatia.)

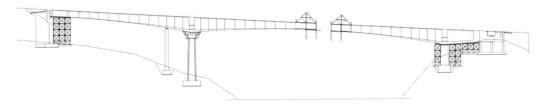

FIGURE 7.14   Construction sequence of Kamačnik Bridge. (Courtesy of the Faculty of Civil Engineering, University of Zagreb, Croatia.)

under the foundation base (Radić et al. 2003a). Construction of the Kamačnik Bridge was completed in 2003 (Figure 7.15).

7.3.7 Rječina Bridges

The first (north) Rječina Bridge on the Rijeka bypass was constructed in the 1980s. Along with this, preliminary works were executed for future construction of a parallel structure, which is postponed until the construction of a second Rijeka Bypass carriageway. The first bridge was designed by the IPZ Design Bureau from Zagreb, for the Croatian Roads Company, and was built by the Hidroelektra Construction Company. The second (south) parallel bridge was designed by experts from the Faculty of Civil Engineering in Zagreb for the Rijeka–Zagreb Motorway Concession Company, who has managed the bypass since 1998. The south bridge was completed in 2009, built by two contractors: Hidroelektra Construction Company from Zagreb and Konstruktor from Split.

The bridge crosses the Rječina River canyon approximately 100 m above the water level. A rigid frame structure with inclined legs was selected for this site after careful economic and aesthetic considerations.

FIGURE 7.15 Overview of Kamačnik Bridge during construction. (Courtesy of the Faculty of Civil Engineering, University of Zagreb, Croatia.)

FIGURE 7.16 Overview of the Rječina Bridges. (Courtesy of Institute IGH, Croatia.)

The north bridge (the older one) consists of a 98.4 m main span and two side spans of 45 m. The effective span of this bridge, considering the base of the slant legs, amounts to 131.65 m. The new south bridge is similar in appearance to the north structure. The terrain required an increase in span length, with the main span of 108.5 m and side spans of 50 m. The distance between the bases of inclined struts amounts to 146 m (Figure 7.16).

Differences in the design of the new bridge in relation to its original design dating back to the late 1970s generally arise from the increased loadings as defined by the contemporary structural design codes. The new design was not only related to new traffic loads, but also to wind actions and seismic effects. Additionally, a more robust concrete structure was designed to meet stricter durability requirements. There are significant differences in construction procedures between the south and north bridges. The first (north) Rječina Bridge was constructed of match-cast precast segments using a launching girder. The second (south) bridge was constructed using the cantilever construction method with in situ concreting (Šavor et al. 2007).

7.4 Steel and Composite Girder Bridges

7.4.1 Belišće Bridge

The bridge over the navigation channel of the Drava River near the town of Belišće was completed in 2000, built for the Croatian Roads Company, the principal operator of state roads. The design was developed at the Faculty of Civil Engineering in Zagreb (Figure 7.17). It consists of the main bridge span, a continuous two-span composite girder structure, and pretensioned concrete approach viaducts. The cross section of the central structure is a composite twin steel girder system and a concrete deck. The fixed bearing is above the central pier. The free profile required for navigation was determined by the longitudinal arrangement with two 95.5 m spans and one pier in the river bed.

The main spans are continuous two-span twin steel plate girders of various heights ranging from 3.3 m at the lateral piers to 5.2 m at the central pier. The upper bearing edge follows the gradient, and the intrados is parabolic. The cross girders are frame-like. Composite construction of the concrete deck

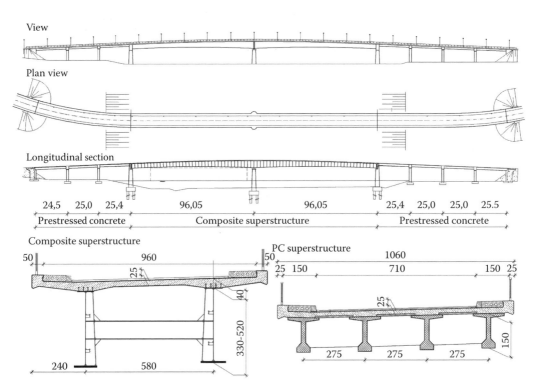

FIGURE 7.17 Sketches of the Belišće Bridge. (Courtesy of the Faculty of Civil Engineering, University of Zagreb, Croatia.)

was achieved by stud connectors. Tensile stresses in the concrete deck plate were decreased by forced deformation of the structure and lifting of bridge ends. The approach viaducts superstructures are made of precast prestressed concrete girders with an additionally concreted deck. Continuity of the structures between the segments was achieved by nonprestressed steel reinforcement (Šavor et al. 2001b).

7.4.2 Mirna Viaduct

The Mirna viaduct is on the Adriatic–Ionian Motorway route connecting the Mediterranean region from Venice, Italy to the Greek port of Kalamata. It was designed and built for the concession company BINA Istra by the Structural Department of Civil Engineering Faculty in Zagreb. The viaduct takes the road over Mirna River valley and basin. The project included a bridge for the right motorway carriageway, completed in 2004, while the other parallel bridge will be constructed at a later stage (Radić et al. 2006).

The bridge design was somewhat restricted by the need to decrease the overall mass as much as possible due to extremely difficult foundation work conditions in the river valley, namely a mud layer. On the other side, the length of the spans had to be increased as much as possible to bring the foundation work to a minimum. The typical span length was defined at 66.5 m, for a total length of 1345.86 m with 22 spans. The final arrangement met all requirements of the soil beneath the bridge, the river, and the canal (Figure 7.18).

The horizontally and vertically curved superstructure is made of two continuous steel plate girders of constant height that are composite with a concrete carriageway deck and crossbeams. The reinforced concrete deck is basically 25 cm thick with haunches at 35 cm above the main girders. External deck edges are increased to thickness of 54 cm and have a safety fence anchored on. The foundations for most piers are on driven steel piles. The pier cross sections are H-shaped, and the central web is placed vertically to the longitudinal bridge axis. The highest pier is 40 m high (Šavor et al. 2006).

The superstructure was erected by launching already-assembled steel girders from both sides of the bridge in the longitudinal grade of the tangent on the bridge axis. After the launching process, the superstructure was lowered on its pot-type bearings. The deck slab was cast on scaffolding, after the launching (Figure 7.19).

7.4.3 Gacka River Bridge

The Gacka Bridge is on the motorway that connects Zagreb with the second-largest Croatian city and one of our most important ports, Split. It is in the immediate vicinity of the town of Otočac, crossing over the Gacka River and the diversion canal used to fill the nearby storage lake with water from the river. The Gacka River has high-quality water along the greater part of its course, placing the complete valley under a special protection regime. Ecological considerations influenced the length, structure, and safety fence elements as well as the drainage system. The bridge design was developed by the Design Bureau IPZ from Zagreb, built in 2004, and is managed by the state motorway company HAC–Croatian Motorway Company (Figure 7.20).

Considering that the structure is on a motorway, two parallel bridges were designed, one for each carriageway. The bridges are in a horizontal curve with a radius of 4000 m and in a vertical concave curve with a radius 25,000 m, with a mean upward grade of ~2.7%. The motorway intersects with the canal and the river under a sharp angle, thus defining the span size since the substructure elements could not be positioned or constructed in the canal or river beds.

Longitudinally, the structure is made of two units divided by an expansion joint. The diversion canal is crossed over by a steel box-type section of variable height, and the bridge over the river is made of steel plate girders. Both sections have an orthotropic steel carriageway deck. The superstructure over the deviation canal is a three-span continuous steel box-type section, with transverse girder frames every 4 m. The spans are 55 m + 90 m + 55 m. The continuous girder in continuation, which crosses the river,

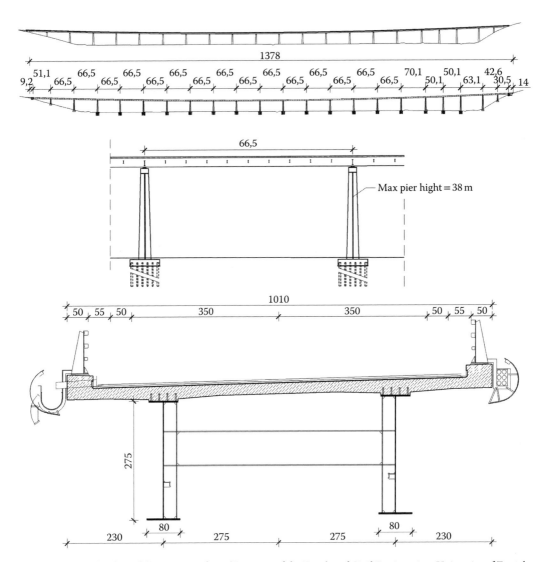

FIGURE 7.18   Sketches of the Mirna viaduct. (Courtesy of the Faculty of Civil Engineering, University of Zagreb, Croatia.)

has five openings with spans of 44 m each and an end span of 32 m. The bridge substructure is made of concrete. All piers and abutments are on bored pile foundations, diameters 150 cm, connected at the top by a head beam (Dumbović et al. 2003).

7.4.4 Limska Draga Viaduct

The viaduct over Limska Draga creek is situated on the north shore of the Istria peninsula, where the coastal road intersects a deep valley. The road alignment at the bridge is at a height of 125 m from the ground, the valley being more than 500 m wide. The bridge was designed by the Design Bureau IPZ from Zagreb. Today it is managed by the concession company Bina–Istra. Works on the concrete bridge parts were performed by the company Viadukt, from Zagreb, and the precast parts of the span superstructure by the company Đuro Đaković, from Slavonski Brod. Bridge construction was completed in 1991 (Dumbović 1995).

The superstructure is a continuous steel box-type section extending over five segments 80 m + 100 m + 160 m + 100 m + 80 m, for a total length of 520 m. The box-type section height in the central

FIGURE 7.19 Overview of the Mirna viaduct. (Courtesy of the Croatian Motorways Limited Company–HAC d.o.o., Croatia.)

FIGURE 7.20 Overview of the Gacka River Bridge. (Courtesy of the Croatian Motorways Limited Company–HAC d.o.o., Croatia.)

segment is 5.5 m, decreasing toward the end segments to 5 m. The central span is somewhat larger than is structurally optimal, but this is because of very unfavorable soil conditions for foundation work at the bottom of the valley. The piers are concrete, hollow box-type cross sections, 800 cm × 250 cm at the top. The wider side, perpendicular to the bridge axis, does not change in height, while the narrower side widens toward the bottom, with an increase of 1:80. The highest pier is 112 m high.

FIGURE 7.21   Overview of the Limska Draga viaduct. (Courtesy of the Croatian Motorways Limited Company–HAC d.o.o., Croatia.)

Bridge girders were erected and assembled on auxiliary steel piers, two in each of the end segments and one in the adjoining segments. The main span was assembled by incremental launching. A total of 22 prefabricated segments were constructed on each side of the bridge, each segment assembled of 8 parts. The assembly work lasted 14 months (Figure 7.21).

7.4.5 Bridge Rehabilitation Projects

This section will present several bridge rehabilitation and repair projects after the Croatian War of Independence (1991–1995), where new assemblies had to replace the ones destroyed during the war. The reconstruction of the damaged bridges may be, and often is, an engineering task even more demanding than the design of a completely new bridge. Major challenges were: (1) all destroyed bridges were designed for the old live load standards and some of them spanned over a motorway too narrow for today's standards; (2) the foundations and pier elements of the old substructures should be used as much as possible during the reconstruction process; and (3) scour considerations for pier foundations that remained standing after bridge superstructures fell into the river, which changed its flow and created scouring problems. In circumstances such as these, most demolished concrete bridge superstructures were replaced with new steel superstructures, while piers were repaired (Radić et al. 2008a).

7.4.5.1 Orašje-Županja Bridge over the Sava River

The road bridge between Orašje and Županja was completed in 1968. This bridge spans the state road over the navigable Sava River. It is managed by the state companies of Croatia and Bosnia and Herzegovina in charge of state roads. The superstructure was divided into three sections: the left bank approach bridge with 10 spans of 33.6 m; the main bridge over the river with three spans, with side spans of 85 m and a central span of 134 m; and the right bank approach with 4 spans of 33.6 m. The main bridge over the navigable part of the river consisted of three continuous spans of two steel plate composite girders of various heights. The concrete deck was prestressed by tendons in the region of hogging bending moments and also by applying special erection measures. Approach bridges were simply supported steel composite girders of constant heights and reinforced concrete slabs.

All the piers and the abutments were concrete. Pile foundations were utilized except for the two middle piers of the main bridge, which were founded on the caisson foundations. The main bridge superstructure totally collapsed during the 1991 war, while the approach bridges were left intact (Šavor et al. 2004b). The new superstructure utilized the existing approach bridges and all the existing foundations (Figure 7.22). This constraint precluded designs in prestressed concrete because the structural weight would have been too large. In addition, the destroyed bridge was designed according to the old bridge code, which prescribed lower live loads than the ones required at present. The height of the main bridge superstructure couldn't be increased because of the navigation requirements. Because of these constraints, the new superstructure of steel orthotropic plate girder was designed and completed in 1998 (Figure 7.23).

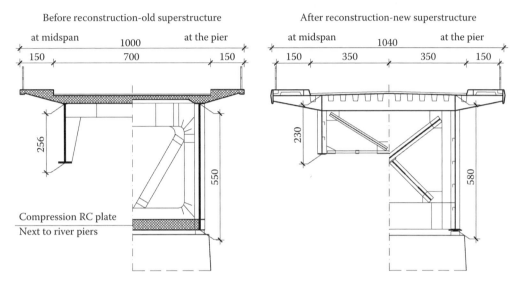

FIGURE 7.22 Cross section of the original superstructure (left) and of the reconstructed superstructure of the main bridge near Orašje (right). (Courtesy of the Faculty of Civil Engineering, University of Zagreb, Croatia.)

FIGURE 7.23 Overview of the Orašje-Županja Bridge over the Sava River. (Courtesy of the Faculty of Civil Engineering, University of Zagreb, Croatia.)

7.4.5.2 Jasenovac Bridge over the Sava River

The bridge over the Sava and Una rivers in Jasenovac connects three geographic units divided by two rivers, the navigable Sava River and a somewhat smaller Una River. It is immediately next to the flow of the Una into the Sava. This bridge, originally built in 1973, carries the state road connecting Croatia and Bosnia and Herzegovina. Reconstruction of the main structure over the Sava River is described next. It was designed by the Civil Engineering Faculty in Zagreb. The bridge is under the management of the state company, Croatian Roads Ltd., Company for Operation, Construction, and Maintenance of State Roads.

The superstructure of the bridge over the Sava River near Jasenovac was destroyed and some piers were severely damaged during the war in 1991. After thorough consideration of possible alternatives, it was concluded that a steel superstructure should be built instead of the original prestressed concrete one. Piers were repaired or reconstructed on existing foundations. The original superstructure was a three-span continuous prestressed concrete box girder, constructed by the free cantilevering method. Foundations were bored piles on a gravely and sandy soil.

The bridge was demolished in such a way that all remaining span structure parts had to be removed. One of the piers in the river was also destroyed up to the pile cap and foundation slab. Inspection showed scouring up to 5 m under the original bottom level and consequently a decreased load-carrying capacity of the foundation. The solution was to keep the basic original outlines and clearances of the original bridge (Figure 7.24).

Reconstruction was planned in two stages: where there was no damage only the bridge equipment was renewed, while the destroyed parts were replaced with a new superstructure. The superstructure is an orthotropic box girder 6 m wide. The profile height above the river piers is somewhat decreased from 5.5 m to 4.5 m in relation to the original superstructure. The structure height at the center and at the ends is equal to the original bridge, 2.5 m. The old piers were either repaired or replaced as well. A new steel superstructure was chosen as the best solution. The extrados of the bridge was kept the same, while the intrados was modified. The cross section of the steel girder is a box with a variable height and an orthotropic deck plate (Figure 7.25).

The original bridge superstructure weight was 2310 tons. The weight of the new steel superstructure is 1100 tons, giving a pile load of the piers in the river of 460 tons. It was concluded that, with adequate rehabilitation of the river bed, the pile loads are acceptable and additional reconstruction of foundations was not necessary.

The new steel superstructure was erected by cantilever method using a floating crane. The length of the assembled parts was about 12 m. The webs were spliced by high-strength bolts while other parts were spliced by welding (Radić et al. 2008a).

FIGURE 7.24 Overview of the Jasenovac Bridge over the Sava River during reconstruction. (Courtesy of the Faculty of Civil Engineering, University of Zagreb, Croatia.)

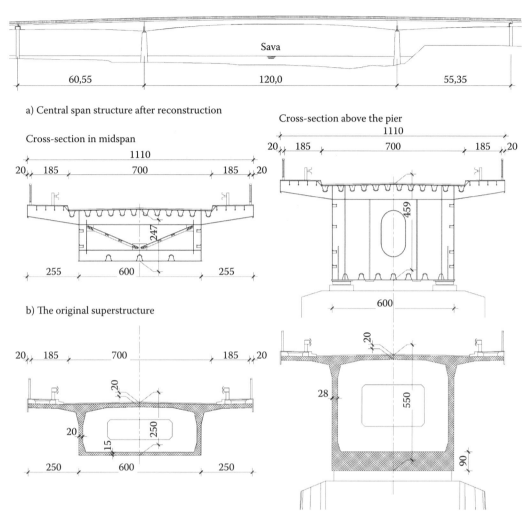

FIGURE 7.25 Sketch of the Jasenovac Bridge over the Sava River: the original and new superstructures. (Courtesy of the Faculty of Civil Engineering, University of Zagreb, Croatia.)

7.5 Arch Bridges

7.5.1 First Modern Arch Bridge: Skradin Bridge

The first modern concept bridge is at the same time an outstanding representation of structural and design achievement. The bridge over the Krka River in the town of Skradin was constructed in place of the old bridge demolished during World War II. The old structure was a through-truss bridge. At a time of great lack of construction material, the great Croatian bridge designer Kruno Tonković designed and built this bridge with very low and slender arches and the carriageway above (Figure 7.26). The arches were made of the remains of the old superstructure, which were pulled out of the river and then filled with concrete, in order to achieve the required load-carrying capacity (Radić et al. 2004). Skradin Bridge even today satisfies the demands of modern traffic loads. A similar concept, with tubular steel arches filled by concrete, was applied in China in 1998 during construction of the arch of the Wangxian–Jangce Bridge, with a record span length of 425 m.

FIGURE 7.26   Overview of Skradin Bridge over the Krka River. (Courtesy of the Faculty of Civil Engineering, University of Zagreb, Croatia.)

7.5.2 Concrete Arch Bridges

Concrete arches were favored for large spans up to the middle of the twentieth century. Then, other structural systems were developed for large spans, which made arches noncompetitive, primarily due to their high construction costs. Heavy falsework was needed for arch construction, which was often more complicated than the arch itself. Four arch bridges, the Šibenik Bridge (246 m span), the Pag Bridge (196 m span), and the Krk Bridge (two arches, spans of 390 m and 244 m) were built from 1966 to 1979. All of them were designed by the same designer, Ilija Stojadinovic, and built by the same construction company, Mostogradnja Company. The most famous is the Krk 1 Bridge, which held the world record in concrete arch span for 20 years.

One of the greatest engineers of all time, Eugene Freyssinet, applied in the 1950s an innovative method of the concrete arch erection by utilizing the free cantilevering method for some arch bridges in the South America. All four arches mentioned above were erected utilizing the free cantilevering method, which was a significant innovation at the time in the field of reinforced concrete arch construction. Ilija Stojadinović perfected this construction method, which did not require heavy scaffolding for large concrete arches. The method was first utilized during the construction of the Šibenik Bridge (Figure 7.27).

All four bridges have a common characteristic: box-type arch cross sections consisting of three cells. The piers are extremely slender and the superstructure comprises precast prestressed concrete girders. These bridges exhibit some common features that influence their durability: minimum dimensions of structural elements and thin concrete covers. They have been inspected, tested, and repaired many times. The results of these inspections confirmed degradation of the structural parts, mostly due to exposure to chlorides from sea water (Radić et al. 2003c).

7.5.2.1 Šibenik Bridge

The Šibenik Bridge, named after the nearby coastal town Šibenik, was the first bridge in the world completely built with the free cantilevering method. It was completed in 1966, and today it is managed by Croatian Roads Ltd., Company for Operation, Construction, and Maintenance of State Roads. It crosses over the Krka River near its delta, where the width reaches 251 m, and the arch span is 246 m (Figure 7.28). It is on the coastal road that runs along the complete Croatian coastline. The arch itself is a hollow box-type, with a sickle-shaped cross section. Near its springing, where it is fixed, it has a height

FIGURE 7.27 Overview of Šibenik Bridge over the Krka River during construction in 1965. (Courtesy of the Faculty of Civil Engineering, University of Zagreb, Croatia.)

FIGURE 7.28 Overview of the Šibenik Bridge. (Courtesy of the Faculty of Civil Engineering, University of Zagreb, Croatia.)

of 2.9 m and at the crown 3.7 m. It is 7.5 m wide, consisting of three segments. The spandrel piers have an octagon-shaped cross section. The carriageway structure consists of a sequence of grillages of prefabricated concrete girders, prestressed transverse girders concreted in situ, and a carriageway deck at the level of the upper flange plate of the longitudinal girders, 13 cm thick.

The arch was concreted on a 29 m mobile steel trussed formwork, fixed at one end to the already constructed arch part, and at the other by skew staying cables. Two staying cables were fastened to each abutment pier top and two at temporary piles over the abutment piers. The first level of staying cables was positioned horizontally at the carriageway level, and the second transversely to the abutments where they are anchored into the concrete block.

7.5.2.2 Pag Bridge

The Pag Bridge over the Fortica Strait provides a fixed road link from the mainland to the island of Pag. The bridge is managed by Croatian Roads Ltd., Company for Operation, Construction, and Maintenance of State Roads. The open spandrel sickle-shaped concrete arch, with a 193 m span, a rise of 28 m, and a rise-to-span ratio of 1:7, was built in 1968. The same equipment and technology was used as for the construction of the Šibenik Bridge.

The Pag Bridge unfortunately deteriorated in a short period of time, due to the influence of a very aggressive maritime environment and also because of inadequate workmanship, poor detailing, and a stringent budget, which contributed to the unsatisfactory quality of works. The worst deterioration was reinforcement corrosion caused by high chloride content, which penetrated into concrete. The critical content of chloride ions in the concrete, which by most standards should not exceed 0.40% of the cement mass, has been measured at the reinforcement level as 1–5%. It should be noted that the salt concentration in Adriatic Sea water is higher than in other seas: the salinity is between 3.5% and 3.8% of water mass. The bridge is situated in an area of very high winds, which form a salty sea spray that reaches all structural elements, depositing salt on all exposed surfaces.

A detailed inspection of the concrete arch in 1988 showed severe cracking near the arch springing and peeling of the concrete cover on approximately 10% of exposed surfaces due to reinforcement corrosion. Arch deflections under test loads were 25% larger than under the same loads in the first test performed prior to bridge opening. These disturbing results called for complete repair of the arch, comprising removal of the damaged concrete cover by a hydro-demolition device, grouting of all visible cracks, and placing shotcrete minimum 4 cm thick strengthened by anchored reinforcement mesh and protected by special long-lasting elastic coating.

Both the superstructure and piers deteriorated even more, inducing serious functional difficulties. More alternative solutions for their repair or replacement were studied, but works on-site were postponed because of war activities. The final choice between the alternative solutions of strengthening the existing superstructure by external prestressing and replacing it with a new steel superstructure was made in favor of the latter option because of the poor state of the concrete superstructure. Piers were to be strengthened by adding 20 cm thick outside reinforced concrete casing to take up all actions, with no loads assigned to the existing heavily corroded piers.

In 1991, the Pag Bridge was additionally damaged by direct aircraft missile hits. Provisional measures had to be taken by erecting heavy scaffolding on the arch to support the superstructure to allow intense heavy road traffic over the bridge, as it remained the only reliable fixed link between south and north Croatia (Šavor et al. 2001a).

A superstructure comprising two steel plate girders connected by an orthotropic deck plate with open stiffeners was selected because of durability requirements, as it allowed accessibility to all load-carrying steel structural elements and also reduced the dead weight, allowing for the live loadings required by relevant Croatian bridge code, which were higher than those for which the bridge was originally designed (Figure 7.29). Two independent continuous superstructures with spans of 5 × 23.30 m + 11.65 m = 128.15 m towards Pag Island and spans of 11.65 m + 4 × 23.30 m = 104.85 m towards the mainland, respectively, were constructed. The last two spans of the superstructure facing the mainland are in a horizontal curve of R = 100 m (Figure 7.30).

7.5.2.3 Krk Bridge

In 1980, the Island of Krk was connected to the mainland by two arch bridges. The first part, with a 390 m span, bypasses the 470 m wide sea channel between the mainland and the Isle of St. Marko, and the other, with a 244 m span, bypasses the channel between the Isle of St. Marko and the Island of Krk (Figure 7.31). The Krk Bridge carries vital pipelines and oil pipelines that exceed the weight of 3.5 tons per one meter of bridge, which is greater than the design traffic live load.

The larger arched girder still holds the record in span size among traditional reinforced concrete structures. The construction work on the arches and piers was done by Mostogradnja, from Belgrade (Serbia), and the pavement structure and finishing works were done by Hidroelektra, from Zagreb. It has changed several management companies during its life span, and in the year 2007 the bridge fell under the concession company ARZ, as a part of a road network with a toll collection regime.

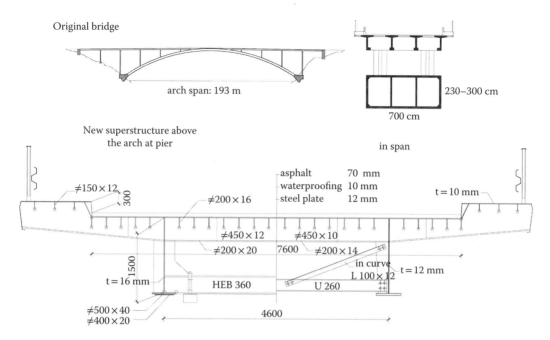

FIGURE 7.29 Sketch of the original bridge and cross section of the new Pag Bridge superstructure, installed during reconstruction in 2001. (Courtesy of the Faculty of Civil Engineering, University of Zagreb, Croatia.)

FIGURE 7.30 Pag Bridge after repair of the arch and reconstruction of the superstructure. (Courtesy of the Faculty of Civil Engineering, University of Zagreb, Croatia.)

The larger arch is supported by a bracing structure with foundations in the sea, and the smaller arch is fixed into the foundations on the rocks. The common characteristic of both arches is the invariability of the outer dimensions along the complete arch length, that is, construction consisting of prescast elements. For example, 86% of the arch with the 244 m span was constructed by assembly of concrete elements, mostly relatively small in size and limited by the assembly equipment capacity to 10 tons of the overall mass. The cross section appearance of the arches is due to its execution by the free cantilevering

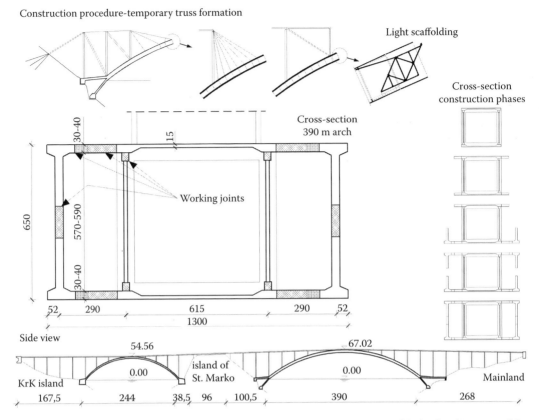

FIGURE 7.31 Sketches of the Krk Bridge, construction of the main arch and layout of the bridge. (Courtesy of the Faculty of Civil Engineering, University of Zagreb, Croatia.)

method, which was brought to its ultimate efficiency for this bridge. The cross section of the bridge arches is constructed in 5 m long sections assembled of precast plates, creating a box, interconnected by concrete jointing. Arch cantilevers were constructed suspended on cables, transferring the force to the ground through ground anchors. The arch piers were constructed parallel to the cantilevers, forming, together with the staying cables, a temporary trussed structure.

The middle unit of the box was constructed first, acting as an arch core, braced at the crown by the temporary steel structure. Using the same mounting procedure, the arch side elements were constructed as well. Some segments were supported by skew staying cables fixed to the truss joint, consisting of previously constructed parts of arch, spandrel posts, and staying cables at the level of the posts and the temporary diagonal.

The larger arch span of the Krk Bridge is smaller than the barrier width to be passed over, that is, the arch does not reach from shore to shore. The structure was extended with frame foundation structures in the sea water, with sloped legs reaching up to 19 m depth, on caissons. The piers are designed with a buttress cross section, that is, as two "T" section posts connected with crossbeams, constructed in a sliding scaffold. The buttress section was intended to enable a greater degree of stiffness with as little mass as possible. Very slender piers end with support beams drawn into the superstructure in a way that the main girders are back cut up to a half of their height.

The carriageway superstructure consists of precast elements. Structurally, it is a sequence of freely placed grillages, consisting of three main girders connected with crossbeams. All the structural parts were designed with minimal statically allowable dimensions so that the carriageway deck is only 15 cm thick. The superstructure girders lean directly on the support beams without the bearings, which soon after completion proved to be a very unfavorable solution; after only 4 years of use, a systematic rehabilitation of these spots was initiated and neoprene bearings were installed.

Damage to structural parts of the bridge occurred very early during its use, which is why the damage cannot be attributed to an aggressive environment as the most influential factor. The main cause for the damage lies in design deficiency and poor construction. Disturbing deterioration of some bridge elements gave rise to much research and thorough rehabilitation of some elements. To date the greatest scope of repair work has been carried out on the section between the Isle of St. Marko and the Island of Krk, whereas in the forthcoming period major operations are taking place on the larger arch bridge.

The lower parts of the smaller arch, below the position of 25 m above sea level, and arch springing, were repaired and protected. Concrete covers were demolished by hydro demolition and renewed with 50 mm of high-quality sprayed mortar. The compressive strength of mortar is above 50 MPa and the adhesion to the base is above 2 MPa. Migration corrosion inhibitors were added to the repair mortar. All reconstructed surfaces were coated with 1.5 mm of elastoplastic polymer coat. On the upper part of the arch, more than 25 m above sea level, and on the lower surface of the span structure, only local repair and protection was applied. Piers were repaired in similar manner.

The arch geometry was tested during bridge exploitation, since the arrangement of internal forces in the cross sections depends on the shape of the arch girder, and even during construction work some deviations from the design were noticed. The original design arch axis of the Krk Bridge was determined by the thrust line adjustment procedure, as an optimum form that minimizes bending in the arch. The first adjustment, the correction of the arch rise by strutting in the crown, was performed in April 1982, where the crown was raised 63 mm. In the following year after the repeated measuring of vertical displacements and the estimate of the foreseen creep-induced deformation had been conducted, additional adjustment was deemed necessary. The arch was once again raised 92 mm.

A survey of the Krk Bridge was conducted in 2006 by classic terrestrial photogrammetry as well as by some new procedures using a laser scanner. It was found that the arch is lower in the crown by 87 cm from the original design, while the vertical deviations at a part of the girder are even greater than 1 m. Horizontal deviations from the designed arch geometry are up to 40 cm.

7.5.2.4 Maslenica Motorway Bridge

The tradition regarding concrete arch bridge construction in Croatia continued through two bridges on the Zagreb–Split Motorway, Maslenica Bridge and Krka Bridge. Whereas in other structures the principle of two separate structures for each carriageway has been applied, in the case of these two arch bridges this principle was abandoned, resulting in one relatively narrow arch carrying two motorway lanes.

Demolition of the Maslenica Bridge on the state road during the 1991 war enhanced the construction of a new bridge at another location. The decision was made that as soon as the conditions allowed, a motorway bridge would be constructed, while the reconstruction of the other bridge, on the state road, followed several years after that. Maslenica Motorway Bridge was designed at the Faculty of Civil Engineering in Zagreb and built by Konstruktor, from Split. It was completed in 1997, and is presently managed by the Croatian Motorway Company.

Two structural solutions prevail in the contemporary design of concrete arch bridges: structural systems with rigid arches and flexible superstructures (standard design solution of most large arch bridges) and structural systems with flexible polygonal arches and rigid superstructures. On the Maslenica Motorway Bridge a wide superstructure is supported on a much narrower arch, resulting in the structural system of a rigid arch and a rigid superstructure.

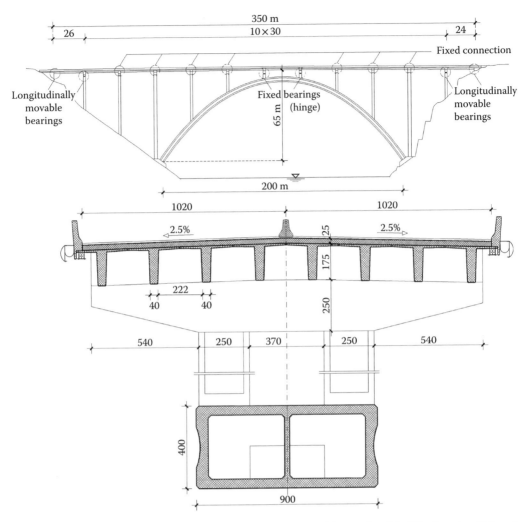

FIGURE 7.32 Sketch of the Maslenica motorway bridge. (Courtesy of the Faculty of Civil Engineering, University of Zagreb, Croatia.)

The Maslenica Motorway Bridge, completed in 1997, belongs to the second generation of bridge design, with the severity of the aggressive maritime environment taken into account. The dimensions of the structural elements were increased in order to avoid reinforcement congestion and increase durability. Structural details and cross sections were simplified in order to minimize construction problems. The bridge lies horizontally in a straight line and the grade line is in an upward curvature of R = 17,500 m, approximately 90 m above the sea level. The overall width of the roadway is 20.4 m, with four traffic lanes.

The superstructure comprises a grillage of eight simple-span precast prestressed girders made continuous over the middle supports by utilizing nonprestressed reinforcement and interconnected by a concrete deck plate 25 cm thick, cast in situ, with cross girders provided only at supports. The superstructure is fixed to all the piers except at the abutments and the piers closest to the abutments, where longitudinally movable bearings have been installed, and the piers nearest to the arch crown, where fixed bearings have been provided. Expansion joints were provided at the abutments only (Figure 7.32).

The main structure is a concrete arch with a span of 200 m and a rise of 65 m, resulting in a rise-to-span ratio f/L = 1/3.08. The arch rib is a fixed double cell box cross section with constant outer dimensions. The overall width of the arch is 9 m and the overall depth is 4 m. The superstructure is continuous over 12 spans of 26 m + 10 × 30 m + 24 m, with the overall length of 350 m. Bridge piers, varying in height from 3.6 m to 67.9 m, consist of two individual columns connected at the top by a head-beam. All piers are box-type cross sections. The bridge is founded on rock. Combined footings are utilized with allowable rock stress of 1.5 MPa.

The construction works started in March 1993. After the constructing of access roads to the bridge piers all foundations were cast. The installation of temporary rock anchors for the free cantilevering erection of the arch followed. Seventy-four anchors were constructed, approximately 25 m long. The installed oscillatory cable crane of 500 m span and 6 ton capacity was used for the transport on the site.

The construction of the arch by free cantilevering was the most demanding. The arch was constructed on traveling formwork carriages with a weight of 55 tons each, in 5.26 m long segments, starting symmetrically from the arch abutments. Piers at the arch abutments were extended by auxiliary steel staying pylons 23 m high to facilitate successive cantilevering. The arch was supported during construction by stays radiating from two levels of the arch abutment piers and from the tops of auxiliary staying pylons, where they were equilibrated by anchor stays connected to rock anchors (Radić et al. 2004). The arch axis was designed and constructed in an elevated position (13.7 cm maximal rise in the crown) so that the designed arch shape would form after long-term creep and shrinkage deformations.

With careful planning, based on 24-hour work days, arch construction was finished in 11 months. Precast prestressed girders were cast near the bridge during the construction of the arch. Spandrel piers and the superstructure above the arch were constructed after both halves of the arch have been closed at the crown in a strictly prescribed order. Just prior to the arch closure a relatively strong earthquake occurred, with the epicenter very close to the bridge site, with no consequences for the bridge structure.

7.5.2.5 Krka Bridge on Zagreb–Split Motorway

The new Adriatic motorway crosses the canyon of the Krka River in the proximity of the entrance to an environmentally valuable and protected area: the Krka River National Park. The location of the bridge is situated near the estuary, in the area where sea water mixes with river water, in a moderately aggressive environment. The bridge was designed at the Structural Department of Civil Engineering Faculty in Zagreb, and built by Konstruktor from Split as the principal contractor and Đuro Đaković from Slavonski Brod as the steelwork subcontractor. The bridge was completed in 2005, and is managed by the Croatian Motorway Company.

The bridge lies horizontally in a straight line, the gradeline being in constant slope of 1.326%, lying 66 m above the water level. The width of the river bed is 190 m, and the width of the canyon is about 390 m. The overall width of the roadway is 21 m, including a median strip of 3 m. The total width of the superstructure is 22.56 m (Radić et al. 2006). The designer offered two alternatives with the arch of the same span and similar shape. The first one envisioned a conventional prestressed concrete superstructure made of precast girders and in situ slab (similar to the Maslenica arch bridge, spanning the sea strait on the same highway), while the second one envisioned a composite superstructure. The arch had considerably smaller dimensions than the second alternative (Šavor et al. 2004a). The steel–concrete composite box girder superstructure was finally chosen.

The rise-to-span ratio of the concrete arch is 0.25 (f/L = 0.25). The arch is fixed of a double cell box cross section with constant outer dimensions of 10 m × 3 m. The arch cross section is constant, except in the proximity of the arch abutments, where the thicknesses of the flanges increase from 50 cm to 60 cm (Figure 7.33). The continuous composite superstructure is supported on longitudinally movable bearings on stiff short piers and abutments, and longitudinally fixed bearings on tall flexible piers. The superstructure steel grillage system consists of two main longitudinal girders spaced at 7.6 m, crossbeams spaced at 4 m, and edge girders. Box-type main girders are 1.7 m high. The concrete deck slab is 25 cm thick.

The superstructure is longitudinally very flexible, which resulted in large horizontal movements in case of a seismic event. For this reason viscous dampers were installed at both ends of the superstructure, transmitting longitudinal forces to massive abutments. The designer proposed circular piers, which would have improved the appearance of the bridge, but the contractor asked for square-shaped piers, in order to facilitate their construction. All piers are of box-type cross sections, except for the shortest ones.

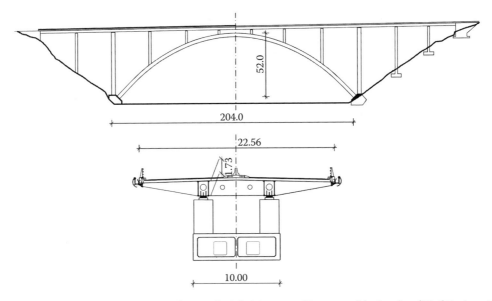

FIGURE 7.33   Sketch of Krka Bridge on the Zagreb–Split Motorway. (Courtesy of the Faculty of Civil Engineering, University of Zagreb, Croatia.)

The optimum arch shape for the Krka Bridge was found by utilizing the equilibrium catenary shapes under concentrated loads applied at the pier locations. They are equal to the total mass of the piers and superstructure, but inverted in direction. The weight of the catenary is equal to the weight of the arch.

The Krka River arch was constructed by free cantilevering, on traveling formwork carriages, in segments 5.25 m long, starting symmetrically from arch abutments. The piers at the arch abutments had to be extended by auxiliary steel staying pylons to facilitate successive cantilevering. The arch was supported during construction by stays equilibrated by anchor stays, connected to rock anchors. The steel grillage was launched from the abutments to the main piers at the arch springing. After the launching procedure, the superstructure was lowered to the permanent bearings. In the next construction stage, precast rein-forced concrete slab elements were placed and interconnected by on-site concreting of longitudinal and transversal joints above shear connectors, thus completing the composite superstructure (Figure 7.34).

7.5.3 Steel Arch Bridges

7.5.3.1 Maslenica Strait Bridge

The steel arch bridge over the Maslenica Strait (Figure 7.35) is a part of the state road connecting the Croatian north and south, and prior to the construction of the Zagreb–Split Motorway, the bridge had very important traffic significance. The original bridge from 1955 was demolished in 1991 during the war, only to be reconstructed in 2002 in a shape very similar to the original but much stronger. The bridge reconstruction design was done by the Design Bureau IPZ from Zagreb, and the actual construction works were done by the company Đuro Đaković from Slavonski Brod. The reconstruction was completed in 2004 and since then the bridge has been managed by the Croatian Roads Ltd., Company for Operation, Construction, and Maintenance of State Roads (Štorga 2005). Inspection and testing of the remaining parts of the demolished bridge showed that the blasting destroyed one of the abutments and four foundations of the piers on the shore, while the remaining foundations were repaired. The whole steel structure was destroyed and could not be reused.

The Maslenica Strait was crossed by a 155 m parabolic double-hinged steel arch span, made of two box-type structures connected with a bracing. The total bridge length was 315 m, with a height of grade

FIGURE 7.34 Overview of the Krka Bridge on the Zagreb–Split Motorway. (Courtesy of the Croatian Motorways Limited Company–HAC d.o.o., Croatia.)

FIGURE 7.35 Reconstruction of the Maslenica Bridge for the state road. (Courtesy of Institute IGH, Croatia.)

line 55 m above the mean sea surface. The carriageway width on the bridge is 7.1 m, with walkways on both sides 1.5 m wide each, making it 1 m wider than the original bridge.

The superstructure consists of continuous steel–concrete composite girders over 17 spans 17.5 m + 3 × 19.7 m + 7 × 17.5 m + 5 × 19.7 m + 17.5 m. It is a grillage structure consisting of steel longitudinal and transverse girders composite with the carriageway deck made of reinforced concrete. The longitudinal supports of the steel superstructure are 1.2 m high welded sheet metal supports, rigidly connected with crossbeams. The piers are steel, quadratic-like cross sections. Portal piers are connected by a beam at the top with a lattice rhomb filling, whose shape corresponds to the fill connecting the arches.

The main load-carrying structure consists of two hinged arched girders, span 155 m, arch rise 41.5 m. The arch axis is shaped as a second-degree parabola running through three determined points. Support hinges are hidden so visually it seems to be a fixed arch. The arches are interconnected by a rhomboid filling of counterbalances without crossbeams. The bridge has shallow foundations, on limestone rock, and all visible parts of the foundations and abutments are lined with dressed stone.

In 1955, when the original bridge was constructed, it was an example of the trends in bridge design at the time, but also an example of good incorporation of a structure into the surrounding environment. Its reconstruction by the same principle preserved the original value of the surroundings.

7.6 Cable-Stayed Bridges

7.6.1 Rijeka Dubrovačka Bridge (Dubrovnik Bridge)

The Rijeka Dubrovačka Bridge, or Dubrovnik Bridge, was the first cable-stayed bridge in Croatia, situated on the western entrance to Dubrovnik, a historical medieval town and one of the most known tourist destinations on the Adriatic Coast in Croatia. It crosses the sea strait, which is approximately 5 km deep. The existing road passes along the seashore so that increased traffic jeopardizes the environment and prevents the development of tourism. The bridge comprises the main cable-stayed bridge with a steel–concrete composite girder superstructure and prestressed concrete approach bridge. The main span of the bridge is 304.5 m. No piers were allowed in the water due to navigation constraints. The vertical navigation clearance was set to 50 m.

Bridge construction started in 1990, before the Croatian War for Independence, when the west approach roads and the west bank abutment were constructed. Construction work continued in 1999, based on a modification of the original design. The main and detailed designs were carried out by a team from the Structural Department of Civil Engineering Faculty in Zagreb, based on a preliminary design by Schambeck and Sporschill from Germany. The bridge was put into service in 2002, and is operated by Croatian Roads Ltd., Company for Operation, Construction, and Maintenance of State Roads.

The overall bridge length between abutment ends is 518 m, while the total bridge width at the larger superstructure part is 14.2 m (the bridge is widened at the west end because of the intersection at the shore, in the immediate vicinity of the bridge). A nonsymmetrical layout with one pylon had to be chosen because the motorway from the west enters the bridge in a sharp curve, so that a more economical three-span cable-stayed structure was not possible. The location of the bridge is in a highly active seismic zone (the design acceleration was prescribed as 0.35 g) and is subject to very strong winds (speeds of up to 50 m/s), which strongly influenced the bridge design.

According to the original design, the bridge had two separate structures, the skew and the curved approach viaduct in prestressed concrete, and the main cable-stayed bridge. The height of the pylon was 163.6 m. The contractor changed the original design in order to simplify the construction procedure and to reduce the maintenance costs. There were two main changes. The prestressed concrete box-type approach viaduct on the right bank was extended 60 m, reducing the suspended span by the same amount, with a hinge between both structures. As a consequence, the height of the pylon was reduced to 141.5 m. The superstructure of the main bridge was changed from a box-type girder to composite girder consisting of two steel girders 2 m high with a 25 cm thick concrete deck plate (Figure 7.36). The designer of the original design opposed the modifications introduced to make the bridge more cost-effective, challenging that they were not a structural improvement. The hinge in the central span should have been avoided in a zone of high seismic risk and because of potential mainte-nance problems.

The prestressed approach structure on the west side starts with an 87.4 m girder span and box-like cross section and continues with a cantilever console into the 60 m main span. The cable-stayed struc-ture with a composite bracing beam is 244 m long at the main opening, continuing with a span of 80.7 m at the end opening. The longitudinal layout of the cables is of a modified fan type with partial suspen-sion. Cable stays in two inclined planes are spaced at 20 m. The composite beam is fixed into the east abutment at its end, the abutment holding nine pairs of anchored tendons (Figure 7.37).

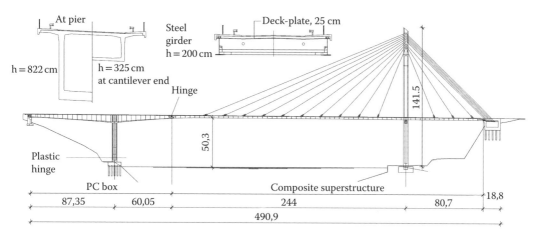

FIGURE 7.36   Sketch of Rijeka Dubrovačka Bridge. (Courtesy of the Faculty of Civil Engineering, University of Zagreb, Croatia.)

FIGURE 7.37   Overview of the Rijeka Dubrovačka Bridge during construction. (Courtesy of the Faculty of Civil Engineering, University of Zagreb, Croatia.)

The A-shaped concrete pylon is of box-type cross section. The designer offered an alternative with cable anchorages in the pylon in steel, but was rejected for economic reasons. Pylon legs are founded on 6 m thick slabs that are interconnected transversally by beams and anchored longitudinally due to unfavorable inclination of rock layers. The prestressed approach viaduct is horizontally partially curved. The superstructure depth varies from 3.2 m at the west bank abutment and cantilever tip to 8.2 m at the fixed connection to the pier.

Two hydraulic dampers of 2000 kN capacity each were installed at the connection of the viaduct to the west bank abutment. The pier base was designed as a plastic hinge during high seismic events. The foundations of the west side pier and east side abutment are on bored piles, connected with thick pile cap. The east bank abutment is a massive box structure with vertically prestressed side walls. The interior is filled with stone. Six pairs of back cable stays are anchored into this abutment, transmitting uplift forces.

The superstructure of the approach viaduct was built by the free cantilevering procedure symmetrically from the pier. Two auxiliary concrete columns were used to stabilize the construction in the 87.4 m span. Two different erection procedures were used for construction of the main bridge superstructure. The steel grillage of the 80.7 m side span and 33 m long adjoining part of the main span was erected by incremental launching from the east bank abutment. The mass of the structure to be launched at the end of the procedure reached 508.2 tons. The erection procedure was completed in 4 months. The combination of the two procedures was chosen in order to avoid auxiliary piers in the side span over an inhabited area.

The remaining 211 m long steel grillage of the main span was erected in 20 m long segments by free cantilevering. A 90 ton capacity derrick crane was used to lift the segments from a barge floating on the sea under the bridge. Concreting of the deck plate followed two segments behind the current derrick crane position. The concrete pile and pier on the west bank were constructed in climbing formwork, using tower cranes. The minimum attained concrete grade for all structural parts was C45/55, basically due to the very low prescribed water-to-cement ratio of 0.35.

After three years in service, the bridge has experienced vibrations of stay cables under certain wind conditions. Measurements and analysis of field observations showed that rain- and wind-induced stay cable vibrations occurred. To mitigate the potential damage due to extremely violent vibration movements of the stay cables and the superstructure under a heavy, damp snow storm, adjustable cable dampers were installed on the bridge to raise the low structural damping of the most-affected stay cables (Radić et al. 2003b).

7.6.2 Homeland Bridge

The Homeland Bridge (Domovinski Most), across Sava River in Zagreb, was the first extradosed bridge in Croatia. The main structure is externally prestressed concrete box girder, with cables extending outside the span structure via short pylons. Thus, even though bridge looks like cable-stayed structure, it is in fact designed as a prestressed concrete girder structure (Radić et al. 2006). Compared to cable-stayed bridges, the height of the main tower in extradosed bridges is lower. In the case of the Homeland Bridge the pylon height is 1/7.5 of the main span.

The bridge carries the main wastewater and potable water pipelines over the Sava River; it also provides a connection over the Zagreb Bypass to the new airport passenger terminal. The design for the construction permit was made by the Croatian Institute for Bridges and Structures (HIMK), and the implementation design was made by the contractor Industrogradnja Company. The bridge is managed by the Zagreb Roads Company–Zagrebačke Ceste. An attractive and modern bridge was selected after winning the competition in 1998. This solution was favored due to its aesthetic advantage. The Homeland Bridge crosses the Sava River bed in a single 120 m long span, and the same girder depth of 3.5 m is kept along the entire 840 m long structure (Figure 7.38).

The bridge superstructure is a continuous post-tensioned concrete box girder with expansion joints placed only at the abutments. The box girder, with four vertical webs and two lateral inclined plates, is 34 m wide as it has to accommodate a double-track tram line, separate two-lane carriageways for each direction, and cycling lanes and pedestrian walkways on both sides. Eight pairs of external tendons on each of four pylons were used. The pylons are 16 m high and fixed to the deck. In addition to vertical suspension, central span box girder is post-tensioned in both the vertical and transverse directions. Because the piers are short and stiff, longitudinal restraint was anticipated only at a pier at the riverbank. Consequently, the effect of seismic forces on the structure had to be carefully considered. The seismic design was based on ground acceleration of 0.19 g. Finally, it was decided to install a set of eight hydraulic viscous dampers into each abutment.

The main extradosed span was erected by the free cantilevering technique in 3 m long sections. Segments were concreted on a 35 m wide form traveler, which was an elaborate structure itself and required careful anchoring. Side spans 60 m long were constructed on scaffolding. The bridge was completed in 2006 (Figure 7.39).

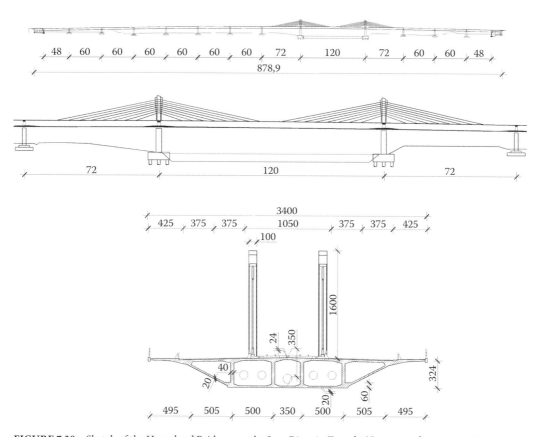

FIGURE 7.38 Sketch of the Homeland Bridge over the Sava River in Zagreb. (Courtesy of Institute IGH, Croatia.)

FIGURE 7.39 Overview of the Domovinski (Homeland) Bridge. (Courtesy of Institute IGH, Croatia.)

7.7 Suspension Bridges

7.7.1 Pedestrian Bridge in Osijek

The pedestrian bridge in Osijek is a suspension bridge with a 209.5 m span, which was chosen in order to fulfill the requirements of river traffic in the proximity of Osijek winter harbor. Steel pylons with four legs support the parabolic main cables, anchored into concrete anchor blocks. The stiffening girder consists of a thin concrete plate, prestressed by two cables underneath, in order to increase the overall stiffness of the structure. The bridge was designed for pedestrian traffic and also enables the crossing of emergency vehicles weighing up to 25 kN (Figure 7.40).

7.7.2 Martinska Ves Bridge over the Sava River Near Sisak

The pedestrian bridge over the Sava River at Martinska Ves was originally designed for pedestrian traffic, but vehicles up to 5 tons are allowed to pass over the bridge in unidirectional traffic regulated by traffic lights. The design was carried out by the team from the Structural Department of Civil Engineering Faculty in Zagreb. The bridge was put into service in 2002, and is managed by the Regional Roads Authority from the town of Sisak. The bridge layout was chosen according to the constraints of river traffic: a 145 m span suspension bridge, 4 m wide (Figure 7.41).

The main cable consists of two pairs of coil locked ropes, each 241.3 m long, 80 mm in diameter, with a breaking force of 6390 kN. They were used instead of the originally designed two 110 mm diameter cables because of fabrication technology and transport requirements. The sag of the cable is f = 16 m (L/f = 9.06). Cables are anchored into massive anchor blocks, founded on 12 drilled piles.

The base part of the concrete pylons is a massive block 10.7 m × 2 m in plane, 6 m high. The pylon above the base part is 22.2 m high, with a cast-steel tower saddle installed on the top. Pylon foundations consist of six drilled piles, 1.5 m in diameter, 12 to 18 m deep. The concrete suspended beam (stiffening girder) is 45 cm thick. The beams are interconnected by a slab, 19 cm thick minimum. The stiffening girder is made of precast segments 3.6 m long, interconnected by wet joints, cast in situ, 40 cm long (Šavor et al. 2001c).

A cable crane was used for assembly of segments instead of a floating platform. The segments were assembled starting from the middle of the span and proceeding toward the pylons symmetrically. Individual girder segments were connected by continuous steel overhangs (Figure 7.42).

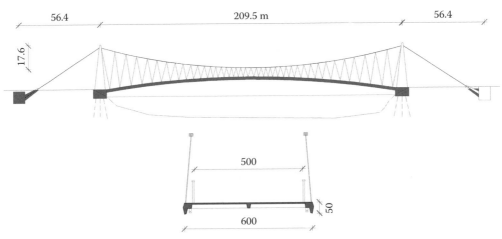

FIGURE 7.40 Longitudinal layout and typical cross section of the pedestrian bridge in Osijek. (Courtesy of the Faculty of Civil Engineering, University of Zagreb, Croatia.)

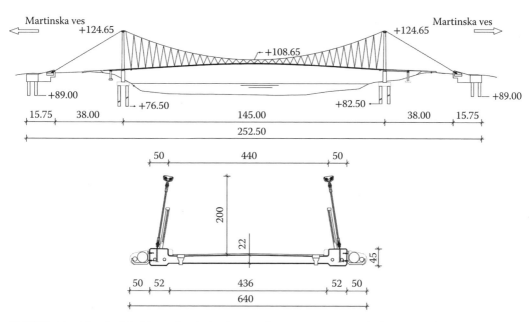

FIGURE 7.41 Sketch of the Martinska Ves Bridge over the Sava River near Sisak. (Courtesy of the Faculty of Civil Engineering, University of Zagreb, Croatia.)

FIGURE 7.42 Overview of the Martinska Ves Bridge. (Courtesy of the Faculty of Civil Engineering, University of Zagreb, Croatia.)

7.8 Bridges under Construction

The bridge across the sea strait between the Croatian mainland and Pelješac peninsula is a major new investment in the region. The southern part of Croatia, including the city of Dubrovnik, is separated from the rest of the state by a small coastal stretch belonging to the state of Bosnia and Herzegovina. The idea of fixed road link to connect the whole territory of Croatia, without having to twice cross the borders, has been studied for more than a decade. Upon analysis of several alternatives, the bridge design developed at the Faculty of Civil Engineering, University of Zagreb was chosen as the best possible solution. An international tender for construction works was published and the contract was awarded to a consortium of Croatian companies, led by the Konstruktor Company from Split. Construction works started in 2008. The investor for this project is the company Hrvatske Ceste (Croatian Roads Ltd.).

The bridge location is in a highly seismic zone with the design ground acceleration of 0.41 g and on extremely bad soil. The four-lane road bridge consists of two approach bridges and the main cable-stayed bridge with a continuous steel trapezoidal box superstructure. The seabed at the bridge alignment is almost level 27 m below the sea level, with the stratigraphic pattern a series of subhorizontal layers and irregular top of the rock along the bridge. The depth to bedrock is variable along the bridge alignment, between 75 m and 102 m in the central portion of the crossing and 39 m close to the mainland coast. The crossing is approximately 2140 m wide at sea level and 2380 m at the gradient line level. The navigation channel is to be at least 400 m wide with a vertical clearance of 55 m.

The bridge site lies in a zone of large seismicity in the vicinity of active seismic faults, with seven significant earthquakes of magnitude M > 6 within 100 km of the site occurring in the past century. The bridge site is also open to high winds with maximum average 10-minute wind speeds of 33.4 m/s and wind gust speeds of 47.1 m/s, both for the return period of 50 years. Hence, the main challenges for the bridge design were the high bridge alignment at approximately 90 m elevation, adverse soil conditions, and high seismicity of the site. Because of these constraints, it was evident that the total bridge dead weight, and especially the dead weight of the superstructure, should be reduced as much as possible, which was accomplished by adopting a steel orthotropic box superstructure. Relatively long spans were also utilized to limit the number of expensive foundations (Radić et al. 2008b).

The four lane roadway width is 2 × 8 m = 16 m, with two lanes in each direction separated by a 3 m wide median strip, so that the total width between safety barriers amounts to 20 m. Survey walkways 1.11 m wide are positioned outside safety barriers, resulting in an overall bridge width of 23 m between cornices. Semicircular cornices 1.05 m deep and 0.81 m wide, which also function as wind deflectors, are installed at transverse bridge edges.

The superstructure is a continuous trapezoidal steel box over 17 spans, with the overall length of 72 m + 96 m + 4 × 120 m + (120 m + 150 m + 568 m + 150 m + 120 m) + 4 × 120 m + 96 m + 72 m = 2404 m. The height of the superstructure in the bridge axis is 3 m for the main cable-stayed bridge and 5 m for the approach spans, with the smooth transition between these heights at cable-stayed bridge ends (Figure 7.43). The cable-stayed bridge is symmetrical, with the main span of 568 m over the navigation channel and extended to both sides over two spans of 120 m and 150 m, respectively. The longitudinal layout of the stay cables is of a modified fan type with partial suspension. Cable stays in two inclined planes are spaced at 20 m in the bridge direction.

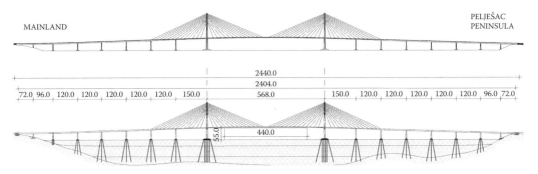

FIGURE 7.43 Longitudinal layout of the future Pelješac Bridge. (Courtesy of the Faculty of Civil Engineering, University of Zagreb, Croatia.)

The streamlined deck cross section has been designed to reduce wind actions on the bridge and to provide high torsional rigidity. Lateral webs of the box girder where cable stays are anchored are designed with sufficient thickness with no longitudinal stiffeners. All the other plates, the deck plate, the bottom plate, inner webs, and inclined bottom plates, are stiffened by longitudinal trough-type closed trapezoidal stiffeners. The cable-stayed bridge deck plate thickness is 12 mm in the fast lanes, 14 mm in the slow lanes, and 20 mm in the edge beams (Figures 7.44 and 7.45).

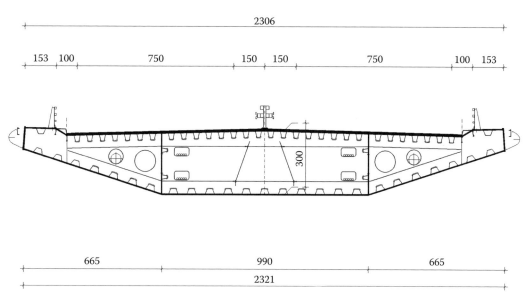

FIGURE 7.44 Typical cross sections at cable stay anchoring of the main cable-stayed bridge. (Courtesy of the Faculty of Civil Engineering, University of Zagreb, Croatia.)

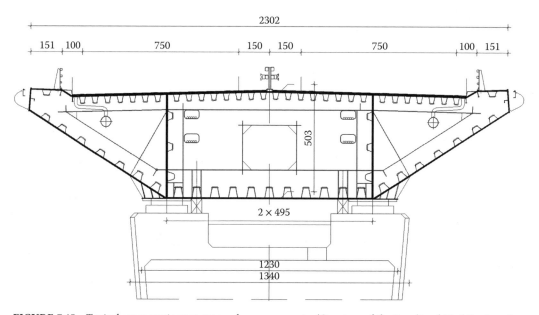

FIGURE 7.45 Typical cross sections at approach span supports. (Courtesy of the Faculty of Civil Engineering, University of Zagreb, Croatia.)

The pylons are diamond-shaped, with a total height of 176 m above the sea. The lower pylon parts, including the prestressed crossbeams, are constructed in concrete, while only the top 52 m, where the cable stays are anchored, are of composite cross section. The main bearing element is a steel box, embedded in a U-shaped concrete outer shell.

All 14 piers are of box-type cross section with 0.50 m thick walls and with variable outer dimensions. The piers are tapered in both the longitudinal and transversal directions. The width of all piers varies linearly in the longitudinal direction from minimum 3.9 m at pier tops with an inclination of 1:60. Piers in the sea strait are founded on a group of 12 driven steel tubular piles, fixed at their tops to 4.5 m thick rectangular concrete pile caps 14 m × 20 m at the sea level. Tubular piles of 2 m diameter are approximately 106 m long. Six outer edge piles are battered at a 10° angle, four corner piles are battered at a 23.4° angle, and only two piles in the middle are vertical. The top 55 m long pile parts are filled with reinforced concrete to increase their stiffness. Pylons are founded on a group of 38 driven steel tubular piles, fixed at their tops to minimum 8 m thick rectangular concrete pile cap 37.5 m × 45 m at sea level. Tubular piles of 2.5 m diameter are approximately 108 m long. These foundations were the most cost-effective solution in terms of constructability and amount of materials used.

Seismic analysis of the Peljesac Bridge was a difficult task. Nonlinear modal time history analysis, based on superposition of Ritz vectors, but with coupled modal equations, was utilized for seismic design. Real earthquake accelerograms for earthquakes in the vicinity (e.g., Bar, Ulcinj, Ston) and also the El Centro earthquake, as well as 72 artificial accelerograms, determined by seismic study for the specific bridge location, were analyzed. The solution for the seismic response of the bridge, situated in a zone of high seismicity and founded on extremely poor soil, was found by combining dampers and flexible pile foundations.

All vertical actions from the superstructure on the substructure are taken by structural bearings, except for uplifting actions, which are taken by prestressing tendons, so that the resulting action remains compressive. The basic idea behind the adopted damper layout was to install them on all supports with rigid foundations, both abutments, both pylons, and piers S2, S3, and S17, to mitigate the seismic energy input and minimize structural damage. Seventeen dampers are to be installed in total (Figure 7.46).

FIGURE 7.46 **(See color insert.)** Simulation overview of the future Peljesac Bridge. (Courtesy of the Faculty of Civil Engineering, University of Zagreb, Croatia.)

7.9 Bridge Management

7.9.1 HRMOS Bridge Management System

The Croatian regulations mandate that bridge inspections be performed at regular intervals, at least every two years. The Croatian Act on Public Roads forms the basis of construction and maintenance of public roads through its construction and maintenance strategy and program. The Maintenance Program refers to a planning period of four years and is generated through yearly construction and maintenance plans adopted by the companies and authorities in charge of maintenance of individual road categories. The yearly maintenance plan includes the situation existing at the beginning of the planning period, determines the level of maintenance priorities and planned investments, as well as an overview of the situation expected at the end of the planning period.

A bridge management system (BMS) named HRMOS is applied to bridges on state roads by National Road authorities. It is based on the Danish system DANBRO, and has been in use since 1995. HRMOS was developed in order to solve the following issues: maintaining bridges in an optimum manner, prioritization of projects, achieving the best results with available resources, making of 5-year budget plan, and ensuring traffic safety. It contains the rating system indicating the relative health of bridge elements, as well as future maintenance needs on an integer value scale from 0 to 5, where 0 stands for no damages and 5 stands for heavy damage.

Each bridge is divided into major parts for condition assessment. The standardized description of bridge parts contains a menu of 13 elements: roadway, expansion joint, sidewalk, railing, embankment, wing walls, abutments, piers, bearings, deck, girders, and approach ramps. Elements are rated separately, as well as the complete bridge. HRMOS BMS comprises the following modules: bridge inventory, general inspection, special inspection, routine inspection, prioritization of repairs, optimization, budgeting and evaluation of results, design (including cost estimates), tender, monitoring work progress, accountancy, administration, and monitoring the results (Bleiziffer and Radić 2004).

One of the major problems in the HRMOS BMS is to ensure uniform and objective condition assessment. This still depends largely on the experience and judgment of the inspector. The other problem lies in the fact that condition ratings assigned during routine bridge inspections, on the basis of visual inspection, do not identify the deterioration process at work or the extent of deterioration. They describe the severity of damage, which is related to the scale of actions, from minor maintenance to replacement of inspected element.

7.9.2 Structure Management System for Motorways

Seven distinct types of structures can be considered and analyzed in the scope of the Structure Management System that has been conceived for management of the national motorway network. The development of a uniform system, encompassing all motorway structures, has been motivated by the fact that funding for the maintenance of all structures found along the motorway system is an essential element of the long-term and annual maintenance plans. The system, which is being implemented by Croatian Motorways Limited Company–HAC d.o.o. (the main operator of highway network in Croatia), recognizes the following types of structures: bridges, tunnels, pavements, drainage systems, geotechnical structures, road furniture, and buildings.

The basic documents of the Structure Management System are:

1. Structure booklet: The main document about every individual structure; it contains information that is needed for the inspection, maintenance, and management of structures.
2. Manual (standard) for the inspection of structures.
3. Manual (standard) for the evaluation of structures.

The following reports are the system deliverables:

1. Inventory of structures
2. Condition of structures
3. Work required
4. Costs
5. Prediction of future condition and maintenance costs

BMS puts special emphasis on the objectivity of the inspection results. It comprises inventory module, documentation of inspection results, condition assessment algorithms, and priority ranking of maintenance activities with a bill of quantities.

7.9.3 Deterioration Prediction Models

In research related to bridge maintenance, emphasis is given to deterioration prediction models. Croatia is a land of great diversity with a continental climate in the inland and a hot Mediterranean climate and distinctive Bora winds along the coast. Chloride attack followed by cracking, delamination, splitting, and peeling of concrete is identified as major deterioration mechanism (Radić et al. 2003c).

Condition ratings assigned during routine bridge inspection indicate the relative health of elements. In order to predict future needs for maintenance and repair, ratings should be related to the stages of the service life, which correspond to specific maintenance actions. Such ratings are to be derived from measurable attributes, which are specific to elements and to deterioration mechanisms. Thus, relevant deterioration mechanisms must be assumed for each element before the detailed inspection procedure, which is to be performed in order to reveal latent defects.

Future deterioration prediction should use the same scale of condition states. Forecasts rely on the data on deterioration of existing bridges and on physical modeling of the deterioration process, given in the dimension of the time. Stochastic models that use the homogenous Markov process with a continuous parameter (time) are particularly suitable for the available condition data, because they use periods of time as input parameters in order to obtain probabilistic predictions. Forecasts are expressed as a distribution of the bridge elements (or bridges) among condition states. Such a procedure enables us to investigate the optimization of lifetime maintenance costs.

Reinforcement corrosion is one of the most frequent deterioration processes in concrete structures, and yet, it is still not completely understood. However, simplified models have been developed for the prediction of the duration of service life phases (initiation and propagation). These models are deterministic, producing only one predicted time for one set of input parameters.

Among a set of nominally similar bridges the rate of deterioration and the need for maintenance vary markedly from bridge to bridge, so that stochastic modeling is required. Among the first stochastic models was one for forecasting future ratings based on Markov chain theory. An attempt to apply the Markov chain model to the data on Croatian bridges failed, because of difficulties related to the calculation of transition probabilities. It was found that inspections were not performed in regular periods. Since Markov chain operates with discrete periods of time, the data set, which is considerably large for the observed group of bridges, could not have been used. A new model using the homogenous Markov process is therefore being developed. While Markov chains operate with discrete parameters (time intervals of 5 years, for example), the homogenous Markov process has a continuous parameter (time). For example, for an element of a concrete bridge close to the sea, which is initially protected with some kind of coating, the process of deterioration can be modeled with four condition states (stages of deterioration). Time periods should be estimated or calculated, so that one can draw the trajectory of the deterioration process (Figure 7.47).

The trajectory shows the expected time the process spends in a certain condition state. Assuming that the concrete element is covered with protective coating, which is expected to last for 10 years, this period is assigned the state "protected." After that, chlorides penetrate the concrete cover until their

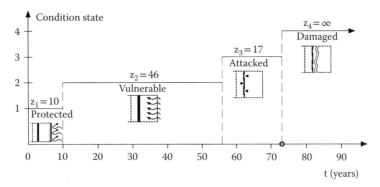

FIGURE 7.47 Example trajectory of the deterioration process.

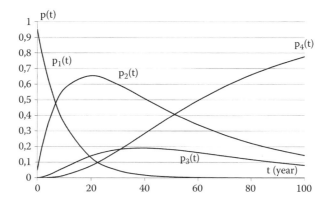

FIGURE 7.48 Probabilities of the process being in a certain condition state in the particular moment t.

concentration reaches the threshold value at the level of the reinforcement. The length of the propagation period, denoted with state "vulnerable," can be calculated using one of the existing models. (For this example it is set to be 30 years.) The next state is "attacked," which means that reinforcement corrosion advances until a certain extent, which is considered to be the limit state (e.g., 10% reduction of the cross section). According to another physical model, the duration of that period should be 15 years. At last the process enters the final state "damaged," in which, theoretically, it remains indefinitely. We can assume that the application of the protective coating was not completely successful, so that 10% of the exposed surface of elements remained unprotected. The results of the model based on the homogenous Markov process with continuous parameters are shown in Figure 7.48.

It is seen that the probability of state 1 "protected" prevails in first 5 years, and after that state 2 "vulnerable" dominates until 37 years from the start of the process, when state 4 "damaged" becomes the most probable. The probability of state 3 "attacked" is lower than 20% through the whole service life. In case of repair, another model should be derived, with new parameters and perhaps with a different number of possible states. Such predictions present valuable aid in creating an optimal maintenance program that states which elements of which bridges should be maintained by specified methods each year such that the lifetime maintenance cost of each bridge is minimized.

References

Bleiziffer, J. and Radić, J. 2004. Bridge management systems. In *Durability and Maintenance of Concrete Structures*, ed. J. Radić, 575–582. Zagreb: SECON HDGK.

Crnjak, M., Puž, G., Marić, A. and Čleković, V. 2008. *Croatian motorways*. Zagreb: Hrvatske Autoceste d.o.o.

Dumbović, I. et al. 2003. Gacka River bridge near Otočac. In *New Technologies in Croatian Civil Engineering* [in Croatian], ed. J. Radić, 151–158. Zagreb: HDGK.

Dumbović, I. 1995. Steel bridge Limska Draga. In *Proceedings of Colloquium on Steel Construction* [in Croatian], ed. J. Radić, 381–396. Zagreb: DHGK.

Friedl, M. and Runjić, A. 2001. Drežnik Viaduct—the longest viaduct in Croatia. In *5th General Assembly of Croatian Society of Structural Engineers* [in Croatian], ed. J. Radić, 157–163. Zagreb: HDGK.

Radić, J. 2004. Croatian achievements in structural engineering. In *Durability and Maintenance of Concrete Structures*, ed. J.Radić, 9–18. Zagreb: SECON HDGK.

Radić, J. 2006. Croatian achievements in bridge engineering. In *Bridges*, ed. J. Radić, 13–26. Dubrovnik: SECON HDGK.

Radić, J., Bleiziffer, J., Žderić, Ž., Šavor, Z. and Tkalčić, D. 2006. *Concrete structures in Croatia 2002–2006*. Zagreb: Croatian Member Group and CSSE.

Radić, J., Puž, G. and Gukov, I. 2003a. Croatian experience in design of long-span concrete bridges. In *Proceedings of International Symposium on System-Based Vision for Strategic and Creative Design*, 941–945. Lisse: Swets & Zeitlinger (A.A. Balkema).

Radić, J., Puž, G. and Žderić, Ž. 2008a. *A time to break down and a time to build up: Reconstruction of bridges in Croatia*. Zagreb: Croatian Academy of Sciences and Arts (HAZU), Faculty of Civil Engineering in Zagreb and IGH.

Radić, J., Šavor, Z., Hrelja, G., Mujkanović, N., Vlašić, A. and Franetović, M. 2008b. Design of Pelješac Bridge. In *Networks for Sustainable Environment and High Quality of Life*, ed. J. Radić and J. Bleiziffer, 31–38. Zagreb: SECON.

Radić, J., Šavor, Z., Hrelja, G. and Puž, G. 2003b. Bridge across Rijeka Dubrovačka. In *Proceedings of International Symposium on New Dimensions in Bridges*, 333–340. Singapore: CI-Premier.

Radić, J., Šavor, Z., Mandić, A. and Kindij, A. 2008c. Implementation of structural Eurocodes to Croatian engineering practice. In *Networks for Sustainable Environment and High Quality of Life*, ed. J. Radić and J. Bleiziffer, 231–238. Zagreb: SECON.

Radić, J., Šavor, Z. and Puž, G. 2003c. Extreme wind and salt influence on Adriatic bridges. In *Structural Engineering International*, Zurich: IABSE.

Radić, J., Žderić, Ž. and Puž, G. 2004. Construction methods for reinforced concrete arch bridges. In *DAAAM International Scientific Book*, ed. B. Katalinić, 519–536. Wiena: DAAAM International.

Šavor, Z., Gukov, I., Bleizzifer, J. and Hrelja, G. 2007. Twin bridges across Rječina River canyon. In *Concrete Structures—Stimulators of Development*, ed. J. Radić, 83–92. Zagreb: SECON.

Šavor, Z., Hrelja, G. and Mujkanović, N. 2006. Design and construction of Mirna viaduct. In *Bridges*, ed. J. Radić, 107–118. Dubrovnik: SECON HDGK.

Šavor, Z., Radić, J. and Puž, G. 2004a. Krka River Bridge near Skradin. In *Proceedings of International Symposium on Arch Bridges IV*, 558–559. Barcelona: International Center for Numerical Methods in Engineering.

Šavor, Z., Radić, J. and Puž, G. 2004b. Reconstruction of bridges on the Sava and Drava Rivers. In *Proceedings of International Symposium on Bridges in the Danube Basin*, 197–208. Novi Sad: Euro Gardi Group.

Šavor, Z., Mujkanović, N. and Puž, G. 2001a. The Pag Bridge renovation—A case study. In *Proceedings of International Symposium on Failures of Concrete Structures II*, ed. E. Javor, 189–195. Bratislava: Expertcentrum.

Šavor, Z., Mujkanović, N. and Hrelja, G. 2001b. Bridge over Drava River near Belišće. In *5th General Assembly of Croatian Society of Structural Engineers* [in Croatian], ed. J. Radić, 189–194. Zagreb: HDGK.

Šavor, Z., Mujkanović, N. and Hrelja, G. 2001c. Bridge over the river Sava at Martinska Ves. In *5th General Assembly of Croatian Society of Structural Engineers* [in Croatian], ed. J. Radić, 195–206. Zagreb: HDGK.

Štorga, S. 2005. The reconstructed Old Maslenica Bridge. In *Proceedings of First Assembly of Croatian Bridgebuilders* [in Croatian], ed. J. Radić, 95–106. Zagreb: SECON HDGK.

Vlašić, A., Radić, J. and Puž, G. 2006. Bridges on the Zagreb–Split Motorway. In *Proceedings of the 2nd CCC Congress*, 238–245. Hradec Kralove: Czech Concrete Society.

8

Bridge Engineering in the Czech Republic

Jiri Strasky
Brno University
of Technology

8.1 Introduction

8.1.1 Geographic Characteristics

The Czech Republic's central European landscape is dominated by the Bohemian Massif, which rises to heights of 900 m above sea level. This ring of mountains encircles a large elevated basin, the Bohemian Plateau. The principal rivers are the Elbe, the Vltava, the Morava, and the Odra. The Czech Republic is a small country that occupies 78,865 km² with 10.5 million inhabitants. The Czech Republic's central position within Europe has always been a natural crossroads for continental trade routes, which later became tracks, roads, and railways, and in recent decades have advanced to freeways and international railway corridors [1].

The Czech Republic has a dense network of railroads and roads. The construction of the first freeway started as early as 1939, however, it stopped during World War II and continued as late as the 1960s. So far 1111 km of freeways has been completed, about 40% of the planned network.

8.1.2 Historical Development

Building bridges has a long tradition in the Czech regions. During all eras and under various social conditions Czech engineers have designed and constructed bridges in proportion to the Czech landscape—small and gentle forms, which have not overpowered it but complemented it; bridges whose forms express their basic function of the safe and economical conveying of traffic over natural and man-made obstacles. Many stone arch bridges can be still found in our regions, from medieval times, the Renaissance, the Baroque era, and even more recent eras, among them the world-famous Charles Bridge (Figure 8.1) across the Vltava River and the oldest remaining bridge in Bohemia across the Otava River in Písek.

At the same time, Czech engineers have always tried to design modern structures erected by progressive technologies. They participated in the expansion and development of steel and concrete structures. They actively developed prestressed concrete, not only because it allowed them to overcome larger spans, but also because it represents a creative way of thinking and designing.

The roots of Czech contemporary engineering and bridge building date back to the beginning of the nineteenth century, when the Czech Corporate Polytechnic Institute was established in Prague. At this time the historical lands of the Czech Republic, consisting of Bohemia, Moravia, and Silesia, began to become similar to the prominent countries of Europe in regard to industry, technology, and education, which was clearly manifested in the advancement in structures and technologies of transportation, including bridge structures. Many of these excel to this day in their unique and daring designs.

Changes in constitutional law and the political makeup of Europe after World Wars I and II significantly influenced the social and economical circumstances of the Czech Republic, including its transportation infrastructure. Fast expansion of motor transport in the second half of the twentieth century brought about construction of new types of roads: highways, expressways, and city ring roads.

8.1.3 Chain Bridges

The first suspended chain road bridge in continental Europe was erected in 1824 in the city of Straznice across one of the branches of the Morava River. Other chain bridge constructions soon

FIGURE 8.1 Charles Bridge across the Vltava River in Prague. (Courtesy of Strasky, Husty, and Partners.)

followed—near Zatec, in Podebrady, Loket, Strakonice, and across the Vltava River in Prague. The only remaining bridge structure of this type in the Czech Republic still used as a road bridge is the bridge across the Luznice River in Stadlec, which bridged the Vltava River in Podolsko since 1848. In 1960, the bridge was dismantled and transferred to a new location as a technical monument of our bridge engineering, and has been registered as a national cultural heritage landmark since 1974 (Figure 8.2).

8.1.4 Steel Bridges

Steel bridge constructions in the Czech regions at the end of the nineteen and beginning of the twentieth centuries was linked mainly to the construction of new railroads and roads in the Austria–Hungarian Empire, many of which are used to this day. These were more or less riveted plate girder bridges; in the case of longer spans they were arch or truss types, very often of variable cross sections with upper or lower parabolic chords. Some of them are still in use and after necessary reconstructions they have became important technical monuments.

A unique construction of its kind in central Europe is the Stephen Bridge across the Labe River in Obristvi, dated 1912. It is a triple flange truss bridge with hinges and a 44.5 m span. Notable truss bridges were also built across the Labe in 1910 in Roudnice nad Labem and in Litomerice. The best-known one is the Czech's Art Nouveau arch bridge (Cechuv Most) bridging the Vltava River in Prague and dating back to 1908, with a 59.2 m main span between the stone pillars.

Welding technology brought a breakthrough in steel bridge construction, and was used in our country—among the first in the world—in 1930 during the construction of a truss bridge in the Skoda Works in Plzen, and a year later on an arch bridge across Radbuza in Doudleby. In 1935 the bridge across the Elbe River in Usti nad Labem was opened for traffic, with a span at its biggest section of 123.6 m. Welding technology significantly influenced the design and manufacture of steel bridges after World War II. This technology extended into all types of steel bridges and gradually enabled design and manufacture of geometrically challenging shapes and structures. A truly unique steel structure is the Zdakov Bridge described in Section 8.5.2.

FIGURE 8.2 Stadlec Bridge across the Luznice River. (Courtesy of Milan Vaisar.)

8.1.5 Concrete Bridges

8.1.5.1 Plain Concrete Bridges

The first concrete bridge on our territory was an arch over the Rokytka Stream in Prague-Liben dated 1896, with a span of over 10 m. Other bridges were later built, often three-hinged made of unreinforced concrete, the most significant of which are the bridges over the River Jizera in Perimov from 1912, with a 45 m span, and the Prague Bridges Hlavek's (1911) and Manes's (1914). Among the last unreinforced concrete arch bridges built was the Libensky Bridge in Prague, erected in 1924–1928, with a 48 m span at its biggest section.

8.1.5.2 Reinforced Concrete Bridges

Construction of reinforced concrete arch bridges supporting upper decks started in 1903 with construction of the three-span bridge, each 22.4 m long, across the Becva River in Prerov. In 1928 an arch bridge across the deep Lužnice River Valley near Bechyne was completed. The arch has a span of 90 m and its rise was almost 50 m. Reinforced concrete slab and girder structures began to appear in the same period and were generally designed for smaller spans; multispan bridges were often designed as statically determined structures with hinges. This concept was also used in our biggest reinforced concrete girder bridge, spanning the Vltava River at Vestec with spans 36.85 m + 3 × 52.5 m + 36.85 m, built in 1936.

In 1939 several reinforced concrete arch bridges with upper or lower decks were erected. The first one, formed by a slender arch of span length 77 m supporting a deck slab, was built across the Luznice River in the city of Tabor. During construction of the first Czechoslovak Freeway from Prague to Romania, the first arch bridges with spans exceeding 100 m were built in Borovsko, Senohraby, and Pist. The construction of these bridges started in 1939, however they were completed after World War II, in 1950.

Beside the freeway two other bridges—the biggest in their category—were erected across the Vltava River: the Dr. Edvard Benes Bridge in Stechovice, with concrete box arches and a suspended deck with 114 m span, and the deck arch in Podolsko—the biggest Czech reinforced concrete bridge, with a 150 m arch span (Section 8.5.1).

8.1.5.3 Prestressed Concrete Bridges

The first using of prestressed concrete in the Czech Republic dated as early as 1947. For the construction of a bridge near Koberovice, pretensioned precast girders composite with a cast-in-place deck slab were used. The development of the prestressed concrete bridges was similar to development in other countries. For short spans precast girder bridges have been used; for longer spans cast-in-place or precast segmental structures erected span by span or in balanced cantilevers have been used.

8.2 Design Practice

The development of the bridge industry was influenced by the political situation in the country. The socialist government strongly supported precast concrete structures that were often used in bridges for which either cast-in-place or steel structures would be more appropriate. After the Velvet Revolution in 1989 the open market allowed engineers to design and built economical structures of all structural systems, for which optimum technologies have been used. For typical bridges concrete proved to be the most economical structural material. Steel structures are mainly designed for bridges built in poor geotechnical conditions or where limited clearance requires slender structures. It is also used for long spans. As is evident from bridges described throughout this book, numerous modern technologies are used in the construction of bridges.

The development of the design approach, codes, and specifications followed the development in surrounding countries. Unlike concrete structures used for buildings, concrete bridges are designed using allowable stresses. In the design of prestressed concrete bridges both allowable stresses and the ultimate capacity of the structural members are checked. For steel structures, load factor design is used. The development of the design load also followed the development of design loads in surrounding countries. However, while common heavy trucks were represented by the weight of military tanks of 60 tons in the so-called west countries, the Czech Republic used a weight of 80 tons. The uniform load of $9\,kN/m^2$ was situated on a strip 3 m wide, and the remaining area is loaded by a uniform load $3.5\,kN/m^2$. Since April 2010 all new bridges have to be designed for loads given by the Eurocodes using the load factor design approach.

8.3 Concrete Girder Bridges

For shorter-span bridges either precast concrete girders with a composite deck slab or slabs are widely used. Long bridges are usually progressively cast in a formwork that is supported by an underslung or overhead movable gantry. The most economic bridges have double-T cross sections. Spine girders with large overhangs are also widely used. For longer-span box girder structures either cast-in-place or precast segmental decks (superstructures) are used. If the bridges are lead well above the grade, structures formed by a single box girder with large overhangs that are supported by precast struts are designed. Since 1958 balanced cantilever construction has been widely used for many bridges with spans longer than 50 m. Typical solutions are evident from the examples described next.

8.3.1 Bridge across Ludina Creek

The bridge across Ludina Creek is situated on the D47 freeway between the cities of Lipnik nad Becvou and Belotin. The twin bridge, with a total length of 320 m, is formed by a continuous girder of 10 spans of lengths from 25 m to 32 m. The superstructure of each bridge has a double-T cross section formed by two girders and a deck slab. The width of the deck is 17.1 m, the depth is 2.1 m, and the longitudinal girder spacing is 8.6 m. The low crossbeams are designed only above the end abutments (Figure 8.3).

FIGURE 8.3 Bridge across Ludina Creek. (Courtesy of Strasky, Husty, and Partners.)

The superstructure is supported by the slender piers of the octagonal cross section. The piers are founded on drill shafts. The fix bearings are on three central piers; all remaining bearings are movable in the longitudinal direction. The superstructure, which was designed as a partially prestressed structure, was post-tensioned both in the longitudinal and transverse directions of the bridge. The deck of the viaduct was cast span by span in a formwork supported by an underslung launching gantry. The gantry was formed by three steel beams situated outside and between the concrete girders. The joint between progressively cast spans is situated at a distance 7 m from the piers. In the joints only one-half of the longitudinal tendons were post-tensioned and coupled. The second half of the tendons were temporarily bent above the erected span and subsequently placed in the next span. The bridge opened to traffic in fall 2008.

8.3.2 Viaduct Kninice, D8 Freeway

Viaduct Kninice, which is situated between a village, Žďárek, and a tunnel, Libouchec, is formed by two parallel bridges with deck lengths of 1027 m and 1077 m. The elevation of the freeway is in a longitudinal slope of 4.5%; the corresponding height difference between the abutments is 48 m. Since the viaduct continues in to the tunnel, the transverse gap between the bridges varies from 0.90 to 9 m. The superstructures of both bridges are formed by continuous structures with a typical span of 42 m (see Figure 8.4). The superstructures are formed by a spine girder with large overhangs. The spine girder has a solid cross section of variable depth from 2.6 m at the supports to 1.4 m at midspan. The superstructure, which was designed as a partially prestressed structure, was post-tensioned both in the longitudinal and transverse directions of the bridge.

The superstructure of the viaduct was cast span by span in a formwork supported by an underslung launching gantry. The gantry is formed by two steel girders of the box section. The girders are supported at the front end by the piers and at the rear end the girders are suspended on the cantilever of the already cast deck. The joint between the progressively cast spans is situated at a distance 8.5 m from the piers. In the joint between the progressively cast spans, only one-half of the longitudinal tendons were post-tensioned and coupled. The second half of the tendons was coupled in floating couplers situated at a distance of 2.6 m from the joint.

FIGURE 8.4   Viaduct Kninice, D8 Freeway. (Courtesy of Strasky, Husty, and Partners.)

The piers are formed by slender columns up to 25.2 m high. The piers have a constant depth of 2.4 m, and their width varies from 3 to 3.91 m. The deck is supported by a pair of pot bearings on each pier. Longitudinal horizontal forces are resisted by three pairs of fixed bearings situated on three central piers. The piers are founded on drill shafts. The bridge was built at a speed of 10 days for one entire span. The bridge opened to traffic in fall 2006.

8.3.3 Viaduct 210, D47 Freeway, Section 06

Between the villages of Bilov and Butovice the D47 freeway crosses a local highway and two creeks on a twin viaduct with total lengths of 638.95 m and 645.57 m. The bridge axis is horizontally curved in plan curvature and in a vertical sag alignment. The superstructure of the left bridge has 16 spans of length of 25 m + 36 m + 13 × 42 m + 29.95 m, the deck of the right bridge has also 16 spans of length of 32.22 m + 13 × 42 m + 36 m + 29.35 m. Due to the skew crossing of the highway and creeks the piers of the bridges are mutually staggered.

The superstructure, of a constant depth of 2.4 m and width of 15.05 m, has a double-T cross section formed by two girders and a deck slab. The girders are indirectly supported by pier cross-beams placed on a couple of pot bearings situated at a transverse distance of 2.4 m on narrow piers. The 2.5 m wide crossbeams cast in advance formed pier tables that supported the launching gantry (Figure 8.5). The piers are fixed in footings supported by drilled piles 0.90 m in diameter with lengths from 16.5 m to 21 m.

The superstructure was progressively cast in sections of lengths that corresponded to the lengths of the spans. The construction joints were situated at a distance of 6 m from the piers. The superstructures were prestressed by continuous tendons with the layout corresponding to the course of dead load bending moments. Only one-half of the tendons were anchored and coupled in construction joints. The second half of the tendons was coupled by floating couplers situated at a distance of 3 m from the joints. Construction of the bridge started in spring 2007 and was completed in fall 2009.

FIGURE 8.5 Viaduct 210, D47 Freeway. (Courtesy of Strasky, Husty, and Partners.)

8.3.4 Bridge across the Uhlava River Valley, D5 Freeway

Close to city of Plzen the D5 freeway crosses the Uhlava River Valley on a twin viaduct with a total length of 430 m. The bridge alignment is straight and in a longitudinal slope. Each bridge is formed by a continuous box girder of nine spans of lengths 35 m + 4 × 50 m + 54 m + 58 m + 46 m + 35 m. The girder is assembled of precast, match cast segments of depth 3 m. The length of the typical segments is 2.2 m; the length of the pier and abutment segments is 1.8 m.

The girder was progressively erected in symmetrical cantilevers by a launching gantry (Figure 8.6). During erection, typical internal cantilever tendons were post-tensioned, and after casting the midspan joints the internal span tendons and external continuity tendons from 18 0.6 inch monstrands were post-tensioned. The piers of the X cross section were provided with widened pier caps that enabled placing hydraulic jacks supporting the erected cantilevers. The bridge was completed in 2007.

8.3.5 Bridge across Rybny Creek Valley

The highway bridge across Rybny Creek Valley was built close to the border between the Czech Republic and Germany. The 356 m long bridge, which is situated up to 52 m above the terrain, is formed by a continuous box girder of seven spans of length from 34 m to 58 m (Figure 8.7). The 30.5 m wide superstructure is formed by a relatively narrow box girder with large overhangs that was incrementally launched from the lower abutment. The beam is supported by narrow I-shaped piers. On four tall intermediate piers, the girder is supported by concrete hinges; on the lower side piers the girder is supported by a couple of the unidirectional pot bearings. At the abutments, the box girder is indirectly supported by wide end crossbeams that are supported by two side multidirectional bearings; the lateral forces are resisted by a central guide bearing.

The superstructure is formed by a relatively narrow box girder with large transversally prestressed overhangs. At distances of 4 m, the overhangs are supported by single struts of dimensions 0.40 m × 0.50 m. The box girder above the piers is stiffened by additionally cast diaphragms. During launching the box girder was progressively post-tensioned by coupled straight tendons uniformly distributed at the top and bottom slabs and at the webs. After launching, the girder was prestressed by continuous external tendons anchored at the end crossbeams.

FIGURE 8.6   Bridge across the Uhlava River, D5 Freeway. (Courtesy SMP CZ.)

FIGURE 8.7   Bridge across Rybny Creek, D8 Freeway. (Courtesy of Strasky, Husty, and Partners.)

The piers were cast in a slip form. The superstructure was cast in the casting yard situated beyond the northern abutment, in two phases. First, the bottom slab and the webs were cast and consequently longitudinally prestressed. Then this section was launched into the second position closer to the abutment, where the precast struts were erected and the top slab was cast. After their transverse and longitudinal prestressing, the structure was launched. A typical bridge segment was 30 m long and it was completed and launched in 10 days.

During erection the bridge superstructure was equipped with a steel launching nose 35 m long. The hydraulic launching units located at the northern abutment towed the bridge forward by means of cables that were anchored to the superstructure using vertical steel pins. The weight of the superstructure was almost 20,000 tons. The tall piers were stiffened by external cables anchored to the abutments. The bridge opened to traffic in fall 2006.

8.3.6 Nusle Valley Bridge, Prague

The bridge across the Nusle Valley carries both a city expressway and a subway [2]. The bridge axis is straight and in a constant slope of 0.65%. The bridge is 485 m long, 26.5 m wide, and is 42.5 m above the ground (Figure 8.8). It is formed by a continuous single-cell box girder of five spans of lengths of 68.25 m + 3 × 115 m + 68.25 m that is frame connected with slender piers. The subway is situated inside the box girder (Figure 8.9).

The box girder is 26 m wide and 6.4 m deep. The overhangs are stiffened by ribs that directly support the sidewalks. While the top slab has constant dimensions, both the bottom slab and webs have variable thicknesses. Since the top surface of the bottom slab is at a constant height, its different depth is achieved by a reducing the slab's thickness in the space between the webs. The deck is prestressed in the longitudinal and transverse directions; the shear stresses are reduced by vertical prestressing.

The piers are formed by four slender walls that gape both in the longitudinal and transverse directions of the bridge. The walls have a constant thickness of 1.2 m; their width is variable. Due to the subway the pier diaphragms had to be omitted. Therefore, the piers' walls are partially situated outside the girder webs. The piers are fixed in footings supported by drilled shafts of 2.4 m diameter and lengths up to 13.5 m.

FIGURE 8.8   Nusle Valley Bridge, Prague.

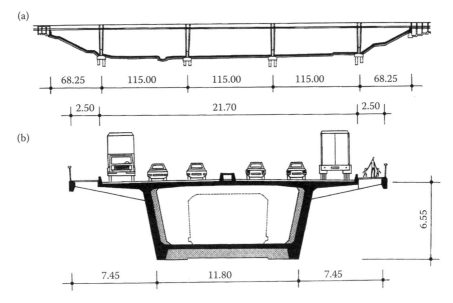

FIGURE 8.9   Nusle Valley Bridge, Prague: (a) elevation, (b) cross section.

The side spans were cast on falsework, and the remaining spans were progressively cast in segments of length from 2 m to 3.5 m in cantilevers. The spans adjacent to the side span were cast in one side cantilevers, and the the central spans in the symmetrical cantilever started from wide pier tables. The piers were stiffened there by temporary towers. The bridge opened to traffic in 1973.

8.4 Steel Girder Bridges

For spans from 30 m to 100 m, steel composite girder bridges are usually designed. For shorter spans, bridges of I-shaped composite plate girders are usually used. For longer spans, steel composite box girders are employed. Where it is appropriate, a composite slab is also used at bottom flanges above the supports of the continuous girders.

The possibilities for steel composite girder bridges are demonstrated by examples of two bridges built on the D47 freeway close to the city of Ostrava. In this area the consequences of the mining activity have to be considered. The design of these bridges was influenced by two opposing requirements. On the one hand, the structures had to be sufficiently stiff to be able to resist the design load; on the other hand, the structure had to be sufficiently flexible to be able to resist the effects of the mining subsidence.

Steel orthotropic decks are used only for long spans or in special circumstances, like replacing an old structure with a wider structure or for movable bridges. An orthotropic deck was also used in a construction of the Vysocina freeway bridge, with a maximum span length of 135 m.

8.4.1 Rudna Bridge, D47 Freeway

The twin Rudna Bridge crosses local highways, a pond, and a railway (Figure 8.10). Due to the skew crossing, the westbound and eastbound bridge sections have different spans and span lengths. The westbound bridge has 11 spans of lengths from 33.45 m to 70 m; its width is from 15 m to 24.78 m. The eastbound bridge has 12 spans of the length from 28.45 m to 69 m; its width is from 14.5 m to 22.88 m.

The prevailing part of the deck of both bridges is formed by two I-shaped plate girders connected by a transversely prestressed concrete deck slab. Above the intermediate supports the plate girders are connected with crossbeams that transfer the load from the girders into two bearings situated on the narrow piers. In the portions of the bridges where the deck is wider, additional longitudinal plate girders are inserted between the edge plate girders. The integrity of the bridges, with the deck of the open cross section, is provided by the concrete deck slab. In the transverse direction the deck slab was designed as a partially prestressed concrete member, and in the longitudinal direction as a reinforced concrete member.

Typical spans were erected span by span using the heavy lifting technique (Figure 8.11). The composite slab was cast using two movable travelers. The first one was used for casting the span portion of the deck, the second one was used for casting the deck above supports. The bridge opened to traffic in fall 2008.

8.4.2 Odra I Bridge, D47 Freeway

This twin bridge, with a total length of 402 m, crosses in a very skew angle of 57° the Odra River and local roads (Figure 8.12). The bridge has five spans with lengths from 49 m to 102 m. Each bridge is formed by one cell box girder. On the intermediate supports the girder is supported by single bearings,

FIGURE 8.10 Rudna Bridge, D47 Freeway. (Courtesy of Strasky, Husty, and Partners.)

FIGURE 8.11   Rudna Bridge, D47 Freeway, shown under construction. (Courtesy of Strasky, Husty, and Partners.)

FIGURE 8.12   Odra I Bridge, D47 Freeway. (Courtesy of Strasky, Husty, and Partners.)

and the torsion is resisted at the end abutments. This means that for torsion effects the span length is 402 m. Since the relative different rotations of the supports caused by the effects of subsidence decrease with the length of the bridge, the points where the rotations are transferred into the deck were designed at the longest possible distance—at the end abutments. That is why the deck is supported by single bearings situated in the bridge axis on intermediate supports.

The integrity of the box girder structures is guaranteed by a concrete deck slab. Since the reduction of the stiffness of the deck slab reduces the torsional stiffness of the box girder, the designer tried to eliminate the cracks. Therefore, the deck is post-tensioned both in the transverse and longitudinal directions of the bridge. The transverse post-tensioning is provided by traditional internal tendons; the longitudinal post-tensioning is provided by external cables situated inside the box. The cables are continuous across the whole length of the superstructure and are bent above the intermediate supports and at the internal diaphragms situated close to the quarters of the spans. The process of construction and a level of post-tensioning were designed in such a way that the tension stresses in the deck slab are smaller than the allowable ones.

The superstructure is assembled from the steel girder of the trough section and the transversely post-tensioned concrete deck slab. Bottom plates above the intermediate supports are stiffened by a concrete slab. The end diaphragms are also composite with a steel T-section. The steel structure was erected in a progressive cantilever from one abutment to the other (Figure 8.13). Static effects in the erected cantilever were reduced by temporary supports. The segments, with lengths up to 24 m, were erected wby a special crane that moved along the already erected structure. Then the composite concrete of the end diaphragms and concrete slabs above intermediate supports were cast. The bridge opened to traffic in fall 2004.

8.4.3 Vysocina Bridge, D1 Freeway

The twin Vysocina Bridge crosses the deep valley of the Oslava River near the city of Velke Mezirici (Figure 8.14). The bridge, with a total length of 425 m, has four spans of 80 m + 110 m + 135 m + 100 m. The superstructure of each bridge is formed by a continuous single-cell box girder of a variable depth from 3 m to 4.1 m. The width of the deck is 14.025 m; the width of the box is 6 m. The girder is stiffened by transverse floor beams and truss stiffeners situated at a distance of 2.5 m. The fix bearings are situated on central tall piers, and the remaining bearings are movable in the longitudinal direction of the bridge.

FIGURE 8.13 Odra I Bridge, D47 Freeway, shown under construction. (Courtesy of Strasky, Husty and Partners.)

FIGURE 8.14   Vysocina Bridge, D1 Freeway. (Courtesy of Antonin Pechal.)

The substructure consists of end abutments formed by hollow boxes and three pairs of H-shaped piers. The piers have a constant width of 6 m and they narrow in the longitudinal direction of the bridge from 6 m to 3 m. The piers were cast in a slip form.

The steel structure was erected in a progressive cantilever from one abutment to the other. Static effects in the erected cantilever were reduced by two temporary supports situated in each span. Both bridges were erected simultaneously by a special crane that moved along the already erected structure. Both bridges were divided into 39 segments with a maximum weight of 80 tons. In each erection step the right segment was erected first, and after its connection to the already erected structure, the left segment was placed. The joints between the segments have a combined arrangement. The top and bottom chords are welded while the webs are connected by means of high-strength bolts. The bridge opened to traffic in fall 1978.

8.5 Arch Bridges: Stone, Concrete, and Steel

Numerous arch structures both from steel and concrete were built during the first half of the last century. While only a few arch bridges were built in the second part, at present numerous structures have been built or are in the design stage. The most significant structures are described next. A hybrid structure system and hybrid structural members were also used in a design of a small arch bridge in 1996. Since then this structural system has been utilized in several new bridges.

8.5.1 Podolsko Bridge across the Vltava River Valley

This bridge, with a total length of 510.5 m, carries a local highway across the deep valley of the Vltava River (Figure 8.15) [2]. The bridge axis is straight and horizontal. The bridge is formed by a main arch span crossing the river and arch approaches situated on both sides. The span length of the main arch is 150 m; its rise is 41.8 m. The span length of the approach two hinge arches is 35.65 m; their rise is 10.35 m. Similar deck arches that eliminates tall piers are also designed in the main span.

FIGURE 8.15   Podolsko Bridge. (Courtesy of Milan Vaisar.)

The deck, with a width of 8.7 m and depth of 0.75 m, is formed by five girders stiffened by crossbeams and a deck slab. The shape of the main arch is a compromise between exact funicular shape and parabolic shape. It is formed by a compound curve formed by third-degree parabolas with a common tangent at the connection with the deck arches. The main arch has a solid rectangular cross section. The arch depth of 2 m is constant; the width is variable from 7.5 m at the arch crown to 9.5 m at the arch spring.

The approach and deck arches are formed by circular segments of a constant depth of 0.15 m that are stiffened by ribs 0.50 m deep. The arch's width is 7.26 m. The main arch and piers are fixed into spread footings supported by sound granite. The bridge was progressively cast on falsework. First, the main arch was cast, then the piers and the approach and deck arches and after that the deck. The bridge was completed in 1941.

8.5.2 Zdakov Bridge across the Vltava River

This bridge, with a total length of 542.91 m, crosses the Vltava River near the city of Zdakov (Figure 8.16) [3]. At the crossing the river is widened by the Orlik dam, which was built during the construction of the bridge. The bridge is formed by a two-hinge arch with a span of 330 m and rise of 42.5 m that is supported by 24.8 m long concrete cantilevers protruding from the arch footings. The deck's typical span length is 23.4 m; its width is 13 m.

The arch is formed by two arch ribs of the steel box section measuring 1 m × 5 m with 1.4 m wide top and bottom flanges. The transverse distance of the ribs is 12 m. The arch ribs are transversely stiffened by a truss bracing. The deck is formed by two steel plate girders stiffened by floor beams that support the concrete deck slab. The piers above the arch footings are formed by concrete frames resisting the wind load, and all other columns are formed by steel pipes with diameters from 0.3 m to 1 m.

The bridge was erected in two cantilevers from the abutments to the middle of the bridge. Both the deck and arches were erected by a special crane that moved along the already erected deck. The static effects in the erected structure were reduced by four temporary towers and by diagonal members that connected the plate girders with the arch ribs. The erection of the bridge started in 1963 and was completed in 1965.

FIGURE 8.16   Zdakov Bridge. (Courtesy of Milan Vaisar.)

FIGURE 8.17   Litol Bridge. (Courtesy of Antonin Pechal.)

8.5.3 Litol Bridge across the Elbe River

This bridge, with a total length of 198.4 m, crosses the Labe River close to the city of Litol (Figure 8.17). The bridge axis is straight and in a crest elevation. The bridge of three spans of 36 m + 128 m + 36 m is formed by a continuous I-girder that is stiffened by an arch in the main span. The 15.73 m wide deck is formed by two plate girders 2.20 m deep that are connected by floor beams and a composite concrete deck slab. The girders protrude 1.1 m above the carriageway and separate the 9.5 m wide highway from sidewalks.

The arches have a steel box section with a constant width of 0.85 m and depth of 1.4 m. The suspenders are formed by steel pipes. The intermediate piers are fixed in footings supported by drilled piles. The steel structure was assembled on the right bank and incrementally launched across the river. The span length was reduced by two temporary towers situated in the river bad. The bridge was completed in 2001.

8.5.4 Rajhrad Bridge, R52 Expressway

Close to a small city of Rajhrad a local communication crosses in the skew angle at a height of approximate 13 m a new expressway (Figure 8.18). The bridge is formed by an arch that supports the prestressed concrete deck. The arch rib is CFST (concrete-filled steel tube) formed by a steel pipe with an external

FIGURE 8.18 Rajhrad Bridge, R52 Expressway. (Courtesy of Strasky, Husty, and Partners.)

diameter of 900 mm and thickness of 30 mm in-filled with concrete. The span of the arch is 67.5 m; its rise is 8.05 m. The steel tube is internally stiffened by ring diaphragms. The arch supports the deck via triangular steel struts. The post-tensioned concrete deck is formed by two edge girders and a deck slab. The edge girders, with the internal shape of New Jersey barriers, have a constant depth of 1.68 m. At the ends, the girders are stiffened to allow anchorage to the longitudinal PT tendons.

The steel arch was erected from 12 m long steel segments that were connected by full penetration welds. Then the triangular-shaped steel struts were welded to the arch and the concrete fill of the struts and arch was cast. The deck's falsework was then erected and reinforcing and prestressing steel were placed. The deck was cast as one unit from one end crossbeam to another. When the concrete reached sufficient strength, the deck and end struts were post-tensioned and the arch was jacked against the footings. The bridge was completed in 1997.

8.6 Cable-Stayed Bridges

While suspension structures were built in the nineteenth century, at present, suspension structures are mainly used for pedestrian bridges. On the other hand, seven cable-stayed bridges have been built since 1987. Five bridges have concrete decks, one has a steel deck, and one is suspended on low pylons and combines a concrete and steel deck. The first cable-stayed bridge was erected across the Elbe River near the city of Podebrady in 1987. A recently completed bridge across the Odra River is formed by a twin structure whose decks are mutually connected in the suspended spans.

8.6.1 Elbe River Bridge, D11 Freeway

The Elbe River Bridge near the city of Podebrady carries the Prague-Hradec Kralove Freeway (Figure 8.19). The freeway axis is straight in a longitudinal gradient of 0.6%. The bridge consists of three spans, 61.6 m + 123.2 m + 61.6 m in length. The superstructure is 31.8 m wide, suspended in the bridge axis on two single pylons 28 m high. The stay cables, arranged in a semi-fan symmetrical to the tower, are anchored in the deck at intervals of 2.2 m and in the pylon at intervals of 0.50 m. The deck is supported by pot bearings placed on the low abutments and massive piers founded on wall diaphragms 18.2 m long.

The superstructure consists of a precast concrete box girder spine and overhangs constructed of precast concrete struts and cast-in-place concrete deck slabs (Figure 8.20). The spine box girders consist

FIGURE 8.19 Elbe River Bridge. (Courtesy of Strasky, Husty, and Partners.)

FIGURE 8.20 Elbe River Bridge under construction. (Courtesy of Strasky, Husty, and Partners.)

of precast match cast concrete segments. The segments are stiffened by prestressed ties transferring the cable force into their bottom corners. The decks of both bridges are prestressed longitudinally and transversally by internal cables.

The single steel column pylons of constant shape are fixed into the superstructure. A central wall, positioned with the bridge axis, divides the pylon into two cells; the stay cables were anchored on this wall in the upper part of the pylon, while the lower part of the pylon was made composite by concrete in-filled from below upwards. The deck is suspended on 21 stays on each side of the pylon (2x2x21) arranged in a semi-fan symmetrical to the pylon. Each stay is formed by two cables of (15 or 18) 0.6 inch strands grouted in the steel tubes.

The side spans were erected in two progressive cantilevers from the abutments, and the static effects in the erected cantilevers were reduced by temporary supports. After the erection of the steel pylon, its lower part was concreted. The segments of the middle span were erected in cantilevers by a special crane that shifted the segment in front of the face of the assembled structure and turned it. After the epoxy resin was applied, the segment was lowered and connected to the erected structure by post-tensioning of the prestressing rods and cables. The erection of the stay cables followed two segments behind the erection of the segments of the main span. The cables were tensioned simultaneously in pairs symmetrical to the pylon. During the erection of the segments the precast struts were erected and cast-in-place deck slab was cast (Figure 8.20). The bridge was completed in 1988.

8.6.2 Odra River Bridge, D47 Freeway

Near the city of Ostrava the D47 freeway crosses the Odra River and Antosovice Lake on a twin bridge with a total length of 589 m (Figure 8.21). The bridge crosses the river in a skew angle of 54°. The freeway's axis is in a plan curvature of 1500 m that transits into a straight line and is in a crest elevation with a radius of 20.000 m. The span length varies from 24.5 m to 105 m. The main span bridging the Odra River is suspended on a 46.81 m high single pylon. Since the stay cables have a symmetrical arrangement, the back stays are anchored in two adjacent spans situated on the land between the river and lake. The stay cables have a semiradial arrangement; in the deck they are anchored at a distance of 6.07 m, at the pylon they are anchored at a distance of 1.2 m.

The decks are formed by two cell concrete box girders of the depth of 2.2 m without traditional overhangs (Figure 8.22). The bottom slab of both cells is inclined and is curved in the middle of the girder. In the suspended spans the box girders are mutually connected by a top slab cast between the girders and by individual struts situated at a distance of 6.07 m. The stay cables are anchored at anchor blocks situated at the connected slab.

The single pylon is formed by the steel core of the octagonal cross section that is composite with a concrete cover. Inside its top portion the stay cables are anchored; the bottom part is filled with concrete. The pylon has a constant depth of 3 m; the width below the deck is 4.1 m, and above the deck is 2.4 m. The pylon was designed from high-strength concrete of a characteristic cube strength of 75 MPa. The stays are of the VSL SSI 2000 system. The stays are assembled from 55 to 91 0.6 inch strands. The stays' dead ends are at the pylon, and the active anchorages are at the deck.

FIGURE 8.21 Odra River Bridge. (Courtesy of Strasky, Husty, and Partners.)

FIGURE 8.22 Image (visualization) of the structure near the pylon of the Odra River Bridge. (Courtesy of Strasky, Husty, and Partners.)

The contractor has decided to cast the deck span by span in two formworks suspended on two movable scaffoldings. With respect to the span length of the movable scaffoldings, temporary piers had to be built in the suspended spans. During construction, the movable scaffolding was shored up by pier segments that were supported by temporarily fixed bearings. First, the northbound bridge was progressively erected using the first movable scaffolding; after completion of the first six spans a progressive erection of the southbound bridge began. As soon as the spans adjacent to the pylon were cast, the pylon's steel core was erected and concrete fill and cover was progressively cast. Simultaneously, the concrete struts between the girders were erected and the top slab between the girders was cast and transversally pre-stressed. After that, the stay cables were erected and tensioned. Then the temporary piers were removed. The construction of the bridge started in 2005 and was completed in fall 2007.

8.7 Pedestrian Bridges

In recent years many pedestrian bridges have been built. They used different structural systems and materials: timber, steel, and concrete. The following structures received international recognition.

8.7.1 Pedestrian Bridge across D8 Freeway Near Chlumec (The Cat's Eyes Bridge)

The pedestrian bridge for which the name "The Cat's Eyes Bridge" is used is situated in an area influenced by the effects of mining. Therefore, a three-span structure with significantly variable depth and a hinge at the middle of the bridge is used (Figure 8.23). The length of the main span is 43.5 m; the length of the side spans is 24.75 m. The structure is formed by two Vierandel girders connected by perpendicular elliptical members. These members substitute traditional vertical members and a transverse connection. The elliptical members are also mutually connected by horizontal members serving as floor beams supporting the sidewalk. The depth of the girders varies from 0.40 at the abutments and midspan to 5.4 m at the intermediate supports. The top and bottom chords are formed by pipes 0.406 m in diameter; the elliptical members are formed by pipes 0.194 m in diameter. The bridge was built in 2006.

FIGURE 8.23 **(See color insert.)** The Cat's Eyes Bridge. (Courtesy of Vaclav Mach.)

8.7.2 DS-L Stress Ribbon Pedestrian Bridges: Vltava River Bridge in Prague

Between 1978 and 1985 seven stress ribbon bridges of similar arrangements were built (Figure 8.24). All of these bridges were assembled of the same precast segments and have similar structural arrangements. The bridges have one (Figure 8.24), two, or three spans of span lengths up to 102 m.

The decks of all the bridges are assembled of precast segments 3 m long, 3.8 m wide, and 0.30 m deep. During erection the segments were suspended on bearing tendons situated at troughs, and after the casting of the joints between the segments, the deck was post-tensioned by prestressing the tendons situated in the deck slab. The bearing and prestressing tendons are formed by six 0.6 inch strands. The number of tendons depends on the span length and the sag. The wearing surface of the segments is formed by a 10 mm thick layer of epoxy concrete.

This structural arrangement is demonstrated by the bridge across the Vltava River in Prague–Troja. The bridge, with a total length of 261.2 m, crosses the Vltava River in the northern suburb of Prague–Troja. It connects the Prague Zoo and Troja Chateau with sports facilities situated on Emperor Island and with Stromovka Park (Figure 8.25).

The bridge has three spans of 85.5 m + 96 m + 67.5 m; the sags at midspan are 1.34 m, 1.69 m, and 0.84 m, respectively. The stressed ribbon is formed by precast segments and by cast-in-place saddles (pier tables) frame connected with intermediate piers. At the bottom of the piers concrete hinges, which allow rotation in the longitudinal direction of the bridge, were designed. The horizontal force from the stress ribbon is resisted by wall diaphragms and micropiles.

After the casting of the end abutment the solid segments were placed on the neoprene pads situated on the front portions of the abutments. Then the first half of the bearing tendons was pulled across the river and tensioned to the design stress. The tendons were supported by steel saddles situated on the piers. The segments were then erected by a mobile crane. The segments were suspended on bearing tendons and shifted along them into the design position. After all segments were erected, the second half of the bearing tendons was pulled and tensioned to the design stress. In this way the structure reached the design shape. The steel tubes that form the ducts in the joints between the segments were then placed and the prestressing tendons were pulled through the deck.

FIGURE 8.24 Svratka River Bridge. (Courtesy of Strasky, Husty, and Partners.)

FIGURE 8.25 Vltava River Bridge. (Courtesy of Strasky, Husty, and Partners.)

The reinforcing steel of the troughs and saddles was then placed and the joints, troughs, and saddles were cast. The saddles were cast in formworks that were suspended on the already erected segments and were supported by the piers. The design assumptions and quality of the workmanship were also checked by a static and dynamic loading test (Figure 8.25). The bridge was completed in 1984. In 2002 the pedestrian bridge was totally flooded. Careful examination of the bridge after the flood confirmed that the structure was safe, without any structural damage.

A disadvantage of the classical stress-ribbon type structure is the need to resist very large horizontal forces at the abutments, which determines the economy of that solution in many cases. For that reason, a new system that combines the stress-ribbon form with arches has been developed.

The stress ribbon is supported or suspended on arches. The structures form a self-anchoring system where the horizontal force from the stress ribbon is transferred by inclined concrete struts to the foundation, where it is balanced against the horizontal component of the arch. Four structures of this type have been built thus far. The most interesting is a small bridge across the Svratka River in Brno.

8.7.3 Stress Ribbon and Arch Pedestrian Bridges: Bridge across the Svratka River in Brno

This bridge connects a newly developed area (Spielberk Office Center) with an old city center (Figure 8.26). An old multispan arch bridge with piers in the river is situated close to the bridge. It was evident that the new bridge should also be formed by an arch structure, however, with a bold span without piers in the river bed. Due to poor geotechnical conditions a traditional arch structure that requires resisting of a large horizontal force would be too expensive. Therefore, the self anchored stress ribbon and arch structure represented a logical solution.

The abutments are supported by pairs of drilled shafts. The rear shafts are stressed by tension forces, and the front shafts are stressed by compression forces. This coupling of forces balances a coupling of tension and compression forces originating in the stress ribbon and arch. The arch span L = 42.9 m, its rise f = 2.65 m, and the rise-to-span ratio f/L = 1/16.19. The 43.5 m long stress ribbon is assembled of segments of 1.5 m. In the middle portion of the bridge the stress ribbon is supported by low spandrel walls. The stress ribbon is carried and prestressed by four internal tendons of 12 0.6 inch diameter monostrands grouted in polyethylene ducts. The stress ribbon and the arch were made from high-strength concrete of the characteristic strength of 80 MPa.

The arch was assembled from two arch segments temporarily suspended on erection cables anchored at the end abutments. After the midspan joint was cast, the deck segments were erected. After casting the joints between the deck segments, the cables were tensioned up to the design stress and, as a result, the deck was prestressed. The bridge was completed in 2007.

FIGURE 8.26 Svratka River Bridge. (Courtesy of Strasky, Husty, and Partners.)

8.7.4 Vranov Lake Bridge

This suspension bridge, built in 1993, is located in a beautiful, wooded recreation area where Lake Vranov was created by a dam in the 1930s (Figure 8.27). The structure was also designed to carry water and gas lines. A very slender superstructure with a depth of only 0.40 m is suspended on two inclined suspension cables of three spans 30 m + 252 m + 30 m. The cables are deviated in steel saddles situated at the diaphragms of the concrete pylons and anchored in anchor blocks. The pull from the cables is transferred into the ground by rock anchors. The anchor blocks and the abutments are mutually connected by prestressed concrete ties.

To stiffen the structure for the effects of the wind load the superstructure is widened from midspan toward the pylons. The superstructure is assembled of precast segments (Figure 8.28) that are suspended at its outer edges on hangers that are perpendicular to the longitudinal axis. The segments of the double-T cross section are only 0.40 m deep. Steel pipe conduits for gas and water lines were placed on the outer, but not mutually connected, overhangs. The superstructure was post-tensioned by four internal tendons that were led through the whole deck and anchored at the end segments. The vertical and horizontal curvatures stabilize the structure by stiffening the external cables situated within the edges of the deck; the cables pass across the expansion joints and are anchored at the end abutments. The deck is supported on both ends by two multidirectional pot bearings situated on the pylons' diaphragms. The horizontal force due to wind is transferred by steel shear keys.

The main cables are formed by two sets of 108 strands 15.5 mm in diameter grouted in steel pipes. The hangers are formed from solid steel rods of 30 mm diameter pin connected with the deck and main suspension cables. The inclined pylons have an "A" shape with curved legs connected by the top and bottom diaphragms. During the erection of the structure, the pylons were supported by pins; after the erection the pylons were cast in the footings. The bridge forms a partly self-anchored system in which the arched deck is suspended on the cables and is flexibly connected with the abutments, which in turn are mutually connected with the anchor blocks by prestressed concrete tie rods. The bridge was completed in 1993.

FIGURE 8.27 Vranov Lake Bridge. (Courtesy of Strasky, Husty, and Partners.)

FIGURE 8.28 Vranov Lake Bridge under construction. (Courtesy of Strasky, Husty, and Partners.)

References

1. Vaisar, M. 2008. *Road bridges and footbridges*. Prague: MJV ProConsult–Milan Vaisar.
2. Klimeš, J. and Zůda, K. 1968. *Concrete bridges*. Praha: SNTL.
3. Faltus, F. 1971. *Truss, arch and suspension steel bridges*. Praha: Academia.

9

Bridge Engineering in Denmark

Niels Jørgen Gimsing
*Technical University
of Denmark*

9.1 Introduction

Denmark consists of the peninsula Jylland (Jutland), extending north from Germany, and 78 inhabited islands (plus more than 200 small uninhabited islands). The capital of the country, Copenhagen, is situated on the largest island, Sjælland (Zealand). The Danish Straits, Lillebælt (Little Belt), Storebælt (Great Belt), and Øresund, form the only natural waterways between the Baltic Sea and the North Sea. In area Denmark is the smallest of the Nordic countries (Denmark, Finland, Iceland, Norway, and Sweden), but it has the second-largest population, although it is only 5.5 million.

Denmark is the southernmost of the Nordic countries and the only one with a land boundary to continental Europe: the 68 km long border with Germany. With its location between the Scandinavian peninsula and central Europe, Denmark has always formed a transit zone not only in relation to traffic but also to cultural exchange. The language spoken in Denmark is Danish but English is spoken by the vast majority of the population as a second language.

9.1.1 Early Bridges

In prehistoric times, minor bridges were probably built of wood, but they all disappeared long ago and nothing is left to indicate the location and form of these bridges. In the Middle Ages wooden bridges were built in large numbers across brooks of moderate widths, but none of these bridges have survived to our time. The remains of a few medieval stone bridges have been located in Jutland and it has been determined that the technique in constructing the vaults was similar to that used to build the vast number of village churches in the twelfth and thirteenth centuries.

Up until around 1770 the majority of road bridges were still built of wood, but after that time natural stone was used to build more reliable and durable bridges, forming a part of a national highway network. The stone bridges, which were all of moderate size, were generally composed of rough-hewed stone beams and piers, as illustrated by the Immervad Bridge from 1786, shown in Figure 9.1. However, a stone bridge of a more refined design was built in 1744 as access to the Christiansborg Castle in Copenhagen (Figure 9.2). With its total length of 30 m, it was at its completion the longest stone bridge in Denmark.

9.1.2 Start of Major Bridge Construction

The first highway bridges across the straits separating the different parts of the island kingdom were constructed in the 1860s as pontoon bridges with movable navigation spans.

FIGURE 9.1   Immervad Bridge in Jutland.

FIGURE 9.2   Marmorbroen (the Marble Bridge) at the Christiansborg Castle in Copenhagen.

More permanent bridges on fixed piers and with steel superstructures (Figure 9.3) appeared sporadically in the second half of the nineteenth century as part of the new railway network, but it was not until the 1930s that a real effort was made to substitute the many ferry routes by bridges.

From 1930 to 1940 11 bridges were built across straits with widths from a few hundred meters to more than 3 km. Five of the bridges were for both road and rail traffic, five for road traffic only, and one for railway only. The most notable bridges were the high-level bridges: the Lillebælt Bridge from 1935 (Figure 9.4) and the Storstrøm Bridge from 1937 (Figure 9.5). With its total length of 3.2 km, the Storstrøm Bridge was at completion the longest bridge in Europe, and it remained the longest road and rail bridge for a period of 60 years until it was surpassed by the West Bridge of the Storebælt Link in Denmark. The ambitious bridge construction program of the 1930s was abruptly interrupted at the outbreak of World War II in 1939. However, two more bridges that were already under construction were opened to traffic in 1942 and 1943, respectively.

The bridges from the 1930s to early 1940s were generally built with steel superstructures in the form of riveted plate girders, trusses, or arches. In a few bridges the approach spans were made as concrete arches, for example in the Queen Alexandrine Bridge, shown in Figure 9.6. After World War II, due to lack of resources during the postwar period, the bridge building program was not revived until 1952, when the small Munkholm Bridge (Figure 9.7) was opened to traffic. It was a fine little bridge in a gentle Danish landscape and it was greatly appreciated by the public, because it gave a clear indication of the fact that the standstill in bridge building was over and a bright future was ahead.

FIGURE 9.3   Railway bridge from 1862 across the Gudenaa.

FIGURE 9.4   The Lillebælt Bridge from 1935.

FIGURE 9.5   The Storstrøm Bridge from 1937. (Courtesy of DTU.)

FIGURE 9.6   The Queen Alexandrine Bridge.

FIGURE 9.7   **(See color insert.)** The Munkholm Bridge from 1952. (Courtesy of DTU.)

A route comprised of three strait-crossing bridges between Funen and Langeland had been on the drawing board already in the late 1930s but was postponed by the war. In the 1950s construction was started, but based on revised designs. The prewar designs had concrete arch approach spans leading up to a larger main span in steel (similar to the Queen Alexandrine Bridge in Figure 9.6), but the Langeland Bridge that was actually constructed in the 1960s had post-tensioned concrete box girder approach spans and the main span was a tied arch in concrete (Figure 9.8). Developments in post-tensioned concrete made it possible to apply a haunched concrete box girder for the main span of the Svendborgsund Bridge (Figure 9.9), the last of the three strait-crossing bridges in the Funen to Langeland link.

FIGURE 9.8 The Langeland Bridge.

FIGURE 9.9 The Svendborgsund Bridge. (Courtesy of DTU.)

FIGURE 9.10 The Lillebælt Suspension Bridge from 1970. (Courtesy of DTU.)

From 1970 to the mid-1980s a number of major bridges were built across straits as part of the new domestic expressway network. Some of these bridges were actually built as parallel bridges across the straits where bridges with narrow two-lane roadways were built in the 1930s. The second Lillebælt Bridge was opened to traffic in 1970 and it was the first major suspension bridge in Denmark (Figure 9.10). In contrast to the other postwar bridges it had a deck made of steel to save weight in the 600 m main span.

The second Lillebælt Bridge was followed by the Vejlefjord Bridge (Figure 9.11), a multispan concrete bridge with a superstructure composed of haunched post-tensioned concrete box girders.

As a supplement to the Storstrøm Bridge from 1937 the Farø Bridge was completed in 1985. The tender designs of the Farø Bridge, a northern bridge with a multispan concrete box girder and a similar southern bridge but with a main span designed as a cable-stayed bridge, were prepared by the owner, the Danish Road Directorate. However, the successful contractor had based his bid on an alternative design with a superstructure in steel (Figure 9.12). That proved to be competitive mainly because the 80 m full-span box girder units could be fabricated at a shipyard and erected in one piece.

With the bridges built in the 1930s the main parts of Denmark were linked together in two units, one comprised of the peninsula Jutland and the second-largest island Funen, and the other of the main island Zealand and islands to the south (Figure 9.13). It is obvious that one bridge was missing to link the entire country together: a bridge across Storebælt between Zealand and Funen.

Building a bridge crossing Storebælt was a task of quite a different magnitude than the other bridges of the 1930s as the width of Storebælt from coast to coast is 18 km—almost six times the length of the Storstrøm Bridge completed in 1937. Furthermore, the Storebælt Bridge would cross the international navigation channel from the Baltic Sea to the North Sea, so it had to be built to allow passage of the largest ocean-going vessels. The first realistic plans (Figure 9.14) to construct a Storebælt Bridge were presented in the late 1930s and preparatory works such as geotechnical investigations were planned to start in 1940, but the outbreak of World War II made it impossible to proceed with the plans at that time.

FIGURE 9.11   The Vejlefjord Bridge. (Courtesy of DTU.)

FIGURE 9.12   The Farø Bridges, southern crossing.

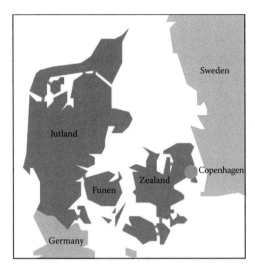

FIGURE 9.13 Strait-crossing bridges built in Denmark up until World War II.

FIGURE 9.14 Design from 1936 for a Storebælt Bridge.

9.1.3 World-Class Bridges across Storebælt and Øresund

In 1987 the Danish Parliament made the final decision to start the construction of a fixed traffic link across Storebælt. Due to the location of a small island, Sprogø, in the middle of Storebælt, the link was to be composed of two bridges and a tunnel. Across the Eastern Channel of the international navigation channel, the link consists of a high-level expressway bridge and a bored railway tunnel, whereas the Western Channel is crossed by a low-level bridge carrying both road and rail traffic.

At completion in 1997 the 6.6 km long West Bridge of the Storebælt Link (Figure 9.15) was the longest road and rail bridge in Europe, a record it held for only three years until it was surpassed by the Øresund Bridge. The most impressive part of the Storebælt Link is undoubtedly the 6.8 km long East Bridge, with a 1624 m main span suspension bridge (Figure 9.16). That span was at the completion in June 1998 the second-longest span in the world, surpassed only by the span of the Akashi Kaikyo Bridge in Japan.

Overlapping with the construction of the Storebælt Bridges was the construction of the Øresund Bridge between Denmark and Sweden. The superstructure is designed as a double-deck truss with two railway tracks on the lower deck and a four-lane expressway on the upper deck (Figure 9.17). The total length of the bridge is 7.8 km, making it the longest bridge in the world for both road and rail traffic. At completion in 2000 the main span of 490 m was also the longest in the world for a cable-stayed bridge carrying both road and rail traffic.

FIGURE 9.15   The West Bridge of the Storebælt Link. (Courtesy of DTU.)

FIGURE 9.16   The main span of the Storebælt East Bridge. (Courtesy of DTU.)

FIGURE 9.17   The Øresund Bridge. (Courtesy of DTU.)

9.2 Design Practice

Structural codes used in Denmark have since the beginning of the twenty-first century been based on Eurocodes supplemented by National Annexes (or National Application Documents) and printed in both Danish and English. National Annexes form a part of the Eurocode complex as choices are open for the member countries to specify their own National Determined Parameters (NDP). The National Annexes contain:

- A table of contents that indicate in which paragraphs national choices are open and in which paragraphs national choices have been made
- Paragraphs numbered corresponding to the Eurocode, where national choices have been made

NDP comprise specified values where alternatives are given in the Eurocodes, specific geographical and climatic data, and accommodation of national safety levels. For ordinary road and pedestrian bridges the Danish Road Directorate has prepared specific rules in the document "Road and Pedestrian Bridges—Load and Design Rules." The document is in its entirety printed only in Danish, but an appendix containing the relevant National Appendix to "Eurocode 1, Basis of Design and Actions on Structures, Part 3: Traffic Loads on Bridges" is also printed in English.

For the largest Danish bridges, the Storebælt Link and the Øresund Link, special structural codes were prepared in documents denoted "Design Basis" for the Storebælt bridges and "Design Requirements" for the Øresund Bridge. For the Øresund Bridge the specifications were based on the Eurocodes. It was one of the very first applications of these codes, which were to a large extent only available in draft form when the design work was initiated. Most importantly, the Eurocode-based design solved the problem related to the fact that the existing Danish and Swedish codes differed in many aspects, so it would in any case have been necessary to prepare a new set of design rules. Because the Øresund Bridge design specifications were applied to a bridge between two countries, the usual National Application Documents (NAD) were substituted by a special Project Application Document (PAD).

9.3 Inclined Wall Frame Bridges

In Denmark, short- to medium-span highway bridges are almost exclusively built as cast-in-place reinforced concrete bridges. This allows the bridges to be designed as monolithic structures without any movable joints. A very popular design for underpass structures is the "inclined wall frame bridge" shown in Figure 9.18. Here the inclined retaining walls are rigidly joined to the deck slab to form a highly efficient framed structure. Besides its structural advantages, the bridge is characterized also by improved traffic safety, as the concrete walls are moved away from the traffic lanes (compared to bridges with vertical retaining walls). Furthermore the inclined walls improve the view beyond the bridge and reduce the tunnel impression.

Inclined wall frame bridges are generally built with spans up to 12 m in normal reinforced concrete and with a deck thickness of typically 0.6 m and wall thicknesses of 0.5 m.

The bridge type shown in Figure 9.18 was developed by the engineers Gimsing & Madsen, Ltd. in the early 1970s.

In contrast to the extensive use of concrete in road bridges, pedestrian bridges are in most cases made with steel superstructures. As an example, Figure 9.19 shows a pedestrian bridge in Hørsholm, north of Copenhagen. The bridge has a superstructure made of a haunched steel box girder with a trapezoidal cross section. The length of the steel girder is 36 m and the width of the deck 3 m. It was fabricated at a steel fabrication shop in northern Jutland and transported in one piece to the bridge site 400 km away. After arrival at the site the prefabricated steel superstructure was lifted by a mobile crane, turned, and lowered onto the abutments, where a moment rigid connection was made by anchor bolts at the rear ends of the girder. Despite the fact that the girder depth is only 0.6 m at midspan, the bridge shows no tendency to vibrate under dynamic loading from pedestrians moving across the bridge.

FIGURE 9.18   Underpass structure built as a monolithic frame. (Courtesy of Gimsing & Madsen, Ltd.)

FIGURE 9.19   Pedestrian bridge at Hørsholm. (Courtesy of Gimsing & Madsen, Ltd.)

In road and rail bridges with spans of more than 80 m, both steel and concrete superstructures are used, but for bridges with spans of more than 150 m only steel superstructures are found. Since the late 1960s steel bridges have been built with all joints welded both in the shop and on site. This is in contrast to many other countries where field joints are more commonly bolted.

Using welded joints, it is possible to get steel box girders with airtight interiors. Fully welded joints were used for the first time in the Little Belt suspension bridge to achieve very efficient corrosion protection of all interior steel surfaces by adding a dehumidification plant inside the steel box and keeping the relative air humidity below 40% so that no corrosion will form on the steel surfaces.

The principle of dehumidifying the air inside box girders was developed by the engineers C. Ostenfeld & W. Jønsson (later renamed COWIconsult) during the design of the Lillebælt suspension bridge, and the system has here been successfully in operation for more than 40 years. Today dehumidification is applied all over the world as an efficient tool to protect interior steel surfaces against corrosion. Also, a large number of box girder bridges originally built without dehumidification have been retrofitted to avoid the complicated process of interior maintenance of steel surfaces.

9.4 Girder Bridges

Bridges with the main girders designed as plate girders (concrete ribs) or box girders were only used in minor bridges up to the mid-1930s, but after World War II they became the most commonly used bridge type for spans up to well over 100 m.

9.4.1 Storstrøm Bridge

The 3.2 km long Storstrøm Bridge constructed in the 1930s is characterized by very long approach spans leading up to the three navigation spans of 102 m, 136 m, and 102 m. Therefore, the approach spans constitute almost 90% of the total bridge length. The approach spans have alternating spans of 58 m and 62 m and the superstructure is statically determinate, with two hinges in every second span. The two main girders are riveted plate girders topped with crossbeams and stringers to support the separated concrete deck slabs under the railway tracks and roadway, respectively. As was common in prewar design practice there was no composite action between the concrete slabs and the underlying steel girders (Figure 9.20).

Although the Storstrøm Bridge was built on bridge technology belonging to the past, in one aspect it introduced a concept that later proved to be very efficient during construction of multispan bridges across water: the concept of fabricating and assembling full-span girder units transported and erected by a large purpose-built floating crane (Figure 9.21).

9.4.2 Funder Valley Bridge

The bridge across the Funder Valley in Jutland near the town of Silkeborg is the longest bridge over land in Denmark. The total length is 730 m and the number of spans is nine, with seven intermediate spans of 85 m and two shorter end spans. The superstructure consists of two independent concrete box girders each 3.5 m deep and with a 14.2 m wide deck slab on top (Figure 9.22). The pier shafts, with a box-shaped cross section and heights of up to 30 m above ground, were cast in situ using jump forms.

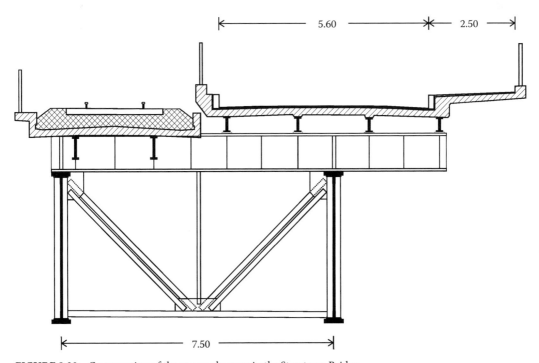

FIGURE 9.20 Cross section of the approach spans in the Storstrøm Bridge.

FIGURE 9.21   A full-span erection unit lifted into place by the floating crane "Stærkodder."

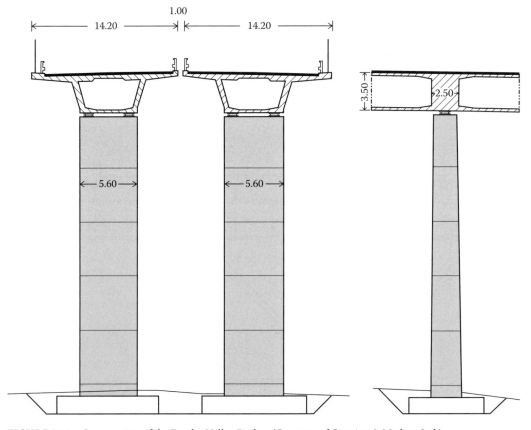

FIGURE 9.22   Cross section of the Funder Valley Bridge. (Courtesy of Gimsing & Madsen Ltd.)

FIGURE 9.23 Temporary pier for midspan support during push launching of the superstructure of the Funder Valley Bridge. (courtesy of Gimsing & Madsen, Ltd.)

The superstructure was erected by push launching with the box girders supported not only on the permanent piers but also on temporary piers at midspan. The temporary piers were also made of concrete but with two columns interconnected by several crossbeams (Figure 9.23). The available construction time allowed one box girder to be launched from one abutment to the other before launching the second box girder. That allowed a large degree of repetitive use of temporary construction equipment; for example, the temporary pier shafts were moved sideways after completion of the first box girder launching to provide support during the second launching.

9.4.3 Storebælt West Bridge

The West Bridge of the Storebælt Link is a low-level bridge with a vertical clearance of 18 m in the central spans, which allows a passage of vessels of moderate size, for example small coasters, fishing boats, and leisure boats, whereas larger vessels will have to pass the navigation channel under the high-level East Bridge. The total length from abutment to abutment is 6618 m, subdivided in six continuous expansion sections with lengths between 1047.6 m and 1158.1 m. All intermediate spans are 110.5 m long, whereas the spans next to the expansion joints are 81.8 m long.

The superstructure in concrete consists of two independent box girders, one for expressway traffic with a deck width of 24.1 m and one for the railway with a deck width of 13.25 m (Figure 9.24). The depth of the haunched roadway girder varies between 3.78 m and 7.34 m and that of the railway girder between 5.13 m and 8.70 m. The two independent superstructures are supported through pot bearings onto two individual pier shafts on each pier. The hollow pier shafts have a constant cross section to 3.5 m below the water surface where they are resting on a common caisson with a width of 29.4 m. In the railway girder the expansion joints are supplemented by a total of 14 hydraulic buffers that allow slow longitudinal movements due to temperature change but are able to transfer short-term loads due to braking of trains.

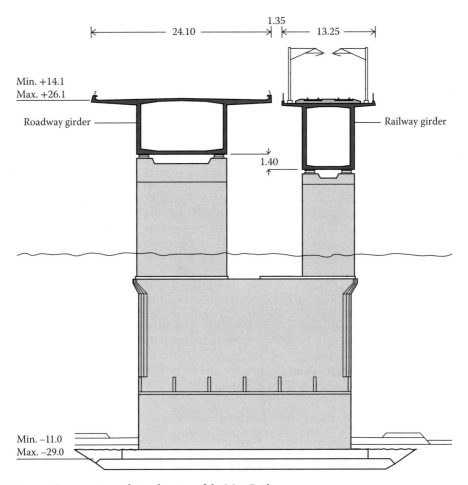

FIGURE 9.24   Cross section and pier elevation of the West Bridge.

The construction of the West Bridge was based on an alternative design by the contractor. The main feature of that design was the use of a large, purpose-built floating crane called "Svanen," with a lifting capacity of 6000 metric tons. That allowed the superstructure erection units to comprise full-span precast elements (from midspan to midspan) as the weight of a road girder element was 5500 metric tons and a rail girder element 4100 metric tons (Figure 9.25). Most of the precast caissons also had weights of less than 6000 metric tons and could be picked up by "Svanen" from a load-out pier, but the largest caissons, with weights of up to 7200 metric tons, had to be completed while the caisson was resting on the sea bottom near the load-out pier to activate the buoyancy of the submerged part of the caisson.

9.4.4 Storebælt East Bridge Approach Spans

The East Bridge of the Storebælt Link is a high-level bridge with a vertical clearance of 65 m in the main span across the international navigation channel from the North Sea to the Baltic Sea. The total length from abutment to abutment is 6790 m, subdivided into the western approach spans with a length of 1556 m, the main suspension bridge with a length of 2694 m, and the eastern approach spans with a length of 2518 m (Figure 9.26). Add to this 11 m on top of each of the two anchor blocks between the approach and main spans.

FIGURE 9.25 Full-span precast road girder element lowered onto the pier top by the floating crane "Svanen." (Courtesy of DTU.)

FIGURE 9.26 The 2518 m long continuous box girder forming the eastern approach spans of the East Bridge. (Courtesy of DTU.)

The superstructures of the approach spans are continuous steel box girders with all intermediate spans 193 m long. The steel box girders have a constant depth of 7.1 m and a deck width of 25.1 m. The cross section of the two-cell box girder is trapezoidal, with the outer webs attached along the edges of the orthotropic deck plate (Figure 9.27).

A total quantity of 48,900 metric tons of steel S420 with a yield point of 420 MPa was required for the box girders of the approach spans. The stiffened panels were fabricated in Italy and subsequently transported by ship to an assembly yard in Portugal to be preassembled into 40 m long units. From Portugal the subassembled units were transported to a second assembly yard in northern Denmark to be joined into full-span units each weighing approximately 2400 metric tons. The final transport to the bridge site took place on a barge and the lift onto the piers was performed by two cranes, one positioned at the end of the previous span and the other a floating crane, as shown in Figure 9.28.

FIGURE 9.27   Road girder element for the East Bridge approach spans. (Courtesy of DTU.)

FIGURE 9.28   Prefabricated box girder element 193 m long in the process of being lifted onto the permanent bearings on top of the pier shafts (courtesy of DTU).

As for all major box girder bridges in Denmark built after 1970, the East Bridge girders have all interior surfaces corrosion protected by dehumidification, and the layout of the trapezoidal box is actually to a large extent governed by the desire to have all complex plate joints between stiffeners, diaphragms, deck plate, webs, and bottom plate inside the box so that surface treatment by painting shall only take place on the plane surfaces of the exterior of the box.

With a depth of 7.1 m and a span of 193 m, the span-to-depth ratio of the approach spans is 27—a value that indicates a certain wind sensitivity. That was confirmed both analytically and during wind tunnel testing, where vertical vortex-induced vibrations were observed for wind speeds that could be experienced under the service conditions. The vibrations would not be critical for the safety of the structure, but they could give rise to user discomfort. It was therefore decided to install a number of tuned mass dampers inside the box girders to eliminate all vibrations experienced from cross wind speeds below 25 m/sec. Vibrations appearing for wind speeds above 25 m/sec were of no concern as the bridge is closed to all traffic when cross wind speeds above 25 m/sec are measured.

9.5 Arch Bridges

Arch bridges were widely used in the strait-crossing bridges built in the 1930s and into the 1940s, whereas they have been only scarcely used in recent years.

9.5.1 Queen Alexandrine Bridge

The most prominent arch bridge from the first half of the twentieth century is the Queen Alexandrine Bridge across the strait separating Zealand from the small island Møn. The 125 m long span across the navigation channel is a latticed steel arch positioned partly above the roadway, whereas the 10 approach spans consist of concrete arches positioned below the bridge deck. The length of the different approach spans varies from 47 to 64 m in such a way that the horizontal thrust from dead load is balanced at all intermediate piers. As a consequence the substructure only had to be designed for the differential thrust from nonuniform traffic load.

The superstructure of all the approach spans was cast in situ on a full-span formwork that could be floated from one span to the next and after length adjustment be reused. The two arch ribs under the deck are made with a solid cross section (Figure 9.29).

9.5.2 Ågade Bridge

Danish bridges are traditionally designed by giving priority to structural efficiency, but with due respect to aesthetic value by choosing clean and harmonic shapes for the structural elements. However, the

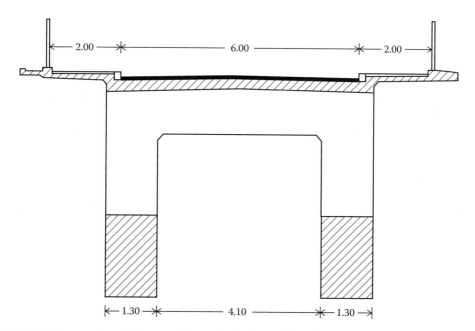

FIGURE 9.29 Cross section of the approach spans in the Queen Alexandrine Bridge.

FIGURE 9.30 The Ågade Bridge in Copenhagen. (Courtesy of DTU.)

recent trend on the international scene to let the design of bridges be determined by architectural competitions judged by evaluation committees without genuine structural competence has also reached Danish shores.

The Ågade Bridge in Copenhagen can be seen as an example of this trend. Here the deck of the bridge, for bicycles and pedestrians, is supported by a single arch in a plane leaning 45° outwards from one of the edges of the deck. That induces a substantial amount of (unnecessary) torsion and out-of-plane bending. As a result the material consumption is excessive, as illustrated by the fact that the steel weight per square meter of deck area in the 65 m long arch span is 25% higher than in the 193 m long spans of the approach bridges (with full roadway loading) in the Storebælt East Bridge. In affluent communities it is of course fully acceptable to spend extra amounts of money to create a "landmark" bridge, and in the case of the Ågade Bridge it was the intention that the numerous people passing under the bridge should experience it as a gateway to the town (Figure 9.30).

9.6 Truss Bridges

At the end of the nineteenth century and the beginning of the twentieth century, truss bridges of steel were the preferred solution for medium-span railway bridges, and for a few road bridges. Most of these bridges have been replaced by new bridges and been demolished, but a few of the old railway truss bridges have survived as bridges for pedestrians and bicycles (Figure 9.31). The largest truss bridge from the first half of the twentieth century is the first Lillebælt Bridge, shown in Figure 9.4. The five-span truss, with a total length of 825 m, has a 220 m main span where the truss depth is 24 m.

9.6.1 Øresund Bridge Approach Spans

The Øresund Bridge between Denmark and Sweden is a high-level bridge with a vertical clearance of 55 m in the navigation span. The total length from abutment to abutment is 7840 m, subdivided into the western approach spans with a length of 3014 m comprised of two expansion sections, the main cable-stayed bridge with a continuous length of 1092 m, and the eastern approach spans with a length of 3739 m

FIGURE 9.31 Old railway bridge across the Gudenaa at Randers converted to a bridge for pedestrians and bicycles.

FIGURE 9.32 The 3739 m long eastern approach spans of the Øresund Bridge. (Courtesy of DTU.)

comprised of three expansion sections. Including the two expansion joints at the abutments, the total number of expansion joints in the 7.8 km long superstructure is only seven. The approach spans have superstructures made as continuous double-deck steel trusses with composite action to the upper deck slab (Figure 9.32). All intermediate spans are 140 m long. The total width of the upper deck slab is 24.5 m and the width between the exterior faces of the steel trusses is 15.0 m (Figure 9.33).

The truss members for the approach spans are made of steel S460 (80%) and S355 (20%) in a total quantity of 65,600 metric tons. The truss elements were fabricated in Cadiz in southern Spain, where the upper deck slab was also cast. From Cadiz to a work site area in Malmö close to the bridge site, the 140 m long truss elements were transported in pairs on a barge. At the work site area the prefabricated, channel-shaped containment girders for the railway tracks were erected on top of the lower crossbeams between the two main trusses.

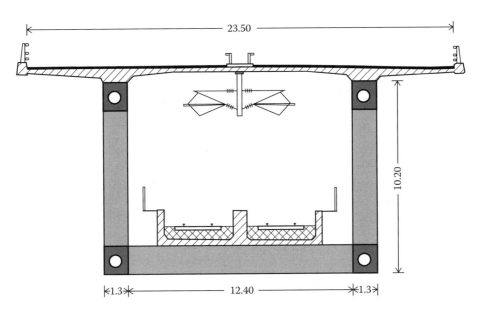

FIGURE 9.33   Cross section of the Øresund approach spans.

FIGURE 9.34   The floating crane "Svanen" in the process of lifting a 140 m long approach span truss element into position.

The final transport and erection of the 140 m full-span elements were performed by the floating crane "Svanen" that had been modified since it was initially applied during the construction of the West Bridge of the Storebælt Link (Figure 9.34). The lifting capacity of "Svanen" was increased to 9500 metric tons and the lifting height was also substantially increased. All truss members and cross-beams in the Øresund Bridge deck are box shaped with all interior surfaces corrosion protected by dehumidification.

9.7 Cable-Stayed Bridges

9.7.1 Farø Bridge, Southern Crossing

The second bridge link between Sjælland and the southern island Falster is the Farø Bridge, completed in 1985. The link consists of the 1696 m long northern bridge between Sjælland and the small island of Farø and the 1726 m long southern bridge between Farø and Falster. Both bridges have continuous superstructures from abutment to abutment with decks designed as wide box girders in steel. Most of the spans are 80 m long but across the navigation channel between Farø and Falster a cable-stayed bridge with a 290 m long main span is found.

The deck of the cable-stayed bridge is supported by a central fan-shaped cable system supported on a diamond-shaped pylon in concrete. The pylon shape was chosen to avoid the additional width of the central reserve required if a vertical pylon in the cable plane should have been used. The resulting diamond shape for the Farø Bridge actually marks the first application of this pylon configuration that has subsequently been widely used all over the world (Figure 9.35).

The 290 m long main span is flanked on either side by 100 m long side spans. Because the fans are of equal length in the main span and side span, the side-span pier is not positioned at the end of the side-span fan but at the deck anchorage of the second stay from the top. That efficiently activates the three stays at the end of the side-span fan as backstays so a special heavy single backstay is avoided (Figure 9.36). At the pylons the wide box forming the deck of the bridge is supported in the lateral direction and against twist but is free to move in the vertical direction. To achieve such a supporting condition it was necessary to connect the deck to the pylon through cross-connected hydraulic cylinders.

The steel superstructure of the Farø Bridge is designed as a wide box in a trapezoidal shape (Figure 9.37). The depth is 3.5 m and the width 22.4 m. In the approach spans the trapezoidal box has only one cell, but in the cable-stayed bridge a central vertical web is added to allow an efficient anchoring of the stay cables. As is the case for all major steel box girder bridges, the interior of the Farø Bridge box is corrosion protected by dehumidification, a feature that proved to be a deciding factor in the choice of a superstructure in steel rather than concrete.

FIGURE 9.35 The Farø Bridge with its diamond-shaped pylons under construction.

FIGURE 9.36   The main span of the Farø Bridge.

FIGURE 9.37   Box-shaped main girder for the Farø Bridge (approach spans).

9.7.2 Øresund Bridge Main Spans

The Øresund main spans at the navigation channel have a total continuous length from expansion joint to expansion joint of 1102 m. The continuous truss is supported by harp-shaped cable systems except for the regions close to the expansion joints. The side-span supports closest to the pylons provide vertical support inside the side-span harps, which has a pronounced influence on the stiffness of the system (Figure 9.38).

The cross section in the main span is basically similar to that of the approach spans, that is, it has an upper roadway deck made as a transversally prestressed concrete slab acting compositely with the top chords of two vertical steel trusses. In contrast to the approach spans, the lower railway deck is made entirely in steel in the form of a shallow multicell box (Figure 9.39). The steel trusses for the cable-stayed bridge were fabricated in a Swedish shipyard and made mainly of S420 in a quantity of 15,200 metric tons.

Because the pylons are free-standing posts above the roadway, the cable planes have to be vertical and positioned some distance from the edges of the roadway slab. To transfer the load from the main trusses to the cable anchorages, triangular brackets or "outriggers" are added outside the trusses. To make it possible to connect the outriggers efficiently to the main trusses, the geometry of the diagonal bracing was changed from the approach span, where all diagonals were of the same length, to a system with long diagonals in the direction of the stay cables and short diagonals in between so that the node distance at the bottom chords

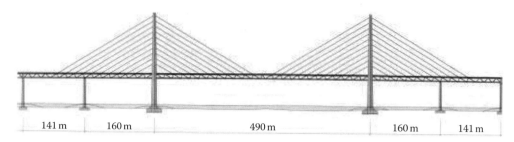

FIGURE 9.38 Øresund Bridge main spans.

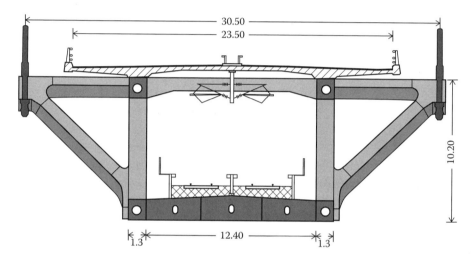

FIGURE 9.39 Cross section of the main span in the Øresund Bridge.

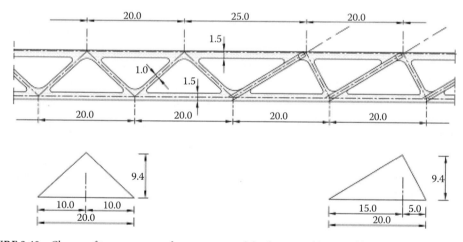

FIGURE 9.40 Change of truss geometry from an approach bridge to a cable-stayed bridge.

was kept constant (Figure 9.40). The stay consists of two cables (one on top of the other) each comprised of approximately 70 seven-wire strands. The total amount of cable steel amounts to 2150 metric tons.

With the overall design leading to vertical cable planes outside the deck, it was possible to let the pylons consist of vertical free-standing posts above the deck (Figure 9.41). By ensuring that the centroid of the pylon was positioned in the vertical cable planes, the cross section of the pylon was subjected to pure uniform compression from dead load and vertical traffic load acting on the deck.

FIGURE 9.41   Øresund Bridge pylon.

9.8 Suspension Bridges

9.8.1 Second Lillebælt Bridge

The second bridge across Lillebælt was opened to traffic in 1970 as part of the regional expressway system. It also provided a much-needed improvement of the traffic capacity offered by the narrow dual-lane roadway on the first bridge from 1935. While the first bridge was built close to the location where the width of the strait is minimum, for the second bridge the desire to have a more smooth alignment resulted in a location where the width of Lillebælt is about 50% larger than the minimum. In order to improve navigational conditions the new bridge was built with a main span of 600 m, almost three times the main span of 220 m in the 1935 truss bridge.

At the time when the Second Lillebælt Bridge was in the detailed design phase it was only realistic to consider a suspension bridge for a span of 600 m. Cable-stayed bridges had at that time only been built with a span of up to about 300 m, so it would be a major step to go to twice that span. The Second Lillebælt Bridge was initially designed with a stiffening truss in accordance with the American tradition, but after having learned about the box girder design introduced for the Severn Bridge it was decided to also use this concept for the Danish suspension bridge. Thus the Second Lillebælt Bridge became the second suspension bridge in the world to have stiffening girder formed as a streamlined box (Figure 9.42). The Second Lillebælt Bridge also became the second major suspension bridge (span >500 m) with pylons made of concrete.

9.8.2 Storebælt East Bridge Main Spans

The main bridge across the international navigation channel through Storebælt is a suspension bridge with a total length of 2694 m: a main span of 1624 m flanked by two side spans each 535 m. At completion in 1998 the main span was the second longest in the world, surpassed only by the span of the Akashi Kaikyo Bridge in Japan.

The steel box girder forming the deck of the suspension bridge is 4.34 m deep and 31 m wide (Figure 9.43). It is made of S355 as the strength requirements made it superfluous to use S420 as in the approach spans of the bridge. The total quantity of structural steel used for the stiffening box girder is

FIGURE 9.42 Second Lillebælt Bridge.

FIGURE 9.43 Streamlined box girder of the Storebælt East Bridge.

29,700 metric tons. The box girder is supported vertically at the anchor blocks and by the vertical hangers from the main cables. At the pylons the deck is only supported in the lateral direction but is free to move vertically. At midspan the main cables are fixed to the box girder through a long clamp. Together with installation of large hydraulic buffers between the box girder and the anchor blocks, the central clamp improves the stiffness under short-term asymmetric load and increases the frequency of asymmetric vibration modes. In the case of long-term movements, for example from temperature changes, the longitudinal buffers do not provide resistance (Figure 9.44).

As the box girder is continuous over the full length of 2694 m and no vertical support was needed, it became unnecessary to add a crossbeam between the two pylon legs immediately below the deck. The omission of the crossbeam below the deck also clearly illustrates the fact that the deck in a suspension

FIGURE 9.44   Central clamp between main cable and deck.

FIGURE 9.45   Storebælt East Bridge pylon.

bridge is supported by the cable system and does not carry the global load by bending to the pylons. Without the crossbeam at deck level it was possible to position the lower deck beam at midheight of the 254 m high pylons (Figure 9.45).

During the conceptual design of the East Bridge emphasis was laid on arriving at a clean and pleasing appearance as seen in the configuration of the pylons. However, the anchor blocks were also treated aesthetically to appear more elegant than is found in many of the existing massive-looking anchor blocks. With its location at mid-sea the designers attempted to give the East Bridge anchor blocks a more transparent look (Figure 9.46).

FIGURE 9.46 Anchor block of the Storebælt East Bridge.

9.9 Summary

Beginning in the early 1930s a large number of bridges have been built across the many straits separating the different regions of the Danish island kingdom. Among these are some of the world's largest bridges constructed during the twentieth century.

In the first decennium of the twenty-first century activities within the construction of major bridges have been at a halt in Denmark, but a new bridge-building era might be experienced soon, as the Storstrøm Bridge from 1937 shall be replaced by a new bridge with a double track railway and a two lane road of present standard. Also, a third Lillebælt Bridge, or a Kattegat Bridge directly between Sjælland and Jylland are under consideration.

References

Gimsing, Niels J. (editor), 1998. *The Great Belt Publications: West Bridge*, A/S Great Belt, Copenhagen.

Gimsing, Niels J. (editor), 1998. *The Great Belt Publications: East Bridge*, A/S Great Belt, Copenhagen.

Gimsing, Niels J. (editor), 2000. *The Øresund Technical Publications: The Bridge*, Øresundsbro konsortiet, Copenhagen.

10

Bridge Engineering in Finland

Esko Järvenpää
WSP Finland Ltd

10.1 Introduction to Bridges in Finland

10.1.1 Old Stone Bridges and Wooden Bridges

Finland began to build roads during the late Middle Ages. The first important bridges were built over the rivers and straits of Finnish lakes when the main road from the Russian border to west cost was built. The road was called the King's Road, and wood was the main material for the bridges because of its availability and low costs. There were also skilled carpenters just for bridge construction. The problem with wooden bridges was the short life span of the materials and work amount needed often for repair of the damages. One of the most interesting wooden bridges was built in Oulu in 1867 (Figure 10.1). The spans of the bridge were 38.1 m + 45.7 m + 38.1 m. The bridges served until the year 1950.

Stone bridges were much more expensive. Only the most important bridges were built as stone bridges. The oldest existing stone bridge is from the year 1777 in Espoo. The construction of stone bridges required skilled workmanship. Some beautiful stone bridges are still in usage (Figure 10.2).

10.1.2 Iron and Steel Bridges

The first railway in Finland was built from Helsinki to Hämeenlinna and opened for traffic in the year 1862. The first iron bridge was built at that time. The knowledge of how to build steel bridges came from England. The first major steel bridge was built over the river Kymi in 1870. The bridge exists today as a historic bridge. Road bridges were also built using steel. Many of those bridges were destroyed during the Second World War.

10.1.3 Concrete Bridges

The first reinforced concrete bridge was built in the 1910s. The usage of concrete bridges quickly became common. One remarkable concrete bridge was the Färjsund Bridge, with a main arch span of 130 m, completed in 1937.

FIGURE 10.1   Aunes Bridge is one of best examples of stone bridges. (Courtesy of Kristiina Hakovirta.)

FIGURE 10.2   Historic wooden bridge of the Oulu River. (Courtesy of Pohjois Pohjanmaa Museum/Uuno Laukka Pictures.)

10.2 Design Practice

In the past, Finland has not had any special bridge codes or standards. The earlier Roads Administration and the States Railways have published their own design practice guides for bridges on how to adopt the official codes and accepted standards. The design work of private consultants has been checked mainly by the same state authorities.

The latest level of traffic loads before the Eurocodes is from 1970. The loading level of bridges based on the Eurocodes has not increased remarkably compared to the earlier loading level. The changes to the Eurocodes have resulted in new codes and practices.

The Eurocodes consist of design standards for load-bearing structures. The Eurocodes are prepared by the European Commission–mandated European Committee for Standardization (CEN).

The Eurocodes currently consist of 58 parts. The Eurocodes determine the reliability of the bridge for various kinds of loads such as traffic, snow, wind, heat, accident, and crane loads. Various building materials have their own detailed instructions. The Eurocodes and National Eurocodes have replaced the earlier codes for load-bearing structures. The Eurocode system aims to harmonize the design of structures and to remove barriers to competition in the European Union member states. Responsibility

TABLE 10.1 Longest Spans of Bridges in Finland

Bridge	Span	Location	Completed
Cable-Stayed Bridges			
Raippaluoto Bridge	250 m	Mustasaari	1997
Kärkistensalmi Bridge	240 m	Korpilahti	1997
Suspension Bridges			
Kirjalansalmi Bridge	220 m	Parainen	1963
Sääksmäki Bridge	155 m	Valkeakoski	1963
Steel and Concrete Composite Bridges			
Saimaa Bridge	140 m	Puumala	1995
Vehmersalmi Bridge	130 m	Valkeakoski	2002
Prestressed Concrete Bridges			
Norströmmen Bridge	135 m	Nauvo	1986
Toijansalmi Bridge	120 m	Taipalsaari	2004
Pedestrian Bridges			
Ylistö Bridge	112 m	Jyväskylä	1993
Laukko Bridge	100 m	Tampere	2010

for the Eurocodes for bridges is carried by the Transport and Communications Ministry. The Eurocodes for bridges were introduced in accordance with the Transport Agency's plan in June 2010.

10.3 Bridges in Finland

Finland's road network was built mainly in the 1950s and in the 1960s, a new era for concrete bridge and post-tensioned bridges started. Today many of those bridges are under repair and rehabilitation work. The number of bridges in country is about 15,000 on the roads and highways under the state organization. The number of railway bridges is about 2300. When adding the bridges of municipalities and cities the total number of bridges is more than 20,000. The longest bridge spans are listed in Table 10.1.

10.3.1 Concrete Girder Bridges

Concrete is the most common construction material in bridges. A typical bridge is a continuous slab bridge up to a span length of 20 m with a massive cross section. Cast-in-situ post-tensioned TT-type girder covers 20–40 m spans. Some big open TT-girder bridges have been built with up to 100 m spans. Concrete box girder bridges have been built with up to 135 m spans using the cast-in-situ free cantilevering method (Figure 10.3).

Examples of modern concrete railway bridges are the bridges on the new direct high-speed railway line between Helsinki and Lahti, which opened for traffic in 2006. The biggest of those bridges is the Luhdanmäki Railway Bridge (Figure 10.4). The bridge is a continuous post-tensioned concrete box girder bridge for two tracks. The total length of the bridge is 548 m and the regular span lengths are 45 m. The main focus in design was constructability and economic construction. The bridge is founded on composite steel pipe piles and spread footing. The box girder was cast in situ and post-tensioned in several stages.

10.3.2 Steel Girder Bridges

Because of steel production in Finland and the availability of skilled steel workers, steel bridges are popular in Finland. The economy of composite plate girder bridges against concrete has been obvious since 1980. The construction method of launching the steel girder first and concreting the deck slab

FIGURE 10.3   Norsströmmen Bridge, designed by Y-suunnittelu Oy. (Courtesy of Torsti Salonen.)

FIGURE 10.4   Luhdanmäki Railway Bridge, designed by WSP Finland Oy. (Courtesy of Esko Leppäluoto.)

supported by the girders fits well with Finnish climatic conditions. Composite structures with only mild reinforcing at tension areas of the concrete slab have showed be a sound structural system.

10.3.2.1 Saimaa Bridge

The Saimaa Bridge (Figure 10.5) is an innovative solution of composite girder bridges built in Finland. The total length of the bridge is 781 m and the main span has a span length of 140 m. The effective width of the deck is 14 m. The bridge has an elevator from the harbor to the bridge deck level for pedestrians and bicyclists. The open two-plate girders working as a composite structure with the deck slab also have a composite concrete bottom slab at the main span supports.

FIGURE 10.5 Saimaa Bridge, designed by Pontek Oy. (Courtesy of Vastavalo Oy.)

FIGURE 10.6 Vehmersalmi Bridge, designed by WSP Finland Oy, aesthetic design by Jussi Tervaoja.

10.3.2.2 Vehmersalmi Bridge

Vehmersalmi Bridge is another example of a bigger composite bridge in Finland (Figure 10.6). The total length of the bridge is 400 m. The main span has a length of 130 m. The bridge is an unsymmetrical composite girder bridge.

10.3.3 Arch Bridges

The concrete arch Inkulansalmi Bridge (Figure 10.7) is one of the most beautiful examples of how a bridge can fit into the landscape. The span of the arch is 76 m and the effective width is 8.5 m.

FIGURE 10.7   Inkulansalmi Bridge in a beautiful lake landscape, designed by WSP Finland. (Courtesy of Vastavalo Oy.)

FIGURE 10.8   **(See color insert.)** Tornio River Bridge, built in 1939. (Courtesy of Pekka Pulkkinen.)

10.3.4 Truss Bridges

Truss bridges have been a solution for railway bridges and bigger road bridges over the stronger rivers in the country. One example is the Tornio River Bridge (Figure 10.8), which was completed in 1939. The bridge was the biggest road bridge at the time. The main span has a length of 108 m. The bridge has been renovated. The concrete deck has been changed to an orthotropic deck and the pedestrian lanes have been widened outside the truss girders.

10.3.5 Cable-Stayed Bridges

10.3.5.1 Lumberjack's Candle Bridge, Finland

The Lumberjack's Candle Bridge (Figures 10.9 through 10.11) and the following Tähtiniemi Bridge are the results of bridge design competitions. The Association of Finnish Civil Engineers has published rules for bridge design competitions and these rules have been used in most high-profile bridge design

FIGURE 10.9 Lumberjack's Candle Bridge at night, designed by WSP Finland Oy.

FIGURE 10.10 Lumberjack's Candle Bridge. (Courtesy of Esko Järvenpää.)

FIGURE 10.11 Another view of the Lumberjack's Candle Bridge. (Courtesy of Antti karjalainen.)

competitions in Finland. Early in the 1990s cable-stayed bridges got a good image in Finland. The opening of the Lumberjack's Candle Bridge near the Arctic Circle aroused a lot of publicity with the beauty of its form. The bridge was the first cable-stayed bridge built in Finland and was one of the first composite cable-stayed bridges ever constructed. The structure celebrated its 20th anniversary in September 2009.

The Finnish Road Administration and the city of Rovaniemi had concluded a need for a bridge over the Kemijoki River. The city of Rovaniemi wanted to have a landmark structure with a symbolic historical image and arranged a design competition in order to find the best solution for the bridge. The competition jury was unanimous about the winning designer and recommended the proposal for the final bridge design. Initially the citizens protested the bridge, but when the bridge was opened the objections ceased. The additional costs compared to a conventional bridge were easy to accept.

The structure is an asymmetric cable-stayed bridge with a composite steel and concrete superstructure and continuous spans of 41 m + 42 m + 126 m + 42 m + 42 m. On the town center side the bridge continues with a frame bridge over the shore road, separated from the main bridge by an expansion joint.

The cross section comprises a composite box with bottom plate and webs of steel and a concrete deck slab. The box is 8.8 m wide and the webs are 2.4 m deep. The deck slab has wide cantilevers supported by steel struts at 5.6 m intervals. The deck slab is post-tensioned longitudinally and transversely. The segments of the steel structure were welded together behind the abutment and launched over the river. Two auxiliary supports were needed to support the steel structure in the main span for launching and casting of the deck slab. The concrete deck slab was cast in segments each 11.2 m long using two sets of formwork equipment. The design of the composite superstructure with the deck slab cast in segments required development of a new computer program not commercially available at that time following the development of the structural properties of the box girder together with shrinkage and creep of concrete.

The effective width of the deck structure is 22.5 m. The main span is supported by eight pairs of cables. Passive anchorages are located at the lower end of the cable inside the girder box of the superstructure. The cable spacing is 16 m at the deck level. The back span cables in six pairs have passive anchorages at the top of the tower and active anchorages at the counterweight abutment.

The concrete tower rises 47 m above the surface of the bridge deck and is capped with two candle columns each having a diameter of 2.3 m. The cables are anchored at the top of the tower inside a steel box within the concrete section. Access to the top of the tower is by means of open air ladders between the tower legs. It is important to consider the appearance of a bridge at all times. Night time views of the Lumberjack's Candle Bridge have been enhanced by carefully selected illumination of the tower, cables, and the fascia beam on both of sides of the bridge.

10.3.5.2 Tähtiniemi Bridge

After the successful completion of the Lumberjack's Candle Bridge more than 10 cable-stayed crossings have been built in Finland. The Tähtiniemi cable-stayed bridge (Figures 10.12 and 10.13) was designed to blend into a beautiful lake landscape. The Tähtiniemi Bridge is located outside the city of Heinola in Lake Finland. The winning proposal from the competition was a single pylon cable-stayed bridge with a total length of 924 m. The bridge was chosen because of its clear structural system and aesthetic harmony with the beautiful landscape.

The main span is 165 m in length and the width of the deck varies from 22 m to 30 m. The top of the tower is 105 m above the water level. The superstructure is a traditional composite structure comprised of two open plate girders and crossbeams at 7 m intervals with a secondary longitudinal girder in the middle.

The steel girders were launched from both abutments and joined by welding. The launching operation was done with the help of pontoons to support the cantilever stages.

The steel grade of flanges was changed during construction to utilize thermomechanically manufactured high-strength steel with a yield strength of 420 MPa. The casting of the deck slab was done using movable scaffolding units. The section length was up to 24 m.

FIGURE 10.12   Tähtiniemi Bridge, designed by WSP Finland Oy.

FIGURE 10.13   Another view of the Tähtiniemi Bridge.

In the competition the pylon was designed to be a steel tower but was later changed to concrete, and was constructed using a climbing formwork. The concrete surface of the tower has been finished with a green-colored coating. One leg of the tower has an elevator.

Difficult foundation conditions at four of the piers and the tower required the use of steel tubular piles. The other supports are founded on soil or bed rock. The steel tubular piles were designed as composite structures. The diameter of each pile is 813 mm with a thickness of 16 mm. Reinforcement is used only at the upper sections of the piles.

10.3.5.3 Saame Bridge, Norway and Finland

The Saame Bridge (Figures 10.14 and 10.15) is located over the border river of Teno between Norway and Finland and it is one of the most northern cable-stayed bridges in the world. The river is an important salmon river, noted for the cleanliness of its water. In the design of the bridge the sensitive nature of northern Lapland and its landscape was taken into consideration. The Saame Bridge has the spans of 35 m + 155 m + 75 m + 35 m. The effective width of the deck is 10.5 m.

The weather conditions dictated design requirements, construction schedule, and construction methods. The ultimate temperature change was from −45° to +35° Celsius degrees. In the winter there are

FIGURE 10.14   Saame Bridge, designed by WSP Finland Oy.

FIGURE 10.15   Another view of the Saame Bridge.

51 days when the sun does not rise at all and during the summer there are 71 days when the sun shines day and night. The change in the weather and nature is rapid. The ice floes during spring can be dangerous for the bridge piers. The ice floes are estimated to be up to 75 m wide. The crushing force against the vertical piers is minimized using the conical shape of the piers on the same level of possible floating ice. The conical shape changes the crushing of ice to bending of ice, thereby reducing ice loads.

The bridge is located far from mainstream construction industry activity. Therefore, it was important to consider the construction methodology throughout the design and recognize the constraints arising from the remoteness of the site. The river channel is sensitive to erosion, so the bridge was founded on steel tubular piles with a diameter of 610 mm and a length of about 25 m. The ultimate load-bearing capacity was calculated to be 5.5 mN. The tubes are filled with concrete and the structural behavior of the piles has been analyzed as a composite structure.

The superstructure of the bridge is of composite construction. The prefabricated deck slab was installed on steel girders and joint cast over the top flanges of the steel beams. The pylons of the bridge are also composite structures. Tapering steel tubes have been filled with concrete up to the cable level. The steel structure was launched over the river with the help of auxiliary steel tube columns in the main span and in the 75 m span. The columns were stabilized with pretensioned horizontal strands to the abutments and pylon superstructure to control horizontal friction forces during launching. Half of the deck element was installed on the steel girders before the first round of stressing of the cables. The active anchorages are situated at the top of the pylons. The cable stressing was controlled by monitoring the length of cables and the level of the deck in elevation, as well as measuring the cable forces.

10.3.5.4 Crusell Bridge

The Crusell Bridge (Figures 10.16 and 10.17) is an example of a beautiful city bridge. The bridge is transversally a steel and concrete composite structure and longitudinally a concrete girder with two main girders supported by stay cables. The steel towers are inclined backwards. The spans of the bridge are 92 m + 51.5 m. The effective width of the bridge is 24.8 m. The short back span needed additional concrete to the cross section in order to balance the weight of the main span.

FIGURE 10.16   Crusell Bridge, designed by WSP Finland Oy.

FIGURE 10.17   Crusell Bridge cross section.

10.3.6 Wooden Bridges in Finland

Wooden bridges built after the Second World War were popular in the country. Laminated wooden girder standard bridges have not been competitive against cast-in-situ concrete bridges and prefabricated concrete bridges. The standard bridge type was design for spans from 4 m up to 20 m and for widths of 4.5 m, 6 m, and 6.5 m. The latest developments have shown that wooden bridges will have a chance take their place in bridges in the future. The following example is the result of a bridge design competition arranged according to the competition rules of the Association of Finnish Civil Engineers.

10.3.6.1 Vihantasalmi Bridge

The Vihantasalmi Bridge (Figure 10.18) is one of the largest wooden highway bridges in the world. The spans of the bridge are 21 m + 42 m + 42 m + 42 m + 21 m. The effective width of the bridge is 14 m including the sidewalk of 3 m for bicyclists and pedestrians. The structure is a three-span king-post truss. The load-bearing elements are made of glued laminated timber and the deck is a wood–concrete composite structure. The two side spans of the bridge are simply supported wood–concrete composite structures.

The main girders of the deck consist of a pair of glued laminated beams with dimensions of 1.35 m × 0.265 m. The hangers of the main girders are located at the midpoint and quarter points of the spans. In all the spans the wooden beams are connected to the concrete deck with shear connectors provided by bars glued into the wood and with shear keys in the wooden beams. The main bearing compression diagonals have two laminated members, having a cross section of 0.99 m × 0.265 m. The members are connected together with nailed weatherproof plywood sheets.

FIGURE 10.18 Vihantasalmi Bridge, designed by Inststo Timo Rantakokko & Co. Oy (Ponvia OY). (Courtesy of Torsti Salonen.)

Second-order nonlinear theory was used for the structural analysis of the main load-bearing system. The strength and stiffness properties of the shear connectors were determined in static and fatigue shear loading tests. All wooden members in the bridge are pressure impregnated and wooden members above the deck are protected against solar radiation and excessive moisture.

10.3.7 Pedestrian Bridges

The traffic culture in Finland has special features because of the climatic conditions and the fact that it is a sparsely inhabited country. Bicycling and walking is a common habit in country. Many cities have developed a light traffic culture and also separated vehicle traffic and light traffic. This culture has led to the development of many bicycle and pedestrian bridges. Three examples of pedestrian bridges are shown in Figures 10.19 through 10.21.

FIGURE 10.19 Iirislahti Pedestrian Bridge, designed by WSP Finland Oy. (Courtesy of Atte Mikkonen.)

FIGURE 10.20 Pornaistenniemi Pedestrian Bridge, designed by WSP Finland Oy.

FIGURE 10.21   Ylistö Pedestrian Bridge, designed by WSP Finland Oy.

10.3.8 Foundation Conditions

The foundation conditions in the country vary widely. Rock bed and spread foot foundations are typical, as well as different piled foundations. Driven concrete pile foundations are typical for bridges. The latest development in large-diameter piles has produced several new systems. Cast-in-situ bored pile up to 1.5 m diameter has a new competitor from driven steel tube piles. The steel tube pile works structurally as a composite structure with concrete and minimum reinforcing. Many remarkable bridges have composite steel tube pile foundations.

10.4 Bridge Management System in Finland

The Finnish National Road Administration has developed a computer-based bridge management system for bridge policy, long-term planning, and programming of investments on the network level. The system helps bridge engineers in the preparation of annual work programs. The system applies probabilistic deterioration models to find a condition distribution of bridges that minimizes maintenance and rehabilitation costs for the existing bridge stock and establishes deterministic repair and reconstruction indexes to organize for annual programming of bridge repair works.

The inspection of the bridges has been organized and controlled by the Finnish Transport Agency. Only skilled persons who have passed the qualification criteria and test arranged by the Finnish Transport Agency are allowed to do bridge inspection works in Finland.

In practice every bridge is inspected annually and every fifth year a more detailed inspection is carried out. Before starting repair design and following repair works, a special inspection is carried out.

11

Bridge Engineering in France

Jean-Armand Calgaro
La Defense

11.1 Introduction

11.1.1 Historical Development

Stones, wood, and ropes were probably the first elements used to build bridges. Bridge construction started approximately 4000 to 5000 years ago. Bridges symbolize the beginning of civilization, the first human conquest over hostile territory, the determination to encounter others, and the means to wage or to end wars. In historical times, the most exceptional bridges were built by Romans. The Pont du Gard in southern France stands as a lasting example of their engineering talent. This majestic three-tiered aqueduct, built in 18 BC, is approximately 49 m tall by 273 m in length. But many other beautiful bridges of smaller dimensions were built by Romans, like the Julien Bridge over the Coulon River near Apt, in southern France (Figure 11.1).

Timber bridges were built for several centuries, but their service life was limited by wars and fire. Timber bridges were no longer built after 1850. Between the eleventh and fourteenth centuries,

FIGURE 11.1   Julien bridge over the Coulon River near Apt in southern France. (Courtesy of Jean-Armand Calgaro, JAC.)

expanding trade spurred the development of a modern network of roads linking provinces and cities, markets, and pilgrimage centers. Often built by religious brotherhoods, bridges represented major financial assets for local lords who collected taxes and tolls from them. Fortified on both ends, they often housed several chapels. The Avignon Bridge is probably one of the most world-famous bridges in France. Stretching nearly 900 m, this bridge is a historic monument steeped in legend. Nobody knows exactly when it was designed, and by whom (the historical date is 1176), but it was destroyed at the end of the thirteenth century. The Avignon Bridge as we know it today with its four arches was completed around 1350, but abandoned in 1669 after multiple floods of the River Rhône. The arches stand as enduring symbols of the papal city.

The world's great cities—Paris, London, and Florence, among others—naturally flourished along rivers that were essential to the transport of merchandise. Waterways brought prosperity, while bridges ensured a sense of community. London Bridge, the Pont-Neuf, the Notre-Dame Bridge, and the Ponte Vecchio are not mere bridges, they are streets, market squares, and living areas; in short, proof that the city stands unified. Of the 38 bridges spanning the Seine River in Paris, the Pont Neuf, despite its name, is not only one of the oldest bridges in the city—it is as much an emblem of the French capital as the Eiffel Tower or Notre-Dame Cathedral. Connecting the right and left banks of the city with its historic center, the "Ile de la Cité," the masonry structure is formed by two successive bridges, one with five arches, the other seven.

The need for a new bridge to take some of the pressure off the overcrowded Notre-Dame Bridge dates to the early sixteenth century. The project was developed slowly, however, and the first stone was not laid until May 31, 1578, in a pompous ceremony attended by Henry III and his court. Designed by Baptiste Androuet du Cerceau and constructed under Guillaume Marchand, the bridge was just about completed in 1607 when Henry IV crossed it on horseback. Flanked by his powerful minister, the Duke of Sully, Henry IV turned Paris into one of the great European capitals (Figure 11.2).

The French Engineering School was founded during the seventeenth century: Jean-Baptiste Colbert organized the development and maintenance of roads and bridges and simultaneously concentrated the decision power. In particular, he created the Corps des Ponts et Chaussées. Technical aspects of bridge design became more important than architectural aspects and many innovations allowed the construction of increased span lengths. For example, the construction of the Pont Royal (1685–1688) over the Seine River by Jules Hardouin-Mansart and Jacques IV Gabriel was an engineering feat.

Born in France into a Swiss family, Jean-Rodolphe Perronet (1708–1794) is considered one of the most famous bridge builders of the eighteenth century. The Concorde Bridge (Figure 11.3) was his last work. The bridge opened in 1791 and is still used today.

FIGURE 11.2   The Pont Neuf over the Seine River. (Courtesy of JAC.)

FIGURE 11.3   The Concorde Bridge over the Seine River (Courtesy of JAC.)

Although metal bridges have existed for centuries, the real Iron Age began with the Industrial Revolution. In 1779, Abraham Darby III built the first iron bridge spanning the Severn, but the material was not strong enough. The design and construction of metal bridges was developed quickly with the use of steel of increasing quality. Gustave Eiffel's projects are striking examples of the innovative use of steel (Figure 11.4).

A series of innovations in concrete technology revolutionized bridge construction. The first, in 1850, was reinforced concrete, but the major breakthrough was the invention of prestressed concrete in 1928 by the engineer Eugène Freyssinet. Combining steel and concrete meant that large cable-stayed bridges and suspension bridges could be built. The birth of the French post-tensioning industry could be said to date from 1939, when Freyssinet developed the conical friction anchorage and the double-acting jack, and industry began to supply steel with the appropriate properties. The origin of prestressed concrete is,

FIGURE 11.4 The Garabit viaduct, designed by Gustave Eiffel. (Courtesy of JAC.)

however, much older. As long ago as 1908, Freyssinet built an experimental arch at Moulins with a 50 m span and 2 m rise. During the Second World War, because of the shortage of primary materials, only three prestressed structures were built by post-tensioning using the Freyssinet system. These are the Luzancy Bridge over the River Marne (1941–1945), a 54 m span portal frame bridge built from prefabricated segments with mortar filled joints and post-tensioned in three directions, and two other solid slab bridges.

The technique of prestressed concrete really took off after the end of the Second World War in spite of the absence of formal design rules. The construction of five portal frame bridges over the Marne River (d'Anet, Changis, Esbly, Trilbardou, and Ussy) between 1947 and 1950 raised the image of French civil engineering at the time and enhanced the reputation of prestressed concrete construction. These five structures, each with a 74 m span, are of a design similar to that of the Luzancy Bridge.

From 1946 to the 1980s, the technique of internal post-tensioning of concrete was predominantly used. External prestressing, however, made a short appearance in the 1950s with the construction of four bridges, three of which are still in service. The design and construction of motorways and other highways from 1955 onwards led to an increase in the use of prestressing, with the development of standardized bridges suitable for incorporation into the motorway system: slab bridges, pretensioned precast girder bridges, post-tensioned prefabricated beams connected by crossbeams, and a general concrete slab cast in situ. Among the structures built at that time, the Saint-Waast Bridge at Valenciennes (1947–1951) is worthy of mention. It has a span of 63.82 m, which is the longest of this type. For bridges of span length ranging from 35 m to 50 m, structures of other types have been built, mostly using temporary falsework (portal frame bridges, bow-string arch bridges, ribbed slab or voided slab bridges, and bridges with ribs).

In the 1980s, the French Roads Directorate introduced a policy of stimulating innovation under the strict supervision of the Central Technical Departments of the Ministry, in order to avoid new ideas being put forward without design and other control facilities. This policy took shape with the construction of innovative bridges (space frame structures in prestressed concrete, composite triangular and prestressed concrete structures, composite structures using corrugated webs and prestressed slabs, etc.). Examples include:

- The Val Maupré viaduct at Charolles, completed in 1988. Its superstructure cross section is of a triangular shape with a composite structure with corrugated webs post-tensioned by external tendons.
- The Sylans and Glacières viaducts were built with high-strength concrete (Class C65) precast segments launched into position and are of similar design to the Bubiyan Bridge in Kuwait.

- The three-dimensional superstructure of the bridge on the Roize River (1987–1990) was made of steel with a high-strength concrete deck slab (Class C80), externally post-tensioned.
- The three viaducts at Boulonnais (north of France, near Calais; 1995–1997) are made of precast segment double decks supported by a steel truss.

In the span range 40–200 m, box girder bridges have been found to be most competitive and to perform best. Initially the top slabs of box girders were not very wide, and decks were constructed using several boxes connected at the level of the top slab, which was itself prestressed transversely in the majority of cases. Progressively, box girders became wider and transverse prestressing became much less common. For wide-deck top slabs, multicell box girders have replaced multiple box girders. Today, in the majority of structures a single box girder, often with inclined webs, is used. For the widest bridge decks, struts or transverse ribs are used to reduce the thickness of the cantilever parts of the top slabs.

Depending on the method of construction (balanced cantilever or incremental launching) the box girders are either concreted in situ in long lengths, or concreted in situ or even prefabricated in short segments. In certain cases the beam cross section is constructed using several operations (for example, first the bottom slab, then the webs, and finally the top slab). All these large bridges are therefore built using successive operations and their classification depends on their characteristics and method of construction.

Beyond 300–400 m span length, the most common bridge type is the cable-stayed bridge. Originally, these bridges were designed with a steel or composite steel-concrete deck with a limited number of stays of high load-carrying capacity. Then concrete decks appeared with several kinds of suspension arrangements and an increasing number of stays analogous to prestressing cables in more traditional bridge types. The 320 m long Brotonne Bridge, designed by Jean Muller and Jacques Mathivat in 1977 in Normandy, is the first prestressed concrete cable-stayed bridge (Figure 11.5). The Iroise Bridge on the Elorn River near Brest (1991–1995) has a central suspension system. A part of the center span of 400 m is formed like a box girder with struts of lightweight concrete. The side spans are partially made of lightweight concrete and were launched into position. Nearly 20 years later, the concrete and steel Normandy Bridge (1990–1995) became the longest cable-stayed bridge in the world, with a roadway measuring 856 m and a double suspension system. A part of the 856 m center span is formed by a steel orthotropic deck box girder (Figure 11.6). In 1996, the Tatara Bridge in Japan, with a main span of 890 m, broke this record.

The longest bridges are suspension bridges, and the current record holder is the 1991 m Aikashi Kaikyo Bridge in Japan, completed in 1998. In France, the design of very-long-span suspension bridges

FIGURE 11.5 The Brotonne Bridge over the Seine River. (Courtesy of Campenon-Bernard.)

FIGURE 11.6   The Normandy Bridge. (Courtesy of JAC.)

FIGURE 11.7   The Tancarville suspension bridge over the Seine River, main span length 608 m. (Courtesy of JAC.)

is not useful because obstacles are rather narrow. Nevertheless, let us mention the Tancarville Bridge over the Seine River, built from 1955 to 1959 (Figure 11.7) and the Aquitaine Bridge over the Garonne River, built from 1960 to 1967 (Figure 11.8). The suspension cables of both of these bridges were replaced in 1998–1999 and 2004, respectively.

There are few arch bridges in France due to geographical conditions and technical traditions. The Chateaubriand Bridge over the Rance River (1988–1990) is an elegant example. It has a 261 m span length constructed in high-performance concrete (Class C60), built using temporary staying wires and temporary piers with a twin girder steel composite deck (Figure 11.9).

11.1.2 Categories of Modern Bridges

The design of bridges is permanently evolving for several reasons: the use of materials of increasing and strictly controlled performances, the development of faster and more precise construction methods, the creation of innovative forms bringing new solutions to problems raised by obstacles created by

FIGURE 11.8 The Aquitaine suspension bridge over the Garonne River, main span length 393.7 m. (Courtesy of JAC.)

FIGURE 11.9 Chateaubriand Bridge over the Rance River. (Courtesy of Service d'études sur les Transports, les Routes et leurs Aménagements.)

sometimes exceptional dimensions, and calculation techniques allowing the use of very sophisticated models of behavior. The design process of an individual bridge requires expertise from the designer to identify the most economical solutions, make the best use of the properties of the materials available, and limit possible hazards during execution without ignoring aesthetic and environmental aspects.

A good knowledge of the main types of structures, the area of their domain of use, and the specific methods of sizing is needed to undertake basic studies of a bridge at a given location. But a bridge is not only a common civil engineering work—it is built with the aim of providing a service for which society requires a high level of quality, safety, and reliability. With regard to the structural resistance, this level of quality is normally guaranteed by rules specified in standards and by rules of the art, forming a technical reference framework. Currently, in France as well as in all other Member States of the European Union, this reference framework is mainly based on European design (Eurocodes), execution, product, materials, test standards, and on European technical approvals. However, only application of these standards and rules is not sufficient; the structural analysis of exceptional bridges needs more and more expertise, beyond the traditional standardization, investigating dynamic behavior under the influence of variable actions like wind and earthquakes with probabilistic models. The safety and the comfort of the users are also taken into account, through an appropriate choice of fittings satisfying the specified requirements and by adopting detailing likely to guarantee the expected durability.

In general a bridge is a construction work in rise, built in situ, carrying traffic in a way to overcome a natural or artificial obstacle, such as a river, valley, road, railroad, channel (canal), and so on. The carried roadway can be a highway (road bridge), a pedestrian way (footbridge), railway tracks (railway bridge), or, in a few cases, a bridge carrying a canal, for example the historical Briare Bridge. Bridges may be classified according to various criteria: the constitutive materials, the construction process, or the mechanical functioning. This last criterion is adopted in the following classification: beam/girder bridges, arch bridges, and cable-stayed bridges.

Beam/girder bridges include all types of horizontal structures resistant to actions and action effects mainly by their flexural behavior, with the reactions of supports being vertical or almost vertical. The bridge deck or superstructure is, generally, a linear structure with independent or continuous spans or, exceptionally, cantilever parts. This linear structure is composed of main beams, parallel to the axis of the bridge deck, connected in the transverse direction by crossbeams on supports and, if needed, within spans. In general, a concrete deck slab carries the road traffic. Solid slab bridges may be classified in this category because the reactions at the supports are mainly vertical and the model of calculation of the longitudinal action effects is the same as for a beam. However, in the transverse direction, several structural arrangements are possible. The slab can be

- Full, generally of constant thickness, with or without side cantilever parts
- Composed with longitudinal holes in the concrete mass to limit the self-weight
- Composed of longitudinal ribs—it can be simply ribbed (a single wide rib with side cantilever parts), or multi-ribbed with an intermediate slab between the ribs

According to span lengths, the thickness of the slab can be constant or variable in the longitudinal direction. In the case of bow string (tie arch) bridges, the horizontal force transmitted by the arches is balanced by the tensile force in the horizontal deck so that the reactions at the supports are vertical. Figure 11.10 shows a traditional bow string made of reinforced concrete.

In arch bridges, the arch ribs work predominantly in compression, and the reactions at the supports are inclined. Such structures are possible only if they can be supported on a resistant rock. Under this condition, the domain of the use of arch bridges is vast (up to 500 m, with an arch rib of concrete-filled steel tube as is used in China). There are few arch bridges in France; other types of structures are more appropriate to cross important obstacles. In particular, cable-stayed bridges are now very commonly designed. They are generally long-span flexible structures. Suspension and cable-stayed bridges have a very different structural functioning. Suspension bridges are bridges where the main carrying system is a system of cables through which the reactions of the bridge deck are transmitted by suspenders. These

FIGURE 11.10 Traditional reinforced concrete bow string bridge. (Courtesy of JAC.)

metallic suspenders (cables) pass at the top of the pylons and are anchored in very heavy abutments. These bridges are mostly three-span bridges; the side spans are generally suspended spans, and are sometimes independent spans.

In cable-stayed bridges the main carrying structure is composed of beams supported by rectilinear oblique cables called stays. These cables are placed either in a single sheet in the axis of the bridge deck or in two side sheets and are arranged in a harp (parallel stays) or are fan shaped (convergent stays). The reactions of the supports are vertical and the structure works the same as bridges with beams subjected to flexion, because of the horizontal component of the tension of the stays. The use of cable-stayed bridges seems to be expanding, minimized only by that of suspension bridges, which remain the only type of structure possible for very long-range bridges (beyond 1500 m).

11.1.3 Data for the Design of Bridges

The design of a bridge can be undertaken only when all the data concerning the crossing are collected. The information required to begin the study appropriately is discussed next. In the open countryside, the construction of a bridge is generally part of a road project. If the project does not include exceptional civil engineering works, the financial weight of bridges is normally rather low compared to that of earthworks. On the other hand, if a large breach or large river needs to be crossed, the design of a bridge or viaduct must be carefully examined. Good cooperation has to be established between specialists of infrastructures (roads, railroads, etc.) and specialists of civil engineering works. In urban zones, the constraints of environment often govern the design.

The geometrical characteristics must be carefully considered. They depend mainly on the nature of the carried way, but can be slightly adjusted to simplify the project of the bridge, or to improve its mechanical functioning or even to offer more freedom in the choice of the type of work. The questions on the skew and horizontal curvature must be examined carefully. As a general rule, large bridges must, as far as possible, be designed straight: a horizontal curvature, even moderate, complicates the execution and may infer a mechanical functioning that can more or less deviate from the usual models of structural analysis. Finally, the length of the bridge must be considered. The progress accomplished in the execution of earthworks upsets the data when comparing the costs of a bridge and an embankment and, in the absence of major constraints of aesthetic or hydraulic nature, the embankment usually constitutes the least-expensive solution.

A visit to the future location of the bridge by the engineer is an essential stage of the project. The main information to be collected on the spot includes:

- Topography: It is useful to have a topographic account and a plan view of the site indicating the possibilities of access, as well as the available areas for the installation of the construction site, storages, and so on.
- Hydrology: In the case of the crossing of a river, the frequency and importance of floods, solid debit, and possible carriage of floating objects susceptible to striking piers must be known. Apart from impacts, the greatest danger is scour effects. It is advisable to estimate the potential height of scour in the neighborhood of the supports and to limit as much as possible the number of supports in the aquatic site.
- Geological and geotechnical data: These data, which concern the nature and properties of the ground and the foundation, without forgetting the knowledge of the level of the groundwater, are very important. Their collection constitutes a decisive stage for the choice of the type of foundation. An insufficient study can mean modifications of the project or a very expensive extension of the already executed foundations will be required if the ground does not have the expected properties. Geotechnical tests are generally rather expensive and the designer has to organize the tests according to the size and the importance of the construction works. The designer has to make them at first for the envisaged location of the supports and collect the test results, which would already have been made in the neighborhood.

Natural actions to which the bridge is susceptible must be applied; besides a high water level, other natural actions that may act on a bridge include direct actions of wind, whose force can be increased in deep valleys; snow and atmospheric ice; earthquakes; in the case of the crossing of an estuary or an arm of the sea, indirect actions such as those from sea sprays; and, in a general way, physic-chemical actions of the environment. Towards actions of environment, it is necessary to adopt appropriate detailing (cover of reinforcement steel, choice of a high-performance concrete, etc.).

To avoid any omission, it is advisable to establish in advance a list of the data required to undertake the study. This list includes the site map; the profile across, taking into account possibly future extensions; the profile in length; operating expenses, normal and exceptional; the free heights and openings to be reserved (road, railroad, waterway); the architectural quality; the construction constraints, which can be of a very varied nature; the relative cost of the work and materials; availability of aggregates and cement; and so on.

11.2 Design Practice

11.2.1 Structural Eurocodes

The first European Directive on public procurement was published in 1971, but its practical application concerning the calculation of civil engineering works was very difficult. This was mainly due to a clause forbidding, for a public tender, the rejection of a proposal for the reason that the tender was based on design standards in force in a country different from the country where the construction works was to be built. For that reason, it was decided in 1976 to develop European structural design standards, mainly based on studies carried out by scientific international associations, which could be widely recognized for the judgement of tenders.

In the early 1980s, the first documents, called Eurocodes, were published as provisional standards under the responsibility of the Commission of European Communities. After lengthy international inquiries and after the adoption of the Unique Act (1986), it was decided to transfer the development of the Eurocodes to CEN (the European Committee for Standardization) and to link them to the Construction Product Directive (CPD). The transfer took place in 1990 and CEN decided to publish the Eurocodes first as provisional European standards (ENVs), and then as European standards (ENs).

The Foreword of each Eurocode mentions that the Member States of the EU and European Free Trade Association (EFTA) recognize that the Eurocodes serve as reference documents for the following purposes:

- As a means to prove compliance of building and civil engineering works with the essential requirements of Council Directive 89/106/EEC, particularly Essential Requirement 1, "Mechanical Resistance and Stability," and Essential Requirement 2, "Safety in Case of Fire"
- As a basis for specifying contracts for construction works and related engineering services
- As a framework for drawing up harmonized technical specifications for construction products (ENs and ETAs)

In fact, the Eurocodes were also developed to improve the functioning of the single market for products and engineering services by removing obstacles arising from different nationally codified practices for the assessment of structural reliability, and to improve the competitiveness of the European construction industry and the professionals and industries connected to it in countries outside the European Union.

On March 31, 2010, all Member States of the European Union adopted the Eurocodes as official standards for the design and verification of buildings and civil engineering works. In fact, in France, the Eurocodes had already been in use for several years for the design of new bridges because the French standards were no longer maintained after the decision to develop European standards. The Structural Eurocode program comprises the standards shown in Table 11.1, generally consisting of a number of parts.

The Eurocodes are intended for the design of new construction works using the most traditional materials (reinforced and prestressed concrete, steel, steel and concrete composite construction, timber,

TABLE 11.1 The Eurocodes Program

EN 1990	Eurocode	Basis of Structural Design
EN 1991	Eurocode 1	Actions on Structures
EN 1992	Eurocode 2	Design of Concrete Structures
EN 1993	Eurocode 3	Design of Steel Structures
EN 1994	Eurocode 4	Design of Composite Steel and Concrete Structures
EN 1995	Eurocode 5	Design of Timber Structures
EN 1996	Eurocode 6	Design of Masonry Structures
EN 1997	Eurocode 7	Geotechnical Design
EN 1998	Eurocode 8	Design of Structures for Earthquake Resistance
EN 1999	Eurocode 9	Design of Aluminium Structures

masonry, and aluminium). But EN 1990, "Basis of Structural Design," is also applicable for the structural evaluation of existing constructions, in developing designs for repairs and rehabilitation or in assessing changes of use. This applies in particular to the strengthening of existing bridges. Of course, additional or amended provisions may have to be adopted for the individual project.

The verification rules in all Eurocodes are based on the limit-state design philosophy using the partial factors method. In the case of bridges, most of accidental situations leading to catastrophic failures are due to gross errors during execution, impacts during normal use, or uncontrolled scour effects. Such risks may be avoided, or their consequences may be strictly limited, by adopting appropriate design and execution measures (e.g., stabilizing devices) and by appropriate quality control procedures. During its service life, the collapse of a bridge may be the consequence of:

- A possible accidental situation (e.g., exceptional scour near foundations).
- Impact (e.g., lorry, ship, or train collision on a bridge pier or deck, or even an impact as a consequence of a natural phenomenon).
- Development of fatigue cracks in a structure designed with a poor redundancy (e.g., cracks in a welded joint in one of the two girders of a composite steel–concrete bridge deck) or failure of cables due to fatigue; concerning this question, the design Eurocodes establish a distinction between damage tolerant and not tolerant structures.
- Brittle behavior of some construction materials (e.g., brittle fracture of steel at low temperatures). This type of risk is very limited in the case of recent or new bridges, but it may be very real in the case of old bridges.
- Deterioration of materials (e.g., corrosion of reinforcement and cables, deterioration of concrete, etc.).

Bridges are public works, for which public authorities may have responsibilities as owners and also for the issue of national regulations on authorized traffic (especially on vehicle loads) and for delivery and control dispensations when relevant, for example for abnormally heavy vehicles. One major requirement is the design service life. A design service life of 100 years is commonly agreed upon for bridges by experts and relevant authorities, but the meaning of this value needs some clarification.

First, it is uneconomical and unpractical to design all parts of a bridge for the same service life. In particular, structural bearings, expansion joints, coatings, or any industrial product cannot be designed or executed for such a long service life. And, in the case of road restraint systems, the concept of design service life is not really relevant.

By the end of the design service/working life, irreversible serviceability limit states should not be too much exceeded, considering a reasonable program of maintenance and limited repair. Of course, the design working life may be used directly in some fatigue verifications for steel members, but more and more frequently, requirements concerning, for example, the penetration of chlorides into concrete or the rate of carbonation after x years are defined in the project specification of the bridge. Finally, the design of a bridge is not only a matter of architecture or of calculation: it has to be considered as a living form which needs care.

11.2.2 Traffic Loads on Bridges

11.2.2.1 General

Part 2 of Eurocode 1, "Actions on Structures" (EN 1991–2), defines load models (vertical and horizontal forces) for road bridges, footbridges, and railway bridges. It also defines accidental design situations over the deck. The collision forces on bridge piers and decks due to road or rail traffic are defined in Part 1-7 of Eurocode 1 (EN 1991-1-7). Collision forces dues to ship collisions against bridge piers are also defined in Part 1-7 of Eurocode 1. All load models defined in EN 1991-2 are generally applicable for the design of new bridges including piers, abutments, upstand walls, wing walls and flank walls, and so on, and their foundations. But specific rules need to be defined in some cases, for example for bridges receiving simultaneous road and rail traffic, for masonry arch bridges, buried structures, retaining walls, and tunnels.

For normal conditions of use of bridges, traffic actions are considered as free (within some limits) variable multicomponent actions, which means that a well-identified type of traffic produces vertical and horizontal and static and dynamic forces. In order to facilitate the combinations of actions, EN 1991-2 defines the concept of "group of loads" for all types of bridges: a group of loads is taken in combinations of actions as a unique variable action.

11.2.2.2 Load Models for Road Bridges

For road bridges, EN 1991-2 defines the following four load models of vertical loads (noted LM1 to LM4) to be used for the verification of all limit states except fatigue, and five load models for fatigue verifications (noted FLM1 to FLM5):

1. A main load model (LM1), including concentrated loads (tandem systems) and uniformly distributed loads and applicable to all bridges
2. A model consisting of a single axle with two wheels (LM2), in addition to the previous one (LM1) for the verification of short structural members (3 m to 7 m)
3. A model made up of a set of special vehicles intended to take into account the effects of exceptional convoys (LM3)
4. A model corresponding to the loading of the surface of the bridge with a uniformly distributed load of 5 kN/m², corresponding to the effects (dynamic amplification included) of a crowd (LM4)

The magnitude of load models LM1 and LM2 may be adjusted at the national level depending on the expected traffic volume and the composition on the bridge to be designed. The basic characteristic level corresponds to a return period of 1000 years, which means a probability of being exceeded by 5% in 50 years or 10% in 100 years. The frequent level corresponds to a return period of one week and the quasi-permanent values are generally equal to zero for traffic loads. Nevertheless, for road bridges that support heavy and continuous traffic, a quasi-permanent value different from zero may be appropriate. For bridges with intense traffic and located in seismic areas, it is recommended to adopt a quasi-permanent value equal to 20% of the characteristic value.

For the application of the various load models, the basic concept is the division of the carriageway into notional lanes. The width w of the carriageway is measured between the inner limits of vehicle restraint systems or between curbs. The carriageway width w is divided into the greatest possible integer number n_l of notional lanes; the normal width of a notional lane is $w_l = 3$ m, except for a carriageway width such that 5.4 m $\leq w <$ 6 m (two lanes). The difference between the carriageway width and the width of all notional lanes is the width of the "remaining area." Where the carriageway on a bridge deck is physically divided into two parts separated by a central reservation, then each part, including all hard shoulders or strips, is separately divided into notional lanes if the parts are separated by a permanent road restraint system, and the whole carriageway, central reservation included, is divided into notional lanes if the parts are separated by a temporary road restraint system.

Load models LM1 and LM2 have been defined and calibrated in order to give effects as close as possible to "extrapolated target effects" (adjusted to the selected return periods) determined from effects due to measured actual traffic. The load models are to be applied on notional lanes, which are not physical lanes, and the numbering of the notional lanes depends on the conditions of application of the load model with the purpose of getting, in all cases, the most critical effect. In other words, there is no *physical* numbering of the notional lanes. Nevertheless, the location and numbering of notional lanes shall be determined in accordance with the principles specified in the Eurocodes. Main characteristic load model 1 (LM1) is described next.

The main characteristic model, LM1, is represented in Figure 11.11. It has been selected and calibrated to cover the most common traffic effects with an appropriate reliability margin. Scientific studies have been performed, based on real traffic data and various theoretical developments. After identification of the notional lanes on the carriageway, these lanes are loaded by

- A uniformly distributed load
- A tandem system including two axles

A maximum of three notional lanes are loaded with a single tandem system per lane, which means that, for an individual project or at the national level, it can be decided to use only one (not recommended) or two tandem systems. For the assessment of general effects, the tandem systems are assumed to travel centrally along the axes of the relevant notional lanes. The characteristic value of each axle load of a tandem system located in lane number i is noted $\alpha_{Q,i}Q_{ik}$, and the two wheels forming the axle transmit the same load $\alpha_{Q,i}Q_{ik}/2$. The characteristic value of the uniformly distributed load is noted $\alpha_{q,i}q_{ik}$ on lane i and $\alpha_{q,r}q_{rk}$ on the remaining area. $\alpha_{Q,i}$, $\alpha_{q,i}$, $\alpha_{q,r}$ are adjustment factors intended to take into account the various types of traffic on bridges.

Uniformly distributed loads are to be applied only in the unfavorable parts of the influence area, longitudinally and transversally. This means, for example in the transverse direction, that the uniformly distributed load may be applied on a width less than the normal width of a notional lane (Figure 11.12). The lanes are numbered 1, 2, 3, and so on in such a way that the lane giving the most unfavorable effect

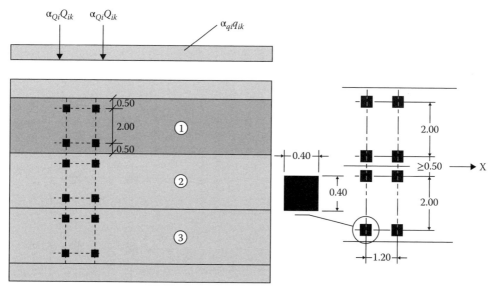

FIGURE 11.11 Load model 1.

is lane 1, the lane giving the second most unfavorable effect is lane 2, and so on. The characteristic values of the loads (basic values) are given in Table 11.2, which derives from Table 4.2 of EN 1991-2. They correspond to heavy long-distance international traffic and the dynamic effects are included.

EN 1991-2 recommends that the factor α_{Q1} shall not be less than 0.8, and the value 0.9 was considered for small roads. It results from a combination of a low density and of a rather favourable distribution of the individual loads. Horizontal forces include braking or acceleration forces and centrifugal forces. The braking or acceleration forces are represented by a longitudinal force, applied at the surfacing level of the carriageway, with a limited characteristic value of 900 kN, and it is calculated as a fraction of the total maximum vertical loads due to LM1 applied to lane 1.

EN 1991-2 defines the characteristic value of a transverse force, noted Q_{tk}, applicable at the finished carriageway level in a direction perpendicular to its axis. This force is given as a function of the horizontal radius of the carriageway centerline.

Concerning fatigue verifications, EN 1991-2 defines five load models called FLM1 to FLM5. These models were defined for various uses. The main fatigue model is FLM3 (Figure 11.13), intended for common verifications without performing any damage calculation. It consists of four axles of 120 kN, each axle having got two wheels with square contact areas of 0.40×0.40 m².

The basic idea was originally to select a fatigue "single vehicle" so that, assuming a conventional number of crossings of the bridge by this vehicle (e.g., 2×10^6), and after a numerical adaptation with appropriate factors, it led to the same damage as real traffic during the designed service life of the bridge. Thus,

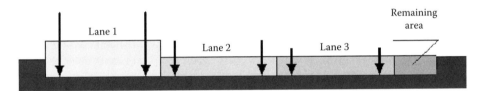

FIGURE 11.12   Example of application of LM1 in the transverse direction.

TABLE 11.2   Load Model 1 Basic Characteristic Values

Location	Tandem system Axle loads Q_{ik} (kN)	Uniformly distributed load system q_{ik} (or q_{tk}) (kN/m²)
Lane 1	300	9
Lane 2	200	2.5
Lane 3	100	2.5
Other lanes	0	2.5
Remaining area (q_{rk})	0	2.5

Note: The contact surface of wheels is a square of 0.40 m × 0.40 m.

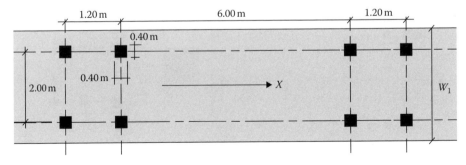

FIGURE 11.13   Definition of FLM3; W_1 is the lane width and X is the bridge longitudinal axis.

the designer calculates the extreme stresses (maximum and minimum) resulting from FLM3 in order to evaluate a stress range $\Delta\sigma_{FLM} = |Max\sigma_{FLM} - Min\sigma_{FLM}|$. This stress range is then multiplied by a dynamic magnification factor φ_{fat} taking account of the carriageway roughness and a load factor λ_e, which gives an equivalent stress range: $\Delta\sigma_{fat} = \lambda_e \varphi_{fat} \Delta\sigma_{FLM}$. This stress range $\Delta\sigma_{fat}$ is compared to the value $\Delta\sigma_c$ of the (Stress range-Number of cycles) S-N curve, corresponding to 2×10^6 applications (Figure 11.14).

For the assessment of the expected annual traffic volume EN 1991-2 gives indicative numbers of heavy vehicles expected per year and per slow lane. For the assessment of action effects:

- The fatigue load models are positioned centrally on the appropriate notional lanes defined in the project specification for general effects.
- The fatigue load models are positioned centrally on the notional lanes assumed to be located anywhere on the carriageway and moreover (e.g. for orthotropic decks), a statistical distribution of the transverse location of the vehicles within the notional lanes may be taken into account.

FLM1 to 4 include dynamic load amplification appropriate for pavements of good quality. It is recommended to apply to all loads an additional amplification factor $\Delta\varphi_{fat}$ near the expansion joints, given by the following formula:

$$\Delta\varphi_{fat} = 1.30\left(1 - \frac{D}{26}\right) \quad \Delta\varphi_{fat} \geq 1$$

where D is the distance (m) of the cross section under consideration from the expansion joint.

For accidental design situations and actions, EN 1991-2 gives rules concerning

- Vehicle collision with bridge piers, soffit of bridge, or decks (Figure 11.15)
- The presence of heavy wheels or vehicles on footways
- Vehicle collision with curbs, vehicle parapets, and structural components

EN 1991-1-7, which was drafted after EN 1991-2, gives more detailed rules concerning collision forces from vehicles under the bridge, covering impact forces on piers and other supporting members, and impact on decks. It also gives rules concerning impact from ships on bridge piers.

The presence of heavy wheels or vehicles on footways is an accidental design situation and needs to be taken into account for all bridges where footways are not protected by a rigid road restraint system. The accidental action is due to one axle load from the tandem system corresponding to notional lane 2, that is, $\alpha_{Q2}Q_{2k} = 200\alpha_{Q2}$, to be applied and oriented on the unprotected parts of the deck so as to give

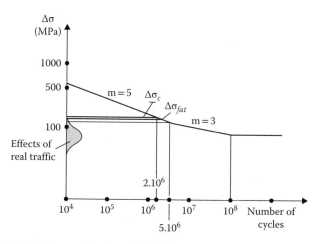

FIGURE 11.14 Principle of fatigue verification with FLM3.

FIGURE 11.15 Example of impact on a bridge deck. (Courtesy of LCPC.)

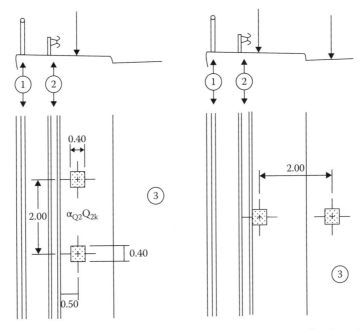

FIGURE 11.16 Examples showing locations of loads from vehicles on footways and cycle tracks of road bridges: (1) pedestrian parapet (or vehicle parapet if a safety barrier is not provided); (2) safety barrier; (3) carriageway.

the most adverse effect. The design situations to be taken into account are defined by the designer in agreement with the client. Figure 11.16 shows two examples of accidental design situations. Of course, vehicle collision forces on unprotected structural members above or beside the carriageway levels need to be taken into account, for example for bridges with lateral lattice girders (Figures 11.17 and 11.18).

11.2.2.3 Load Models for Footbridges

Eurocode 1 Part 2 (chapter 5) defines and describes traffic loads applicable to footways, cycle tracks, and footbridges during permanent and transient design situations. Static loads due to pedestrians or cycles are very light compared to loads due to vehicle road or railway traffic. Therefore, long-span footbridges are very slender or flexible structures, especially when designed with innovative architectural ideas.

FIGURE 11.17 Example of bridge with protection of lateral girders. (Courtesy of JAC.)

FIGURE 11.18 Example of accidental situation on a suspension bridge. (Courtesy of LCPC.)

Dynamic stability, in connection with structural flexibility due to wind actions and footbridge–pedestrian interaction, has been strongly highlighted. When crossing a footbridge, people can walk in a number of ways, run, jump, or dance. On footbridges, people's walking movements such as running, jumping, or dancing may increase vibrations which are not yet correctly covered by design standards. The number and location of people likely to be simultaneously on the bridge deck depend on the bridge under consideration, but also on external circumstances, more or less linked to its location; these parameters are commonly highly random and even uncertain. Some accidental situations like vandalism may occur. During such situations, the structural behavior can be strongly modified. These situations are not explicitly considered in the Eurocodes, but simulations based on appropriate dynamic load models may be performed.

At present, the Eurocode gives only static load models for pedestrian and cycle loads, and some general rules dealing with vibrational aspects. The field of application of these static load models is only slightly limited by the footbridge width, and a value of 6 m is suggested, but this value is rather conventional. In fact, various human activities may take place on wide footbridges and special analysis may be needed for specific projects. A dynamic analysis needs to be performed in order to determine if the consideration of static load models is sufficient.

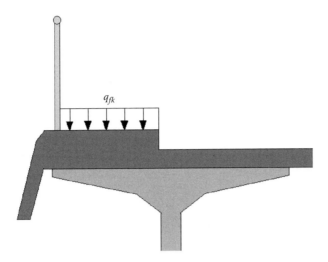

FIGURE 11.19   Pedestrian load on a footway or cycle track (recommended value 5 kN/m²).

Three static independent models of vertical loads which are not intended to be used for fatigue verifications are defined in the Eurocode:

- A vertical uniformly distributed load q_{fk}, applicable to footways, cycle tracks, and footbridges
- A concentrated load Q_{fwk}, applicable to footways, cycle tracks, and footbridges
- A load representing a service vehicle Q_{serv}, applicable only to footbridges as a "normal" or an "accidental" load

In addition, horizontal forces are defined, accidental design situations are evoked, and, for road bridges, load models on embankments are defined. However, loads on access steps are not defined. The recommended value for the characteristic value of the uniformly distributed load is equal to $q_{fk} = 5$ kN/m² (Figure 11.19). The characteristic value $q_{fk} = 5$ kN/m² represents a physical maximum load including a limited dynamic magnification (five heavy persons per square meter).

The Eurocode leaves the choice of the characteristic value at the national level or for the individual project, but gives the following recommendations:

- For footbridges carrying (regularly or not) a continuous dense crowd (e.g., near the exit of a stadium or an exhibition hall), a characteristic value $q_{fk} = 5$ kN/m² may be specified.
- For bridges not carrying a continuous dense crowd a reduced value for long-span footbridges may be adopted. The recommended value for q_{fk} is then

$$q_{fk} = 2{,}0 + \frac{120}{L+30} \quad \text{kN/m}^2$$

$$q_{fk} \geq 2{,}5 \text{ kN/m}^2; \quad q_{fk} \leq 5{,}0 \text{ kN/m}^2$$

where L is the loaded length in [m].

The consideration of concentrated loads is needed to check the local resistance of a footbridge. In general, loads on footbridges may differ depending on their location and on the possible traffic flow of some vehicles. Three cases are envisaged by the Eurocode:

1. Permanent provisions are made to prevent access of all vehicles to the footbridge. A concentrated load Q_{fwk} of 10 kN acting on a square surface of sides 0.10 m is recommended to check the resistance with regard to local effects due, for example, to small equipment for maintenance of the footbridge. The concentrated load does not act simultaneously with the uniformly distributed load.

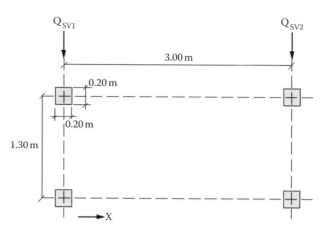

FIGURE 11.20 Model for accidental presence of a vehicle on a footbridge deck: X is the bridge axis direction; $Q_{sv1} = 80$ kN; $Q_{sv2} = 40$ kN.

2. The presence of a "heavy" vehicle on the footbridge is not normally foreseeable but no permanent obstacle prevents this presence. The Eurocode recommends strongly to apply the accidental presence (accidental design situation) of a vehicle on the bridge deck as shown in Figure 11.20. This model may be adjusted at the national level or for the individual project.

3. A "heavy" vehicle, Q_{serv}, for maintenance, emergencies (e.g., ambulance, fire), or other services is foreseen to be driven onto the footbridge deck. Its characteristics (axle weight and spacing, contact area of wheels, etc.), the dynamic amplification, and all other appropriate loading rules may be defined for the individual project or at the national level. In the absence of available information, the vehicle load specified in the second case (Figure 11.20) may be used as the service vehicle (characteristic load). The concentrated load Q_{fwk} shall not act simultaneously with this load model. Where relevant, several service vehicles, mutually exclusive, may be defined for the individual project.

No horizontal forces are associated with the uniformly distributed load on footways. However, for footbridges, the Eurocode recommends

- A horizontal force equal to 10% of the total vertical load shall be applied together with the uniformly distributed load.
- A horizontal force equal to 60% of the total weight of the service vehicle shall be applied together with the service vehicle.

EN 1991-2 does not define dynamic load models of pedestrians. It only highlights the need to define appropriate dynamic models of pedestrian loads and comfort criteria, and gives a few recommendations intended to introduce the general comfort requirements evoked in Annex A2 to EN 1990, "Basis of Structural Design (Application for Bridges)." In the absence of significant response of the bridge, a pedestrian normally walking exerts on it the following simultaneous periodic forces:

- Vertical, with a frequency that can range between 1 and 3 Hz
- Horizontal, with a frequency that can range between 0.5 and 1.5 Hz

Groups of joggers may cross a footbridge with a frequency of 3 Hz.

In general, it seems accepted that the use of three dynamic models may be appropriate as follows:

1. A model for a single pedestrian
2. A model for a group of pedestrians, for example from 10 to 15
3. A model for a dense crowd

11.2.2.4 Load Models for Railway Bridges

The design of railway bridges is, of course, very different from the design of a road bridge or a footbridge. Moreover, the design includes different aspects depending on the nature of the railway line: normal traffic or high-speed traffic. Figure 11.21 shows a typical cross section of a bridge for high-speed trains, with two lateral main girders (welded plates). The deck plate is a concrete slab with embedded cross girders.

For railway bridges, EN 1991-2 defines the following load models:

- Vertical traffic actions (based on International Union of Railways (UIC) Codes 700[2], 702[1], 776-1[3])
- LM 71
- LM SW/0
- LM SW/2
- Load model HSLM (high-speed load model to be used where required by the Technical Specification for Interoperability of High-Speed Traffic in accordance with the relevant EU directive and/or the relevant authority, based on UIC Code 776-2[4])
- Load model unloaded train for checking lateral stability in conjunction with the leading lateral wind actions on the bridge
- Load effects from real trains (where required by the relevant authority)
- Centrifugal forces
- Traction and braking
- Nosing
- Longitudinal forces (based on UIC Code 774-3[5] for load effects generated by the interaction between track and structure)
- Load effects generated by the interaction between train, track, and structure to variable actions, in particular speed (based on UIC Code 776-2[4])
- Live load surcharge horizontal earth pressure
- Aerodynamic actions (slipstream effects from passing rail traffic and so on, based on UIC Code 779-1[6])

Detailed discussion of all these actions and load models is beyond the scope of this chapter. The following section is devoted to the main loading system for vertical forces, which is load model 71 (LM71). This load model represents the static effect of vertical loading due to normal rail traffic. The load arrangement and the characteristic values for vertical loads are shown in Figure 11.22.

The characteristic values given in Figure 11.22 are to be multiplied by a factor α on lines carrying rail traffic that is heavier or lighter than normal rail traffic. When multiplied by the factor α, the loads are

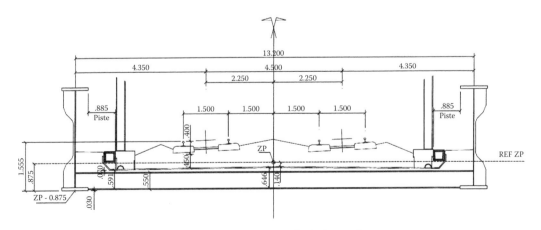

FIGURE 11.21 Example of cross section of a modern bridge for high-speed trains.

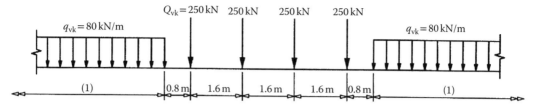

FIGURE 11.22　LM71 and characteristic values for vertical loads.

called classified vertical loads. This factor α shall be selected in the following set: 0.75, 0.83, 0.91, 1.00, 1.10, 1.21, 1.33, 1.46. For international lines, α = 1.33 is the recommended value because it takes into account the gradual increase of axle loads from 250 kN today up to 300 kN in the next decades.

To consider bridge–track interaction, the permissible additional rail stresses and deformations are calibrated on the existing practice. Theoretically, this is a serviceability limit state (SLS) for the bridge and an ultimate limit state (ULS) for the rail. However, as the given permissible rail stresses and deformations were obtained by deterministic design methods and calibrated using existing practice, the calculations for interaction should not be carried out with α = 1.33 but always with α = 1.0. The axle loads of 300 kN will only be introduced over the next hundred years and the future track characteristics are not known at this time. Calculations with α = 1.0 have enough reserves so that in the future no additional expansion devices will be necessary for the bridges calculated with α = 1.0 today. The other models for vertical loads are given in Figure 11.23.

Load Model	q_{vk} (kN/m)	a (m)	c (m)
SW/0	133	15.0	5.3
SW/2	150	25.0	7.0

Key

Universal Train	Number of Intermediate Coaches N	Coach Length D (m)	Bogie Axle Spacing d (m)	Point Force P (kN)
A1	18	18	2.0	170
A2	17	19	3.5	200
A3	16	20	2.0	180
A4	15	21	3.0	190
A5	14	22	2.0	170
A6	13	23	2.0	180
A7	13	24	2.0	190
A8	12	25	2.5	190
A9	11	26	2.0	210
A10	11	27	2.0	210

Fatigue verifications shall be performed with LM71 and α = 1.0. Fatigue is a real problem for railway bridges. In order to get optimal life-cycle costs to achieve the intended design life, all important structural members shall be designed for fatigue to an acceptable level of probability that their performance will be satisfactory throughout their intended design life. Due to the long design working life of bridges (100 years in general) it is necessary to take into account long-term considerations. The Eurocodes define severe permissible deflections. For the assessment of the deflections, α = 1.0 remains the recommended

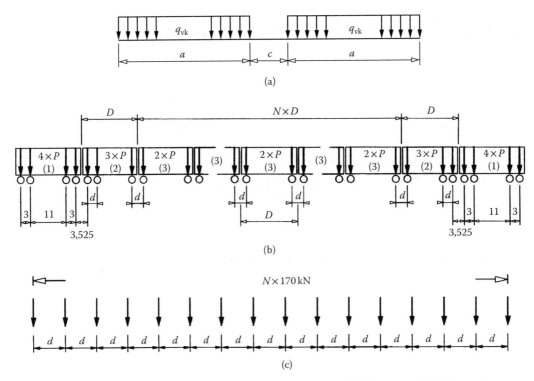

FIGURE 11.23 Definition of load models SW/0, SW/2, HSLM, and A & B (HSLM-A and HSLM-B together represent the dynamic load effects of articulated, conventional, and regular high-speed passenger trains in accordance with requirements from the European Technical Specification for Interoperability). (a) Load models SW/0 and SW/2 (heavy traffic); (b) HSLM-A; (c) HSLM-B. HSLM-B comprises N number point forces of 170 kN at uniform spacing d (m) where N and d are defined in the Eurocode. (1) Power car (leading and trailing power cars identical); (2) end coach (leading and trailing end coaches identical); (3) intermediate coach.

value with LM71 (and SW/0 if relevant), even if $\alpha = 1.33$ is adopted for ULS design. In Annex 2 to EN 1990, "Basis of Structural Design," only minimum conditions for bridge deformations are given. If these conditions are the governing factors in the design of a bridge, this would lead to bridges with insufficient stiffness for track maintenance at the ends of the bridges.

The twist of the bridge deck shall be calculated taking into account the characteristic values of LM71 as well as SW/0 or SW/2 as appropriate, multiplied by specific factors Φ and α (and load model HSLM for speeds over 200 km/h), including centrifugal effects. Twist shall be checked on the approach to the bridge, across the bridge, and for the departure from the bridge. This is an important condition for rail traffic safety. Therefore, the value $\alpha = 1.33$ has to be taken with LM71 or SW/0. The maximum twist t (mm/3m) of a track gauge s (m) of 1.435 m (distance between inner sides of rails) measured over a length of 3 m (Figure 11.24) should not exceed the values given in Table 11.3.

The total track twist is that due to any twist which may be present in the track when the bridge is not subject to rail traffic actions (for example in a transition curve), plus the track twist due to the total deformation of the bridge resulting from rail traffic actions. The total track twist shall not exceed t_T with a recommended value of 7.5 mm/3m.

The vertical traffic loads applied to the bridge cause the deck to bend, resulting in a vertical displacement of every point on the surface of the deck. In general, maximum displacement occurs at the point in the middle of the deck, or at midspan. This displacement is known as the deflection of the deck. For speeds up to 200 km/h the deflections are calculated with LM71 (or SW/0) multiplied by Φ, without

TABLE 11.3 Limiting Values of Deck Twist

Speed Range V (km/h)	Maximum Twist t (mm/3m)
$V \leq 120$	$t \leq t_1$
$120 < V \leq 200$	$t \leq t_2$
$V > 200$	$t \leq t_3$

Note: The recommended values for the set of t are: $t_1 = 4.5$, $t_2 = 3.0$, $t_3 = 1.5$.

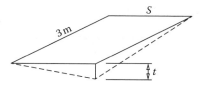

FIGURE 11.24 Definition of deck twist.

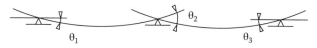

FIGURE 11.25 Angular rotation of two adjacent decks.

dynamic study. For all structure configurations loaded with the classified characteristic vertical loading in accordance with the Eurocode, the maximum total vertical deflection measured along any track due to rail traffic actions should not exceed $L/600$, but the quality of the track maintenance is not taken into account.

The deflection of the deck also causes rotation of the ends of the deck. For a succession of simple beams (Figure 11.25), the permissible values for deflections may therefore be reduced to avoid the permissible total relative rotation between the adjacent ends of two decks being doubled.

The deflection of the deck under traffic loads causes the end of the deck behind the support structures to lift. This lift must be limited to

$$V \leq 160 \text{ km/h} \quad \leq 3 \text{ mm}$$

$$160 < V \leq 200 \text{ km/h} \quad \leq 2 \text{ mm}$$

taking LM 71 (SW/0) multiplied by Φ, with $\alpha = 1.0$.
Last but not least, two accidental design situations are defined in the Eurocode. They are shown in Figure 11.26.

Key

(1) Max. 1.5s or less if against wall.
(2) Track gauge *s*.
(3) For ballasted decks the point forces may be assumed to be distributed on a square of side 450 mm at the top of the deck.

Key

(1) Load acting on edge of structure.
(2) Track gauge *s*.

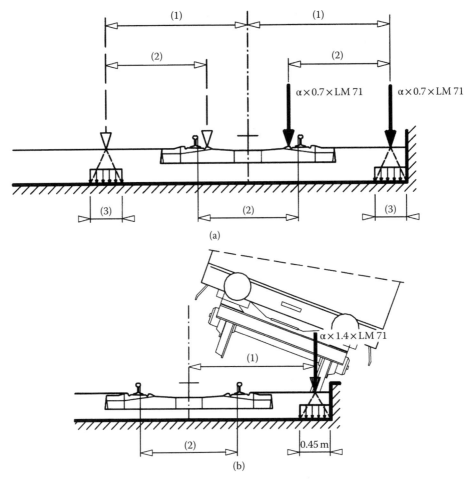

FIGURE 11.26 Accidental design situations for the deck of railway bridges. (a) Design situation I, equivalent load Q_{A1d} and q_{A1d} and (b) design situation II, equivalent load q_{A2d}.

11.2.3 Actions Other Than Traffic Loads

This section is concerned with the determination of nontraffic actions applicable to bridges during persistent design situations. The material in this section is covered in the following parts of Eurocode 1 (EN 1991), "Actions on Structures":

- EN 1991-1-1, General actions—Densities, self-weight, imposed loads for buildings
- EN 1991-1-3, General actions—Snow loads
- EN 1991-1-4, General actions—Wind actions
- EN 1991-1-5, General actions—Thermal actions
- EN 1990 – "Basis of Structural Design," Annex A2 "Application for Bridges"

11.2.3.1 Self-Weight of the Structure and Other Permanent Actions

In accordance with Eurocode 1 Part 1-1, the self-weight of a bridge includes the structure, structural elements and products, and nonstructural elements (fixed services and bridge furniture), as well as the weight of earth and ballast. Examples of fixed services are cables, pipes, and service ducts (generally located within footways, sometimes within the deck structure). Examples of bridge furniture are waterproofing, surfacing and other coatings, traffic restraint systems (safety barriers, vehicle and pedestrian parapets), acoustic and anti-wind screens, and ballast on railway bridges.

The total self-weight of structural and nonstructural members is normally taken in combinations of actions as a single action. Then, the variability of *G* may be neglected if *G* does not vary significantly during the design working life of the structure and its coefficient of variation is small (less than 10%). G_k should then be taken equal to the mean value.

11.2.3.2 Snow Loads

The field of application of Eurocode 1 Part 1-3, "Snow Loads," does not include special aspects of snow loading, for example snow loads on bridges. Hence, Eurocode 1 Part 1-3 is normally not applicable to bridge design for persistent design situations. During execution, rules are defined where snow loading may have significant effects (chapter 3 of the Eurocode). However, there is no reason to exclude snow loads on bridges, particularly roofed bridges (Figure 11.27) for persistent design situations.

For road and railway bridges in normal climatic zones:

- Significant snow loads and traffic loads cannot generally act simultaneously.
- The effects of the characteristic value of snow loads on a bridge deck are far less important than those of the characteristic value of traffic loads.

Concerning snow loads on the roof of a roofed bridge, the characteristic value is determined exactly in the same way as for a building roof. The combination of snow loads and traffic loads may be defined at the national level or directly for the individual project.

11.2.3.3 Wind Actions on Bridges

Section 8 of Eurocode 1 Part 1-4 (EN 1991-1-4) gives rules for the determination of quasi-static effects of natural wind actions (wind effects due to trains along the rail track are defined in Eurocode 1 Part 2) for the structural design of bridges (decks and piers). These rules are applicable to bridges having no span greater than 200 m, the height of the deck above ground being less than 200 m, and not subject to aerodynamic phenomena. Eurocode 1 Part 1-4 indicates that for normal road and railway bridge decks of less than 40 m span, a dynamic response procedure is generally not needed. This part of Eurocode 1 is applicable to single bridge decks with one or more spans of classical cross section (slab bridges, girder bridges, box girders, truss bridges, etc.) and constant depth. Examples are given in Figure 11.28.

FIGURE 11.27 Example of roofed bridge in the Alps. (Courtesy of SETRA.)

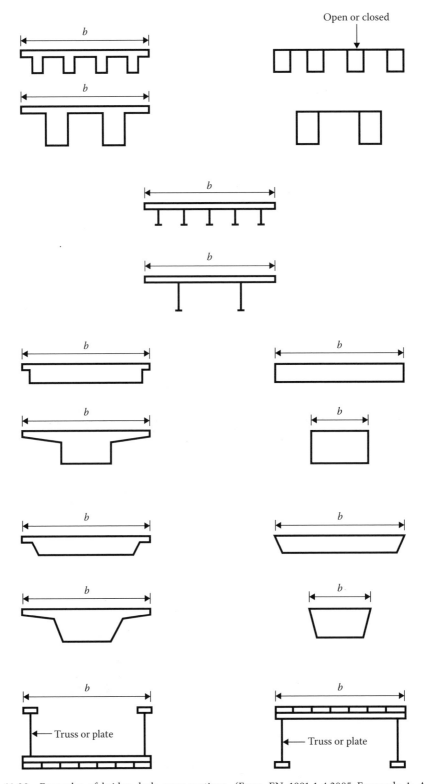

FIGURE 11.28 Examples of bridge deck cross sections. (From EN 1991-1-4:2005 Eurocode 1: Actions on structures, Part 1-4: General actions, Wind actions.)

Arch, suspension or cable-stayed, roofed, moving bridges, and bridges including multiple or significantly curved decks are normally excluded from the field of application of the Eurocode, but the general procedure is applicable with some additional rules which may be defined in the National Annex or for the individual project. For skew bridges the rules given in Section 8 of the Eurocode may be considered approximations whose acceptability depends on the skew angle.

11.2.3.4 Thermal Actions

Eurocode 1 Part 1-5 (EN 1991-1-5) defines the thermal actions to be taken into account for bridges. For the calculation of these actions, the thermal expansion coefficient of materials is needed. For example, for traditional steel and concrete, it is $\alpha_T = 12(10)^{-6}$, but values for other materials are given by the Eurocode. Eurocode 1 Part 1-5 distinguishes three types of bridge decks:

1. Steel deck: Steel box girder, steel truss or plate girder
2. Composite deck
3. Concrete deck: Concrete slab, concrete beam, concrete box girder

The thermal effects in bridge decks are represented by the distribution of the temperature resulting from the sum of four terms, shown in Figure 11.29.

The extreme characteristic values of the uniform temperature component are given in the national temperature map. These values are based on a return period of 50 years, but formulae are given in Annex A, derived from the Gumbel law (law of extreme values of type I) for the assessment of extreme temperatures based on a different return period.

11.2.3.5 Actions Caused by Water (Q_{wa})

Groundwater is considered to belong to the family of geotechnical actions. Eurocode 1 Part 1-6 gives rules for the determination of

- (Quasi-static) actions exerted by currents on immersed structures
- (Quasi-static) actions due to accumulation of debris against immersed structures

These actions are not specific for transient design situations, but they may have dominant effects on auxiliary structures during execution. Forces due to wave actions are treated in ISO/DIS 21650. Water and wave actions due to earthquakes (tsunamis) are not treated in the Eurocodes.

11.2.3.5.1 Actions Exerted by Currents on Immersed Structures

First, the determination of the water depth of a river should take into account an appropriate scour depth. Usually, a distinction is made between the general and local scour depths. The general scour

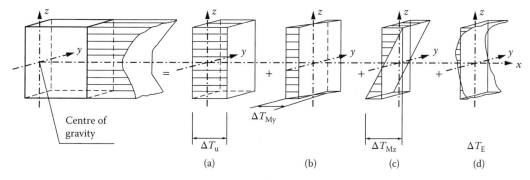

FIGURE 11.29 Diagrammatic representation of constituent components of a temperature profile. (a) Component of the uniform temperature; (b) and (c) components of the temperature linearly variable according to two axes contained in the plan of the section; (d) a residual component.

depth is the scour depth due to river flow, independent of the presence of an obstacle (scour depth depends on the flood magnitude). The local scour depth (Figure 11.30) is the scour depth due to water vortices in the vicinity of an obstacle such as a bridge pier.

Actions caused by water, including dynamic effects, where relevant, exerted by currents on immersed structures are represented by a force to be applied perpendicularly to the contact areas (Figure 11.31). The magnitude of the total horizontal force F_{wa} (N) exerted by currents on the vertical surface is given by the following formula:

$$F_{wa} = \frac{1}{2} k \rho_{wa} h b v_{wa}^2$$

where

v_{wa} is the mean speed of the water averaged over the depth, in m/s
ρ_{wa} is the density of water, in kg/m3
h is the water depth, but not including local scour depth, in m
b is the width of the object, in m
k is the shape factor
$k = 1.44$ for an object of square or rectangular horizontal cross section
$k = 0.70$ for an object of circular horizontal cross section

In general, the force due to water current is not critical with regard to the stability of the bridge piers, but it may be significant for the stability of cofferdams.

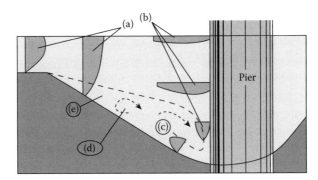

FIGURE 11.30 Local scour near a bridge pier. (a) Representation of horizontal water velocities; (b) representation of vertical water velocities; (c) vortex; (d) small secondary vortex; (e) dead water.

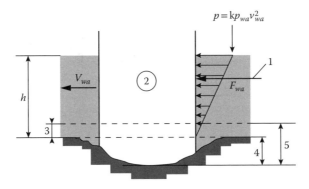

FIGURE 11.31 Pressure and force due to currents. (1) Current pressure (p); (2) object; (3) general scour depth; (4) local scour depth; (5) total scour depth.

11.2.3.5.2 *Actions Due to Accumulation of Debris against Immersed Structures*

In some rivers, an accumulation of debris against immersed structures is possible, and the phenomenon may occur regularly. Eurocode 1 Part 1-6 recommends the effects of such accumulation be represented by a force F_{deb} (N), calculated for a rectangular object (e.g., a cofferdam), for example, from the following expression:

$$F_{deb} = k_{deb} A_{deb} v_{wa}^2$$

where

k_{deb} is a debris density parameter; the recommended value is $k_{deb} = 666$ kg/m³
v_{wa} is the mean speed of the water averaged over the depth, in m/s
A_{deb} is the area of obstruction presented by the trapped debris and falsework, in m²

11.2.3.6 Construction Loads (Q_c)

A construction load is a load that can be present due to execution activities, but is not present when the execution activities are completed. For consistency with this definition, it has been considered that construction loads be classified as variable actions. A construction load may have vertical as well as horizontal components, and static as well as dynamic effects. In general, construction loads are very varied. To take them easily in to account, six sets have been defined in Eurocode 1 Part 1-6 and models are proposed for some of them. These sets are described in Table 11.4. The designer has to identify the construction loads for the design of an individual bridge. However, some heavy loads will only be known after the contractor, who will design the construction loads for the individual project, is selected.

After the identification of the construction loads for the individual project, these loads may be represented in the appropriate design situations, either as one single variable action or, where appropriate, different types of construction loads may be grouped and applied as a single variable action. Single and/ or grouping of construction loads should be considered to act simultaneously with nonconstruction loads as appropriate. In a general manner, construction loads are represented by the symbol Q_c.

11.2.3.7 Representation of Other Actions

Eurocode 1 Part 1-6 highlights some aspects concerning the following actions, which are already defined in other parts of Eurocode 1, due to the construction phase:

- Actions on structural and nonstructural members during handling.
- Geotechnical actions.
- Actions due to prestressing; if prestressing forces during the execution stage should be taken into account as permanent actions, the loads on the structure from stressing jacks during prestressing activities should be classified as variable actions for the design of the anchor region. This rule is an innovation, and means that the maximum prestressing force should be multiplied by a partial factor (probably 1.35) for a verification of the reinforcement at the ultimate limit state of the anchor region.
- Predeformations.
- Temperature, shrinkage, and hydration effects. The effects are due to the time difference between casting one concrete element and another element that has already hardened. In general, the limit state to be checked is the prevention of unacceptable cracks or crack widths, especially in the case of composite steel and concrete structures. Attention is also paid to possible restraints from the effects of friction of bearings.
- Snow loads. As shown in Figure 11.32, snow loads may become a dominant action for bridges during execution when located on mountain routes; indeed, they may remain several months (in winter) without any human intervention and accumulation of snow may lead to problems of static equilibrium.

Annex A2 to Eurocode 1 Part 1-6 gives the following rules. Snow loads on bridges during execution take account of the relevant return period. When daily removal of snow (also during weekends and

TABLE 11.4   Representation of Construction Loads (Q_c)

| Construction Loads (Q_c) | | | | |
| Actions | | | | |
Type	Symbol	Description	Representation	Notes and Remarks
Personnel and hand tools	Q_{ca}	Working personnel, staff and visitors, possibly with hand tools or other small site equipment	Modeled as a uniformly distributed load q_{ca} and applied to obtain the most unfavorable effects	Note 1: The characteristic value $q_{ca,k}$ of the uniformly distributed load may be defined in the National Annex or for the individual project. Note 2: The recommended value is 1.0 kN/m².
Storage of movable items	Q_{cb}	Storage of movable items, e.g., building and construction materials, precast elements, equipment	Modeled as free actions and should be represented as appropriate by a uniformly distributed load q_{cb} and a concentrated load F_{cb}	Note 3: The characteristic values of the uniformly distributed load and the concentrated load may be defined in the National Annex or for the individual project. For bridges, the following values are recommended minimum values: $q_{cb,k} = 0.2$ kN/m² and $F_{cb,k} = 100$ kN, where $F_{cb,k}$ may be applied over a nominal area for detailed design.
Nonpermanent equipment	Q_{cc}	Nonpermanent equipment in position for use during execution, either static (e.g., formwork panels, scaffolding, falsework, machinery, containers) or movement (e.g., traveling forms, launching girders and nose, counterweights)	Modeled as free actions and should be represented as appropriate by a uniformly distributed load q_{cc}	Note 4: These loads may be defined for the individual project using information given by the supplier. Unless more accurate information is available, the loads may be modeled by a uniformly distributed load with a recommended minimum characteristic value of $q_{cc,k} = 0,5$ kN/m². A range of CEN design codes are available, for example, see EN 12811 and for formwork and falsework design see EN 12812.
Movable heavy machinery and equipment	Q_{cd}	Movable heavy machinery and equipment, usually wheeled or tracked, (e.g., cranes, lifts, vehicles, lift trucks, power installations, jacks, heavy lifting devices)	Unless specified should be modeled on information given in the relevant parts of EN 1991.	
Accumulation of waste materials	Q_{ce}	Accumulation of waste materials (e.g., surplus construction materials, excavated soil, or demolition materials)	Taken into account by considering possible mass effects on horizontal, inclined, and vertical elements (such as walls)	Note 5: These loads may vary significantly and over short time periods, depending on types of materials, climatic conditions, build-up rates, and clearance rates, for example.
Loads from parts of a structure in a temporary state	Q_{cf}	Loads from parts of a structure in a temporary state (under execution) before the final design actions take effect (e.g., loads from lifting operations)	Taken into account and modeled according to the planned execution sequences, including the consequences of those sequences (e.g., loads and reverse load effects due to particular processes of construction, such as assemblage)	

FIGURE 11.32 Snow loads on a bridge deck in winter, during execution, in the Alps. (Courtesy of Joël Raoul.)

holidays) is required for the project and safety measures for removal are provided, the snow load should be reduced compared to the value specified for the final stage; the recommended value during execution is 30% of the value for permanent design situations. But for the verification of static equilibrium, and where justified by climatic conditions and anticipated duration of the construction phase, the snow load should be assumed to be uniformly distributed in the areas giving unfavorable action effects with a recommended characteristic value equal to 75% of the value for permanent design situations.

- Actions due to atmospheric icing; these include mainly loads by ice on water (floating ice), or icing of cables or other structural parts of masts and towers. Eurocode 1 Part 1-6 refers mainly to the ISO 12494 standard.
- Accidental actions. In accordance with EN 1991-1-6, accidental actions such as impact from construction vehicles, cranes, building equipment, or materials in transit (e.g., skip of fresh concrete), and/or local failure of final or temporary supports, including dynamic effects, that may result in collapse of load-bearing structural members, shall be taken into account, where relevant.

It is the responsibility of the designer to select the accidental design situations and the design values of accidental actions during construction, depending on the type of bridge under construction. The most critical accidental actions are

- The loss of stability of a bridge deck during launching due to an exit from temporary bearings
- The fall of equipment (for example a traveling form during its displacement), including the dynamic effects (Figure 11.33)
- The fall of structural elements (for example the fall of a precast segment before the final prestressing is active), including dynamic effects (Figure 11.34)
- The fall of a crane

In general the dynamic effects may be considered by a dynamic amplification factor for which the recommended value is equal to 2. This implies that the action effect of the fall (for example of the traveling

FIGURE 11.33   Fall of a traveling form. (Courtesy of JAC.)

FIGURE 11.34   Fall of a precast segment.

form) is equivalent to a force equal and opposite to its self-weight. For seismic actions, Eurocode 1 Part 1-6 mentions that the design values of ground acceleration and the importance factor γ_I need to be defined for the individual project, if they are not defined at the national level through a national regulation or in the National Annex of the relevant Eurocodes. Nevertheless, a project specification for very short-term phases or local effects is generally irrelevant.

11.3 Reinforced and Prestressed Concrete Bridges

11.3.1 Closed-Frame and Open-Frame Reinforced Concrete Bridges

The two major types, closed- and open-frame reinforced concrete bridges, are usually used for short spans. The closed-frame reinforced concrete bridge looks like a box (Figure 11.35) with its bottom surface resting on the foundation soil. It is completed by retaining walls intended to keep in the earth constituting the embankment of the formation. This type of structure has been most widely used to bridge narrow roads or railway lines (trams), since it combines strength with ease of execution. It is suitable for bridging any road with a skewed opening of less than 12 m. It may be advantageous to accept a fill on certain frames; a structure having a 2 m to 3 m embankment is generally more economical than a structure providing more clearance than is needed, and is also more aesthetic (Figure 11.36).

The frame can be designed with a shallow foundation and a soil of poor quality; the mean pressure on the soil is in the order of 0.10 MPa. The recommended thickness of the top slab is 1/25 of the skewed opening. The skew angle, under common conditions, ranges from 60 to 90 degrees. In France, about 200 bridges of this type are built every year.

The open-frame reinforced concrete bridge is most often founded on spread footings. It is completed by retaining walls intended to retain the earth constituting the embankment of the formation. The open portal frame is used for skew openings ranging from 10 m to 22 m (Figure 11.37). A structure on spread

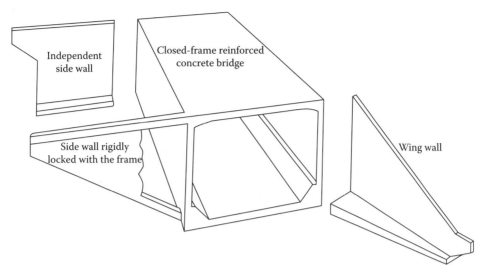

FIGURE 11.35 General morphology of a closed-frame reinforced concrete bridge.

FIGURE 11.36 Example of a closed-frame reinforced concrete bridge. (Courtesy of JAC.)

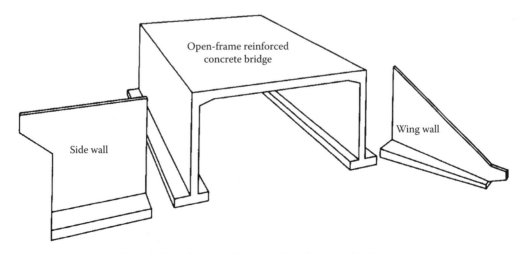

FIGURE 11.37 General morphology of an open-frame reinforced concrete bridge.

FIGURE 11.38 Example of open-frame reinforced concrete bridge over a tram line. (Courtesy of JAC.)

footings requires a foundation soil of good quality (i.e., a bearing capacity greater than 0.25 MPa); if this is not the case, a piled foundation is needed. The recommended slab thickness is 1/25 of the skewed opening. The skew angle, in common cases, is in the range 60 to 90 degrees (the skew angle is usually defined as the angle between the axis of support relative to a line normal to the longitudinal axis of the bridge, i.e., a 0 degree skew denotes a rectangular bridge). In France, about 150 bridges of this type are built every year (Figure 11.38).

11.3.2 Portal and Slab Bridges

The portal bridge is a structure with integral supports and a prestressed concrete deck (solid slab, voided slab or box girder—Figure 11.39). This type of structure, which enables bridging of relatively wide gaps without intermediate supports, leaving a substantial clearance over a great width, has aesthetic advantages that may lead to its use at certain judiciously chosen locations. It may, for example, be better than a conventional structure in the case of a highway or a motorway, leaving an unobstructed view and creating a kind of structural signal interrupting the monotony of the road. However, this structure requires a soil of good bearing capacity in case of shallow foundations, and its skew angle is limited to between 70 and 90 degrees.

Reinforced concrete slabs are generally of constant thickness, with or without side overhangs, with simply supported or continuous spans, and of moderate skew. Reinforced concrete slabs are used to span roads or motorways when the longest skew span does not exceed 18 m; however, this value may be easily increased if needed, depending on economic conditions. The depth-to-span ratio, which depends on the number of spans and the ratio between the spans, is in the range 1/22 for a simply supported span to 1/28 for a continuous slab of at least three spans. These values are given for a solid slab of rectangular cross section, or with small side overhangs; in the case of wide overhangs the thickness may be increased by about 10%. About 150 bridges of this type are built every year (Figure 11.40).

Prestressed concrete slabs with constant thickness are most commonly used for medium long-span bridges. They are prestressed longitudinally and reinforced transversely, which may or may not include side overhangs. The spans may be simply supported or continuous. The alignment, rectilinear or with a slight curvature in plan, should however hold a moderate skew for the structure (Figure 11.41). The range of use is large, extending from simply spanning a road or motorway to spanning interchanges or rivers. The depth-to-span ratio ranges from 1/25 for a simply supported span to 1/33 for a continuous slab of at least three spans. These values are given for a slab of rectangular cross section or with small side overhangs; in the case of wide overhangs, the thickness may be increased by about 10%. In the most

FIGURE 11.39 Example of portal bridge on a French motorway. (Courtesy of JAC.)

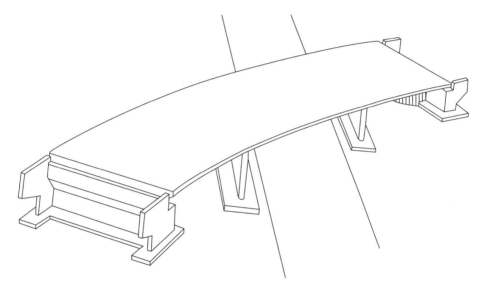

FIGURE 11.40 General view of a curved slab bridge.

FIGURE 11.41 Example of prestressed concrete slab bridge. (Courtesy of JAC.)

common cases, the number of spans is limited to 6 and the skew angle to 45 degrees; the horizontal curvature is limited in such a way that the ratio of the longest span to the radius of curvature is not more than 0.30. In France, about 300 bridges of this type are built every year.

There are two types of classical prestressed concrete slab bridges: hollow and ribbed continuous slabs. A hollow slab structure is characterized by the presence of longitudinal voids in the concrete, yielding an appreciable saving of dead load; this results in longer spans than with a solid slab, making it possible to build slender decks leaving a more unobstructed view (Figure 11.42). The use of the hollow slab is rather extensive. It can span as much as 35 m. In particular, this type of structure gives a means of spanning motorways of all widths without a support on the central reserve of the motorway. The slab thickness may be constant or variable, depending on the span. The constant slab thickness can be used for spans up to about 25 m. For longer spans, the thickness must be increased with the aid of haunches in the vicinity of the intermediate supports (Figure 11.43).

For constant thickness slabs, the depth-to-span ratio ranges from 1/22 for a simply supported span to 1/30 for a continuous bridge over at least three spans. For variable thickness slabs, it ranges from 1/20 on the middle support to 1/30 into the spans in the case of two spans, and from 1/24 on the intermediate supports to 1/42 into the spans in the case of three or more spans.

The ribbed slab is a solid slab with wide side overhangs. It has advantages comparable to those of the hollow slab and a more attractive appearance as well. The single-rib slab, which has a single rib, has the same range of use as the hollow slab. A multiribbed slab (at least two ribs) has the same range of use as the hollow slab, but is used primarily for bridges of more than 14 m width (Figures 11.44 and 11.45).

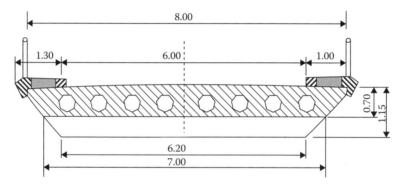

FIGURE 11.42 Typical cross section of a hollow slab bridge deck.

FIGURE 11.43 Example of hollow slab bridge deck. (Courtesy of JAC.)

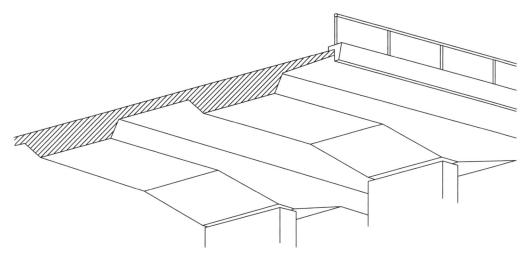

FIGURE 11.44 Morphology of a ribbed slab.

FIGURE 11.45 Example of a slab bridge with a single rib. (Courtesy of JAC.)

11.3.3 Pre- and Post-Tensioned Girders

Precasting was discovered almost at the beginning of the twentieth century, very soon after reinforced concrete was born. After a wavering start in the 1930s, followed by a near-total stoppage during the Second World War, the need of housing in particular, and heavy constructions of all natures, caused an irreversible development of concrete precasting in Europe and then in the United States and Asia. In the field of bridges, precasting applications started soon after the Second World War. From then on, bridges were designed and built with independent girders with prestressed or post-tensioned beams bound together by a concrete slab (Figure 11.46). A typical bridge commonly used in France has simply supported precast pretensioned concrete beams with a cast-in-situ concrete deck slab. The beams are not cross-braced, except at their ends where there are butt ties.

This structure should not be considered a systematic replacement for conventional standard cast-in-situ structures; on the other hand, it is highly advantageous in particular cases in which prefabrication can be benefited from (e.g., excessive cost of scaffolding, site located near the precast plant, or special requirements as regards completion times) and if the precast plant normally produces beams of the

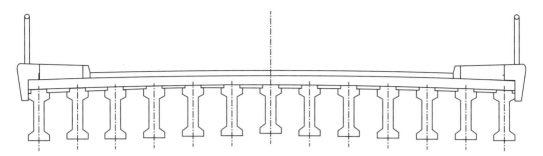

FIGURE 11.46   Typical cross section of a precast, pretentioned, prestressed, concrete beam bridge.

FIGURE 11.47   Transportation of a precast, pretensioned, prestressed, concrete beam. (Courtesy of Fombeton.)

required quality. Its normal range of use with spans from 10 m to 25 m makes it a particularly attractive solution for spanning roads for which traffic cannot be stopped, railway lines, and some rivers. The usual depth-to-span ratio is about 1/20. Factory precasting of the beams generally requires steam curing of the concrete, with a view to acceleration in the fabrication cycles in order to achieve faster investment redemption. This process of steam curing has many effects on the properties of the materials.

The range of use, which owes something to the advantages afforded by the use of the metal framework, is extensive, since spans of more than 100 m are conceivable with bridges of this type; however, spans of about 35 to 50 m are the range of normal use for this type of structure. The depth-to-span ratio ranges from 1/25 to 1/30 in the case of continuous beams of constant height. The skew angle possible without introducing special problems is between 55 and 90 degrees in the case of a narrow structure ($\leq$ 8 m); if the structure is wide, a special study should be performed regarding the effects the skew (Figure 11.47).

Viaducts with post-tensioned prestressed beams are made up of simply supported spans, each consisting of a certain number of prefabricated beams with bottom flanges and of constant height, prestressed by tendons, with or without crossbeams, and linked by cast-in-situ slabs made of reinforced or prestressed concrete (Figure 11.48). Viaducts built in this way may have a large number of spans. Coupling the spans makes it possible to space the expansion joints as much as 100 m apart for a greater user comfort. This type of structure is especially well suited to the spanning of isolated nonstandard obstacles, such as a series of closely spaced roads, railway lines or canals for which a series of isolated bridges would not be competitive, or zones in open terrain if embankments are not feasible (on peaty ground, for example). Developed for rectilinear, possibly skewed bridges, this structure can also be

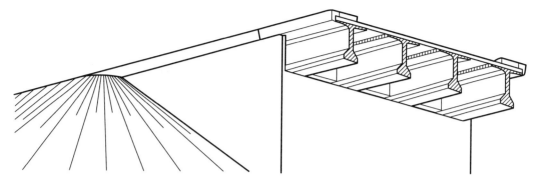

FIGURE 11.48 General morphology of a bridge deck with post-tensioned prestressed beams.

adapted to curved bridges—the beams then form a broken line and the deck overhang, of variable width, matches the general curvature of the alignment.

The spans range from 25 m to 45 m, in exceptional cases to as much as 50 m. An economic beam depth-to-span ratio is 1/17, which requires a rather substantial deck height; reducing the depth-to-span ratio to 1/22 increases the cost by about 20%. It should, however, be noted that viaducts are quite often high enough over ground or water for clearance problems, and in these circumstances a relatively thick deck may well tend to enhance the appearance of the structures.

11.4 Segmental Concrete Bridges

In the span range of 40–200 m, box girder bridges are the most competitive structures. Decks have also been designed and constructed in the past with ribbed beams, but only for span lengths not exceeding 80 m. At their origin, box girders were of limited dimensions, and wide decks were constructed using several boxes connected at the level of the top deck slab, which was often prestressed transversely. Progressively, box girders became wider and transverse prestressing became much less common. For wide decks, multicell box girders have replaced multiple box girders.

Today, in the majority of structures a single box girder, often with inclined webs, is used. For the widest bridges, struts or transverse ribs are incorporated. These members are intended to reduce the thickness of the cantilever slabs and to ensure their resistance to bending moment. Depending on the method of construction (formwork launching girder, balanced cantilever, or incremental launching) the box girders are either concreted in situ in long segments, or concreted in situ or prefabricated in short segments. In certain cases the deck cross section is constructed in several steps (for example, first the bottom slab, then the webs, and finally the top deck slabs). All these large bridges are therefore built in steps and their classification depends on their characteristics and method of construction.

11.4.1 Prestressed Concrete Bridges Built by the Cantilever Method

This method of construction was the first generally adopted in France for large prestressed concrete bridges. Three generations of this type of construction have been developed.

- First-generation structures (hinged at midspan): The first bridge constructed in France by balanced cantilevering was the Chasey Bridge over the Ain River, built in 1957 by the company GTM, which built many other bridges of this type. All these structures showed the same defect after several years: excessive deflection of the cantilevers, due, in particular, to underestimated effects of creep and shrinkage.
- Second-generation structures (continuous): In 1961–1962, the bridges of Goncelin over the Isère River and Lacroix-Falgarde over the Ariège River, designed by STUP (which became later

Freyssinet International), were the first bridges with continuous decks. These were followed by numerous others. During this period, the first concrete bridge decks made of precast segments with matched glued joints appeared.

- Third-generation structures: The design assumptions and rules regarding the evolution of forces arising from long-term deformations and the effects of temperature gradients were found wrong. In 1972, the French authorities launched a program of inspection and repair of numerous structures, in particular bridges built by the cantilever method. Simultaneously, the design rules were improved and from 1975 a new generation of segmentally constructed bridges appeared.
- External and combined internal-external prestressing: From 1983, in the majority of bridges built by cantilever method most of the internal continuity tendons were replaced by continuous external tendons which extended over several spans. The tendons in the cantilevers remained inside the concrete structure except for the Sermenaz viaduct near the city of Lyon. Two remarkable bridges may be mentioned: the Ile de Ré Bridge (1987–1988) which is 3840 m long (total length) and has a deck formed of prefabricated segments (Figure 11.49) and the Pierre Pflimlin Bridge over the Rhine River (Figure 11.50). The Ile de Ré Bridge (Re Island Bridge) is now rather old. It was completed in 1988. The main spans (24) are 110 m length and it was the first concrete bridge made of what was called at that time "high-performance concrete."

The Pierre-Pflimlin Bridge over the Rhine River between France and Germany was built in 2000–2002. The owners are the French Republic, the German Land Baden-Württemberg, and the German Ministry of Transportation. Its main span is 205 m and was designed to resist very strong earthquakes. When the bridge deck was fixed to the piers, a solution to ensure the longitudinal flexibility of the piers to shape them as a double thin wall.

The Tulle viaduct (Figure 11.51) was built in 2003 on Motorway A89 and belongs to the company Autoroutes du Sud de la France. It has three main spans of about 180 m and its height above the valley floor or water (Corrèze River) is 150 m.

As already said, the superstructures of bridges built by the cantilever method are box girders. While in the past sections usually possessed more than two webs as soon as the upper slab width exceeded

FIGURE 11.49 The Ile de Ré Bridge under construction. (Courtesy of Bouygues.)

FIGURE 11.50   Pierre-Pflimlin Bridge over the Rhine River. (Courtesy LCPC.)

FIGURE 11.51   The Tulle viaduct; note the double piers. (Courtesy M. Capelli.)

13–14 m, the current trend is in the single cell box. For the widest bridge decks, the transverse resistance of the upper slab and even of the bottom slab can be assured by various means: an appropriate thickness with a prestressing in the transverse direction or a transversely reinforced concrete upper slab by reinforced concrete ribs (Figure 11.52). Such a design is appropriate for deck width in the range 20 to 30 m. In the same range of widths, another type of design consists of supporting the upper slab by braces, which may be made of concrete or, more generally, made of steel members (Figure 11.53).

Figure 11.54 shows an example of a bridge with concrete braces (this bridge, the Scardon viaduct, was constructed by the incremental launching method). This viaduct was built in 1995–1997 and carries Motorway A16.

During the last decades, the design of very wide bridge decks significantly evolved and, nowadays, two main materials of the construction industry, concrete and steel, are associated with it. Research was

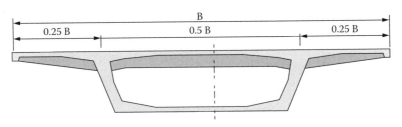

FIGURE 11.52   Example of box girder with an upper slab stiffened by reinforced concrete ribs.

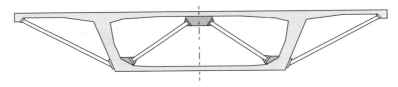

FIGURE 11.53   Example of cross section of a box girder with braces.

FIGURE 11.54   Scardon viaduct on Motorway A16 (north of France). (Courtesy of SETRA.)

undertaken in several directions with the main objective being to reduce the self-weight of webs (it is practically impossible to reduce the upper slab thickness, which supports traffic loads directly, as well as the lower slab thickness, which has to resist larger compression forces) and to simplify the methods of construction or connections. A major idea was to replace the traditional webs by lattice works, at first in concrete (Figure 11.55) then, more and more systematically, by using metallic tubes (Figure 11.56).

Concerning the tendons profiles in concrete bridge decks, two families of cables may be roughly identified: one family to ensure the deck resistance (bending moment, shear force) during construction and the second family to ensure the continuity and resistance of the bridge deck at the final stage. Figure 11.57 shows the principles generally adopted for the design of the tendons profiles.

In order to illustrate the most recent trends in the design of road bridges, three examples are proposed: the Abra Bridge over the Taravo River (Corsica) (Figure 11.58); the Bras de la Plaine viaduct (La Réunion Island) (Figure 11.59); and the Grande Ravine viaduct at La Réunion Island (Figure 11.60). The Abra Bridge was constructed between 2006 and 2008. Its center span is 96.1 m long. The design of the webs was selected for aesthetical reasons and to reduce their self-weight. The bridge is located on a highway.

FIGURE 11.55 Sylans viaduct on Motorway A40. (Courtesy of Bouygues.)

(a)

(b)

FIGURE 11.56 Echingen viaduct on Motorway A16: (a) arrangement of prestressing cables and (b) design of deck segments. (Courtesy of JAC.)

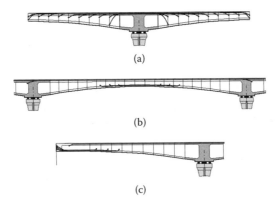

(a)

(b)

(c)

FIGURE 11.57 Tendons profiles in a bridge deck built by the cantilever method: (a) first phase; (b) second phase (continuity tendons); (c) second phase (continuity tendons in end spans).

FIGURE 11.58 Abra Bridge over the Taravo River (Corsica). Under construction. (Courtesy of Razel.)

FIGURE 11.59 Bras de la Plaine viaduct (La Réunion Island). (Courtesy of Jean-Muller International.)

FIGURE 11.60   Grande Ravine viaduct at La Réunion Island. (Courtesy SETEC.)

The Bras de la Plaine viaduct was built in 2000–2001 by the cantilever method, but it looks like an arch bridge of about 280 m (the total length is 305 m). The owner is the local authority (Conseil Général de la Réunion). The Grande Ravine viaduct is a very slender bridge built by the cantilever method in 2006–2009. The owner is the regional authority. The main span is 140 m and its total length is 288 m. The Grande Ravine is 300 m wide and 170 m deep.

11.4.2 Concrete Bridges Built by the Incremental Launching Method

This method of construction, deriving directly from the launching method for steel bridges, was developed in France in the 1970s and has been widely used throughout the world. In some countries, such as Germany, Austria, Italy, and France, this technology is now considered traditional. This method of construction gives rise to some specific design problems.

The principle of the incremental launching method consists of building the superstructure segments in a casting area located behind one of the abutments of the bridge. Each segment is cast in contact with the previous one, and connected by prestress to the section of deck already built. Once a segment is completed, the entire structure is moved forward on a distance equal to the length of this segment, releasing the formwork for the construction of the following segment. When the deck reaches its final position, it is jacked up, and the temporary slides are replaced with permanent bearings. Such a method has several advantages:

- It eliminates the traditional scaffolding required for supporting the formwork. This is particularly valuable for projects with high piers, in urban areas, or spanning over water, roads, or railways.
- Most of the construction operations take place in the same location, which may be organized as a precasting yard. The work is thus more easily supervised, leading to a high quality of workmanship.
- The sequence of the construction stages is repetitive. After a rather short training period, the work on the site becomes generally very efficient.
- The method requires only a low investment in specific equipment such as a launching nose, launching jacks, conventional jacks, temporary slides, and guides.
- The number of construction joints, which always constitute weak sections in the structure, especially if prestressing couplers are used, is reduced to a minimum. Moreover, they can be located in zones where the bending moments are reduced.

These advantages should not mask the fact that the method requires close supervision of the construction. The conformity of the soffit with its theoretical shape is particularly important if the structure is not to be subjected to unacceptable forces. For the same reason, the jacking of the deck for replacing the provisional bearings by the permanent bearings must be carried out carefully.

During launching, the superstructure is extended forwards by a temporary structure, generally made of steel and called a launching nose. The purpose of the nose is to reduce the large cantilever bending moments which would otherwise occur in the superstructure before each pier is reached. Before the superstructure cantilevers are too far, the nose bears on the next pier. The vertical reaction generated in this way reduces the cantilever bending moment (Figure 11.61).

The Meaux viaduct was built in 2001–2005, crosses the Marne River, and carries Motorway A140 near Paris. Its owner is the French Republic. The superstructure plus the launching nose make up a continuous beam of variable properties. By a careful choice of the stiffness of the nose as a function of the bridge spans and of the section properties of the superstructure, it is possible to keep the stresses in the concrete below the allowable limits at all launching stages.

A typical section of the superstructure is subjected to bending moments of both signs during launching that are of variable magnitude. Therefore, a deflected profile is not appropriate; if the eccentricity is favorable for a given position, it is unfavorable for another. As a consequence, the prestress cables must be straight during launching. The current design principles of prestress are shown in Figure 11.62.

The design of cross sections of bridge superstructures built by the incremental launching method follows the same principles as bridges built by the balanced cantilever method. The most difficult problems appear with very wide bridge decks. Figure 11.63 shows the cross section of the Meaux viaduct previously mentioned. The cross section of the Meaux viaduct is a 4/5 m deep composite deck box girder with steel tube webs designed to enhance prestressing efficiency and save self-weight. Figure 11.64 shows this elegant structure after completion. The incremental launching method is generally used without temporary supports for the construction of bridges, with typical spans between 40 and 60–70 m. For projects with spans longer than 70 m, intermediate temporary supports are generally required (Figure 11.65).

FIGURE 11.61 Meaux viaduct during launching with the launching nose. (Courtesy of Razel.)

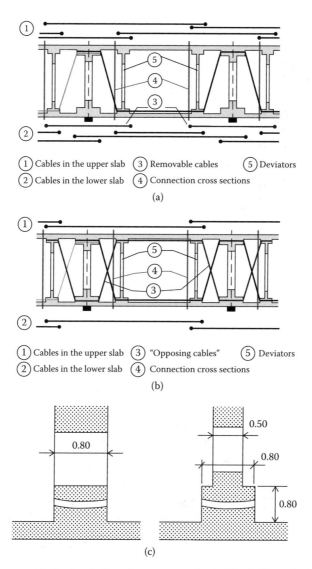

FIGURE 11.62 Various aspects for the design of prestressing in a bridge built by the incremental launching method. (a) Design of prestressing cables without opposing cables for the launching phase; (b) design of prestressing cables with opposing cables for the launching phase; (c) examples of deviators for prestressing cables.

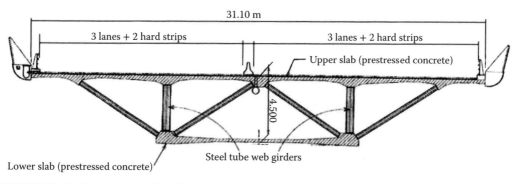

FIGURE 11.63 Cross section of the Meaux viaduct.

FIGURE 11.64 The Meaux viaduct, general view. (Courtesy of Razel.)

FIGURE 11.65 Launching of the Millau Bridge with intermediate temporary supports. (Courtsey of SETRA.)

11.5 Steel and Steel–Concrete Composite Bridges

11.5.1 General

Steel bridges benefit from continuous progress accomplished in the manufacturing of steels of high, regular, and guaranteed mechanical properties, the improvement of connection techniques, methods of assembly, and from the techniques of corrosion protection, and are rather competitive compared to concrete bridges. Today, high-performance steels are an essential part of many market segments, such as pipelines, offshore platforms, pressure equipment, bridges, and so on. For all these markets, the steel industry has continuously developed steels showing higher strength and enhanced fabrication properties. This explains why steel construction has indisputable advantages resulting from the excellent strength-to-weight ratio of the material and from the best conditions for the fabrication of components of a high quality and that are easy to assemble.

The evolution of economic conditions leads engineers to simplify the designs of structures and even to exclude certain types of structures. This strong tendency to widely simplify the design of composite steel–concrete girder bridges with concrete deck slabs connected to steel girders (I-beams or box girders) is widely encouraged. But structural steel may be used to bridge a range of spans from short to very long (15–1500 m) carrying traffic loads with a minimum dead weight.

11.5.2 Steel–Concrete Composite Girder Bridges

Typical cross sections of composite steel-concrete girder bridges with concrete slabs are shown in Figure 11.66. Sometimes, deck slabs are prestressed in the transverse direction. The connection between the slab and the steel beams is assured by shear connectors mainly constituted by angles or vertical shear stud connectors welded on the top flanges of beams. In some cases, the bridge superstructure may be a steel box girder with a top concrete slab, in particular for long span lengths and wide decks (Figure 11.67). For railway bridges, the cross section is often a braced box girder (Figures 11.68 through 11.70).

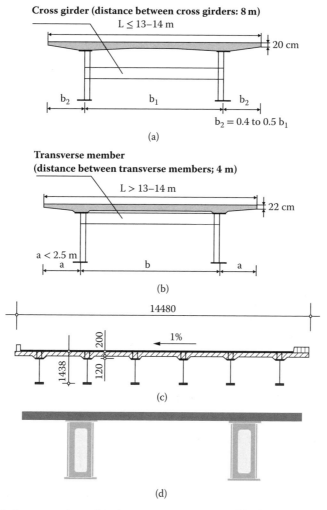

FIGURE 11.66 Typical cross sections of steel–concrete composite road bridge superstructures: (a) example of bridge cross section with cross girders; (b) example of bridge cross section with transverse members; (c) example of bridge cross section with rolled-steel girders; (d) example of possible evolution of a bridge cross section with box girders and without transverse members.

FIGURE 11.67   Box girder of the Verrières viaduct. (Courtesy of SETRA.)

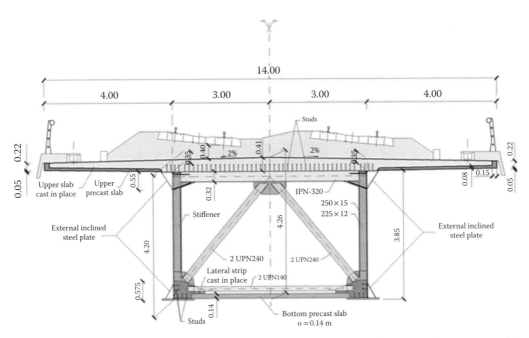

FIGURE 11.68   Example of railway bridge superstructure (braced cross section) with a concrete bottom precast slab.

11.5.3 Orthotropic Girder Bridges

Orthotropic plates are constituted by a continuous steel plate of 12 or 14 mm minimum thickness, stiffened in the two perpendicular directions by closed trapezoidal ribs and by regularly spaced stiffeners (Figure 11.71). The basic advantage of orthotropic plates compared to concrete slabs, besides the speed of their assemblage, lies in their lightness, but they are relatively expensive and their proper use is reserved for long-span bridges (where the gain of self-weight is particularly interesting), or for medium-span bridges, when the size conditions require a very slender superstructure, and for temporary steel viaducts, which are not often used today (Figure 11.72).

FIGURE 11.69 Viaduct over the Moselle River, High Speed Trains (TGV) East, total length 1510 m. (Courtesy of French National Railway Corporation, SNCF.)

FIGURE 11.70 The Claye Souilly railway viaduct, crossing the Beuvronne Valley and the Ourcq Canal (415.5 m). (Courtesy of SNCF.)

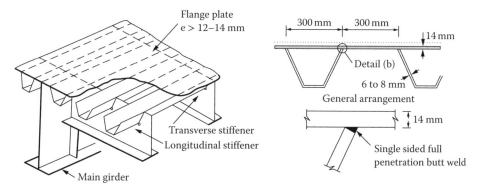

FIGURE 11.71 Principle of orthotropic plates.

FIGURE 11.72   Example of orthotropic deck. (Courtesy of JAC.)

11.5.4 Superstructure Design Principles

Bridge superstructures with beams (I-beams, box girders, etc.) are the most commonly used because they cover a wide range of span lengths (up to 160 m for a simply supported span in the case of the Cheviré Bridge over the Loire River near Nantes, shown in Figure 11.73). I-beams may be placed below the top deck slab or laterally over the top slab when problems of free height are met for the crossing of the obstacle. Sometimes, but very seldom, truss girders may be used laterally or over the top slab.

The usual range of span lengths for steel–concrete composite road bridges is 30 m to 120 m for continuous spans and 25 m to 90 m for simply supported spans. Before 1970, steel–concrete composite girder bridge were of the multibeam type. Today they are rather systematically designed with only two beams because their construction is very easy. However, multibeam bridge decks remain interesting in some cases. If the bridge deck is rather wide (more than about 13 m), the most common solution consists in connecting a concrete top slab to the steel beams in the longitudinal direction (thickness 20 cm to 24 cm) and cross girders in the transverse direction with a spacing of approximately 4 m, in order to obtain a composite bridge.

In other cases, the concrete top slab (usually 25 cm thickness) is connected to the beams, the distance between beams in the transverse direction being about 0.55 times the width of the upper slab. The beams are connected with transverse cross girders. In order to decrease the weight of the concrete top slab, it may be prestressed in the transverse direction (in general, single strands T15S positioned every 20 cm to 60 cm) when the slab width exceeds 17–18 m (connectors are then of the friction type).

11.5.4.1 Steel-Concrete Composite Box Girder Bridges

Box girders are the best solution to ensure a better rigidity to bridge superstructures, particularly in case of horizontally curved alignments or intermediate supports of small dimensions in the transverse direction (in urban areas, between railway tracks, etc.). A rigid section with regard to the torsional moment can be necessary also to facilitate the launching of long spans.

Box girders also have a good resistance to corrosion: the outside surfaces are smooth, which is favorable to prevent the accumulation of water, and the internal surfaces are in a protected atmosphere, possibly with the presence of devices absorbing humidity, as in the case of very big bridges (for example, the Normandy Bridge). Furthermore, their aspect is rather fine due to clear volumes. However, they are in

FIGURE 11.73 The Cheviré Bridge over the Loire River near Nantes (Courtesy of SETRA).

general more expensive than I-beams because they require a larger quantity of steel and a more complex manufacturing process. They are competitive only if a limited number of members can be completely prefabricated in factories and then carried on site; the dimension in the transverse direction of prefabricated members is limited to 5 m when transported by road.

The deck slab is made of reinforced concrete and is connected to the webs of the box girder. If the deck width is important, it may be technically interesting to design a multibox girder in order to reduce the thickness of bottom flanges. The depth-to-span ratio of a steel–concrete box girder depends on the top slab width; the order of magnitude is around 1/36 for a superstructure of constant height and 12 m wide.

11.5.4.2 U-Shaped Girder Bridges

U-shaped girder bridges are mainly used for railway bridges for span lengths in the range of 6 m to 25 m for beams of limited height and spans up to 50–60 m for high beams. The first ones are often designed to replace old bridge decks with twin beams. The second are always designed with welded beams interdependent with a bridge deck with small steel girders incorporated in a concrete slab (Figure 11.74). Their aspect may be rather poor if it is not very carefully handled; for that reason, the maintenance footpaths are sometimes set outside the beams to reduce their monotony by the shaded areas they can create.

11.5.4.3 Steel–Concrete Composite I-Beam Bridges

Steel–concrete composite I-beam bridges may easily be suited for a moderate skew or a moderate curvature. In the first case, the cross girders or the transverse members are in general perpendicular to the main longitudinal beams, except in the case of end cross girders, to avoid complicated joints. In the second case, it is possible to design straight girders and adjust the geometry of the top slab, or it is possible to use horizontally curved girders which can be easily manufactured if the radius of curvature is constant.

The control of the concrete top slab cracking, which is a natural phenomenon, should be perfectly organized and should follow some rules of the art concerning mainly the concrete quality (compactness, moisture content, components intended to increase the concrete workability, etc.) and the construction process (continuous pouring of the concrete slab for short spans or pouring step by step with appropriate support subsidence). Concrete cracking is not only a consequence of negative bending moments, but is also due to total shrinkage, which is composed of two elements, the drying shrinkage and the autogenous shrinkage. The drying shrinkage develops slowly, since it is a function of the migration of

FIGURE 11.74   Crossing of A104 Motorway at Pomponne (length 75 m, two spans of 37.25 and 37.75 m). (Courtesy of SNCF.)

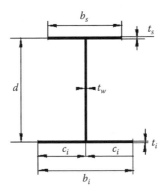

FIGURE 11.75   Notation for the design of I-beams for road bridges

the water through the hardened concrete. The autogenous shrinkage strain develops during hardening of the concrete; the major part therefore develops in the early days after casting. The autogenous shrinkage depends on the concrete strength. When the construction period is rather short, it is possible to use precast slab members with cast-in-place joints.

The depth-to-span ratio of steel–concrete I-beams for road bridges is about 1/22 to 1/25 for simple spans and 1/28 for continuous spans of constant height. If the height of the beams is variable, the depth-to-span ratio is about 1/27 over piers and 1/30 for current cross sections, but it can reach easily 1/35, even 1/40 if needed. With the notation defined in Figure 11.75, the width b_s of the top flange and the width b_i of the bottom flange may be taken from the following formulae, for $L(m)$ in the range 30 to 100 m:

$$b_s(mm) = \frac{1}{7}(40L + 1600) \qquad b_i(mm) = 10L + 200$$

The web thickness, t_w, is generally equal to 12 mm or more to limit the number of its stiffeners, and it is recommended that the average value of the shear stress due to the shear force V calculated at the serviceability limit state is such that

$$\frac{V}{dt_w} \leq 100 \quad MPa$$

In order to avoid any risk of buckling, the width-to-thickness ratio of a plate is limited by the following formula:

$$\frac{b}{t} \leq 0.9 \sqrt{\frac{E}{f_y}}$$

where

E is the steel elasticity modulus
f_y is the yielding stress of steel

The data for the design of railway bridges depend widely on the nature of the carried traffic: normal traffic, heavy traffic, high-speed traffic, and so on. The problems of rigidity and dynamic behavior are determining factors for bridges located on lines for heavy or very-high-speed traffic.

11.5.4.4 Orthotropic Deck Bridges

Steel orthotropic decks are in general used for very-long-span bridges (spans of more than 120 m), slender bridges, movable bridges, and in all cases where the speed of construction is the most important factor (Figure 11.76). For a three-span symmetric bridge without specific constraints, the length of access

FIGURE 11.76 **(See color insert.)** Example of a bascule bridge. (Courtesy of SETRA.)

spans is 0.40 to 0.60 times the center span. The choice between the variable depth and the constant depth of the deck depends mainly on natural constraints of the project or on considerations of aesthetics. Mechanically, the variable depth is fully appropriate only beyond a main span of 150 m length.

Orthotropic deck plates are designed either with two beams, or with a simple box (two webs). In the first case, it is recommended to adopt a cross section with distant beams so as to benefit from the membrane effect of the deck plate. For box girders, it may be convenient to design the cross section with wide lateral cantilevers to save the volume of steel members between webs. With regard to the average depth-to-span ratio, there is practically no difference between a box girder and an I-beam. In statically determined spans, this ratio about 1/30. For a continuous bridge, the average ratio is also about 1/30 with a constant depth. It is in the range 1/25–1/30 over piers and in the range 1/40–1/50 at midspan.

11.5.4.5 Fatigue in Steel Bridges

Road or rail traffic loads develop stress variations in bridge decks (magnitude, frequency), which may potentially create fatigue damage. Load effects due to the wheels or axles of vehicles in the members of a bridge deck have a dynamic character due to the suspended masses, to the irregularity of the road or of the railway track, and to the dynamic response of the bridge deck. Generally, the dynamic magnification is the highest near discontinuities of the surface where traffic loads are applied, for example near expansion joints at the ends of a road bridge.

Orthotropic decks of road bridges are particularly concerned with the effects of fatigue. Bridges designed and built in the 1950s to 1960s proved their durability over several decades. The orthotropic box girders lead to considerable weight economy, but the effects of fatigue were sometimes underestimated in the past. For example, in some bridges where the welded joints between transverse frames and stiffeners of the webs were not correctly designed with regard to fatigue problems, cracks were discovered, in particular where, due to the presence of rigid stiffeners or diaphragms, the flexibility of the top orthotropic deck plate under the influence of traffic loads was subject to a strong discontinuity, and hence was subjected to dynamic actions amplifying the repeated local stress variations.

Today, steel bridges are systematically designed for fatigue via a direct or indirect damage calculation by means of the Palmgren–Miner rule using a histogram of the variations of real stresses (recorded in situ) or based on a simplified method using an appropriately calibrated unique lorry so that a single crossing of the bridge by this lorry leads to a stress variation representing the effects of real traffic. For railway bridges, a series of fatigue load models is given in an annex of Eurocode 1 Part 2. In the practice, the standards specify a classification of details sensitive to fatigue (Figure 11.77) in reference to curves, called Wöhler curves, connecting every range of stress variation to a number of cycles generating a rupture with safety margins, which are not always apparent. The fatigue considerations are mainly focused on the design of the main beams of short or medium span lengths.

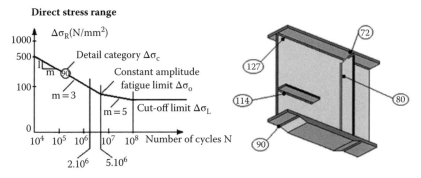

FIGURE 11.77   Example of classification of details for steel structures.

11.6 Arch and Portal Bridges

11.6.1 General

An arch bridge is primarily a vertically curved beam, carrying a bridge deck, with a curved gravity axis generally of circular or parabolic geometry with oblique support reactions requiring a foundation ground of good bearing capacity. The modern arch bridge is generally made of reinforced, sometimes slightly prestressed, concrete or steel. Its general dimensions are characterized by its span length L, measured between its springings, and its rise f, which represents the distance between the line joining the springings and the highest point of the gravity axis, as shown in Figure 11.78. The mean value of the ratio L/f is nearby 6 and in the range 5 to 8.

11.6.2 Concrete Arch Bridges

Two families of bridges need to be distinguished: deck arch bridges with a deck over the carrying arch (arch with upper bridge deck) and through-arch bridges with a deck placed below the arch (arch with suspended or intermediate bridge deck). The former are the only ones that can be constructed by the segmental method (Figure 11.79).

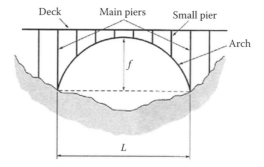

FIGURE 11.78 Various parts of an arch bridge.

FIGURE 11.79 Trellins Bridge on the Isère River near Grenoble (1984, main span 138 m). (Courtesy of SETRA.)

Arches sometimes have fixed ends to limit their deformation due to concrete creep. However, in the case of long-span arch bridges, it is possible to use jacks to compensate, during construction and in the maintenance program after construction, for the effects of deformations of the arch (immediate deformations and then progressive deformations due to creep and shrinkage). An arch rib structure is mostly:

- A concrete box with an architectural design of its webs for spans larger than 150 m and a depth H with a ratio of about $L/H \cong 60$
- Designed with beams connected with crossbeams for common spans from 100–150 m
- A ribbed upper slab for shorter spans less than 100 m

The bridge deck is a reinforced or prestressed concrete slab with or without ribs or a bridge deck with prestressed girders of 15 m to 40 m of span length (depending on the distance between piers). But a steel–concrete composite bridge may be a selected deck (see the Chateaubriand bridge, Figure 11.80) or a box girder set up by the incremental launching method, as in the case of the Trellins Bridge (Figure 11.79) although, in that case, the construction technique entailed a dissymmetry of internal efforts during construction the control of which turned out to be difficult and lead to an expensive construction process.

In the past, arch bridges were built on a curve, a real temporary, mostly wooden bridge on which the real bridge was constructed (Figure 11.81). After the Second World War, the cost of curved falseworks

FIGURE 11.80   Chateaubriand Bridge over the Rance River (see also Figure 11.9).

FIGURE 11.81   Example of timber curved falsework. (Courtesy of Ph. Ramondenc.)

became prohibitive and the technique of arch bridges disappeared for many years until the use of temporary stays and construction with segments created a new interest in the catalog of solutions for the crossing of wide obstacles. However, arch bridges are still not very popular in the French environment.

11.6.3 Steel Arch Bridges

With the development of cities in modern countries, many footbridges have been designed during recent years. There are several examples in France, but two of them are remarkable: the Léopold Sédar-Senghor footbridge (Figure 11.82) and the Simone de Beauvoir footbridge (Figure 11.83). The owners of the Léopold Sédar-Senghor footbridge are two French ministries (Ministry of Environment and Ministry of Culture). Its main span is 106 m, its total length is 140 m, and its deck width varies from 11 m to 15 m. This footbridge is well known because on this bridge the dynamic effects of pedestrian traffic (vertical and horizontal vibrations) were discovered for the first time in France (similar problems were discovered with the Millenium footbridge in London). The Simone de Beauvoir footbridge belongs to Paris City Hall. Completed in 2006, its main span is 190 m long, its total length is 304 m, and its width is 12 m. It was designed in accordance with the Eurocodes.

FIGURE 11.82   Léopold Sédar-Senghor footbridge over the Seine River, Paris, 1999. (Courtesy of Mr. Ramondenc.)

FIGURE 11.83   Simone de Beauvoir footbridge over the Seine River, Paris. (Courtesy of JAC.)

FIGURE 11.84   The Mornas Bridge over the Rhône River (HSL South of France, 1999, length 121.4 m). (Courtesy of SNCF, photo J.-J. d'Angelo.)

FIGURE 11.85   The Garde-Adhémar Bridge over the Rhône River (HSL South of France, 2000, length 325 m). (Courtesy of SNCF, photo J.-J. d'Angelo.)

Other steel arch bridges built recently are often intermediate between arch bridges and bow strings. Figures 11.84 and 11.85 show the Mornas Bridge and the Garde-Adhemar Bridge. These two bridges were designed by SNCF for the high-speed railway line in the south of France.

11.6.4 Portal Bridges

Like arch bridges, portal bridges can be an elegant solution to crossing deep valleys. The deck of modern portal bridges is in general a three-span continuous girder made of prestressed concrete supported by two inclined piers (Figure 11.86), but various types of decks may be envisaged. An interesting bridge with double inclined legs is shown in Figure 11.87. This bridge was built a few years ago, crosses the Grand Canal du Havre for the A29 Motorway, and is not far from the Normandy Bridge.

FIGURE 11.86 Example of portal bridge (bridge over the Truyère River at Garabit on the A75 Motorway). (Courtesy of SETRA.)

FIGURE 11.87 Portal bridge carrying the A29 Motorway and crossing the Grand Canal du Havre. (Courtesy of JAC.)

The bridge over the Truyère River at Garabit was completed in 1993 and carries the A75 Motorway. The owner is the French Republic, Ministry of Environment. Its total length is 308 m, and the main span length is 144 m with a deck width of 20.5 m. The Grand Canal Bridge at Le Havre was completed in 1994, carrying the A29 Motorway. It was built by Cimolai Costruzioni Metalliche and the owner is Scetauroute. Its total length is 1410 m and the main span length is 275 m.

11.7 Cable-Stayed Bridges

11.7.1 General

The idea to support a beam with inclined ropes or chains hanging from a mast is very old—the Egyptians used such stays for their sailing ships. But the double role of stays supporting the bridge deck by vertical tension component and compressing the bridge deck by horizontal compression was used for the first time in the case of the aqueduct of Tempul, in Spain, built in 1926 by E. Torroja. In 1946, the famous French engineer Albert Caquot developed these ideas to build the first modern cable-stayed road bridge with a reinforced concrete deck over a canal near the hydroelectric plant of Donzère-Mondragon. The design and construction of modern cable-stayed bridges quickly developed after the Second World War, at first with steel decks and then, thanks to R. Morandi, with concrete decks.

Originally, all cable-stayed bridges were designed with stiff decks and a limited number of stays of high resistance. The major evolution came from the development of multistay systems: these systems allowed better control of efforts with the help of advanced computer software, and allowed an application to the design of bridges with prestressed concrete decks easily built by the cantilever method. The multistay system has aesthetic properties in the transparency which it confers to the structure. And finally, more stay cables allow more slender girders.

The application of cable-stayed bridges has been expanding and overtaking the use of suspension bridges for very long span lengths; the limitation in the increase of their main span length is only a matter of aerodynamic stability (Figure 11.88). The three most important cable-stayed bridges in France are the Brotonne Bridge, the Normandy Bridge, and the Millau viaduct. Two other bridges should be added: the Rion-Antirion Bridge, designed and constructed by a French company in Greece, and the recent Térénez Bridge.

The Brotonne Bridge, crossing the Seine River downstream from Rouen, opened to traffic in 1977. The deck is a prestressed concrete bow girder 3.97 m deep and 19.2 m wide. The center span length is 320 m long. It was designed by J. Muller and J. Mathivat. The Normandy Bridge crosses the Seine River near its estuary. It is a multi-cable-stayed bridge built between 1989 and 1995. Its owner is the Chambre de Commerce et d'Industrie du Havre and it carries the French Motorway A29. The main span length is 856 m, including a part of prestressed concrete cross section (116 m), a part of steel cross section (624 m), and another part of concrete cross section (116 m) (Figure 11.89).

The Millau viaduct (Figure 11.90) was built in 2001–2004. It is located near Millau in south of France. It carries the French Motorway A75 and is a multispan cable-stayed bridge, multicable with a fan arrangement. The conceding authority is the French Republic. Its most important features are that the highest pylon is 343 m and the main span lengths are 342 m. The superstructure is a steel box girder (Figure 11.91) with a total length of 2460 m.

The Rion-Antirion Bridge (Harilaos Trikoupis Bridge) was designed and built in Greece between 1999 and 2004 by the French company VINCI Construction. It is a multiple-span cable-stayed bridge located between Rion, Achaea, West Greece and Antirion, Aitolia-Acernania, West Greece. It belongs to the Republic of Greece. Its total length is 2880 m and includes three main spans 560 m long and two side spans 286 m long. The superstructure is a steel–concrete composite girder of 27.2 m width

FIGURE 11.88 Millau viaduct—example of aerodynamic study in a laboratory for checking stability during the launching phase. (Courtesy of CSTB).

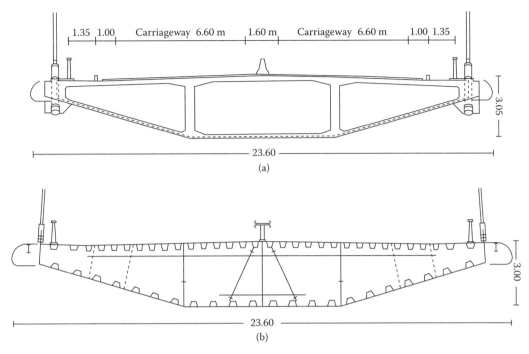

FIGURE 11.89 Cross sections of the Normandy Bridge: (a) cross section of the prestressed concrete part of the deck and (b) cross section of the steel part of the deck.

FIGURE 11.90 General view of the Millau viaduct (Courtesy of JAC.)

and 2.82 m depth. The specific design difficulty for this bridge was due to environmental characteristics: wind, earthquakes, tectonic movements, sea depth, ship collision, and so on (Figure 11.92). The Térénez road bridge, located in Bretagne, crosses the Aulne River. It opened to traffic in 2011. It is a cable-stayed bridge with a curved superstructure and a semi-fan arrangement of stays. Its main span is 285 m long (Figure 11.93).

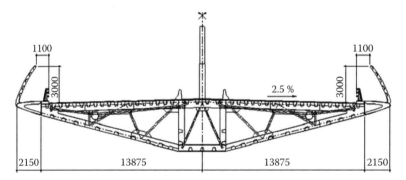

FIGURE 11.91 Cross section of the Millau viaduct.

FIGURE 11.92 View of the Harilaos Trikoupis Bridge in Greece, steel–concrete composite deck during execution. (Courtesy of JAC.)

FIGURE 11.93 Térénez Bridge over the Aulne River. (Courtesy of SETRA, photo Gérard Forquet.)

11.7.2 Design of the Stay System

Symmetric three-span bridges constitute the most numerous family of cable-stayed bridges. The ratio between the main and side spans, which has a significant influence on the stress variations in the side span and backstay cables, is nearby 0.40 in most cases. The shape of the pylons is essentially governed by the width of the bridge deck, the height of the deck above ground or water level, and the choice of the cable system (cables in a single plane or in two planes).

The choice of a cable system, which directly influences the design of the pylons, is not only made on a rational basis. It is the result of technical as well as aesthetic arguments. Considering the complexity of the problem and the variety of possible solutions, only some general ideas are expressed. The solution of cables in a single plane is possible if the traffic way includes a central reservation where a pylon with a unique mast may be erected. In that case, the bridge superstructure needs a proper torsional stiffness to resist the torsional moments due to an unsymmetrical loading; as a consequence, the superstructure deck should not be too wide (less than 20 m, to give an order of magnitude). Bridges with single-plane stays are of high aesthetic quality, compared to a double plane, and avoid any nonaesthetic optical crossing of cables. The presence of a slender central mast also confers to the bridge an interesting feeling of lightness. Cables in two planes are an appropriate system:

- For medium-span bridges with decks of moderate width (for example, less than 15 m), the pylons are two independent vertical masts, and the cables are in two quasi-vertical planes with a fan-shaped, semi-fan-shaped, or even a harp-shaped configuration. For wide decks, the pylons can be two vertical masts connected by a transverse member or, more frequently, A-shaped towers.
- For long-span bridges to benefit from the natural rigidity due to two planes of cables (to resist asymmetrical loads as well as to ensure aerodynamic stability, which becomes determining in that case), the pylon, of a large height, is almost systematically in the shape of a reversed Y, and stays are anchored in the vertical part of the Y (Figure 11.94).

A harp-shaped cable-stay system is very harmonious and is especially convenient for stays in a single plane or in two planes. A semi-fan shape is the most common system for medium- to very-long-span bridges with A- or Y-shaped towers.

11.7.3 Design of the Bridge Superstructure

With the almost systematic adoption of the multistay system (for economic reasons and to facilitate maintenance, i.e., the replacement of one or several stays is rather easy), the requirements of rigidity with regard to bending moment in the vertical plane of the bridge superstructure do not govern the design. This evolution has facilitated the development of bridges with concrete or steel–concrete composite superstructures. As a general rule, the design of the bridge superstructure is governed by the resistance to transverse bending, by the resistance to concentrated forces in the zone of stay anchors and, in the case of bridge superstructures with axial compression, by the limitation of torsional deformations due to unsymmetrical loading. For very-long-span bridges, the geometry of the bridge superstructure deck depends also on conditions of aerodynamic stability.

The use of concrete superstructures (self-weight in the range 10 to 15 kN/m²) can probably be extended to about 500 m. Steel and concrete composite superstructures (self-weight in the range 6.5 to 8.5 kN/m²) should be an interesting solution in a range of span lengths from 300 m to 600 m. Orthotropic steel superstructures (self-weight in the range 2.5 to 3.5 kN/m²) allow the longest span lengths (probably up to 1500 m or 1800 m).

There is no mathematical rule to design the connection of the bridge superstructure to the towers. A fixed liaison is reserved for big bridges with concrete superstructures with axial suspensions, whereas the vertical simple support or the total suspension (with blocking of the horizontal movement) are adopted in the case of bridge superstructures with cables in two planes. In France, some short-span

FIGURE 11.94   Harilaos Trikoupis Bridge in Greece, view of the pylons from the highway. (Courtesy of JAC.)

FIGURE 11.95   Example of a slab bridge transformed into a cable-stayed bridge (Motorway A6). (Courtesy of JAC.)

solid slab bridges over motorways have been in the past converted into cable-stayed bridges to widen the motorway platform (removal of intermediate piers). Figure 11.95 gives an example of such a slab bridge converted into a cable-stayed bridge. This kind of technique was used a few times, but it turned out to be expensive. The best solution is probably to replace bridges that do not have the appropriate dimensions for the evolution of traffic and needs.

11.8 Suspension Bridges

The most common suspension bridges are symmetric three-span bridges with continuous parabolic suspension cables anchored in independent massive anchorages. In the past many suspension bridges were built. The biggest French modern suspension bridge is the Tancarville Bridge (see Figure 11.7). The bridge superstructure is hung on, in an almost continuous way, by suspenders on a pair of parabolic suspension cables. Under the influence of direct vertical loads, the deck tightens the suspenders and these suspenders transfer the loads to the suspension cables by tightening them in turn. For obvious reasons of limitation of self-weight, and considering the range of span lengths covered by this type of construction work, the superstructure is always made of steel. The superstructure is often a truss or of tubular shape and its aerodynamic stability is carefully studied.

The cost of suspension bridges depends widely on that of the massive abutments for the anchoring of the suspension cables, which can be gigantic if the foundation ground is not good rock. To reduce this cost, some engineers had the idea to anchor suspension cables directly on the bridge superstructure at its ends. However, such a design is not convenient for two major reasons: first, the bridge superstructure is significantly compressed by the anchored cables, which is not beneficial for a steel superstructure. Secondly, while in bridges with massive abutments the step-by-step erection of the bridge superstructure deck is relatively easy by using the suspension cables, a general scaffolding is needed to support the bridge superstructure during its preparation until the anchoring of cables or the use of temporary external anchoring cables, which may be used as elevated cableways until the end of construction.

In France, there are few suspension bridges due to the smaller dimensions of crossings and because cable-stayed bridges are financially more interesting. Nevertheless, some suspension bridges have been designed for aesthetic reasons, like the Chavanon viaduct (Figure 11.96). The Chavanon viaduct was built in 1997–2000 over the Chavanon Valley in the center of France. It is a monocable suspension bridge, and the superstructure is a steel–concrete composite box girder and carries the French Motorway A 89. The owner is the company Autoroutes du Sud de la France, and the main span is 300 m (total length 360 m).

FIGURE 11.96 The Chavanon viaduct. (Courtesy of EGIS.)

Abbreviations

CPD: Construction Products Directive 89/106/EEC, as amended by CE Marking Directive 93/68/EEC. It is now replaced by the Construction Products Regulation, adopted by the European Parliament and the Council on January 18, 2011 and implemented in 2013.

PPD: Public Procurement Directive. The Guidance Paper L of the Commission refers to the Council Directive 93/37/EEC of June 14, 1993 concerning the coordination of procedures for the award of public works contracts.

EN: European standard.

EN: Eurocode Version of Eurocode approved by CEN as a European standard.

NDP: Nationally determined parameter.

CEN: European Standardization Organization (Comité Européen de Normalisation).

NSB: National Standards Body (CEN Member).

EOTA: European Organization for Technical Approval (article 9.2 of the CPD).

ETA: European Technical Approval.

JRC: Joint Research Center, ISPRA, Italy.

LCPC: Central Laboratory of Bridges and Roads (Laboratoire Central des Ponts et Chaussées). The name has been recently changed to IFSTTAR, French Institute of Science and Technology for Transport, Development and Networks (Institut Français des Sciences et Technologies des Transports, de l'Aménagement et des Réseaux).

SETRA: Technical Department for Transport, Roads and Bridges Engineering and Road Safety, French Ministry of Ecology, Sustainable Development and Energy (Service d'études sur les Transports, les Routes et leurs Aménagements).

UIC: International Union of Railways (Union Internationale des Chemins de Fer).

SLS: Serviceability limit state.

ULS: Ultimate limit state.

References

The Eurocode suite

EN 1990:2002 Eurocode, Basis of structural design
EN 1990:2002/A1:2005 Eurocode, Basis of structural design, Application for bridges (Annex A2)

Eurocode 1: Actions on Structures

EN 1991-1-1:2002 Eurocode 1: Actions on structures, Part 1-1: General actions, Densities, self-weight, imposed loads for buildings
EN 1991-1-2:2002 Eurocode 1: Actions on structures, Part 1-2: General actions, Actions on structures exposed to fire
EN 1991-1-3:2003 Eurocode 1: Actions on structures, Part 1-3: General actions, Snow loads
EN 1991-1-4:2005 Eurocode 1: Actions on structures, Part 1-4: General actions, Wind actions
EN 1991-1-5:2003 Eurocode 1: Actions on structures, Part 1-5: General actions, Thermal actions
EN 1991-1-6:2005 Eurocode 1: Actions on structures, Part 1-6: General actions, Actions during execution
EN 1991-1-7:2006 Eurocode 1: Actions on structures, Part 1-7: General actions, Accidental actions
EN 1991-2:2003 Eurocode 1: Actions on structures, Part 2: Traffic loads on bridges
EN 1991-3:2006 Eurocode 1: Actions on structures, Part 3: Actions induced by cranes and machinery
EN 1991-4: 2006 Eurocode 1: Actions on structures, Part 4: Silos and tanks

Eurocode 2: Design of Concrete Structures

EN 1992-1-1:2004 Eurocode 2: Design of concrete structures, Part 1-1: General rules and rules for buildings

EN 1992-1-2:2004 Eurocode 2: Design of concrete structures, Part 1-2: General rules, Structural fire design

EN 1992-2:2005 Eurocode 2: Design of concrete structures, Part 2: Concrete bridges, Design and detailing rules

EN 1992-3:2006 Eurocode 2: Design of concrete structures, Part 3: Liquid retaining and containment structures

Eurocode 3: Design of Steel Structures

EN 1993-1-1:2005 Eurocode 3: Design of steel structures, Part 1-1: General rules and rules for buildings

EN 1993-1-2:2005 Eurocode 3: Design of steel structures, Part 1-2: General rules, Structural fire design

EN 1993-1-3:2006 Eurocode 3: Design of steel structures, Part 1-3: General rules, Supplementary rules for cold-formed members and sheeting

EN 1993-1-4:2006 Eurocode 3: Design of steel structures, Part 1-4: General rules, Supplementary rules for stainless steels

EN 1993-1-5:2006 Eurocode 3: Design of steel structures, Part 1-5: General rules, Plated structural elements

EN 1993-1-6:2007 Eurocode 3: Design of steel structures, Part 1-6: Strength and stability of shell structures

EN 1993-1-7:2007 Eurocode 3: Design of steel structures, Part 1-7: Strength and stability of planar plated structures subject to out of plane loading

EN 1993-1-8:2005 Eurocode 3: Design of steel structures, Part 1-8: Design of joints

EN 1993-1-9:2005 Eurocode 3: Design of steel structures, Part 1-9: Fatigue

EN 1993-1-10:2005 Eurocode 3: Design of steel structures, Part 1-10: Material toughness and through-thickness properties

EN 1993-1-11:2006 Eurocode 3: Design of steel structures, Part 1-11: Design of structures with tension components

EN 1993-1-12:2007 Eurocode 3: Design of steel structures, Part 1-12: General, High strength steels

EN 1993-2:2006 Eurocode 3: Design of steel structures, Part 2: Steel bridges

EN 1993-3-1:2006 Eurocode 3: Design of steel structures, Part 3-1: Towers, masts and chimneys, Towers and masts

EN 1993-3-2:2006 Eurocode 3: Design of steel structures - Part 3-2: Towers, masts and chimneys, Chimneys

EN 1993-4-1:2007 Eurocode 3: Design of steel structures, Part 4-1: Silos

EN 1993-4-2:2007 Eurocode 3: Design of steel structures, Part 4-2: Tanks

EN 1993-4-3:2007 Eurocode 3: Design of steel structures, Part 4-3: Pipelines

EN 1993-5:2007 Eurocode 3: Design of steel structures, Part 5: Piling

EN 1993-6:2007 Eurocode 3: Design of steel structures, Part 6: Crane supporting structures

Eurocode 4: Design of Composite Steel and Concrete Structures

EN 1994-1-1:2004 Eurocode 4: Design of composite steel and concrete structures, Part 1-1: General rules and rules for buildings

EN 1994-1-2:2005 Eurocode 4: Design of composite steel and concrete structures, Part 1-2: General rules, Structural fire design

EN 1994-2:2005 Eurocode 4: Design of composite steel and concrete structures, Part 2: General rules and rules for bridges

Eurocode 5: Design of Timber Structures

EN 1995-1-1:2004 Eurocode 5: Design of timber structures, Part 1-1: General, Common rules and rules for buildings
EN 1995-1-2:2004 Eurocode 5: Design of timber structures, Part 1-2: General, Structural fire design
EN 1995-2:2004 Eurocode 5: Design of timber structures, Part 2: Bridges

Eurocode 6: Design of Masonry Structures

EN 1996-1-1:2005 Eurocode 6: Design of masonry structures, Part 1-1: General rules for reinforced and unreinforced masonry structures
EN 1996-1-2:2005 Eurocode 6: Design of masonry structures, Part 1-2: General rules, Structural fire design
EN 1996-2:2006 Eurocode 6: Design of masonry structures, Part 2: Design considerations, selection of materials and execution of masonry
EN 1996-3:2006 Eurocode 6: Design of masonry structures, Part 3: Simplified calculation methods for unreinforced masonry structures

Eurocode 7: Geotechnical Design

EN 1997-1:2004 Eurocode 7: Geotechnical design, Part 1: General rules
EN 1997-2:2007 Eurocode 7: Geotechnical design, Part 2: Ground investigation and testing

Eurocode8: Design of Structures for Earthquake Resistance

EN 1998-1:2004 Eurocode 8: Design of structures for earthquake resistance, Part 1: General rules, seismic actions and rules for buildings
EN 1998-2:2005 Eurocode 8: Design of structures for earthquake resistance, Part 2: Bridges
EN 1998-3:2005 Eurocode 8: Design of structures for earthquake resistance, Part 3: Assessment and retrofitting of buildings
EN 1998-4:2006 Eurocode 8: Design of structures for earthquake resistance, Part 4: Silos, tanks and pipelines
EN 1998-5:2004 Eurocode 8: Design of structures for earthquake resistance, Part 5: Foundations, retaining structures and geotechnical aspects
EN 1998-6:2005 Eurocode 8: Design of structures for earthquake resistance, Part 6: Towers, masts and chimneys

Eurocode 9: Design of Aluminum Structures

EN 1999-1-1:2007 Eurocode 9: Design of aluminium structures, Part 1-1: General structural rules
EN 1999-1-2:2007 Eurocode 9: Design of aluminium structures, Part 1-2: Structural fire design
EN 1999-1-3:2007 Eurocode 9: Design of aluminium structures, Part 1-3: Structures susceptible to fatigue
EN 1999-1-4:2007 Eurocode 9: Design of aluminium structures, Part 1-4: Cold-formed structural sheeting
EN 1999-1-5:2007 Eurocode 9: Design of aluminium structures, Part 1-5: Shell structures

Further Reading

General

Calgaro, J. A. June 26–28, 2006. "Recent trends in the design of bridges, The French experience," (Keynote Lecture), *First International Conference on Advances in Bridge Engineering*, Brunel University West London.

Calgaro, J. A. June 21–24, 2009. "Assessment of existing bridges: General aspects and safety principles," (Keynote Lecture), *2009 Conference on Protection of Historical Buildings*, Rome, Italy.

Gaumy, J., Miquel, P., Autissier, I., Blanc, G., Léotard, Ph., Sella, Ph., Rykiel, S., Orsenna, E. and Lemoine, B. 1998. *Views and Visions: Art of Major Constructions*. Le Patio Editeur, Saints, France.

Calibration of traffic loads

Bruls, A. 1996. "Resistance of Bridges carrying road traffic, Load models, Assessment of Existing Bridges," Doctoral thesis, University of Liège, Belgium.

Bruls, A., Calgaro, J. A., Mathieu, H. and Prat, M. March 27–29, 1996. "ENV 1991, Part 3: Traffic loads on bridges, The main models of traffic loads on road bridges, background studies," *IABSE Colloquium - Basis of Design and Actions on Structures*, Delft, Netherlands.

Bruls, A., Croce, P., Sanpaolesi, L. and Sedlacek, G. March 27–29, 1996. "ENV 1991, Part 3: Traffic loads on bridges, Calibration of load models for road bridges," *IABSE Colloquium, Basis of Design and Actions on Structures*, Delft, Netherlands.

Calgaro, J. A., Tschumi, M. and Gulvanessian, H. 2010. *Designers' Guide to Eurocode 1: Actions on Bridges, EN 1991-2, EN 1991-1-1, -1-3 to -1-7 and EN 1990 Annex A2*, Telford, London.

Cantieni, R. 1992. *Dynamic Behavior of Highway Bridges Under the Passage of Heavy Vehicles*. EMPA (Swiss Federal Laboratories for Materials Testing and Research), Dübendorf.

Conti, E. and Fouré, B. 1994. *External Prestressing in Structures*. French Association for Bridges and Structural design [Association française des Ponts Charpentes].

Croce, P. March 27–29, 1996. "Vehicle interactions and fatigue assessment of bridges," *IABSE Colloquium, Basis of Design and Actions on Structures*, Delft, Netherlands.

Flint, A. R. and Jacob, B. March 27–29, 1996. "Extreme traffic loads on road bridges and target values of their effects for code calibration," *IABSE Colloquium, Basis of Design and Actions on Structures*, Delft, Netherlands.

Gandil, J., Tschumi, M. A., Delorme, F. and Voignier, P. March 27–29, 1996. "Railway traffic actions and combinations with other variable actions," *IABSE Colloquium, Basis of Design and Actions on Structures*, Delft, Netherlands.

Grundmann, H., Kreuzinger, H. and Schneider, M. 1993. "Vibration tests for pedestrian bridges," *Civil* 68: 215–225.

Jacob, B. and Kretz, T. March 27–29, 1996. "Calibration of bridge fatigue loads under real traffic conditions," *IABSE Colloquium, Basis of Design and Actions on Structures*, Delft, Netherlands.

Mathieu, H., Calgaro, J. A. and Prat, M. 1989. *Final Report to the Commission of the European Communities on Contract Nr. PRS/89/7750/MI 15, Concerning Development of Models of Traffic Loading and Rules for the Specification of* Bridge Loads.

Mathieu, H., Calgaro, J. A. and Prat, M. 1991. *Final Report to the Commission of the European Communities on contract Nr. PRS/90/7750/RN/46, Concerning Development of Models of Traffic Loading and Rules for the Specification of* Bridge Loads.

Merzenich, G. and Sedlacek, G. 1995. *Background Document to Eurocode 1, Part 3, Traffic Loads on Road Bridges*. Federal Ministry road traffic, Research for construction of Roads and road circulation, Document 711.

Prat, M. 1997. "The use of the road traffic measurements in bridge engineering—WAVE (Weighing in motion of Axles and Vehicles for Europe)," *Proceedings of the Mid-Term Seminar*, Delft.

Prat, M. and Jacob, B. 1992. "Local load effects on road bridges," *Third International Symposium on Heavy Vehicle Weights and Dimensions*, June 28 - July 2, Cambridge, UK.

Ricketts, N. J. and Page, J. 1997. "Traffic data for highway bridge loading," Transportation Research Laboratory *Report* 251.

Rolf, F. H. and Snijder, H. H. March 27–29, 1996. "Comparative research to establish load factors for railway bridges," *IABSE Colloquium, Basis of Design and Actions on Structures*, Delft, Netherlands.

SIA: Swiss Society of Engineers and Architects [Société suisse des Ingénieurs et des Architectes] 261, SN 505 261. 2003. *Actions on Structures*, Zürich.

TRL. 1999. "Post-tensioned concrete bridges," Highways Agency (TRL), SETRA, LCPC. Thomas Telford.

UIC Code 700. November 2004. *Classification of lines. Resulting load limits for wagons*, 10th edition.

UIC Code 702. March 2003. *Static loading diagrams to be taken into consideration for the design of rail carrying structures on lines used by international services*, 3rd edition.

Vrouwenvelder, A. and Waarts, P. H. 1991. "Traffic loads on bridges: Simulation, extrapolation and sensitivity studies," TNO Building and Construction Research, Report b-91-0477.

Books on bridges

CEVM. 2005. The Millau Viaduct—La viaduc de Millau, *The Millau Viaduct Portfolio* (English/French). Compagnie Eiffage du Viaduc de Millau, Montreuil, France.

Désveaux D. 2008. The bet of impossible: the Tamarins road in La Réunion Island. *Le pari de l'impossible: La route des Tamarins à La Réunion*. Presses des Ponts et Chaussées Paris, France.

Montens, S. 2001. The most beautiful bridges in France. Bonneton.

Relevant websites

http://en.structurae.de/structures
http://eurocodes.jrc.ec.europa.eu
http://www.eurocodes.co.uk

12

Bridge Engineering in Greece

Stamatios
Stathopoulos
DOMI S.A.

12.1 Introduction

12.1.1 Geographical Characteristics

Greece achieved its independence from the Ottoman Empire in 1829. During the second half of the nineteenth and the first half of the twentieth century, the liberation of almost all Greek territories was gradually completed. Greece was involved in both World Wars; during World War II it was occupied by Germany, Italy, and Bulgaria, resulting in the total destruction of its infrastructure. In 1981 Greece joined the European Union and in 2001 it became the twelfth member of the European Economic and Monetary Union.

Greece is located in southeastern Europe, bordering Albania (282 km), Fyrom (246 km), Bulgaria (494 km), and Turkey (206 km); for its major part it is bordered by the Mediterranean and more specifically by the Ionian and Aegean Seas, with a coastline of 13,676 km. The total area of Greece amounts to 131,957 km² (land 130,647 km²; water 1,310 km²).

With respect to the natural hazards, the country is characterized by high seismicity and locally by intense tectonic movements. Many severe earthquakes have occurred during the second half of the twentieth century with a magnitude of up to 7.5 degrees on the Richter scale; among them two (1981 and 1999) have afflicted the capital of Athens and one (1978) affected the second-largest city of the country, Thessalonica. In earlier years, earthquakes up to 8.2 on the Richter scale have been reported.

The population numbers approximately 10.8 million people with a growth rate equal to 0.127% (2009 estimation); 61% of the total population is urban. To a great degree Greece is mountainous; the bulky mountain chain of Pindus, which is the end of the Alps in the Balkans, stretches from north to south. Large rivers exist mainly in northern and northeastern Greece (Aliakmon, Axios, Strimonas, Nestos, Evros) and less in western Greece (Acheloos, Arachthos, etc.).

The road network of the country is today satisfactory, amounting to ~117,500 km, of which about 108,000 km are paved. Six motorways (Figure 12.1) are included in the road network; some of them are already completed while the remaining ones are being improved, with an estimated completion year of 2015. On the contrary, the rail network (Figure 12.1) is limited. It is about 2500 km in length, of which almost 1000 km are of old technology; the remaining 1500 km are continuously being improved and electrified (~750 km).

12.1.2 Historical Development

12.1.2.1 Ancient Greek Period (Twelfth Century BC–31 BC)

The lack of large rivers and the city/state (polis) system which characterized ancient Greece did not encourage the growth of bridge construction, at least to the same extent that temple construction was developed; it is, however, certain that small- and medium-sized bridges existed, about which many historical reports as well as ruins exist. In Crete, small stone culverts and small bridges of the Minoan period have been detected, such as the one on the southern side of the palace of Knossos on the river Elysian, constructed between 2000 BC and 1450 BC (three spans 3.1 m/3.25 m/2.3 m, free height 5 m up to 5.5 m) and small posterior bridges like the one in Eleftherna (total length 9 m, span 3.85 m, width 5.35 m, and height 4.6 m, constructed in the fourth century BC) (Figure 12.2).

Peloponnese, the powerful Mycenaean state, developed its road network, constructing bridges over the various streams of its territory with small stone culverts. Almost 20 such small structures have been located, some of which exist today in good condition, such as the bridge (culvert) of Kazarma (length 12 m, width 5 m, height 6–7 m, constructed ~1200 BC) (Figure 12.3). The ruins of these bridges testify to the know-how and the constructional capacity of the Mycenaean bridge constructors.

The common characteristic of the Minoan and Mycenaean bridges was corbelling (Figure 12.4) with coarsely processed bulky stones (first version of the current cantilevering) and the cover of the free span with continuously converging walls (abutments). The gap between the walls, provided that it was not null (e.g., Eleftherna), was covered by a horizontal stone beam (e.g., Drakonera Bridge, northwest of the citadel of Mycenae) or by a wedge-like stone (e.g., Kazarma Bridge). This type does not appear to be developed further in classic Greek years and was possibly abandoned.

In the classic years three main bridge construction systems particularly were developed:

- Bridges with stone piers and decks with wooden beams: The piers were constructed with big local limestones, of thickness up to 2 m, perfectly sculpted and structurally joined, while the superstructures were wooden beams (Figure 12.5). A representative bridge of this type, which partially exists now, is the Mavrozoumaina Bridge (Figure 12.6), near ancient Messina, at the southwestern end of Peloponnese. The bridge was constructed at the junction of two rivers in the fourth century BC. It is exceptionally interesting from a technical point of view, shaped like a "Y" in the ground plan, with arm length almost 20 m and clear spans between the piers 5–7 m. The whole structure reveals exceptional experience in the art of stonemasonry. The present-day superstructure consists of nine semicircular arches of the Roman type; it was constructed on the old footings

FIGURE 12.1 Road and railway network of Greece. (Courtesy of ADT-OMEGA SA.)

FIGURE 12.2   Bridge in Eleftherna of Crete, with boulders of limestone; the gap at the crown is null [1], [2], [30]. (Courtesy of G. D. Makris.)

FIGURE 12.3   Downstream view of Kazarma Bridge, raw limestones [1], [2], [30]. (Courtesy of G. D. Makris.)

much later during the Ottoman period and is rather clumsy. The old deck must have consisted of wooden beams, at least at one part of it. This type of bridge is related to the most important bridge on the Euphrates (near Babylon), constructed by Queen Semiramis or King Nebuchadnezzar; it is very likely that the know-how came to the Greeks from the Assyrians.

• Bridges with stone piers and decks with stone beams: This type of bridge was built over small- and intermediate-width waterways, with small free span between the piers due to the material of the deck. A representative example of such a bridge was that in Asia Minor Assus, on the northern coast of the Adramyttian Gulf (Figure 12.7). It is dated fourth century BC and constitutes the most representative sample of Greek bridges in pre-Roman years. The existing runes, of length 52 m,

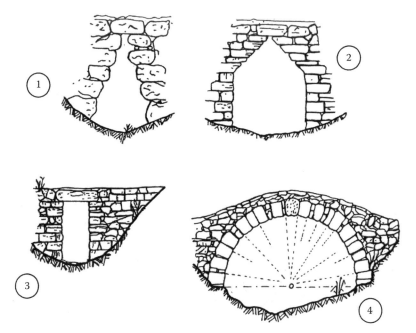

FIGURE 12.4 Basic types of stone bridges [1], [2]: (1) with continuously converging piers with the corbelling system, with or without gap at the crown; (2) with constant piers up to a specific height and null key (Eleftherna); (3) with constant piers and cover of the span with horizontal beams (Mavrozoumaina); (4) with arched construction (Samothrace, Eretria). (Courtesy of G. D. Makris.)

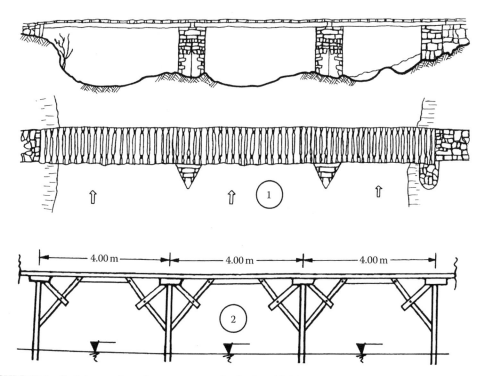

FIGURE 12.5 Basic types of wooden or stone wooden bridges [1], [2]: (1) one or many spans, with stone piers and wooden beams (Assus); (2) one or many totally wooden spans for the bridging of big rivers (Amphipolis). (Courtesy of G. D. Makris.)

FIGURE 12.6 General view of the triple bridge of Mavrozoumaina according to gravure [1]. The inlaid picture shows the still-existing parts of the ancient piers (fourth century BC). (Courtesy of G. D. Makris.)

FIGURE 12.7 Assus Bridge [4], [5], [30]. (Courtesy of Technical Chamber of Greece.)

bridge lightly skew (8°) the paved riverbed. The deck of the bridge was composed of four to six simply supported stone beams of width 0.44–0.46 m (three still exist at the first span and four at the second span), height 0.34 m and length 2.70–3.73 m connected between them through wooden small shear connectors. It was seated on 15 piers (six of which still exist), constructed in stone. The piers had an intense hydrodynamic (rhomboid) form and dimensions of 3.6 m length, 1 m thickness, and over 2.6 m height. A span of the bridge (in the center) was probably bridged with wooden beams, which could be removed more easily in case of hostilities.

- Bridges totally wooden, seated on wooden piles: This type of bridge was implemented on large rivers, where there was a need to reduce the number of the piers with an increase of the free span between them. A representative sample of such bridges is the Amphipolis Bridge (Figure 12.8), in the estuaries of the Strimonas River, of length 275 m. The bridge is reported in Thucydides (d. 103), on the occasion of the battle between the Spartans and Athenians that took place in the region in 422 BC, without however any technical description. The excavations revealed 77 oak piles, of circular and square sections, which had been vertically driven in the sandy ground of the left bank of the river. The peaks of the piles were sharp (obviously in order to facilitate the driving) and often had iron noses (similar to what was found in the Roman Bridge on the Rhine). The piles were driven with percussion, aided by a dropped weight from a small height; 35–40 piers existed in total. The deck consisted of wooden piers, bridging the spans between the piers and could be easily dismantled during wars.

Other bridges also existed on the Strimonas river, from the era of the Greco-Persian wars, whose ruins no longer exist, according to Herodotus, who reports that the army of Xerxes passed by the bridges of Strimonas. The end of the classic bridge of Amphipolis is not known; we know that a new arched stone bridge was constructed at the same location in the first century AD, during the era of the Roman Empire.

Beyond the aforementioned three systems, an impressive arched three-center bridge was built in Rhodes Island (Figure 12.9). For unknown reasons, the Greeks hesitated in widely implementing this technique, which a little later was perfected by the Romans and allowed the bridging of large spans. It should be noted that, in parallel with fixed bridges, the technique of crossing large waterways with off-hand means, mainly for military reasons, was exceptionally developed.

Herodotus (VII 36) [29] certified that the Persian armies passed the Bosporus and Hellespont at least twice (Darius in the expedition to Dacia, 513–512 BC, and Xerxes in the expedition to Greece, 481–480 BC) on bridges consisting of boats in contact, anchored into the sea ground and connected between them with ropes. The Greeks Mandrocles and Arpalus are reported as the architects and engineers of these bridges. Both structures can be characterized as the first links of sea straits in the history (the length of the Hellespont bridge is ~1580 m).

FIGURE 12.8 Wooden piles of the ancient bridge on the Strimonas River in Amphipolis, as they are today [1], [6]. (Courtesy of G. D. Makris.)

FIGURE 12.9 The arched three-center bridge in Rhodes (end of fourth century BC [1]). (Courtesy of G. D. Makris.)

Later, on the crusade of Alexander the Great, in his course to India he was obliged to pass through the large and impetuous rivers of Afghanistan and Pakistan under martial conditions. Arrianus reports the construction of a navigable bridge on Oxos River, the length of six stadiums (~1100 m), as well as a technique using wooden barges, aided by inflated airbags so that they become unsinkable; according to an old, unconfirmed Italian manuscript, Alexander also constructed a fixed wooden bridge on the Euphrates with nails and iron chains.

12.1.2.2 Roman and Byzantine Period (31 BC–1453 AD)

The Roman Peace (pax romana) and the military needs of the empire made provision for or imposed the construction of a large number of roads and bridges. The bridges of the Roman period are mainly stone, constructed with the classic semicircular arch. These bridges were to a great degree the product of Greek craftsmen and designers, such as Apollodorus of Damascus, the official architect of Emperors Trajan and Adrian, who constructed inter alia the large Roman Bridge on the Danube (105 AD) and probably the Pantheon of Rome.

A more characteristic example of bridge construction in the Roman period is the multiarched bridge in Kleidi, which bridged the Loudias River (Figure 12.10). According to the testimonies of observers, the bridge had a total length of 190 m, with 8–10 arches, only one of which exists today. The bridge was built with special limestones, the so called "tholitis",irreproachably sculpted and assembled. The free span of each arch was 17 m, by no means negligible for this period and comparable to the spans of big Roman bridges (25–30 m). It is obvious that the construction of such a bridge required an adequate and well-designed formwork, capable of receiving the weight of the stones during the construction phase; the formwork was formed for one span and was reused for the subsequent ones.

Among the remaining bridges of this period are the multiarched bridge in the strait of Leucas Island, of supposed length of 700 m, the one-arched bridge in Xirokampi of Sparta (it still exists in a very good condition, see Figure 12.11) with a semicircular arch of diameter 7.5 m, the four-arched bridge in Eleusinian Cephisus (still in existence) constructed with skillfully sculpted Piraeus tufa and Roman pozzolan as mortar, and others. Cast stonemasonry, a type of early cement (opus caementicium) in the core of the structures and the use of pozzolan as an additive allowed the growth of large free spans (e.g., the bridge in Kleidi) as masonry units, orthogonal stones (opus quadratum), or bricks (opus testaceum) were used.

FIGURE 12.10 The still-existing arch of the multiarched Roman bridge on the Loudias River. (Courtesy of the Center of Study of Stone Bridges.)

FIGURE 12.11 One-arched stone Roman bridge in Xirokampi of Sparta [1], [2]. (Courtesy of G. D. Makris.)

During the successive Byzantine Empire, the administrative center departed from classic Greece, which was demoted to a province, and moved east to Constantinople and Asia Minor. The Byzantines continued the growth of bridges, based on knowledge of the Roman period. In the early Byzantine period (fourth to seventh centuries AD), bridge construction was included in the sector of public works. There are some reports from the historian Procopius on issues of bridge construction or repairs during the period of Justinian Emperor (527–565 AD). In later years, bridge construction was connected with private initiative, mainly under the form of donations. There are reports on bridges that were built with social-religious criteria.

The Byzantine bridges were usually wooden or stone arched. The piers were strengthened with noses of triangular or circular section (starlings), so that they were protected from the pressure of waters. Near the bridges, towers, hostels, and chapels were often built. The following still-existing bridges deserve particular mention: the multiarched bridge on the Sangarius River of Asia Minor (Figure 12.12), built during the Justinian period and completed within three years (559–562); the bridge on Afrin River in Syria (Figure 12.13); and the Karamaraga Bridge in Cappadocia (Figure 12.14). The bridges of Monemvassia and Karytaina in Peloponnese follow (Figure 12.15).

FIGURE 12.12 The Justinian bridge on the Sangarius River [7]. (Sketch by A. Nikitas.)

FIGURE 12.13 The early Byzantine bridge on the Afrin River in Syria [2], [8]. (Courtesy of L. Evert.)

FIGURE 12.14 The early Byzantine Karamaraga Bridge in Cappadocia of Asia Minor (Fifth to sixth century AD). [3], [7] (Sketch by A. Nikitas.)

FIGURE 12.15 The medieval bridge of Karytaina (1440 AD) is an Assus bridge; when necessary, it could be taken immediately out of operation with the removal of the wooden deck, which appears at the right end. (Courtesy of G. D. Makris.)

12.1.2.3 Ottoman Period (1453 AD–1828 AD)

Following the fall of Constantinople, Greece was, until its liberation (completed in the beginning of the twentieth century), a part of the wider Ottoman Empire. The needs of internal trade in the Empire (mainly with the Danubian regions) imposed the development of the road network and thus the construction of bridges, mainly in the hard mountainous regions. The bridges were financed by religious resources (the Orthodox Church had relative autonomy in the frame of the Empire) or donations of Greek and Turkish notables in the region. During this period, and especially in the second half of the eighteenth century and the entire nineteenth century, bridge construction blossomed impressively. The common characteristic of all the bridges of this period is their construction with stone, a material scattered throughout mountainous Greece without the need of production and transport, and therefore a cheap material.

The applied technique was exclusively the technique of the arch; the style is Greek with the implementation of semicircular (Roman) or lowered elliptic arches, contrary to the Islamic sharp headed ones (without doubt there were also bridges like those in Zerma, of Kyra in Gjirokastër, in Tsipiani, etc.). Usually one-span arched bridges were built. The span of the bridge varied between 20 m and 35 m, with largest recorded span 45 m at the Korakos Bridge on Acheloos River. Their height, due to the semicircular arch, was particularly large, up to 20 m, while their width was rather narrow (3–4 m), wide enough for a carriage. An important number of one-span arched bridges in northwestern Greece still exist today, the most impressive ones being the Konitsa (Figure 12.16) and Pyli Bridges (Figure 12.17).

On the large rivers, multiarched bridges were also constructed, the most characteristic ones being the Arta Bridge on the Arachthos River (four arches, Figure 12.18), the Spanos Bridge on the Venetikos

FIGURE 12.16 Konitsa Bridge on the Aoos River (1871), span 37 m. (Courtesy of DOMI SA.)

FIGURE 12.17 Pyli Bridge on the Portaikos River (1515), span 28 m. (Courtesy of DOMI SA.)

FIGURE 12.18   Four-arched Arta Bridge on the Arachthos River (1610), maximum span 14.3 m, total length 145 m. (Courtesy of G. D. Makris.)

FIGURE 12.19   Five-arched Spanos Bridge on the Venetikos River (1846), maximum span 14.3 m, total length 84 m. (Courtesy of DOMI SA.)

River (five arches, Figure 12.19), the bridge on the Kompsatos River (three arches, Figure 12.20), and the Kalogerikos Bridge (three arches, Figure 12.21). A big central span and smaller side spans were usually formed, gradually decreasing to the edges of the bridges. Almost always, at the axes of the piers, relieving openings were formed for static as well as hydraulic reasons. The main technical characteristics of these bridges, common for all, are as follows:

- The large arches in northwestern Greece were mainly built with two equitant series of stones, sculpted on all their surfaces; occasionally, for aesthetic reasons, the overlying side stuck out from the subjacent one (there is, however, a bridge with three layers of stones). On the contrary, in north-eastern Greece arches were formed with one row of stones. At the crown of the arch a well-sculpted feathered stone was embedded, which restored the continuity and lent the static function of the arch. The two fronts of the subjacent internal arch were connected with incorporated iron ties.
- At the lower parts of the arch, two side walls with stones sculpted only externally were built on the extrados, while the interior of the bridge was filled with loose material like earth or stones (in

FIGURE 12.20 Three-arched bridge on the Kompsatos River (early Ottoman period), maximum span 21.5 m. (Courtesy of DOMI SA.)

FIGURE 12.21 Three-arched Kalogerikos Bridge. (Courtesy of DOMI SA.)

contrast to the Roman arches, where the interior was massive). At the high parts of the bridge, around the crown, the arch appeared. Over the embankment and the prominent part of the arch a narrow corridor was paved with transverse terraces of width 1.5–2 m. At the edges of the deck low parapets or earlier individual standing stones were formed for safety reasons.

- In the mountainous or semi-mountainous areas, the arches were founded on the rock; in the soft flat grounds, they were founded on wooden piles, usually oak or chestnut, following the model of the ancient bridge of Amphipolis. In general, the foundation was the biggest problem in the construction of these bridges, particularly in flat, soft soils; bridges repeatedly collapsed even after their completion.
- The necessary temporary scaffoldings were seated mainly on the stone piers and were anchored at their base; often, they were seated directly on the bed, with great danger of being carried away by

the flow of the river or receding under the weight of the stonemasonry. The removal of scaffoldings was the most crucial phase in the construction of the bridge; many bridges collapsed during the unmolding.

- A powerful lime mortar was used, strengthened with eggs, straw, and goat hair, which changed in the foundation, with the addition of graded tile, to a hydraulic type.
- The time frame of construction was usually short, limited to the summer months leading up to the first rains. Construction was performed on two fronts, with populous independent crews that interchanged periodically for the achievement of a homogeneous result.
- At the bases of the piers starlings were constructed for the reduction of the hydrodynamic pressure in front and spinning behind.

During that period, construction of 431 stone bridges has been recorded, of which 155 have unfortunately been destroyed.

12.1.2.4 Modern Greek Period (1828 AD–2010 AD)

Following the progressive liberation of Greece and its independence from the Ottoman Empire (1821–1920) an increase occurred in the construction of steel railway bridges, which were exceptionally advanced for their period. Some of them exist today and are in operation despite being more than 100 years old and the relevant fatigue of the material as well as increases in the weight and speed of the trains. Seventy-five bridges with spans longer than 15 m exist now in the entire Greek state. The majority of these bridges are truss bridges and a smaller portion are beam bridges; on the whole, the bridges were riveted. Their material was of very good quality, homogeneous, without internal faults, and could easily be classified in class S235 (fracture elongation over 20%, ultimate stress of 40 MPa). Their abutments were stone, while their piers were stone or steel. Those bridges are usually simply supported with spans up to 80 m or continuous with spans of 50–60 m. Sporadically, double overhanging beams with spans up to 120 m, three-hinged arches with spans up to 80 m, and tied arches were constructed.

Certain bridges had a total length over 300 m, spans up to 120 m, and piers of heights up to 60 m. Included among the largest bridges are the bridges of Corinth Canal (Figure 12.22), span 81.4 m; the Achladocampos Bridge, total length of 253 m (36.2 m + 54.2 m + 72.2 m + 54.2 m + 36.2 m), continuous, with pier height over 40 m (Figure 12.23); the railway bridge on the Alpheios River, total length 6 × 51.6 m (Figure 12.24); the Gorgopotamos Bridge (Figure 12.25), total length 7 × 29.12 m; the Asopos Bridge (Figure 12.26) total

FIGURE 12.22 Initial bridge of Corinth Canal (1893), span 81.4 m [5]. (Courtesy of Technical Chamber of Greece.)

FIGURE 12.23   Achladocampos Bridge, total length 253 m, maximum span 72.2 m [5]. (Courtesy of Technical Chamber of Greece.)

FIGURE 12.24   Alpheios Bridge, total length 6 × 51.6 m [5]. (Courtesy of Technical Chamber of Greece.)

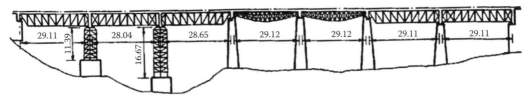

FIGURE 12.25   Gorgopotamos Bridge, total length 7 × 29.12 m [5]. (Courtesy of Technical Chamber of Greece.)

length 180 m (5 × 25.3 m + 1 × 80 m); and the Bralos Bridge (Figures 12.27 and 12.28) total length ~325 m (16 m + 42 m + 52.5 m + 60 m + 120 m + 33 m).

After World War II and its severe destructions, the construction of modern bridges began in Greece. Reinforced and prestressed concrete was widely used, while steel, due to its cost and lack of domestic industry, passed unnoticed, and appeared again after 2000 and the incorporation of the country into the European Union. Today in Greece, all the structural systems of bridges have been used, with the exception of the classic suspended bridges (there is only a minimum number of pedestrian bridges); the most representative samples of these bridges are presented in the following sections.

FIGURE 12.26   Asopos Bridge, total length 180 m, maximum span 80 m (three-hinged arch) [5]. (Courtesy of Technical Chamber of Greece.)

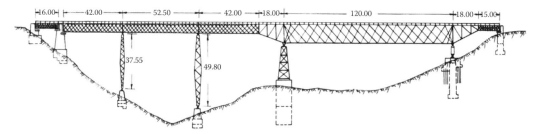

FIGURE 12.27   Initial railway Bralos Bridge (construction year 1905), maximum span 120 m [5]. (Courtesy of Technical Chamber of Greece.)

FIGURE 12.28   Rocking steel pier of height 50 m of the old Bralos Bridge (see Figure 12.27) [5]. (Courtesy of Technical Chamber of Greece.)

12.2 Design and Construction Practice

12.2.1 Materials

12.2.1.1 Concrete

The material that was predominantly used after World War II was concrete, in its different forms (plain, reinforced, prestressed). Plain concrete was applied in small culverts, with spans less than 6 m, while reinforced concrete was used in small-span road bridges less than 20 m. Prestressed concrete was mainly used for longer spans. In Greece, concrete precisely followed the German standards and their evolution (DIN 1045, 1048, 4225, 4227). Initially and up to 1972, the so-called B series was applied (B160, B225, B300, B450, B600). The classification of the concrete was based on the mean average of three cubic specimens 20 cm × 20 cm × 20 cm, without any reference to the standard deviation.

From 1972 until 1997 the new classes of German standards were applied, at first the Bn series and shortly afterwards the revised B series (B15, B25, B35, B45, B55); in this series the classification criterion was the smallest value of three cubic strengths (cubes 20 × 20 × 20), with a mean diversion less than 5 MPa, which was considered the characteristic strength. From 1997 onwards concrete followed the Greek Standard of Concrete Technology, which practically coincides with EC 206. Classes C8/10, C12/15, C15/20, C20/25, C25/30, C30/37, C35/45, C40/50, C45/55, and C50/60 are used (the first number refers to the cylinder strength and the second to the cubic strength). Since 2010, EC 206 has been fully applied.

Greek concrete used mainly limestone aggregates, crushed or collected; less often, harder aggregates like granite and gneiss were used. Light aggregates were never used. Almost exclusively, cement produced in Greece was of the Portland type with or without additives such as pozzolan, of mean strength at 28 days of 35 MPa or 45 MPa. Today, Greek cements are in accordance with the European prototype EN 197-1.

12.2.1.2 Reinforcing Steel

Imported as well as Greek steel was used for the construction of technical works. Initially, smooth steel St I, of maximum diameter $\varnothing$ 26 and yield strength 220 MPa was used; shortly afterwards the application of ribbed steel (type St III and St IV) with yield strengths 400/420, respectively, and 500 MPa began, compatible with the older German standards. Today, weldable steel of high ductility (B500C) and occasionally steel of low ductility (B500A; only for grids) are used, according to the European and Greek standards of steel technology.

12.2.1.3 Prestressing Steel

Immediately after World War II, prestressing tendons consisting of $\varnothing$ 6, $\varnothing$ 7, and $\varnothing$ 8 wires of a class up to 1470/1670 (cold formed) were used; to a much smaller extent, Dywidag bars $\varnothing$ 26.5 and $\varnothing$ 32, of a class up to 885/1080, were also used. The material was totally imported. To a great extent the prestress systems Freyssinet and Morandi were used, and to a lesser extent the Polensky & Zöllner, Leoba, CCL, and other systems were used. The allowable prestressing force of the tendons did not exceed 1200 kN per tendon. Every system complied with the requirements and checks imposed by its approval certificate; concerning design matters, the German standard DIN 4227 was generally applied. The tendons were preassembled, usually in situ, and were usually manually placed, while less often a crane was used; due to this, their length and weight were crucial to their design.

From the 1980s onwards, strands 0.5 inch and 0.6 inch were gradually used, of a class up to 1670/1860, as well as newer prestressing systems (VSL, Alga, Freyssinet, etc.). Today the tendons are completely formed with strands, which are wired after concreting using special machines, thus allowing the implementation of great lengths (e.g., the maximum tendon length at Metsovo Bridge was 250 m). Usually tendons of 19 strands and secondarily of 4, 15, and 22 strands are applied (the maximum number of strands in a Greek bridge is 31).

External prestressing was never applied up to now, due to the fact that it was not permitted by the German standards. The new codes, however (Eurocodes, DIN FB), not only allow but also favor it; thus, its use is soon expected. Couplers were used relatively early on bridges with cantilevering construction (Dywidag-type tendons). From the 1980s they began to be used in other prestressing systems, although not to a great extent, due to the posterior wiring of the strands; today they have been limited to structures with progressive construction (launched bridges).

12.2.1.4 Structural Steel

The steel used in Greece today is totally imported from abroad. It is of quality S235, S275, and mainly S355, according to EC 10025. According to EN 1998-2, the earthquake damping ratio for structural steel is taken equal to 2% for welded and 4% for bolted structures. The anticorrosion protection of large steel elements is realized using appropriate systems of epoxy and polyurethane-base paint of thickness 200 to 500 mm, depending, according to EN 12944, on the environmental corrosiveness, the expected lifetime, and the acceptable breakdown level. The method of deep hot galvanizing is only applied to small-size elements.

12.2.1.5 Stay Cables

At the three (up to now) Greek cable stayed road bridges, stay cables with parallel individual galvanized strands, protected with cement grout or with simple protection of grease without cohesion inside PVC pipes, have been used. At the smaller pedestrian bridges wire ropes of closed type or bars of MC Alloy type have been used. At the arched bridge of Tsakona, currently under construction, galvanized locked coil strands $\emptyset$ 80 up to $\emptyset$ 100 in diameter have been taken into account, with guaranteed ultimate strength (F_{GUTS}) of 6384 to 10096 kN.

12.2.1.6 Masonry

The application of masonry has totally disappeared, even for small culverts, having been replaced by plain or reinforced concrete. In the first years after World War II, it was applied for a short period to the piers and retaining walls of railway bridges. Its ultimate strength, measured in a cube 50 cm × 50 cm × 50 cm after hardening for 28 days, ranged between 25 MPa and 8 MPa, according to DIN 1055/55 (depending on the stones and the type of the stone masonry). The allowable compression stress was 1/5 of the ultimate compression stress. Tensile stresses were allowed only for hyperstatic structures, if they did not exceed 1/5 of the relevant compression and at the most 0.50 MPa.

12.2.1.7 Bearings

Almost exclusively (at least until 2000), elastomeric bearings were used in Greek bridges, simple (type 1 of the German standard 4141, part 14) as well as anchored (type 2, 3, 4, or 5 of the same standard). The reason for the wide use of these bearings, apart, of course, from the low cost and the easy installation, was the ability to create a simple, but very effective, seismic isolation of the deck. The bearings, usually placed between the pier and the deck, increase the basic eigenperiod of the bridge strongly through their shear deformation, thus reducing its excitation (it is noted that the bearings do not absorb seismic energy, because their whole behavior is totally elastic). The bearings followed, in their design and production, DIN 4141 Part 14, constructed from synthetic elastic (chloroprene), with embedded steel plates. The bearings were designed under the service actions following DIN 4141 and under the seismic actions following the Greek guidelines E39/99.

The Greek experience with the use of elastomeric bearings has been up to now very good concerning their behavior against earthquakes (no failures have been recorded despite frequent earthquakes) as well as aging (old bearings, which were deposited from demolishing bridges and checked in the laboratory, showed very good behavior after operation greater than 20 years).

Along with the above low seismic damping bearings, also high damping bearings were used (in bridge strengthening); the lack of experience and some doubts concerning their long-term effectiveness did not allow their further appliance. In railway and large road bridges pot bearings are used, fixed or sliding, as well as special ordered shear keys. Recently, spherical friction pendulum bearings, of simple or triple curvature have begun to be used, which (apart from undertaking vertical loads) increase the eigenperiod of the deck and simultaneously absorb seismic energy through friction.

12.2.2 Specifications and Codes

12.2.2.1 General

Following World War II the American standards AASHO 1944 were applied for a short period; from the 1950s until now, all Greek bridges are designed according to German standards in general (e.g., [9], [10], [11], [12]), with the only exception of DIN 4149, which refers to earthquakes. Instead of DIN 4149, the relevant Greek standards and guidelines are used, and for the most part the guideline E39/99 [13]. Since 2007, the Eurocodes mainly and the new German standards (FB) have been gradually applied.

12.2.2.2 Road Bridges

Road bridges actions are described in DIN 1072 and EN 1991. Additionally, the French guidelines for collision loads onto safety barriers of heavy type BN4 are applied, which take into account the implementation of a horizontal load $P = 300$ kN and a moment $M = 200$ kN/m to the base of the barrier post. For structures with high requirements, special actions and design statements are used, such as:

- Collapse of one or two successive stays in cable-stayed bridges, without further damage
- Replacement of one stay in cable-stayed bridges under simultaneous presence of 50% of live loads
- Ship collision loads on the piers depending on the kinetic energy of the ship
- Special very heavy trucks
- Potential differential settlement between successive piers equal to 50% of the maximum expected settlement of each pier
- Angular foundation rotation equal to 1/500 rad

The analysis and dimensioning of the road bridges was done according to DIN 1072 (road bridges actions), 1075 (design of concrete bridges), 1045 (plain and reinforced concrete), 4227 (prestressed concrete), 1054 (foundations), and 4014 (bored piles). Especially for steel and composite bridges DIN 18800, 18809, and so on were applied. Since 2007, Eurocodes EN 1990, 1991, 1992, 1993, 1994, 1995, 1996, 1997, and 1998 and DIN FB 101, 102, 103, and 104 have been gradually applied.

12.2.2.3 Railway Bridges

The design of railway bridges was effected according to German code DS 804, followed by DIN 1045, 4227, 1054, 4014, 18800, and 18809. The recent Greek bridges are designed for load model UIC 71 with train speeds up to 220 km/h. Load model UIC 71 foresees a locomotive of four axles, of total weight 1000 kN and length 6.4 m, accompanied by a linear load 80 kN/m at both its sides (Figure 12.29). These loads should be totally or partially imposed, since the action effect under checking is disfavored (Figures 12.30 and 12.31). The loads are multiplied by the classification coefficient v, which in the Greek network is generally taken equal to 1.25.

Continuous bridges are also checked with the train of type SW, which is applied on their whole as it is (Figure 12.32). In case of heavy trains, the bridges are also checked for the train SSW, which is also

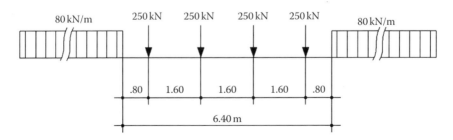

FIGURE 12.29   Load model UIC 71.

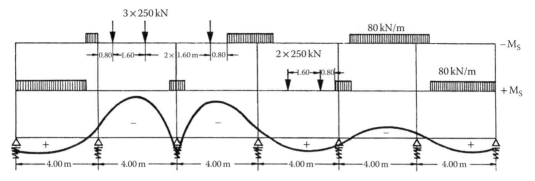

FIGURE 12.30 Partial loading (UIC 71) for the achievement of hogging moment.

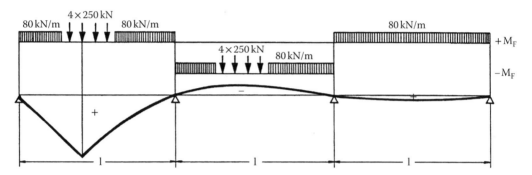

FIGURE 12.31 Partial loading (UIC 71) for the achievement of sacking moment.

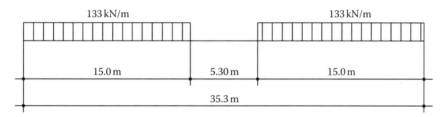

FIGURE 12.32 Train SW for continuous bridges.

applied on their whole as it is (Figure 12.33). The above loads are multiplied by the dynamic coefficient φ, depending on the effective span l'; this coefficient comes up to

$$\varphi = 1.67 \text{ for } l^* \leq 3.61\text{m}$$

$$\phi = \frac{1.44}{\sqrt{l^*} - 0.2} + 0.82 \text{ for } 3.61\text{m} < l^* \leq 65\text{m}$$

$$\varphi = 1.00 \text{ for } l^* \geq 65\text{m}$$

12.2.2.4 Pedestrian Bridges

All Greek pedestrian bridges have been designed with the uniform surface load

$$p = 5.00 \geq 5.50 - 0.05l \geq 4.00\text{kN/m}^2 \tag{12.1}$$

where l is the theoretical length of the span.

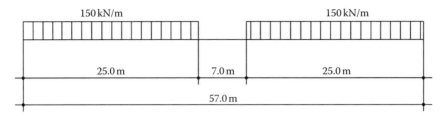

FIGURE 12.33   Heavy train SSW.

For pedestrian bridges up to 10 m a uniform surface load equal to $p = 5$ kN/m^2 is taken into account. The vibrations of pedestrian bridges, for which there was no reference in the standard, were checked with the fundamental eigenperiod of the deck as a criterion, which should be outside some dangerous ranges.

12.2.2.5 Earthquake

The seismic design of Greek bridges is governed by the Greek earthquake standard EAK 2000 and guideline E39/99, most parts of which have been incorporated into EN 1998-2. According to the above codes, the country is divided into three seismic zones (I, II, and III), each one of which is characterized by the design ground acceleration (0.16 g/0.24 g/0.36 g); this acceleration is considered to have a probability of being exceeded by 10% in the next 50 years (return period of the seismic event 475 years).

Depending on the importance of the bridge (financial, technical, circulatory) the above acceleration can be increased, that is, multiplied by the importance factor $\gamma = 1.30$ (return period of the seismic event 950 years). It is also increased by 25% if there is an active seismic fault in the area of the structure. Depending on the soil class the response spectrum of the structure is also chosen; there are four soil classes: A (rock, $T_1 = 0.10$, $T_2 = 0.40$ s), B (sand gravel soils or hard clay, $T_1 = 0.15$, $T_2 = 0.60$ s), C (soft clays, $T_1 = 0.20$, $T_2 = 0.80$ s), and D (unstable soils, $T_1 = 0.20$, $T_2 = 1.2$ s). Figure 12.34 shows the typical response spectra.

Depending on the available ductility of the structure, the elastic spectra are turned into design spectra, with the introduction of the so-called ductility factor q ranging between 1 (elastic design) and 3.5 (plastic design). The choice of q value depends on the ability of the structure to absorb the seismic energy in predefined positions (where plastic hinges are formed) and the extent of the acceptable damage. In parallel to the horizontal ground acceleration, a vertical ground acceleration is taken into account, equal to 0.70 A, having the same design spectrum but without implementation of any ductility factor ($q = 1$) (Figure 12.34).

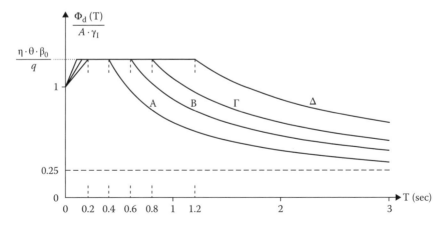

FIGURE 12.34   Design response spectrum.

$$0 \leq T < T_1 : \Phi_d(T) = \gamma \times A \times \left[1 + \frac{T}{T_1} \left(\frac{\eta \times \theta \times \beta_0}{q} - 1 \right) \right] \tag{12.2}$$

$$T_1 \leq T \leq T_2 : \Phi_d(T) = \gamma \times A \times \frac{\eta \times \theta \times \beta_0}{q} \tag{12.3}$$

$$T_2 < T : \Phi_d\ T = \gamma \times A \times \frac{\eta \times \theta \times \beta_0}{q} \times \left(\frac{T_2}{T} \right)^{2/3} \geq 0.25 \times \gamma \times A \tag{12.4}$$

where

$A = \alpha \times g$ is the maximum horizontal ground acceleration
g is the gravity acceleration
γ is the bridge importance factor
q is the bridge behavior coefficient (ductility factor)
η is the corrective coefficient for damping percentage $\neq 5\%$
θ is the foundation influence coefficient equal in general to 1.00
T_1 and T_2 is the characteristic periods of the spectrum depending on the soil class
β_0 is the amplification factor equal in general to 2.5

The corrective damping coefficient is calculated by the formula

$$\eta = \sqrt{\frac{7}{2 + \zeta}} \geq 0.7 \tag{12.5}$$

where $\zeta(\%)$ is the viscous damping (depending on the material, 2–6%).

12.2.2.6 Tectonic Displacements

Some Greek bridges have been designed taking into account tectonic displacements. The most characteristic of them is Rion–Antirion Bridge, which has been designed against relative displacement of all of its piers (every two) of ±2 m in all possible directions. In general, tectonic displacements are not simultaneous with earthquakes, appearing close to the end of the seismic event; according to tender documents, only a part of them, ranging from 30–50%, is combined with the earthquake.

12.2.3 Design Methodology

12.2.3.1 Service Actions

According to the German standards, bridges were designed under the service actions, based on an elastic analysis; further redistribution of the stress, even to a reduced level, was not allowed. In progressive constructions (cantilevering, launching, etc.) the analysis of the internal forces takes into account the time development of the structure and the creep and shrinkage effects. Initially, the design of the cross sections was performed on the basis of the allowable stresses; following the evolution of the German standards since 1972, the cross sections have been checked at the serviceability limit state (SLS) and the ultimate limit state (ULS).

At the SLS it was verified that the developed stresses did not exceed the allowable ones (at least in prestressed structures), as well as that the cracking of the sections remained very small (usually within the limitation of the steel stresses under the permanent loads plus 50% of the live ones in a range of 160–200 MPa, depending on the diameter and the spacing between the bars). At the ultimate state, the cross sections were checked with overall strength methods, with the limitation of the limit steel strain to 5% and the concrete one to 3.5%.

Three groups of actions were distinguished: group (H), which included permanent and traffic loads; group (Z), which included secondary service loads (temperature, wind, settlements etc.); and group (A), which included accidental loads except earthquake (collision, etc.). The load combinations (H), (HZ), and (HA) were relative. A global safety factor equal to 1.75 was applied for combinations of (H) type, $1.75 \times 0.90 = 1.575$ for combinations of (HZ) type, and 1.0 for combinations of type (HA). The reduction of the stiffness of the sections due to cracking was taken equal to ~60%.

Since 2007 the design of bridges based on the Eurocodes or DIN FB has been gradually applied; from current experience, the dimensioning of the sections is not seriously affected by the change in the standards.

12.2.3.2 Earthquake Actions

In the first years after World War II, earthquake was treated in a static manner as an additional horizontal inertia force, equal to the product of mass times ground acceleration, ranging, depending on the soil class, between 0.04 g and 0.16 g. Under the earthquake combinations, increases of the allowable concrete stresses equal to 20%, of steel equal to 20%, and of soil equal to 50% were allowed. Since the 1980s, the philosophy of the earthquake design has totally changed. The capacity design philosophy was adopted, according to the New Zealand school, with dynamic spectral analysis as the basic tool.

According to this philosophy, which has been totally adopted by EN 1998, controlled damage in a limited number of usually accessible locations of the structure are allowed under the earthquake design. These regions (plastic hinges) are suitably reinforced with confinement, so that they become ductile, preventing brittle fracture. In these regions, due to confinement, it is possible that concrete strain of 0.01–0.012 is developed, which allows local plastic rotation, thus relieving seismic stress.

Structure analysis is usually performed linearly, according to the design spectrum, which is produced from the typical elastic spectrum for each soil class (for structural damping $\zeta = 5\%$) and the ductility factor, ranging between 1 and 3.5 according to the structural system and its ability to absorb energy. The design moment in the predefined plastic hinges is equal to the one resulting from linear spectral analysis. The regions among the plastic hinges are designed using capacity design forces, which are the maximum actions the specific cross section can develop. It generally holds

$$M_{o,h} = \gamma_o \times M_{Rd,h} \qquad (12.6)$$

where

γ_o is the factor of overstrength of the plastic hinge equal to 1.4.

$M_{Rd,h}$ is the moment resistance according to the actual reinforcement and the dimensions of the cross section.

The maximum value of the capacity design forces is defined as the relevant elastic one. Shear design is based on the capacity stress, with the maximum limit being the elastic one. With the above technique, the development of seismic damages in the predefined positions (plastic hinges) is ensured. The analytical seismic displacement d_E is equal to the displacement d_{Eo} of the equivalent linear analysis multiplied by q ($d_E = q \cdot d_{Eo}$). A set of semi-empirical rules controls the dimensioning and the structural design of cross sections so that local ductility is ensured. In bridges of special seismic requirements, usually seismically isolated, time history analysis is used based on a set of suitably chosen accelerograms (see also Section 12.11).

12.2.3.3 Laboratory and Loading Tests

As a rule in bridges of special requirements, laboratory tests of materials and structural elements, as well as loading tests of completed elements or the whole of the structure, are anticipated. Testing is aimed at the verification of basic design assumptions and the improvement of the analysis model. Among these tests, the most characteristic ones are

- Dampers testing: As normal practice, a prototype unit of seismic dampers, in natural scale or smaller, is tested in the laboratory; for example, the dampers of Rion–Antirion Bridge were checked in the laboratory of University of San Diego. As a rule, the constitutive law of their operation is tested in the laboratory.
- Tests of aerodynamic tunnel: The cross sections of the Rion–Antirion and Euripus stay-cable bridges were tested concerning their aerodynamic behavior and the potential appearance of aeroelastic instability in the Central Laboratory of Public Works (LCPC) of France and the laboratory of the Stuttgart University.
- Centrifuge model and seismic table tests: The foundation model of the Rion–Antirion pylons based on soil improvement with steel inclusions was meticulously checked by centrifuge model tests in LCPC, in order to verify and improve its theoretical base. Tests of a smaller extent have also been realized in the seismic table of National Technical University of Athens (NTUA).
- Stay cables tests: The stays of the Rion–Antirion and Euripus bridges were experimentally tested as structural elements in their whole, including the anchorage head in the laboratories of Swiss Federal Laboratories for Materials Testing and Research (EMPA) and French Central Laboratory for Bridges and Pavements (LCPC).
- Material tests: Testing of materials incorporated in the structure, such as concrete, prestressing steel, couplers, and so on is an obligatory routine.

Loading tests are done basically on piles for the verification of their ultimate capacity and their behavior under vertical and horizontal loading. The loading takes place on a service or test pile; in the second case the loading can reach failure. Test loads up to 12 MN vertically and 1.2 MN horizontally have often been applied; piles of diameter up to $\varnothing$ 1.8 m have been tested. Apart from this, test loadings, mainly for the estimation of the existing strength in old bridges or the verification of the structural behavior in new ones, are not performed very often.

The loading tests on the Corinth Canal road bridge, the old railway bridges of Bralos (over 100 years old), and the new Euripus cable-stayed road bridge should be mentioned. The loadings were done with trucks or locomotives, standing or moving. The behavior of the structure was recorded and compared to the calculated expected behavior, providing information for the improvement of the analysis model.

12.2.4 Construction Techniques

12.2.4.1 In Situ (Cast-in-Place) Construction

A large number of Greek bridges, especially urban ones, were constructed with a conventional formwork on scaffoldings; concrete was poured on whole bridges or span by span, with or without the use of reinforcement bar couplers. With this method, spans up to 60 m and bridges up to 300 m have been constructed.

12.2.4.2 Prefabrication

The industrialization of construction very soon made necessary the use of precast beams, the construction of which is usually done in situ, with the implementation of conventional prestressing or a prestressing bed in the factory (recently, small, portable on-site beds have been used, which allow the production of beams up to ~38 m long with direct bond in situ). With prefabrication, beams up to 44 m long and 1200 kN weight have been constructed. Beam placement is done using mobile cranes (if the height of the deck is less than 20–25 m) or a special traveler moving longitudinally and transversally along the bridge. The prefabrication method, despite the increased maintenance cost, is very popular in Greece due to the easy construction and low initial cost. It is noted that the beams are designed and placed as simply supported beams on (usually) elastomeric bearings; for seismic reasons, however, partial continuity with the so-called continuity slabs in the axes of the piers is ensured.

12.2.4.3 Cantilevering

Another method with wide implementation is cantilevering; very large bridges have been constructed with this method (with a central span up to 235 m and length of over 1000 m), especially in mountainous, not easily accessible areas. The usual length of the individual segments amounts to 5 m and their maximum weight to ~2500 kN. Prestressing is totally realized using conventional tendons (usual units 19T15), wired after the concreting; the maximum tendon length in a Greek bridge is 250 m (Metsovo Bridge).

The usual production cycle of one pair of segments is one week, although in some cases a shorter period may be obtained. The construction of the cantilevers is especially based on the precamber design and in situ measurement of the deflection deformations; great experience in this specific sector has already been developed, so that the cantilevers are usually met with a gap less than 20 mm.

12.2.4.4 Launching

The implementation of incremental launching of bridges (prestressed as well as composite, road and railway) began in 1997. More than 10 prestressed decks and approximately 10 steel decks have been launched, with a maximum launching length of 520 m (railway bridges) and maximum span of 55 m in prestressed and 77.5 m in composite structures.

12.2.4.5 Composite Girder Bridges

In 1981 the first composite Greek road bridge was constructed in Athens, but it was not a popular form, mainly due to the high cost of steel compared to concrete. Since 1995, and especially after 2000, this technology strongly extended, mainly due to the speed and ease of construction and the low dead weight of the deck. The beams are placed with launching or mobile cranes and the traffic slab is concreted in situ on preslabs or trapezoidal steel sheets. The connection of the steel beams and the concrete slab is exclusively realized with Nelson-type shear studs, of quality S235 J2 G3 + C450 ($f_{yk} \geq 355$ MPa, $f_u \geq 450$ MPa, $A_s \geq 15\%$). Anticorrosion protection is provided with epoxy and polyurethane paintings in three or four layers, of total thickness 200–500 μm (depending on the corrosiveness of the environment). The steel plates are imported from abroad, while the fabrication takes place mainly in Greek factories. The development of international transportation allowed the import of ready-made large elements from abroad (Italy, Egypt, and Czech Republic).

12.2.4.6 Cable-Stayed Bridges

Another technology with remarkable implementation in Greece is cable-stayed bridges. In 1993 the first Greek road bridge of Euripus, with a total span of 395 m and center span of 215 m, was completed. The bridge was constructed using the method of free cantilevering, aided by a self-moving carrier especially designed for this bridge. On the contrary, in the Rion–Antirion Bridge ($\Sigma L = 2252$ m, max $L = 560$ m), where the deck was composite, the steel segments were transported using floating means and were placed with a crane seated on the deck (derrick), aided by the next stay.

12.3 Concrete Girder Bridges

12.3.1 General

Concrete bridges, reinforced or prestressed, constitute the overwhelming majority of Greek bridges. The abundant feedstock (aggregates) and the powerful Greek cement industry (in contrast to the lack of steel production units) were the definitive factors that imposed concrete as the exclusive bridge material and led to the development of strong construction companies. The first reinforced concrete bridge was the railway bridge of Vrichonas (a simply supported tied-arched bridge), constructed in 1917 (Figure 12.35).

FIGURE 12.35 Vrichonas Bridge, the first reinforced concrete bridge in Greece (1917). (Courtesy of the Municipal Center for Historical Research and Documentation of Volos, Zimeris' Photographical Archive.)

The bridge, after repairs, also serves the present-day needs of a lightweight local train. The first pre-stressed bridge, with a span of 20 m, was constructed on Kifisos River in Athens in 1954 (designed by Th. Tasios). Since then, a large number of bridges of all types have been designed and constructed, with the most important being the cable-stayed Rion–Antirion Bridge (Figure 12.36).

12.3.2 Cast-in-Place Bridges

12.3.2.1 General

Interchange bridges with geometrically difficult plan views have mainly been constructed using this technique. Bridges with simply supported slabs in national and provincial roads, with or without voids, with a deck at a small distance from the ground, up to 20 m, were also designed; to decrease the height and the cost of the scaffoldings, embankments up to 10 m high are usually constructed. The deck concrete is poured in one phase for short bridges or step by step for longer ones to decrease the equipment cost. Using this method, bridges up to 300 m long with maximum spans of 60 m have been constructed.

12.3.2.2 Examples

12.3.2.2.1 Petrou Ralli Bridge

This is a continuous prestressed concrete box girder bridge of five spans of 52 m, with a total length of 260 m (Figure 12.37). The bridge is curved in plan ($R = 100, 350, 700$ m). The superstructure at the major part of the bridge is twin with a width of 15.6 m, with the remaining part split into two individual sections of width 8.6 m each (Figure 12.38). The typical section consists of boxes of a total height 2.1 m. The superstructure was constructed, due to traffic reasons, in three stages. Elastomeric bearings are used to provide a satisfactory seismic isolation to the structure. The piers are supported on pile groups $\varnothing$ 120 and $\varnothing$ 150. The bridge was designed by DOMI SA and constructed by Hellenic Technodomiki SA, and was completed in 1988.

12.3.2.2.2 Koukaki Interchange Bridges

These are two horizontally curved bridges of an approximate length 200 m each and a deck width 10.55 m (Figure 12.39). For traffic reasons during construction each bridge was divided into three sub-bridges, from which the first one is continuous for four spans and the other two continuous for two

FIGURE 12.36 **(See color insert.)** Rion–Antirion Bridge. (Courtesy of Jacques Combault and GEFYRA SA.)

FIGURE 12.37 Petrou Ralli interchange bridge in Athens. (Courtesy of DOMI SA.)

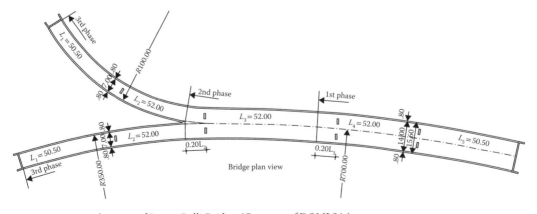

FIGURE 12.38 Plan view of Petrou Ralli Bridge. (Courtesy of DOMI SA.)

FIGURE 12.39 Koukaki interchange bridge in Athens. (Courtesy of DOMI SA.)

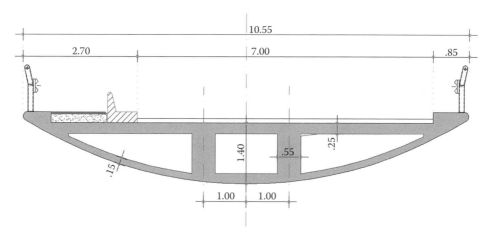

FIGURE 12.40 Typical cross section of Koukaki interchange bridge. (Courtesy of DOMI SA.)

spans. Three of the total six sub-bridges are conventionally reinforced and the other three longitudinally prestressed. Transverse deck prestressing was implemented in only one bridge. The typical section is a three-cell box, with a thin-wall curved bottom flange (Figure 12.40). The superstructure is supported on elastomeric bearings on prismatic columns. The foundation consists of pile groups. The bridges successfully withstood a very strong earthquake in Athens in 1999 without any damage. They were designed by DOMI SA and constructed by Hellenic Technodomiki SA, and were completed in 1985.

12.3.2.2.3 Kaisariani Bridge

This is one of the bridges in the first interchange in Athens. The bridge is curved ($R = 250$ m) and skew in plan, of a total length ~80 m, with six traffic lanes and two sidewalks, having a longitudinal slope of 4% and a transverse slope of 6%. The very difficult geometry of the deck imposed the in situ construction.

The superstructure (Figure 12.41) consists of two continuous prestressed horizontally curved beams of section 2.00/1.25 m, spaced at 13 m, and a prestressed slab of variable section with a central span 13 m

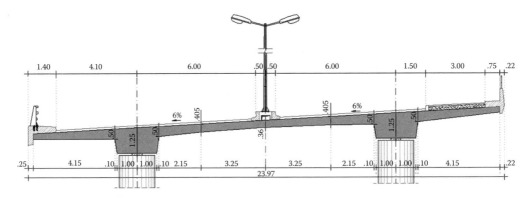

FIGURE 12.41 Cross section of Kaisariani Bridge in Athens. (Courtesy of DOMI SA.)

and two cantilevers 5.5 m. Crossbeams exist only over the end supports, resulting in a decrease in the degree of the transverse distribution but also the unfavorable effects of the skewness at the area of the acute angle of the bridge. The superstructure sits on elastomeric bearings on two buttressed piers, skew to the axis of the bridge, of length ~50 m and with six circular columns ∅ 1.3. Spread footings seated on the rocky subsoil of the area constitute the foundation.

The bridge successfully withstood a very strong earthquake in Athens in 1999 without any damage. Despite that, in order to improve the seismic resistance of the bridges after the 1999 Athens earthquake, the columns were strengthened by jackets of self-compacted concrete of thickness 0.15 m and the bearings were replaced by new anchored, more enforced, elastomeric ones. It should be noted that despite 20 years of life and exposure to the environment, the old bearings were in very good condition. The bridge was designed by DOMI SA and constructed by Hellenic Technodomiki SA, and was completed in 1983.

12.3.2.2.4 Posidonos Interchange Bridge

This bridge is one of 11 road bridges, 3 underpasses, and 1 pedestrian bridge in a very complex interchange. The bridge runs partially over Kifisos River and is quasi-straight in a length of 179 m and circular in the remaining 209 m. It consists of a continuous hollow slab of 20 m width and 1.5 m thickness with circular voids of 1 m diameter, supported on 13 V-shaped piers and 2 abutments through elastomeric bearings. The piers in the part over the river transfer their vertical load to a wall founded on bored piles of 1.8 m diameter, while the horizontal loads are transferred through horizontal beams lying on the ground level on retaining piles of 1.2 m diameter, bored outside the side walls of the underlying river bed. The remaining piers are founded on bored piles through pile caps. High-strength concrete B35/DIN 1045 for the slab and B25/DIN 1045 for the piers was used, while prestressing was carried out through tendons of 19 ∅ 15 strands. The slab was cast in situ on formwork supported on temporary steel bridges, permitting construction under simultaneous traffic. It was designed by Dr. Bairaktaris and Associates Ltd., constructed by Mochlos SA, and completed in 2004.

12.3.3 Precast Bridges

12.3.3.1 General

Precast bridges are by far the most popular type of bridge in Greece up, mainly due to the low cost and short construction time. They are implemented in spans of small and medium length (up to 44 m). The typical superstructure cross section consists of precast prestressed beams spaced at 3–4 m and connected between them by the deck slab. Beams are usually placed at both edges of the section, so that they facilitate the construction of the deck slab, which is poured on preslabs with a shear connection to them or on galvanized trapezoidal steel sheets of thickness 1–1.5 mm. Crossbeams are placed only over the supports; the deck slab thickness is about 30–35 cm. The beams are usually precast in situ for transportation reasons; however, beams precast in factories are also used. Prestressing is conventional or in a concreting bed.

The concrete classes usually applied are C30/37 up to C40/50 according to the Eurocodes or B35 and B45 according to the old DIN standards; however, class C50/60 has also been implemented. The deck slab is usually continuous in order to reduce the number of expansion joints and improve the seismic behavior of the superstructure. The maximum length of a continuous part of the bridge is around 250 m. The deck slab is designed locally over the supports as a potential plastic hinge.

The total weight of the beams amounts to ~120 tons. The beams are placed by means of a special traveler or by mobile cranes. The mean material consumption is 0.65–0.70 m³/m² (concrete), 20–25 kg/m² (prestressed steel), and ~120 kg/m³ (reinforced steel). Beams are typically seated on anchored (due to seismic actions) replaceable elastomeric bearings on the piers. Bridges with lengths up to 780 m have been constructed with the application of the precast method.

12.3.3.2 Examples

12.3.3.2.1 Ravine Bridge in Kavala

This is a dual road bridge with total branch lengths 270 m (seven spans) and 345 m (nine spans) (Figure 12.42). The horizontal alignment presents a very small radius of 200 m and the vertical alignment exhibits a varying gradient along the bridge with values up to 5.7%. The lateral gradient of the cross section of the deck is also variable, ranging from 2% to 8%. The superstructure cross section consists of four prestressed I-beams per branch, placed on the piers by using a traveler, connected between them via an in situ concrete slab. The height of the superstructure is 3.1 m while the theoretical span of the beams ranges from 22.75 m to 43.5 m. The beams are simply supported on elastomeric bearings with seismic links between the superstructure and the pier caps preventing the fall of the deck during an earthquake stronger than specified for the bridge. The piers have a significant maximum height of 63.65 m; they are hollow box-type cross sections with external dimensions of 4 m × 4 m and wall thickness of 0.40 m, while their foundation is cylindrical rock sockets 6 m in diameter. It was designed by Odomichaniki SA, constructed by Ergas SA, and completed in 2001.

FIGURE 12.42 Kavala Ravine Bridge on the Egnatia Motorway. (Courtesy of Odomichaniki SA.)

12.3.3.2.2 Nestos Bridge

This is a twin road river bridge in northeastern Greece, at section 14.1.2 of the Egnatia Motorway. The bridge has a total length of 456 m (Figure 12.43). Each branch consists of 12 spans of 38 m; every 6 spans are connected over the supports with the so-called continuity slabs, forming two independent sub-bridges of 228 m. The typical cross section has a total width of 2 m × 14.45 m; it is formed with five precast beams constructed on site in a concreting bed, connected between them with the deck slab with a total thickness of 0.30 m, poured on preslabs. The precast I-beams are seated on the piers through anchored elastomeric bearings.

The placement of the beams was realized with mobile cranes; the weight of each beam was 1000 kN. The main design challenges were the very poor foundation ground, subject to liquefaction in case of strong earthquakes, and the very high seismic ground acceleration of 0.40 g. Stone piles 20 m deep and ∅ 80 and a seismic isolation system were used to meet geotechnical and earthquake requirements (see also Section 12.11). It was designed by DOMI SA, constructed by IONIOS SA, and completed in 2009.

12.3.3.2.3 Panaghia Bridge

This is a twin road valley bridge in northwestern Greece, at section 4.1.1 of the Egnatia Motorway. The bridge has total lengths of 300 m and 337 m for the left and right branches, respectively. It consists of simply supported spans of 37.5 m (Figure 12.44). The spans are connected between them with the deck slab, forming thus independent sub-bridges (two in the left branch and three in the right one). The superstructure consists of precast prestressed beams, constructed in situ and conventionally prestressed. The beams are seated on hollow circular piers through elastomeric bearings. The placement of the beams has been realized with mobile cranes; the weight of each beam was 930 kN.

The special characteristic of the bridge is its foundation within a creeping area with a depth up to 13 m. Small-diameter shafts ranging from 5 m to 8 m and of a 35 m depth are used to control the produced pressures from the creeping mass on the shafts. Future lateral anchorages of the shafts, if needed, have already been designed. It was designed by DOMI SA/LAP GmbH, constructed by AttiKat SA, and completed in 2009.

FIGURE 12.43 Nestos Bridge on the Egnatia Motorway. (Courtesy of DOMI SA.)

FIGURE 12.44   Panaghia Bridge within a creeping area on the Egnatia Motorway. (Courtesy of DOMI SA.)

12.3.4 Launched Bridges

12.3.4.1 General

This very interesting construction method has a short implementation history in Greece. The first implementation was in 1997 for the Lissos Bridge; ever since, more than 10 road and railway superstructures have been constructed with maximum length ~520 m and maximum span 56.3 m. In one of these, an auxiliary suspension system was used. The maximum longitudinal slope was 5%.

The typical superstructure cross section is a box with a ratio of the top and bottom section moduli $W_o/W_u \approx 2$. Straight tendons are commonly used (double at the upper flange compared to the bottom), serving mainly the launching phase, and complementarily additional curved ones, arranged within the webs of the box and wired after pouring by using special machines, serving the additional and traffic loads. External tendons have not been yet used. The basic prestressing tendons are provided with couplers at every two segments, while a strict rule that no more than 50% of the tendons may be coupled at the same section is followed.

The launching is realized with special launchers; the temporary seat of the deck (superstructure) is realized on sliding teflon slabs. A segment of length 20–25 m is usually produced per week. The concrete classes are C30/37 and C35/45/DIN 1045. The mean material consumption comes up to 0.75 m³/m² (concrete), 40–50 kg/m² (prestressing steel), and 130 kg/m³ (concrete steel). The seat of the deck on the piers is realized with anchored (due to seismic actions) elastomeric bearings.

12.3.4.2 Examples

12.3.4.2.1 Lissos Bridge

The 433 m long Lissos River road bridge, located on the 687 km long Egnatia Motorway in northeastern Greece, represents the first bridge in Greece built by the incremental launching method (Figure 12.45). The bridge consists of two independent branches, each one having a total width of 14.4 m. The superstructure is a straight 11-span continuous single-cell box of a constant depth, equal to a 2.75 m structure with spans of length 29.56 m + 3 × 37.05 m + 6 × 44.35 m + 26.5 m, with a minor slope of 0.70%. The nose, attached to the front of the deck for the reduction of the cantilever moments during launching,

FIGURE 12.45   Lissos Bridge, first implementation of the launching method in Greece (1997).

was a 29.2 m long steel beam structure. Prestressing has been realized mainly by straight tendons (16 at the top slab and 8 at the bottom slab) and a few curved tendons (four) inside the webs. The launching was performed through special launchers seated on the abutments. It was designed by DOMI SA, constructed by joint venture J&P/Hellenic Technodomiki SA, and completed in 1997.

12.3.4.2.2 T11 Bridge on Athens Ring Road

This bridge is located on the Athens Ring Road. It consists of two independent branches of a total width equal to 12.03 m each one (Figure 12.46). The superstructure follows a horizontal curve with $R = 1000$ m and a slope of 1.4%. The bridge has a total length of 411.36 m with nine spans of 35.93 m + 44.86 m + 45.12 m + 51.2 m + 56.3 m + 51.21 m + 45.13 m + 45.11 m + 36.5 m. The span arrangement and in particular the central span were determined by the railway track and roadway locations. The available construction equipment imposed the finally applied cross section of a single-cell box type with a constant depth equal to 2.75 m, thus resulting in a depth-to-span ratio of 1/20.5. To safely launch the superstructure above the long central span, it was necessary to install a temporary cable-stayed system. It was designed by DOMI SA, construction Ellaktor SA, and completed in 2002.

12.3.4.2.3 SG12 Bridge in Lianokladi

This is a railway double valley bridge in central Greece (Figure 12.47). The superstructure is a horizontally curved continuous single-cell box girder with five spans (36 m + 3 × 45 m + 36 m). The bridge width is 13.9 m. The piers have a concave orthogonal section, founded on bored piles. The bridge was constructed with the launching method, with the aid of a steel nose of length 29 m. Due to the high seismicity of the area an extended seismic isolation was implemented, consisting of friction pendulum sliding bearings and viscous hydraulic dampers. It was designed by Kanon SA, constructed by Aktor SA, and completed in 2009.

12.3.4.2.4 SG10 Bridge in Lianokladi

Bridge SG10 of the double high-speed railway in central Greece ensures the passage of the railway track over a valley. The total length of the bridge is 522 m, with a typical span length of 45 m (Figure 12.48). The bridge comprises eleven piers (36 m + 10 × 45 m + 36 m) with a maximum pier height of 45 m. The superstructure features a continuous single box girder of concrete B45/DIN 1045 both longitudinally

FIGURE 12.46 TE11 bridge on the Athens Ring Road. (Courtesy of DOMI SA.)

FIGURE 12.47 SG12 railway bridge in Lianokladi. (Courtesy of Kanon SA.)

FIGURE 12.48 SG10 railway bridge in Lianokladi. (Courtesy of TTA SA.)

and transversally prestressed with tendons 19T15 Super, 18T15 Super, and 4T15 Super (transverse prestressing). The middle piers are made of concrete B35 having a rectangular hollow box section; the foundation is achieved on bored piles of diameter $D = 1.5$ m.

The superstructure is seismically isolated. In particular, viscous hydraulic dampers were installed in the longitudinal direction of the bridge at the abutments and in the transverse direction at each one of the middle piers. Friction pendulum sliding bearings were also installed at the nine central piers. At the remaining two piers spherical sliding/friction bearings were used. Expansion joints are provided only at the abutments. The bridge was constructed by the incremental launching method. The design of the bridge was based on a nonlinear dynamic time history analysis using real accelerograms in three directions. Also, due to the fact that the bridge lies on a circular arc of radius 825 m, a special nonlinear analysis was conducted against the centrifugal forces developed during the passage of the trains. The bridge was designed by TTA and Associates, constructed by Michaniki SA, and completed in 2009.

12.3.5 Moving Formwork Bridges

12.3.5.1 General

Wide experience in this construction method does not exist in Greece, despite the fact that the first bridge of this type was constructed in 1970. For a long period, until the end of 1990s, no more similar bridges were constructed. Today over 10 road and railway bridges of the moving formwork type, among them one with very interesting technical challenges (Kristallopigi Bridge), are constructed or are under construction. Bridges of maximum length up to 850 m and maximum span 55 m have been constructed with the moving formwork method. The usual span of the bridges constructed with this method is about 50 m. Box type is widely used for the superstructure section; however, double-T sections are also used.

12.3.5.2 Examples

12.3.5.2.1 Achladokampos Bridge

This was the first Greek bridge constructed with the moving formwork method and also the first railway prestressed bridge. The bridge consists of 11 independent girders (10×26.5 m + 1×14.5 m, total 279.5 m) simply supported on two abutments and 10 box-type piers of variable heights (Figure 12.49). The girder cross section is of a double-I type, without cross girders between supports. High-strength concrete B300 and B450 (according to the old German codes) and Freyssinet prestressed tendons of type 12 Ø 7 were used. The beams rest on the piers through elastomeric bearings. The superstructure is fully prestressed, thus avoiding any risk of fatigue and cracking.

FIGURE 12.49 Achladocampos railway bridge, first implementation of the moving formwork method in Greece (1970) [14]. (Courtesy of N. Chroneas.)

The cross section of the piers varies, with height up to 60 m. B300 type concrete was used for their construction, with the exception of the lower part of the highest piers, where B450 was used. The piers were erected using slip forms. Analysis of the piers was performed using the nonlinear second order theory, taking into account deformations from all possible sources (train load, wind, dynamic impact, random eccentricity etc.). As for earthquakes, a time history analysis, using as input the "El Centro 1940 N-S Component" suitably adapted, was carried out. The bridge was designed by Pagonis–Chroneas–Kinatos, constructed by Pantechniki SA, and completed in 1970.

12.3.5.2.2 Kristallopigi Bridge

This is a twin road valley bridge of the Egnatia Motorway in northwestern Greece. The left branch of the bridge has a total length of 638.2 m, consisting of 12 spans (44.17 m + 10 × 54.99 m + 44.17 m), and the right branch a total length of 847.8 m, consisting of 16 spans (43.2 m + 14 × 54.38 m + 43.25 m) (Figure 12.50). Both branches are curved in plan. The superstructure is a continuous prestressed concrete box girder section. The piers, of a small height in general, have a hollow rectangular section. Due to seismic activity, five central spans are fixed to the piers while the others are bored on the piers through elastomeric bearings, sliding in the longitudinal direction, blocked in the transverse one.

The main design challenge of the bridge was to reach, in the longitudinal direction, an optimum equilibrium between two conflicting requirements, namely:

• Maximize the seismic resistance by increasing the number of piers monolithically connected to the superstructure.
• Minimize longitudinal restraints by reducing the above number.

After a theoretical investigation, it was found that five piers was the best choice. The bridge was constructed using a moving formwork. It was designed by Denco SA, constructed by Michaniki SA, and completed in 2004.

12.3.5.2.3 SG16 Bridge in Lianokladi

This is a double railway valley bridge in central Greece. Its total length is 657 m (Figure 12.51). The superstructure is a continuous single-cell box girder with 15 spans (36 m + 13 × 45 m + 36 m) and a width of 13.9 m. The piers have a concave orthogonal section, founded on bored piles. The bridge was constructed with the moving formwork method. Due to the high seismicity of the area an extended seismic isolation has been implemented, consisting of spherical sliding/friction bearings and viscous hydraulic dampers. The bridge was designed by Kanon SA, constructed by Aktor SA, and completed in 2010.

FIGURE 12.50   Kristallopigi Bridge on the Egnatia Motorway. (Courtesy of DENCO SA.)

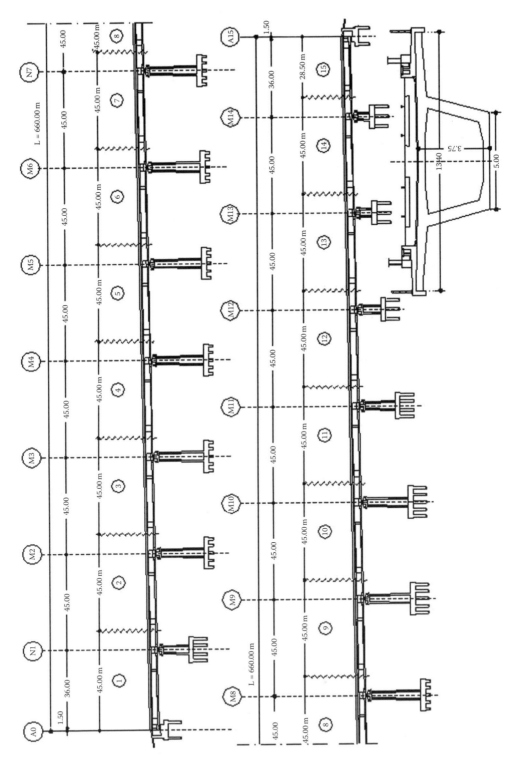

FIGURE 12.51 Longitudinal and cross-section of SG16 bridge in Lianokladi. (Courtesy of Kanon SA.)

FIGURE 12.52 T12 railway bridge on the Axios River. (Courtesy of MeteSysm SA.)

12.3.5.2.4 T12 Bridge on the Axios River

This is a railway double bridge on the Axios River in northern Greece with a total length of 800 m (Figure 2.52). The bridge superstructure is a continuous 18-span (40 m + 16 × 45 m + 40 m) prestressed concrete single-cell box girder of a constant depth of 3.6 m and width of 13.9 m. The bridge piers are circular hollow columns with an outside diameter of 4.5 m and wall thickness of 50 cm, resting on bored piles. The maximum pier height is 40 m. The bridge was constructed using the moving formwork method.

The bridge is located in an area of relatively high seismicity (design ground acceleration 0.24 g). In order to reduce the seismic response, the bridge was provided with an isolation system. The box girder is connected to each pier by two lead-rubber bearings (LRB) 1200 mm × 1200 mm with a lead core Ø 250 mm. Overall system damping is further increased by two longitudinal single-action fluid viscous dampers (F_{max} = 4500 kN) at each abutment. The isolation system makes the retention of seismic forces and displacement of the bridge within acceptable limits possible. In the longitudinal direction an effective damping close to 40% was obtained and the seismic displacement of the deck was reduced to 140 mm.

The bridge is equipped with tailor-made, adjustable expansion joints at the abutments, designed to accommodate total displacements up to 500 mm. Relative transverse displacement of the superstructure at the abutments is blocked by longitudinally sliding shear keys with a capacity of 4000 kN. It was designed by MeteSysm SA, constructed by Michaniki SA, and completed in 2008.

12.3.6 Cantilevering Bridges

12.3.6.1 General

Cantilevering has become a widely common construction method during recent years, used to build a great number of bridges, especially large-span bridges. Already by the end of 1960s, Tatarna Bridge, with a central span of 196 m, had been designed and constructed. Since then, several bridges have been constructed, two of which (Votonosi and Metsovo) have very large central spans (230 m and 235 m, respectively). The maximum length that has been constructed with this method is 1036 m (Arachthos Bridge).

The superstructure in all Greek bridges is a typical box girder with vertical walls. The height of the sections comes up to 13.5 m, with wall thickness 50–60 cm and bottom slab thickness up to 2.5 m. The ratio l/h at the supports ranges between 17 and 20 and at the key between 40 and 50. The usual concrete class is B45 (DIN 1045) and occasionally B55, while the class of prestressing steel is St 1570/1760 and more rarely 1670/1860. Almost all Greek bridges are monolithically connected to the pylons (due to earthquakes). The junction of the cantilevers at the keys is always monolithic.

The tendons are in general arranged in the top deck slab and more rarely in the webs, stressed successively during the construction of the segments; continuity tendons are obligatorily arranged in the bottom slab of the central and the side spans, stressed after the completion of the deck. The tendons are always wired after concreting through special launching machines. The construction joints are usually coarse without any special treatment; in some special cases corrugated joints have been formed. The length of the segments is 5 m with a fresh concrete weight up to 2.5 mN. Overpass-type travelers with a dead weight 600–850 kN (including the formwork) are used in Greece. The usual ratio between the side and the central span is 0.60:1. The last part of the side spans is usually cast in situ; in extremely inaccessible terrains, the side span may be totally constructed by using the traveler, with the aid of a temporary pier. Sometimes the deck line is restored and the stress of the structure is redistributed through jacking at the locations of the abutments.

A fundamental point for successful construction is precamber design; precamber is computed based on specific assumptions as to the materials and their behavior, such as creep and shrinkage, the environmental conditions and the construction time schedule. These assumptions are continuously checked during construction and the design is adequately adjusted. The current practice in precamber management in Greece is at a very high level; thus, the cantilevers meet at the key with a gap of 10–20 mm.

The key is usually constructed by using the traveler. Its length varies between 2 m and 4 m. After the completion of the two last segments, any difference in the levels is recorded and corrected, if required, with ballasts or with the narthex method. The two segments are fixed between them with steel beams (e.g., HEB 600) and tied by Dywidag bars into the deck, which inhibit their relevant vertical movement. For the longitudinal fixation of the cantilevers, horizontal steel beams are used, acting as spacers, in combination with the temporary stressing of 2–4 tendons at the bottom flange. After the total fixing of the two opposite segments, the key is concreted, the steel beams are removed, and the temporary tendons are relieved so that the continuity tendons are stressed.

12.3.6.2 Examples

12.3.6.2.1 Tatarna Bridge

This was the first Greek bridge constructed with the cantilevering method (1966–1970). In that period, it was the second-longest bridge span-wise in the world (197 m), after the avant-garde Bendorf Bridge (206 m). The bridge has three spans (98 m + 197 m + 155 m) (Figure 12.53). One of these is on a circular arc of a small radius, while the remaining spans are straight. Its typical cross section is a box with vertical walls of width 1.5 m + 8 m + 1.5 m = 11 m. The piers are rectangular hollow, spread founded on a rocky ground. The bridge was designed based on the know-how of that period, namely with a hinge at the key and prestressing by Dywidag bars Ø 32 St 80/105 with couplers at each segment; the length of the segments was 3.5 m. It was designed by A. Ikonomou, constructed by Odon and Odostromaton/P. Zografidis, and completed in 1970.

12.3.6.2.2 Servia Bridge

This bridge is over an artificial lake. It has a total length of 1339.2 m with a main span of 100 m (41.15 m + 22 × 42.30 m + 2 × 71.15 m + 2 × 100 m + 25.51 m) (Figure 12.54). The bridge was constructed with a mixed system (precast beams and cantilevering). Soil conditions vary significantly along the bridge from soft clay to limestone. In the region of silty clay the foundation bed was formed by removing the top 5 m of the original soil and replacing it with gravel, while in the case of the higher piers that happen

FIGURE 12.53 Tatarna road bridge, first implementation of the cantilevering method in Greece. (Courtesy of P. Zografidis.)

FIGURE 12.54 Servia road bridge over the homonymous artificial lake. (Courtesy of DOMI SA.)

to rest on limestone, the footings were anchored to the rock by prestressed anchors in order to ensure additional stability against overturning.

The hollow piers, ranging in height from 20 m to 55 m, were constructed by using slip forms. The superstructure was constructed partly in situ, as in the case of the cantilever construction, and partly by assembling 40 m long precast prestressed beams. The long spans of 100 m were constructed by creating 30 m cantilevers projecting from either side of adjacent piers and by using a standard 40 m precast beam to span the resulting intermediate gap. Whenever precast beams were used, the deck was formed by precast prestressed slabs and cast in situ concrete. It was designed by R. Morandi, constructed by joint venture Xekte SA–Skapaneus SA, and completed in 1975.

12.3.6.2.3 Bridge over the Corinth Canal

This bridge is one of the most important Greek bridges as it runs over the famous man-made Corinth Canal. It is the first railway bridge in Greece constructed with partial seismic isolation, by use of lead rubber bearings at the abutments and piers and longitudinal hydraulic shock absorbers at the abutments. It has three spans of 60 m + 110 m + 60 m, totaling 230 m in length (Figure 12.55). Its central span bridges the Corinth Canal, leaving a free height of 52 m for ship passage. The width of the deck is 12.7 m.

The superstructure is a continuous prestressed box girder of concrete B45/DIN 4227. The center span was constructed using the free cantilever method while the two side spans were cast in situ. The foundation is a combination of deep concrete shafts and shallow pile groups working as anchors to avoid excessive loading imposed on the banks of the channel. The bridge structure was designed to withstand design earthquake acceleration of 0.78 g while the isolation system was designed for the maximum credible earthquake of 0.975 g.

Provisions were made so that the bridge could also sustain tectonic movements of 0.25 m between its first abutment and pier due to a fault crossing between them. Frame-type special structures were designed at the two abutments in order to prevent uplift during extreme loading conditions. End and intermediate anchorage blocks were also designed to enable future external prestressing in case of retrofit or upgrade of the load capacity of the bridge. It was designed by Odomichaniki SA, constructed by Michaniki SA, and completed in 2005.

12.3.6.2.4 Arachthos Bridge

The bridge is one of the longest bridges ever constructed in Greece. It is a continuous frame composed of six spans of 142 m and two end spans of 92 m, totaling 1036 m (Figure 12.56). The maximum pier height is about 80 m. The bridge is located in an area of a future artificial lake with a water depth of 60 m. The first to sixth piers are connected monolithically into the superstructure while the seventh one and the two abutments are provided with special spherical bearings sliding in the longitudinal direction and a concrete block preventing the lateral movement of the deck. The structure is equipped with fluid viscous seismic dampers, acting in the longitudinal direction, with a maximum force of 2.7 mN

FIGURE 12.55 Corinth Canal railway bridge. (Courtesy of Odomichaniki SA.)

FIGURE 12.56   Arachthos road bridge on the Egnatia Motorway. (Courtesy of TTA SA.)

and a total design displacement of 1410 mm. The piers are designed as twin blades, with a variable cross section 1.80 × 9.00 m to 1.80 × 7.00 m; they are supported on a pile cap over a group of bored piles ∅ 1.5 to a depth of 10 m to 20 m. The design of the bridge is governed by earthquakes; with a fundamental period of 2.9 s and a strong external damping, the seismic behavior of the bridge was absolutely acceptable as far as the forces and the movements were concerned. The seismic response of the bridge was assessed by a nonlinear time history analysis using eight appropriate pairs of accelerograms and the American Association of State Highway and Transportation Officials (AASHTO) "Guide Specifications for Seismic Isolation Design," Interim 2000. It was designed by T. Tsiknias and Associates SA, constructed by AEGEK SA, and completed in 2009.

12.3.6.2.5 Votonosi Bridge

Votonosi Bridge is a twin road bridge in northwestern Greece, at section 3.2 of the Egnatia Motorway, with a total length of 490 m (Figure 12.57). The superstructure is a continuous three-span (130 m + 230 m + 130 m) prestressed concrete single cell box girder, rigidly connected to the piers and simply supported at the abutments, using two free sliding bearings as well as one longitudinally movable shear key.

In plan, the geometry of both branches consists of a clothoid ($A = 300$) and a circular arc ($R = 1000$ m). The piers of both branches, which define the central span, are located extremely close to the river slope. Their foundation consists of shafts ∅ 10 with effective depths of 20 and 25 m into good rock layers. The cross section of the piers is constant along the height. The dimensions of the hollow rectangular section are 5 m × 7 m ($t = 0.60$) for three of them and 6 m × 7 m ($t = 0.75$) for the highest (~52.7 m).

The box girder has a constant external width of 7 m. The height of the box decreases from 13.5 m at the pier to 5.5 m at the key. The thickness of the web takes three values (0.60, 0.525, and 0.45, constant in each section). The top deck slab has a thickness of 0.30 m, increased to 0.50 m near the web. The bottom slab has a variable thickness decreasing from 1.3 m (pier) to 0.30 m (key). The box girder is prestressed along its longitudinal axis. All tendons are arranged in the top deck slab, except for a small number in the vicinity of the piers, which are placed in the webs. The maximum length of the tendons is about 200 m. The prestressing tendons consist of 19 superstrands of diameter 15.7 mm (19T15), wired after concreting. Expansion joints with a capacity of 400 mm are placed at both ends of the bridge. Concrete B45/DIN

FIGURE 12.57 Votonosi road bridge on the Egnatia Motorway [22]. (Courtesy of DOMI SA.)

4227 (river aggregate), steel S500s, and prestressing steel 1570/1770 were used for the construction of the deck. It was designed by DOMI SA, constructed by Michaniki SA, and completed in 2005 [22], [24].

12.3.6.2.6 Metsovo Bridge

The Metsovo Bridge, situated in an area of outstanding natural beauty, within a seismically active zone with tectonic faults, forms a part of the Egnatia Motorway in northwestern Greece (Figure 12.58). The structure is a four-span frame, with a total length of approximately 537 m and a maximum span of 235 m; its height from the river bed measures 150 m approximately (Figure 12.59). The bridge has the longest span in Greece among cantilever bridges and one of the longest worldwide.

The typical cross section has a total width of 13.95 m. In the longitudinal direction both branches present a constant slope of 2.60%. The superstructure and piers M2 and M3 are monolithically connected to a three-span frame, supported at the first span by another pier (M1), on which the superstructure moves through sliding pot bearings, free in all directions. At the abutments the superstructure is free to rotate around the horizontal transverse axis and the vertical axis and moves longitudinally via sliding pot bearings. Regarding the foundation ground, the area base consists of thick-bedded sandstones of flysch with siltstone and (locally) limestone intercalations.

The first pier (M1), of approximately 42 m height at the left branch and 32 m at the right one, has an octagonal cross section with overall dimensions 4.5 m × 2.5 m. Its foundation is a spread footing on the rock, anchored against overturning through permanent prestressed anchors. The second pier (M2), of approximately 110 m height, is a box section, with variable dimensions for the decrease of the weight as well as for aesthetic reasons. The external dimensions begin from 12 m × 9.3 m at the base of the pier and end at 8 m × 8 m at the constant part of the section. The thickness of the longitudinal walls varies between 1.5 m and 0.70 m, while the thickness of the transverse walls varies between 1.35 and 0.70 m. The piers are founded on large shafts of effective circular section ∅ 12 m and depth $L = 25$ m. The third pier (M3), of approximately 32 m height, is twin with two walls of dimensions 2.3 m × 8 m, constant at its full height. The clear distance between the walls is equal to 1.5 m. The piers are fixed on large shafts of effective circular section ∅ 12 m and depth $L = 15$ m. The length of the pier has been artificially increased through the design of a preshaft for decreasing its rigidity.

The superstructure is a single cell box section, of concrete class B45/DIN 4227 (locally B55), monolithically connected to piers M2 and M3. At pier M2, the section has a height of 13 m, while at pier M3, its

FIGURE 12.58 **(See color insert.)** Metsovo road bridge on the Egnatia Motorway [23]. (Courtesy of DOMI SA.)

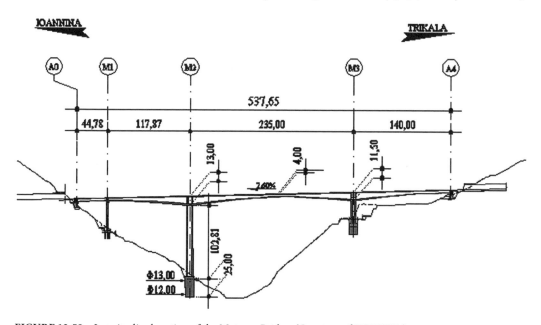

FIGURE 12.59 Longitudinal section of the Metsovo Bridge. (Courtesy of DOMI SA.)

height is 11.5 m. The box has a width of 7 m, while the total width of the deck is 13.95 m. The height of the cross section varies parabolically, decreased to 4 m at the key of the central span, as well as at the keys of the abutments at the side spans. The box girder is prestressed in the longitudinal direction through tendons of 19 superstrands of diameter 15.7 mm (19T15). The length of the tendons varies between 15 m and 250 m. Due to the length of the last span, the installation of a temporary steel pier was required, which was removed after the completion of the deck. An expansion joint of 440 mm is placed at abutment A0 and of 330 mm at abutment A4.

A very interesting and crucial point in the construction of the superstructure was the jacking of the free cantilevers at the permanent pier M1 by 175 mm and at the temporary one by 150 mm for the adjustment of the geometry and the reduction of the negative moment of the cantilevers. The bridge was designed by DOMI SA/LAP GmbH, constructed by Ellaktor SA, and completed in 2009 [23], [24].

12.3.6.2.7 Bridge G2 of the Egnatia Motorway

This is a twin road bridge in northwestern Greece at section 4.1.1 of the Egnatia Motorway, with a total length of ~350 m (Figure 12.60). The bridge superstructure is a continuous three-span (95 m + 160 m + 95 m) prestressed concrete box girder, rigidly connected to the piers and simply supported at each abutment, using two free sliding bearings as well as one movable shear key, for the block in the transverse direction. The girder cross section consists of a single cell box with diaphragms at each support. The depth of the cross section varies parabolically from 3.5 m at the center of the middle span and the abutments to 9 m at the front of the piers, resulting in a slenderness ratio of 1/45.7 (key) to 1/17.8 (supports). The deck width of each branch is 14.2 m. High-class concrete C35/45 was used for the superstructure. Prestressing in the longitudinal direction was applied by Ø 15.7 mm superstrands.

The bridge piers are rectangular hollow columns with outer dimensions 6 m × 7.2 m and wall thickness 0.80 m. The concrete class used was C30/37/DIN 4227. The height of the piers is around 60 m; they were erected using both sliding (M1) and climbing (M2) formworks. The piers are founded on circular concrete shafts. The diameter of each shaft is 10 m and their depth varies from 18 m to 25 m. The concrete class used was C20/25 except for the upper 3 m, where the class was increased to C30/37.

The viaduct is located in an area of moderate seismicity (design ground acceleration 0.16 g). The importance factor used was $\gamma_I = 1.3$. The behavior factor (q-factor) considered was $q = 3.5$ in the longitudinal as well as in the transverse direction. The bridge was designed by MeteSysm SA, constructed by AttiKat SA, and completed in 2008.

12.3.6.2.8 Greveniotikos Bridge

This is a twin bridge at section 4.1.5 of the Egnatia Motorway, crossing a valley. Each branch is a continuous frame composed of eight 100 m spans and two 60 m end spans for a total length of 920 m (Figure 12.61). The maximum pier height is about 45 m. The three central piers are monolithically connected into the superstructure. The two adjacent piers are equipped with teflon (PTFE) bearings sliding in the longitudinal direction and with lockup devices (four at each pier head, with a force capacity of 1500 kN) in order to distribute the seismic force to a larger number of piers. The four remaining piers and both abutments are equipped with longitudinal PTFE sliders without seismic restraints. Finally, all noncontinuous piers and both abutments are equipped with steel shear keys preventing transverse displacements of the superstructure. They are designed to resist a horizontal force in the order of 10 mN. Piers are supported on a pile cap over a group of bored piles 1.5 m in diameter and 20 m deep. The bridge was designed by TTA and Associates Ltd, constructed by Empedos SA/Pantechniki SA, and completed in 2006.

FIGURE 12.60　G2 road bridge on the Egnatia Motorway. (Courtesy of MeteSysm SA.)

FIGURE 12.61 Greveniotikos road bridge on the Egnatia Motorway. (Courtesy of TTA SA.)

12.4 Steel and Composite Girder Bridges

12.4.1 General

At approximately the end of the nineteenth and the beginning of the twentieth century, an extended application of steel bridges is noted in the railway network of the country. Most of these bridges were truss bridges and the remaining bridges were I-beams; all bridges were riveted. The materials used, checked in existing bridges still in service, has been proved to be of very good quality, homogeneous, without any internal defects; such a material could be easily classified as steel grade S235 of the current standards. The most common structural system was that of simply supported beams, with spans up to 80 m; however, the rather daring, for that period, continuous system was also commonly applied, with usual spans ranging between 50 m and 60 m. There were also sporadic applications of double overhanging beams with main spans up to 120 m, three-pinned arches with spans up to 80 m, and tied arches. Following that initial period, the production of steel bridges in Greece was significantly reduced, and practically came to an end after World War II, with the exception of the bridges on the Corinth Canal (1949, 1972) and three interchange overpasses in Athens (1980), one of which is still in service.

Steel bridges have recently returned in praxis, with the new form of composite bridges, mainly applied in road projects. The abolition of taxes, the development of transportation, and increased flexibility and reduced construction time were factors that favored their development. It is estimated that more than 110,000 m^2 of constructed composite decks exist today in Greece. The common characteristic of all Greek decks is that they are of a girder type; the decks consist of multiple built-up beams placed in their final position by the use of mobile cranes or by launching. From a structural point of view, statically indeterminate systems have prevailed, mainly continuous beams. Concrete deck is poured on load-bearing preslabs or trapezoidal steel sheeting; in a few cases specialized concreting travelers were deployed.

The design of these bridges has been performed mainly according to the German standard DIN 18800-1/81 and its amendments; the new bridges are being designed according to the European standard EN 1994 or the corresponding German version DIN FB 104. In general, steel grade S355 is used, with subgrades J_2G_3, J_R, J_0 according to EN 10025, Nelson-type shear studs, 10.9 class prestressed bolts, and B35 concrete according to DIN 1045 or C30/37 concrete according to EN 206.

12.4.2 Examples

12.4.2.1 Potidea Bridge

Potidea Bridge spans the homonymous canal at the beginning of Cassandra Peninsula. It is continuous, with one branch of width 11.5 m, a total length of 165.5 m, and three spans (44 m + 77.5 m + 44 m) (Figure 12.62). It is supported on piers of reinforced concrete founded on bored piles. It was the first composite bridge in Greece erected with the launching method. The deck is supported on the piers through elastomeric bearings. It consists of three main built-up I cross section steel girders of a constant height 2.4 m, with variable width of the top and bottom flanges. Each main beam comprises 13 segments connected between them with prestressed bolted connections. The main beams are connected between them with transverse trusses or massive diaphragms, while on the top flange horizontal bracing is used, necessary for the overall stability during erection.

FIGURE 12.62   Potidea road bridge over the homonymous canal at the beginning of Cassandra Peninsula. (Courtesy of DOMI SA.)

The steel beams are welded I-shaped sections. The thickness of the webs ranges between 20 mm and 25 mm. In addition, the dimensions of the flanges range from 750 mm × 25 mm to 880 mm × 80 mm (top flange), and from 880 mm × 25 mm to 880 mm × 80 mm (bottom flange). The secondary elements of the structure (diaphragms, bracing) are standard angle cross sections. The deck was launched on steel rollers placed on the abutments and the piers. For the reduction of the cantilever moment during launching, a light truss steel nose was placed in the front end. The maximum speed during launching was about 1 m/min. The maximum pulling force applied by the winch to overcome the friction forces was about ~5.50% of the total weight of the steel structure.

The deck slab, of thickness 0.30 m and concrete B35/DIN 1045, was poured on prefabricated slabs 0.10 m thick. To facilitate construction, some of the preslabs were placed on the steel structure from the beginning and were launched together. The slab is connected with the steel beams through Nelson-type shear studs. Due to the marine environment, an especially strict painting system was specified, with a guaranteed life expectancy of 20 years. Four layers were applied, with a total thickness of 550 μm, with glass fibers (inorganic zinc silicate, intermediate epoxy layers, external layer of polyurethane). It was designed by DOMI SA, constructed by Hellenic Technodomiki SA/Cimolai SA, and completed in 2001.

12.4.2.2 Bridge on Kymis Avenue

The twin road bridge on Kymis Avenue crosses the Kifissos River just outside Athens (Figure 12.63). The bridge is continuous, with a total length of 300 m, six spans (40 m + 50 m + 60 m + 60 m + 50 m + 40 m), and two independent branches of 11.80 m width each (Figure 12.64). Its superstructure is composite welded I-section with inclined webs, supported through elastomeric bearings on piers of rectangular hollow box cross sections, founded on bored piles, laterally braced by truss vertical diaphragms and lateral braces with angles on the top and bottom flange. The total height of the cross section, including the asphalt layers, is 3.1 m.

The steel beams are welded I-shaped sections. The thickness of the webs ranges between 20 mm and 27 mm. In addition, the dimensions of the flanges range from 850 mm × 27 mm to 850 mm × 99 mm (top flange) and from 900 mm × 40 mm to 900 mm × 90 mm (bottom flange). The I-girders were launched on steel rollers placed on the abutments and the piers. After the completion of launching and the removal of the rollers, the deck was gradually lowered by 60 cm and placed on the permanent

FIGURE 12.63   Kymis Avenue Bridge. (Courtesy of DOMI SA.)

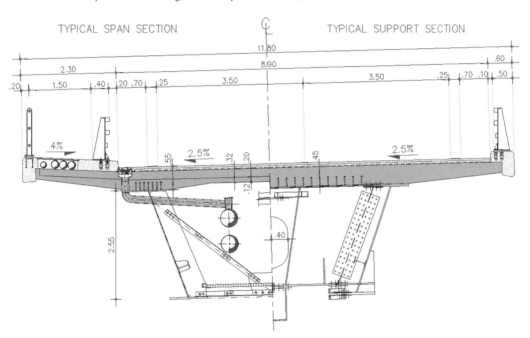

FIGURE 12.64   Typical cross section of Kymis Avenue Bridge. (Courtesy of DOMI SA.)

bearings; at the same time, it was slightly rotated around the longitudinal axis in order to achieve a transverse gradient of 2.5%. The average speed during launching was 12–13 m/h, while the average launching length was 50 m.

The deck slab, of variable thickness 0.32–0.47 m and concrete B35/DIN 1045, was poured on prefabricated slabs 0.12 m thick; it is connected with the steel beams through Nelson-type shear studs placed in pockets along the bridge. The painting system included three layers of paint (two epoxy layers and one polyurethane) of a total thickness of 240 μm. The bridge was designed by DOMI SA, constructed by Aktor SA/Cimolai SA, and completed in 2004.

12.4.2.3 Kifissos Avenue Bridge

This bridge has a total length of ~1400 m, with a theoretical span of 32.9 m over the Kifissos River in Athens (Figures 12.65 and 12.66). The bridge was formed as a composite bridge in order to reduce the construction time and the dead loads (due to foundation problems). The whole bridge was divided into 18 sub-bridges, with a maximum length of ~100 m. Each sub-bridge was formed with two partially prestressed concrete main beams placed on the banks of the river, parallel to its axis, steel cross girders every ~2.5 m, and a cast-in-situ deck slab with an effective thickness of 0.20 m. The typical sub-bridge consists of three spans of 25 m bridged by the main longitudinal prestressed beams of a rectangular cross section 2.05 m × 2 m, of concrete B35/DIN 1045, supported on elastomeric bearings. The connection between the steel crossbeams and the longitudinal concrete crossbeams was achieved by the use of prestressed anchors of 10.9 class and starters.

The steel girders are welded I-shaped sections. The thickness of the webs ranges between 16 mm and 20 mm. The dimensions of the flanges are constant 600 mm × 40 mm for the whole length. The secondary elements (diaphragms, bracing) are standard angles cross sections. The prestressed main beams

FIGURE 12.65   Kifissos Avenue Bridge. (Courtesy of DOMI SA.)

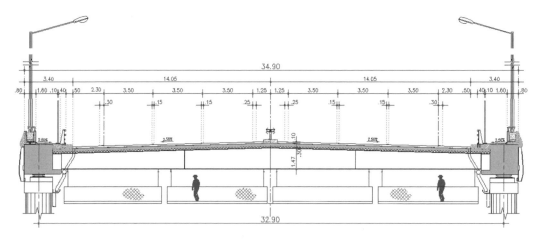

FIGURE 12.66   Typical cross section of Kifissos Avenue Bridge. (Courtesy of DOMI SA.)

parallel to the river were concreted on scaffoldings; before their connection to the steel cross girders, they were fixed against overturning using temporary bearings. The steel beams were put in place by the use of mobile cranes for a part of the structure, and by an overhead crane moving on the upper flange of the main beams for the rest. The deck slab, of concrete B35/DIN 1045 and an effective thickness of 20 cm, was poured on trapezoidal steel sheets.

The connection of the slab to the steel beams was achieved by the use of Nelson-type shear studs ∅ 22/200. The painting system included four layers of paint (three layers of epoxy base 75/150/150 and a surface one of polyurethane base 60 μm), with a total thickness of 435 μm. The bridge was designed by DOMI SA, constructed by Aegek SA/Aktor SA, and completed in 2004.

12.4.2.4 Railway Bridge in Athens

This is a twin railway bridge in Athens over the Kifissos River and the local roads (Figure 12.67). The bridge, with a total length of 436 m, is divided into four independent continuous sub-bridges (118 m + 115 m + 79 m + 78 m), with a maximum span of 34.5 m (Figure 12.68). The composite deck is supported on fixed or unidirectional sliding pot bearings, seated on concrete circular columns. The column foundation consists of piles and pile caps. The steel structures were erected by the use of mobile cranes.

The total width of the deck is 21.6 m (10.7 m per branch, with a 0.20 m gap between them), and the total height 3.1 m. The cross section is composite with two steel beams of 1.9 m height, placed with a distance of 5 m between them, connected with lateral bracing at the top and bottom flanges, as well as with truss diaphragms every 5 m and full crossbeams over the supports. The steel beams are welded I-shaped

FIGURE 12.67   Railway bridge in Athens. (Courtesy of DOMI SA.)

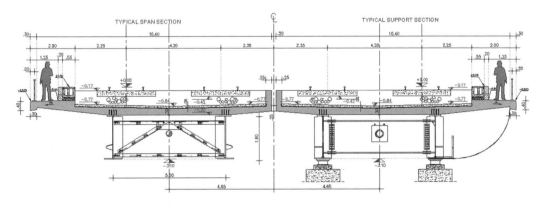

FIGURE 12.68   Typical cross section of the railway bridge in Athens. (Courtesy of DOMI SA.)

FIGURE 12.69   G4 road bridge on the Egnatia Motorway. (Courtesy of TTA SA.)

sections. The thickness of the webs ranges between 20 mm and 32 mm. In addition, the dimensions of the flanges range from 400 mm × 30 mm to 900 mm × 80 mm (top flange) and from 750 mm × 55 mm to 1000 mm × 100 mm (bottom flange). The secondary elements (diaphragms, bracing) are standard angles cross sections.

The deck slab, of concrete B35/DIN 1045, has a slightly variable thickness, with a minimum of 0.30 m. It was concreted totally in situ; for part of the project a suspended traveler was used, while that remaining was concreted on conventional formwork upon scaffolding. The painting system included three layers of paint (two epoxy layers and one polyurethane), with a total thickness of 295 μm. It was designed by DOMI SA, constructed by Aktor SA/Cimolai SA, and completed in 2004.

12.4.2.5 G4 Bridge

The G4 Bridge is a twin road bridge in northwestern Greece, at section 1.1.2 of the Egnatia Motorway, with a total length of 187.75 m (54.75 m + 65 m + 68 m) (Figure 12.69). The bridge comprises two piers with maximum pier height 75 m. The superstructure is composite continuous consisting of four welded steel I-beams. A concrete deck on reinforced concrete preslabs was placed on the steel beams before the deck launching. The piers are made of concrete B35/DIN 1045 and they have rectangular hollow box cross sections. The foundation of the bridge is achieved through rectangular wells. The superstructure is seismically isolated on the piers and abutments through lead rubber bearings. The bridge was designed by TTA and Associates Ltd, constructed by Pantechniki SA, and completed in 2008.

12.5 Arch Bridges

12.5.1 General

During the post-Ottoman period and the first period of the new Greek state a large number of masonry arched bridges were constructed, based mostly on experience and less on the scientific knowledge of the workmen (see also Section 12.1). These bridges, even if they were simple in their conception, contributed to the development of general experience. By the end of the nineteenth century a steel railway three-pinned arch of span 80 m and a steel tied arch had been constructed; unfortunately, these arches were destroyed during World War II. From then on and until recently, a few arched bridges of small spans, mainly of reinforced concrete, were constructed in the country; the construction cost of the necessary formwork soon proved prohibitive and this type has gradually disappeared. Recently steel arches have come again in praxis; two railway bridges and one road bridge are already under construction, while there are also a bridge for light traffic and electromechanical (E/M) pipes and three pedestrian bridges in the works.

12.5.2 Examples

12.5.2.1 Tsakona Bridge

This is an important road bridge of four traffic lanes with a total length of 490 m and a main span of 300 m (Figure 12.70); the main span consists of two parallel fixed vertical arches with a fully independent suspended composite deck. The rise of the arch reaches 45 m. A 40 m part of the arches is of concrete B45, while the rest, 260 m, consists of two steel arches of S355 J2 G3 class (according to EC 10025),

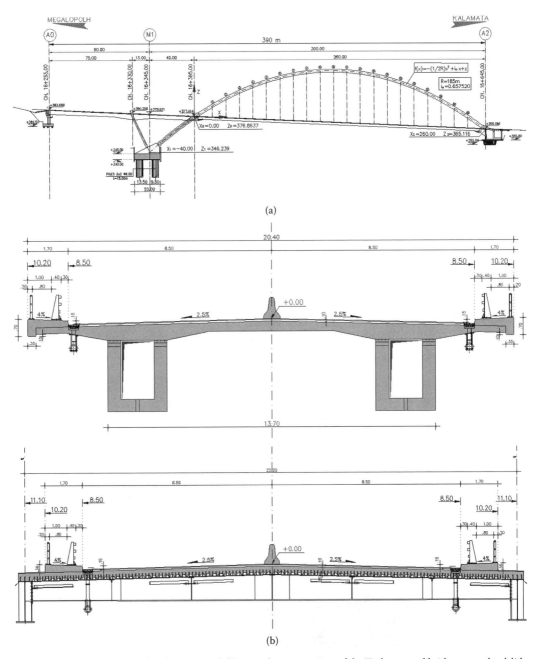

FIGURE 12.70 (a) Longitudinal section and (b) typical cross sections of the Tsakona road bridge over a landslide area. (Courtesy of DOMI SA.)

connected between them with a lateral K-type bracing. The bridge crosses an already landslide-prone mountainous area, whose slopes can underlie to tectonic displacements of 0.20 m.

The arch is a welded steel box section with external dimensions 1600 mm × 2800 mm and plates of thickness 70–120 mm at the flanges and 20–40 mm at the webs, accessible all along. The arches are connected between them with a lateral K-type bracing, consisting of pipes. The deck is hung from the arches with galvanized locked coil strands ⌀ 80–100. It is seated on the approach deck and the right abutment with four anchored elastomeric bearings 900 mm × 900 mm × 198 mm and eight dampers of 500 kN. The project is under construction; the erection of the arch is designed using the heavy lifting method with auxiliary towers. The bridge is designed by DOMI SA and is being constructed by the Joint Venture TERNA SA - ALPINE SA; expected completion year: 2013.

12.5.2.2 Bridge over the Corinth Canal

This is a steel arch bridge for light traffic and pipes over the Corinth Canal (Figure 12.71). The bridge has a span of 110 m and a rise of 18 m. The free height over the sea level is 55 m. The two parabolic arches of the pipe cross section are tied, supported on caps of bored pile groups; they are inclined, in contact between them at the crown. The superstructure consists of a truss box girder suspended from the arches through hangers at both sides, with a spacing of 5.5 m. The arches and the superstructure, consisting of welded segments, were prefabricated in the factory and connected on site by prestressed bolts. Additional external welding was applied to the internal connections between the elements of the arches. It was designed by D. Bairaktaris and Associates Ltd., constructed by Dorikos SA, and completed in 1998.

12.5.2.3 Ikonio Bridge

This is a single railway bridge, very close to Athens, consisting of seven simply supported horizontally curved spans with a total length of 270 m (25 m + 25 m + 110 m + 25 m + 30 m + 30 m + 25 m) (Figure 12.72). The main span of 110 m crossing a local landslide is an arch, while the remaining spans are composite box beams. There are two vertical tied arches, of parabolic form with a rise of 18 m; the axial distance between them reaches 12.8 m. Their section is box-shaped orthogonal, of constant width

FIGURE 12.71   Arch bridge over the Corinth Canal. (Courtesy of D. Bairaktaris.)

FIGURE 12.72 Ikonio railway bridge near Athens during construction. (Courtesy of A. Chrysinas.)

1000 mm and variable height 1200–2700 mm. The suspension of the deck from the arches is designed with 20 (2 × 10) hangers, of solid circular section Ø 130 at a spacing of 9.6 m. The ties of the arches, connected monolithically to them at their ends, have an I-section of height 3000 mm and flange width 850/1250.

The superstructure is welded T-shaped beams spaced at 2400 mm and with heights of 1100 mm. The steel orthotropic deck thickness is 25 mm, stiffened through longitudinal rib stiffeners every 500 mm. The arches are connected between them with box section struts of width 1000 mm and diagonals Ø 298.5/t = 12.5 mm.

The arch rests on elastomeric bearings on the box piers. The horizontal loads at the longitudinal direction are undertaken by four shock transmission units of Maurer type MHD 1000/1000, which act as seismic isolators. In the transverse direction, the bridge is supported through two shear keys. The behavior of the aforementioned units is viscous with a constitutive law $F = C \times v^{\alpha}$, where $C = 1000$ kNs/m and $\alpha = 0.02$; the choice of the exponent allows the activation of the units under earthquake and braking, but it renders them inactive under temperature actions. The approach spans rest on bored piles Ø 120 and Ø 150. The assembly of the arch was realized adjacent to its final position; after assembly, the arch was shifted by jacks to its final position. It was designed by N. Loukatos SA, constructed by Ellaktor SA, and completed in 2010.

12.6 Truss Bridges

Post–World War II there were not many new truss bridges constructed in Greece. Among those that were, the Aliakmonas road bridge on the Larissa–Kozani road, consisting of three simply supported spans of 50.4 m (Figure 12.73) and most of all the road and railway bridge of the Corinth Canal (Figure 12.74) with a span of ~80 m, dominate. The latest bridges, despite the fact that they were very near to the seismic focus of a strong earthquake in 1981, have not suffered major damages; the Aliakmonas road bridge, which has been repaired (replacement of bearings) and slightly improved, still operates.

Many of the old railway truss bridges have also been repaired and improved (or reconstructed), and some of them still operate due to very good initial construction and excellent maintenance. Among them the most important are the Papadia Bridge, total length of 277 m (52 m + 3 × 58 m + 54 m,

FIGURE 12.73   Aliakmonas road bridge. (Courtesy of the Technical Chamber of Greece.)

FIGURE 12.74   Old railway bridge on the Corinth Canal. (Courtesy of the Technical Chamber of Greece.)

reconstruction year 1949, Figure 12.75); Gorgopotamos Bridge, total length of 194 m (5 × 28 m + 2 × 27 m, initial construction year 1906, reconstruction year 1948, Figure 12.76); the Ekkara Bridges, total lengths of 120 m (6 × 20 m) and 80 m (4 × 20 m, construction year 1906, Figure 12.77); the Gallikos Bridge, total length of 122.4 m (2 × 61.2 m, construction year 1896, Figure 12.78); and the Axios Bridge, total length of 286 m (8 × 32 m + 30 m) and ~590 m (17 × 34.7 m, construction years 1888 and 1931, Figure 12.79).

FIGURE 12.75 Papadia railway bridge. (Courtesy of the Technical Chamber of Greece.)

FIGURE 12.76 Gorgopotamos railway bridge. (Courtesy of the Technical Chamber of Greece.)

FIGURE 12.77 Ekkara railway bridge. (Courtesy of DOMI SA.)

FIGURE 12.78   Railway bridge on the Gallikos River. (Courtesy of the Technical Chamber of Greece.)

FIGURE 12.79   Railway bridge on the Axios River. (Courtesy of the Technical Chamber of Greece.)

12.7 Cable-Stayed Bridges

12.7.1 General

Cable-stayed bridges started to be constructed in Greece in the 1980s. Today, three road and five pedestrian cable-stayed bridges exist, with the most important of all being the Rion–Antirion Bridge. The largest free span for a road bridge measures 560 m and for a pedestrian bridge 70.7 m. A composite superstructure has been applied, with a prefabricated or cast-in-situ concrete slab, and a massive concrete deck as well. As far as stay cables are concerned, the following have been applied:

- Parallel individual strands ∅ 5/8 inch, with PVC sleeve and protection with grease, inside a hard polyethilenium duct with external helix
- Parallel individual galvanized strands ∅ 5/8 inch, inside a hard PVC duct wrapped externally with plastic tape against solar radiation and the inside protected with cement grouting

- Galvanized locked coil strands
- Dywidag-type bars, with couplers, without any external protection

At the two largest bridges (Rion–Antirion and Euripus) it was necessary, due to vibrations of the cables, to connect them transversally with slightly prestressed tendons; especially at the Rion–Antirion Bridge the addition of external damping at the base of certain cables was required. The behavior of Greek cable-stayed bridges is very good, so far.

12.7.2 Examples

12.7.2.1 Rion–Antirion Bridge

12.7.2.1.1 General

The Rion–Antirion Bridge crosses the Gulf of Corinth near Patras in western Greece. It consists of an impressive multicable-stayed span bridge, 2252 m long, connected to the land by two approaches. An exceptional combination of physical conditions made the project quite unusual: high water depth, deep strata of weak soil, strong seismic activity, and fault displacements; in addition, a risk of heavy ship collision had to be taken into account. To make the bridge feasible, innovative techniques were developed: the strength of the in situ soil was improved by means of inclusions, and the bridge deck is suspended on its full length, and therefore isolated as much as is possible.

Due to the high water depth, construction of the main bridge faced major difficulties. In relation to this, foundation works, including dredging and steel pipe driving, but also precise laying of the required gravel bed under the pylon bases, formed an impressive work package requiring unusual skills and equipment. To achieve this task, the latest techniques available in the construction of concrete off-shore oil drilling platforms and large cable-stayed bridges were used.

The seabed of the stretch presents fairly steep slopes on each side and a long horizontal plateau at a depth of 60 m to 70 m. No bedrock was encountered during soil investigations down to a depth of 100 m. Based on a geological study, it was considered that the thickness of sediments is greater than 500 m. General trends identified through soils surveys are as follows:

- A cohesionless layer is present at mudline level consisting of sand and gravel to a thickness of 4 m to 7 m, except in one location (near the Antirion side), where its thickness reaches 25 m.
- Underneath this layer, the soil profile, rather erratic and heterogeneous, presents strata of sand, silty sand, and silty clay.
- Below 30 m, the soils are more homogeneous and mainly consist of clays or silty clays.

In view of the nature of the soils, liquefaction does not appear to be a problem except on the north shore, where the first 20 m are susceptible to liquefaction.

The seismic actions in the form of a response spectrum correspond to a 2000 year return period at seabed level (Figure 12.80). The peak ground acceleration is equal to 0.48 g and the maximum spectral acceleration is equal to 1.2 g between 0.20 and 1 s. Furthermore, the Peloponnese is drifting away from mainland Greece by a few millimeters per annum. For that reason, contractual specifications required the bridge to accommodate possible fault movements of up to 2 m in any direction, horizontally and/or vertically, between two adjacent supports. In addition, the bridge pylons must be capable of withstanding the impact from an 180,000 dwt tanker sailing at 16 knots.

12.7.2.1.2 Description

Connected to the land by two approaches, 392 m long on the left side and 239 m long on the right side, the cable-stayed bridge (Figure 12.81) consists of three central spans 560 m long and two side spans 286 m long. The four pylons of the main bridge simply rest on the seabed through a large concrete substructure foundation, 90 m in diameter and 65 m high at the deepest location (Figure 12.82).

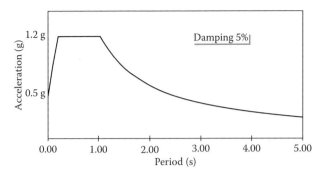

FIGURE 12.80   Rion–Antirion bridge: design response spectrum. (Courtesy of J. Combault–GEFYRA SA.)

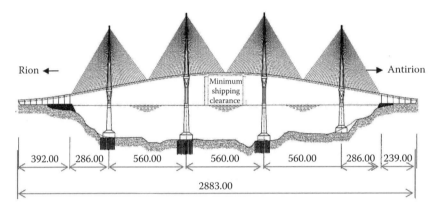

FIGURE 12.81   Rion–Antirion bridge: bridge elevation. (Courtesy of J. Combault–GEFYRA SA.)

FIGURE 12.82   Rion–Antirion bridge: foundation and inclusions. (Courtesy of J. Combault–GEFYRA SA.)

To provide sufficient shear strength to the top 20 m of soils, which are rather heterogeneous and of low mechanical characteristics, the upper soil layer of the seabed is reinforced by inclusions to resist large seismic forces arising from structural inertia forces and hydrodynamic water pressures. The inclusions are hollow steel pipes, 25 m to 30 m long and 2 m in diameter, driven into the upper layer at a regular spacing of 7 m to 8 m (depending on the piers). About 150 to 200 pipes were driven in at each pier location. They are topped by a 3 m thick, properly leveled gravel layer, on which the foundations rest. Due to the presence of a thick gravel layer, these inclusions are not required under the last pylon.

12.7.2.1.3 Pylons

The pylon bases consist of a 1 m thick bottom slab and 32 peripheral cells enclosed in a 9 m high perimeter wall, covered by a top slab slightly sloping up to a conical shaft. For the deepest pier, this shaft, 38 m in diameter at the bottom, 27 m at the top, rises 65 m over the gravel bed up to 3 m above sea level. These huge bases support, through vertical octagonal pylon shafts 24 m wide and nearly 29 m high, a 15.8 m high pyramidal head which spreads to form the 40.5 m wide square base of four concrete legs.

Rigidly embedded in the head to form a monolithic structure, the four legs (4 m × 4 m), made of high-strength concrete, are 78 m high; they converge at their tops to give the rigidity necessary to support asymmetrical service loads and seismic forces. They are topped by a pylon head, 35 m high, comprised of a steel core embedded in two concrete walls, where stay cables are anchored. From sea bottom to pylon top, the pylons are up to 230 m high (Figure 12.83).

12.7.2.1.4 Superstructure and Cables

The superstructure is a composite steel–concrete structure, 27.2 m wide, made of a concrete slab, 25 to 35 cm thick, connected to twin longitudinal steel I-girders, 2.2 m high, braced every 4 m by transverse crossbeams (Figure 12.84). It is fully suspended from 8 sets of 23 pairs of cables and continuous over its total length of 2252 m, with expansion joints at both ends.

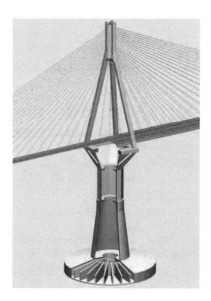

FIGURE 12.83 Rion–Antirion bridge: global view of a pylon. (Courtesy of J. Combault–GEFYRA SA.)

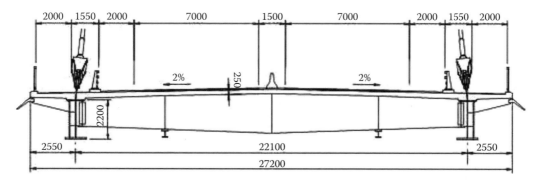

FIGURE 12.84 Rion–Antirion bridge: typical deck cross section. (Courtesy of J. Combault–GEFYRA SA.)

In the longitudinal direction, the superstructure is free to accommodate all thermal and tectonic movements, while the joints are designed to accommodate 2.5 m displacements under service conditions and movements up to 5 m under an extreme seismic event; in the transverse direction it is connected to each pylon with four hydraulic dampers of 3500 kN capacity each and a horizontal steel strut of 10,000 kN capacity (Figure 12.85). The stay cables are arranged in two inclined planes according to a semi-fan shape. They are made of 43 to 73 parallel galvanized strands individually protected.

12.7.2.1.5 Main Bridge Concept and Design Philosophy

The predominant loading for the design of the structure was seismic loading. The impact from the 180,000 DWT tanker, equivalent to a static horizontal force of 280 mN at sea level, generates horizontal forces and overturning moments at the soil–pylon base interface smaller than the seismic forces, and which only necessitate a local strengthening of the pylons in the impact zone. Due to environmental conditions, the primary major concerns were the soil bearing capacity and the feasibility of the required foundations.

After a thorough theoretical investigation a shallow foundation was chosen as the most satisfactory solution as long as it was feasible to significantly improve the top 20 m of soils. This was achieved by means of steel inclusions without any connection between them and the caisson raft, which allows for the pylon bases to partially uplift or slide with respect to the gravel bed. This type of soil reinforcement was quite innovative and necessitated extensive numerical studies and centrifuge model tests that validated the concept.

The other major points of concern were the large tectonic displacements and the large seismic forces to be resisted by the structure. Taking into account that the pylon bases could move on the gravel bed, it was found that the best way to minimize the problem was to make the pylons monolithic and the cable-stayed superstructure continuous, fully suspended, and therefore isolated as much as possible. Thus, the superstructure will behave like a pendulum in the transverse direction during a severe seismic event, its lateral movements being buffered and limited by the hydraulic dampers located at each pylon, while it is kept in place during the strongest winds by a horizontal steel strut connected to each pylon that is intended to break only during a seismic event of low occurrence (over 350 year return period).

These features of the project significantly reduce seismic forces in the superstructure and allow the bridge to accommodate fault movements between adjacent piers due to its global structural flexibility. According to the capacity design principle, the structure will only be the subject of "controlled damage" under an extreme seismic event at a limited number of well-identified locations:

- The pylon bases may slightly slide on the gravel bed and partially uplift.
- Plastic hinges may form in the pylon legs.
- The wind-stabilizing struts may fail and thus make the dampers free to operate both in tension and compression and able to dissipate a substantial amount of energy.

FIGURE 12.85   Rion–Antirion bridge: fully suspended deck, concept and connection to the pylons. (Courtesy of J. Combault–GEFYRA SA.)

The dynamic response of the structure was estimated by using artificial and natural accelerograms matching the design spectrum. This analysis took into account large displacements, hysteretic behavior of materials, nonlinear viscous behavior of the energy dissipation devices, sliding and uplifting elements at the raft–soil interface, and a geological model for the soil–structure interaction.

In addition to the finite element analysis, a nonlinear 3D push-over analysis was performed for the four leg pylons in order to estimate their available capacity after the formation of plastic hinges and to confirm their ductile behavior. Finally, the dynamic relative movement between the superstructure and a pylon during an extreme seismic event on the order of 3.5 m, with velocities up to 1.6 m/sec, a prototype test for the dampers, and special concrete confinement tests on the high-strength concrete used for the pylon legs, were performed at University of California at San Diego.

12.7.2.1.6 Construction

Pylon bases were built in two stages near Antirion: the footings were cast first in a 230 m × 100 m dry dock and the conical shafts were completed in a wet dock. In the dry dock, two cellular pylon footings were cast at a time (Figure 12.86). In fact, two different levels in the dock provided 12 m of water for the first footing and 8 m for the other.

When the first footing, including a 3.2 m lift of the conical shaft, was complete, the dock was flooded and the 17 m tall structure was towed out to the wet dock located 1 km away (Figure 12.87). At the wet dock, where the water depth reaches 50 m, the pylon base remained afloat (Figure 12.88) and was kept

FIGURE 12.86 Rion–Antirion bridge: pylon bases, works in the dry dock. (Courtesy of J. Combault–GEFYRA SA.)

FIGURE 12.87 Rion–Antirion bridge: from the dry dock to the wet dock. (Courtesy of J. Combault–GEFYRA SA.)

FIGURE 12.88   Rion–Antirion bridge: pylon base at the wet dock, progressing towards the top of the cone. (Courtesy of J. Combault–GEFYRA SA.)

FIGURE 12.89   Rion–Antirion bridge: driving the inclusions. (Courtesy of J. Combault–GEFYRA SA.)

in position by three big chains, two anchored in the sea and one on land. Cells in the base were used to keep the pylon base perfectly vertical through a differential ballasting system controlled by computer.

After completion of the conical shaft, the pylon base was towed to its final position and immersed on the reinforced soil. Meanwhile, the seabed was prepared at the future locations of pylon bases (Figure 12.89). The process of dredging the seabed, driving 200 inclusions, and placing and leveling the gravel layer on the top, with a depth of water reaching 65 m, was a major marine operation which necessitated special equipment and procedures. After being immersed at their final position, the pylon bases were filled with water to accelerate settlements, which were significant (between 0.20 and 0.30 m). This preloading was maintained during pylon shaft and capital construction, thus allowing a correction for potential differential settlements before erecting the pylon legs.

The huge pyramidal capitals are key elements of the pylon structure, as they withstand tremendous forces coming from the pylon legs. During a major seismic event, three legs can be in tension, while all vertical loads are transferred to the fourth one. For that reason, these capitals are very heavily reinforced (up to 700 kg/m^3 concrete) and prestressed (Figure 12.90). Their construction was probably the most strenuous operation of the project. The pylon legs during construction required a heavy temporary

FIGURE 12.90 Rion–Antirion bridge: pylon capital and pylon legs under construction. (Courtesy of J. Combault–GEFYRA SA.)

bracing in order to allow them to resist earthquakes. This bracing could be removed once the legs were connected together at their tops. The steel core of the pylon head was made of two elements which were placed at their final location by a huge floating crane able to reach a height of 170 m above sea level (Figure 12.91).

The composite superstructure was constructed in 12 m long segments, prefabricated in the yard, including their concrete slabs. The segments were placed at their final location by a floating crane (Figure 12.92) and bolted to the previously assembled segments using the classical balanced cantilever erection method. Only small joints providing enough space for an appropriate steel reinforcement overlap had to be cast in place. The superstructure was erected from two pylons simultaneously. Five to seven deck segments were put in place each week. In total the deck erection took 13 months.

Roughly 210,000 m³ of concrete, 57,000 tons of reinforcing steel, 28,000 tons of structural steel and reinforcing steel, and 3,800 tons of stay cables were required. The $800 million project was managed by a private build-operate-transfer scheme, led by the French company Vinci. Design was conducted by the joint venture of Vinci Construction Grands Projets/Ingerop/DOMI SA and Geodynamique et Structures. The independent engineer was Buckland and Taylor (Vancouver) supported by the Greek firm Denco SA (Athens) and the supervisor was Fabermaunsell Ltd. (United Kingdom). The construction was realized by a joint venture between the French company Vinci (the leader) and five Greek construction companies. Completed in August 2004, the Rion–Antirion Bridge opened to traffic 4 months before the contractual deadline ([15] to [18]).

12.7.2.2 Euripus Bridge

The Euripus Bridge crosses the Euboean Channel, 80 km north of Athens. It consists of a multicable-stayed bridge, 395 m long, connected to the land by two approaches, 143.5 and 156 m long (total length 694.5 m) (Figure 12.93). It was the first cable-stayed bridge constructed in Greece; with a concrete deck slab thickness of 45 cm in the central span of 215 m, it has perhaps the most slender deck in the world (Figure 12.94).

Apart from the normative traffic loads, specific actions had to be taken into account, such as the relatively high seismicity of the area (peak ground acceleration 0.20 g, ground velocity 0.17 m/s, ground displacement 5 cm), the tectonic movement between the two shores (0.20 m in every direction), the strong wind, the very difficult foundation conditions of one pylon (mainland), and the ship impact of a tanker or a freighter of 15,000 DWT sailing at 10 knots.

The central bridge consists of three spans, 90 m + 215 m + 90 m, supported on two piers and suspended from two pylons. The support at the piers permits movement along the longitudinal axis while blocking it transversally. On the contrary, the connection between the deck and the two pylons is monolithic. The stay cables are arranged into two vertical planes according to a semi-fan shape, anchored in the deck slab every 5.9 m. The deck slab is massive, with a constant depth 0.45 m along the whole length, except for the pylon area where the depth increases linearly to 0.75 m. It is prestressed transversally by

FIGURE 12.91 Rion–Antirion bridge: placing the steel core of the pylon head. (Courtesy of J. Combault–GEFYRA SA.)

FIGURE 12.92 Rion–Antirion bridge: placing the 12 m long segments. (Courtesy of J. Combault–GEFYRA SA.)

FIGURE 12.93 Euripus Bridge crossing the Euboean Channel, 80 km north of Athens [25]. (Courtesy of DOMI SA.)

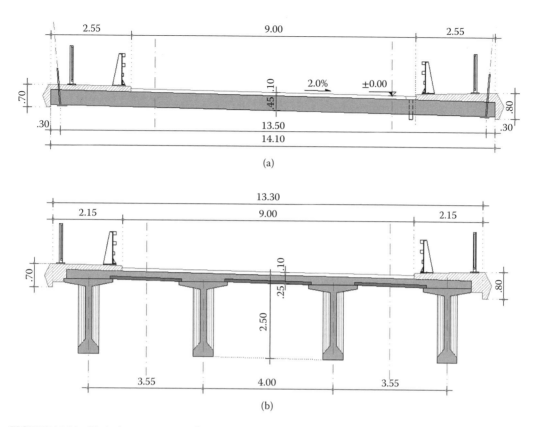

FIGURE 12.94 Typical cross sections of Euripus Bridge: (a) main bridge, (b) approach bridges [25]. (Courtesy of DOMI SA.)

four strand ∅ 5/8 inch tendons at a spacing of 0.59 m. Longitudinal prestress of the slab is foreseen only locally at the central area of the bridge, where the deck axial force equals zero, as well as at the area of the approach piers, where the horizontal force induced by the back stay cables is not sufficient to cover the tensile stresses resulting from high local bending moments.

The concrete is of grade B40/DIN 1045 for the deck slab (extra class, specified only for this project), the reinforcing steel is grade St 420, and the prestressing steel is grade St 1670/1860. The cables consist of seven 21 individual parallel galvanized strands of 0.6″ nominal diameter of grade 1670/1860 with a bearing capacity between 1820 and 5200 kN. Prestressing of the cables was made from the open pylon head. The anchor heads permit an adjustment of the cables up to 80 mm. The strands are arranged within a PVC pipe, filled with cement grout, and protected externally against solar radiation by plastic tapes.

The twin pylons are made of concrete grade B45/DIN 1045, and their dimensions vary from 4 m × 4 m × 0.50 m (at the base) to 2.5 m × 2.5 m × 0.40 m (at the top). The columns of each pylon are connected transversally through two double concrete girders, one below the deck and the other under the pylon top. Regarding the pylon head, an internal steel structure was designed to bear the cable forces and undertake their horizontal components, transferring (in parallel) the cable forces to the external mantle of concrete (acting as a composite structure). The twin pylons are founded on a pile cap at level +1.00. Their heights are 88.35 m for the left and 84.26 m for the right. Each pylon is founded on a group of 20 bored piles ∅ 1.2 m, connected by a pile cap 5 m high. They are founded in an ophiolithic complex (M5 pylon) or in a limestone layer (M6 pylon); their lengths are 30 and 15 m respectively.

Support of the side spans on the piers was realized by using hinged steel holding down devices. Owing to the tensile reaction force applied by the bridge to the piers, these are prestressed vertically and anchored into the soil by micropiles. The approach viaducts consist of four spans of 35.875 m long at the mainland coast and 39 m long at the island coast. They are formed of precast prestressed I cross section beams 2.25 m high, with 0.20 m web thickness and an in situ casted prestressed deck slab 0.25 m thick. The deck slab continues over the piers, connecting the individual spans. Crossbeams are designed only over supports in order to simplify construction.

The twin piers are designed as frames with columns of hollow rectangular cross sections, 2.5 m × 2.2 m × 0.30 m, and girders with massive cross sections, 2.2 m × 2.5 m. The foundation of the piers is shallow, with a local soil improvement by lean concrete (where necessary). In order to increase safety against sliding and overturning, lean concrete was placed between the footing and the surrounding rock. The bridge was designed by DOMI SA/Schlaich–Bergermann–Partner, constructed by Hellenic Technodomiki SA, and completed in 1993 [25].

12.7.2.3 Pallini Bridge

This is a one-span road bridge of 58.3 m over the Ring Road of Athens (Figure 12.95). The superstructure, of variable width 13.4–16.5 m, is composite with two steel main beams (S355 J_2 G_3) at the edges, 22 steel crossbeams at a spacing of 2.5 m, and a concrete deck with active thickness 0.20 m. The suspension system comprises three pairs of cables, with bearing capacity (F_{GUTS}) 5035 and 11130 kN. The pylon has an "Λ" shape, while the tensile external leg is made of prestressed concrete and the compressed internal leg is made of structural steel. The pylon is based on a box-shaped abutment, partially filled with gravel for stability reasons. The pylon head, on which cables are anchored, is constructed with high-strength concrete. Due to traffic restrictions, the deck was erected behind the right abutment and launched at its final position by using two temporary piers.

12.7.3 Cable-Stayed Pedestrian Bridges

During recent years, some interesting pedestrian bridges have been constructed in Athens. Among them, the steel bridge designed by Calatrava (Figure 12.96), the composite pedestrian bridge designed by DOMI SA (Figure 12.97), and the steel pedestrian bridge at Piraeus port designed by TTA (Figure 12.98) are the most important.

FIGURE 12.95 Pallini Bridge over the Ring Road of Athens. (Courtesy of DOMI SA.)

FIGURE 12.96 Calatrava pedestrian steel bridge. (Courtesy of DOMI SA.)

FIGURE 12.97 Composite pedestrian bridge in Athens. (Courtesy of DOMI SA.)

FIGURE 12.98 Pedestrian bridge at Piraeus port. (Courtesy of G. Assimomytis.)

FIGURE 12.99   Suspension bridge over the Pinios River [4]. (Courtesy of the Technical Chamber of Greece.)

12.8 Suspension Bridges

In the past, simple suspension bridges were constructed in mountainous areas, as pedestrian and animal crossings. There is a reference [4] to the Tsimovo Bridge on Arachthos River, of a 50 m span, suspended by a cast iron chain, with a wooden deck (constructed in 1930 and destroyed during World War II) and the Gogos Bridge on Kalarytikos River in the same area, of a 30 m span (constructed in 1928). During the postwar period, in 1961, a 94 m suspension pedestrian bridge was constructed over the Pinios River, with three spans (20 m + 54 m + 20 m), prestressed concrete deck in the side spans, steel beams in the central span (Figure 12.99), and main cables consisted of steel plates.

Attempts to construct large suspension bridges have failed. During the tender for the Rion–Antirion Bridge, two solutions were proposed dealing with a suspension bridge with a span exceeding 1000 m, which did not proceed due to special problems in the foundation of the pylons (a cable-stayed bridge was constructed instead). Also, on the Egnatia Motorway, near the village of Metsovo, a suspension bridge of excellent originality and aesthetics was proposed (designer Ove Arup), with a 551.5 m span, which did not proceed due to problems in anchoring the main cables and the high cost (a cantilever bridge was constructed instead).

12.9 Movable Bridges

There is no tradition in Greece for constructing and managing movable bridges, however there is a bridge with an important history and significant technical originality: the old Euripus Bridge. The bridge is located within the city of Chalcis crossing the Euripus Strait, 40 m wide and 8.5 m deep today.

It is most likely that the channel width at the crossing location was much larger in antiquity than today, given that Hesiodus (eighth century BC) [26] mentions that he went across from Boeotia to Euboea by ship. According to the descriptions of Diodorus Siculus [27], it seems that in 411 BC, people from Chalcis embanked the channel for defense reasons and left only a small canal for navigation, totally controlled by towers on either side and over which they constructed the first movable wooden bridge, which was repaired a little later in 378 BC.

Historians from later historical periods (e.g., the era of Alexander the Great, Hellenistic era, Roman era, Byzantine era), also confirm the existence of the bridge. Indeed, according to some written testimonies of the historian Procopius, there was a wooden bridge at this point, normally operating, which served road traffic as well as navigation [28]. The main features of the bridge (with reserve) are shown in Figures 12.100 and 12.101.

After Constantinople was taken by the Crusaders (1204 AD) and the establishment of Venetian rule in Greece, the channel was noticeably widened, a fort was built in its center, and two bridges were constructed at both sides of the fort; that bridge maintained its features during the Ottoman period as well. Since the re-establishment of the Greek state (1828), several attempts to construct a more modern bridge have been made; the last bridge, constructed in 1960, still exists (with some improvements made in the year 2000) and serves its purpose satisfactorily. It is a movable bridge with two independent longitudinally rolled cantilevering decks, of a 40 m span and 11 m width (1.5 m + 8 m + 1.5 m; rolling drawer type) (Figures 12.102 through 12.104).

In the bridge's closed position, the two cantilevers are hydraulically hinged in the center of the channel and fixed inside the hollow abutments, forming a bilaterally fixed girder articulated at the center. In order to open the bridge and set the channel free for navigation, the two cantilevers are hydraulically lowered by 0.63 m and thereafter mechanically rolled on tracks into box section tunnels of reinforced concrete (abutments), which also form the channel's quay walls.

The superstructure consists of two steel trusses, with a 21 m cantilever and a back end of 8.3 m. Their main support, at a distance of 1 m from the side of the quay wall, is realized onto two jacks of 3000 kN

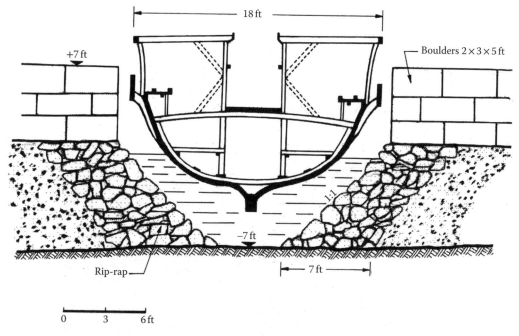

FIGURE 12.100 Assumed minimum cross section of the Euripus Strait for the passage of a trireme ($b = 5.5$ m). (Courtesy of Th. Tassios.)

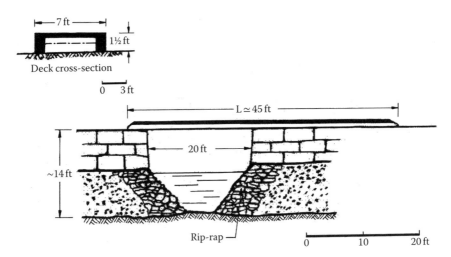

FIGURE 12.101   Assumed longitudinal section of the span of the Euripus Strait. (Courtesy of Th. Tassios.)

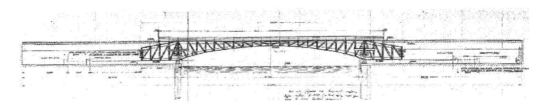

FIGURE 12.102   Longitudinal section of the rolling Euripus Bridge. (Courtesy of Th. Tassios.)

FIGURE 12.103   Completed Euripus Bridge. (Courtesy of Th. Tassios.)

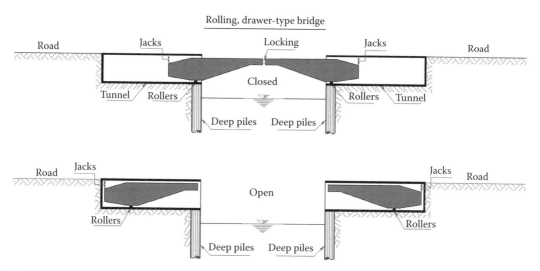

FIGURE 12.104 Sketch of the rolling drawer-type bridge. (Courtesy of Th. Tassios.)

FIGURE 12.105 Operation stages of the submersible bridge of the Corinth Canal. (Courtesy of DOMI SA.)

and the secondary, at a distance of 9.4 m from the same edge, onto two jacks of 1000 kN. The spread (shallow) foundation of the box section tunnels is assisted at the front, under the main support point of the girder, by Benoto-type piles ∅ 150, of 20 m length. The structural design is by Th. Tassios, the E/M design is by E. Mallakis, and it was constructed by a joint venture between Bio SA/Edok SA.

Beyond the aforementioned bridge there is also another submersible bridge at the east entrance of Corinth Canal that serves light vehicles and pedestrians (Figure 12.105). The span of the bridge is up to 38.6 m and its width is up to 9 m. The bridge sinks into the sea during the passage of the ships and is lifted back to its permanent position for the passage of pedestrians and vehicles. The deck of the bridge is formed with four steel beams of HEM 1000, crossbeams of HEA 200 (St S235) at each 1 m, and a wooden floor of coniferous wood of section 300/40.

There is another floating steel bridge between Lefkas Island and the continent. The bridge has a length of 70 m, a total width of 11.8 m (1.4 m + 9 m + 1.4 m), and a bearing capacity of 600 kN. The bridge is divided in three pieces, one central piece 50 m long and two side catapults with a length of 10 m each. For serving small ships, the two catapults are hydraulically lifted, while for the passage of large ships the whole bridge is rotated around the central vertical axis.

12.10 Foundation Techniques

The morphology of the Greek territory is multifarious and in some mountainous areas very unfavorable, resulting in the application of almost all known foundation types. The choice of foundation depends on the local topography, geotechnical conditions, and seismic actions. The most simple and cheap type of foundation is spread footings of constant or variable thickness. This type is applied mainly in rocky, semi-rocky, and well-compacted sandy-gravely soils.

Foundations with vertical bored piles of diameter $\varnothing$ 100, $\varnothing$ 120, and $\varnothing$ 150, and in some special cases $\varnothing$ 180, are widespread. The custom length of the piles is up to 40 m. The piles are constructed in dry conditions in the case of cohesive soils with stable walls and bentonite in the case of soft soils; in extremely soft soils, casing is indispensable. Piles are usually designed as fully elastic elements where predefined plastic hinges are not permitted; the pile heads are, however, reinforced as potential plastic hinges at their connection with the pile cap. The same rule is also applied on both sides of an interface between two different geotechnical layers (in a length of $\pm 2\varnothing$). The reinforcement of the piles is at least 1% of the geometric pile cross section up to 6 $\varnothing$ depth from the pile cap and 0.50% deeper. The longitudinal bars are surrounded by a continuous spire of minimum $\varnothing$ 10 diameter, which is significantly increased in the area of potential plastic hinges.

Until recently, the design of piles followed the German standards DIN 1054 and DIN 4014, as well as Greek guidelines for the seismic design of bridges. Their bearing capacities were analytically calculated based on the aforementioned standards or by using international standards and were frequently verified through loading tests. From 2007 on, design follows the Eurocodes or DIN FB. During the last 15 years, shafts of massive circular cross sections, and in some cases of hollow rectangular cross sections, especially in inaccessible mountainous areas, have usually been applied. This type of foundation offers increased reliability regarding the fixing of high piers with significant horizontal loads and overturning moments, and in addition it easily adjusts in inaccessible mountainous areas. Such shafts with 13 m diameter and depth up to 35 m have already been designed and constructed. Their excavation is achieved with the use of explosives, with controlled predrilling in the perimetric zone in order to reduce the rock mass disturbance in the internal part of the excavation. The immediate support measures consist of shotcrete, soil nails, and prestressed anchors; in highly loose soils a perimetric shield is constructed with micropiles connected between them by a shotcrete wall.

The advantage of shafts is the ability to view and control the foundation level. Their reinforcement is perimetrically arranged, while the concreting is progressively executed at heights of 4 m to 5 m, thus maintaining the concrete temperature below 60°C. In order to achieve this goal, sometimes freezing and maintenance of the concrete with a water flow of 12°C to 16°C are required. Low-class concrete (e.g., C20/25) is typically used except for the highest segment, which is formed as a pile cap of strongly reinforced high-class concrete (e.g., C35/45).

In some foundations micropiles were used either as inclusions in order to increase the bearing capacity of the soil, or as anchors in order to avoid the overturning of the piers under seismic loads (Euripus Bridge). The micropile diameter that is usually applied ranges from 20 cm to 30 cm, with a reinforcement of concrete or structural steel (pipes or I-sections). The micropile grout, which transfers the tensile force to the surrounding soil and at the same time protects the steel element, is inserted into the hole by gravity or pressure. Prestressed permanent anchors of massive circular sections with steel bars of class 1080/1230 and double protection corrosion systems have also been used instead of micropiles (Metsovo Bridge). In special cases, ground improvement techniques with steel pipes (inclusions; Rion–Antirion Bridge) or stone columns (Nestos Bridge) have been used, but the most frequent technique is soil replacement with sandy-gravely material.

12.11 Seismic Isolation and Energy Dispassion

12.11.1 General

The high seismicity of Greece has always been a problem in the design of bridges, which has led to a continuous increase in the seismic protection requirements. The introduction in the 1970s of compact elastomeric bearings released the seismic response of the superstructure from the ground movement, resulting in significant reduction of the inertia forces, but a simultaneous displacement increase. In terms of spectral analysis, the existence of the elastomeric bearings increased the fundamental period of the oscillated superstructure, moving the structure to the right in the graph of the response spectrum; the elastomeric bearings behave practically fully elastic, acting as isolators of the seismic shock without absorbing any seismic energy.

By increasing the size of the structures, it appeared that dealing with the seismic behavior of large bridges with high piers by using only rubber bearings was not satisfactory because of the significant displacements of the superstructure and the consequent effects (second-order effects, expensive expansion joints etc.). Thus, the necessity for additional absorption of the seismic energy, which can be achieved only through specific devices (dampers) became obvious. By the end of 1990s and especially after 2000, a systematic implementation of such dampers began; the revision of the Greek Earthquake Standard (2000) assisted in the introduction of new modern concepts on earthquake engineering (response spectra, spectral, time history and push-over analysis, natural and artificial accelerographs, etc.) to a great extent.

From the globally proven seismic isolation systems, hydraulic viscous dampers were widely applied; elastomeric bearings with lead cores (LRB), sliding/friction bearings, and elastomeric bearings with high damping were used to a lesser extent. Mechanical-type dampers were not yet implemented. Seismic isolation systems are designed according to the Greek Earthquake Standard and the Greek seismic isolation guidelines for bridges, which have been incorporated in EN 1998-2. According to these guidelines, for the design and implementation of a seismic isolation system, the following are required:

- A detailed time history analysis based on accelerographs (natural, artificial, or semi-artificial) compatible to the nominal design spectrum
- The carrying out of laboratory tests for each prototype isolator
- Assessment of the change in the basic parameters of the isolators over time
- Accessibility for inspection, maintenance, and replacement (if necessary)

Essential requirements and compliance criteria can be summarized as follows:

- In a seismically isolated bridge the basic design requirements of conventional bridges must be satisfied.
- The seismic response of the structure must remain practically elastic entirely.
- Increased reliability is required in the strength and integrity of the isolation system.
- The seismic isolation system must be self-restoring in all horizontal directions.
- The seismic isolation system must also be able to respond to strong aftershocks.

It is estimated that seismic isolation has been implemented in Greece in more than 15 bridges, some of which are listed below.

12.11.2 Examples

12.11.2.1 Rion–Antirion Bridge

The Rion–Antirion Bridge (Figure 12.36) was designed to withstand an extreme seismic event (ground peak acceleration 0.48 g and tectonic movements between adjacent piers up to 2 m) corresponding to a return period of 2000 years. The protection of the main stay-cable bridge comprises fuse restraints and viscous dampers acting in parallel, connecting the superstructure to the piers. The restrainers were designed to work as a rigid link in order to withstand high wind loads. Under the specified earthquake (350-year return period), the fuse restrainers are designed to fail, leaving the viscous dampers free to dissipate the energy induced by the earthquake into the structure.

Four viscous dampers (F_{max} 3500 kN, stroke ± 1750 mm) and one fuse restrainer (F_{max} 10500 kN) were installed at each pylon, while at the transition piers two viscous dampers (F_{max} 3500 kN, stroke ± 2600 mm) and one fuse restrainer (F_{max} 340 kN) were designed. The seismic isolation system in the northern approach bridge, consisting of 6 spans of 38.26 m with prefabricated beams, comprises 96 elastomeric bearings 400 mm × 500 mm × 120 mm, 32 longitudinal dampers 80/400 is the type of damper (capacity 800 kN, stroke 2 × 200 mm) and 24 transverse 120/400 (F_{max} 1200 kN, stroke ± 200 mm). The constitutive law of all dampers is of type $F = c \times v^{\alpha}$, where $\alpha = 0.15$ The seismic isolation system was designed by Vinci Construction Grands Projects/Ingerop (France), DOMI SA (Greece), and the dampers were manufactured by FIP (Italy) [21].

12.11.2.2 Arachthos Bridge

In the Arachthos Bridge (Figure 12.56), owing to the large mass of the structure and the very high piers, in order to absorb a large part of the induced seismic energy, two seismic dampers of large capacity were placed in each of the two individual branches of the bridge, between the superstructure and right abutment at the height of the center of gravity of the cross section. The dampers were hydraulic (viscous), operating with silicon fluid with the following technical data:

- Max force 2715 kN, stroke 1410 mm, max velocity 0.364 m/s
- Axial stiffness 78,000 kN/m and damping $c = 3,159$ kN/(m/s)$^{\alpha}$ where $\alpha = 0.15$
- Ambient air temperature range of -10° to +60°

The weight of each unit together with its anchor plates is 8600 kg. A very strong abutment was designed to resist the concentrated force of the dampers The seismic isolation system was designed by TTA SA (Greece), and the dampers were manufactured by FIP (Italy) [19].

12.11.2.3 Nestos Bridge

The Nestos Bridge (Figure 12.43) was designed with extensive use of seismic isolation, in order to reduce the large inertia forces of the superstructure, due to the very high seismic design ground acceleration and the poor local soil conditions (extensive soil improvement by stone columns has been carried out in the foundation area). The isolation system of the bridge consists of 240 elastomeric bearings (which give to the low decks a satisfactory fundamental eigenperiod from 1.45 s to 1.75 s, serving in parallel in undertaking the vertical loads) and 24 longitudinal and 48 transverse dampers.

All dampers are hydraulic (viscous), operating with silicon fluid silicone, fully accessible for inspection and replacement (if necessary) and able to react to strong aftershocks, which can occur a short time after the main shock. The technical specifications of the system are as follows:

- Elastomeric bearings 500 mm × 600 mm, elastic thickness 154 to 187 mm ($Gb = 1$ to 1.5 MPa)
- Constitutive law of all dampers: $F = C \times v^{\alpha}$
- Longitudinal dampers OTP 150/700
  - Max force 1500 kN, stroke ± 350 mm, max velocity 0.90 m/s
  - Damping C = 1400 kN/(m/s)$^{\alpha}$ where $\alpha = 0.15$

- Transverse dampers OTP 75/600
  - Max force 750 kN, stroke ± 300 mm, max velocity 0.90 m/s
  - Damping C = 700 kN/(m/s)$^\alpha$ where α = 0.15
- The overall effective damping of the structure was in the order of 30%.

The seismic isolation system was designed by DOMI SA (Greece), and the dampers were manufactured by FIP (Italy) [20].

12.11.2.4 Bridge SG12

The seismic isolation and energy dispassion system of the SG12 railway bridge (Figure 12.47) comprises friction pendulum bearings and hydraulic viscous dampers, the use of which increases the fundamental period of the structure and the overall damping of the system, respectively, and subsequently reduces the seismic forces and displacements of the structure. Friction pendulum bearings and longitudinal viscous dampers are provided at all piers and abutments. In the transverse direction, the superstructure is fixed on the abutments, whereas dampers are provided at all piers. The transverse force is transferred to the abutments through vertically installed sliding bearings, allowing movement in the longitudinal direction; the bearings are located at the sides, approximately at the level of the shear center of the cross section, in order to avoid the development of an important overturning moment at the level of the viscous dampers and possible uplift at the bearings.

The dampers act as shock transmission units; as such, they remain inactive in case of gradually imposed forces (e.g., temperature effect), but they transfer the breaking and centrifugal forces to the substructure, limiting the displacements of the superstructure to 10 mm approximately, whereas they have a nonlinear viscous behavior in case of an earthquake. For the design of the isolation system of the bridge a nonlinear time history analysis was performed. Seven different design accelerograms were applied to two models, assigning different values for the friction and damping coefficient of the bearings and dampers respectively, in order to obtain the most onerous displacements and design forces for the system. The seismic isolation system was designed by Kanon, and the dampers and bearings were constructed by EPS.

12.12 Bridge Maintenance

12.12.1 Monitoring

Over the last decade, the systematic instrumentation of major bridges also began in Greece, with the objective target of monitoring the actual actions on the structure and its responses in order to prevent unexpected situations and facilitate maintenance. Particular attention has been paid to the recording of seismic events, both at the free area and the individual structural elements of the bridge (e.g., the Rion–Antirion and Euripus bridges), carried out by accelerometers. The same accelerometers, with appropriate adjustment, record the response of those elements to other dynamic actions (i.e., wind, traffic loads). Furthermore, anemometers are settled in order to measure speed and wind direction (e.g., in the Rion–Antirion Bridge for recording speeds up to 50 m/s), jointmeters for measuring the real movements in expansion joints, thermometers and hygrometers for reflecting environmental conditions, thermometers embedded in the structural members for measuring the actual temperature of the structure, and so on.

The micro-movements of the piers in areas with soil creep phenomena (Panaghia Bridge) or wider tectonic movements (Rion–Antirion Bridge) are continuously monitored with geodetic methods. In addition to these, in cable-stayed bridges special measurements are carried out in order to specify the actual force and natural frequencies of the stays. The management and evaluation of such information is the next issue to be resolved; in the Rion–Antirion Bridge, which is a concession project, the data is managed by the concessionaire. In the bridges of the Egnatia Motorway, which belong to a company of the public sector, the information is collected and evaluated by a specific department of the company itself. The same applies to the railway bridges of the railway network, controlled by companies of the public sector. The other bridges are controlled by the Ministry of Public Works.

12.12.2 Inspection and Maintenance

Particular emphasis has been paid to bridge inspection and maintenance in the last 15 years. Future maintenance criteria have been also considered in the design of the structures. Furthermore, specific structural forms are required by the design guidelines in order to facilitate access to and inspection of the crucial areas of the structure. The design guidelines for Greek bridges require the existence of corridors beneath the expansion joints, as well as an appropriate area for the inspection and replacement of the bearings, dampers, and so on. According to the same guidelines all hollow piers must be accessible for inspection. Access is generally provided at the head of the pier (it is not desirable to create openings at their base, where there is a possibility of plastic hinge formation in the case of a very strong seismic event). Access along the piers is served by staircases (usually steel ones) and, in special cases, by a lift (e.g., in the Rion–Antirion and Metsovo bridges).

Superstructures are generally inspected using a special vehicle. Internal inspection of superstructures with box cross sections is performed from one end to the other via manholes at the diaphragms. In special cases (e.g., the Rion–Antirion, Euripus, and Kifissos bridges) a special suspended vehicle, manually or electrically operated, is provided. The inspection of bridges is primarily visual, accompanied by nondestructive material tests. Assessment of the degree of oxidation and in particular of the tendency towards oxidation is foreseen for reinforcement steel. For concrete the following are usually required:

- Integrity checks by ultrasonic or electromagnetic waves
- Strength assessment by impact tests and core specimens
- Chloride control
- Carbonation thickness assessment

Based on the inspection results and other data recorded by the instruments, a maintenance procedure is planned. Maintenance is distinguished as regular and exceptional (after a catastrophic event, e.g., an earthquake). Regular maintenance has usually three levels: an initial level of routine maintenance (e.g., annual) limited to a simple visual inspection, a mid-term level (e.g., 3 years), and a long-term level (e.g., 7 years). Important bridges have their own maintenance programs, which are based on the special requirements of such structures; in current bridges an ongoing, mutually common maintenance program is implemented.

12.13 Prospects

Road bridges built over the last 15 years, supplemented by those in the five new major highway projects, which are already being built and will be completed by 2015, largely cover the road needs of Greece. This is not the case for railway bridges, where, despite progress, there are still deficiencies, which are expected to be covered in the next decade. A need will also exist for urban interchanges and pedestrian bridges, due to the continuous increase of traffic volume, not only in Athens but also in the other major cities.

The major challenge for all Greek bridges will be systematic maintenance and improvement in order to meet today's increased traffic loads and more particularly to cover updated seismic requirements. This improvement is expected to be based on modern standards such as the Eurocodes and new materials (e.g., fiber-reinforced polymer[FRP]); it is also certain that such improvement will be governed by a new design philosophy, which will refer to credibility, durability, aesthetics, and adaptation of the structure to the environment. An important role in this effort is the instrumentation of bridges and the systematic analysis and evaluation of their measurements. The significant experience in repair and strengthening gained by Greek engineers after the major earthquakes of the last 30 years will be valuable in the process of bridge improvement.

References

1. Makris, G. D. *The Bridges in Ancient Greece*, Aiolos Editions, Athens 2004.
2. Kathimerini, *Special Oblation to the Stone Bridges of Greece*, 13.02.2000.
3. Teperoglou, P. Lecture in the Open University of Berlin, 1997, unpublished.
4. Kosteas, A. *Historical Bridge Development*, Technical Chamber of Greece editions, Athens 1941.
5. Kosteas, A. *Steel Bridges*, Technical Chamber of Greece editions, Athens 1978.
6. Lazaridis, A. *Amphipolis, Ministry of Civilization* editions, Athens 1993.
7. Mango, C. *Byzantine Architecture*, New York, 1985.
8. Evert, L. *Syria: Greek Resonances*, Adam Editions, 1993.
9. DIN 1075/59, 73, 81. Concrete bridges; design and construction.
10. DIN 1072/59, 72, 85. Road bridges; design loads.
11. DIN 4227/53, 79, 88, 95. Prestressed concrete structures; design and construction.
12. DIN 1045/43, 59, 72, 88, 96. Reinforced concrete structures; design and construction.
13. *Guidelines for Designing Bridges in Seismic Regions*, Greek Ministry of Public Works editions, November 1999.
14. *Hellenic Outstanding Structures*, 7th FIP Congress, Technical Chamber of Greece editions 1974.
15. Combault, J., Morand, P. and Pecker, A. 2000. "Structural Response of the Rion–Antirion Bridge." *12th World Conference on Earthquake Engineering*. Auckland, New Zealand 2000.
16. Combault, J. 2009.The Rion–Antirion Bridge. "What Made the Bridge Feasible?" *3rd Greek Conference on Earthquake Engineering and Engineering Seismology*. Athens, Greece 2009.
17. Pecker, A. 2003. "A Seismic Foundation Design Process: Lessons Learned from Two Major Projects, The Vasco da Gama and the Rion–Antirion Bridges." *ACI International Conference on Seismic Bridge Design and Retrofit*. La Jolla, California 2003.
18. Teyssandier, J. P., Combault, J. and Morand, P. 2000. "The Rion–Antirion Bridge Design and Construction." *12th World Conference on earthquake Engineering*. Auckland, New Zealand 2000.
19. FIP Industriale. Technical Report on the Arachthos Bridge Viscous Dampers, September 2006.
20. FIP Industriale. Technical Report on the Nestos Bridge Viscous Dampers, June 2009.
21. FIP Industriale. Technical Report on the Rion Bridge Viscous Dampers, May 2002.
22. Stathopoulos, S. et al. "Votonosi Bridge in Greece," *Fib Symposium on Segmental Construction in Concrete*, New Delhi 2004
23. Stathopoulos, S. et al. "Metsovo Bridge, Greece," SEI, Issue 1 2010.
24. Tzaveas, Th. and Gavaise, E. "Planning, Design and Construction Challenges of Major Bridges in Egnatia Motorway," SEI, Issue 1 2010.
25. Stathopoulos, S. "Evripos Bridge," *8th Greek Concrete Conference*, Xanthi 1987.
26. Hesiodus, *Works and Days*, Editions Zitros, 2005.
27. Diodorus Sicilus, *Library*, Editions Kaktos.
28. Procopius, *Buildings*, Georgiadis Editions, Athens, 1996, section 4.3.20.
29. Herodotus, *Histories*, Cactus Editions, Athens, 1992.
30. Georgakopoulos, K. *Ancient Greek Exact Scientists*, Georgiadis editions, Athens, 1995.

13

Bridge Engineering in Macedonia

Tihomir Nikolovski
FAKOM AD-Skopje

Dragan Ivanov
University Sts Cyril &
Methodius, Skopje

13.1 Introduction

The Republic of Macedonia is a small country in the central part of the Balkan Peninsula. To the north, south, east, and west, it borders on Serbia and Kosovo, Greece, Bulgaria, and Albania, respectively. Its territory covers an area of about 25,713 km² and it has about 2 million inhabitants. Macedonia is a mainly mountainous country with several fertile valleys: Pelagonia, Tikves, Strumicko Pole, Ovce Pole, Skopsko Pole, and Polog. The biggest river is Vardar, with a total length of 420 km, which rises in West Macedonia, near Gostivar, passes the border with Greece in the vicinity of Gevgelija, and empties into the Aegean Sea near Thessaloniki. The Greek name for the Vardar is Axios. Its bigger tributaries are Treska, Lepenec, Pcinja, Bregalnica, and Crna Reka.

13.1.1 Road Network

The Republic of Macedonia has a well-conceptualized network of national and regional roads which cover well its territory and enable good transportation with neighboring countries and beyond, and with Southeast and Central Europe as well as the Near East. Two important European transportation corridors pass through Macedonia: the East–West (EC-8) corridor from Burgas to Durres and the North–South (EC-10) corridor from Central Europe to the south of Greece (Figure 13.1).

The roads in the Republic of Macedonia are classified into three categories: national roads connecting its territory with neighboring countries, regional roads connecting adjacent municipalities according to the territorial division of the state, and local roads within the frames of a single municipality. The national and the regional roads are administered by the Agency for State Roads of the Republic of Macedonia. The total length of the national roads amounts to 934 km, of which 244 km are highways and 3620 km are regional roads. According to data from 2007, there are a total of 990 bridges on the national and regional roads in Macedonia, of which 43% are on the national roads. Their total length is over 36 km, of which about 21 km are national roads. Only 8.5% of all the bridges are constructed in compliance with the Yugoslav Regulations on Engineering Norms for Loading of Road Bridges (1991), which is practically identical to the German DIN 1072.

Bridges constructed prior to 1991 and bridges on the highways constructed up to 1980 are designed and constructed according to the "old" Temporary Technical Regulations for Loading of Road Bridges PTP-5 and the Temporary Technical Regulations for Concrete and Reinforced Concrete PTP-3, both dated 1949. For these bridges, the M25 loading scheme was used as a rule. These regulations are no longer effective now.

Most Macedonian bridges are reinforced concrete structures with a plate or beam structural system. The percentage of steel and stone bridges is less than 4%. Since Macedonia is a small country and does not have large rivers, deep and wide walleyes, or many natural obstacles, the average length of bridges is only about 36 m, while that of bridges on the national roads is around 47 m. Bridges with a length of over 200 m and spans larger than 50 m are scarce. All the bridges of the current network of national and regional roads were built after 1950 (98%), which reflects the conditions prevailing until and immediately after the Second World War. The most intensive period of construction was between 1960 and 1980, when over 62% of the bridges on national roads and 75% of the bridges on regional roads were built. They are now about 40 years old.

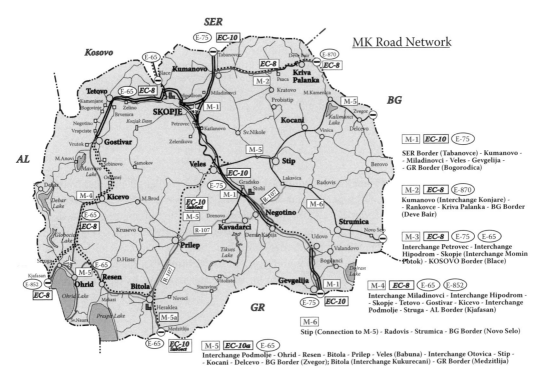

FIGURE 13.1   Road network of the Republic of Macedonia.

With the field surveys and analyses of the conditions of the bridges on national road M-2 from Kumanovo to Deve Bair performed in 2000 and national road M-1 from Kumanovo to Gevgelija performed in 2005, it was concluded that extensive measures for repair of damages mostly caused by aggressive freezing and thawing effects and spraying with salt in winter, as well as strengthening to enable sustaining of increased loads according to the Regulation of 1991, were necessary. In this context, a total of 69 bridges (48 bridges on M-1, 17 bridges on M-2, and 4 bridges on M-3, with a total length of about 4800 m) were reconstructed by donation of NATO in the course of 2000 and during 2007–2008.

With the Spatial Plan of the Republic of Macedonia, three road corridors with technical and serviceability characteristics compatible with the European highway system are planned to have been established by 2020:

- North–south (EC-10, E-65, M-1): Tabanovce (Serbian border)–Kumanovo–Veles–Gevgelija–Bogorodica (Greek border)
- East–west (EC-8, E-870, M-2, and M-4): Deve Bair (Bulgarian border)–Kriva Palanka–Kumanovo–Skopje–Tetovo–Struga–Qafe Tane (Albanian border) and Skopje–Blace section (Kosovo border)
- East–west (M-5): Zvegor (Bulgarian border)–Delchevo–Stip–Veles–Bitola–Ohrid–Qafe Tane (Albanian border) and the section Bitola–Medzitlija (Greek border)

13.1.2 Railroad Network

The railroad network in the Republic of Macedonia mainly coincides with the European transportation corridors passing through its territory. These are the railway lines of the EC-10 corridor: Tabanovce (Serbian border)–Skopje–Veles–Gevgelija (Greek border), with the Skopje–Blace section (Kosovo border, which runs further to Pristine and Belgrade), as well as the railway line within the EC-10A corridor, Veles–Bitola–Kremenica (Greek border). The railway lines within corridor EC-8 are not completed: Kumanovo–Beljakovce–Kriva Palanka–Gjueshevo (under construction from Beljakovce toward Bulgaria, with frequent disruptions) and Skopje–Gostivar–Zajas–Kicevo. The Spatial Plan of the Republic of Macedonia anticipates construction of the section running from Kicevo to the Albanian border by the year 2020, by which time the railway line along the EC-8 corridor from Burgas (Bulgaria) to Durres (Albania) will be completed. Another railway line between Veles and Kocani represents a section of the future east–west corridor.

Within the Macedonian railway, there are around 700 km open track lines, about 220 km station track lines, and approximately 230 km industrial track lines. Over 150 bridges with small and medium spans exist within the Macedonian railway network. The biggest bridges, with spans of 50 m or more, cross Vardar River or deep valleys. All the railway bridges constructed prior to 1970 are steel structures. Most of the bridges on the main railway line (EC-10 corridor) were built or reconstructed in the periods 1922 to 1928 and 1946 to 1950 after their destruction in the course of the First and the Second World Wars and, in addition, during the period 1970 to 1975. Figure 13.2 shows the new railway lines anticipated with the Spatial Plan of the Republic of Macedonia or that are planned to be constructed after 2020.

All the bridges within the Macedonian railway network are designed for loads in compliance with the regulations of the International Union of Railways (loading schemes UIC 776-1 and UIC 702). The managing and the maintenance of the railway bridges in the Republic of Macedonia are responsibility of the Public Enterprise Infrastructure of Macedonian Railways.

13.1.3 Historical Development

The first recorded beginnings of construction of bridges in the present territory of the Republic of Macedonia date back to the period of Roman rule. The most representative example is the Old Stone Bridge in Skopje, with 14 arches and a length of approximately 210 meters (Figure 13.3), which was constructed at the time of Justinianus I (527–565 AD) or even before that, according to some scientific

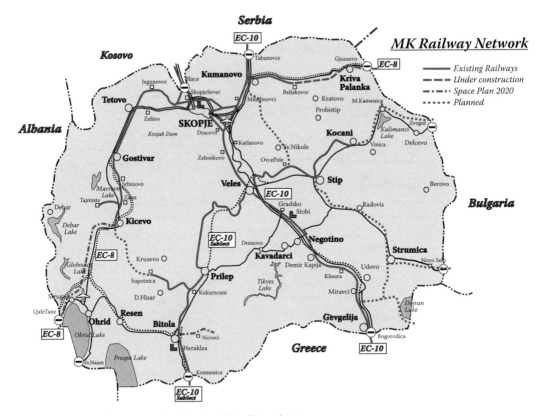

FIGURE 13.2 Railway network of the Republic of Macedonia.

FIGURE 13.3 The Old Stone Bridge in Skopje (dating back to the sixth century AD or earlier).

assumptions. Its construction was carried out by use of hewn stone called travertine, while the joints between the stone blocks were poured over with lead and burnt lime (quicklime) mortar and eggs. In the tradition of the people living in this area, there have been preserved characteristic rules of construction of timber bridges that are assumed to date back to the period of the Roman rule, however, no material evidence has been preserved.

From Medieval times, there are examples of beautifully constructed stone bridges in Kratovo, in Eastern Macedonia, a town that represented an important mining center for silver and gold (Figure 13.4) during Roman times and particularly in the period of Ottoman rule. Another example of a Medieval

(a)   (b)

FIGURE 13.4 Two of the old Kratovo stone bridges (fifteenth century or earlier). (a) Grofcanski bridge, (b) Radin bridge.

FIGURE 13.5 The "Deer's Leap" Bridge over Garska River (approximately sixteenth century).

bridge is the small bridge over Garska River in the beautiful Radika River valley in Western Macedonia, referred to as the "Deer's Leap," associated with a beautiful legend (Figure 13.5).

According to historic data, the traffic situation was catastrophic until the middle of the nineteenth century. Goods were transported by caravans consisting of at least 50 camels, horses, and mules, with an armed escort, while the transport of passengers by wagons was possible only in the plains and was a real trouble. Mail was transported by horses that were replaced at stations at an inter distance of 25 km, while the average speed was about 6 km/h. The transport of mail was organized every 15 days, so, for example, from Bitola to Belgrade, it traveled 9 days during summer and 11–12 days in winter. The transport of goods and passengers was even longer.

Due to this situation, in the second half of the nineteenth century, the Ottoman Empire started to take measures for construction of a railway line. In the period 1872–1873, the Thessaloniki–Miravci–Krivolak–Veles–Skopje railway, referred to as the Macedonian railway, was constructed, while the Skopje–Pristine railway line was constructed in 1874. In the beginning, the average speed of the trains was 25–30 km/h, which increased to 40 km/h after 10 years. When in 1882 the direct Thessaloniki–Skopje–Belgrade–Budapest–Vienna railway line was constructed, the journey lasted about 55 hours.

At first, bridges were constructed of wood. Then, steel bridges started to be built, ordered from Dortmund, Westphalia. Along the Thessaloniki–Skopje railway line, over 50 bridges were constructed with an opening larger than 10 m, of which 6 were bridges over Vardar River with a total length exceeding 1000 m. The completed bridges, with a span of up to 10 m, were put in place as delivered.

During the Balkan Wars of 1912–1913 waged in the territory of Macedonia, all the bridges over Vardar River and a large number of smaller bridges were destroyed and replaced by temporary wooden structures. The First World War and the breakthrough of the Macedonian front by the Triple Alliance/Entente brought even heavier consequences. In that period, bridges and temporary bridges were passed at a low speed or even without locomotive, by dragging the railway cars from one side to the other. Final renovation of the railways took place in the period from 1920 to 1933, and according to some sources, it continued until the beginning of the Second World War. New bridge designs were elaborated with axle loads of 18 tons, while steel bridge structures were procured from Czechoslovakia.

The Second World War caused even heavier consequences for railway bridges. Almost all the bridges with larger spans and truss structures (Vardar, Crna Reka, Bosava) were blown up and demolished, along with several locomotives on them. The repair of these bridges started later in 1944 and lasted until the midst of 1946. Later, in the 1970s, complete reconstruction of the railway line of the EC-10 corridor from Tabanovce to Gevgelija began as well as construction of new railway lines with modern bridges: Gostivar–Kicevo with the Zajas–Tajmiste arm, Gradsko–Sivec and Bakarno Gumno–Sopotnica. The construction of the Beljakovce–Gjueshevo railway line, on which mainly prestressed concrete bridges with precast main girders (Figure 13.6) are anticipated, has been going on since 1995. With many interruptions and decelerate execution of works due to lack of investment, this railway line is still under construction.

One of the first modern road bridges, the so-called New Bridge in the central city area of Skopje, was built in 1938–1939. It was designed by the young engineer and later doyen of Yugoslav structural engineering and academician Gjorgje Lazarevic from Belgrade. The bridge is a reinforced concrete cantilever bridge system (gerber beam structural system). To date, it has been in excellent condition (Figure 13.7).

Intensive construction of the road infrastructure and bridges started in the 1960s when the present road network of the Republic of Macedonia was established along with the current European corridors. In a period of about 20 years, around 680 bridges of different structural systems (deck, beam, arch, prestressed, and so on; around 75% of the bridges within the road network of the Republic of Macedonia)

FIGURE 13.6   One of the bridges on the Kumanovo–Gjueshevo railway under construction.

FIGURE 13.7 The New Bridge in Skopje (1938–1939).

were built. The second wave of intensive construction is expected to take place in the period up to 2020 with the construction of modern highways along the east–west corridors and finalization of the highway network.

13.2 Design Practice

13.2.1 Design Philosophy and Specifications

For the design of all road and railway bridges, linear elastic analysis is used. In practice, there is a difference between proportioning of reinforced concrete and steel bridges. As early as 1971, the concept of ultimate limit state and serviceability limit state was introduced for reinforced concrete structures. For the ultimate limit state, both linear and nonlinear method are used with partial loading factors for dead and live loads, while for the serviceability limit state, the linear method is usually used. In the computation, two serviceability limit states are controlled: the state of deformations and the state of cracks under the effect of live loads.

A further leap forward in the application of limit states in the design of reinforced concrete structures was made in 1987 with the passing of the Regulations on Engineering Norms for Concrete and Reinforced Concrete (PBAB-87) (YOS 1987a), which is based on the CEB-FIB Model Code.

For steel structures, Technical Regulations for Bearing Steel Structures were passed in 1964. These regulations are based on the concept of allowable working stresses with global safety factors for different groups of loads: main loads, main loads + additional loads, and main loads + exceptional loads.

In 1986, with the passing of the group of stability standards MKS U.E7.81:1986 to MKS U.E7.121:1986 (YOS 1986a), harmonized with the Recommendations of the European Convention for Construction in Steel (ECCS), the concept of limit states has implicitly been introduced through global load factors that are identical in their numerical value with the global safety factors (1.50, 1.33, and 1.2). In the same period, the application of the standards to welded steel structures (MKS U.E7.150:1987) (YOS 1987d) and the standards for joints of steel structures with non-preloaded and preloaded bolts (MKS U.E7.140:1987 and MKS U.E7.145:1987) (YOS 1987b and 1987c) began. Note that all Macedonian standards issued by 1991 with a designation MKS are identical with former Yugoslav standards having the designation JUS. Since 2009, Eurocodes EN 1990:2002 to EN 1998:2005 have started to be used in the design of reinforced concrete and steel structures in the direction of harmonization and acceptance of the European standards in the Macedonian technical regulations.

13.2.2 Loads

Due to the application of two different technical regulations, the road bridges in Macedonia can be divided into two categories:

1. Bridges designed according to the Temporary Technical Regulations (PTP-5). This category includes bridges constructed from 1949 to 1991 and bridges on motorways constructed from 1949 to 1980. According to these regulations, four loading schemes are anticipated: (1) commercial vehicles of 13 tons with uniformly distributed loads, (2) 20 ton steamrollers, (3) caterpillars of 30 and 60 tons, and (4) heavy vehicles with trailers M-25 with a weight of 84 tons. As a rule, loading schemes 1 and 4 hold for the main girders (Figure 13.8).

2. Bridges designed according to the Regulations for Loading of Road Bridges of 1991 (YOS 1991). This category includes bridges constructed in the period after 1991 and bridges constructed on motorways after 1980. According to these regulations, the load scheme referring to hybrid vehicles V600 + V300 and uniformly distributed loads (Figure 13.9) is anticipated for the design of bridges on motorways. For the design of bridges on the remaining national roads, scheme V600 is used. Scheme V300 + V300 is used for bridges on regional roads, while scheme V300 is used for bridges on local roads. The second category of road bridges also includes a total of 68 bridges designed according to PTP-5, but reconstructed and strengthened by carbon fiber–reinforced polymers (CFRP) materials in the elapsed period in accordance with the Regulations of 1991.

The Regulations for Loading of Road Bridges of 1991 are practically identical to German standard DIN 1072. The magnitudes of the dynamic coefficients according to PTP-5 and the Regulations of 1991 are presented in Figure 13.10. In the range of spans between 7 m and 26.7 m, the dynamic coefficients pursuant to the Regulations are negligibly higher, while for spans larger than 50 m, they are lower for more than 10%. The loads for which all the railway bridges are designed comply with the documents of

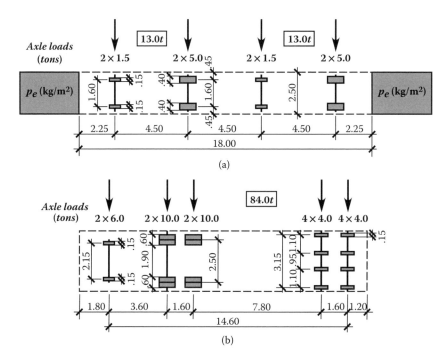

FIGURE 13.8 Loading schemes according to PTP-5 of 1949. (a) Commercial vehicles with UDL, (b) heavy vehicle with trailer M-25.

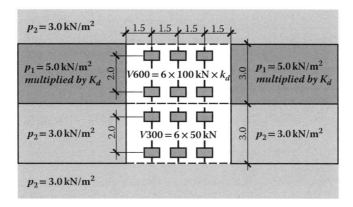

FIGURE 13.9 Loading scheme V600 + V300 according to the Regulations of 1991.

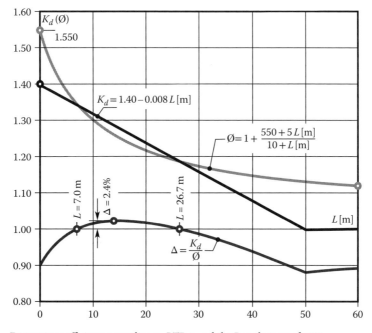

FIGURE 13.10 Dynamic coefficients according to PTP-5 and the Regulations of 1991.

the International Union of Railways. Starting in 1964, all bridges have also been designed for the effect of seismic loads according to the Temporary Regulations and after 1987, according to the Regulations on Seismic Design of Engineering Structures.

13.3 Concrete Girder Bridges

13.3.1 Kalimanci Bridge

At km 39 + 676 on national road M-3, near the Kalimanci Dam and artificial lake, a bridge–viaduct is constructed with spans 3 m × 33 m = 99 m and precast main girders made of prestressed concrete (Figure 13.11). The design documentation was elaborated by Granit Project–Skopje, while the bridge was built by Granit–Delcevo in 1995. Each span of the bridge superstructure is composed of four main

FIGURE 13.11   View of the Kalimanci Bridge. (Courtesy of Balkan Consulting–Skopje.)

prestressed concrete girders, concreted in a metal formwork on the plateau in the immediate vicinity of the bridge. The prestressing of the main girders was done by 13 cables composed of four tendons Ø 15.2 mm according to the IMS-Belgrade system. The assemblage of the main girders was done by means of a launching truss girder. The deck, along with the transverse girders placed in each third part of the span, is concreted in situ upon reinforced concrete omnia slabs.

The intermediate piers are precast and consist of 21 and 23 prefabricated segments, with a double-T cross section and a height of 100 cm each. The segments were manufactured in the Granite Prefabrication Plant in Delcevo in the winter and were transported and assembled in situ, reducing construction time to a minimum. The connection between the segments is realized by a precast standard reinforcement placed in special openings additionally poured over with fine-grained concrete. The upper parts of the intermediate piers end with a supporting beam. After its concreting, vertical prestressing of the intermediate piers was done by 4T15 cables placed in special openings through the segments. The tightening and the protection of the cables was performed using an injection mixture at the lower side for the foundation, whereas fixed anchors were used for the upper end.

Neoprene bearings were used for supporting the main girders over the abutments and intermediate piers. The Kalimanci Bridge was the first bridge in the Republic of Macedonia with precast prestressed piers. The bridge is owned by the Republic of Macedonia, while construction was organized by the Agency for State Roads of the Republic of Macedonia, which is also responsible for management and maintenance. In the vicinity of this bridge, there is another bridge constructed with precast prestressed intermediate piers over five spans.

13.3.2 Arazliska Bridge on National Road M-1 (E-65)

The Arazliska Bridge, the structure with the longest cast-in-place reinforced concrete superstructure in Macedonia, located in Demir Kapija–Udovo section of national road M-1, has spans 18 m + 10 × 21.5 m + 18 m and a total length of 251 m (Figure 13.12). The bridge was designed by the Bureau for Studies and Design at the Technical Faculty in Skopje in 1962. The bridge was constructed by CC Beton–Skopje in 1963.

The bridge superstructure consists of main girders, transverse girders, and a deck. The main and the transverse girders have a constant structural height, whereas the connection with the deck is strengthened by means of horizontal haunches. The deck is constructed with bearing reinforcements in two

FIGURE 13.12 View of the Arazliska Bridge during strengthening works. (Courtesy of Sintek AD–Skopje.)

directions monolithically connected with the main and the transverse girders. Reinforced concrete girders are supported on the expansion bearings in the abutments and firmly connected to the intermediate piers with diameter of Ø 100 cm. Frame action is introduced by the effect of braking forces, seismic forces, temperature variations, and shrinkage of concrete. The abutments are constructed as lightweight. The bridge is founded in favorable geotechnical conditions. The construction of the bridge was carried out by means of a wooden formwork and a steel scaffold in accordance with the then-common practice for construction of bridges.

In 2005, a detailed inspection was carried out to determine the conditions of the structure in the context of necessary strengthening in compliance with the Regulations for Loading of Road Bridges of 1991 and NATO standards. After more than 40 years of serviceability, pronounced carbonization of the concrete and extensive damage to the areas of the expansion joints were detected. Strengthening was done by increasing the concrete cross sections with self compacting concrete and additional steel reinforcement or with carbon strips and carbon meshes, while the damages were repaired and protective coatings were applied. Within the frame of the repair works, the New Jersey concrete barriers were arranged and the expansion joints were replaced.

All of the strengthening and repair works were done without disruption of traffic in the course of 2008, and were financed by donation of NATO. Their realization was headed by NAMSA–Luxembourg. The main contractor was Cesas Company from Turkey, while the subcontractors were Sintek–Skopje (carbon materials and protection) and Granit–Skopje (concrete and other works). The bridge is owned by the Republic of Macedonia, while management and maintenance of the bridge is the responsibility of the Agency for State Roads of the Republic of Macedonia.

13.3.3 Goce Delcev Bridge in Skopje

The Goce Delcev Bridge crosses Vardar River at one of the busiest boulevards in Skopje—Ilinden Boulevard (Figure 13.13). This bridge is characteristic for its structural solution and the geometrical shape of the cross section (Figure 13.14). The bridge was designed at the Institute for Design at the Civil Engineering Faculty in Belgrade, while construction was carried out by CC Pelagonija from Skopje in 1971.

The bridge consists of two parallel and identical prestressed concrete continuous frame structures with spans of 24 m + 88 m + 24 m. The width of each structure is 17.8 m, with three traffic carriageways, pedestrian and bicycle pathways on one side, and a total length of 136 m. The superstructure represents part of a revolving hyperboloid, while the cross section represents a circular sector consisting of an upper, flat plate and a lower plate as part of a circular ring, connected with three longitudinal ribs to obtain a box cross section composed of four chambers. The height of the superstructure varies from 2.08 m in the middle of the central span to 2.79 over the intermediate piers.

FIGURE 13.13   General view of the Goce Delcev Bridge in Skopje.

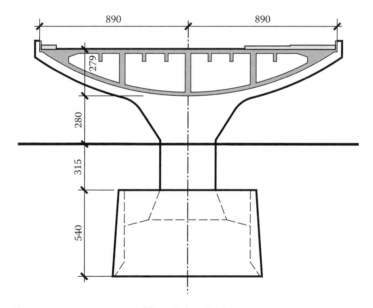

FIGURE 13.14   Characteristic cross section of Goce Delcev Bridge.

The intermediate piers are reinforced concrete walls with variable cross sections continuing through the structure as supporting diaphragms. The intermediate piers are founded in caissons.

The superstructure was prestressed using the IMS–Belgrade system. Three types of cables were formed with patented wire 6 Ø 7 mm, 12 Ø 7 mm, and 16 Ø 7 mm. Prestressing was performed by Mostogradnja Company from Belgrade. Construction was carried out by means of a carefully manufactured wooden formwork and combined wood–steel scaffold.

Since 1971, according to the results from field measurements at the beginning of its service life, an intensive increase of deflections in the central span has been detected. In a period of about 4 years these deflections reached 25 cm. The main reason for this is excessive creeping of the concrete and probably the redistribution of effects in the bridge superstructure due to the loosening of the abutments. In 2002, by donation of the government of France, the bridge was repaired according to a project by Freyssinet (France), which also organized the works. The repair consisted of adding external 19T15 cables to the three longitudinal ribs. By including additional prestressing force, some excessive deformations were eliminated and the stress state and safety during serviceability was improved. The bridge is the property of the Municipality of Skopje, which is also responsible for management and maintenance.

13.3.4 Orman Bridge on Skopje Northern Bypass

This bridge is located at km 18 + 338 on the Skopje Northern Bypass, part of the European corridor EC-8. The bridge is over a railway line, local roads, and larger agricultural lands. It consists of two independent structures, each with 17 spans and a length of 36 m + 14 × 37 m + 36 m = 629 m (Figure 13.15). The bridge's longitudinal axis is in a long transitional curvature with a transverse slope of the pavement ranging from 2.5% to 6%. The total width of each of the bridge structures is 12.25 m. Each of the structures consists of four precast prestressed reinforce concrete main girders with a length of 35 m and a height of 2.2 m.

The main girders were constructed by use of metal formwork on a plateau in the immediate vicinity of the bridge. The prestressing was done according to the Freyssinet system with five 12T15 cables. The assemblage of the main girders was done by a launching truss girder with a capacity of 120 tons. The deck structure, with a total thickness of 7 cm + 15 cm = 22 cm, along with the transverse girders placed in each third part of the span, was concreted in situ upon reinforced concrete omnia slabs with a thickness of 7 cm. The central piers are constructed as reinforced concrete walls with a width of 6 m, depth of 1.2 m to 1.8 m, and maximum height of up to 28 m. In the upper part, they end with a supporting beam. The abutments and the wings are constructed as reinforced concrete retaining walls. The main girders are supported via neoprene bearings.

The bridge was designed by Balkan Consulting–Skopje in accordance with the Regulations of 1991, while construction was carried out by CC Beton–Skopje in 2007 and 2008. The owner of the bridge is the Republic of Macedonia, while construction was organized by the Agency for State Roads of the Republic of Macedonia, which is also responsible for management and maintenance. With its total length of 629 m, this bridge is the longest road bridge in the Republic of Macedonia.

13.3.5 Zdunje Bridge

With the artificial lake for the new hydroelectric power plant Kozjak, near Skopje, all the existing local roads were to be flooded and a number of villages in the Treska River valley would be cut off. Therefore, the main design for Kozjak anticipated construction of a bridge at the village of Zdunje on a regional local road that passes through the dam crest, to connect this region of Porecje with Skopje. Zdunje Bridge (Figure 13.16) is a prestressed concrete structure with spans of 50 m + 3 × 86 m + 50 m = 358 m, fixed at the intermediate piers S3 and S4 and freely supported at the abutments S1 and S6 and intermediate piers S2 and S5. The bridge alignment is in a longitudinal inclination of 1.5%.

FIGURE 13.15 Orman Bridge on the Skopje Northern Bypass. (Courtesy of Balkan Consulting–Skopje.)

FIGURE 13.16   Zdunje Bridge. (Courtesy of Beton AD–Skopje.)

FIGURE 13.17   Construction of the Zdunje Bridge. (Courtesy of Beton AD–Skopje.)

The superstructure of the bridge is a box cross section with a height of 2.5 m in the spans and 4.3 m over the intermediate piers, varied following parabolic law, and a width of 4.5 m. The upper deck slab has a constant thickness of 22 cm, whereas the bottom soffit slab has a variable thickness ranging from 18 cm in the spans to 45 cm over the intermediate piers. The thickness of the box walls varies in a cascade manner between 24 cm (spans) and 36 cm (intermediate piers).

The intermediate piers have a box cross section with two chambers, with proportions of 2.9 m × 6.2 m at the top and end with a supporting beam with a height of 2 m. In the direction of the bridge axis, the width of each intermediate pier is increased by a ratio of 1/85.2 so that the highest pier, S3, is proportioned 4.4/6.2 m at the contact with the foundation. The thickness of the walls is constant along the height (30 cm). The height of the piers follows the configuration of the valley, from about 30 m (pier S5) to 63.9 m (pier S3). The intermediate piers were constructed by means of a lifting formwork, with concreting in segments with a height of 3.1 m. The foundation of the intermediate piers represents solitary footings with foundation depths ranging from 4 m to 7 m in the rock, while the largest foundation (pier S3) is proportioned 10 m × 13 m × 4 m.

The Zdunje Bridge was constructed using the cantilever method (Figure 13.17). After concreting the base segments over the intermediate piers with a length of 2 × 4.8 m upon a stable scaffold, the "cages" were mounted and concreting of segments with lengths of 4.65 m, eight at each cantilever "arm," continued. Finally, the "plugs" in the middle of the central spans were concreted, while at the end spans, the structure was closed with segments with a length of 8 m. The prestressing of the structure was performed following a strictly defined procedure. Prestressing cables 4T15.2 mm and 7T15.2 mm with low relaxation were used.

The bridge was designed by the Institute for Studies and Design of CC Beton–Skopje in cooperation with Mostogradnja (Belgrade, Serbia) in accordance with the Regulations of 1991, while the construction of the bridge was carried out by the CC Beton–Skopje and Mostogradnja in 2002. All organization of construction and investments was provided by Elektrostopanstvo–Skopje (Electric Power Company of the Republic of Macedonia). Zdunje Bridge was the first bridge in the Republic of Macedonia constructed using the cantilever method.

13.3.6 Skopje Railway Station and Transportation Center

After the disastrous earthquake of 1963, the Urban Plan of the City of Skopje of 1966 had foreseen dislocation of both the railway lines through the city's central area and the passenger railway station that had already been destructed. Construction of the new passenger railway station began in parallel with the construction of the new railway lines and bridges of the Skopje railway system. The new passenger railway station is a complex bridge structure with its tracks elevated 10.5 m above the surrounding terrain and streets. It consists of the northern approach structure with double track and a bridge over the Vardar River; a central part with a passenger railway station with 10 tracks, six being in active use; and the southern approach structure. The total length of the complex structure is about 1400 m, and the central part is 418.6 m (Figure 13.18).

The central part consists of six independent structures, the two end ones carrying one track each and the four middle ones carrying two tracks each (Figure 13.19), with a total width of 105.62 m. In the longitudinal direction, each structure consists of six prestressed reinforced concrete bridges with massive slab-type cross sections of modified trapezium shapes. The spans of the bridges (in the north–south direction) are 2 × 18.2 m; 6 × 18.2 m; 4 × 18.2 m; 3 × 36.4 m; 2 × 18.2 m; and 2 × 27.3 m. The bridges with spans of 36.4 m and 27.3 m have cross sections with variable heights of 1.2 m + 0.75 m = 1.95 m in the middle of the midspan and above the end piers and 1.85 m + 0.75 m = 2.6 m above the middle piers. The bridges with spans of 18.2 m are of constant height of 1.2 m + 0.75 m = 1.95 m. The section width of the end structures (one truck) is 10.33 m while that of the middle structures (double truck) is 15.86 m.

FIGURE 13.18 The new passenger railway station and Transportation Center in Skopje. (Photo by Alan Grant.)

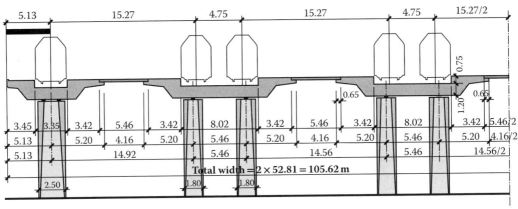

FIGURE 13.19 Cross section of the structures of the new passenger railway station.

FIGURE 13.20 A view of the Transportation Center in Skopje.

Reinforced concrete piers were used with modified rectangular sections of 2.5 m × 1.8 m for the end structures and 1.8 m × 1.8 m for the middle ones, with a height of 8.7 m to 11 m over the terrain. Solitary foundations were used, common for the piers of the middle structures. Below the platforms, incorporated in the new railway station structures, are the premises of the Transportation Post Office, the International and Intercity Bus Station, a parking lot, and other structures (Figure 13.20).

The design of the Transportation Center was entrusted to Kenzo Tange from Tokyo, Japan in 1969. Construction commenced in 1971 and the structures were completed in 1976. It was put in use in 1981 after completion of the new railway lines and bridges of Skopje railway system. The bridge structures were constructed by CC Beton, CC Pelagonija, and CC Mavrovo from Skopje, while prestressing was completed by Mostogradnja using the IMS Belgrade prestressing system. The Transportation Center is a property of the Republic of Macedonia, while maintenance of the bridge structures is entrusted to the Public Enterprise Infrastructure of the Macedonian Railways.

13.4 Steel Girder Bridges

13.4.1 Melnicki Bridge Near Debar

The Melnicki Bridge over the Spilje artificial lake is located at the exit of the wonderful Radika River canyon in Western Macedonia, on the regional road Mavrovo–Debar (Figure 13.21). The bridge is a composite structure with steel main girders and a prestressed reinforced concrete deck. The superstructure of the bridge consists of two separate parts: in a straight alignment, with spans of 68 m + 2 × 83 m + 68 m,

FIGURE 13.21   Melnicki Bridge near Debar.

and in a curvature with a radius of 70 m, with spans of 2 × 35 m. The straight part consists of two main girders with a constant height and prestressed reinforced concrete precast concrete slabs with a thickness of 22 cm. The curved part of the superstructure is a closed steel box cross section with a cast-in-place concrete deck. The height of the steel girders is 2.5 m and the total height of the composite cross section is 2.8 m. The width of the carriageway is 6 m, while the total width of the bridge is 8 m.

The superstructure is supported on two abutments (A and G) on the lake banks and five intermediate piers in the lake (C to F) with heights of up to 47 m. Within abutment A on the Mavrovo side, a characteristic retaining block was built. The straight part of the structure ends over lake pier E, over which an expansion joint is placed. The connection of the straight part with the piers is realized with a fixed-end bearing, whereas the curved structure has free bearings over pier E and fixed bearings on piers F and G.

The bridge is founded in a very specific natural geological medium characterized by complex properties and manifestations. Namely, the location and the wider surroundings of the bridge mostly consist of gypsum rocks of a considerable thickness, while the area downstream from the bridge on the right bank of Radika River is known as one of the richest findings of gypsum of the finest quality in Europe. According to geological data, in the Debar area, and most likely in the region of the bridge there is an active regional fault starting in Bulgaria, running through Macedonia, and continuing to the territory of Albania.

The bridge was built in 1970 according to the design of academician Nikola Hajdin from Belgrade. The steel superstructure was constructed by Metalna (Maribor, Slovenia), while the concreting works (foundations, piers, deck) were done by CC Pelagonija–Skopje. The bridge was funded by the Electric Power Company of Macedonia. Today, the bridge is managed by the Agency for State Roads of the Republic of Macedonia.

13.5 Arch Bridges

13.5.1 The Stone Bridge in Skopje

The old Stone Bridge over the Vardar River is located in the central part of the city of Skopje, between Makedonija Square and the Old Market Place (Stara Carsija). It is among the most important cultural monuments in the Balkans (Figure 13.22). The total length of the bridge is about 210 m, while the deck width of the bridge according to the original shape was probably about 5.5 m. Since 1972, it has exclusively served as a pedestrian bridge. The basis of the primary disposition are 14 openings of different lengths (from 4.05 m to 13.48 m), all of which are vaulted with semicircular vaults resting on massive river piers with a width amounting to 5.6 m. For the construction of the bridge, travertine was used and was connected by being poured over with lead and burnt lime mortar and eggs. The stone piers were founded on horizontally placed oak trunks that were previously boiled in tar.

FIGURE 13.22   **(Color insert)** The Stone Bridge in Skopje.

Considering the location of the bridge and the authentic historic facts as well as the numerous structural details that are visible even today, the origin of the bridge can be placed with certainty at the time of the Byzantine emperor Justinianus I (527–565 AD), who is believed, based on relevant historic data, to have come from the old Byzantine town of Taurezium (present village of Taor) in the surrounding of Skopje. Material evidence discovered during architectonic and archaeological investigations performed after the Skopje earthquake of 1963 identified the bridge with the archaeological findings at the Skopje Fortress (opus quadratum, the walling pattern and the forms of the foundations of the river piers built with recognizable elements from the Roman practice of bridge construction), which most probably originate from the period of its reconstruction after the earthquake of 518 AD. Therefore, it is assumed that the bridge could be even older.

In the course of its long existence, the bridge has been repaired many times. There is no historic data on possible repairs until the midst of the fifteenth century. The construction of the "bows" of the river piers (1444–1446), the renovation of four vaults and the construction of the "mihrab" (the guard post) in 1579 after the catastrophic earthquake of 1555, and the repair works in 1736 performed by Christian protomasters Gigo, Nikola, and Tanas, who incorporated many elements of Turkish bridge construction, were all recorded during the time of the Ottoman Empire. Some historic sources assume that the bridge was built at the time of the Ottoman Empire upon foundations dating back to the Roman period or, at least, that the bridge was completely reconstructed at that time. At the end of the nineteenth century, after the catastrophic flood in 1896–1897, quaysides were constructed in the Vardar River bed, while the end openings on the left bank were partially buried. In 1909, the original stone fence was removed and the bridge was widened by addition of steel pedestrian cantilevers on both sides.

After the catastrophic flood of 1962 and the catastrophic earthquake of 1963, ample repair works were carried out for the bridge within the project for the regulation of the Vardar River: repair and stabilization of the foundation, injection of the piers and the front walls, and unearthing of the buried openings on the right bank. Finally, between 1992 and 2007, a complete reconstruction was done: the steel cantilevers were dismantled and the original stone fence was returned, and four vaults were rebuilt with travertine, by which the bridge was given its original appearance. As a cultural monument of the highest rank, the bridge is under permanent care of the Institute for Protection of Cultural-Historic Heritage of the Republic of Macedonia.

13.5.2 Vault Bridge M1-57D between Skopje and Veles

Vault bridge M1-57D is located at km 52 + 917 on the south lane of national highway M-1 (E-75, EC-10) between Skopje and Veles. The bridge was designed in 1961 by the Technical Faculty in Skopje, while its construction was carried out by the construction company Ramiz Sadiku from Pristine, Kosovo in

FIGURE 13.23   View of the M1-57D Bridge after reconstruction. (Courtesy of Sintek AD–Skopje.)

1963. The bridge was designed and constructed in compliance with the then-valid Yugoslav regulations PTP-3 and PTP-5.

The main span of the bridge is a vault fixed at both ends, with a span $L = 54$ m divided in nine bays of 6 m each, and with a rise of $f = 17$ m. The vault has a constant width of 8.75 m and variable thickness of 1 m at the footings and 0.60 m at the apex. The access structures extend through three bays of 6 m each on the Skopje side, and five bays of 6 m each on the Veles side. The total length of the bridge amounts to about 104 m. With its elegant outline, adaptation to the terrain, and fit into the environment, the bridge is one of the most beautiful in Macedonia (Figure 13.23).

The width of the bridge carriageway is 2×3.8 m = 7.6 m, while the total width is 9.2 m. The longitudinal axis of the bridge superstructure is in left curvature with a radius of $R = 200$ m and with an inclination of 3.2% toward Veles. The transverse slope of the deck structure is 7.7%.

The abutments are lightweight concrete structures, composed of three parallel walls with a covering and inclined transitional plate. The piers of the access structures are founded on solitary footings, while the vault is founded on a strip foundation along its entire width.

In the course of 2007, within the project for strengthening bridges by donations from NATO, detailed repair, reconstruction, and strengthening of the bridge were carried out according to the Regulations of 1991 and NATO standards. The owner of the bridge is the Republic of Macedonia, while the Agency for State Roads is responsible for management and maintenance.

13.5.3 Arch Bridge M1-68D between Skopje and Veles

Arch bridge M1-68D is located at km 62 + 996 on the south lane of national highway M1 (E-75, EC-10) between Skopje and Veles, near the Otovica artificial lake (Velesko Ezero). The bridge was designed in 1961 by the Technical Faculty in Skopje and was constructed by the construction company Ramiz Sadiku from Pristine, Kosovo in 1963. The bridge was designed and constructed according to the then-valid Yugoslav regulations PTP-3 and PTP-5.

The main span of the bridge is a fixed arch with a span of $L = 58$ m divided in 10 bays and with a rise of $f = 16.3$ m. The arches have a variable cross section of 70/205 cm at the footings and 70/120 cm at the apex. The access structures consist of six bays on the Skopje side and five bays on the Veles side, all with spans of 6 m. The total structural length of the bridge amounts to 126 m. The bridge fits well into the environment (Figure 13.24). The width of the bridge carriageway is 2×3.85 m = 7.7 m, while the whole width amounts to 9.2 m. The longitudinal axis of the bridge deck is in right curvature with a radius

FIGURE 13.24   View of the M1-68D Bridge between Skopje and Veles.

of $R = 250$ m, while the transverse slope is 7.0%. The abutments are lightweight concrete structures, composed of three parallel walls with a covering and inclined transitional plate. The piers of the access structures are founded on solitary footings, while arches are founded on joint foundations.

In the course of 2007, within the project for strengthening bridges by donations from NATO, detailed repair, reconstruction, and strengthening of the bridge were carried out according to the Regulations of 1991 and NATO standards. The owner of the bridge is the Republic of Macedonia, while the Agency for State Roads is responsible for management and maintenance.

13.6 Truss Bridges

13.6.1 Srbinovo Railway Bridge

The Srbinovo Bridge is located at km 79 + 169 on the Skopje–Kicevo railway, bridging the deep valley of Zelezna Reka (Iron River) in the vicinity of a village named Srbinovo. The bridge consists of three structures: a reinforced concrete access structure 3 × 18 m = 54 m, a steel truss structure 52.64 m + 65.8 m + 52.64 m = 171.08 m, and a reinforced concrete access structure 5 × 18 m = 90 m. Such an illogical solution is the result of the original planning to construct a reinforced concrete vault with a span of about 170 m in the central part, which was later abandoned. This section describes the steel truss structure only (Figures 13.25 and 13.26).

The superstructure consists of two continuous trusses with 8 bays at the end spans (8 × 6.58 m = 52.64 m) and 10 bays in the central span (10 × 6.58 m = 65.8 m). The structural height of the trusses is 6.6 m, while the inter distance is 4.5 m. At the level of the upper and lower chords of the trusses, there are horizontal wind bracings, whereas over the supports there are vertical bracings of crossed diagonals with strengthened portal connections.

The longitudinal girders are continuous beams with a welded H cross section, with a span of 6.58 m and a height of 750 mm, placed directly over the transverse girders. The bracings against lateral impacts were placed at the level of the upper chord of longitudinal girders. The transverse girders were constructed with a welded H cross section with a height of 950 mm and a span of 4.5 m and are leveled with the upper chord of the main truss girders. In addition, at the level of the upper chord of the transverse

FIGURE 13.25   The Srbinovo railway bridge. (Courtesy of the Faculty of Civil Engineering, UKIM, Skopje.)

FIGURE 13.26   Trial loading during testing of the Srbinovo Bridge. (Courtesy of the Faculty of Civil Engineering, UKIM, Skopje.)

girders, horizontal bracings were placed ending with a braking bracing. By means of special devices, the braking force is transferred to the reinforced concrete access structures and further to their abutments. Such construction enables all of the supports of the steel structure of the bridge to be constructed as longitudinally movable.

The trusses are supported on high reinforced concrete piers with a box cross section and thickness of the walls of 35 cm, which end, in the upper part, with a supporting beam with a height of 2 m. The second intermediate pier has the tallest height (about 60 m). The longitudinal axis of the bridge is in a straight line, while the alignment is in inclination of 21% toward Kicevo. The entire steel structure of the bridge is constructed of steel St 37.2 according to DIN (C.0363 according to the Macedonian regulations), as riveted.

The bridge was designed by the Institute for Studies and Design of the Yugoslav Railways from Belgrade, for loading scheme UIC 24.5/8, while the steel superstructure was constructed and mounted by MIN Nis in 1968. The Skopje–Kicevo railway line has been constantly in operation since 1970. The owner of the bridge is the Republic of Macedonia, while management and maintenance of the bridge is responsibility of the PE Infrastructure of Macedonian Railways.

13.7 Reconstruction of Old Highway Bridges Using CFRP Materials

13.7.1 Pilot Project M-2

Since 1999, the use of a number of roads in the Republic of Macedonia has been characterized by intensive military and civil transport of goods and equipment by use of heavy transportation vehicles with trailers. The age of the bridges on these roads, which were designed according to PTP-5, and, first and foremost, their condition, were the main reasons that the Macedonian road infrastructure authorities and NATO paid greater attention to the security and safety aspects of civil and military road traffic in Macedonia.

The initiative was first taken by NATO because of the need for transportation of equipment from Bulgaria to Macedonia using heavy transportation vehicles. Upon inspection of the bridges on the national road M-2 (EC-8, E-870), running from Deve Bair (Bulgarian border) to Kumanovo with a length of about 70 km, and the performed numerical analysis of bearing capacity regarding increased loads, it was concluded that 17 out of 28 bridges along that section needed to be strengthened. After elaboration of the projects, the strengthening works began in early 2001. Due to the urgency of heavy transportation, after certain dilemmas and consultation considering the experiences of individual countries in the region (Greece, Slovenia), it was decided that the bridges would be strengthened by application of CFRP—carbon strips and carbon wraps.

Work on 17 bridges with a total length of about 800 m was executed within 4 months by NATO donation. About 12 km of carbon strips of different proportions and around 1000 m of carbon wraps were incorporated into the bridges. In the course of the strengthening works and after their accomplishment, trial testing of three characteristic bridges with different structural systems (girder bridge; simple beam system, girder bridge; and continuous beam system and continuous slab bridge) was carried out in two phases, nonstrengthened bridge and strengthened bridge, following a specific procedure. Almost at the same time, two more bridges on the north lane of the Kumanovo–Tabanovce (Serbian border) highway and one bridge in the vicinity of Demir Kapija (Figure 13.27) were strengthened as well.

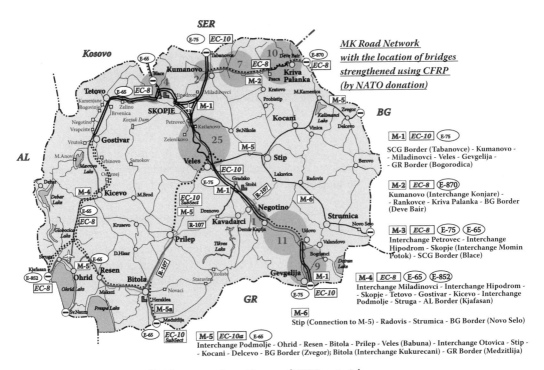

FIGURE 13.27 Locations of bridges strengthened by use of CFRP materials.

The realization of Project M-2 (E-870) enabled key lessons on the process of design by use of CFRP materials for strengthening concrete structures, methods to incorporate CFRP materials, testing by trial loading for verification of the design model and entire computations, and the organization and cooperation of the FRP team in the case of such projects. An internal review of Project M-2 (E-870), or the so-called CFRP Pilot Project, completed four years later by representatives of NATO and University of Skopje, finally resolved all the dilemmas regarding the justification of application of CFRP materials.

13.7.2 Project M-1

The second phase of strengthening the bridges in Macedonia started again by NATO donation in 2005. After detailed inspection of the conditions of the bridges, performed using a special vehicle called "Barin" (Figure 13.28), previous analyses, and elaboration of the projects on reconstruction and strengthening, 29 bridges along Blace (Kosovo border)–Skopje section (4 bridges) and Katlanovo–Veles section (25 bridges) and other 20 bridges along the Demir Kapija–Udovo–Gevgelija section were reconstructed and strengthened in two separate lots. The realization of the project was headed by NAMSA–Luxembourg. The contractor that performed the works was Cesas from Turkey, while the subcontractors were Sintek–Skopje (carbon materials and protection) and Granit–Skopje (concrete and other works). The project was realized in 2007–2008.

With the performed works on strengthening and reconstruction of the bridges, the north–south (EC-10, E-65, M-1) and east–west (EC-8, E-870, M-2) transportation corridors between Deve Bair and Skopje became completely passable for heavy transportation vehicles according to the Regulations of 1991 and DIN 1072. Also, in the course of the construction of the Skopje–Tetovo highway (1998–2002), carbon materials were used to strengthen an additional 10 existing bridges on the road alignment so that the EC-8 was extended through Skopje and up to Gostivar.

13.7.3 Strengthening Design

During bridge inspection, numerous defects of the same or similar character as those detected on the bridges on national road M-2 were observed, particularly on the south lane of the highway between Katlanovo and Veles, passing through the shady side of the Pcinja River gorge. Depending on the age of the bridges (i.e., the year of construction), the greatest number of flaws, in addition to concrete

FIGURE 13.28   Inspection of M1-65D Bridge by means of a special vehicle called "Barin."

carbonization, were a result of the effect of frost and salt used for winter maintenance of roads. Generally, it can be concluded that the main reasons for these defects are the low-quality (or nonexistent) water insulation and the low-quality drainage of atmospheric water from the bridges.

A large number of bridges constructed through the end of the 1970s were not water insulated because of the consideration that the asphalt pavement itself provides sufficient and efficient protection. In addition, and despite the evident repairs done, the weak points of the greatest number of bridges are the expansion joints, the bearings, the settlement of the embankments and the access ramps of the bridges, the pot holes in the pavements, and so on, which can mostly be attributed to the above-stated aggressive effects and penetration of water through the structures.

Structural computations and strengthening design were done in several steps:

1. 3D modeling by use of existing dimensions, characteristics of the cross sections, and properties of the materials, obtained by field inspection, measurements, and laboratory testing of materials.
2. Computer analyses for definition of cross sections forces, first according to PTP-5, then according to the Regulations of 1991 (DIN 1072), and finally according to NATO standards. In order to obtain the envelope of the maximum internal forces, all the stated loading schemes were slid along the longitudinal axis of the bridge with steps of maximum 1/20 of the span, for three different positions: one symmetric in respect to the longitudinal axis and two asymmetric, with the external wheel immediately next to the curbstone on the left and right sidewalks, respectively. Larger deviations with respect to the original design values obtained by manual computation were observed particularly in bridges with high piers, at the transverse girders, bridges in curvature, and so on.
3. Computation of maximum strengthening factors. Each strengthening factor represents a ratio between the maximum effects obtained with the load schemes according to the Regulations of 1991 (or NATO standards) and PTP-5.
4. Computation of strengthening. Strengthening was carried out by adding carbon strips in the tensile zone, carbon wraps in the shear zone, an additional concrete layer in the compressed zone, and finally coating the cross sections with additional self-compacting concrete and adding additional steel reinforcement. In the cases when an additional layer or coating of the cross sections with concrete was also done, due to the change in the stiffness of the elements, corrections were made in the 3D model and steps 1, 2, and 3 were iterated. To define the cross section of the carbon strips, the producer's original software was used. It is based on the ultimate limit state (ULS) and was checked by a number of "manual" computations done in the preliminary phase. Additionally, consideration was given to the development of strains in the existing reinforcement (Figure 13.29) and to the history of loading.

Notice from the above list that strengthening of the structural members is carried out by achieving composite action (by bonding) of the FRP reinforcement, or applying an additional concrete layer on these members. This kind of composite action is "subsequent," which means that, in any such structural members, there already exists an established, initial (residual) state of strains and stresses. Consequently, each such strengthening will be effective only for the influences of those loadings which, at the moment of their placement, are not acting on that element. Thus, the sequence of execution of works becomes an important parameter. The best solution is to apply any kind of strengthening after removal of all permanent loadings that can be removed (asphalt layer, concrete cover layer, curb stones, sidewalks, guard rails, etc.).

In relation to the "classic" composite steel–concrete structures, strengthening by using carbon strips has a particular characteristic, namely, the carbon materials have no reserve to large plastic deformations, since the maximum elongation at the break amounts to about 1.7%. Thus, the maximum ultimate bending resistance of a strengthened cross section is reached at the break of the carbon strips, at the already-commenced plastic yielding of the steel reinforcement, however prior to the failure of the concrete. In addition, it is necessary to observe one of the basic principles for properly designed reinforced concrete members—that the flexural mode of failure with the crushing of the concrete in compression should not appear before the yielding of the reinforcing steel (or FRP reinforcement).

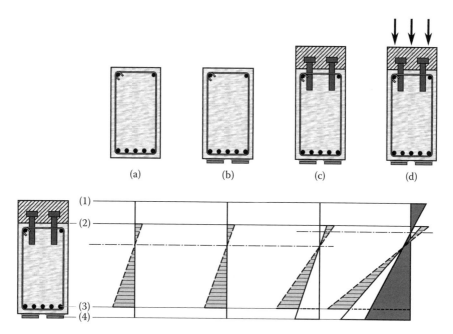

FIGURE 13.29 Strain state of cross section strengthened by FRP reinforcement and an additional concrete layer. (a) Initial (residual) strains due to dead loads; (b) strain after placement of carbon strips; (c) strain after placement of additional concrete layer; (d) total strains in service conditions.

13.7.4 Construction of CFRP Strengthening

Prior to the announcement of the tender and the selection of the most favorable bidder for Pilot Project M-2, a several-day seminar that included practical in situ training was organized at the Faculty of Civil Engineering in Skopje. Through this seminar and passing of an examination, around 30 engineers (designers, site engineers, and supervisors) and over 20 highly qualified workers were awarded licenses. Strengthening construction by use of CFRP materials consisted of the following operations, which are elaborated in the details of the Technical Specifications:

1. Cleaning of the concrete areas by quartzite sand blasting, or alternatively by water jetting, poking, bush hammering, or diamond grinding, then vacuuming the cleaned areas.
2. Leveling the cleaned areas with reparation epoxy mortar.
3. Cleaning the carbon strips.
4. Bonding the carbon strips with two-component epoxy resin (Figure 13.30).
5. Checking the quality of incorporation (gluing) of the carbon strips ("pull-off" test). The most reliable results are obtained if individual carbon strips are cut into pieces of a somewhat greater length (10–15 cm) and incorporated as such. After curing the glue (3–7 days), the strips are cut to the designed length and a "pull-off" test is performed on the remaining parts (Figure 13.31).
6. Checking the quality of glue. Samples are taken at the beginning and end of each daily application, for each bridge separately.
7. Protection of the incorporated carbon strips.

13.7.5 Testing of Bridges by Trial Live Loads

It is undisputable that, in the application of any new procedure for the execution of works, testing by trial live loads is necessary for a final check of all the project phases. In the considered case, the objectives of the tests are the following: (1) verifying the mathematical model and its correlation with the

FIGURE 13.30 Strengthening construction steps. (a) Bridge B35, preparation of carbon strips; (b) completed strengthening of bridge M1-B2N slab-type structure; (c) application of carbon strips on bridge M2-B18; (d) noncarbon strengthening using the same method but steel strips instead of carbon (M1-B112 Bridge).

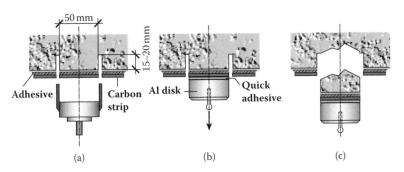

FIGURE 13.31 "Pull-off" test to check the quality of the placement of the carbon strips.

physical model of the bridge; (2) checking and confirming the effectiveness of the adopted procedure of strengthening by use of carbon materials; (3) checking the possibility of incorporation of carbon materials without disrupting regular traffic; and (4) verifying the performance quality of the strengthening works, that is, the final meeting of the criteria for the demanded bearing capacity in compliance with the Regulations of 1991 and the NATO standards.

To achieve these objectives, the bridges were tested in two phases: (1) in their existing state, prior to strengthening, and (2) upon completion of the strengthening works. In both phases, the stresses and

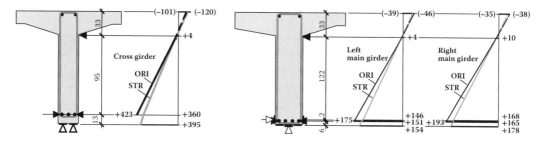

FIGURE 13.32 Strains measured in the transverse and the main girders of bridge M2-B37. ORI: non-strengthened state, STR: strengthened state.

strains were measured by (1) the same trial load and completely identical positions of loads in the individual testing phases, and (2) at the same points of the most critical elements and cross sections. After testing in phase 1, prior to strengthening, the strain gauges applied to the bridge members were carefully protected and then they were again used for the tests in phase 2.

For experts who are familiar with testing structures and composite structures, the main issue is the development and proportionality of the strains in the "composite" cross section, composed of the "initial" cross section and the carbon strips after strengthening. Three representative bridges with different structural systems were tested: reinforced concrete slab structure (continuous beam system over two spans), reinforced concrete girder structure (simple beam system), and another reinforced concrete girder structure (system of continuous beam over four spans).

Undoubtedly the most important conclusion drawn from the tests refers to the proportionality of the strains measured in the strengthened state. The strains in the carbon strips of the main and the transverse girders deviate from the straight line connecting the measured strains in the reinforcement and the concrete ± 3%. Hence, it can be considered that the carbon strips participate, completely and along with the original cross section, in sustaining all the live loads and give rise to a proper composite effect (Figure 13.32).

It is particularly important to point out that all the strengthening works were done under regular traffic, without limitations or halts, although the frequency of the traffic during the execution of works can be estimated to have been low to moderate, with a relatively smaller number of heavy vehicles with high axial pressure. However, the applied adhesive for the carbon strip bonding cured properly, without disturbances, and created a sound connection between the substrate (concrete) and the carbon strips, which agrees well with the laboratory tests performed at EMPA–Zurich (EMPA 2002) and casts another light upon the possibilities for bridges strengthening under service conditions, without disturbing traffic. As to the measured deformations, for instance in the case of bridge M2-B36, strengthened with carbon strips and an additional concrete layer on the upper side (Figure 13.33), proper behavior and considerable "appeasement" is evident as a result of the increased bearing capacity and stiffness of the composite cross section.

13.8 Future Bridges

The Program for Construction of Macedonian Highways includes European corridors EC-10 (from Tabanovce on the Serbian border down to Bogorodica on the Greek border), then the EC-10A subsection (from Veles through Babuna to Medzitlija on the Greek border) as well as the EC-8 corridor (from Deve Bair on the Bulgarian border to Qafe Tane on the Albanian border). Modern highways with two separated lanes will be constructed from Tabanovce to Demir Kapija, from Udovo to Bogorodica, and from Skopje to Gostivar, as well as the Skopje Northern Bypass, for which an upgrade is planned with a junction to the highway running to Kosovo. The main designs for all the highway solutions from Deve Bair to Kumanovo and from Gostivar to Qafe Tane, as well as the main design for the road from Veles to Prilep through Babuna, have already been prepared (Figure 13.1).

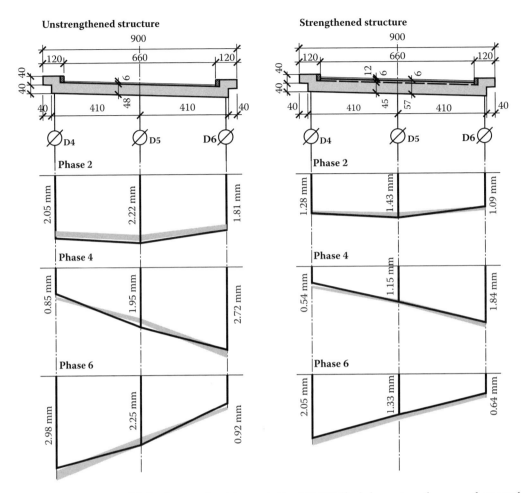

FIGURE 13.33 Measured deflections on the slab-type bridge M2-B36 Shaded are given the areas of expected deflections.

13.8.1 Susicka Bridge

At km 7 + 842 on the future Gostivar–Qafe Tane highway, over the deep valley of the Susicka River, construction of a prestressed reinforced concrete bridge with spans of 115.0 + 150.0 + 115.0 = 380.0 m is anticipated (Figure 13.34). The bridge consists of two identical independent structures with a trapezoidal box cross section and a height that varies according to parabolic law from 3.5 m in the spans to 8 m over the intermediate piers, while it is constant at the ends of the structures, in the first and the third span, within a length of 37.5 m. The thickness of the cross section of the box girders is also variable: the deck plate from 25 cm in the spans to 80 cm over the central piers and the lower plate from 40 cm to 160 cm, and from 50 cm to 100 cm at the sides. The total width of each individual structure is 11.75 m. Both structures are in a transverse slope of 2.5% and a longitudinal inclination of 3%.

The abutments are designed as reinforced concrete with parallel wing walls. The central piers are box type, with a variable cross section and heights of 59.5 m and 63.5 m, with extension at the upper end by which the first basic segment is initially formed for the construction of the superstructure by application of the cantilever method. The foundation of both intermediate piers is anticipated to be carried out upon circular foundation with diameter of 13 m and height of 15 m.

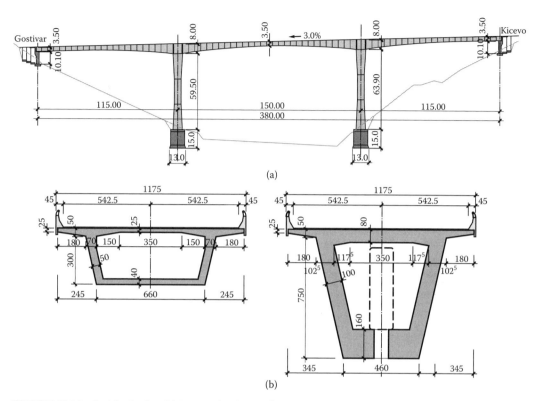

FIGURE 13.34 Susicka Bridge. (a) Longitudinal view; (b) cross section in the middle span and over the central piers. (Courtesy of Balkan Consulting–Skopje.)

The construction of the superstructure will start with the mounting of the cantilever platforms and successive concreting of the segments, symmetrically with respect to the pier axis so that, in the course of the construction, the structural system will be presented by symmetric cantilever girders, while after the complete concreting of the superstructure, closing of the "plugs," and tightening of the cables, the structural system will transform into a continuous frame. The connection between the concreted segments will be realized by cables and standard reinforcement, while the tightening of the cables will be done after achieving the necessary concrete strength. Prestressing will be done by cables composed of tendons with Ø 15.2 mm made of steel with characteristics 1570/1770 and low relaxation. According to the main design, the end parts of the superstructure will be constructed by use of a standard scaffold and formwork.

When the construction of the Susicka Bridge is completed, it will be the bridge with the largest spans in the Republic of Macedonia. The bridge is designed by Balkan Consulting–Skopje. The bridge will be in ownership of the Republic of Macedonia. All the activities that have so far been taken and future activities on this bridge will be realized through the Agency for State Roads of the Republic of Macedonia.

Acknowledgments

The authors wish to express their appreciation and sincere gratitude to all persons and organizations who supported the preparation of this manuscript, in particular Mr. Blaze Panov from CC Beton AD, Mr. Emil Nikolov from Balkan Consulting, Mr. Dobre Tasevski from Sintek AD, Prof. Petar Cvetanovski from the Faculty of Civil Engineering, UKIM, Skopje, and Mr. Nikola Dimitrovski from the Railway Regulatory Agency of the Republic of Macedonia. Their kind and explicit agreement and permission to use valuable data sources and photos contributes a lot to the presentation of the material.

References

EMPA. 2002. *Report No 418–931E: Bonding of CarboDur CFRP plates under oscillating load. Static testing of prestressed narrow slabs strengthened with CarboDur,* February, Swiss Federal Laboratories for Materials Science and Technology, ETH Zürich, Switzerland.

YOS. 1986a. *Group of Standards on Stability of Steel Structures* (JUS U.E7.081/86; JUS U.E7.086/86; JUS U.E7.091/86; JUS U.E7.096/86; JUS U.E7.101/86; JUS U.E7.111/86; JUS U.E7.121/86), Yugoslav Organization for Standardization, Official Gazette of SFRYu 21/1986, Belgrade, Serbia.

YOS. 1987a. *Regulations on Engineering Norms for Concrete and Reinforced Concrete* (PBAB-87), Yugoslav Organization for Standardization, Official Gazette of SFRYu No 11/1987, Belgrade, Serbia.

YOS. 1987b. *Connections Made with High-Strength Bolts in Steel Structures. Technical Requirements (JUS U.E7.140/87),* Yugoslav Organization for Standardization, Official Gazette of SFRYu 5/1987, Belgrade, Serbia.

YOS. 1987c. *Steel Structures Connected with Rivets and Bolts. Technical Requirements* (JUS U.E7.145/87), Yugoslav Organization for Standardization, Official Gazette of SFRYu No 17/1987, Belgrade, Serbia.

YOS. 1987d. *Welded Steel Structures* (JUS U.E7.150/87), Yugoslav Organization for Standardization, Belgrade, 1987, Official Gazette of SFRYu No 17/1987, Belgrade, Serbia.

YOS. 1991. *Regulations on Engineering Norms for Loading of Road Bridges,* Yugoslav Organization for Standardization, Official Gazette of SFRYu No 1/199, Belgrade, Serbia.

Bibliography

ACI Committee 440. 2002. *Design and Construction of Externally Bonded FRP Systems for Strengthening Concrete Structures,* ACI 440.2R-02, American Concrrete Institute, Farmington Hills, MI.

Bitovski, K. et al. 1973. *History of Railways in Macedonia 1873–1973,* Skopje Railway Transport Organization, Skopje, Macedonia.

Crawford, K. and Nikolovski, T. 2006. "The Design and Application of CFRP Composite Materials to Strengthen 19 Concrete Highway Bridges in the Republic of Macedonia," *2nd International FIB Congress 2006,* Paper 0442, Naples, Italy.

Crawford, K. and Nikolovski, T. 2007. "The Application of ACI 440 for FRP System Design to Strengthen 19 Concrete Highway Bridges on European Corridor 8," *8th International Symposium on Fiber Reinforced Polymer Reinforcement for Concrete Structures*
(8-FRPRCS 2007), paper 230, University of Patras, Patras, Greece.

Folic, R. 2013. "Chapter 16 Bridge Engineering in Serbia," *Handbook of International Bridge Engineering,* Ed. Chen, W. F. and Duan, L., CRC Press, Boca Raton, FL .

Gojkovic, M. 1989. *Old Stone Bridges,* Naucna Knjiga, Belgrade, Serbia. pp. 127–130.

Ivanov, D. and Nikolovski, T. 2003. "Old Bridges on Macedonian Roads. Conditions, Need and Methods for Their Strengthening", *Conference on Contemporary Civil Engineering Practice,* March, Novi Sad, Serbia.

Ivanov, D., Nikolovski, T., et al. 2000. *Design for Strengthening of Bridges on National Road M-2 (E-870) between Kumanovo and Deve Bair,* Vol. 1 and 2, Faculty of Civil Engineering, Skopje, Working Group TMCA/Bridges, Sept. , Skopje, Macedonia.

Nikolovski, T., et al. 2006. "Melnicki Bridge at Debar. About the Influences of a Geological Phenomenon," *International Conference GNP 2006,* Zabljak, Montenegro, pp. 173–180.

Nikolovski T., Todoroski, G. et al. 2002. "Trial testing of three bridges on National road M-2 (E-870) between Kumanovo and Deve Bair strengthened using CFRP systems (Carbon Fiber Reinforced Polymers)," *VSU Conference,* Sofia, Bulgaria.

Paskalov T., Nagulj, V. and et al. 2003. "Bridge over Lake Kozjak at Zdunje Village", *10th International Symposium of MASE,* Paper PR-3, Sept., Ohrid, Macedonia.

SRTO. 1981. *Skopje Railway Station, Monograph on the occasion of putting into operation of Skopje Transport Center,* Skopje Railway Transport Organization, Skopje , Macedonia.

YOS. 1986b. *Regulations on Engineering Norms for Steel Structures,* Yugoslav Organization for Standardization, Official gazette of SFRYu No 61/1986, Belgrade, Serbia.

YOS. 1986c. *Regulations on Engineering Norms for Design and Construction of Civil Engineering Structures in Seismic Regions,* Draft Proposal commonly used in practice, Yugoslav Organization for Standardization, Belgrade, Serbia.

14

Bridge Engineering in Poland

Jan Biliszczuk
*Wrocław University
of Technology*

Jan Bień
*Wrocław University
of Technology*

Wojciech Barcik
*Mosty-Wrocław Research &
Design Office*

Paweł Hawryszków
*Wrocław University
of Technology*

Maciej Hildebrand
*Wrocław University
of Technology*

14.1 Introduction

14.1.1 Geographical Characteristics

Poland as a state was established in 966 AD. Its territory and frontiers have changed throughout 1000 years of history. Within its contemporary boundaries Poland boasts not only Polish accomplishments, but also historical structures of great value to bridge engineering erected by German engineers.

FIGURE 14.1   Map of Poland with major cities and rivers.

Other achievements of Polish engineers can also be found in the current territories of Russia, Lithuania, Belarus, and Ukraine.

The total area of Poland is 312,679 km² (120,726 mi²), and it has a population of over 38 million people. Poland is a lowland country rather rich in woodland areas. Mountainous regions comprise only 9% of the country's territory. The major rivers posing communication obstacles are the Vistula, Odra, Bug, Narew, Warta, Pilica, Dunajec, and others (Figure 14.1).

Construction of bridges in the middle and lower courses of the major rivers in Poland has always been difficult due to the geological shape of the river valleys requiring relatively deep foundations as well as due to flooding and ice drift common in Central Europe. The basic building materials used in Poland from the early Middle Ages to the early 1800s were wood, universally available, and stone, geographically restricted. Brick was used occasionally. Rapid development of bridge engineering in Poland began in the mid-nineteenth century with the introduction of railway transport and modernization of the road network.

14.1.2 Historical Development

14.1.2.1 Timber Bridges

In the village of Biskupin located in central Poland, the oldest known traces of bridge engineering, dating back to the Pre-Slavic period (737 BC, a year determined by means of dendrochronology), were found. These are the remnants of a 120 m long timber bridge which linked an ancient settlement located on an island on the Lake of Biskupin with the shore. Over 1700 years later, Mieszko I, the first historical ruler of Poland, built a fortified settlement on the island of Ostrów Lednicki, located on Lednickie Lake 50 km southwest from Biskupin. Two timber bridges, 438 m and 187 m long, were used to link the eastern and western parts of the island with the shores of the lake (Figure 14.2). The structures were from 4.1 m to 4.5 m wide and their spans were from 4 m to 4.5 m long.

The structures, made of oak timber, consisted of girder spans placed on supports made up of a girt and two groups of piles (Figure 14.2b). Since the maximum depth of the lake reached 8–10 m, some of the piles were 12–14 m long. All components were connected by means of carpentry, without applying any metal elements. On the basis of dendrochronological examination of the excavated piles and given other historical records, it was established that the bridges existed from 963 to 1038.

The greatest Polish achievement in the preindustrial era was the construction of a timber bridge over the Vistula River in Warsaw (Chwaściński 1997; Mistewicz 1991) in the sixteenth century

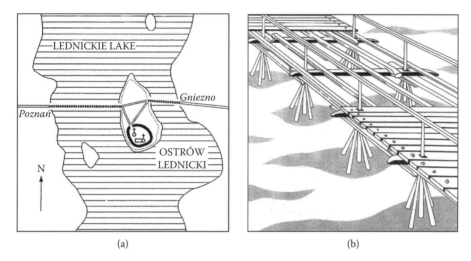

FIGURE 14.2 Reconstruction of the timber bridges on Lednickie Lake (963–1038): (a) location, (b) structure.

FIGURE 14.3 First timber bridge over the Vistula River in Warsaw (1573–1603).

(Figure 14.3). The Polish monarch Zygmunt August (Sigismund Augustus) commissioned the first bridge over the Vistula River in Warsaw, the construction of which started in 1568. According to old pictures from Warsaw the bridge was 500 m long and consisted of 23 timber spans. The 18 main spans, each 23 m long, were constructed as triangle truss superstructures with timber bars and iron joint connections. Five shorter spans (approximately 10 m long) were movable to enable shipping, which was intensive at that time. The bridge was opened in 1573 and operated until 1603, when drifting ice destroyed the structure.

Floating timber bridges were constructed for military purposes. One of the most popular bridges of this type was built during the war between the Kingdom of Poland and the State of the Teutonic Order in 1409–1410. Before the military operation began, the "prefabricated" bridge elements had been hauled on boats down the Vistula River and mounted at the destination during half a day. The bridge made it possible for the united Polish–Lithuanian army under the command of the Polish King Władysław Jagiełło to concentrate on the right bank of the river and, as a consequence, contributed to the victory in the Battle of Grunwald (Tannenberg) on July 15, 1410.

14.1.2.2 Masonry Bridges

Since the thirteenth century, stone masonry bridges have been constructed in southern Poland (Jankowski 1973). Many of these structures survived until modern times. The bridge over the Młynówka River in Kłodzko, built in 1390, is the oldest masonry bridge in Poland and has been in service for over 600 years (Figure 14.4). The 52.2 m long bridge consists of four skew (irregular) stone arch spans. Its vaults are 6.25 m wide and 0.50 m thick.

From the mid-nineteenth century to World War I (1914–1918), the construction of masonry bridges in Poland was very common. It was a time when railways were built across the country. There were about 5000 stone and brick masonry bridges erected. Figure 14.5 presents the longest stone bridge in Poland. It was built over the Bóbr River in Bolesławiec in 1844–1846 (Biliszczuk et al. 2000). Nowadays, the modernized bridge is operated by the PKP Polish Railway Lines Company.

14.1.2.3 Iron Bridges

In Poland, iron was introduced into bridge engineering already in the eighteenth century (Biliszczuk et al. 2009; Rabiega and Biliszczuk 1996). The first iron bridge in continental Europe was built in the Silesia region (see Figure 14.1). In the late 1700s, two steelworks functioned in Silesia, which significantly contributed to the development of iron bridges in this part of Europe: Małapanew Ironworks in Ozimek, established in 1755, and Royal Iron Foundry in Gliwice, established in 1796 (now Technical Equipment Plant). These ironworks produced over a dozen of the first iron bridges located in Silesia and in Germany.

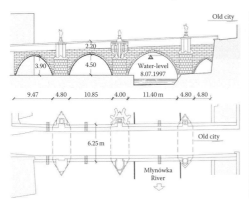

FIGURE 14.4   Stone masonry bridge over the Młynówka River in Kłodzko (1390), the oldest bridge in Poland still in use.

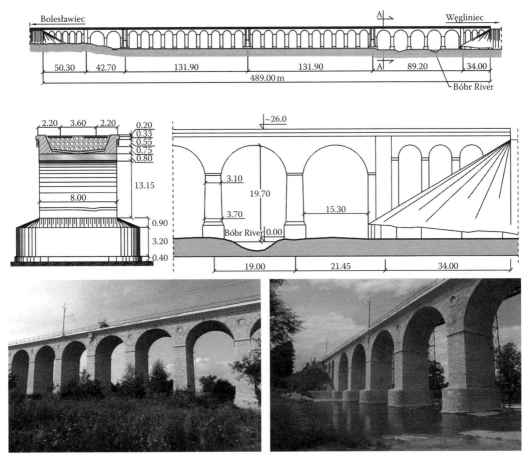

FIGURE 14.5 Stone railway bridge over the Bóbr River in Bolesławiec along the Wrocław–Berlin line (1846).

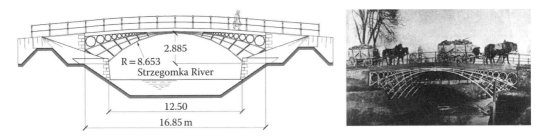

FIGURE 14.6 The bridge over the Strzegomka River in Łażany (1796–1945).

The bridge over the Strzegomka River in Łażany—the first iron bridge on the European continent (Biliszczuk et al. 2009; Pasternak et al. 1996)—was, at the beginning, supposed to be a stone structure. However, inspired by the iron bridges built in England (e.g., Coalbrookdale in 1779) and in America, the investor Nicolas August W. von Burghauss decided to erect an innovative iron structure. In 1793 he commissioned the Małapanew Ironworks in Ozimek, the most highly developed steelworks in Silesia, to produce the bridge. The project of the bridge and the technology of its production (casting, transport of the elements, assembly, and erection) was managed by English engineer John Baildon, under the supervision of Count Reden (the owner of the ironworks in Ozimek). The structure (Figure 14.6) was a one-span bridge. The span consisted of five iron arches, each 16.85 m long and 2.885 m high, with spacing at 1.35 m. Iron arch girders were joined by bolts in the middle of the span. The arches were braced by appropriate transverse elements. The deck was made of iron slabs 50 mm thick, 0.50 m wide, and 5.8 m long, which equals the width of the bridge. On the edges of the spans, there were cornice slabs placed, 0.36 m high, to which decorative railings were fixed. The deck surfacing was made up of paving stone laid down on a pit run gravel layer.

The 46 ton structure was cast in 1794–1795. In the fall of 1795, it was transported to its destination. Construction of the structure took 10 weeks in the spring, and the bridge opened to traffic on May 20, 1796. It served until 1945, when it was destroyed during World War II. The preserved elements are exhibited at Wrocław University of Technology.

The footbridge in Opatówek was built in 1824 in Józef Zajączek's palace and park complex (Biliszczuk and Hildebrand 2000). The structure (Figure 14.7) is a one-span arch structure consisting of four main girders cast in iron. Each of the girders is made up of three segments joined by bolts. Originally, the deck was probably built up of timber planks, which have been replaced by reinforced concrete slabs. The massive supports are made of stone. The span between the supports extends to 10.3 m and the total length of the superstructure is 13.8 m. The total width of the deck is 3.5 m. The main girders are adorned with ornaments cast as wholes. Figure 14.7 shows the current condition of the footbridge (right) as well as its original shape (left).

In 1827, a suspension bridge was built over the Mała Panew River (Rabiega and Biliszczuk 1996). The designer and builder of the structure was Karl Schottelius. The main span between pylons is 31.4 m long. The bridge is 6.28 m wide. The open-work pylons form an ornate portal for the gate to the bridge (Figure 14.8). There are four main tendons suspended on each pylon. The tendons consist of elements connected by bolts and cover plates. The tendons are arranged along the catenary and anchored in an anchorage foundation below the ground level. The hangers, fixed to the tendons, are mono round rods carrying the deck. The road surface, built up of three layers of timber planks, was placed on crossbeams formed of rolled I-beams.

The era of iron bridges lasted from 1779 to the late 1830s. Technological development in metallurgy led to massive low-cost production of steel, a material of high tensile strength and of high resistance to cracking, which limited the use of cast iron.

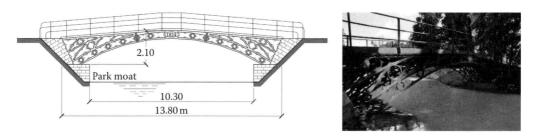

FIGURE 14.7   Bridge over the park moat in Opatówek (1824), the oldest iron bridge in Poland still open for pedestrians.

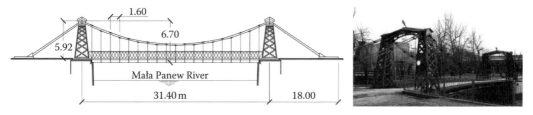

FIGURE 14.8   Bridge over the Mała Panew River in Ozimek (1827).

14.1.2.4 Steel Bridges

The two oldest steel bridges are located over the Vistula River in Tczew and the Nogat River in Malbork. The bridges, constructed by the Prussian engineer Carl Lentze along the Eastern Railway in Tczew and Malbork (1850–1857), are significant to world bridge engineering (Jankowski 1973). They were designed as tubular (box) structures modeled on the Britannia Tubular Bridge (1846–1850) constructed by George Stephenson. As opposed to Stephenson, Lentze used a lighter truss construction, which better resists wind forces. The superstructure was formed of three two-span space trusses with parallel chords and a dense system of diagonal members. The distances between the axes of the piers are 6 × 130.88 m. The total length of the bridge is 785.28 m. The girders are 11.8 m high and the clearance between the truss webs of the box is 6.5 m.

The structural system of the girders was formed of a dense riveted truss of flat bars. The supports, adorned with towers, were made of stone. The foundation was also made of stone and protected by steel sheet pile walls. The bridge over the Nogat River in Malbork had identical structural solution and consisted of one two-span segment. The structures in Tczew and Malbork were designed for rail and road traffic. Road traffic was withheld when trains crossed the bridge. Both structures were severely damaged during World War II. In Tczew, only three out of six spans have survived (Figure 14.9).

The first durable bridge structure over the Vistula River in Warsaw was constructed by famous Polish engineer Stanisław Kierbedź in 1859–1864 (Jankowski 1973). It was a steel truss bridge with a total length of 475 m. The spans were 79 m long. The structure was similar to that of the bridge in Tczew. The superstructure was divided into three segments, each in the form of a continuous two-span beam. The clearance between the 9.1 m high girders was 10.5 m. The bridge was a steel riveted structure. A tram track with groove rails was laid down on the bridge. The sidewalks were placed outside the box girder. Figure 14.10 shows a general view of the structure. During World War I, one of the piers was blown up. Afterwards the bridge was reconstructed. The bridge was not reconstructed after being destroyed in World War II. Its substructure, settled on steel caissons, is still in use.

FIGURE 14.9 Carl Lentze's bridge over the Vistula River in Tczew (1857).

FIGURE 14.10 Kierbedź's bridge over the Vistula River in Warsaw (1864–1945).

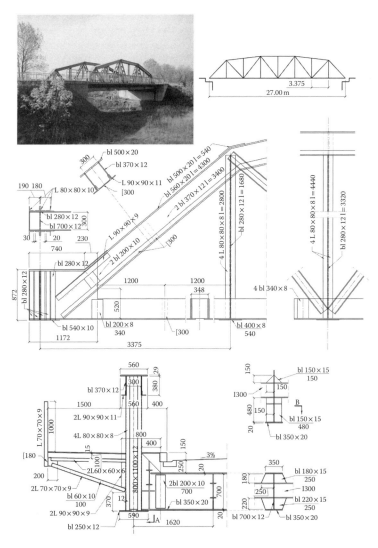

FIGURE 14.11 The first welded steel road bridge in the world erected over the Słudwia River near Łowicz (1929). Today it is exhibited as a historical structure.

Construction of the first welded steel road bridge in the world (Figure 14.11) was a significant achieve-ment of Polish bridge engineering (Pancewicz 1989; Mistewicz 1991). The design process was preceded by theoretical studies focused on tests of welded joints and on small steel structures connected by means of gas and electric welding. The laboratory results assured the initiator of these tests, Professor Stefan Bryła – precursor of using welding in house and ship building in Poland, that it was possible to use such joining methods also for steel bridges. The test results together with the engineering intuition of Bryła were the basis on which the road bridge over the Słudwia River near Łowicz (Figure 14.11) was fabri-cated by means of arc welding in 1928. The erection of this bridge resulted in faster development of steel structures in civil engineering. The welded steel bridge was completed in December 1928 and the bridge on the Słudwia River was opened to traffic in August 1929. It is worth mentioning that it was the first welded steel bridge in Europe and the first road bridge of this type in the world.

The bridge consists of two main truss beams with a straight bottom chord and parabolic top chord, as presented in Figure 14.11. The effective span of the truss beams was 27 m, the effective height at mid-span was 4.3 m, and the distance between the bottom truss nodes, that is, the lengths of the stringers, amounted to 3.375 m. The width of the bridge between main truss beams equaled 6.76 m and clear-ance between truss girders was 6.2 m. On both sides of the bridge a 1.5 m wide sidewalk was provided. The bridge was made of steel of strength 370–420 MPa and minimum unit elongation 20%. The main truss beams are connected with crossbeams and horizontal braces located at the bottom chord, without braces at the top chord. In 1977 the bridge was closed and is today exhibited as a technical monument.

14.1.2.5 Concrete Bridges

The first Portland cement plant in Poland was established in Grodziec in 1857 (Jankowski 1973). Cement produced in that plant was used to construct railway bridges along the lines Warszawa–Bydgoszcz (1861–1862) and Kraków–Lwów (1858–1860). At first, concrete was used exclusively in elements sub-jected to compression, that is, in the foundations of the piers and in the vaults of small spans. The first reinforced concrete bridge in Poland was a pedestrian footbridge in the form of an arch located in the courtyard of Lviv University of Technology (now in the territory of Ukraine). It was erected in 1894 according to a project by Professor Maksymilian Thullie and still exists. The span length of the foot-bridge is 11.05 m and its width is 2 m.

The development of reinforced concrete bridges was very rapid. Already in 1910 the Sosnowski and Zacharewicz company built a reinforced concrete arch road bridge crossing the Wisłoka River, over 160 m long. The bridge consisted of five spans of 2 × 30.5 m + 38 m + 2 × 30.5 m. Prestressed concrete was first used in Poland after World War II (Biliszczuk 1997a, Biliszczuk 1997b). There is, however, a prestressed concrete bridge along the A4 highway (Wrocław–Opole) that is chronologically the fourth bridge of this type in the world (Figure 14.12). It was built in 1941–1942 by German company Wayss und Freytag on the basis of a license granted by E. Freyssinet (Biliszczuk 1997a, Biliszczuk 1997b). The bridge crosses the Nysa Kłodzka River in Sarny Wielkie, Silesia. It consists of one span of 42.2 m. The structure of the span (under one roadway of highway) is formed of seven prefabricated prestressed concrete beams. The beams are spaced at 1.7 m. The reinforced concrete deck slab is 0.15 m thick. The span carrying one lane of highway is 12.23 m wide. The bridge, remaining in good condition, is still in operation. Table 14.1 summarizes the introduction of various technologies used in prestressed concrete structures in Poland in chronological order (Biliszczuk 1997).

14.1.3 Bridge Infrastructure

14.1.3.1 Road Bridges

Road bridge infrastructure in Poland (Mistewicz 1991) consists of about 30,000 bridge structures (bridges, viaducts, footbridges, tunnels, retaining walls) with a total length of over 540,000 m, as well as over 100,000 road culverts. The length of the public roads is almost 380,000 km, and the roads are

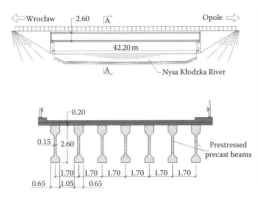

FIGURE 14.12   Bridge over the Nysa Kłodzka River along the A4 highway, one of the first prestressed concrete bridges in the world (1942).

TABLE 14.1   Development of Prestressed Concrete Structures in Poland

No.	New Technique	Year of First Implementation in Poland	Inventors in Poland	Implementing Company	Location in Poland
1	Casting in the whole formwork	1953	T. Kluz	KPRM	Bridge at Stary Młyn near Końskie
2	Prestressed precast beams	1954	J. Zieliński Z. Czerski	PPRM	Bridge at Village Kujan
3	Cantilevered casting	1963	M. Wolff	PPRM	Bernardyński Bridge in Bydgoszcz
4	Post-tensioned rigid nodes between neighboring spans	1967	S. Filipiuk	KPRM	Bridge over the Vistula River at Annopol
5	Cantilevered assembly of precast units	1969	M. Wolff	PPRM	Pomorski Bridge in Bydgoszcz
6	Multiuse falsework implemented for span-by-span method	1979	S. Jendrzejek, J. Kasperek, A. Skrzypek	KPRM	Viaduct in the city of Chorzów
7	Longitudinal launching	1987	S. Jendrzejek	KPRM	Bridge over the Soła River in Oświęcim

divided into the following categories: national roads (5% of the total length), regional roads (8%), county roads (34%), and communal roads (53% of the total length of public roads). The General Directorate for National Roads and Motorways in Warsaw (www.gddkia.gov.pl) is responsible for the management of the national roads, including bridge structures. Other categories of roads are managed by local authorities. The locations of the main highways and the most important express roads are shown in Figure 14.13.

The majority of the existing road bridge superstructures (87%) are constructed of concrete, either reinforced concrete (RC) or prestressed concrete (PC), as shown in Figure 14.14. Other bridges are steel structures (8% of all bridges) as well as stone and brick masonry structures (2.6%). The last group (2.4%) consists of composite structures constructed of two or more materials (steel/RC, PC/RC, etc.). The average length of a road bridge structure in Poland is about 20 m and the average distance between bridge structures located on the road network is about 13 km. Road bridges in Poland are relatively young: 37.6% of the structures were built during the last 20 years, 32.7% are 20–50 years old, and 29.7% of all bridge structures are more than 50 years old (Figure 14.15).

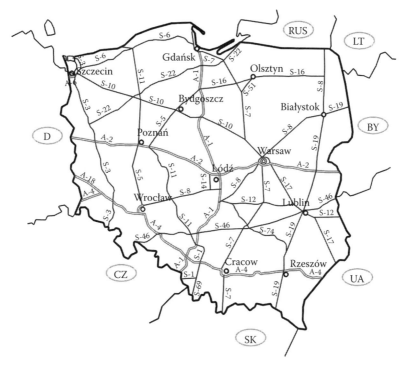

FIGURE 14.13 Main highways and express roads in Poland (existing, under construction and planned).

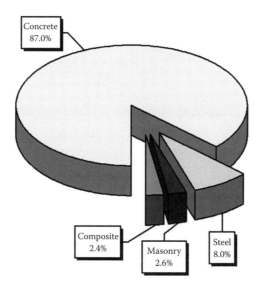

FIGURE 14.14 Road bridges in Poland—material of superstructure (Bień 2010).

14.1.3.2 Railway Bridges

The total length of railway lines in Poland is almost 20,000 km, and about 8,000 bridge structures and over 24,000 culverts are in operation. The entire railway network, including bridge structures, is managed by the PKP Polish Railway Lines Company (www.plk-sa.pl). The most important national railway lines in Poland are shown in Figure 14.16 on the map implemented in the Railway Bridge

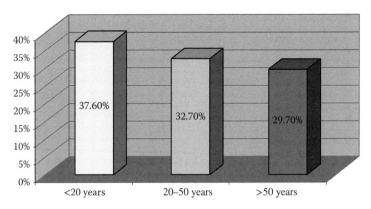

FIGURE 14.15   Age profile of road bridges in Poland (Bień 2010).

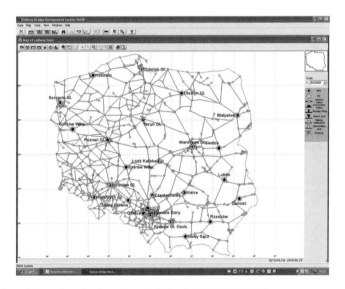

FIGURE 14.16   Map of the national railway lines in Poland (RBMS "SMOK").

Management System (SMOK). A percentage comparison of basic structural materials used for construction of railway bridges is presented in Figure 14.17. Superstructures are dominated by steel (42% of all bridges) as well as by RC and PC (36.8%). Other important groups are masonry (12%) and composite structures (9.2%).

Materials applied in the construction of railway bridge substructures are also presented in Figure 14.17. The domination of concrete (plain and RC) is very clear (53.9% of all structures). Masonry supports are used in 30.5% of the bridge structures and composite supports in 15%. Use of steel supports is limited to 0.6%. The average length of a railway bridge in Poland is about 30 m and average distance between bridge structures located along the railway lines is about 2.5 km. Development of the railway network started in Poland in the first half of nineteenth century, so a large number of bridge structures are relatively old (Figure 14.18). Almost 45% of bridges are over 100 years old and only 3.6% of the structures were constructed during the last 20 years.

14.1.3.3 The Longest Bridge Spans in Poland

The longest bridge spans in Poland are listed in Table 14.2. The locations of the main rivers and cities in Poland are shown in Figures 14.1 and 14.13.

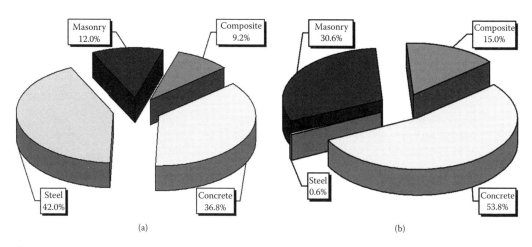

FIGURE 14.17 Material of railway bridges in Poland: (a) superstructures, (b) substructures (Bień 2010).

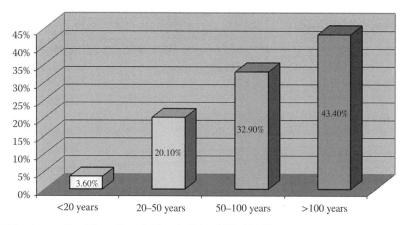

FIGURE 14.18 Age profile of the railway bridges in Poland (Bień 2010).

TABLE 14.2 Bridges with the Longest Spans in Poland (as of January 2013)

No.	Location, Obstacle	Longest Span (m)	Superstructure Material	Year of Construction
Girder Bridges with Solid Web Girders				
1	Grudziądz (Highway A1), Vistula River	180.00	PC	2011
2	Warsaw (North Bridge), Vistula River	160.00	Steel + RC	2012
3	Knybawa (National Road No. 22), Vistula River	142.40	Steel	1941/1950*
4	Kędzierzyn Koźle (National Road No. 40), Odra River	140.00	PC	2010
5	Wyszków (National Road No. 8), Bug River	136.00	PC	2008
Girder bridges with truss girders				
1	Toruń (Piłsudskiego Bridge), Vistula River	130.00	Steel	1934/1950*
2	Tczew (road bridge), Vistula River	128.60	Steel	1857/1959*
3	Tczew (railway bridge), Vistula River	128.60	Steel	1891/1971*
4	Płock (Legionów Piłsudskiego Bridge), Vistula River	110.40	Steel	1937/1950*
5	Puławy (Mościckiego Bridge), Vistula River	110.00	Steel	1934/1949*

(Continued)

TABLE 14.2 (*Continued*)   Bridges with the Longest Spans in Poland (as of January 2013)

No.	Location, Obstacle	Longest Span (m)	Superstructure Material	Year of Construction
Arch bridges				
1	Warsaw (Krasińskiego Bridge), Vistula River	277.09	Steel + RC	Planned
2	Toruń (National Road No. 15), Vistula River	270.00	Steel	Under construction
3	Puławy (National Road No. 12), Vistula River	212.00	Steel + RC	2008
4	Kraków (Kotlarski Bridge), Vistula River	166.00	Steel	2001
5	Wolin (National Road No. 3), Dziwna River	165.00	Steel + RC	2003
Cable-stayed bridges				
1	Płock (Solidarności Bridge), Vistula River	375.00	Steel	2005
2	Wrocław (Highway A8), Odra River	256.00	PC	2011
3	Warsaw (Siekierkowski Bridge), Vistula River	250.00	Steel + RC	2002
4	Gdańsk (III Millennium Bridge), Martwa Vistula River	230.00	Steel + RC	2001
5	Kwidzyn (National Road No. 90), Vistula River	204.00	PC	Under construction
Suspension bridges				
1	Witryłów Footbridge, San River	150.00	Steel	2010
2	Wrocław (Grunwaldzki Bridge), Odra River	114.00	Steel	1910/1947*

*Year of reconstruction after World War II.

14.2 Design Practice

The bridge design process in Poland is based on the following legal grounds:

1. The Polish Construction Law enacted by Polish Parliament; in accordance with this legal act construction permits are issued by administration, and the project should meet all the requirements, No. 89, item 414
2. The ordinance of Minister of Transport and Maritime Economy of May 30, 2000 on the technical requirements for road facilities and their locations, No. 63, item 735
3. The ordinance of Minister of Transport and Maritime Economy of September 10, 1998 on the technical requirements for railway facilities and their locations, No. 151, item 987
4. Technical requirements for railway structures PKP–D2, Polish State Railways Bulletin, 2001
5. A system of bridge design standards issued by the Polish Committee for Standardization

The legal acts in 2, 3, and 4 determine the technical and operation requirements for bridge structures that should be satisfied in every project. These requirements pertain to:

- Clear spans considering the flow of flood waters; important bridges are designed for 0.3% or 0.5% probability of flood waters
- The number and width of the bridge lanes
- The bridge load class
- The type of equipment elements, etc.

Poland applies the following national bridge design standards:

- PN-S-10030:1985. Bridges. Loads.
- PN-S-10052:1982. Bridges. Steel structures. Design.
- PN-S-10042:1991. Bridges. Concrete, reinforced concrete and prestressed concrete structures. Design.
- PN-S-10082:1993. Bridges. Timber structures. Design.

The essence of designing contained in these standards lies in the assessment of the degree to which a structure is endangered when it reaches the borderline state. The ultimate and serviceability limit states have been distinguished. The ultimate limit states are:

- Exhaustion of load capacity in the critical sections of construction; in the Polish standard system the load capacity is tested by means of the separate factors of the safety method (LRFD - load and resistance factors design)
- Loss of global and local stability
- Exhaustion of load capacity due to fatigue

The serviceability limit states are:

- Deformations hindering the serviceability of the structure
- Unacceptable vibration

It has been established that from April 30, 2010 bridge structures will be designed according to the Eurocode system supplemented by Polish annexes. It is permissible to design structures according to different standard systems, in accordance with the technical specifications defined for each structure.

14.3 Girder Bridges

In Poland, there are over 30 girder bridges with span lengths over 100 m. The structures with the longest spans are presented in Table 14.2. These are steel bridges with steel orthotropic decks or composite concrete decks as well as prestressed concrete bridges erected by the cantilever method. The largest or most interesting girder bridges are presented in Sections 14.3.1 through 14.3.3.

14.3.1 Bridge over the Bug River in Wyszków

This bridge superstructure (Figure 14.19) is a nine-span continuous prestressed concrete box girder. The total length of the structure is 600 m. C45/50 class concrete was used. The piers were settled on driven 1.5 m diameter piles of various lengths. The spans over the river and the adjacent spans were built by the cantilever method, whereas the approach spans over the flood area were constructed using scaffolding.

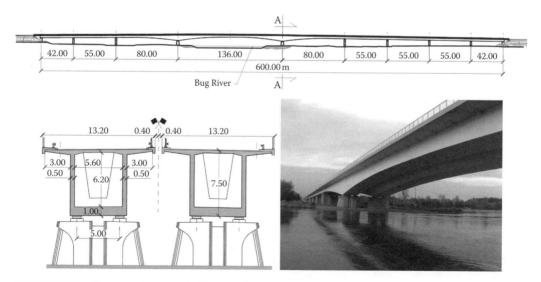

FIGURE 14.19 Prestressed concrete bridge over the Bug River in Wyszków along the S8 expressway (2008).

All of the segments were then connected by prestressing tendons to form a continuous beam. The bridge was designed by Profil and Transmost, Warsaw (Krzysztof Nagórko and Wojciech Kujawski) and constructed by Mosty Łódź, Łódź, Poland, from 2006 to 2008. The owner is the General Directorate for National Roads and Motorways.

14.3.2 Bridges along the Castle Route in Szczecin

The total length of the Castle Route in Szczecin is 2200 m, 1830 m (83%) of which runs along bridges and flyovers. The biggest bridges are located across the Parnica Canal (Figure 14.20, bottom right). The four-span superstructure of the northern bridge is 75.634 m + 135.314 m + 87.593 m + 57.871 m = 356.412 m long. The height of the structure varies from 2.024 m to 4.3 m. The southern bridge consists of three spans of 75.292 m + 134.670 m + 75.295 m = 285.257 m. The superstructure is 2.644–4.30 m high. The construction of the spans is formed by a curved, steel continuous beam that is a box section with two chambers. The route was finished in 1987.

The bridge was designed by BPBK Gdańsk (Henryk Żółtowski) and constructed by PBTK Trakt Szczecin, Szczecin, Poland. The owner is the city of Szczecin.

14.3.3 Bridge over the Vistula River in Wyszogród

This bridge is 1200 m long and consists of two segments separated by expansion joints (Pilujski 1998). The approach spans above the flood area are 50 m + 10 × 60 m long, for a total length of 650 m. The river spans are 75 m + 4 × 100 m + 75 m long. The total length of this part of the bridge is 550 m. The width of the bridge is 12.37 m. It is a steel composite structure. The spans consist of two steel girders connected with traverse beams and an RC deck slab. The approach span girders are constant height of 2.55 m, whereas the height of the girders over the river varies from 2.55 m to 5.37 m. The piers are supported by reinforced concrete driven piles 1.5 m in diameter. Both parts of the bridge were built by assembling prefabricated steel elements off-site and launching them onto the piers. Launching of the spans with variable heights required special alignment construction and temporary suspension (Figure 14.21, left). The reinforced concrete deck slab was poured on a movable scaffolding. The bridge was constructed from 1996 to 1999.

FIGURE 14.20   View of the Castle Route in Szczecin and the main bridges (1987).

FIGURE 14.21 Bridge over the Vistula River in Wyszogród along National Road No. 50 (1999).

The bridge was designed by Transprojekt Warszawa, Warsaw (Witold Doboszyński) and constructed by Mostostal Kraków, Cracow, Poland. The owner is the General Directorate for National Roads and Motorways.

14.4 Truss Bridges

The majority of the old road and railway bridges with spans longer than 50 m in Poland are truss bridges. As examples two structures are presented next: the old bridge in Płock built in 1938 (Figure 14.22) and the contemporary bridge in Brok erected in 1995 (Figure 14.23).

14.4.1 Bridge over the Vistula River in Płock

This bridge is a four-girder truss structure of variable height carrying one railway track, roadway, and a sidewalk. The total length is 690.4 m. The spans of the bridge are 75 m + 2 × 84 m + 92 m + 2 × 110.4 m + 92 m long. The maximum height of the truss is 19.8 m. The bridge was constructed from 1936 to 1938 (Jankowski 1973). The bridge was designed by the Office of Vehicular Roads Department Warsaw (Andrzej Pszenicki and Eugeniusz Hildebrandt) and constructed by K. Rudzki i S-ka, Mińsk Mazowiecki, Poland. The owner is the city of Płock.

14.4.2 Steel Bridge over the Bug River in Brok

The bridge is a six-span continuous structure consisting of two truss girders joined with a deck slab, which is partly longitudinally prestressed. The spans are 58 m + 3 × 69 m + 88 m + 49 m long and the total length of the bridge is 402 m (Bąk et al. 1995). The structure is 13.12 m wide. The abutments and piers in the flood areas are supported by Franki piles. The piers on the riverbed are supported by concrete piles 1.2 m in diameter. The structure was erected by assembling the steel elements on erection stays. The bridge was built from 1993 to 1995. The bridge was designed by Pomost Warszawa, Warsaw (Jerzy Bąk) and constructed by Mostostal Warszawa, Warsaw, Poland. The owner is the General Directorate for National Roads and Motorways.

14.5 Arch and Frame Bridges

Arch bridges in Poland were widespread in the period when railway tracks were laid down and later, when it was necessary to build spans of greater lengths. One historical and several contemporary structures are presented next.

FIGURE 14.22 (**See color insert.**) Road and railway truss bridge over the Vistula River in Płock called the Marshal Józef Piłsudski's Legions Bridge (1938).

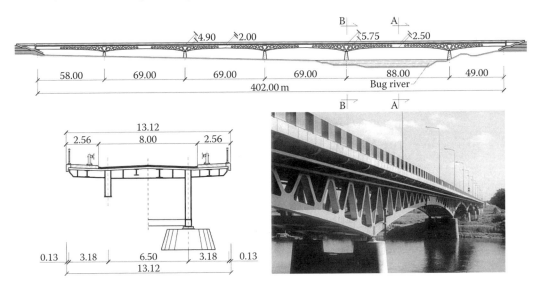

FIGURE 14.23 Bridge over the Bug River in Brok (1995).

14.5.1 Zwierzyniecki Bridge in Wrocław

The Zwierzyniecki Bridge (Figure 14.24) in Wrocław was constructed from 1895 to 1897. It is an aesthetically interesting steel arch structure 60.63 m long. The distance between arch truss girders is 12.54 m and the length of the cantilevers under the sidewalks is 4.63 m. Arch girders are connected over the roadway by truss bracing. The bridge has stone masonry abutments. The bridge was assembled on scaffolding next to the existing crossing; after this was demounted, the new span was launched transversely to the target position. The width of the roadway is 10 m. There are two tramway tracks laid down on the deck. The owner of the bridge is the city of Wrocław.

14.5.2 The Rainbow Overpass

The so-called Rainbow Overpass (Figure 14.25) is an overpass consisting of two pairs of concrete-filled steel tube arches and a suspended deck strengthened by two arch members. Structure is constituted by

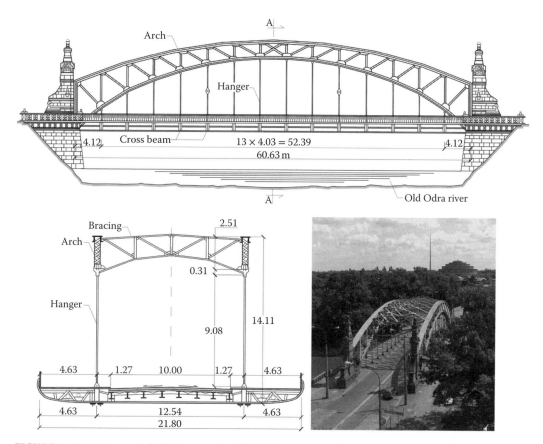

FIGURE 14.24 Zwierzyniecki Bridge over the Old Odra River in Wrocław (1897).

a reinforced concrete grating which consists of two outside grid and traverse beams with a reinforced concrete slab, minimum 0.27 m thick. Prestressing tendons are used in the outside beams along the whole span. The total length of the structure is 76.3 m and the width is 14.08 m. The overpass was built in 2006.

The bridge was designed by Transprojekt Gdański, Gdańsk (Ewa Kordek) and constructed by a consortium of Zakład Robót Mostowych Mostmar Marcin and Grzegorz Marcinków, PP-WBI Complex Projekt 2, Mosty Śląsk. The owner is the General Directorate for National Roads and Motorways.

14.5.3 Castle Bridge in Rzeszów

The Castle Bridge in Rzeszów was built in 2002. It is a five-span composite steel box structure (Figure 14.26). The spans are 2 × 44 m + 3 × 28 m long. Two river spans are supported on three reinforced concrete fixed arches; their theoretical span is 50 m. The arches are supported on diaphragm walls, 16 m deep. This was the first significant use of self-compacting concrete in bridge construction in Poland. The bridge was designed by design office Promost Consulting, Rzeszów (Tomasz Siwowski) and constructed by Mosty Łódź, Łódź, Poland. The owner is the city of Rzeszów.

14.5.4 National Road No. 69 Bridge in Milówka

This 662.53 m concrete bridge carries National Road No. 69 across the valley near Milówka in southern Poland. The structure consists of 12 spans, including 3 arch spans of 103.84 m each (Figure 14.27). The bridge is 12.74 m wide. The arch spans were built using large prefabricated elements up to 36 tons placed on temporary supports. Gaps between the prefabricated elements were filled with concrete, obtaining

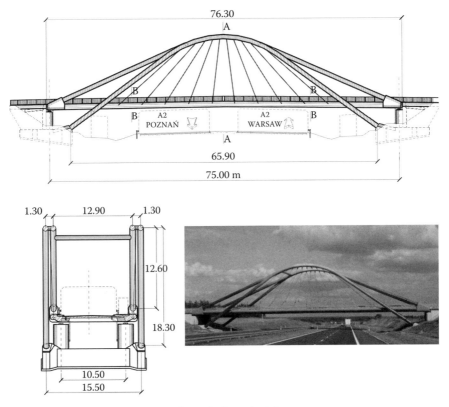

FIGURE 14.25   The Rainbow Overpass over the A2 highway (2006).

continuous structures. The deck on the arch spans and the beam part of the overpass were made by means of the step-by-step method using movable scaffolding. The structure was designed by Stähler + Knoppik Ingenieursgesellschaft GmbH, Germany and bulit by Skanska, Zakład Robót Mostowych, as well as Mostmar Marcin and Grzegorz Marcinków from 2004 to 2006. The owner is the General Directorate for National Roads and Motorways.

14.5.5 Bridge over the Narew River in Ostrołęka

The bridge consists of four spans of 32 m + 110 m + 2 × 32 m (Łagoda and Łagoda 1998). The main span is a steel tie arch 110 m long. A steel–concrete composite deck attached to the arch acts as a tension member (Figure 14.28). The rise of the arch is 21 m. The main span was assembled on the shore and then launched onto the piers by means of a barge. The bridge was designed by the Road and Bridge Research Institute, Warsaw (Marek Łagoda) and built by Beton-Stal Ostrołęka, Poland from 1993 to 1995. The owner is the city of Ostrołęka.

14.5.6 Steel Bridge over the Dziwna River at Wolin

This bridge (Figure 14.29) consists of three types of structures divided into five parts connected with expansion joints. The total length of the structure is 1112.4 m (Filipiuk et al. 2005; Żółtowski and Topolewicz 2007). The main span is a steel tie arch 165 m long. The span consists of two inclined

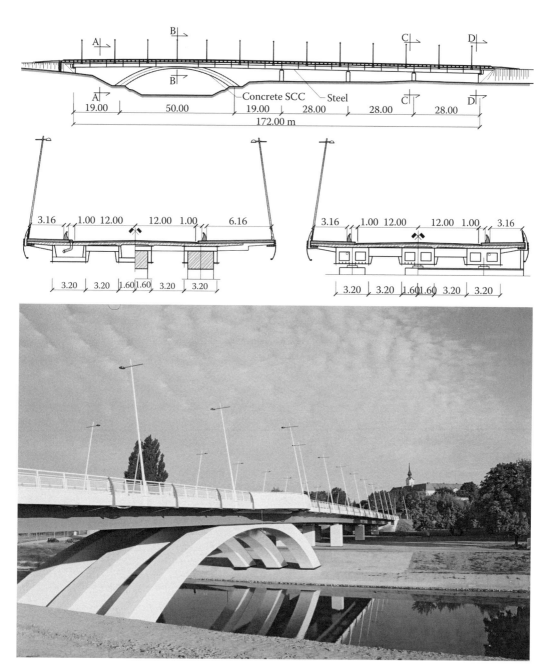

FIGURE 14.26 The Castle Bridge in Rzeszów (2002).

steel box arches with a rise of 24 m and a composite steel–reinforced concrete deck. The arches are box girders measuring 1.8 m × 1 m connected by seven pipe braces. The deck binding the arches is formed of two outside beams, one central stringer, and traverse beams. The traverse beams stick out beyond the outline of the bridge and the hangers are attached to them. The Freyssinet's closed ropes constitute the suspension. The thrust of the arch is transferred by six external cables of size 37L15. The prestressing force is greater than the thrust of the arch and causes prestressing of the reinforced

FIGURE 14.27   Bridge along National Road No. 69 in Milówka (2006).

FIGURE 14.28   **(See color insert.)** Bridge over the Narew River in Ostrołęka (1995).

concrete deck slab. The arch span was assembled using temporary supports. The bridge opened to traffic in 2003. The bridge was designed by Transprojekt Gdański, Gdańsk (Krzysztof Topolewicz) and built by Mostostal Warszawa, Warsaw, Poland. The owner is the General Directorate for National Roads and Motorways.

14.5.7 Kotlarski Bridge in Cracow

The superstructure of the 166 m long main span is a steel structure (Figure 14.30) (Majcherczyk et al. 2002). It consists of four arch girders and a steel orthotropic deck. The girders are formed of parabolic upper arches and of circular lower arches with various rises. In the midspan the upper arches are 13.258 m

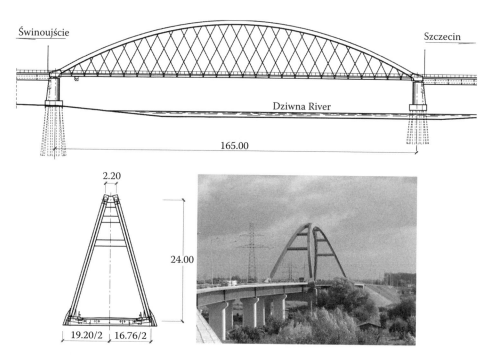

FIGURE 14.29   Steel bridge over the Dziwna River in Wolin along the S3 expressway (2003).

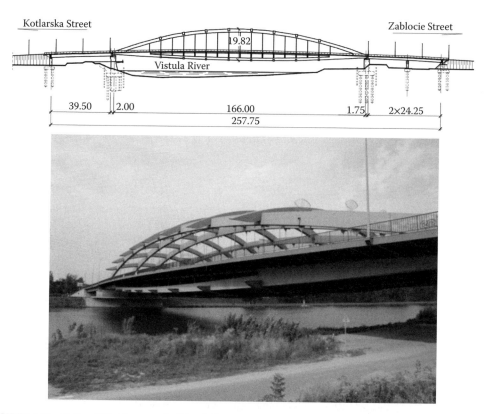

FIGURE 14.30   Kotlarski Bridge over the Vistula River in Cracow (2001).

high (inside girders) and 10.258 m high (outside girders). The spaces between the axes of upper and lower arches are 18.05 m (inner arches) and 14.05 m (outer arches). There are two tram tracks and two roadways on the bridge. Pedestrian and bicycle lanes are located outside the arches. The span is 36.84 m wide.

The assembly of the elements was carried out by means of launching along a circular track by means of three temporary supports. The lower part of the span—the lower arches integrated with the deck—was launched. The upper arches were assembled on the launched deck placed on temporary supports. The hangers were made of Macalloy rods. Steel consumption amounted to 690 kg/m². The high consumption of steel resulted from the necessity to comply with the architect's vision. The bridge was designed by Transprojekt Kraków, Cracow, architecture was designed by Witold Gawłowski, structural solutions were resolved by Bogdan Majcherczyk, and the bridge was built by Mostostal Kraków, Cracow, Poland from 2000 to 2001. The owner is the city of Cracow.

14.5.8 Bridge over the Vistula River in Puławy

The total length of this structure is 1038.2 m (Figure 14.31) and its width is 22.3 m. The main span of 212 m consists of two arches inclined at a 10-degree angle from vertical. On the piers, the arches turn into struts which direct the thrust to the deck. The deck is underslung by hangers, formed of 82 mm diameter rods, attached to it every 12 m. Each hanger consists of four rods. The bridge foundations are supported on concrete piles 1.2 m or 1.5 m in diameter.

The bridge was erected by assembling the large construction elements using temporary supports. All steel elements were welded together. The bridge was designed by Pomost Warszawa, Warsaw (Krzysztof Grej) and built by a consortium of Mosty Łódź, Łódź, Poland; Hermann Kirchner Bauunternehmung GmbH, Germany; and Vistal, Gdynia, Poland, from 2005 to 2008. Scientific consulting was done by Wrocław University of Technology, Wrocław, Poland. The owner is the General Directorate for National Roads and Motorways.

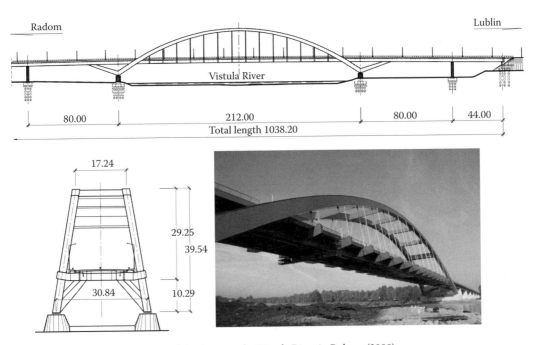

FIGURE 14.31 Steel composite arch bridge over the Vistula River in Puławy (2008).

14.6 Cable-Stayed Bridges

The first cable-stayed bridge in Poland was erected in 1959. It was a pedestrian footbridge over the Dunajec River in Tylmanowa (see Section 14.8.1). The majority of the structures presented next were built in the last 15 years.

14.6.1 Bridge over the Vistula River in Gdańsk along the Henryk Sucharski Route

The cable-stayed bridge over the Vistula River in Gdańsk is a six-span continuous structure with a total length of 372 m (Figure 14.32). The main span, 230 m long, is supported by cables attached to the 97.04 m high pylon (an inverted Y-shaped pylon), which is balanced by opposing cables anchored in the side spans and in the abutment. The deck is 21.31 m wide. It is a double girder steel construction combined with a reinforced concrete deck slab located on the edges, in which the cables are anchored. In the cross section the cables supporting the spans form two planes of support meeting in the pylon. The main span was erected by the cantilever method. The structure was built from 1999 to 2001.

The bridge was designed by Biuro Projektów Budownictwa Komunalnego, Gdańsk (Krzysztof Wąchalski) and built by Demathieu et Bard, France and Mosty Łódź, Łódź, Poland. The owner is the city of Gdańsk. Scientific consulting was done by Wrocław University of Technology, Wrocław, Poland.

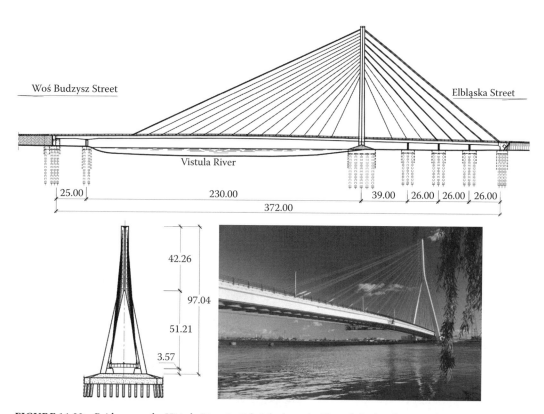

FIGURE 14.32 Bridge over the Vistula River in Gdańsk along the Henryk Sucharski Route (2001).

14.6.2 Bridge over the Vistula River in Warsaw along the Siekierkowska Route

The bridge over the Vistula River in Warsaw along the Siekierkowska Route (Figure 14.33) (Biliszczuk and Barcik 2009) is the third-longest cable-stayed bridge in Poland, with a span length of 250 m. The total length of the bridge is 826.5 m. The whole structure consists of the western flyover, 521 m long, the main bridge, 500 m long, and the eastern flyover, 57 m long. The main bridge consists of five spans and it is a two-beam steel structure, combined with the reinforced concrete deck slab, with spans of 48 m + 77 m + 250 m + 77 m + 48 m. The main span is supported symmetrically to two 90.07 m high H-shaped pylons balanced by opposing cables anchored in the side spans. In the cross section the cables supporting the spans form two vertical planes of support. In the side view they form a symmetrical fan system. The structure was built from 2000 to 2002.

The bridge was designed by Transprojekt Gdański, Gdańsk (Stefan Filipiuk) and built by Mostostal Warszawa, Warsaw, Poland. The owner is the city of Warsaw. Scientific consulting was done by Wrocław University of Technology, Wrocław, Poland.

14.6.3 Millennium Bridge over the Odra River in Wrocław

The total length of this bridge over the Odra River in Wrocław along the downtown ring road is 972 m (Figure 14.34). The structure consists of the western flyover, 325 m long, the main bridge, 290 m long, and the eastern bridge 357 m long. The main bridge is formed of three spans and it is a two-beam concrete structure with spans of 68.5 m + 153 m + 68.5 m. The main span and the side spans are supported symmetrically to the 50 m high pylons. In the cross section the cables supporting the spans form two vertical planes of support. In the side view they form a symmetrical fan system. The bridge was designed by BBR Polska, Warsaw (Piotr Wanecki) and built by Skanska, Poland. The owner is the city of Wrocław.

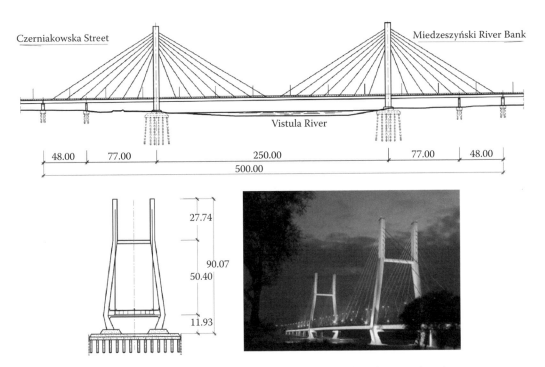

FIGURE 14.33   Bridge over the Vistula River in Warsaw along the Siekierkowska Route (2002).

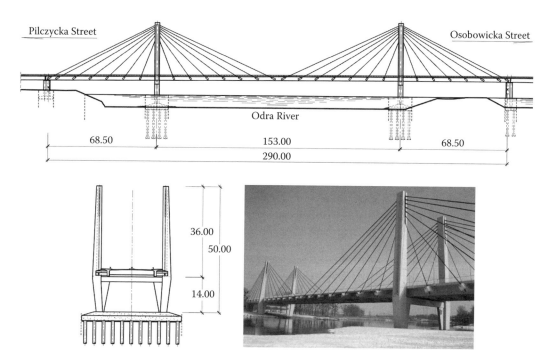

Pilczycka Street

Osobowicka Street

Odra River

68.50   153.00   68.50

290.00

36.00
50.00

14.00

FIGURE 14.34   Millennium Bridge over the Odra River in Wrocław (2004).

14.6.4 Bridge over the Vistula River in Płock

The bridge over the Vistula River in Płock (Figure 14.35) is the longest cable-stayed bridge in Poland, with a span length of 375 m (Biliszczuk and Barcik 2009). It is also the fifth-largest bridge in terms of total length. The structure is 1712 m long and consists of two southern flyovers, each 292.5 m long, the cable-stayed bridge, 615 m long, and the northern flyover, 512 m long. The main bridge is a five-span steel box structure with spans of 2×60 m $+ 375$ m $+ 2 \times 60$ m. The main span is supported symmetrically to two 63.7 m high column pylons, fixed in the span structure, balanced by opposing cables anchored in the side spans. The supporting cables form a fan system. In the cross section they form one central plane of support. The main bridge was erected by the cantilever method. The structure was constructed from 2003 to 2005.

The bridge was designed by Budoplan, Płock (Nicola Hajdin, Bratislav Stipanic) and constructed by Mosty Łódź, Łódź, Poland and Płockie Przedsiębiorstwo Robót Mostowych, Płock, Poland. The owner is the city of Płock. Scientific consulting was done by Wrocław University of Technology, Wrocław, Poland.

14.6.5 Bridge over the Odra River in Wrocław

The bridge over the Odra River in Wrocław along the A8 motorway is the fourth-longest bridge in Poland and the second-longest in terms of the length of the spans (Biliszczuk J., Onysyk J., Barcik W., et al. 2007). The bridge is 1742 m long and it is divided into three independent subsequent structures, the southern flyover, 611.05 m long, the 612 m long cable-stayed bridge, and the northern flyover, 521.05 m long. The main bridge is a four-span continuous structure with spans of 49 m $+ 2 \times 256$ m $+ 49$ m and with two parallel concrete decks, one for each roadway of the motorway (Figure 14.36). The total width of the spans is 38.58 m. The main spans are supported by cables attached to the 122 m high trapezoidal pylon, centrally situated. They are balanced by the side beam spans. In the cross section the cables supporting the spans (20 pairs for each span) form four surfaces of support (due to two separated decks)

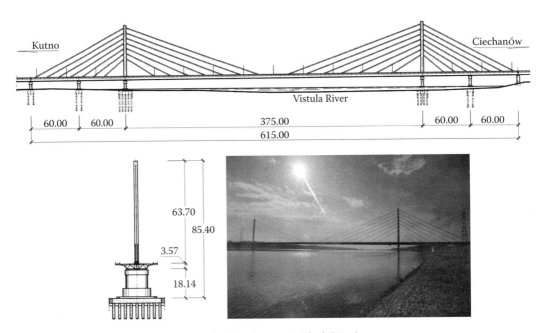

FIGURE 14.35　Cable-stayed bridge over the Vistula River in Płock (2005).

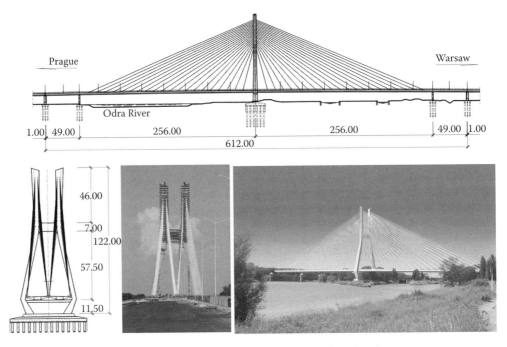

FIGURE 14.36　Bridge over the Odra River in Wrocław along the A8 highway (2011).

meeting in the arms of the pylon. In the side view they form a symmetrical fan system. The cables are anchored every 12 m and every 1.8 m in the upper arms of the pylon.

The bridge was designed by Research & Design Office Mosty-Wrocław, Wrocław (Jan Biliszczuk, Wojciech Barcik, and Jerzy Onysyk) and built by Mostostal Warszawa and Acciona, Warsaw, Poland. The owner is the General Directorate for National Roads and Motorways.

14.6.6 Overpass WD-22 over the A4 Motorway

Figure 14.37 presents a viaduct over the A4 motorway (Biliszczuk J., Onysyk J., Barcik W., et al. 2005). This is a prestressed concrete deck structure with steel pylons. The span lengths are 45.47 m and 45.28 m and the height of the steel part of the pylon is 15.3 m. The steel pylons are fixed to concrete bases located in a central reservation. The bridge was designed by Research & Design Office Mosty-Wrocław, Wrocław (Jan Biliszczuk and Wojciech Barcik) and constructed by DTP Terrassement, Poland. The owner is the General Directorate for National Roads and Motorways.

14.6.7 Overpass over the Klucz Interchange

This overpass (Figure 14.38) is situated at the S3 expressway and A6 motorway interchange. It is a landmark steel cable-stayed construction. The total length of the viaduct is 61.5 m and its width is 8.8 m. The deck was designed as a steel grid with a reinforced concrete slab. The 26 m high pylon is a welded box structure. The overpass was designed by Transprojekt Gdański, Gdańsk (Tadeusz Stefanowski) and erected by Hermann Kirchner GmbH et al. in 2009. The owner is the General Directorate for National Roads and Motorways.

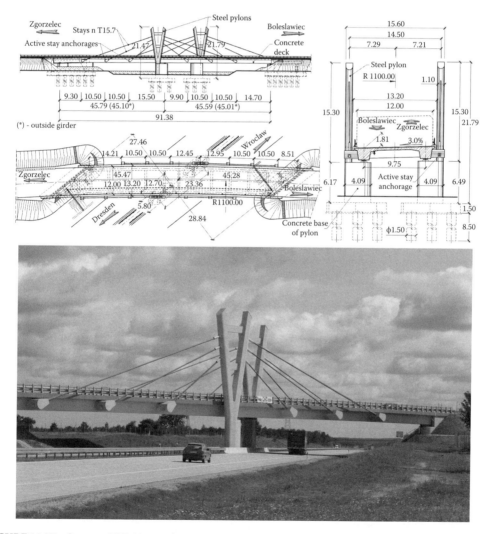

FIGURE 14.37   Overpass WD-22 over the A4 motorway (2008).

FIGURE 14.38   Overpass over the Klucz interchange on the S-3 expressway (2009).

14.7 Suspension Bridges

There is only one important suspension bridge in Poland: Grunwaldzki Bridge over the Odra River in Wrocław. Grunwaldzki Bridge, called "Kaiserbrücke" before 1945, is the only road and tramway suspension bridge in Poland. The bridge was designed by R. Weyrauch and M. Mayer in 1905. The construction works started in 1908 and were completed in September 1910. Construction consumed 1976 tons of cast steel (Thomas), 289.5 tons of cast iron, 1.4 tons of copper, and 7160 m³ of concrete. Beuchelt u. Sohn Company from Zielona Góra was the contractor of the bridge.

The pylons were built of stone face bricks with stone slab facing. The deck is comprised of two truss-structure edge beams. The suspension bands were made of metal sheets. All joints were riveted. The span length between the axes of pylons on which the suspension bands are suspended is 126.6 m and the distance between the axes of their anchor blocks is 172.6 m. A painting by C. Denner showing the bridge right after it was built in 1910 is presented in top of Figure 14.39. The bottom picture shows the current view of the structure. The pictures display changes both in the bridge and in the neighborhood that took place during the last 100 years. The owner is the city of Wrocław.

14.8 Pedestrian Bridges

Dozens of pedestrian bridges (footbridges) have been built in Poland in recent years. Some information about design and aesthetic solutions can be found in Biliszczuk et al. 2002 and Biliszczuk et al. 2005. It seems that the main goal of making the highways more attractive and promotion of the areas was achieved. These footbridges are recognizable from a distance by motorway users and serve as landmarks.

14.8.1 Cable-Stayed Footbridge over the Dunajec River in Tylmanowa

The footbridge in Tylmanowa (Figure 14.40) was the first cable-stayed structure in Poland, built in 1959. The deck is made of steel with wooden pavement. The span is 78 m long. The bridge was designed and erected by Mostostal Kraków (Józef Szulc and Władysław Główczak). The owner is the country borough of Tylmanowa.

14.8.2 Footbridges over the A4 Highway

A few representative pedestrian bridges built on the A4 (Wrocław–Katowice) highway include cable-stayed footbridges A(Mo)037 and KP-15 and the arch footbridge B(Mo)027. The pylon of the

FIGURE 14.39 Grunwaldzki suspension bridge over the Odra River in Wrocław (1910). The top picture presents the bridge in 1910; the bottom picture shows contemporary view of the bridge.

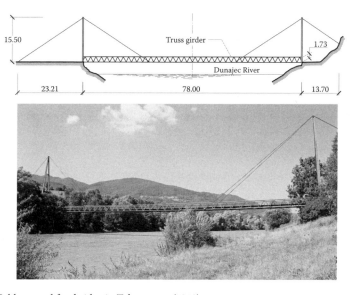

FIGURE 14.40 Cable-stayed footbridge in Tylmanowa (1959).

A(Mo)037 (Figure 14.41) footbridge is "A" shaped. The deck, made of prestressed concrete, is supported by stays anchored to the steel pylon. The pavement of the deck is situated lower than the upper part of the Z-shaped edge beams. The stays are anchored outside the girder just below the upper flange (Biliszczuk et al. 2005). The arch footbridge B(Mo)027 (Figure 14.42) has a similar deck. The plane of the single steel arch is not parallel to the axis of the deck. The arch was erected on independent concrete foundations. The bridge was designed by Research & Design Office Mosty-Wrocław, Wrocław (Jan Biliszczuk, Czesław Machelski, Jerzy Onysyk, and Przemysław Prabucki) and constructed by Mota Engil. The owner is the General Directorate for National Roads and Motorways.

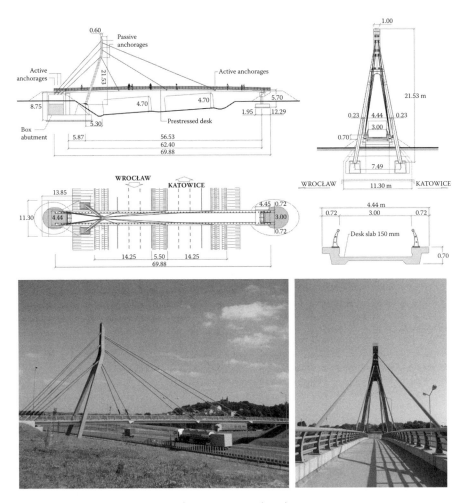

FIGURE 14.41   Footbridge A(Mo)037 over the A4 motorway (2000).

Footbridge KP-15 (Figure 14.43) (Biliszczuk et al. 2005) was designed as a double-span cable-stayed structure with a prestressed concrete deck and a steel pylon. The span lengths are 54.5 m + 7.02 m, and the height of the pylon is 23.66 m. The deck is made of C40/50 concrete. It is prestressed at both ends of the structure. The deck is fixed to the P2 abutment and supported by fixed bearings at the pylon and unidirectional bearings at the P1 abutment. The deck consists of two lateral edge beams connected by a slab. Each beam has a depth of 0.88 m and both are prestressed by two 13T15S cables. The main span is supported by 14 3T15S cables to the pylon. The pylon is stabilized by 14 backstays anchored in P2 abutment. Passive anchorages of the stays were placed in the pylon, whereas active ones are located in the deck and abutment P2. Ground anchors were applied to anchor the P2 abutment to the bed rock. The deck was made by means of scaffoldings. The steel tubes of the pylon were bent and then welded together. The pylon was fixed in the foundation. The bridge was designed by Research & Design Office Mosty-Wrocław, Wrocław (Jan Biliszczuk and Wojciech Barcik) and constructed by Prinż, Katowice, Poland. The owner is the General Directorate for National Roads and Motorways.

14.8.3 Footbridge over the A4 Highway in Katowice

The structural system of the footbridge is a steel arch fixed in the abutments (Figure 14.44). The span is 73 m. The 0.25 m thick concrete deck was connected to the arch in five points using the tension bars in polyethylene sheath. Two 0.60 m high beams are formed in the cross section. The bridge was designed by

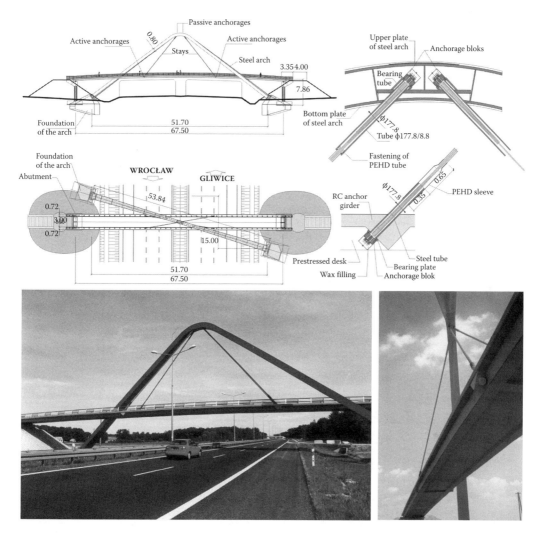

FIGURE 14.42 Footbridge B(Mo)027 over the A4 motorway (2000).

Mosty Katowice, Katowice (Krzysztof Markowicz) and built by Mosty Łódź, Łódź, Poland. The owner is the General Directorate for National Roads and Motorways.

14.8.4 Footbridge over S11 Expressway

The main span of the structure consists of a welded steel box arch and polymer composite deck. The inclined arch is circular with the radius equal to 21 m. The theoretical length of the arch (chord length) is 40 m. The inclination of the arch is 17 degrees measured from the vertical axis (Figure 14.45). The steel box arch was filled by concrete to the first hanger in order to improve the dynamic behavior of the structure. The bridge girder is a horizontally curved box structure. Double-sided cantilevers were welded to the main girder. The configuration of the cantilevers is radial. Hangers are anchored in every second cantilever. The bridge was designed by Transprojekt Warszawa, Warsaw (architecture by Bartłomiej Grotte, structure by Henryk Zobel) and built by Mota Engil. The owner is the General Directorate for National Roads and Motorways.

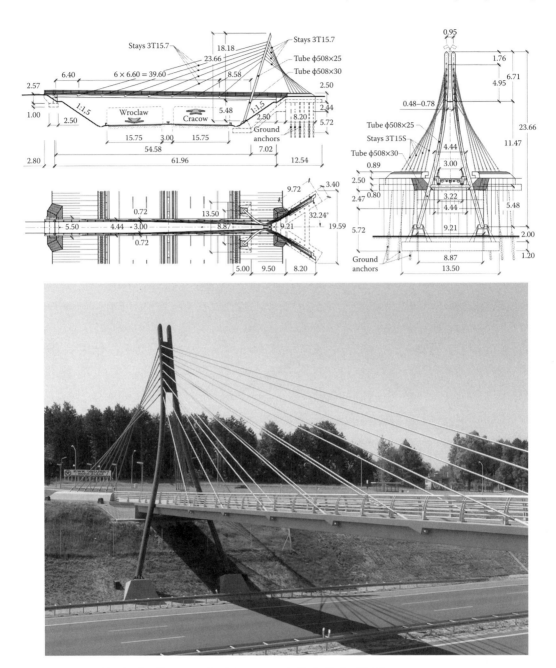

FIGURE 14.43   Footbridge KP-15 over the A4 motorway in Ruda Śląska (2004).

14.8.5 Footbridge in Sromowce Niżne

The footbridge in Sromowce Niżne was designed as a cable-stayed structure (Figure 14.46). The main span is 90 m long, whereas the approach spans are each 10.5 m long. The total length of the structure is 149.95 m (Biliszczuk et al. 2007). The wooden deck is attached to a steel pylon with five pairs of stays manufactured by Pfeifer. Full locked cables of Ø 40 mm or Ø 28 mm and 1570 MPa ultimate strength are used in the main span and tensioned rods of 860 MPa Ø 60 mm or Ø 52 mm type are used for the

FIGURE 14.44 Footbridge over the A4 motorway in Katowice (2007).

FIGURE 14.45 Footbridge over the S11 expressway (2008).

back stays. The deck of the footbridge is constructed with glued laminated wood braced by steel semiframes and wind bracing. The girders were designed using GL32 class pinewood with a rectangular cross section of 1.6 m × 0.30 m. The width of the deck is 2.5 m, and the total width is 3.5 m. The height of the deck is 1.87 m. The total length of the wooden girders is 112 m. The bridge was designed by Research & Design Office Mosty-Wrocław, Wrocław (Jan Biliszczuk, Paweł Hawryszków, and Mariusz Sułkowski) and built by Schmees & Lühn Polska and Remost PHU Dębica. The owner is the Commune of Czorsztyn.

14.8.6 Footbridge over the Vistula River in Cracow

This footbridge consists of three parallel steel structures (Figure 14.47). The middle part is dedicated to tourist traffic due to the excellent view of the city. The length of the structure is 127 m. The footbridge is expected to be constructed by 2015. The bridge was designed by Lewicki & Łatak (Kazimierz Łatak) for the architecture and Research & Design Office Mosty-Wrocław (Jan Biliszczuk, Tomasz Kamiński, and Robert Toczkiewicz) for the structure.

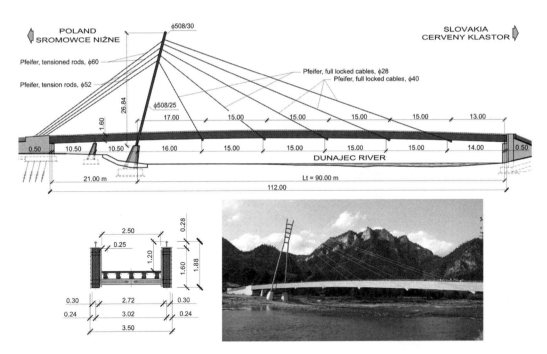

FIGURE 14.46   Design drawing and photo of the footbridge over the Dunajec River (2006).

FIGURE 14.47   Design of a steel footbridge over the Vistula River in Cracow. (Computer image courtesy of Lewicki & Łatak.)

14.9 Bridge Maintenance

14.9.1 Bridge Management Systems

Management of road bridges is divided between the General Directorate of National Roads and Motorways, with 16 regional divisions, and local governments (over 2800 offices on all levels). Owners of each category of the roads are responsible for management, maintenance, and operation of the bridge structures located along the roads. Management activities are supported by computer-based bridge management systems (BMS) (www.gddkia.gov.pl) in all offices of the General Directorate of National Roads and Motorways as well as in the majority of the transportation departments of local governments. The most important groups of information collected in the BMS database are presented in Figure 14.48. The management process is based on systematic assessment of bridge condition by means of results from bridge inspections and tests of bridge structures. Before opening a bridge for traffic all road structures with a theoretical span

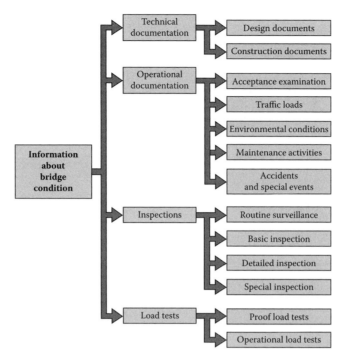

FIGURE 14.48 Main groups of information collected in the BMS database.

length exceeding 20 m and all railway bridges with a span length greater than 10 m are tested under proof loads. The system of bridge inspections consists of routine surveillance performed from the ground level by road or track staff every 3 months, basic inspections required for all bridges once a year executed as visual investigations, detailed inspections performed every 5 years by highly experienced staff, and special inspections with the application of advanced equipment in the case of problems in condition assessment.

PKP Polish Railway Lines Company (www.plk-sa.pl), with eight regional divisions, is responsible for management, maintenance, and operation of all railway bridges. Since the late 1990s the management process has been supported by the computer-based railway BMS (RBMS) (Bień and Rewiński 1996) implemented in all offices in the whole country. The general functional scheme of the system is presented in Figure 14.49. The management system is equipped with a database and knowledge base, which enables implementation of expert systems (Bień and Rawa 2004) supporting decision processes.

Computer-aided BMS are used for permanent analysis of bridge conditions and safety of the transportation systems as well as for optimization of maintenance planning and allocation of funds. For condition assessment, advanced non-destructive testing methods (Helmerich et al. 2007; Olofsson et al. 2005) as well as sophisticated dynamic tests (Bień et al. 2002, 2004) are used. The most important large bridge structures are also equipped with individually designed monitoring systems for permanent observation of the structure.

14.9.2 Bridge Monitoring Systems

On a worldwide scale, modern structures are becoming more and more complicated, sophisticated, daring, and sometimes amazing—and there are a lot of bridges among them. Very often their spans are long, slender, and supported by cables. For this reason, it was proposed to observe select bridges using a modern approach, by employing a procedure of permanent data acquisition concerning the state and behavior of the structure. Such a procedure is often called a monitoring system. There are three big bridges in Poland that are equipped with systems of permanent data acquisition. They were built in the beginning of the twenty-first century across the Vistula River, as well as the Odra River. One is a cable-stayed bridge in Płock with a main span of 375 m, the second is an arch bridge in Puławy

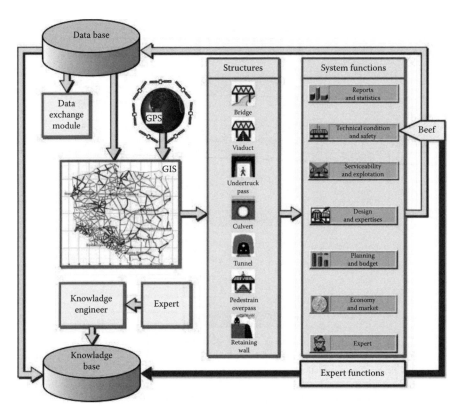

FIGURE 14.49   General scheme of the RBMS implemented in Poland.

with the main span of 212 m, and the third is a cable-stayed bridge in Wrocław. Some remaining bigger structures were in the past or will be in the future equipped with monitoring systems.

The monitoring system of Płock Bridge was expected to gain information on the behavior of the structure regarding force in the stays, inclination of the pylons, strains of the structure, and wind load. The following gauges were employed (Figure 14.50): load cells (8 sensors in selected stays), inclinometers (2 sensors in pylons), strain gauges (10 sensors in the main span girder and one pylon), and anemometers (2 sensors, one in the midspan, 19 m above water level, and one on the top of a pylon, 83 m above water level). Furthermore, there is a data acquisition unit involving a computer, amplifiers, a power unit designed to ensure uninterrupted supply, and other devices/accessories. The conclusions from the operation on the monitoring system may be divided into three groups:

1. Remarks concerning the state of the structure and its safety
2. Knowledge-building observations on the behavior and features of cable-stayed structures
3. Data on real environmental effects on the bridges

Attention was focused on forces in the stays and the inclination of the pylons, understood to be crucial parameters defining structural safety. The conclusion drawn after the first period of data acquisition is reassuring. The observed behavior of the bridge is not unexpected and the measured parameters are within the foreseen limits. What is interesting is that environmental effects can lead to serious loads and extraordinary behavior of large cable-stayed bridges. For instance, the force in a stay induced by trucks passing over the deck was smaller than the force induced by gusts of wind in the evening of the same day (Figures 14.51 and 14.52). Furthermore, the movement of the pylons during strong wind can be bigger

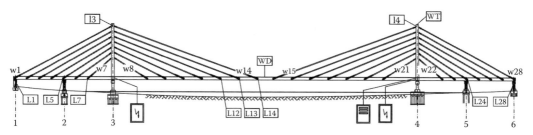

FIGURE 14.50 The diagram of the Płock bridge. The gauges L1–L28 are load cells, I3–I4 are inclinometers, and WD and WT are anemometers. The strain gauges are spread in the bottom of pylon 4 and the superstructure. There are also three temperature sensors.

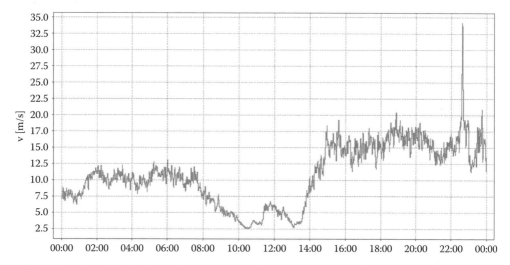

FIGURE 14.51 The wind velocity on January 18, 2007. The maximum value in the evening reached 34.14 m/s.

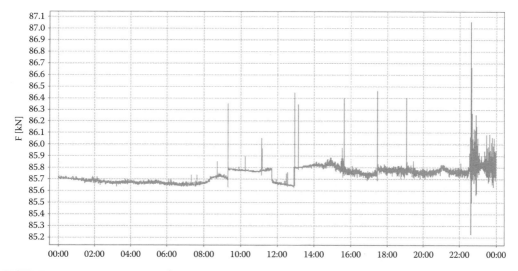

FIGURE 14.52 Force in one strand on January 18, 2007. The maximum value in the evening corresponds to the maximum wind speed. The other peaks represent the passages of trucks.

than during the ordinary service of a structure. Over the whole of 2006, the maximum rotation of the top of one pylon (total range) was lower than 0.05 degrees, but on one day a much bigger rotation was observed. The maximum rotation was about 0.3 degrees, on the evening of January 18, 2007.

The system that was installed inside the bridge in Płock is a very useful tool for gathering important information on the loads generated by environmental effects and the loads induced by automobile traffic on a cable-stayed bridge. Moreover, it is a very good opportunity to assess the real range of forces in stays due to different types of loads over long periods. The data gathered in the system can be very useful to verify the standards and design criteria for cable-stayed bridges. Furthermore, it can be used to asses the probability of simultaneity of occurrence of different types of loads.

14.9.3 Professional Education in Bridge Engineering

Education and research focused on bridge engineering are conducted in Poland mainly by the technical universities in Warsaw, Wrocław, Cracow, Poznań, Gdańsk, Rzeszów, and Gliwice. The universities offer study on a BSc level (7–8 semesters) and MSc level (3–4 semesters) as well as post-graduate studies, including PhD studies (8–10 semesters). Wrocław University of Technology (www.pwr.wroc.pl) and Rzeszów University of Technology (www.prz.edu.pl) also manage special training for bridge inspectors required for professional authorization. The academic centers systematically organize short-term trainings and courses for bridge engineers as a part of the Continuous Education System.

Professional certification is required in Poland for independent bridge design and for construction works management. To receive certification, it is necessary to have a few years experience in bridge engineering under the supervision of a professional engineer, as well as to pass special exams arranged by the Polish Chamber of Civil Engineers (www.piib.org.pl). The professional bridge engineer with long-term experience can receive the title of bridge expert.

Bridge engineers can be members of the Polish Society of Bridge Engineers and the Polish Association of Engineers & Technicians of Transportation. They are usually members of the Polish Chamber of Civil Engineers, which has over 100,000 engineers and technicians with building qualifications in all specialties.

References

Bąk J., Grej K., Sałach W. 1995. Road Bridge Crossing Bug River in Brok (in Polish). *Inżynieria i Budownictwo* 4.

Bień J. 2010. *Defects and Diagnostic of Bridge Structures* (in Polish). Transport and Communication Publishers, Warsaw, 417 pp.

Bień J., Gładysz M., Rawa P. 2002. *Vibration Tests of Bridge Structures*. First International Conference on Bridge Maintenance, Safety and Management, IABMAS. Barcelona, Spain.

Bień J., Krzyżanowski J., Rawa P., et al. 2004. Dynamic Load Tests in Bridge Management. *Archives of Civil and Mechanical Engineering* 2:63–78.

Bień J., Rawa P. 2004. Hybrid Knowledge Representation in BMS. *Archives of Civil and Mechanical Engineering* 1:41–55.

Bień J., Rewiński S. 1996. SMOK—Railway Bridge Management System (in Polish). *Inżynieria i Budownictwo* 3:180–84.

Biliszczuk J. 1997a. On the Oldest Prestressed Concrete Bridge in Poland (in Polish). *Inżynieria i Budownictwo* 6:285–86.

Biliszczuk J. 1997b. The History of the Development of Prestressed Concrete Bridges in Poland (in Polish). *Inżynieria i Budownictwo* 9:427–31.

Biliszczuk J., Barcik W. 2009. Steel Bridges in Poland, State of the Art and Properties (in Polish). *Inżynieria i Budownictwo* 1–2:36–42.

Biliszczuk J., Barcik W., Hawryszków P., et al. 2005. *New Footbridges in Poland*. Footbridge 2005, Second International Conference, Venice.

Biliszczuk J., Budych L., Kubiak Z., et al. 2000. Historic Large Stone Masonry Railway Bridges in Silesia (in Polish). *Inżynieria i Budownictwo* 9:477–80.

Biliszczuk J., Hawryszków P., Maury A., et al. 2007. *A Cable-Stayed Footbridge Made of Glued Laminated Wood: Design, Erection and Experimental Investigations.* Improving Infrastructure Worldwide, IABSE Symposium, Weimar.

Biliszczuk J., Hildebrand M. 2000. On Historic Bridge from the Early Years of the 19th Century in Opatówek (in Polish). *Inżynieria i Budownictwo* 9:475–76.

Biliszczuk J., Hildebrand M., Rabiega J., et al. 2009. The Oldest Iron Bridges in Poland (1796–1827) (in Polish). *Inżynieria i Budownictwo* 9:487–90.

Biliszczuk J., Machelski Cz., Onysyk J., et al. 2002. *Examples of New Built Footbridges in Poland.* Proceedings of the International Conference on the Design and Dynamic Behavior of Footbridges, Paris, France.

Biliszczuk J., Onysyk J., Barcik W., et al. 2005. *Bridge Structures as Landmarks along Polish Motorways.* Fib Symposium: Keep Concrete Attractive, Budapest.

Biliszczuk J., Onysyk J., Barcik W., et al. 2007. *Bridges along Highway Ring of Wrocław.* Improving Infrastructure Worldwide, IABSE Symposium, Weimar.

Chwaściński B. 1997. *Bridges over Vistula River and Their Builders* (in Polish). Fundacja Rozwoju Nauki w zakresie Inżynierii Lądowej im. A. i Z. Wasiutyńskich.

Filipiuk S. 2005. *Bridge over Dziwna River in Wolin along Bypass of Wolin National Road No. 3.* Qax Manufaktura Artystyczna.

Helmerich R., Bień J., Cruz P. 2007. A Guideline for Inspection and Condition Assessment Including the NDT-Toolbox. In *Sustainable Bridges: Assessment for Future Traffic Demands and Longer Lives*, eds. J. Bień, L. Elfgren, and J. Olofsson, Lower Silesian Educational Publishers, Wrocław, Poland.

Jankowski J. 1973. *Bridges in Poland and Polish Bridge Engineers.* Ossolineum.

Łagoda G., Łagoda M. 1998. The New Cable-Stayed Bridge Over the Narew River in Ostrołęka (in Polish). *Inżynieria i Budownictwo* 5:253–56.

Majcherczyk B., Mendera Z., Pilujski B. 2002. The New Kotlarski Bridge in Cracow: The Longest Arch Bridge in Poland (in Polish). *Inżynieria i Budownictwo* 3–4:125–30.

Mistewicz M. 1991. *Road Bridges in Poland.* General Directorate of Public Roads, Bridge Division.

Olofsson J., Elfgren L., Bell B., et al. 2005. Assessment of European Railway Bridges for Future Traffic Demands and Longer Lives—EC Project Sustainable Bridges. *Structure and Infrastructure Engineering* 2:93–100.

Pancewicz Z. 1989. *The First Welded Road Bridge in the World: 60 Years of the First Welded Bridge in the World.* WKŁ.

Pasternak H., Rabiega J., Biliszczuk J. 1996. 200 Jahre eiserne Brücken auf dem europäischen Kontinent—auf Spurensuche in Schlesien und der Lausitz. *Stahlbau* 12:542–46.

Pilujski B. 1998. Assembly of Steel Road Bridge Construction over the Vistula River in Wyszków (in Polish). *Inżynieria i Budownictwo* 5:231–5.

Rabiega J., Biliszczuk J. 1996. Two Hundred Years of Iron Bridges in Silesia Region (in Polish). *Inżynieria i Budownictwo* 3:139–42.

Ramm W. 2004. Witness of the Past: The Old Bridge over the Vistula River in Tczew. Technische Universität Kaiserslautern.

Żółtowski K., Topolewicz K. 2007. Arch Bridge above Dziwna River in Wolin—Design and Realization after 3 Years of Operation (in Polish). *Inżynieria i Budownictwo* 1:27–30.

Relevant Websites

www.gddkia.gov.pl
www.piib.org.pl
www.plk-sa.pl
www.portal.prz.edu.pl
www.pwr.wroc.pl

15

Bridge Engineering in Russia

Simon A. Blank
Consulting Engineer

Vadim A. Seliverstov
Giprotransmost

15.1 Introduction

Bridge design and construction practice in the former USSR, especially Russia, is not well known by foreign engineers. Many advanced structural theories and construction practices have been established. In view of the global economy, the opportunities to use such advanced theories and practices especially exist with the collapse of the Iron Curtain. In 1931, Franklin D. Roosevelt said "There can be little doubt that in many ways the story of bridge building is the story of civilization. By it, we can readily measure progress in each particular country." The development of bridge engineering is based on previous experiences and historical aspects. Certainly the Russian experience in bridge engineering has its own specifics.

15.2 Historical Evolution

15.2.1 Masonry and Timber Bridges

The most widespread types of bridges in past times were timber and masonry bridges. Because there were plenty of natural wood resources in ancient Russia, timber bridges were solely built up to the end of the fifteenth century. For centuries masonry bridges (Figure 15.1) were built on territories of such former republics of the USSR as Georgia and Armenia. From different sources it is known that the oldest masonry bridges in Armenia and Georgia were built in about the fourth to sixth centuries. One of the remaining old masonry bridges in Armenia is the Sanainsky Bridge over the River Debeda-chai, built in 1234. The Red Bridge over the River Chram in Georgia was constructed in the eleventh century.

Probably the first masonry bridges built in Russia were in the city of Moscow. The oldest, constructed in 1516, was the Troitsky arch masonry bridge near the Troitsky Gate of the Kremlin. The largest masonry bridge over the Moskva River was named Bolshoi Kamenny and was designed by Yacobson and Kristler. Construction started in 1643, but after two years it was halted because of Kristler's death. Only in 1672 did construction continue by an unknown Russian master and was mostly completed in 1689. Finally, the Russian czar Peter the First completed construction of this bridge. Bolshoi Kamenny Bridge is a seven-span structure with a total length of 140 m and width of 22 m (Belyaev 1945). This bridge was rebuilt twice (in 1857 and 1939). In general, a masonry bridge cannot compete with bridges of other materials due to cost and duration of construction.

Ivan Kulibin (Russian mechanics engineer, 1735–1818) designed a timber arch bridge over the Neva River with a span of about 300 m, illustrating one attempt in the search for efficient structural form (Evgrafov and Bogdanov 1996). He tested a 1/10 scale model bridge to investigate the adequacy of the member cross sections of the arch and found a strong new structural system. However, due to unknown reasons the bridge was not built.

15.2.2 Iron and Steel Bridges

Extensive progress in bridge construction in the beginning of the nineteenth century was influenced by overall industry development. A number of cast iron arch bridges for roadway and railway traffic were built. At about the same time construction of steel suspension bridges was started in St. Petersburg. In 1824 the Panteleimonovsky Bridge over the Fontanka River, with a span of about 40 m, was built. In 1825 the pedestrian Potchtamsky Bridge over the Moika river in Bankovsky and the Lion's river over the Ekaterininsky kanal were constructed, and the Egiption Bridge over the Fontanka river, with a span of 38.4m was constructed in 1827.

The largest suspension bridge built in the nineteenth century was a chain suspension bridge over the Dnepr River in Kiev (Ukraine), with total length 710 m, including six spans by 68.3 m and one (134.1 m) constructed in 1853. In 1851–1853 two similar suspension bridges were constructed over the Velikaya River in Ostrov with spans of 93.2 m. Development in suspension bridge systems was based on the invention of wire cables. In Russia, one of the first suspension bridges using wire cables was built in

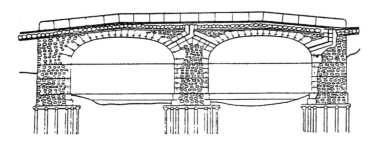

FIGURE 15.1 Typical masonry bridge (1786).

1836 near the Brest–Litovsk fortress. This bridge crossed the West Bug River, with a span of about 89 m. However, suspension bridges of the first half of the nineteenth century, due to a lack of understanding structural performance, resulted in inadequate stiffness in both the vertical and horizontal directions. This appeared to be the main reason for a series of catastrophes with suspension bridges in different countries. Any occurrence of catastrophes in Russia was not noted.

To reduce the flexibility of suspension bridges, first timber and then steel stiffening trusses were applied. However, this innovation improved the performance only partly and the development of beam bridges became inevitable in the second half of the nineteenth century. The first I-beam bridges in Russia were the Semenovsky Bridge over the Fontanka River in St. Petersburg, constructed in 1857, and the bridge over the Neman River in Kovno on the St. Petersburg–Warsaw railway, constructed in 1861. However, I-beams for large spans proved to be very heavy and this led to a more wide application of truss systems.

The construction of the railway line in 1847–1851 between St. Petersburg and Moscow required a large number of bridges. Zhuravsky modified the structural system of timber truss implemented by Howe in 1840 to include continuous systems. These bridges over the Volga, Volhov, and some other rivers had relatively large spans; for example, the bridge over the Msta River had a 61.2 m span. This structural system (known in Russia as the Howe–Zhuravsky system) was widely used for bridge construction up to the mid-twentieth century.

Steel truss bridges at first were structurally similar to types of timber bridges, such as plank trusses or lattice trusses. A distinguished double track railway bridge with a deck truss system, designed by Kerbedz, was constructed in 1857 over the Luga River on the St. Petersburg–Warsaw railway line (Figure 15.2). The bridge consists of two continuous spans (each 55.3 m). Π-shaped cross sections were used for chord members, and angles for diagonals. Each track was carried by a separate superstructure that includes two planes of trusses spaced at 2.25 m. The other remarkable truss bridge built in 1861 was the Borodinsky Bridge over the Moskva River in Moscow, with a span length of 42.7 m. The cornerstone of the 1860s was the introduction of caisson foundations for bridge substructures.

Until the 1880s, steel superstructures were fabricated of wrought steel. Cast steel bridge superstructures appeared in Russia in 1883. After the 1890s, wrought steel was not used for superstructures. In 1884, Belelubsky established the first standard designs for steel superstructures, covering a span range of 54.87–109.25 m. For spans exceeding 87.78 m, polygonal trusses were designed. A typical superstructure with a 87.78 m span is shown in Figure 15.3. Developments in structural theory and technology advances in the steel industry expanded the capabilities of shop fabrication of steel structures and formed a basis for further simplification of truss systems and increase of panel sizes. This improvement resulted in an application of triangular-type trusses. By the end of the nineteenth century, a tendency toward transition from lattice truss to triangular truss was outlined and Proskuryakov initiated usage of a riveted triangular truss system in bridge superstructures. The first riveted triangular truss bridge in Russia was constructed in 1887 on the Romny–Krmenchug railway line.

FIGURE 15.2 Bridge over the Luga River (1857).

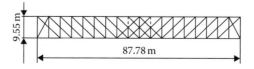

FIGURE 15.3 Standard truss superstructure.

Many beam bridges built in the mid-nineteenth century were the continuous span type. Continuous bridge systems have economic advantages, but they are sensitive to pier settlement and have bigger movement due to temperature changes. To take these aspects into account was a complicated task at that time. In order to transfer a continuous system to a statically determined cantilever system, hinges were arranged within the spans. This led to a new direction in bridge construction. The first cantilever steel railway bridge, over the Sula River, designed by Proskuryakov, was built in 1888. The first steel cantilever highway bridge, over the Dnepr River in Smolensk, was constructed in 1898. A steel bridge using the cantilever system over the Dnestr River, a combined railway and highway bridge with a span of 102 m, was designed by Boguslavsky and built in 1894. In 1908, a steel bridge over the River Dnepr near Kichkas, carrying railway and highway traffic and with a record span of 190 m, was constructed. New techniques developed for the construction of deep foundations, and an increase of live loads on bridges made the use of continuous systems more feasible compared to other systems.

In the mid-nineteenth century arch bridges were normally constructed of cast iron, but from the 1880s, steel arch bridges started to dominate over cast iron bridges. The first steel arch bridges were designed as fixed arches. Hinged arch bridges appeared later and became more widely used. The need for application of arches in flat areas led to the creation of depressed through-arch bridges.

15.3 Modern Development

The twentieth century has been remarkable in quick spread of new materials (reinforced concrete, prestressed concrete, and high-strength steel), new structural forms (cable-stayed bridges), and new construction techniques (segmental construction) in bridge engineering. The steel depressed through-arch truss railway bridge over the Moskva River built in 1904 is shown in Figure 15.4. To reduce thrust, arches of the cantilever system were used. For example, the Kirovsky Bridge over the Neva River, with spans of 97 m, was built in St. Petersburg in 1902. Building new railway lines required construction of many long multispan bridges. In 1932, two distinguished arch steel bridges designed by Streletsky were constructed over the Old and New Dnepr River (Ukraine) with main spans of 224 m and 140 m, respectively (Figure 15.5).

The first reinforced concrete structures in Russian bridge construction practice were culverts at the Moscow–Kazan railway line (1892). In the early twentieth century, the use of reinforced concrete was limited to small bridges with spans of up to 6 m. In 1903, the road ribbed arch bridge over the Kaslagach River was built (Figure 15.6). This bridge had a total length of 30.73 m and a length of arch span of 17 m. In 1904, the road bridge over the Kazarmen River was constructed. The bridge

FIGURE 15.4　Steel depressed arch railway bridge over the Moskva River.

FIGURE 15.5 Steel arch bridge over the Old and New Dnepr River (Ukraine) with a span of 224 m.

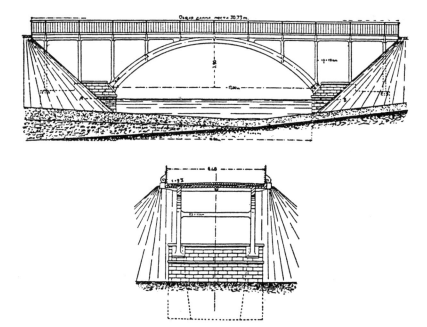

FIGURE 15.6 Reinforced concrete ribbed arch bridge over the Kaslagash River (1903).

had a total length of 298.2 m and comprised 13 reinforced concrete arch spans of 21.3 m, with box section ribs.

The existing transportation infrastructure of Russia is less developed compared with other European countries. The average density of railway and highway mileage is about five times less than that of the United States. A tremendous effort in renovation of the transport infrastructure was made in Moscow and St. Petersburg cities in the 1990s. This effort began with reconstruction of the outer Moscow 109 km Ring Road (MKAD), which was originally built as a dual two-lane motorway in 1962. Reconstruction of the Ring Road into a dual four-lane motorway was completed in the fall of 1998 (Figure 15.7). The project comprised reconstruction and new building of 69 highway bridges and overpasses, 2 railway bridges, 41 pedestrian bridges, and 12 pedestrian tunnels (Perevoznikov and Seliverstov 1999).

The next development in Moscow was the Third Ring Road. This 37 km inner city motorway was completed in 2005. Today, most bridge construction is oriented toward solving transportation problems within the city zone. Besides some of the most critical locations (bottlenecks of the city transportation system), the main bridge construction has started at the first section of one of the most ambitious projects: the Fourth Ring Road, another 74 km inner city ring road. The first section of the Ring Road around St. Petersburg was opened to traffic in 2008. This 80.5 km ring road has four lanes in each direction. Along with the Ring Road a so-called West Speed Diameter motorway is currently under construction. This 47 km road is designed as a dual four-lane motorway. Bridges make up 55% of the whole

FIGURE 15.7   Interchange of Moscow Ring Road (MKAD).

length of the motorway and comprise 14 interchanges; bridges must have navigation clearances of 55 m and 35 m. Construction is planned to be completed in 2011.

In the last few years, some major bridge crossings over the Volga and Kama rivers were built and form an important part of the country's transportation system. Major bridge crossings are under construction in Vladivostok. Also, to improve the transportation system of Russia, a long-term transport strategy up to year 2030 was recently accepted by the government.

15.3.1 Standardization of Superstructures

An overview of the number and scale of bridges constructed in Russia has shown that about 70% of railway bridges have span lengths less than 33.6 m and 80% of highway bridges less than 42 m. Medium and small bridges are predominant in construction practice and standard structural solutions have been developed and used efficiently, using standardized design features for modern bridges. The current existing standardization covers the design of superstructures for certain bridge types. For railway bridges, standard designs are applicable to spans from 69 m to 132 m. These are reinforced concrete superstructures of slab, stringer types, box girder, steel superstructures of slab, composite superstructures, steel superstructures of through-plate girder, deck truss, and through-truss types. For highway bridges, standard designs cover a span range from 12 m to 147 m. These are reinforced concrete superstructures of voided slab, stringer, channel (Π shape) girder, solid web girder, box girder types, composite superstructures of steel web girder types, and steel superstructures of web and box girder types.

Modern highway bridges with spans up to 33 m and railway bridges up to 27.6 m are normally constructed with precast concrete simple beams. For highway bridges, continuous superstructures of solid web girder precast concrete segments are normally used for spans of 42 m to 63 m, and box girder type precast box segments are used for spans of 63 m to 84 m. The weight of precast segments should not exceed 60 tons and should meet the requirements of railway and highway transportation clearances. For railway bridges, steel box superstructures of full-span (33.6 m), shop-fabricated segments have become the most widespread in current practice.

The present situation in Russia is characterized by a relative increase in the scale of application of steel superstructures for bridges. After the 1990s, a large number of highway steel bridges were constructed. Construction was primarily based on modularization of superstructure elements: shop-fabricated segments with lengths of up to 21 m. Many of them were built in the city of Moscow, and on and over the Moscow Ring Road; also bridges were built over the Oka River in Nizhni Novgorod, over the Belaya River in Ufa, and many others. A number of steel and composite highways bridges with spans from 60–150 m are currently under design or construction. In construction of bridges over the Moskva, Dnepr, Oka, Volga, Irtish, and Ob rivers, on the peripheral highway around Ankara (Turkey), in Moscow, St. Petersburg (the Ring Roads), Sochi, and some other cities, continuous steel superstructures of web and box girder types with permanent structure depths are typically used. The superstructures are assembled with modularized, shop-fabricated elements, which are welded at a shop or construction site to form a complete cross section configuration. Erection is normally preceded by incremental launching or cantilever segmental construction methods.

Extensive use of steel bridges in Russia is due to their optimum structural solutions, which account for the interaction of fabrication technology and erection techniques. Efficiency is proven by high-quality welded connections, which allowed the erection of large prefabricated segments and reduction in quantities of works and the construction period. A low maintenance cost that can be predicted with sufficient accuracy is also an advantage. However, superstructures of prestressed reinforced concrete in some cases require essential and frequently unpredictable expenses to ensure their capacity and durability.

15.3.2 Features of Substructures

Construction of bridge piers in Russia has been mainly oriented toward the use of precast concrete segments in combination with cast-in-place concrete. The usual practice is to construct piers with columns of uniform rectangular or circular sections fixed at the bottom to the foundations. For highway bridges with spans up to 33 m, full-height precast rectangular columns of 50 cm × 80 cm are standard. For longer-span bridges, the use of precast contour segments forming the outer shape of the piers and cast-in-place methods has become more widespread. A typical practice is the use of driven precast concrete piles for the pile foundation. Standard types of precast concrete piles are of square sections 0.35 m × 0.35 m, 0.40 m × 0.40 m, and 0.45 m × 0.45 m and a circular hollow section 0.60 m in diameter. Also, in the last two decades, CIDH (cast-in-place drilled hole) piles of 0.80–1.7 m in diameter have been widely used for foundations.

An increasing tendency to use pile shafts has been seen, especially in urban areas when superstructures bear directly on piles extended above the ground level (as columns). Efficiency was reached by implementing piles of square sections 35 cm × 35 cm and 40 cm × 40 cm and a circular section 80 cm in diameter. Bored and cast-in-place piles of large diameters ranging from 1.6–3 m, drilled to a depth of up to 50 m, and steel casings are widely employed in bridge foundations. In foundations for bridges over rivers and reservoirs, bored piles using a nonwithdrawable steel casing within the zone of change in the water level and scour depth are normally used. These foundation types were applied for the construction of bridges over the Oka River on the peripheral road around Nizhni Novgorod, over the Volga River in Kineshma, over the Ob River in Barnaul, over the Volga River in Ulyanovsk, and some others.

Open abutments are the most commonly used type for highway bridges. Typical shapes are bank seats, bank seats on piles, and buried skeleton (spill-through). Wall abutments and bank seats on piles are the types of substructures mainly used for railroad bridges. Wing walls are typically constructed back from abutment structures and parallel to the road.

15.4 Design Theory and Methods

15.4.1 Codes and Specifications

The Russian Bridge Code SNIP 2.05.03-84 (SNIP 1996) was first published in 1984, amended in 1991, and reissued in 1996. In Russia the new system of construction codes was adopted in 1995. In accordance with this system bridges must be designed to satisfy the requirements of the bridge code, regional codes, and the standards of enterprises or companies. The standards introduce new requirements resolving inconsistencies found in the bridge code. The bridge code covers design of new bridges and rehabilitation of existing bridges and culverts for highways, railways, tramways, metro lines, and combined highway–railway bridges. The requirements specified are for the locations of structures in all climatic conditions in the area of the former USSR, and for seismic regions up to magnitude 9 (ground acceleration of 0.4 g) on the MSK-64 scale. The bridge code has seven main sections: (1) general provisions, (2) loads, (3) concrete and reinforced concrete structures, (4) steel structures, (5) composite structures, (6) timber structures, and (7) foundations.

In 1995, the Moscow City Department of Transportation developed and adapted "Additional Requirements for the Design and Construction of Bridges and Overpasses on the Moscow Ring Highway" to supplement the bridge code for the design of highway widening and rehabilitation of more than 50 bridge structures on the Moscow Ring Highway. The live load is increased by 27% in the "Additional Requirements." Further, in 1999 the new standard MGSN 5.02-99 (MGSN 1999) was published on the basis of the "Additional Requirements." This new standard specifies an increased live load and abnormal loading and reflects the need to improve the reliability and durability of bridge structures. In 2008 the application of the increased live load was widened from the regional Moscow city area to all of Russia (GOST P 52748 2007).

15.4.2 Concepts and Philosophy

In the former Soviet Union, the ultimate strength design method (strength method) was adopted for the design of bridges and culverts in 1962. Three limit states (1) strength at ultimate load, (2) deformation at service load, and (3) crack width at service load, were specified in the bridge design standard, a predecessor of the current bridge code. Later, limit states 2 and 3 were combined into one group. The state standard GOST 27751-88, "Reliability of Constructions and Foundations" (GOST 1989), specifies two limit states: strength and serviceability. The first limit state is related to structural failure, such as loss of stability of the structure or its parts, structural collapse of any character (ductile, brittle, fatigue), and development of a mechanism in a structural system due to material yielding or shear at the connections. The second limit is related to cracking (crack width), deflections of the structure and foundations, and vibration of the structure.

The main principles for the design of bridges are specified in the Building Codes and Regulations, "Bridges and Culverts," SNIP 2.05.03-84 (SNIP 1996). The ultimate strength is obtained from specified material strengths (e.g., concrete at maximum strength and usually the steel yielding). In general bridge structures should satisfy the ultimate strength limit in the following format:

$$\gamma_d S_d + \gamma_1(1+\mu)S_1\eta \le F(m_1, m_2, \gamma_n, \gamma_m, R_n, A) \tag{15.1}$$

where S_d and S_1 are force effects due to dead load and live load, respectively; γ_d and γ_1 are overload coefficients; μ is the dynamic factor; η is the load combination factor; F is a function determining the limit state of the structure; m_1 is the general working condition factor accounting for possible deviations of the constructed structure from design dimensions and geometrical form; m_2 is a coefficient characterizing uncertainties of structure behavior under load and inaccuracy of calculations; γ_n is the coefficient of material homogeneity; γ_m is the working condition factor of the material; R_n is the nominal resistance of the material; and A is the geometrical characteristic of the structure element.

The serviceability limit state requirement is:

$$f \le \Delta \tag{15.2}$$

where f is the design deformation or displacement and Δ is the ultimate allowable deformation or displacement.

The analysis of bridge superstructures is normally implemented using three-dimensional analysis models. Simplified two-dimensional models considering interaction between the elements are also used.

15.4.3 Concrete Structure Design

Concrete structures are designed for both limit states. The load effects of statically indeterminate concrete bridge structures are usually obtained with consideration of inelastic deformation and cracking in concrete. Proper consideration is given to redistribution of effects due to creep and shrinkage of concrete, force adjustment (if any), cracking, and prestressing which are applied using coefficients of reliability for loads equal to 1.1 or 0.9.

The analyses of the strength limit state include calculations for strength and stability at the conditions of operation, prestressing, transportation, storing, and erection. The fatigue analysis of bridge structures is made for operation conditions. The analyses of the serviceability limit state comprise calculations for the same conditions as indicated for the strength limit state. The bridge code stipulates five categories of requirements to crack resistance: no cracks; allowing a small probability of crack formation (width opening up to 0.015 mm) due to live load actions on the condition that closing the cracks perpendicular to the longitudinal axis of the element under the dead load is assured; and allowing the opening of cracks after the passing of live load over the bridge within the limitations of crack width openings 0.15 mm, 0.20 mm, and 0.30 mm, respectively. The bridge code also specifies that the ultimate elastic deflections of super-structures shall not exceed $L/600$ for railroad bridges and $L/400$ for highway bridges. The new standard (MGSN 1999) provides a more strict limit of $L/600$ for the deflections of highway bridges in Moscow.

15.4.4 Steel Structure Design

Steel members are analyzed for both groups of limit states. Load effects in elements of steel bridge structures are usually determined using elastic small deformation theory. Geometrical nonlinearity must be accounted for in the calculation of systems in which such an account causes a change in effects and displacements more than 5%. The strength limit states for steel members are limited to member strength, fatigue, general stability, and local buckling. Calculations for fatigue are obligatory for railroad and highway bridges. In calculations for strength and stability to the strength limit state for steel super-structures, the code requires consideration of physical nonlinearity in the elastoplastic stage. Maximum residual tensile strain is assumed as 0.0006, and shear strain is equal to 0.00105.

Net sections are used for the strength design of high-strength bolt (friction) connections and the cross sections for fatigue, stability, and stiffness design. A development of limited plastic deformations of steel is allowed for flexural members in the strength limit state. The principles for design of steel bridges with consideration for plastic deformations are reviewed in Potapkin 1984. Stability design checks combined flexure and torsion forms of stability loss, flange, and web buckling. A composite bridge superstructure is normally based on a hypothesis of plane sections. Elastic deformations are considered in calculations of effects occurring in elements of statically indeterminate systems as well as in calculations of strength and stability, fatigue, crack control, and ordinates of camber.

15.4.5 Stability Design

The piers and superstructures of bridges must be checked for stability with respect to overturning and sliding under the action of load combinations. Sliding stability is checked with reference to the horizontal plane. Working condition factors of more than unity should also be applied for overturning and driving and less than unity for resisting forces.

15.4.6 Hydraulic Considerations

Bridge design criteria against floods is specified in the bridge code (SNIP 1996). The selection of design flood frequency is based on highway and railway classifications depending on the category of railway or highway on which the bridge is located. The probability of exceedance is normally taken in the range of 1–2% for railways and 1–3% for highways. In addition, bridges on railways of the first and second categories are required to be checked for a flood with a probability of exceedance of 0.33%. A new approach to the selection of the design flood frequency was suggested by several studies (Perevoznikov, Ivanova, and Seliverstov 1997; Perevoznikov and Seliverstov 2001). The new design criteria is based on importance classification of bridges and fits the requirements of the basic standard GOST 27751 (see Tables 15.1 and 15.2). The new design approach requires two design cases to be analyzed: stream action on bridges of basic flood design (regular conditions of operation) and extreme flood (extraordinary conditions of operation when design flood is exceeded), and the designs should be reviewed for the extent of probable damage.

15.4.7 Temporary Structures

The current design criteria for temporary structures used for bridge construction is set forth in the guideline VSN (Department Building Norms) 136-78 (VSN 1978, 1984). This departmental standard was developed mainly in addition to the bridge code and also some other codes related to bridge construction. VSN 136-78 is a single volume first published in 1978 and amended in 1984. These guidelines cover the design of various types of temporary structures (sheet piling, cofferdams, temporary piers, falsework etc.) and devices required for the construction of permanent bridge structures. It specifies loads and overload coefficients, working condition factors to be used in the design, and requirements for the design of concrete and reinforced concrete, steel, and timber temporary structures. It also provides special requirements for devices and units of general purpose, construction of foundations, forms of cast-in-place structures, and erection of steel and composite superstructures.

TABLE 15.1 Design Flood Probability of Exceedance Related to Bridge Importance Classification

	Bridge Crossing										
	Railways						Highways in Towns				
Design Case	Probability of Exceedance Depending on Important Class of a Structure, %					Design Case	Probability of Exceedance Depending on Important Class of a Structure, %				
	I	II	III	IV	V		I	II	III	IV	V
Basic	0.33	1	2	5	10	Basic	0.33	1	2	5	10
Extreme	0.1	0.33	1	3	5	Extreme	0.2	0.33	1	3	5

TABLE 15.2 Importance Classification

Importance Level	Flood Hazard Exposure Class	Description of Structures
I	I	Major highway and combined bridges, bridges on railways of I and II categories (main transportation system); bridges on roads (irrespective of their categories) damage of which would cause nonrecoverable ecological consequences and also long-term interruption of traffic with losses exceeding the initial construction cost.
	II	Bridges on railways of III and IV categories (main transportation system), highways and roads of I–III categories.
II	III	Bridges on railways of IV and V category (approach railroads), railroads on internal links of enterprises, highways of IV–V categories. All bridges on highways not classified in classes I–III.
III	IV	Temporary and short-term bridge structures with an operation period of up to 10 years.
	V	Temporary structures used for bridge construction (rehabilitation protection).

Temporary structures are designed to the two limit states, similar to the principles established for permanent bridge structures. Meanwhile, the overload coefficients and working condition factors have lower values compared to those of permanent structures. A study has shown that guideline VSN 136-78 requires revision, and therefore initial recommendations to improve the specified requirements and some other aspects have been reviewed (Seliverstov 1997). VSN 136-78 specifies a 10-year frequency flood. Also, on the basis of technical-economical justification, up to a 2-year return period may be taken in the design, but in this case special measures for high water discharge and passing of ice are required. The methods for providing technical hydrologic justification for reliable functioning of temporary structures are reviewed in the guideline (Perevoznikov, Ivanova, and Seliverstov 1997). To widen the existing structures, the range of the design flood return period for temporary structures in the direction of lower and higher probabilities of exceedance is also recommended (Perevoznikov, Ivanova, and Seliverstov 1997).

15.5 Inspection and Test Techniques

The techniques discussed herein reflect current requirements set forth in SNIP 3.06.07-86, the rules of inspection and testing for bridges and culverts (SNIP 1988). This standard covers inspection and test procedures of constructed (new) or rehabilitated (old) bridges. This standard is also applicable for inspection and tests of structures currently under operation or for bridges designed for special loads such as pipelines, canals, and others. Inspection and testing of bridges are implemented to determine conditions and to investigate the behavior of structures. These works are implemented by special test organizations, contractors, or operation agencies.

All newly constructed bridges are inspected before opening to traffic. The main intention of inspection is to verify that a bridge meets the design requirements and the requirements on quality of works specified by SNIP 3.06.04-91 (SNIP 1992). Inspection is carried out by means of technical check up, control measurements, and instrumentation of bridge structural parts. If required, inspection may additionally comprise nondestructive control of materials quality, investigation into the strength characteristics of materials by lab tests, setting up long-term instrumentation, and so on. Results obtained during inspection are then compared to allowable tolerances for fabrication and erection specified in SNIP 3.06.04-91 (SNIP 1992). If tolerances or other standard requirements are breached, the influence of these noted deviations on bridge load-carrying capacity and the service state are estimated.

Prior to structure testing various details should be precisely defined on the basis of inspection results. Load tests require an elaboration safety procedure to protect both the structure and human life. Maximum load, taking into account the design criteria and existing structural deviations, needs to be established. The position of the structure (before testing is started) for future identification of changes needs to be recorded. For the purpose of dynamic load tests the conditions of loads passing over the bridge are evaluated.

15.5.1 Static Load Tests

During static load testing, displacements and deformations of the structure and its parts, stresses in typical sections of elements, and local deformations (crack opening, displacements at connections, etc.) are measured. Moreover, depending on the types of structures, field conditions, and the testing purpose, measurements of angle strain and load effects in the stays or struts may be executed. For static load tests a bridge is loaded by locomotives, rolling stock of railways, metro or tramway trains, or trucks as live loads. In cases when separate bridge elements are tested or the stiffness of the structure needs to be determined, jacks, winches, or other individual loads may be needed.

Load effects in members obtained from the test should not exceed the effects of live load considered in the design with account for the overload factor of the unit and the value of the dynamic factor taken in the design. At the same time, load effects in members obtained from the test shall not be less than 70% of that due to design live loading. The weight characteristics of transportation units used for tests should be measured with an accuracy of at least 5%. When testing railroad, metro, or tramway bridges or bridges for heavy trucks, load effects in a member normally should not be less than those due to the most heavy live loads passing over a given bridge. The quantities of static load tests depend on the bridge length and complexity of structures. Superstructures of longer spans are usually tested in detail. In multispan bridges with similar equal spans, only one superstructure is tested in detail; other superstructures are tested on the basis of a reduced program, and thus only deflections are measured.

During testing the live loads are positioned on the deck in such a manner that maximum load effects (within the limits outlined above) occur in the member being tested. The duration of the test at each position shall be determined by a stabilization of readings at measuring devices. Observed deformation increments within a period of 5 minutes should not exceed 5%. In order to improve the accuracy of measurements, the duration of loading, unloading, and taking readings shall be minimized as much as possible. Residual deformations in the structure shall be determined on the basis of the first test loading results. Loading of structures by test load will normally be repeated. The number of repeated loadings is established considering the results that are obtained from the first loading.

15.5.2 Dynamic Load Tests

Dynamic load tests are performed in order to evaluate the dynamic influence (impact factor) of actual moving vehicles and to determine the main dynamic characteristics of the structure (free oscillation frequency and oscillation form, dynamic stiffness, and characteristics of damping oscillation). During dynamic load testing, overall structure displacements and deformations are measured (e.g., midspan deflections, displacements of the superstructure end installed over movable bearings), and also in special cases displacements and stresses in individual members of structure are measured.

Heavy vehicles that may really pass over the bridge shall be used in dynamic load tests to determine the dynamic characteristics of structures and moving impact. Vibration, wind, and other loads may be used for dynamic tests. To investigate oscillations excited by moving vehicles, the trucks are required to pass over the bridge at different speeds, starting from 10–15 km/h. The behavior of the structure within a range of typical speeds can thus be determined. At least 10 heats of trucks at various speeds are recommended, and those heats should be repeated when increased dynamic impact is noted. In some cases when motorway bridges are tested, to increase the influence of moving vehicles (e.g., to ascertain dynamic characteristics of the structure), a special measure may be applied. This measure is to imitate the deck surface roughness, for example by laying planks (4 cm thick) perpendicular to the roadway spaced at the same distance as the distance between the truck wheels.

When testing pedestrian bridges, excitation of structure free oscillations is made by throwing down a load or single pedestrians or a group of pedestrians walking or running over bridges. Throwing down a load (e.g., castings having a weight of 0.3–2 tons) creates the impact load on a roadway surface, typically from 0.5–2 m high. The location of test load application shall coincide with the section where maximum deflections occur (midspan, cantilever end). To protect the roadway surface a sand layer or protective decking is placed. The load is dropped down several times and every time adjusts the height. The results of these tests give diagrams of free oscillations of superstructures. Load effects in structural members during the test execution shall not exceed those calculated in the design, as stipulated in the section above.

15.5.3 Running-In of Bridges under Loads

To reveal adequate behavior of a structure under the heaviest operational loads, running-in of bridges is conducted. Running-in of railroad and metro bridges is implemented under heavy trains, while bridges designed for AB highway loading are run-in by heavy trailers. Visual observations of the structure's behavior under load are performed. Midspan deflection may also be measured by simple means such as leveling. At least 12 load passes (shuttle type) with different speeds are recommended in the running-in procedures of railroad and metro bridges. The first two or three passes shall be performed at a low speed of 5–10 km/h; if deflection measurements are required, the trains are stopped. Positioning trailers over marginal lanes with 10 m spacing between the back and front wheels of adjacent units is recommended in running-in of bridges designed for AB highway loading of two or more lanes. It is recommended that single trailers pass over free lanes at a speed of 10–40 km/h, and at least 5 passes are normally taken. When visual observations are completed, trailers are moved to another marginal lane and single trailers pass over the lane, which is set free. For running-in of single lane bridges the passes of single trailers are used only.

15.6 Steel and Composite Bridges

In recent decades, there has been further development in the design and construction of steel bridges in Russia. New systems provide a higher level of standardization and reduce construction costs.

15.6.1 Superstructures for Railway Bridges

Most railroad bridges are steel composite bridges with single-track superstructures. Standard super-structures (Popov et al. 1998) are applicable to spans from 18.2 m to 154 m (steel girder spans 18.2 m to 33.6; composite girder spans 18.2 m to 55 m; deck truss spans 44 m to 66 m; and through-truss spans 33 m to 154 m). Figures 15.8 and 15.9 show typical railroad bridges. A steel box girder, as shown in Figure 15.10, has a span of 33.6 m. Similar box girders can be assembled by connecting two prefabri-cated units. These units are shop welded and field bolted with high-strength friction-type bolts (Monov and Seliverstov 1997; Popov et al. 1998). For truss systems, the height of trusses is from 8.5 m to 24 m and the length of panels is 5.5 m to 11 m. Two types of bridge deck, ballasted deck with a road bed of reinforced concrete or corrosion-resistant steel and ballastless deck with a track over wooden ties or reinforced concrete slabs, are usually used. For longer than 154 m or double-track bridges, a special design is required.

FIGURE 15.8 Typical through-truss bridge of 44 m span on the Baikal–Amour railway bridge line.

FIGURE 15.9 **(See color insert.)** Deck girder composite bridge of 55 m span over the Mulmuta River (Siberia).

FIGURE 15.10 Steel box superstructure of 33.6 m span for railway bridges.

15.6.2 Superstructures for Highway Bridges

Standard steel superstructures covering spans from 42 m to 147 m are based on modularization of elements: 10.5 m length box segments, 21 m double-T segments, and 10.5 m orthotropic deck segments. Typical standard cross sections are shown in Figure 15.11. The main technological features of shop production of steel bridge structures have been maintained and refined. Automatic double arc-welding machines are used in 90% of shop fabrications.

15.6.3 Construction Techniques

Steel and composite superstructures for railway bridges are usually erected by cantilever cranes (Figure 15.12) with capacities up to 130 tons and boom cranes with larger capacities. The superstructures of 55 m span bridges may be erected by the incremental launching method using a nose. When

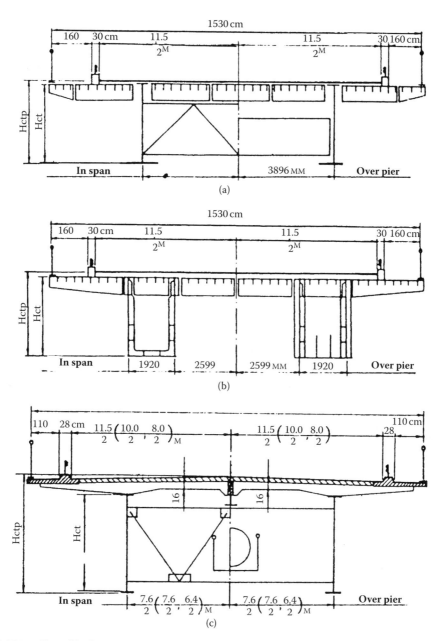

FIGURE 15.11 Typical highway superstructure cross sections: (a) steel plate girder; (b) box girder; (c) composite girder.

a cantilever crane is applied for erection of superstructures of 55 m span, a temporary pier is required. For truss superstructures, cantilever and semicantilever erection methods are widely implemented. Figure 15.13 shows the cantilever erection of a bridge over the Lena River (spans of 110 m + 132 m + 110 m) using derrick cranes.

For highway bridges, the incremental launching method has been the main erection method for plate girder and box girder bridges since 1970, although this method has been known in Russia for a long time. Cantilever and semicantilever methods are also used for girder bridges. Floating-in is the most effective method to erect superstructures over waterway when a large quantity of assembled superstructure

FIGURE 15.12   Erection of 33.6 m steel box superstructure by boom cranes.

FIGURE 15.13   Cantilever erection of typical truss bridge using derrick cranes.

segments are required. The application of standard pontoons simplified the assembly of the erection floating system. Equipping floating temporary piers by an air leveling system and other special equipment made this erection method more reliable and technological.

15.6.4 Typical Girder Bridges

15.6.4.1 Pavelesky Railroad Overhead

The design of overpasses in Moscow was a challenge to engineers. It requires low construction depth and minimum interruption of traffic flow. Figure 15.14 shows the Paveletsky Overhead built in 1996 over the widened Moscow Ring Road. The bridge has a horizontal curve of $R = 800$ m and carries a triple track line. The steel superstructure (Figure 15.15) was designed with a low construction depth to meet the specified 5.5 m highway clearance and to maintain the existing track level. The new four-span (11 m + 30 m + 30 m + 11 m) skewed structure was designed as simple steel double tee girders with a depth of 1.75 m. An orthotropic plate deck 20 mm thick with inverted T-ribs was used. The greatest innovation of this bridge is the combination of a crossbeam (on which the longitudinal ribs of the orthotropic deck are beared) and a vertical stiffener of the main girder connected to the bottom flange forming a diaphragm spaced at 3 m. A deck cover sheet in contact with the ballast was protected by a metallized varnish coating.

FIGURE 15.14 Paveletsky railroad overpass.

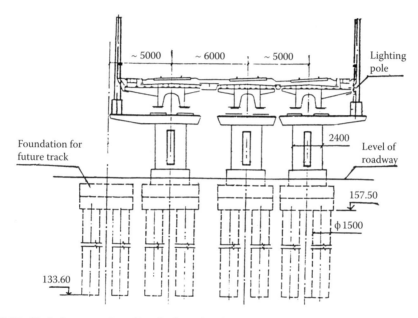

FIGURE 15.15 Typical cross section of Paveletsky railroad overpass (all dimensions in mm).

15.6.4.2 Moskva River Bridge

Figure 15.16 shows the Moskva River Bridge in the Moscow region, built in 1983. The superstructure is a continuous three-span (51.2 m + 96 m + 51.2 m) twin box girder (depth of 2.53 m) with orthotropic deck. The bridge carries two lanes of traffic in each direction and sidewalks of 3 m. In the transverse direction boxes are braced, connected by cross frames at 9 m. Piers are of reinforced concrete Y-shaped frames. Foundation piles are 40 cm × 40 cm sections and driven to a depth of 16 m. The box girders were launched using temporary piers (Figure 15.17) from the right bank of the Moskva River. Four sliding devices were installed on the top of each pier. The speed of launching reached 2.7 mi/h.

15.6.4.3 Oka River Bridge

Oka River Bridge, near Nizhny Novgorod, consists of twin box sections (Figure 15.18) formed with two L-shaped elements with an orthotropic deck and opened to traffic in 1993. This 988 m long steel bridge is over the Oka River and has a span arrangement of 2 × 85 m + 5 × 126 m + 2 × 84 m.

FIGURE 15.16   Moskva River Bridge in Krilatskoe.

FIGURE 15.17   Superstructure launching for the Moskva River Bridge in Krilatskoe.

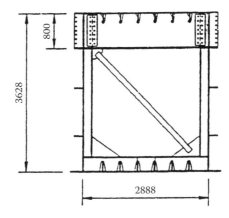

FIGURE 15.18   Cross section of main box girder formed of two L-shaped elements (all dimensions in mm).

15.6.4.4 Ural River Bridge

The Ural River Bridge near Uralsk was is completed in 1998. This five-span (84 m + 3 × 105 m + 84 m) continuous composite girder bridge (depth of 3.6 m) lies on a gradient of 2.6 m and carries single lane of highway with an overall width of 14.8 m. Traditionally, when precast deck slabs were used for bridge construction, these precast elements were fabricated with the provision of holes for shear connectors

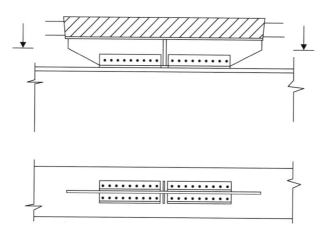

FIGURE 15.19 Connection details of precast reinforced concrete deck and steel girders.

that were later filled with concrete. This solution has several disadvantages. To improve the practice, a new type of joint between structural members was introduced. The use of steel embeds in precast slabs allows them to connect to the main girders by means of angles and high-strength bolts. Precast concrete deck segments are connected with steel girders by high strength friction bolts (Figure 15.19). Connection details are presented in Monov and Seliverstov (1997).

15.6.4.5 Chusovaya River (Perm–Beresniki) Highway Bridge

The Chusovaya River Bridge, with a length of 1504 m, lies on convex vertical curves in a radius of 8,000 m and 25,000 m and was completed in 1997. The superstructure comprises two continuous steel composite girders (4 × 84 m + 84 m + 126 m + 5 × 147 m + 126 m + 84 m). One main problem with bridge construction was the construction of pier foundations within the river under complex geological and hydrological conditions. All river piers were constructed under protection of sheet pilling with the use of scows. Piers were constructed using floating cranes. The superstructure segments, with lengths of 94.8 m, 84 m, and 99.7 m, were assembled on the right bank and then slipped to the river and placed over the floating pier.

15.6.4.6 Donbassky Bridge Project in Moscow

The existing tram and highway bridges located in the tight urban area of Moscow were built in 1913 and 1945, respectively. Both bridges were severely deteriorated and classified as structurally deficient. Based on the investigation inspection staged replacement of old bridges while maintaining uninterrupted traffic flow was recommended. The entire project consists of the reconstruction of a 586.5 m portion of the Warshavskoe city road including a new highway bridge of 150 m length, construction of a new tram bridge and approaches of 830 m total length, and a 156 m long bridge for service cables.

The Donbassky reconstruction bridge project resulted in a highly complex mixed-use multidisciplinary project consisting of five major components: construction of a new steel tram bridge, construction of a new steel highway bridge (left and right carriageways), construction of a new steel bridge for service cables, demolition of the old tram bridge, and demolition of the old highway bridge. The situation before reconstruction is shown in Figure 15.20.

15.6.4.6.1 Ground Conditions and Substructure Design

The bridge site is characterized by complex geological conditions mainly represented by variable interlayers of sand and sandy clays. The design geotechnical parameters were calculated on the basis of a probability of failure strength of the material (soil) equal to 0.98. The preliminary design presented two types of

FIGURE 15.20   Intersection of railway, city road, and tramway (left).

foundations for piers and abutments: driven piles with square section 35 cm × 35 cm and bored piles 1.2 m in diameter. Compared to driven piles, bored piles create less noise during construction and have minimum influence on the stability of the railway subgrade. Therefore, bored piles were recommended for the detailed design.

15.6.4.6.2 Design Criteria

The bridges were designed according to Russian Bridge Code 2.05.03-84 (SNIP 1996) and the Moscow city norms for the design of bridge structures (MGSN 1998). The highway bridge was designed to a specified highway loading composed of a two-axle load of 14 tons per axle with 1.5 m spacing between the axles and a uniformly distributed load of 1.4 ton/m. It was also specified that bridge structures should carry a special live load (a four-axle load of 20 tons per axle with 1.2 m spacing between the axles). The tram bridge was designed for tram loading composed of trains with 8.5 m spacing between the axles of the nearest trains. Each train was composed of four four-axle carriages of 7.5 tons per axle.

The thermal range to be accommodated is 70°C in accordance with the requirements of SNIP 23-01-99 (SNIP 2000). The concrete grades specified for the main components of the bridges are B35 for the tram superstructure deck, B30 for pier columns and retaining walls, and B25 for pile caps and piles. Low-alloy structural steel of grades 10XCHD and 15XCHD, which conform to the Russian standard (GOST 1992), was specified for the girders. The yield strength depends on the thickness of the rolled steel and varies from 330 MPa to 390 MPa.

An environmental impact assessment for the operation and construction stages proved the ecological safety. Based on this assessment a combination of noise protection measures were recommended. This combination included planting of trees, construction of a noise protection barrier, and replacing existing windows in buildings located near the construction site.

15.6.4.6.3 Construction Sequence

Due to the tight urban area congested with underground facilities of different purposes, the project had to be implemented in a specific sequence. First the new tram bridge had to be built, and then the tram traffic was relocated to this new bridge and the old existing tram bridge was demolished. Then construction of a new bridge for service cables commenced in order to relocate the existing service cables from the construction area. When demolition of the old existing tram bridge was completed the area was free to build the new highway bridge (left carriageway) alongside the existing old highway bridge. Further, when highway traffic was relocated to the constructed new highway bridge, demolition of the old existing highway bridge began. Upon demolition the new highway bridge (right carriageway) was constructed and finally the design scheme was implemented: four traffic lanes in each direction.

FIGURE 15.21   Steel superstructure of the Donbassky Bridge.

The final design of the Donbassky Bridge components was completed progressively in separate design packages as construction proceeded. The first design package covered the tram bridge. The bridge crossing is composed of one continuous two-span structure and two continuous four-span structures. The bridge scheme of the steel orthotropic system is based on spans 47 m + 60 m + 98 m + 70 m. The single steel box girder is 2.6 m high (Figure 15.21). Diaphragms are provided over the piers and the south abutment. The two bridge schemes of the composite system are based on spans 2 × 33 m and 38 m + 2 × 50 m + 38 m. The single steel U-shape girder is 2.1 m high, acting compositely with a concrete deck (Figure 15.21). Diaphragms are provided over the piers and the north abutment. Steel transverse floor beams were used to support the 220 mm thick cast-in-place concrete deck.

A composite action of the reinforced concrete deck with a steel beam is provided by shear connectors (shear studs Ø 22) welded on the top flange. The use of angles as shear connectors was the typical practice in Russia up to the end of the twentieth century. Wide application of cast-in-place or monolithic concrete for bridge superstructures began in the 1990s. Shear studs have technical advantages over the other types of connectors; modern technology allows completely mechanization of the application process and labor consumption reduction for reinforcement placement in a slab. The first application of stud welding technology in Russia was based on the use of Köco (Germany) welding equipment and materials (studs). Based on complex studies of the compatibility of German studs welded to Russian steels, the special standard STP 015-2001 (STP 2001) was worked out in 2001. Today more than 2 million pieces of studs produced by Köco have been supplied to Russia for bridge construction.

The tram tracks were set on a ballast bed over the top concrete or orthotropic deck, depending on the superstructure type. The structural peculiarity of a ballasted superstructure deck required a special type of expansion joint. Movements on the order of 200 mm occurred at the junction between the steel and composite superstructures and the south abutment, requiring multiple steel beam and flexible seal joints. The Maurer system of expansion joints was installed at the level of the top deck, allowing drainage water to pass over the joint. In the author's knowledge this was the first application in the world of Maurer system expansion joints with a capacity of 200 mm movements for a tram (railway) bridge.

To optimize the use of ground space and for easier installation in the vicinity of an area extremely congested by underground utilities, groups of six to eight bored piles 1.2 m in diameter, cast in situ, were selected for the pier and abutment foundations. The piles penetrate to a depth of 12–18 m. The pile caps for piers 2 and 3 comprise a uniform reinforced concrete section of 6.5 m × 4.3 m × 2 m deep and for

FIGURE 15.22   Composite superstructure of the Donbassky Bridge (no concrete deck).

piers 4 through 10, 8.3 m × 4.3 m × 2 m deep supporting the single pier column. The piers are cast-in-place reinforced concrete. A number of distinctive architectural features were incorporated in the final design at the direction of the project architect. These features included handrails, granite inserts into the piers, and coloristic solutions (Figure 15.22).

15.6.4.6.4 Highway Bridge (Final Design)

To satisfy the planned construction schedule, the contractor suggested changing the original composite design to a steel superstructure with an orthotropic deck. These changes were acceptable to the engineer and agreed to by the owner. These modifications meant that a complete redesign of the superstructure was required. The layout of the piers and abutments remained unchanged although some modifications were required. The redesigned bridge crossing is a continuous three-span structure, with a span arrangement of 42 m + 66 m + 42 m. The steel superstructure consists of three longitudinal rectangular cross section box girders at the pier sections and six I-girders at the midspan section. The girders depth is 2 m. Transverse beams are spaced at 3 m.

15.6.4.6.5 Bridge for Telephone Cables (Final Design)

This bridge is a continuous three-span structure carrying about 150 cables. The bridge scheme is based on spans 45 m + 66 m + 45 m. Since the steel structure has no skew, the detailing is simplified compared to the steel structure of the highway bridge, where extraordinary skew is adopted in the design. The steel superstructure is composed of a single U-shaped girder 2 m high. The polycarbonate roof system was adopted in the final design to improve the overall appearance and reduce dead weight.

15.6.4.6.6 Construction Aspects

Underground utilities presented a serious obstacle for foundation construction; over the years all kinds of subsurface utilities have been clustered up as a result of repeated diversion works by different utility companies. Therefore, preparation works started with relocation of underground utilities to free the space for bridge foundation construction. The tram bridge—the first project component—was completed in March 2003, and comprised building a new 520 m long bridge structure, building two approach fills with retaining walls, removing 2520 linear meters of existing and temporary tram track, and laying 1760 linear meters of new tram track.

Russian steel bridge construction is mainly oriented toward superstructures composed of modules fabricated at specialized shops. This technology is based on modularization of steel superstructures and orthotropic deck elements: fabricated shop units, with lengths of up to 21 m. The superstructures are assembled with modularized shop-fabricated elements, which are welded at the shop or construction site to form a complete cross section configuration. For detailed discussion of this structural technological system refer to Popov et al. (1998). The same structural technological system was adopted for the steel bridge structures of all project components.

For the tram bridge, shop-fabricated steel girders were delivered to the construction site by road (Figure 15.23). A further steel superstructure section was formed with these segments over the temporary framework (Figure 15.24). Mobile cranes were used for the erection. The connections of the main girder steel segments are welded. The tracks were set in the usual manner, on a bed of ballast over

FIGURE 15.23　Delivery of steel segments to the site.

FIGURE 15.24　Erection of steel superstructure.

the orthotropic or cast-in-place concrete slab, depending on the superstructure type. For the highway bridge the steel superstructure was assembled with shop-fabricated elements in the area behind the abutment. The erection of the superstructure was implemented by incremental launching (Figure 15.25). Launching proceeded from one side using noses. The peculiarity of launching was that two noses of different lengths were attached to the superstructure composed of three main girders. No temporary piers were required in the spans. The highway bridge (both carriageways)—the third project component—was completed in June 2004, and comprised building a new 150 m long bridge structure and building two approach fills 586.5 m long.

For the telephone cable bridge the steel superstructure was preassembled with shop-fabricated elements in the area behind the abutment and then launched into its final position (Figure 15.26).

FIGURE 15.25   Launching by the incremental method (left carriageway).

FIGURE 15.26   Launching the first side span using a special temporary pier.

Launching was implemented using a special temporary pier composed of a plane structure founded between the railway lines and a frame structure founded outside the railway lines. This temporary pier allowed placement of a sliding device approximately in the middle of the main 66 m span. The bridge for telephone cables—the second project component—was completed in May 2003, and comprised building a new 150 m long bridge structure and relocation of underground conduits in the approached area.

This project is a key part of ongoing initiatives to improve the flow of traffic throughout the Moscow metropolitan area. The constructed crossings have increased the average speed of motor traffic through this section of the internal city road from 15 km/h up to 50 km/h.

15.6.4.7 Railroad Bridge over Bolshaya Tulskaya Street

The complex reconstruction project of Bolshaya Tulskaya Street in Moscow consists of widening of the street itself, rebuilding of the deteriorated railway bridge, relocation of the tram line, and creation of various underground facilities. The main freeway (street), a double track tram line, and pedestrian walkways had to be bridged. The existing railroad grade, plan location of the rail tracks (no relocation is possible), and highway clearance requirements over the street limited the available structural depth of the superstructure. The existing street is skewed 76° relative to the axis of the bridge and could not be changed.

To integrate the bridge into the town surroundings eight different concepts have been studied. These studies showed that a bow string truss system and asymmetric solid slab deck were the most appropriate and satisfied the abovementioned design constraints (Figure 15.27). This bridge comprises a central single-span steel structure and one side single-span concrete tapered structure. To alleviate the detailing problem of the skewed configuration, the central superstructure was designed straight in plan. Upon completion in mid-2006 the bridge carried three railway tracks. To satisfy the requirements of street widening and tram passage, the central span length is 97.6 m, a record in Russian railroad bridge construction practice for this system type.

The central superstructure comprises twin steel box girders 1.8 m high and 0.652 m wide. These two main girders are interconnected by transverse beams spaced at 2.033 m. Along with the transverse beams, two stringers in the cross section, spaced at 1.7 m, support the precast concrete slabs of the ballastless deck, which carries a single track line. Structural steel with a 490 MPa of ultimate strength and 340 MPa of yield strength was employed (GOST 1992). The side superstructure has a complex tapered

FIGURE 15.27 General view of railroad overcross bridge.

configuration spanning a pedestrian walkway. The average span of 10.8 m is solid reinforced concrete deck slab 0.8 m high and carries ballasted track line. Grade B40 concrete was specified for the deck.

All three steel superstructures of the bow string truss system were preassembled from shop-fabricated segments up to 21 m long to form a complete configuration. The steel superstructures were fabricated in two specialized shops located in the Russian cities Kurgan and Voronezh. The off-site fabricated segments were transported by rail and automobile trailers to the bridge site. One of the main design features was to simplify fabrication and minimize on-site assembly of the superstructure from prefabricated elements. A lower chord is divided into five box segments which vary in length from 18.3 m to 20.3 m and in mass from 16.2 tons to 20.9 tons. To simplify the assembly of the on-site joints of crossbeams to main girders, vertical connection elements were designed T-shaped and fixed to webs of the box segments in the shop (Figure 15.28).

Each superstructure was assembled on temporary jig supports behind the abutment by means of two mobile boom cranes with capacities of 120 tons and 160 tons. In accordance with the adopted sequence of two-staged construction the assembly of two trusses was implemented simultaneously at the first stage. Thus, using the off-site prefabricated segments, 1240 tons of steelwork was assembled in about 4 months. Finally, the superstructure was launched out into the span within 5 days.

In Russian railroad bridge construction practice ballastless decks formed of precast slabs are the most widespread (Monov and Seliverstov 1997). Standard structural solutions were developed in the 1960s, including precast slabs, panels that have a standard configuration and dimensions that cover a range of spacing between the main (longitudinal) girders from 1.8 m to 2.2 m. The precast concrete panels with nonprestressed reinforcements are produced at specialized shops. The concrete grade specified for these panels is B40. The lifespan of precast panels did not exceed 10 years due to cracks. To improve the reliability of ballastless decks, fiber reinforced concrete was developed to prolong the structure lifespan. Based on experimental investigations the optimum steel fiber quantity to be used for the production technology has been recommended in the range of 60–100 kg/m^3. Full-scale tests on fatigue and crack resistance performed by the Federal Scientific Research Institute of Railway Transport (VNIIZT) and the Scientific Research Institute of Transport Construction (TSNIIS) showed that the crack resistance of samples with steel fiber was improved by 1.5 times compared to samples without steel fiber. Also, crack resistance in the transverse direction was increased by 2–3 times and impact strength improved by 8–10 times. Tension strength under flexure was increased by 2–3 times. Application of steel fiber for precast

FIGURE 15.28   Connection details of crossbeams to main girders and ballastless deck.

slabs of ballastless decks is a new technology that put into practice durable structural components. Furthermore, the use of steel fiber for slabs of ballastless decks in railway bridge construction allows reduction of bridge maintenance costs significantly. The bridge project was completed in 2006.

15.6.4.8 Severyanin Overpass

The existing overpass accommodates 6 lanes of traffic and crosses 10 rail tracks and local roads with 2 lanes of traffic. The overall width is 31.7 m, including two sidewalks 3.1 m wide. The intensity of automobile and railway traffic is classified as high. The overpass has an overall length of 1048 m and comprises three main sections—two side and one central. The city side section has a length of 104 m, the suburban section 108 m, and the central 340 m. The central section crosses the railway with a span arrangement of 2×34.1 m + 3×34.1 m + 2×34.1 m. This superstructure comprises three continuous skewed (32°) steel structures, each structure composed of eight steel I-girders which support a 20 cm thick concrete deck, acting noncompositely. Approaching superstructures are composed of Π-shaped reinforced concrete girders with a span of 6 m. These Π-shaped precast blocks are 0.9 m deep and vary in width from 1.6–1.94 m. Besides typical expansion joints to reflect specific details of the bridge deck, the expansion joint was installed in the longitudinal direction.

Widening the main street carriageway called for reconstruction of the existing overpass. Since the architectural view of the overall structure had to be unchanged, the main design concept was to widen the bridge carriageway and at the same time narrow an existing footway. The reconstructed overpass accommodates four lanes of traffic in each direction and has a central reserve of 2 m and footway of 1 m at each side (Figure 15.29). The design widening was achieved by the increased width of the new concrete monolithic deck, which reached 34 m. The superstructure was upgraded by partially interfering with structural function: the steel portion was strengthened and the existing concrete deck was demolished and a new monolithic concrete deck over the main steel girders was placed using a new type of shear connection. Some of the existing piers were used in the final bridge configuration. Controlled design loads have allowed the specific structural function of the existing piers on low-capacity soils to be maintained.

The approach fills were reconstructed also. Since the main carriageway had to be widened, construction of retaining walls was required in order to maintain the existing roadways along the overpass and to provide new design turns. This specific design task had its own challenges related to the stability of the existing approach fill. Since the existing overpass was built in 1953, all design calculations were based on the method of allowable stresses which formed the basis of the old bridge code in Russia. To assess the condition of the structure and justify a design approach, a thorough investigation, including visual observations of structures to allow for fixing of the actual defects, instrumental studies of the material properties of concrete and steel, assessment of degradation processes, analyses of the main characteristics of the structure in terms of load capacity and safety of traffic passage, and review

FIGURE 15.29 General view of reconstructed overpass.

of probable strategies for existing structure rehabilitation conforming to modern requirements, was implemented at the end of 2001.

Based on the results of this complex investigation it was concluded that the main load-carrying members of the overpass were in fair condition and the load capacity of the bridge satisfied A11 class live load. However, the code currently in use (MGSN 1998) specifies A14 class live load, which is 29% higher than the A11 class live load. To satisfy the requirements of further long-term operation of at least 70 years, and to increase the load capacity conforming to the higher A14 class live load, the following measures were recommended:

- Partial strengthening of the existing steel structure
- Replacement of the corroding steel members
- Introducing a composite action of a reinforced concrete deck with a steel superstructure
- Improving traffic safety

Widening of the carriageway, approach fills, and a need for additional roadway for traffic passage underneath the overpass determined structural–technological solutions for reconstruction. The upgraded central section of the overpass is a continuous seven-span structure with a span arrangement as described previously. The cross girders in the areas nearest to the former expansion joints were completely replaced. The existing precast concrete deck was replaced with a new concrete monolithic deck. A composite action of the steel girders and concrete deck is achieved by a combination of linear shear connectors and typical studs (Figure 15.30). With this new structural solution the existing expansion joint in longitudinal direction is eliminated.

To build an additional road passage with a width of 16 m underneath the existing overpass, the existing concrete piers and superstructures from pier 1 to pier 6 were demolished. In their place a new bridge part with a tapered configuration in plan was constructed. This trapezoidal configuration varies in length from 36.8 m to 80.2 m. This reconstructed side section of overpass is a continuous three-span structure with a span arrangement of 11.3 m + 2 × 34.1 m measured along the maximum span length. The composite superstructure comprises eight main steel girders of permanent depth 2.13 m and variable length, acting compositely with a concrete monolithic deck. This bridge project was completed in 2006.

FIGURE 15.30 Typical detail of shear connectors.

15.6.5 Development of Construction Techniques

15.6.5.1 Launching Method

Launching techniques have been steadily refined during recent years. A typical erection design may comprise various measures assuring the safety of superstructure launching: a nose fixed to the super-structure end, removal of some orthotropic deck panels from the superstructure, additional strengthening, or a combination of these measures. Normally when the launching span exceeds 100 m, temporary piers may be built. However, depending on the local conditions (e.g., deep water), this might be costly. The use of a temporary strut tower on top of the launched superstructure reduces erection effects, but its application has some disadvantages, mainly installation of special jacking equipment for the adjustment of cable tensioning depending on the stage of launching (e.g., in the span or at the over-pier section). The development of a special strut-frame system installed below the launched superstructure alleviates most of the difficulties. This temporary system ensures a design range of vibration at all stages of launching, reduces bending moments and reaction value at the cantilever root, and eliminates deflection of the superstructure end reaching the pier.

This innovative system has been employed for the erection of several bridges over the Volga River and proved its efficiency. Furthermore, additional study showed that the described temporary strut-frame system may be effective for launching spans of up to 170 m. This system has recently been used for the erection of a steel superstructure over the famous Russian Volga River in Volgograd (Figure 15.31). This large bridge has a total length of 1.2 km. It is an urban bridge with a span arrangement of 84 m + 3 × 126 m + 3 × 155 m + 126 m + 84 m and accommodates dual three lanes. Each group of three lanes is 14.25 m wide and has a single sidewalk of 1.5 m wide. The first section of motorway carrying three lanes opened to traffic in September 2009.

15.6.5.2 Erection by Rotation

The pedestrian bridge crossing over the Moscva River is a continuous three-span structure, with a span arrangement of 53 m + 105 m + 53 m (Figure 15.32). The superstructure comprises a single steel box girder with an orthotropic deck 12.5 m wide. The girder is variable in depth, 4.27 m at the over-pier section and 1.63 m in the middle of the central span. The steel superstructure box section is formed with L-shaped prefabricated segments. To avoid the difficulties of using scaffolding for the side spans and a

FIGURE 15.31 Launching of superstructure using the strut-frame system.

FIGURE 15.32   General view of the pedestrian bridge crossing over the Moscva River.

FIGURE 15.33   Erection by rotation.

huge floating crane for the main span, a new erection scheme was developed. The idea was to assemble two superstructure halves at artificial fill platforms constructed along the river banks. Furthermore, each assembled half had to be rotated around intermediate piers into the final position and locked by a welded joint in the center of the river span.

To assemble the superstructure, sand fills along the existing embankments had to be formed. The length of these fills was about 150 m and the width was minimized in order to allow scaffolding erection and technological roadways for the delivery of the superstructure elements. Although constructed from both river banks the fills resulted in some river contraction; a ship navigation passage of about 50 m was provided. Scaffolding was erected along embankments (Figure 15.33). The assembly of the superstructure, with elements of up to 35 tons, was implemented by boom cranes with a capacity of 60–100 tons. Starting with over-pier segments, the assembly continued by balanced erection.

Each superstructure half comprised nine box segments, however for the superstructure half at pier 3 only eight segments were erected. This is because of lack of room to rotate (see right side of Figure 15.33); accordingly, the weights of the halves reached 634 tons and 617 tons respectively. Upon assemblage, additional loads were placed at the ends of each half. This measure assured a value of coefficient of stability of at least 1.4 during rotation and locking operations. The rotation system comprises 12 columns of steel tubes 530 mm in diameter installed around the intermediate piers; a rotation steel circle 14 m in diameter; a polished steel sheet placed on top of the rotation circle, with bracings in the horizontal plane of the rotation circle; a centering device protecting displacement of over-pier segment during rotation; sliding devices taking the role of means of transportation during rotation; and a jacking system composed of two jacks (capacity 185 tons, stroke 1.12 m).

Prior to rotation of the superstructure, stress control gauges were installed at the rotation circle. An additional load composed of reinforced concrete panels was placed at the end of side spans. Then the assembled superstructure was jacked up by 200 mm. Jig supports were removed and the superstructure half was lowered over sliding devices. The structure was fixed to the sliding devices to prevent horizontal displacement. A further centering device was mounted. Plywood pads and a strip of ftoroplast were placed in the gaps between the sliding devices and rotation circle. The pushing jacks were fixed to the leading sliding devices and a sample test rotation (about 0.5 m) was made. One rotation cycle included a working jack stroke of 800–850 mm, removal of fixings to the rotation circle, return of stroke, and fixing again. Normally this cycle was implemented within 15 minutes. Less than 24 hours were needed for the whole rotation process. Compared to traditional methods of erection the quantity required for temporary steelwork was reduced by 400 tons. The main works related to pier construction and superstructure erection were completed within one year. During construction an uninterrupted passage of vehicles and ships was provided.

15.7 Concrete Bridges

About 90% of modern bridges are concrete bridges. Recent decades have been characterized by intensive development of standardized precast concrete bridge elements. The main operational requirements and various conditions of construction are taken into account by existing standardization. This allows a flexible approach to solving architectural and planning tasks and construction of bridges of various span lengths and clearances. Precast concrete bridges, both railway and roadway, have been built in many environments ranging from urban to rural. The general characteristics of the common types of concrete bridge structures are provided next.

15.7.1 Superstructures for Railway Bridges

For railway bridges, standardized shapes for girders and slabs have been widely used for span lengths of 2.95–27.6 m. The simplest type of bridge superstructure is the deck slab, which may be solid or voided. Standard structures are designed to carry live load C14 (single track) and may be located on curved sections of alignment with a radius of 300 m and more. Typical nonprestressed concrete slabs (Figure 15.34a) with structural depths from 0.65 m to 1.35 m are applicable for spans ranging 2.95 m to 16.5 m. Nonprestressed precast T-beams (Figure 15.34b) with depths from 1.25 m to 1.75 m are used for spans ranging from 9.3 m to 16.5 m. Prestressed concrete T-beams (Figure 15.34c) with depths from 1.75 m to 2.6 m are applicable for spans from 16.5 m to 27.6 m. Single track bridges consist of two precast full segments which are connected at diaphragms by welding of steel joint straps and then pouring concrete.

All three types of precast superstructure segments are normally fabricated at a shop and transported to the construction site. The waterproofing system is shop applied. Longitudinal gaps between the segments are covered by steel plates. All superstructure segments over 11.5 m in length are placed on steel bearings. Superstructures may be connected into a partially continuous system, thus allowing adjustment of the horizontal forces transferred to piers.

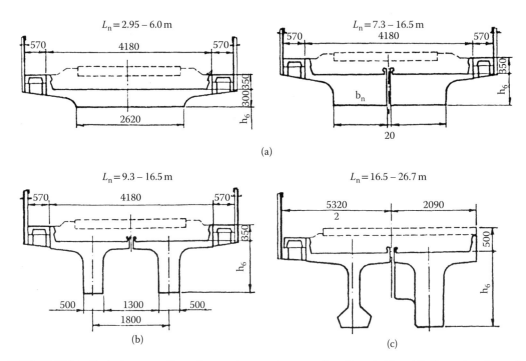

FIGURE 15.34 Typical cross sections of concrete superstructures for railway bridges (all dimensions in mm): (a) nonprestressed reinforced concrete slab; (b) nonprestressed reinforced concrete T-beam; (c) prestressed concrete T-beam.

15.7.2 Superstructures for Roadway Bridges

More than 80% of bridges on the federal highway networks have span lengths not exceeding 33 m. Figure 15.35 shows typical standard cross sections for roadway bridges. Nonprestressed concrete void slabs with structural depths from 0.6 m to 0.75 m are applicable for spans ranging 12 m to 18 m. Nonprestressed precast T-beams with depths from 0.9 m to 1.05 m are used for spans ranging from 11.1 m to 17.8 m. Prestressed concrete T-beams with depths from 1.2 m to 1.7 m are applicable for spans from 20.3 m to 32.3 m. Figure 15.36 shows a construction site of precast concrete beams on the Moscow Ring Road.

Fabrication of pretensioned standard beams is conducted at a number of specialized shops. However, some technological difficulties have been noted: operations related to installation of reinforcements (space units) into molds of a complex configuration (e.g., bulbous bottom of the girder section), and placing of concrete into and taking off the beam from these molds. To improve the fabrication procedure, a special shape of the beam web (see Figure 15.35; last cross section) has been developed. The standard design of a beam with a 27 m length has been elaborated. Precast prestressed T-beams, which are the most widespread in the current construction practice, have structural depth from 1.2 m to 1.7 m with typical top width of 1.8 m and weight of one beam from 32.3 tons to 59 tons.

Where special transport facilities are not available, transportation limitations exist, or manufacturing facilities such as stressing strands are expensive, precast in segments post-tensioned beams are beneficial. The designs of standard T-beams have been worked out considering both systems: post-tensioning and pretensioning. Standard T-beams have been designed precast in segments with subsequent post-tensioning for span lengths of 24 m, 33 m, and 42 m. However, the use of 42 m T-beams is very rare in practice.

Due to transportation limitations or restrictions or other reasons a rational alternative for post-tensioned beams is beams with transverse joints. In this case the beam consists of segments of limited

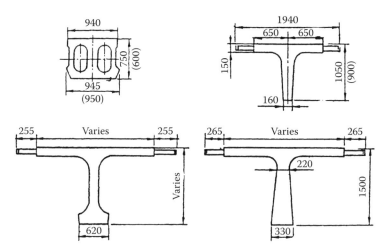

FIGURE 15.35 Standard precast beams for highway bridges (all dimensions in mm).

FIGURE 15.36 Typical section of overhead on the Moscow Ring Road.

length and weight (up to 11.8 tons) that may be transported by usual means. Such segments may be precast either on site or in short lengths at the factory with subsequent post-tensioning at the site. The joints between segments may be implemented: by filling the joint gap of 20–30 mm thickness with concrete (thin joint) or by placing concrete minimum thickness 70 mm (thick joint) or epoxy glue with a 5 mm thickness.

To reduce expansion joints and improve road conditions, the partially continuous system—simple beams at the erection stage and continuous beams at the service stage—has been widely used for superstructure spans up to 33 m. The girders are connected by casting the deck slab over support the locations, and slabs are connected by welding steel straps on the top of deck. Figure 15.37 shows a typical continuous bridge scheme composed of standard precast slabs or T-beams. Figure 15.38 shows a standard design for continuous superstructures with double-T beams. The overall width of precast segments may reach 20 m, but length is limited by 3 m due to transportation constraints. Ducts for prestressing cables are placed in the web only.

In 1990, another standard design for box girder continuous systems was initiated. A typical box section is shown in Figure 15.39. Each segment has 1.4 m at the bent and 2.2 m within the spans, with

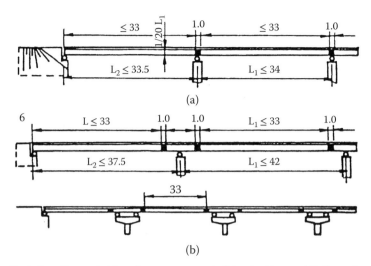

FIGURE 15.37   Typical continuous bridge schemes composed of standard beams (all dimensions in m).

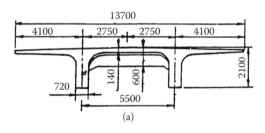

FIGURE 15.38   Typical cross sections of standard solid web girders (all dimensions in mm).

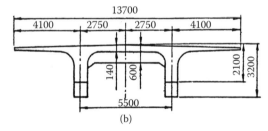

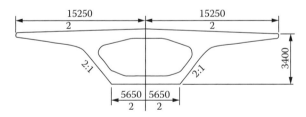

FIGURE 15.39   Typical cross section of a precast box segment (all dimensions in mm).

weight up to 62 tons. Standard precast segments superstructures can have spans 63 m, 71.8 m, 84 m, and 92.8 m. Due to financial difficulties the design works on standard box segmental superstructures have not been completed. However, the general idea has been implemented at a number of bridges. Since the new standard (GOST 2008) specified the increased live load, superstructure standard designs cannot be used without adjustments at present.

15.7.3 Construction Techniques

Precast full-span beam segments are fabricated at specialized shops on the basis of standard designs. Erections of solid girder segments up to 33 m are conducted by mobile boom cranes of various capacities, gantry cranes, and launching gantries. Special scaffoldings have been designed to erect solid girders of 24–63 m (Figure 15.40). When large-span bridges are constructed, different cranes may be used simultaneously. At low water areas of rivers, gantry cranes may be applied, but at deep water areas, SPK-65 cranes or others may be used. A typical erection scheme of a precast cantilever bridge over the Volga River built in 1970 is shown in Figure 15.41.

Although concrete bridge superstructures are constructed mainly by precast segments, in recent years cast-in-place superstructure construction has been reviewed on a new technological level. By this method, construction of the superstructure is organized in the area behind the abutment on the approaches to the bridge. The successive portion of the superstructure is cast against the proceeding segment and prestressed to it before proceeding with erection by incremental launching. Cast-in-place concrete main girders and modified precast concrete decks are usually used. Traditionally, when precast deck slabs were used for bridge construction, these precast elements were fabricated with provision of holes for shear connectors that were later filled with concrete. This solution has several disadvantages. To improve the practice, a new type of joint between structural members was introduced. The use of steel embeds in precast slabs allows them to connect to the main girders by means of angles and high-strength bolts (Monov and Seliverstov 1997).

15.7.4 Typical Bridges

15.7.4.1 Komarovka Bridge

A railway bridge over the Komarovka River was built in the 1990s in Ussuryisk (in the far east of Russia). The 106.85 m long bridge has a span arrangement of 6 × 16.5 m (Figure 15.42). The super-structure is reinforced concrete beams of standard design. The intermediate piers are of cast-in-place

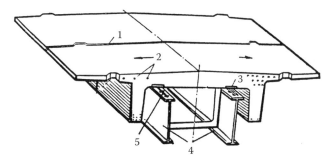

FIGURE 15.40 Position of precast segments of solid web girders over special scaffoldings. 1: Glue joints; 2: Ducts; 3: Embeds; 4: Movable special scaffoldings; 5: Rails for segment moving.

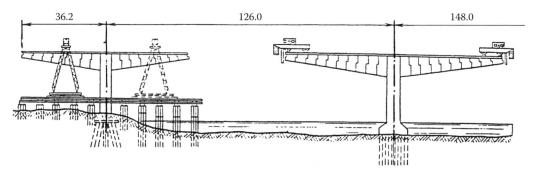

FIGURE 15.41 Typical erection scheme of the bridge over the Volga River (all dimensions in m).

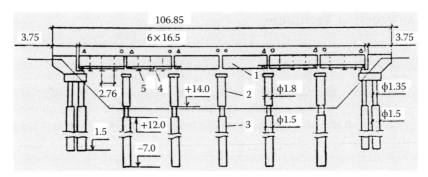

FIGURE 15.42 General arrangement of the bridge over the Komarovka River. 1: Reinforced concrete superstructure; 2: Cast-in-place pier; 3: Bored pile; 4: Special girt; 5: Key to limit sideways movement (all dimensions in m).

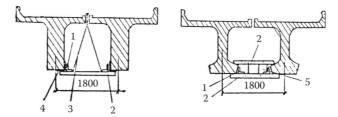

FIGURE 15.43 Special girt details: (left) nonprestressed and (right) prestressed girder. 1: Girt; 2: Crosstie; 3: Anchor element; 4: Noftlen pad; 5: Jointing plank (all dimensions in mm).

reinforced concrete. The abutments are spill-through type of precast elements and cast-in-place concrete. Foundations are on bored 1.5 m diameter piles. Steel casings 1.35 m in diameter were placed in the top portion of the piles. Separate connections of the superstructure into the two systems reduced the temperature forces in the girt. Structurally the girt (Figure 15.43) consists of two angles of 125 mm × 125 mm × 10 mm, which are jointed to the bearing nodes by means of gussets installed between the bottom flange of the superstructure and the sole plate of the fixed bearings. The girt is attached to each gusset plate by high-strength bolts. To reduce temperature stresses the girts were fixed to the superstructure and embeds over the abutments at the temperature of 0°C.

15.7.4.2 Kashira Oka River Bridge

In 1995, a bridge with a total length of 1.96 km crossing over the Oka River near Kashira was constructed. The main spans over the navigation channel are 44.1 m + 5 × 85.5 m + 42.12 m. The bridge carries three traffic lanes in each direction. The seven-span continuous superstructures are precast concrete box segments with depths of 3.4 m and widths of 16 m constructed using the cantilever method with further locking in the middle of the spans. A typical cross section is shown in Figure 15.44. The precast segments vary from 1.5 m to 1.98 m and are governed by the capacity of the erection equipment, limited to 60 tons. The piers are cast in place, slipformed, and the foundations are on bored piles 1.5 m and 1.7 m in diameter.

15.7.4.3 Frame Bridge with Slender Legs

The development of structural forms and erection techniques for prestressed concrete led to the construction of fixed rigid frames for bridges. Compared to continuous span frame bridge systems, this system is less commonly used. To form a frame system, a precast superstructure and pier elements are concreted at the over-pier section (1 m along the bridge) and at the deck section (0.36 m wide in the transverse direction). Figure 15.45 shows a typical frame bridge with slender legs. The design foundations need an individual approach depending on the site's geological conditions. The simple forms of the precast elements, low mass, clear erection scheme, and aesthetic appearance are the main advantages of this bridge system.

15.7.4.4 Buisky Perevoz Vyatka River Bridge

This highway bridge (a nonconventional structure), as shown in Figure 15.46, opened to traffic in 1985. The bridge is a cantilever frame system with a suspended span of 32.3 m. The superstructure is a single-box rectangular box girder with a depth of 3.75 m and width of 8.66 m. The overall deck width is 10 m. The river piers are cast-in-place concrete and the foundations are on bored 1.5 m piles penetrated to a depth of 16 m. One of the piers is founded on a caisson placed to a depth of 8 m. The 32 mm diameter bars of the pile reinforcement were stressed with 50 kN force per bar for better crack resistance.

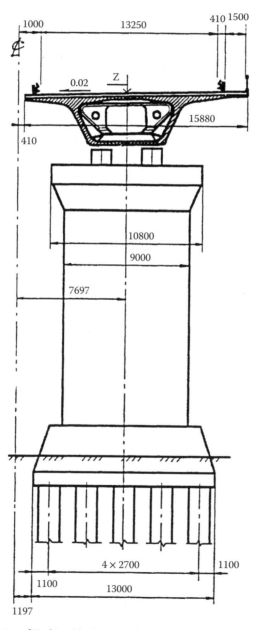

FIGURE 15.44 Typical section of Kashira Oka River Bridge (all dimensions in mm).

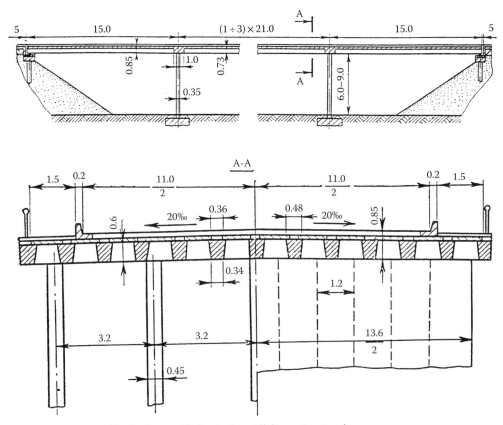

FIGURE 15.45 Typical bridge frame with slender legs (all dimensions in m).

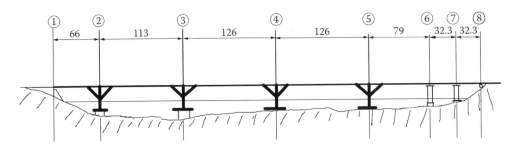

FIGURE 15.46 General scheme of Buisky Perevoz Vyatka River Bridge (all dimensions in m).

15.7.4.5 Penza Sura River Bridge

This highway bridge comprises a precast prestressed concrete frame of a two-hinge system (Figure 15.47) and was built in 1975. The superstructure is three boxes with variable sections (Figure 15.48) with a total of 66 prestressing strands. The inclined legs of the frames have a box section of 25 m × 1.5 m at the top and a solid section 1.45 m × 0.7 m at the bottom. The legs of each frame are reinforced by 12 prestressing strands. Each strand diameter is 5 mm consists of 48 wires. The piers are cast-in-place concrete, and the foundations are on driven hollow precast concrete piles of 0.6 m. Each frame structure was erected of 60% precast segments. These segments are 5.6 m wide and 2.7–3.3 m long. Each frame system superstructure was erected of 12 segments in a strict sequence, starting from the center of the span to the piers. The span segments were placed into design position, glued, and prestressed. The leg segments are

FIGURE 15.47   Penza Sura River Bridge.

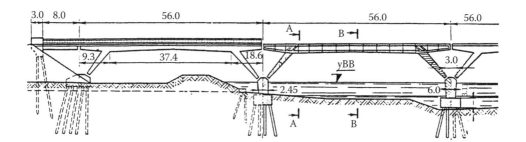

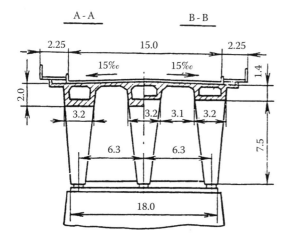

FIGURE 15.48   Typical cross section of Penza Sura River Bridge (all dimensions in m).

connected with span segments by cast-in-place concrete. A special sequence, as shown in Figure 15.49, for tensioning the strands was established for this bridge.

15.7.4.6 Moscow Moskva River Arch Bridge

This cantilever arch bridge (Figure 15.50) over the Moskva River on the Moscow Ring Road was built in 1962. The bridge is a three-span structure with an arrangement of 48.65 m + 98 m + 48.64 m. The overall road width is 21 m and the sidewalks are 1.5 m on each side. The half arches are connected by a tie of 10 prestressing strands at the level of the roadway. Erection of the superstructure was implemented on steel scaffoldings. The arches are erected of precast elements weighting from 10 tons to 20 tons.

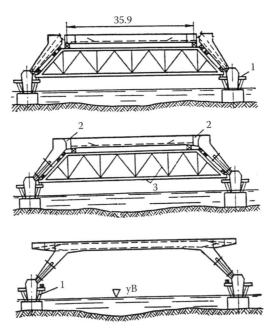

FIGURE 15.49 General sequence of frame structure erection (all dimensions in m). 1: Temporary steel cantilevers; 2: Cast-in-situ joint; 3: Temporary steel truss.

FIGURE 15.50 Moscow Moskva River arch bridge.

15.7.5 Monolithic Concrete Bridges

Russian concrete bridge superstructures have been mainly constructed by precast (prefabricated) concrete girders since the 1960s. Therefore, the technology for concrete highway superstructure construction in Russia was predominantly based on employment of prefabricated shop units, thus speeding up the erection process. In the early 1990s, various studies proved the cast-in-place or monolithic superstructures to be more beneficial and durable compared to precast concrete. The durability issue is one of the important aspects for bridges and especially for bridges in urban areas where traffic congestion and the requirement to not interrupt traffic flow do not allow easy access to build a new structure or repair or replace the deteriorated existing one. The mentioned constraints have called for development of efficient structural–technological system (Gadaev, Perevoznikov, and Seliverstov 2002). This structural–technological system comprises installation of formworks (Figure 15.51), placing of reinforcement (Figure 15.52), concreting works (Figure 15.53), and launching of the structure. Launching of the main structure into the span is realized with a jacking-pushing device, which comprises sliding devices, jacks, and an informational coordinating system.

Application of the new structural–technological system began in 1997. Efficiency with this system has been achieved on a number of built bridges. For concrete overcross bridges, constructed within the frame of the large-scale transportation project "Reconstruction of the Moscow Ring Road," launching of monolithic concrete superstructures was successfully implemented. An overview of this large-scale project (Figure 15.54) was given by Perevoznikov and Seliverstov (1999). Specific features of the superstructure launching used for the construction of monolithic concrete overcross bridges over Ring Road around Moscow are discussed in the next sections.

FIGURE 15.51 Formworks.

FIGURE 15.52 Reinforcement placement works.

FIGURE 15.53 Concreting works.

FIGURE 15.54 Overcross bridge at km 11 of the Moscow Ring Road (MKAD).

During construction of overcross bridges two types of mechanisms are used for launching superstructure segments or sections. This determines some variability in execution of works. Launching is implemented by superstructure segments with lengths of 18–25 m; the produced section is moved over sliding bearings with the use of special pads. One surface of these pads is Teflon, with a friction coefficient of 0.02, and the other is rubber, with a friction coefficient of 0.7. When the concrete has gained 70% of its strength the superstructure section is prestressed and jointed to the previous section; half of prestressing reinforcement is connected at the joint, and the other half is passed through. Temporary and permanent piers are equipped with side curbs, which allow maintenance of the design position of the superstructure in plan.

The length of the superstructure concrete casting segment is chosen so that the superstructure joints in the final design configuration will be in the zone of minimum moments. According to this principle, the length of the first and last launching superstructure segments are considered to be 0.65 of the side span lengths, and the length of the intermediate launching superstructure segments to be 0.5 of the central span length. Naturally, the other factors, which influence the choice of superstructure launching length, are also taken into account (e.g., in this case the quantity of concrete casting was limited to 300–350 m³). The launching system for produced superstructure segments varies from one bridge to another. Two types of launching systems are used in practice. One type uses pushing of a superstructure special clamping system (when a pulling force is applied to the rear of the latest-cast segment). The second type uses a combined vertical/horizontal jacking unit normally placed at the first pier.

An example of a superstructure launching over Moscow Ring Road (with uninterrupted traffic flow) at 11 km is shown in Figure 15.55. In this case the pushing device consisted of a system of combined vertical and horizontal jacks installed at one of the abutments. Horizontal jacks were fixed to the vertical ones and allowed the completed superstructure section to be moved into the span towards the permanent piers of the bridge. Having a horizontal jack stroke of 250 mm, the time required for one cycle of pushing does not exceed 1.5 minutes, thus giving a speed of superstructure section launch of about 10 m/h. One superstructure segment (structural module) was cast in about one section per week. The average speed of construction reached for bridges built in 1997–1998 as part of the Moscow Ring Road reconstruction project is given in Table 15.3 below. Some specific design parameters on built superstructures, including relative consumption of concrete and prestressed and nonprestressed reinforcement quantity per square meter of the bridge, are given in Table 15.4.

Construction technologies adopted in the former USSR and Russian Federation for launching reinforced concrete prestressed superstructures (precast or monolithic) were very labor-consuming (4.5 man-hour/m³ for precast concrete and 2.83 man-hour/m³ for monolithic concrete), which resulted in enlargement of construction period. Based on recent experience, the average rate of construction is 2.6 linear meters of superstructure per day. The developed structural–technological system is five times less labor-consuming than the existing analogues in Russia.

The structural–technological system for erection of monolithic concrete superstructures by the launching method has been employed on a number of bridges. This system allows execution of works in a tight space, satisfies the requirements of environmental protection, and has proved its efficiency in specific urban conditions of Moscow. For bridge span lengths of 15–20 m, monolithic concrete superstructures are considered to be feasible, and also can be advantageous in application for interchanges and crossings, having the complex configurations of bridge superstructures in plan and profile.

FIGURE 15.55 Launching of superstructure over the Moscow Ring Road.

TABLE 15.3   Average Speed of Superstructure Construction

Name of Bridge	Length of Bridge (m)	Number of Concrete Casting Sections	Average Length of Segment (m)	Average Speed of Superstructure Construction (m/day)
Solomei Neris	174	8	21.75	3.10
Shelkovskoe Shosse	110	5	22.00	3.14
Novoryazanskoe Shosse	130	7	18.50	2.64
Reytovo	136	9	15.10	2.15

TABLE 15.4   Design Parameters of Built Superstructures

Name of Bridge (Type of Superstructure Cross Section)	Span Arrangement (m)	Relative Span (m)	Concrete Consumption (m³/m²)	Prestressed Reinforcement Consumption (kg/m²)	Nonprestressed Reinforcement Consumption (kg/m²)
Solomei Neris (plate)	22.5 + 2 × 30 + 2 × 22 + 25 + 22.5	25.3	0.855	57	100
Shelkovskoe Shosse (plate)	22.5 + 2 × 30 + 22.5	26.8	0.946	79	97
Novoryazanskoe Shosse (box)	25 + 2 × 39 + 25	33.5	0.68	30.4	118.7
Reytovo (plate-ribbed)	23.5 + 20.1 + 2 × 31 + 21 + 20.1	25.4	0.877	39.,5	108

15.8 Cable-Stayed Bridges

The first cable-stayed bridges were constructed in the former USSR during the period 1932–1936. A cable-stayed highway bridge with a span of 80 m, designed by Kriltsov over the Magna River (former Georgian SSR), was constructed in 1932. The bridges over the Surhob River, with a span of 120.2 m, and over the Narin River, with a span of 132 m, were constructed in 1934 and 1936, respectively. Figure 15.56 shows a general view of the Narin River Bridge. The stiffening girder of the steel truss system was adopted in these bridges.

The modern period of Russian experience in cable-stayed bridge construction may be characterized by the following projects: the Dnepr River Bridge in Kiev (1962), the Moscow Dnepr River Bridge in Kiev (1976), Cherepovets Scheksna River Bridge (1980), Riga Daugava River Bridge (1981), and the Dnepr River South Bridge in Kiev (1991). In the past two decades a number of cable-stayed bridges of different scales have been built, over the Moscva and Neva rivers, over the Ob River near Surgut, and near the city of Murom. The two largest projects—Bosphorus (the Eastern Strait) and Golden Horn Bay in Vladivostok (far east of Russia)—are currently under construction.

15.8.1 Kiev Dnepr River Bridge

The first concrete cable-stayed bridge (Figure 15.57), crossing over the harbor of the Dnepr River in Kiev, was constructed in 1962. The three-span cable-stayed system has spans of 65.85 m + 144 m + 65.85 m. The bridge carries highway traffic on a roadway 7 m wide and has 1.5 m wide sidewalks on each side. The superstructure comprises two main II-shaped prestressed concrete beams 1.5 m deep, 1.4 m wide, and spaced at 9.6 m. The cable arrangement is a radiating shape. The stays are composed of strands of 73 mm and 55 mm in diameters. The towers are cast-in-place reinforced concrete structures.

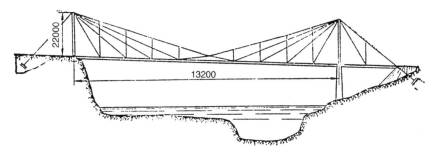

FIGURE 15.56 Narin River Bridge (all dimensions in mm).

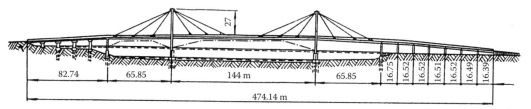

FIGURE 15.57 Kiev Dnepr River Bridge.

15.8.2 Moscow Dnepr River Bridge

In 1976, a cable-stayed bridge was constructed in Moscow. The bridge carries six lanes of traffic and five large-diameter pipes below the deck, and has an overall width of 31 m. The three-span continuous structure has a span arrangement of 84.5 m + 300 m + 63 m. The stiffening girder comprises twin steel box beams with an orthotropic deck fabricated of 10XCND low-alloyed steel. To meet transportation clearances the depth of the girder was limited to 3.6 m. In the cross section the main beams are 5.5 m wide with a distance between inner the webs of adjacent girders equal to 20.2 m and diaphragms spaced at 12.5 m. The stiffening girder has a fixed connection to the abutment and is movable at the pylon and intermediate pier. An A-shaped single reinforced concrete pylon 125 m high has box section legs. Each stay is formed from 91 parallel galvanized wires (diameter 5 mm). The stays have a hexagonal section of 55 mm × 48 cm and are installed in two inclined planes.

15.8.3 Serebryanyi Bor Bridge in Moscow

To arrange a traffic link between the inner city ring road and the peripheral motorway MKAD, a new bridge crossing over the Moskva River was built. The bridge crosses the river at an angle of 15° (almost parallel to the river). The specifics of the route alignment and navigation clearance requirements called for a cable-stayed system with arch pylons (Figure 15.58). The overall length of the bridge crossing is 1.4 km comprising three sections. The bridge accommodates four lanes of traffic in each direction. The central cable-stayed section has a span arrangement of 2 × 105 m + 410 m + 2 × 105 m, with a total length of 830 m. The arch pylon has a 138 m span and a height of 102 m. The stiffening girder comprises two steel main box beams with an orthotropic deck, with a permanent depth of 3.16 m.

The bridge has 72 stays which are fan arranged at the pylon and linear arranged along both edges of the deck. Each stay consists of 27–47 strands. The number of wires per strand is seven. Anchorage of the Freyssinet system is provided at the deck and the pylon. The whole superstructure is protected against longitudinal vibration by large Maurer dampers with a capacity of 400 tons (longitudinal force). The bridge has a complicated space configuration and three-dimensional analysis is the only viable tool for understanding structural behavior and forming a basis for design. Based on numerical analysis results

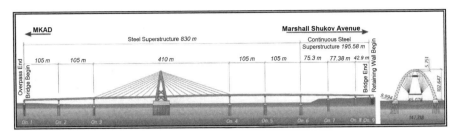

FIGURE 15.58 General arrangement of the Serebryan Bor Bridge.

FIGURE 15.59 Construction of arch pylon.

the stresses at the arch members do not exceed 300 MPa. Special attention had been paid to a search of rational forms and material distribution in the structural elements. The overall and local strengths of the diaphragms, orthotropic deck, and joints of the cable fixing were determined on the basis of this scheme.

The adopted method of erection for the cable-stayed bridge segment called for three-dimensional analysis of the system for each construction stage. To guarantee permanent geodetic control at the stages of superstructure erection and stay stressing, the effects and displacements of control markings at the stiffening girder and arch were determined. This data assured control of accuracy for the elements of the cable-stayed system, stressing the value of stays and levels of position for the elements in the combined arch cable-stayed system.

The arch pylon (Figures 15.59 and 15.60) was constructed by cantilever erection from two sides. Up to the middle of the arch pylon height assembly was implemented with elements, having a mass of 8–10 tons, by a boom crane with a capacity of 300 tons. Finally, two arch halves had to be locked at the crown and the main task was reaching a high accuracy. In reality, the displacement of one arch half against another at the locking point did not exceed 20 mm.

15.8.4 Surgut Ob River Bridge

The new bridge crossing over the Ob River near the city of Surgut is 2110 m long and has only one tower. Its central span of 408 m is the longest for a single-tower cable-stayed bridge. The bridge opened to traffic in September 2000 and carries road traffic. The general scheme of the bridge is shown in Figure 15.61.

FIGURE 15.60 Erection of observation platform.

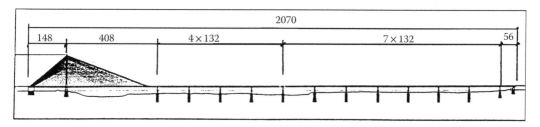

FIGURE 15.61 Surgut Ob River Bridge (all dimensions in m).

The bridge has an overall width of 15.2 m and will allow two lanes of traffic (Figure 15.62). It is located in profile on a convex curve with a radius of 120,000 m. The superstructure is a single steel box girder with an orthotropic deck (Figure 15.63). A single pylon 146 m high was constructed in the bottom portion of precast segments, forming an outer shape and cast-in-place concrete core, and in the upper portion of the two parallel steel towers (transverse section) with struts creating a frame. The intermediate piers are constructed of precast segments with cast-in-place concrete cores and with foundations on bored piles with steel casings 1420 mm in diameter. The abutments are cast-in-place concrete with foundations on reinforced concrete piles of hollow sections 0.6 m in diameter and filled by concrete.

15.8.5 Zolotoi Rog Bay Bridge

The construction of a new crossing over Zolotoi Rog (Golden Horn) Bay in the city of Vladivostok was completed in 2012. The Bay separates Vladivostok into two pieces. The idea for this bridge is very ancient. A postcard in the city museum with a picture of the old bridge is dated 1906. The total length of the crossing is 2.1 km and the cable-stayed bridge itself is 1388 m, with a span arrangement of 49.98 m + 2 × 90 m + 100 m + 737 m + 100 m + 2 × 90 m + 41.94 m. The bridge accommodates two dual lanes, and the navigation clearance is 64.25 m. The bridge comprises two identical V-shaped pylons, with a height of 221.16 m, which are the main feature of this bridge (Figure 15.64). The main task in the pylon design was to eliminate bending moments at the legs from the dead load. To compensate the bending moment at the legs in the transverse direction, the connections of the cables to the pylons are placed with eccentricity to the neutral axis of the pylon leg. Therefore, the weight of the stiffening girder "runs" the pylon in the opposite direction of its self-weight. The bridge was completed in 2012 and links the

FIGURE 15.62   Surgut Ob River Bridge and the railway steel truss bridge nearby.

FIGURE 15.63   Surgut Ob River Bridge box girder.

federal motorway with Russky Island and will be a part of the route for host delegates of the Asia-Pacific Economic Cooperation (APEC) Summit.

15.8.6 Eastern Bosphorus Strait Bridge

This bridge crossing between the mainland and Russky Island across the Eastern Bosphorus Strait has a total length of 3.1 km. The cable-stayed portion is 1885 km in length, with a span arrangement of 60 m + 72 m + 3 × 84 m + 1104 m + 3 × 84 m + 72 m + 60 m and a record central span. The bridge accommodates a dual two-lane motorway and the navigation clearance is 70 m. The region is

FIGURE 15.64 Zolotoi Rog Bridge under construction.

FIGURE 15.65 Eastern Bosphorus Strait Bridge under construction.

characterized by complex climatic conditions: the temperature ranges from −40°C to +40°C, with gust wind speeds up to 36 m/s, wave heights up to 6 m, and ice thickness up to 0.7 m.

The bridge has two identical concrete A-shaped towers 320.9 m high. Each tower is supported by 120 bored piles penetrating to a depth of more than 70 m. The superstructure passes through each tower between the legs (Figure 15.65). The superstructure in the central span is designed of steel and the side spans are reinforced concrete. The central span steel superstructure has an aerodynamic section to compensate for the load of gust winds. The configuration of this cross section is based on aerodynamic analysis and optimized on the results of an experimental scale test of the bridge model. The defining feature of the Eastern Bosphorus Strait Bridge is cables that minimize wind load by 30% using a compact polyethylene sheath system (PSS) where the compact configuration of the strands requires casings smaller in diameter. The PSS cables include strands 15.7 mm in diameter. Each cable comprises from 13 to 79 strands. The bridge was opened to traffic in 2012.

15.9 Prospects

The strategy for 2010–2030 transport system development plans includes construction of modern high-speed motorways with a budget of about $2000 billion It is supposed that the total length of the road network will reach 1350 thousands km. Preparations for the 2012 APEC Summit called for priority investment for the Southern Primorye Coastal Zone infrastructure. Investment into the transportation system plays the main role in the development plans. Because the 2012 APEC Summit was hosted on Russky Island, the improvement of the transportation system of the city of Vladivostok has priority, with a budget of $663 billion. The construction of two major cable-stayed bridges plays an important role in this program. Another major investment is improvement of the transportation system for the Olympic Games in Sochi. The current bridge construction activity within Sochi requires a dual two-lane carriageway comprising 13 bridges and overpasses with a total length of 4.8 km, 3 interchanges, and 6 tunnels with a length of 4 km, with a budget of $2 billion.

The significant demands for the rehabilitation and strengthening of existing bridge structures are increasing every year. The above large-scale plans determine the future directions of bridge design and construction practice.

- Harmonization and modification of national standards considering mostly Eurocode standards; consideration of interactions of structural solutions with technological processes, including aesthetic, ecological, and operational requirements and speed construction
- Development of new structure forms such as precast or cast-in-place reinforced concrete and prestressing concrete, steel, and composite structures for piers and superstructures to improve reliability and durability
- Redesign of standard structures considering practical experiences in engineering, fabrications, and erection practices
- Unification of shop-fabricated steel elements ready for erection to a maximum dimension fitting the transportation requirements, and improvement of precast concrete decks; improvement of corrosion protection systems to a lifespan up to 12 years
- Development of relevant mobile equipment and practical considerations of cast-in-place concrete bridges

References

Belyaev, A. V. 1945. *Moskvoretsky's Bridges*, Chap. 1 (in Russian). Academy of Science of the USSR, Moscow, Leningrad.

Evgrafov, G. K., Bogdanov, N. N. 1996. *Design of Bridges*, Chap. 1 (in Russian). Transport, Moscow.

Gadaev, N., Perevoznikov, B., Seliverstov, V. 2002. Construction technology for monolithic concrete bridges in Moscow megacity. *Proceedings of the 1st Fib Congress 2002*, Vol. 2. Japan Prestressed Concrete Engineering Association and Japan Concrete Institute, Osaka, Japan, pp. 439–446.

GOST. 1988. 9238-83 (State Standard), *Construction and Rolling Stock Clearance Diagrams for the USSR Railways of 1520 mm Gauge* (in Russian). Gosstroy of USSR, Moscow.

GOST. 1989. 27751-88 (State Standard), *Reliability of Constructions and Foundations, Principal Rules of Calculations* (in Russian). Gosstroy of USSR, Moscow.

GOST. 1992. 6713-91 (State Standard), *Low Alloyed Structural Rolled Stock for Bridge Building* (in Russian). Gosstroy, Moscow.

GOST. 1997. 26775-97 (State Standard), *Clearances of Navigable Bridge Spans in the Inland Waterways, Norms and Technical Requirements* (in Russian). Gosstroy, Moscow.

GOST. 2008. P-52748, *Automobile Roads of General Use. Standard Loads, Loading Systems and Clearance Approaches* (in Russian). Standardinform, Moscow.

Kriltsov, E. I., Popov, O. A., Fainstein, I. S. 1974. *Modern Reinforced Concrete Bridges* (in Russian). Transport, Moscow.

MGSN. 1998. 5.02-99 (Moscow City Building Norms), *Design of Town Bridge Structures* (in Russian). Moscow City Government, Moscow.

Monov, B., Seliverstov, V. 1997. Erection of composite bridges with precast deck slabs. *International Conference, Composite Construction-Conventional and Innovative*, September 16–18, 1997, *Innsbruck, Austria*, pp. 531–536.

Perevoznikov, B. F., Ivanova, E. N., Seliverstov, V. A. 1997. Specified requirements and recommendations to improve the process of design of temporary structures and methods of hydrologic justification of their functioning (in Russian). *Information Agency 7*, pp. 1–46.

Perevoznikov, B. F., Seliverstov, V. A. 1999. Reconstruction of the Moscow Ring Road. *Structural Engineering International* 9(2):137–142.

Perevoznikov, B., Seliverstov, V. A. 2001. Reliability of bridges: Hydraulic aspects, *International Conference Malta 2001: Safety, Risk and Reliability-Trends in Engineering, Conference Report*. IABSE, Zurich, pp. 209–214.

Popov, O. A., Chemerinsky, O. I., Seliverstov, V. A. 1997. Launching construction of bridges: The Russian experience. *Proceedings of International Conference on New Technologies in Structural Engineering*, Lisbon, Portugal, July 2–5.

Popov, O. A., Monov, B., Kornoukhov, G., Seliverstov, V. 1998. Standard structural solutions in steel bridge design. *Proceedings of 2nd World Conference on Steel in Construction*, San Sebastian, Spain, 11–13 May 1998. The Steel Construction Institute, Elsevier, London, UK, pp. 60–61.

Potapkin, A. A. 1984. *Design of Steel Bridges with Consideration for Plastic Deformation* (in Russian). Transport, Moscow.

Seliverstov, V. A. 1997. Specified requirements to determine forces from hydrologic and meteorological factors for design of temporary structures (in Russian). *Information Agency 8*, pp. 8–31.

SNIP. 1988. 3.06.07-86 (Building Norms and Regulations), *Bridges and Culverts, Rules of Inspection and Testing* (in Russian). Gosstroy of USSR, Moscow.

SNIP. 1992. 3.06.04-91 (Building Norms and Regulations), *Bridges and Culverts* (in Russian). Gosstroy of USSR, Moscow.

SNIP. 1996. 2.05.03-84 (Building Norms and Regulations), *Bridges and Culverts* (in Russian). Minstroy of Russia, Moscow.

SNIP. 2000. 23-01-99 (Building Norms and Regulations), *Construction Climatology* (in Russian). Gosstroy of Russia, Moscow.

STP. 2001. 015-2001, Standard of Enterprise. *Technology of Shear Studs Application in Bridge Structures Using Import Materials* (in Russian). Corporation "Transstroy", Moscow.

VSN. 1978 and 1984. 136-78 (Departmental Building Norms), *Guidelines for Design of Temporary Structures and Devices for Construction of Bridges* (in Russian). Mintransstroy, Moscow.

Zhuravov, L. N., Chemerinsky, O. I., Seliverstov, V. A. 1996. Launching steel bridges in Russia. *Structural Engineering International* 6(3):183–186.

FIGURE 1.1   Canada's oldest bridge, the Percy Covered Bridge, Powerscourt, Québec.

FIGURE 1.16   North Arm Bridge. (Photo courtesy of MMM Group.)

FIGURE 2.35   Bixby Creek Bridge. (Courtesy of the California DOT.)

FIGURE 2.48 Golden Gate Bridge. (Courtesy of Lian Duan.)

FIGURE 3.11 Overview of General San Martín Internacional Bridge, between Fray Bentos (Uruguay) and Puerto Unzué (Argentina).

FIGURE 3.25 The Nuestra Señora del Rosario Bridge, viewed from the Rosario side. Protections against vessel collisions can be seen.

FIGURE 4.11 Side view of the Rio-Niterói Bridge main span.

FIGURE 4.21 Aerial view of the Tancredo Neves Bridge.

FIGURE 5.7 Old Bridge in Mostar.

FIGURE 6.54 Rowing Canal Bridge. (Courtesy of A. Georgiev.)

FIGURE 7.2 Overview of Krk Bridge. (Courtesy of the Institute IGH, Croatia.)

FIGURE 7.46 Simulation overview of the future Pelješac Bridge. (Courtesy of the Faculty of Civil Engineering, University of Zagreb, Croatia.)

FIGURE 8.23 The Cat's Eyes Bridge. (Courtesy of Vaclav Mach.)

FIGURE 9.7 The Munkholm Bridge from 1952. (Courtesy of DTU.)

FIGURE 10.8   Tornio River Bridge, built in 1939. (Courtesy of Pekka Pulkkinen.)

FIGURE 11.76   Example of a bascule bridge. (Courtesy of SETRA.)

FIGURE 12.36 Rion–Antirion Bridge. (Courtesy of Jacques Combault and GEFYRA SA.)

FIGURE 12.58 Metsovo road bridge on the Egnatia Motorway [23]. (Courtesy of DOMI SA.)

FIGURE 13.22 The Stone Bridge in Skopje.

FIGURE 14.22   Road and railway truss bridge over the Vistula River in Płock called the Marshal Józef Piłsudski's Legions Bridge (1938).

FIGURE 14.28   Bridge over the Narew River in Ostrołęka (1995).

FIGURE 15.9   Deck girder composite bridge of 55 m span over the Mulmuta River (Siberia).

FIGURE 16.1 The Sava River in Belgrade, view from Belgrade Fortress.

FIGURE 16.59 Construction of the pylon. (Folić and Misulic 2010.)

FIGURE 18.29 Gülburnu Bridge.

FIGURE 20.8 China's Expressway Network (From China Ministry of Transport, http://glcx.moc.gov
.cn/roadMap.jsp; also Wikipedia on Answers.com: Expressways of China, http://www.answers.com/topic/
china-national-highways.)

FIGURE 20.22 Shanxi Dan River Bridge. (From China Ministry of Transport, http://www.moc.gov.cn/huihuang60/difangzhuanti/shanxi/jinrixinmao/200907/W020090721562697210807.jpg.)

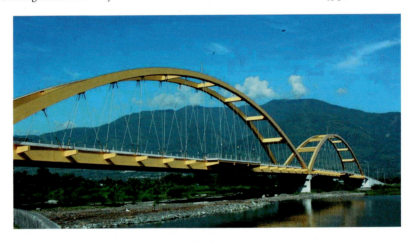

FIGURE 21.56 Palu IV Bridge (300 m), Palu, Central Sulawesi.

FIGURE 21.101 The Suramadu Bridge (5438 m), over the Madura Strait (June 2009).

FIGURE 22.1   Dezful Bridge in Dezful.

FIGURE 22.34   Eighth Bridge in Ahwaz. (HCE 2011).

FIGURE 23.37   Tokyo Gate Bridge (Tokyo). (Courtesy of JBA.)

FIGURE 23.40   Akashi Kaikyo Bridge (Hyogo).

FIGURE 24.38   Night view of the Gaoping Creek Bridge (1999). (Adapted from Taiwan Area National Expressway Engineering Bureau 2003.)

FIGURE 25.11   Rama III Bridge.

FIGURE 25.20 Rama VIII Bridge.

FIGURE 28.1 Siduhe Bridge.

FIGURE 28.3 Pont Du Gard stone aqueduct.

FIGURE 28.26 Balinghe cable-stayed bridge proposal.

Bridge Engineering in Serbia

Radomir Folić
Faculty of Technical
Sciences

16.1 Introduction

16.1.1 Geographical Characteristics

16.1.1.1 General

The Republic of Serbia, has a population of about 7.3 million (not including Kosovo), and Kosovo has around 1.8 million (Statistical Office of the Republic of Serbia [RZS]). It is a landlocked country located at the crossroads of Central and Southeastern Europe, covering the southern lowlands of the Carpathian Basin and the central part of the Balkans. It can be classified as a Dinaric, Balkan,

Carpathian, Pannonian, and Danubian state. The province of Vojvodina covers the northern third of the country, and is entirely located within the Central European Pannonian Plain. The easternmost tip of Serbia extends into the Wallachian Plain.

The northeastern border of the country is determined by the Carpathian mountain range, which runs through the whole of Central Europe. The Southern Carpathians meet the Balkan Mountains following the course of the Velika Morava, a 500 km long river. The Midžor peak is the highest point in eastern Serbia, at 2156 m. In the southeast, the Balkan Mountains meet the Rhodope Mountains. The Šar Mountains of Kosovo form the border with Albania, with one of the highest peaks in the region, Djeravica, reaching 2656 m at its peak. The Dinaric Alps of Serbia follow the flow of the Drina River, overlooking the Dinaric peaks on the opposite shore in Bosnia and Herzegovina.

Serbia borders Hungary to the north; Romania and Bulgaria to the east; the Republic of Macedonia to the south; and Croatia, Bosnia and Herzegovina, and Montenegro to the west; its border with Albania is disputed. Despite the country's small size, the European Union's largest river and tributary to the Black Sea, the Danube, passes through Serbia with 21% of its overall length, joined by its biggest tributaries, the Sava and Tisza rivers. Serbia's capital city, Belgrade, is among the most populous in Southeastern Europe.

After their settlement in the Balkans, Serbs formed a medieval kingdom that evolved into a Serbian Empire, which reached its peak in the fourteenth century. By the sixteenth century Serbian lands were conquered and occupied by the Ottomans, at times interrupted by the Hapsburgs. In the early 1800s, the Serbian revolution established the country as the region's first constitutional monarchy which subsequently expanded its territory, pioneering the abolition of feudalism and serfdom in Southeastern Europe. The former Hapsburg crown land of Vojvodina joined Serbia in 1918. Decimated as a result of World War I, the country united with other South Slavic peoples into a Yugoslav state that would exist in several formations up until 2006, when Serbia once again became independent (Statistical Office of the Republic of Serbia [RZS]).

In February 2008, the parliament of Kosovo, Serbia's southern province with an ethnic Albanian majority, declared independence. The response from the international community has been mixed. Serbia regards Kosovo as its autonomous province governed by UNMIK, a UN mission. Serbia is a member of the United Nations and Council of Europe.

16.1.1.2 Water

Spanning over 588 km across Serbia, the Danube River is the largest source of fresh water. Other fresh water rivers are the Sava (Figure 16.1), Morava, Tisza, and Timok. The Drina River flows into the Black Sea. The largest artificial reservoir, Djerdap, locally known as Djerdap Lake, is located on the river Danube.

FIGURE 16.1   (See color insert.) The Sava River in Belgrade, view from Belgrade Fortress.

TABLE 16.1 Rivers that Run through Serbia and Other Countries

	River	Km in Serbia	Total Length (km)	Number of Countries
1	Danube	588	2783	9
2	Great Morava	493	493	1
3	Ibar	250	272	2
4	Drina	220	346	3
5	Sava	206	945	4
6	Timok	202	202	1
7	Tisa	168	966	4
8	Nišava	151	218	2
9	Tamiš	118	359	2
10	Bega	75	244	2

The Danube originates in the Black Forest in Germany as the much smaller Brigach and Breg rivers, which join at the German town of Donaueschingen to become the Danube. The river then flows eastwards for a distance of some 2800 km, passing through four Central and Eastern European capitals, before emptying into the Black Sea via the Danube Delta in Romania and Ukraine. Known to history as one of the long-standing frontiers of the Roman Empire, the river flows through or forms a part of the borders of 10 countries: Germany (7.5%), Austria (10.3%), Slovakia (5.8%), Hungary (11.7%), Croatia (4.5%), Serbia (10.3%), Romania (28.9%), Bulgaria (5.2%), Moldova (0.017%), and Ukraine (3.8%) (Table 16.1).

The Sava is a river in Southern Europe, a right side tributary of the Danube River at Belgrade. It is 945 km long. It flows through four countries: Slovenia, Croatia, Bosnia and Herzegovina (making its northern border), and Serbia. Through the Danube, it belongs to the Black Sea drainage basin, and represents the Danube's longest right tributary and second-longest of all, after Tisza, as well as the richest with water, by far.

The Tisza or Tisa is one of the main rivers of Central Europe. It rises in Ukraine, and is formed near Rakhiv by the junction of headwaters of White Tisa, whose source is in the Chornohora Mountains, and Black Tisa, which springs in the Gorgany range. It flows roughly along the Romanian border and enters Hungary at Tiszabecs; downstream, it marks the Slovak–Hungarian border, passes through Hungary, and falls into the Danube in central Vojvodina of Serbia. There, it forms the boundary between the regions of Bačka and Banat. The river also forms short portions of the border between Hungary and Ukraine and between Hungary and Serbia.

The Ibar is a river in Montenegro, Serbia and Kosovo, with a total length of 276 km (171 mi). It starts in eastern Montenegro and, after passing through Kosovo, flows into the Zapadna Morava, central Serbia, near Kraljevo. It belongs to the Black Sea drainage basin. Its own drainage area is 8059 km² (3112 mi²), and average discharge at the mouth is 60 m³/s. It is not navigable. The Nišava is a river in Bulgaria and Serbia, a right tributary, and with a length of 218 km also the longest river of the Southern Morava.

The Great Morava is the final section of the Morava, a major river system in Serbia. The Great Morava is created by the confluence of the Southern Morava and the Western Morava located near the small town of Stalac, a major railway junction in central Serbia. From there to its confluence with the Danube northeast of the city of Smederevo, the Great Morava is 185 km long. With its longer branch, the Western Morava, it is 493 km long. The Southern Morava, which represents the natural headwaters of the Morava, used to be longer than the Western Morava, but due to the regulations of river bed and melioration, today it is shorter. Regulations were made on all three Morava rivers, and they all used to be much longer; the total Morava was over 600 km long. Today, the most distant water source in the Morava watershed is the source of the river Ibar, the right and longest tributary of the Western Morava, originating in Montenegro, which gives the Ibar–Western Morava–Great Morava river system a length of 550 km, making it the longest waterway in the Balkan Peninsula (Statistical Office of the Republic of Serbia [RZS]).

16.1.1.3 Bridges

There are 2638 bridges on the roadways of the Republic of Serbia (without Kosovo), with a total surface area of 800,000 m², of which 215 bridges are on highways, 996 bridges are on first-order state-owned roadways, and 1427 bridges are on second-order state-owned roadways. The unredeemed value of all bridges is estimated at $1 billion. The bridges are of different ages, shapes, construction, and materials (timber, stone, concrete, prestressed concrete, steel) as well as of different static systems, spans, and lengths, ranging from 5 to 2212 m (the length of the bridge over the Danube near Beška). There are 11 large bridges: 7 bridges over the Danube and 5 bridges over the Sava River. Due to longstanding insufficient investments into the maintenance and reconstruction of bridges, they are in dissatisfying condition.

Serbia is often called "the tie between East and West," which usually refers to the Morava valley, since it is the easiest route between Greece and Asia Minor on the one hand and the rest of Europe on the other. This beneficial position was the main cause of its painful past. Serbia has well-developed roadway, railway, aerial, and water traffic. Due to many geographic advantages, it is expected that the future development of traffic in Serbia will be faster and more extensive. The main traffic junction in the country is its capital, Belgrade. There are two pan-European traffic corridors passing through Serbia: the roadway–railway corridor 10, with its arms B and C, and the waterway corridor 7.

The roadway network of the Republic of Serbia consists of public roads of first and second order (former mainline, regional, and local roadways). Public roads of the first order are Serbia's basic roadway network, consisting of 30 roadway directions with a total length of 5.525 km. A separate category of this first-order roadway network is the category of highways and semihighways. Within the framework of the first-order roadway network, 2150 km of the roads in Serbia belong to the category of European roadway network, the so-called E roads (CIP 2010).

16.1.1.4 The Roadway Network of Serbia

The value of the first- and second-order state-owned roadway networks of the Republic of Serbia is estimated at $13 billion. Serbia is a European country with moderate population density and a well-developed roadway network. The roadway network altitude ranges between 30 m (Negotin) and 1700 m (Golija). It is estimated that 40% of the total length of roadway network is situated at altitudes over 600 m. The asphalt carriageways of the first- and second-order state-owned roadways were constructed between 1962 and 1985 and many have kept the elements of the old roadway, that is, the asphalt was placed over the existing broken stone. In the same period, the most important routes were built according to the projects, so these sections consist of better elements of longitudinal and transversal profile (CIP 2010).

The 40,845 km long roadway network of the Republic of Serbia (Figure 16.2) consists of 5,525 km of first-order state-owned roadways, 11,540 km of second-order state-owned roadways, and 23,780 km of local roadways. The roadway network consists of 498 km of highways and 136 km of semihighways with tolls. Unpaved roadways comprise two-fifths of the entire roadway network. Thirty-two percent of roadways are over 20 years old and only 14% of roadways are less than 10 years old. Due to lack of investments into the maintenance and reconstruction of roadways over many years, the current condition of the roadway network is dissatisfactory.

E–A class roadways in Serbia are the following:

- Road E65: (Montenegro)–Ribariće–Kosovska Mitrovica–Priština–Đeneral Janković–(Macedonia).
- Road E70: (Croatia)–Batrovci–Beograd–Vršac–Vatin–(Romania). Where the E70 overlaps with Corridor 10 (section Batrovci–Beograd), there is a modern highway.
- Road E75: (Hungary)–Horgoš–Subotica–Novi Sad–Inđija–Beograd–Pojate–Niš–Grdelica–Mijatovac–(Macedonia). There is a modern highway on section Novi Sad–Inđija–Beograd–Niš–Grdelica. Except for a smaller part from Batajnica to Zemun, the road overlaps with Corridor 10 (branch B).

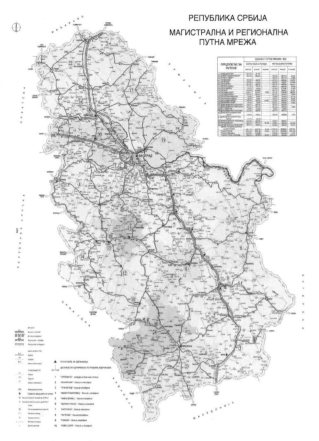

РЕПУБЛИКА СРБИЈА

МАГИСТРАЛНА И РЕГИОНАЛНА
ПУТНА МРЕЖА

FIGURE 16.2   Roadway network map of Serbia.

- Road E80: (Bosna-Herzegovina)–Kotroman–Požega–Pojate–Niš–Gradina–(Bulgaria). Where the road overlaps with Corridor 10 (Niš–Gradina) a modern highway was planned, with the section from Niš to Niška Banja already finished.

E–B class roadways in Serbia are the following:

- Road E662: Subotica–Sombor–Bogojevo–(Croatia)
- Road E761: Pojate–Paraćin–Zaječar–Vrka Čuka–(Bulgaria)
- Road E763: Beograd–Požega–Nova Varoš–Gostun–(Montenegro), with a modern highway on the Beograd–Požega section
- Road E771: (Romania)–Kladovo–Zaječar–Niš
- Road E851: (Albania)–Prizren–Priština–Niš

There are two Pan-European traffic corridors passing through Serbia: Corridor 10, with its branches B and C, and waterway Corridor 7 (CIP 2010).

16.1.1.5 Railway Traffic

The total length of the railway network (Figure 16.3) in Serbia is 3808 km (data from 2008), of which 1196 km is electrified. This is regarding the railroads of standard track width. There are also railways with narrow tracks, which are nowadays out of use or used for specific purposes (a tourist railway, known as "Šarganska osmica" [Shargan eight]). Also, there is a small percentage of railways with two

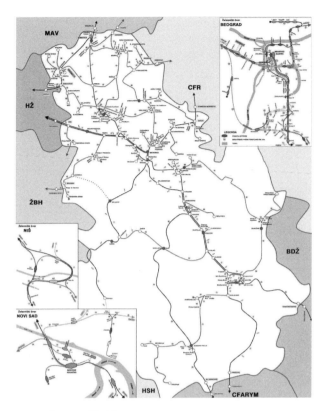

FIGURE 16.3   Railway network map of Serbia.

tracks. The only a two-track railway is the Belgrade–Inđija–Sremska Mitrovica–Croatian border, whose length will be increased from Belgrade to Niš.

The main railway are directed from Belgrade:

- West toward Sremska Mitrovica and the Croatian border
- North toward Novi Sad, Subotica, and the Hungarian border
- Northeast toward Pančevo, Vršac, and the Romanian border
- Southeast toward Niš, Pirot, and the Bulgarian border
- South toward Niš, Leskovac, and the Macedonian border
- Southwest toward Valjevo, Užice, and the Montenegro border
- South toward Lapovo, Kraljevo, Kosovo Polje, Đeneral Jankovic, and the Macedonian border

The largest railway junctions are Belgrade and Niš, with five railway lines each. Junctions of smaller importance are Novi Sad, Subotica, Inđija, Požega, Stalać, Kraljevo, and Kosovo Polje.

16.1.2 Historical Development

Since the Roman Empire, stone and wood bridges were mainly built in the area of current Serbia, which was followed by significant achievements. Remarkable accomplishments were also achieved by bridge builders from Serbia beyond the borders of current Serbia. The bridge over Đurđevića Tare (1938–1940) (Trojanović 1968, 1974), constructed 150 m above the water level, with its main arch span of 116 m, was designed by M. Trojanović, a professor of concrete bridges in Belgrade. Significant accomplishments were achieved by Serbian designers and contractors after World War II. Bridges designed and constructed by Mostogradnja construction company from Belgrade are obvious examples. They are primarily concrete arch bridges connecting the mainland and the island of Pag, and connecting the mainland and the island of Krk (Sram 2002) on the Adriatic highway over Sibenik Bay (Muller 2000). Thus, it is important to mention the continuous girder bridge over the rivers Sava and Una near Jasenovac, which connects Croatia and Bosnia and Herzegovina (Sram 2002).

Although some of the mentioned bridges are part of the architectural heritage, their rich legacy has been addressed only marginally in histories of architecture. During their long existence, the old stone bridges have undergone many repairs and modifications, so today they are preserved in an altered form (Gojkovic 1989). Wars happened more often in this part of Europe than others and resulted in destruction, primarily of superstructures, but often of the entire structures. Hence, the history of engineering is filled with construction, destruction, renewal, and reconstruction. As to Serbia, regrettably this has also been the case in the near past. In 1999, during 78 days of a NATO air raid, 62 bridges were destroyed or severely damaged (Vojinovic and Folić 2002). Therefore, a separate section (Section 16.9) is included that relates to this subject.

It is well-known that Roman engineers achieved major accomplishments, mainly in structures made of stone. Over the period of the Roman Empire, masonry bridges were built in Serbia, but most of them were destroyed and thus they are not preserved today. The largest bridge in the Roman Empire was the Trajan Bridge (Figure 16.4), built on the Danube near Kladovo between 103 and 104 AD linking this part of Serbia with Romania. It was made of masonry columns and a wooden superstructure, with 21 spans, each of 33 m. Some of the remains of Trajan Bridge indicate the use of so-called Roman concrete. Its bonding properties were obtained by adding flour of ground bricks and volcanic ash to lime plaster, thereby achieving hydraulic properties (underwater bonding). The fall of Roman Empire in the Middle Ages resulted in a serious setback in architectural progress. However, accomplishments in building masonry bridges in that period are extremely significant.

During Ottoman Empire, stone and wooden bridges were built to connect the territories known today as Serbia and Romania. Thus, starting from the twelfth century, radical changes were made to the Belgrade Fortress by reconstruction of the entrance gate and the bridge over the city trench, as compared to the Byzantine period. In the sixteenth century, a wooden bridge was built by the Turks, supported by masonry columns of bricks and stone, a part of which was movable. A masonry stone bridge was erected instead of the wooden bridge over the river Kubršnica in Smederevska Palanka in 1730. The bridge consists of four spans of 8.6 m + 8.2 m + 8.5 m + 3.4 m. It was damaged in 1987. The 54 m long Kasapic Bridge in Užice, with five spans (7.4 m + 0.7 m + 10.4 m + 8.2 m + 5.3 m) was destroyed in 1945. The bridge over the river Đetinja in Užice was built in 1628 and destroyed during World War II. There is another bridge on the river Uvac with a stone arch spanning 15.4 m. The 28.1 m long and 3.15 m wide bridge in Ljubovađa is arched with an aperture of 15.4 m. The bridge was reconstructed in 1987 (Gojkovic 1989).

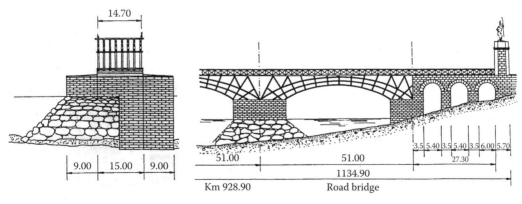

FIGURE 16.4 Trajan Bridge reconstruction. (From Ivanyi, M., and B. Stipanic, Bridges on the Danube, Bridges over the Danube-Catalogue, First & Sec. Intern. Conf. Ed., 1992, 1995, and 2008.)

In Prizren, the city bridge 25.1 m long, with three spans of 4.2 m + 9.88 m + 2.62 m was built in the sixteenth century. It was seriously damaged in 1979 by high waters, and rehabilitated in 1982. A similar bridge in the canyon of Prizrenska Bistrica in the vicinity of Prizren was also rehabilitated. In the vicinity of Djakovica, three stone bridges were built: Terzija Bridge (Figure 16.5), Shvanjska Bridge, and Tabakska Bridge. They were in use until the 1950s, when they suffered large-scale deformations due to the huge traffic flow. However, the structural deficiencies of these bridges were primarily caused by unequal settlement of the foundations. An example of this is Terzijski Bridge, which was damaged by high waters. In 1961 huge deformations occurred at both the foundation of the third column and the column itself. This 192.8 m long bridge over the river Erenik has 11 spans of different dimensions ranging from 10.5 to 12.8 m. To secure traffic, a new reinforced concrete bridge was built in its vicinity. The bridge was rehabilitated between 1984 and 1988. Similar bridges are the Vojinovic Bridge near Vucitrn and the bridge over the river Istok near the village Zac Budisava (Gojkovic 1989). Depending on geotechnical conditions, these bridge superstructures were supported either directly on wooden beams stiffened on arch joints or on wooden boxes. This is why the foundation was the most frequent cause of damage to these structures.

The largest and the most beautiful bridges of that period are in Mostar, over the river Neretva as well as in Visegrad, over the river Drina (in Bosnia and Herzegovina). Many bridges have disappeared in time but there are a few that have been preserved or reconstructed in their original form as invaluable monuments of culture. Two basic construction materials used for old bridges are stone and lime mortar. Stones that could be easily cut were used for suitable stone blocks. The use of lime mortar with the addition of fine ground brick, and of "tera rosa," which provided some hydraulic properties to the lime mortar, was found on many structures. Stone arches were erected at both sides from the end towards the central part of the bridge. Many bridges suffered extensive damage due to increased traffic loads, the influence of urbanization, atmospheric and climate impact, and as a consequence of low-quality building materials or deformation on the foundation. When only lime plaster was used, larger joints occurred, and they were later filled with lead. Culverts and bridges made of brick were built in Vojvodina, but rarely in other parts of Serbia. Combined with wooden constructions, they were built also within city fortresses, an example of which is the above-mentioned Belgrade Fortress.

Timber bridges were built on forest roads of girder systems, hanging trusses, and systems of truss frames. The trusses were rarely made using the Howe system, and they were also used as beams for stiffening suspension footbridges with small spans. The durability of timber bridges is a limiting factor of their use, so they were often replaced by reinforced concrete bridges and sometimes by steel bridges. In the first half of the twentieth century, wooden scaffolds were used for building both beam and arch concrete bridges. Later, by introduction of more efficient construction technologies, cantilever, launching, and segmental methods were used. The first permanent timber bridge in Vojvodina, which was a truss beam, was built in 1873 over the river Tisa near Senta (Vukmanovic 2007). It was intended for pedestrian and cart traffic. The bridge perished in fire and its remains were demolished in 1902.

People have been crossing Danube from Backa to Srem from time immemorial. Aerial ferries were used initially; they were followed by pontoon bridges and since the nineteenth century road bridges have

FIGURE 16.5   Terzija Bridge after Reconstruction in 1988, upstream view. (Gojkovic 1989.)

been built (Folić, Radonjanin, and Malesev 2000). The first permanent lattice steel bridge on concrete pillars (four in the river bed and two on the banks) was intended for railway traffic. This bridge was the longest-lasting of all bridges over the Danube in Novi Sad—it lasted from 1883 to 1941.

Pontoon bridges were the first to connect the banks of the river Danube. The first pontoon bridge was constructed by the Turks in 1526. After the forces of the Ottoman Empire were driven out of Petrovaradin in 1688, the construction of strategically important pontoon bridges began. The first two bridges had a significant strategic role in the victory of E. Savoy in 1716. The pontoon bridge connecting Petrovaradin Fortress and the former Petrovaradin trench, known from numerous engravings, dates back to 1788 when it was moved from Sremska Kamenica. The central part of the bridge was movable and it was moved aside when ships and boats were passing. Crossing was free of charge for citizens of Novi Sad and Petrovaradin. Shortly before ice appeared along the Danube, the bridge was disassembled and kept in winter storage until spring. Pontoon elements were 10 cords long and 2 cords wide and were made only of oak, with sheet metal tops (Vukmanovic 2007).

The first two iron bridges in Serbia were constructed in 1883 over the Danube in Novi Sad (supported by six concrete columns) and 1884 over the Sava in Belgrade, respectively. Since the Franz Joseph Bridge in Novi Sad was a railway bridge with an additional pedestrian footway, pontoons and ferries were used for roadway traffic. The roadway–railway bridge near Bogojevo over the Danube was constructed in 1911. Traffic was bustling and the ferry was used for crossing the Danube between Bogojevo and Erdut in 1193 and later on in 1372, 1480, and 1526. Railway traffic between Sombor and Osijek was established over a large ferry on metal thrust bearings. That was the practice until 1911, when a roadway–railway bridge was constructed there (Vukmanovic 2007).

The first permanent metal bridge for roadway traffic was constructed in 1910 on the river Tisa near Senta. During the war in 1915 a roadway bridge was constructed over the Danube near Novi Sad. It was destroyed by floating ice in February 1924 as a consequence of inadequate ice blasting. In the same year, a bridge of almost 500 m length was constructed over the Tisa near Zabalj, with its metal construction supported by timber columns. It was in service until 1922. Also in 1915, a roadway bridge was constructed over the river Tisa near Titel for war purposes, which was in service until 1927. In the same year, two identical parallel bridges were constructed, a roadway and a railway bridge. After World War II, it was renewed as a dual bridge, but the roadway bridge was built from presetressed concrete, and the railway from steel.

Modern bridge construction over the Danube began in 1928, with the construction of a roadway bridge—one of the most beautiful bridges in Europe at the time, in Novi Sad. Named the Bridge of Prince Tomislav, it was mounted without supporting yokes in the river bed. Unfortunately, the bridge was demolished in 1941, and the new, Tito Bridge was erected on its pillars (later named Varadin Bridge) for only 160 days; it was the first steel bridge constructed in Europe after World War II (Figure 16.6). The bridge was opened to traffic in January 1946 and was destroyed during the NATO campaign on April 1, 1999 (Folić, Radonjanin, and Malesev 2000).

There are 37 steel truss bridges over the Danube. The Varadin Bridge, connecting the central part of Novi Sad with Petrovaradin, was erected on the well-preserved piers of the Prince Tomislav Bridge, from May 1945 to January 1946. It was mainly built of the remains of the old railway bridge. Until 1962 it was used both for road and railway traffic, and afterwards only for roadway and pedestrian traffic. The truss was a gerber beam girder system (with a span of 87 m + 130 m + 87 m). The main truss had parallel chords and a K-web, with a middle span of 130 m and two cantilevers of 36 m. The lower chord of the main truss and the bank piers supported two trusses (with spans of 51 m) with their upper chords (Figure 16.6). The total length of the bridge was 345 m, and the width of the deck was 5.5 m. The height of the bridge over the river was 7 m. This height limited river traffic on the Danube. According to the Danube Convention, the required navigation channel height and width were 9.5 m and 150 m respectively. Therefore, plans for the reconstruction on the bridge were designed to raise the vertical clearance by 2.4 m.

Early in the morning of April 1, 1999, Varadin Bridge was destroyed (Folić, Radonjanin, and Malesev 2000). The first missile hit and broke through the deck structure close to the right river pier. The second

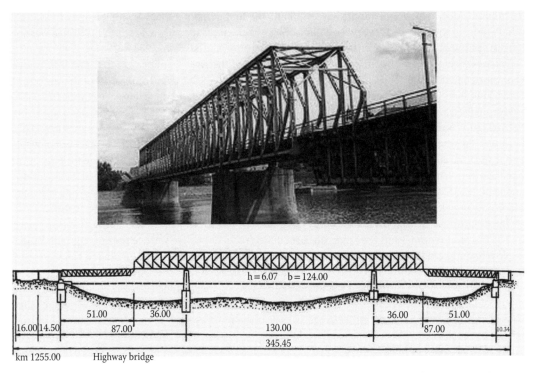

FIGURE 16.6   Varadin Bridge on the Danube in Novi Sad (destroyed in 1999). (Ivanyi and Stipanic 1998.)

missile hit the support part of the structure on the left river pier, which led to the breaking of the main truss girders on both river piers and the loss of the stability of the span structure, that is, the destruction of the whole bridge. The river and bank piers were basically undamaged.

Peace and wars alternated, bridges were demolished and renewed, and a large number of small and big bridges were under construction. This happened especially at the beginning and end of World War II. Also, after the war old bridges were renewed and new bridges were constructed, often by using the steel from demolished bridges. However, the Franz Joseph Bridge was not renewed, although its three concrete columns were preserved. After inspecting the strength of concrete and of the pillar foundations in March 2003, it was proved that its quality was satisfactory. In 1999, the columns were used to fix steel ropes for anchoring the floating bridge on barges. It was the 135th bridge on that crossing, counting from the first in 1526 (Vukmanovic 2007).

The first reinforced concrete bridge of 5 m span in Serbia was built in 1903 over the river Vukodraža in the vicinity of Valjevo. The wars between 1912 and 1918 caused a temporary setback in bridge construction. From 1920 onwards, extensive construction of both steel and concrete bridges resumed. Until World War II, large steel bridges were designed and constructed, mostly by foreign companies. Owing to Petar Micić, our eminent structural engineer and professor of metal structures, further design and construction of steel bridges after the war was taken over by Serbian experts (Darijevic, Nenadic, and Certic 1999).

16.2 Design

16.2.1 Design Philosophy and Regulations

Elastic linear analysis of the structures at normal working loads and working stress design was carried out. The strength of the structural member is assessed by imposing a factor of safety between the maximum stress at working loads and the critical stress. Ultimate limit state (strength method) and serviceability limit state were adopted in 1971 for concrete structures. For the ultimate limit state, both linear and nonlinear methods are used, and for the serviceability limit state usually the linear method is used. There are two serviceability limit states: deformation under service load and cracks at service load. The Rules for Concrete Structures in 1987 (Regulations for Concrete Structures and Reinforced Concrete Structures 1987) provided further advancement. The documents from 1971 and 1987 were based on Model Code CEB-FIP. In recent years the design methodology from EN 1990:2002 to EN 1998:2005 was introduced.

Regarding steel bridges, the Regulations for Steel Structures (1986), which consider different structures and bridges, are used in bridge design and protection. For the stability of compression members and arches, as well as for determining shear lag effects, corresponding standards from the group of Standards for Stability of Steel Structures (JUS U.E7.081/86, for instance) are used. Bearing and hinges of steel structures are used according to JUS U.E7.131/80, connections with rivet and screw are used according to JUS U.E7.145/87 and JUS U.E7.140/85, and welded steel structures are used according to JUS U.E7.150/87.

For concrete constructions, a particular code applies, and the bridges' foundation engineering is regulated by a unique code for all objects beginning in 1991. Since 1964 seismic design was considered for bridge structures according to the temporary code, and in 1987 the Rules for Engineering Objects were applied. For the design, construction, and reconstruction of railway bridges and culverts, the Regulations on Engineering Norms for Bridge Maintenance (1992) are used. In accordance with the standard for inspection of bridges (Regulations on Engineering Norms for Bridge Maintenance 1992), regular inspections must be performed before bridges are approved for service and followed by regular control during their service. Evaluation of results is based on drawing comparisons between the measured and calculated deflection, as well as on analysis of permanent deflection and cracks after unloading.

16.2.2 Loads

The Temporary Code (PTP-5) was applied in former Yugoslavia for traffic road load from 1949 to 1991. Two main categories of bridges were built. The first were designed according to Code PTP-5 (1949–1991), and the others according to the new codes of 1991 (the Regulations for Determining Intensity of Loads on Bridges), similar to DIN 1072. The bridge members' effects were calculated, depending on road category, with hybrid vehicle V300 or V600, that is, V600 + V300 for highway bridges with the vehicle configuration shown in Figure 16.7. V600 + V300 were applied for highway schemes; V600 for first-class road schemes; V300 + V300 for second-class schemes; and V300 for local road schemes. All bridges designed prior to 1991 do not satisfy the conditions of the Rules of 1991, so their strengthening is highly recommended. Furthermore, traffic loads according to EC-1 are more unfavorable than those according to our codes, and bridges will be in need of further strengthening once Eurocode-EN 1991 is brought to power in Serbia.

According to the codes, the loads of railway (Rules for Determining Intensity of Loads 1992) and roadway (Regulations of 1991) bridges are different. Loads have been subdivided into basic (permanent load and traffic load), additional (braking, concussions, wind), and special/uncommon (interruption of cords, loading during construction, seismic forces) loads.

16.2.3 Theory

The design of short- and middle-span bridges is based on the assumptions of first-order linear elastic theory (small deformation theory), and for large-span bridges a second-order theory has been applied. Currently, beside national codes, bridge structures are designed in accordance with international codes and standards such as Model Code CEB-FIP 1993, EN 1990, 1991, and 1992 to 1998, ACI 318, RILEM, and so on. The minimum requirements to be fulfilled are stated in national codes

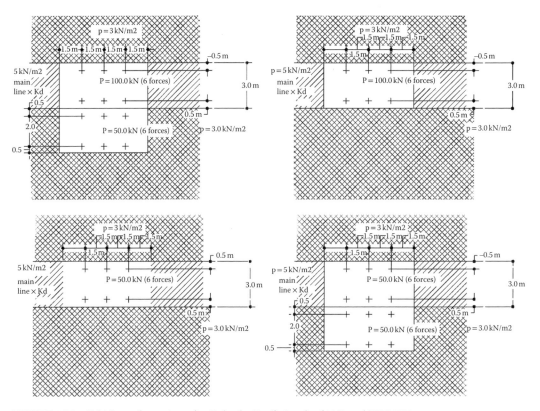

FIGURE 16.7 Vehicle configuration, after Rules for Traffic Loads of 1991 and DIN 1072.

and standards. History and tradition have influenced codes and standards differ considerably from country to country. The modern design concept of concrete structure durability was developed mainly within CEB-FIP (Folić 1992), based on the consistent deterioration mechanisms of engineering models.

16.2.4 Fabrication and Standardization in Concrete (Prefabricated Bridges)

Between 1966 and 1969, 500 bridges with spans up to 20 m and widths of 4–10 m were constructed. Prefabricated prestressed girders 1 m wide were placed next to each other over the prepared bearings and tightened in the transverse direction (Figure 16.8) (Želalić 1969). The safety measures against slipping between the prefabricated elements proved to be significant. They were also used for bridges with multiple fields that substituted deteriorated wooden bridges and bridges over the Danube–Tisza–Danube Canal. These standard procedures have been used both for new and existing bridges, of all lengths between 4 m and 10 m, with or without pedestrian walkways between 50 cm and 125 cm width, an integral part of prefabricated girders or separate prefabricated girders.

The cross section of prefabricated bridges is usually formed by mounting the prefabricated longitudinal girders and concreting a plate over and between them. This approach does not require in situ concreting after assembling, but the ideal mutual incumbency of precast members is important for prestressing in the transverse direction. The girders are concreted next to each other, allocated the same way as in the completed object. Cavities for the transverse cables in all girders are kept at the same height

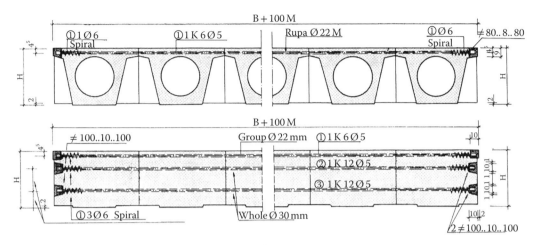

FIGURE 16.8 Detail of precast elements for bridges. (Želalić 1969.)

and at the designed mutual distances. The girders for spans of up to 3 m are standardized solid plate for spans between 6 m and 12 m and hollow plate for spans of between 13 m and 20 m. Prestressed box girders were designed of 52–70 cm depth, depending on width and span.

The partition of elements enables the heaviest elements to weigh up to 5 tons, thereby allowing the use of a single 6 ton crane. Besides the superstructure, the standardization also included the substructure. Thus, these standard bridges are constructed with several fields. Inspection under test loading indicated satisfactory behavior of the superstructure. In order to achieve the steadiest possible force distribution over the bridge length, the distance between transverse cables should be within 2 m.

16.3 Concrete Girder Bridges

Conventional methods for construction of reinforced or prestressed concrete girder bridge systems require adequate scaffoldings which could have, particularly in cases of specific physical obstacles (long length to be crossed, deep rivers or valleys, lakes), a major impact upon project cost-effectiveness, especially where deep foundations must be built. Researchers and builders have decided to leave conventional scaffolds and construct bridges of girders such that girders are prefabricated on the bank or in special workshops, transported to the site, and put in place by some adequate method (Trojanovic 1968). An example of this approach is the construction of the Roadway-Railway Bridge (later named Žeželj Bridge, after its designer).

Gaining reputation through successful applications in bridge engineering, prestressed concrete slowly pushed away ordinary reinforced concrete in girder (beam) bridges in the 1950s and the latter practically disappeared, at least from long-span bridge designs. The first prestressed bridge in Yugoslavia was constructed in 1952, over the Samoilska River, on the Kraljevo–Čačak roadway. Its typical cross section is presented in Figure 16.9. This trend eventually happened in Serbia, though with some delay. However, in the hands of connoisseurs, reinforced concrete beam bridges can efficiently compete with those of prestressed concrete under certain conditions and circumstances. This position is best of all illustrated by the design and construction of a road bridge over the river Nišava in Niš (Figure 16.10), on the Belgrade–Niš–Skopje highway, the so-called Jagodinmalska Bridge (1959–1960).

The slender bridge superstructure was achieved because the girder structure had small moments in the midspan of the girder. With an appropriate arrangement of moments of inertia in the girder and

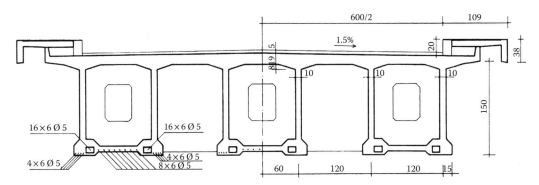

FIGURE 16.9 Cross section of the bridge on the river Samoilska on the Kraljevo–Čačak roadway.

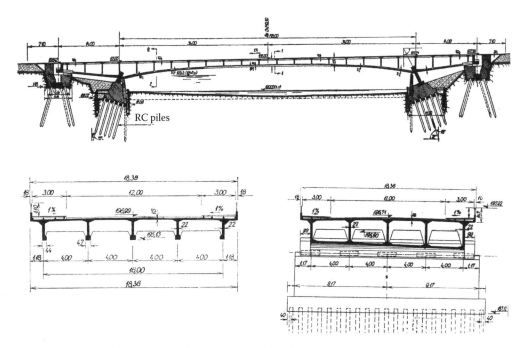

FIGURE 16.10 Road bridge over the river Nišava in Niš. (Trojanovic 1968.)

mass redistribution, the bridge dead load was significantly relieved. The static parameters were then further influenced by turning the classical continuous girder into a continuous system with adjusted relatively small reaction forces at the end supports. One of the inner supports is an inclined pendulum (Figure 16.11).

The end support in the selected continuous three-span system does not have any natural reaction but has assigned reactions of desired magnitudes induced by hydraulic presses and directly influencing the static values in each span. Reactions in the middle supports (fixed bearing and inclined pendulum) follow the well-chosen angular direction. The angular reactions of the middle piers induce moments in the middle span similar to horizontal thrusts in an arch system.

A railway bridge was built over Yuri Gagarin Street on the Belgrade–Stara Pazova–Šid railroad (Figure 16.12). The bridge is a continuous concrete beam girder with four spans with a total length

FIGURE 16.11 Bridge over the river Nišava in Niš, 1960. (Trojanovic 1968.)

FIGURE 16.12 Yuri Gagarin Bridge on the Belgrade–Stara Pazova–Šid railroad. (CIP 2010; Mihajlovic 2010.)

of 188 m (40 m + 2 × 54 m + 40 m). The concrete beam's cross section is a trapezoidal closed box of constant height of $H = 3$ m.

Another type of bridge structure, approach structures to new bridges over big rivers, have been built in urban areas. Their shapes in plane must quite often be adapted to complex requirements (horizontal curves, counter curves). An example from the national practice is the approach structure on the railway bridge over the Sava River in Belgrade. Crossing the Bulevar V. Misica and the Belgrade Fair Site, it consists of a prestressed concrete continuous girder with six spans of 75.5 m, the longest spans on bridge structures of this kind.

16.3.1 Access Structures of the Railway Bridge over the Sava River in Belgrade

The total length of the railway bridge over the river Sava between the Novi Beograd and Prokop railway stations is 1928 m (Figure 16.13). It consists of the part over the river and access spans on the left and the right side. On the right bank, the structure is made of prestressed reinforced concrete with spans of 63.03 m + 75.5 m + 63.03 m, and in the extension of the common reinforced concrete construction with spans of 35 m + 2 × 38.5 m + 35 m (CIP 2010; Mihajlovic, 2010).

16.3.2 Roadway Bridge in Vladicin Han

A roadway bridge was built in Vladicin Han over the river Southern Morava as a beam structure of prestressed reinforced concrete (class 45), 9 spans and 10 piers of concrete class 30. It is composed of two continuous structures 89.8 m + 1 m + 134.2 m = 225 m (Figure 16.14). It consists of two parts. The structure above the regulated bed of the Southern Morava River is 24.9 m + 40 m + 24.9 m = 89.8 m length. The structure in the town, over the single truck railway line Niš–Presevo, spans 20.6 m + 3 × 21 m + 24.9 m = 134.2 m. The cross section is a quasi-box section structure 12.9 m wide (Mrkonjic and Vlajic 2009).

FIGURE 16.13   Access structures of the railway bridge over the Sava River in Belgrade. (CIP 2010; Mihajlovic 2010.)

FIGURE 16.14   Roadway bridge in Vladicin Han. (Mrkonjic and Vlajic 2009.)

16.3.3 Concrete Bridge over the River Southern Morava and Railway Bridge at the Entry of Grdelicka Klisura

A reinforced concrete bridge, 60 m span over the river with an approach span 7 × 16.8 m + 18.8 m, total length $L = 196.4$ m, was built over the Southern Morava (Figure 16.15). It is 11.5 m wide. The first span at the Niš end is 16.8 m long and it is followed by a reinforced concrete arch span of 60 m and a 10.15 m rise. Then, the 18.8 m long span is followed by another 18.8 m span and again with three spans 16.8 m each.

A railway bridge was built over the river Suvaja on the Beograd–Bar railroad (Figure 16.16). Made of prestressed concrete on five spans of 25 m each, the total length of this bridge is 125 m (CIP 2010; Mihajlovic 2010).

16.3.4 Beška Bridge

The exceptional importance of this bridge lies in the fact that it is situated on the main European road-way E-75. At the same time it is a remarkable architectural achievement. In the course of its construction, state-of-the-art structural solutions and up-to-date building methods were applied, resulting in an imposing bridge structure, which once again proved the high quality of Serbian bridge constructors and builders. The bridge was built from 1971 to 1975. Its total width is 14.4 m, and it has three traffic lanes

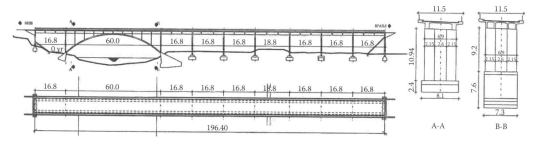

FIGURE 16.15   Concrete bridge on road M-1 over the Southern Morava and over the railway line at the entry of Grdelicka Gorge (partly destroyed in 1999). (CIP 2010; Mihajlovic 2010.)

FIGURE 16.16   Bridge over the river Suvaja. (CIP 2010; Mihajlovic 2010.)

with a total width of 11 m and two pedestrian sidewalks with widths of 1.7 m each. The total length of the bridge is 2250 m and the slope in the longitudinal direction is 2.3%.

The bridge span over the water is a three-span continuous prestressed concrete beam of 105 m + 210 m + 105 m and two cantilevers on both ends with lengths of 15 m each (Figure 16.17). The bridge girder has variable box cross sections, that is, the width of the vertical walls (ribs) ranges from 20 cm to 40 cm, whereas the depth of the lower plate ranges from 40 cm to 80 cm. The top slab carries the traffic load and its thickness ranges from 24 cm to 28 cm. The depth of the box section is 11 m in the middle river pylons and 6 m in the middle of the central span (Sram 2002), respectively. The phases of construction are shown in Figure 16.18.

The length of the viaduct on the Novi Sad side is 1575 m (35 × 45m) and 225 m (5 × 45m) on the Belgrade side. The main supports of the viaduct, with a span of 45 m, are made of prestressed concrete. Different types of cables were used for prestressing. The main cables consist of 36 wires 7 mm in diameter, with an initial force capacity of 1800 kN. The foundation was done on caissons and Franki piles.

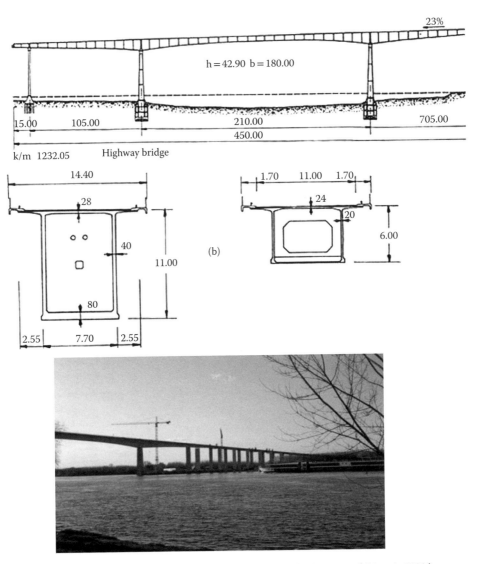

FIGURE 16.17 Longitudinal section of Beška Bridge over the Danube. (Ivanyi and Stipanic 1998.)

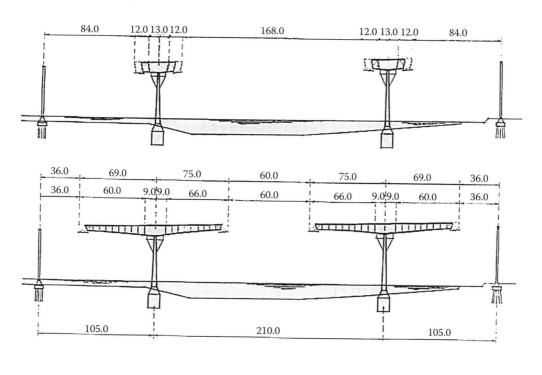

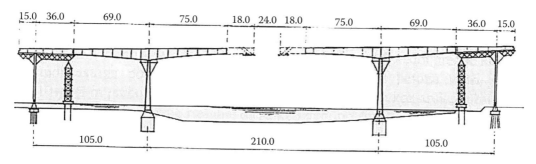

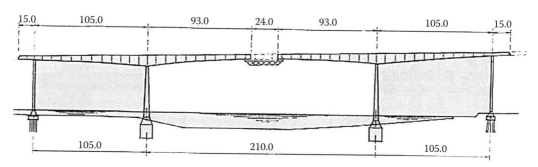

FIGURE 16.18 Phases of construction Beška Bridge over the Danube. (Sram 2002.)

The two middle river piers, with heights of 40 m, have variable box cross sections. These pylons were concreted in sections of 4 m each, using movable formwork and hydraulic cranes. Other pylons, with constant box cross sections 12 m to 45 m high, were concreted by means of slip form. The largest part of the bridge structure (approximately 70%) was built using the cantilever method, with symmetrical and parallel advance from the middle river pylons. The designer was B. Žeželj of the Institute for Testing of Materials (IMS), Belgrade. The main construction company was Mostogradnja (Belgrade).

The bridge was first hit at 5:25 a.m. on March 31, 1999. During this strike only the fence of the bridge was damaged, although the actual target was the right river pylon in the largest span. In the next strike, at 3:00 a.m. on April 21, 1999, the first span of the approach structure on the Novi Sad side was torn down.

16.4 Steel Girder Bridges

16.4.1 Beam and Frame Structure Girders and Orthotropic Decks

According to the Vienna Convention the first connection with Serbia was established when the railway bridge over the river Sava was finished in 1884. The bridge was a truss structure with large spans for that period (CIP 2010; Perisic 1999): 86.1 m + 3 × 96 m + 86.1 m. Its structure was destroyed four times and reconstructed afterwards, but in 1945 the German army blew it up.

One of the higher bridges that was constructed in the postwar period was the Varadin Bridge. It was a temporary bridge for combined road and railway traffic that was reconstructed for light roadway traffic in 1961 (Figure 16.6).

After the war, the Pancevo Bridge, the bridge near Bogojevo, and the bridge over the river Sava near Ostruznica were also reconstructed. From that time on, bridges were built by engineers from Serbia, providing room for development of the metal industry. Riveted structures withdrew, leaving room for welded structures that led to new structural systems. Some time later a road-railway bridge was constructed over the river Morava in Ljubicevo. The river was bridged with three semiparabolic truss simple beams with a 61.62 m span. High-strength bolts for connecting prefabricated steel members were introduced. A family of prefabricated and mountable-dismountable bridges for emergency measures was designed (spans 36 m to 60 m). Later, bridges with spans of 72 m and even of 102 m were designed as well (Mihajlovic and Shiftmiler 2001).

The application of welding procedures has enabled the use of orthotropic plate on steel bridges (e.g., the bridge on Brankova Street over the river Sava in Belgrade). The bridge is a continuous steel plate girder with spans of 75 m + 261 m + 75 m. The 12 m wide roadway and two footways of 3 m width each are on the upper band. The bottom side of the main girders cross section is opened, with two vertical ribs at a distance of 12.1 m. The lower belt is curvilinear, with 4.5 m height in the middle and 9.6 m height over the central beams. With a central span of 261 m, this bridge held a word record at the time it was built (Figure 16.19).

The bridge was constructed by several serbian companies in cooperation with the German firm MAN. Due to the experience gained through the building procedure, a large number of orthotropic deck bridges were constructed in Serbia from 1953 to 1956. This bridge was reconstructed between 1973 and 1978, and three new traffic lanes were added with a total width of 10.5 m. For the new construction, the central beams of the existing bridge were used. The bridge has longitudinal spans of 81.5 m + 161 m + 81.5 m. The endmost supports are anchored by ties into massive reinforced concrete blocks. The construction is mounted 3.5 m away, transversally pulled across, and placed into the designed position (Buđevac et al. 1999).

A railway bridge was built over the river Ribnica on the Beograd–Bar railroad (Figure 16.20). The construction consists of six apertures with a total length of 168 m (36.5 m + 2 × 37 m + 50.5 m + 60 m + 50 m; a combination of three apertures, continuous girder, and three apertures of the composite construction).

The road bridge over the Sava River at Ostruznica near Belgrade lay on the Dobanovci–Bubanj Potok highway section. The steel structure over the river is a continuous box girder of an orthotropic deck system, 13.6 m wide, with spans of $L = 101.45$ m + 198 m + 2 × 99 m + 90.45 m = 587.9 m (Figure 16.21). The approach structures are prefabricated beams made continuous on the site; the left is 593.35 m long

FIGURE 16.19 The road bridge over the river Sava on Brankova Street, Belgrade. (Buđevac et al. 1999.)

FIGURE 16.20 Railway bridge over the river Ribnica. (CIP 2010; Mihajlovic 2010.)

FIGURE 16.21 View of bridge structure over the river Sava near Ostruznica. (Buđevac et al. 1999.)

with 16 openings 35.2 m to 40 m and the right is 608.35 m long with openings 30.25 m to 30.4 m. The total bridge length is 1789 m. The foundation is supported on bored HW piles of 1500 mm diameter. The bridge was designed by S. Cvetkovic and D. Dragojevic of Mostogradnja, who is also the contractor.

16.4.2 Roadway Bridge over the Danube in Novi Sad–Varadin Rainbow

At the location where the Varadin truss bridge across the Danube in Novi Sad was destroyed in 1999, a new steel bridge was erected (Figure 16.22). It was designed as a city bridge for roadway, pedestrian, and bicycle traffic. Its total length is 357.2 m and the width is 14.2 m (7 m + 2 × 3.6 m). Its main construction consists of a steel orthotropic continuous girder of three spans 87 m + 130 m + 87 m (Figures 16.22 and 16.23). The approach span on the left bank is a simply supported composite beam span of 35.6 m, and on the right bank is a simply supported prestressed concrete beam span of 14.6 m. The bridge was designed to be able to adopt tramway traffic as well (Nenedic, Djukic, and Ladjinovic 2001). The two piers in the river bed were heightened 4.5 m to secure the navigation height of 9.5 m. The main girder is made of two boxes with a trapezoidal cross section with inclined ribs (Figure 16.24). The height of the vertical sheet metal of the steel deck is 2.9 mm, which together with the upper and lower layers and the asphalt (60 mm) yields a total height in the middle of the span of 2997 mm (Figure 16.24). The orthotropic deck on the roadway and the footway is 12 mm and 15 mm thick, respectively.

A detailed static and dynamic analysis was conducted in order to construct the bridge. The orthotropic deck was treated as the grid of the longitudinal and transverse girders, together with the deck, which lean over the ribs of the box. Based on the influential lines, extreme influences were obtained for the proper position of a heavy vehicle. Results of the planar and spatial model were in good agreement. The structure was analyzed regarding the influence of the wind and earthquake. The analysis was performed on a 3D model with lumped mass, with the natural undamned vibrations that were previously identified. Influences of earthquake were identified by multimodal spectral analysis and time history analysis (from the Montenegro Earthquake of 1979). Geometrical characteristics are shown in Figures 16.23 through 16.25.

The superstructure was assembled in three parts (111.3 m + 81.4 m + 111.3 m) on a plateau on the right bank and erected in three parts using the floating-in method. The lengths were identified according to the position of zero points in the three-span continuous girder's central field (Figure 16.23). Through special runways on the bank and partially over the river, these assemblies were transversely

FIGURE 16.22   Varadin Rainbow Bridge in Novi Sad. (CIP 2010; Mihajlovic 2010.)

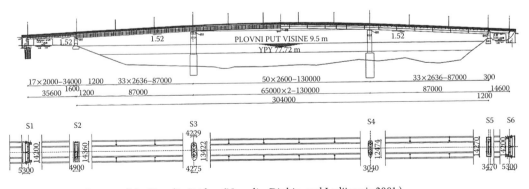

FIGURE 16.23   Layout of the Varadin Bridge. (Nenedic, Djukic, and Ladjinovic 2001.)

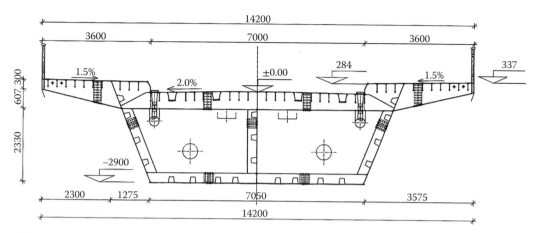

FIGURE 16.24 Main bridge structure. (Nenedic, Djukic, and Ladjinovic 2001.)

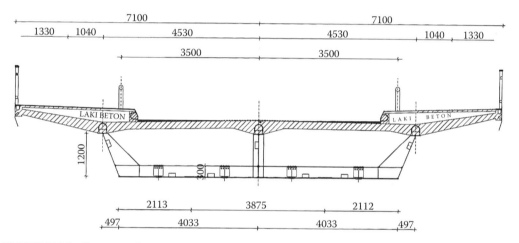

FIGURE 16.25 The approaching composite structure. (Nenedic, Djukic, and Ladjinovic 2001.)

pulled over the vessels (barges). On the barges, steel yokes were set up with hydraulic jacks on their tip, which enabled lifting the parts from the runway and placing them onto the barges. Thus, the assemblies (parts) were transported by vessels to the river piers, lifted to the appropriate height, and laterally moved into the designed position. The side parts were mounted first (beams with cantilever), and the central piece in the end, followed by the system's interconnection and placing the superstructure on the beds (Nenedic, Djukic, and Ladjinovic 2001). Quantities of 2.028 tons of steel, 2.3 tons of prestressing cables, 40.6 tons of reinforcement, and 316 m³ of concrete were used.

The bridge was load tested with approximately 77% of its design load, according to the standard U.M1.046. The bridge satisfied all the requirements for load-carrying capacity, as well as the criteria for pedestrian traffic. Work on the design began in March 2000, and it was completed on October 13 of that year. The bridge was designed by Institute CIP (G. Nenadić and Lj. Đukić); the contractor was Mostogradnja; and the testing institute was K. Savić of Belgrade.

16.4.3 Other Steel Girder Bridges

A railway bridge was built near Bistrica, over the accumulation lake of the hydroelectric power station Potpeć, on the Beograd–Bar railroad. Its structure consists of five steel frame constructions with a total length of 215 m (Figure 16.26).

FIGURE 16.26   Railway bridge over the accumulation lake near Bistrica. (CIP 2010; Mihajlovic 2010.)

The Gazela Bridge over the Sava River in Belgrade is a combination of beam and arch (Figure 16.27). It was built from 1966 to 1970 and represents a unique solution in its conception and aesthetics. The static system of the bridge, the shallow frame, is 332 m long. The bridge "jumps over the river" in a leap, and thus became associated with the graceful gazelle and got its name. The beam is supported at a distance of 63 m from upright towards the center by the hinge onto the inclined piers, which are positioned at a 290-degree angle with the horizontal line. The bridge follows a highway profile 2 × 10.5 m, with two pedestrian lanes 3 m wide, for a total width of 27.5 m. The arrangement of the spans is 41.65 m + 249.92 m + 40.3 m and two simple beams on a flood span of 66.8 m. The superstructure is an orthotropic box girder with a maximum height at middle span of 21.8 m above water. The total weight of the steel structure is 6050 tons (Buđevac and Stipanic 2007).

The middle part of the structure was erected by the system of simple mounting with the help of a floating crane, whereas for the shore inclined piers a steel scaffold and truck cranes with capacities of 60 tons and jibs 30 m long were used. The assembly elements are 15 m to 25 m long and weigh 6 to 5 tons. The inclined columns are supported on the caisson of the outer dimensions 40.5 m × 24.6 m, which can withstand great horizontal forces and moments. The caissons were filled with 9,200 m^3 of reinforced concrete, 13,000 m^3 concrete as filling, and 720 tons of reinforcement. The massive piers at the place of crossing from steel to concrete structure, were founded on Franki piles and designed by M. Djurić. The contractor was Mostogradnja.

The bridge over the Danube at Bezdan (Figure 16.28) has a total length of 683.2 m and a maximum span of 169.6 m above the river surface. Both approach structures are of prestressed concrete; the left is 2 × 30.6 m and the right is 3 × 30.6 m. The bridge width is 11.2 m. The foundation was built on φ 150 cm bored HW piles. The pile caps were done in the open pit behind a sheet piling of steel Larsen pile. The bridge was designed by B. Tripalo and D. Certic of Mostogradnja, who was also the contractor.

The highway bridge Bogojevo–Erdut has two lines, a roadway width of 7.5m, and a total width of 12 m. The main beam is a continuous orthotropic deck plate girder of partly open, partly box section. V-shaped transverse bracings are spaced at a distance of 6 m. The foundation structures are two abutments and six piers of reinforced concrete, faced by stone on steel, caissons, and bored HW-type piles, 120 cm in diameter (Figure 16.29). The bridge, like the other bridges, was static and dynamic loads were tested with tracks of 47 tons each, with measurement displacement, strains, deflection, tangent angles, natural frequency, and amplitude of vibrations. It opened to the traffic in 1980.

FIGURE 16.27 Gazela Bridge over the river Sava in Belgrade. (Buđevac et al. 1999.)

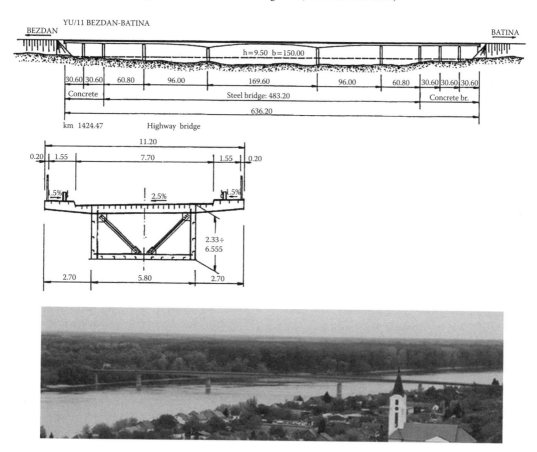

FIGURE 16.28 Longitudinal and cross section of the Bezdan–Batina Bridge. (Ivanyi and Stipanic 1998.)

An original bridge over the Drina River at Loznica (designed by G. Nenadic and H. Eric) has a total length of 286 m (48.5 m + 3 × 63 m + 48.5 m) and a width of 1.75 m + 8 m + 1.75 m = 11.5 m. It was erected from 1973–1975. The combined action of different levels of bearings and cable prestressing keeps the bridge deck in a permanent compressive state. The foundations lie on HW piles, Ø 150 cm (Darijevic, Nenadic, and Certic 1999; Perisic 1999).

YU/10 BOGOJEVO-ERDUT

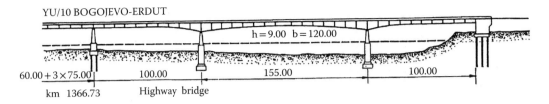

60.00 + 3 × 75.00   100.00   155.00   100.00

km 1366.73   Highway bridge

FIGURE 16.29 Longitudinal section of the Bogojevo–Erdut road bridge. (Ivanyi and Stipanic 1998.)

The bridge over the Danube at Backa Palanka has a total length of 725 m and a maximum span of 160 m above the river surface (Figure 16.30). The bridge is divided into two sections that visually form a whole. The section of six spans (60 m) on the left bank is a simply supported composite beam, and the section with three main river openings, spans 100 m + 160 m + 100 m, is a continuous beam. The bridge superstructure is a box section 5940 mm wide. The box depth of the composite girder, including the pre-stressed concrete deck, is constant, while the depth of the continuous girders of the orthotropic deck in the side spans is increased from 3000 mm to 4500 mm. In the central span the box has constant height of 4500 mm. The bridge is 725 m long and 10.5 m wide. All 10 piers are founded on bored (HW) piles, 150 cm in diameter, 12 m to 15.5 m long. The caps on majority of them are expanded to 250 cm and are made against sheeting Larsen piles. The bridge was designed by B. Tripalo, Đ. Cupurdija, and M. Rančić, and the contractor and construction company was Mostogradnja.

A bridge was built over the Danube at Smederevo (designed by B. Tripalo and D. Certic) with a total length of 1231 m and the maximum span of 253.7 m above the river surface. The bridge is designed as an upstream carriageway of a potential highway (8.5 m wide with stop lanes and footways 80 cm wide on other side). The river bridge structure is steel and the approaching structures are reinforced concrete. The central structure spans are 108.8 m + 171.2 m + 108.8 m = 388.8 m. The right-hand structure is constant in height with four 97.5 m spans, while the left structure has five spans, 89.6 m each. The foundations are placed on HW piles with caps.

The steel bridge over the Sava in Sabac has a total length of 628 m with the biggest span of 160 m above the river surface. The bridge is 12.4 m wide. The foundations lie on bored HW piles with pile caps fabricated in an open pit using steel Larsen sheeting piles. It was designed by D. Simic, I. Stojadinovic, and V. Njagulj of Mostogradnja, who was also the contractor. A bridge was also built over the Sava River at Sremska Mitrovica (designed by D. Simic and N. Kokanovic) with a total length of 739.3 m and the biggest span of 180 m above the river surface.

The Smederevo–Kovin highway bridge over the Danube is designed as an upstream carriageway of a potential highway (Figure 16.31). The approach structures are reinforced concrete, and the river bridge part is steel. The steel parts of the bridge are three continuous beams of variable heights on 13 piers. The central spans are 108.8 m + 171.2 m + 108.8 m = 388.8 m. The flood bridges consist of the following: left, 5 × 89.6 m (steel) + 3 × 32 m (concrete); right, 4 × 97.6 m (steel) + 3 × 32 m (concrete). The roadway width is 8.5 m, and total width is 12 m. The bridge was built in 1976. It was designed by B. Tripalo and D. Certic of Mostogradnja, who was also the contractor.

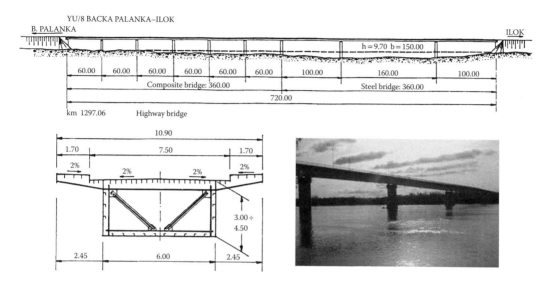

FIGURE 16.30 Longitudinal and cross section of the Backa Palanka–Ilok (Croatia) Bridge. (Ivanyi and Stipanic 1998.)

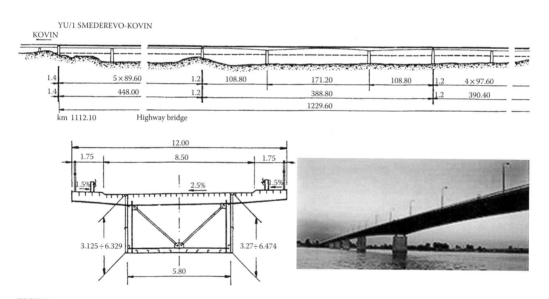

FIGURE 16.31 Longitudinal and cross section of Smederevo–Kovin Bridge. (Ivanyi and Stipanic 1998.)

16.5 Truss Bridges

The Cerovo railway bridge was built over the river Southern Morava in the vicinity of the villages Supovac and Cerovo (Figure 16.32) in 1989, on the second track of the Beograd–Niš railway, with a total length of 156 m (3 × 52 m) (CIP 2010; Mihajlovic 2010).

The Krivelja railway bridge was erected on Mala Krsna–Bor–Rasputnica railroad (Figure 16.33) in 1969. Its superstructure is continuous steel trusses on seven piers with total length of 435.6 m (52.8 m + 5 × 66 m + 528 m) (CIP 2010; Mihajlovic 2010).

FIGURE 16.32   Cerovo Bridge over the Southern Morava River. (CIP 2010; Mihajlovic 2010.)

FIGURE 16.33   Krivelja Bridge. (CIP 2010; Mihajlovic 2010.)

16.5.1 Pancevo Bridge

A roadway and railway bridge was erected on the Beograd–Pancevo railway line. Its main span is a continuous steel truss with a total length of 809 m, with spans of 161.3 m + 3 × 162.144 m + 161.3 m. Its approach spans consist of eight spans of girders of 32 m = 256 m. The total width of the bridge is 29.94 m. In the flood (approach) spans, the roadway traffic is directed through composite constructions with spans of $L = 32$ m (Pancevo side). The truss height is $H = 18$ m.

The bridge was intended for railway traffic with two tracks. The roadway traffic unfolds on two tracks in both directions that were placed onto composite cantilever on both sides of the bridge (upstream and downstream). The approach spans to the railroad portion are prestressed continuous concrete girders. The roadway track on both sides is 7 m wide, and the footways 1.5 m each (Figures 16.34 and 16.35). The composite superstructure deck was prestressed in order to avoid the tension in the concrete deck (Belgrade side). The railway of the bridge was built in 1961 and in 1964 roadway was finished.

FIGURE 16.34 Pancevo Bridge. (CIP 2010; Mihajlovic 2010.)

FIGURE 16.35 Another view of Pancevo Bridge. (CIP 2010; Mihajlovic 2010.)

16.5.2 Other Truss Bridges

The Sremska Rača Bridge is a 400 m long and 10.9 m wide roadway bridge over the river Sava that was constructed with the aim to separate the railway and roadway traffic. Its superstructure consists of trusses, continuous orthotropic steel plate girders with spans of 125 m + 150 m + 125 m (Figure 16.36).

The single track Vrelo railway bridge was erected on the Beograd–Bar railroad. The construction consists of six apertures of composite constructions with a total length of 232.5 m (2 × 36 m + 3 × 41.5 m + 36 m). The Uvac railway bridge was erected over the river Uvac on the Beograd–Bar railroad (Figure 16.37). Its construction is continuous truss steel beam with five apertures with a total length of 303.6 m (52.8 m + 3 × 66 m + 52.8 m).

FIGURE 16.36 Sremska Rača Bridge. (CIP 2010; Mihajlovic 2010.)

FIGURE 16.37 Railway bridge over the river Uvac. (CIP 2010; Mihajlovic 2010.)

YU/9 BOGOJEVO-ERDUT

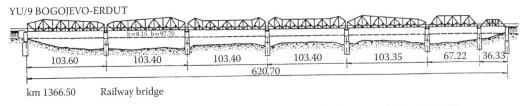

km 1366.50   Railway bridge

FIGURE 16.38 Longitudinal section of Bogojevo-Erdut railway bridge. (Ivanyi and Stipanic 1998.)

The railway bridge Bogojevo–Erdut (Figure 16.38) was built of seven single-span trusses. The first bridge was built in 1911, containing six single-span trusses on two abutments and five piers. It was destroyed in 1941. The second structure, which is still in operation, is a combined highway and railway bridge, built on the original piers and abutments. One side span was divided into two, thus, a new pier was built there. Now the bridge carries only a railway line, as a new highway bridge was built nearby. There are seven spans with a total length of 620.7 m. Load testing under static and dynamic loads for heavy railway traffic was performed in 1974.

16.6 Arch Bridges

16.6.1 Concrete Arch Bridges

A railway bridge was built over the river Djetinja on the Beograd–Bar railroad. This is a continuous reinforced concrete bridge with a 56 m span over the arch, and 108.5 m total length (Figure 16.39). A railway viaduct was also erected on the Beograd–Bar railroad (Figure 16.40). The construction consists of seven restrained reinforced concrete arches with a total length of 126.4 m (18.7 m + 5 × 17.8 m + 18.7 m).

16.6.1.1 Road Bridge over the River Tisza at Titel

A road bridge made of prestressed and reinforced concrete crosses over the river Tisza at Titel and is placed alongside the railway steel bridge built on the existing piers of the road–rail bridge destroyed in World War II. Both bridges have similar geometric characteristics and the same static system. The main bridge structure above the river is a three-span continuous prestressed girder 50 m + 154 m + 50 m, strengthened in the central span by a slender reinforced concrete arch above the bridge deck, known as Langer's girder. The approach structures on the right and left banks are also made of prestressed concrete, 2 × 35 m in each span. The longitudinal and cross sections are shown in Figure 16.41.

FIGURE 16.39 Railway bridge over the river Djetinja. (CIP 2010; Mihajlovic 2010.)

FIGURE 16.40 Railway viaduct on the Belgrade–Bar railway. (CIP 2010; Mihajlovic 2010.)

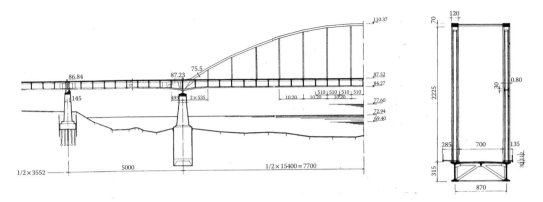

FIGURE 16.41 Longitudinal and cross sections of the road bridge over the river Tisza. (Trojanovic 1968.)

In order to compensate the effects of concrete shrinkage and creep by lightening and simplifying the scaffolds and reducing the quantity of shuttering, the bridge structure was constructed in several stages before the final system was established. The prestressing cables were guided along the outer sides of the main girder web plates and dragged with special devices upwards or downwards to cater to the changes in the static system in phases until the system was established.

The original concept for the design and construction of this bridge was one of the supreme achievements in bridge engineering in the world at the time of its construction. With its 145 m long span, this bridge holds the span record of its time for this type of structure. The designer was B. Žeželj, and the contractor was Pionir Company, Belgrade (Trojanovic 1974).

16.6.1.2 Žeželj Roadway-Railway Bridge over the Danube in Novi Sad

This bridge, called the Žeželj Bridge, was constructed in 1961. It is a bridge of excellent design parameters, with a complex foundation and method of construction. The bridge is 466 m long and consists of two arches (Figure 16.42) and two inundation structures. The spans of arches are 211 m and 165.75 m, both with a 1/6.5 ratio of rise to span of arch. The inundation structure has a box cross section. The total width of the roadway is 20.15 m. The distance between arches is 15.05 m. The railway profile is 4.4 m, and the roadway profile is 9 m.

The arches have a box cross section. The roadway, by means of hangers in the greater portion, is hung on arches, while a minor portion is supported by means of rectangular reinforced concrete columns on arches. The arches are connected by truss bracing, which is made of precast prestressed elements of a T cross section. All structural elements were prestressed by IMS system cables of 6 φ 7 mm. Two inundation and one river reinforced concrete pier are founded by caissons. The left abutment pier is founded at dimension rate 20 × 25 m. The middle pier receives asymmetrical compression, so that it is placed asymmetrically in relation to the pier's axis. The foundation of the end pier is a shallow foundation on fine sand.

The reinforced concrete caissons for the abutment piers were cast in situ and sunk to the designed levels. In order to safely take up the arch thrust and prevent horizontal moving of the arch support, movable reinforced concrete walls were foreseen behind the abutments. Between the backsides of the abutments and the masonry there were hydraulic jacks to activate soil resistance (Trojanovic 1968).

The scaffolds in the bigger navigable opening had a few supports because of the depth and velocity of the river Danube, ice occurrence, and an obligation to provide a 100 m navigation clearance during construction. The clearance was provided by means of a concrete arch scaffold, 108 m in span, formed of prefabricated elements and freely erected starting from the supports and working towards the crown. The side sections in the bigger opening have steel truss strutting structures. At that time, the applied designs and technology were new and bold and enriched the national and world practice in bridge engineering.

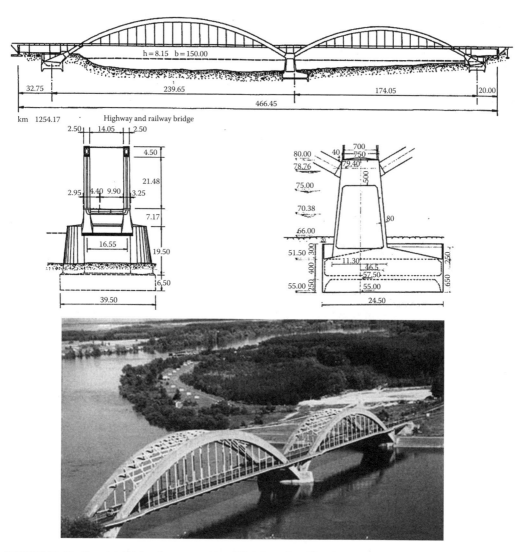

FIGURE 16.42 Longitudinal and cross sections of Žeželj Bridge. (Žeželj and Petrovic 1992.)

16.6.1.3 Reconstruction of Žeželj Bridge

From the beginning of its service the structure was monitored by undertaking necessary interventions. In 1965 there was a serious settlement (to some 30 cm) with longitudinal displacement of the arch top (some 11 cm) of the inundation pier at the right side. This was a consequence of a high river level, which initiated sand dislocation at the right pier and caused the settlement. Protection of this pier from further erosion was made by means of a cofferdam made of stone fill and bags full of sand placed upstream from the pier. This deformation in the foundation reflected upon the structure as well, such that a set of cracks occurred on the wall panels and reinforced concrete slabs over caisson of the right pier, which were later repaired.

The monitoring of the bridge structure's behavior was organized along with necessary registration of phenomena that could have influenced the stability and durability of the structure. From a survey carried out over a long period until 1989, it was decided that the settlement of the abutment piers and the size of their spreading were in allowable limits that provided stability and safety for the bridge.

16.6.2 Steel Arch Bridges

The railway bridge made of steel over the river Tisza in Titel was built in 1959 (Figure 16.43). This bridge has similar geometric characteristics and the same static system as the road bridge placed on the same piers. The bridge has a three-span continuous beam of $L = 50$ m + 154 m + 50 m with the central span strengthened with an arch (Langer's system) and a width of $B = 5$ m. The end of the arch is eccentrically connected with the beam. The height of the main girder is 3200 mm and slightly increases at the central supports. The arch rib height is $h = 600$ mm while the rise is 24.2 m. The arches are connected with rhombic bracing. The bridge was designed by M. Djuric and Metalna Company, Maribor, and the contractor was Pionir Belgrade.

The road bridge over the Gazivode water storage is a two-hinged truss arch, 195 m in span with a carriageway 7.5 m wide. With the approach structures on the left bank (6.8 m + 2 × 15 m + 1.6 m = 38.4 m) and right bank (1.6 m + 18.9 m + 2 × 20.3 m + 19.75 m + 9.4 m = 90.25 m) the total length of the bridge is 323.65 m (Figure 16.44). This bridge, with its main span of 195 m, is the biggest steel arch bridge in Serbia. The total steel weight is 832 tons. The author was G. Nenadic, and the designers were S. Cvetkovic for the steel and V. Nagulj for the concrete structure (Buđevac and Stipanic 2007).

FIGURE 16.43   Two bridges over the Tisa near Titel (left: steel; right: prestressed reinforced concrete). (Trojanović 1968.)

FIGURE 16.44   Gazivode Bridge over the Gazivode water storage. (Buđevac et al. 1999.)

16.7 Cable-Stayed Bridges

16.7.1 The Pedestrian Bridge of St. Irinej

This pedestrian bridge (authors and designers G. Sreckovic and D. Isailovic) crosses the river Sava to connect the towns of Macvanska Mitrovica and Sremska Mitrovica, and attracts attention with its simplicity and elegance (Figure 16.45). The bridge consists of an approach structure of a simple beam system of 35 m span and the main bridge with an inclined (stayed) cable of span 35 m + 192.5 m + 35 m. The typical bridge width is 5.5 m and increases to 6.5 m in the pylon zones since the pylons are positioned in the midplane of bridge. The stiffening beam consists of a steel trapezoidal box girder that is composite with the concrete deck slab. The bridge pylon is a solid reinforced concrete square section, fixed into the pylon piers. The total pylon height from the fixing point to the cap top is 38.9 m. They protrude through the stiffening beam and are completely separated from it. It is a daring bridge design. The aspect coefficient of the bridge is 35.

16.7.2 Railway Bridge across the River Sava in Belgrade

This cable-stayed railway bridge (designed by N. Hajdin and Lj. Jevtovic) was designed and built within the framework of a new conception of the railway complex, so-called Prokop (Figure 16.46). The entire bridge consists of three parts: approach bridges on the left- and right-hand side of the river, made up of continuous presteressed concrete girders with spans of 45 m to 75.5 m and reinforced concrete with spans of 35 m to 38.5 m and a central cable-stayed steel structure of 557.94 m total length, with a main river span of 253.7 m. A navigable profile including a possible bridge pier with two spans of 100 × 7.5 m was designed.

The bridge was built from 1975 to 1979. The pylons are placed in the main girder centers above the columns and exceed the main girders by 52.49 m. The pylon width is constant at 1.94 m, with the second dimension decreasing at a 5% rate. The connection of pylons with the main girders is stiff. They are intercrossly connected between the anchored places of the upper and lower clusters of the steel ropes. The pylon column was supported on reinforced concrete caissons and of all other columns on HW-type piles, 1500 mm in diameter, with length about 17 m. The main girders are two orthotropic box girders of constant depth and width. This railway bridge system was the second in the world of this type, and the first to carry two railway tracks. Constructed with the continuous girder system, the bridge was designed with two tacks with two pairs of inclined ties on each end of the bridge, with 52.49 m high portal-like pylons fixed into the main girders.

Two main box girders are interconnected by an orthotropic plate. The orthotropic deck plate is sunk 90 cm from the main girder upper belt, so that a canal is formed for the ballast on which the double railway track is laid. Service footways are situated outside the main girders. The main girders are under the

FIGURE 16.45 The pedestrian bridge of St. Irinej. (Sreckovic 2009.)

FIGURE 16.46   The railway cable-stayed bridge over the Sava River in Belgrade. (Buđevac and Stipanic 2007.)

footways. Under the footway floors, which can be lifted, there are electric cable ducts. Sewage pipes are placed under the orthotropic plate and are served by one pathway along the entire bridge. The pathways for an inspection cart with telescopic pulling of the lateral parts are positioned near the lower belt, on the interior side of the main girders. The main contractor was Mostogradnja.

16.7.3 Most Slobode (Bridge of Freedom)

Completed in 1981, this is the only bridge in former Yugoslavia with a full highway profile. Aesthetically, the bridge excellently fits into the environment—the river, its banks, and the city. The modern bridge construction—a bridge with angular cables of 351 m main span—was a record span of the time for this type of bridge. The downstream snapshot in Figure 16.47 illustrates three bridges in Novi Sad with different structural systems. Bridge of Freedom is the last one in the figure.

The road bridge over the Danube in Novi Sad (designed by N. Hajdin and G. Nenadic) is similar to the Bridge of Freedom. The total bridge length is 1312 m. The cable-stayed spans are 591 m long (2 × 60 m + 351 m + 2 × 60 m), which was the world record for bridges of this type at the time the bridge was completed. The towers and cable stays are located in the intermediate bridge plane (Figure 16.48). The stiffening beam is a trapezoidal box. The bridge is 27.68 m wide.

The orthotropic box girder ensures necessary stiffness in torsion in the event of asymmetric loads. The bridge towers, also a box cross section, rectangular at the bridge axis, are fixed into the main beam and are 58.7 m high. The 351 m long main span is taken up with three groups of cables consisting of parallel wires that start at the pylons and take up the main girder at 54 m + 48 m + 8 m, spacing symmetrically on either side. The cables are anchored into the towers at 34.7 m, 44.7 m, and 54.7 m levels. Each group contains four cables forming a stay, the mutual distances of cables being 760 mm in both directions.

Though Novi Sad is nearly 400 km away from Kosovo, these three bridges (Figure 16.47) were destroyed in the NATO attack on Serbia. Varadin Bridge was destroyed on April 1, 1999, which resulted not only in interruption of the traffic between the center of Novi Sad and Petrovaradin, but also in complete obstruction of traffic stream on the Danube. Demolition, cleaning, and reconstruction of Sloboda Bridge was presented at the 2004 International Conference on Bridges across the Danube in Novi Sad. A part of bridge structure (main structure) is shown in Figure 16.48. The undamaged part, partially damaged part, and parts of the bridge requiring replacement are presented in Figures 16.49 and 16.50.

16.7.4 The New Roadway Bridge across the Sava in Belgrade

A new cable-stayed roadway bridge will pass over the lower tip of Ada Ciganlija Island and will carry six traffic lanes and a double-track rail line, as well as two lanes for the use of pedestrians and cyclists (Figure 16.51). The route from New Belgrade overpasses, by approaches, the shipyard and comes to the main bridge; it crosses the "winter boat storage" bay (~130 m), left bank area (~170 m), the Sava River

FIGURE 16.47 View of Novi Sad bridges before the NATO air strike in 1999. (Folić, Radonjanin, and Malesev 2001b.)

YU/7 NOVI SAD, "SLOBODA" BRIDGE

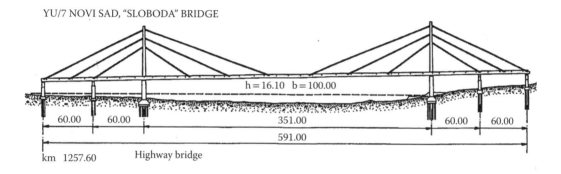

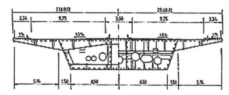

FIGURE 16.48 Elevation and cross section of Sloboda Bridge. (Ivanyi and Stipanic 1998.)

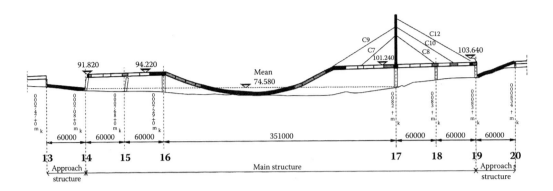

Legend: white - undamaged parts of the bridge
        gray - partially damaged parts of the bridge
        black - parts of the bridge requiring replacement

FIGURE 16.49 Bridge structure after bombing. (Hajdin and Stipanic 2004.)

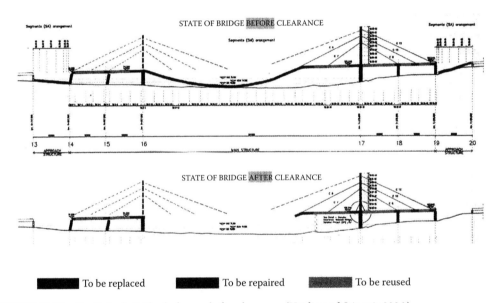

FIGURE 16.50 The Sloboda Bridge before and after clearance. (Hajdin and Stipanic 2006.)

(~350 m), the lower tip of Ada Ciganlija Island (~50 m), and Cukaricki Bay (~180 m) (Figures 16.51 and 16.52). The bridge and associated roads will form the first section of the Inner City Semi-Ring Road. The conceptual design was done by Ponting Maribor, DDC Ljubljana, and CPV Novi Sad (designer Viktor Markelj). A cross section of the bridge in span according to the preliminary design is shown in Figure 16.53.

The Sava Bridge is a six-span continuous superstructure with an overall length of 929 m between the deck expansion joints (Hajdin and Stipanic 2007). The main support system is a single pylon asymmetric cable-stayed structure, with a main span of ~376 m and a back span of 200 m. The side spans (~69 m + 108 m + 2 × 81 m) are continuous, with the main bridge section having a cable-stayed structure. The pylon is 200 m high. The piers are founded on sets of bored piles.

FIGURE 16.51 Photo visualization of the bridge across the Sava River at lower tip of Ada Ciganlija Island. (Hajdin and Stipanic 2007.)

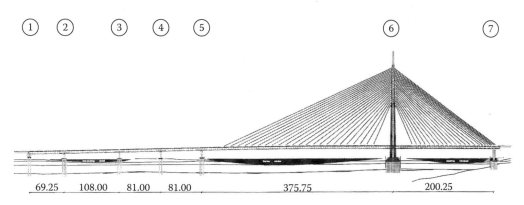

FIGURE 16.52 The bridge across the Sava River at the lower tip of Ada Ciganlija Island. Numbers are showing axes (no. 6 is an axe of the pylon, and other numbers are showing pier axes). (Hajdin and Stipanic 2007.)

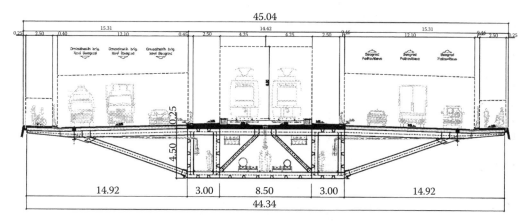

FIGURE 16.53 Cross section of the bridge in span according to the preliminary design. (Hajdin and Stipanic 2007.)

Construction of the bridge was extensive in 2010 and is illustrated in Figures 16.54 through 16.59, taken in November 2010. The concrete back span is 200 m long with a self-weight (g1) of around 100 tons/m. It was constructed using the push launching technique. The 54 m × 45 m concrete launching platform was made at the right river bank to enable the production of 18 m long segments which were then pushed over the set of Teflon bearings positioned on temporary piers with the 50 m distance from the permanent pier. To eliminate big cantilevers, a 25 m long steel nose was used to help the structure reach the supports, thus enabling rational design of the bridge girder.

FIGURE 16.54 Push launching of the back span. (Folić and Misulic 2010, www.savabridge.com)

FIGURE 16.55 Temporary supports for the back span. (Folić and Misulic 2010.)

The connection of the bridge structure was carried out on August 8, 2011, as shown in Figures 16.56 and 16.57, and the bridge was opened to traffic at midnight on New Year's Eve 2012. So, the traffic started on January 1, 2012. The finished bridge with south approaches is shown in Figure 16.58.

A set of four temporary piers were constructed in the bay, all on piled foundations. The piers are made of precast concrete elements which are tied together with 6 m long Dywidag tie rods. The pier heads were

FIGURE 16.56 Connection of Bridge structure. (www.savabridge.com, accessed April 2013.)

FIGURE 16.57 Bridge "closure" panorama. (www.savabridge.com, accessed April 2013.)

FIGURE 16.58 Ada bridge with the south approaches. (www.savabridge.com, accessed April 2013.)

FIGURE 16.59   **(See color insert.)** Construction of the pylon. (Folić and Misulic 2010.)

made out of two "rock" elements which enabled some movement of the launched bridge superstructure. Steel guide girders were also positioned on the sides. To make the system stable in the horizontal direction, all temporary piers were connected with a system of post-tensioned cables. The practicality of this design is to dismantle temporary structures easily after the permanent works are completed.

The main feature of this bridge is its 200 m high cone-shaped pylon. The first 175 m are made of prestressed concrete and the last 25 m are stainless steel with a brushed finish. The concrete part was constructed using a 4.5 m high DOKA self-climbing jumping formwork, which is adjusted to a new shape in every cycle. The high-strength concrete was poured by a tower crane using a 3 m³ concrete bucket.

16.8 Suspension Bridges

In the period between World War I and II, a large number of railway and road steel bridges were built over wide rivers such as the Danube, the Sava, the Drina, the Morava, and the Ibar. Some of them had main spans of 120 m to 150 m. The largest bridges of the period were Pančevo Bridge over the Danube, on Belgrade–Pančevo road, and the King Alexander red-tram suspension bridge over the river Sava in Belgrade (Figure 16.60). This suspension bridge was constructed between 1930 and 1934. It resembles the bridge over the river Rhine in Cologne (Buđevac and Stipanic 2007).

The construction of the King Alexander Bridge began in May of 1930 and was completed in 1934. The spans were 75 m + 261 m + 75 m (the suspension part), with a total length of 474 m and width of 3 m + 12 m + 3 m = 18 m. The bridge was destroyed in April 1941, at the beginning of World War II. In addition, many bridges (mainly made of steel) were destroyed by the withdrawing German army. Reconstruction of bridges began in 1945 in order to meet postwar traffic requirements. Initially, using the material from the previously destroyed bridges, provisional bridges were constructed in order to secure railway traffic. The two new bridges over the rivers Ibar and Ribnica near Kraljevo were the first composite bridges in Serbia.

The bridge over the river Tisza between Kanjiza and Knezevac is shown in the Figure 16.61. (designed by Lj. Jevtovic, M. Joksimovic, and M. Hiba). The bridge is a classic suspension bridge with a 154.5 m long central span for road traffic (Figures 16.61 and 16.62). The carriageway is 7 m wide. A stiffening beam (height of 1.2 m) was hanged on steel rope with ties 10.3 m distance. Ropes are 4 φ 87mm under each main girder and translate over the pylons and anchorage in special blocks at a distance of 67 m. The reinforced concrete composite deck was connected with the vertical sheet metal of the main girder.

FIGURE 16.60 King Alexander Bridge. (Buđevac and Stipanic 2007.)

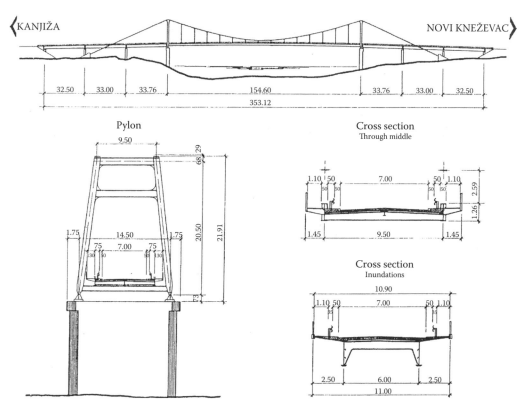

FIGURE 16.61 Suspension bridge over the river Tisza between Kanjiza and Knezevac. (Buđevac and Stipanic 2007.)

A suspension bridge for transmitting pipeline over the Danube in Smederevo, with a length of $L = 859.74$ m + 480.78 m + 462 m = 182.52 m was built in 1972, to carry a pipeline (four pipes φ 219.1 mm) across the Danube river (Figures 16.63 and 16.64). Ada, an island in the Dunavac river branch, is part of the space structure. The suspension structure over the Danube has a span of $L = 193.02$ m + 479.7 m + 193.02 m = 859.74 m. The main support cable (2 φ 81mm) is parabolic, rising 48 m. The pylons are

FIGURE 16.62 Another view of the bridge over the river Tisza between Kanjiza and Knezevac.

FIGURE 16.63 Suspension bridge over the Danube in Smederevo. (Buđevac et al. 1999.)

YU/2 SMEDEREVO

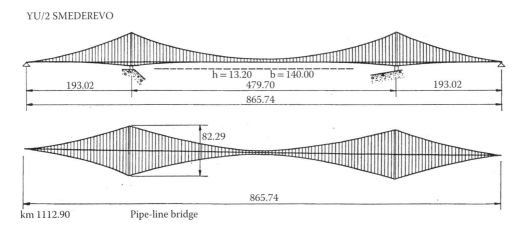

FIGURE 16.64 Layout of suspension bridge transmitting pipeline. (Ivanyi and Stipanic 1998.)

50 m high. The pipes lie in frames suspended on hangers and the strand. The bridge was designed by M. Milosavljevic and Z. Pavlovic, and the contractor was Monting Zagreb.

One of the more valuable pedestrian suspension bridges over the river Drava in Osijek (Croatia) was designed by B. Tripalo and was built by Mostogradnja. It is a hanging concrete slab system without stiffening beams but with curved prestressed tendons underneath the slab. Aesthetically, it is a very successful achievement.

16.9 Destruction and Damage in NATO Bombardment and Renewal of Bridges

16.9.1 Destruction and Damage

Civil wars in former Yugoslavia led to its disintegration and the establishment of five new countries. The war ended by the bombardment of the Federal Republic of Yugoslavia over the period from March 24 to June 10, 1999 by allied NATO forces. Events of this kind are of interest in almost all countries of the world. During 78 days of bombardment NATO destroyed or partly damaged numerous and various structures, including 62 bridges: 45 road bridges, 16 railway bridges and 1 road-rail bridge were completely or partly destroyed.

Renewal of structures that were destroyed or damaged during the war is of interest in selected expert circles to deal with planning, design, and construction. The volume and type of damage to structures during the air raids as well as their efficient and quick renewal are of interest to the expert public. Table 16.2 presents a survey of the damage and renewal of bridges. The quick and efficient reconstruction of damaged bridges in extremely difficult circumstances resulted from exceptional tradition, organization, and knowledge of modern construction technologies and also the vitality of building companies in former Yugoslavia.

The destruction of the Varadin Bridge resulted in the immediate halt of the traffic on the Danube, to the detriment of numerous European countries, particularly in the Danube Basin region. It should be noted that European shipping companies suffered considerable financial loss. The damage inflicted was not only economic in nature, but to a large extent ecological as well (Folić, Radonjanin, and Malesev 2000).

There are a large number of bridges designed and constructed by engineers and companies from Serbia. Very important bridges were built after World War I and particularly after World War II, such as:

- Bridge over the river Tara, span 116 m (1938–41)
- King Alexander Bridge in Belgrade, spans 78 m + 261 m + 78 m (1932)
- New Pančevo Bridge over the river Sava in Belgrade, spans 5 × 162 m
- Gazelle Bridge over the Sava in Belgrade, span 250 m
- Bridge over the river Tisa near Titel, Langer's system with middle span 154 m
- Bridge over the Danube near Bezdan; the largest span is 169.6 m long
- Bridge over the Sava near Sremska Mitrovica; the largest span is 180 m long
- Cable-stayed railway bridge over the Sava in Belgrade, main span 253.7 m, the first of its kind in the world
- Railway and road concrete Žeželj Bridge made by the arch system over the Danube in Novi Sad, arch spans 211 m and 165.75 m
- Freedom Bridge, made of steel with oblique ties in Novi Sad, main span 351 m
- Bridge over the Danube near Beška, total length 2250 m with the main structure spans 15 m + 105 m + 210 m + 105 m + 15 m
- Concrete bridge near Sibenik, span 246 m; now in Croatia
- Mainland Bridge in Krk, arches with spans 390 m and 244 m; now in Croatia

TABLE 16.2 Survey of Damaged Bridges

No.	Type of Bridge	Bridge Name	Date of Damage (1999)
1	Road	Steel bridge Varadin over the Danube on road M-22, first section Novi Sad–Petrovaradin	April 1
2	Road	Concrete bridge Beška over the Danube on road M-22, section Novi Sad–Belgrade	April 1, April 29
3	Road and railway	Concrete bridge Žeželj over the Danube on road M-22, first section Novi Sad–Petrovaradin	April 25, May 21
4	Road	Steel bridge Freedom over the Danube on road M-22, first section Novi Sad–Sremska Kamenica	April 3
5	Road	Steel bridge near Backa Palanka over the Danube on road M-17, section Novi Sad–Croatia	April 2, April 27
6	Road	New concrete bridge over the river Ibar in Biljanovac on road M-22, section Kraljevo–Novi Pazar	April 5
7	Road	Concrete bridge over the river Toplica near Kursumlija on road M-25, section Niš–Pristina	April 20, June 15, June 18
8	Road	Concrete bridge over the Southern Morava and over the railway line at the entrance to Grdelica Gorge, on road M-1, section Niš–Vranje	April 13, April 30
9	Road	Concrete bridge over the railway line on road M-25, section Kursumlija–Pristina	April 13, April 30
10	Road	Concrete bridge Kosanica over the river Kosanica, on road M-25, section Kursumlija–Pristina	April 13, April 18, April 29
11	Road	Steel bridge over the Western Morava near Jaska, on road P-102, section Krusevac–Kragujevac	April 16
12	Road	Steel bridge over the Danube, on road M-24, section Smederevo–Kovin	April 16
13	Road	Concrete bridge over the river Ibar at Biljanovac	April 16
14	Road	Concrete bridge over the river Bistrica on road P-115, section Priboj–Bistrica	April 20, May 2
15	Road	Steel bridge over the river Sava near Ostruznica, on the bypass around Belgrade	April 30
16	Road	Steel bridge over the Western Morava near Trstenik on road P-219	May 1
17	Road	Concrete bridge over the Southern Morava near Vranjski Priboj, on road M-1, section V. Han–Vranje	May 2
18	Road	Concrete bridge over the river Lim in Prijepolje	May 3
19	Road	Concrete bridge over the river Velika Morava near Mijatovac, on highway E-75, section Belgrade–Niš	May 9
20	Road	Concrete bridge over the river Nišava of the town of Niš on road P-124	May 9
21	Road	Concrete bridge over the railway line and cross connection with Horgos on road Subotica–Horgos	May 13
22	Road	Steel bridge over the Danube–Tisa–Danube Canal on road P-118 through Vrbas	May 18
23	Road	Steel bridge over the river Velika Morava on road P-220, section Varvarin–Cicevac	May 28
24	Road	Concrete bridge over the railway line near the village Veliko Orasje on the highway E-75, section Belgrade–Niš	May 11

Source: Vojinović, B., Folić, R. 2002. *Jubilee Sc. Conf. 60 Years University of ACEG*, Sofia, Vol. 4, pp. 35–45.

Bridges made of prestressed concrete according to the original building technologies are especially important. Some of them were once world record holders in their categories. Serbian bridge builders were among the best in the world. The merit belongs to the Belgrade construction school (DJ. Lazarevic, M. Trojanovic, M. Djuric, N. Hajdin), to structural engineer Branko Žeželj from the IMS Institute, and to the structural engineers of Mostogradnja, especially to Ilija Stojadinovic.

16.9.2 Organization and Resources Engaged in Renewal of Bridges

The Direction of Renewal of the Country, an organization for renewal of destroyed and damaged bridges, was formed on April 4, 1999. The reconstruction works on damaged bridges began on June 14, 1999, with the bridge near Beška, and this bridge was finished on July 20, 1999. The program of renewal was divided into two phases:

1. From April 1999 to November 1999, phase I covered the rehabilitation and construction of bridges on transportation systems of high priority in the Republic of Serbia. In this phase of renewal, 4 railroad bridges and 28 road bridges were completed.
2. From November 1999 to July 2000, phase II covered the rehabilitation and construction of all 58 damaged bridges except 2 bridges in Novi Sad (the Sloboda Bridge and the road-rail bridge, for which alternative solutions were built instead) (Vojinovic and Folić 2002).

As early as October 11, 1999, the following five bridges were completed on highway E-75 from Belgrade to Niš:

1. Jasenica, rehabilitation of the right concrete bridge
2. Veliko Orasje, rehabilitation of the right concrete bridge
3. Mijatovac, rehabilitation of the left concrete bridge
4. Trupalska joint, rehabilitation of the right bridge
5. River Nišava, rehabilitation of the right concrete bridge over the Nišava south of Trupale

Consequently, permanent traffic flow was practically re-established on that section of E-75 less than four months from the beginning of reconstruction.

The renewal of damaged structures was financed from real sources (budget, donations) and was carried out during the period of sanctions by the UN against the Federal Republic of Yugoslavia. At that time there was a serious deficiency of basic building materials—cement, structural steel, and so on—so the building companies were supplied with these materials by the Direction for Reconstruction of the Country (2000). Table 16.3 summarizes companies engaged in the design and execution of works on most structures. Experts from the Traffic Institute CIP, the Institute for Roads, the Kirilo Savic Institute, and the IMS Institute supervised the reconstruction.

16.9.3 Technology of Reconstruction of Bridges

Technological solutions for the rehabilitation of bridges were selected according to expert investigations that provided a basis for the design and reconstruction of each bridge. Works for both concrete and steel bridges included the following:

- Removal of parts or/and complete bearing structures
- Replacement of removed parts and complete bearing structures with new ones
- Execution of final works and traffic signalization

TABLE 16.3 Survey of Companies Engaged in Phase 1 of Bridge Reconstruction

Company Name	Design	Construction	Testing
Construction Co. Mostogradnja–Belgrade	7	11	–
Institute of Traffic CIP–Belgrade	6	–	8
Road Institute–Belgrade	10	–	–
Institute Kirilo Savić–Belgrade	–	–	8
Brvenik–Raška (construction company)	1	3	–
Goša–S. Palanka (steel structure company)	–	4	–

Source: Vojinović, B., Folić, R. 2002. *Jubilee Sc. Conf. 60 Years University of ACEG*, Sofia, Vol. 4, pp. 35–45.

Varadin Bridge in Novi Sad was reconstructed through the extension of its pillars and a total replacement of the superstructure (Figure 16.6 and 16.22).

Several damaged bridges are presented in detail in Mrkonjic and Vlajic 2009, and the paper Vojinovic and Folić 2002, along with some that were destroyed. Some of damaged bridges are shown in Figures 16.65 through 16.75.

FIGURE 16.65   Destroyed concrete bridge over the Southern Morava at the entry of Grdelicka Gorgev. (Data from Mrkonjic and Vlajic, *Bridges, Library, Reconstruction,* Book No. 3, Trect milentjum, Belgrade, 2009.)

FIGURE 16.66   Damaged steel bridge Lesak on km 163 + 644 of the Kraljevo–Kosovo Polje rail line. (Data from Mrkonjic and Vlajic, *Bridges, Library, Reconstruction,* Book No. 3, Trect milentjum, Belgrade, 2009.)

FIGURE 16.67   Sloboda Bridge in Novi Sad after the NATO air strike, April 1999. (Data from Folić, Radonjanin, and Malesev, *Journal Road and Traffic,* 1, 3–20, 1999.)

FIGURE 16.68 Sloboda Bridge in Novi Sad. (Data from Folić, Radonjanin, and Malesev, *Proceedings of the 9th International Symposium of MASE,* Ohrid, BP-5/1-BP5/16, 2001a.)

FIGURE 16.69 The view of the destroyed pylon and the broken link between the pylon and deck of Sloboda Bridge. (Data from Folić, Radonjanin, and Malesev, *Proceedings of the 9th International Symposium of MASE,* Ohrid, BP-5/1-BP5/16, 2001a.)

FIGURE 16.70 Damaged Žeželj Bridge. (Data from Folić, Radonjanin, and Malesev, *Proceedings of the 4th International Conference on the Bridges across the Danube,* Bratislava, 349–358, 2001b.)

FIGURE 16.71   Pier of Žeželj Bridge in Novi Sad. (Folić, Radonjanin, and Malesev 2000a.)

FIGURE 16.72   Repair and temporary structure after damaged Žeželj Bridge in Novi Sad. (Folić, Radonjanin, and Malesev 2000a.)

FIGURE 16.73   Support of the main girder after damaged Žeželj Bridge in Novi Sad. (Folić, Radonjanin, and Malesev 2000a.)

FIGURE 16.74 Torn down span of the approach bridge structure Beska. (Folić, Radonjanin, and Malesev 2001a.)

FIGURE 16.75 Debris of Varadin Bridge. (Folić, Radonjanin, and Malesev 1999a.)

The Grdelicka Gorge Bridge deck in the first span and the reinforced concrete arch together with the piers were destroyed. The first three fields towards Vranje were completely demolished and the fourth was damaged. All the piers preceding the destroyed span were also destroyed. The entire construction preceding the fourth pier was completely rebuilt, including the main arch.

Several repaired and renewed bridges are presented in detail in Perisic 1999 and Mrkonjic and Vlajic 2009. Some of these are shown in Figures 16.76 through 16.79.

In the city of Novi Sad three permanent bridges were destroyed, so that between April 26 and September 15, 1999 the banks of the Danube were connected only by ferries, boats, and small vessels. Besides the repair and renewal of some bridges, several bridges were rebuilt as temporary solutions. A mountable-dismountable bridge was erected instead of the Žeželj Bridge. Varadin Bridge was replaced by a floating (pontoon) bridge, which was in operation from September 1999 until October 2005, when the reconstruction of the Sloboda Bridge was finished. About 45,000 vehicles were crossing the bridges every day before they were demolished. The problem of crossing the Danube was alleviated by the construction of a temporary railway-roadway mountable-dismountable bridge which opened to traffic on May 29, 2000.

FIGURE 16.76   Bridge of Renewal. (Mrkonjic and Vlajic 2009.)

FIGURE 16.77   Bridge in Trstenik. (Vojinovic and Folić 2002.)

FIGURE 16.78   Bridge in Biljanovac. (Vojinovic and Folić 2002.)

16.9.3.1 Temporary Mountable-Dismountable Railway-Roadway Bridge in Novi Sad

Constructed on the Belgrade–Novi Sad–Hungarian border railway, this bridge is designed for mixed alternate railway-roadway traffic (Figures 16.80 through 16.82). Its construction consists of six steel truss structures of a simple beam system (4 × 72 m + 102 m + 36 m). The bridge's total length is 433.5 m.

FIGURE 16.79 Kokin Brod Bridge. (Mrkonjic and Vlajic 2009.)

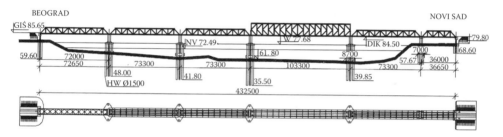

FIGURE 16.80 Temporary railway-roadway mountable-dismountable bridge in Novi Sad. (Mihajlovic and Shiftmiler 2001.)

FIGURE 16.81 Temporary railway-roadway bridge. (Mihajlovic and Shiftmiler 2001.)

The bridge is axially situated 75 m upstream from the demolished Žeželj Bridge. The roadway track is 4.5 m wide, with unidirectional/alternating traffic. The waterway is 94.5 m wide and 6.82 m high. The trusses were made of a small number of different elements to form a single-floor truss with a maximum span of 72 m (6 m high) and two-floor truss of 102 m span (12 m high). The bridge is 5.3 m wide. The length of a single field is 6 m and the roadway board and main girders are designed for a 300 kN roadway vehicle (Mihajlovic and Shiftmiler 2001).

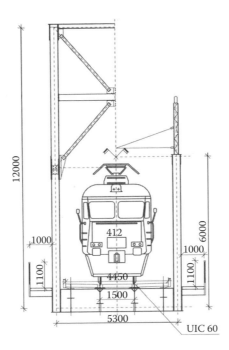

FIGURE 16.82 Cross section of the temporary railway-roadway bridge. (Mihajlovic and Shiftmiler 2001.)

FIGURE 16.83 Erection of truss in main span. (Mihajlovic and Shiftmiler 2001.)

The truss construction, together with the railway, was mounted on a huge platform on the right bank. The platform was equipped with two slipways entering the river. The full-size trusses were erected using steel yokes and hydraulic jacks on special barges (Figure 16.83). All trusses were premounted and slipped over the water where they were accepted by four barges and pulled to the place where they were installed. The trusses were lifted and lowered by means of yokes and prestressing hydraulic jacks. The largest piece in the erection procedure was 102 m long, 12 m high, and weighed 450 tons.

The bridge is supported on 4 river piers and 3 piers on the bank, on 35 bored HW piles, ϕ 1500 mm, 20 m to 42 m long (Figure 16.84). The piers were made during the maximum water level. They are batteries of six drilled piles (ϕ 1500mm) stabbed into the bottom of the river. The piles are stiffened with

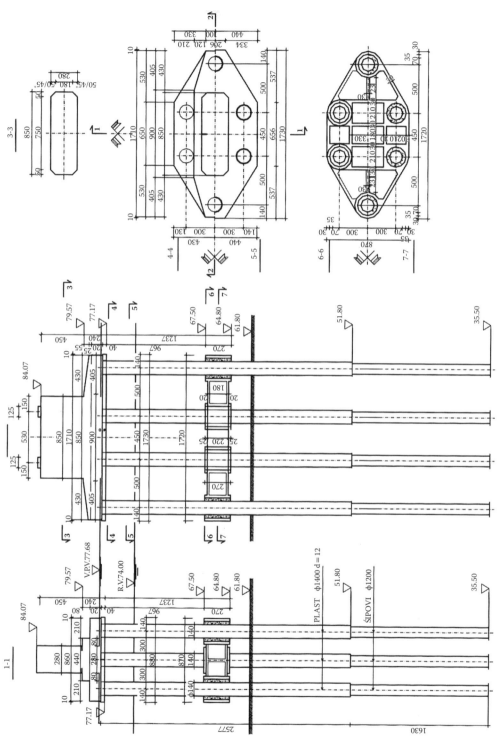

FIGURE 16.84 Pier of the bridge. (Mihajlovic and Shiftmiler 2001.)

prefabricated reinforced concrete diaphragm. Over the concrete piles prefabricated reinforced concrete plates were mounted as the helmet beams' formwork. The four river piers were made in such manner, and the bank piers were made on site through the stabbed piles. The work was completed in 100 days. The bridge was designed by CIP Belgrade; the contractor was Mostogradnja. (Mihajlovic and Shiftmiler 2001).

16.9.3.2 Road-Pedestrian Bridge over the Danube at Novi Sad

In order to resume road and pedestrian traffic between Novi Sad and Petrovaradin a temporary bridge was built on barges upstream of the destroyed Petrovaradnin Bridge. The bridge consists of two traffic lines 3 m wide each, for B-300 standard-type vehicles, two footways 1.5 m wide each, and 20 cm railings (9.4 m total width). The total length of the bridge, including the transition bridges, is 369.6 m. The bridge structure rests on the abutment piers, two barges installed in the longitudinal direction on the left and right banks, and four barges installed in the transverse direction to the stream, each barge 71 m long. The adopted position of the barges, at right angles to the river flow, caused strong forces, which were received with connection anchors and the piers of the former F. Josef railway bridge.

The bridge structure consists of three parts. The first is the structure on the navigation channel, while the other two parts are the connecting structure between the floating part and the abutments. The approach structure consists of transition bridges, with steel orthotropic decks (5.454 x 5.0 m). The part of the bridge that rests on the barges was formed of a grid with transverse and longitudinal girders, while bridge deck on the top of them was formed of precast reinforced concrete slabs.

The longitudinal barges are interconnected by rigid bolts jointed to the barge structure. The hinges are distributed so as to allow rotation in the vertical direction along the length of the bridge. Lateral titling is possible only to the extent permitted by the tensional stiffness of the barges and the whole system. The parts of bridge are movable so that ships and ice can pass through the gap when opened. By opening the barges, a 140 m wide navigation gap is formed. The bridge was designed by D. Lukic of Mostogradnja and the contractor was Construction Company Belgrade. Trial loading of the bridge was done by Institute CIP Belgrade. The bridge was completed within 60 days and opened to traffic on September 15, 1999.

16.10 Bridge Management System and Bridge Evaluation

Bridges are inspected regularly to ensure that they perform properly. If a discord appears, it is carefully studied since it could influence future actions. This applies to expansion joints, anchorage plates, hinges, bearings and supporters, and so on. The development of information systems for managing the structures has been enabled by the development of software for creating a relational data base (Folić, Vojinovic, and Vugrincic 1998). The information system offers the required information on the basis of available data. The precision of the information is in direct proportion with the precision of the input data.

The information systems for managing structures can mostly function with the use of computer software. In Serbia, three information systems have been developed within the information systems for roads:

1. Database for Roads (including sliding)
2. Database for Bridges (BMS)
3. Database for Traffic

The Bridge Management System (BMS) was developed 20 years ago. It contains data for some 3200 bridges on the network of primary roads in Serbia, with two groups of data:

1. Permanent data (road number, chain age, name of the obstacle, name of the bridge, data about the designer and contractor, technical data on the bridge, photos, sketches, etc.)
2. Variable data regarding the state of the bridge used to determine the "rating" of the bridge, that is, its priority in maintenance

Most checklists include a description of the structure, condition of the structure, nature of loading and detrimental elements, original condition of the structure, materials used in construction, construction practices, and the initial physical properties of the hardened concrete. The Regulations of 1992 cover in detail the technical standards for utilization and regular maintenance of bridges on primary, regional, and town routes. They include the following:

- Contents of the data to be registered (data bank)
- Kinds of inspections (control, regular, main, and extraordinary), as well as their content regarding the elements of the bridges
- Specification of works on regular maintenance, considering cleaning the bridge structure, servicing the used expansion devices, and other necessary works on the bridge, depending on the type of structure and the conditions of usage

Maintenance includes inspection, observation, and registration of changes and their condition, maintenance activities, and measures for eliminating all defects and damage. For the observation of the technical condition of structures on regional and main roads, the following types of inspections are performed (Regulations for Maintenance 1992):

- Control inspections
- Regular inspections
- Detailed inspections
- Special inspections
- Extraordinary inspections

The aim of regular inspections is to control the state of the structure directly and keep traffic safety satisfactory. Control inspections are done by a road inspector at least once a month. In addition to monitoring deficiencies in due time, it is necessary to study all available documentation in order to define, in the first place, the causes of the damage. This begins with an initial visual inspection, for which it is necessary to provide safe access to the damaged structure. During this inspection the structural condition category is determined, depending on damage properties and seriousness. On the basis of the visual inspection, the scope of detailed inspection and testing methods are determined.

A detailed inspection aims to obtain information on optimal scope in order to assess the condition of the structure and the possibility of further use with necessary interventions. Detailed inspections of bridges should be done at least every two years. During the inspection all elements of the bridge are included, especially the following:

- Foundations of middle and end piers
- Middle piers, abutments (end piers), and bearings
- Mean and transverse girders
- Bracing
- Deck and cantilever
- Reinforcement or steel structure corrosion
- Humidity insulation and pavement
- Expansion joints
- Waterway or the area under the bridge
- Slopes, approach slab, and embankment
- Footpaths, parapets, curbs, and fences
- Drainage system
- Utilities, devices
- Traffic signs on the bridge

For bridges with a 15 m span or larger permanent control is conducted according to the regulations and the regulations system. During special inspections, special equipment and measuring instruments are used in order to check the actual degree of damage, especially at structures for which regular or occasional inspection has established it is in danger of falling down or ruining, which require urgent repairs. A special inspection is conducted according to a previously planned program. Extraordinary inspections are conducted according to regular inspection rules after an unexpected event, as well as before and after the transportation of an extraordinary load that might endanger the capacity or function of the structure. In the case of damage, the equipment for the inspection is the same as during special inspections.

If the level of reliability cannot be preserved with regular maintenance, the owner should initiate adequate research works in line with the enacted Regulations for Maintenance (1992), followed by the repair or reconstruction of the bridge. The program and scope of the research works, as well as the resulting interventions, depend to a large degree on the subjective opinion of the expert team to whom the job is entrusted. They primarily concern the priority, type, and scope of intervention measures, as well as the corresponding costs. Such problems are successfully resolved by the development and application of information systems for managing the structures.

The variable data comprise the empirical evaluation of the condition (b) of 26 elements of the bridge (including the position of the bridge on the route and the traffic loading), classified into 6 groups of so-called factors of importance, and the following 5 gradations are adopted for the evaluation of the condition of the elements: dissatisfactory, bad, insufficient, acceptable, and good (in extreme cases, "does not exist" and "dangerous" can also be used). The numerical values of these gradations (b_i) range between 1 and 20 (in extreme cases 0 and 100). The management of a structure should be formed in such a way as to provide uninterrupted use of the structure within a time interval, with minimal investments in behavior monitoring and repairs. Therefore, it is necessary to define requirements regarding the condition of the structure and its components. On the basis of the recorded range and type of damage, it is possible to assess the costs of particular repairs, considering priorities.

The described BMS requires further elaboration, above all by introducing financial indices, which enable the owner to plan funds for maintenance, and a system of evaluating the condition of bridges in connection with rehabilitation works. Evaluation criteria for the determination of bridge adequacy are structural and functional. Technical regulations for bridge structure maintenance have existed since 1992. If it is necessary to carry out revitalization measures of bridges, optimum solutions are obtained on the basis of research works, in line with a special procedure. The successful and technically complex repair of bridge structures in Serbia proved its high technological level of bridge construction.

16.11 Future Bridge Activities

It has been shown that the sufficient durability of bridges is not only assured by their design according to the regulations and standards, especially if the continuity over the beam bridge support is not secured. A good concept is important as well as a design recognizing previous experiences and feedback information regarding maintenance and BMS. In order to increase the durability of a bridge's superstructure and reduce maintenance costs, integral bridges are increasingly built.

Characteristic developmental trends are based on the progress of the theory of structure and computerization, which provide the possibility of inclusion of the entire design process and better insight into the behavior of bridge structures, and increase the reliability, durability, and service life. The construction–ground interaction should be more effectively included in the inertial and kinematic components, which is especially important with integral bridges.

The contribution of designers and contractors from Serbia to the development and advance of engineering technology remains significant, but further work on respecting local conditions is certainly of great importance. Therefore, the construction of monolithic and precast concrete prestressed bridges has been developed. Cable-stayed bridges are increasingly designed, both for large- and medium-sized spans, as well as for smaller spans. There is a trend toward combining suspension bridges and cable-stayed bridges to master extremely large spans.

In addition to new bridge construction, the application of external prestressing is gaining importance in repairing and strengthening existing superstructures. Further work on the advancement of maintenance and strengthening constructions is appreciated, as well as the application of composite materials. Thus, there are ongoing works for designing bridges with the possibility of altering certain parts.

Acknowledgments

I would like to express my sincere gratitude to the Institute of Traffic in Belgrade for the photographs and material they have provided for this work. Especially, I am thankful to their structural engineer Sinisa Mihajlovic (CIP) and to Dr. Bratislav Stipanic (Faculty of Civil Engineering) for the photographs and dates they kindly provided. Also, I would like to thank Dragana Glavardanov for helping me out with formatting this manuscript. This work has been undertaken as part of Project No. 16018 and No. 36043, partly financially supported by the Ministry of Education and Sciences of Serbia.

References

Buđevac, D., Stipanic, B. 2007. *Metal bridges*. Građevinska knjiga, Belgrade, Serbia (in Serbian).

Buđevac, D. et al. 1999. *Metal structures*. Građevinska knjiga, Belgrade, Serbia (in Serbian).

CIP. 2010. Bridges designed by CIP. Prepared by S. Mihajlovic, collection of Bridge Figures Designed by Institute for Traffic and Design (CIP) Belgrade, Serbia.

Eurocodes. EN 1990 to 1999; EN 2002 to 2006. CEN, Brussels.

Folić, R. 1992. Some Problems of maintenance and repair of prestressed concrete bridges—two examples of Bridges over the Danube river. *Int. Conf. Bridges over the Danube*, Budapest, Hungary, II/149–158.

Folić, R., Vojinović, B., Vugrinčić, V. 1998. Some experiences of maintenance and repair of reinforced and presstresed concrete bridges. *Proceedings of the DCM of Bridges across the Danube* (ed. K. Zilch et al.), Bauingineur, Springer VDI Verlag, Dusseldorf, Germany, 49–70.

Folić, R., Radonjanin, V., Malešev, M. 1999. Damage and destruction of Bridges over the Danube in Novi Sad and near Beška 1999. *Journal Road and Traffic*, No. 1, pp. 3–20 (in Serbian).

Folić, R., Radonjanin, V., Malešev, M. 2000. Novi Sad Bridges and their destruction in April 1999. *Institute IMS Proceedings*, Belgrade, No. 1, 7–20.

Folić, R., Radonjanin, V., Malešev, M. 2001a. Review of destruction and renewal of some important structures in Novi Sad and its vicinity. *Proceedings of the 9th International Simposium of MASE*, Ohrid, pp. BP-5/1–BP-5/16.

Folić, R., Radonjanin, V., Malešev, M. 2001b. Damage and destruction of Bridges over the Danube in Novi Sad and near Beška in spring 1999. *Proceedings of the 4th International Conference on Bridges across the Danube*, Bratislava, pp. 349–358.

Folić, R. 2007. Some aspects of deterioration and maintenance of concrete bridges. In *Bridges in the Danube Basin*, ed. M. Ivanyi, R. Bancila, 6th International Conference on Bridges across the Danube 2007, Muegyetemi Kiado, Budapest, Hungary, pp. 377–388.

Gojković, M. 1989. *Old Stone Bridges*. Naučna knjiga, Belgrade (in Serbian).

Hajdin, N., Stipanic, B. 2004. Sloboda bridge reconstruction project preleminary design and design for tender. *Proceedings. of the 5th International Conference on Bridges across the Danube*, Novi Sad, 2004, Vol. 1, pp. 53–60.

Hajdin, N., Stipanic, B. 2006. Reconstruction of the Cable-Stayed Bridge over the Danube in Novi Sad. *Bridges Int. Conf.* (ed. J. Radic), *Dubrovnik*, pp. 293–290.

Hajdin, N., Stipanic, B. 2007. Project of new roadway Bridge across Sava River in Belgrade. *Proc. 6th Int. Conf. Bridges in Danube Basin*, Budapest, September 12–14, pp. 107–115.

Ivanyi, M., Stipanic, B. Bridges on the Danube. 1992, 1995 & 2008. Bridges over the Danube-Catalogue, First & Sec. Intern. Conf. Ed. (Part Bridges in Yugoslavia), Budapest, 1992; Buchurest-Romania, 1995; and Regensburg-Germany, 1998.

JUS U.E7.081/86; JUS U.E7.086/86; JUS U.E7.091/86; JUS U.E7.096/86; JUS U.E7.101/86; JUS U.E7.111/86; JUS U.E7.121/86. Standards for stability steel bearing structures (in Serbian).

JUS U.E7.131/80. Bearing and hinges bearing steel structures.

JUS U.E7.145/87, JUS U.E7.140/85. Connections with rivets and screws.

JUS U.E7.150/87. Welded steel bearing structures.

Mihajlovic, S., Shiftmiler, Z. 2001. The temporary assembly-type railway-road Bridge over the river Danube at Novi Sad. *Conf. Contemporary Civil Eng. Practice, Soc. CE N. Sad and ICE & Arch.*, May 10, pp. 39–54 (in Serbian).

Mrkonjic, M., Vlajic, L. J. 2009. *Bridges, Library, Reconstruction*, Book No. 3, Treci milenijum, Belgrade.

Muller, J. M. 2000. Design practice in Europe. In *Bridge Engineering Handbook*, ed. W. Chen and L. Duan, chapter 64, CRC Press, Boca Raton, FL.

Nenedić, G., Djukić, L. J., Ladjinović, D. J. 2001. Structural design of Varadin's Rainbow Bridge. *Conf. Contemporary Civil Eng. Practice, Soc. CE N. Sad and ICE & Arch.*, May 10–11, pp. 55–70 (in Serbian).

Perisic, Z., ed. 1999. *Proceedings of the International Conference Bridge Rehabilitation and Reconstruction*, CIP, Belgrade, Serbia Belgrade.

Reconstruction–Renewal. 2000. Direction for Reconstruction of the Country, April, Belgrade (in Serbian).

Regulations on engineering norms for bearing steel structures. 1986. *Official Yu Gazette No. 61* (in Serbian).

Regulations on engineering norms for the design of engineering structures in seismic regions, 1987 (in Serbian).

Regulations on engineering norms for concrete and reinforced concrete. 1987. *Official Yu Gazette No. 11* (in Serbian).

Regulations on engineering norms for determination intensity of loads on bridges. 1991. *Official Yu Gazette No. 1* (in Serbian).

Regulations on engineering norms for design, construction, and rehabilitation railway bridges and culverts. 1992. *Official Yu Gazette No. 4* (in Serbian).

Regulations on engineering norms for maintenance of bridges. 1992. *Official Yu Gazette No. 20* (in Serbian).

Rule for determining intensity of loads and categorization railway line. 1992. *Official Yu Gazette No. 23* (in Serbian).

Šram, S. 2002. *Construction of bridges—concrete bridges*. Golden Marketing, Zagreb, pp. 448–460 (in Croatian).

Statistical Office of the Republic of Serbia- RZS 2012. http://webrzs.stat.gov.rs/WebSite/Public/PageView. aspx?pKey=2

Temporary Code 1964, Temporary code for building in seismic regions. *Official Gazette No. 39/64* (in Serbian).

Trojanović, M. 1968, 1974. *Reinforced and prestressed concrete bridges*. ZIUS, Belgrade (in Serbian).

Vojinović, B., Folić, R. 2002. Renewal of bridges damaged in NATO bombardment of FR Yugoslavia in 1999. *Jubilee Sc. Conf. 60 Years University of ACEG*, Sofia, Vol. 4, pp. 35–45.

Vukmanovic, V. 2007. *Bridges on the Danube*. Prometej, Novi Sad (in Serbian).

Želalić, P. 1969. Five hundred prefabricated bridges with spans up to 20 m. In *Special publication: Prestressed concrete structures*, Izgradnja, Belgrade, Serbia, Belgrade, pp. 261–272 (in Serbian), Izgradnja.

Žeželj, B., Petrović, B. 1992. A concrete Bridge over the river Danube at Novi Sad. *Int. Conf. Bridges on the Danube*, Vol. 1, Budapest, Hungary, pp. 247–252.

17

Bridge Engineering in the Slovak Republic

Ivan Baláž
*Slovak University
of Technology*

17.1 Introduction

17.1.1 Geographical Characteristics

Slovakia became an independent country on January 1, 1993, after peaceful separating of Czechoslovakia into Czech Republic and the Slovak Republic. It is located in Central Europe, bordering Hungary (677 km), Poland (444 km), Czech Republic (252 km), Ukraine (97 km), and Austria (91 km). The total area of Slovakia is 49,035 km². The population is about 5.43 million people. Slovakia is a member state of the United Nations (1993), European Union (2004), and the European Economic and Monetary Union (2009), among others. The official language is Slovak, very similar to the Czech language, both members of the Indo-European Slavic language family.

17.1.2 Historical Development

17.1.2.1 Historical Stone Bridges

The oldest and the most interesting stone bridges in the Slovak Republic are

- St. Gotthard Bridge in the village Leles (Figure 17.1) was built in the fourteenth or fifteenth century over original trough of Tisa River. The bridge is 30 m long with a 5 m wide roadway and is no longer in service. The span of the arch is 2.5 m. The bridge obtained its name in the eighteenth century. A statue of St. Gotthard was removed in 1848. The bridge was reconstructed in 1994.

(a)

(b)

FIGURE 17.1 St. Gotthard Bridge in village Leles (fourteenth or fifteenth century): (a) overview, (b) single arch. (Courtesy of The Monument Boards of the Slovak Republic.)

- Gothic Arch Bridge in the village Dravce (Figure 17.2) was built with sandstone in the fifteenth century. The earliest reference to the village Dravce dates from 1263. The bridge over Bicír Stream originally carried the main road and served as the gateway to the village. The bridge is 8 m long and 10.5 m wide. The arch, with a span of 3.6 m, is 3 m high and 0.4 m thick. The bridge is still in service today for the local road only.
- The bridge in Spišský Hrhov over Lodzina Stream is shown in Figure 17.3. The original bridge may have been built between the sixteenth and seventeenth centuries and rebuilt in the Baroque period, but with signs of the Renaissance in 1803–1808 under the supervision of Andrej Probstner. The bridge, with four sandstone arches, is 39.93 m long and 9.95 m wide. The arches have spans from 5.49 m to 5.79 m. The third arch (over the stream) has the highest clearance at 3.9 m. The piers are 2 m thick. The bridge still serves as a local road.

FIGURE 17.2 Gothic Arch Bridge in Dravce over Bicír Stream (fifteenth century). (Courtesy of The Monument Boards of the Slovak Republic.)

FIGURE 17.3 Stone bridge in Spišský Hrhov over Lodzina Stream (1803–1808). (Courtesy of The Monument Boards of the Slovak Republic.)

- Turkish bridge over Drevenica Stream (Figure 17.4) in Nová Ves nad Žitavou was probably based on the tradition arches built at the end of the sixteenth century by Turkish conquerors. The bridge, with three arches, was built from quarry stone in the end of the eighteenth century in the Late Baroque period.
- Baroque bridge over Sikenica Stream (Figure 17.5) in Bátovce was built according to a very bad readable inscription on the stone plate at the top of the middle arch in about 1700. It has three semicircle arches. There is a statue of St. John of Nepomuk on the bridge made of the same stone material. St. John of Nepomuk is a national saint of the Czech Republic, a protector from floods and patron saint of bridges.

FIGURE 17.4   Turkish bridge over Drevenica Stream in Nová Ves nad Žitavou (end of eighteenth century). (Courtesy of P. Paulík.)

(a)   (b)

FIGURE 17.5   Baroque bridge over Sikenica Stream in Bátovce (about 1700): (a) overview, (b) original inscription giving the year. (Courtesy of The Monument Boards of the Slovak Republic.)

- The magnificient bridge with three elliptic arches in Kráľová pri Senci has baroque and secession elements (Figure 17.6). Designed by Ján Nagymihály, it was built by Italian builders in 1904. Its predecessor bridge was from the eighteenth century. The bridge is 40.09 m long, 6.8 m high, and 6.02 m wide. The road is 4.7 m wide.

17.1.2.2 Historical Timber Bridges

The oldest timber bridges (Dutko and Ferjenčík 1965) are:

- The remains of a pile bridge found in Dievčenský Hrad–Leányvár (Girls' Castle) in the village of Iža near Komárno in the former Roman fortress of Brigetio (Figure 17.12). It is part of the ruins of the Roman fortification, which was built around the first or second century as a part of Limes Romanus.
- Timber bridge over Váh River in the Roman military camp Eleutheropolis at Hlohovec.
- Pontoon bridges over Danube River, mentioned for the first time in the seventh century.

The oldest covered bridge may be seen in a painting by J. Willenberg from 1599 (Dutko and Ferjenčík 1965). The bridge over Hron River was located at Zvolen (Figure 17.7). Another one, which was in Zvolen

FIGURE 17.6 Park bridge in Kráľová pri Senci built in 1904. Its predecessor bridge was from the eighteenth century. (Courtesy of P. Paulík.)

FIGURE 17.7 Covered bridge in Zvolen over the Hron River, seen in a painting by J. Willenberg from 1599. (Courtesy of The Monument Boards of the Slovak Republic.)

FIGURE 17.8   Combined system of covered timber truss bridge: Kluknava Bridge over Hornád River (1831–1981) (Courtesy of The Monument Boards of the Slovak Republic.)

over Slatina River, may be seen in an engraving from the seventeenth century made according to a painting by Gérard Bouttats.

Data collected in 1954 found that there were 25 timber bridges in service in 1925 and 9 bridges in 1954. All 29 bridges were built in period 1831–1900 and demounted in 1925–1981, including 10 bridges destroyed during World War II. Ten bridges were built over Hnilec River. The bridges used statical systems: 14 bridges with the queen post system, 11 bridges with the Howe–Zhuravsky system, 3 with combined systems, and 1 from 1599 (Figure 17.7) with an unknown system. Kluknava Bridge (Figure 17.8), 27.02 m long, was demounted in 1981 and reconstructed in 1981–1984, and its deck was repaired in 2003–2004.

17.1.2.3 Historical Cast Iron Bridges

The first cast iron bridge (Figures 17.9 and 17.10) in Middle Europe was built in 1810 in Hronec (Rohniz, Kisgaram) near Podbrezová during the time of the Hungarian Kingdom (Nesvadba and Plevák 1982). There were seven blast furnaces on Hronec in 1740. The rolling mill of the plate originated in Hronec in 1814. The bridge had an arch span of 3.01 m and consisted of 36 components, including 5 arch ribs. The bridge had a length of 4.87 m, width of 4.845 m, and height of 1.31 m. The chemical composition of the cast iron sample taken from the rib was C 3.87, Mn 2.96, Si 1.23, P 0.776, and S 0.018, and from the foundation plate was C 3.55, Mn 2.00, Si 1.39, P 0.640, and S 0.030. The stress at the failure of a sample with diameter $d = 6$ mm obtained in a mechanical tension test was 187.3 MPa. The bridge served 152 years until 1962. Only part of the bridge has survived, consisting of three arch ribs. It is exposed in the front of the local foundry. Bridge details are shown in Figure 17.10.

The second cast iron bridge in Hronec was built probably at the same time as the first one. The second bridge, with four arch ribs, was covered by stone. It was 5 m long and 3.5 m wide. After repeated repairs it was removed in 1942. Its remaining parts disappeared. The third cast iron bridge (Figure 17.11) in Hronec over Čierny Hron River was built in 1815. The bridge, with five arch ribs, was 10.2 m long and 4.9 m wide. A rib consisted of two half-arches connected in the middle of the span. It was destroyed by German soldiers in 1945. The ribs had complicated laced mode decoration, which was very similar to the

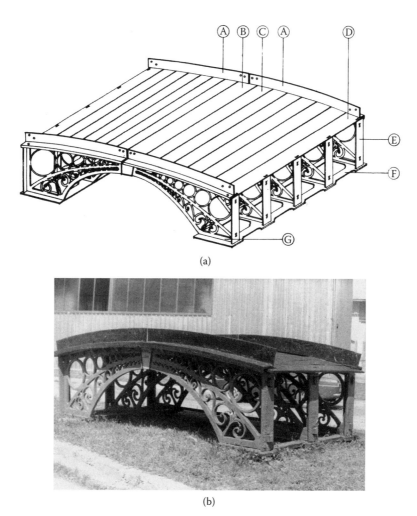

(a)

(b)

FIGURE 17.9 The first cast iron bridge in Hronec (1810–1962): (a) original bridge with elements A-G, (b) remaining bridge. (Courtesy of M. Plevák and O. Nesvadba.)

(a)

(b)

FIGURE 17.10 Details of the first cast iron bridge in Hronec (1810–1962): (a) half arch, (b) arch crown. (Courtesy of M. Plevák and O. Nesvadba.)

FIGURE 17.11   The third cast iron bridge in Hronec over Čierny Hron River (1815–1945).

FIGURE 17.12   Location of Bratislava (Carnutum, Gerulata) and Komárno (Brigetio) at Danube River, which defines Limes Romanus (the fortified border of the Roman Empire).

famous bridge over Strzegonka River in Łażany from 1796 (see Figure 14.6 in Chapter 14). This may be explained by the fact that founder specialists from Prussian Silesia came to Hronec in 1807 to improve the foundry technology. The third iron bridge was mentioned among other remarkable world bridges in the paper by Gender Stort (1935).

17.1.2.4 Bridges over Danube River in Bratislava, Capital of Slovakia

During the Roman period (27 BC–476 AD) temporary boat bridges were built mainly for military use in Bratislava and surroundings of the Danube River (Figure 17.12). There were only fords, pile bridges (Figure 17.13), and pile bridges combined with pontoons in the period up until 1676 (Table 17.1).

FIGURE 17.13 Bratislava before 1676: pile timber bridge. Copper plate published in Gabriel Bodenehr, Europens Pracht und Macht Augsburg 1720 (1729–1730).

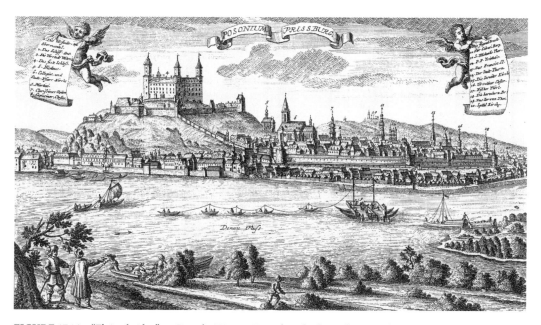

FIGURE 17.14 "Flying bridge" on Danube River in Bratislava (eighteenth century).

From 1676–1825 *die fliegende* Brücken ("flying bridges") were built and used. It was an anchoraged system of pontoons connected by cables to a ship or ferry. In the period between February 9, 1563, and September 9, 1830, altogether 19 coronation ceremonies were organized in Bratislava. At these occasions temporary timber pontoon bridges were built. The bridges were created from several connected pontoons on which timber deck was laid (Figure 17.14, Table 17.1). Current situation see on Figure 17.19.

TABLE 17.1   Bridge Historical Development in Bratislava

Year	Bridge Type/Name/Structure	Designed By/Built on the Order of Ruler	Main Features of Bridge/Comments
PERIOD OF ROMAN EMPIRE: 27 BC–476 AD			
81–96	Temporary military pontoon bridge	Roman Emperor T. Flavius Domitianus (*51; reign 81–96)	Allegedly built near Bratislava
172	Timber military pontoon bridge	Roman Emperor Marcus Aurelius (*121; reign 161–180)	Built on Slovak territory
PERIOD 1000–1676: Holy Roman Empire (962–1806), named Holy Roman Empire of German Nation after 1450, Hungarian Kingdom (1000–1918)			
976	Simple timber bridge	Destroyed by flood in Bratislava	
1252	Stone toll-house water tower	Built at the first known Bratislava Danube ford under the castle; former structure belonged to the Roman fortification system	
1271–1273	Two temporary military bridges	Czech King Přemysl Otakar II (*1232, †1278)	Near water tower
1407	Bridge	Emperor Zigmund Luxemburg (*1368, †1437)	King alone paid big expenses
1430	Second bridge	Emperor Zigmund Luxemburg; bridge builder Gutzel from Vienna	
1439	Bridge	King Albrecht II Habsburg (*1397, †1439), by deed of donation issued in Vienna, gave to Bratislava town the right of ferryage and right of wet toll under the condition that the town would build and maintain a bridge	
1468	Bridge supported only by piles, not by pontoons	King Matthias Corvinus; Edmund Fleck, the bridge builder, came from Vienna. Earl Ernö János Csáktornayi from Zvolen was responsible for bridge construction.	Located in the front of Fishing Gate in prolongation of Bridge Street (River Street); the bridge led over Gereidt Island, and consisted of three parts
1487	Bridge supported only by piles	King Matthias Corvinus (*1443, †1490) sent order to Bratislava Earl Lidvai Bánffy to allow the city to build a bridge	Located in the front of Weidrits Gate (Fishing Square), the bridge led over Danube Island and its middle part was a hoist bridge; lasted 3 years
1493	The same type as the bridge from 1439	Located in the front of Fishing Gate; demounted in 1496 and never renovated	
1510	Bridge	Protected by boats and rangers on horses against Turkish soldiers	
1563	Temporary timber pontoon bridge	Emperor Maximilian II (*1527, †1576)	Built at the occasion of coronation
1608	Temporary timber pontoon bridge	Emperor Matthew II (*1557, †1619)	Built at the occasion of coronation
1618	Temporary timber pontoon bridge	Emperor Matthew II ordered to build the bridge again	
1620	Water tower	Emperor's General Dampière ordered destruction of the Water Tower during the time of the Earls Uprising led by Gábor Bethlen (*1580, †1629)	
1626	Iced bridge (straw laid on ice and poured over with water)	The last crossing, called "Eiswegmachen," over the frozen Danube River, occurred on January 26—a coachman, carriage and pair of horses broke the ice and drowned	
1638	Temporary bridge supported by pontoons	Mary Anne of Spain, wife of Emperor Ferdinand III (*1608, †1657)	Built at the occasion of coronation, seen on a copperplate
1655	Timber girder bridge	Emperor Leopold I (*1640, †1705)	Built at the occasion of coronation, seen on a copperplate
PERIOD 1676–1825: Holy Roman Empire (962–1806), named Holy Roman Empire of German Nation after 1450, Hungarian Kingdom (1000–1918), Austrian Empire–Kaisertum Österreich (r. 1804–1918)			
1687	Bridge	Joseph I (*1678, †1711)	Built at the occasion of coronation

1712	Bridge	Built at the occasion of coronation of Charles VI (III) (*1685, †1740)	Bridge was dedicated by Emperor to the city and was not demounted after coronation as usual
1727	Masonry bridge	The oldest bridge in Bratislava, successor of a hoist bridge	
1735	Bridges to watermills on Danube River	City council decided on March 3 that people may use these bridges	
1741	Bridge supported by pontoons	City council decided to built a bridge at the occasion of the coronation of Maria Theresa (*1717, †1780), the only female ruler in the Habsburg dominions; it is not seen on the map from 1742 by surveyor Samuel Mikovíny (1700–1750)	
1742	Bridge supported by piles	Only bridge seen on Mikovíny's map, a former one was probably demounted after coronation	
1776	"Flying bridge" (Figure 17.14)	City council built advanced "flying bridge" designed by inventor Johann Wolfgang Ritter (Knight) von Kempelen de Pázmánd, born in Bratislava; the year of the first use of the "flying bridge" is not known	
1786	"Flying bridge"	Bridge is able to carry 700 people or 70 cattle or 16 couches; it is the second largest source of money for the city (Korabinsky 1786).	
1790	Bridge supported by pontoons	Emperor Leopold II (*1747, †1792)	Built at the occasion of coronation
1796	Bridge	Vienna demanded to build the bridge	Built at the occasion of Parliament meeting
1809	Bridge supported by pontoons	Napoleon Bonaparte (*1769, †1821) demanded in a letter to his general that bridges be built at Bratislava and Devin; this bridge, built by French soldiers, served until January 1, 1810	
1811	"Flying bridge"	This bridge has been in service since April 18, 1811	
1815	"Flying bridge"	Engraving shows chapel with statue of St. John of Nepomuk, patron of sailors and bridge builders	
1818	Beginning of steam navigation on Danube	Steamer Caroline, inventor Anton Bernard	The first navigation on Danube River occured on September 2
PERIOD 1825–1890: Hungarian Kingdom (1000–1918), Austrian Empire (1804–1918), Austro–Hungarian Monarchy (1867–1918)			
1825	Karolinen–Brücke (Caroline Bridge), timber pontoon bridge	Empress Caroline Charlotte Augusta of Bavaria (*1792, †1873)	Built at the occasion of coronation; bridge served 65 years till 1889 (Figure 17.15); demounted during winter's periods
1830	Beginning of regular navigation on Danube		The navigation of steamer Franz I. occured on June 17
1840–1872	The first railway bridge; horse railway; pressure to built permanent bridge	The first railway in the Hungarian Kingdom; it served between Bratislava and Sväty Jur from September 27, 1840 to October 10, 1872, when use of steam locomotives began; length 15.2 km	
1848	The first railway tunnel	First in Hungarian Kingdom, opened August 19 in Bratislava; length 703.6 m, width 6.4–6.95 m	
PERIOD 1844-1937: Temporary military timber pontoon bridges			
1844–1937	Temporary military pontoon bridges (Figure 17.16)	Designer Karl Ritter (Freiherr; Knight) von Birago (*1792, †1845); Birago's excellent "Militärbrückentrains" (military bridge wagon trains) were used in 1844–1937 (Birago 1839)	

(Continued)

TABLE 17.1 (Continued) Bridge Historical Development in Bratislava

Year	Bridge Type/Name /Structure	Designed By/Built on the Order of Ruler	Main Features of Bridge/Comments
		PERIOD 1890–1945: Austro–Hungarian Monarchy (1867–1918), Czechoslovakia (1918–1938), Slovak state (1939–1945)	
1870–1890	Regulation of Danube River	Together with reinforcing of river banks created conditions for building the first permanent bridge	
1890	Emperor Franz Joseph Bridge; the first permanent bridge; road steel through-truss bridge	Opened by Emperor Franz Joseph I (*1830, †1916) on December 30	During the Slovak State it was renamed General M. R. Štefánik Bridge; it served 55 years until 1945 (Figures 17.17 and 17.18)
1891	Railroad steel through-truss bridge	Opened in September or November	Laid on the same piers as the road bridge; it served 55 years until 1945 (Figure 17.17)
		PERIOD AFTER 1945: Czechoslovakia (1945–1992), the Slovak Republic (from 1993)	
1945	Temporary military pontoon bridge located near castle	Built by Red Army	It served 9.5 months from April 5, 1945 to January 24, 1946
1945	Temporary military pontoon bridge located near Winter Port	Built by Red Army	
1945	Temporary timber pile bridge	Built by Red Army	It served 2 months from August 1945 to October 1945
1946	Old Bridge (original name Red Army Bridge), steel through-truss road bridge planned for 15 years	Built by Red Army using the standardized Austrian Roth–Waagner bridge system; demountable provisional truss arrangement developed and used for the first time in 1915 by R. Ph. Waagner–L. & J. Bíró and A. Kurz; it served traffic until 2010 (see Section 17.4.1)	
1950	Old Bridge, steel diamond-type through-truss railway bridge	Laid on the same piers as the road bridge, it served 35 years until 1985 (see Section 17.4.2); total bridge length is 460.07 m	
1972	SNP Bridge, (called New Bridge in the period 1993–2012) steel cable-stayed bridge	Structural engineer was Árpád Tesár, architect was Ján Lacko, and their teams (Baláž 2012); total length 431.8 m, three-span continuous girder: 74.8 m + 303 m + 54 m (see Section 17.6.1)	
1983, 1985	Port Bridge, steel Warren through-truss bridge with double deck	Motorway on upper deck and railway on lower deck. It has served public traffic since December 1983, when the railway and half profile of the motoray opened. The second profile of the motorawy was opened after completing all approaches. Total length 460.8 m, four-span continuous girder: 102.4 m + 204.8 m + 64 m + 89.6 m (see Section 17.4.4)	
1991	Lafranconi Bridge, concrete prestressed girder bridge	Total length 761 m, seven-span continuous girder: 83 m + 174 m + 172 m + 4 × 83 m (see Section 17.3.2.2)	
2005	Apollo Bridge, steel arch bridge	Project author was Miroslav Maťaščík; total length 517.5 m, six-span continuous girder: 52.5 m + 2 × 61 m + 63 m + 231 m − 49 m (see Section 17.5.1)	

*born, † died

Note: Older, German, Hungarian or Latin names for Bratislava: Prešporok, Prešpurk, Požoň, Pressburg, Presburg, Poszony, Posonium, Wilson city. Photographs of all 392 bridges over Danube River may be found in Träger, H., Tóth, E. (eds.). 2010. *Danube Bridges: From the Black Forest to the Black Sea.* Photographs by P. Gyukics. Yuki Studio, Budapest. Photogrphs and information about more than 250 Slovak bridges see in (Paulík 2012).

FIGURE 17.15 Karolinen–Brücke Bridge (Caroline Bridge) over Danube River in Bratislava (1825–1889).

FIGURE 17.16 Temporary military pontoon bridge on Danube River near Bratislava before 1912.

FIGURE 17.17 Emperor Franz Joseph Bridge over Danube River in Bratislava, railway bridge and roadway bridge (1890–1945).

FIGURE 17.18 Emperor Franz Joseph Bridge over Danube River in Bratislava. Railway bridge; road bridge is behind it on the same piers (1890–1945).

FIGURE 17.19 Bridges over Danube River in Bratislava. Downstream from the left: Lafranconi Bridge (1991), SNP Bridge (1972), old road (1945), and railway (1950) bridges on the same piers, Apollo Bridge (2005), Port Bridge (1983, 1985) under an airplane, not seen. (Courtesy of M. Maťaščík.)

17.1.3 Modern Bridges Development

17.1.3.1 Road Network and Roadway Bridges

The road network, consisting of 17,985 km of motorways and 25,942 km of urban roads, is shown in Figure 17.20. During World War II in 1944–1945, 2036 roadway bridges with a total length of about 31 km were destroyed. There was a shortage of structural materials in the period after World War II, except timber. Therefore, long-term temporary bridges were built from timber and military steel assemblies of Bailey bridge types with planned service lives of 15 years. Tables 17.2 and 17.3 list roadway bridge history from 1800 to January 1, 2011. Table 17.4 lists milestones in roadway bridge development.

There were a total of 7698 roadway bridges in 2000 and 7821 bridges in 2010 in Slovakia. Roadway bridges are classified into eight grades: perfect, very good, good, satisfactory, wrong, bad, emergency, and undefined, based on their conditions. Table 17.3 shows that the amount of bridges rated satisfactory and above is 75% to 80%.

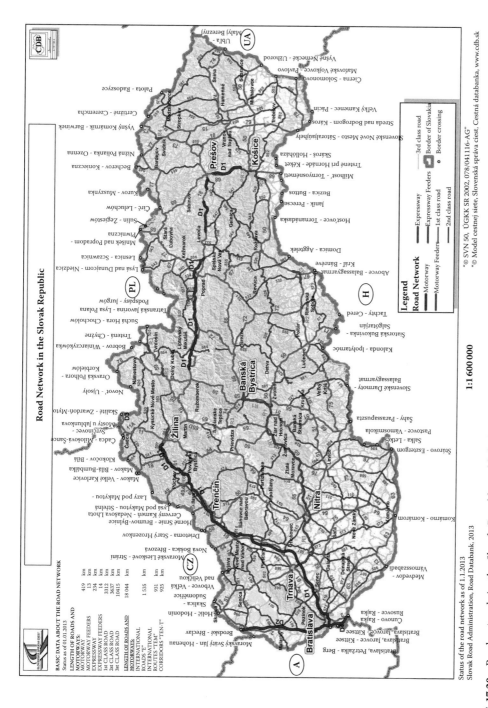

FIGURE 17.20 Road network in the Slovak Republic published in 2013 by Slovak Road Administration (a non-profit organisation established by the Ministry of Transport, Posts and Telecommunications of the Slovak Republic). Updateated editions may be downloaded here: http://www.cdb.sk/files/img/zakladne-mapy-cs/anglicke/road-network_sr.jpg.

TABLE 17.2   Number of Roadway Bridges According to Year of Construction and Structural Material

Structural Material of Bridges	1800 and before	1801–1850	1851–1900	1901–1910	1911–1920	1921–1930	1931–1940	1941–1950	1951–1960	1961–1970	1971–1980	1981–1990	1991–2000	After 2001	Undefined	Number of Bridges
Wood								1							0	1
Stone	2	2	109	24	5	22	26	23	4		2	1	2		0	222
Stone and concrete			1												0	1
Steel			2	1		1	2	5	3	8	19	29	33	11	3	116
Steel plate girders			2	2	2	8	18	22	47	24	9	3	1	2	0	140
Steel truss girders			2	2	1			2	4	2	2	1	1	1	0	18
Steel–concrete composite					1	1	1	3	7	1		1	1		0	16
Precast prestressed concrete			2*			1*	2*	8	137	515	391	287	179	48	12	1576
Precast reinforced concrete			11*	3	1	3	8	24	225	827	471	231	93	21	2	1919
Prestressed concrete							1*	1	2	19	37	31	54	53	11	205
Plain concrete			9	5	2	8	68	57	56	20	4	1	1	3	0	234
Plain concrete and prefabricated reinforced concrete							1								0	1
Brick	1		11	1	2	2	4	3	4	2			1		0	31
Reinforced concrete	1*		18*	14	22	248	647	811	1010	263	62	23	30	47	9	3201
Reinforced concrete and precast prestressed concrete							3								0	3
Others		1						1	1			1		3	5	10
Unknown									1		1			5	227	127
Number of bridges	4	3	167	52	36	294	781	961	1501	1681	998	608	396	195	269	7821

*Some data before 1900 may be incorrect or not updated when bridge was repaired.

Source: Published by Slovak Road Administration in 2011.

TABLE 17.3   Number of Roadway Bridges Classified in Seven Grades of Bridge States

	Grade of Bridge State								
	I	II	III	IV	V	VI	VII	0*	Total Number of Bridges
Year	Perfect	Very Good	Good	Satisfactory	Wrong	Bad	Emergency	Undefined	
2000	431	1040	3675	1824	460	162	18	88	7698
2001	446	982	3636	1912	455	169	16	80	7696
2002	307	911	3604	1988	442	164	15	13	7444
2003	448	925	3682	2154	468	178	13	0	7868
2004	292	815	3574	2036	461	180	17	11	7386
2005	465	830	3574	2137	477	181	18	22	7704
2006	427	837	3556	2129	475	183	9	91	7707
2007	480	837	3514	2169	484	172	8	133	7797
2008	471	830	3477	2199	486	170	9	123	7765
2009	501	833	3475	2198	489	167	10	139	7812
2010	504	825	3468	2209	489	173	11	142	7821

*The Central Register of Technical Roads (ÚTECK) contains incomplete information about the object.

Source: Published by Slovak Road Administration in 2011.

TABLE 17.4 Milestones in Roadway Bridge Development

Year/Period	Bridge/Name//Type/Location	Bridge Parameters/Main Features	Comments
1890	Emperor Franz Joseph Bridge	See Table 17.1	
1891	Reinforced concrete bridge over Bystrica River in Krásno nad Kysucou	Monier system; length 36.4 m, openings 2 × 16.45 m; height of reinforced concrete arch 0.25 m; three stone supports (Figure 17.21)	
1892	Alžbeta (Elisabeth) Bridge over Danube River links Komárno (Slovakia) and Komárom (Hungary); steel Pratt through-truss bridge	Through-type hogging Pratt truss consisting of four single spans 102 m long; total length is 412.5 m; roadway of 2.56 m + 5.6 m + 2.59 m = 10.78 m	Replaced a former pontoon bridge (1586–1892); designed by János Feketeházy; rebuilt in 1951 and reconstructed in 1980
1895	Mária Valéria Bridge over Danube River (Figure 17.22) links Štúrovo (Slovakia) and Esztergom (Hungary); steel crescent-type Pratt through-truss bridge	Five simply supported truss girders: 83.5 m + 102 m + 119 m + 102 m + 83.5 m + 13 m; roadway widths: 2.55 m + 5.72 m + 2.55 m	Replaced former floating bridge; designers, technical inspectors were Aurél Czekelius, Gyula Pischinger, and Ákos Éltető; destroyed in 1919 and in 1944, reconstructed in 2001; see also year 2001 in this table
1908	Reinforced concrete bridge in Kamenica nad Hronom	Three deck arches, main span cca 42 m, total bridge length cca 130m	Designers: Aladár Sebestény Kovács, and József Kovács, chancellor of TU Budapest. Destroyed in 1945
1933	Colonnade bridge over Váh River in Piešťany; concrete bridge (Figure 17.23)	Seven spans: 3 × 20 m + 28 m + 3 × 20 m; two three-span continuous with middle span suspended	Covered bridge with shops on the bridge; destroyed in 1945; reconstructed
1934 (1932–1936)	Three steel modified Warren through-truss bridges over Váh River channel near Dolné Kočkovce, near Beluša and near Ladce	Three simply supported steel truss bridges with spans 46 m, 46.6 m, and 48.8 m (Figure 17.24)	First fully welded truss bridges built in the period of building of the first Váh River cascade, Dolné Kočkovce-Ladce
1935	Steel Vierendeel arch bridge at Michalovce over Laborec River	Span 53.405 m; distance of main girders: 8 m; the arch rise: 7.865 m	First fully welded bridge, fabricated in Škoda Plzeň
1942	Steel modified Warren through-truss links Medvedov (Slovakia) and Vámosszabadi (Hungary), over Danube River	Three-span continuous truss girder: 114 m + 133 m + 114 m; roadway widths 3.2 m + 6 m + 3.2 m	Destroyed in 1944 and 1945, rebuilt in 1973
1945	Timber bridges	Short period after World War II	
1945	Temporary military steel assemblies of Bailey bridge types	Short period after World War II	Temporary bridges with planned life 10 years
1946	Old Bridge (original name Red Army Bridge) over Danube River in Bratislava	See Section 17.4.1 and Table 17.1	
1947	Steel–concrete composite bridge at Púchov over Váh River	Five-span continuous steel plate girder with reinforced concrete slab: 37.4 m + 3 × 37.6 m + 37.4 m	Welded steel–concrete composite bridge; haunched steel plate girders at piers
1949	Steel–concrete composite bridge, Velká Bytča, over Váh River	Five-span continuous girder; haunched concrete slab is connected with five main steel girders of depth 2.27 m	First large road steel–concrete composite bridge; steel consumption 765 tons

(Continued)

TABLE 17.4 (Continued)   Milestones in Roadway Bridge Development

Year/Period	Bridge/Name//Type/Location	Bridge Parameters/Main Features	Comments
1955	Concrete arch bridge, Komárno, over Váh River	Clear span 112.5 m, ratio rise/span = 8.5/112.5	Reinforced concrete bridge (Figure 17.25)
1956	Two concrete bridges over Oravica River in Trstená		First two prefabricated bridges
1958	Concrete bridge over Nitra River in Nové Zámky	22 m + 28 m + 22 m	First continuous prefabricated bridge
1958	Steel single-cell box girder over water cushion of waterwork Krpeľany	Bridge span 58 m, height of the cross section 2.75 m; steel consumption 132 tons	First thin-walled box girder bridge with orthotropic deck; project of concern Vítkovice, Inc. office in Bratislava
1960	Steel Warren through-truss bridge over branch of Danube River in Medveďov	Steel truss bridge with concrete slab 150 mm thick connected with steel cross girders; bridge span 102.6 m	Rigid bottom truss chord with distance of bottom chord joints 17.1 m; all elements are welded in fabrication shop; on-site connections are welded except riveted connections of intermediate diagonals
1961	Prestressed concrete bridge over Váh River in Sereď	Six spans of continuous arches 6 × 55.2 m; length 330 m	
1960	Beginning of construction of segmental bridges with larger spans made from prestressed concrete and erected by progressive cantilever method		
1963–1987	Fifteen concrete bridges built using this method	Maximum spans: 61.2–90 m; lengths: 79 m–634.04 m	Built by Doprastav, Inc. in Bratislava
1963	Concrete bridge over Váh River in Nové Mesto nad Váhom	Three-span continuous bridge: 42.8 m + 70 m + 43.2 m; width of structure 11.3 m; two-cell box girder; nonuniform depth of box girder from 1.8–3.3 m	Only the middle span was built by cantilever method; it replaced three-span steel bridge (43.7 m + 70 m + 43.7 m) destroyed in 1945
1964	Concrete bridge over Váh River in Hlohovec	Three-span frame with hinge in the middle field: 56 m + 80 m + 56 m; difficult foundation conditions; about 80% was built by cantilever method	There was a Roman military bridge in this place in the past; it replaced a former steel truss bridge destroyed in 1945; first use of large-diameter Benoto piles, d = 1.35 m
1965	Concrete bridge over Váh River in Kolárovo	Frame structure with five spans: 35.1 m + 3 × 61.8 m + 35.1 m; total length 256 m; single-cell box girder with nonuniform depth from 1.5–3.11 m	The largest structure at that time built by the cantilever method; the end of the period of frame structures with a hinge in the middle
1963	Beginning of construction of segmental bridges with spans 50–80 m made from prestressed concrete prefabricates erected by cantilever method		

Year	Bridge	Details	Notes
1964–1986	Thirteen concrete bridges built by this method	Frame structure with middle hinge; maximum spans: 60–77 m; lengths: 78.5–1038 m	Built by Inžinierske Stavby, Inc. in Košice
1964	Concrete frame two-cell box girder bridge with middle hinge over Ondava River in Sirnik	Three spans: 30.15 m + 60 m + 30.15 m; length 122 m; width 11.5 m; twin box girders connected with reinforced concrete panel	First segmental bridge from prestressed concrete prefabricates erected by the balanced cantilever method; built by Inžinierské Stavby, Inc. in Košice
1965	Concrete bridge over Váh River in Kolárovo	See Section 17.3.2.1	
1967	Steel bascule bridge to the island of Red Fleet; over Danube River branch in Komárno	Two arms 25 m long each; width of the road 7 m, two 2.4 m wide sidewalks; navigation opening 40 m	One of two bascule bridges (Figures 17.72 and 17.73) in Slovakia; project of Dopravoprojekt, Inc. in Bratislava
1972	SNP Bridge (called New Bridge in the period 1993–2012) over Danube River in Bratislava	See Section 17.6.1 and Table 17.1	
1983	Concrete viaduct in Podtureň	See Section 17.3.1.1	
1983, 1985	Port Bridge over Danube River in Bratislava	See Section 17.4.3 and Table 17.1	
1986	Concrete bridge over Dovalovec; motorway D1	Two independent parallel prefabricated bridges; spans 39.5 m + 6 × 75 m + 39.5 m; total length 533 m; uniform depth of two single-cell box girders 3.5 m; width 15.14 m; built by Inžinierske Stavby, Inc. in Košice	
1987, 1988	Two parallel concrete bridges over Váh River in Sereď (Figure 17.26)	Cast-in-place segmental concrete bridge erected by the cantilever method; spans 46 m + 2 × 90 m + 46 m + 12 × 30 m; total bridge length 634.04 m; the second bridge opened in 1988	
1991	Lafranconi Bridge over Danube River in Bratislava	See Section 17.3.2.2 and Table 17.1	
2001	Mária Valéria Bridge over Danube River in Komárno; steel crescent-type Pratt through-truss bridge	See also year 1895 in this table; reconstruction of the bridge destroyed in 1944 was realized after 57 years using money from the European Union	
2005	Apollo Bridge over Danube River in Bratislava	See Section 17.5.1 and Table 17.1	
2007	Viaduct Skalité	See Section 17.3.3	
2010	Viaduct Studené	See Section 17.5.2	
2010	Extradosed urban viaduct in Považská Bystrica	See Section 17.3.2.3	
2010	Extradosed concrete bridge (Figure 17.27) on motorway R1 over Hron River between Žiar nad Hronom and Ladomerská Vieska	Continuous monolithic concrete structure; spans 26 m + 36 m + 50 m + 80 m + 45 m; reinforced concrete pylon is 8.45 m high; total length of the structure is 239.9 m; width of structure 26.13 m. Project of Dopravoprojekt, Inc. Bratislava, division Zvolen, center Liptovský Mikuláš.	

FIGURE 17.21 Reinforced concrete bridge (Monier system) over Bystrica River in Krásno nad Kysucou (1891). Length 36.4 m, openings 2 × 16.4 m. (Courtesy of P. Paulík.)

FIGURE 17.22 Mária Valéria Bridge over Danube River connecting Slovak Štúrovo with Hungarian Ostrihom (1895–1919). Destroyed in World War I in 1919 and in World War II in 1944. Built again in 2001 in its original shape.

FIGURE 17.23 Colonnade bridge over Váh River in the world-renowned spa Piešťany (1933).

FIGURE 17.24 One of the first fully welded bridges in Slovakia built over Váh River channel near Dolné Kočkovce (1934). (Courtesy of P. Filadelfi, Slovak Water Management Company.)

FIGURE 17.25 Concrete arch bridge over Váh River in Komárno (1955). (Courtesy of P. Paulík.)

FIGURE 17.26 Two parallel bridges over Váh River in Sereď (1987, 1988). (Courtesy of P. Paulík.)

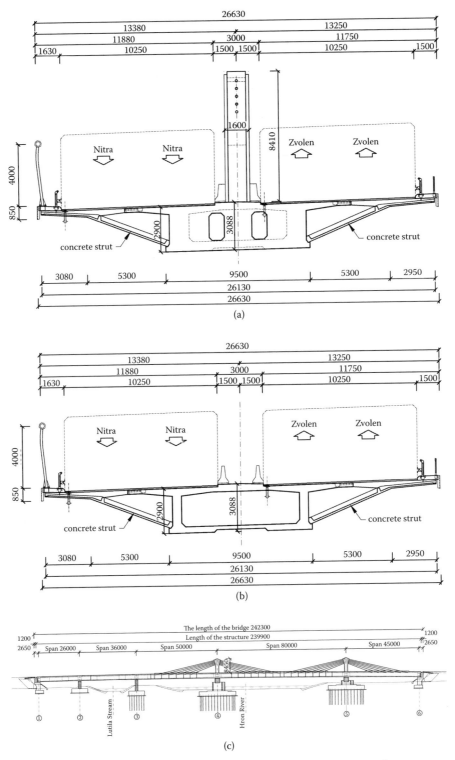

FIGURE 17.27 Extradosed concrete bridge on motorway R1 over Hron River between Žiar nad Hronom and Ladomerská Vieska (2010): (a) typical section at pier, (b) typical section at midspan, (c) elevation, (d) overview. (Courtesy of J. Guoth, Dopravoprojekt, Inc.)

(d)

FIGURE 17.27 **(*Continued*)**

17.1.3.2 Rail Network and Railway Bridges

The current railway network of the Slovak Republic, with a total length of 6876 km, is shown in Figure 17.28. Table 17.5 lists milestones of railway bridge development in modern history. As of 2011 there were 2282 railway bridges with a total length of 50.3 km, including 456 steel bridges and 1826 concrete bridges. About 25% of bridges were built before 1933 and 95% of bridges are in good or sufficient condition. It should be noted that 798 bridges, 72% of the total 1111 bridges, were destroyed during World War II, from 1944 to 1945.

17.1.3.3 Longest Bridge Spans in the Slovak Republic

The longest bridge spans in Slovakia with record span lengths in various types of static schemes are listed in Table 17.6.

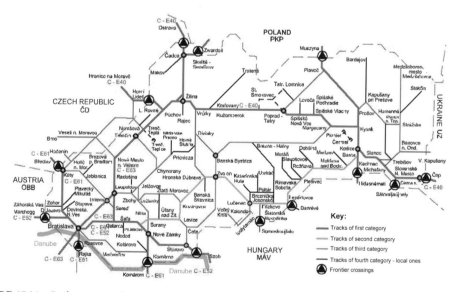

FIGURE 17.28 Railway network in the Slovak Republic. Thickness and level of black color of lines mean first, second, third and fourth category of railway tracks, respectively. (Courtesy of Železnice Slovenskej Republiky Railways of the Slovak Republic, state-owned railway infrastructure company).

TABLE 17.5 Milestones in Railway Bridge Development

Year/Period	Bridge/Name/Location	Bridge Parameters/Main Features	Comments
Railway bridges built in the 1840s and 1850s were mostly stone arches with clearance between piers less than 20 m or timber lattice bridges with spans up to 40 m (Figure 17.30)			
1848	Timber lattice bridge plus 12 brick arches, viaduct over Morava River in Marchegg	Two timber spans 2 × 43 m plus 12 brick arches with clear spans of 15 m; total length 474 m; from 1868 replaced by similar but steel underslung parallel-chord lattice truss bridge	First railway bridge in Monarchy; destroyed in 1866 by Austrian army; built again in 1868 (Figure 17.29), destroyed in 1945; in 1961 replaced by steel semithrough-truss bridge; see 1961 in this table
1848	Red Bridge made from red bricks in Bratislava	Nine arches: 10.7 m + 7 × 18.86 m + 15.75 m	Two tracks, from 1881, reconstructed in 1935, destroyed on April 4, 1945; see also 1948 in this table
1858	First railway lattice bridges made of wrought iron in Hungarian monarchy were built over Ipeľ River and Hron River between Štúrovo and Szob by designer Ruppert Figure 17.30		
1871	Lattice railway bridge over Váh River in Turany made of wrought iron (Figure 17.31)	Span arrangement 2 × 35.1 m + 43.6 m	On Žilina–Poprad track; construction began on December 15, 1869
In the 1870s lattice bridges were replaced by truss bridges made of wrought iron or later made of steel. For smaller spans up to 25 m plate girder bridges were used. After 1880 no timber bridges were built on the main tracks.			
1887	**Temporary Eiffel bridge systems used by the army were replaced by temporary Kohn bridge systems after this year.**		
In the 1880s efforts to save material led to through-truss bridges with broken upper chords or underslung truss bridges with broken bottom chords. 1883: bridges over Váh River in Trenčín, Púchov and Nosice; 1888: bridge over Vlára River on the Trenčianska Teplá–Veselí nad Moravou track.			
1891	Bratislava railroad bridge	See railway Old Bridge in Table 17.1	
1895	In this year, mild Martin's steel replaced wrought iron in railway bridges.		
1896	Viaduct Pod Dielom on the cog railway track Tisovec–Pohronská Polhora, built in 1893–1896; steel underslung Pratt truss girder (Figure 17.32)	Stone arches at bridge ends and five-span steel trusses simply supported by stone towers 31 m high; clear span 25 m; curved in plane with radius 250 m	Abt's cog railway (Swiss inventor Carl Roman Abt (1850–1933), patent from 1882) was used to overcome 48% slope; destroyed in 1945; replaced in 1946 by unique temporary timber trestle with three stories; repaired in 1952 (fire), 1955, and 1959; see also 1959
1896	Čertov (Devil's) Viaduct in Devil Valley on the cog railway track Tisovec–Pohronská Polhora, built in 1893–1896; steel underslung Pratt truss girder (Figure 17.33)	Three spans simply supported by high stone towers 20 m high	Abt's cog railway was used to overcome 35% slope; destroyed in 1945

Year	Description	Notes
1900	After 1900 temporary Roth-Wagner bridge systems were used. The bridge systems were used for spans up to 84 m, exceptionally up to 114 m. At the turn of century reinforced concrete appeared in railways, but only in structures with very small spans.	
1909	Friendship Bridge over Danube River in Komárno; through-truss bridge	Blown up in World War II in March 1945; rebuilt in 1954; see also 1954
1912	Viaduct Handlová on the Horná Štubňa–Prievidza track; underslung Pratt truss girder with parabolic curved bottom chord	Six spans: 2×26 m + 3×31 m + 26 m; 20 m high; curved in the plan with R = 275 m; Destroyed in 1945
Circa 1924	Viaduct Dobrá Niva made of plain concrete	83 m long and 14 m high; On the Zvolen–Krupina track, built in 1923–1925

1930s: Progress in reinforced concrete technology made possible the period of building of mountain railways on the tracks Červená Skala–Margecany (1931–1936), Banská Bystrica–Horná Štubňa (1936–1940), Slavošovce–Lubeník (1940–1941). Steel bridges dominated: Until 30 m long spans these bridges were designed as through, half-through, but mostly as deck plate girders, using in the later period prefabricated elements. For longer spans underslung or through-truss girders were used.

Year	Description	Notes
1933	Telgártsky Viaduct; reinforced concrete and stone arches (Figure 17.34)	86.2 m long; 22 m high; middle arch from reinforced concrete with clear span 31.88 m; two plus two side stone arches with clear span 9 m; First railway reinforced arch bridge; on the Červená Skala–Margecany track, built in 1931–1936
Circa 1933	Chmarošský Viaduct over Chmaroška Stream in Gregová Valley (Figure 17.35); stone arches	113.6 m long, 18 m high; 9 stone arches with clear spans 10 m
Circa 1937	Steel through-truss bridge over Váh River in Púchov; the largest truss bridge from this period	Five-span truss bridge for two tracks with main span 73 m long and with parabolic upper chord; the other four spans had parallel upper and bottom chords and spans from 32–40 m; on the Púchov–Horní Lideč track (1935–1937)
1939	Viaduct at Ulanka (Figure 17.36); four reinforced concrete arches	19 m + 55 m + 2×19 m; maximum height 42 m; On the Banská Bystrica–Horná Štubňa track, built in 1936–1940 (also with four unitized steel viaducts: Kostiviarsky, Čierna Voda, Glózy, and Na Vode)
1940	Viaduct Čierna Voda; steel underslung plate girder	Five spans 25 m long simply supported by high stone towers
1941	Viaduct between Kopráš and Mníšany, parts of village Magnezitovce, (Figure 17.37); concrete and reinforced concrete arches	Five arches, middle arch has a span about 53 m long; on the Slavošovce–Lubeník track, one of three so-called Gemer connections, which were started in 1940–1941 but never finished; After Vienna arbitrage in 1938, a part of Slovak territory was annexed by Hungary; this was the reason for building the Gemer connections, which were never finished after returning the territory after World War II

(*Continued*)

TABLE 17.5 (*Continued*)   Milestones in Railway Bridge Development

Year/Period	Bridge/Name/Location	Bridge Parameters/Main Features	Comments
1942–1943	Four viaducts near Hanušovce nad Topľou on the Prešov–Kapušany–Vranov nad Topľou–Strážovce track (1939–1943); several spans of simply supported steel modified Warren underslung trusses with broken bottom chords	1) Bridge length 99.8 m with a span arrangement 3 × 30 m; 2) curved in plain; bridge length 395.05 m with a span arrangement of one stone arch and nine spans (30.5 m + 8 × 42.4 m; the highest pillar 28 m high (Figure 17.38); 3) bridge length 283.36 m with a span arrangement of two stone arches + 4 × 52.2 m + two stone arches (Figure 17.39); and 4) bridge length 202.95 m with a span arrangement 34 m + 3 × 40 m + 34 m. Destroyed in 1945, reconstructed in 1946–1948.	
Before 1945	Before 1945, all concrete bridges were made from reinforced concrete.		
1945	Emergency or temporary timber bridges built in short period after World War II (Figure 17.40)	Usually built parallel to original bridge axis at 10 m – 12 m distance	Temporary bridges with planned life 10 years; shortage of steel temporary assemblies
1945	Steel temporary assemblies built in a short period after World War II (Figure 17.40)		Shortage of steel temporary assemblies
1948	Red Bridge in Bratislava; steel underslung plate girders	Two parallel bridges structures on the same piers; continuous three-span bridges (3 × 32 m) polygonally curved in plan girders	Designed by A. Tesár; the bridge replaced damaged original Red Bridge; see 1848 in this table
1948–1950	Period of replacing temporary bridges with permanent ones. The only solutions were steel bridges and monolithic concrete bridges.		
After 1950	After 1950, a shortage of steel materials opened the door to design of prestressed concrete and prefabrication bridges.		
1950–1961	Older steel railway bridges were fabricated and constructed using rivet connections.		
1949	Steel viaduct on the Youth Track; modified Warren underslung truss with broken bottom chord	Arrangement of spans in section 3 of the Hronská Dúbrava–Banská Štiavnica track (1948–1949): 3 × 42.4 m	
1949	Concrete viaduct with five arches on the Youth Track	Arrangement of spans in section 5 of the Hronská Dúbrava–Banská Štiavnica track (1948–1949): 2 × 8 m + 3 × 10 m	

Year			
1949	Concrete viaduct with five arches on the Youth Track	Arrangement of spans in section 6 of the Hronská Dúbrava–Banská Štiavnica track (1948–1949): 5 × 16 m	
1950	Old Bridge over Danube River in Bratislava; steel diamond-type through-truss bridge	Laid on the same piers as road bridge, it served 35 years until 1985 and was not in use in the following years (see Section 17.4.2); total bridge length is 460.07 m	
1954	Friendship Bridge over Danube River in Komárno; steel modified Warren through-truss bridge	Single-span plus four-span continuous riveted truss girders; 81.84 m + (102.9 m + 2 × 103.32 m + 102.9 m); it crosses borders between Slovakia and Hungary	The longest railway bridge in Slovakia at 495.34 m long; it replaced a bridge from 1909 destroyed in 1945; built together with Hungary
1959	Viaduct Pod Dielom on the Tisovec–Pohronská Polhora track; underslung Pratt truss bridge	Designed by A. Tesár; five-span continuous polygonally arranged in plane trusses with spans 5 × 28 m	First steel railway truss fully on-site welded in Czechoslovakia; replacement of temporary timber trestle from 1946; see also 1896
1961	The first two railway bridges (see next two items) with welding used in the fabrication shop and high-strength friction-grip (HSFG) bolts used on site. Until 1972 on-site friction connections were used for another 28 bridges (truss girders, plate girders, and girders stiffened by arch).		
1961	Bridge over Rajčianka River in Žilina; steel skew modified Warren semithrough-truss bridge (U-frame)	Two independent bridge structures for each track; span of main girder 44.16 m; chords and diagonals have closed profiles; weight of one steel bridge structure: 142.5 tons	First time used in railway bridges in Czechoslovakia: elements welded together in a fabrication shop, with in situ friction connections using HSFG bolts
1961	Marchegg Bridge over Morava River; steel modified Warren semithrough-truss bridge (U-frame)	Continuous two-span bridge, 2 × 43 m; friction connections of both chords and diagonals; weight of the steel bridge: 160.3 tons	First time used in railway bridges in Czechoslovakia: elements welded together in a fabrication shop, with in situ friction connections using HSFG bolts
1962–1966	Development of prefabrication in railway bridge engineering		
1966	Precast prestressed concrete structure, Jaklovce Viaduct	First railway bridge in Europe from prestressed concrete prefabricates erected without scaffold; spans 7 × 18 m; length 119 m	Built by Inžinierske Stavby, Inc. in Košice from fabricated KT-18 beams post-tensioned on site, erected by launching using special erecting TMS bridge (TMS = Ťažká mostná sústava, Heavy Bridge System)
1967	Two concrete bridges on the Margecany–Jaklovce track	They differ only in height of piers; three-span frame: 30.5 m + 55 m + 30.5 m; single-span box girder with nonuniform depth from 2.17–3.94 m	First two segmental bridges in Europe built by Inžinierske Stavby, Inc. in Košice

(Continued)

TABLE 17.5 (Continued)   Milestones in Railway Bridge Development

Year/Period	Bridge/Name/Location	Bridge Parameters/Main Features	Comments
1970	Bridge on the Čadca–Skalité pri Čiernom track; underslung plate girders with broken bottom flange	Bridge span is 25.44 m; reconstruction of the previous bridge	First use of timber ties supported by a central steel longitudinal bar; the steel bar was attached directly to the flange of main plate girder
1973	Steel–concrete composite bridge on the track between Devínske Jazero–Stupava over motorway D2	One track; single-cell box girder with 205 mm thick monolithic reinforced slab; three spans: 14.45 m + 28.9 m + 14.45 m; steel consumption 103.6 tons	First two railway steel–concrete composite bridges; project concern Vitkovice, Inc., office in Bratislava
1973	Steel–concrete composite bridge on Zohor–Plavecký Mikuláš track over a state road	One track; single-cell box girder with 220 mm thick monolithic reinforced slab; three spans: 11 m + 22 m + 11 m; steel consumption 88.5 tons	First two railway steel–concrete composite bridges; project concern Vitkovice, Inc., office in Bratislava
1973	Steel Langer-type through bridge over Vlára River on the Trenčianska Teplá–Vlára track	Bridge span is 63 m	First steel plate girder stiffened by an arch (Langer girder)
1973	Steel box girder bridge in Trnovec nad Váhom at Šaľa over Váh River on the Bratislava–Štúrovo track	In spans 1 and 2 there are original modified Warren underslung trusses with broken bottom chords from 1956; in other spans there are simply supported fully welded box girders with 31.2 m long spans (spans 3 to 8) and 30.56 m long spans (spans 9 to 14).	An effort to save timber ties led to using first time rails directly attached to the steel box girder (which was too noisy). The total length is 445.79 m; a parallel bridge on the same piers is a modified Warren underslung truss with a broken bottom chord with 14 simply supported spans.
1977	Steel through-truss bridge over Ipeľ River in Ipeľský Sokolec		First steel through-truss bridge with rails directly attached to steel structure (this solution was too noisy)
1979	Blue Bridge in Bratislava; steel modified Warren through-truss bridge	Truss bridge with open bridge deck; span 56 m; bridge elements were welded in a fabrication shop, with on-site friction connections using HSFG bolts	One of the larger bridges with timber ties supported by a central steel longitudinal bar
1980	Skew steel plate girder bridge, Bratislava–Pálenisko	Two-span continuous girder: 2 × 26.28 m; bottom orthotropic deck with 15 mm thick steel plate and 50 mm thick concrete layer	First bridge with ballasted deck
1984	Bridge over Váh River in Trenčín; steel modified Warren through-truss bridge	Set of simply supported girders with arrangement of spans: 4 × 61.62 m	Rails directly attached to steel structure (this solution was too noisy)
1983	Port Bridge	See Section 17.4.3 and Table 17.1	
1983	Steel–concrete composite railway approach spans of Port Bridge in Bratislava	The largest steel–concrete composite structure in Slovakia	
2000	Two parallel Langer-type through bridges over Váh River near Žilina	Span 112 m, total height 19.79 m, spacing of main giders 7.1 m; the girder is 2.86 m high; plate thickness of the steel deck is 14 mm; opened on December 5, 2000. (Figure 17.41)	

TABLE 17.6 Bridges with the Longest Spans in the Slovak Republic (state in Sepetmber 2013)

No.	Bridge, Location, Obstacle	Longest Span (m)	Superstructure Material; Railway, Motorway or Roadway	Open in Year
I. Solid-Web Girders				
1	Lafranconi over Danube River in Bratislava	174	PC; motorway	1981
2	Roadway bridge over Orava River in Dolný Kubín	90	PC; roadway	1983
3	Two parallel roadway bridges over Váh River in Sereď (the second one was opened in 1988)	90	PC; roadway	1987
4	Roadway bridge, Nové Zámky–Komoča	88.5	PC; roadway	1968
5	Motorway bridge over Morava River on motorway D2	80	PC; motorway	1980
6	Roadway bridge in Tomášov	78	PC; roadway	1967
II. Truss Girders				
1	Port Bridge over Danube River in Bratislava	204.8	Steel; railway and motorway	1983 1985
2	Roadway bridge over Danube River in Medveďov	133	Roadway	1942 1973*
3	Roadway bridge over Váh River in Komárno	115	RC; roadway	1913 1945†
4	Friendship Bridge over Danube River in Komárno	(101.76) 103.32	Steel; railway	1909 1954*
5	Roadway bridge over Danube River distributary in Medveďov	102.6	Steel, roadway	1960
6	Elisabeth Bridge over Danube River in Komárno	102	Steel; roadway	1892 1951*
7	Old Bridge over Danube River in Bratislava; two parallel bridges, roadway and railway	92.24	Steel; roadway Steel; railway	1890 1946*
8	Roadway bridge over Bodrog River in Viničky	92	Steel; roadway	
III. Arch Bridges, Langer-Type Bridges				
1	Apollo Bridge over Danube River in Bratislava	231	Steel; roadway	2005
2	Roadway bridge over Váh River in Komárno (replaced the bridge in table part II.3)	(112.5)	RC; roadway	1955*
3	Two parallel Langer-type bridges over Váh River near Žilina	112	Steel; railway	2000

(Continued)

TABLE 17.6 (Continued) Bridges with the Longest Spans in the Slovak Republic (state in Sepetmber 2013)

No.	Bridge, Location, Obstacle	Longest Span (m)	Superstructure Material; Railway, Motorway or Roadway	Open in Year
4	Roadway bridge over Orava River in Dlhá nad Oravou	80 (78)	RC; roadway	1924 1958*
4	Viaduct on motorway D1 between Studenec and Jablonov	80	Steel; motorway	2007
IV. Cable-Stayed Bridges (C-sB) and Footbridges (C-sF), Extradosed Bridges (ExB)				
1	C-sB: SNP bridge (called New Bridge in the period 1993–2012) over Danube River in Bratislava	303	Steel; roadway	1972
2	ExB: Urban viaduct in Považská Bystrica	122	PC; motorway	2010
3	ExB: Bridge between Žiar nad Hronom and Ladomerská Vieska	80	PC; motorway	2010
4	C-sF: Footbridge over Hron River in Žarnovica	65	Steel	2006
5	C-sF: Footbridge over Váh River in Stankovany	60.4	Steel	1992
V. Suspension Pipelines and Footbridges, Prestressed Space Cable Structure (PSCS)				
1	Pipeline and footbridge in Winter Port on Danube River in Bratislava	156	Steel (PSCS)	1965
2	Footbridge over Orava River in Dolný Kubín	100	Steel (PSCS)	1969
3	Footbridge over Hron River in Voznica	77.4	Steel	?
4	Pipeline bridge over Váh River in Rybarpole	75.9	Steel (PSCS)	1965

*Year of reconstruction or rebuilding after destruction in World War II.

†Year in which the bridge was destroyed.

Note: PC = prestressed concrete; numbers in parentheses indicate clear spans. PSCS = prestressed space cable structure, RC = reinforced concrete.

FIGURE 17.29   Steel lattice railway bridge in Marchegg over Morava River (1868–1945).

FIGURE 17.30   Timber lattice railway bridge over Hornád River on the track Kysak–Prešov (about 1869–1881). It was replaced by a steel bridge; span arrangement 2 × 25 m. (Courtesy of MDC, ŽSR-VVÚŽ in Bratislava. Museum Documenation Center, Railways of the Slovak Republic, and Research and Development Institute of Railways of the Slovak Republic)

FIGURE 17.31   Steel lattice railway bridge over Váh River in Turany (1871) with a span arrangement 2 × 35.1 m + 43.6 m on the track Žilina–Poprad with beginning of building on December 15, 1869. (Courtesy of MDC, ŽSR-VVÚŽ in Bratislava. Museum Documenation Center, Railways of the Slovak Republic, and Research and Development Institute of Railways of the Slovak Republic.)

FIGURE 17.32   Viaduct Pod Dielom (under Mountain Diel) near railway station Tisovec–Bánovo on the Tisovec–Pohronská Polhora–Brezno track (1896–1945). (Courtesy of Považské Museum in Žilina.)

FIGURE 17.33   Devil's Viaduct over Devil's Valley on the Brezno–Tisovec–Jesenské track (1896–1945). (Courtesy of The Monument Boards of the Slovak Republic.)

FIGURE 17.34 Telgártsky Viaduct on the Červená Skala–Margecany track (1933). (Courtesy of P. Páteček.)

FIGURE 17.35 Chmarošský Viaduct over Chmaroška Stream in Gregová Valley on the Červená Skala–Margecany track (1936). (Courtesy of P. Páteček.)

FIGURE 17.36   Viaduct at Uľanka (1939). (Courtesy of Ľ. Turčina.)

FIGURE 17.37   Viaduct in Magnezitovce, between Kopráš and Mníšany (1941). (Courtesy of P. Páteček.)

FIGURE 17.38 The highest stone pier of the second viaduct in Hanušovce nad Topľou (1943). (Courtesy of P. Páteček.)

FIGURE 17.39 The third viaduct in Hanušovce nad Topľou (1943). (Courtesy of P. Páteček.)

FIGURE 17.40 First train on temporary bridge structure over Váh River near Strečno (September 13, 1945). (Courtesy of MDC, ŽSR-VVÚŽ in Bratislava. Museum Documenation Center, Railways of the Slovak Republic, and Research and Development Institute of Railways of the Slovak Republic.)

FIGURE 17.41 Two parallel Langer-type bridges over Váh River near Žilina (2000). (Courtesy of P. Paulík.)

17.2 Design Practice

The older standards were based on the allowable stress design (ASD) method. The ASD method is a traditional and intuitive method based on elastic theory, adopted for the design of structures with the objective to provide adequate assurance that the structure will support the applied working loads. In this philosophy, the stresses produced in a member by the applied loads must not exceed a specified allowable stress derived from the yield strength by dividing by an appropriate safety factor. The newer standards are based on the limit state design (LSD) method, which is called in the United States the load and resistance factor design (LRFD) method. In this philosophy, the design value of action effect E_d is compared with the design value of resistance R_d (the characteristic value of material property divided by the appropriate partial factor for material property).

A. Brebera, from the Research Insitute of Civil Engineering Production in Prague, Šletr from Research Institute of Iron Metallurgy in Prague and A. Mrázik from the Slovak Academy of Sciences in Bratislava obtained 1,053,394 values of yield stress and 984,391 values of tensile strength of steel made in Czechoslovakia from extensive tests in the periods 1956–1959 and 1966–1988. A. Mrázik and J. Djubek from the Slovak Academy of Sciences in Bratislava elaborated on these tests the "Framing Guidelines for the Analysis of Steel Structures According to Limit States," published on June 20, 1960. The first structure recalculated using these guidelines was a hangar of Ruzyně airport in Prague in 1960. This recalculation, which was done on the recommendation of F. Faltus, led to the saving of the steel material and showed that the quality of foreign steel was not better than Czechoslovak steel of grade 52. The first publication in Czechoslovakia relating to limit states and theory of reliability was the proceedings edited by K. Havelka in 1964 (Havelka 1964).

The LSD method was used in the Soviet standard SniP II-C.3-62, "Steel structures—Designing standards" in 1962. This standard was the base for common Council for Mutual Economic Assistance (RVHP) RS 161-64 "Steel structures—Designing standards" from 1964, approved by standing committee in 1965 (Mrázik 2010). Consequently Czechoslovak standard ČSN 73 0031 "Analysis of civil engineering structures and foundations." Basic requirements were approved on December 28, 1961, and effective from January 1, 1963. This standard was followed by publishing of a set of standards based on the LSD method, including those for the design of structures made of different structural materials: ČSN 73 1001 "Foundation of structures: Foundation soil under flat foundations" (approved March 30, 1966 and effective from July 1, 1967), ČSN 73 0035 "Loading of building structures" (approved March 30, 1967 and effective from March 1, 1968), ČSN 73 1401 "Design of steel structures" (approved June 15, 1966 and effective from January 1, 1969), ČSN 73 6203 "Actions on bridges" (approved March 13, 1968 and effective from July 1, 1969), ČSN 73 1201 "Design of concrete structures" (approved October 25, 1967 and effective from January 1, 1970), ČSN 73 2603 "Execution of steel bridge structures" (approved December 18, 1968 and effective from January 1, 1970), ČSN 73 1701 "Design of timber structures" (approved July 30, 1969, and effective from January 1, 1971), and ČSN 73 62005 "Design of steel bridges" (approved December 17, 1969, and effective from January 1, 1972), and so on.

Development of some Czechoslovak and Slovak national standards may be seen in Table 17.7.

17.2.1 Loading on Roadway Bridges

Regulations from 1904 and 1919 (Table 17.7) specified different sizes of vehicles, and different axle spacing for loading classes I, II, and III. Loading values increases were as follows: a steam roller in 1919 had 180 kN, and a plough machine in 1919 had 220 kN for loading class I. Already in the 1904 regulations there was uniform loading with value 4.6 kN/m^2. The impact factor was specified for the first time in the 1931 standard and formulae for it were effective until 1987. Uniform regulation from 1937 specified for all loading classes I, II, and III a uniform shape and dimension of vehicles, which differed by loading value. For class I a steamroller had 240 kN and uniform loading had 5 kN/m^2. The decree from 1945 had loading classes A and B and an ideal vehicle with 600 kN. The Temporary Regulations from 1951 and standard from 1953 were used in the design of bridges for a long time. They specified loading classes A, B, and C and five types of ideal vehicles, line loading and tabulated values of an impact factor depending on the size of the span. Similar standards ČSN 73 6203 from 1969 and 1976 put into practice loading classes A and B, the synchronic effect of ideal vehicle and uniform loading, and coefficients of loading combinations.

TABLE 17.7 Standards and Methods for Design of Bridges in Slovak Territory

Effective Date	Content	Standard, Regulation, Guide, Decree; Design Method: ASD or LSD
August 28, 1904	Bridges	New bridge regulation published by Austrian Ministry of Railways
September 15, 1918	Bridges	Amendment to regulation from June 15, 1911 about execution of structures of roadway bridges made of plain concrete
April 10, 1919	Bridges	Modified amendment version in the Czech language published by the Ministry of Public Works of the Czechoslovak Republic
1923	Bridges	Draft of Czechoslovak bridge regulation, V/1923, No. 21, No. 23
1929	Steel structures	ČSN 1051 "Design of steel structures"
1931	Bridges	ČSN 1090–1931 "Design of bridge structures," effective from 1936
1937	Bridges	ČSN 1230–1937 "Uniform bridge regulations Part I: Design of bridges"
September 15, 1945		Decree of the Ministry of Transport No. 128/4–II/7
June 1950	Steel structures	ČSN 05110–1949, "Design of steel structures," ASD
1950	Loading of structures	ČSN 1050–1950 "Loading of structures"
1951	Bridges	"Regulations for railway and roadway bridges Part I: Guideline for design of bridges"
May 1, 1953	Loading of bridges	ČSN 73 6202–1953 "Loading and analysis of bridges Parts I–V"
April 6, 1954		Directive of the Ministry of Civil Engineering
1960	Bridges	Guideline for designing of bridges, amendment
January 1, 1960	Prestressed concrete structures	Guideline for design of structures made of prestressed concrete
January 1, 1963	Basic requirements	ČSN 73 0031 "Analysis of civil engineering structures and foundations." Basic requirements
1964		Directive of the Ministry of Civil Engineering
July 1, 1967	Foundation soil	ČSN 73 1001 "Foundation of structures Foundation soil under flat foundations," LSD
1968		Decree of the Ministry of Transport
March 1, 1968	Loading of buildings	ČSN 73 0035 "Loading of building structures"
January 1, 1969	Steel structures	ČSN 73 1401 "Design of steel structures," LSD
July 1, 1969	Actions on bridges	ČSN 73 6203 "Actions on bridges," LSD
January 1, 1970	Concrete structures	ČSN 73 1201 "Design of concrete structures," LSD
January 1, 1970	Execution of steel bridges	ČSN 73 2603 "Execution of steel bridge structures"
April 1, 1970	Prestressed concrete structures	ČSN 73 1251 "Design of structures made of prestressed concrete," ASD
January 1, 1971	Timber structures	ČSN 73 1701 "Design of wood structures," LSD
January 1, 1972	Steel bridges	ČSN 73 6205 "Design of steel bridge structures," LSD
June 1, 1972	Concrete bridges	ČSN 73 6206 "Design of concrete and reinforced concrete bridge structures," ASD
July 1, 1976	Actions on bridges	ČSN 73 6203 "Actions on bridges," amendment a 4-5/1976
January 1, 1977	Bridge terminology	ČSN 73 6200 "Bridges: Terminology"

January 1, 1978	Steel structures	ČSN 73 1401 "Design of steel structures," LSD
February 1, 1978	Loading of buildings	ČSN 73 0035 "Loading of building structures"
1980	Steel–concrete composite bridges	Guideline "Structures of steel composite bridges made of I-profiles"
July 1, 1970	Bridge register	ON 73 6220 "Register of bridges on motorways," expressways and urban roads, amendment a–4/1983
August 1, 1984	Timber structures	ČSN 73 1701 "Design of wood structures," LSD
March 1, 1986	Steel structures	ČSN 73 1401 "Design of steel structures," LSD
January 1, 1987	Steel bridges	ČSN 73 6205 "Design of steel bridge structures," LSD
August 1, 1987	Assessment of CE structures	ČSN 73 0038 "Design and assessment of civil engineering structures during rebuilding"
September 1, 1987	Actions on bridges	ČSN 73 6203 "Actions on bridges"
May 1, 1988	Actions on structures	ČSN 73 0035 "Actions on structures"
October 1, 1988	Concrete structures	ČSN 73 1201 "Design of concrete structures," LSD
February 1, 1990	Bridge carrying capacity	ON 73 6232 "Recalculations of steel railway bridges and calculation of their load-carrying capacity"
1990	Timber structures	Amedment a–9/1990 to ČSN 73-1701, LSD

Since April 1, 2010, only Eurocodes STN EN 1991 together with the Slovak National Annexes may be used for the design of buildings and civil engineering structures in Slovakia.

March 2004	Actions on bridges	STN EN 1991-2 "Actions on structures. Part 2: Traffic loads on bridges"
April 2005	Timber bridges	STN EN 1995-2 "Design of timber structures. Part 2: Timber bridges"
June 2006	Concrete bridges	STN EN 1992-2 "Design of concrete structures. Part 2: Concrete bridges"
June 2006	Steel–concrete composite structures	STN EN 1994-2 "Design of composite steel and concrete structures. Part 2: General rules and rules for bridges"
June 2006	Earthquake resistant bridges	STN EN 1998-2 "Design of structures for earthquake resistance. Part 2: Bridges"
April 2007	Steel bridges	STN EN 1993-2 "Design of steel structures. Part 2: Steel bridges"

Note: CE = Civil Engineering

17.2.2 Loading on Railway Bridges

Regulations effective from 1867 specified for main track bridges three, three-axle tank engines with 13 tons per axle, and for local track bridges three, four-axle tank engines with 10 tons per axle. The increasing weight of locomotives led to modification of loadings beginning in 1893. For main tracks two four-axle tank engines with 16 tons per axle (6.34 tons/m) together with train wagons with 2.8 tons/m were specified. For local tracks three, three-axle tank engines with 12 tons per axle or three four-axle tank engines with 10 tons per axle, together with train wagons with 2.8 tons/m, were specified. These loadings were increased for bridges with spans smaller than 15 m, to take into account dynamic effects. Further loading increasing were realized in 1907. For main tracks tank engines with 6.8 tons/m, together with 4 tons/m for train wagons, were used. For local tracks four-axle tank engines with 12 tons per axle, together with train wagons with 4 tons/m, were used. From September 1, 1920, the new loadings were specified. The bridges should carry ideal engine with six axles and 20 tons/axle connected with a four-axle tank with 16 tons/axle. These values were modified in 1937 by standard ČSN 1230, which specified 25 tons/axle of ideal engine for all important bridges.

Under German influence the Slovak state modified loadings on November 26, 1941, so that bridges would be designed for three types of loading trains. The heaviest engine had 13.67 tons/m and train wagons 8 tons/m. The new loading scheme was specified in ČSN 73 6202 from 1953, with loading trains of type A (12 axles with 24 tons/axle), type B (20 tons/axle), and type C (18 tons/axle). Group of four axles had 1.25 times greater axle loads than trains. This scheme was replaced in 1987 by UIC/71 loading, with 80 kN/m together with train wagons with four axles and 25 kN/axle. The current traffic loads on bridges are specified in Eurocode STN EN 1991-2 (Table 17.7).

Standards for materials were less often changed than standards for design of structures; they were continuously complemented, and they are not listed in Table 17.7. Basic material properties may be found in primary standards ČSN 73 1201, ČSN 73 1401, ČSN 73 1701, and so on. The many other standards, for example ČSN 73 6209 "Loading tests of bridges," ČSN 73 2089 "Guideline for design of steel-concrete composite girders," ON 73 6212 "Design of timber bridge structures," and ČSN 73 2601 "Construction of steel structures" are also not in Table 17.7. After the peaceful separation of Czechoslovakia into Czech and the Slovak Republics on January 1, 1993, the former Czecholsovak standards became Czech standards (ČSN) and Slovak standards (STN). The new STN standards published after 1993 were no longer identical with ČSN. For example, STN 73 1401, effective from March 1998, differs from ČSN 73 1401, also effective from March 1998.

In the period from November 1998 to June 2004 the European prestandards ENV 199i (i = 1, 2...9) were implemented into the system of Slovak standards. Sixty parts and subparts of European prestandards and two amendments denoted STN P ENV 199i (4451 total pages) could be used in the design of buildings and civil engineering structures parallel to relevant national standards, provided that they were used together with their Slovak National Application Documents and that the complete package of prestandards STN P ENV 199i was available for designed structures. In the period January 2001 to May 2007 the European standards EN 199i (i = 0, 1... 9) were implemented into the system of Slovak standards. Fifty-eight parts and subparts of European standards and several amendments and corrigenda denoted STN EN 199i (5027 total pages) could be used in the design of civil engineering structures parallel to relevant national standards, provided they were used together with their Slovak National Annexes defining National Determined Parameters and that the complete package of standards STN EN 199i was available for the designed structure. In the transition period it was possible to use all three systems of standards, STN, STN P ENV, and STN EN, but the particular standards of different systems could not be mixed. Since April 1, 2010, only Eurocodes STN EN 199i together with Slovak National Annexes may be used for the design of buildings and civil engineering structures in Slovakia.

A system of foreign standards may be used in Slovakia for the design of a structure, but it must be verified that reliability of the design is at least the same as the reliability obtained according to Eurocodes STN EN 199i. From the set of 58 parts and subparts of Eurocodes STN EN 199i the parts "-2 are devoted to bridges, namely STN EN 1991-2 (actions on bridges), STN EN 1992-2 (concrete bridges), STN EN 1993-2 (steel bridges), STN EN 1994-2 (steel–concrete composite bridges), STN EN 1995-2 (timber bridges) and STN EN 1998-2 (earthquake resistance of bridges). The parts -2 refer to other parts of Eurocodes and for design of bridges a package of Eurocodes parts must be used.

17.3 Girder Bridges

17.3.1 Precast Segmental Concrete Bridges Erected by the Cantilever Method

Inžinierske Stavby, Inc. of Košice built 13 segmental bridges made from prestressed concrete prefabricates in the period 1964–1986 (Table 17.4).

17.3.1.1 Viaduct Podtureň

Viaduct Podtureň (Figure 17.42) is one of three bridge structures: the viaduct at the village of Podtureň (1983; length 1038 m, two parallel independent single-cell box girders), Belá Bridge over Belá River (1986; 305 m, two-cell box girder), and Dovalovec Bridge at the village of Dovalovo (1986; 533 m, two-cell box girder) located on motorway D1. The segmental Podtureň Bridge over Váh River was made from prestressed concrete prefabricates erected by the cantilever method. It was built in 1983 by Inžinierske Stavby, Inc. of Košice. In 1986 the longest bridge in Slovakia had a total length of 1038 m and an arrangement of 17 spans, 30 m + 58 m + 2 × 64 m + 58 m + 64 m 2 × 70 m + 67 m + 61 m + 58 m + 61 m + 67 m + 3 × 70 m + 36 m. The structure is 13.4 m wide. The segments were 2 m, 2.5 m, and 3 m long with weights from 60 tons to 76 tons. The concrete was grade B-500.

17.3.2 Cast-in-Place Segmental Concrete Bridges Erected by the Progressive Cantilever Method

Slovakia is among the pioneers of segmental concrete bridge construction. The first prestressed concrete bridge built by the progressive cantilever method was constructed in 1960. More than 30 segmental concrete bridges have been constructed during the past 50 years.

17.3.2.1 Bridge over Váh River in Kolárovo

This bridge has spans 35.1 m + 3 × 61.68 m + 35.1 m, as shown in Figure 17.43. The foundation is on open caissons lowered to a depth of about 15 m below the river bed. The piers have circular cross sections 3.2 m in diameter restrained to the concrete filling of the caisson. The superstructure is a single-cell box cross section with a variable depth between 1.5 m and 3.1 m with short cantilevers on both sides, prestressed longitudinally and transversally. This rigid frame with midspan hinges belongs to the very lastest class of bridges of this type with typical hinges. Problems with deformations of the ends of the cantilevers caused by shrinkage and creep of concrete were the main reason for midspan hinges being abandoned for continuous superstructures without hinges. This bridge opened to traffic in 1965.

17.3.2.2 Lafranconi Bridge over Danube River in Bratislava

The crossing of the Danube River, 300 m wide at this site, offered an extraordinary opportunity to Slovak bridge designers and bridge builders. The bridge (Figures 17.44 and 17.45) consists of two identical, mutually independent parallel prestressed concrete bridge structures of an overall length of 761 m with spans of 83 m + 174 m + 172 m + 4 × 83 m. The cross section of the bridge superstructure is a box girder with cantilever slabs on both the top and bottom levels. The lower cantilever slabs are used for pedestrians and bicycles. The box girder has a variable depth from a maximum of 11.0 m above the pier in the Danube River bed to a minimum of 4.7 m constant depth. The segments are 3.5 m or 5 m long. The total number of segments for the whole bridge is 228. The bridge is an example in which the external prestressing tendons were used probably the first time in the world in this type of bridge construction.

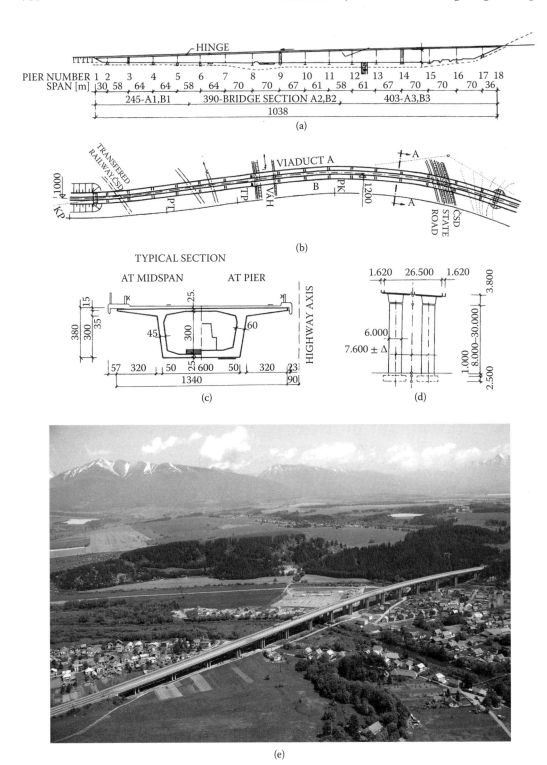

FIGURE 17.42 Viaduct Podtureň (1983): (a) longitudinal section and lengths, (b) elevation, (c) cross section (cm), (d) cross section A-A, (e) birdview. Three continuous girders have lengths 245 m, 390 m, and 403 m. A hinge is located between piers 5 and 6 and between piers 11 and 12. (Courtesy of D. Turanský, Dopravoprojekt, Inc.)

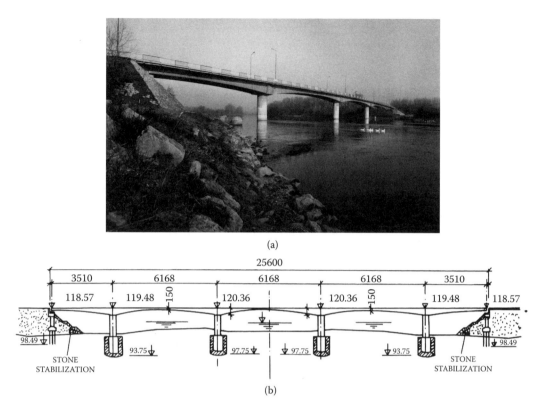

FIGURE 17.43 Bridge over Váh River in Kolárovo (1965) (a) overview, (b) elevation. (Courtesy of G. Tevec, Doprastav, Inc.)

They are placed inside the box girder void and anchored in the blocks near the crossbeam closely below the top slab. The main pier in the river bed has spread footing on the bedrock. The spread foundation was constructed by using a double sheet-pile cofferdam, assembled on shore and floated into position. All other piers have deep foundations consisting of diaphragm walls.

The designers agreed that the overall character of the territory and the proximity of Bratislava Castle Hill and the cable-stayed steel bridge make it preferable to design the bridge without distinct vertical features. The biggest mass of concrete is concentrated in the superstructure, 11 m deep above the river bed pier, which is relatively distant from an observer's position on either river bank. The contours of the structure gradually recede into the part with a constant depth cross section. The horizontal line of the pedestrian path cantilever makes the structure optically more slender. The shaping of the pier in the river bed emphasizes the dominant position and the rhythm of all other supports corresponds with the depth of the superstructure. This architectural concept has also governed the structural design, which has used strict, structurally advantageous symmetry. The adopted asymmetrical solution meant an exacting technical problem both for designer and for the contractor consisting in the jointing of two markedly asymmetrical cantilevers 120 m and 50 m long. The bridge was completed in 1991. The owner of the bridge is NDS, Inc. (National Motorway Company, Inc.).

17.3.2.3 Multispan Extradosed Urban Motorway Viaduct in Považská Bystrica

The motorway extradosed bridge (Figures 17.46 and 17.47) over the urban area of the town of Považská Bystrica, railway track, and Váh River is the latest evidence of the high quality of Slovak bridge designers and builders. The total bridge length is 968 m. The superstructure is a continuous single-cell box girder with 10 spans of 34.16 m + 48.8 m + 70.76 m + 6 × 122 m + 68 m. The main bridge (868 m) was constructed by the cantilever method. The concrete deck is suspended on seven low pylons situated in

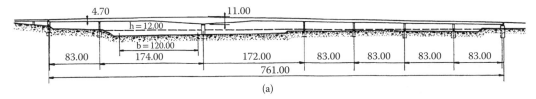

(a)

(b)

FIGURE 17.44   Lafranconi Bridge in Bratislava over Danube River (1991) (a) elevation, (b) birdview. (Courtesy of D. Turanský, Dopravoprojekt, Inc.)

the bridge axis. The box girder of the deck has large overhangs supported by precast V-shaped struts. In addition to the statical function of inclined V-struts, they also create an interesting appearance for the bridge. This is the first motorway bridge in Slovakia with the superstructure carrying the full motorway profile. The superstructure is 30.6 m wide with a variable depth (from 4.6 m up to 6 m) and extended cantilevers. The bridge was simultaneously cast in 2 × 7 symmetrical cantilevers originating from H-shaped piers (Figure 17.46). There is a system of external extradosed stay cables situated within the structure center line, which is routed through seven pylons with heights of 14 m. The fan of external cables consists of eight radially arranged stay cables.

The superstructure is supported by standalone, built up piers that vary in height from 23 m to 33 m. The piers consist of two hexagonal walls finished with four square columns on the top of piers. The stability of the balanced cantilevers under construction is ensured by temporarily fixing the deck at the piers with vertically placed bars. When all cantilevers beams were linked, the joint was released by removing the temporary prestressing. The pier shape and elements, which were temporarily built in it, very effectively (and fancily as well) enabled neuralgic problems with regard to the superstructure construction to be dealt with, namely the superstructure starter section (hammerhead) construction method and the stabilization of the balanced cantilever during the superstructure construction. The implemented shape of the piers enabled construction of the hammerhead as well as the whole balanced cantilever without the need for auxiliary supporting towers. The piers were designed in such way that no further provisional supports were needed during construction of the balanced cantilevers. The 30.6 m wide cross section with precast struts was cast in one shift without expressive interruption.

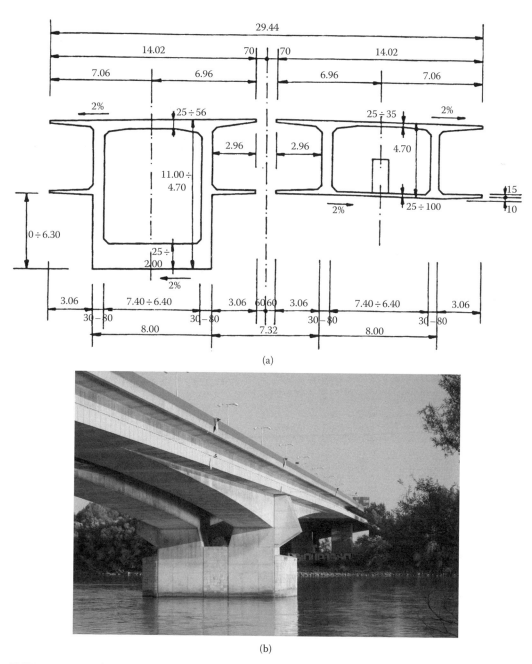

(a)

(b)

FIGURE 17.45 Lafranconi Bridge in Bratislava over Danube River (1991) (a) typical section at pier and at mid-span, (b) overview. (Courtesy of P. Paulík.)

The short construction time required the use of 14 form travelers on the all high piers. The use of the new fast split system with overlapping rear parts of form travelers allowed the length of the hammer head to be shortened, with the possibility of immediately continuing the construction of the cantilevers on both sides. The installation of the concrete precast struts, with a mass of 3 tons, was carried out by a light crane and chain hoist. The anchoring of the stays was solved by opening the upper deck. After launching the form traveler, the opening was used for concreting anchoring block with units for spreading the tensile forces from the stays.

A full-size test segment was constructed before commencement of the construction works on the bridge superstructure. The concrete parameters were tested and the heat of hydratation was measured during and after concreting. A loading test of the segment was also performed. The bridge was built within 22 months and opened to traffic in May 2010 (Figure 17.47). The owner is NDS, Inc.; the designer was a joint venture of firms Alfa 04, Inc. in Bratislava, Slovakia and Strásky, Husty and Partners, Ltd. in Brno, Czech Republic; and the contractor was a joint venture of Doprastav, Inc. in Bratislava, Slovakia

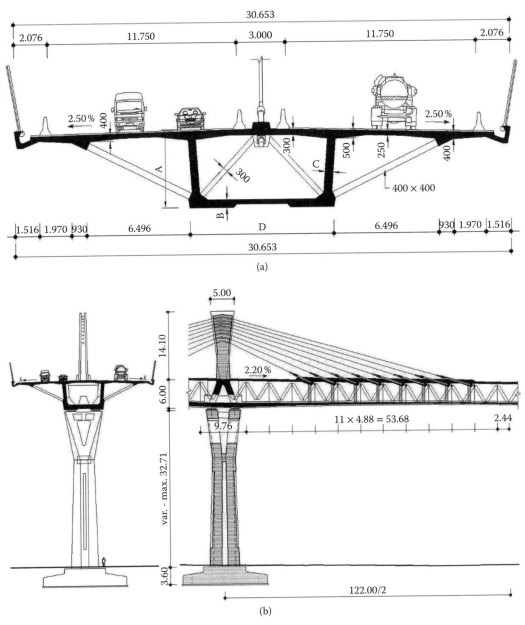

FIGURE 17.46 Multispan extradosed urban motorway viaduct in Považská Bystrica (2010) (a) typical section, (b) tower, cross-section and side view, (c) under construction. The total length is 968 m, in 10 spans. (Courtesy of M. Maťaščík, Alfa 04, Inc.)

(c)

FIGURE 17.46 (*Continued*)

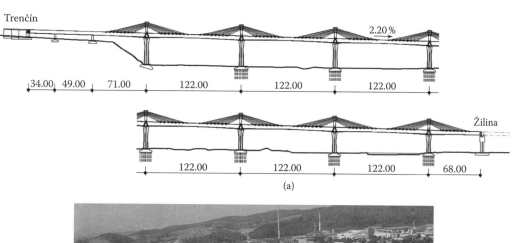

Trenčín

2.20 %

34.00 49.00 71.00 122.00 122.00 122.00

Žilina

122.00 122.00 122.00 68.00

(a)

(b)

FIGURE 17.47 Multispan extradosed urban motorway viaduct in Považská Bystrica (2010) (a) elevation, (b) birdview. The total length is 968 m, in 10 spans. (Courtesy of M. Maťaščík, Alfa 04, Inc.)

and Skanska CZ, Inc. Division 77 Bridges in Brno, Czech Republic. More details may be found in Strásky, Maťaščík, Novák, and Táborská 2011.

17.3.3 Skalité Viaduct

There are four steel bridges with steel–concrete composite decks located on motorway D3 between the village of Skalité and Slovak–Polish border Skalité. Only a half profile of motorway D3 is finished in this part of motorway (Figures 17.48 and 17.49). This part of motorway should be finished in 2018. The bridges have the following total lengths and arrangements of spans. D201: 36 m + 7 × 60 m + 36 m = 492 m (Figure 17.50); D202: 45 m + 60 m + 45 m = 150 m; D203: 45 m + 3 × 60 m + 45 m = 270 m; D204: 40 m + 6 × 60 m + 40 m = 440 m. The contractor and owner is NDS, Inc.; the general designer was Geoconsult, Ltd.; and the designer of the steel structures was Distler-Šuppa, Ltd.

SECTION A-A
M 1:50
Section at pier

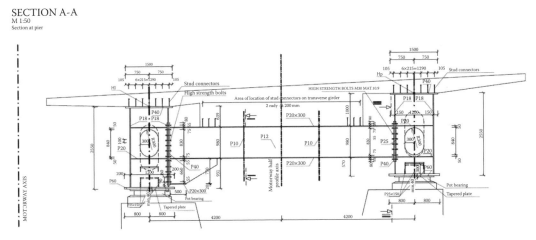

FIGURE 17.48 Cross section of the steel bridge of Viaduct Skalité over a pier, creating a finished half profile of motorway D3. (Courtesy of Distler-Šuppa, Ltd.)

SECTION B-B
M 1:50
SECTION AT MIDSPAN

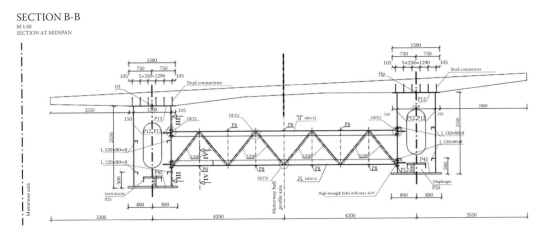

FIGURE 17.49 Cross section of the steel bridge of Viaduct Skalité in the middle of the span, creating a finished half profile of motorway D3. (Courtesy of Distler-Šuppa, Ltd.)

FIGURE 17.50 Skalité Viaduct (2007). (Courtesy of M. Paulini, Váhostav, Inc.)

17.4 Truss Bridges

17.4.1 Old Bridge (Red Army Bridge) over Danube River in Bratislava (Roadway Bridge)

This roadway bridge (Figure 17.51) was built on the piers of Emperor Franz Joseph Bridge, which was destroyed by German soldiers on April 2 and 3, 1945. The bridge was built by 600 soldiers of the Red Army, 450 German war prisoners, and 300 Czechoslovak specialists in 5 months and 17 days between August 4, 1945 and January 24, 1946. The standardized Austrian Roth-Waagner bridge system was used for reconstruction of spans 2 to 5. This demountable provisional truss arrangement was developed and used for the first time in 1915 by Vienna company R. Ph. Waagner–L. & J. Biró & A. Kurz. Spans 1, 6, and 7 are Warren trusses.

The bridge was reconstructed in 1986 and served until 2010. The bridge has a total length of 461.07 m and a span arrangement of 32.2 m + 2 × 75 m + 91.5 m + 75 m + 75.18 m + 32.22 m. Spans 1 and 2 are simply supported. Spans 3 to 5 are continuous truss. Spans 6 and 7 are also continuous truss. The two main trusses are spaced 7 m apart. The height of trusses is 8.28 m (only 5.435 m in span 1 and in spans 6 and 7, where there are Warren trusses). The bridge is 9.77 m wide and the roadway is 5.96 m wide, carrying two traffic lanes. The bridge abutments and piers are made of concrete and supported by caisson foundations. Fix bearings are N-Neotopflager rocker, neoprene, and cast iron and expansion bearings are NGe roller, neoprene, and cast iron.

17.4.2 Old Bridge over Danube River in Bratislava (Railway Bridge)

The railway bridge was built in 1950 on the same piers as the roadway Old Bridge (Figure 17.52). Spans 1 and 7 are Warren trusses. Spans 2 to 6 are double Warren trusses (diamond truss systems). The bridge, with one track, served until 1985 and is now out of operation. The total bridge length is 460.07 m. Two main trusses are spaced 5 m and 8.86 m high (only 4.83 m in spans 1 and 7). All spans are simply supported with a span arrangement of 32.31 m + 75.93 m + 75.64 m + 92.24 m + 75.71 m + 75.82 m + 32.42 m.

17.4.3 Port Bridge (Heroes of Dukla Pass Bridge) over Danube River in Bratislava

This bridge is a double deck steel truss bridge (Figure 17.53). The upper deck has a width of 29.4 m and carries three traffic lanes in each direction. The lower deck carries electrified double-track trains. The main span of the navigation channel has a span length of 204.8 m and clearance of 10 m over the

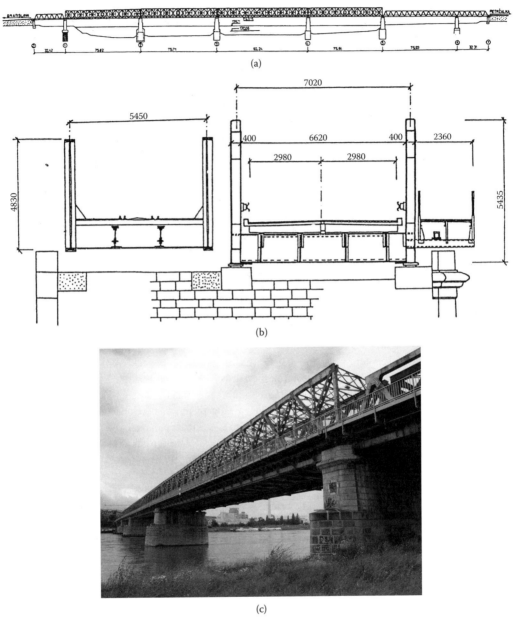

(a)

(b)

(c)

FIGURE 17.51 Old Bridge (Red Army Bridge) over Danube River in Bratislava (roadway bridge; 1946) (a) elevation, (b) typical section, (c) overview. Closed for individual vehicle transport in 2008 and for all vehicles in May 17, 2010. Deck bridge demounting started August 13, 2010. Now it is used only by pedestrians. The owner of the bridge is the city of Bratislava. Attacks of architects and wishes of conservationists continuing till 2013 have prevented to accept and to realise several projects of Old bridges (road and railway) replacement. In August 7, 2013 it was decided that the reconstruction will be in 2013–2015. (Courtesy of I. Baláž.)

maximum shipping water level. The bridge was designed by Dopravoprojekt Inc., Bratislava, Vítkovice Bratislava, SUDOP Bratislava, Hydrostav Bratislava, and Doprastav Bratislava. The bridge was built by Doprastav Bratislava, Vítkovice Ostrava, Hutní Montáže Ostrava, Hydrostav Bratislava, and other firms from 1977 to 1985. It has partly served public traffic since December 1983, when one of carriageways was completed. The owner of the bridge is NDS, Inc.

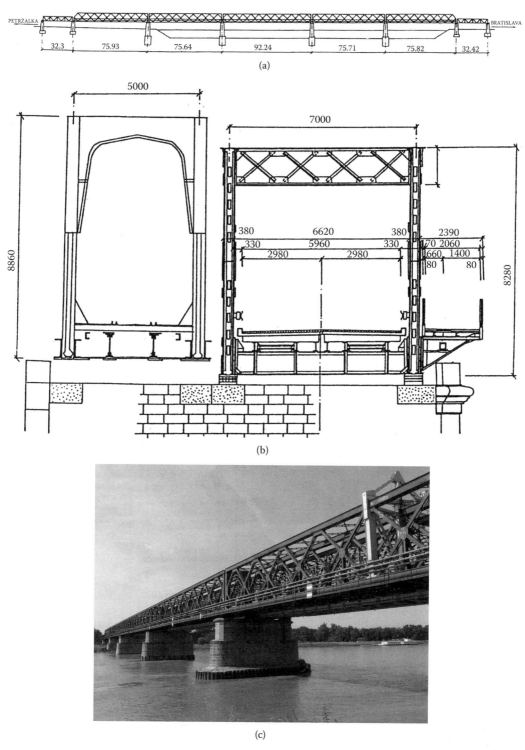

FIGURE 17.52 Old Bridge over Danube River in Bratislava (railway bridge; 1950) (a) elevation, (b) typical section, (c) overview. Out of operation since 1983. The silver cantilever structure seen in the picture was designed to help the neighboring roadway bridge. Due to protest of Slovak Society of Steel structures and shortage of money the erection of another four cantilevers was stopped. The owner of the bridge is the city of Bratislava. (Courtesy of I. Baláž.)

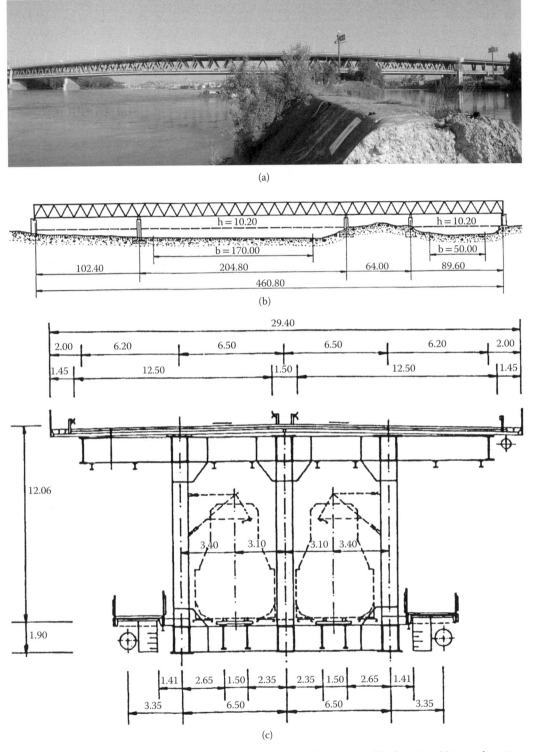

FIGURE 17.53 Port Bridge over Danube in Bratislava (1985) (a) overview, (b) elevation, (c) typical section, (d) inside-view, (e) erection, (f) under construction. (Courtesy of L. Nagy, Vítkovice office, later Dopravoprojekt, Inc., now retired and I. Baláž.)

(d)

(e)

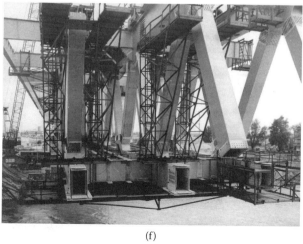

(f)

FIGURE 17.53 *(Continued)*

The bridge has a total length of 460.8 m with a span arrangement of 102.4 m + 204.8 m + 64 m + 89.6 m. It consists of three main steel continuous trusses (Figure 17.53) framed together in the transverse direction by the crossbeams of the carriageway and the railway cross girders. The main trusses are spaced 6.5 m apart and the theoretical depth is 11.7 m. The bay length of the Warren-type truss is 12.8 m. For all field connections of chords and diagonals high-strength friction grip bolts M24, M30, and M36 of steel grade 10.9 were used. The chords and diagonals, with box-shaped cross sections, are made of steel plates of grade 11523 for thickness < 20 mm (f_y = 360 MPa) and grade 11503 for thickness = 20–60 mm (f_y = 350 MPa). The bridge was erected by derrick and Demag cranes.

The carriageway is designed as a steel–concrete composite plate girder connected by shear connectors only to the crossbeams, formed of two standard section I-340 at a distance of 1.83 m. The concrete slab, with total thickness 150 mm, is supported along the bridge on PTFE (Polytetrafluoroethylene) sliding bearings and was designed as a "swimming" deck. Each main truss sits on PTFE disc-type bearings. The fixed bearing laying in the center line of pier 3 is bolted to chord of the middle truss. All other bearings are PTFE sliding disc type and enable free longitudinal expansion of the structure.

The superstructure is supported by five massive concrete piers. Pier 2 was founded on an artificial island using large-diameter bored piles. The other four piers are founded on shallow foundations.

The consumption of steel was as follows: main steel structure 7,630 tons; roadway 2,415 tons; railway 1,215 tons; HSFG bolts 305 tons; pedestrian and bicycle pavements 335 tons; bridge facilities 1,300 tons; for a total of 13,200 tons.

The port bridge was put into service for the railway in December 1983, when the main inspection and loading were performed, including velocity tests on the bridge and adjacent track lines. The loading tests gave results that were in good agreement with theoretical values and the requirements of the Czechoslovak standard ČSN 73 6209 (Table 17.7). The obtained values were natural frequencies from 0.4 to 0.8 Hz, impact factor 1.008, and logarithmic decrement from 0.10 to 0.14. Special requirements for future bridge assessment and maintenance were included in the design proposals. The bridge has been properly inspected and maintained since it was finished in 1985.

17.5 Arch Bridges

17.5.1 Apollo Bridge in Bratislava

The design preparation for the Apollo Bridge (Figure 17.54 to 17.57) started in 1999. Construction began on February 1, 2003 and the bridge opened to traffic on September 5, 2005. The Apollo Bridge crosses the Danube River not far from the Bratislava historical center. The total length of bridging is 840 m. The

FIGURE 17.54 Erection of Apollo Bridge in Bratislava over Danube River (2005). Upstream view, Old Bridge (called before Red Army Bridge) and Bridge (called New Bridge in the period 1993–2012) Bridge in the background. (Courtesy of I. Baláž.)

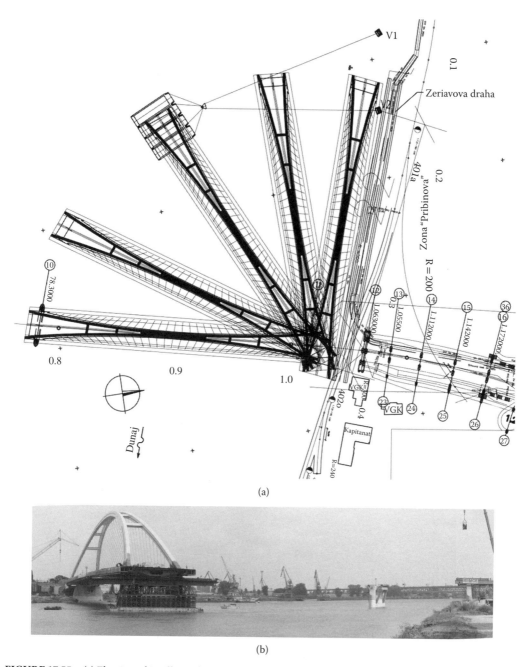

(a)

(b)

FIGURE 17.55 (a) Floating of Apollo Bridge in Bratislava over Danube River (2005) (a) bridge rotation, (b) bridge floating. Downstream view, Port Bridge in the background. (Courtesy of I. Baláž.)

length of the steel bridge structure is 517.5 m with a span arrangement of 52.5 m + 2 × 61 m + 63 m + 231 m + 49 m. The height of the steel arch over the main span is 36 m. The number of cable stays is 2 × 33 pcs-type DYNAGrip 12(0.62). At both sides of the main bridge structure there are approaching flyover structures with lengths of 141.5 m and 195 m made of cast-in-place prestressed concrete. The roadway width is 2 × 8 m and the width between parapets is 19 m–31 m. The clearance of the navigation channel is 210 m width and 10 m height above the maximum level of Danube River.

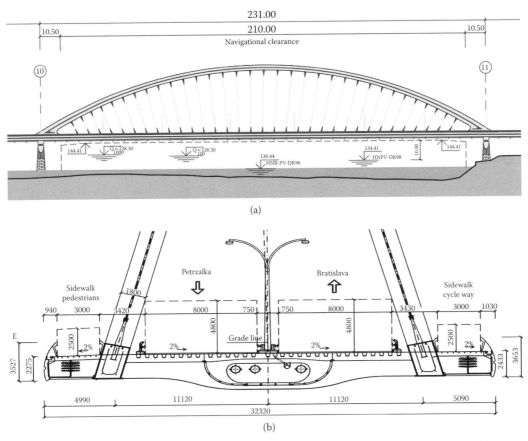

FIGURE 17.56 Cross section of Apollo Bridge in Bratislava over Danube River (2005) (a) elevation, (b) typical section. (Courtesy of M. Maťaščík, Alfa 04, Inc.)

FIGURE 17.57 Apollo Bridge in Bratislava over Danube River (2005). (Courtesy of M. Maťaščík, Alfa 04, Inc.)

The load-carrying system of the main bridge structure is designed as a six-span continuous steel structure with two main girders. The main girders within the main span are suspended by a system of hangers on two mutually inclined arches so that the girders have the function of a tie of the arches. The girders within the main span protrude over the deck in the form of a parapet. A greater depth of the girder is statically more efficient and enables the girders to be more comfortably passable during both the construction and operation stages. The parapet of the bridge also serves as a security element, preventing the free fall of vehicles into the space of the pedestrian and bicycle ways as well as from hitting the hangers.

A dominant visual element of the bridge soffit is the system of the main cross girders, which have a nonstandard shape design. The main cross girders carry an orthotropic steel bridge deck and a longitudinal tubular body in which the lines of service on the bridge are placed. The high-voltage and telecommunication cables are placed in the space of the cantilevered elements of the sidewalks at both sides of the longitudinal main girders.

Due to the skewness of the bridge and the curved lines of the bridge in plan as well as its curved vertical alignment and the inclination of both arches, the geometrical shape of the bridge is very sophisticated. The arches are formed as unsymmetrical sections of ellipse and some of the other construction parts of the bridge are designed as "buckled surfaces." The design of the steel bridge structure is characterized as an all-welded steel structure using weldable fine-grain structural steels ranging in thickness from 8 mm to 40 mm. The bridge deck is made of normalized hot-rolled steel of grade S355 NL, while the main girders and arches are made of thermomechanical rolled steels of grades S355 ML and S420 ML. For the thin-walled and/or less loaded members, steel of grades S355 K2 G3 and S235 J2 G3 were accepted. The total weight of the steel structure is 7850 tons.

Parts of the main girders above the piers are strengthened at their bottom flanges with concrete, giving the effect of a composite beam. All surfaces of the steel structure were painted in a combination of light gray and dark blue. At night, the bridge can be illuminated so that the most interesting features of its appearance will be highlighted.

The main bridge span was constructed by a very interesting and nonstandard method. The steel structure of the main span was erected on the left bank of the river so that one of its bearings was placed on pier 11. The other bridge bearings were supported by a temporary steel work. The opposite end of the steel structure was launched on a group of barges anchored to the bank of the river. The entire structure of the main span was then swung around the bearing placed on pier 11 so that the free opposite end of the main span could be floated to pier 10, which was founded in the river bed. Then the bridge end was launched on pier 10 and placed into its final position above the Danube River.

The Apollo Bridge (known as the Košická Bridge during the construction phase) was honored by the European Steel Design Awards in 2005 for outstanding design in steel construction. The award was given by the European Convention for Constructional Steelwork. The Apollo Bridge was the only European project named one of five finalists for the 2006 Outstanding Civil Engineering Achievement Award (OPAL Award) by the American Society of Civil Engineers. The owner of the bridge is the city of Bratislava. The project author was Miroslav Maťaščík. The consultants were Dopravoprojekt, Inc., Bratislava and PIO Keramoprojekt, Inc., Trenčín. The contractors were Doprastav Inc., Bratislava (lead partner) and MCE Stahl und Maschinenbau GmbH Co. Linz (partner). The project manager was Alexander Menyhárt, Doprastav, Inc., Bratislava.

17.5.2 Studenec Viaduct

The Studenec Viaduct has seven steel arches with a steel–concrete composite deck and is located on motorway D1 between the villages of Jablonov and Studenec (Figure 17.58). A half profile of motorway D1 was finished in this part of motorway on June 7, 2010. The whole profile was finished on December 14, 2012. The total length of the viaduct is 700 m, with a span arrangement of 100 m made of prestressed concrete I-profiles and a 500 m long steel bridge with seven arches (60 m + 70 m + 3 × 80 m + 70 m + 60 m) + 100 m made of prestressed concrete I-profiles. The contractor and owner is NDS, Inc.; the general designer was Geoconsult, Ltd.; the designer of the steel structures was Distler-Šuppa, Ltd.; and the designer of the reinforced concrete deck was Projkon, Ltd.

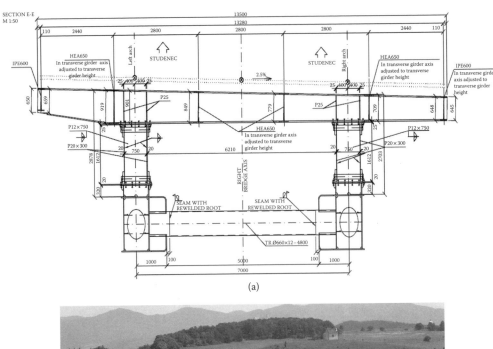

(a)

(b)

FIGURE 17.58 Studenec Viaduct (2010) (a) typical section, (b) overview. Finished half profile of motorway D1 between Jablonov and Studenec. (Courtesy of Distler-Šuppa, Ltd. and F. Tomek, Doprastav, Inc.)

17.6 Cable-Stayed Bridges

17.6.1 SNP Bridge (Bridge of Slovak National Uprising Called New Bridge in the Period 1993–2012)

SNP Bridge is the second permanent bridge over Danube River in Bratislava, the capital of the Slovak Republic. It is a steel cable-stayed bridge with a single backward inclined steel tower and a single cable plane. The asymmetrical position of the inclined 84.6 m high A-shaped tower, crowned by a circular restaurant with 32 m in diameter, creates a natural balance to the famous Bratislava Castle and the St. Martin Dome (Figure 17.59). The restaurant UFO (previously called Bystrica; capacity 120 people) and the public observation deck on its roof offer a nice view of Bratislava (Figure 17.60). SNP Bridge is the only bridge registered in the World Federation of Great Towers, where the key criterion is to have a public observation deck. The bridge certainly looks distinctive and elegant.

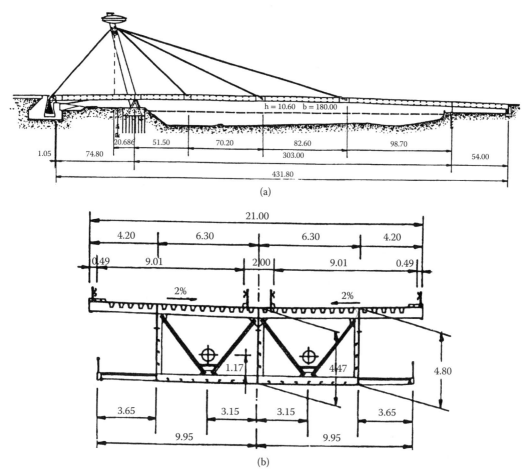

(a)

(b)

(c)

FIGURE 17.59 SNP Bridge (called New Bridge in the period 1993–2012) over Danube River in Bratislava (1972) (a) elevation, (b) typical section, (c) overview. The main span is 303 m. Upstream view; Lafranconi Bridge is in the background. (Courtesy of E. Chladný and I. Baláž.)

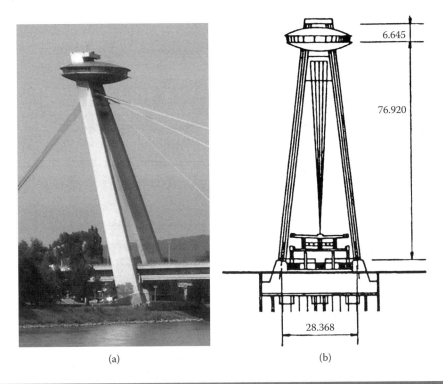

(a)   (b)

(c)

FIGURE 17.60 SNP Bridge. (a) and (b) Steel 84.6 m high A-shaped fully welded steel tower crowned by UFO restaurant with public observation deck at the top. Dimensions of leg cross section: 2.5 m × 8.5 m at the fixed base and 3.6 m at the top, (c) end span with length 54 m between framed pendulum wall and Bratislava bridge end. (Courtesy of E. Chladný and I. Baláž.)

When it was constructed in 1969, the bridge, with main span 303 m, was the world record holder in the category of cable-stayed bridges. The bridge was the fourth-longest cable-stayed bridge span when it opened to traffic on August 29, 1972 (Tesár 1970, Zvara 1971, Tesár 1973). It was the winner of the competition Structure of the Twentieth Century in Slovakia, organized in 2001. The original name of the bridge was Slovak National Uprising Bridge (Most Slovenského Národného Povstania). It was renamed, like other Bratislava Danube bridges, after the collapse of the communist regime in the beginning of the 1990s. The name "New Bridge" was valid only in the period 1993–2012. Before and after this period the name is SNP Bridge. The bridge was designed by a team (members see Baláž, 2012) from Slovak University of Technology (STU) in Bratislava led by Professors Arpád Tesár (structural engineer) and Ján Lacko (architect). The general designer was Dopravoprojekt, Inc., Bratislava; the general contractor was Doprastav Bratislava; the complete design solution was by Project-construction office VŽKG n.p. (Vítkovícke Železárny Klementa Gotwalda, narodný podnik = Vítkovice Iron Works Klement Gotwald, national enterprise), branch Bratislava in cooperation with the Department of Metal and Timber Structures, FCE (Faculty of Civil Engineering), STU (Slovenská technická univerzita = Slovak University of Technology in Bratislava, Slovakia); fabrication of the steel structure was by Vítkovice Iron Works; the bridge was erected by Hutní Montáže, Ostrava (Metal Works Erection company in Ostrava, Czech Republic); ČSPD (Czechoslovak Danube River Navigation company in Bratislava, Slovakia) Bratislava was in charge of the floating; and loading tests were performed by TSUS (Technický a skúšobný ústav stavebný = Building Testing and Research Institute) in Bratislava in cooperation with FCE STU, VÚIS Bratislava (Výskumný ústav inžinierskych stavieb = Research Institute of Civil Engineering Works in Bratislava, Slovakia) and Geodetic Institute Bratislava.

The bridge has three spans of 74.8 m + 303 m + 54 m with a total length of 431.8 m (Figure 17.59). Three cables support the 4.6 m high steel orthotropic box girder, dividing the main span into four segments of 51.5 m + 70.2 m + 82.6 m + 98.7 m = 303 m. The main span is 80.2% of the total length of two stayed spans (74.8 m + 303 m). The reason for the major span being longer than usual is that the bridge has a single back stay anchored to the abutment rather than several back stays distributed along the side span. The navigation opening was prescribed to be 10 m × 180 m. The original design of the bridge support system was improved by using a triangular steel structure located between two tower legs to support the girder. This arrangement resulted in vertical loading (165 MN) of the foundation ground because the sum of the horizontal reaction of the girder and the horizontal reactions of the tower legs became zero inside of the concrete foundation block.

The bridge superstructure consists of a steel orthotropic two-cell box girder supported by a single-plane fan cable system. Nevertheless, along the length 86 m + 54 m = 140 m, the cross section was originally a single-cell box steel girder because the middle web of the box girder was omitted from the point of the last, most-inclined cable anchorage. In 1991 the bridge was strengthened by adding a Warren-type truss as a middle web to make a two-cell box girder along the whole length of the bridge (Chladný and Baláž 1993). Six-meter spacing of the "nodes" and a diagonal area of 12,000 mm^2 provide fictitious wall shear thickness $t_s = 2.73$ mm. Adding a truss middle web along the length 140 m has a positive influence on increasing the effective breadth when taking into account shear lag phenomenon caused by variable loading at the support in the form of a framed pendulum wall on the Bratislava side. Cables in one vertical plane along the bridge middle axis are used. The bridge has four traffic lanes.

There are two pedestrian walk ways of 3 m on both sides of the bottom flange. The steel box girder has a width of 21 m. The orthotropic deck plate is mainly of 12 mm thick and increased to 16 mm, 20 mm, 25 mm, and 30 mm at the areas of the cable anchorages. The bottom flange plate thickness varies between 12 mm, 14 mm, 16 mm, 18 mm, 20 mm, and 22 mm, and three vertical web plates are 12 mm thick. Cross frames are spaced at 3 m. The system of longitudinal stiffeners is as follows: closed trapezoidal ribs with 600 mm spacing made of 6 mm and 8 mm thick plates are used in the bridge deck, while the three vertical web plates (spacing 6.3 m, height 4.6 m) and the bottom flange of the box girder are stiffened by L-profiles.

The tower has two hollow legs. There is a lift in the left leg when looking from the Bratislava side and there is an emergence staircase with 430 stairs in the right leg. St 52 steel with yield stress 360 MPa and ultimate tensile strength 510 MPa was used for all structures. The total steel used for the superstructure was 6500 tons, including 1600 tons for the tower and 620 tons for the cables. The tower legs are inclined 17° from the vertical plane and fixed to the concrete block foundation. The fix support of the box girder is located between the tower legs and attached to the concrete block foundation 7 m × 20.5 m × 40.5 m. The concrete block is supported by 56 piles with dimensions 3 m × 0.6 m. The load-carrying capacity of the piles is between 10.2 MN and 12.5 MN. The anchoring block was constructed in an open pit protected against water infiltration by steel sheet piling, which reached the clay layer. It is made of 12,000 m³ concrete transmitting a tension force of 129 MN by its own weight.

The steel superstructure was fully welded and fabricated in the factory into seven sections. The first section tower was erected from 6 m length parts with the help of slanting temporary supports. Then a 133.2 m part of the box girder was completed on the scaffolding in the first field on the right bank and pushed over the river with a 53 m cantilever. Then this part of the box girder was laid down on bearings and stayed by cables I and II. Sections 4, 5, and 6 (lengths from 72 m to 78 m), with weights approximately 800 kN, were mounted on the right river bank 300 m downwards. They were floated to the axis of bridge, lifted, supported by a temporary pier, welded to the previous piece of girder, and stayed by cables. During box girder erection 3 intermediate temporary piers (A, B, and C) were used. The seventh section (the last part of box girder) was completed on the left river bank on scaffolding. Consequent vertical movements caused by removing the temporary piers and controlled by telescopic lifting jacks located on permanent supports enabled the desired level of stresses in the cables and box girder to be achieved. Static and dynamic loading tests by moving vehicles and impulses of rocket motors were performed during July and August 1972 and achieved 78% of the total loading specified by the design codes. Good agreement between theoretical and test results was found (Harasymiv 1973). See also Baláž, Chladný (June 1994 and September 1994). The bridge has been inspected and maintained in accordance with code ON 73 62221, once every three years. The shape of the bridge is verified twice a year by surveying. The owner of the bridge is the city of Bratislava.

17.7 Footbridges and Pipeline Bridges

17.7.1 Steel Footbridges and Pipeline Bridges

There are several suspension footbridges and pipeline bridges in Slovakia (Figures 17.61 to 17.69). The first prestressed cable bridge in Czechoslovakia was the pipeline bridge in Kralupy over Vltava River, built in 1965. The Czechoslovak patent was awarded to Arpád Tesár from Slovak University of

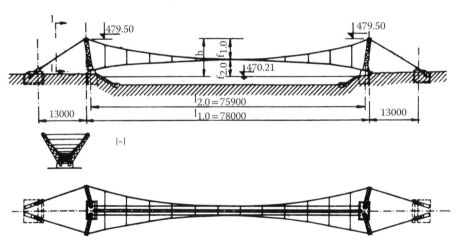

FIGURE 17.61 Combined presstressed pipeline and footbridge over Váh River in Rybarpole (1965). (Courtesy of E. Chladný.)

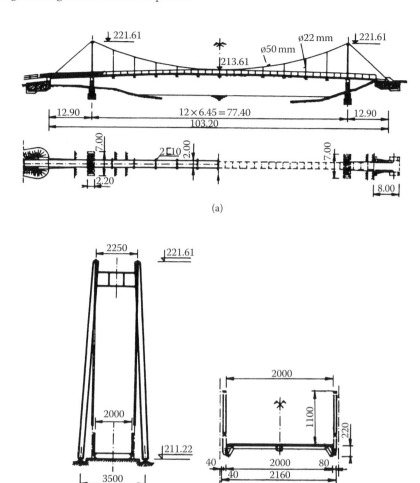

FIGURE 17.62 Footbridge over Hron River in Voznica (a) elevation, (b) tower, (c) typical scetion. (Courtesy of E. Chladný.)

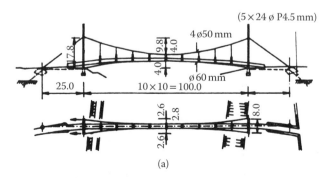

FIGURE 17.63 Footbridge over Orava River in Dolný Kubín (1969) (a) elevation, (b) tower, (c) overview. (Courtesy of E. Chladný and P. Paulík.)

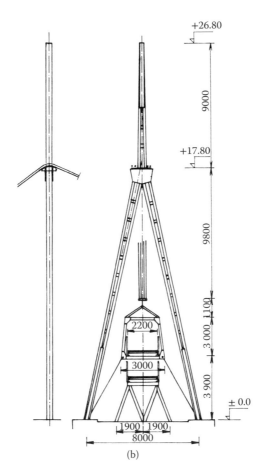

(b)

(c)

FIGURE 17.63   (*Continued*)

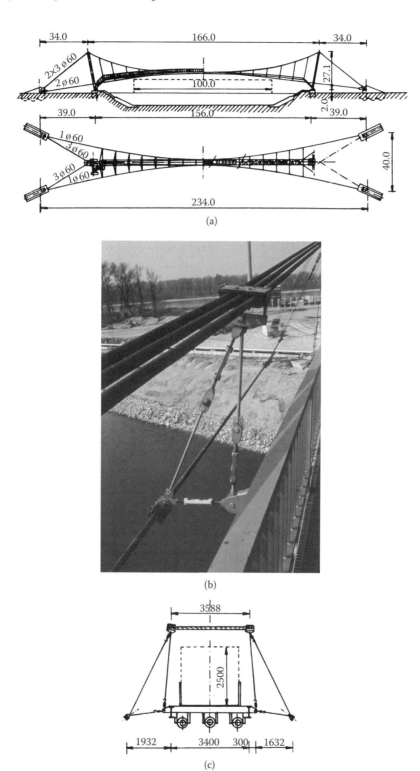

FIGURE 17.64 The largest suspension bridge in Slovakia: combined pipeline and footbridge in Winter Port at Danube River in Bratislava–Pálenisko (1981). Prestressed space structure: (a) elevation (m), (b) cable connection details, (c) typical section (mm). (d) "inside" view, (e) column base, (f) anchoring block. (Courtesy of E. Chladný and I. Baláž.)

(d)

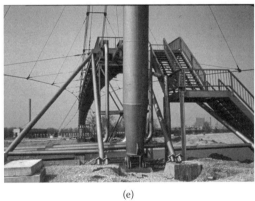

(e)

(f)

FIGURE 17.64   (*Continued*)

FIGURE 17.65 Cable-stayed footbridge over Hron River in Žarnovica, direction Lukavica (2006). Two spans are 40 m and 65 m long. The "tennis rocket" tower is 17.85 m high. (Courtesy of A. Zsigmond, Dopravoprojekt, Inc.)

FIGURE 17.66 Footbridge over Domanižanka River in Považská Bystrica. (Courtesy of P. Paulík.)

FIGURE 17.67 Cable-stayed footbridge over Dunajec River in Červený Kláštor. (Courtesy of P. Paulík.)

FIGURE 17.68   Suspension footbridge over Poprad River in Stará Ľubovňa. (Courtesy of P. Paulík.)

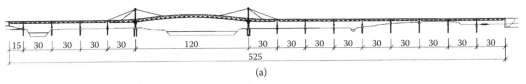

(a)

(b)

FIGURE 17.69   Cyclebridge Freedom links Devínska Nová Ves (Slovakia) with Schlosshof (Austria) over Morava River (a) elevation, (b) overview during construction. Three span cable-stayed truss girder 30 m + 120 m +30 m, the bridge length 525 m, opened on September 22, 2012. In the Slovak public voting the name Chuck Norris Bridge was winner, which was refused by Austrain side preferring Maria Theresa Bridge (Courtesy of Z. Agócs).

Technology in Bratislava for this bridge. Several footbridges were built based on this concept, for example the footbridge in Dolný Kubín (Figure 17.63). More details about Slovak suspension footbridges and pipeline bridges may be found in the book (Chladný, Paulíny, Poliaček, Vébr, Zvara 1989.) and in Agócs, Chladný, and Brodniansky (1999).

17.7.2 Timber Footbridges

The timber bridge over the Danube River distributary is shown in Figure 17.70 The bridge is made of Robinia pseudoaccacia, is 86 m long and 2.25 m wide, and leads to the floating boat mill. Figure 17.71 shows the timber Colonnaded Footbridge, built in 1994.

FIGURE 17.70   Timber footbridge in Kolárovo, Danube River distributary (1992). (Courtesy of P. Paulík.)

FIGURE 17.71   Timber Colonnaded Footbridge over Orava River in Dolný Kubín (1994). (Courtesy of P. Paulík.)

17.8 Movable Bridges

Until 1985, the only bascule bridge in Slovakia was the bridge in Komárno (Figure 17.72). Later another steel bascule bridge was built as a part of Čunovo Dam on Danube River near Bratislava (Figure 17.73).

FIGURE 17.72   Steel bascule bridge to the island of Red Fleet in Gábor Steiner shipyard in Komárno (1967). Two arms made of single-cell box girder with length 25 m, navigation opening is 40 m, roadway is 7 m wide. (Courtesy of P. Paulík and I. Baláž.)

FIGURE 17.73 Steel bascule bridge in Čunovo Dam near Bratislava on Danube River (1996). Repaired in 2009, navigation channel is 2 × 12.4 m wide. (Courtesy of F. Škvarka, Steel OK Levice, Ltd.)

17.9 Bridge Maintenance

17.9.1 Maintenance of Roadway Bridges

The regulations "TP 9B/2005 Inspections, maintenance and repairs of roadway communications—bridges" were approved by the Ministry of Transport, Construction, and Regional Development of the Slovak Republic. According to these specifications there are ordinary, main, extraordinary, and control bridge inspections. Results of all inspections shall be recorded.

- Ordinary inspections are done once a year in the spring.
- Main inspections are done at least once every 4 years depending on the condition of the bridge structure. In the frame of the main inspection a bridge is evaluated by using seven damage grades of elements, components, or the whole structure of the bridge:

- Perfect state, without any visible or hidden apparent damage
- Very good state, with only appearance imperfections not influencing the load-carrying capacity of the bridge
- Good state, with greater imperfections not influencing the load-carrying capacity of the bridge
- Satisfactory state, with damages that have no influence on the momentary load-carrying capacity of the bridge, but may influence it in the future
- Wrong state, with damages that have a negative influence on the load-carrying capacity of the bridge and which are eliminable without exchange of damaged components
- Bad state, with damages that have a negative influence on the load-carrying capacity of the bridge and which are not eliminable without exchange of damaged components or without completion of missing components
- Emergency state, with damages that influence the load-carrying capacity of the bridge to such extent that immediate repair is necessary to stave off threatened disaster

On the basis of bridge state evaluation and its remaining life and safety, the sequence of urgent repairs and reconstruction of the bridge are recommended.

- Extraordinary inspections are executed by the owner or administrator to obtain the immediate qualtitative state of the bridge. It is recommended to invite the relevant roadway administrative body. These inspections are sporadically executed, mainly in cases when safety is jeopardized.
- Control inspections are executed by the relevant roadway administrative body in the frame of enforcement of state expert supervision or by a superior body.

Maintenance is executed by relevant administrators in accordance with the Roads Act 135/1961 (Law on the roads) and Act 278/1993 (Law on state property administration) and relevant regulations. Maintenance is divided into structural and nonstructural and is done anually depending on necessity and climatic conditions. The owners report the shortage of resources for maintenance. Repair, rehabilitation, and reconstruction of bridges are executed on the basis of the bridge state evaluation, its remaining life, and its safety and with respect to available resources.

Presently, the SSC (Slovak Road Administration) is implementing a new application of a computer program for central evidence for the technical data of bridges, results of inspections with the consequent evaluation of the bridge state and elaboration of the sequence of urgency repairs, and reconstruction or rebuilding of the bridge. The information should serve to increase the effectiveness and quality of the process of management, maintenance, and rehabilitation of bridges. The state of roadway bridges in the Slovak Republic is continuously decreasing (Table 17.3). If this problem cannot be solved in an appropriate way and resources cannot be increased in the near future, the technical state and serviceability of the roadway network will not satisfy the standard transportation conditions.

17.9.2 Maintenance of Railway Bridges

Binding rules Railways of the Slovak Republic (ŽSR) S5 "Administration of railways bridge structures," which include culverts, temporary bridge structures, footbridges, and wagon weigh bridges, replaced former federal Czechoslovak rules S5 "Administration of bridge structures" and have been effective since April 1, 2009. In the frame of these rules an administrator shall perform the following supervisory activities:

- Permanent supervisory activities are part of the track control performed visually between one and seven times per week without keeping regular records.
- Ordinary inspections are performed anually by a direct administrator, who records data into a register book of ordinary inspections.

- Detailed inspections are revisions of all bridges with spans greater than 10 m once every 3 years. The results of independent evaluation of the superstructure and substructure are registered in a revision report using grades 1 (good state), 2 (sufficient state), and 3 (unsufficient state).
- Control inspections are performed in special cases or if grade 3 is assigned to the structure.
- Extraordinary inspections are performed in cases of accident, track distortion, or special behavior of the structure.
- Special observations are assured by an administrator when necessary.

The monitoring of selected special structures is performed by ŽSR in cooperation with external contractors. Maintenance is performed on the basis of the above-described system of supervisory activities by the executive units of ŽSR, the bridge districts, or by outside contractors. Reconstructions are performed when the bridge structure is evaluated by grade 3 (unsufficient state) or when space capacity and/or space arrangement of the bridge are unsufficient. Works on smaller bridges are performed by the Bridge District in Bratislava or the Bridge District in Košice. Works of a greater extent are performed by outside contractors selected in a transparent selection procedure.

Acknowledgments

This chapter would not have been completed without great help of different kinds. I appreciate the help and recommendations I received from Bartolomej Šechný, Jaroslav Guoth, Arpád Zsigmond, and Dušan Turanský, all from Dopravoprojekt, Inc.; Ladislav Nagy, originally in the Vítkovice office, later in Dopravoprojekt, Inc., now retired; Miroslav Maťaščík, Alfa 04, Inc.; Pavel Distler and František Šuppa, Distler-Šuppa, Ltd.; Samuel Jelínek, Marianna Králiková, and Alica Szebényiová, PhD., SSC (Slovak Road Administration); Ján Husák, ŽSR (Railways of the Slovak Republic); Michal Tunega, Museum Documemation Center, Railways of the Slovak Republic, and Research and Development Institute of Railways of the Slovak Republic); Jiří Kubáček, Ministry of Transport, Construction, and Regional Development of the Slovak Republic; Viera Plávková, PhD, and Ľuboslav Škoviera, RNDr., The Monument Boards of the Slovak Republic; Pavol Filadelfi, Vladimír Holčík, and Boris Rakšányi from Slovak Water Management Company; Peter Páteček, Železničné.info (a magazine about railways in Slovakia); Ľudovít Turčina from Ružomberok, Peter Šimko from Považské Museum in Žilina, and Peter Paulík, PhD., Faculty of Civil Engineering, Slovak University of Technology in Bratislava, who was ready to travel on a motorcycle to any bridge to take a photograph and check bridge data. They were all very nice and always ready to help, for which I am very grateful.

I wish to especially acknowledge much work done by Prof. Emeritus Eugen Chladný, PhD., concerning steel roadway and railway bridges, and late Assoc. Prof. Jozef Zvara concerning concrete roadway and railway bridges, both from Slovak University of Technology in Bratislava in the book (Chladný, Paulíny, Poliaček, Vébr, Zvara 1989.) Finally I would like to thank to Gabriel Tevec, Doprastav Inc., Bratislava, who wrote the text of Section 17.3.2.

References

Agócs, Z., Chladný, E., Brodniansky, J. 1999. Hängeseil and Schrägseilkonstruktionen von Fußgänger und Rohrleitungsbrücken. (Suspension and cable-stayed footbridges and pipeline bridges). *Stahlbau* (Steel Construction) No. 1, pp. 51–55.

Baláž, I. 2003. History of record bridges (in Slovak). *Eurostav* No. 6, Bratislava, Slovakia, pp. 12–16.

Baláž, I. 2005. Bridges over Danube River in Bratislava. Part 1: Historical bridges (years 0–1890) (in Slovak). *Eurostav* No. 4, pp. 23–27.

Baláž, I. 2005. Bridges over Danube River in Bratislava. Part 2: Modern bridges (years 1890–2005) (in Slovak). *Eurostav* No. 6, pp. 60–64.

Baláž, I. 2012. Forty years of SNP Bridge in Bratislava. *Eurostav* No. 7–8, pp. 16–19.

Baláž, I. 2013. Timber bridges with record spans. *Eurostav* No. 5, pp. 62–63.

Baláž, I. 2013. Timber bridges with record spans 2. *Eurostav* No. 7–8, pp. 50–51.

Baláž, I., Dajun, D. September 5–8, 2006. Cable-stayed bridges. *Proceedings of International Colloquium on Stability and Ductility of Steel Structures*, Lisboa, pp. 1083–1090.

Baláž, I., Virola, J. 2000. Bridges with record spans and curious (in Slovak). *ASB* VII, No. 4, pp. 26–28.

Baláž, I., Chladný, E. June, 1994. The load carrying capacity of cable-stayed steel bridges with box cross-section. *Proceedings of Slovak-US American Bridge Conference*, Bratislava, Slovakia, pp. 271–276.

Baláž, I., Chladný, E. September, 20–22, 1994. Some remarks to the analytical evaluation and strengthening of the cable-stayed bridge across the Danube in Bratislava. *Proceedings of International Bridge Conference*, Warsaw '94, Poland.

Birago, K. Ritter von. 1839. *Untersuchungen über europäischen Militärbrückentrains und Versuch einer verbesserten, alles Vorderungen entsprechenden Militärbrückeneinrichtung*, Wien. Investigations of European military bridge wagon trains and test of improved military bridge equipment matching to all requirements.

Búci, B., Choma, Š., Hrnčiar, Ľ., Šefčík, T., Zvara, J. 1994. The Lafranconi Bridge across the Danube River in Bratislava. *Inženýrske stavby (Civil Engineering Works)*, No. 2–3, pp. 73–81.

Búci, L. 1981. Viaduct Podtureň (in Slovak). *Inženýrske Stavby (Civil Engineering Works)*, No. 3, pp. 117–126.

Chladný, E., Baláž, I. 1993. Inspection, evaluation, and strengthening of the SNP bridge in Bratislava. Bridge management 2. Inspection, maintenance, assessment, and repair. *Proceedings of the Second International Conference on Bridge Management*, Thomas Telford, London, pp. 407–417.

Chladný, E., Paulíny, L., Poliaček, L., Vébr, V., Zvara, J. 1989. Civil Engineering in Slovakia in 1945–1985. Transportation structures. Alfa Bratislava, Slovakia.

Dutko, P., Ferjenčík, P. 1965. Timber structures in Czechoslovak Socialist Republic (in Slovak). *Proceedings of Scientific Works of Faculty of Civil Engineering* SVŠT in Bratislava.

Ferjenčík, P., Dutko, P. September 4–6, 1984. Timber structures in Czechoslovakia (in Slovak). *Proceedings of the Third International Symposium on Timber in Civil Engineering Structures II*, Kočovce, STU (Slovak University of Technology) Bratislava, Czechoslovakia.

Gender Stort, E. A. 1935. Small span bridges. Historical note—general considerations. *L'ossature Metallique* 6, pp. 301–310.

Harasymiv, V. 1973. Loading test of steel SNP Bridge over Danube in Bratislava (in Slovak). *Inženýrske stavby (Civil Engineering Works)*, No. 7, pp. 321–331.

Havelka, K. (ed.). 1964. *Theory of Calculations of Civil Engineering Structures and Basis of Limit State* (in Slovak). Slovak Academy of Sciences, Bratislava, Slovakia.

Ivanyi, M., Bancila, R., Stipanic, B., Chladný, E., Ramberger, G., Kupfer, H., Nather, F. (eds.). September 11–15, 1995. *Bridges on the Danube*. Edited for Second International Conference, Bucharest, Romania.

Korabinský, M. J. 1786. Geographisch-historisches und Produkten-Lexikon von Ungarn *(Geographic-Historical and Economical Lexicon of Hungary)*. Pressburg.

Kubáček, J. et al. 2007. *History of Railways on Slovak territory* (in Slovak). Second edition. OTA, Košice, Slovakia.

Majdúch, D. et al. 1989. *Handbook for Calculation of Load-Carrying Capacity of Older Bridges* (in Slovak). Faculty of Civil Engineering, Slovak University of Technology. Bratislava, Slovakia.

Maurenz, J., Kubáček, J., Golda, B. 2010. *History of Railways in Slovakia and in Ruthenia on Contemporary Viewcards and Photographs* (in Slovak). Publisher Ružolíci Chrochtík (Snorting piggy with pink face). Prague, Czech Republic.

Mrázik, A. 2010. Memoirs on the preparation of transition of design of steel structures from allowable stress design to design based on limit states. *Proceedings of 36th Meeting of Specialists in the Field of Steel Structures*, Žilina, Slovakia, pp.151–154.

Nesvadba, O., Plevák, M. 1982. The oldest cast iron bridges in Slovakia (in Slovak). Supervisor: P. Ferjenčík. Student research work, Department of Metal and Timber Structures, Faculty of Civil Engineering, STU Bratislava, Czechoslovakia.

Paulík, P. 2012. Bridges on Slovak territory. History and the present time of more than 250 the most beautiful and the most interesting Slovak bridges. JAGA, Bratislava, Slovakia.

Šefčík, T., Tevec, G. 1990. Motorway bridge over Danube River near Lafranconi in Bratislava (in Slovak). *Inženýrske Stavby (Civil Engineering Works)*, No. 3–4, pp. 136–143.

SSC (Slovak Road Administration). 2001. Data overview about road communications in the Slovak Republic. Bratislava, Slovakia.

Strásky, J., Maťaščík, M., Novák, R., Táborská, K. 2011. Urban viaduct in Považská Bystrica. *Structural Engineering International* 21(3), pp. 356–359.

Tesár et al. 1970. Design of new road bridge over Danube River in Bratislava. *Proceedings of Scientific Works of Faculty of Civil Engineering*, Slovak University of Technology in Bratislava, pp. 7–88.

Tesár, A. 1973. Most SNP cez Dunaj v Bratislave. SNP Bridge over Danube River in Bratislava. (in Slovak). *Inženýrske Stavby (Civil Engineering Works)*, No. 7, pp. 291–321.

Träger, H., Tóth, E. (eds.). 2010. *Danube Bridges: From the Black Forest to the Black Sea*. Photographs by P. Gyukics. Yuki Studio. Budapest.

Zvara, J. 1971. Tests of piles of tower foundation of SNP Bridge over Danube River in Bratislava. *Proceedings of Scientific Works of Faculty of Civil Engineering*, Slovak University of Technology in Bratislava. Published by Department of Metal and Timber Structures, Department of Concrete Structures and Bridges and Department of Structural Mechanics, pp. 317–350.

18
Bridge Engineering in Turkey

Cetin Yilmaz
Middle East Technical University

Alp Caner
Middle East Technical University

Ahmet Turer
Middle East Technical University

18.1 Introduction

In 1923, at the time of the establishment of the Republic of Turkey, the total length of the roads was around 18,000 km with 94 bridges. In 1929, an early version of the Highway Department was established within the Ministry of Public Works. In the 1950s, bridge and road construction were accelerated and the total length of highways reached 62,000 km. The total number of bridges reached 5860 with a total length of 255 km. In Asia Minor, Anatolia was home to many civilizations prior to the establishment of Turkey. In Sections 18.1.1 and 18.1.2 the development of roads and bridges will be presented.

18.1.1 Historical Anatolian Bridges from the Middle Ages to 1900

The first known civilization, the Hittites, ruled over Anatolia between 1900 and 1200 BC. The Hittites built some major cities and trade routes in Anatolia as shown in Figure 18.1 (KGM 2007). The Kingdom of Urartu was established around 900 BC close to the Van Lake in the eastern part of Anatolia. They

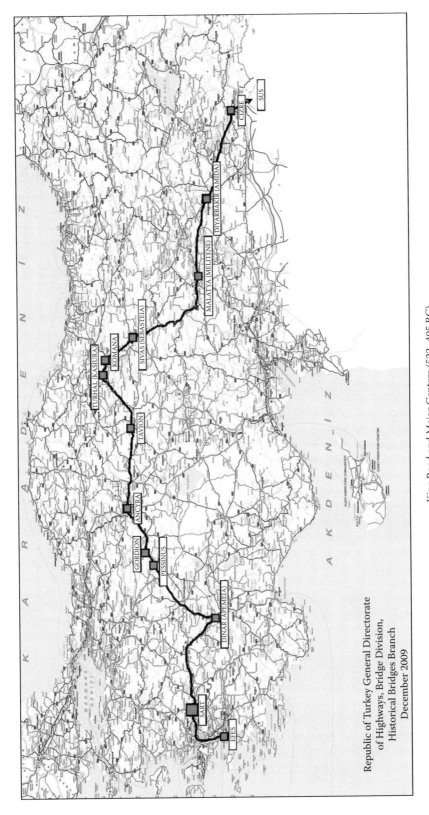

King Road and Major Centers (522–405 BC)

Persian King DARIUS (DARA) Captured SART, the Capital of Lidya, and built a road between his capital SART famous historian Herodotus called road the KING road

Republic of Turkey General Directorate
of Highways, Bridge Division,
Historical Bridges Branch
December 2009

FIGURE 18.1   Major Hittite cities and trade routes.

built 3.9 m to 5.4 m wide roads. On the transportation network system, the rest stations were located about 20 km to 30 km apart from each other. The road connecting the city of Susa to the Aegean coast was about 2165 km long, as shown in Figure 18.2 (KGM 2007). Around 330 BC. Alexander the Great invaded Anatolia, and the southern route was constructed, as shown in Figure 18.3. Alexander the Great used this route to proceed to India (KGM 2007).

About 80 bridges can be identified from the time of the Roman Empire between 130 BC and 476 AD, as shown in Figure 18.4. The Romans also placed signs on the roads indicating the travel distance between cities (KGM 2007). Byzantine Empire bridges were built between 395 and 1453 AD. The locations of the 10 major bridges are shown in Figure 18.5 (KGM 2007).

The Seljuk Empire, which ruled Anatolia between 1040 and 1308, built some major bridges mostly in the southern part of the country close to the capital city of Konya, as depicted in Figure 18.6 (KGM 2007). The Seljuk Empire built "kervansaray" hostels on the roads for tradesman. These hostels were also used to accommodate military forces. The Malabadi Stone Arch Bridge from the time of the Seljuk Empire was the longest arch bridge of its era when it was was built around 1147; a recent picture of the bridge is shown in Figure 18.7. The bridge is 7.8 m wide and the main span is around 40 m long (Ilter 1978). The total length of the bridge is around 150 m.

During the Ottoman Empire, there were 413 registered bridges; the locations of some 1059 bridges are shown in Figure 18.8 (KGM 2007). The Ottomans ruled Anatolia from 1453 to 1923. Some pictures of major bridges from the Ottoman Empire period are shown in Figures 18.9 to 18.11. Many other bridges were built in the Balkans and the Middle East during this period.

18.1.2 Historical Development of Contemporary Bridges in Turkey (1923 to Present)

Common construction techniques and examples of the most well-known bridges in Turkey are presented in this section. Not many bridges were constructed between 1923 and 1950. An arch bridge construction in 1938 is shown in Figure 18.12. During the construction, the reinforced concrete arch was supported with temporary cables connected to temporary towers. Short-span bridges were constructed using timber elements. Short-span bridges were typically used for one-way traffic at those times, which resulted in control passes.

In early 1950, the General Directorate of Highways (KGM), equivalent to the U.S. Federal Highway Administration, was established. From the 1950s to the current time, a significant increase in construction of new bridges has been observed. The total number of highway bridges in Turkey was estimated to be around 5860 in the year 2008 (www.kgm.gov.tr). Out of this total, the number of cast-in-place cantilever segmental post-tensioned bridges is less than 10, including the ones under construction or design. A brief history of the development of bridge technology in Turkey is described next based on observations made during the inspection of bridges between the cities of Canakkale and Bursa (Caner et al. 2008).

Most Turkish standard highway bridges were designed using reinforced concrete technology in early 1950. Cast-in-place reinforced concrete girder slab bridges were made continuous over the piers. In these bridges, either the girders sit directly on top of the piers over a thin steel plate or on a series of reinforced concrete rocker bearings. Reinforcement was provided in concrete rocker bearings, connecting them to the superstructure and pier. The maximum span length of these bridges is 30 m, as shown in Figure 18.13. Solid shear wall type piers were preferred around the 1950s.

In the 1980s, superstructures were still formed from cast-in-place reinforced concrete girder slab but there was a significant change in the selection of the pier. The multicolumn piers became more slender compared to the 1950s piers, as shown in Figure 18.14. In the mid-1980s, one of the few cast-in-place cantilever post-tensioned segmental bridges in Turkey was constructed connecting the city of Elazığ to Malatya, as shown in Figure 18.15. The total length of the dam bridge was around 284 m.

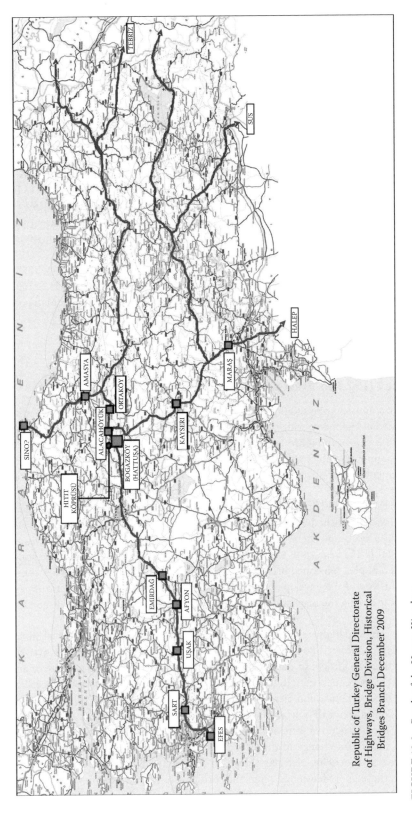

Republic of Turkey General Directorate
of Highways, Bridge Division, Historical
Bridges Branch December 2009

FIGURE 18.2 Roads of the Urartu Kingdom.

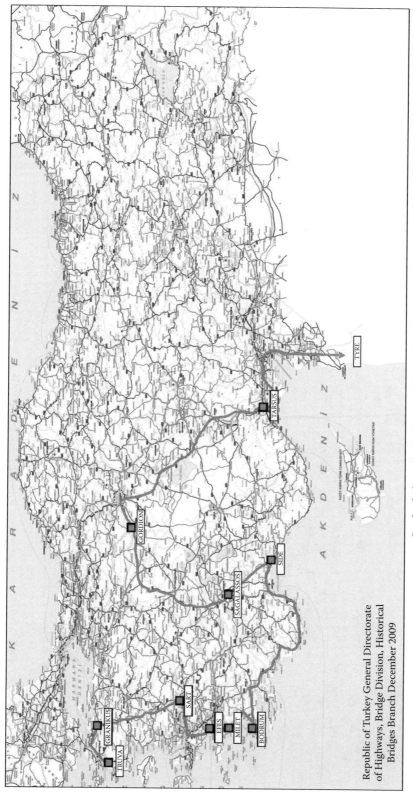

Republic of Turkey General Directorate
of Highways, Bridge Division, Historical
Bridges Branch December 2009

Roads of Alexander the Great and Major Cities (334–323 BC)

FIGURE 18.3 Roads used by Alexander the Great.

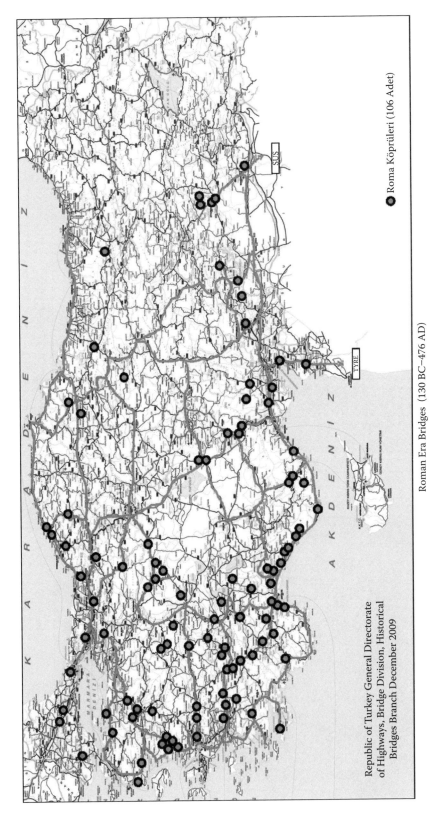

Republic of Turkey General Directorate
of Highways, Bridge Division, Historical
Bridges Branch December 2009

Roman Era Bridges (130 BC–476 AD)

FIGURE 18.4 Identified bridges of the Roman Empire built between 130 BC and 476 AD.

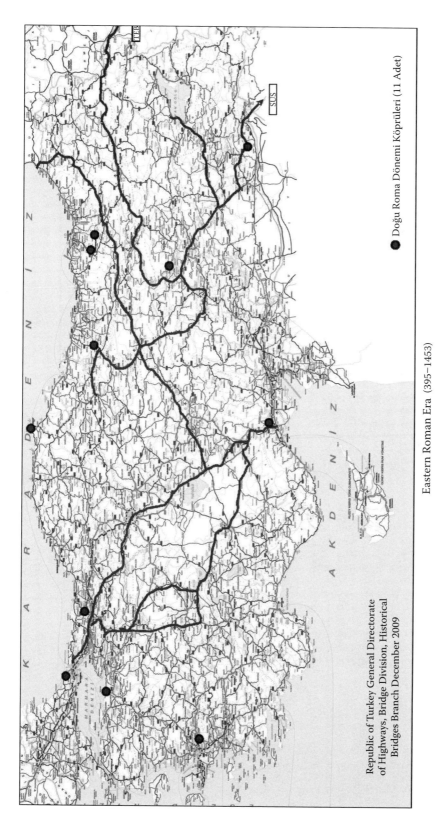

Republic of Turkey General Directorate
of Highways, Bridge Division, Historical
Bridges Branch December 2009

Eastern Roman Era (395–1453)

● Doğu Roma Dönemi Köprüleri (11 Adet)

FIGURE 18.5 Identified major Byzantine Empire bridges built between 395 and 1453 AD.

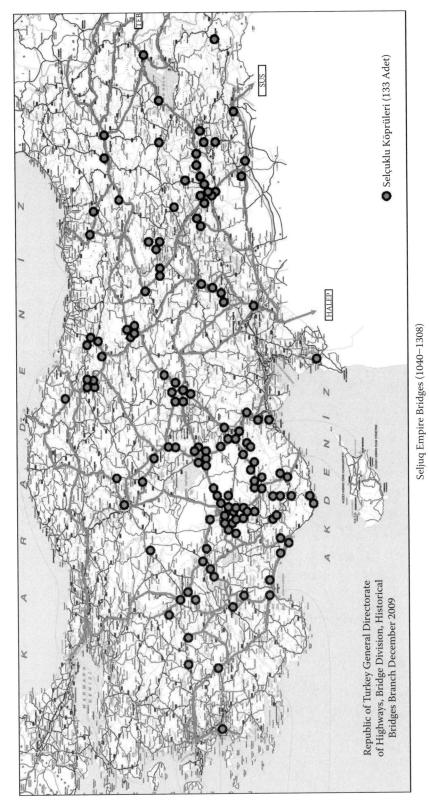

Republic of Turkey General Directorate
of Highways, Bridge Division, Historical
Bridges Branch December 2009

Seljuq Empire Bridges (1040–1308)

● Selçuklu Köprüleri (133 Adet)

FIGURE 18.6 Major Seljuk Empire bridges built between 1040 and 1308.

FIGURE 18.7   Malabadi Bridge (1147 AD).

Around mid the-1990s, the superstructure type was switched to precast prestressed girder slab from post-tensioned concrete girder slab. The girders were spaced a certain distance from each other and the deck was formed over a stay-in-place steel form. The piers of these bridges were changed to a different geometry from the earlier versions, as shown in Figure 18.16. Elastomeric bearings were placed between the girders and the pier cap. These bearings remained in place by the gravity loads of the superstructure elements.

Around the year 2000, the superstructures were still formed from precast prestressed girders; the only difference was that the girders were spaced closely and the cast-in-place slab was formed over the top flanges of these girders, as shown in Figure 18.17. Different types of bridge construction are shown in Figure 18.18. This figure indicates that the most commonly used bridge type in Turkey is single-span beam-type reinforced concrete bridges. Cast-in-place cantilever bridges are becoming more popular due to the recent transfer of knowledge in this field. However, current bridge design specification in Turkey does not yet cover cast-in-place cantilever bridges.

Bridges in Turkey are mainly divided into highway and railway bridges. Bridges within city limits are owned by municipalities and evaluated as a third category. According to the KGM 2000 census, there are about 5000 interstate highway bridges with about 186 km total length (Figure 18.18). The dominant bridge type in Turkey is the simply supported reinforced concrete bridge, with about 2800 bridges and a total length of 82 km. Steel-type bridges in highways are quite rare. On the other hand, the steel-type bridge is the most common one in railways.

18.2 Design Practice

18.2.1 Highway Bridge Design Specifications

The first bridge specification adapted for the design of structures in Turkey was based on German engineering design practice. In 1953, the reinforced concrete design guidelines developed by the Turkish Association of Bridges and Structures were used in the design of bridges. The same guidelines were also utilized in the design of building codes. In the design specifications of the early 1970s, seismic issues were included in the body of the text. In the same years, the American Association of State Highway and Transportation Officials (AASHTO) bridge specifications did not recognize the importance of seismic events.

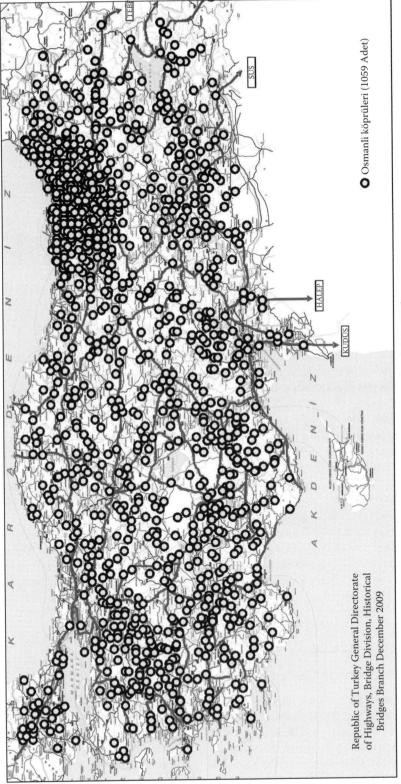

Republic of Turkey General Directorate
of Highways, Bridge Division, Historical
Bridges Branch December 2009

Ottoman Empire Bridges (1300–1923)

FIGURE 18.8		Ottoman Empire bridges (1300 to 1923).

FIGURE 18.9 Uzun Bridge (1266 m long) with 174 arches, built by Sultan Murat the 2nd (1430s).

FIGURE 18.10 Saray Bridge (60 m long), built by Sultan Suleiman the magnificent (1553).

FIGURE 18.11 Prewar picture of Mostar Bridge in the Balkans (1566), built by the architect Hayrettin, a student of the great architect and builder Sinan (destroyed during the Bosnian War around the mid-1990s and reconstructed after the war).

FIGURE 18.12 Construction of an arch bridge, 1938.

FIGURE 18.13 Reinforced concrete bridge from the 1950s.

FIGURE 18.14 Piers of reinforced concrete girder slab bridge piers from the 1980s.

FIGURE 18.15 One of the first cast-in-place cantilever post-tensioned segmental bridges, Kömürhan.

In the 1980s, for new express highway projects and motorways, project-specific seismic design criteria were developed and used. The Turkish seismic map (Figure 18.19) was developed to determine the seismic zone of the bridge, which would be utilized to find the expected ground acceleration magnitude for earthquake analysis. The five seismic zones defined in Turkey were similar to the seismic zones defined in the AASHTO bridge specifications. For instance, seismic zone 1 in Turkey corresponds to

FIGURE 18.16   Precast prestressed girder slab bridges around the mid 1990s.

FIGURE 18.17   Precast prestressed girder slab bridges around the 2000s.

seismic zone D in the U.S. and seismic zone 4 in Turkey corresponds to seismic zone A in the U.S. The seismic design mostly follows the minimum requirements of the AASHTO bridge design specifications (2002). The design earthquake return period was taken as 475 years. In some special designs, bridges were designed for larger earthquakes such as those with a 1000-year return period. The peak ground acceleration data for a 1000-year return period earthquake used in a design was developed at the Middle East Technical University around the mid-1990s.

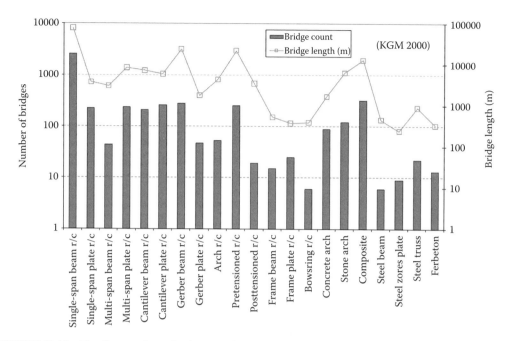

FIGURE 18.18 Distribution of state bridges, number and lengths according to bridge type. (Courtesy of KGM–Census 2000.)

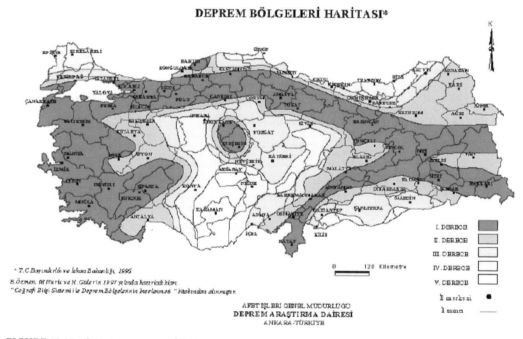

FIGURE 18.19 Seismic zone map of Turkey.

A modified version of the AASHTO specifications was used in the design of gravity and wind load combinations. The major difference between the AASHTO specifications and the Turkish specifications was the truck load. The truck load used in Turkish specifications (Figure 18.20) was much heavier than the ones used in U.S. engineering practice. The load resistance and factor design method (LRFD) would be the new type of design method in Turkey in the near future.

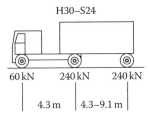

FIGURE 18.20   Turkish truck loading for spans ≤ 40 m.

18.2.2 Railroad Bridge Design Specifications

Similar to highway bridges, the early railroad bridges were designed according to German design specifications. The design train loading was based on German codes and was accepted as the S1950 train. Until the last decade, not much investment was done for the development of railroads. In design of the first high-speed rail bridges between Ankara and Eskisehir, European norms were utilized to determine passenger comfort levels at different speeds of the train. The superstructures of these bridges were much stiffer than the bridges of conventional lines. The seismic design was still based on the modified AASHTO requirements.

18.2.3 Seismic Design and 1999 Izmit Earthquake

The estimated length of fault rupture during the 1999 Izmit earthquake was about 170 km, of which 50 km to 60 km was believed to be extended under the sea bed. Only one overpass collapsed due to excessive seismic relative movements of the fault line. The fault line crossed the end span of the bridge, and the fault line offset was observed to be 4 m. The other 32 similar overpasses in the vicinity of the fault line had minor damage. The seismic design of these bridges was based on a modified version of the AASHTO Standard Specifications. The seismic performance of the bridges was observed to be satisfactory compared to the seismic performance of the buildings in that area.

18.2.4 Seismic Isolation

Use of a friction pendulum and lead core bearings has become very popular in the seismic design of critical bridges. In new designs with seismic isolation systems, engineering design decisions are usually made to have an essentially elastic bridge pier and foundation system. A new type of seismic isolation, steel ball rubber bearing, was developed at Middle East Technical University. The test results indicated that the bearings have a similar energy dissipation capacity compared to other systems. The bearings can be manufactured at a very low cost compared to other systems and the manufacturing processes do not involve complex steps.

18.3 Prestressed Concrete Bridges

Prestressed technology has been utilized in an increasing number of bridge projects within the last 20 years in Turkey. Some of the remarkable prestressed concrete bridges are described in this section.

18.3.1 Bolu Viaducts

The Gümüşova–Gerede Highway Project is a part of the Anatolian Motorway, which connects Istanbul and Ankara, the two largest cities in Turkey (Figure 18.21). The Bolu Mountain Pass is an important part of Gümüşova–Gerede Highway Project, which aims at meeting the demand of national and

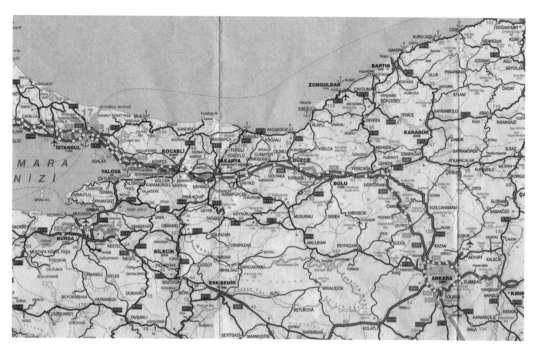

FIGURE 18.21 Location of Bolu Viaduct.

international transportation on the Edirne–Istanbul–Ankara route, the main artery of the Turkish highway network.

The Bolu Mountain Pass, also known as Stretch 2 of the Gümüşova–Gerede Highway Project, is 25.6 km long and starts at km 30 of the Gümüşova–Gerede Highway. It goes east along the Asarsuyu River, passes through Bolu Mountain with a tunnel, and ends at Yumrukaya. The total cost of this part of the project is $570.5 million, and the contractor for the project is a joint venture between an Italian company, Astaldi S.p.A., and a Turkish company, Bayindir Construction. The Bolu Mountain Pass is composed of two viaducts and two long tunnels; one of the two viaducts was almost completed in November 1999 (Figure 18.22) when the second Duzce Earthquake, which occurred on November 12, 1999, caused extensive damage to the viaduct.

The highway is curved in plan, has a vertical slope of approximately 4% toward Istanbul, and consists of two parallel bridges, each having a separate traffic direction. The number of spans is 59 and 58 in the Ankara and Istanbul bridges, respectively. The total length of the bridge is 2313.03 m and 2273.75 m for the Ankara and Istanbul carriageways, respectively. There are tall reinforced concrete piers with large and nearly rectangular hollow cross sections (Figure 18.23). The total number of piers is 58 for the left bridge and 57 for the right bridge. Piers are numbered starting from the Düzce end in the direction of Bolu from 1 to 58 and 1 to 57 for the left and right bridges, respectively. Pier lengths vary from 10 m to 49 m in height.

All piers rest on massive and monolithic column footings supported on 12 cast-in drilled hole friction piles with a diameter of 1.8 m. The depth of the piles ranges from 20 m to 30 m. The typical size of the footing is 3 m deep, 18.7 m long in the longitudinal direction, and 16 m wide in the transverse direction. Pier spacing is 39.2 m from center to center and the superstructure is made continuous over 10 spans of the section by means of 1.5 m long and 24 cm thick span connecting slabs. At the end piers of these 392 m long segments, suitable expansion joints are provided to the deck, which terminate with longitudinal cantilevers. A typical cross section, pretensioned beams, and pier cap drawing is given in Figure 18.24.

FIGURE 18.22   General view of Bolu Viaduct during the construction stage (1995).

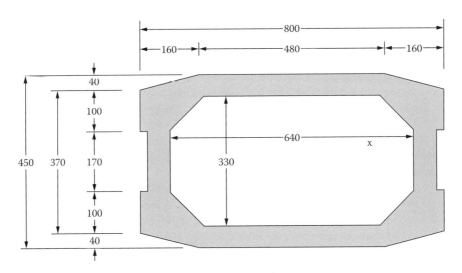

FIGURE 18.23   Cross section of a typical pier (dimensions in cm).

Initially, energy dissipating units (EDUs) were placed between the girders and pier cap to dissipate energy in case of a large earthquake (Figure 18.25). During the 1999 earthquake, the EDUs were heavily damaged and dislocated. The pier locations were changed during the earthquake as a surface crack passed between piers 45 and 47; therefore, substantial retrofit and strengthening were carried out. The bridge deck was lifted and shifted to the most appropriate position and EDUs were replaced by friction pendulum-type seismic isolators. The pier caps were retrofitted and spans were made continuous for groups of 10 spans. Bolu Viaduct is currently operational and one of the most important structures on the Anatolian Motorway.

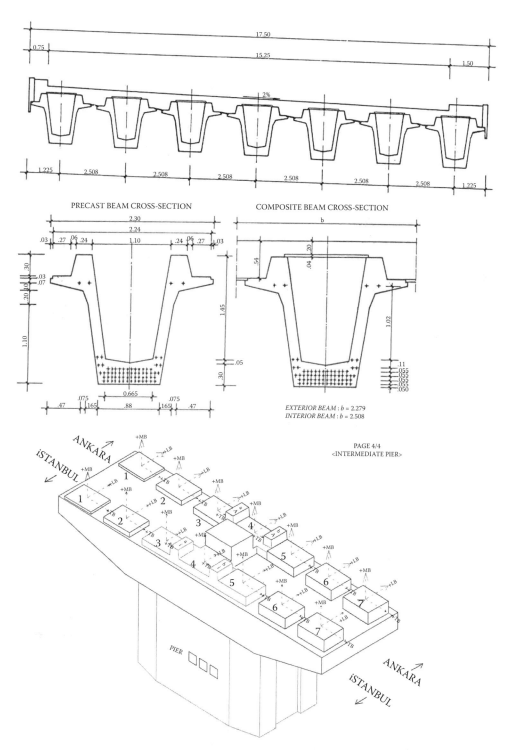

FIGURE 18.24 Drawings and details of Bolu Viaduct.

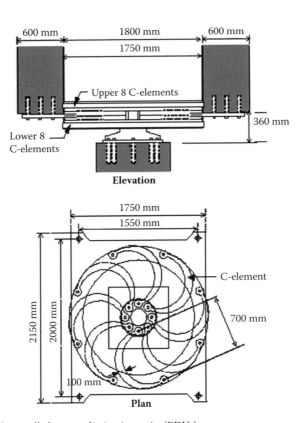

FIGURE 18.25   Initially installed energy dissipating units (EDUs).

18.4 Segmental Concrete Bridges

Segmental post-tensioned bridges are not very common in Turkey. The cast-in-place cantilever balanced method is typically used in such construction. The best-known bridges are Imrahor Bridge in Ankara, Gulburnu Bridge in Giresun, Beylerbeyi Bridge in Malatya, and Akarsin Viaduct and Ortakoy Bridge in Artvin.

18.4.1 Cast-in-Place Cantilever Balanced Segmental Bridges

18.4.1.1 Ortakoy Bridge

The two-lane 12 meter wide Ortakoy Viaduct connecting the cities of Artvin and Erzurum has a total length of 156 m with two equal spans. The typical girder cross sections at the abutment and the pier are shown in Figures 18.26 and 18.27. These sections are selected based purely on engineering judgment relying on some basic hand computations. The section depth is decreased from pier to abutment in a half parabolic shape. The typical length of the segment is around 4.5 m, except the ones close to pier and the abutment.

It is known that as the span length increases, the live load moment contribution in the total design moment becomes less effective compared to the moments induced by the dead load of the structure. The number of tendons was initially assessed for the construction stage at the cantilever position of the bridge just before the last segment connected to the abutment segment, almost at a straight position.

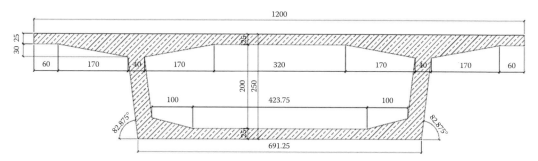

FIGURE 18.26 Girder section at abutment.

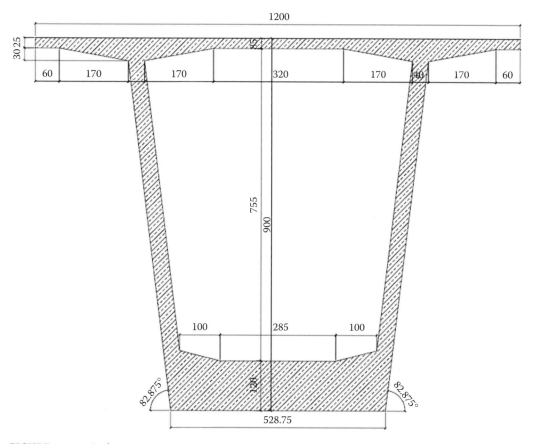

FIGURE 18.27 Girder section at pier.

For the selected tendon layout and sections, a construction stage analysis, including the time-dependent effects, was performed using a computer model (LARSA-4D) as shown in Figure 18.28. The construction of each segment had a typical seven-day cycle. In a typical seven-day cycle, the 90 ton traveler form moved to its new position and after the concrete was placed and tendons were stressed, the cycle was completed. The assessment made at the preliminary design was determined to have minor changes in the final design stage. Over the pier segment, 44 ducts each having 19 tendons were used. The cantilever construction almost reached to the abutment at the straight position. Principal stresses were checked at the webs to determine if there would be excessive cracking at the webs.

FIGURE 18.28   Construction stage model.

FIGURE 18.29   (**See color insert.**) Gülburnu Bridge.

The bridge was designed based on the minimum requirements of the AASHTO-LRFD (2007) and (Euro-International Concrete Committee) CEB-FIP (International Federation of Prestressing) speci-fications (1990). The bridge was designed for the HS30-24 truck. The transverse design of the box section revealed that transverse post-tensioning was needed at the top flange to minimize cracking, which can affect the durability of the bridge and the long-term performance. The need for transverse post-tensioning at the top flange was determined through moving load and temperature analysis. The magnitude of live load and the distance between the webs were the governing factors in the design of transverse post-tensioning.

18.4.1.2 Gülburnu Bridge

This three-span bridge, with a total length of 330 m (82.5 m + 165 m + 82.5 m), is located on the Karadeniz Coastal Road between Giresun and Espiye, as shown in Figure 18.29. The 14.5 m wide bridge can accommodate two lanes of traffic. The superstructure height is 8.5 m over the piers and 3.5 m at the ends. The cast-in-place segmental concrete had a concrete compressive strength of 40 MPa at 28 days. The bridge was designed based on the methods suggested in the AASHTO-LRFD (2004) and CEB-FIP specifications (1990). The bridge was designed for live load truck HS30-24. In the construction stage analysis LARSA software (2006) and in the seismic analysis SAP2000 software (CSI 2000) was utilized.

18.5 Steel Girder Bridges

Steel girder bridges are not common in Turkey. Steel girders are typically used to span more than 40 m of clear span length. The construction of a composite steel girder bridge in Mekece is shown in Figure 18.30. The composite effect between the steel girder and the concrete deck is achieved by the I-profile shear studs, as shown in Figure 18.31. Each 58.5 m long span has five 3 m deep girders connected by cross braces at a certain spacing. The allowable tension stress is equal to 140 MPa. The 25 cm thick reinforced concrete deck has a 28-day concrete compressive strength of 30 MPa.

FIGURE 18.30 Construction of a steel girder bridge.

FIGURE 18.31 I-profile shear studs over the top steel flange.

18.6 Arch Bridges

18.6.1 Ayvacik Eynel Arch Bridge

This bridge connects the villages that will be separated by the water body of the Ugurlu Dam. The total length of the bridge is 366 m, of which 168 m is the length of the main span. The steel arch is in the form of a box with internal stiffeners. The total width of the bridge is set to 12 m, accommodating two lanes and two pedestrian walkways. The truck live load is selected to be HS20-S16, which is lighter than the design truck loads used on the highways. The bridge was designed based on the minimum requirements of the AASHTO standards (2002). The fatigue life of the bridge was also evaluated in the computations for a total 500,000 cycles of truck loading. Corrosion protection and maintenance guidelines were included in the design. During construction the water body was detoured, as shown in Figure 18.32. The bridge was completed and opened to traffic in March 2009.

18.6.2 Dikmen Valley Arch Bridge

Dikmen Valley Arch Bridge is constructed in a private residential area. The 180 m long bridge connects the two sides of the valley. The bottom arch, formed from steel box sections, has a separate foundation than the superstructure above. The construction of the Dikmen Valley Arch Bridge is shown in Figure 18.33. The bridge carries two lanes of traffic and was constructed in 120 days using 1000 tons of steel and opened to traffic in 2008.

(a)

(b)

FIGURE 18.32   Samsun Ayvacik Bridge (a) after construction and (b) during construction.

FIGURE 18.33   Dikmen Valley Arch Bridge.

18.7 Truss Bridges

18.7.1 Karakaya Bridge

The Karakaya Bridge, 2030 m in length, was built over a dam reservoir close to the city of Malatya in the late 1980s. The main construction phases were foundation works, pier construction, and steel truss erection. The 64 m long steel truss superstructure was supported with hollow concrete bridge piers spaced about 70 m from each other, as shown in Figure 18.34. Each steel span was weighed about 366 tons. The yield strength used in the design of the steel elements was 235 MPa. The steel trusses were built on the ground and lifted to their positions, as shown in Figure 18.35. The lift speed was 6 m/h. Each pier was constructed using a traveling form. The longest pier was about 58 m long. The foundation was formed on piles. The total cost of the bridge was around 4.5 million Turkish liras.

FIGURE 18.34 Karakaya Bridge around the mid-1980s.

FIGURE 18.35 Construction stages of Karakaya Bridge.

FIGURE 18.36   Typical high-speed train viaduct.

18.8 High-Speed Train Bridges

The high-speed rail project began in 2003. The project aims to connect Ankara to Istanbul via Eskisehir. Ankara–Eskisehir high-speed train services started in March 2009. When the 533 km long project is completed, the total number of viaducts will reach up to 21. There will also be 34 river, 2 highway, and 4 railroad bridges on the network. The typical span length of a bridge or viaduct ranges from 30 m to 34 m, as shown in Figure 18.36. In the design of these bridges, a set of related European Norms (EN), current as of 2012, are utilized. Bridge–rail interaction and passenger comfort are critical service load design checks. The superstructure rigidities of these bridges are relatively larger than those of conventional railroad bridges in order to satisfy passenger comfort levels.

The columns are typically stiffer compared to those of highway bridges. Over the pier caps, special fixed supports are used at one end of the span only to prevent excessive longitudinal displacements of the span due to braking of incoming and acceleration of outgoing trains. The precast I-girders manufactured at a different site were shipped to the bridge and placed side by side. The manufacturing of the precast prestressed I-girders is shown in Figure 18.37.

18.9 Cable-Supported Pedestrian Bridges

A relatively new trend in Turkey is building pedestrian bridges using cable stays. The aesthetics of the bridge is thought to be the primary driving force for building many similar bridges, especially in the capital city of Ankara. In addition to the nice architectural pattern, omitting the middle supporting pier enhances traffic safety in the case of an accident. Urban areas have traffic congestion problems, and while building new underpasses and over connections makes the traffic flow faster, it also divides regions of the city with fast-flowing traffic roads. Such an action would necessitate building tens, even hundreds of pedestrian bridges in the city.

Two almost identical pedestrian bridges located on Esenboga Road are shown in Figure 18.38. Measurement of cable forces is necessary in order to obtain even distribution of forces in the cables and maintain the correct camber of the bridge under service. The cable force measurements are typically made using two techniques, either by (1) static methods—bending the cable in the transverse direction and obtaining tension indirectly as a function of the force applied to bend the cable and amount

FIGURE 18.37 Precast prestressed I-girder manufacturing plant.

FIGURE 18.38 General view of Esenboga road pedestrian bridges.

FIGURE 18.39   Static and dynamic tests conducted on Esenboga road pedestrian bridge cables.

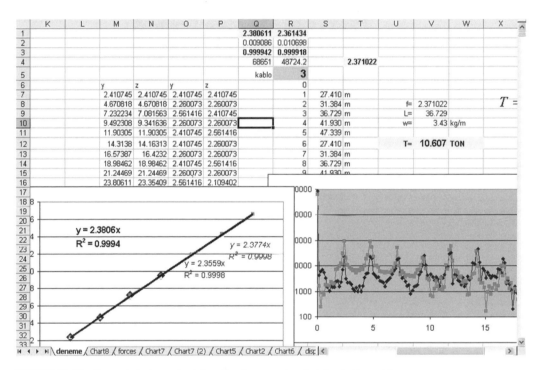

FIGURE 18.40   Dynamic tests conducted on Esenboga road pedestrian bridge cables.

of transverse deformation (Figure 18.39), or (2) measuring the vibration frequencies of the cable and converting the vibration frequencies into axial tension in the cable, akin to the vibration characteristics of a guitar string (Figure 18.40).

18.10 Major Istanbul Bridges

18.10.1 Bosphorus Bridge

The Bosphorus Bridge is located on the major Istanbul belt highway, close to the Marmara Sea at the shortest distance on the Bosphorus (Figure 18.41). The bridge has a total length of 1560 m and a main span of 1074 m. The bridge has extraordinary shaped cables that are inverse

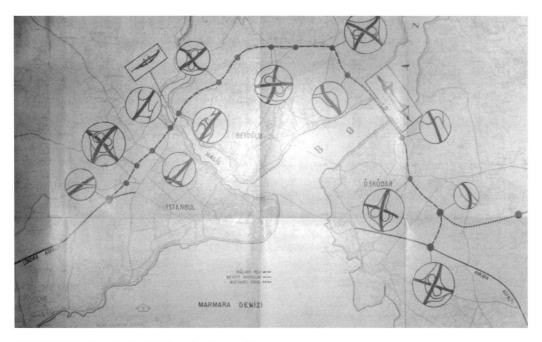

FIGURE 18.41 Istanbul belt highway bridges and intersections.

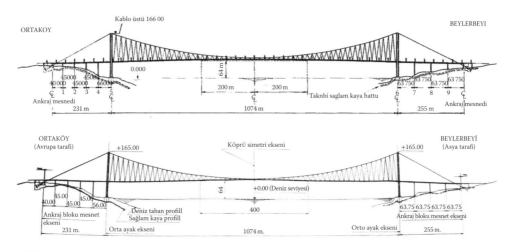

FIGURE 18.42 Bosphorus Bridge elevation drawing.

"V" shaped and located at 59 positions along the length of the main span (Figure 18.42). The Bosphorus Bridge has a width of 25 m which carries total of six lanes of traffic, three lanes in each direction (Figure 18.43). The bridge opened to traffic in October 30, 1973, for the fiftieth anniversary of the Turkish Republic.

18.10.2 Fatih Sultan Mehmet Bridge

The Fatih Sultan Mehmet Bridge is located on the Bosphorus Strait and is the second bridge constructed between Europe and Asia, as shown in Figure 18.44. The bridge has 1090 m of main span and the total length between anchorages is 1510 m (Figure 18.45). The bridge carries four lanes of traffic in each

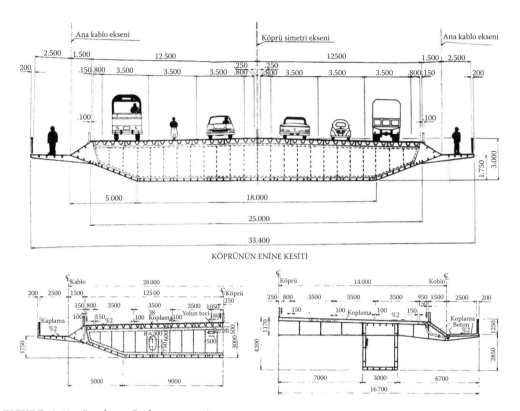

FIGURE 18.43 Bosphorus Bridge cross section.

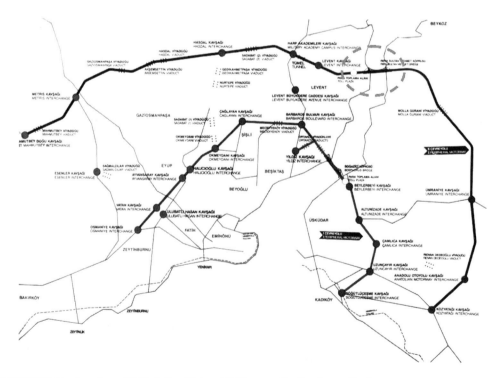

FIGURE 18.44 Location of Fatih Sultan Mehmet Bridge over the Bosphorus Strait.

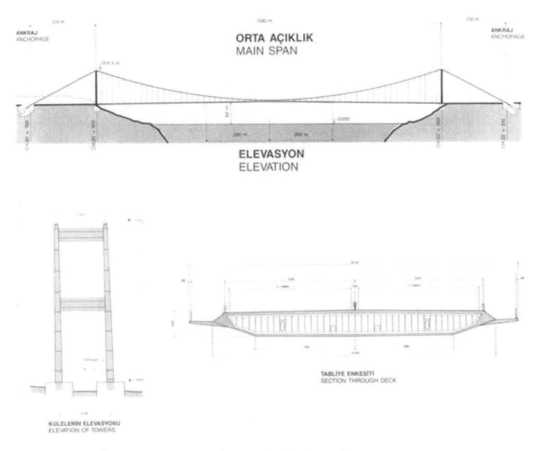

FIGURE 18.45   Elevation, cross section, and towers of Fatih Sultan Mehmet Bridge.

direction and the sidewalks, for a total width of 39.4 m. The towers are 107.1 m high above the footing and have a variable cross section from 5 m × 4 m at the base to 3 m × 4 m at the top. The bridge has a clearance of 64 m from the sea level, which is also equal to the clearance of Bosphorus Bridge, located about 6 km towards Marmara Sea.

Construction began on December 4, 1985, and was completed in 908 days, on May 29, 1988—192 days earlier than its scheduled date. When the bridge was constructed, it was the sixth longest bridge in its category in the world, after Humber (1410 m; United Kingdom), Verrazano-Narrows (1298 m; United States), Golden Gate (1280 m; United States), Mackinac (1158 m; United States), and Monami-Bisanseto (1100 m; Japan).

18.10.3 Halic Bridge

The Halic (Golden Horn) Bridge is a part of the 22 km long Istanbul belt highway. The Golden Horn separates the European part of Istanbul into two. The Galata and Ataturk bridges over the Golden Horn became insufficient for the increasing traffic demands of Istanbul, and Halic Bridge construction was completed in 1974 as the third bridge on the Golden Horn (Figure 18.46). The bridge project was prepared by Turkish and Japanese engineers. The bridge has eight spans of continuous steel girders and five shorter spans of precast/pretensioned concrete girders for a total length of 995 m, and is located 22 m above the sea level (Figure 18.47). Halic Bridge originally carried total of six lanes of traffic (Figure 18.48) but had to be extended to 10 lanes by adding two lanes in each direction.

FIGURE 18.46   General view of the Halic Bridge (www.turkiyehaberci.com).

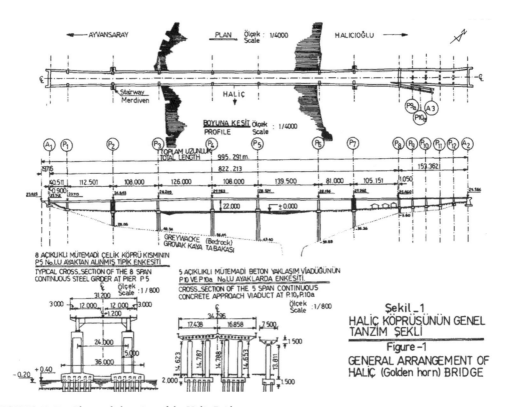

FIGURE 18.47   Plan and elevation of the Halic Bridge.

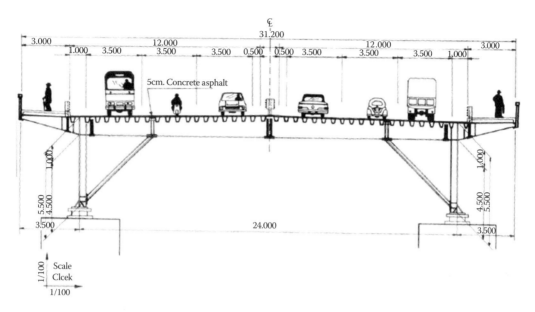

FIGURE 18.48   Cross section of the Halic Bridge.

18.11 Bridge Maintenance

18.11.1 Bridge Health Monitoring Systems

Bridge health monitoring (BHM) studies in Turkey gained momentum in the last decade of the twentieth century. Monitoring of bridges around the world has become possible with the advancement of computer and sensor technologies. As the bridges grew longer, taller, and became more challenging, monitoring started to become an integral part of the bridge structure. Bridge monitoring studies in Turkey can be basically grouped under three categories: (1) important bridges that require continuous attention to observe any signs of deterioration or damage, (2) a standard and descriptive subgroup of ordinary bridges that are members of a large population, in order to understand their performance over time, and (3) bridges with serious problems that necessitate monitoring studies.

Early monitoring attempts were made on Bosphorus Bridge, which is one of the most important modern bridges in Turkey. Natural vibration frequencies and mode shapes of the 1073 m long main suspension span were measured at different occasions. An initial study for long-term monitoring of the Bosphorus Bridge was conducted by the State Highway Department, Istanbul district (KGM–17th division) by placing 10 temperature transducers to monitor temperature changes over time. Currently, a comprehensive monitoring study has been initiated, including GPS based deflection sensors, strain gages, cable force measurement, tilt sensors, and so on.

Recent studies on monitoring of the Bolu Viaduct are being conducted jointly by Middle East Technical University (METU) and the State Highway Department (KGM) as a part of the TUBITAK 104I108 project. The pilot study involves placement of 16 strain gages, 2 accelerometers, a relative humidity probe, 3 temperature gages, and 3 LVDTs (Linear Variable Differential Transformer) to measure horizontal movement in the longitudinal and transverse directions between the deck and the selected pier 48 (Figure 18.49). The third LVDT measures changes in the expansion joint between spans 50 and 51. The viaduct experienced damage during the Izmit (August 17, 1999; Mw = 7.5) and Duzce earthquakes (November 12, 1999; Mw = 7.2), with a surface crack passing between piers 45 and 47 (Figure 18.50). The monitoring study started in 2008. The vibration characteristics, such as natural vibration modes and damping ratios, are also obtained and recorded. Analytical model simulations and actual measurements

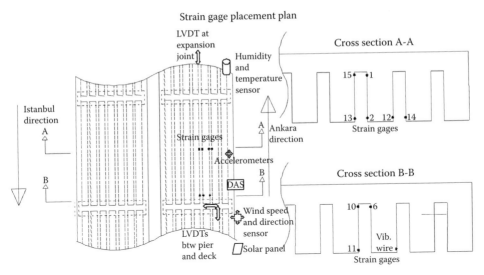

FIGURE 18.49   General view of the Bolu Viaduct structural health monitoring system.

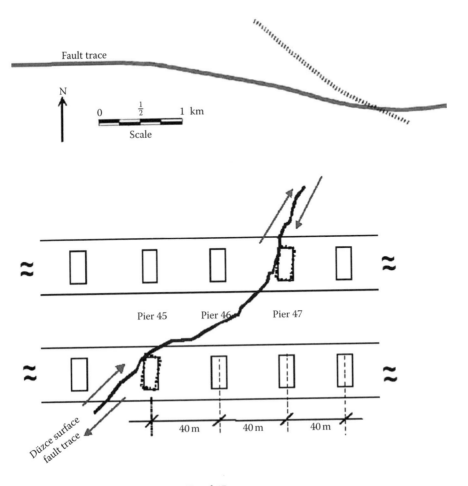

FIGURE 18.50   Surface crack between piers 45 and 47.

(Satellite view)

FIGURE 18.51 General view and location of Andirap Bridge.

are compared to identify structural parameters. The initial studies show temperature-dependent movement of the deck, which is placed over pendulum-type seismic isolators.

Another example related to bridge health monitoring is the Andirap Bridge, which is a 38 m long arch bridge with an 18 m long approach span and is located near Adana (Figure 18.51). The bridge gained an important role when a hydroelectric power plant dam was constructed in close proximity. All of the dam excavation and fill material had to pass over the Andirap Bridge, which was not initially designed for such heavy (50 tons/truck) and frequent (about 400 trucks/day) traffic. Although the bridge was constructed in 2001, diagnostic tests were carried out to determine its allowable load capacity. Strengthening of some members promised an overall capacity increase; however, long-term health monitoring was necessary to obtain its performance and intervene if residual deformations were observed. A total of 10 vibrating wire strain gages, 11 temperature sensors, and 2 LVDTs were installed on the bridge. The monitoring study started in early 2009 and has been collecting data since then (Figure 18.52). The strain readings at the arch, columns, beams, and deck show temperature-dependent measurement, which return back to original values as the climate slowly goes into the winter season, when the measurements were first initiated.

Bridge monitoring studies are also conducted on pedestrian bridges. Two such examples were carried out for the METU and Esenboga pedestrian bridges in Ankara. Any bridge monitoring study can be performed using continuous long-term measurement systems installed on structural members or discrete sets of measurements taken from the structure with certain time intervals (such as every year). Measurement instruments have been installed on structural parts of METU pedestrian bridge in the factory to capture demand on the structural members during transportation, erection, and service conditions (Figures 18.53 and 18.54). Experimental and analytical results were compared in order to conduct structural identification (St-Id). The results indicated that construction stresses reach about 120%

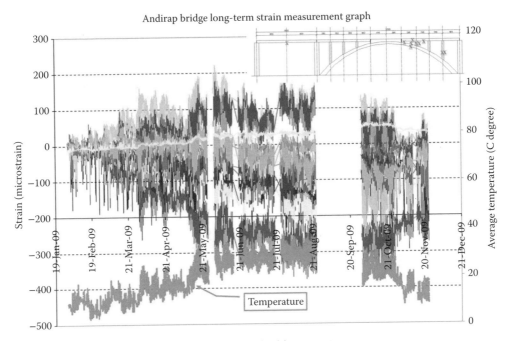

FIGURE 18.52   Data collected from the Andirap Bridge health monitoring system.

FIGURE 18.53   METU pedestrian bridge.

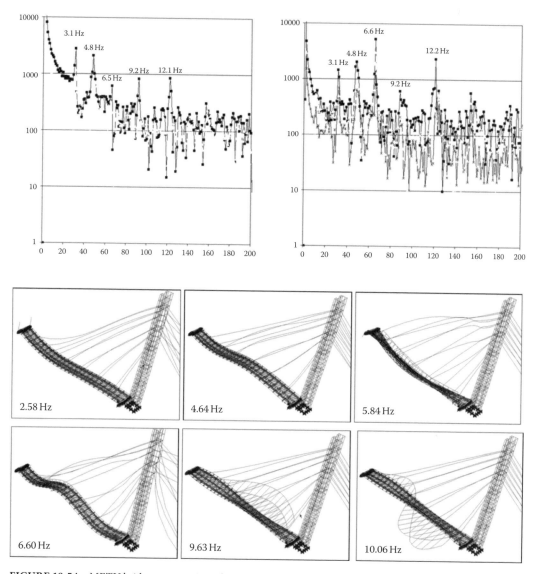

FIGURE 18.54 METU bridge; comparison of analytical and experimental modal frequencies.

of the service conditions during the first three months of monitoring, but obviously do not exceed the allowable limits of the structure.

Pedestrian bridges may also be constructed using cable stays, such as the METU and Esenboga pedestrian bridges (Figure 18.55). The measurement of cable forces were carried out by obtaining cable vibration frequencies (f_n) and converting them into axial forces T (in Newtons) using Equation 18.1:

$$T = \frac{4^* w^* (f_n^* L)^2}{n^{2\,*}g} \tag{18.1}$$

where w is the unit weight of the cable (N/m); n and f_n are the nth resonant vibration and corresponding frequency (Hz) of the cable, respectively; L is the length (m); and g is the gravitation acceleration (m/s^2). Other units may also be used provided that the units are compatible with each other.

FIGURE 18.55   Esenboga pedestrian bridges.

Comparison of cable forces against symmetric counterparts as well as other cables has shown that the existing cable forces do not match the design loads. Although the cables are overdesigned, large differences in cable forces may lead to fatigue or failure of secondary members such as connecting plates. The cable forces were corrected using turn buckles located at the cable and bridge connection locations.

18.11.2 Bridge Inspection

The current Turkish practice of bridge inspection is based on a condition rating of bridges with four levels. The four main levels of the condition of any structure are failed (grade A), seriously deteriorated (grade B), minor deterioration (grade C) and good condition (grade D). An improved bridge inspection grading card is proposed as shown in Figure 18.56. The inspection grading card includes explanations for the different condition rates at the bottom. The main elements of the structure that will be evaluated are listed so that the inspection can be made easily and systematically: main body components, earth retaining components, and serviceability components. Each of the components is examined with the sub-group elements and properties that are mentioned in the checklist shown in Figure 18.56. This grouping for the components is done according to the function of the elements within the structure and concerning the similarities of the elements and their effects on the structure (Caner et al. 2008).

With the primary concern of how seriously or suddenly bridge components can cause the structural performance to deteriorate or how long the structure can survive without them, some coefficients are assigned to the components in order to differentiate the ones that need more attention than others. To this end, 1.0 is assigned to the main components and earth retaining components, whereas 0.5 is set for the serviceability components. These weight coefficients are determined based on knowledge gathered through engineering experience in practical life. The critical level of structural stability or integrity is determined by the condition of the main components and the earth retaining components. Serviceability components should also be considered with great care as they are the elements affecting the passenger comfort level of the structure. The safety items are selected to have a larger weight coefficient than the comfort level items. Nondestructive testing of structural elements, condition grading, and taking pictures can be integrated into each other to minimize errors in the inspection grading. In order to avoid any subjectivity in grading, the inspection grades were re-evaluated by re-examining the pictures.

Bridge Name: ULUBAT SAğ (ESKİ)
Construction Year : 1951

MANI BODY COMPONENTS					EARTH RETAINING COMPONENTS	
Deck		**Support**			**Abutment**	
Cracks	4	Metallic (M) /Elastomeric (E)		(M)	Deformation	5
Concrete disintegration	3	Main damage		4	Cracks	6
Apparent reinforcement	5	Support bed		4	Concrete disintegration	6
Holes and cavities	5	Loss of elements (M)		5	Exposed reinforcement	7
Water leakage	4	Anchorage (M)		5	Holes and cavities	6
Overall Grade	4.20	Surface arrangement (E)		9	Water & debris abrasion	9
Beams		Deformation (E)		9	Scour in foundation	9
Steel (S) / Concrete (C)	C	Overall grade		4.50	Overall grade	6.00
Deformation	5	**Piers**			**Approach fill**	
Cracks	4	Deformation		6	Settlement and Slump	5
Rusting (S)	9	Cracks		6	Erosion on road platform	5
Bolts and Rivets (S)	9	Concrete disintegration		6	Erosion in fill	5
Welding (S)	9	Apparent reinforcement		6	Overall grade	5.00
Concrete disintegration (C)	3	Holes and cavities		5	**Stabilization**	
Apparent reinforcement (C)	3	Water and debris abrasion		5	Settlement and Slump	6
Holes and cavities (C)	4	Scour in foundation		6	Erosion	5
Overall grade	3.80	Overall grade		5.71	Scour in bed level	9
					Overall grade	5.50

SERVICEABILITY COMPONENTS

Coating		Border-railing			Expansion joint	
Waving	4	**Border (B) /Railing (R)**		**R**	Noise	4
Tire tracks	5	Cracks in the concrete (B)		9	Water leakage	4
Cracks	4	Concrete disintegration (B)		9	Deformation	4
Holes and cavities	5	Apparent reinforcement (B)		9	Holes and cavities	4
Overall grade	4.50	Deformation in railing (R)		5	Loss of elements	5
Drainage		Rusting in railing (R)		4	Loss of function	5
Pipe damage	6	Deficiency in railing (R)		5	Overall grade	4.33
Blockage	6	Overall grade		4.67		
Cleanout	9					
Overall grade	6.00					

ITEM	GRADE		Condition rating	Description
MAIN BODY COMPENENTS	4.55			
EARTH RETAINING COMPONENTS	5.50		1	Totally deteriorated or failed
SERVICEABILITY COMPONENTS	4.88		2	A grade between 1 and 3
CUMULATIVE WEIGHTED GRADE	5.00		3	Serious deterioration, not functioning as originally designed
			4	A grade between 3 and 5
			5	Minor deterioration but functioning as originally designed
			6	A grade between 5 and 7
			7	New condition
			8	Not applicable
			9	Unknown

FIGURE 18.56 Condition rating and proposed inspection card.

Bridges on route D200, connecting Bursa to Canakkale, were inspected by this new system. At the investigated bridges, cracks were observed at the beams, decks, columns, and abutments. Bridges with supports at mid-third of the spans were damaged and had almost complete loss of function, as illustrated in Figure 18.57. Beam cracks were due to either shear or flexure or concrete disintegration. Some exposed reinforcement was also observed at beams, as shown in Figure 18.58. Some of pier cracks were observed to be map cracks that indicate an alkali-silica reaction or some other chemical reaction. In a number of slender piers, buckling of columns was observed. Some of the abutment cracks were considerably big, that is, in the order of centimeters. Cracks were measured by a handheld crack meter.

In evaluation of some bridges, a rebar locator was used to sense the spacing of stirrups and longitudinal bars. In some other cases, because of concrete spalling, the size and spacing of rebars were easily detected by vision inspection. The visible rebars were corroded. The concrete compressive strength of bridges

FIGURE 18.57   Karadere Bridge, deteriorated support.

FIGURE 18.58   Ulubat Sag, exposed reinforcements.

was usually measured at the piers using a portable electronic Schmidt hammer. During nondestructive tests, ten hits were applied to obtain statistical information from the computer. It was determined that on average, the concrete compressive strength was about two times greater than the 28-day minimum strength specified on the inventory card of the bridges.

Most of the aging bridges have concrete disintegration, water leakage, debris accumulation, loss of function at expansion joints, serviceability problems, cracks at structural elements, and corroded exposed reinforcement at piers and beams. The bridges with supports at mid-third lengths of the spans have major problems facing loss of function. The compressive strengths of concrete at the inspected bridges are found to be satisfactory.

A method for bridge life expectancy assessment for bridges with no regular inspections was developed by Caner et al. (2008). The deterioration in the main components is higher than in earth retaining components and serviceability components. When the deterioration rate for the main components was projected to grade 3, the not functioning as designed grade, the average life of bridges was estimated to be around 80 years, which is similar to those in OECD (Organization for Economic Cooperation and Development) countries.

References

AASHTO. 2007. *AASHTO LRFD Bridge Design Specifications*, 4th Edition, American Association of State Highway and Transportation Officials, Washington, D.C. (1st Edition, 1994; 2nd Edition, 1998; 3rd Edition, 2004.)

AASHTO. 2002. *Standard Specifications for Highway Bridges*, 17th Edition, Association of State Highway and Transportation Officials, Washington, D.C. (12th Edition, 1977; 13th Edition, 1983; 14th Edition, 1989; 15th Edition, 1992; 16th Edition, 1997.)

Caner, A., Yanmaz, A. M., Yakut, A., Avsar, O. and Yilmaz, T. 2008. "Service life assessment of existing highway bridges with no planned regular inspections," *Journal of Performance of Constructed Facilities*, ASCE, April 22(2):108–114.

CEB. 1990. *CEB-FIP Model Code for Concrete Structures*, Comite Euro-International de Beton, Lewis Brooks, New Malden, UK.

CSI. 2000. *SAP 2000*, Computers and Structures, Inc., Berkeley, CA.

Ilter, F. 1978. *Turkish Anatolian Bridges until Ottomans*, (in Turkish), General Directorate of Highways (KGM). Ankara, Turkey.

KGM. 2007. "History of Highways," (in Turkish) Pelin Publishing, Ankara, Turkey.

LARSA. 2006. *LARSA 4D User's Manual*, LARSA, Inc. Melville, NY.

Further Readings

Demirezer T. 2007. "Turkey high speed rail projects and structures", *1st Bridge and Viaducts Symposium Proceedings*, (in Turkish), November, Antalya, Turkey.

Harputoglu, Z. Celebi N., and Tulumtas F. 2007. "Gulburnu Bridge", *1st Bridge and Viaducts Symposium Proceedings*, (in Turkish), November, Antalya, Turkey.

Turker, I. L., Ahmad, A. and Ozdemir H. 2009. "Samsun Ayvacik- Eynel Bridge" *The third National Steel Structures Symposium*, (in Turkish), Gaziantep, Turkey.

Yuksel I. A.S. 1986. *Karakaya Dam Lake, Firat railway bridge*, ERAS Publication. (in Turkish).

19

Bridge Engineering in Ukraine

Mykhailo Korniev
JSC "Mostobud"

19.1 Introduction

19.1.1 Geographical Characteristics

Ukraine, the largest European country by territory (Figure 19.1), has a population of almost 47 million people. The country is bisected in the north–south direction by the Dnipro River. Along its length of about 1000 km many hydroelectric power stations have been constructed, with dams creating artificial lakes several kilometers wide. Large cities and industrial centers such as Kyiv, Cherkasy, Kremenchuk, Dniprodzerzhynsk, Dnipropetrovsk, Zaporizhia, and Kherson are located along the Dnipro River. Bridges crossing the Dnipro play a vital role in the Ukraine transportation system. The Ukrainian bridge construction industry, much affected by economic crisis in the early 1990s, is currently in the process of revival. At the present, three major bridges over the Dnipro River are being constructed, two in the city of Kyiv and one in Zaporizhia City.

19.1.2 Historical Development

Permanent bridges across the Dnipro River were not built until the middle of the nineteenth century, replacing the ferries or floating bridges used previously. Wooden or stone bridges were used across smaller rivers, and some stone viaducts built for approaches to castles were quite imposing (Figure 19.2).

FIGURE 19.1 Map of Ukraine.

FIGURE 19.2 Bridge over the Smotrych River in Kamenets–Podilsky.

Novoplanivsky Bridge over the Smotrych River in Kamenets–Podilsky was built in 1874. This bridge has a length of 136 m and height of 38 m. It has seven spans of 19.4 m. Another famous bridge in Kamenets–Podilsky is the Castle Bridge, at present similar to a dam, but initially it was stone bridge with spans of 7.8 m and width of 3.9 m. Scientists believe that this bridge was built in the second century by Roman troops. The Castle Bridge has been rebuilt many times [17],[20],[9].

The first iron bridge over the rivers was built in railway construction in the second half of the nineteenth century. Many stone arch railway viaducts over the valleys were built in the Carpathian Mountains in the Western part of the country. Those bridges are still in service (Figure 19.3). For long-span railway bridges across the Dnipro, iron closed-mesh lattice trusses were used. Two of the structures of this type, in Kyiv (1870) and in Dnipropetrovsk (then Katerynoslav; 1884), are shown in Figures 19.4 and 19.5, respectively. The bridge in Kyiv had 14 simple spans of 76 m with a total length of 1068 m. It was at that time the longest bridge in Europe. Individual truss spans were assembled on the bank and floated to their locations

FIGURE 19.3   Railway viaduct in the Carpathian Mountains in Vorokhta, 1894; main span 65 m.

FIGURE 19.4   Bridge in Kyiv across the Dnipro, 1870.

on the piers. The total weight of iron for this structure was about 4000 tons, and the cost of construction was 3.2 million rubles. Every day seven trains crossed the bridge at a speed of 30 km/hr.

The two-level railway bridge in Dnipropetrovsk (then Katerynoslav; 1884), with 15 spans at 83 m and a total length of 1250 m, was even longer (Figure 19.5). Its total weight was 9525 tons, and the cost was 4 million rubles. The bridge was destroyed twice, in 1918 and 1941, respectively and was rebuilt twice in 1920 and 1943, and then in 1955 it was completely rebuilt [4].

The first major bridge for vehicular and pedestrian use was built in 1853: the chain suspension Mykolaivskyj Bridge over the Dnipro in Kyiv. This bridge is 790.9 m long, with 68.58 m + 4 × 134.1 m + 68.58 m fixed spans and one swing span of 15.24 m (Figure 19.6). The width of the deck is 16 m, with a vehicular roadway 10 m wide. The iron structure, weighing 1600 tons, was fabricated in Birmingham, England, delivered by steamships to Odesa, and carted by oxen carts to Kyiv. The bridge was designed for a live load of 520 kg/m². Tolls were charged for bridge crossing: 6 copecks for a cow, 9 for a horse. With the advent of automobiles, 15 copecks were charged for a car. In 1912, a tram line was laid on the bridge [19]. This bridge was designed by the British engineer Charles Blacker Vignoles. A silver bridge model was exhibited in London in 1854.

FIGURE 19.5 Bridge in Dnipropetrovsk across the Dnipro, 1884.

FIGURE 19.6 Mykolaivskyj suspension bridge in Kyiv across the Dnipro, 1853.

FIGURE 19.7 Kichkas bridge across the Dnipro, 1902.

The Kichkas Bridge over the Dnipro near Zaporizhia was built in 1902 (Figure 19.7). This bridge was designed by famous bridge designer L. D. Proskuryakov (author of works in structural mechanics and teacher of E. O. Paton). The bridge had span of 190 m and arch height of 20 m, with a total length of 336 m. On the top level was two railway tracks (railway traffic on the bridge opened in 1908) and on the bottom level by two sides of the bridge were pedestrian sidewalks. In 1920, the central part of the span was exploded and was recovered in 1921. Due to construction of the Dnipro hydro electric power plant and its reservoir the Kichkas Bridge was disassembled in 1931 [2].

In 1920, during the civil war, one of the spans of the Mykolaivskyj Bridge in Kyiv was blown up and the link chain was broken, which caused its total failure (Figure 19.8). In 1925, a new bridge was built utilizing the existing piers, designed by E. Paton, a professor at the Kyiv Polytechnic Institute (Figure 19.9). Well-known scholars S. Timoshenko, author of *Strength of Materials* (1918), and Ihor Sikorsky, designer of helicopters who later emigrated to the United States, were graduates of this institute.

Bridge construction activities were much intensified prior to World War I across all of Ukraine, which was then divided between the Russian empire and the Austro-Hungarian monarchy. Many steel bridges of all types were built during this period. The twin tied arch bridge over the Rusanivskyj Strait in Kyiv, constructed in 1906 and designed by M. Belelubskyj with spans of 100 m, is an example (Figure 19.10) [20]. However, most of these bridges were destroyed during World War I from 1914 to 1918 and the subsequent civil war from 1918 to 1921.

FIGURE 19.8 Mykolaivskyj bridge failure, 1920.

FIGURE 19.9 The rebuilt Mykolaivskyj bridge on old piers according to E. Paton's project, 1925.

FIGURE 19.10 Rusanivskyj Strait bridge, 1906.

19.2 Design Practice

19.2.1 Bridge Design Specifications and Design Philosophy

After World War II, all of the Ukrainian territories were incorporated into the former Soviet Union, and the bridge design and construction was governed by the relevant USSR standards. The latest edition of the standards, SNiP 2.05.03-84, was published in 1984. It consists of seven chapters: general provisions, loading, reinforced concrete bridges, steel bridges, composite bridges, wood bridges, and foundations. Bridge standards are based on the load and resistance factor design principle. The bridge design norm State Construction Norms (DBN) B.2.3-14: 2006 [5] is currently used in Ukraine.

According to the 1984 standards, bridges are designed in accordance with the design philosophy of limit state design. The design equation in the Ukrainian Norms (UN) is defined by the inequality

$$E_d \leq R_d \tag{19.1}$$

where E_d denotes the effects of loading multiplied by load factors (safety factors based on stipulated probabilistic considerations of variability of loading, probability of simultaneous occurrence of various loadings, lack of design accuracy, importance of the structure, etc.) and R_d denotes structure resistance, with a specified probability of material quality defects. The recommended load factors are essentially similar to these of the Eurocode.

The UN stipulates two groups of limit states:

Group I: Catastrophic failure of the structure due to loss of strength, instability or cracking
Group II: Serviceability limitation which may cause service complications or service interruption

Ukraine, aspiring to membership in the European Union, since 1995 a member of the Council of Europe and, since 2008, a member of the World Trade Organization, is actively engaged in export and import trade, with the Ukrainian bridge design and construction industry competitively engaged in fabrication and construction of bridges abroad. International commercial activities would be further facilitated by upgrading the national transportation network and harmonizing Ukrainian industrial and bridge design norms with those of the neighboring European Union members. For this purpose the Ukrainian Society of Civil Engineers, in collaboration with the Ukrainian Group of the International Association of Structural Engineering (IABSE), has set up a task force for study of the Eurocodes with the aim of gradual adaptation of these norms for Ukrainian use. As a first step, a Ukrainian translation of the basic Eurocode 1, "Basis of design and actions on structures," (ENV 1991-I) [7] was prepared in 1997, and appropriate revisions of the Ukrainian Norms [5] are being studied.

19.2.2 Loadings for Highway and Railway Bridges

The norms determine rules of bridge loading with permanent and live loads. The norms provide characteristic values of traffic loads for railway and motor transport, metro and pavement loads, and pedestrian values as well as values of safety and dynamic factors. Design loading for highway bridges currently specified in the Ukrainian Norms provide for two vehicle load models: heavy single-vehicle NK or AK lanes. The NK heavy single-vehicle, with a weight of 981 kN, is used without other live loadings in the most unfavorable positions on the bridge, including safety lanes.

The AK loading consists of two tandem axles of $P = 10K$ kN, combined with a uniformly distributed lane loading $q = K$ kN/m (Figure 19.11), where K is loading class. For new bridges $K = 15$ kN is used, and a heavy vehicle. Two tandems, with a reduction factor of 0.85, are used for calculation of negative moments in continuous spans. AK loading is used for the first and second lanes and is subject to reduction for additional lanes.

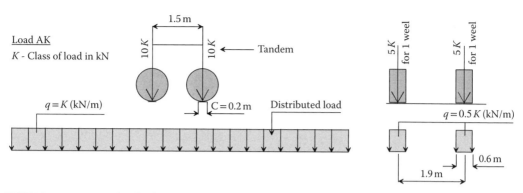

FIGURE 19.11 Design live load AK for bridges.

For comparison, the tandem axle load stipulated for the design is 150 kN (with $K = 15$) in the UN, 300 kN in the Eurocode, and 110 kN in the American Association of State Highway and Transportation Officials (AASHTO) load resistance and factor design method (LRFD) specifications. The magnitudes of q are 15.27 and 9.3 kN/m, respectively, for these specifications. Design loading for railway bridges is CK. For permanent railway bridges use CK14 loading, which is performed in table form depending on bridge length, table values was obtained from local loadings of $24.5K = 343$ kN and uniformly distributed loading of $9.81K = 137$ kN/m. Longitudinal traction and braking forces and transverse loading are also specified [1],[5],[18],[13].

19.3 Concrete Girder Bridges

In the former USSR, modular precast prestressed concrete bridges were widely used because industrially fabricated components were more economical and could be erected expeditiously, regardless of weather conditions. Typical prefabricated components included deck panels with spans up to 12 m and beams of 18–33 m spans (Figure 19.12). For longer spans up to 45 m precast prestressed double girder elements, shown in Figure 19.13a, were used, weighing up to 60 tons (590 kN). Box girder segments (Figure 19.13b) were used for spans up to 90 m, with cables consisting of 12 or 19 strands of seven wires used for prestressing. The joints between the segments were filled with epoxy compounds.

Such box girder segments were used on the concrete approach spans of the South Bridge in Kyiv (see Section 19.7) with spans of 50 m + 7 × 79 m + 60 m. The bridge cross section consists of three box segments (Figure 19.14), with vehicular traffic carried by the two outer boxes and the Metro tracks by the central box. The segments, fabricated at the Dnipopetrovsk reinforced concrete plant, were subassembled into longitudinal units. Individual segments have constant exterior dimensions: depth of 3.4 m, width at the top of 11.2 m, and length along the bridge 1.98 m and 1.50 m, for regular (in span) and basic (on pier) segments, respectively. Changes in the bottom flange and the web thickness of the boxes were obtained by variation of the interior dimensions (Figure 19.15). The weight of the segments varied from 40–54 tons (394–531 kN).

Erection of preassembled units of five segments proceeded by cantilever method starting from the supporting piers. Subsequent erection assemblies were added in a balanced manner. This operation required special equipment for assuring placement in the correct position, working platforms for erection personnel, and accommodation of equipment for moving forward and prestressing of assemblies. Prestressing (Figure 19.16) was accomplished by 12-strand cables exerting a force of 180 tons (1770 kN). Cables were inserted in 90 mm diameter openings, which were subsequently filled with cement grout.

The Kaydakskyj precast concrete bridge over the Dnipro in Dnipropetrovsk (Figure 19.17) was built in 1982. It crosses the river island and consists of two portions over the two river channels with the spans of 3 × 43 m + 73.5 m + 2 × 105 m + 73.5 m + 42 m and 42 m + 12 × 63 m + 43 m, with a total bridge length of 1732 m [4]. Another major bridge of this type, the Antonivskyj Bridge over the Dnipro at Kherson (Figure 19.18), was built in 1985 with spans of 35 m + 5 × 42 m + 70 m + 2 × 105 m + 70 m + 16 × 42 m + 35 m and a total length of 1302 m.

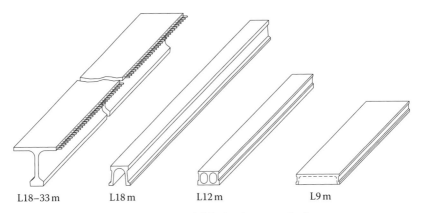

FIGURE 19.12 Typical precast concrete girders and slabs for short-span bridge structures.

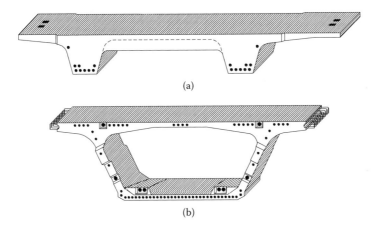

FIGURE 19.13 Typical precast concrete segments for long-span bridges. (a) Double girder. (b) Box girder.

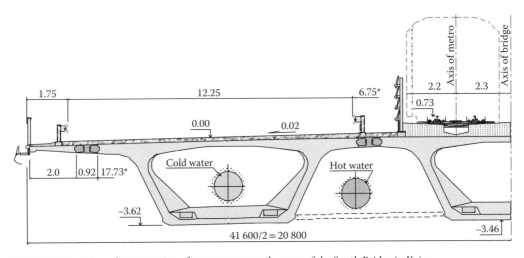

FIGURE 19.14 Typical cross section of concrete approach spans of the South Bridge in Kyiv.

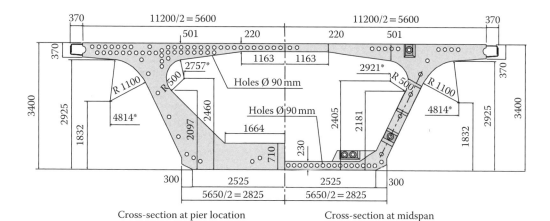

FIGURE 19.15 Cross section of modular concrete segments used in the South Bridge in Kyiv.

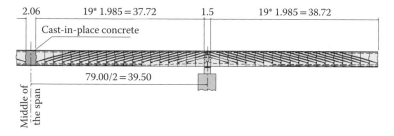

FIGURE 19.16 Layout of prestressing strands in the concrete approach part of the South Bridge in Kyiv.

FIGURE 19.17 Kaydakskyj bridge in Dnipropetrovsk, 1982.

FIGURE 19.18 Antonivskyj bridge in Kherson, 1985.

19.4 Steel Girder Bridges

19.4.1 Bridges with Concrete Decks

The spans of steel girder bridges with concrete decks are typically used for bridge spans in the range of 80 m to 120 m. A typical bridge of this kind is the Paton Bridge in Kyiv, built in 1953 (Figure 19.19). It is a continuous girder structure with span lengths of 17.5 m + 9 × 58 m + 4 × 87 m + 11 × 58 m + 17.5 m and a total length of 1543 m. The vehicular roadway width is 21 m (Figure 19.20). Concrete caisson foundations were used. The concrete deck is supported by transverse beams with regular spacing of 2.9 m, which is supported by four I-shaped girders. It is of interest to note that all girder splices are welded. Automatic welding equipment was used for welding of the vertical walls. All equipment was developed by the Electric Welding Institute in Kyiv. The institute was established by the outstanding bridge engineer E. Paton at the Kyiv Polytechnic Institute [6].

19.4.2 Bridges with Orthotropic Decks

Orthotropic deck bridges were introduced in the former USSR in the 1970s. The first major structure using such decks was a cable-stayed bridge in Kyiv that opened in 1976. Its spans are 84.5 m + 300 m + 63.5 m, and the roadway width is 31.4 m (see Section 19.7). An open-rib orthotropic deck was used as the roadway and the top flange of the stiffening girders. Following the example of orthotropic decks being

FIGURE 19.19   The Paton Bridge in Kyiv across the Dnipro, 1953.

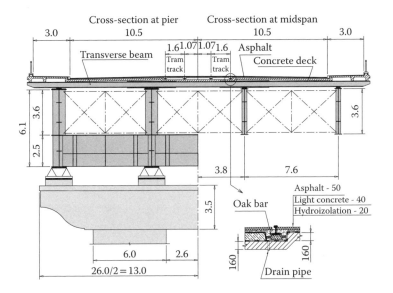

FIGURE 19.20   Cross section of the Paton Bridge over the Dnipro in Kyiv.

built at that time in Germany, where this deck system was developed, the deck plate thickness was 12 mm and the flat bar longitudinal ribs were 180 mm deep and 12 mm to 25 mm thick spanning between crossbeams spaced at 2.5 m (Figure 19.21). The design methods and fabrication technologies developed for this project served as guides for subsequent structures of this type in Ukraine and in other republics of the USSR, such as the cable-stayed bridge in Riga, Latvia.

The Prypiat Bridge near Chernobyl (1987) (Figure 19.22) is typical of several such orthotropic deck structures with open ribs built subsequently in Ukraine. The bridge, with spans of 62 m + 84.6 m + 3 × 26 m + 84.6 m + 62 m, consists of two parallel structures. The bridge has an 11.5 m width roadway supported by box girders 3.2 m deep. The weight of the steel superstructure is 430 kg/m². Girders of a similar design were also applied for several bridges in Ukraine. The design of the steel orthotropic decks of these bridges utilized the former Soviet SNiP specifications, as well as the relevant provisions of the German DIN norms and the American AASHTO bridge design specifications.

New Ukrainian structural design norms (DNB) are under development based on the latest progress in this field. Revised specifications for the design and construction of orthotropic decks will aim at harmonizing their recommendations with the relevant provisions of the Eurocode and the AASHTO LRFD

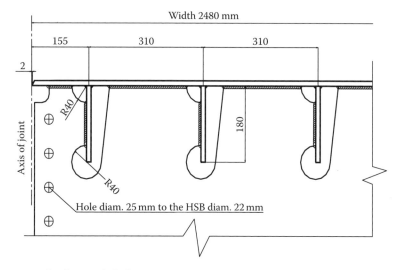

FIGURE 19.21 Details of open-rib deck.

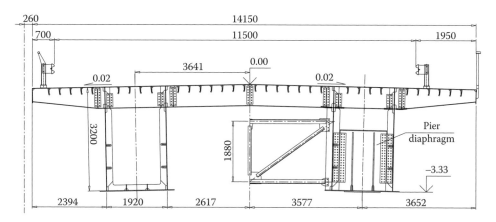

FIGURE 19.22 Cross section of the Prypiat Bridge near Chernobyl.

specifications, which are currently undergoing significant rethinking. One of the problems now being considered is how to establish a reliable rational method of assuring safety against cracking of steel orthotropic decks. According to the formerly prevalent approach, this could be achieved by limiting the "stress ranges" to which various deck details were subjected. However, based on accumulated world-wide experiences with orthotropic deck bridges, cracking is caused mainly by the effects of fabrication, with the applied stress fluctuation contributing to crack propagation being of secondary importance. It has also been shown that classical elastic fatigue theory based on stress fluctuation is not applicable to details at welds.

Based on these findings, empirical provisions for highway bridges relying on recommendations for structural details and weld preparation, not requiring numerical analysis, have been stipulated in the Eurocode (EN 1993-2: 2004 (E), Art 9.1.2). A similar trend is reflected in the AASHTO LRFD bridge design specifications of 2007, where, in the 2009 Interim Edition, the "Detail Categories" for ortho-tropic decks stipulating "allowable stress ranges" have been omitted. Introduction of similar rules in the Ukrainian design norms is now under consideration. The old SNiP norms do not contain provisions for orthotropic decks with closed ribs. Now in Ukraine the new bridge design rules (DBN) are prepared with complete recommendations for the design of orthotropic decks [1],[5],[18].

Typical details of orthotropic decks with open ribs are shown in Figure 19.21. Figure 19.23 shows two methods of solving the problem of rib misalignment at the field splices. Traditionally, field splices were made by high-strength bolting. The problem of horizontal rib misalignment was solved either by the rib-to-deck weld not completed in the shop to permit adjustment (Figure 19.23a) or by using shims under the rib splice plates (Figure 19.23b). Deck plate field spices are made by two methods (Figure 19.24). The traditional way consists of automatic multipass submerged arc welding

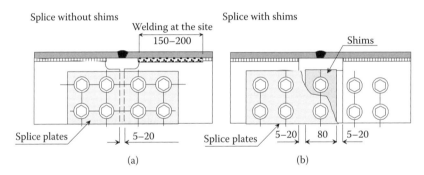

FIGURE 19.23 Typical field splices of open ribs: (a) with not completed weld; (b) with shims.

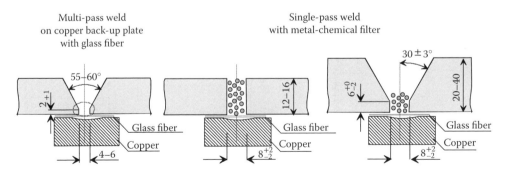

FIGURE 19.24 Welded field splices of deck plate.

made after manual welding of the weld root. The plate edges are beveled 27° ±2°. A copper back-up plate with 18 mm minimum thickness overlaid by glass fiber is used. Tests on welds made by this method show good mechanical properties and fine-grained structure of the weld zone. However, this method is labor intensive.

The other method uses metal-chemical filler (deformed 2 mm welding wire and chemical additive) and the weld is made on glass fiber and copper backup automatically in one pass. For plate thicknesses greater than 16 mm the edges are beveled. Instead of glass, fiberglass could be used. This method is much faster, however the weld quality is lower because of larger heat input.

In some cases other types of rib splices are used. Rib splices are usually made at quarter points of their spans. Figure 19.25a shows rib splices on the highway on top of the dam at the Dniprohes hydroelectric power station. Figure 19.25b shows closed-rib splices of the Dnipropetrovsk Bridge. Closed ribs for the Dnipropetrovsk Bridge were fabricated on cold bending presses.

19.4.3 Composite Steel and Concrete Bridges

Composite steel and concrete bridges are widely used in Ukraine for spans with ranges of 40–60 m. Under current economic conditions, steel bridges are uneconomical for short spans, while concrete bridges with spans exceeding 35–40 m require the use of costly equipment. Cast-in-place concrete decks of composite bridges do not require the use of temporary support from the ground. In composite bridges, welded shear stud connectors (Figure 19.26) are used.

Figures 19.27 and 19.28 show a continuous composite trestle at the Trukhaniv Island and Rybalsky Peninsula in Kyiv (presently under construction) serving vehicular traffic on the top level and Metro rail traffic on the bottom level. The spans of this horizontally curved trestle range from 30 m to 36 m. The designed traffic capacity is 60,000 vehicles per day. Concrete of 50 MPa and steel with yield strength of 390 MPa were used. Welded shear connectors and spherical bearings and modular expansion joints were used [15].

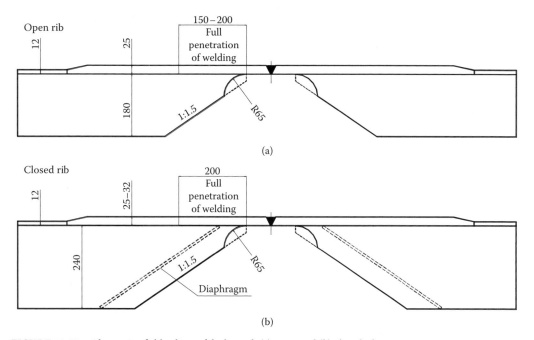

FIGURE 19.25 Alternative field splices of decks with (a) open and (b) closed ribs.

FIGURE 19.26   Metro beam (for viaduct on the Turkhaniv Island in Kyiv) with stud shear connector.

FIGURE 19.27   Viaduct in Kyiv at Trukhaniv Island.

Another composite bridge in Kyiv is the Harbor Bridge, completed in 2007 (Figures 19.29 and 19.30) with spans of 35 m + 45 m + 60 m + 75 m + 60 m + 45 m + 35 m. Its shape was determined by architectural considerations. Concrete of 50 MPa and steel with yield strength of 390 MPa, as well as welded shear connectors, were used. Spherical bearings and modular expansion joints were also used. The wearing surface used a waterproofing system, which consisted of three-ply (primer, membrane, and bond coat) sprayed polymer eliminator and mastic-asphalt with a thickness of 8 cm [8],[15].

Several composite overpass structures with center spans of 42 m were built on the main Kyiv–Odesa auto route (Figures 19.31 and 19.32). The depth of the steel girders at the midspan is 1.1 m. Elastomeric bearings were used. Figures 19.33 and 19.34 show a horizontally curved interchange in the city of Zaporizhia with spans of 40–48 m (presently under construction). The total length of

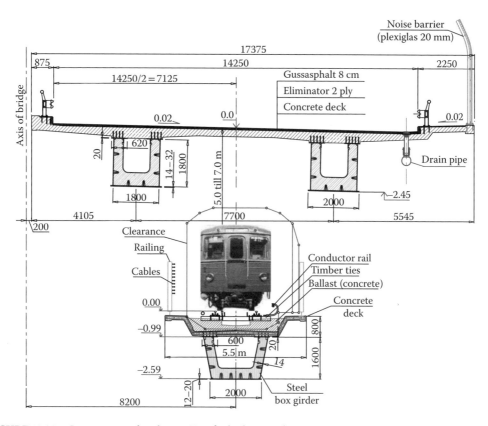

FIGURE 19.28 Cross section of viaduct in Kyiv for highway and metro use.

FIGURE 19.29 Overall view of the Harbor Bridge in Kyiv.

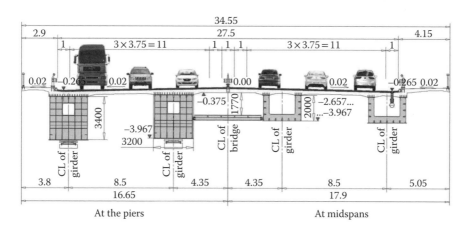

FIGURE 19.30 Cross section of the Harbor Bridge in Kyiv.

FIGURE 19.31   Typical overpass of the Kyiv–Odesa autoroute, spans 8 m + 42 m + 8 m, 2004.

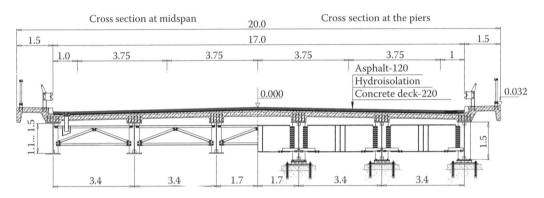

FIGURE 19.32   Cross section of overpass on the Kyiv–Odesa autoroute.

FIGURE 19.33   Horizontally curved composite viaduct with spans of 40–48 m in Zaporizhia (under construction, photo 2009).

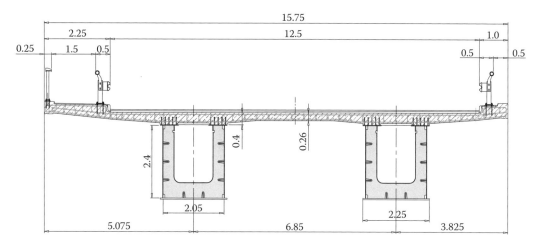

FIGURE 19.34 Cross section of the viaduct in Zaporizhia.

the bridges with viaducts is 8.2 km. Concrete of 50 MPa and steel with yield strength of 390 MPa were used. Welded shear connectors, spherical bearings, and modular expansion joints were also used [16].

19.5 Arch and Frame Bridges

19.5.1 Concrete Arches

Concrete arch bridges are generally used for railway crossings of major rivers. Figures 19.35 and 19.36 show the Merefa–Kherson railway bridge over the Dnipro River in Dnipropetrovsk. It was built in 1932 and then reconstructed after World War II in 1948. The two main arch spans are 2 × 106 m, the approach deck arch spans are 55 m, and total length of bridge is 1610 m. The bridge carries a single track. The material is cast-in-place monolith concrete. The steel trusses of the main spans were built in 1932 and in 1948 concrete arches were erected in their place [17],[12],[11].

The bridge across the Dnipro River at Zaporizhia (Figures 19.37 to 19.43) was built in 1952. It carries two railway tracks on the gravel ballast at the upper level and a narrow vehicular deck at the lower level. The bridge over the main Dnipro navigational channel consists of four 140 m arch spans; the arch over the Old Dnipro has a clear span of 228 m [2].

19.5.2 Steel Arches

Figures 19.44 and 19.45 show the steel arch portion of the Darnytsky railway bridge consisting of three spans of 110 m. The highway arch bridge over the Old Dnipro in Zaporizhia was built in 1974 (Figures 19.46 to 19.48); the span of the arch is 196 m, total length of the crossing is 320 m [3].

19.5.3 Steel Frames

A typical structure of this type, a highway bridge over the Smotrych River in Kamenets–Podilsky, is shown in Figure 19.49. The span of the main frame is 149 m, total length of crossing is 378 m, and the roadway is 14 m wide. Steel frame bridges are not widely used in Ukraine [9].

FIGURE 19.35   Merefa–Kherson railway bridge across the Dnipro River in Dnipropetrovsk, 1932.

FIGURE 19.36   Merefa–Kherson railway bridge, concrete arches of approach, 1932.

FIGURE 19.37   Arch bridge crossing the Old Dnipro, Zaporizhia, 1952.

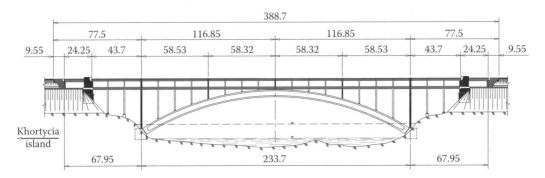

FIGURE 19.38   Elevation of the bridge over the Old Dnipro, Zaporizhia.

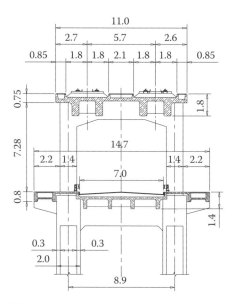

FIGURE 19.39 Cross-section of the bridge over the Dnipro, Zaporizhia.

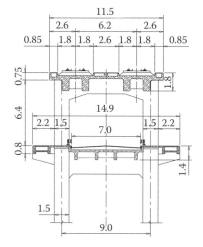

FIGURE 19.40 Cross-section of the bridge over the Old Dnipro, Zaporizhia.

FIGURE 19.41 Arch bridge across the Dnipro, Zaporizhia, 1952.

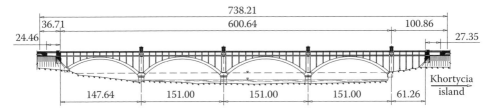

FIGURE 19.42 Elevation of the arch bridge across the Dnipro in Zaporizhia.

FIGURE 19.43 Darnytsky railway bridge crossing the Dnipro, Kyiv (approach), 1952.

FIGURE 19.44 Main arch spans of the Darnytsky railway bridge across the Dnipro, Kyiv, 1952.

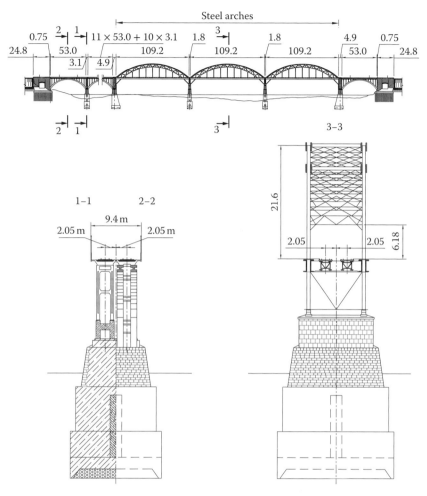

FIGURE 19.45 Elevation and cross sections of the Darnytsky railway bridge across the Dnipro, Kyiv, 1952.

FIGURE 19.46 Highway bridge across the Old Dnipro in Zaporizhia, 1974.

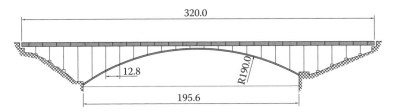

FIGURE 19.47   Elevation of the highway bridge across the Old Dnipro in Zaporizhia.

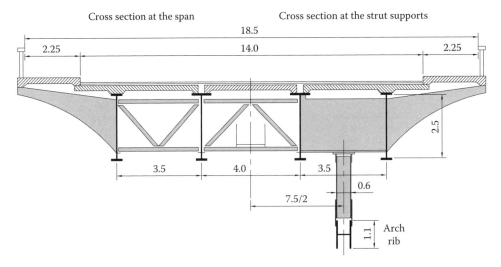

FIGURE 19.48   Cross section of the highway bridge across the Old Dnipro in Zaporizhia.

FIGURE 19.49   Frame bridge crossing the Smotrytch River in Kamenets–Podilsky, 1973.

19.6 Truss Bridges

Trusses are used primarily for railway bridges. In the former USSR, standard designs were available for spans of 33–127.4 m for through trusses, and for spans of 44–66 m for deck trusses (Figures 19.50 and 19.51). The main types of track fastening are shown in Figure 19.52. For the lattice members, typical H-members with conjunction by high-strength bolts were used. Some truss bridges are shown on Figures 19.53 to 19.58. A new steel truss bridge for metro and highway crossing of the Desenka River in Kyiv was built in 2008 (Figures 19.59 to 19.62). An orthotropic deck plate with closed ribs was used.

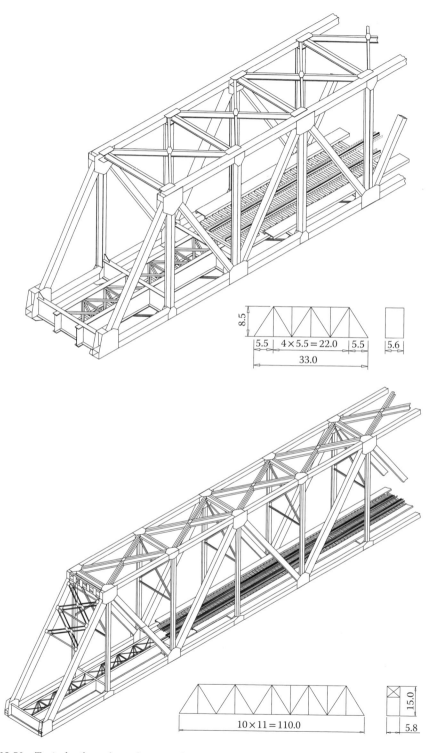

FIGURE 19.50 Typical railway through trusses for spans 33–110 m.

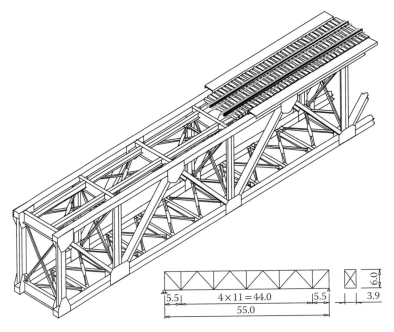

FIGURE 19.51   Typical railway deck trusses for spans 44–66 m.

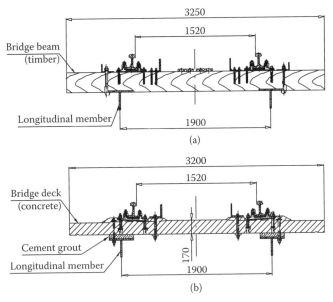

FIGURE 19.52   Track fastening: (a) to timber bridge beams; (b) to a ballast-free concrete deck.

Construction used steel with a yield strength of 390 MPa. Spherical bearings and modular expansion joints were used [15].

19.7 Cable-Stayed Bridges

The first cable-stayed bridge in Ukraine (Figure 19.63) was a concrete highway initially, later converted to a pedestrian bridge, crossing the Dnipro Harbor in Kyiv and built in 1963. The main span is 140 m with a roadway width of 8 m. Stay cables were made of locked coil strands [20].

FIGURE 19.53 Petrovsky railway bridge crossing the Dnipro River in Kyiv, built in 1929; rebuilt in 1945.

FIGURE 19.54 Combined railway and highway bridge crossing the Dnipro River in Kremenchuk with spans of 82 m.

FIGURE 19.55 Combined railway and highway bridge crossing the Dnipro River in Kremenchuk.

FIGURE 19.56 Movable span of the combined railway and highway bridge (foreground), 1955, and railway bridge (background), 1977, across the Dnipro River in Dnipropetrovsk.

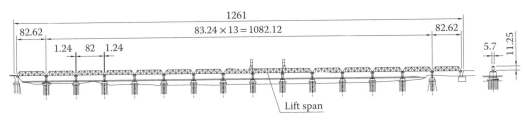

FIGURE 19.57 Elevation of the railway bridge over the Dnipro River in Dnipropetrovsk.

FIGURE 19.58 Highway bridge crossing of the Dnipro River in Cherkasy.

FIGURE 19.59   Bridge crossing the Desenka River in Kyiv, 2008.

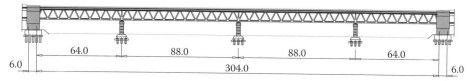

FIGURE 19.60   Elevation of the Desenka River bridge.

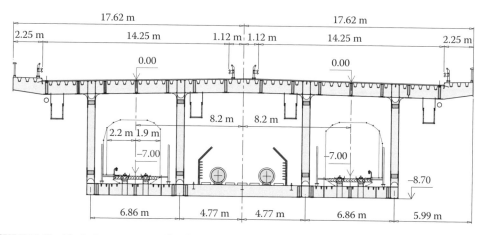

FIGURE 19.61   Typical cross section of bridge across the Desenka River in Kyiv.

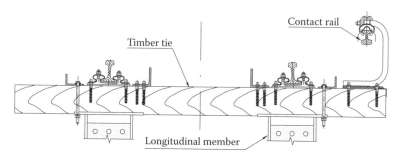

FIGURE 19.62   Rail fastening (bridge across the Desenka River in Kyiv).

FIGURE 19.63   Cable-stayed bridge over Dnipro Harbor in Kyiv, 1963.

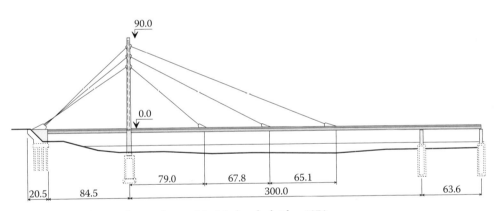

FIGURE 19.64   Cable-stayed part scheme of the Moskovsky bridge, 1976.

The Moskovsky cable-stayed bridge in Kyiv (Figures 19.64 through 19.67) crossing the Dnipro also opened to vehicle traffic in 1976. This bridge has an orthotropic deck roadway (Figure 19.65) with the spans of 84.5 m + 300 m + 63.5 = 448 m. The pylon was made of concrete. Stay cables were made of parallel wire strands.

The new major bridge crossing the Dnipro in Kyiv is the South Bridge, completed in 1992 (Figure 19.68 to 19.70). This crossing, serving vehicular traffic and the metropolitan transit trains, is over 3 km long. The cable-stayed bridge crossing the navigation waterway has an orthotropic deck with a main span of 271 m. The roadway is 41.6 m wide and accommodates six traffic lanes and two metropolitan transit tracks located in the center of the bridge (Figure 19.71). In addition, the bridge also carries four large-diameter water supply ducts. The total bridge loading is equivalent to 22 vehicular traffic lanes [13],[20]. The cable-stayed part has the spans of 80 m + 90 m + 271 m + 63 m + 63 m and consists of two parts: steel 80 m + 90 m + 268 m and concrete 60 m + 63 m. The concrete part of the bridge counterbalances

Regular cross section Cross section at the cable-stay anchorage

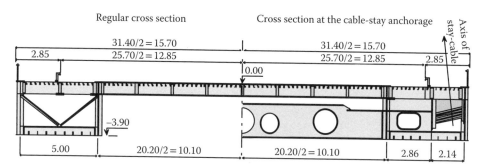

FIGURE 19.65 Cross section of the cable-stayed portion of the Moskovsky bridge, Kyiv.

FIGURE 19.66 The Moskovsky bridge, 1976.

the steel part; therefore it is in addition loaded with cast iron and is drawn by ropes for support. The regular weight of these parts is equal to 230 t/m (Figure 19.71). A typical cross section of the steel part with the spans 268 m + 90 m + 80.5 m is shown in Figure 19.72.

The orthotropic deck on the South Bridge is similar to the one used on the 1976 bridge, with a minimal 12 mm deck thickness and flat bar ribs. Under the rail transit tracks the transverse and longitudinal stiffeners were placed on opposite faces of the deck plate to avoid stress concentrations at rib intersections (Figure 19.73). The rails are mounted on timber ties supported by 20 × 400 mm longitudinal stiffeners. These were cut in the field after completion of the stay-supported span to compensate for construction inaccuracies and to ensure that the design longitudinal profile of the track was within a 10 mm tolerance. The cover plate to which the ties were fastened was subsequently welded to the ribs. A special automatic welding machine for these overhead welds was designed for this project by the Paton Welding Institute in Kyiv. The 16 mm fillet welds were obtained in one pass by this gas-shielded welding procedure. Legs of 130 m concrete pylon were placed by height on a dividing strip between the traffic lanes and two metropolitan transit tracks. The pylon was designed by optimizing the bending moment achieved by adjusting the support elevations of the girders and the initial length of the cable stays [14],[15].

FIGURE 19.67   Testing the Moskovsky bridge.

FIGURE 19.68   The South Bridge in Kyiv, 1992.

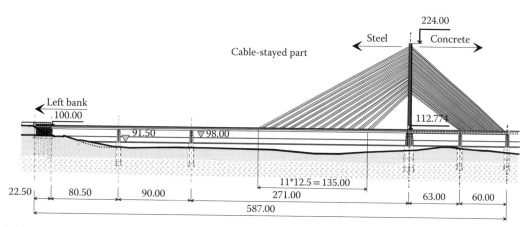

FIGURE 19.69 Elevation of the South Bridge in Kyiv.

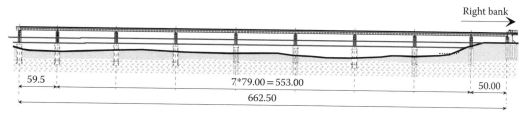

FIGURE 19.70 Elevation of the concrete approach of the South Bridge in Kyiv.

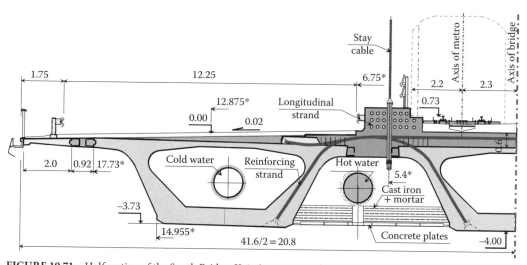

FIGURE 19.71 Half section of the South Bridge, Kyiv (concrete part).

A cable-stayed bridge in Odesa (Figure 19.74) was completed in 1998. This bridge crosses the railway lines in Odesa Sea Port. The main span is 150 m. Both the roadway deck and the pylon are made of steel. Cables were made of locked coil strands. The deck plate thickness of this structure was increased up to 14 mm and no cope holes in the crossbeams at the tops of the open ribs were provided. A thicker deck plate thickness was used due to unsatisfactory performance of the wearing surfacing on previously built bridges; avoidance of the copes in the web was aimed at improvement of the weld quality at these locations [20].

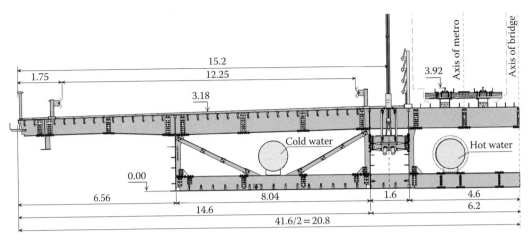

FIGURE 19.72 Half section of the South Bridge, Kyiv (steel part).

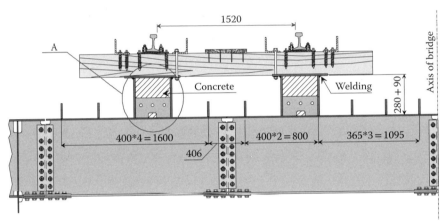

FIGURE 19.73 South Bridge – supports of rail transit tracks.

FIGURE 19.74 Odesa Bridge, 1998.

19.8 Pedestrian Bridges

A steel chain pedestrian bridge (Figure 19.75) over the Dnipro River to the beach in Kyiv was built in 1957. The chain scheme is 60 m + 180 m + 60 m. The pedestrian roadway width is 8 m. Steel beams were used as chains. A steel pedestrian girder bridge (Figure 19.76) over the Desna River in Chernihiv was built with the scheme 52.8 m + 136.5 m + 52.8 m = 242.1 m in 1990. The pedestrian walkway is 7 m wide. A pedestrian suspension bridge (Figure 19.77) over the Ros River in Bohuslav was built in 1992. The span is 90 m. The pedestrian walkway width is 2 m.

19.9 Bridges Currently in the Design or Construction Stage

19.9.1 Podilskyj Bridge Crossing in Kyiv

Kyiv is the capital of Ukraine, with a population of 3 million people. The four existing bridges serving the daily traffic of 270,000 vehicles are overloaded and the city cannot afford to close any of them for badly needed rehabilitation. All the Dnipro bridges must have long clear spans to satisfy the river navigation requirements. Bridge construction in Ukraine has been revived after the deep crisis in the beginning of the 1990s. At present, construction of three bridges across the Dnipro River are underway, two in Kiev and one in Zaporizhia. These bridges are being constructed under conditions of money, equipment, and qualified personnel deficiencies. Rehabilitation of the bridge has already begun but

FIGURE 19.75   Pedestrian bridge crossing the Dnipro River to the beach in Kyiv, 1957.

FIGURE 19.76   Bridge over the Desna River in Chernihiv, 1990.

FIGURE 19.77   Pedestrian bridge over the Ros River in Bohuslav.

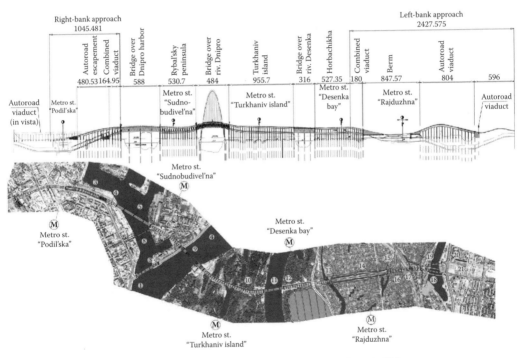

FIGURE 19.78   Plan and overview of Podilsky bridge crossing the Dnipro River in Kyiv.

achievement of the former level will need time. At the same time, the opening of borders has allowed practical application of all the best developments of the world bridge community in design.

The Podilskyj bridge crossing the Dnipro River (Figure 19.78) in Kyiv includes three big two-level bridges and 4 kilometers of trestles. One of the bridges is a combined steel arch bridge (Figures 19.79 through 19.81) with a main span of 344 m. This will be the longest steel arch span in Europe [20],[8],[15].

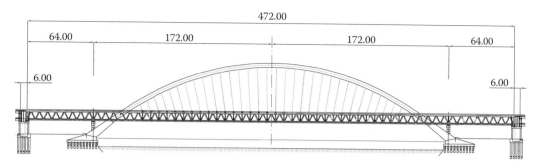

FIGURE 19.79 Podilskyj arch bridge crossing the Dnipro River in Kyiv.

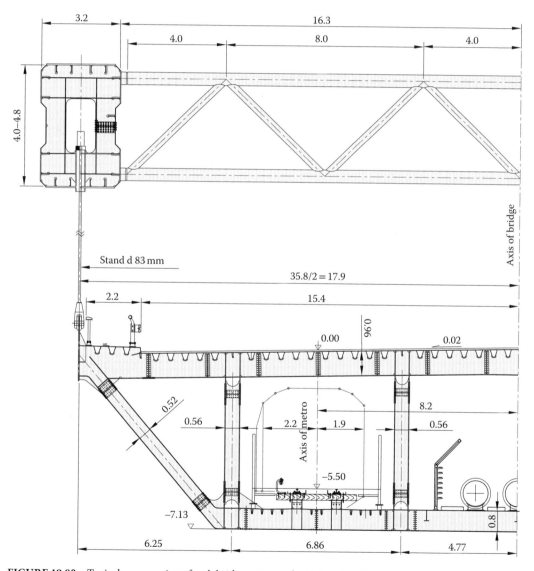

FIGURE 19.80 Typical cross section of arch bridge crossing the Dnipro in Kyiv.

FIGURE 19.81 Arch bridge crossing the Dnipro in Kyiv (erection of the arch rib).

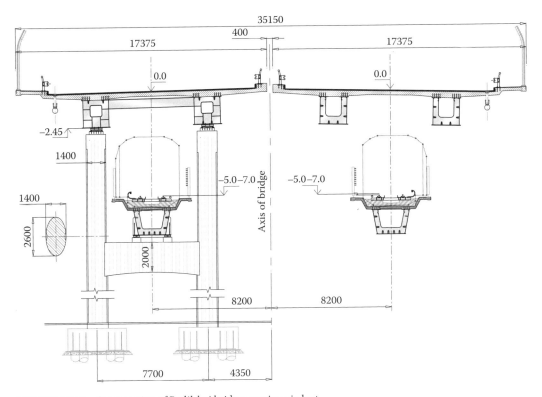

FIGURE 19.82 Cross section of Podilskyj bridge crossing viaduct.

Composite steel and concrete viaducts (Figures 19.82 through 19.84) are used for the roadway at the top level, and for metro traffic at the bottom level. The Podilskyj bridge is expected to be completed and open to traffic by 2014, but the project has suffered poor funding.

19.9.2 Bridge Crossing in Zaporizhia

The second big bridge crossing is now under construction in Zaporizhia. The bridge crossing includes two big bridges (Figures 19.85 through 19.88) over the Dnipro and the Old Dnipro. The bridge is expected to be completed and open to traffic by 2012 [16],[15].

FIGURE 19.83 Highway and metro viaduct on Dnipro Island in Kyiv, 2007.

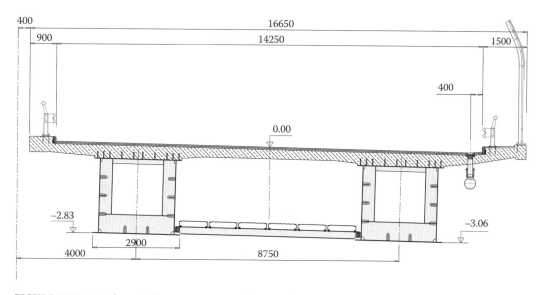

FIGURE 19.84 Roadway viaduct cross section with spans of 42–48 m.

19.9.3 Combined Railway and Highway Bridge Crossing in Kyiv

The second big bridge crossing in Kyiv is under construction in the south of the city near the old railway bridge (Figures 19.89 through 19.92). The bridge partially opened to traffic in 2010, and full completion was in 2011. Two bridges are shown in Figure 19.89. In the foreground we can see the old railway bridge with three 110 m spans; in the background of the picture there is a new bridge [10].

FIGURE 19.85 Cable-stayed bridge across the Dnipro in Zaporizhia with spans 80 × 4 m + 260 m + 80 m = 660 m.

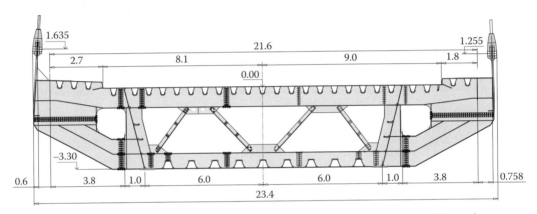

FIGURE 19.86 Cable-stayed bridge crossing the Dnipro River in Zaporizhia, cross section of the girder.

FIGURE 19.87 Cable-stayed bridge crossing the Dnipro in Zaporizhia, erecting the pylon, 2009.

FIGURE 19.88 Frame bridge across the Old Dnipro River in Zaporizhia with a span of 264 m.

FIGURE 19.89 Combined railway and highway bridge crossing the Dnipro River in Kyiv (in the background).

FIGURE 19.90 Erection of the combined railway and highway bridge crossing the Dnipro River in Kyiv, summer 2009.

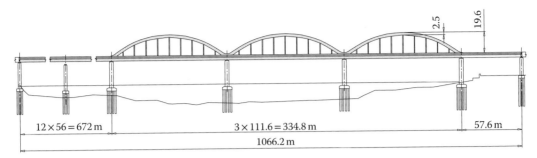

FIGURE 19.91   Elevation of the combined railway and highway bridge crossing the Dnipro River in Kyiv.

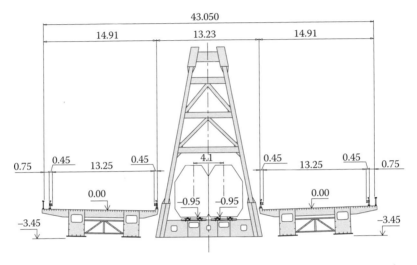

FIGURE 19.92   Cross section at midspan of the combined railway and highway bridge crossing the Dnipro River in Kyiv.

References

1. AASHTO. 2007. *AASHTO LRFD Bridge Design Specifications*, 4th Edition, American Association of State Highway and Transportation Officials, Washington, DC.
2. Adelberg, L. 1980. *Bridges of Zaporizhzhya*.
3. Barenboim, I.Y., Teslenko, L.M., Skachkov, I.A., Lugovoi, N.V., Karasik, M.E. 1974. Express-Information. Construction of a metal arch road bridge over the Old Dnipro in Zaporizhzhia. CININTI Orgtransstroy, Moscow.
4. Information Agency "The City Site" Bridges of Dnipropetrovsk. http://gorod.dp.ua/history/article_ru.php?article=56 (accessed January, 2010).
5. National Transport University DBN B.2.3–14. 2006. Transport constructions. Bridges and pipes. Design rules. Ministry of Construction, Architecture and Housing, Kyiv.
6. Dekhtyar, A.S., Zhoga, I.V. 2008. Bridge name Paton in Kiev—to the history of design. *Transport Construction of Ukraine* 6(14). pp. 14–15.
7. ENV 1991-1. 1997. *Eurocode 1: Basis of Design and Actions on Structures. Part 1: Basis of Design*.
8. Fuks, G.B. 2006. Continues construction of the Podilskyj bridge crossing. *Transport Construction of Ukraine* 1(1/4). p. 21.

9. Hlotova, I.A. 2006. Journey in Ukraine—Bridges of Kamenets–Podilsky. *Transport Construction of Ukraine* 3(3). pp. 24–25.

10. Karmanova, I. 2008. Kyrpa Bridge: Feasibility of four years of successful construction. *Transport Construction of Ukraine* 6(14). pp. 17–19.

11. Kienya, M.A. 1933. *Transport Construction* 2–3.

12. Kolokolov, N.M. 1933. *Transport Construction* 2–3.

13. Korniyev, M.M. 2003. *Steel Bridges: Theoretical and Practical Handbook for Their Design* (in Russian). VIPOL, Kyiv.

14. Korniyiv, M., Fuks, G. 1994. The South Bridge, Kyiv, Ukraine. *Structural Engineering International* 4/94, pp. 223–225.

15. Korniev, M. 2009. Features and examples: Bridge construction in Ukraine. *Bridge Construction* February 2009.

16. Panasyuk, I.A., Korniyev, M.M. 2006. Construction of the bridge crossing in Zaporizhya. *Transport Construction of Ukraine* 4(4). pp. 11–15.

17. Plamenytska, O., Plamenytska, E. 1999. Kamyanets-Podilsky. *Remembrances of Ukraine* 4, p. 41.

18. SniP 2.05.03-84. 1985. Building norms and rules. Bridges and pipes. State Building Committee of the USSR, Moscow.

19. International Database and Gallery of Structures. 2002 Tsar Nicholas I Suspension Bridge http://en.structurae.de/structures/data/index.cfm?ID=s0005655 (accessed January, 2010).

20. Ukrainian bridges. 2003. Magazine A.C.C. 6(47). pp. 41–59.

20
Bridge Engineering in China

Quan Qin
Tsinghua University

Gang Mei
China Highway Planning &
Design Institute
Consultants, Inc.

Gongyi Xu
China Railway Major
Bridge Reconnaissance and
Design Institute, Co., Ltd.

20.1 Introduction

China has a population of 1.43 billion as of 2013, a total area of 9.641 million km², including 6,530 sea islands with area of 80,000 km², and coastlines of 14,500 km. China ranges from mostly plateaus and mountains in the west to lower lands in the east. In the west, the north has a great alluvial plain, and the south has a vast calcareous table land traversed by hill ranges of moderate elevation, and the Himalayas, containing Earth's highest point, Mount Qomolangma. The northwest also has high plateaus with more arid desert landscapes such as the Takla-Makan and the Gobi Desert. In the east, along the shores of the

Yellow Sea and the East China Sea there are extensive and densely populated alluvial plains. On the edges of the Inner Mongolian plateau in the north, grasslands can be seen. Southern China is dominated by hills and low mountain ranges. In the central-east are the deltas of China's two major rivers, the Yellow River and the Yangtze River. The principal rivers flow from west to east, including the Yangtze (central), Yellow River (north-central), and Heilong River (northeast), and sometimes toward the south (including the Pearl River), with most Chinese rivers emptying into the Pacific Ocean. Plateaus, mountains, and hill ranges account for 66% of China's area, and 34% of the area are plains. Figure 20.1 shows a map of China.

The economy of the People's Republic of China is a rapidly developing and influential market economy. As of 2013, China is the second-largest economy in the world after the United States, with a nominal GDP of $8.3 trillion when measured in exchange-rate terms. It is the second-largest in the world, after that of the United States with a GDP of $11.3 trillion, when measured on a purchasing power parity (PPP) basis. China has had the fastest-growing major economy for the past 34 years with an average annual GDP growth rate above 9.9%. China's per capita income has likewise grown at an average annual rate of more than 8% over the last three decades, drastically reducing poverty.

The earliest reference to bridges in Chinese history is the Jü Bridge dating from the Shang Dynasty (eleventh century BC). Ancient Chinese bridges can be classified under four categories: beam, arch, suspension, and floating bridges. In the period from the Zhou Dynasty through the Qin and Han Dynasties (1100 BC to 220 AD) Chinese bridges with timber beams supported by stone piers were dominant. Even though China erected thousands of ancient bridges, very few of them survived until now. The oldest bridge in China, which survives today and is well-preserved, is a single-span segmental stone arch bridge known as the Anji Bridge or Zhaozhou Bridge (Figure 20.2). It is located in Zhao County, Hebei Province. It was completed in the first year of Daye's reign of the Sui Dynasty (605 AD). It is composed of 28 individual arch ribs 0.33 m thick boned transversely, 37.03 m in span, rising 7.23 m above the chord line, and averaging 9 m in width, narrower in the upper part and wider in the lower.

The longest ancient stone bridge is the Anping Bridge (Figure 20.3) across a sea bay located in Jinjiang, Fujian Province. As a stone beam bridge, its construction was completed in the twenty-first year of the

FIGURE 20.1 China's topography. (From China State Bureau of Surveying and Mapping, http://www.sbsm.gov.cn/contentimage/zgds.jpg; http://www.chinatouristmaps.com/china-maps/topography-of-china/detailed-map.html))

FIGURE 20.2 Anji Bridge. (From Zhao County Government, Hebei Province, http://www.zhaoxian.gov.cn/admin/eWebEditor/UploadFile/2007225155929607.jpg.)

FIGURE 20.3 Anping Bridge. (From Jinjiang County Government, Zhejiang Province, http://www.jinjiang.gov.cn/main/download/2005/02/01/20050201/6750.jpg.)

Shaoxing reign of the Song Dynasty (1151 AD). It is 3 m to 3.8 m wide, 2255 m long, a national record for over 700 years, and has 5 pavilions and 361 stone piers composed of rectangular blocks of stone in the orthogonal directions assembled without mortar. The piers have three different shapes of the cross section: rectangular, rectangular with the upstream starlings only, and rectangular with both upstream and downstream starlings for the main navigation channels. Its deck consists of large stone slabs 5 m to 11 m long, 0.6 m to 1 m wide, and 0.5 m to 1 m deep, and the maximum weight of a single slab is 250 kN.

The Lugou Bridge (Figure 20.4), Beijing, also known as the Marco Polo Bridge, was constructed in the Jin Dynasty and completed in the third year of the Mingchang reign of the Jin Dynasty (1192 AD). It is 212.2 m long, 9.3 m wide, and has 11 semicircular stone arches, ranging from 11.4 m to 13.45 m in span. The piers are from 6.5 m to 7.9 m wide; their pointed cutwaters upstream are inlaid with triangular iron bars, while the downstream sides are square in shape but without two angles. The parapets are divided into 269 sections with columns in between, each column crowned with a carved lion. When the bridge was first erected, all the lions were alike and very simple, but through the ages they were replaced each time by better ones, more delicately carved and different in style. Now, each lion has an individual posture, and the lion cubs are even more fascinating. They are playing around their parents, clinging to the breast, squatting on the shoulder, nestling at the feet, or licking the face. These exquisite sculptures on the bridge and on the ornamental columns, which show the practical application of the aesthetic principle of unity and variation, have become a scene of attraction.

In 1937, a railway–highway bipurpose bridge (Figure 20.5) with a total length of 1453 m, the longest span being 67 m, was constructed over the Qiantang River. It might be the first modern bridge designed by Chinese engineers. Generally speaking, the construction of modern bridges in China started relatively late. Before the 1950s, many bridges were designed and constructed by foreigners. Most highway bridges were made of wood. After the 1950s, China's bridge construction entered a new era. In 1957, the Wuhan Yangtze River Bridge (Figure 20.6), the first bridge across the Yangtze River, was built. It is a highway–railway bipurpose double-deck steel truss bridge with nine main spans of 128 m and a total length of 1155 m. The trusses were erected by the cantilever method and pipe foundations were driven by vibrators for the first time. It was designed by the Major Bridge Reconnaissance and Design Institute Co., Ltd. and constructed by the China Railway Major Bridge Engineering Group Co., Ltd. Since then, modern bridges have made giant

FIGURE 20.4   Lugou Bridge. (Courtesy of Quan Qin.)

FIGURE 20.5 Qiantang River Bridge. (Courtesy of Zhao Wei.)

FIGURE 20.6 Wuhan Yangtze River Bridge. (Courtesy of Gongyi Xu.)

strides. Nanjing Yangtze River Bridge (Figure 20.7) is also a highway and railway double-deck continuous steel truss bridge in Nanjing, Jiangsu Province. It was completed in December 1968.

Before 1978, China constructed 53,000 km of highway including 1,313 km of high-grade highway (expressway). Up to 1993, China built 33,000 km of expressways. During the period from 1993 to 2008, China built five north–south and seven east–west national expressways of 37,300 km. In 2003, completed investment in highway construction was 350 billion RMB and 219 key highway projects progressed. By the end of 2004, the total length of highways open to traffic reached 1.871 million km, including 34,300 km of expressways up to advanced modern transportation standard, ranking second in the world. The nation's highway density has now reached 19.5 km per 100 km². With the completion in 2008 of 'the five north-south and the seven east-west national arterial highways', totaling 35,000 km, Beijing and Shanghai have been linked by major highways, chiefly expressways, to the capitals of all provinces and autonomous regions of China, creating highway connections between over 200 cities. Up to the end of 2012, the China's highway network has a total length of 4.237 million km, including 131.100 km expressways.

FIGURE 20.7   Nanjing Yangtze River Bridge. (Courtesy of Gongyi Xu.)

A new network, termed the 7918 Network (also known as National Trunk Highway System (NTHS)), which began use on a nationwide level beginning late 2009 and early 2010. It is composed of 7 radial expressways leaving Beijing (Numbered 1xx, red lines in Figure 20.8); 9 vertical expressways going north to south (Numbered 2xx, blue lines in Figure 20.8) and 18 horizontal expressways head west to east (Numbered 3xx, green lines in Figure 20.8)

100 Series:

CNH 101: Beijing–Shenyang (Liaoning), 879 km
CNH 102: Beijing–Harbin (Heilongjiang), 1,311 km
CNH 103: Beijing–Tanggu (Tianjin), 149 km
CNH 104: Beijing–Fuzhou (Fujian), 2,387 km
CNH 105: Beijing–Zhuhai (Guangdong), 2,653 km
CNH 106: Beijing–Guangzhou (Guangdong), 2,505 km
CNH 107: Beijing–Shenzhen (Guangdong), 2,509 km
CNH 108: Beijing–Kunming (Yunnan), 3,356 km
CNH 109: Beijing–Lhasa (Tibet), 3,855 km
CNH 110: Beijing–Yinchuan (Ningxia), 1,135 km
CNH 111: Beijing–Jiagedaqi (Heilongjiang), 2,123 km
CNH 112: Gaobeidian–Tianjin–Tangshan–Xuanhua–Gaobeidian ring route, 1,228 km

200 Series:

CNH 201: Hegang (Heilongjiang)–Dalian (Liaoning), 1,946 km
CNH 202: Heihe (Heilongjiang)–Dalian (Liaoning), 1,818 km
CNH 203: Mingshui (Heilongjiang)–Shenyang (Liaoning), 720 km
CNH 204: Yantai (Shandong)–Shanghai, 1,031 km
CNH 205: Shanhaiguan (Hebei)–Guangzhou (Guangdong), 3,160 km
CNH 206: Yantai (Shandong)–Shantou (Guangdong), 2,302 km
CNH 207: Xilinhot (Inner Mongolia)–Hai'an (Guangdong), 3,738 km
CNH 208: Erenhot (Inner Mongolia)–Changzhi (Shanxi), 990 km
CNH 209: Hohhot (Inner Mongolia)–Beihai (Guangxi), 3,435 km
CNH 210: Baotou (Inner Mongolia)–Nanning (Guangxi), 3,097 km
CNH 211: Yinchuan (Ningxia)–Xi'an (Shaanxi), 645 km
CNH 212: Lanzhou Gansu–Chongqing, 1,195 km
CNH 213: Lanzhou (Gansu)–Mohan (Yunnan), 2,827 km

FIGURE 20.8 (**See color insert.**) China's Expressway Network (From China Ministry of Transport, http://glcx
.moc.gov.cn/roadMap.jsp; also Wikipedia on Answers.com: Expressways of China, http://www.answers.com/topic/
china-national-highways.)

CNH 214: Xining (Qinghai)–Jinghong (Yunnan), 3,345 km

CNH 215: HongliuRMB (Gansu)–Golmud (Qinghai), 591 km

CNH 216: Altay (Xinjiang)–Baluntai (Xinjiang), 853 km

CNH 217: Altay (Xinjiang)–Kuche (Xinjiang), 1,023 km

CNH 218: Huocheng (Xinjiang)–Ruoqiang (Xinjiang), 1,073 km

CNH 219: Yecheng (Xinjiang)–Lhatse (Tibet), 2,279 km

CNH 220: Dongying (Shandong)–Zhengzhou (Henan), 570 km

CNH 221: Harbin (Heilongjiang)–Tongjiang (Heilongjiang), 662 km

CNH 222: Harbin (Heilongjiang)–Yichun (Heilongjiang), 358 km

CNH 223: Haikou (Hainan)–Sanya (Hainan), 320 km

CNH 224: Haikou (Hainan)–Central Sanya (Hainan), 293 km

CNH 225: Haikou (Hainan)–Western Sanya (Hainan), 427 km

CNH 227: Xining (Qinghai)–Zhangye (Gansu), 338 km

300 Series:

CNH 301: Suifenhe (Heilongjiang)–Manzhouli (Inner Mongolia, China–Russia–Mongolia
border), 1,680 km

CNH 302: Hunchun (Jilin)–Ulanhot (Inner Mongolia), 1,028 km

CNH 303: Ji'an (Jilin)–Xilinguole (Inner Mongolia), 1,263 km

CNH 304: Dandong (Liaoning)–Holingol (Inner Mongolia), 889 km

CNH 305: Zhuanghe (Liaoning)–Linxi (Inner Mongolia), 816 km

CNH 306: Suizhong (Liaoning)–Hexigten Qi (Inner Mongolia), 480 km

CNH 307: Qikou (Hebei)–Yinchuan (Ningxia), 1,351 km

CNH 308: Qingdao (Shandong)–Shijiazhuang (Hebei), 786 km

CNH 309: Rongcheng (Shandong)–Lanzhou (Gansu), 2,372 km

CNH 310: Lianyungang Jiangsu)–Tianshui (Gansu), 1,395 km

CNH 311: Xuzhou (Jiangsu)–Xixia (Henan), 738 km

CNH 312: Shanghai–Huoch/Yining (Xinjiang), 4,967 km

CHN 313: Anxi (Gansu)–Ruoqiang (Xinjiang), 821 km

CNH 314: Urumqi (Xinjiang)–Khunjerab Pass (Xinjiang), 1,948 km

CNH 315: Xining (Qinghai)–Kashi (Xinjiang), 3,048 km

CNH 316: Fuzhou (Fujian)–Lanzhou (Gansu), 2,678 km

CNH 317: Chengdu (Sichuan)–Naqu (Tibet), 2,028 km

CNH 318: Shanghai–Zhangmu (Tibet, China–Nepal border), 5,334 km

CNH 319: Xiamen (Fujian)–Chengdu (Sichuan), 3,027 km

CNH 320: Shanghai- Ruili (Yunnan, China–Burma border), 3,748 km

CNH 321: Guangzhou (Guangdong)–Chengdu (Sichuan), 2,168 km

CNH 322: Hengyang (Hunan)–Friendship Gate (Guangxi, China–Vietnam border), 1,119 km

CNH 323: Ruijin (Jiangxi)–Lincang (Yunnan, China–Burma border), 2,926 km

CNH 324: Fuzhou (Fujian)–Kunming (Yunnan), 2,583 km

CNH 325: Guangzhou (Guangdong)–Nanning (Guangxi), 831 km

CNH 326: Xiushan (Chongqing)–Hekou (Yunnan, China–Vietnam border), 1,674 km

CNH 327: Heze (Shandong)–Lianyungang Jiangsu), 421 km

CNH 328: Nanjing (Jiangsu)–Hai'an County (Jiangsu), 295 km

CNH 329: Hangzhou (Zhejiang)–Putuo District (Zhejiang), 292 km

CNH 330: Shouchang (Zhejiang)–Wenzhou (Zhejiang), 331 km

Meanwhile, 590,000 highway bridges with a total length of 250,000 km have been constructed. Among them, 165 long-span bridges are across the Yangtze River, including rigid frame, arch, cable-stayed, and suspension bridges. Since the 1990s 72 long-span bridges with spans longer than 400 m have been built in China, especially since 2005, when three sea-crossing bridges with total lengths longer than 20 km each have been built.

In a period from 1876, when the first railway was completed, to 1949, China built 20,000 km railways. Up to 1998, China had railways of 66,000 km including 20,000 km of multi-track railways and 10,000 km of electrified railways. In 2007, China had 46,888 railway bridges with a total length of 3,355 km, including 5,459 bridges longer than 100 m, 826 bridges longer than 500 m, and 4 bridges longer than 10,000 m. In 2012, China has 980,000 km of railways in operation, including so-called 'four north-south and four east-west' railway lines for passenger traffic of 16,000 km (see Figure 20.9). The 980,000 km of railways with total 140,000 spans of railway bridges includes 9,356 km high-speed railways. The total length of the traditional railway bridges is about 8% of the 970,000 km total length of the traditional railways, and the total length high-speed railway bridges is averagely about 30% of the 9,356 km total length of the high-speed railways with different percentages in between 32% to 94% for different high-speed railways. China plans to have 140,000 km of railways in 2020 with 50% of multi-track railways and 45% electrified railways (www.chnrailway.com).

In China, four national ministries are responsible for bridge construction and management: the China Ministry of Transport (former Ministry of Communications), China Ministry of Railways which is being merged to the China Ministry of Transport, China Ministry of Housing and Urban–Rural Development, and China State Administration of Cultural Heritage. The China Ministry of Transport is responsible for highway bridges, the China Ministry of Railways is responsible for railway bridges, the China Ministry of Housing and Urban–Rural Development is responsible for highway bridges in urban regions, and the China State Administration of Cultural Heritage is responsible for ancient bridges as national heritages. This chapter discusses mostly highway bridges and railway–highway bridges. Table 20.1 lists some important and representative bridges in mainland China and the Hong Kong Special Administrative Region.

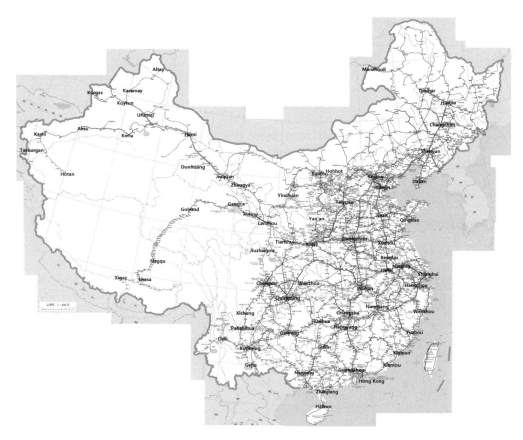

FIGURE 20.9 2013 China railway network. (Quan Qin modified form http://www.china-mor.gov.cn)

TABLE 20.1 Important and Representative Bridges in Mainland China and the Hong Kong Special Administrative Region

Bridge	Year and Designer	Comments
Zhaozhou Bridge	605, Li Chun	Stone arch, span of 37.02 m
Fujian Anping Bridge	1151, Unknown	Stone beam, total length of 2223 m, 362 spans
Lugou Bridge	1192, Unknown	Stone ach, length of 212.2 m with maximum span 13.45 m
Hangzhou Qiantang River Bridge	1937, Design Group of Qiangtang River Bridge	Steel truss, 18 spans of 65.84 m, railway and highway
Wuhan Yangtze River Bridge	1957, Major Bridge Reconnaissance and Design Institute	Steel truss, nine spans of 128 m, railway and highway
Yibin Jinsha River steel truss bridge	1968, Major Bridge Reconnaissance and Design Institute	Steel truss, spanning 112 m + 176 m + 112 m, railway
Nanjing Yangtze River Bridge	1968, Major Bridge Reconnaissance and Design Institute	Steel truss, nine spans of 160 m, railway and highway
Jiujiang Yangtze River Bridge	1992, Major Bridge Reconnaissance and Design Institute	Steel trussed arch bridge, spanning 180 m + 216 m + 180 m, railway
Hubei Huangshi Yangtze River Bridge	1995 China Highway Planning and Design Institute	Continuous segmental prestressed concrete, three spans of 245 m, highway
Wanzhou Yangtze River Bridge	1997, Sichuan Highway Planning and Design Institute	Reinforced concrete arch, world record span of 420 m, highway

(Continued)

TABLE 20.1 (*Continued*)   Important and Representative Bridges in Mainland China and the Hong Kong
Special Administrative Region

Bridge	Year and Designer	Comments
Kap Shui Mun cable-stayed bridge	1997, Greiner International, Ltd.	Concrete box girder of main span 430 m, railway and highway
Tsing Ma suspension bridge	1997, Mott McDonald H.K., Ltd.	Steel truss stiffening girder of span 1377 m, railway and highway
Humen Extra Major Bridge	1997, Guangdong Highway Planning and Design Institute	Segmental continuous prestressed concrete, maximum span of 270 m, highway
Ting Kau cable-stayed bridge	1998, Schlaich, Bergermann and Partners, GmbH	Separate composite girder, main spans of 488 m + 475 m, highway
Jiangyin Yangtze River Bridge	1999, China Highway Planning and Design Institute	World's fifth-longest suspended span of 1385 m, highway
Shanxi Dan River Bridge	2000, First Highway Planning and Design Institute	Stone arch, world record stone span of 146 m, highway
Baishazhou Yangtze River Bridge	2000, Major Bridge Reconnaissance and Design Institute	Cable-stayed span of 618 m, highway
Wuhu Yangtze River Bridge	2001, Major Bridge Reconnaissance and Design Institute	Main cable-stayed span 312 m, width of 21 m, integrated steel truss of 2193.7 m, railway and highway
Nanjing Second Yangtze River Bridge	2001, Highway Planning and Design Institute	Cable-stayed bridge, main span of 628 m, highway
Qingzhou Min River Bridge	2001, Major Bridge Reconnaissance and Design Institute	Cable-stayed bridge, main span of 605 m, highway
Shanghai Lupu Arch Bridge	2003, Shanghai Municipal Engineering Design Institute	Steel box tied arch with deck through, world record span of 550 m, highway
Wushan Yangtze River Bridge	2005, Sichuan Highway Planning and Design Institute	Half-through concrete-filled-steel tubular arch bridge, main span of 492 m, highway
Runyang Yangtze River Bridge	2005, Jiangsu Communication Planning and Design Institute	Span of 1490 m, world's fourth-longest suspended span, highway
East Sea Bridge	2005, Shanghai Municipal Engineering Design Institute	China's third-longest bridge, total length 32.5 km, including a cable-stayed bridge of 420 m span, highway
Shibanpo Double-Line Bridge	2006, Shanghai Municipal Engineering Design Institute	Continuous rigid frame prestressed concrete and steel box girder, world record span of 330 m, highway
Hangzhou Bay Bridge	2008, China Highway Planning and Design Institute	China's second-longest total length 36 km, including two cable-stayed bridges, main span 448 m, highway
Sutong Yangtze River Bridge	2008, Architectural Design and Research Institute of Tongji University	World record cable-stayed span of 1088 m, world record tower height of 300.4 m, highway
Xihoumen suspension bridge	2009, China Highway Planning and Design Institute	Main span of 1650 m, world's second-longest suspension span, highway
Stonecutters cable-stayed bridge	2009, Arup	Main span of 1018 m, world's second-longest cable-stayed span, highway
Hubei Longtan River Bridge	2009, Second Highway Consultants Co., Ltd., CCCC	Continuous segmental prestressed concrete, 3×200 m, world's second-tallest pier height of 179 m, highway
Chaotianmen Yangtze River Bridge	2009, Major Bridge Reconnaissance and Design Institute	Steel half-through three-span continuous truss tie arch, main span 552 m, railway and highway
Shanghai Yangtze River Bridge	2009, Shanghai Municipal Engineering Design Institute	Cable-stayed main span of 730 m, world's widest steel box girders at 51.25 m, railway and highway

TABLE 20.1 (*Continued*) Important and Representative Bridges in Mainland China and the Hong Kong Special Administrative Region

Bridge	Year and Designer	Comments
Edong Yangtze River Bridge	2010, Hubei Highway Planning and Design Institute	Cable-stayed bridge, nine spans with main span 926 m, the third-longest span in the world, highway
Dashengguan Yangtze River Bridge	2011, Major Bridge Reconnaissance and Design Institute	Steel half-through six-span continuous truss tie arch, main span 336 m, railway and highway
Qingdao Jiaozhou Bay Bridge	2011, China Highway Planning and Design Institute	China's longest bridge of 36.4 km, including two cable-stayed bridges and a suspension bridge, highway
Jiaxing-Shaoxing River-Sea-Crossing bridge	2013, China Highway Planning and Design Institute	China's future longest bridge of 70 km, including a 6 tower and 4 stay plane 5 × 335 m span cable-stayed bridge, highway

20.2 Design

20.2.1 Philosophy

Mainland China uses a group of unified common rules for structural design of highway bridges. These rules are included in 10 codes, standards, or guidelines. Since the 1980s the China Ministry of Transport has started preparation work for the reliability-based design—the load and resistance factor design (LRFD)—codes for highway bridges and published two LRFD design codes, a code for design of masonry highway bridges and a code for reinforce concrete (RC) and prestressed concrete (PC) highway bridges, both published in 1985 with the design formulae corresponding to both ultimate and serviceability limit states. The codes for design of steel and timber structures (1986) and for design of foundations of highway bridges (1985) were still based on the working stress design (WSD). All of the remaining codes dealing with loads, wind, and seismic actions work for both LRFD and WSD.

In 1999, China published the national standard GB/T 50283-1999 Unified Standard for Reliability Design of Highway Engineering Structures, where GB stands for "national standard." Most of these codes, except the code for design of steel and timber structures of highway bridges, were revised from 2004 to 2008. This time, the Code for Design of Ground Base and Foundation of Highway Bridges and Culverts has been changed to an LRFD design code. Following are the latest versions of these codes used in China where JTG is the initials of three Chinese characters in the Chinese phonetic alphabet of the "Standards of the Ministry of Transport."

- Unified Standard for Reliability Design of Highway Engineering Structures (China Ministry of Transport, GB/T 50283-1999)
- General Code for Design of Highway Bridges and Culverts (China Ministry of Transport, JTG D60-2004)
- Code for Design of Reinforced Concrete and Prestressed Concrete Highway Bridges and Culverts (China Ministry of Transport, JTG D62-2004)
- Code for Design of Masonry Highway Bridges and Culverts (China Ministry of Transport, JTG D61-2005)
- Code for Design of Steel and Timber Structures of Highway Bridges and Culverts (China Ministry of Transport, JTG 025-86 1986)
- Wind-Resistant Design Specification for Highway Bridges (China Ministry of Transport, JTG/T D60-01-2004)

- Code for Design of Ground Base and Foundation of Highway Bridges and Culverts (China Ministry of Transport, JTG D63-2007)
- Guidelines for Design of Highway Cable-Stayed Bridges (China Ministry of Transport, JTG/T D65-01-2007)
- Guidelines for Seismic Design of Highway Bridges (China Ministry of Transport, JTG/T B02-01-2008)
- Guide to Design and Construction Technology of Road Concrete–Filled Steel Tube (CFST) Highway Bridges (Sichuan Province Highway Planning Survey Design and Research Institute, 2008)

Meanwhile, due to strong influences on the rules of highway traffic management of the central and local governments on highway traffic loads, many investigations and studies on highway loads, such as new traffic load grades, for instance, the super grade I and super grade II, may be added, and changes to the reliability-based design codes still continue.

At the present time, the codes for design of railway bridge structures are still based on the working stress methods, even though design philosophy based on the structural reliability theory has been studied and the first reliability-based railway structural design code, Unified Standard for Reliability Design of Railway Engineering Structures, was published in 1994 with a revised version published in 2009. The following are the latest versions of these codes used in China, where TB stands for "Standards of the Ministry of Railways."

- Unified Standard for Reliability Design of Railway Engineering Structures (China Ministry of Railways, GB 50216-2009)
- Fundamental Code for Design of Railway Bridges and Culverts (China Ministry of Railways, TB 10002.1-2005)
- Code for Design of Concrete and Prestressed Concrete Structures in Railway Bridges and Culverts (China Ministry of Railways, TB 10002.3-2005)
- Code for Design on Subsoil and Foundation of Railway Bridges and Culverts (China Ministry of Railways, TB 10002.5-2005)
- Code for Design of Steel Structures of Railway Bridges (China Ministry of Railways, TB 10002-2005)
- Code for Seismic Design of Railway Engineering (China Ministry of Railways, GB 50111-2006)
- Code for Design of Concrete and Block Masonry Structures of Railway Bridges and Culverts (China Ministry of Railways, TB 10002.4-2005)

20.2.2 Loads

20.2.2.1 Loads on Highway Bridges

For highway and municipal bridges, JTG D60-2004 uses the representative values of the actions on bridges on the basis of a reference period of 100 years to determine the corresponding design values. The variable actions are discussed here. For natural actions, the maximum characteristic values of wind and temperature actions for bridges with the most important category are determined as the values within a return period of 100 years. The wind action (pressure) is transformed from the maximum values of the wind speed, measured at 20 m above the ground and averaged over 10 min in a reference period of 100 years. Both wind and temperature actions are assumed as basic variables with a type I asymptotic extreme value distribution. The maximum characteristic values of seismic actions for bridges in the most important category are determined as the values in a return period of 475 years corresponding to a 10% exceeding probability in 50 years. Some reduction factors on the above maximum values are given for less-important bridges. For all highway bridges, the design flood is not higher than that corresponding to a recurrence period of 300 years.

The characteristic values of vertical highway traffic loads for both bending moment and shear force are determined as a 0.95 fractile of the maximum values of the traffic loads in a reference period of

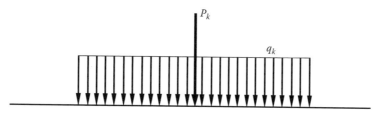

FIGURE 20.10 The traffic lane load. (From China Ministry of Transport, JTG D60-2004.)

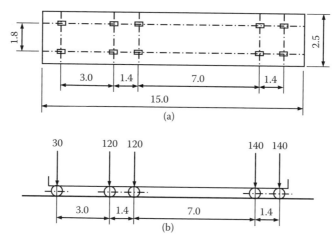

FIGURE 20.11 The traffic vehicle load in kN and m. (a) Elevation, (b) plan. (From China Ministry of Transport, JTG D60-2004.)

100 years, and described as a combination of the design truck and design lane load. The highway traffic loads are based on measured data of vehicle tandems on certain representative highway sections and modeled as some stochastic processes with parent distributions of Weibull distribution for the normal traffic state and a normal distribution for the abnormal (dense) traffic state. JTG D60-2004 stipulates the traffic load as a lane load and a vehicle load (Figures 20.10 and 20.11).

The snow load is not considered for highway bridges.

For the persistent situation of the ultimate limit state, when two or more variable actions are applied to bridges simultaneously, all not-dominating variable effects should be reduced by a factor of a combination value not larger than 0.8. For the serviceability limit state, variable actions should be reduced by a factor of a frequent value not larger than 1.0 for frequent combination or by a factor of a quasi-permanent value not larger than 0.8 for a quasi-permanent combination. The design values of the actions are determined as productions of the above representative values of the actions times the corresponding action partial safety factors and are used for both LRFD and WSD.

20.2.2.2 Variable Loads on Railway Bridges

For variable loads on railway bridges (GB 50216-2009), the design values of the natural loads are based on a return period of 100 years. Wind pressure is transformed from the maximum wind speed measured at 20 m above the ground averaged over 10 min in a reference period of 100 years. The maximum seismic loads for bridges in the most important category are determined as the values in a return period of 475 years corresponding to a 10% exceeding probability in 50 years, with reduction factors on the above maximum values for less important bridges. For all railway bridges, the design flood is not higher than that corresponding to a return period of 300 years.

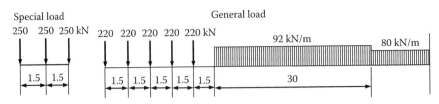

FIGURE 20.12   Medium variable loads for vertical train loads. (From China Ministry of Railways, GB 50216-2009.)

The design train loads (GB 50216-2009) are for passenger trains with speeds not higher than 160 km/h and freight trains with speeds not higher than 120 km/h. The vertical train loads are stipulated as medium-variable loads (Figure 20.12) consisting of a special load and a general load. For railway–highway bipurpose bridges, the train loads are the same as the above, but the highway traffic loads take the actions given by the highway bridge code (JTG D60-2004) reduced by a factor 0.75. For bridges on railways with speeds higher than 200 km/h, a special dynamic analysis on train–bridge interaction is required.

20.2.3 Resistances

For highway bridge design, the characteristic values of the resistances of structural members with different failure modes and different materials are determined as a 0.05 fractile of the probability distributions of the corresponding resistance basic variables. The design values of the resistances are determined as productions of the above characteristic values of the resistances times the corresponding resistance partial safety factors and are used for both of LRFD and WSD. For railway bridge design, the allowable stresses are used to represent the strengths of different materials.

20.3 Segmental Concrete Bridges

The segmental concrete bridge was primarily developed and built in China in the 1960s. This kind of structure is most suitable to be erected by the balanced cantilever construction process, either by cantilever segmental concreting with suspended formwork or by cantilever erection with segments of precast concrete. The first example of cantilever erection is the Wei River Bridge (completed in 1964) in Wuling, Henan Province, while the Liu River Bridge (completed in 1967) in Liuzhou in Guangxi Zhuangzu autonomous region is the first by cantilever casting. The Yangtze River Highway Bridge at Chongqing (completed in 1980), with a main span of 174 m, was regarded as the largest of this kind when it was completed.

Major segmental concrete bridges include the Luoxi Bridge in Guangzhou, Guangdong Province (completed in 1988), which features a 180 m main span. The Huangshi Yangtze River Bridge in Hubei Province has a main span of 245 m. The Humen Bridge side spans in Guangdong Province (completed in 1997) and Chongqing Shibanpo double-line bridge (completed in 2006), which have a 270 m main span and a 330 m span, respectively, were regarded as the largest of this kind in the world.

20.3.1 Hubei Huangshi Yangtze River Bridge

The Hubei Huangshi Yangtze River Bridge (Figure 20.13) is located in Huangshi, Hubei Province, with a total length reaching 2,580.08 m. A 162.5 m + 3 × 245 m + 162.5 m prestressed concrete continuous box girder rigid frame bridge was designed for the main bridge. The deck is 20 m wide, providing 15 m for motor vehicle traffic and 2.5 m on both sides for nonmotor-vehicle traffic. The designed average daily traffic flow (ADT) is 250,000 vehicles. The approach along the Huangshi bank is 840.7 m long, consisting of continuous bridges and simply supported T-girder bridges with continuous decks, while the approach along the Xishui bank is 679.21 m, simply supported T-girder bridges with continuous decks.

FIGURE 20.13   Hubei Huangshi Yangtze River Bridge. (Courtesy of Tao Liu.)

A 28 m diameter double-wall steel cofferdam with a 16.3 m diameter bored piles foundation was employed for the piers of the main span, which provided enough capacity to resist the collision force of ships. The navigation clearances of the bridge are 200 m × 24 m, which allows the navigation of a vessel of 5,000 tons. The bridge was designed by China Highway Planning and Design Institute Consultants. It was constructed by China Communication Construction Co., Ltd., completed in December 1995, rehabilitated with a wider deck of 18 m for motor vehicle traffic with ADT 480,000 vehicles in December 2002, and strengthened for cracking and excessive midspan deflection of the girder.

20.3.2 Humen Bridge Over the Auxiliary Navigation Channel

The Humen Bridge (Figure 20.14), a highway bridge over Pearl River, is on the freeway linking Guangzhou and Zhuhai to Shenzhen. It is composed of bridges of different types. A rigid frame bridge (150 m + 270 m + 150 m) is arranged over the auxiliary navigation channel, with a main span reaching 270 m, a world record for this bridge type. The superstructure of the bridge consists of two separate bridges, each a single-box, single-cell prestressed concrete continuous rigid frame. The 24 m wide deck provides 214.25 m for motor vehicle traffic. Adoption of a 15.24 mm VSL prestress system (a prestress technique created by VSL International Ltd., Berne, Switzerland) makes thinner top slabs and no bottom slabs of the box girder possible, the single-box single-cell thin-wall section offers a greater moment of inertia per unit area, and the depth of the main girder at the supports is 14.8 m (one-eighteenth of the main span) and 5 m (about one-fiftieth of the main span) in midspan. The substructure consists of double thin-wall piers resting upon group pile foundations. The symmetrical cantilever casting method was employed for the erection of the superstructure. The bridge was designed by Guangdong Highway Planning and Design Institute and constructed by Highway Engineering Construction Ltd., Guangdong Province. It opened to traffic in July 1997.

20.3.3 Chongqing Shibanpo Double-Line Bridge

Spanning 86.5 m + 4 × 138 m + 330 m + 132.5 m across the Yangtze River, the continuous prestressed rigid frame Chongqing Shibanpo double-line bridge (Figure 20.15), has a world record main span of 330 m in its category. Its 19 m wide deck is one-way and four lanes of traffic. In order to reduce the

FIGURE 20.14   Humen Bridge over the auxiliary navigation channel. (Courtesy of Gang Mei.)

FIGURE 20.15   Chongqing Shibanpo double-line bridge. (Courtesy of He Huang.)

weight of the prestressed concrete box girder, the center section of the 108 m long main span pre-stressed concrete box girder was changed to a steel box section, instead of lightweight aggregate concrete. The central steel box section is connected to two prestressed concrete box sections of 222 m long each through two joints, each 2.5 m long (Figure 20.16). The steel box section weighs 1400 tons. Each of the connecting joints has 76 Perfobond Leiste (PBL) steel shear boards welded inside the steel box at the upper flange, webs, and bottom. A cross section of the joints is shown in Figure 20.17. Other spans were erected by the balanced cantilever construction process. The bridge was designed by Shanghai Municipal Engineering Design Institute, constructed by Chongqing Bridge Construction Company, and completed in September 2006. Research on the efficiency of the joint between the steel box and prestressed concrete box has been carried out. A system monitoring the joint behavior has been installed in the bridge.

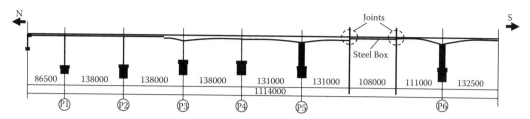

FIGURE 20.16 Central steel box section of the main span of the Chongqing Shibanpo double-line bridge. (Courtesy of Xueshan Liu.)

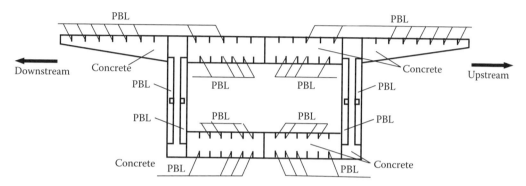

FIGURE 20.17 A cross section of the joints connecting the concrete box to the steel box with PBL boards. (Courtesy of Xueshan Liu.)

FIGURE 20.18 Hubei Longtan River Bridge. (From Transportation Bureau of Yichang City, Hubei Province, http://www.ycjt.gov.cn/UploadFiles/200893101717418.jpg.)

20.3.4 Hubei Longtan River Bridge

This five-span (106 m + 3 × 200 m + 106 m) continuous prestressed concrete rigid frame box girder bridge (Figure 20.18) is across the Longtan River valley, Hubei Province. This bridge is on an expressway linking Shanghai to Chengdu, Sichuan Province. The segmental concrete bridge includes two identical, parallel, side-by-side bridges separated by 12.5 m. The twin bridge (Figure 20.19) connects twin highway tunnel tubes at the two ends. Each of their center reinforced concrete built-up piers of two legs with variable box

FIGURE 20.19 Hubei Longtan River Bridge under construction. (From Xiaogan Construction Information Net, http://www.xgjs.gov.cn/Web/Article/2006/09/29/1051370068.aspx?ArticleID=bb66c367-4eb2-4808-adf0-1aa4fad06090.)

cross sections is 179 m high, the tallest in China, consists of 8000 m³ of concrete and 1000 tons of steel, and cost more than 20 million RMB (yuan). A cast-in-place balanced cantilever construction method with movable form carriers was adopted to build the prestressed concrete box girder. Each span of each bridge had one closure segment. The construction of the closure segments was in high summer and extra axial thrust forces in the longitudinal direction were applied to the girder to compensate for the thermal expansion. The bridge was designed by Second Highway Consultants Co., Ltd. of China Communications Construction Company, Ltd. (CCCC), and its construction was carried out by the China Railway 17th Bureau Group Co., Ltd. and Hubei Road and Bridge Group and completed in 2009.

20.4 Steel Truss Bridges

Steel structures are employed primarily for railway and railway–highway bipurpose bridges. In 1957, in the city of Wuhan, a railway–highway bipurpose bridge with a total length of 1670 m was completed over the Yangtze River, another milestone in China's bridge construction history. The bridge has continuous steel trusses with 128 m main spans. The rivet-connected truss is made of grade No. 3 steel. Its upper deck is 23 m wide with six 3 m wide traffic lanes and two 2.25 m wide sidewalks. A newly developed cylinder shaft 1.55 m in diameter was initially used in the deep foundation. (In 1962, a 5.8 m diameter cylinder shaft foundation was laid in the Ganjiang South Bridge in Nanchang Jiangxi Province.)

In 1968, another such bridge over the Yangtze River—the Nanjing Yangtze River Bridge—came into being. The whole project, including its material, design, and installation, was completed through

Chinese efforts. It is a rivet-connected continuous truss bridge with 160 m main spans. The material used was high-quality steel of 16-Mnq grade. In construction, a deep-water foundation was developed. Open caissons were sunk to a depth of 54.87 m, and pretensioned concrete cylinder shafts 3.6 m in diameter were laid, thus forming a new type of compound foundation. Underwater cleaning was performed at a depth of 65 m.

China's longest steel highway bridge, with a total length of 1390.6 m, was the Beizhen Yellow River Bridge in Binzhou, Shandong Province (1972). Its four main spans are 112 m long each. It has a continuous truss of bolt-connected welded members. The foundation is composed of 1.5 m diameter concrete bored piles, whose penetration depth into the subsoil reaches 107 m, the deepest pile ever drilled in China at that time. A new structure of a field-bolting welded box girder bridge with an orthotropic steel deck was first introduced at the North River Highway Bridge at Mafang, Guangdong Province, which was completed in 1980.

Another attractive and gigantic structure standing over the Yangtze River is the Jiujiang Railway–Highway Bridge, with a total length of 7645 m (railway) and 4460 m (highway), completed in 1992. Chinese-made 15-MnVN grade steel was used and shop-welded steel plates 56 mm thick were bolted on site. The main span reaches 216 m. The continuous steel truss is reinforced by flexible stiffening arch ribs. In laying the foundation, a double-walled sheet piling cofferdam was built, in which a concrete bored pile was cast in place. When erecting the steel beams, a double suspended cable frame took the place of a single one, which is another innovation. Since the 1980s, steel girder or composite girder bridges have been adopted in the construction of long-span or complex-structure city bridges in China. For example, they were applied in Guangzhong Road Flyover and East Yan'an Road Viaduct in Shanghai.

20.4.1 Qiantang River Bridge

The Qiantang River Bridge (Figure 20.5), located in Hangzhou, Zhejiang Province with a total length 1453 m, is a double-deck highway–railway bipurpose steel truss bridge. The 6.1 m wide upper deck supports two highway lanes and two 1.52 m wide sidewalks. The lower deck supports a single-track railway. It consists of 2×14.63 m spans of steel plate girders and 16×65.84 m spans of simply supported parallel chord rivet Warren trusses with a height of 10.7 m and center-to-center distance of 6.1 m, supported by reinforced concrete piers. It was designed by the Design Group of the Qiantang River Bridge, constructed by Corritt Co. (Denmark) and Dormanlong Co. (England), and completed in October 1937.

20.4.2 Wuhan Yangtze River Bridge

The Wuhan Yangtze River Bridge (Figure 20.6), located in Wuhan, Hubei Province with a total length of 1670 m, is a double-deck highway–railway bipurpose steel truss bridge. The 18 m wide upper deck supports six highway lanes and two 2.25 m wide sidewalks. The lower deck supports a double-track railway. Its major part consists of three 128 m span continuous steel trusses supported by reinforced concrete piers and tubular concrete piles. It was designed by the Major Bridge Reconnaissance and Design Institute, constructed by the China Railway Major Bridge Engineering Bureau, and completed in 1957.

20.4.3 Nanjing Yangtze River Bridge

The Nanjing Yangtze River Bridge (Figure 20.7) is a highway–railway double-deck continuous steel truss bridge in Nanjing, Jiangsu Province across the Yangtze River. The upper deck provides four lanes of highway traffic, which are 15 m wide, plus two sidewalks each 2.25 m wide, and the lower deck carries two railway tracks. The main bridge is 1576 m long. If the approaches are taken into account, the length of the railway bridge reaches 6772 m and the highway bridge is 4588 m long.

Ten spans were arranged for the main bridge, including a 128 m simply supported steel truss side span and nine 160 m continuous steel truss spans, continued every three spans. The main truss is a parallel chord rhombic truss with a reinforcing bottom chord. It was erected by cantilever assembling.

Considering the complex geologic conditions at the bridge site, different types of foundations were used, including heavy concrete caissons with a depth freeway of penetration reaching 54.87 m for areas with shallow water and deep coverings and a floating-type steel caisson combined with pipe column foundations, which was used for the first time at a deep water site. The bridge was designed and constructed by the China Railway Major Bridge Engineering Group Co., Ltd. It was completed in December 1968.

20.4.4 Yibin Jinsha River Bridge

Located in Yibin, Sichuan Province, this single-track railway bridge has a total length of 1053 m (Figure 20.20). Its major part consists of continuous rivet steel trusses spanning 112 m + 176 m + 112 m. The trusses are 20 m tall and with a center-to-center distance of 8 m. The superstructure was erected by cantilever assembling at both banks and closed at the midspan. It was designed by the Major Bridge Reconnaissance and Design Institute, constructed by the China Railway Major Bridge Engineering Bureau, and completed in October 1968.

20.4.5 Jiujiang Yangtze River Bridge

Jiujiang Yangtze River Bridge (Figure 20.21) located in Jiujiang, Jiangxi Province, is a double-deck highway–railway bipurpose bridge. The upper deck supports four 14 m wide highway lanes and two 2 m wide sidewalks and the lower deck supports a double-track railway. It has a total length of 7675 m for railway and 4460 m for highway. Its 1806 m long major part includes a combined welded and bolted steel trussed arch spanning 180 m + 216 m + 180 m at the center, two 3 × 160 m spanned continuous steel trusses at the north, and two 3 × 162 m spanned continuous steel trusses at the south. The heights of the truss are 16 m at the midspans and 32 m at the piers. The arch rises are 24 m, 32 m, and 24 m, respectively. It first used 15 MnVNq steel (yield strength 420 MPa) for the superstructure. It was designed by the Major Bridge Reconnaissance and Design Institute, constructed by the China Railway Major Bridge Engineering Bureau and completed in May 1992 at a cost of 788 million RMB.

FIGURE 20.20 Yibin Jinsha River Bridge. (From Xiang et al. 1993. *Bridges in China*. Tongji University Press, Shanghai.)

FIGURE 20.21 Jiujiang Yangtze River Bridge. (From China Ministry of Transport, http://www.moc.gov.cn/2006/05zhishi/qiaoliangzs/W020050731548581258297.jpg.)

20.5 Arch Bridges

Of all types of bridges in China, the arch bridge plays a leading role in variety and magnitude. Statistics from all sources available show that about 60% of highway bridges are arch bridges. China is renowned for its mountains with an abundant supply of stone. Stone has been used as the main construction material for arch bridges. The Dan River Bridge (Figure 20.22) in Shanxi Province, for instance, with a span of 146 m, is the longest stone arch bridge in the world. However, reinforced concrete arch bridges are also widely used in modern highways.

Most of the arches used in China fall into five categories: box arch, two-way curved arch, ribbed arch, trussed arch, and rigid framed arch. The majority of these structures are deck bridges with wide clearance, and it costs less to build such bridges. The box arch is especially suitable for long-span bridges. The longest stone arch ever built in China until 1991 was the Wu River Bridge in Beijing, Sichuan Province, whose span is as long as 120 m. The Wanzhou Yangtze River Bridge in Wanzhou, Chongqing City, with a spectacular span of 420 m, set a world record in concrete arch literature. A unique and successful improvement of the reinforced concrete arch, the two-way curved arch structure, which originated in Wuxi, Jiangsu Province, has found wide application all over the country, because of its advantages of saving labor and falsework. The largest span of this type goes to the 150 m span Qianhe River Bridge in Henan Province, built in 1969. This trussed arch with light deadweight performs effectively on soft subsoil foundations. It has been adopted to improve the composite action between the rib and the spandrel.

On the basis of the truss theory, a light and congruous reinforced concrete arch bridge has been gradually developed for short and medium spans. Through prestressing and with the application of the cantilevering erection process, a special type of bridge known as a "cantilever composite trussed arch bridge" has come into use. An example of this type is the 330 m span Jiangjie River Bridge (1995, Figure 20.23) in Guizhou Province. The Yongning Yong River Bridge (1996, Figure 20.24), located in Guangxi Province, is a half-through ribbed arch bridge with a span of 312 m, the longest of its kind at that time.

With a simplified spandrel construction, the rigid framed arch bridge has a much better stress distribution on the main rib by means of inclined struts, which transfer to the springing point the force induced by the live load on the critical position. In the city of Wuxi, Jiangsu Province, three such bridges with a span of 100 m each were erected in succession across the Great Canal. Many bridges, quite a number of which are ribbed arch bridges, have been built either with tied arches or with Langer girders. The Wangcang Bridge in Sichuan Province and the Gaoming Bridge in Guangdong Province are both steel tubular arch bridges. The former has a 115 m prestressed tied arch, while the latter has a 110 m half-through fixed rib arch. A few steel arch bridges and slant-legged rigid frame bridges have also been constructed.

FIGURE 20.22 **(See color insert.)** Shanxi Dan River Bridge. (From China Ministry of Transport, http://www .moc.gov.cn/huihuang60/difangzhuanti/shanxi/jinrixinmao/200907/W020090721562697210807.jpg.)

FIGURE 20.23 Guizhou Jiangjie River Bridge. (From Wengan City Government, Guizhou Province, http://zsj .wengan.gov.cn/uploadfile/jpg/2009-4/2009424143755644.jpg.)

FIGURE 20.24 Yongning Yong River Bridge. (Courtesy of Lianyou Li.)

FIGURE 20.25 Arch bridge erection by the method of overall rotation. (From Zhuang, W.L. and Huang, D. 2001. *Proceedings of the 3rd International Arch Bridge Conference* [ed. C. Abdunur], Press of The National Institute of Bridges and Roads, Paris.)

In building arch bridges of short and medium spans, precast ribs are used to serve as temporary falsework, and sometimes a cantilever paving process is used. Large-span arch bridges are segmented transversely and longitudinally. With precast ribs, a bridge can be erected without scaffolding, its components being assembled complemented by cast-in-place concrete. Also, successful experience has been accumulated on arch bridge erection, particularly by the method of overall rotation without any auxiliary falsework or support (Figure 20.25). Along with the construction of reinforced concrete arch bridges, research on the following topics has been carried out: optimum arch axis locus, redistribution of internal forces between concrete and reinforcement caused by concrete creep, analytical approaches to continuous arches, and distribution of load between the filled concrete and steel tube in the concrete-filled steel tubular arches.

20.5.1 Shanxi Dan River Highway Bridge

This 413.7 m long highway stone arch bridge (Figure 20.22) has spans 2 × 30 m + 146 m + 5 × 30 m, and is and located in Jin City, Shanxi Province. The main span is 146 m long and is the largest stone arch in the world. Its rise is 32.444 m. The bridge deck is 24.8 m wide and 80.6 m tall. Some special light material was developed and filled in the interior of the spandrel on the subarches on the main open spandrel arch. The bridge was designed by the First Highway Survey and Design Institute and its construction using steel falsework was completed in July 2000. Its construction was carried out by the Jincheng City branch of the Shanxi Highway Administration Bureau and China Railway Seventeenth Group Co., Ltd.

20.5.2 Wanzhou Yangtze River Bridge

This bridge is located in Huangniu Kong, 7 km upstream from Wanzhou, Chongqing, and is an important structure on the No. 318 national highway (Figure 20.26). It is 864.12 m long. A reinforced concrete box arch with a rise-to-span ratio of 1:5 offers a single span of 420 m. Its preliminary design showed that the arch bridge type has less cost than the cable-stayed bridge type. Steel pipes were used to form stiffening arch skeletons before the erection of the main arch; there are 14 spans of 30 m prestressed concrete. Simply supported T-girders make up the spandrel structure, while 13 spans of the same girders are used for the approaches. The continuous deck is 24 m wide, providing two 7.75 m lanes for motor vehicle traffic and two 3 m wide sidewalks. A longitudinal slope of 1% is arranged from the midspan to either side

FIGURE 20.26   Wanzhou Yangtze River Bridge. (Courtesy of Yang Bin.)

with a radius of the vertical curve of 5000 m, while the cross slope is 2%. The bridge was designed by the Highway Survey and Design Institute of Sichuan Province and constructed by the Highway Engineering Company of Sichuan Province. It was completed in 1997 at a cost of 168 million RMB.

20.5.3 Chonqing Wushan Yangtze River Bridge

This 612.2 m long highway bridge, also named the Wu Gorge Bridge (Figure 20.27) is a half-through concrete-filled-steel tubular arch bridge with a world record main span of 492 m, across the Yangtze River in the eastern part of Chongqing. Its deck is 19 m wide and carries four two-way lanes of traffic. It is the first concrete-filled-steel tubular arch bridge across the Yangtze River. The total length of the bridge and its approaches is 8 km and the total cost was 196 million RMB. The diameter of the steel tubes of the arch is the largest in the world. Its steel arches were erected section by section by using suspension cables and slight expansion concrete was filled into the steel tubes by pressure. The bridge was designed by the Sichuan Provincial Highway Planning, Survey, Design and Research Institute, constructed by the Sichuan Road and Bridge Engineering Group Co., Ltd., and was completed in January 2005.

20.5.4 Shanghai Lupu Bridge

This half-through steel tie arch highway bridge has spans 100 m + 550 m + 100 m with a world record main span of 550 m in its category, and is located in Shanghai across the Huangpu River (Figure 20.28). Its steel orthotropic deck is 28.7 m wide, provides six two-way traffic lanes for ADT 60,000 motor vehicles and four sightseeing sidewalks 2.25 m wide and 750 m long. It has a 46 m clearance under the deck. Its two welded steel box arches with cross sections of 9 m × 5 m, the largest in the world, are 100 m tall, and made of S355N and S355N-Z25 steel. It has 28 pairs of steel hangers. The tie cables of the arches consist of 427 steel wires 7 mm in diameter and 762 m long.

The bridge used 46,000 tons of steel. The total length of the arch bridge and approaches is 3900 m. Stayed cables supported by temporary towers were used for erecting 27 sections of each steel box arch section by section. The complete top surfaces of the two box arches are fully decorated as areas for travelers' sightseeing. Considerations on aesthetics, the surroundings, and the environment led to the arch bridge alternative with a cost of 2.2 billion RMB, higher than that of cable-stayed bridge alternatives. The bridge was designed by Shanghai Municipal Engineering Design Institute, and its construction was carried out by Shanghai Construction Engineering Group and completed in 2003.

FIGURE 20.27 Chongqing Wushan Yangtze River Bridge. (Courtesy of Lianyou Li.)

FIGURE 20.28 Shanghai Lupu Bridge. (Courtesy of Quan Qin.)

20.5.5 Chongqing Chaotianmen Yangtze River Bridge

Chongqing Chaotianmen Yangtze River Bridge (Figure 20.29) has a total length of the main bridge and both west and east approaches of 1741 m, including the 932 m long main bridge with a span layout of 190 m + 552 m + 190 m to adopt a half-through continuous steel truss tie arch with tie girders in a double-level traffic arrangement and the 314 m long west and 495 m long east approaches, both continuous PC box girder bridges. The full width of the main bridge is 36.5 m. The height from arch top to the middle supports is 142 m, the lower chord is in quadratic parabola with a rise of 128 m, and the ratio of rise to span is 1:4.31. The N-type truss was adopted for the main truss with central depth of 14 m; the depths at the middle support and at the end support are 73.13 m and 11.83 m, respectively.

FIGURE 20.29   Chongqing Chaotianmen Yangtze River Bridge. (Courtesy of Gongyi Xu.)

The 31 m wide upper deck carries six lanes two ways and a pedestrian lane on each side with the lane arrangement of three two-way lanes with lane widths of 3.75 m, and two side pedestrian lanes with total widths of 2.5 m, and the lower deck carries dual municipal light rails in the middle, with a boundary of rail traffic with clear width ≥ 9.2 m, clear height over the rail top ≥ 6.5 m, and two lanes on each side with a clear road height ≥ 5 m. The span of this bridge (552 m) became a new record.

For the main truss, Q420qD, Q370D and Q345qD steel was adopted; for the deck system, Q345qD was mainly adopted and Q370qD was partially adopted; for the members in the joining system, Q345qD was mainly used. The design traffic loads include the highway load of a first-grade municipal main road with a design velocity of 60 km/h and railway load of two-way light rails with a distance between lines of 4.2 m, design velocity of 80 km/h, and allowable maximum of 100 km/h.

The main bridge was designed by the Major Bridge Reconnaissance and Design Institute and constructed by the Second Navigational Engineering Bureau Co., Ltd. of CCCC. Bridge construction started on December 29, 2004, and opened to traffic on April 29, 2009, at a cost of 3 billion RMB.

20.5.6 Nanjing Dashengguan Yangtze River Bridge

Nanjing Dashengguan Bridge (Figure 20.30) is a key engineering project in the high-speed railway linking Beijing to Shanghai. With a total length of 9270 m, the Dashengguan Yangtze River Bridge includes a six-span continuous steel truss arch bridge with a total length of 1272m as its main bridge in a span arrangement of 108 m + 192 m + 336 m + 336 m + 192 m + 108 m. The rise of the arch is 54.3 m. The height of the arch rib changes from 12 m high in the crown to 56.8 m high in the spring section. The center to center distance of the two truss ribs is 15 m. The six-track railway bridge is across the Yangtze River and supports double tracks for the Beijing–Shanghai high-speed railway with a design speed of 300 km/h, double tracks for the Shanghai–Wuhan–Chengdu railway with a design speed of 200 km/h for a grade I railway passenger train, and double tracks for the joint of Nanjing municipal railway with a design speed of 80 km/h for class B vehicle loads. The project commenced in 2006 and the complete bridge was joined on September 28, 2009. In 2011, this bridge opened to electric multiple unit high speed trains. The Major Bridge Reconnaissance and Design Institute carried out the entire design of the bridge and the Major Bridge Engineering Group Co., Ltd. under China Railway Engineering Cooperation (CREC) constructed the bridge. Its total cost is 4.49 billion RMB.

FIGURE 20.30   Nanjing Dashengguan Yangtze River Bridge. (Courtesy of Gongyi Xu.)

20.6 Cable-Stayed Bridges

Cable-stayed bridges were first introduced into China in the early 1960s. Two trial cable-stayed bridges, the Xinwu Bridge, with a main span of 54 m in Shanghai, and the Tangxi Bridge, with a span of 75.8 m in Yuyang, Sichuan Province, are both reinforced concrete superstructures and were completed in 1975.

In 1977, the construction of long-span cable-stayed bridges began. The Jinan Bridge across the Yellow River, with a main span of 220 m, was completed in 1982. In the 1980s, the construction of cable-stayed bridges developed rapidly over a wide area in China. More than 30 bridges of various types were built around China. Among them, the Yong River Bridge in Tianjin has a main span of 260 m, and the Dongying Bridge in Shandong Province has span of 288 m, China's first steel cable-stayed bridge. In addition, the Haiying Bridge in Guangzhou has a 35 m wide deck, a single cable plane, and double thin-walled pylon piers; the Jiujiang Bridge in Nanhai of Guangdong Province was erected by a floating crane with a capacity of 5,000 kN; the Shimen Bridge in Chongqing has an asymmetrical single cable plane arrangement and a 230 m cantilever cast in place; and the attractive-looking Xiang River North Bridge in Changsha, Hunan Province, was completed in 1990 with a light traveling formwork. All are representative of this period with their respective features.

At the beginning of the 1990s, with the completion of the Nanpu Bridge in Shanghai in 1991, a new high tide of construction of cable-stayed bridges began in China. Now, more than 20 cable-stayed bridges with spans of over 400 m have been completed, and a large number of long-span cable-stayed bridges are under design and construction. The most outstanding is the Sutong Yangtze River Bridge, with a main span of 1088 m, a steel orthotropic deck cable-stayed bridge in Jiangsu Province.

20.6.1 Hong Kong Kap Shui Mun Cable-Stayed Bridge

This double-decked cable-stayed composite box girder bridge is a bipurpose highway–railway bridge with spans of 2 × 80 m + 430 m + 2 × 80 m that leads to asymmetrical cable systems (Figure 20.31). A column in each of the back spans of the cable-stayed bridge makes four 80 m spans to add to the 430 m main span, which made the bridge the longest cable-stayed bridge in the world before 2000 that transports both highway and railway traffic. A 70 m long approach brings the total length of the bridge to 820 m. Across the main marine channel, Kap Shui Mun, the central 387 m of the main span uses steel composite steel–concrete box girder. The back spans and the remaining main span are concrete girders.

FIGURE 20.31 Hong Kong Kap Shui Mun Bridge. (From Environment Protection Department, Hong Kong Government, http://www.epd.gov.hk/epd/misc/ehk04/english/hk/images/pic04.jpg.)

Using the lighter steel cross section in the majority of the main span serves to equalize the horizontal forces on the towers and balance the bridge.

Its deck is 32.5 m wide and 7.46 m deep with an upper deck of six highway lanes, and a lower deck for both two highway lanes and two railway lines. Because the lower deck carries both rail and vehicles, the cross section was designed as a Vierendeel truss. Its two H-shaped reinforced concrete towers are made of grade 50/20 concrete. The navigation clearance of 47 m is part of the reason that the towers are 150 m tall. It has 176 stay cables. The 503 m Ma Wan Viaduct connects the bridge to the Tsing Ma Bridge, thus forming the Lantau Link that was built to provide access to the new airport. The 500 tonne units of the deck girder were floated out on a barge, lifted into position, and the cable stays connected. It was originally designed by Kennedy and Donkin Leonhardt, Andräund Partner, Maunsell Group URS, and Greiner Woodward Clyde. Its construction was carried out by Fugro Hong Kong, Ltd., Hitachi Zosen Inc., Hsin Chong Group, Kumagai Gumi (Hong Kong) Ltd., Maeda Corporation, and Yokogawa Bridge Corporation and completed in 1997. The cost of the Kap Shui Mun bridge with the Ma Wan Viaduct was HK$1643 million.

Since opening to traffic, the bridge is maintained by Tsing Ma Management, Ltd. under contract to the Highways Department of the Government of Hong Kong. Along with the Tsing Ma Bridge and Ting Kau Bridge, it has been closely monitored by the Wind and Structural Health Monitoring System (WASHMS). More detailed information on WASHMS is given in the section on the Tsing Ma Bridge.

20.6.2 Hong Kong Ting Kau Cable-Stayed Bridge

This highway triple tower bridge adjacent to Tsing Ma Bridge (Figure 20.32) is a 1177 m long cable-stayed bridge that spans from the northwest of Tsing Yi Island and Tuen Mun Road. This toll-free bridge has six highway traffic lanes and four spans: 127 m + 448 m + 475 m + 127 m. The bridge is part of Route 3, connecting northwest New Territories with Hong Kong Island, and carrying the heaviest traffic volume of the bridges, with many container trucks traveling to and from mainland China and the Hong Kong container port. The Ting Kau Bridge and approach viaduct are 1875 m long.

Ting Kau Bridge was the world's first major four-span cable-stayed bridge. This meant that the central tower had to be stabilized longitudinally; the problem was solved using the longest (465 m) cable stays then used in a bridge. The design of this bridge involves special features such as single leg towers, which are stabilized by transverse cables just like the masts of a sailboat. Three reinforcement concrete towers were designed to withstand extreme wind and typhoon conditions, with a Ting Kau tower height of 173.3 m, main tower height of 201.55 m, and Tsing Yi tower height of 164.3 m, located on the Ting

FIGURE 20.32 Hong Kong Ting Kau Bridge. (Courtesy of Quan Qin.)

Kau headland, on a reclaimed island in Rambler Channel (which spans 900 m), and on the northwest Tsing Yi shoreline, respectively. The bridge uses 384 stay cables.

The arrangement of separate 18.77 m wide composite decks, each with steel girders, steel crossbeams, and precast concrete panels on both sides of the three towers contributes to the slender appearance of the bridge while acting favorably under heavy wind and typhoon loads. Each deck carries three traffic lanes and a hard shoulder. Its design and construction cost was HK$1.94 billion. It was designed by Binnie Consultants, Ltd. and Schlaich, Bergermann and Partners, GmbH. Its construction was carried out by Ting Kau Contractors joint venture, which designed and built Ting Kau Bridge between 1994 and 1998. The joint venture comprised lead partners Cubiertas Y Mzov and Entrecanales Y Tavora, both of Spain (now both part of Acciona SA); Germany's Ed Züblin; Australia's Downer and Co.; and Hong Kong's Paul Y. Along with the Tsing Ma Bridge and Kap Shui Mun Bridge, it has been closely monitored by the Wind and Structural Health Monitoring System (WASHMS); see the section on the Tsing Ma Bridge for details.

20.6.3 Wuhan Baishazhou Yangtze River Highway Bridge

This cable-stayed highway bridge, also called the Wuhan Third Yangtze River Bridge (Figure 20.33), has two A-frame towers, a two-plane cable system with 96 pairs of stayed cables, and combined steel box and prestressed concrete box girders. Its steel main bridge has spans of 50 m + 180 m + 618 m + 180 m + 50 m. The integrated steel girder is 906.72 m long and is connected by high-strength bolts section by section. The total length, including the main bridge and approaches, is 3586.38 m. Its deck is 26.5 m wide and carries two-way six lanes of traffic for an ADT of 50,000 motor vehicles with design vehicle speed 80 km/h. The two diamond-shaped concrete main towers are 175 m tall and support a pair of two double fans of the modified fan system. The construction cost was 1100 million RMB. It was designed by the Major Bridge Reconnaissance and Design Institute Co., Ltd. and the Wuhan Municipal Engineering Design and Research Institute Co., Ltd., and was constructed by the 2nd Harbor Engineering Co., Ltd. of CCCC and the China Railway Major Bridge Engineering Group Co., Ltd. The bridge was completed in 2000.

20.6.4 Wuhu Yangtze River Bridge

The Wuhu Yangtze River cable-stayed bridge, with a total length of 10,521 m for highway and 5,681 m for railway, is the longest bipurpose highway–railway cable-stayed bridge in the world. Its main spans (Figure 20.34) have an arrangement of 120 m + 8 × 144 m + 180 m + 312 m + 180 m + 2 × 120 m. It is across the Yangtze River and located in Wuhu, Anhui Province. The main bridge has a double-deck composite concrete slab and welded steel truss structures with the upper deck carrying two-way and four highway lanes and the lower deck carrying two track railways.

The 21.7 m wide and 14.6 m deep welding steel truss consists of three, three-span continuous trusses, one two-span continuous truss, and a three-span cable-stayed truss, and is made of 14MnNbq (SM490C)

FIGURE 20.33   Wuhan Baishazhou Yangtze River Bridge. (Courtesy of Gongyi Xu.)

FIGURE 20.34   Wuhu Yangtze River Bridge. (Courtesy of Gongyi Xu.)

steel. Precast concrete panels at the top of the truss are connected by bolts to the top chords of the truss. The two concrete H-type towers are only 84.2 m tall as the height of the bridge towers is constrained by related airspace requirements. A pair of two vertical double fans of the modified fan system has 32 pairs of stay cables. Its navigation clearance is 24 m. It was designed by the Major Bridge Reconnaissance and Design Institute Co., Ltd. and constructed by the China Railway Major Bridge Engineering Group Co., Ltd. The cost of this bridge was 590 million RMB. The expressway was completed in 2000.

20.6.5 Qingzhou Min River Bridge

This cable-stayed highway bridge (Figure 20.35) has spans of 40 m + 250 m + 605 m + 250 m + 40 m and a total length of 1185 m. The 29 m wide concrete deck provides two-way and six lanes of traffic and is composited with two steel I-section plate girders. The main span of 605 m was the longest of the cable-stayed composite girder bridges in the world at the time of construction. The two diamond-shaped concrete main towers are 175 m tall and support a pair of two double fans of the modified fan system with 72 pairs of stay cables. Its navigation clearance is 43 m. The cable anchorage structures are welded

FIGURE 20.35 Qingzhou Min River Bridge. (Courtesy of Gongyi Xu.)

on the top flange of the longitudinal girder. The foundations of the two towers are steel pipe piles 2 m in diameter and bored piles 3 m in diameter, respectively. This bridge was designed by the Major Bridge Reconnaissance and Design Institute and its construction was carried out by the Shanghai Construction Engineering Group and completed in 2001 at a cost of 65 million RMB.

20.6.6 Nanjing Second Yangtze River Bridge (South Bridge)

The 21.4 km long Nanjing Second Yangtze River Bridge (Figure 20.36) includes the South Bridge, North Bridge, South Approach, North Approach, and Baguazhou Approach, at a total cost of 3 billion RMB. This section describes only the South Bridge. Its 1238 m long streamlined steel box girder has spans of 58.5 m + 246.5 m + 628 m + 246.5 m + 58.5 m with an orthotropic steel deck and a navigation clearance of 35 m. The 37.2 m wide and 3.144 m deep superstructure carries two-way and six lanes of 100 km/h automobile traffic. A total of 93 standard deck modules, each 15 m long and about 270 tons in weight, were transported to the site by barges and erected into the deck position. Its two inverted Y-shaped concrete towers, with a height of 195.41 m, are deeply seated on bedrock under water. 60 m tall coffer-dams were built to construct the foundations of 42 concrete bored piles 3 m in diameter and 5130 m^3 of concrete caps at its top. It has 160 stay cables 160 to 313 m long. It was designed by the China Highway Planning and Design Institute Consultants, built by the Shandong Highway and Bridge Construction Corporation, and completed in 2001 at a cost of US$410 million. The orthotropic steel deck box girders were fabricated by the Baoji Bridge Girder Plant.

20.6.7 Sutong Yangtze River Cable-Stayed Bridge

The Sutong Bridge, linking Nantong to Suzhou, Jiangsu Province, has a total length of 32.4 km, including the north linking viaduct, main spans, and south linking viaduct. With a span of 1088 m, the main bridge (Figure 20.37) across the Yangtze River is the cable-stayed bridge with the longest main span in the world. The main spans over the river are 2 × 100 m + 300 m + 1088 m + 300 m + 2 × 100 m. With the two approaches, the main bridge is 8206 m long across the 6 km wide river channel of the Yangtze River. The 270 m thick soft soil layer of the river banks is not suitable to build a suspension bridge. The 40 m wide and 4 m deep streamline closed steel box girder with an orthotropic deck provides two-way and six lanes of highway traffic at 100 km/h, and 891 m × 62 m navigation clearances.

The bridge has 113 piers, 92 of which are in the water. The two inverted Y-shaped steel fiber concrete towers of the main bridge are 306 m tall and thus the second-tallest in the world. Their foundations have 131 bored piles 2.5 m to 2.8 m in diameter and 120 m long, and 50,000 m^3 concrete caps of 114 m × 48 m × 9 m dimensions, the largest and deepest foundations in the world. Two 118 m × 52 m × 17 m cofferdams

FIGURE 20.36   Nanjing Second Yangtze River Bridge. (Courtesy of Nanjing Yangtze River Second Bridge Co., Ltd.)

FIGURE 20.37   Sutong Yangtze River Bridge. (From You, Q.Z. 2007. *Proceedings of the 5th International Conference on Current and Future Trends in Bridge Design, Construction and Maintenance*, Beijing, hold by the Institution of Civil Engineers.)

were built for their construction. Its two double-fan systems have 136 pairs of stay cables with dimples, including the longest four cables, which are 577 m long, the longest in the world. Figure 20.37 shows the bridge on the auxiliary navigation channels.

The bridge was designed for a scour depth of 30 m, a wind speed of 75 m/s, a functional evaluate earthquake of return period 1000 years, and a safety evaluate earthquake of 2500 years. The construction cost was 6.45 billion RMB. The 165th bridge across the Yangtze River was designed by China Highway Planning and Design Institute Consultants, the Architectural Design and Research Institute of Tongji University, and Jiangsu Province Communications Planning and Design Institute. It was constructed by China Harbour Engineering Company Group, China Railway Shanhaiguan Bridge Group Co., Jiangsu Fasten Nippon Steel Cable Co., Ltd., Second Highway Engineering Bureau of CCCC, and Second Navigational Engineering Bureau of CCCC, and opened to traffic on May 25, 2008.

20.6.8 The East Sea (Donghai) Bridge

The East Sea Bridge, also named the Donghai Bridge (Figure 20.38) is a strategic 32.5 km highway bridge across the sea, the longest in the world before 2008 and the third-longest in the world now, linking mainland Shanghai with the Small Yangshan Port in the East Sea just outside Shanghai, which is expected to become the biggest deep-water port in northeast Asia. Providing a clearance of 40 m, its 31.5 m wide deck carries six lanes of traffic at a speed of 80 km/h. It has the following four parts:

1. A 45 km long approach embankment section linking to the Lu-Mu expressway
2. A 26 km-long viaduct of 76 spans of 30 m prestressed concrete single-cell box girders, with every five spans continued
3. A 56 km long section across the sea linking the new embankment of the Luchao Port to Dawugui Island, including:
   a. A 29 km long subsection on shallow sea water of 26 × 50 m span prestressed concrete single-cell box girders with every five spans continued and supported by 112 bored piles 1600 mm in diameter and 198 tube piles 1200 mm in diameter
   b. A "not navigation channel" subsection consisting of eight spans of 50 m long prestressed concrete single-cell box girders and a number of 60 m or 70 m long precast standard concrete girder spans
   c. An auxiliary navigation channel subsection of 12 spans with an arrangement of 2 × 160 m + 2 × 120 m + 2 × 140 m continuous prestressed concrete box girders
   d. A main navigation channel subsection including a one-plane stay-system cable-stayed bridge, the main cable-stayed bridge, with steel–concrete composite girder spans of 73 m + 132 m + 420 m + 132 m + 73 m and two 140.2 m tall inverted Y-shaped concrete towers each supported by 76 bored piles 2.5 m in diameter and 110 m long
4. A 3.5 km long approach connecting Dawugui Island to the Small Yangshan Port, including:
   a. A 1.22 km long, 41 m tall, and 55 m wide (at the top) sea embankment section
   b. A two-plane stay-system cable-stayed bridge with a composite steel and concrete girder with a main span of 322 m and two 100 m tall H-shaped concrete towers supported by 70 piles 2.5 m in diameter and continuous concrete approaches of 50 m spans

FIGURE 20.38 East Sea Bridge. (From Shanyang Town Government, Jinshan County, Shanghai, http://shanyang .jinshan.gov.cn/UpLoadFile/UploadFile/200710812226241.jpg.)

The bridge was designed by the Shanghai Municipal Engineering Design Institute, constructed by the China Railway Major Bridge Engineering Group Co., Ltd. and completed in 2005 at a cost of 10 million RMB.

20.6.9 Hangzhou Bay Bridge

This 36 km long over-the-sea highway bridge (Figure 20.39), is across the Hangzhou Bay in an S-shape and links Jiaxing to Ningbo, Zhejiang Province. Its 33 m wide deck provides two ways and six lanes of highway traffic at 100 km/h speed. Now, this bridge is the second-longest bridge in China, with a total length of 36 km. It has two cable-stayed bridges: the North Bridge across the north navigation channel and the South Bridge across the south navigation channel. The North Bridge (Figure 20.39) has spans of 70 m + 160 m + 448 m + 160 m + 70 m. Its streamline flat steel orthotropic deck box girder is 38.2 m wide and 3.5 m deep. It has two diamond-shaped concrete towers 178.8 m tall and 56 pairs of stay cables in two stay planes. It provides navigation clearances of 325 m × 47 m. The South Bridge (Figure 20.40) has spans of 318 m + 160 m +100 m, one A-shaped concrete tower 194.3 m tall, and two stay planes.

FIGURE 20.39   Hangzhou Bay Bridge and its North Bridge. (Courtesy of Tao Liu.)

FIGURE 20.40   South Bridge of the Hangzhou Bay Bridge. (Courtesy of Tao Liu.)

Its steel orthotropic deck box girder provides navigation clearances of 125 m × 31 m. All approaches have continuous prestressed concrete box girders with spans 30 m to 80 m and a 12,000 m² platform for traffic service, sightseeing, and rescue.

It was designed by a joint venture of the China Highway Planning and Design Institute Consultants, the Major Bridge Reconnaissance and Design Institute, and the Third Highway Survey and Design Institute, constructed by the 2nd Harbor Engineering Co., Ltd. of CCCC, and opened to traffic in 2008 at a cost of 14 billion RMB.

20.6.10 Hong Kong Stonecutters Cable-Stayed Bridge

This highway bridge, also called the Angchuanzhou Bridge (Figure 20.41), with a main span of 1018 m, is the world's second-longest spanning cable-stayed bridge after the Sutong Bridge, with spans of 289 m + 1018 m + 289 m. The deck itself is made of steel in the main span and concrete in the side spans. It is the centerpiece of the new Route 8, which improves access between Hong Kong International Airport and the urban areas of West Kowloon, as well as providing enhanced links to one of the busiest container ports in the world.

Its steel main span is supported by two 295 m tall single pole towers with 51 m wide decks (including a 14 m gap between the two carriageways) split into two streamlined boxes connected by cross girders and carrying a dual three-lane expressway. The two towers are in concrete until level +175m and in composite construction consisting of an inner concrete ring with a stainless steel (grade 1.4462 to BSEN10088) skin with a shot peened surface finish for the top 120 m to limit wind vibrations of the stay cables. The deck allows a navigation clearance of 73.5 m over the full entrance to the container port. The tower pedestals have a size of 24 m × 18 m. The cables consist of parallel 7 mm galvanized wire with HiAm anchorages, and they range in size from 187 to 421 wires. The longest stay cables supporting the bridge superstructure are more than 670 m long. The concept design of this bridge was by Dissing + Weitling Arkitektfirma, Flint & Neill Partnership, Halcrow Group, Shanghai Construction Engineering Group, and Maeda Corporation. A group led by Ove Arup & Partners with COWIA/S as the main subconsultant carried out the further design.

In October 2002, a 50 m mast was installed at the site to measure the speed, direction, and turbulence of winds in the area. Readings, which were continued until at least 2004, were transmitted in real time to an offsite location for wind analysis. In particular, the stability of the 509 m long cantilevers during construction required special consideration in the design. Its construction was carried out by Hitachi Zosen Inc., Hsin Chong Group and Yokogawa Bridge Corporation and completed in 2009. The construction cost of the bridge was HK$2.76 billion.

FIGURE 20.41 Hong Kong Stonecutters Bridge. (From Wikipedia, http://en.wikipedia.org/wiki/Stonecutters_Bridge.)

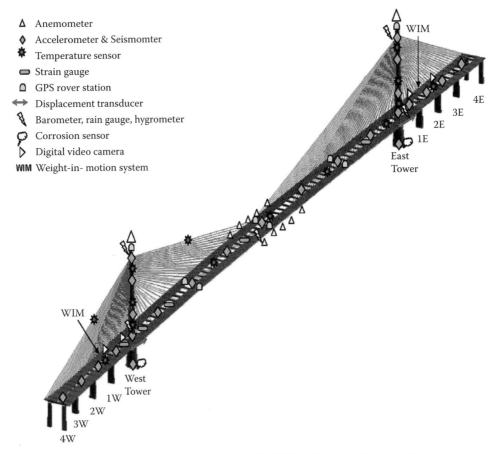

△ Anemometer
◊ Accelerometer & Seismomter
✳ Temperature sensor
▭ Strain gauge
⌂ GPS rover station
↔ Displacement transducer
⚲ Barometer, rain gauge, hygrometer
○ Corrosion sensor
▷ Digital video camera
WIM Weight-in- motion system

FIGURE 20.42 Layout of sensors of Stonecutters Bridge WASHMS. (Courtesy of Kai-Yuen Wong.)

An independent wind and structural health monitoring system was installed on the Stonecutters Bridge to monitor the wind and structural integrity, durability, and reliability of the bridges. Like the Tsing Ma Bridge WASHMS, the Stonecutters Bridge WASHMS also has four different levels of operation: sensory systems, data acquisition systems, local centralized computer systems, and a global central computer system. Its sensor system has 48 accelerometers and seismometers, 40 anemometers, 388 temperature sensors, 28 temperature and relative humidity sensors, 20 GPS sensors, 34 displacement sensors, 515 strain gauges, 12 weigh-in-motion sensors, 16 digital video cameras, and 30 corrosion cells. Some sensors were used for measuring temperature and strain inside the stay cables by inserting several fiber optic strain and temperature chains inside the stay cables. The sensors were installed inside the stay cable during the fabrication process by the stay cable manufacturer. Figure 20.42 shows the layout of the sensor system of the Stonecutters Bridge WASHMS.

20.6.11 Shanghai Yangtze River Bridge

The Shanghai–Chongming across the Yangtze River tunnel and bridge project, of total length 25 km, located in the east of Shanghai, connecting Wuhaogou, Pudong, at the south and Chenjiazhen, Chongming Island at the north, includes an 8.9 km long tunnel with an inside diameter of 13.7 m as its north part and the 10.2 km long Shanghai Yangtze River Bridge as its south part. The part of the Shanghai Yangtze River Bridge (Figure 20.43) that crosses the river is of total length 9974 m and consists of a prestressed concrete continuous girder spanning 6 × 21 m + 15 × 30 m +15 × 50 m + 23 × 70 m, a steel–concrete composite

FIGURE 20.43 The Shanghai Yangtze River Bridge. (From China Central People's Government Net, http://www.gov.cn/jrzg/images/images/00123f3793250c239e3501.jpg.)

continuous girder spanning 85 m + 5 × 105 m + 90 m, a cable-stayed bridge with two single-column pylons and separated twin orthotropic steel box girders spanning 92 m + 258 m + 730 m + 258 m + 92 m, a steel–concrete composite continuous girder spanning 90 m + 5 × 105 m + 85 m, a prestressed concrete continuous girder spanning 9 × 70 m, a prefabricated prestressed concrete segmental continuous girder spanning 32 × 60 m, a prestressed concrete continuous girder spanning 80 m + 144 m + 144 m + 80 m, and a prestressed concrete continuous girder spanning 14 × 50 m + 17 × 30 m.

The orthotropic steel box girders of the cable-stayed bridge, with cross sections of 4 m height and 20.75 m width for each girder, are spaced at 10 m and connected by transverse beams. The girders, with a total width of 51.25 m, a world record, support six highway lanes with a design speed of 100 km/h and two 4.15 m wide hard shoulders of roads for a future two-track municipal railway with a 90 km/h design speed. The two inverse Y-shaped concrete pylons for earthquakes with a 3000-year return period are 216 m tall and support the girders through 96 pairs of stay cables. The bridge, completed in October 2009, was designed by the Shanghai Municipal Engineering Design Institute and built by the Second Navigational Engineering Bureau Co., Ltd. of the CCCC and Shanghai Foundation Engineering Co. The Shanghai Yangtze River Bridge cost about 12.6 billion RMB.

20.6.12 Edong Yangtze River Bridge

The Edong Yangtze River Bridge (Figure 20.44) located in Huangshi, Hubei, is a two-pylon cable-stayed highway bridge with a major bridge of nine spans with an arrangement of 3 × 67.5 m + 72.5 m + 926 m + 72.5 m + 3 × 67.5 m. The 926 m main span is the third-longest cable-stayed span in the world. The total length with approaches is 15.149 km. The orthotropic steel box girder of the main span connects to the two continuous prestressed concrete box girders of the side spans through C55 steel fiber concrete joints starting from 12.5 m in the main span. The girders support a 38 m wide continuous deck with six traffic lanes. The two double leg prestressed concrete pylons are 236.5 m tall. It was designed by the Hubei Highway Planning and Design Institute and constructed by the 2nd Harbor Engineering Co., Ltd. and the 2nd Highway Engineering Company, both under the CCCC, at a cost of 3.193 billion RMB. The steel girders were fabricated by the 15th Metallurgical Construction Co., Ltd. of the China Nonferrous Metal Mining (Group) Co., Ltd. The Edong Yangtze River Bridge has been completed in September 2010.

20.6.13 Qingdao Jiaozhou Bay Bridge

The Qingdao Jiaozhou Bay Bridge has a Y-shaped plan with the east end at Yangjiawan, in the downtown area of the city of Qingdao, the west end at Hongshiya, Huangdao, and the north end at Hongdao. Its 35 m wide superstructure has separated twin decks supporting two separated ways each for three traffic

FIGURE 20.44   Edong Yangtze River Bridge. (From Hubei Province Transportation Department, http://www
.hbjt.gov.cn/Theme/Upload/20091103111565625483.jpg.)

FIGURE 20.45   Cangkou cable-stayed bridge portion of the Qingdao Jiaozhou Bay Bridge. (From Qingdao
Development and Reform Commission, http://www.qddpc.gov.cn/uploadfile/2007116171558659.jpg.)

lanes with 80 km/h speed and a total length of 36.4 km, the longest total length of bridge now in China.
Its across-sea section is 26.75 km long, and includes three orthotropic steel box girder bridges crossing
three navigation channels: the Cangkou cable-stayed bridge (Figure 20.45) with a main span of 260 m
and two concrete pylons, the Hongdao cable-stayed bridge with two main spans of 120 m and a single
concrete pylon, and the Daguhe self-anchored suspension bridge with two main spans of 260 m and a
single concrete pylon 160 m tall. Other sections of the bridge consist of prestressed concrete box girder
sections 50 m or 60 m long.

The bridge was designed by the China Highway Planning and Design Institute Consultants. It is
being constructed by the 2nd and 3rd Highway Engineering Company and 4th Harbor Engineering Co.,
Ltd., under the CCCC, the China Road and Bridge Corporation, Major Bridge Engineering Group Co.,
Ltd., and China Railway 9th, 14th, and 15th Bureau Group Co., Ltd. under the CREC, and so on. It was
opened to traffic in 2011. The total cost was 9.08 billion RMB.

20.7 Suspension Bridges

The construction of modern suspension bridges in China started in the 1960s. Some flexible suspension bridges with spans less than 200 m were built in the mountain areas of southwestern China,
the Chaoyang Bridge in Chongqing being the most famous one. However, the Dazi Bridge in Tibet,
completed in 1984, has a span of 500 m. The upsurge of transportation engineering construction in
the 1990s has been leading to a new stage of modern suspension bridges. The Shantou Bay Bridge

in Shantou, Guangdong Province, was completed in 1995, with a 452 m concrete stiffening girder. The Humen Pearl River Bridge, a steel box girder suspension bridge with a main span of 888 m, was completed in 1997. The Jiangying Yangtze River Bridge, with a main span of 1385 m, and the Runyang Yangtze River bridge, with a main span of 1490 m, were built in 1999 and 2005, respectively, and the Xihoumen across-sea suspension bridge, with a main span of 1650 m, was completed in 2009. In addition, quite a few self-anchored suspension bridges have been constructed in China in the recent years.

20.7.1 Hong Kong Tsing Ma Bridge

The Tsing Ma Bridge (Figure 20.46), a highway–railway bipurpose suspension bridge on the freeway between the new airport and the urban district linking Tsing Yi Island and Mawan Island in Hong Kong, is the world's longest of its kind. It has a suspended main span of 1377 m, a suspended side span of 355.5 m, and continuous girder side spans of 4 × 72 m. The two concrete towers, with a two-legged and four portal beam form design, are 206 m above sea level. The legs were constructed with high-strength concrete of 50 MPa (concrete grade 50/20) strength. One tower is located on Wok Tai Wan on the Tsing Yi side and the other on a man-made island 120 m from the coast of Ma Wan Island. Both towers are founded on relatively shallow bedrock and built by using a slipform system in continuous operation.

The design of its main steel stiffening girder was mainly based on consideration of its aerodynamic stability under a design 1-minute mean wind speed of 95 m/s and a truss type was finally adopted. The steel crossbeams of the truss deck are Vierendeel truss. With two expansion joints set at the two anchorages, the integrated steel truss girder of the six spans is 2160 m long. Its 41 m wide and 7.232 m deep truss superstructure carries six lanes of highway traffic, with three lanes in each direction on the upper deck and three carriageways on the lower deck including the central one for two rail tracks and two sheltered highway lanes for maintenance access and as backup for traffic when particularly severe typhoons strike Hong Kong. There is no sidewalk on the bridge. A total of 96 standard deck modules, each 18 m long and about 480 tons in weight, were brought to the site by specially designed barges and erected into the deck position by a pair of strandjack gantries that could maneuver along the main cable.

The main cables, 1.1 m in diameter, were constructed by an aerial spinning process and are of 80 × 368 + 11 × 360 steel wires, each galvanized wire having a diameter of 5.35 mm. The bridge has 94 pairs of hangers. The design gap of the Tsing Yi expansion joint is 89 cm wide. Two large gravity anchorages are located at both ends of the bridge. They are massive concrete structures deeply seated on bedrock on the sides of Tsing Yi and Ma Wan Islands. The total weight of concrete used in the Tsing Yi anchorage is 200,000 tons and the Ma Wan anchorage is 250,000 tons. The bridge was designed by Mott MacDonald. The construction of the bridge was carried out by an Anglo Japanese joint venture of Trafalgar House Construction (Asia), Ltd. (part of the Kvaerner Group of Norway), Costain Civil Engineering, Ltd. of Britain, and Mitsui and Co., Ltd. of Japan, and the bridge opened to traffic in 1997. It cost HK$7.2 billion.

FIGURE 20.46 Hong Kong Tsing Ma Bridge. (From Hong Kong Observation, Hong Kong Government, http://www.hko.gov.hk/education/edu06nature/images/ele-bridge-2.jpg.)

Like the Ting Kau Bridge and Kap Shui Mun Bridge, the three bridges are closely monitored by an integrated WASHMS. Surveillance cameras were also installed over the bridges to record traffic conditions. WASHMS is the first sophisticated bridge structural health monitoring system, with a cost of US$1.3 million, used by the Hong Kong Highways Department to ensure road user comfort and safety of the three bridges, and was designed by Flint, Neill, and Partnership.

In order to oversee the integrity, durability, and reliability of the bridges, WASHMS has four different levels of operation: the sensory system, data acquisition system, local centralized computer system, and global central computer system. The sensory system consists of approximately 900 sensors including 29 GPS stations and their relevant interfacing units. With more than 350 sensors on the Tsing Ma Bridge, 350 on Ting Kau Bridge, and 200 on Kap Shui Mun Bridge, the structural behavior of the bridges is measured 24 hours a day, 7 days a week. The sensors include accelerometers, straingauges, displacement transducers, level sensing stations, anemometers, temperature sensors, and dynamic weight-in-motion sensors. They measure everything from tarmac temperature and strains in the structural members to wind speed, the deflection and rotation of the cables, and any movement of the bridge decks and towers. These sensors also work as an early warning system for the bridges, providing the essential information that helps the Highways Department to accurately monitor the general health conditions of the bridges.

The information from these sensors is transmitted to the data acquisition outstation units. There are three data acquisition outstation units on the Tsing Ma Bridge, three on Ting Kau, and two on Kap Shui Mun. The computing powerhouse for these systems is in the administrative building used by the Highways Department in Tsing Yi. The local central computer system provides data collection control, post-processing, transmission, and storage. The global system is used for data acquisition and analysis, assessing the physical conditions and structural functions of the bridges and for integration and manipulation of the data acquisition, analysis, and assessing processes.

20.7.2 Jiangyin Yangtze River Bridge

Jiangyin Yangtze River Bridge (Figure 20.47) is located on the planned north–south principal highway system in the coastal area between Jiangyin and Jinjiang in Jiangsu Province. It is a large suspension bridge with a central span of 1385 m, and was the first bridge designed with a span in excess of 1000 m. Its total length reaches nearly 3 km. The bridge carries six lanes of highway traffic, while median and emergency parking strips are also considered, with two 1.5 m wide pedestrian walks on the central span.

Flat steel orthotropic deck box girders with wind fairing were adopted, whose depth and width are 3 m and 37.7 m, respectively. The two main cables, with sag-to-span ratios of 1:10.5, are composed of 169

FIGURE 20.47   Jiangyin Yangtze River Bridge. (From Jiangsu Province Transportation Department, http://www .jscd.gov.cn/art/2007/9/1/art_913_51570.html.)

strands of 127 galvanized high-strength wires 5.35 mm in diameter, and were built by the prefabricated parallel wire strand (PPWS) method. The bridge provides a vertical navigation clearance of 50 m. Two 196 m tall towers are reinforced concrete structures. The northern tower, located in shallow water outside the north bank, rests on a 96 bored pile foundation constructed by the sand island method, whereas the southern tower is on the rock stratum of the bank. The north anchorage is a large massive concrete gravity anchorage 69 m long, 51 m wide, and 58 m tall and is located on soft soil ground.

The bridge was designed by the China Highway Planning and Design Institute Consultants, Major Bridge Reconnaissance and Design Institute, and the Jiangsu Province Communications Planning and Design Institute and constructed by Dorman Long Technology Ltd., Kvaerner Cleveland Bridge and Engineering SdnBhd, Second Navigational Engineering Bureau Co., Ltd. of China Communication Construction Co., Ltd. and Shanghai Foundation Engineering Co. This bridge cost 337 million RMB and was erected in 1999.

20.7.3 South Bridge of Runyang Yangtze River Suspension Bridge

This 7.1 km long highway bridge across the Yangtze River linking Zhenjiang and Yangzhou, Jiangsu Province, provides six lanes of highway traffic in two directions (Figure 20.48). With a total cost of 5.78 billion RMB, it includes the North Bridge, a viaduct, and the South Bridge. The South Bridge is a single-span suspension bridge with 470 m long continuous spans for the two side spans. Its one 490 m main span, supported by 91 pairs of hangers, is the fourth-longest span in the world. Its 38.7 m wide and 3 m deep flat steel orthotropic deck box girder provides navigation clearances of 700 m × 50 m.

Its two concrete main towers are 215.58 m tall. Each main cable is made of 184 strands of 127 steel wires 5.3 mm in diameter by the PPWS method. The bridge has two massive concrete gravity anchorages. The north anchorage is made of 60,000 m³ concrete with pit sizes of 69 m × 50 m × 50 m, the largest in China. Its south anchorage has the sizes of 69 m × 51 m × 29 m and is made of 102,000 m³ of concrete. The ground freezing method was used for the pit of the south anchorage. It was designed by the Jiangsu Communication Planning and Design Institute and its construction was carried by China Communication Construction Co., Ltd. and Dorman Long Technology, Ltd. and was completed in 2005 at a cost of US$700 million.

20.7.4 Xihoumen Suspension Bridge

This highway suspension bridge is a part of 46.5 km long highway project linking the Zhoushan Archipelago, Zhejiang Province, with a total cost of 13.1 Billion RMB. This highway bridge across the sea has spans of 578 m + 1650 m + 485 m, with a main span of 1650 m, the second-longest span in the

FIGURE 20.48 The South Bridge of Runyang Yangtze River Bridge. (Courtesy of Hao Huang.)

FIGURE 20.49   The Xihoumen Suspension Bridge. (Courtesy of Xiaodong Wang.)

world, and links Taoyaomen Hill on Cezi Island and Jintang Island (Figure 20.49). The 578 m long side span is also a suspended span, but the 485 m long side span is a six-span continuous girder. The steel stiffening girder of the two suspended spans has separated double streamline flat closed box sections with a total width of 36 m and depth of 3.5 m. Each of the steel boxes is 16.24 m wide with a space between the two boxes and is connected by steel crossbeams. The orthotropic deck provides four lanes of highway traffic in two directions with a design speed of 80 km/h.

The bridge is across the Xihoumen navigation channel and provides navigation clearances of 630 m × 49.5 m. Its two H-shaped concrete towers are 211 m tall. Each of its two main cables consists of 169 strands of 127 steel 1770 MPa grade wires 5.25 mm in diameter and was erected by the PPWS method. Each of the towers is supported by 12 piles 2.8 m in diameter with a cap of 22.8 m × 16.8 m × 7 m. The bridge was designed by the China Highway Planning and Design Institute Consultants, and its construction was carried out by the Sichuan Road and Bridge Company and Second Navigational Engineering Bureau Co., Ltd. of CCCC. The steel box girder was fabricated by the Baoji Bridge Girder Plant. It opened to traffic on December 25, 2009, at a cost of 2.48 billion RMB.

20.8 Future Bridges

Nationwide highway construction in China continues. More large-scale bridge projects are being constructed or are planned.

20.8.1 The Jiaxing-Shaoxing River-Sea-Crossing bridge

This highway links Jiaxing to Shaoxing, Zhejiang Province across the Hangzhou Bay, with a total length of 69.46 km including a major sea-crossing bridge of 10.14 km, with a north approach of 42.95 km and a south approach of 16.3 km. Its major part includes a multispan cable-stayed bridge (Figure 20.50) which has six 175 m tall single-column concrete towers made of C50 concrete. Its separated streamline flat steel box girder, including spans of 5 × 480 m, is 2680 m long, 55.6 m wide, and supports eight traffic lanes with a speed of 100 km/h, the future longest and widest in the world, and will provide navigation clearances of 335 m × 32.5 m. The girder has rigid expansion joints which were used for the first time in the world. Each tower supports four overlapping fan stay planes. The bridge was designed by the China Highway Planning and Design Institute Consultants and its construction started in December 2008 and is planned to open to traffic in June, 2013. The section crossing the sea will cost 6.5 billion RMB. The expected cost of the complete project is 13.98 billion RMB.

FIGURE 20.50 The Jiaxing-Shaoxing river-sea-crossing bridge. (Courtesy of Tao Liu.)

References

China Ministry of Railways. 2005. TB 10002.1-2005. *Fundamental Code for Design of Railway Bridges and Culverts.* China Railway Publishing House, Beijing (in Chinese).

China Ministry of Railways. 2005. TB10002.3-2005. *Code for Design of Concrete and Prestressed Concrete Structures in Railway Bridges and Culverts.* China Railway Publishing House, Beijing (in Chinese).

China Ministry of Railways. 2005. TB10002.4-2005. *Code for Design of Concrete and Block Masonry Structures of Railway Bridges and Culverts.* China Railway Publishing House, Beijing (in Chinese).

China Ministry of Railways. 2005. TB10002.5-2005. *Code for Design on Subsoil and Foundations of Railway Bridges and Culverts.* China Railway Publishing House, Beijing (in Chinese).

China Ministry of Railways. 2005. TB10002-2005. *Code for Design of Steel Structures of Railway Bridges.* China Railway Publishing House, Beijing (in Chinese).

China Ministry of Railways. 2006. GB50111-2006. *Code for Seismic Design of Railway Engineering.* China Planning Press (in Chinese).

China Ministry of Railways. 2008. China's medium-to-long-term plan to construct a national railway net (modified in 2008), http://www.china-mor.gov.cn/tllwjs/tlwgh_6.html (This website recently does not work and is being merged into www.moc.gov.cn, the site of China ministry of transport.)

China Ministry of Railways. 2009. GB50216-2009. *Unified Standard for Reliability Design of Railway Engineering Structures.* China Railway Publishing House, Beijing (in Chinese).

China Ministry of Transport. 1986. JTJ 025-86. *Code for Design of Steel and Timber Structures of Highway Bridges and Culverts.* China Communications Press, Beijing (in Chinese).

China Ministry of Transport. 1999. GB/T 50283-1999. *Unified Standard for Reliability Design of Highway Engineering Structures.* China Planning Press, Beijing (in Chinese).

China Ministry of Transport. 2004. JTG D60-2004. *General Code for Design of Highway Bridges and Culverts.* China Communications Press, Beijing (in Chinese).

China Ministry of Transport. 2004. JTG D62-2004. *Code for Design of Highway Reinforced Concrete and Prestressed Concrete Bridges and Culverts.* China Communications Press, Beijing (in Chinese).

China Ministry of Transport. 2004. JTG/T D60-01-2004. *Wind-Resistant Design Specification for Highway Bridges.* Wuhan University of Technology Press, Wuhan (in Chinese).

China Ministry of Transport. 2005. JTG D61-2005. *Code for Design of Highway Masonry Bridges and Culverts.* China Communications Press, Beijing (in Chinese).

China Ministry of Transport, 2005, The National Highway Network Planning (in Chinese), www.moc.gov.cn

China Ministry of Transport. 2007. JTG D63-2007. *Code for Design of Ground Base and Foundations of Highway Bridges and Culverts.* China Communications Press, Beijing (in Chinese).

China Ministry of Transport. 2007. JTG/T D65-01-2007. *Guidelines for Design of Highway Cable-Stayed Bridges.* China Communications Press, Beijing (in Chinese).

China Ministry of Transport. 2008. JTG/T B02-01-2008. *Guidelines for Seismic Design of Highway Bridges.* China Communications Press, Beijing (in Chinese).

China Ministry of Transport. 2012. Development statistics bulletin '2012 highway and waterway transportation industry' (in Chinese).

Xiang, H.F. et al Ed. 1993. *Bridges in China.* Tongji University Press, Shanghai.

Li, W.Q., Fan, W. et al. 2001. The Wanxian Bridge: The world's longest concrete arch span. In *Proceedings of the 3rd International Arch Bridge Conference* (ed. C. Abdunur), Press of the National Institute of Bridges and Roads, Paris.

Sichuan Province Highway Planning Survey Design and Research Institute. 2008. *Guide to Design and Construction Technology of Road Steel Tube Concrete Bridges.* China Communications Press, Beijing (in Chinese).

You, Q.Z. 2007. A thousand-meter span cable-stayed bridge across Yangtze River—Sutong Bridge. In *Proceedings of 5th International Conference on Current and Future Trends in Bridge Design, Construction and Maintenance,* Institution of Civil Engineering, Beijing.

Zhuang, W.L. and Huang, D. 2001. The 360 meter CFST arch of the Yajisha Bridge. In *Proceedings of the 3rd International Arch Bridge Conference* (ed. C. Abdunur), Press of the National Institute of Bridges and Roads, Paris.

Relevant Websites

http://glcx.moc.gov.cn China Highway Information service,

http://www.0499.cn"www.0499.cn Lingshi Civil Engineering Network, China,

http://www.ccccltd.com.cn China Communications Construction Company Ltd. (CCCC),

http://www.chinahighway.com China Highway Network,

http://www.china-mor.gov.cn China Ministry of Railways, which is being merged to the website of China Ministry of Transport,

http://www.chnrailway.com China Railway Network,

http://www.cndaoqiao.com China Highway and Bridge—Construction Technology Network,

http://www.gov.cn The Central People's Government, PRC,

http://www.hyd.gov.hk Highways Department of Hong Kong Special Administration Region, China,

http://www.jscd.gov.cn Jiangsu Province Communication Department, China,

http://www.moc.gov.cn China Ministry of Transport,

http://www.rbtmm.com China Specialty Network for Management of Highway Bridges and Tunnels,

http://www.raildoor.com China Railway Forum,

http://www.sbsm.gov.cn National Administration of Surveying, Mapping and Geoinformation,

http://www.sinobridge.net China Bridge Network,

http://www.stats.gov.cn National Bureau of Statistics of China,

http://www.zjt.gov.cn Zhejiang Province Communication Department, China,

http://www. ztjs.net.cn China Railway Construction Group Co., Ltd.,

http://www. ztmbec.com China Railway Major Bridge Engineering Group Co., Ltd.,

21

Bridge Engineering in Indonesia

Wiryanto
   Dewobroto
Universitas Pelita
Harapan Tangerang

Lanny Hidayat
Department of Public
Works Cibinong

Herry Vaza
Department of
Public Works

951

21.1 Introduction

21.1.1 Geographical Characteristics

Indonesia is the world's largest archipelago country, covers around 17,000 islands, and lies to the north of the equator, between 6° and 11° south latitude and 97° and 114° east longitude (Figure 21.1). Not only is Indonesia located between two great continents, Asia and Australia, but it is also located between two great oceans, the Indian and Pacific Oceans. The accumulation of land and water is around 1.9 million km² or approximately 5.8 times that of Malaysia, which is only around 0.3 million km². However, Indonesia is still only around 0.2 times the size of the United States of America, which is around 9.8 million km². Indonesia is the largest among its neighboring countries in Southeast Asia, such as Singapore, Malaysia, the Philippines, Papua New Guinea, and East Timor.

The population in Indonesia is 222 million based on the Badan Pusat Statistik Republik Indonesia (Statistics Indonesia; http://www.bps.go.id) of 2006. Over half the population lives on Java Island, which is only around 7% of the total area. The population density is not spread evenly. In general, the population is concentrated on five major islands: Sumatra (474 km²), Java (132 km²), Kalimantan, the third world's largest island (540 km²), Sulawesi (189 km²), and Papua New Guinea (422 km²). Most of these main islands, except Kalimantan, are located along earthquake rings. The contours of the regions are varied, consisting of mountains, valleys, and many rivers which divide one region from another. This means land transportation requires bridges to connect one roadway to another.

21.1.2 Bridge Development

According to the road laws of the Republic of Indonesia (Number 38, 2004), a road, as part of the national transportation system, plays an important role in particularly supporting not only the

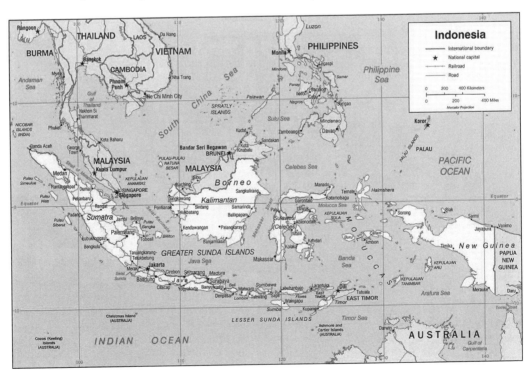

FIGURE 21.1   Map of the Republic of Indonesia. (From http://www.surftrip.com.)

economic, social, and cultural sectors but the environment as well. Roads are developed by means of a regional expansion approach so interregional development balance and equal distribution can be achieved, thus forming and affirming national unity and national defense and security, as well as forming a spatial structure within the framework of realizing national development's goals. Therefore, the government must provide excellent quality, useful, and sustainable road infrastructure, which is needed to support peace, justice, and democracy, and to improve the people's welfare in Indonesia.

At present, road transportation is the main mode of transportation in Indonesia and far exceeds the volume of other available modes of transportation. About 90% of all goods and more than 95% of all passengers in Indonesia are transported by road. Accordingly, the road networks in Indonesia have been expanded from 85,000 km in 1971 to around 437,759 km by the end of 2008 (BPS Statistics Indonesia). Along these roads, there are approximately 88,000 bridges and other types of crossing methods with a total length of almost 1000 km. Apart from that, 28,000 bridges are located along national and provincial roads and another 60,000 bridges are located along local and urban roads.

From a strategic point of view, it is clear that bridges have an important role in the operation and function of road networks and involve a large initial investment. To accelerate road infrastructure developments, particularly in bridge construction, the government's policies are directed towards superstructure standardization by providing a component stock of bridge standard spans as well as technical construction standard drawings that can be constructed in the fields. The main purpose of bridge superstructure standardization is to guarantee that the quality of the product fulfills the requirements as specified to ease construction works. This reason for standardization is not a surprise because, in reality, there are currently 88,000 bridges, which mostly cross small rivers.

Out of all existing bridges along the national and provincial road links, the number of bridges that cross rivers with a width of more than 100 m is less than 2%. Although there are not so many rivers with big channels compared to other neighboring countries, Indonesia has also adopted advanced bridge structural technology as indicated by the application of prestressed concrete structures, cable-supported structures, and more challenging architectural demands for crossing big rivers as well as for urban infrastructure developments. For bridge spans up to 60 m, superstructure constructions in Indonesia usually follow the Bina Marga bridge standards; for long spans over 60 m, designs are made specifically for the systems of different structures or the existing standard system is modified, resulting in nonstandard bridge constructions.

Construction of nonstandard bridges in Indonesia started in the 1960s; the first implementation was the Ampera Bridge, a continuous steel girder and constructed from 1962 to 1965 in Palembang, South Sumatra. The construction of the Danau Bingkuang Bridge in Riau also introduced the construction of continuous composite steel girders from 1968 to 1970. Special bridge constructions have continued to develop with the use of various types of constructions, including continuous concrete box bridges, continuous steel girders, concrete arches, steel arches, suspension bridges, and cable-stayed bridges, among others. These types of bridges are designed to be economical bridge spans and special attention is paid to the aesthetic aspects and compatibility with the surrounding environments.

The trends for choosing proper bridge types are closely linked to the geographical characteristics and the needs of local traffic. Bridges on the island of Java were generally designed to serve dense traffic, crossing rivers or river banks that are relatively short, resulting in wide bridges with short to medium spans. As for some areas of Borneo and some parts of Sumatra, there are many great rivers with heavy water traffic, but the road traffic is not as heavy as in Java. This calls for the availability of bridges with long spans and heights sufficient to accommodate water traffic. Table 21.1 presents a list of nonstandard bridges in Indonesia.

TABLE 21.1   List of Nonstandard Bridges in Indonesia

No.	Name	Location	Main Span (m)	Total Length (m)	Year Built	Data
1	Continuous Concrete Bridges					
	Rantau Berangin	Riau	121	201	1972–1974	√
	Rajamandala	West Java	132	222	1972–1979	√
	Serayu Kesugihan	Central Java	128	274	1978–1985	
	Mojokerto	East Java	62	230	1975–1977	
	Arakundo	Aceh	96	210	1987–1990	√
	Tonton–Nipah	Riau Isles	160	420	1995–1998	√
	Setoko–Rempang	Riau Isles	145	365	1994–1997	√
	Siti Nurbaya	West Sumatra	76	156	1995–2002	
	Tukad Bangkung	Bali	120	360	2006	√
	Air Teluk II	South Sumatra	104	214	2006	√
	Perawang	Riau	180	1473	2007	√
2	Continuous Steel Bridges					
	Ampera	South Sumatra	75	1100	1962–1965	√
	Danau Bingkuang	Riau	120	200	1968–1970	√
	Siak I	Riau	52	350	1975–1977	√
	Kapuas Timpah	Central Kalimantan	105	255	2010	√
3	Concrete Arch Bridges					
	Karebbe	South Sulawesi	60	60	1996	√
	Serayu Cindaga	Central Java	90	214	1993–1998	√
	Rempang–Galang	Riau Isles	245	385	1995–1998	√
	Besuk Koboan	East Java	80	125	2000	√
	Pangkep	South Sulawesi	60	84	2006	√
	Bajulmati	East Java	60	90	2007	√
4	Steel Arch Bridges					
	Malo	East Java	128	178	2007	
	Ogan Pelengkung	South Sumatra	100	100	2008	
	Kahayan Hulu	Central Kalimantan	80	160	2007	
	Kahayan	Central Kalimantan	150	640	1995–2000	√
	Rumbai Jaya	Riau	150	780	2003	√
	Barito Hulu	Central Kalimantan	150	561	2008	√
	Martadipura	East Kalimantan	200	560	2004	√
	Mahulu	East Kalimantan	200	800	2003	
	Palu IV	Central Sulawesi	125	300	2006	√
	Rumpiang	South Kalimantan	200	754	2008	√
	Batanghari II	Jambi	150	1351	2009	√
	Teluk Masjid	Riau	250	1650	2012 (construction finished)	√
	Siak III	Riau	120	518	2011 (construction finished)	
	Pela	East Kalimantan	150	420	2010	√
	Sei Tayan	West Kalimantan	200	1420	Under construction	√
5	Cable-Stayed Bridges					
	Teuku Fisabilillah	Riau Isles	350	642	1998	√

TABLE 21.1 (*Continued*) List of Nonstandard Bridges in Indonesia

No.	Name	Location	Main Span (m)	Total Length (m)	Year Built	Data
	Pasupati	West Java	106	2282	2005	√
	Grand Wisata Overpass	West Java	81	81	2007	√
	Siak Indrapura	Riau	200	1196	2007	√
	Siak IV	Riau	156	699	Under construction	√
	Sukarno	North Sulawesi	120	622	Under construction	√
	Melak	East Kalimantan	340	680	Under construction	√
	Galalapoka	Maluku	150	1065	Under construction	√
6	Suspension Bridges					
	Old Bantar	Yogyakarta	80	176	1932	√
	Memberamo	Papua	235	235	1996	√
	Barito	South Kalimantan	240	1082	1997	√
	Kartanegara	East Kalimantan	270	714	2001	√
	Balang/Teluk Balikpapan	East Kalimantan	708	1344	Design	√
7	Special Bridges Projects					
	Kelok 9	West Sumatra	90	5 km	2013	√
	North Java Flyover	Java	Varies	Varies	Under construction	√
	Suramadu	East Java	434	5438	June 10, 2009	√
	Bali Strait	East Java–Bali	2000	2000	Preliminary study	√
	Sunda Strait	Banten	2.2 km	29 km	Preliminary design	√

21.2 Design Practices

21.2.1 General Information

Prior to the 1970s, the Department of Public Works had guidelines called "Peraturan Beban PU" as the load regulations for bridge design. Now, we call it load regulation "PU Lama." In 1970 the Directorate General of Highways (Ditjen Bina Marga) issued regulations about more advanced loading specifications for highway bridges (Peraturan Muatan untuk Djembatan Djalan Raya, No. 12/1970), as a formal reference for bridge designs and construction in Indonesia. From 1971 to the 1990s, there was no significant change in the regulations. Even if there had been, it would be limited to efforts to improve the existing regulations, such as improving the regulations of earthquake specifications for highways and bridges, or improving the existing load regulations (No. 12/1970) to become codes for load designs of highway bridges (Tata Cara Perencanaan Pembebanan Jembatan Jalan Raya, SNI-03-1725-1989).

In 1989, there was collaboration between Indonesia and Australia to produce complete bridge design codes. This collaboration lasted for a fairly long time so that in 1992, not less than 17 modules, popularly known as Bridge Management Systems 1992 (BMS-92), were created. Those modules were relatively complete because they covered all activities in bridge management, starting from managerial activities and bridge operations, including bridge design codes and how to use the manual. The manual, which also showed how to use the modules, could in fact become a practical guideline to choosing and determining construction types. This really simplified preliminary bridge designs. Since the scopes and substance of the discussions were very broad, BMS-92 made it possible for designers to conduct their activities in designing bridges, particularly ones that extended up to 200 m.

BMS-92 adopted modern design concepts by applying the limit-state design concept. A limit-state design is a term used to describe a design approach in which all of the ways a structure may fail are taken into account. Failure is defined as any state that makes a design objective infeasible (i.e., it will not work for its intended purpose). Such failures are usually grouped into two main categories (or limit states): ultimate limit state (ULS) and serviceability limit state (SLS). This analytical approach was obviously different from its predecessor, which commonly utilized working stress designs, which were based on the elasticity theory. Although these modern design concepts seemed more complex at first, the limit-state design philosophy is a more rational approach than the working stress design approach. A design produced by application of limit-state principles will be more economical and will result in bridges of more uniform capacities or strength reserves. Therefore, the designs tend to be more efficient and reliable. Half of the manuals of BMS-92 have been transformed to Indonesia's Standard (SNI) since 2001, especially for the regulations of bridge designs, such as stress requirements for concrete and steel.

21.2.2 Code References

The current bridge design codes in Indonesia are as follows:

- Bridge design code BMS-92 with revisions as follows:
  - Part 2, Bridge Loads (SK.SNI T-02-2005), referring to Kepmen PU No. 498/KPTS/M/2005
  - Part 6, Reinforced Concrete Designs for Bridges (SK.SNI T-12-2004), referring to Kepmen PU No. 260/KPTS/M/2004
  - Part 7, Steel Designs for Bridges (SK.SNI T-03-2005) referring to Kepmen PU No. 498/KPTS/M/2005
- Standard Designs on Earthquake Strength for Bridges (Revised SNI 03-2883-1992)
- Bridge Design Manual BMS-92
- Bridge Design Criteria (referring to the letter of the Directorate General of Highways No. UM 0103-Db/242, March 21, 2008)

The above Bina Marga codes include highway and pedestrian bridge designs. Long-span bridges (more than 200 m long) and bridges with special structures or that are built with new materials and methods will be treated as special bridges.

21.2.3 Bridges Loads

21.2.3.1 General Information

Bridge design loads, which cover permanent, traffic, and environmentally induced loads, should refer to the load regulations for bridges (SK.SNI T-02-2005), which is in compliance with Kepmen PU No. 498/KPTS/M/2005, which was a revision of part 2 of BMS-92.

21.2.3.2 Design Vehicle Load

In considering vehicle actions, the vehicle load has three components: a vertical component, a braking component, and a centrifugal component (for horizontally curved bridges). The traffic loads for the design of bridges consist of lane load D and truck load T. The D lane load applies to the entire width of the vehicle's lane with the intensity of the load by 100% for the number of lanes (n) and 50% for the rest of the vehicle floor space available. This load is used to calculate the effects on the bridge equivalent to a convoy of actual vehicles. The total number of the D lane load that works depends on the width of the vehicle's own lane. The T truck load is a heavy vehicle with three axes which is placed in several positions in the traffic lane design. Each axis consists of two loading contact areas which are meant to be a simulation of the influence of the heavy vehicle's wheels. The T load is used more for the calculation of the strength capacity of the deck system of the bridge. In general, the D load will govern the design of medium- to long-span bridges, whereas the T load is used for short spans and deck systems.

21.2.3.3 D Lane Load

The D lane load (Figure 21.2) consists of

1. A uniform distributed load (UDL) of intensity q kPa, where q depends on the total loaded length L as follows:

$$L \leq 30 \text{ m} \quad q = 9.0 \text{ kPa} \tag{21.1}$$

$$L > 30 \text{ m} \quad q = 9.0 \times \left(0.5 + \frac{15}{L} \right) \text{ kPa} \tag{21.2}$$

The UDL may be applied in broken lengths to maximize its effects. In this case L is the sum of the individual lengths of the broken load. The D lane load is positioned perpendicular to the direction of traffic.

2. A knife edge load (KEL) or line load of p kN/m, placed in any position along the bridge perpendicular to the traffic direction:

$$q = 49.0 \text{ kN} / \text{m} \tag{21.3}$$

In continuous spans, the KEL is placed in the same lateral position perpendicular to the traffic direction in two spans to maximize the negative bending moments.

21.2.3.4 T Truck Load

The T truck load consists of a semitrailer truck vehicle that has a structure and axis weight as shown in the Figure 21.3.

Only one truck shall be placed in any design traffic lane for the full length of the bridge. The T truck shall be placed centrally in the traffic lane. The maximum number of the design traffic lanes is given in Table 21.2. These lanes are placed anywhere between the curbs.

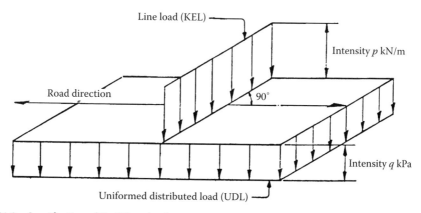

FIGURE 21.2 Specification of the D lane load.

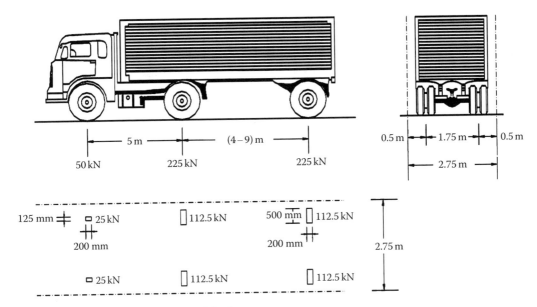

FIGURE 21.3   Specification of the T truck load.

TABLE 21.2   Number of the Design Traffic Lanes

Bridge Type	Bridge Roadway Width (m)	Number of Design Traffic Lanes
Single lane	4–5	1
Two-way, no median	5.5–8.25	2
	11.25–15	4
Multiple roadway	10–12.9	3
	11.25–15	4
	15.1–18.75	5
	18.8–22.5	6

TABLE 21.3   Dynamic Allowance for the D Lane KEL

Equivalent Span L_E (m)	DLA (for both limit states)
$L_E \leq 50$	0.4
$50 < L_E < 90$	$0.525 - 0.0025L_E$
$L_E \geq 90$	0.3

21.2.3.5 Dynamic Load Allowance

A dynamic load allowance (DLA) is applied to the D lane KEL and the T truck to simulate the impact of the moving vehicles on the bridge structure. The DLA applies equally to the SLS and ULS to all parts of the structure to the foundation. For the T truck, the DLA is 0.3. For the D lane, the KEL is listed in Table 21.3.

For a simple span $L_E =$ actual span length; for a continuous span, use $L_E = \sqrt{L_{av} \cdot L_{max}}$, where L_{av} is an average span length of the continuous spans and L_{max} is the maximum span length of the continuous spans.

21.2.3.6 Braking Forces

Braking and acceleration effects of the traffic shall be considered as longitudinal forces. This force is independent of the bridge width but is influenced by the length of the bridge structures.

21.2.3.7 Collision Loads on Bridge Support System

The bridge support system in traffic areas shall be designed to resist accidental collisions or be provided with a special protective barrier.

21.3 Standard Superstructures

21.3.1 General

For small spans without specific requirements, the Bina Marga bridge standard is commonly selected based on the economic span and the conditions of the water traffic beneath it. Some design standards for typical bridge superstructures (Directorate General of Highways 2005) are discussed next.

21.3.2 Short-Span Bridges (1–25 m)

The available standardization of short span bridges is

- Reinforced concrete rectangular box culverts, with spans of 1–10 m (see Figure 21.5)
- Pretensioned precast concrete flat slabs, with spans of 5–12 m (see Figure 21.6)
- Pretensioned precast concrete voided slabs, with spans of 5–16 m (see Figure 21.7)
- Steel composite girders, with spans of 8–20 m
- Reinforced concrete tee girders, with spans of 5–25 m (see Figure 21.8)

The square culvert or concrete box culvert is the simplest option for bridges, based on the idea of how to utilize the water channels across the street (the sewer) to serve as well as a bridge in supporting traffic loads. Because of its square shape, the bottom part of the foundation and the top part of the bridge deck are identical. The upper side of the concrete box culvert can be directly used as traffic lanes but if the river has a steep river bank, earth fills can first be added on it.

The Patukki II Bridge in Sulawesi (Figure 21.4) is only a 6 m span and therefore adopts the box culvert single-cell type. For a bridge with a larger span, double- or triple-cell box culverts can be adopted. Figure 21.5 shows a detail of a double-type box culvert. A box culvert structure is relatively simple in

FIGURE 21.4 The Patukiki II Bridge (6 m), Sulawesi.

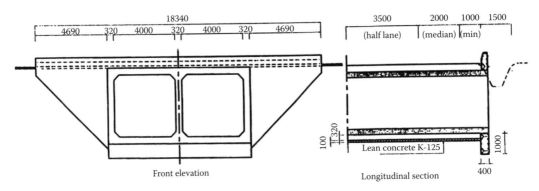

FIGURE 21.5   Double-type box culvert (example of 8 m span).

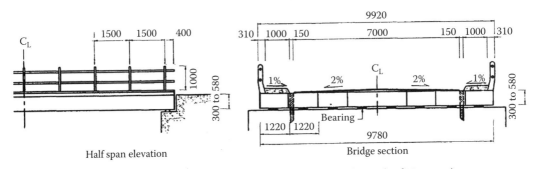

FIGURE 21.6   Precast prestressed concrete solid slab for spans of 5–12 m (example of 12 m span).

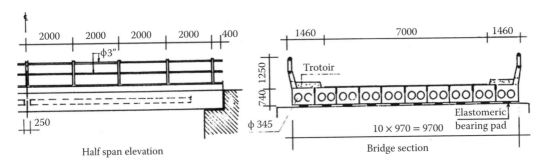

FIGURE 21.7   Precast prestressed concrete voided slab for spans of 5–16 m (example of 16 m span).

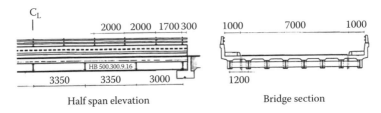

FIGURE 21.8   Composite steel bridge for a span of 8–20 m (example of 20 m span).

design, but the construction work at the bottom side is relatively more complex compared to the upper side. If the length of the span is adequate, the river is relatively shallow, and the soil conditions underneath are good enough and easy to dry, then construction is relatively easy.

These systems are often found in constructing bridges over irrigation system, drainages, or creeks, which are quite prevalent in Indonesia. If the river conditions do not allow building of a unified bottom structure, such as for a box culvert, then the bridge needs special substructures and to use a super-structure that follows the reference of the Bina Marga standard. The use of precast prestressed concrete (PC) elements for bridge superstructures would certainly facilitate and accelerate bridge construction. However, precast elements are only economical if they are in large quantities and transportation is available. If this is not possible, then the use of composite steel girders or cast-in-situ reinforced concrete girders can become an alternative.

21.3.3 Prestressed Concrete Girders for Spans of 20–40 m

21.3.3.1 General

Concrete structures are popularly used on various development projects in Indonesia, including bridge projects, mainly because aggregate and sand, which are the main materials for concrete structures, are easy to get and are affordable. Another reason is that currently there are more cement factories than steel factories in Indonesia. If the conditions make it possible, concrete bridges, especially using the prestressed concrete structure system, are the best choice for bridges, specifically for the middle spans. Further, the future maintenance costs of concrete bridges are more economical compared to those of steel bridges.

21.3.3.2 Bina Marga Standard of Prestressed Concrete Girders

The Directorate General of Highways (Direktorat Jenderal Bina Marga) provides standard designs of prestressed concrete I-girder bridges for spans of 20 m to 40 m (Figure 21.9) with a traffic lane width of 7 m, footway width of 2×1 m, and distance between the two outer edges of the back of 9.92 m.

21.3.3.3 Tol Cipularang Bridges

As evidenced, concrete bridges, especially prestressed concrete bridges, are popular in Indonesia; there-fore, the bridges of the Cipularang toll road project (phase II) will be shown. This is a special project within the framework of the Indonesian government in preparing for the Asian-African Summit (2005) in Bandung. All of the bridges built in the project are prestressed concrete girder bridges.

The Cipularang toll road project (Cikampek–Purwakarta–Padalarang) constitutes a toll road which connects the cities of Jakarta and Bandung, particularly the new road segments from Cikampek–Purwakarta until Padalarang. The construction was divided into two phases: phase 1, Cikampek–Sadang and Padalarang–Cikamuning (17.5 km) and phase 2, Sadang–Cikamuning (41 km). The construction of

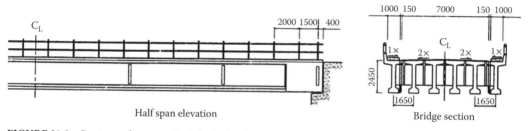

FIGURE 21.9 Prestressed concrete I-girder bridge for a span of 20–40 m (example of 35 m span).

phase 1 was 17.5 km along a relatively flat area and there are no large rivers, so there was no construction of long-span bridges. On the other hand, the construction of phase 2 was 41 km and the area covers some mountains with some steep rivers, so many long and tall bridges were required. Information on the bridges is given in Table 21.4.

Although the terrain is challenging, by cutting through the mountains and steep rivers and due to the anticipated Asian-African Summit in Bandung in April 2005, the project was successfully completed only in one year. With this new toll road, the distance between Jakarta and Bandung, which was previously traveled in about 3 to 4 hours, is now shortened to 1.5 to 2 hours, especially for a normal trip (if there are no traffic jams).

The Cikao Bridge (Figure 21.10) is the first bridge on the segment of phase II of the Cipularang toll road project from Jakarta to Bandung. The Cikao Bridge configuration is a simply supported standard prestressed concrete I-girder bridge. The Ciujung Bridge (Figure 21.11), which is located on the Plered–Cikalong Wetan segment, is a total length of 500 m. The bridge is the second bridge on the segment of phase II of the Cipularang toll road project from Jakarta to Bandung. The Ciujung Bridge configuration is a simply supported standard prestressed concrete I-girder bridge.

The Cisomang Bridge (Figure 21.12) is the third bridge on the segment of phase II of the Cipularang toll road project from Jakarta to Bandung. The superstructure configuration is a combination of simply supported and continuous beams using a precast prestressed concrete bulb T-girder system. The continuous beam structure was achieved by post-tensioning precast segments and embedded directly to the pier head. The integral system of the bridge should be adopted due to high pier characteristics and the requirement that the structure be seismic resistant (Imran et al. 2005).

The Cikubang Bridge (Figure 21.13) is one of the bridges on phase II of the Cipularang toll road, and is interesting in terms of location and method of construction. The bridge, with its highest piers of 59 m and total length of 520 m, is located in the village of Cikubang, Cikalongwetan district, Bandung

TABLE 21.4   Bridges in Phase II of the Cipularang Toll Road Project

Bridge		PC I-Girder		Pier		Location
Name	L (m)	Number	L (m)	Number	h (m)	(km)
Cikao	80	28	35, 45	1	14	81 + 200
Ciujung	500	154	40	11	5–30	95 + 225
Cisomang	252	98	25, 40	6	7–46	100 + 695
Cikubang	520	156	26, 40, 42.8	12	5–59	109 + 535
Cipada	707	192 + 12	40	16	5–35	111 + 804

FIGURE 21.10   The Cikao Bridge (80 m) at km 81 + 200 of the Cipularang toll road, West Java.

FIGURE 21.11   The Ciujung Bridge (500 m) at km 95 + 225 of the Cipularang toll road, West Java.

FIGURE 21.12   The Cisomang Bridge (252 m) at km 100 + 695 of the Cipularang toll road, West Java.

FIGURE 21.13   The Cikubang Bridge (520 m), under construction (2004–2005).

regency. The bridge crosses a valley, the Cikubang River, and railways. The bridge carries 2 × 2 lanes of traffic with a radius of horizontal curvature of 1200 m. The structural system of Cikubang Bridge is similar to Cisomang Bridge except for the type of precast concrete girder. The Cikubang Bridge uses standard prestressed concrete I-girders. A similar system is also used in the Cipada bridge (see Figure 21.14).

21.3.4 Steel Bridges for Spans of 40–60 m

21.3.4.1 General

The most popular bridge type in Indonesia is the steel bridge. From 1970 to 1990, steel bridges with a total length of about 237 km were imported from various sources, for instance, the United Kingdom (Callender Hamilton, Compact Bailey), the Netherlands (Hollandia Kloos), Australia (Transfield and Trans Bakrie), and so on. At that time, Indonesia was expected to reach a certain target of road links to open isolated areas with many kind of conditions; therefore, the faster way was to construct standard steel truss bridges. By 2000, the use of steel bridges was declining, although demand was still high in areas with limited transportation access because, after all, steel bridges are superior in terms of ease of construction and uniformed achievements in quality.

21.3.4.2 Bina Marga Standard of Steel Trusses

The Department of Public Works, through the Directorate General of Highways (Ditjen Bina Marga), provides standard designs for steel truss bridge spans of 40 m to 60 m and some up to 80 m and 100 m. The quality of steel material for the main structure of the bridge is SM 490/BJ 55 ($fu = 550$ MPa and $fy = 410$ MPa) and for others is SS400/BJ 50 ($fu = 510$ MPa and $fy = 290$ MPa). This means that the strength of the steel material for the bridge is higher than typical steel material used for the building construction, which is generally BJ 37 ($fu = 370$ MPa and $fy = 240$ MPa). Figure 21.15 shows the Binamarga standard steel bridge.

One of the main reasons for using a steel truss bridge is due to the high strength-to-weight ratio, so that the structure is relatively lighter, which results in smaller foundations. Another special feature is that steel truss elements are prefabricated in workshops and assembled on the project site by bolting small segments together. Standardized elements and products have made the segments easy to put

FIGURE 21.14   The Cipada Bridge (707 m) at km 111 + 804 of the Cipularang toll road, West Java.

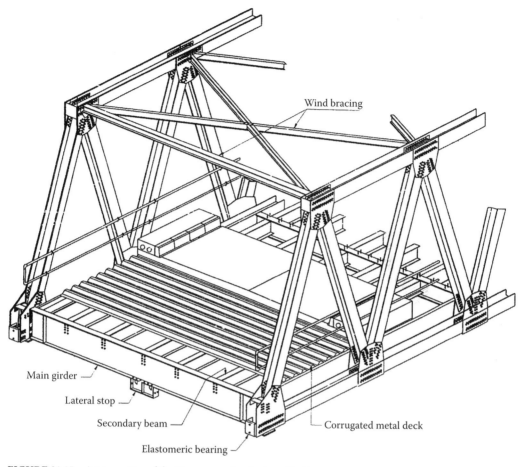

Wind bracing

Main girder

Lateral stop

Secondary beam

Elastomeric bearing

Corrugated metal deck

FIGURE 21.15   A perspective of the Binamarga standard steel bridge.

together with high quality control and high precision. The typical erection of structural steel segments to form a bridge on-site is to place a falsework (temporary support) in the middle of the river while the superstructure is being assembled, as shown in Figure 21.16. After the bridge is erected and before pouring the concrete deck, the falsework is removed. This allows the superstructure to deflect as designed when the deck is poured.

The biggest advantage of this method is that there is no need for additional anchor spans linking the kit or kentledge (counterweight), which the piece-by-piece cantilever method employs. In addition, there is no need for any heavy lifting equipment as the heaviest component is only ± 1.5 tons. However, it is a labor intensive method with a minimum amount of lifting equipment required. On many sites, the existing bridge can be used as the basis for the falsework support to reduce construction costs. In general, a falsework pier or trestle is set up under each cross girder, with a space of about 5 m. Since the segments of the steel truss elements are prefabricated in workshops and they are only assembled by using bolts, the construction speed is relatively faster when compared with that of a concrete bridge.

The use of the falsework system will certainly create a problem if the river is too deep or there is heavy traffic in the river, or if floods may suddenly occur in the wet season. Construction seasons should also be considered carefully. When using a falsework is infeasible, a standard steel truss bridge can be used as a construction tool on the site, namely by using the method of piece-by-piece cantilever construction, as shown in Figure 21.17. To use this method, it is necessary to have the following tools: additional anchor

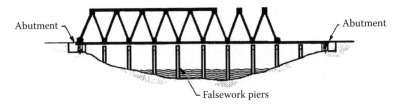

FIGURE 21.16 Erection of the falsework.

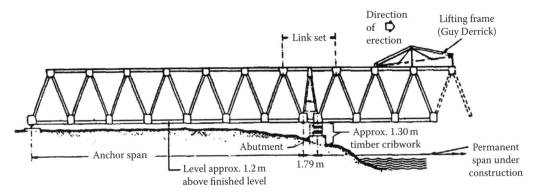

FIGURE 21.17 Piece-by-piece cantilever construction of steel truss bridges.

spans, linking kits, and a kentledge (counterweight). Since it is necessary to have a steel truss bridge to serve as the anchor span, there are two structural frameworks of the bridge; it is economical to construct a bridge consisting of two or more spans. It is thus appropriate to be used over a deep river or over a river with heavy water traffic.

21.3.4.3 Other Standards for Steel Bridges

Although the Directorate General of Highways (Ditjen Bina Marga) has its own standards for steel truss bridges, each with several benefits, there are also various standards for steel truss bridges which vary depending on the origin country. Table 21.5 lists a number of steel bridges by country, together with their construction periods, with a total length of 296.7 km.

From Table 21.5, note that the most widely used type of standard steel truss bridge is the Australian (Transfield and Trans Bakrie) and was built from 1984 to 1993. This type of bridge is about 35% of the total number of standard steel bridges constructed until the year 2007. So, it is quite understandable why Australia was interested in helping Indonesia produce the most comprehensive bridge regulations, BMS-92. The Australian (Transfield and Trans Bakrie) system (Figure 21.18) provides bridge spans with a range of 35 m to 60 m of through-truss designs. The permanent spans are supplied in three classes, A, B, and C, which only differ in roadway width and curb/footway configurations. The spans in all classes have composite reinforced concrete decks. This bridging system is planned to have low maintenance characteristics. To this end, all of the steelwork and bolts are galvanized and the bearings are elastomeric.

The Australian system has more advantages over the Bina Marga system, especially in the construction methods, in which the single-span launching (SSL) method (Figure 21.19) can be used. With this method of erection, the truss span is completely assembled on one bank and rolled out into position using an anchor and kentledge (counterweight). No falsework is required within the crossing since the span is designed to be fully cantilever. The SSL method is suitable for a single span or the first span

TABLE 21.5 Construction of Steel Truss Bridges in Indonesia

No.	Country/Bridge System	Year	Production (m)
1	Britain (Callender–Hamilton bridge)	1970–1980	17,269
2	Britain (compact Bailey bridge)	1986	12,108
3	Britain (compact Bailey bridge)	1990	5,100
4	Holland/the Netherlands (Hollandia Kloos)	1979–1993	32,130
5	Australia (Transfield and Trans Bakrie)	1984–1993	104,480
6	Indonesia (Marubeni, KBI and Bakrie IBRD, Loan No. 2717–IND)	1985–1987	11,175
7	Indonesia (BTU/Credit Export)	1996	26,740
8	Austria (Wagner Biro)	1987–1998	56,495
9	Spain (Centurion and PT. Wika)	1998–2002	8,055
10	Indonesia (KBI/OECF, Loan IP–444)	1999	4,500
11	Indonesia (Trans Bakrie and DSD/OECF, Loan IP 444)	2001	3,715
12	Indonesia (Trans Bakrie/IBRD, Loan No. 4643–IND)	2002	3,926
13	Indonesia (Bukaka and WBI/APBN)	2003	2,045
14	Indonesia (Bukaka and WBI/APBN)	2004	1,765
15	Indonesia (Bukaka and WBI/APBN)	2005	1,465
16	Indonesia (WBI WBBB/IBRD, Loan No.4744 IND)	2005	1,230
17	Indonesia (KBI, WBI, and BTU/APBN)	2006	2,230
18	Indonesia (KBI and BTU/APBN)	2007	2,235

Source: Vaza, H. 2008. *Indonesia Bridges: Now and Later.* Directorate of Technical Affairs, Directorate General of Highways, Department of Public Works (in Bahasa Indonesian).

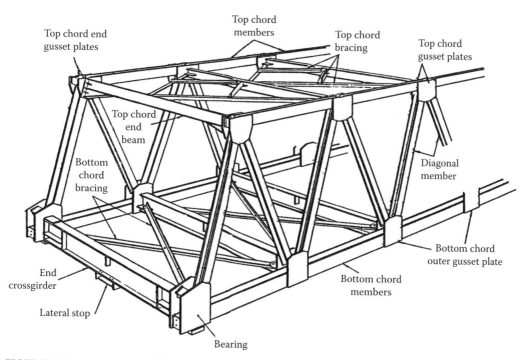

FIGURE 21.18 A perspective of the Australian (Transfield and Trans Bakrie) steel bridge.

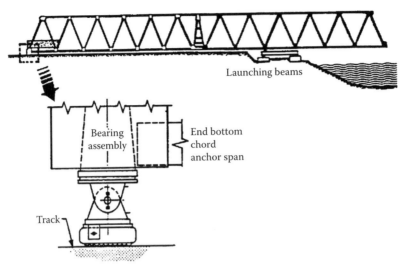

FIGURE 21.19   A single-span launching method and its device.

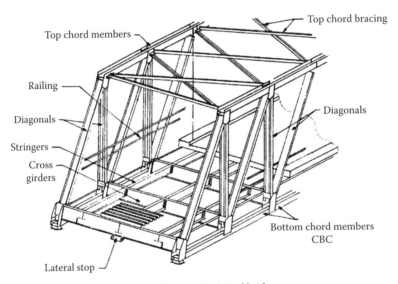

FIGURE 21.20   A perspective of the Austrian (Wagner Biro) steel bridge.

of a multispan bridge. It is particularly suited to a bridge site of one span which cannot be erected on falsework. Not all bridge sites are suitable for this system, because a longer assembly area is required on the bank, from which the launching process is carried out, compared to the piece-by-piece cantilever method, where no assembly area is needed on the river bank other than what has been specified previously for the assembly of the anchor span. The additional area required for the SSL method is due to the need for rolling tracks, which must be constructed to accommodate both the main span and anchor span. The area required on the river bank will depend on the length of the main span and anchor span plus the work area surrounding the spans.

The second most used type of standard steel truss bridges is the Austrian (Wagner Biro) system, which covers approximately 19% of the total number of standard steel truss bridges built so far. The Austrian system of truss bridges (Figure 21.20) comprises precision-made standard steel components which are assembled by bolting together to form a bridge span of a continuous truss design in a range of 35 m to

60 m. The various spans available are classes A, B, and C, which differ in roadway widths and curb/footway configurations. The spans in all classes have concrete decks supported by corrugated trapezoidal steel sheets, supplied as part of the bridge system. The system has been designed to permit progressive assembly by cantilever working from one bank, without the use of a falsework on the river. Other methods of assembly and erection, such as part cantilever or erection on a falsework, are feasible. The Austrian bridge system is designed to be of low maintenance, so all steel work and bolts are galvanized.

The third most used type of standard steel truss bridges is the Dutch (Hollandia Kloos) system (Figure 21.21), which covers approximately 11% of the total number of bridges. Its unique features, if compared with the other standard steel-framed bridges, are the availability of 40 m to 105 m bridge spans, which is approximately 175% longer than existing standard steel truss bridges. In addition, just like the other standard steel truss bridges, which require low maintenance, all the steel components and bolts are galvanized.

21.3.4.4 Steel Truss Standards for Crossing Wide Rivers

The maximum span of a standard steel truss is generally 60 m. However, a standard steel truss bridge is also often used on bridges over wide rivers, especially if bridge piers can be built on the river. There are a lot of Australian (Transfield & Trans Bakrie) steel truss bridges built along the provincial roads on the island of Java and some also on other islands too. The following bridge projects will illustrate their application.

21.3.4.4.1 Bantar II Bridge

The Bantar II Bridge is located on the Progo River, Bantar village, Wates, between the roads of Yogyakarta and Purworejo. The bridge total length is 226 m, composed of four spans (51.4 m + 2 × 61.6 m + 51.4 m) with a width of 9 m (1 m + 7 m + 1 m). The maximum span length is 61.6 m. This bridge is the one that has taken over the role of the old Bantar Bridge, a suspension bridge inherited from the Dutch (before the independence of Indonesia). The image in Figure 21.22 was taken from the old Bantar bridge. At the time the photo was taken, the Bantar III Bridge (prestressed concrete girder), located between the old Bantar Bridge and the Bantar II Bridge, was not built. Currently the old Bantar Bridge, Bantar II Bridge, and Bantar III Bridge are lined up in parallel in that location.

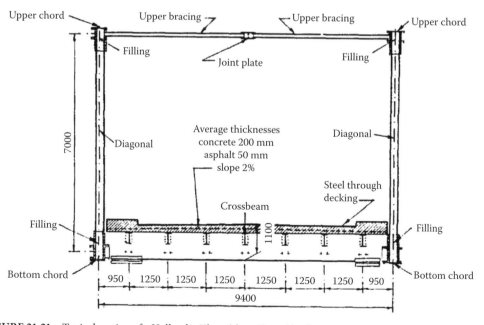

FIGURE 21.21 Typical section of a Hollandia Kloos (class A) steel bridge.

FIGURE 21.22   Bantar II Bridge (226 m), Yogyakarta.

FIGURE 21.23   Koto Panjang Bridge (295 m), Riau.

21.3.4.4.2 Koto Panjang Bridge

The Koto Panjang Bridge, in Kampar regency, Riau, Sumatra, was built over a lake that formed as a result of a barrier of the Kampar River for the construction of a dam for a hydropowered electricity source for Koto Panjang (Figure 21.23). The bridge total length is 295 m, composed of five spans with a width of 6 m. The maximum span length is 60 m.

21.3.4.4.3 Berbak Bridge

The Berbak Bridge, located at a distance of 40 km from the city of Muarosabak and 70 km from the city of Jambi, was precisely situated in the district of Berbak, Tanjung Jabung Timur regency, Jambi, Sumatra. The steel truss bridge (Figure 21.24), with a length of 369 m and a width of 9 m, crosses the Batanghari River. Construction started in 2003 and was completed in 2007. This bridge is a support bridge for the Batanghari II Bridge in the direction of East Jambi province, providing public access to residents in Nipang Panjang, Rantau Rasau, and Berbak when they travel to the city of Jambi. The bridge is also to support the growth of the port of Muara Sabak, where loading and unloading containers is a big business.

21.3.4.4.4 Batanghari I Bridge

The Batanghari I Bridge (Figure 21.25) is located on the Batanghari River, in the village of Aur Duri, district of Telanaipura, five kilometers west of the Jambi city center, Sumatra. The bridge is the only link connecting the economic and social interests from Jambi province to Riau via the eastern traffic lines.

21.3.4.4.5 Mahakam Bridge

The Mahakam Bridge (Figure 21.26), which is also called the Mahkota I Bridge, is a bridge which connects the off-shore Samarinda region with the city of Samarinda, Kalimantan. The Mahakam Bridge was

FIGURE 21.24 Berbak Bridge (369 m), Jambi.

FIGURE 21.25 Batanghari I Bridge, Jambi.

FIGURE 21.26 Mahakam Bridge (400 m), Samarinda.

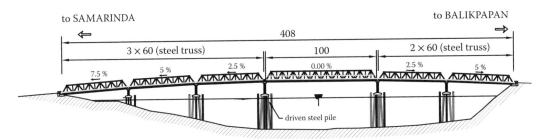

FIGURE 21.27   The Hollandia Kloos truss elevation of Mahakam Bridge.

FIGURE 21.28   Collision due to the RM 352 pontoon (Kompas, 24/01/2010).

built in 1987 and was inaugurated by President Soeharto. This bridge utilized a Dutch (Hollandia Kloos) type of steel frames, having a frame length of 60 m and 100 m, a total length of 400 m, and a total width of 10.3 m (Figure 21.27).

Although a standard steel truss bridge can also be used on a wide river, the 100 m span of the standard steel truss bridge is still too narrow for heavy water traffic. This will put the bridge piers at a high risk of vessel collisions. On Saturday, January 23, 2010, the Mahakam Bridge was hit by an RM 352 pontoon (Figure 21.28). Fortunately, the bridge survived without collapse. A nonstandard bridge with a larger span is required to avoid vessel collisions.

21.4 Concrete Continuous Bridges

21.4.1 Rantau Berangin Bridge (198 m), Riau (1974)

The Rantau Berangin Bridge is a modern prestressed concrete box girder bridge constructed in 1972–1974 by the cantilever segmental system (freie vorbau) and which was the first bridge of its kind to be built in Indonesia (Figure 21.29). It crosses the Batanghari River, Riau Province, Sumatra. The total length of the bridge is 201 m, the main span is 121 m, and the side spans (right and left) are 40 m.

The bridge designer was NV IBIS from the Netherlands. Prof. Dr. (HC) Ir. Roosseno Soerjohadikoesoemo, a professor of the Institute of Technology Bandung who was later known as the father of concrete structures in Indonesia, was actively involved as an expert advisor to monitor the implementation of the project and at the same time to learn the new techniques. Thus, the success of the bridge construction project of the Rantau Berangin Bridge not only produced the physical

FIGURE 21.29 The Rantau Berangin Bridge (201 m), the first cantilever segmental concrete bridge in Indonesia.

FIGURE 21.30 Rajamandala Bridge (222 m), Cianjur, West Java.

construction, but also became an important milestone in the transferring technology from other nations. This is, of course, very important for the self-reliance of the nation, particularly in the field of construction engineering.

The process of transferring the technology from the project, which was led by Professor Roosseno and also Lanneke Tristanto (now a research professor at the Department of Public Works). The construction of similar bridges in several other places in Indonesia without the need to involve foreign experts later proved how well the the knowledge and modern technologies has been absorbed. The bridges were fully constructed by Indonesian experts.

21.4.2 Rajamandala Bridge (222 m), Cianjur, West Java (1979)

The Rajamandala Bridge (Figure 21.30) is located on the Citarum River, Cianjur regency, West Java. If observed carefully, it is seemingly similar in shape to the Rantau Berangin Bridge in Riau. The only difference is in the bridge pier (hollow box cross section), which is higher. It could have indeed happened that way because the designer of this bridge was Professor Roosseno, who had previously been actively involved as an expert advisor in the construction of the Rantau Berangin Bridge. The bridge has a total length of 222 m with a central span of 132 m and side spans of 45 m (right and left). The bridge has a width of 9 m to carry two lanes of highway traffic. Since the successful construction of this bridge, Indonesian experts became more confident in engaging independently in large-span bridge projects, especially involving the construction of prestressed concrete bridges.

21.4.3 Arakundo Bridge (210 m), East Aceh, Aceh (1990)

The Arakundo Bridge (Figure 21.31) is located on the segment of the roads of Lhokseumawe–Langsa at km 342 + 580, East Aceh regency, Aceh province. The bridge was built between September 12, 1987 and July 31, 1990. The length of the bridge is 57 m + 96 m + 57 m = 210 m; the width is 1.5 m + 7 m + 1.5 m = 10 m. This superstructure is a continuous cast-in-place prestressed concrete box girder, constructed by the balanced cantilever method. The bridge was designed by Rendall Parkman, England and PT Indah Karya, Indonesia. This bridge was part of the Arakundo–Jambo Aye Irrigation and Flood Control Project that got a loan from the Saudi Fund for Development (SFD), with the purpose of developing irrigation, flood control, and the transportation roads surrounding the Jambo Aye River in Arakundo.

21.4.4 Tonton–Nipah Bridge (420 m), Riau Isles (1997)

The Tonton–Nipah Bridge (Figure 21.32), which connects the islands of Tonton and Nipah, is one of the few bridges built together in Batam and is also called the Barelang Bridge. The bridge was built to promote tourism and industry on the island of Batam. The Tonton–Nipah Bridge is a cantilever segmental prestressed concrete bridge with a total length of 420 m and a main span of 160 m. This bridge is one of the Barelang bridges.

FIGURE 21.31 Arakundo Bridge (210 m), Aceh.

FIGURE 21.32 Tonton–Nipah Bridge (420 m), Riau Isles.

21.4.5 Setoko–Rempang Bridge (365 m), Riau Isles (1997)

The Setoko–Rempang Bridge (Figure 21.33), which connects Setoko and Rempang islands, is one of the Barelang bridges. The bridge is a concrete box girder which was constructed by means of a balanced cantilever. The designer was LAPI ITB. The bridge, with a width of 18 m, a total length of 365 m, and a greatest span of 145 m, was the second-longest balanced cantilever bridge built in Indonesia at the time of its construction in 1997. There were two technological innovations for this bridge. The first was the marine foundation work and the creation of the traveler form module. The results of these innovations provided many benefits: being lighter, being stronger in carrying loads, being more easily manufactured and operated, having smaller effects of deflections, and resulting in faster molding cycle times per segment.

21.4.6 Tukad Bangkung Bridge (360 m), Badung, Bali Island (2006)

The Tukad Bangkung Bridge (Figure 21.34) is located at the village of Plaga, Petang district, Badung regency, Bali. It opened to the public on December 19, 2006. The bridge, which connects three districts, Badung, Bangli, and Buleleng, was the longest bridge in at the time. (The superstructure is

FIGURE 21.33 The Setoko–Rempang Bridge (365 m), Riau Isles.

FIGURE 21.34 Tukad Bangkung Bridge (360 m), Bali.

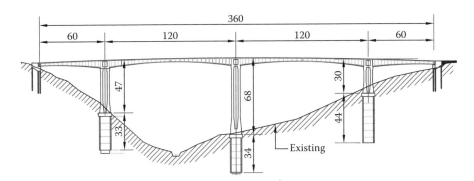

FIGURE 21.35   Elevation of the Tukad Bangkung Bridge.

a continuous prestressed concrete single-cell box girder built by balanced cantilever construction. The total length of the bridge is 360 m, with a long span of 120 m and a width of 9.6 m. The box girder is supported by double prestressed concrete piers 47–68 m high fixed connected on φ 9 m caisson foundations 33–44 m deep. The abutments are supported on concrete piles with a diameter of 60 cm (see Figure 21.35).

21.4.7 Air Teluk II Bridge (214 m), Sekayu, South Sumatra (2006)

The Air Teluk II Bridge (Figure 21.36), Sekayu, South Sumatra, is a concrete box girder balanced cantilever bridge, with spans of 104 m + 55 m + 55 m and a total length of 214 m. It was completed in 2006. Within a distance of 100 m from the bridge, there is an old concrete arch bridge built in the era of the Dutch colonialization. In order to maintain the historical value of the old bridge, therefore, an additional arch steel ornament, which does not function as a structural element, has been installed in the new Air Teluk II Bridge.

21.4.8 Perawang Bridge (1473 m), Siak, Riau (2007)

The Perawang Bridge is located in Maredan, Tualang district, Siak regency, Riau. The distance is approximately 80 km from the city of Pakanbaru. In this city, there is the second-largest paper company in Indonesia, PT Indah Kiat Pulp Paper. The bridge is a type of box balanced cantilever bridge

FIGURE 21.36   Air Teluk II Bridge (569 m), Sekayu, South Sumatra.

FIGURE 21.37 Site view of the Perawang Bridge (1473 m), Maredan, Riau.

with 14 piers and 4 abutments. The total span is 1473.4 m, consisting of the main span of 180 m, side spans of 101 m each, approach spans of 254 m each, slabs-on-pile of 321.7 m and 261.7 m, and a width of 12.7 m (Figure 21.37). The Perawang Bridge forms an alternative path from the eastern path of Sumatra, which is an alternative road connecting the towns of Simpang Lago, Perawang, and the city of Minas. This is a new path that is shorter because the local residents do not need to go around the city of Pakanbaru.

21.5 Steel Continuous Bridges

21.5.1 Ampera Bridge (1100 m), Palembang, South Sumatra (1965)

The Ampera Bridge (Figure 21.38) over the Musi River is an icon of the city of Palembang. It was constructed between April 1962 and May 1965 on a 350 m width of Musi river to connect the area of Plaju and Palembang. It was proposed as vertical lift bridge in Indonesia. The total length of the bridge is about 1100 m (including approach). The main bridge consists of spans of 22.5 m + 58.5 m + 58.5 m + 75 m + 58.5 m + 58.5 m + 22.5 m = 354 m. The bridge width is 22 m, including four lane of carriageway 4 × 3.5 m = 14 m, a bicycle way on both sides 2 × 1.75 m = 3.5 m, and sidewalks on both sides 2 × 2.25 m = 4.5 m.

The side spans comprise two-span continuous plate girders of 58.5 m, and two simple span plate girders of 22.5 m for each side. The approach bridge at the Palembang side consists of 27 m + 30 m = 57 m, and at the Plaju side is 27 m + 6 × 30 m + 27 m = 234 m. The tower of the bridge is 78 m high. The central part of the superstructure can be lifted up to allow large ships to pass (see inset in Figure 21.39). It consists of a simple plate girder 75 m long, with a weight of about 100 tons. However, this special mechanism no longer works. In fact, it was the only bridge in Indonesia bridge history with such a mechanism.

21.5.2 Danau Bingkuang Bridge (120 m), Kampar, Riau (1970)

The Danau Bingkuang Bridge (Figure 21.40) is over the Kampar River, Kampar regency, Riau. The bridge connects the cities of Pekanbaru and Bangkinang. The bridge is a part of the early construction of nonstandard bridges in Indonesia in the era of the 1960s. The Danau Bingkuang Bridge consists of three spans of continuous steel trusses with a configuration of 40 m + 120 m + 40 m. The bridge has a width of 9 m to carry two lanes of highway traffic.

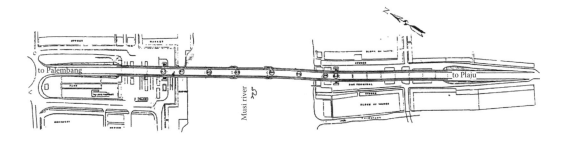

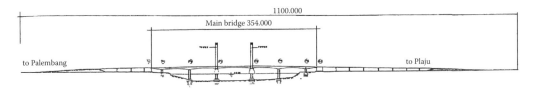

FIGURE 21.38 Plan and elevation of Ampera Bridge, Palembang, South Sumatra.

FIGURE 21.39 Ampera's main bridge (inset: the lifting of the central span).

FIGURE 21.40 Danau Bingkuang Bridge (120 m), Kampar, Riau.

21.5.3 Siak I Bridge (349 m), Riau (1977)

The Siak I Bridge (Figure 21.41 for the elevation and Figure 21.42 for the site view), better known as the Leighton Bridge, has a length of 349 m and connects the city of Pekanbaru with the coastal Rumbai district. This bridge is the first permanent bridge over the Siak River, was a contribution of PT Chevron Pacific Indonesia (formerly known as PT Caltex Pacific Indonesia), and construction was carried out by the Leighton Company from Australia between 1975 and 1977. The current capacity of the bridge is no longer sufficient to support the mobility of the local residents. Frequent traffic jams on the bridge have worsened the situation. As a result, more bridges have been built over the Siak River, notably the Siak II, Siak III, and Siak IV bridges. The inset figure in Figure 21.42 depicts a convoy of vehicles across the Siak River through the Poton Bridge, which is limited in its use. With the construction of the Siak I Bridge with a larger capacity, the development of the Riau region was significantly influenced.

21.5.4 Kapuas Timpah Bridge (255 m), Central Kalimantan (2010)

The Kapuas Timpah Bridge (Figure 21.43) over the Kapuas River is located in Lungkuh Layang, Timpah district, Kapuas regency, Central Kalimantan province. The bridge connects the road from Palangkaraya to Buntok. The bridge was constructed using a continuous steel truss that consist of three spans (62.5 m + 105 m + 62.5 m) and one side of approach composite steel girder 25 m long. The total length of the bridge is 255 m and the width is 9 m (1 m + 7 m + 1 m). Construction began in 2006/2007 and it opened to the public in April 2010.

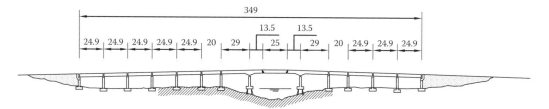

FIGURE 21.41 Elevation of the Siak I Bridge, Riau.

FIGURE 21.42 Siak I Bridge (349 m), Riau.

FIGURE 21.43   Kapuas Timpah Bridge (255 m), Central Kalimantan.

21.6 Arch Bridges

21.6.1 Concrete Arch Bridges

21.6.1.1 Karebbe Bridge (60 m), East Luwu, South Sulawesi (1996)

The Karebbe Bridge (Figure 21.44) is located in the road segment of Malili–BTS Sultra, km 574, Malili district, capital city of the East Luwu regency, South Sulawesi, Indonesia, about 565 km off Makassar. The bridge is a fixed-arch type structure, with total length 60 m and crown span ratio 1:5. The thickness of the arch at the support is 90 cm and at the crown (top center) is 60 m. For the construction of the bridge, the full shoring system was used.

21.6.1.2 New Serayu Cindaga Bridge (214 m), Banyumas, Central Java (1996)

The New Serayu Cindaga Bridge (Figure 21.45) is located in Banyumas District, Central Java and has a main span of 90 m; each side span is 31 m and has a width of 9 m. The foundation consists of prestressed concrete piles. The bridge was built from 1993 to 1998.

21.6.1.3 Rempang–Galang Bridge (385 m), Riau Isles (1998)

The Rempang–Galang Bridge (Figure 21.46), which is now called the Tuanku Tambusai Bridge, is part of the Barelang bridges, located on the island of Batam, Riau Islands province. This bridge has the form of a concrete arch bridge, the longest ever built in Indonesia, with a total length of 11 × 35 m = 385 m, with a bow span of 245 m, symmetrical side spans on either side of 35 m, and wide decks of 18 m. The bridge was built from 1995 to 1998.

21.6.1.4 Besuk Koboan Bridge (125 m), Lumajang, East Java (2000)

The Besuk Koboan Bridge (Figure 21.47), in Lumajang, East Java has a main span of 80 m and side spans of 20 m and 25 m, and was completed in 2000.

21.6.1.5 Pangkep Bridge (86 m), Pangkep, South Sulawesi (2006)

The Pangkep Bridge (Figure 21.48) is located in Pangkep regency, South Sulawesi. The bridge is in the road connecting Bungoro and the Maros region. The Pangkep Bridge has a length of 86 m, consisting of three spans (12 m + 60 m + 12 m) and width of 10 m.

FIGURE 21.44   Karebbe Bridge (60 m), East Luwu, South Sulawesi.

FIGURE 21.45   New (foreground) and old Serayu Cindaga Bridges, Banyumas, Central Java.

FIGURE 21.46   Tuanku Tambusai Bridge (385 m), Batam, Riau Isles.

FIGURE 21.47   Besuk Koboan Bridge (125 m), Lumajang, East Java.

FIGURE 21.48   Pangkep Bridge (86 m), Pangkep, South Sulawesi.

21.6.1.6 Bajulmati Bridge (90 m), Malang, East Java (2007)

The Bajulmati Bridge (Figure 21.49) is located in the Malang regency, precisely in the road segment of the southern coast line of East Java (662 km). Its construction was to support the improvement of the natural resource sectors and tourism, which have not been optimally explored in the southern region of East Java. Therefore, a unique form has been chosen for this bridge, namely the construction of a reinforced concrete suspension bridge with a single arch pylon as a supporter of the steel cable located in the middle. The Bajulmati Bridge has a length of 90 m, consisting of three spans (15 m + 60 m + 15 m) with a width of 15 m. With the Bajulmati Bridge, tourists going from Surabaya to tourist resorts around Malang are expected to continue their journeys to the Kondangmerak and Sendangbiru beaches, which have not been optimally explored, and at the same time they can save a distance of 80 km.

21.6.2 Steel Arch Bridges

21.6.2.1 Kahayan Bridge (640 m), Palangkaraya, Central Kalimantan (2000)

The Kahayan Bridge (Figure 21.50) is located on the segment of Palangkaraya–Buntok streets, crossing the Kahayan river, Palangkaraya, Central Kalimantan. The bridge was built from 1995 to 2002 to

FIGURE 21.49   Bajulmati Bridge (90 m), Malang, East Java.

FIGURE 21.50   Kahayan Bridge (640 m), Palangkaraya, Central Kalimantan.

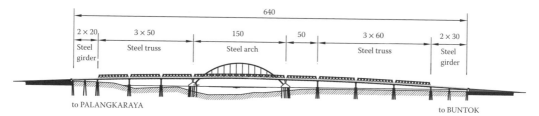

FIGURE 21.51   Elevation of the Kahayan Bridge, Palangkaraya, Central Kalimantan.

connect the city of Palangkaraya to the South Barito and North Barito regencies. The bridge has a length of 640 m and a width of 9 m (7 m wide for vehicles and a 1 m footway on both sides), consisting of 12 spans with a particular span of 150 m at the river cruise line (Figure 21.51). The Kahayan River has shipping traffic which requires a free space of 14 m from the highest water surface and 18 m from the normal water surface.

21.6.2.2 Rumbai Jaya Bridge (710 m), Indragiri Hilir, Riau (2004)

The Rumbai Jaya Bridge (Figure 21.52), which crosses over the Indragiri River in Indragiri Hilir, on the segment of Pekanbaru–Tembilahan streets, is the longest bridge in the area of the Riau mainland.

FIGURE 21.52   Rumbai Jaya Bridge (710 m), Indragiri Hilir, Riau.

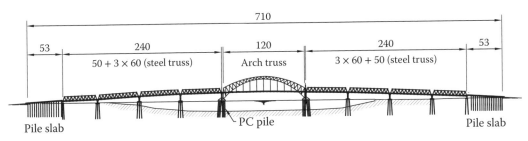

FIGURE 21.53   Elevation of the Rumbai Jaya Bridge, Indragiri Hilir, Riau.

FIGURE 21.54   Martadipura Bridge (569 m), Kotabangun, East Kalimantan.

The Rumbai Jaya Bridge has a total length of 710 m, with its largest span of 120 m and a width of 7 m (Figure 21.53). Its structural system is that of a steel truss arch bridge. The bridge was inaugurated by President Megawati on March 13, 2004.

21.6.2.3 Martadipura Bridge (569 m), Kotabangun, East Kalimantan (2004)

The Martadipura Bridge (Figure 21.54) in Kotabangun, East Kalimantan is the third bridge crossing the Mahakam River. The first is the Mahakam I Bridge in Samarinda and the second is the Kartanegara Bridge in Tenggarong. The Martadipura Bridge is the second bridge built of its kind in Indonesia, after the Rumbai Bridge in Riau, and the longest span ever built, with a main span of 200 m. The bridge width is 9 m and its total length is 569 m. Its main vertical clearance is 15 m. That superstructure consists of steel truss (SM 490 YB) with an arch height of 36 m (see Figure 21.55). The deck employs reinforced concrete K-350 U-40 and a substructure steel pipe with a diameter of 1000 mm. The approach spans are H-beam composite steel girders with span lengths of 2 × 6 × 30 m (see Figure 21.55) and its substructure is in the form of a steel pipe with a diameter of 600 mm.

21.6.2.4 Palu IV Bridge (300 m), Palu, Central Sulawesi (2006)

The Palu IV Bridge (Figure 21.56) spans the Talise Bay, linking the East Palu district and West Palu, in the city of Palu, Central Sulawesi. The bridge is located in the segment of the Palu Gulf coastal ring road. The bridge is also a landmark of the city of Palu because, philosophically, the curved shapes represent the two major mountains flanking the city. The bridge was inaugurated by President Susilo Bambang Yudhoyono in May 2006. The total length of the bridge is 300 m, with two main spans of 125 m in the form of twin steel squares, and a width of 7 m for vehicles. The links on the right and left are composite girders with measurements of 25 m (see Figure 21.57).

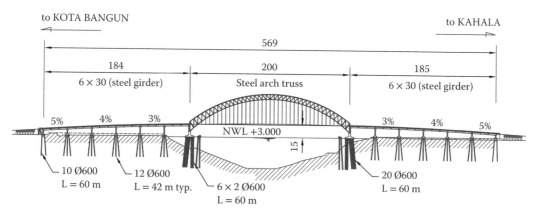

FIGURE 21.55   Elevation of the Martadipura Bridge, Kotabangun, East Kalimantan.

FIGURE 21.56   (**See color insert.**) Palu IV Bridge (300 m), Palu, Central Sulawesi.

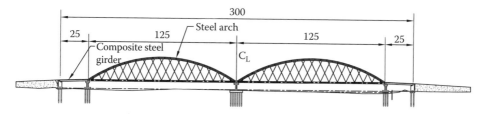

FIGURE 21.57   Elevation of the Palu IV Bridge, Palu, Central Sulawesi.

21.6.2.5 Barito Hulu Bridge (561 m), Puruk Cahu, Central Kalimantan (2008)

The Barito Hulu Bridge (Figure 21.58), which is now called the Merdeka Bridge, crosses over the Barito River, Puruk Cahu city, Murung Raya (Mura) regency, Central Kalimantan. This bridge connects the Puruk Cahu region with Muara Teweh, North Barito (Barut) regency and the surrounding areas. The total bridge length is 561 m and its width is 9 m, with a configuration as shown in Figure 21.59. The bridge was designed by PT Perencana Jaya, Jakarta, and construction took approximately five years (2003–2008). The main structure is a steel arch bridge truss, with spans of 62 m + 153 m + 62 m. At first glance, it looks like the Batanghari II Bridge, on the Batanghari River, Jambi, Sumatra.

21.6.2.6 Rumpiang Bridge (754 m), Barito Kuala, South Kalimantan (2008)

The Rumpiang Bridge (Figure 21.60) is located above the Barito River, Barito Kuala regency, South Kalimantan province. The construction of the bridge began on December 1, 2003, and was completed on April 25, 2008. The total length of the bridge is 754 m with a width of 9 m (see the elevation in Figure 21.61). The configuration is that of a main bridge flanked by approach bridges, which consist of several composite steel girders.

Next, with the advantages of standard steel truss bridges in general, an arc shape pointing upward means the bridge can be easily erected by applying piece-by-piece cantilever construction, without the need for the tools of complex constructions except for an additional temporary tower at the end of the bridge to put the pull cables to produce a cantilever effect, as seen in Figure 21.62 of the erection of

FIGURE 21.58   Merdeka Bridge (561 m), Puruk Cahu, Central Kalimantan.

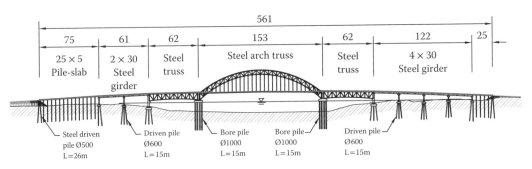

FIGURE 21.59   Elevation of the Merdeka Bridge, Puruk Cahu, Central Kalimantan.

FIGURE 21.60 Rumpiang Bridge (754 m), Barito Kuala, South Kalimantan.

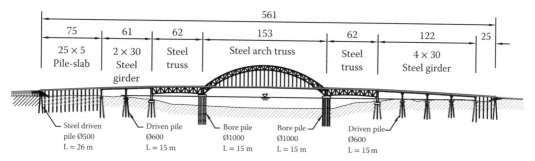

FIGURE 21.61 Elevation of the Rumpiang Bridge, Barito Kuala, South Kalimantan.

FIGURE 21.62 Erection of the main span of the bridge.

the main span on the bridge. After the steel truss arch on the top has been completed, then the road body underneath the bridge may be installed. The final result is depicted in Figure 21.62.

21.6.2.7 Mahulu Bridge (800 m), Samarinda, East Kalimantan (2008)

The Mahakam Ulu (Mahulu) Bridge (Figure 21.63) crosses the Mahakam River, which connects the village of Loa Buah and the Kujang River with Sengkotek regency, East Kalimantan. The bridge is one of five other bridges across the Mahakam River: the Martadipura Bridge, the Kertanagara Bridge, the Mahakam Ulu Bridge, and the Mahkota I and Mahkota II bridges (which were under construction as of April 2013). This bridge length is 800 m with a central span of 200 m and a width of 11 m. The distance between the bridge and the water surface is 18 m. Bridge construction was carried out in two parts, notably the bridge approaching from Loa Janan (six spans, 240 m) and from Loa Buah (nine spans, 360 m) using prestressed concrete I-girders, and its main span is a steel arch (200 m).

FIGURE 21.63   Mahulu Bridge (800 m), East Kalimantan.

FIGURE 21.64   Batanghari II Bridge (1351 m), Sijenjang, Jambi.

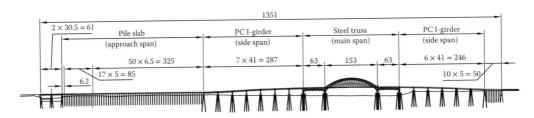

FIGURE 21.65   Elevation of the Batanghari II Bridge, Jambi, Sumatra.

21.6.2.8 Batanghari II Bridge (1351 m), Sijenjang, Jambi (2009)

The Batanghari II Bridge (Figure 21.64) crosses over the Batanghari River in the Sijenjang regency, Jambi, Sumatra. The location is approximately 6 miles east of the downtown area. The total length is 1351 m with a width of 9 m, including a 1 m footway on the right and left sides of the bridge (see Figure 21.65). The Batanghari II Bridge is also safe for shipping traffic on the Batanghari River. The distance or height between the water surface and the bridge floor (clearance) at times of flooding is 15 m and at low tide is 17.5 m.

For the province of Jambi, the Batanghari II Bridge provides a significant contribution in terms of transportation as well as in enhancing business relations between the regions, namely between the city of Jambi and the regency of Jambi Muaro, including the Eastern Tanjung Jabung regency. This bridge has provided economic access to most parts of Muaro Jabung, Jambi, and Eastern Tanjung Jabung, which have been constrained in that the two areas have been separated by the Batanghari River so that different water transport systems should be utilized.

21.6.2.9 Pela Bridge (420 m), Kota Bangun, East Kalimantan (2010)

The Pela Bridge (Figure 21.66) crosses over the Mahakam River connecting the road of Kahala to Pelabaru. It is located in the village of Pela, in Kota Bangun district, Kutai Kartanegara Regency, East Kalimantan Province. Pela Bridge's superstructure is a through-deck steel box arch bridge with simple supported restraints. The total length is 420 m, composed of two 45 m steel girders, a 45 m side truss, a 150 m steel arch as the main span, a 45 m side truss, and two 45 m steel girders (see Figure 21.67).

The arch is formed by a prismatic box member along the span (1 m × 1.2 m) except for the origin side, which is bigger (1.6 m × 1.2 m). The bridge's horizontal beam is straight and strengthened by 8 prestressing tendons, each composed by 19 strands with 0.5 inch diameters. These prestressed strands allow the bridge's superstructure to develop horizontal force optimally since most of the horizontal force due to dead load and life load are directly diminished by the prestressed force. Hence, the bridge's foundation bears only a small horizontal force. The bridge's camber is formed by several cross girders attached to the straight horizontal beam.

FIGURE 21.66 Pela Bridge (420 m), Kutai Kartanegara, East Kalimantan.

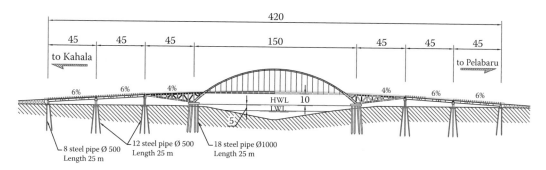

FIGURE 21.67 Elevation of the Pela Bridge, Kutai Kartanegara, East Kalimantan.

21.6.2.10 Sei Tayan Bridge (1420 m), Tayan, West Kalimantan

The Sei Tayan Bridge, which is under construction in the lower Tayan district, has been designed to cross the Kapuas River to connect the cities of Tayak and Piasak, in the Sanggau regency, approximately 112 km from the city of Pontianak. The bridge crosses the island of Tayan, which is relatively small in size (58 ha) with about 2100 residents. Bridge construction is intended to replace the ferry transportation that functions to serve the local people's activities. When the Sei Tayan Bridge is finished, it will be part of a road segment of the South Kalimantan ring, which connects West Kalimantan and Central Kalimantan.

The bridge consists of two parts, with a total length of 1420 m and a width of 11.5 m (2 × 2 lanes). The first bridge can be accessed from the northern part of Tayan to the island with a length of 280 m (see Figure 21.68). The second bridge, which is the main bridge, extends from Tayan Island towards Piasak with a distance of 1074 m. The main bridge is a continuous truss arch with a span of 350 m, which has been designed to accommodate shipping traffic navigations (see Figure 21.69). A detailed description of the second bridge is given in Table 21.6 and Figure 21.70.

FIGURE 21.68 An artist's impression of the first segment of the Sei Tayan Bridge (280 m), Pontianak, West Kalimantan.

FIGURE 21.69 An artist's impression of the second segment of the Sei Tayan Bridge (1074 m), Pontianak, West Kalimantan.

TABLE 21.6   Second Bridge of the Sei Tayan Project

Segment	Superstructure	Length of Span
2A	Steel composite girders	120 m (3 × 40 m)
2B	Continuous steel composite girders	600 m (4 × 45 m + 4 × 60 m + 4 × 45 m)
2C	Continuous steel arch truss	350 m (75 m + 200 m + 75 m)
2D	Pile slab	70 m (14 × 5 m)

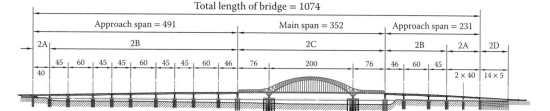

FIGURE 21.70   The second segment of the Sei Tayan Bridge, Pontianak, West Kalimantan.

21.7 Cable-Stayed Bridges

21.7.1 Teuku Fisabilillah Bridge (642 m), Riau Isles (1998)

The Teuku Fisabilillah Bridge (Figure 21.71) is part of the Barelang bridges, a chain of six bridges of various types, and connects the islands of Batam, Rempang, and Galang. The bridge is an icon of the local area and is a popular tourist destination. The full stretch of the six bridges reaches a total of 2 km. Traveling from the first bridge to the last is about 50 km and takes about 50 minutes. The construction of the bridges started in 1992, and this bridge was named Teuku Fisabilillah, after the fifteenth to eighteenth century rulers of the Melayu–Riau Kingdom.

The Teuku Fisabilillah Bridge is the first cable-stayed bridge built in this country and it was the longest cable-stayed bridge span in the southern hemisphere at the time of its completion. The total length of the bridge is 642 m with a 350 m main span and 2 × 146 m side spans. The bridge is supported by a concrete pylon 124 m high. The width is 21.5 m and the vertical clearance below the main span is 38 m. The VSL-200-SSI-type stays were installed in two planes of a semi-fan arrangement with their backstays tied to 20,000 ton weight abutments. The pylon legs stand on concrete pile caps with diameters of 24 m and 12.5 m thickness and are also supported by 2 × 30 pieces of 40 m deep concrete bore piles with diameters of 1 m.

21.7.2 Pasupati Bridge (2282 m), Bandung, West Java (2005)

The Pasupati Bridge is the incoming elevated highway in central Bandung, West Java, Indonesia. The bridge connects the western and eastern parts of the city of Bandung, lies in the northern part of the Bandung downtown, and passes through the Cikapundung valley. The 2282.4 m long elevated road consists of a 1278.7 m long west viaduct followed by a 303.5 m long main cable-stayed bridge and a 700.2 m long east viaduct (VSL Brochure, 2013). This bridge resolved the traffic congestions in North Bandung. Above the Cikapundung valley, a cable-stayed structure is particularly used so that the bridge also serves as an icon of the city. The designers were Sir William Halcrow and Partners Ltd. (UK), Inco (Kuwait), and PT Indec & Lapi ITB (Indonesia).

The Surapati Bridge can be divided into two main parts, a viaduct and a main cable-stayed bridge.

The viaduct (Figure 21.72) is a bridge road over the highway between Pasteur Street and the Cikapundung valley. The 44.5 m typical span viaduct is formed of a 2.95 m long precast segmental, three-cell, concrete box girder. The 21.53 m wide single deck is designed to accommodate a two-way

FIGURE 21.71   Teuku Fisabilillah Bridge (642 m), Bantam, Riau.

FIGURE 21.72   Precast segmental concrete box girder of the Surapati Viaduct.

road with two traffic lanes in each direction. The viaduct was built by launching gantry using the segmental balanced cantilever method and is supported by 48 Y-shaped piers.

The main span cable-stayed asymmetrical single pylon (Figure 21.73) stands over the Cikapundung valley with a single pylon at a height of 52 m off the ground, or about 38 m of the bridge deck. The pylon height of the bridge is suited to the regulations of Husein Sastranegara Airport, which is nearby. The span between the pylons is 55 m on the west side/back span and 106 m on the east side/main span. The width of the superstructure is 3 × 3.5 m for each direction (see Figure 21.74).

21.7.3 Grand Wisata Overpass (81 m), Bekasi, West Java (2007)

The Grand Wisata cable-stayed overpass is located in eastern Bekasi, around 20 km off Jakarta (see Figure 21.75). The bridge opened to traffic in July 2007. The overpass has a span length of 81 m, and was designed mainly for aesthetic purposes. It consists of a single pylon with two inclined concrete columns interconnected by an arch at the top and precast prestressed concrete girders overcrossing an expressway.

Due to the three-dimensional inclination and nonprism shapes of the pylon's sections, a full support system was used for concrete pouring on the 40 m high pylon (see Figure 21.76). A careful design of the scaffolding was considered, particularly to resist the lateral forces induced by the pylon's self-weight

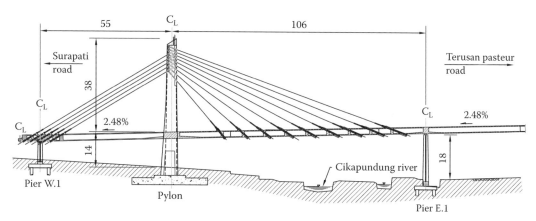

FIGURE 21.73 Elevation of the Surapati main bridge span, Bandung, West Java.

FIGURE 21.74 The Surapati main bridge span (161 m), Bandung, West Java.

FIGURE 21.75 Grand Wisata Overpass (81 m), Bekasi, West Java.

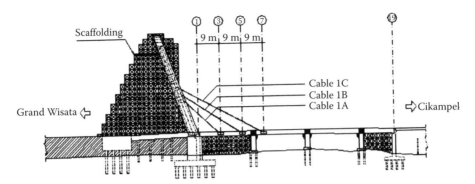

FIGURE 21.76 A fully supported scaffolding system for the pylon construction. (From Supartono, F.X. 2009. *Structural Engineering International* 19(1).)

FIGURE 21.77 Siak Indrapura Bridge (1196 m), Siak, Riau.

during concreting. Meanwhile, a step-by-step deflection analysis was performed to accommodate the pylon's deformation during the construction of the overpass. Due to nonprism hollow shapes of pylon's sections, in particular for the top arch connection, special care was taken on the concrete compactness. A high-performance and self-compacting concrete of 60 MPa cylindrical strength, with a slump flow criteria of 65–75 cm, was used for the pylon. This represents the highest grade of self-compacting concrete produced in Indonesia for cast-in-place concrete.

The deck system consists of a longitudinal main girder and crossbeams that are interconnected with its longitudinal ribs (secondary beams). The deck slab consists of a precast half slab 100 mm thick and a cast-in-place top slab 120 mm thick. Due to the limited capacity of the temporary support in the median of the expressway, the erection of the precast girder was combined with partial stressing of the stayed cables before completely pouring the concrete top slabs, in order to decrease the dead weight being loaded on the temporary support. The stayed cables are unbonded high-strength epoxy-coated strands with a diameter of 0.6 in (15.2 mm) and are protected by HDPE (high-density polyethylene) pipes. This type of cable-stayed bridge was used for the first time in Indonesia.

21.7.4 Siak Indrapura Bridge (1196 m), Siak, Riau (2007)

The Siak Indrapura Bridge, an icon of the region, lies on the Siak River, in Siak regency, Riau, Sumatra (Figure 21.77). This bridge is important for the development of the area because it connects the northern and southern regencies. This bridge has unique features and may even be the only one in Indonesia. The 77 m high bridge pylon has a slightly curved "A" shape. Its peak has become a diorama on the north side of the pier and a restaurant on the south side of the pier. The total length of the bridge is 1196 m (2 × 150 m + 8 × 51 m + 2 × 94 m + 200 m), whereas the width of the bridge is 16.95 m to accommodate the four 2.75 m vehicle lanes and two 2.25 m footways. The pylon legs sit on an individual pile cap of 12 m × 24 m × 4m and are supported by 24 steel pipe piles.

The Siak Bridge is similar to the Batam–Tonton Bridge in Batam (Sucipto and Lontoh 2003). However, several modifications were made to overcome specific problems in different locations. The Siak cable-stayed bridge was built to replace the existing ferry transportation on the river. The bridge provides a 24 m freeboard that enables vessels to pass through.

21.7.5 Siak IV Bridge (699 m), Pekanbaru, Riau

The city of Pekanbaru is divided by the Siak River into two areas. This bridge was built to accommodate rapid urban development in the region (see Figures 21.78 and 21.79). Except for highway traffic, the bridge will be used as an icon of Pekanbaru, particularly to welcome National Sports Competition (PON) XVIII in 2012. Hence, the cable-stayed bridge system was selected. Because the Siak Indrapura Bridge adopted the same system, the Siak IV Bridge was made differently by using a tilted single pier so that the bridge becomes asymmetric. It was designed so that the height of the bridge deck is 14 m and its length is 699 m. Its main stretch is that of a cable-stayed system with a length of 156 m. Its width is 20 m and the slope of the elongated part is 3.5–5%. The bridge has four lanes, with each direction consisting of two lanes. The bridge is expected to be completed by 2014.

21.7.6 Sukarno Bridge (622 m), Manado, North Sulawesi

The Sukarno Bridge (Figure 21.80) is part of the Manado ring road in Manado and is expected to accelerate the development of cities in North Sulawesi. It is a cable-stayed bridge, with a main span of 240 m and a free vertical height of 15 m. The bridge is expected to be completed by the end of 2013.

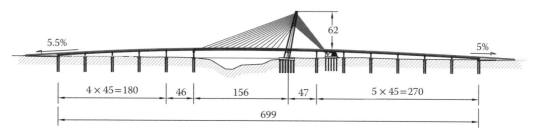

FIGURE 21.78 Elevation of the Siak IV Bridge, Pekanbaru, Riau.

FIGURE 21.79 An artist's impression of the Siak IV Bridge (699 m), Pekanbaru, Riau.

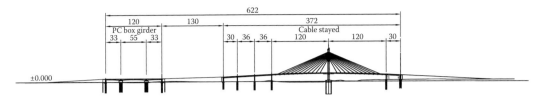

FIGURE 21.80   Sukarno Bridge, Manado, North Sulawesi.

FIGURE 21.81   An artist's impression of the Melak Bridge (680 m), West Kutai, East Kalimantan.

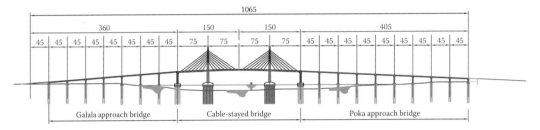

FIGURE 21.82   Elevation of the Merah Putih Bridge, Ambon, Maluku.

21.7.7 Melak Bridge (680 m), West Kutai, East Kalimantan

The Melak Bridge is located in West Kutai, East Kalimantan, and is a double-pylon cable-stayed bridge with three spans of 170 m + 340 m + 170 m. The bridge deck also represents an open crossed section with twin side girders of prestressed concrete with a width of 14.2 m and a height of 2.4 m. The pylons have a slightly curved "A" shape with a height of about 108 m. Construction of the bridge began in 2010.

21.7.8 Merah Putih Bridge (1065 m), Ambon, Maluku (under construction)

The Merah Putih Bridge (Figure 21.82) crosses over Ambon Bay, connecting Galala, in the Sirimau subdistrict, with Poka, in the Ambon Bay subdistrict. The bridge will serve as a shortcut from Galala and Poka (around the Pattimura International Airport) to Ambon. The total length of the bridge is 1065 m. The main span is a cable-stayed system consisting of three spans of 75 m + 150 m + 75 m with prestressed floor systems and a width of 22.3 m. The pylon height is 110 m above the pile cap. The approach bridge is a prestressed concrete I-girder. The bridge is expected to be completed by 2014.

21.8 Suspension Bridges

21.8.1 Old Bantar Bridge (176 m), Kulon Progo, Jogyakarta (1932)

The Old Bantar Bridge, a suspension bridge, was the first bridge to cross the Progo River in the village of Bantar, Kulon Progo regency, Jogjakarta (Figure 21.83). The bridge was built a long time ago during Dutch colonial times. The total length of the bridge is 176 m, consisting of three continuous spans with a main span of 80 m and a width of 7 m. The road segment above it belongs to the southern path of Java Island, which is an alternative national road to serve the southern region of Java Island. Due to the increased traffic volume from year to year and also the age factor, the bridge is no longer used for highway traffic. Only light traffic, such as motorcycles and the like, is allowed to pass over the bridge. Furthermore, at the same place, next to the Old Bantar Bridge, two parallel bridges have been built side by side, namely the Bantar II and Bantar III bridges. Both of the two newly built bridges are actively used for traffic. The Old Bantar Bridge is still retained for its historic value and artistic shape.

21.8.2 Mamberamo Bridge (235 m), Papua (1996)

The Mamberamo Bridge, over the Mamberamo River in Papua, was built in 1996 as part of a road development project in an agricultural area (Figures 21.84 and 21.85). Since the road works were incomplete at the time of the construction, the transport of materials to the bridge site was a particular challenge in this project. A suspension bridge with a single span of 235 m using a double cable system was designed to overcome rapid river flow and a rocky river bed. The towers are prefabricated steel frames with site-bolted connections. The main cables are in dual asymmetric arrangements for optimum stiffness. The

FIGURE 21.83 Old Bantar Bridge (176 m), Kulon Progo, Yogyakarta.

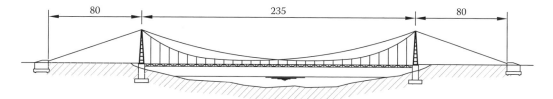

FIGURE 21.84 Elevation of the Mamberamo Bridge (235 m), Papua.

FIGURE 21.85   The Mamberamo Bridge (235 m), Papua.

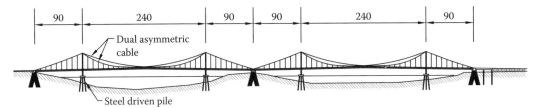

FIGURE 21.86   Elevation of the Barito Bridge (1082 m), Banjarmasin, South Kalimantan.

FIGURE 21.87   Barito Bridge (1082 m), Banjarmasin, South Kalimantan.

total weight of the cables is about 120 tons. The timber deck consists of transverse deck planks with timber curbs to delineate the carriageway. The cables were anchored into rock anchorages.

21.8.3 Barito Bridge (1082 m), Banjarmasin, South Kalimantan (1995)

The Barito Bridge over the Barito River, Banjarmasin, South Kalimantan, is a twin suspension bridge 2 × 90 m + 2 × 240 m + 2 × 90 m (see Figure 21.86), with a total length of about 1082 m. The designers were McMillan, Britton & Kell, the superstructure engineering site works contractor was PT Adhi Karya, and the main subcontractor was Transfield Construction. The engineering supervision was carried out by PT Perencana Jaya, Jakarta. The construction period was from 1993 to 1997. The bridge comprises two 3.5 m traffic lanes and a 1.5 m footway on each side, with a total width of 10 m. The width between the cables is approximately 12 m. A key feature of the design is the dual asymmetric cable arrangement, which provides a 70% increase in bridge stiffness when compared to a conventional single cable arrangement (see Figure 21.87 for the site view).

21.8.4 Kartanegara Bridge (714 m), Kutai Kartanegara, East Kalimantan (2001)

The Kartanegara Bridge (Figure 21.88) in Tenggarong, Kutai Kartanegara, is the second bridge that crosses over the Mahakam River, after the Mahakam I Bridge in the city of Samarinda. Both bridges are located in East Kalimantan. This bridge is part of the Kalimantan central axis lane, which connects the cities of Samarinda and Tenggarong. The main bridge span is 270 m and is the third-longest suspension bridge in Indonesia, after the Mamberamo Bridge (235 m) in Papua and the Barito Bridge (240 m) in South Kalimantan.

The bridge used a single catenary cable with a stiffening truss of modified Bina Marga steel bridge standard of A45 class. The total length of the stiffening truss is 470 m with a lane width of 7 m, equipped with a 1 m footway on both sides of the lane. The entire deck of the bridge is reinforced concrete K-350/U-40 with a wearing course in the form of 40 mm thick Hot Rolled Sheet (HRS) asphalt. The structure of the tower legs constitutes a series of four steel pipes φ 600 mm, with a height of 37 m, supported by the 15 m tall concrete construction. Figure 21.89 shows the process of erection of the stiffening truss segments, which form a bridge with an A45 class standard framework. See Figure 21.90 for site view of the bridge. On November 26, 2011, the bridge had collapsed at the time of maintenance efforts; therefore, it only functioned for ten years after it was completed. The failure of the bridge killed at least 20 people and injured 40.

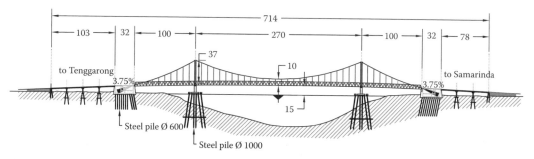

FIGURE 21.88 Elevation of the Kartanegara Bridge, Kutai Kartanegara, East Kalimantan.

FIGURE 21.89 The piece-by-piece erection of the stiffening truss segments.

FIGURE 21.90   Kartanegara Bridge (714 m), Kutai Kartanegara, East Kalimantan.

FIGURE 21.91   An artist's impression of the Pulau Balang Bridge (1344 m), Balikpapan, East Kalimantan.

21.8.5 Balikpapan Bay Bridge (1344 m), East Kalimantan

Of the entire trans-Kalimantan road, Kalimantan Island, with its rivers that are generally in the shape of deep trenches, still has some road segments that are not yet connected directly through the land infrastructure, such as the Balikpapan Bay lane via Balang Island and the road crossing the Kapuas River via Tayan Island. The two lanes have a relatively large stretch of approximately 1000 m to 2000 m. This proposed bridge will connect both sides of the bay, between the city of Balikpapan and the North Penajam Paser regency and will pass Balang Island, such that it is also called the Pulau Balang Bridge (see Figure 21.91 for an artist's impression of the bridge).

The bridge was designed as a suspension bridge for four-lane traffic from both directions, with a main span of 708 m and a width of 22 m (see Figure 21.92). The stiffening system of the suspension bridge is in the form of a galvanized steel truss with a pylon height of approximately 80 m above the elevation of the road. The 20 m bridge elevation was designed to cater to marine navigation below the deck of the

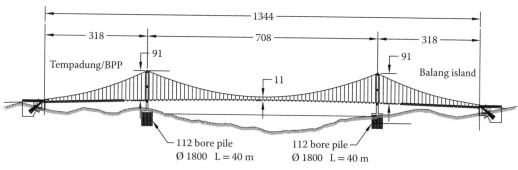

FIGURE 21.92   Main Span Elevation of the Pulau Balang bridge, East Kalimantan.

bridge. When completed, the bridge will be part of the trans-Borneo ring road from the south lane and is expected to support regional transportation and development of the Kariangau Balikpapan industrial area, which is yet to be realized.

21.9 Special Bridge Projects

21.9.1 Road Improvement Projects

21.9.1.1 Kelok-9 Project, Lima Puluh Kota, West Sumatra

The Kelok-9 bridge project is located in the regency of Lima Puluh Kota, West Sumatra Province. This project is part of the Payakumbuh–Batas Riau road improvement project (km 130 + 000 to km 148 + 000), which functions as a connecting road between the central Sumatra lane and the eastern coast of Sumatra. The project consists of a 5 km road and a ±1 km bridge. The condition of the Kelok-9 area reflects its own name, that is, winding or turning. Naturally, the road is very curvy, as shown in Figure 21.93.

With steep and winding road conditions, the smooth flow of the traffic is reduced. Semitrailer trucks or trailers cannot pass this road. The road via the Kelok-9 route is a national road which greatly affects the life of the surrounding communities. As a result, the improvement project has been designed not only to widen the existing road but also to make a new route to accommodate heavy vehicles. In seeking a new route, the route of the old site will be retained. Furthermore, since the Kelok-9 project also requires almost 20% of its entire design to be bridges, surely this bridge will become a new icon for the area.

An estimate of the new road by an artist's description is shown in Figure 21.94, in which the bridge is seen from the city of Bukittinggi. The arch bridge that can be seen most prominently is part of Bridge IV (454 m). The segments of the other bridges, which consist of six segments, together with their locations and information, can be seen in the map in Figure 21.95. Table 21.7 presents a list of detailed configurations of bridges in the Kelok-9 road project.

The elevation of Bridges II, IV, and VI can be seen in Figures 21.96 through 21.98. Because of the difficult conditions of the area, that is, the location in a hilly and steep area, it is sometimes necessary to create a special access road to reach the location and stage constructions are necessary. Bridge VI (Figure 21.98) was built in 2005 and required open road and blast rocks. The level of difficulty can be seen in Figure 21.99. Although the bridge has been completed, the end of the road is still blocked by the hill and needs to be cleared. In order to clear the hill, explosives should be used. However, the intervention of manual workers using heavy equipment is definitely still required. The bridge project is expected to be completed by the end of 2013.

FIGURE 21.93 The existing Kelok-9 road, West Sumatra.

FIGURE 21.94 An artist's impression of the Kelok-9 bridge, West Sumatra.

21.9.1.2 North Java Corridor Flyover Project

Motor vehicle transportation on the road, either for public or private vehicles, is still a major concern of the community, especially in Java. However, the railway network in Java is relatively complete compared to those on other islands of Indonesia. Because the local residents still rely on road traffic, it is very natural that on certain days, particularly national holidays, congestion will always hit some street segments. In facing these conditions, the Directorate General of Highways (Ditjen Bina Marga), Ministry of Public Works, will build a bridge overpass using a Japanese government grant in order

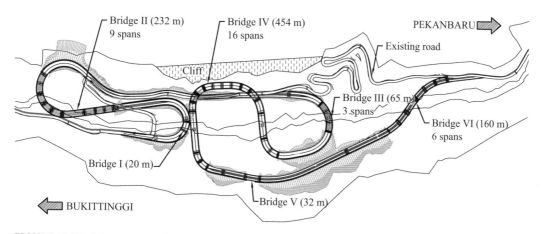

FIGURE 21.95 Map Location of the Kelok-9 bridge, West Sumatra.

TABLE 21.7 The Bridge Segments of the Kelok-9 Project

No.	Station	Abutment/ Pier	Bridge Type	Segment Span (m)	Total Length (m)
I	3 + 018–3 + 038	2	RC box girder	20	20
II	3 + 225–3 + 265	2	PC I-girder	20–20	230
	3 + 265–3 + 365	4	PC box girder	25–50–25	
	3 + 365–3 + 455	5	RC box girder	22–23–23–22	
III*	4 + 103–4 + 168	4	RC box girder	20–25–20	65
IV	4 + 301–4 + 321	2	RC box girder	20	462
	4 + 321–4 + 436	4	PC box girder	30–55–30	
	4 + 436–4 + 506	4	RC box girder	25–21.5–21.5	
	4 + 506–4 + 546	3	PC I-girder	20–20	
	4 + 546–4 + 626	5	RC box girder	20–20–20–20	
	4 + 626–4 + 716	2	Arch bridge	90	
	4 + 716–4 + 756	3	RC box girder	20–20	
V	4 + 844–4 + 876	2	PC I-girder	31	31
VI*	5 + 181–5 + 366	7	PC I-girder	27–29–35–29–29–27	176

*The bridge has already been finished.

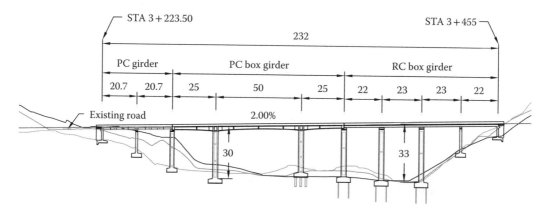

FIGURE 21.96 The Kelok-9 bridge project: Bridge II.

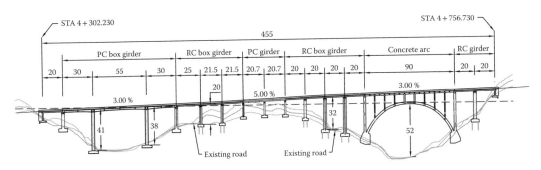

FIGURE 21.97   The Kelok-9 bridge project: Bridge IV.

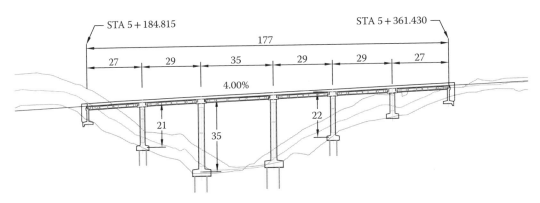

FIGURE 21.98   The Kelok-9 bridge project: Bridge VI.

FIGURE 21.99   Site condition of the Kelok-9 bridge project, West Sumatra.

to enhance the transportation capacity of the North Java Corridor (Figure 21.100). This will provide east–west linkage in northern Java Island to reduce traffic congestion and thereby improve the climate in the Java region. The project includes the construction of flyovers, which will carry both road and railway traffic at six intersections along the North Java Corridor and its alternative routes: Merak, Balaraja, Nagreg, Gebang, Peterongan, and Tanggulangin. In this flyover project, prestressed steel or concrete bridges will be used, where the shapes and lengths will be adapted to the conditions and needs of each location. The quantitative information on the bridges in the project plan can be seen in the Table 21.8.

For the section of the flyover at Tangulangin, the details are not available yet. However, it is already known that its location is in East Java so that it is relatively close to the Peterongan flyover. The Merak and Balaraja flyovers constitute the part of the road segment that connects Merak Port and Jakarta, which carries vehicles from the island of Sumatra to Java Island. The vehicles that generally dominate are those for transporting goods. The Nagrek flyover over the railway line is located in the road segment from Bandung to Malangbong, the part of the road that is often referred to as the "Southern Route" to get to Central Java. The road segments that pass through the northern coast of Java are the national roads connecting Banten, West Java, Central Java, and East Java.

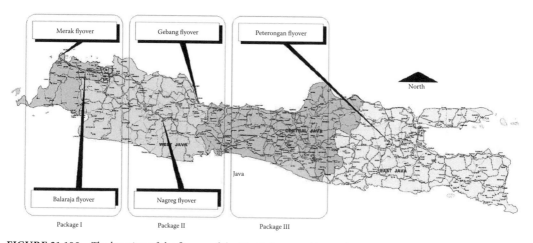

FIGURE 21.100 The location of the flyover of the North Java corridor project.

TABLE 21.8 Flyover Bridges of the North Java Corridor Project

Flyover	Width (m)	Length (m)		
		Total	Steel	Concrete
Merak (Pulorida side)	6.75	285	125	160
Merak (Jakarta side)	9	60	–	60
Merak (terminal exit)	7	70	60	110
Balaraja	13	221	81	140
Nagrek	13	224	104	120
Gebang	9	385	225	160
Peterongan	13	262	82	180
Tanggulangin	13	200	100	100
Total		1807	777	1030

21.9.2 Island-Crossing Bridges

In a speech delivered at the Bandung Institute of Technology (ITB) in 1960, Professor Ir. R. Sedyatmo proposed the idea of permanently connecting the islands of the archipelago. At that time the main topic was a fixed crossing system between Java and Sumatra across the Sunda Strait. Further discussions were also developing as to how connect the island of Java, which is regarded as the center, with the large islands nearby, notably the islands of Bali and Madura.

In 1986, the government of Indonesia launched the Three-Island Linkage project (Tri Nusa Bima Sakti), a study on the connection of the three islands of Bali, Java, and Sumatra, conducted by the Agency for the Assessment and Application of Technology (BPPT) and the Department of Public Works. The purpose of the study was to find the most suitable fixed crossing system between the three islands, whether using a bridge or tunnel system. For the Sunda Strait crossing system between Java and Sumatra, no distinct conclusion was indicated.

Later in 1989, a memorandum of agreement (MoA) between BPPT, the Ministry of Public Works, and the national development planning board of the study was reached, which was later expanded into the "Three-Island Linkage project and main crossing system" to carry out some preliminary studies to connect Sumatra, Java, and Madura/Bali. The implementation of the above studies is discussed next.

21.9.2.1 Suramadu Bridge (5438 m), East Java (2009)

Twenty-three years after the Three-Island Linkage project was launched, only one bridge over the strait of Madura was successfully constructed. Various studies indicated that a bridge over the strait of Madura was the most feasible link to be completed first in terms of funding limitations, engineering capabilities, and experiences. The Suramadu Bridge (Figure 21.101) is not the first inter-island bridge in Indonesia. The first was the Barelang Bridge, a chain of six bridges of various types that connect the islands of Batam, Rempang, and Galang, in the province of Riau Isles, giving the system its name.

The bridge, which spans over the Madura Strait, is located in the northern part of East Java province, and connects Surabaya on the island of Java with Bangkalan on the island of Madura. The name "Suramadu" comes from the special abbreviation of the names of *Sura*baya and *Madu*ra. The successful construction of the Suramadu Bridge can be considered the most significant collaborative project between China and Indonesia, because China was the donor country and, concurrently, provided the designers as well as the technology transfer and major implementation of the bridge construction. The construction contract was signed for the first time in September 2004, whereas the real construction work began in October 2005, and was completed in June 2009. The total length of the bridge is 5438 m (Figure 21.102a) and the navigation channel is 400 m × 35 m. This is the largest and longest bridge ever built in Indonesia until 2009.

FIGURE 21.101 **(See color insert.)** The Suramadu Bridge (5438 m), over the Madura Strait (June 2009).

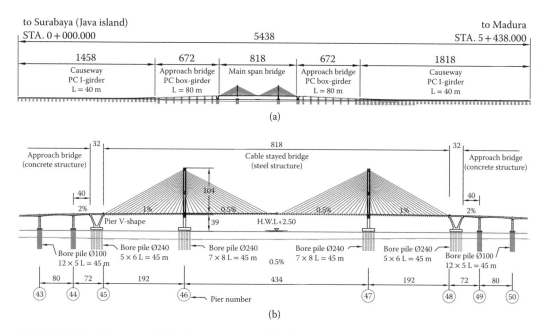

FIGURE 21.102 Suramadu Bridge, between Java and Madura islands (5438 m): (a) full elevation of the bridge, (b) main bridge span.

Because the design and construction of the bridge was contracted by the Consortium of Chinese Contractors (CCC), Chinese, Indonesian, and British standards were utilized as references for the design (Binli and Yiqian 2005). The Indonesian BMS-92 standard is convenient to provide the design of normal bridges with span lengths of less than 200 m. For the Suramadu Bridge, which should be a special large bridge according to the code, BMS-92 is not commonly applicable. A design life of 100 years is assumed in the Chinese standards, 120 years in the BS5400, and 50 years in the BMS-92. Since a design life of 100 years is specified for the Suramadu Bridge, it was designed as a special bridge. As designated in the main contract, the Suramadu Bridge was mainly designed in accordance with the Chinese standards. Allowing for some special features of this project, some specifications in the Indonesian BMS-92 and British BS5400 would be taken as references for checking the reliability of the superstructure.

In fact, to ensure the technical reliability of the Suramadu Bridge, the results of the CCC design needed to be reviewed separately by an independent and competent bridge consultant. In this case, Cowi of Denmark was chosen. The Indonesian government also formed an expert team under the coordination of the National Road V Implementation Task Force (Balai Besar Pelaksanaan Jalan Nasional V), Directorate General of Highways, Department of Public Works, Republic of Indonesia. The expert team referred to selected specialists who were expert academics considered competent to help monitor the design and implementation stages of the project (Ismail et al. 2009).

The Suramadu Bridge consists of a cable-stayed bridge system as the main span structure with a steel composite girder deck, supported by twin pylons with heights of 146 m. The main span configuration is 192 m + 434 m +192 m = 818 m (see Figure 21.102b). The approach bridge at each side is a continuous prestressed concrete box girder bridge built by the cantilever method with a length of 40 m + 7 × 80 m + 40 m = 640 m. The main span and approach bridges are connected with V-shaped concrete piers with a span of 32 m.

The steel structures of the main span of the bridge are composed of the steel main girders, steel floor beams, and stringers. In the cross section of the bridge, two steel main girders are arranged at the outer sides and two stringers at the inner sides (Figure 21.103). For a standard segment, the steel floor beams are set every 4 m along the bridge. The connections for every part of the steel structure segment use high-strength bolts (see Figure 21.104). The main span of the bridge is set up in a floating system,

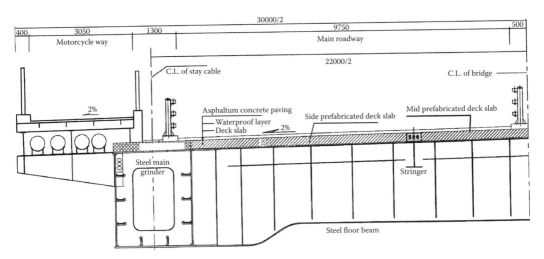

FIGURE 21.103 Half section of the main bridge span.

FIGURE 21.104 The segment of the main bridge span after erection.

hanging on a cable to the pylon. The vertical bearings are set only at the side piers. To limit movements along the bridge, longitudinal earthquake-resistant dampers are set at the pylon towers. In the transverse direction, rubber positive blocks are arranged between the main girders and pylon shafts at the pylon, and concrete stoppers are set on the V-shaped piers to restrain the transverse movements at the bridge ends (see Figure 21.105).

For the process of erection of the steel girder segment of the main bridge span, a special crane was placed at the end of the steel girder. The installation started from the edge (pylon), moving to the center. The crane would then lift the steel girder segment, which was transported from the factory by ship and was finally installed at the end of the bridge (see Figure 21.106). The cable-stayed construction was then installed, and then the crane moved forward. That was the steel girder assembly implementation progress until finally the two ends of the bridge floor were united. The assembly process of the steel girder of the main bridge span started from the pylon first, from the two directions symmetrically. Because there

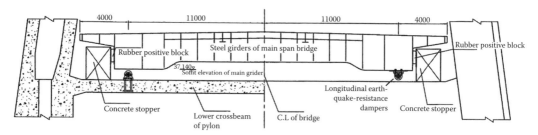

FIGURE 21.105   The floating system of the main bridge span.

FIGURE 21.106   Typical erection of the steel box girder segments.

were two pylons, four cranes were used in parallel (see Figure 21.107). Because the steel girder was used after the bolts and the cables were also secure, the structure worked fully.

At the main span of the bridge, a steel girder is used, while some parts of the approach bridge use concrete construction, namely the prestressed concrete segmental cantilever box girder, where the new structure would work fully when the concrete strength has reached a certain age (see Figure 21.108). The construction of the bridge utilized the balanced cantilever method, while in the beginning, before the ends of the bridge were connected with each other, the bridge behavior during construction was cantilever. Consequently, the cross section of the bridge at the pier was at a greater depth compared with the sectional girder in the center of the span. The cross sections in question can be seen in Figure 21.109.

FIGURE 21.107 Construction of the main bridge span (434 m).

FIGURE 21.108 Construction of an approach bridge (8 × 80 m each side).

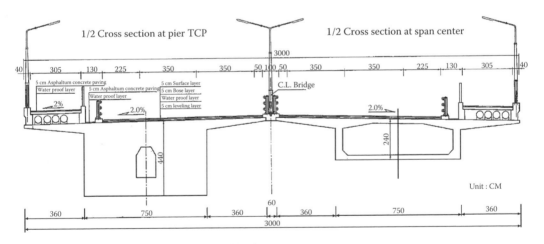

FIGURE 21.109 Typical section of an approach bridge.

Although the span of each structure of the prestressed concrete segmental box girder of the approach bridge is relatively short at 80 m, compared with the main span of the bridge, which is 434 m, and although the building process could also be done in parallel and simultaneously from one end to the other, it was proven that the completion of steel construction was much faster, possibly because the level of difficulty of the lower structure was greater than that of the superstructure.

The Suramadu Bridge has been operated as a toll road, where the users must pay in advance before they pass the bridge. Nevertheless, the character of the traffic crossing the bridge is interesting because it not only consists of four-wheeled vehicles but also two-wheelers. As a result, dedicated lanes for motorcycles are provided on the outer edge of each side of the bridge, as shown in the Figure 21.110. This is, of course, something new and the first time in Indonesia for such a toll road.

21.9.2.2 Bali Strait Bridge (± 2 km), East Java, Bali (Proposal Study)

The Bali Strait Bridge is one of the bridges recommended by the Three-Island Linkage project study. In 1992, there were some private parties who responded: the Scotia International Associates (UK), PT Mitra Trans Balongan (Indonesia), and Brown Beech and Associates Ltd. (UK Consultant). All the three combined themselves to form the Scotia Bali Bridge Co., Ltd. to submit a proposal to the government of Indonesia. In 1996, the minister of public works approved a 1-year study. The result was the third generation of the Bali Strait suspension bridge, designed by Brown Beech & Asscociates, which was a build-operate-transfer (BOT) proposal submitted to the Department of Public Works by the Scotia Bali Bridge Co., Ltd. The bridge has a total length of 2900 m with a main span of 2300 m and has a six-lane bridge floor (see Figure 21.111). The location of the bridge, as proposed by them, is in a narrow area of the strait of Bali, which has a width of 2 km and is located approximately 6 km to the north of the Ketapang–Gilimanuk ferry crossing area.

FIGURE 21.110 A view of the roadway on the Suramadu Bridge.

FIGURE 21.111 Proposal of the Bali Strait Bridge.

The plan to construct a bridge which connects Java and Bali will be an attractive infrastructure project in Indonesia. Its presence will increase the accessibility and economic prospects of the two provinces. The island of Bali is rich in its tourist potential, while the island of Java is a supplier for the basic needs of community life in general. However, since the Bali bombings (in 2002 and 2005), which were carried out by people outside the island, the desire to permanently unite the two provinces with the bridge has no longer resonated strongly.

21.9.2.3 Sunda Strait Bridge (29 km), West Java, Sumatra (Preliminary Design)

Since the disclosure of the idea to build a connection between Java and Sumatra by Professor Ir. R. Sedyatmo in 1960, there have been several attempts to respond to the idea. In 1986, the government of Indonesia launched the Three-Island Linkage project to study how to connect the three main islands: Sumatra, Java, and Bali. In 1989, there was an additional follow-up study, which included the island of Madura. The study was then known as the "Three-Island Linkage and main crossing" project. Twenty years later, this study finally led to the construction of the Suramadu Bridge (2009). Although some studies have also been conducted on the Sunda Strait, a definite conclusion has not been obtained as to whether the link should be in the form of a tunnel, bridge, or ship.

From 1986 to early 1997 there was no further news about the link. In May 1997, Dr. Ir. Wiratman Wangsadinata submitted a study report to BPPT, entitled "The Sunda Strait crossing and its feasibility as a link between Java and Sumatra." The study report was the result of an assignment from the state minister of Indonesia for research and technology to study the feasibility of a bridge crossing between Java and Sumatra. This study indicated that a bridge crossing between Java and Sumatra was feasible, provided that the building techniques for an ultra-long-span suspension bridge similar to the strait of Messina crossing were applied.

Professor Wiratman's proposal was quite attractive because it was supported by conditions which showed that the traffic across the two islands was so high that the current ferry system was not able to meet the growing demands. Even from records in 2002 it was known that more than 19,000 tons of goods per day from Sumatra Island were transported to Java Island through the Lampung Province. Also, more than 25,000 people and 6,000 vehicles, such as trucks, buses, private cars, and motorbikes, passed over the Sunda Strait per day. During the Eid holidays, the number could increase sharply to about 85,000 people and 11,500 vehicles. In these conditions, the capacity of the ferry crossing system was not enough, causing long queues of vehicles and causing extraordinary traffic jams. Consequently, passengers had to wait for hours just to cross the strait. If the ultra-long-span suspension bridge were completed, such terrible situations would not happen.

In May 2005, an idea of building the bridge was supported. PT Wiratman and associates was invited to collaborate with PT Bangungraha Sejahtera Mulia (BSM) to set up a consortium for the infrastructure construction of the Sunda Strait Bridge. Next, the consortium actively lobbied the governor of Banten in 2007, and even the president and the presidential staff in 2008, all of which was positively welcomed. Therefore, in May 2009, PT Wiratman was asked by BSM to carry out a preliminary study on the Sunda Strait Bridge, which was then equipped with bathymetric and topometric surveys (Wiratman 2009). The purpose of this survey was to produce a topographic contour map of the seabed and a topographic profile along the surveyed area.

It will take three more years to conduct the feasibility study before the construction process will begin. The entire project is expected to be completed in 2025. When the bridge is successfully completed, it will be the world's longest suspension bridge, passing about 50 km across the Sunda Strait from the active Krakatau volcano, traversing 29 km in length.

As the bridge is located in the Sunda Strait, which is prone to earthquakes and tsunamis, its construction would include four important phases involving hydrographic, oceanographic, geologic, seismological, climatological, and environmental aspects. Experts say the bridge is technologically feasible but extensive and expensive safety measures are essential to withstand earthquakes. Several quakes measuring more than 7 on the Richter scale have struck the waters of Sumatra and a stronger quake caused a massive tsunami of the west coast in 2004. According to a senior design consultant, Professor Wiratman Wangsadinata, flexible construction materials would be used to protect the bridge against earthquakes of up to magnitude 9, based on the Messina Strait bridge in Italy. Thus, the Sunda Strait Bridge will use technology of the third generation with the following specific advantages, especially on ultra-long-span suspension bridges: a relatively flexible pylon structure and a very light and aerodynamic multiple box girder. As a result, the bridge will have very small wind-drag forces, be insensitive to flutters, and will respond as required to earthquakes.

According to the study, the Sunda Strait Bridge will cross over Sangiang Island between Java and Sumatra and Panjurit Island near Sumatra, the most suitable locations. The bridges themselves consist of two ultra-long-span suspension bridges to span the two very wide and deep valleys of the Sunda Strait, and the overall bridge length is 29 km. The ultra-long-span suspension bridge near Anyer in the east of the strait and the ultra-long-span suspension bridge near Bakauheni in the west have the same dimensions, which were very advantageous for their design and construction. The five sections of the Sunda Strait Bridge are shown in Figure 21.112, indicating a general arrangement of the bridge elevations. Note that the vertical scale is different from the horizontal scale.

The two ultra-long-span suspension bridges are made up of high-strength steel materials, each with a main span 2200 m in length and two side spans 800 m in length. The ultra-long-span

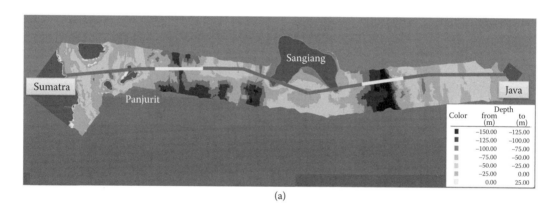

(a)

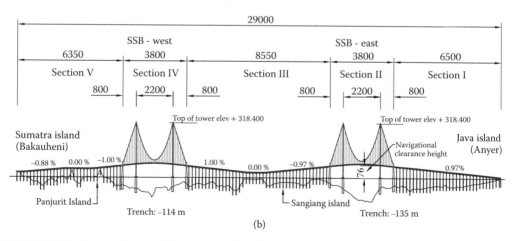

(b)

FIGURE 21.112 Route of the Sunda Strait Bridge (based on Wiratman's concept): (a) the islands connected by the bridge plan; (b) the longitudinal section.

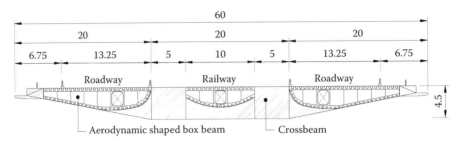

FIGURE 21.113   The ultra-long-span suspension bridge deck (based on Wiratman's concept).

suspension bridge decks are adopted from the concepts of the Messina Strait Bridge, an aerodynamic shaped orthotropic triple box, two carrying roadway traffic and one carrying railway traffic, supported by 4.5 m deep crossbeams spaced at 30 m apart (see Figure 21.113). The deck accommodates six lanes of roadway traffic, one double-track railway, one maintenance path, and a pedestrian way on each side.

The Sunda Strait Bridge, with a total length of 29000 m, is divided into five sections as follows:

Section I: This section consists of 32 balanced cantilever bridges, each of which is 200 m in span length, and 1 transitional balanced cantilever bridge with a span length of 100 m. The total length of Section I is 6500 m.

Section II: This section consists of one ultra-long-span suspension bridge, ULSB East, spanning over a wide and deep sea valley (trough). ULSB East has a main span of 2200 m and two side spans 800 m in length. The total length of Section II is 3800 m.

Section III: These sections consist of 42 balanced cantilever bridges, each of which is 200 m in span length, and 1 transitional balanced cantilever bridge with a span length of 150 m. The total length of Section III is 8550 m.

Section IV: This section consists of one ultra-long-span suspension bridge, ULSB West, spanning over a wide and deep sea valley (trough). ULSB West has a main span of 2200 m and two side spans 800 m in length. The total length of Section IV is 3800 m.

Section V: This section consists of 31 balanced cantilever bridges, each of which is 200 m in span length, and 1 transitional balanced cantilever bridge with a span length of 150 m. The total length of Section V is 6350 m.

The location of the Sunda Strait Bridge is inside the Indonesian Islands Sea Water Channel for international sea vessels; therefore, a minimum navigational clearance height of 75 m is mandatory. This navigational clearance height is taken as the distance measured from the high water level (± 1 m from the sea level mean) up to the soffit of the deck at the pylon location. The steel pylons of the ultra-long-span suspension bridge will reach a height of 318.4 m above the sea level mean. This distance takes into account the deck sloping height, navigational clearance height, and a sag-to-span ratio of 1:10. A prefabricated panel with longitudinal stiffeners is adopted for the pylon's cross section. The foundation for each pylon consists of a hollow with a diameter of 100 m and a caisson with a 15 m wall thickness filled with lean concrete.

The ultra-long suspension bridge system used is like that of the suspension bridge of San Francisco's Golden Gate Bridge. However, the Golden Gate Bridge is a suspension bridge of the first generation, whereas the Sunda Strait Bridge uses third generation technology, almost similar to the Xihoumen Bridge in China, except that the designer only uses a twin box concept without the railway lines. Figure 21.114 shows an artist's impression of the completed Sunda Strait Bridge. Notice that Krakatoa Island can be seen in the distance, and with a clearance height of ±76 m above sea level, the merchant vessels which pass through will seem small.

FIGURE 21.114   An artist's impression of the Sunda Strait Bridge (based on Wiratman's concept).

The plan sustainability of the Sunda Strait Bridge construction has become a powerful issue in the Indonesian media, especially based on the fact that the Suramadu Bridge, in East Java, which is Indonesia's longest bridge (5438 m) is now open to the public. Therefore, the government is confident enough to include it as a project on the Blue Book of the National Development Planning Agency (Bappenas).

The team for the national development of the Sunda Strait Bridge was established under presidential decree (Ref. No. 39 of 2009), dated December 28, 2009. This team is cochaired by the coordinating minister for economic affairs and deputy chairman of the coordinating minister for political, legal, and security affairs. The executive chiefs are the minister of public works and deputy executive chief of the Ministry of Transportation. Although it is still a dream, it is certain that the majority of the Indonesian people hope that the construction of the Sunda Strait ultra-long-span suspension bridge can really come true in the future. Only time will tell whether it is just a dream or a reality.

Acknowledgments

First we would like to express our sincere gratitude to Professor Wiratman Wangsadinata, the director of PT Wiratman and Associates in Jakarta, for the latest data concerning the Sunda Strait Bridge. Second, I would also like to thank Professor Lanneke Tristanto of the Research and Development Center for Roads and Bridges, Department of Public Works of Indonesia, who has provided good comments and feedback. Finally, my special thanks go to my colleague, Agus Santoso, EdD, for the earlier versions of the script translation.

References

Badan Pusat Statistik. 2006. http://www.bps.go.id, last modified 2012.

Binli, G., and L. Yiqian. 2005. *Design Criteria for Suramadu Bridge* (Revision 2), Consortium of Chinese Contractors.

Bridge Management System. 1992. "Bridge Construction Techniques Manual," Directorate General of Highways, Ministry of Public Works Republic of Indonesia and Australian International Development Assistance Bureau.

Directorate General of Highways. 2005. "Reference No. 04/BM/2005: Standard of Drawings for Road and Bridge Works," Department of Public Works Republic of Indonesia (in Bahasa Indonesian).

Imran, I., B. Budiono, K. Adhi, and Rusdiman. 2005. "Bridge Continuous Girder System, Case Study: Cisomang Bridge Design," National Seminar for Engineering Material and Concrete Construction, Civil Engineering Department, Itenas Bandung.

Ismail, et al. 2009. *Suramadu Bridge: The Dream and the Obsession, Overview Books Written by the Team from Induk Pelaksana Kegiatan Jembatan Suramadu*, Surabaya (in Bahasa Indonesian).

Sucipto, B., and S. Lontoh. 2003. "The Riau Cable-Stayed Bridges: Improvements of Construction Methodology of the Batam Cable-Stayed Bridges That Apply to the Siak Cable-Stayed Bridges," *Proceedings of the Ninth East Asia-Pacific Conference on Structural Engineering and Construction,* Bali, Indonesia, December 16–18, 2003.

Supartono, F.X. 2009. "Grand Wisata Cable-Stayed Overpass, Indonesia." *Structural Engineering International,* Volume 19, Number 1, February 2009 , pp. 46–47(2).

Vaza, H. 2008. *Indonesia Bridges: Now and Later.* Directorate of Technical Affairs, Directorate General of Highways, Department of Public Works (in Bahasa Indonesian).

VSL. "Brochure of Pasupati Bridge-Indonesia: Precast Segmental Balanced Cantilever Bridge with Overhead Launching Gantry." http://www.vsl-sg.com/flip/pasupati/files/pasupati-two-pages.pdf, dated April 25, 2013.

Wiratman and Associates. 2009. "The Preliminary Feasibility Study of Sunda Strait Bridge," Report to PT Bangungraha Sejahtera Mulia (BSM) to Fulfill Contract Agreement No. 0021/P-AGN-BSM-JSS/V/2009, dated May 5, 2009.

22

Bridge Engineering in Iran

Shervin Maleki
*Sharif University
of Technology*

22.1 Introduction

Iran has 1,648,000 km² of land area and is home to over 70 million people. Its rugged mountainous landscape has made road building very costly and difficult, requiring tunnels and bridges in most parts. It has 66,000 km of asphalt-paved roadway and over 5,000 km of railway (MRT 2008). There are more than 316,000 roadway and 21,000 railway bridges in the country. Most of the roadway bridges are of short spans. It is estimated that the number of roadway bridges with spans over 10 m is only 6500, and less than half of that have a span over 20 m (MRT 2008). Roadway and bridge building has a long history in Iran. A brief chronological review of this art and technology is given below in three parts: ancient times from the dawn of civilization to the seventh century AD, the Middle Ages and beyond covering bridge construction to the beginning of the twentieth century, and finally the modern era.

22.1.1 Ancient Times

Iran, formerly known as Persia, is a country with over 6000 years of history. It has been home to great civilizations, such as the Elamites (2800 BC) and Medes (600 BC), and empire builders like the Achaemenids (559–330 BC), Parthians (238 BC–226 AD), and Sassanids (226–632 AD). These vast empires built roads and bridges for communication and transportation. The 2700 km long Royal Road was built by King Darius in the fifth century BC. Couriers traveled on this road over seven days using a system of rest areas and a supply of fresh horses. Later, the Parthians completed and maintained the Silk Road in the Persian territory. Little remains from the bridges that were built in that era. The remnants show that the primary construction material in the Achaemenidian period was stone, which was abundant in Iran. The stone was used with mud and lime as mortar. Some metals were also used for connections and splices, mostly lead, poured in a molten state. The ancient Greek historian Herodotus gave the details of some of the engineering achievements of the Achaemenids. He reported that King Darius built the first floating bridge over Bosphorus in 490 BC for his army to cross. Later, King Xerxes constructed a floating bridge across the Helespont in 480 BC during his invasion of Greece.

The construction of masonry arches was perfected during the Sassanid's (226–632 AD) rule. Brick substituted stone as the primary material for arch-making in this era, although the piers were made from stone masonry and lime. The Dezful and Shadervan bridges are believed to have been constructed in this period; they are located in the southwest of Iran. The Dezful Bridge (Figure 22.1), built over the Dez River, has piers almost 1800 years old, but its pointed arches were reconstructed at a later time. The Shadervan Bridge (Figure 22.2), also known as Bande Qaisar (or Caesar's Dam), is believed to have been constructed by the captive Romans after the defeat of Emperor Valerian by Shapur, the king of Persia in 260 AD. The bridge had 44 masonry arches, of which very little remains. The use of semicircular arches in this bridge attests to the influence of Roman architecture.

FIGURE 22.1   **(See color insert.)** Dezful Bridge in Dezful.

FIGURE 22.2   Bande Qaisar Bridge in Shushtar.

FIGURE 22.3 Shapouri Bridge in Khorramabad.

Another bridge built in the same era with 28 stone masonry arches is the Shapouri Bridge near the city of Khorramabad in the west of Iran (Figure 22.3). Only seven masonry arches of this bridge have remained (Mokhlesi 1998). It should be noted that these bridges also functioned as dams to raise the water level for irrigation purposes. The smaller arches at higher elevations over the piers would release the excess water before it could cause collapse of the bridge.

22.1.2 Middle Ages and Beyond

The Islamic Era started in the mid-seventh century in Iran after the Arab invasion. The construction of bridges declined for several hundred years due to political instability in the country. This was exacerbated when the Mongols invaded the country in the thirteenth century. This decline in bridge building and in general construction continued throughout the Middle Ages. It was not until the beginning of the sixteenth century that the country again had a united leadership. This was achieved by the Saffavid Dynasty (1501–1736 AD), who ruled Iran for 200 years. During the Saffavid's rule, Iran underwent notable growth in science and engineering, especially in architecture. The buildings and bridges built in this period are still standing and are considered marvels of Persia.

The city of Isfahan soon became the capital of Persia, with magnificent developments. These structures are a major tourist attraction. The passage of Zayande Rud River through the middle of this city was an incentive for the kings to show their majestic power by building beautiful bridges. One of these is the Siose Pol, a two-story masonry arch bridge consisting of 33 arches in the lower story and spanning 295 m across the river (Figure 22.4). The bridge was built in 1602 and is the longest masonry bridge in terms of total bridge length in the country from before the modern era. The upper level, approximately 14 m wide, carries pedestrian traffic and is protected on each side by more masonry arches acting as parapet walls (Figure 22.5). The foundation and piers are built with stone and mortar, whereas the arches are made from brick masonry. The piers are 3.5 m wide and the clear span of each arch is 5.5 m. Unlike the Roman semicircular arches, the arches in this bridge are pointed or ogival. This was a common form of arch building for bridges in Iran at least from the tenth century on.

Nearly 50 years later another two-story bridge was built upstream from the Siose Pol in Isfahan. It is called the Khaju Bridge and is 110 m long and 20 m wide and has 24 arches (Figure 22.6). The bridge has an octagonal pavilion in the middle of its span for social gatherings. The artwork on the structure includes paintings, ornamental tiling, and gypsum carving, which add to its beauty (Figures 22.7 and 22.8).

FIGURE 22.4   Siose Pol Bridge in Isfahan.

FIGURE 22.5   Siose Pol Bridge passageway.

22.1.3 Modern Era

The modern era of bridge building in Iran started after World War I. By this time Europeans had gathered tremendous amount of experience in building all types of long-span concrete and steel bridges. Meanwhile, in Iran during this period, no advancement in construction materials or methods was noticeable. The Great Powers desire to expand and enter the global market of engineering was ever-increasing after the recession following the war. The newly found oil reserves in Iran made the country

FIGURE 22.6 Khaju Bridge in Isfahan.

FIGURE 22.7 Khaju Bridge artwork on arches.

more desirable for such investments. Although the British were leading in the petroleum sector, in the bridge engineering sector the leading role was almost exclusively reserved for a Danish engineering and construction company by the name of Kampsax. An excellent account of the Kampsax role in the Iranian politics of that era is given in Andersen (2008).

The company was founded by engineers Per Kampmann, Otto Kierulff and Jørgen Saxild in 1917 (Andersen 2008). Kampsax started its work in Iran in 1933 by winning the contract to build the first major railroad in the country, running from the Caspian Sea in the north to the Persian Gulf in the

FIGURE 22.8   Khaju Bridge pavilion.

south. Its previous involvement in the building of the Turkish railroad helped secure this project. Considering the rough terrain of the Iranian Plateau, with its Alborz and Zagros mountain ranges, and numerous rivers initiated from these drainage basins, the project was a huge undertaking and involved many bridges and tunnels (Figure 22.9). Nevertheless, Kampsax finished the project in 5 years with a cost of $9 million (Andersen 2008). The length of this railroad is 1400 km and it passes through 230 tunnels and 4100 bridges. The railroad and its bridges are still in operation and many of its bridges have survived the moderate earthquakes that happen routinely in the country. All sorts of bridges were constructed for this project including steel, plain, and reinforced concrete, brick, and stone masonry. Some of the longest and most beautiful masonry arch bridges in the country belong to this railroad project (Figure 22.10).

This event marked the first exposure of Iranians to modern construction materials and building techniques. A masterpiece of the Kampsax work, constructed at 110 m height, is the Veresk Bridge in the north of Iran, shown in Figure 22.11. It has an arch span of 66 m crossing the mountains and is made of unreinforced concrete. The bridge was completed in 1938.

Following the success of Kampsax, other European countries like Sweden, Germany, England, and later the United States got involved in domestic Iranian projects and have contributed to building many long-span bridges, among which the Sefid and Ghotur bridges are noteworthy. The Sefid Bridge was built in 1936 by a Swedish company over the Karun River in the city of Ahwaz in southwest of Iran. It has a through-arch main span made of steel with the deck hanging from suspension cables (Figure 22.12). The spans consist of two steel arches of 130 m and 136 m length and three deck-type steel arches of 49 m span. The bridge is 10 m wide and is currently operating under one-way traffic. Up until 1971 this was the longest span bridge in the country. In 1971, a joint venture of an American and Austrian company designed and built a 223 m trussed steel arch as part of the western Iranian railroad project that connects to the Turkish railroad. The bridge is called the Ghotur Bridge (Figure 22.13) and has a total length of 448 m. It was the longest arch bridge until a decade ago and is the second-longest arch bridge in the country today. Iranian bridge consulting firms played a more active role in the 1970s. Starting with short-span bridges, eventually over two decades they were able to master the art.

FIGURE 22.9 Iranian railroad arch bridges and tunnels.

FIGURE 22.10 Masonry arch bridge in Lorestan.

Historically, most long-span bridges in Iran cross the Karun River. This river is the longest river (950 km) with the most flow in the country. It initiates from the basins in the Zagros Mountains and passes through Khuzistan province in the southwest of the country before entering the Persian Gulf. The oil-rich province has seen a good share of development in the past decade. Several dams and hydroelectric projects have come to an end recently, requiring new access roads and bridges. The Karun-3 project was one of these achievements (see Section 22.5).

FIGURE 22.11   Veresk Bridge in Firuzkuh.

FIGURE 22.12   Sefid Bridge in Ahwaz.

22.2 Design Practice

Bridge design with modern materials started seriously in the country in the early 1970s. At that time, since the country lacked unified specifications for bridge design, most bridge designers used foreign specifications. The American Association of State Highway and Transportation Officials (AASHTO) Standard Specifications (AASHTO 1996) was the most popular. The first attempt by the government to standardize bridge loading materialized in 1957 by the Ministry of Roads and Transportation (MRT). In this year, a small booklet titled "Technical Specification No. 11" (MRT 1957) was published in which a 45 ton truck with three axles was stipulated as the standard truck loading for highway bridges. In 1995, a complete highway and railroad bridge loading specification was published (MRT 1995). This was further

FIGURE 22.13 Ghotur Bridge near Khoy.

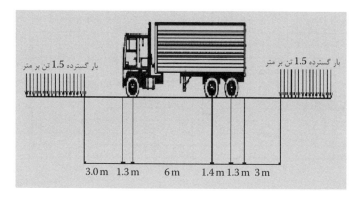

FIGURE 22.14 Design truck in Specification No. 139.

revised in 2000 and is known as "Specification No. 139" (MRT 2000). This specification introduced a 40 ton truck combined with a 1.5 ton/m uniform load as the standard loading, as shown in Figure 22.14. The front axle weighs 8 tons and the two rear axles weigh 16 tons each. The truck is 10 m in length and the wheels in each axle are 2 m apart center to center. The truck occupies a standard lane of 3 m.

The specification also specifies two military loading schemes as overload (extreme live load) criteria. The military loadings consist of a 70 ton tank (Figure 22.15) measuring 3.5 m in length and a 90 ton tractor-trailer 16.16 m in length (Figure 22.16). These vehicles need only be occupying one lane of the bridge without any consideration for the impact factor. The longitudinal distances between the two consecutive tanks are 30 m center to center and for the tractor-trailer is 12 m between the front and back wheels.

A comparison of the Iranian code live loads with that of AASHTO Standard Specifications (AASHTO 1996) is shown in Figure 22.17. The figure shows the maximum moment in a simply supported beam when loaded with one lane of live load. It is seen that the Iranian code loading is much more severe than

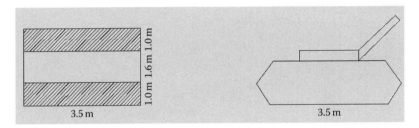

FIGURE 22.15 Design tank in Specification No. 139.

FIGURE 22.16 Design trailer in Specification No. 139.

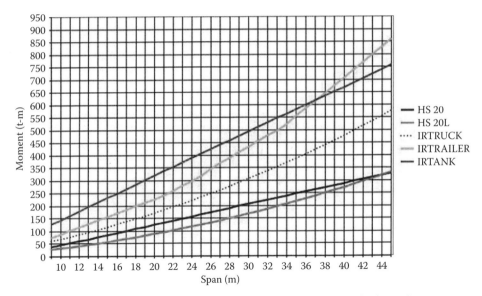

FIGURE 22.17 Moment comparison of Specification No. 139 and AASHTO Standard Specifications.

AASHTO HS 20 truck and lane loadings. Figure 22.18 compares the live loads with AASHTO Load and Resistance Factor Design (LRFD) Bridge Design Specifications (AASHTO 2007). It is seen that the truck loadings of the two codes produce about the same amount of moment. However, the military loading of the Iranian code produces much higher moments. Considering that in the United States each state has its own overload and permit vehicles, this is not unexpected.

The seismic loading and design criteria for bridges is handled separately in specification No. 463. This specification was first published in 1995 and has been modified twice since. The latest version (DTA 2008) came out in 2008 and for the most part follows AASHTO's Division IA (AASHTO 1996) for the seismic design of members. However, the design seismic load and the design response spectra are different and they match that of the Iranian code for seismic design of buildings (MHUD 2001). The seismic design force is obtained from the following equation:

$$F = C \times W$$

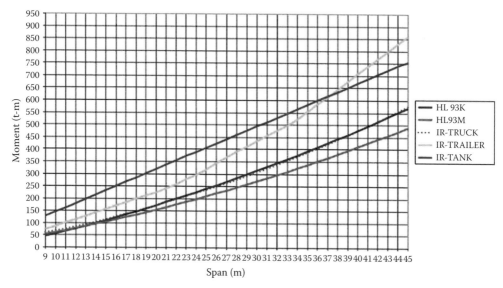

FIGURE 22.18   Moment comparison of Specification No. 139 and AASHTO LRFD specification.

where $C = \dfrac{ABI}{R}$ and W is the weight of the structure. A is the site acceleration, R is the response modification factor and I is the importance factor (equal to 1.2 for most bridges). The factor B is obtained as follows based on the period of the structure and soil type S:

$$B = 1 + S\left(\frac{T}{T_0}\right) \qquad 0 \le T \le T_0$$

$$B = S + 1 \qquad T_0 \le T \le T_S$$

$$B = (S+1)\left(\frac{T_S}{T}\right)^{\frac{2}{3}} \qquad T \ge T_S$$

The plot of the C/A ratio for soil type I ($S = 1.5$, $T_0 = 0.1$, $T_s = 0.4$) and importance factor $I = 1.2$ and $R = 0.8$ (as for connections in bridges) is compared to AASHTO LRFD (AASHTO 2007), C_{sm}/A factor with $R = 0.8$ in Figure 22.19. It is seen that regardless of the acceleration coefficient (A) of the site, for short-period structures ($T < 1$ sec) the seismic demand will be higher according to the Iranian code.

Prior to 1995, most bridges built in Iran lacked details for proper seismic response. However, past earthquakes showed that masonry arch bridges and single-span bridges are not very vulnerable to severe ground shaking. In 2005, the government embarked on nationwide projects for seismic retrofitting of existing bridges. Initially, the U.S. Federal Highway Administration (FHWA) retrofitting guidelines (FHWA 1995) were followed by most consulting engineers. The country's first seismic retrofit guideline for bridges came out in 2010 (DTA 2010). The capital city of Tehran is leading the country in this regard, with over 100 of its bridges being retrofitted. The first phase of these projects identified the weak points of most bridges to be inadequate seat width, transverse reinforcement and longitudinal bar lap splices for concrete piers, and not having restraint in the form of shear keys or something else for the superstructure.

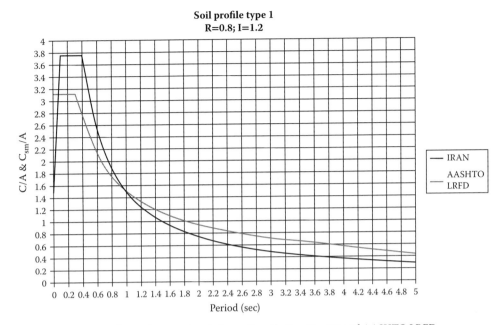

FIGURE 22.19   Seismic demand comparison between Specifcation No. 463 and AASHTO LRFD.

22.3 Concrete Girder Bridges

Concrete slab girder bridges are very popular in Iran for spans less than 30 m (Figure 22.20). Precast I-girders are usually used without prestressing in a composite design with the in-situ poured deck slab of 200 mm thickness. For longer spans concrete box girders are used often. The next section describes one of these long-span concrete box girder bridges.

22.3.1 Seventh Ahwaz Bridge

A record holder on the Karun River is the Seventh Ahwaz Bridge in the city of Ahwaz (Figure 22.21). This prestressed concrete box girder bridge was built in 1999 with the cantilever staged construction method. It has a total length of 490 m and it is 16 m wide. The main span is 140 m and it is currently the longest concrete box girder in the country. The piers are solid walls of concrete with piled foundations. Figure 22.22 shows the use of shear keys at the piers for stability of the superstructure against transverse seismic loads.

22.4 Steel Girder Bridges

Most steel girder bridges in Iran are of the plate girder type. The local steel mills do not produce deep-rolled sections for bridges. Therefore, designers use welded plate girders even for short spans of 12 m. Consequently, the fabrication cost and the susceptibility to corrosion has made steel bridges less popular for short spans.

22.4.1 Steel Plate Girder Bridges

Figure 22.23 shows a typical composite plate girder bridge in the capital city of Tehran over the Chamran expressway. Channel shear connectors are used exclusively in these bridges due to low cost and ease of welding. Elastomeric bearings are also the bearings of choice for most bridges in the

FIGURE 22.20 A typical concrete slab girder bridge.

FIGURE 22.21 Seventh Bridge in Ahwaz.

FIGURE 22.22 Seventh Bridge shear keys.

FIGURE 22.23   Plate girder bridge over the Chamran expressway in Tehran.

FIGURE 22.24   Elastomeric bearings.

country (Figure 22.24). The Sixth Bridge in Ahwaz (Figure 22.25) was built in 2004 using haunched steel composite plate girders. The bridge spans 428 m over the Karun River and has a main span of 75 m. The bridge is 21 m wide and uses multicolumn concrete piers and solid wall abutments (HCE 2011). This is one of the longest plate girder bridges in the country.

22.4.2 Steel Box Girder Bridges

Composite steel box girders are used often for spans greater than 30 m. The box girder is usually fabricated in the shop and lifted into place without any temporary supports. Figure 22.26 shows a typical box girder bridge with a span of 38 m in the city of Marvdasht by the name of Khan Bridge. The bridge replaces the old Khan Bridge built in the sixteenth century (Figure 22.27).

FIGURE 22.25 Sixth Bridge in Ahwaz.

FIGURE 22.26 New Khan Bridge in Marvdasht.

FIGURE 22.27 The old masonry arch Khan Bridge.

22.5 Arch Bridges

The Karun-3 Bridge (Figure 22.28) is the longest span arch bridge in Iran today. It was built in 2004 as part of the Karun-3 dam project. It is 12 m wide and carries two lanes of traffic. The two-hinged arch spans 264 m over the river at an elevation of 250 m. The arch is a trussed arch and supports a deck that has four steel girders (Figure 22.29). A total of 2500 tons of steel was used in this project.

22.6 Cable-Stayed Bridges

Cable-stayed bridges are more recent in Iran. Due to the unavailability of high-strength cables and anchors in the country, this type of system has usually been constructed with the cooperation of foreign companies, such as the Freyssinet Company. The first cable-stayed bridge was built in Tehran in 1997 as

FIGURE 22.28 Karun-3 Bridge in Ahwaz.

FIGURE 22.29 Karun-3 steel arch.

part of the Chamran expressway (Figure 22.30). The 52 m main span is made of hollow cored concrete box girder. The cables were used to create an aesthetically pleasing appearance in the middle of the city, rather than creating a structural advantage.

The capital city of Tehran has seen a tremendous increase in traffic for the past decade. This has led to construction of many new interchanges and overpasses (Figure 22.31). Another cable-stayed steel box girder bridge overpass was built over the Hemmat expressway in Tehran (Figure 22.32). In 2006, the Shushtar Bridge in southwest Iran was completed (Figure 22.33). This bridge has a length of 674 m and is 13.8 m wide. The main span is 150 m in length and uses cable-stayed steel girders. This is the longest cable-stayed span in the country as of today.

FIGURE 22.30 Chamran expressway cable-stayed bridge in Tehran.

FIGURE 22.31 A view of Tehran overpasses.

FIGURE 22.32   Cable-stayed overpass bridge in Tehran.

FIGURE 22.33   Cable-stayed bridge in Shushtar.

22.7 Future Bridges

22.7.1 Eighth Ahwaz Bridge

The Eighth Ahwaz Bridge has just finished construction and has a total span of 642m, width of 21 m, and a main span of 212 m. Figure 22.34 shows an artist's rendering of the proposed bridge (HCE 2011). The bridge, like the other Ahwaz bridges, crosses the Karun River. The design calls for a cable-stayed prestressed concrete box girder system for the main span. Once finished, this will be the longest cable-stayed bridge in the country.

22.7.2 Lake Urumiyeh Bridge

One of the biggest construction projects in the history of the country is still ongoing near Lake Urumiyeh. Phase 1 of the Shahid Kalantari Flyover Bridge on Lake Urumiyeh in northwest Iran opened in 2009. The bridge is part of a highway linking the northwestern cities of Urumiyeh and Tabriz and is the longest overall bridge in Iran, spanning 1.3 km directly over the lake (Figure 22.35). The bridge

FIGURE 22.34   **(See color insert.)** Eighth Bridge in Ahwaz. (HCE 2011).

FIGURE 22.35   Shahid Kalantary Bridge over Lake Urumiyeh.

consists of two decks of traffic lanes 9.5 m wide and a 5 m wide railroad track in the middle. The bridge spans consist of 2 × 50 m, 6 × 55 m, and 12 × 70 m and a main span 100 m long that uses a steel trussed arch girder. The main span provides 12 m of clearance underneath for the passage of ships. The other spans use steel composite girders. Close to 80% of the entire project has been completed to date. Due to harsh water conditions with very high chlorine content and high currents the foundation system of this bridge has been studied for several years by many foreign and domestic engineering firms. The final design called for sleeved piles 80 m in length penetrating 40 m into the lake bed. The piles are 32 in and the sleeves are 40 in in diameter.

References

AASHTO. 1996. *Standard Specifications for Highway Bridges*, 16th Edition, American Association of State Highway and Transportation Officials, Washington, D.C.

AASHTO. 2007. *AASHTO LRFD Bridge Design Specifications*, 4th Edition, American Association of State Highway and Transportation Officials, Washington, D.C.

Andersen, S. 2008. *Building for the Shah: Market Entry, Political Reality and Risks on the Iranian Market, 1933–1939*, Enterprise and Society, vol 9, no. 4, pp. 637–669.

DTA. 2008. *Specification No. 463*, Deputy of Technical Affairs, Office of the President, Islamic Republic of Iran.

DTA. 2010. *Seismic Retrofitting Manual for Bridges*, Deputy of Technical Affairs, Office of the President, Islamic Republic of Iran.

FHWA. 1995. *Seismic Retrofitting Manual for Highway Bridges*, FHWA-RD-94-052, Federal Highway Administration, Washington, D.C.

HCE. 2011. Hexa Consulting Engineers. Tehran, Iran, official website: www.hexa.ir/en/page/projects.

MHUD. 2001. *Design Loads for Buildings, Section 6 of the National Building Code*, Ministry of Housing and Urban Development, Tehran, Iran.

Mokhlesi, M.A. 1998. *Ancient Bridges of Iran*, Iranian Cultural Heritage Organization,Tehran, Iran.

MRT. 1957. *Technical Specification No. 11*, Ministry of Road and Transportation, Tehran, Iran.

MRT. 1995. *Specification No. 139*, Ministry of Road and Transportation, Tehran, Iran.

MRT. 2000. *Specification No. 139*, Revision 1, Ministry of Road and Transportation, Tehran, Iran.

MRT. 2008. *Annual Report*, Ministry of Road and Transportation, Tehran, Iran.

23

Bridge Engineering in Japan

Masatsugu Nagai
*Nagaoka University
of Technology*

Yoshiaki Okui
Saitama University

Yutaka Kawai
Nihon University

Masaaki Yamamoto
Kajima Corporation

Kimio Saito
Kajima Corporation

23.1 Introduction

23.1.1 Geographical Characteristics

The territory of Japan is about 378,000 km², as large as Norway. The population of Japan is about 127.43 million as of February 2010 (provisional estimates), which is about four times that of Canada. The population density of Japan is 337 persons per km², which is 7.2 times the world average. Japan has about 6850 islands; the main islands are Hokkaido, Honshu, Shikoku, and Kyushu. These islands are stretched like a bow from north to south over 3800 km. There are many mountains, and more than 20 exceed 3000 m high, so that plains are few and narrow. In addition, there are many rivers characterized by their relatively short lengths and considerably steep gradients. The climate is generally warm and comfortable; however, Japan has had trouble with many kinds of natural disasters, such as earthquakes, tsunamis, and typhoons. All kinds of infrastructures have suffered serious damage from natural disasters, so that the design specifications of infrastructures have been revised regularly to endure.

There are 11 big cities whose populations are over 1 million as of February 2010 (provisional estimates): Tokyo (8.5 million), Yokohama (3.6 million), Osaka (2.6 million), Nagoya (2.2 million), Sapporo (1.9 million), Kobe (1.5 million), Kyoto (1.5 million), Fukuoka (1.4 million), Kawasaki (1.3 million), Saitama (1.2 million), Hiroshima (1.2 million), and Sendai (1 million). Japan has dense ground transportation infrastructures, both road and railway networks, and they link the major cities. Roads are classified into four categories: national motorways, national roads, prefectural roads, and municipal roads, and the total length of these roads reached 1,196,217 km as of April 2008. There are are 9499 km of expressways or highways, mostly national motorways, whose main routes are shown in Figure 23.1a, as of August 2009. The target length of expressways is 14,000 km.

The total length of bridges on the roads is 9,502,572 m as of April 2008. Steel bridges, prestressed concrete bridges, and reinforced concrete bridges occupy 47.9%, 33.3%, and 12.5%, respectively, as shown in Figure 23.2a. The total number of bridges on the roads is 153,529; steel bridges, prestressed concrete bridges, and reinforced concrete bridges occupy 38.3%, 41.3%, and 16.9%, respectively, as shown in Figure 23.2b. In terms of bridge length, bridges are classified as shown in Figure 23.2c. Bridges over 100 m long make up 12.2% of the total. In terms of bridge type, bridges are classified as shown in Figure 23.2d; girder bridges make up 76% [1].

Four major islands are also covered with an extensive and reliable network of railways. The railway network is mainly operated by 22 private railway companies. Japan Railways (JR Group), the most major private railway company, is the successor of the Japanese National Railways (JNR). The JR Group consists of six regional passenger railway companies and one nationwide freight railway company. Together they operate a nationwide network of urban, regional, and interregional train lines, night trains, and the Shinkansen, known as the bullet train, which is a network of high-speed railway lines shown in Figure 23.1b. The Tokaido Sinkansen connecting Tokyo and Osaka started service in 1964 at the speed of 210 km/h (130 mph). The Sinkansen network has expanded to 2459 km long, linking most major cities on the islands of Honshu and Kyushu at the maximum speed of 300 km/h. JR Tokai plans to start operations of the new maglev Chuo Shinkansen line connecting Tokyo and Nagoya in 2025 and subsequently extend it to Osaka.

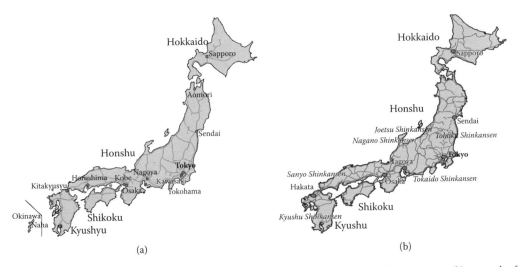

(a) (b)

FIGURE 23.1 Main ground transportation infrastructures in Japan: (a) network of expressways; (b) network of Shinkansen.

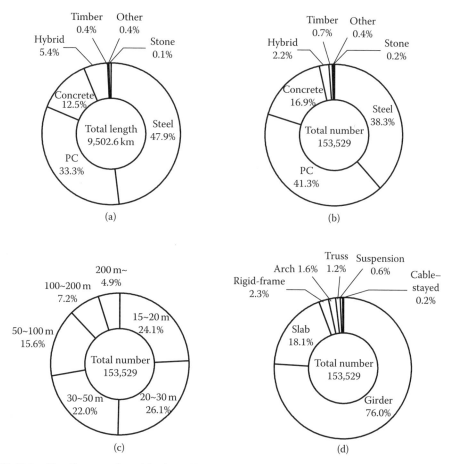

FIGURE 23.2 Classification of road bridges. (a) Bridge material ratio in length; (b) bridge material ratio in number; (c) bridge length ratio in number; and (d) bridge type ratio in number.

23.1.2 Historical Development

In Japan, there are now around 670,000 bridges with spans exceeding 2 m. Among them, as shown in Section 23.1.1, more than 150,000 bridges have spans exceeding 15 m. Figure 23.3 shows the number of bridges constructed in Japan. Bridge construction kept increasing after World War II until the 1970s and then began decreasing.

The oldest steel bridge in Japan is Kurogane Bridge in Nagasaki, Nagasaki prefecture, which was constructed in 1868 by a Japanese ship building company. Since then, most steel bridges have been imported, and since 1910, Japanese companies have fabricated them. Before World War II, only a few welded steel bridges were constructed. In 1939, specifications for welded steel bridges were issued, and were revised in 1957. Since then, welded steel bridges have been preferred and have become commonly used. Regarding the field connection of structural members, since 1965, almost all steel bridges have used high-strength bolt connections instead of rivet connections. Even now, most steel bridges employ the connections for field joints, and field weld connections are not often used in Japan. Steel materials are introduced in Section 23.3, including the recent development of high-performance steel (HPS).

Steel–concrete composite bridge construction started in 1951 in Japan. After issue of the guideline for composite bridge construction in 1960, construction of composite bridges started increasing.

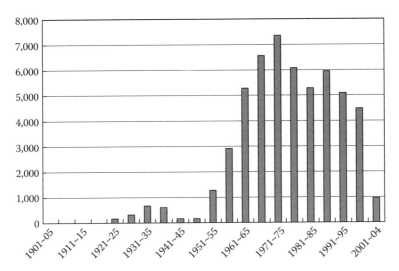

FIGURE 23.3   Number of bridges constructed in Japan.

However, 10 years from the start of composite girder bridge construction, a lot of damage in the slabs appeared. This was due to employment of relatively thin deck slab, low longitudinal reinforcement, no waterproof layer, poor concrete quality, and so on. Since the late 1970s, we seldom employ composite design. In the 1960s, the so-called modern steel plate girder bridge system was established in Japan and since then this system has been preferably employed. It is a noncomposite design, has a relatively thick concrete deck slab supported by multigirders consisting of a thin web plate with many stiffeners, has a complicated transverse stiffening system, and employs high-strength bolt connections. In addition, the allowable stress design (ASD) method (see Section 23.2) has been employed for more than 40 years.

The oldest concrete bridge is Wakasa Bridge in Kobe, constructed in 1903, and the oldest prestressed concrete bridge is Chosei Bridge in Nanao, Ishikawa prefecture, constructed in 1952, which has a span of 3 m. Since the introduction of prestressed concrete bridges, reinforced concrete bridge construction has decreased. However, for span lengths less than 20 m, concrete hollow slab bridges and arch concrete bridges have been employed. Material and the strength of concrete for bridges are presented in Section 23.3, and the design method of concrete bridges is discussed in Section 23.2. From 1980 to 2000, Japanese bridge engineers enjoyed long-span bridge construction, in particular long-span cable-supported bridge construction, in the Honshu–Shikoku islands connecting project, on the Tokyo and Osaka–Kobe bayshore route projects. In these big projects, the world's longest suspension bridge, Akashi Kaikyo Bridge, with a span of 1991 m, and the third-longest cable-stayed bridge, Tatara Bridge, with a span of 890 m, were constructed. Examples of cable-supported bridge construction are introduced in Section 23.7.

Cost reduction for bridge construction became top priority in the early 1990s, when new Tokyo–Nagoya–Osaka Expressway construction began. Responding to this, the steel–concrete hybrid (composite and mixed) structure has been preferably selected from an economical viewpoint. It is a new trend of bridge design in Japan, and examples of this newly developed bridge system, hybrid bridges, are introduced in Section 23.4. Since roadway network construction has reached an advanced stage in Japan, new bridge construction business has been decreasing, as shown in Figure 23.3. On the contrary, old bridges have been increasing. In 20 years, around a half or more existing bridges will exceed 50 years of age. Even now, we sometimes face severe deterioration of old bridges, and sometimes experience collapse. Repair, strengthening, and rebuilding of old bridges must be one of our future major works. Several innovations in technology in Japan are briefly introduced in Sections 23.9 and 23.10.

23.2 Design

23.2.1 Philosophy

There are three major design codes for bridges in Japan:

1. Specifications for Highway Bridges (SHB) (JRA 2002) [2]
2. Design Standards for Railway Structures and Commentary (DSRS) (RTRI 2008) [3]
3. Standard Specifications by Japan Society of Civil Engineers (SS-JSCE) (JSCE 2007) [4]

These design codes are based on different philosophies depending on objectives and historical background. The first two design codes are used in design practice and published by the associated governing legal authorities. On the other hand, SS-JSCE is a model code published by the Japan Society of Civil Engineers (JSCE). It suggests a future direction for design codes in Japan rather than current design practice.

The SHB is composed of five parts:

Part I: General
Part II: Steel Bridges
Part III: Concrete Bridges
Part IV: Substructures
Part V: Seismic Design

The current SHB employs the performance-based ASD method except for seismic design, which has been based on the performance-based limit state design (LSD) method. In principle, the performance-based design method can be described as follows. First, the required performance levels for individual bridge functions must be clarified. Then, the designated performance levels must be achieved. The methods to confirm the performance levels are not specified in the main body of the performance-based design codes. Instead, several appropriate procedures are shown in the commentary. Hence, designers can select an appropriate design procedure from several procedures that meet their purposes. This framework of the performance-based design method offers greater flexibility to designers, and is expected to promote new technologies, such as new materials and innovative structures. In reality, however, it is not so easy to implement the performance-based design codes, because the required performance levels, especially for durability and maintainability, are not clearly defined. In addition, some design equations have been developed on the basis of experimental data employing materials confirming Japan Industrial Standards (JIS). Accordingly, if we do not use materials confirming JIS, the application of some design equations might not be appropriate. To avoid this kind difficulty, some provisions include specification-based descriptions rather than performance-based.

Although the performance-based design codes involve this difficulty, they have been successfully employed in seismic design. In the SHB, Part V: Seismic Design, the target seismic performances are clearly specified, as shown in Table 23.1. Two levels of design earthquake motions are considered: Level 1 and 2. The Level 1 earthquakes are highly probable to occur during the service life of bridges, while the Level 2 earthquakes possess higher intensity and are less probable to occur during the service life. For Level 1 earthquakes, elastic behavior of bridges without damage is required; for Level 2 earthquakes,

TABLE 23.1 Design Earthquake and Required Seismic Performances

Design Earthquake		Bridge of Standard Importance (Bridge Class A)	Bridge of High Importance (Bridge Class B)
Level 1		No damage (seismic performance 1)	
Level 2	Type 1	To prevent fatal damage	To limit damage (seismic performance 2)
	Type 2	(seismic performance 3)	

the extent of acceptable damage is different depending on the importance of the bridge. In this table, a Type 1 earthquake means a plate boundary–type large-scale earthquake, and Type 2 is an inland direct strike–type earthquake, such as the 1995 Hyogoken-Nanbu earthquake. The SHB is scheduled to be revised completely into a performance-based LSD method in 2011.

The DSRS, which is the official design code for railway bridges, has been revised from the ASD to LSD method since 1992. Because the philosophy of seismic design in DSRS is very similar to the SHB, only the design procedure, except for seismic design, will be introduced here. The design verification equations in the DSRS are written in the partial factor format with five partial factors. The structure of the partial factor format is illustrated in Figure 23.4. In this figure, the structural factor γ_i varies depending on the importance of the bridge, and the member factor γ_b depends on the certainty of the capacity prediction equation and the consequences of the corresponding failure mode. The other partial factors, the load factor, structural analysis factor, and material factor, are introduced to deal with the uncertainty of the load, the accuracy of structural analysis, and the material properties, respectively.

The SS-JSCE is based on the performance-based LSD method employing the same partial factor format as the DSRS. In fact, the DSRS has adopted the partial factor format in the SS-JSCE. The SS-JSCE consists of the following six parts:

Part I: General Provision
Part II: Structural Planning
Part III: Design
Part IV: Seismic Design
Part V: Construction
Part VI: Maintenance and Management

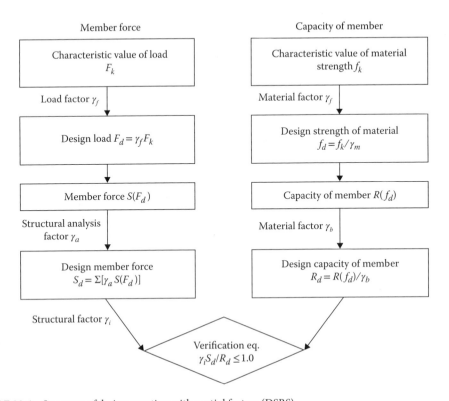

FIGURE 23.4   Structure of design equation with partial factors (DSRS).

The first five parts have been published, and Part VI will be issued in 2013. The English version of Parts I to III will be published within 2011. In SS-JSCE, the following seven performances are required:

1. Safety
2. Serviceability
3. Durability
4. Social and environmental compatibility
5. Maintainability
6. Post-earthquake serviceability
7. Constructability

Most required performances listed above are lucid, except for post-earthquake serviceability, which is considered to be very important due to past earthquake experiences in Japan. In the 1995 Hyogoken-Nanbu earthquake, many parts of the highway viaduct in Kobe area collapsed, and the highway could not be used as a lifeline. If at least emergency vehicles, such as ambulances and fire trucks, could use the highway after an earthquake, large-scale fire and loss of human life could be minimized. Of course, this post-earthquake serviceability is not required for all of bridges, but we require this performance for important "lifeline" bridges.

23.2.2 Load

Design loads for road bridges will be introduced in this section. The SHB (JRA 2002) classifies design loads into four categories:

1. Primary load (referred to as P): dead load (D), live load (L), impact (I), prestress (PS), etc.
2. Secondary load (S): wind load (W), temperature effect (T), earthquake effect (EQ)
3. Particular load corresponding to the primary load (PP): snow load, settlement of support, etc.
4. Particular load corresponding to the secondary load (PA): braking load (BK), load during erection (ER), collision load (CO), etc.

The SHB requires the safety of bridges against the ten load combinations listed in Table 23.2. In this table, the factors multiplied by the allowable stress for the respective load combinations are also given. Since the standard safety factor for steel members is 1.7 in the SHB, the net safety factor for the load combination No. 3 becomes 1.7/1.25 = 1.36, as an example.

It should be noted that only Level 1 earthquakes are considered in this load combination. As mentioned earlier, the SHB specifies two levels of earthquake motions. The Level 1 earthquake is treated in the static linear elastic structural analysis based on the allowable design method, while the Level 2 earthquake motions require nonlinear dynamic analysis using the performance-based LSD method. The intensity of Level 1 earthquake motion is calculated from three acceleration response spectra depending on ground classification, a modification factor for the regional zone, and the damping constant. The acceleration response spectra specified in SHB are established by integrating past practices and statistical analyses of 394 earthquake ground records observed in Japan.

Two levels of live load are specified in SHB, depending on the importance of the bridge: B live load is for bridges on expressways and major traffic routes, and A live load is for those in other loads. Both live loads consist of two distributed loads p_1 and p_2, whose magnitudes are listed in Table 23.3. In this table, D stands for the longitudinal length of load placement for p_1 as shown in Figure 23.5. The load intensity p_1 varies depending on design cross-sectional forces to be calculated, such as bending moment and shear force. The placement of the distributed live load is determined based on the specifications shown in Figure 23.5 and the influence line analysis to determine maximum load effects.

TABLE 23.2   Design Load Combinations and Associated Multiplier Factors for Allowable Stress

No.	Load Combination	Multiplier Factor
1	P+PP	1.00
2	P+PP+T	1.15
3	P+PP+W	1.25
4	P+PP+T+W	1.35
5	P+PP+BK	1.25
6	P+PP+CO	1.70 for steel members, 1.50 for concrete member
7	P*+EQ	1.50
8	W	1.20
9	BK	1.20
10	ER	1.25

Note: P* stands for the primary load P excluding the live load and impact.

TABLE 23.3   Load Intensity of Live Load and Loading Length D

	Loading Length D (m)	p_1 (kN/m²)		p_2 (kN/m²)		
		For Moment	For Shear	$L \leq 80$	$80 < L \leq 130$	$130 < L$
A live load	6	10	12	3.5	4.3−0.01L	3.0
B live load	10					

Note: L = span length (m)

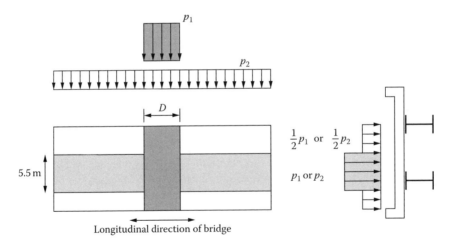

FIGURE 23.5   Placement of live load p_1 and p_2 (SHB).

23.3 Material

23.3.1 Steel

23.3.1.1 Conventional Steel

In Japan, a variety of steels, such as steel plates, tapered steel plates, H-shaped steels, deformed H-shaped steels, circular and rectangular steel pipes, high-strength steel wires, and rods have been used for all types of bridges. Steel plates, which are basically allowed in both highway and railway bridges in Japan, are specified in the JIS. The yield strength levels range from 235 MPa to 700 MPa, which are similar to those used in the United States and Europe. Among them, JIS-SM490Y steel, with a yield stress of 365 MPa (t ≤ 16 mm), 355 MPa (16 mm < t ≤ 40 mm), is the most commonly used material because of its high allowable stress and cost performance. Thermomechanical controlled processing (TMCP) is commonly used in Japan for the production method of higher-strength steels to obtain improved weldability [5]. Use of weathering steel has been increasing because of its anticorrosion performance, which will minimize the lifetime cost of bridges.

23.3.1.2 High-Performance Steel

In Japan, a type of high-performance steel (HPS) has been developed for special use in bridges, called BHS (bridge high-performance steel). BHS is produced by advanced TMCP. BHS was developed to design economical and highly durable bridges, and to fabricate steel members efficiently. BHS has numerous excellent properties, such as higher strength, higher fracture toughness, better weldability, and higher fabrication efficiency when it compared with the aforementioned JIS-SM490Y. BHS also features wide variety of weathering performance [6].

The specifications of the HPS specified in JIS are shown in Table 23.4 compared with the corresponding conventional high-strength steels. Conventional low-alloy weathering steel, JIS-SMA, shows poor performance in environments where airborne salinity is high, such as marine and coastal areas. Recent research revealed a rusting mechanism in the presence of chloride, and advanced weathering steels and coatings have been developed for this purpose [7]. Figure 23.6a shows the first application of 3% Ni advanced weathering steel in the world to a bridge on the Hokuriku Shinkansen line (bullet train

TABLE 23.4 Specifications of HPS and Conventional Structural Steels in Japan

Bridge High-Performance Steels			Conventional High-Strength Steels		
Grade	Thickness (mm)	Yield Strength (MPa)	Grade	Thickness (mm)	Yield Strength (MPa)
JIS G 3140 SBHS500 SBHS500W	6<t≤100	≥500	JIS G 3106 SM570 JIS G 3114 SMA570W	≤16	≥460
				16<t≤40	≥450
				40<t≤75	≥430
				75<t≤100	≥420
JIS G 3140 SBHS700 SBHS700W	6<t≤75	≥700	JIS G 3128 SHY685	6<t≤50	≥685
				50<t≤100	≥665
			JIS G 3102 HT80	8<t≤50	≥685
				50<t≤75	≥665

(a)   (b)

FIGURE 23.6 Application of 3% Ni advanced weathering steel on (a) a bridge on the Hokuriku Shinkansen line and (b) Nahari Bridge.

railway) over Hokuriku Expressway. Figure 23.6b shows Nahari Bridge of the Tosa Kuroshio railway in Kochi prefecture, where 3% Ni advanced weathering steel was applied without any surface treatment.

23.3.2 Concrete

In general, concrete with compressive strength up to 40 MPa is applied to superstructures, while high-strength concrete with compressive strength greater than 40 MPa is adopted in special cases. Recently, the number of such special cases is increasing because concrete-producing technology shows progress. At the same time, some design standards for high-strength concrete have been published in recent years.

23.3.2.1 Conventional Concrete

Ready-mixed concrete with compressive strength up to 60 MPa was defined in JIS-A-5308 in 2003. This definition indicates that high-strength concrete up to 60 MPa is available as a general industrial product in Japan by the progress of processing technology. In major design standards, the application range of concrete strength has been expanded. In accordance with the latest edition of SHB, high-strength concrete up to 80 MPa can be applied to concrete bridges. In the same way, 60 MPa concrete can be utilized according to the latest edition of DSRS.

23.3.2.2 Ultra-High-Strength Concrete

"Guidelines for Design and Construction of High-Strength Concrete for Prestressed Concrete Structures" was published by the Japan Prestressed Concrete Engineering Association in 2008 [8]. The guidelines were established especially for structures utilizing high-strength concrete from 60 MPa to 160 MPa. In addition, a guideline for ultra-high-strength fiber-reinforced concrete (UFC) was also published by the Japan Society of Civil Engineers [9]. The guidelines cover two types of UFC: Ductal® and SUQCEM®. The former was developed in France, and the latter was developed in Japan. The compressive strength of the UFC is 180 MPa. Some examples of high-strength concrete are introduced in Section 23.9.

23.3.2.3 Reinforcing Steels

Reinforcing bars with guaranteed yield strength (f_{sy}) of 295 MPa and 345 MPa are generally applied to both super- and substructures, while a reinforcing bar with f_{sy} up to 490 MPa is defined in JIS-G-3112. Although the application range of SHB is up to 345 MPa, high-strength rebar, f_{sy} up to 685 MPa, and has been applied to some pier columns for highway bridges in order to improve workability in special cases. Furthermore, high-strength rebar, f_{sy} up to 785 MPa, has been applied to railway bridges after confirmation through analyses and experiments. The application range of DSRS is up to 390 MPa.

23.3.3 New Materials

23.3.3.1 Fiber-Reinforced Polymer

For more than 20 years, fiber-reinforced polymer (FRP) has been tried to be applied to infrastructures such as bridges. This is because FRP is not only a light and high-strength material but also a highly durable material. In the 1980s, FRP reinforcements for construction, such as FRP rods, cables, and grids were developed. After that, research to apply FRP to concrete structures became active. However, in the construction field, FRP reinforcement was not easily accepted because the cost-effectiveness of FRP was inferior to such conventional materials as steel. In the early 1990s, research to use FRP reinforcement such as FRP sheets for repairing and/or strengthening existing structures became active. The lightweight properties and workability of FRP sheets have drawn attention because they do not require heavy facilities and human power is enough at work site to retrofit existing structures with FRP sheets. Today, the FRP sheet bonding method is very popular for seismic retrofitting of concrete pier columns and strengthening of concrete superstructures in the rehabilitation field.

Figure 23.7 shows the first application of CFRP (carbon fiber–reinforced polymer) sheets bonding to corroded steel members. The repair work was carried out in 2007 in order to restore the original

FIGURE 23.7   Application of CFRP sheets bonding.

TABLE 23.5   Strength and Cost Performance of Aluminum Alloy

	Aluminum Alloy		Steel
	5083-O	6061-T6	JIS-SM400
Yield Stress'/Unit Weight	4.8	9.2	3.1
Unit Price/Yield Stress*	4.8	2.5	3.3

*The yield strength at 0.2% offset

cross-sectional properties in Asari River Bridge located on the route of the Chuo Expressway, known for heavy traffic. Not only initial cost but also LCC (life cycle cost) are important for long-term maintenance. It is possible to use FRP for improving the durability of structures because FRP itself is highly durable against corrosion. Since the application of FRP to infrastructures for maintenance is rather new technology, its long-term performance will be monitored.

23.3.3.2 Aluminum

The structural aluminum alloy specified in JIS conforms to the international alloy designation system, which is the most widely accepted naming scheme for wrought alloys. Each alloy is given a four-digit number in this naming scheme. The Research Institute of Aluminum Structures in Japan recommends applying 6000 series with the minimum yield strength of 250 MPa for bridge structures because of its good cost performance as shown in Table 23.5. The 6000 series is alloyed with magnesium and silicon, easy to process by machine, and can be precipitation hardened.

23.4 Girder Bridges

23.4.1 Steel Girder Bridges

The first iron plate girder bridge in Japan was Kurogane Bridge in Nagasaki, and is a riveted cast iron bridge completed in 1868. It was reconstructed three times, most recently completed in 1990 as a walkway bridge. Imajuku Bridge (bridge length 9.5 m; width 3.2 m) was completed in Kanagawa in 1931, and was the first welded steel plate girder bridge in Japan. It is the oldest welded highway steel bridge in Japan. In 1952, Hon Kyu Bashi Bridge in Hyogo (bridge length 112.9 m; maximum span length 24.6 m; width 9 m) was completed. It was the first all-welded highway steel plate girder bridge in Japan, and is a five-span cantilever-type plate girder bridge. Since then, many welded steel plate girder bridges and box girder bridges have been constructed.

Steel girder bridges are the most popular bridge type for medium span lengths ranging from 30 m to 200 m, especially in urban highway viaducts. Since the mid-1990s, the concept for optimum design has widely changed. The minimum fabrication cost design, instead of the minimum steel weight design, became considered the optimum bridge design procedure, because of rising labor costs. One of the solutions to the minimum fabrication cost design is twin I-girder bridges. They have several advantages compared with conventional multiple girder bridges, such that construction costs can be reduced due to their structural simplicity. Twin I-girder bridges have been specially developed for the New Tomei Expressway project, headed by the Japan Highway Public Corporation (currently Nippon Expressway Company Limited). Since 1995, more than 800 bridges of this type have been constructed in Japan.

Figure 23.8 shows Warashinagawa Bridge (total bridge length 1005 m; maximum span length 75 m; width 18.01 m). This bridge was completed in 2003 on New Tomei Expressway and the total steel weight is 6938 tons. This was the first twin I-girder bridge in Japan constructed with long-span cast-in-place prestressed concrete slabs. The span length of the slabs is 11 m [10]. When Warashinagawa Bridge was constructed, rapid construction, cost savings, and structural durability were required. To satisfy these demands, various advanced construction technologies, such as a large traveling form and advanced field joint methods for deck slabs, were adopted.

Trans-Tokyo Bay Highway (TTB; total length 15.1 km) was completed in 1997 and connects Kawasaki City in Kanagawa, on the west side of the bay, to Kisarazu City in Chiba, on the east side of the bay. TTB consists of a 4424 m long elevated highway bridge, the Tokyo Bay Aqua-Line Bridge, shown in Figure 23.9, and a 9.6 km long tunnel under Tokyo Bay [11,12]. The bridges are continuous steel box girders whose span lengths decrease gradually from two 240 m central spans to the approach spans. At the maximum span, a set of special tuned dumpers is installed in the box girder to prevent wind-induced vibration. The bridges are set onto the Y-shaped piers, which are covered with titan clad steel plates for low-maintenance corrosion protection system [13]. Figure 23.10 shows one of the recently developed girder bridges in Japan, the Kobe Newtransit concrete-filled tube (CFT) girder bridge (total bridge length 193.3 m; span length 87.9 m; bridge width 7.5 m), completed in Kobe in 2004. This bridge is the guideway girder for the new transportation system and consists of CFT for the lower flange to increase the rigidity of girder and to decrease construction cost [14]. The total steel weight is 550t.

In recent years, steel girder bridges with a particular purpose have been developed. As the traffic volume has increased in urban areas, grade-separated crossings are required by constructing under-passes or overpasses. However, construction works for grade-separated road crossover bridges or flyover bridges over railways in heavy-traffic urban areas have often caused additional serious traffic jams, and

FIGURE 23.8   Warashinagawa Bridge (Shizuoka). (Courtesy of Japan Bridge Association [JBA].)

FIGURE 23.9 Tokyo Bay Aqua-Line Bridge (Chiba). (Courtesy of JBA.)

FIGURE 23.10 Kobe Newtransit CFT girder bridge (Hyogo). (Courtesy of JBA.)

consequently most reconstruction plans have not yet been realized. To solve this problem, several "quick construction methods" for crossover bridges, in which steel plate girders or box girders are used for the main girders, have been developed.

Shin-Koiwa Overpass (length of bridge 557 m; maximum span length 26 m; width of bridge 9 m; total steel weight 2780 tons), shown in Figure 23.11, is one of the typical applications of "quick construction methods." The bridge was completed in 2007. It is a four-span continuous steel orthotropic box girder bridge. The foundations of the steel box piers are steel footings. In this case, the flyover bridge, with a length of about 818 m, was completed in only 108 days. Using a conventional construction method, bridge construction would have taken more than four years. Many benefits were revealed by shortening the construction period, such as eliminating traffic jams and decreasing construction cost.

23.4.2 Concrete Girder Bridges

Construction of reinforced concrete girder bridges in Japan started with the Wakasa Bridge (length of bridge 3.66 m) in 1903, 30 years later than the West. In the same year, Hinooka No.11 Bridge (length of bridge 7.28m; width of bridge 1.6 m) as shown in Figure 23.12, which is the oldest existing concrete bridge in Japan, was also completed in Kyoto. This bridge is a Melan-type concrete bridge, reinforced

FIGURE 23.11   Shin-Koiwa Overpass (Tokyo).

FIGURE 23.12   Hinooka No. 11 Bridge (Kyoto). (Courtesy of A. Kurebayashi of Tokyo Metropolitan Expressway Company Ltd.)

by steel frames. Until prestressed concrete technology was introduced to Japan in the 1950s, a number of concrete girder bridges were constructed as the typical type of concrete bridge, extending the span length of concrete bridges. In 1941, Tokachi Bridge was completed, which had a maximum span length of 41 m, the longest in Japan at that time. This bridge was an 11-span gerber girder bridge, with the largest bridge surface of 6987 m² in the world at that time; however, the bridge was demolished in 1996. New Tokachi Ohasi Bridge, which is a three-span concrete cable-stayed bridge with a center span of 251 m, was completed in 1996.

In the beginning of the 1950s, the construction of prestressed concrete bridges began. The first prestressed concrete bridge in Japan, Chosei Bridge (length of bridge 11.6 m; span length 3.6 m), was completed in Ishikawa in 1952. Figure 23.13 shows Daiichi-Daidogawa Bridge (span length 30 m; width of bridge 4 m), which was completed in Shiga in 1954 and is still in service for the Sigaraki Kougen railway [15]. In 1959, Ranzan Bridge (length of bridge 75 m; span length 51.2 m) was constructed by the cantilever erection method, and extension of the span lengths of prestressed concrete bridges proceeded.

FIGURE 23.13 Daiichi-Daidogawa Bridge (Shiga). (Courtesy of S. Yajima of JR West Japan Consultants Company.)

FIGURE 23.14 Hamana Bridge (Shizuoka).

Figure 23.14 shows Hamana Bridge, which was completed in Shizuoka in 1974. This bridge is a five-span continuous rigid frame–type concrete bridge with a central hinge; the center span length of 240 m was the longest in the world at that time [16]. Ejima Ohasi Bridge, which has a span length of 250 m, was completed in 2005.

Furukawa Viaduct (length of bridge 1475 m; maximum span length 45.5m; width of bridge 15.97–17.27 m), as shown in Figure 23.15, was constructed by a unique method in 2002. This viaduct consists of a PC 9-span continuous bridge and a PC 13-span continuous bridge. The erection method using lightweight U-shaped precast segments with ribs was applied, because precast segments were transported from the production yard to the erection site via an ordinary road. The viaduct was erected using the span-by-span method and slab deck concrete was cast using a precast mold [17].

23.4.3 Hybrid Girder Bridges

Normally, a steel girder whose section is made up of steel plates with different strengths has been called a hybrid girder. While a number of hybrid steel girder bridges have been built as a type of steel girder bridge, steel–concrete composite or mixed girder bridges have been built for several decades. In Japan,

FIGURE 23.15   Furukawa Viaduct (Mie). (Courtesy of Dr. A. Kasuga of Sumitomo Mitsui Construction Co., Ltd.)

FIGURE 23.16   Kinugawa Bridge (Tochigi). (Courtesy of Dr. A. Kasuga of Sumitomo Mitsui Construction Co., Ltd.)

the term "hybrid bridge" has been used as a general term for steel–concrete composite bridges and mixed bridges. Concrete girder bridges with corrugated steel webs became typical composite girder bridges in Japan. In 1993, Sinkai Bridge (span length 30 m) was completed in Niigata, and it was the first application of this type in Japan. In 1998, Hontani Bridge (maximum span length 97.2 m) was completed on the route of an expressway in Gifu. This bridge is a three-span continuous rigid frame–type corrugated steel web bridge. More than 100 bridges of this type have been constructed in Japan since 1998.

Kinugawa Bridge (length of bridge 1005 m; maximum span length 71.9 m; width of bridge 10.7 m), as shown in Figure 23.16, was completed in Tochigi in 2006. This bridge is a 16-span continuous corrugated steel web bridge and was erected by the rapid erection method using a light moving platform. Pre-erected corrugated steel web was used as a part of the platform during the erection. Torisakigawa Bridge (length of bridge 554 m; maximum span length 56 m; width of bridge 11.3 m), as shown in Figure 23.17,

FIGURE 23.17   Torizakigawa Bridge (Hokkaido). (Courtesy of N. Watanabe of Taisei Corporation.)

FIGURE 23.18   Sugitanigawa Bridge (Siga). (Courtesy of H. Kobayashi of P.S. Mitsubishi Construction.)

was completed in Hokkaido in 2007. This bridge is an 11-span continuous corrugated steel web bridge. This bridge was erected by the launching erection method using a corrugated steel web as a launching nose; the erection cost was reduced by this special erection method [18].

Sugitanigawa Bridge (length of bridge 445.4 m; maximum span length 94 m; width of bridge 11.46 m), as shown in Figure 23.18, was completed in Shiga in 2007. This bridge is a six-span continuous rigid frame corrugated steel web bridge. A new platform, in which the concurrent erection of three brocks is possible, was applied [19]. The corrugated steel web is also applied to girders for concrete cable-stayed bridges and extradosed-type cable-stayed bridges (see Sections 7.2.2 and 7.3).

A composite truss bridge was introduced as a type of hybrid bridge aiming at reducing the weight of concrete girders. Several node systems at the panel point between the concrete slabs and steel truss members of a hybrid truss bridge were developed in Japan. Kinokawa Viaduct (length of bridge 268 m; maximum span length 85 m; width of bridge 11.15 m), as shown in Figure 23.19, was completed in Wakayama in 2003. This bridge is a four-span continuous steel and concrete hybrid truss bridge [20]. There are some mixed girder bridges in Japan as well. Several types of connecting systems between steel girders and concrete girders were developed in order to complete mixed girder bridges.

FIGURE 23.19 Kinokawa Viaduct (Wakayama).

23.5 Arch Bridges

23.5.1 Steel Arch Bridges

Steel arch bridges in Japan have been applied to medium- and long-span bridges ranging from 40 m to 300 m. This type of bridge is often selected for aesthetic reasons. The first steel highway arch bridge, Eitaibashi Bridge in Tokyo (bridge length 184.7 m; bridge width 22 m), was constructed in 1926 and it is the oldest existing steel tied arch bridge in Japan. This bridge was reconstructed from a steel truss bridge with wooden decks to the steel arch bridge as part of the Great Kanto Earthquake Reconstruction Project.

Ujina Bridge (total bridge length 550 m; main span length 270 m; bridge width 18.8 m), as shown in Figure 23.20, was completed in 1999 and the total steel weight is 8585 tons. This is a three-span continuous box girder bridge with steel deck slab reinforced with a monochord arch member, and it has the longest girder span length in Japan. This bridge was designed as a symbol of the entrance of the southeastern part of the Hiroshima harbor area [21].

Goshiki Zakura Big Bridge (total bridge length 142.2 m; span length 142.2 m; bridge width 17 m), as shown in Figure 23.21, was completed in 2002 and the total steel weight is 4036 tons. This bridge was the first double-deck-type Nielsen arch expressway bridge in the world without upper lateral bracings. A new electric generation system was attempted in this bridge: electricity is generated by converting the energy created by the vibration of structural members caused by automobiles passing on the bridge. This system is expected to be applied to other bridges of the Metropolitan Expressway Co. Ltd., the owner of the bridges. The system is not yet able to generate a lot of power, but the electricity is enough to provide power for some lighting facilities on the bridge.

The New Kitakyushu Airport Access access bridge (total length 2100 m; main span length 210 m), as shown in Figure 23.22, was completed in 2005, and has a unique configuration. This bridge connects the man-made island airport, New Kitakyushu Airport, and the Kyushu mainland. It is the first balanced-type steel arch bridge with a monochord arch rib in Japan and a ten-span continuous steel box girder bridge [22].

New Saikai Bridge (bridge length 620 m; main bridge 300 m + approach bridge 320 m; bridge width 14 m), as shown in Figure 23.23, is a steel braced–rib half-through arch bridge with a 230 m arch span.

FIGURE 23.20 Ujina Bridge (Hiroshima). (Courtesy of JBA.)

FIGURE 23.21 Goshiki Zakura Big Bridge (Tokyo). (Courtesy of JBA.)

FIGURE 23.22 New Kitakyusyu Airport access bridge (Fukuoka). (Courtesy of JBA.)

FIGURE 23.23   New Saikai Bridge (Nagasaki). (Courtesy of JBA.)

The bridge was designed from an aesthetic viewpoint to harmonize with the existing Saikai Bridge, which marks the beginning of long-span bridge construction in Japan. Each arch rib consists of three steel pipes, which are filled with high-fluidity concrete (CFT) to obtain high local buckling strength and to increase the concrete compressive strength. The quality and performance of the filled concrete have been ensured by a series of full-scale loading tests.

23.5.2 Concrete Arch Bridges

Japan's first concrete arch bridge was Oiwa Bridge (length of bridge 12.6 m), which was completed in Kyoto in 1904. Until 1925, many concrete arch bridges with medium span lengths were constructed. Typical concrete arch bridges include Shijo Ohasi Bridge and Shichijo Ohasi Bridge (length of bridge 112 m; width of bridge 18 m) both completed in 1913, crossing over Kamogawa River in Kyoto. From that time on, the span length of concrete arch bridges has increased. Bandai Bridge (length of bridge 309.8 m; main span length 42.37 m; width of bridge 22 m), as shown in Figure 23.24, was completed in Niigata in 1929. This bridge is a six-span concrete arch bridge and is still in service.

In the 1970s, the cantilever erection method was introduced in the construction of arch bridges and the span length of arch bridges rapidly increased. Figure 23.25 shows Hokawazu Bridge (span length 170 m; width of bridge 10.1 m), completed in 1974. This bridge was erected by the cantilever truss erection method for the first time in the world [23]. Figure 23.26 shows Akayagawa Bridge (span length 116 m; width of bridge 12 m) completed for railway service in Gunma in 1979. This bridge is a reverse Langer arch bridge. Beppu Myoban Bridge, completed in 1989, has an arch span length of 235 m; the span length was the longest in the East at that time. Figure 23.27 shows Ikeda Hesokko Bridge (span length 200 m; width of bridge 10.4 m) completed in Tokushima in 1999. This bridge is a five-span continuous reverse Langer arch bridge and has a peculiar feature of a balanced arch bridge [24,25].

23.5.3 Hybrid Arch Bridges

Figure 23.28 shows Fujikawa Bridge (length of bridge up line 365 m, down line 381 m; maximum span length 265 m; width of bridge 18.05 m) completed on the New Tomei Expressway in Shizuoka in 2005. This bridge is a steel–concrete hybrid arch bridge with concrete arch ribs and a composite twin I-girder bridge supported by concrete pillars installed on the arch ribs [26].

FIGURE 23.24   Bandai Bridge (Niigata).

FIGURE 23.25   Hokawazu Bridge (Saga). (Courtesy of Dr. A. Kasuga of Sumitomo Mitsui Construction Co., Ltd.)

FIGURE 23.26   Akayagawa Bridge (Gunma).

FIGURE 23.27   Ikeda Hesokko Bridge (Tokushima).

FIGURE 23.28   Fujikawa Bridge (Shizuoka). (Courtesy of N. Watanabe of Taisei Corporation.)

23.5.4 Stone Arch Bridges

High-quality stones, which are applicable to structures, have been obtained in the Kyushu area, on the west side of Japan. Hence, stone bridges are seen mainly in the Kyushu area. The longest span of a stone arch bridge in Japan is 90 m. Megane Bridge (*megane* means "eyeglass"), as shown in Figure 23.29, is located in Nagasaki and is the oldest stone bridge in Japan. The bridge, consisting of two arches with a bridge length of 22.8 m, was completed in 1634. Two circles can be seen from the combination of the bridge itself and its reflected shape on the surface of the river water, hence, it has been called the Eyeglass Bridge.

Tsujun Bridge is a waterway bridge connecting two residential areas. It is located in Kumamoto and was constructed in 1854. The bridge length is 47.5 m and the arch span is 28.2 m. Three stone water pipes installed on the bridge carry water. Figure 23.30 shows Tsujun Bridge. Water spray from the crown of the arch can be seen. The main reason for the spray is to remove accumulated sand and mud in the stone waterway, and the water spray can be seen in farmer's off season.

23.5.5 Timber Arch Bridges

Kintai Bridge (bridge length 193.3 m) is a timber bridge consisting of three arch bridges (span length 35.1 m) at the center three spans and two girder bridges (span length 34.8 m) at both end spans, as shown in Figure 23.31. The original Kintai Bridge, designed by Mr. K. Kodama, was constructed in 1673 under a feudal lord. This bridge spanned over Nishiki River to link Iwakuni Castle to a residential area

FIGURE 23.29 Megane Bridge (Nagasaki).

FIGURE 23.30 Tsujun Bridge (Kumamoto).

FIGURE 23.31 Kintai Bridge (Ymaguchi).

of vassals. In the next year, however, the bridge collapsed due to a flood, and was reconstructed within the same year by reinforcing the stone abutments from the original design. Since this reconstruction, the Kintai Bridge was not destroyed by flood for 276 years, although periodical maintenance and minor improvement of structural details were carried out. In 1950, some of the piers were washed away by a flood due to a typhoon, and reconstruction started in 1951 and was completed in 1953. The most recent damage to the bridge by flood occurred in 2005, and it has been restored and hitherto exists as it is.

23.6 Truss Bridges

23.6.1 Steel Truss Bridges

Steel truss bridges have been applied to medium and long span lengths ranging 40 m to 500 m in Japan. Construction of steel truss bridges in Japan started with imported bridges from Europe.

The first truss bridge, Yoshida Bridge in Kanagawa (bridge length 24 m; bridge width 6 m), was completed in 1869. It was designed by an English engineer and made of iron. The bridge has been reconstructed five times. Shinsai-Bashi Bridge in Osaka, which was imported from Germany, was completed in 1873. It was originally made of wrought iron and was reconstructed as a stone arch bridge in 1900.

The first steel truss highway bridge in Japan was Eitai-Bashi Bridge in Tokyo, completed in 1897. As mentioned, the bridge collapsed in the Great Kanto earthquake and was rebuilt as a steel arch bridge in 1926. Third Chikumagawa Bridge on the Hokuriku Shinkansen line (bridge length 309 m; span length 103 m), as shown in Figure 23.32, was completed in 1996 and the total steel weight is 1885 tons. This bridge is a double truck railway bridge and one of the largest railway truss bridges in Japan. This bridge consists of weathering steel and its color matches with the environment.

Takehana No. 3 Bridge, shown in Figure 23.33, was completed in 2000 on Shikoku Island. This bridge is a cable-trussed girder bridge constructed on Tokushima Expressway [27]. This was the first application of a cable-trussed girder bridge to a highway bridge in Japan. The cable-trussed bridge (or reversed cable-stayed girder bridge) consisting of relatively slender steel plate I-girders, a spatial frame-type post arranged beneath the girders at the middle of the span, and external cables anchored at the ends of the girders. Ikitsuki Bridge, shown in Figure 23.34, has the longest center span length in Japan, at 400 m. Shibakura Bridge (bridge length 163 m; span length 160.4 m; bridge width 10.8 m), shown in Figure 23.35, was completed in 2000 and the total steel weight is 1445 tons. This bridge is a simply supported highway truss bridge with the longest span in Japan.

Sky Gate Bridge R, shown in Figure 23.36, is the access bridge to Kansai International Airport completed in 1992 and has a total length of 3750 m. This bridge is the longest truss bridge in the world (structural height from 13.7 m to 17.5 m; bridge width 30 m), and consists of a three-span continuous steel

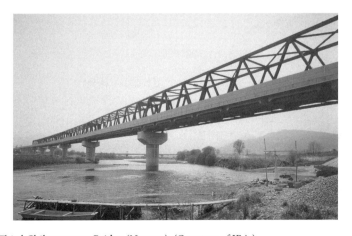

FIGURE 23.32 Third Chikumagawa Bridge (Nagano). (Courtesy of JBA.)

FIGURE 23.33 Takehana No. 3 Bridge (Tokushima). (Courtesy of West Nippon Expressway Company Ltd.)

FIGURE 23.34 Ikitsuki Bridge (Nagasaki). (Courtesy of JBA.)

FIGURE 23.35 Shibakura Bridge (Akita). (Courtesy of JBA.)

FIGURE 23.36   Sky Gate Bridge R (Osaka). (Courtesy of JBA.)

FIGURE 23.37   (**See color insert.**) Tokyo Gate Bridge (Tokyo). (Courtesy of JBA.)

truss, a three-span continuous steel box girder, a two-span continuous steel box girder, and simply supported steel box girders. The bridge carries both roadway and railway access to the airport. The bridge is double-level type and has a six-lane highway on the upper level and a two-track railway on the lower level. Electricity, gas, water service, telephone, and all others utilities in the airport run through the bridge.

The Tokyo Bay Highway has a total length of approximately 8 km, connecting the man-made landfill islands in the Tokyo Port. The highway consists of a submersed tunnel (in service) and a long-span bridge (under construction). Tokyo Gate Bridge, as illustrated in Figure 23.37, is a series of bridges, in which the major bridge is designed as a three-span continuous steel hybrid bridge combined with truss members and single-cell box girder (total length 760 m; main span length 440 m; span length 160 m + 440 m + 160 m). In this bridge, BHS was applied for the first time in Japan together with new advanced construction technologies, for the purpose of decreasing construction cost and increasing workability. The Tokyo Gate Bridge was completed in 2012.

23.6.2 Timber Truss Bridges

Great Karikobozu Bridge (total bridge length 140 m; span length 50 m; bridge width 7 m), shown in Figure 23.38, was completed in Miyazaki in 2002 and the total steel weight is 305 tons. This bridge is one of the world's biggest timber king post–type truss roadway bridges [28]. As Miyazaki prefecture is the

FIGURE 23.38   Great Karikobozu Bridge (Miyazaki).

most productive county of cedar timber in Japan, cedar glue laminated timber was used as the material for the main structural member of this bridge to promote the local forestry industry and preserve the landscape and the environment [29].

23.7 Cable-Supported Bridges

23.7.1 Suspension Bridges

Steel suspension bridges are applied to relatively wide span ranges from 70 m to 1991 m in Japan. Most suspension bridges with small spans are unstiffened pedestrian suspension bridges over deep valleys, and those with long spans are strait bridges. The oldest modern steel suspension bridge in Japan is Mino Bridge in Gifu (span length 116 m; bridge width 3.1 m), completed in 1915. It is a simple span suspension bridge with a stiffening girder and its main towers are made of concrete. This bridge was designated an important national cultural landmark in 2003. Konohana Bridge (bridge length 540 m; main span length 300 m; bridge width 23.5 m), shown in Figure 23.39, was completed in 1990. This bridge is the first monocable self-anchored suspension bridge with diagonal hangers in the world. The total steel weight is 9643 tons.

Akashi Kaikyo Bridge, also known as the Pearl Bridge (bridge length 3911 m; center span length 1991 m, bridge width 30 m), shown in Figure 23.40, was completed in 1998. This bridge is the longest suspension bridge in the world. The total steel weight is 200,000 tons. It links Kobe on the mainland of Honshu to Awaji on Awaji Island by crossing the Akashi Strait, an important international waterway. It is a part of the Honshu–Shikoku Highway and one of the key links of the Honshu–Shikoku Bridge project, which includes three routes across the Seto Inland Sea. The center span was originally 1990 m, however, the 1995 Hyogoken-Nanbu earthquake on January 17, 1995 moved the two towers and the span increased by 1 m.

The bridge was designed as a two-hinged stiffening girder bridge, allowing the structure to withstand winds of 286 km/h, earthquake magnitudes of 8.5 on the Richter scale, and harsh sea currents. The bridge also contains pendulum-type tuned mass dampers (TMD) in the tall towers that are designed to operate at the resonance frequency of the towers to damp forces. The two main supporting towers rise 298 m above sea level, and the bridge can expand up to 2 m a day due to temperature changes. Each concrete anchorage weighs 350,000 tons. The total length of the steel cables is 300,000 km and each cable is 112 cm in diameter and contains 36,830 strands of wires. The wires have been protected by an innovative cable protection method called the dry air injection system (DAIS), which was the first application in the world. Dry air is injected among wires in the cables and water and moisture are removed. Since the opening of the bridge in 1998, the condition of the cables has been monitored and inspected every 10 years.

FIGURE 23.39   Konohana Bridge (Osaka). (Courtesy of JBA.)

FIGURE 23.40   **(See color insert.)** Akashi Kaikyo Bridge (Hyogo).

Kurushima Kaikyo Bridge (total bridge lengths 4 km), shown in Figure 23.41, was completed in 1999. The bridge is also a part of the Honshu–Shikoku Bridge project and consists of three successive suspension bridges with six towers and four anchorages. It spans the 4 km wide Kurushima Strait in Seto Inland Sea. The strait is well known for rapid currents of over 8 knots. The bridge is located at the entrance to Seto Inland Sea and was designed to allow ocean-going vessel passage. The configuration of three consecutive suspension bridges was selected by taking into consideration its superior conservation of the natural environment, navigational safety, and the drivability of vehicles on the highway. The design and construction of the bridges are the culmination of advanced technologies developed and adapted during the design and construction stages of the Seto Bridge and the Akashi Kaikyo Bridge projects, through the cooperative effort of various bridge research and testing programs.

Kokonoe "Yume" Otsurihashi, shown in Figure 23.42, is the longest pedestrian unstiffened suspension bridge in Japan. In order to construct the bridge, with its main span of 390 m and road surface of 173 m above the ground, a wind tunnel test and dynamic analyses to assess the stability of the bridge against wind and pedestrian loads were carried out, and design verification was executed in field excitation tests. The bridge was constructed only for tourism and it has brought considerable economic benefits to the region and contributed to its promotion.

FIGURE 23.41   Kurushima Kaikyo Bridge (Ehime). (Courtesy of JBA.)

FIGURE 23.42   Kokonoe "Yume" Otsurihashi (Oita). (Courtesy of JBA.)

23.7.2 Cable-Stayed Bridges

23.7.2.1 Cable-Stayed Steel Bridges

The first cable-stayed bridge in Japan, Katsuse Bridge in Kanagawa (bridge length 128 m; bridge width 4 m), was completed in 1959. It is a recycled bridge that was originally a suspension bridge. The cable-stayed bridge was constructed using its main towers made of concrete. This bridge, however, was removed due to deterioration in 2010 and the New Katsuse Bridge (bridge length 270 m; bridge width 12.5 m; cable-stayed bridge) was completed in 2006 adjacent to the old bridge.

Meiko Triton Bridge, shown in Figure 23.43, consists of three cable-stayed bridges crossing the port of Nagoya, and was completed in 1998. The three bridges are Meiko East Bridge (bridge length 700 m; main span length 410 m; bridge width 29 m), Meiko Central Bridge (bridge length 1170 m; main span length 590 m; bridge width 29 m), and Meiko West Bridge (bridge length 758 m; main span length 405 m; bridge width 13.8 m). The total steel weights are 21,760 tons, 38,700 tons, and 11,448 tons, respectively.

FIGURE 23.43   Meiko Triton Bridges (Aichi). (Courtesy of JBA.)

FIGURE 23.44   Tatara Bridge (Hirohsima–Ehime). (Courtesy of JBA.)

Meiko Central Bridge has a center span length of 590 m, which was the second-longest cable-stayed bridge in the world at its completion. Meiko West Bridge consists of two cable-stayed bridges, located in parallel, which were completed in 1985 and 1997, respectively. All of their pylons are A-shaped.

Tatara Bridge (bridge length 1480 m; main span length 890 m; bridge width 30.6 m) shown in Figure 23.44, was completed in 1999 and the total steel weight is 37,300 tons [30]. Tatara Bridge forms a part of the 60 km long West–Seto Expressway, which is one of the three routes built by the Honshu–Shikoku Bridge Authority, connecting seven islands, and which is formed by 10 bridges of various different structural typologies. This bridge was the longest cable-stayed bridge in the world when it was completed, supplanting the Normandie Bridge, which has a center span of 856 m. The superstructure is a mixed-type hybrid structure that consists of a steel stiffening girder for the center span and a concrete girder for a part of the side spans.

Hiroshima Nishi Ohashi Bridge (bridge length 476.5 m; main span length 78 m; bridge width 18.7 m), shown in Figure 23.45, was completed in 2001 and the total steel weight is 4529 tons. This bridge is a seven-span continuous cable-stayed bridge and is located at the entrance of the center of Hiroshima, so that the bridge was designed as a symbol of the entrance.

Katsushika Harp Bridge (bridge length 455 m; span length 0.5 m + 134 m + 220 m + 60.5 m; bridge width 23.5 m), shown in Figure 23.46, was completed in 1987 and the total steel weight is 8119 tons.

FIGURE 23.45 Hiroshima Nishi Ohashi Bridge (Hiroshima). (Courtesy of JBA.)

FIGURE 23.46 Katsushika Harp Bridge (Tokyo). (Courtesy of JBA.)

This bridge is a four-span continuous cable-stayed S-curved bridge and was the first horizontally curved cable-stayed bridge in the world. The curve is S-shaped, so that its deck slab has a variable cross slope and its structural members have complex three-dimensional shapes. The towers are different heights; one is 65 m and another is 29 m. The bridge was named the Harp Bridge because of the shape of its 48 stay cables and curved main girders.

23.7.2.2 Cable-Stayed Concrete Bridges

Japan's first concrete cable-stayed bridge is EXPO East Gate Bridge (span length 37.8 m), completed in Osaka in 1969. In the 1970s several pedestrian bridges with short spans were constructed. In 1979, Omotogawa Bridge (span length 85 m; width of bridge 7 m), shown in Figure 23.47, was completed for a railway in Iwate. This bridge is a three-span continuous cable-stayed bridge and the first practical concrete cable-stayed bridge for a railway in Japan; prestressed diagonal tension members were applied [31]. In the 1980s, the span lengths of concrete cable-stayed bridges increased; Yobuko Bridge (span length 250 m) was completed in Saga in 1989.

FIGURE 23.47   Omotogawa Bridge (Iwate).

FIGURE 23.48   Oshiba Bridge (Hiroshima). (Courtesy of Dr. A. Kasuga of Sumitomo Mitsui Construction Co., Ltd.)

FIGURE 23.49   Yabegawa Bridge (Fukuoka). (Courtesy of Dr. A. Kasuga of Sumitomo Mitsui Construction Co., Ltd.)

In the 1990s, a number of concrete cable-stayed bridges were constructed; more than 120 concrete cable-stayed bridges have been completed in Japan. Figure 23.48 shows Oshiba Bridge (length of bridge 410 m; main span length 210 m; width of bridge 5 m), completed in Hiroshima in 1997. This bridge is a three-span continuous cable-stayed bridge and precast segments were applied to the main girder [32]. Figure 23.49 shows Yabegawa Bridge (length of bridge 517 m; center span length 261 m; width of bridge 20.2 m), completed in Fukuoka in 2008. This bridge is a three-span continuous concrete cable-stayed bridge and has the longest span of its kind in Japan [33]. Figure 23.50 shows Yahagigawa Bridge, also known as Toyota Arrow Bridge (length of bridge 820 m; center span length 235 m; width of bridge 43.8 m), which applied corrugated steel web first in the world and was completed in 2004. This bridge is a four-span continuous concrete and steel hybrid cable-stayed bridge [34]. Innovative technology on the post-tensioning cable system, the JSAS® system, was applied [35].

23.7.3 Extradosed-Type Cable-Stayed Bridges

In 1994, the world's first extradosed bridge, Odawara Blue Way Bridge, shown in Figure 23.51, was completed in Kanagawa. This bridge is a three-span continuous extradosed bridge; the span length is 122.3 m [36]. In 1995, Yashiro North Bridge (span length 90 m) and Yashiro South Bridge (span length 122.3 m) were completed. The stiffness of the girder in extradosed bridges is higher than in cable-stayed bridges, and the higher stiffness of the girder is advantageous against the fatigue of stay-cables. Since 2000, extradosed bridge construction has overtaken concrete cable-stayed bridges and more than 50 extradosed bridges have been completed.

FIGURE 23.50 Yahagigawa Bridge (Aichi).

FIGURE 23.51 Odawara Blue Way Bridge (Kanagawa). (Courtesy of Dr. A. Kasuga of Sumitomo Mitsui Construction Co., Ltd.)

Sannohe Bokyo Bridge (span length 180 m; width of bridge 10.25 m), a typical concrete extradosed bridge, was completed in Aomori in 2004. This bridge is a three-span continuous extradosed bridge. Figure 23.52 shows Himi Yume Bridge, with a corrugated steel web (length of bridge 365 m; main span length 180 m; width of bridge 9.75 m), which was completed in 2004. This bridge is a three-span continuous corrugated steel web extradosed bridge [37]. Figure 23.53 shows Sannai-Maruyama Bridge (span length 150 m; width of bridge 13.85 m), which was completed in 2008. This bridge is a four-span continuous extradosed bridge [38].

Figure 23.54 shows a hybrid extradosed bridge, Kiso River Bridge (length of bridge 1145 m; maximum span 275 m) and Ibi River Bridge (length of bridge 1397m; maximum span 271.5 m), which are composed of prestressed concrete girders and steel girders. Both bridges were completed in 2001. Kiso River Bridge is a five-span continuous concrete and steel hybrid bridge; Ibi River Bridge is a six-span continuous concrete and steel hybrid bridge [39].

23.7.4 Stress Ribbon Bridges

Japan's first stress-ribbon bridge was No. 9 Footway Bridge at EXPO'70 (span length 27m) shown in Figure 23.55. The bridge is a pedestrian bridge completed in 1969 for the Osaka International Exposition [40]. Figure 23.56 shows Seiun Bridge (span length 93.8 m), a self-balanced stress ribbon bridge. This bridge was completed in 2004 and has a very unique appearance. In the erection stage, the reaction forces were supported by ground anchors and the bridge was not self-balanced; however, the bridge was shifted to a self-balanced structure at completion. Therefore, ground anchors are not required at completion and stress in the ground is reduced [41].

FIGURE 23.52 Himi Yume Bridge (Nagasaki). (Courtesy of Dr. A. Kasuga of Sumitomo Mitsui Construction Co., Ltd.)

FIGURE 23.53 Sannai-Maruyama Bridge (Aomori). (Courtesy of Dr. H. Akiyama of The Zenitaka Corporation.)

FIGURE 23.54   Kiso River Bridge and Ibi River Bridge (Aichi and Mie).

FIGURE 23.55   No. 9 Footway Bridge at EXPO'70 (Osaka).

FIGURE 23.56   Seiun Bridge (Tokushima). (Courtesy of Dr. A. Kasuga of Sumitomo Mitsui Construction Co., Ltd.)

23.8 Movable Bridges

Yumemai-Ohashi Bridge (total bridge length 940 m; floating bridge length 410m; bridge width 33.8 m), shown in Figure 23.57, was completed in 2001. This bridge was the first floating swing bridge in the world. It connects two man-made landfill islands, Yumeshima Island and Maishima Island in the Port of Osaka. The superstructure is a steel arch with a double arch rib and it has two pontoons with which the bridge can move laterally. It allows the passage of large ships by swinging of the bridge itself in the main waterway of the Port of Osaka. The mooring system of the bridge consists of rubber fenders, steel reaction walls, steel beams, and dolphins, which support the bridge in the horizontal direction. When the bridge swings, the walls are laid down to move the bridge laterally and after swinging they are stood up again. Such a mooring system was the first trial in the world. Since the bridge is supported by huge pontoons, maintenance of the bridge, especially in the splash and tidal zones, becomes extremely difficult after bridge service starts. To avoid frequent maintenance due to corrosion, the titanium clad steel plates are used in the splash and tidal zones [42].

Kachidoki Bridge (total bridge length 246 m; center span length 51.6 m; side span length 86 m; bridge width 26.6 m; completed in 1940) is the one of the most well-known movable bridges in Japan. It carries Harumi-Dori Avenue over Sumida River in Tokyo to link Kyobashi, which used to be a center of commerce, and the old Tokyo landfill bay area. Kachidoki Bridge consists of three spans, as shown in Figure 23.58. The center span is a double-leaf trunnion bascule bridge with a span of 51.6 m from center

FIGURE 23.57 Yumemai-Ohashi Bridge (Osaka). (Courtesy of JBA.)

FIGURE 23.58 Kachidoki Bridge (Tokyo).

to center of the trunnions, which are the axis of rotation for the bascule leaf, and the side spans are steel tied arch bridges with a span of 86 m. From 1940 to 1945, the bridge was opened five times per day as general operation. However, as maritime traffic decreased, the number of openings per day reduced to three times from 1945 to 1963. Finally, the Tokyo metropolitan government decided to stop the bascule opening in order to avoid vehicular traffic congestion; the last opening was on December 29, 1970. Since 2005, a museum for Kachidoki Bridge has been open at a transformer substation beside the bridge. Original drawings and photos of Kachidoki Bridge are exhibited [43].

23.9 New Bridge Technologies

23.9.1 Seismic Design

Earthquake activities in Japan are very high and severe, and every structure has to be designed against earthquakes. The 1995 Hyogoken-Nanbu earthquake caused catastrophic damage to infrastructures, including bridges, in the Kobe area. After the earthquake, the bridge design specifications were changed, and the base isolation system and structural response control devices were applied in the seismic design of bridges. In addition, seismic retrofits of existing bridges were executed. In the base isolation design, the earthquake force on superstructures is reduced by a base isolation bearing, and in the structural response control, earthquake force is absorbed by control equipment such as seismic dampers. On the other hand, in the conventional seismic design, structural members are designed to resist seismic forces.

23.9.1.1 Seismically Isolated Bridges

Many rational and efficient optimal design methods for seismically isolated bridge systems against strong earthquake motion, which consist of superstructures, seismic-isolation bearings, piers, and pile foundations, have been developed and applied to the existing and newly constructed bridges. In Japan, application of seismic isolation bearings is the most economical and popular method for both highway bridges and railway bridges. Replacement of the conventional steel bearings (a set of fixed and movable bearings) with seismic isolation bearings (rubber bearings) has especially been carried out on the existing bridges for seismic retrofitting. Figure 23.59 shows Miyagawa Bridge (length of bridge 105.8 m), which is the first seismically isolated bridge in Japan [44]. The bridge is a three-span continuous girder bridge for road use.

FIGURE 23.59 Miyagawa Bridge (Shizuoka). (Courtesy of Nippon Steel Corporation and OILES Corporation.)

Railway bridges have more restrictions on seismic design than highway bridges, such as the limitation of displacement response of bridges. Figure 23.60 shows Kumegawa Bridge (bridge length 144.2 m: span length 71 m), completed in 2001, which is the first seismic isolation steel railway truss bridge in Japan, solving these difficulties using LRB (lead rubber bearing) for horizontal seismic force dispersion. In this bridge, however, the reduction of seismic force was not taken into account in the structural design, but was considered as the safety margin for the earthquake. Following this bridge, the seismic isolation method has been improved and applied to many railway bridges.

23.9.1.2 Structural Response-Controlled Bridges

Figure 23.61 shows Riverside Senshu Connecting Bridge (length of bridge 30.5 m; maximum span length 26 m; width of bridge 3.5 m), completed in 2007, which is the first structural response controlled bridge in Japan [45]. This bridge is a pedestrian bridge located in Niigata and a three-span continuous rigid frame concrete bridge with HiFleD (high-flexibility and damping) piers. HiFleD piers, which consist of slender columns connected by steel dampers, achieved a structural response control bridge. The superstructure is made of an ultra-high-strength fiber reinforced concrete (UFC) developed in Japan. It is called SUQCEM (super-high-quality cementitious material) and its bending compressive strength is 200 MPa.

23.9.1.3 Seismic Retrofit for Existing Bridges

Since the 1995 Hyogoken-Nanbu earthquake, seismic retrofitting of existing bridges has been hastened. The Minato Bridge (center span 510 m), shown in Figure 23.62, was completed in 1974. This bridge is one of the biggest cantilever truss bridges in the world. After the 1995 Hyogoken-Nanbu earthquake, the

FIGURE 23.60   Kumegawa Bridge (Nagano). (Courtesy of East Japan Railway Company.)

FIGURE 23.61   Riverside Senshu Connecting Bridge (Niigata).

FIGURE 23.62   Minato Bridge (Osaka). (Courtesy of JBA.)

Hanshin Expressway Public Corporation (the present Hanshin Expressway Company Limited) carried out seismic retrofit work on Minato Bridge as a part of a seismic structural reinforcement project in 2003. The earthquake-proof improvement was designed against Level 2 earthquake motion. For motion in the longitudinal direction, the floor frame earthquake-proof system using slip supports on the steel floor beams was adopted. For motion in the transverse direction, the diagonal truss damping system using buckling restraint braces (hysteresis type damper) was adopted.

23.9.2 Bridge Renewal Technology

23.9.2.1 Renewal of Deck Slabs

In a coastal environment, concrete is prone to deterioration due to airborne salt attack. Figure 23.63 shows Okubi River Bridge, which was renewed in 2008. This bridge is composed of a three-span continuous hollowed slab bridge (length of bridge 51.5 m; maximum span 17 m; width of bridge 9.255 m) and a six-span continuous hollowed slab bridge (length of bridge 102.3 m; maximum span 17 m; width of bridge 9.255 m). The bridge, originally completed in 1975, is located near the mouth of Okubi River in Okinawa. At the construction stage, marine sands, from which removal of salt was not enough, were used. The damage of the hollowed slab started after ten years of service and the superstructures dangerously deteriorated. Finally full renewal was decided upon based on analysis of the LCC of the bridge. At construction, precast concrete girders were applied to shorten the construction period. The original bridge was two piers without lateral beams, while in the new bridge, precast lateral beams were set on the piers in order to connect the precast concrete girders. All concrete includes ground granulated blast furnace slag to secure durability against salt attack.

23.9.2.2 Ultra Joint

Due to discontinuity at the joints of bridges, noise occurs, and driving comfortability deteriorates. Pavement near the joints is also damaged. The ultra joint method was developed to solve this problem [46]. In the method, old joints are removed and continuous pavement is made by using a precast engineered cementitious composite (ECC) slab. The ECC slab, which has high flexibility and high crack dispersion, follows deflection of the girders due to live loads and displacement of gap between the girders due to temperature change. Figure 23.64 shows the detail of the ultra joint method. This method was used at the Shibuya line on the Tokyo Metropolitan Expressway. The construction of one joint took 9 hours.

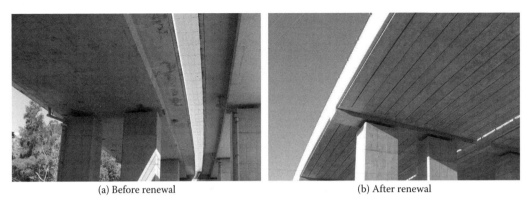

(a) Before renewal   (b) After renewal

FIGURE.23.63   Okubi River Bridge (Okinawa). (Courtesy of T. Yoshikawa of Oriental Shiraishi Corporation.)

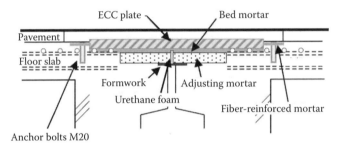

FIGURE 23.64   Detail of ultra joint.

FIGURE 23.65   Miyazu Bridge (Toyama).

23.9.2.3 Jointless Retrofit

Concrete box girders with central hinges were a popular structural type in the early history of prestressed concrete bridges. From 1959 to 1980, more than 90 bridges were constructed. Recently, the damage and function failure of the hinges, and traffic inconveniences due to creep deflection at the hinges have been observed. Jointless retrofit, which removes the central hinge and makes the center span continuous, has been applied to many bridges of this type; the bridge is renewed to a concrete box girder without a central hinge. Jointless retrofit abolishes joint which becomes obstacles to bridge maintenance and improves comfortability in drive [47]. Figure 23.65 shows Miyazu Bridge (length of bridge 259.4 m; center span length 115 m), which was renewed in 2007 by jointless retrofit.

23.9.3 Anticorrosion Technology

23.9.3.1 Dehumidification Systems

A new corrosion protection method different from previous common means has been applied to the inside of the box girder of the Shin-Onomichi Bridge (total length 546 m, center span 215 m, completed in 1999) shown in Figure 23.66. This bridge is a part of the West-Seto Expressway of the Honshu-Shikoku Bridge project, which opened to traffic in 1999 [48]. The previous method of corrosion protection of the interior of a box section used in the Honshu-Shikoku Bridge project was painting with nonorganic zinc-rich paint plus three layers of modified epoxy resin paint (each layer 90 μm thick), and this painting system is specified in the "Painting Standard for Steel Bridges and Structures, HSB," finally amended in April 1990.

The new method used in Shin-Onomichi Bridge is to suppress rust generation by dehumidifying the inside air with drying machines, instead of applying three layers of the modified epoxy resin paint. One drying machine has a capacity to handle the air of 300 m³/h and to remove moisture of 1.6 kg/h, while total air volume inside the box girder reaches 21,400 m³. Five drying machines are arranged to dehumidify the entire area of the bridge. When one of the sensors deployed inside the box girder indicates that the relative humidity is higher than 60%, the dryers begin to work until the humidity at every sensor drops below 40%. At approximately the center of the bridge, the air condition inside the box girder and the operation of the system are monitored. It is found that the inside humidity rises somewhat when rain falls, but it is confirmed that the inside moisture has been approximately kept at the target value and no rust has been observed so far. From now on, the air tightness of the box girder will to be improved, and then better and more economical operation is to be pursued through accumulation of the data.

23.9.3.2 Thermal Spraying (Metal Spraying)

In Japan, since the 1990s, a corrosion protection system using thermal spray coatings has been applied to many steel bridges located in corrosive atmospheres. Thus, this method was specified in the "Manual for Painting and Corrosion Protections for Steel Highway Bridges," revised in 2005 to reduce LCC by preventing corrosion. The manual includes the technical details and execution procedures of thermal spray coatings [49]. This corrosion protection method was applied to Nijubashi Bridge in the Imperial Palace for the first time in Japan in 1963, and after that it was applied to Kanmon Bridge for the first time in a Japanese long-span bridge in 1971.

FIGURE 23.66 Shin-Onomichi Bridge (Hiroshima). (Courtesy of JBA.)

FIGURE 23.67   Uminonakamichi Bridge (Fukuoka). (Courtesy of JBA.)

Thermal-sprayed coatings of zinc, aluminum, and their alloys have been proved to be cost-effective for long-term corrosion protection of steel in various natural environments. For sealants, which are used to fill inherent holes in sprayed coatings and increase the component life, both organic and inorganic types are available. For the top coat, chlorinated rubber paint was used previously, however, fluoro-polymer paint coating has become popular.

Figure 23.67 shows Uminonakamichi Bridge (total bridge length 260 m), completed in 1999. The entire surface of 14,000 m² of arch ribs were coated with thermal spray aluminum alloy coating with a thickness of over 160 μm. After sealing with inorganic-type sealant, two layers of fluoro-polymer paint coating were applied. In a part of 18 km long Fukuoka Expressway Route 5, which consists of continuous steel orthotropic box girders, thermal spray zinc–aluminum coating was applied. The applied surface area reached approximately 200,000 m². It is also notable that colored inorganic sealing was applied for the first time.

23.9.4 Deck Slabs for Plate Girders

23.9.4.1 Fatigue Damage and Repair of Orthotropic Steel Decks

The oldest existing steel bridge with an orthotropic steel deck in Japan is Nakazato Viaduct, which was constructed in 1954. From 1955 to 1965, construction of this type of bridge increased with the development of both design and fabrication technology. Orthotropic steel decks in this period were stiffened with open section ribs, such as steel strips and bulb steel flat bars. In the period of high-growth industries after 1965, this type of bridge was constructed increasingly and their stiffening ribs were shifted to a closed section type, such as an inverted trapezoid-shaped and inverted Y-shaped steel ribs, to obtain increased torsional rigidity [50]. Since the late 1970s, numerous long-span bridges had been constructed in Japan assisted by the technological development of cable-stayed bridges and suspension bridges. Following these technical trends, many steel girder bridges with orthotropic steel decks were applied to medium- and long-span bridges, especially for urban highway bridges, because the heights of the girders and weights of the bridges are limited in many cases of urban highway construction. Consequently, in Japan we have more orthotropic steel decks than other countries in the world.

In recent years, fatigue damage has been observed in orthotropic steel deck bridges 20 years or older that have been subjected to heavy traffic loads since their opening. The observed fatigue damage includes fatigue cracks in the welded joints of the longitudinal ribs with deck plate and closed cross sections, as shown in Figure 23.68. This type of fatigue cracks was not observed previously. Fatigue cracks in welded joints between deck plates and longitudinal ribs can be classified into weld bead cracks, which initiate

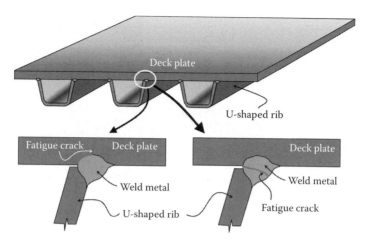

FIGURE 23.68   Fatigue cracking in orthotropic steel decks.

from the root of the welds and run along the weld bead direction, and deck plate cracks, which initiate from the root of the welds and run through the thickness of the deck plates. Urgent repair actions are required for deck plate cracks, because the cracks are found only after they reach the upper surface of the deck plates. This may cause an anomalous condition of the road surface and deck plate cracking and affect the safety of the moving vehicles. Most maintenance manuals for orthotropic steel decks focus mainly on weld bead cracks, so they are by no means adequate. Much effort has been expended to develop efficient ways to inspect and repair a large number of orthotropic steel deck bridges.

23.9.4.2 Steel–Concrete Composite Deck Slabs

Many types of bridge deck systems have been developed and applied to the decks of girder bridges in Japan. Among them, the cast-in-place reinforced concrete (CIP RC) deck is most commonly used. However, the major shortcoming of the CIP RC deck may be its low constructability during erection and low durability. Furthermore, in the design of steel box girder bridges, especially two box girder systems, the spacing of the girders is often controlled by the span length of the RC deck due to its low bending rigidity. For this reason, CIP RC decks with typical spans are usually supported by intermediate beams.

In recent years, stay-in-place steel forms have often been used to enhance the constructability of CIP RC decks for steel girder bridges. The steel forms, which consist of thin steel plates (around 10 mm thick), are generally used as tension members (bottom plates). If a composite action between the bottom plates and concrete is obtained, the bottom plates can partially act as a tensile reinforcement and the thickness of the deck can be reduced. To obtain enough composite action between the bottom plates and concrete, shear connecters can be used for providing the required horizontal shear resistance. Figure 23.69 shows two schematic examples of the typical steel-concrete composite bridge deck system applied in Japan. The left figure shows the deck system consisted of bottom steel plates, hot-rolled T-shapes with deformed flange surface which act as shear connectors and steel reinforcements [51]. The right figure shows another type of deck system that consists of bottom steel plates, perfobond ribs shear connectors, and concrete [52].

23.9.5 Bridge Management System

As mentioned in Section 23.1.2, there are more than 150,000 road bridges with span lengths of 15 m or more in Japan. In the period of rapid growth after the World War II (1955–1973), approximately 35% of total bridges were constructed. Generally speaking, maintenance of these bridges has not been carried out sufficiently because of ignorance and limited budgets. More than 50 years have passed since the beginning of the period; the time to replace or repair many bridges has arrived.

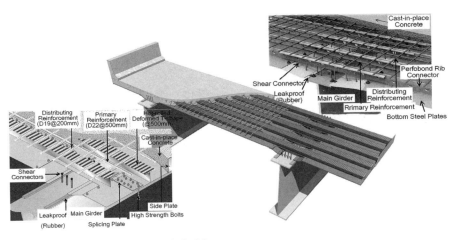

FIGURE 23.69   Steel–concrete composite deck slabs.

Until now, the repair of bridges was executed when degradation of the bridge was clear (so-called breakdown maintenance); however, this strategy often shortens the life span of the bridge and the total repair cost may be more expensive than preventive repair. Recently, the national and local governments' budgets for the maintenance of bridges have been limited because of their financial difficulties; therefore, the asset management of bridges has to be taken into account. It is recognized that the strategy to minimize the life cycle cost is appropriate for this purpose. In order to perform this strategy, many organizations and civil engineers have developed systems to minimize the life cycle cost of bridges by applying so-called preventive maintenance and asset management systems.

The Japanese government has supported local governments, prefectures, and cities, establishing a long-range repair plan of for bridges since 2007, because the extension of bridge service life and reduction of the cost of repair and replacement of bridges are crucial for the policy of bridge maintenance as infrastructures. The Ministry of Land, Infrastructure, Transport, and Tourism established a granted project for planning highway road repairs aiming to prolong their service life in 2007. Several local governments realized the importance of long-range repair plan and developed computer-based systems to apply for asset management of bridges. One of the systems, the Bridge Management System (BMS), is introduced next [53,54]. BMS is owned and maintained by the Regional Planning Institute of Osaka and several local governments started to apply it for asset management of their bridges. The Regional Planning Institute of Osaka intends to spread it among the local governments in Japan.

23.9.6 Bridge Health Monitoring

More than 80% of road bridges in Japan are managed by local governments; the national government manages less than 20% of road bridges. National road bridges, which are managed by the national government, are inspected periodically every 5 years; however, about 10% of prefectures and about 90% of cities did not conduct inspection of roads in 2006. Before local governments plan road repairs, they must understand the condition of existing bridges. Therefore, central government offers technical support such as lectures or guidelines to local governments that do not have inspection manuals. As a way to capture the current state of bridges, structural health monitoring (SHM) is promising.

SHM uses integrated sensing and collects data about the performance and integrity of a structure in relation to the system's expected safety, serviceability, and reliability to provide indicators when anomalies are detected in a structure. The sensing device introduced here is a laser Doppler vibrometer

FIGURE 23.70 Monitoring with LDV.

(LDV), shown in Figure 23.70. This is an optical instrument employing laser technology to measure velocity. In comparison with conventional transducers such as accelerometers, LDV makes it possible to conduct noncontact and long-distance measurement without adding mass or stiffness to an object. In addition, the resolution of velocity is very high, and frequency bandwidth is broad. Furthermore, by attaching a scanning unit on a laser sensor head, measurement of multiple points can be realized. LDVs have been applied to monitoring of steel box girder bridges, concrete viaducts, and cables in long-span bridges [55].

23.10 Future Prospective Bridges

23.10.1 Ultra-High-Strength Concrete Bridges

Ultra-high-strength concretes have been developed in Japan; they have bending compressive strengths ranging from 100 to 200 MPa. Although they are seldom used in actual bridge construction, they will change the figure and design of bridges in the near future. Figure 23.71 shows Akihabara pedestrian deck (length of bridge 63.7 m; maximum span length 33.2 m; width of bridge 8 m), completed in Tokyo in 2006. This bridge is a pedestrian bridge located at Akaihabara in Tokyo and is a curved PC two-span continuous bridge [56]. The superstructure is made of a newly developed low-shrinkage ultra-high-strength concrete called PowerCrete [57]. Because it has a bending compressive strength of 120 MPa, the girder depth is controlled within 1.2 m against the maximum span length of 33.2 m. An ultra-high-strength prestressing strand φ 15.2 mm was used for the first time in the project [58]. This concrete has solved the problem of ultra-high-strength concrete, which has high self-shrinkage. Ultra-high-strength concrete will be more widely used in the construction of bridges in the near future.

23.10.2 FRP Composite Deck Slabs

In Japan, FRP composite deck slabs, which consist of built-in-place formwork and reinforced concrete, have been developed since the 1990s and have been applied for some bridges. The first bridge with FRP composite deck slabs in Japan was Matsukubo Bridge (bridge length 142.5 m; bridge width 9.9 m), completed in 1997. FRP composite slabs are lightweight and high-corrosion resistant, so they have been applied for both newly constructed bridges and rehabilitation of deteriorated bridge decks.

The construction cost of FRP composite deck slabs is expensive; however, the total construction cost of these bridges, including both superstructures and substructures, may be competitive with bridges with RC deck slabs. Figure 23.72 shows Ushioshinmachi Route Bridge, which is a jetty-type bridge

FIGURE 23.71   Akihabara pedestrian deck (Tokyo).

FIGURE 23.72   Ushiosinmachi Route Bridge (Kochi). (Courtesy of JBA.)

spanning an inlet [59]. The sea level of the inlet is very close to the bottom of the deck slabs; therefore, FRP composite deck slabs were adopted because of their superior salt tolerance.

23.10.3 Aluminum Bridges

Recently, life cycle cost has become an important factor in the design of infrastructures, including bridges. Aluminum alloy is almost free of maintenance and lighter than concrete. Because of these characteristics, construction of bridges made of aluminum alloy has started [60]. Kinkei Bridge (span length 20 m) is the first aluminum bridge in Japan; the bridge was completed in Hyogo in 1961. This bridge is a single-span aluminum–concrete composite girder bridge. In the 1990s, more than 10 aluminum pedestrian bridges were completed; since then, several pedestrian bridges have been completed every year. Figure 23.73 shows Jubanminato Bridge (span length 30.9 m), which was completed in Kumamoto in 2002. This bridge is a single-span pony truss pedestrian bridge. In Japan the design load of automobiles has changed from 196 N to 245 N and reinforcement of concrete deck slabs and steel girders has

FIGURE 23.73   Jubanminato Bridge (Kumamoto). (Courtesy of JBA.)

hastened. Also, seismic reinforcement of piers has hastened since the 1995 Hyogoken-Nanbu earthquake. Aluminum deck slabs, which are lighter than concrete slabs, have been used for seismic reinforcement of existing bridges.

Acknowledgments

The authors would like to express our sincere gratitude to Mr. A. Kobayashi (Nippon Steel Materials Co., Ltd.) and Dr. T. Miyashita (Nagaoka University of Technology) for their contributions to Sections 23.3.3.1 and 23.9.6. The authors also would like to thank those who furnished the many photographs in this chapter.

References

1. Planning Division, Road Bureau, Ministry of Land, Infrastructure and Transport. 2009. *Annual Report on Road Statistics 2009* (in Japanese).
2. JRA. 2002. *Specifications for Highway Bridges: Part II Steel Bridges.* Japan Road Association, Maruzen, Tokyo, Japan (in Japanese).
3. RTRI. 2008. *Design Standards for Railway Structures and Commentary (Steel–Concrete Hybrid Structures).* Railway Technical Research Institute, Maruzen, Tokyo, Japan.
4. JSCE. 2007. Japan Society of Civil Engineers (JSCE). 2007. *Standard Specifications for Steel and Composite Structures,* Japan Society of Civil Engineers, Tokyo, Japan (in Japanese).
5. Nishimura, K., K. Matsui and N. Tsumura. 2005. High-Performance Steel Plates for Bridge Construction—High-Strength Steel Plates with Excellent Weldability Realizing Advanced Design for Rationalized Fabrication of Bridges. *JFE Technical Report* No. 5, pp. 30–36 (in Japanese).
6. Kodama, T. 2000. Weathering Steel in Coastal Atmospheres. *Corrosion Engineering* Vol. 49, No. 1, pp. 3–9.
7. Kiahara, H., M. Tanaka, H. Yasunami, et al. 2004. 3% Ni-Advanced Weathering Steel and Its Applicability Assessing Method. *Nippon Steel Technical Report* [*Shinnittetu Giho*] No. 380, pp. 28–32 (in Japanese).
8. JPCEA. 2008. *Guidelines for Design and Construction of High-Strength Concrete for Prestressed Concrete Structures,* Japan Prestressed Concrete Engineering Association, Tokyo, Japan (in Japanese).
9. JSCE. 2004. *Guideline for Ultra-High-Strength Fiber Reinforced Concrete.* Japan Society of Civil Engineers. Tokyo, Japan (in Japanese).

10. Uehara, T., H. Matumoto, K. Fujii and T. Kasai. 2002. Construction of Steel Girders and Long-Span Cast-in-Place PC Slab in 2nd Tomei Expressway Warashinagawa Bridge. *Miyaji Technical Report* No. 18, pp. 18–43 (in Japanese).

11. Shinohara, H. 1991. Design and Construction of Trans-Tokyo Bay Highway Bridge. *Proceedings of IABSE Symposium*, Leningrad, USSR.

12. Uchida, K. and H. Asakura 1998. Trans-Tokyo Bay Highway. *Structural Engineering International* Vol. 8, No. 1, pp. 7–9.

13. Nakamura, S. 2000. Durability of Titanium-Clad Steel Plates as an Anti-Corrosion System. *Structural Engineering International* Vol. 10, No. 4, pp. 262–265.

14. Onishi, E., O. Kawabata, Y. Tsumura, et al. 2005. New Style Bridge: First Steel Deck CFT Girder Bridge in Japan. *K.H.I. Technical Review* No. 157, pp. 10–15 (in Japanese).

15. Sugawara, M. 2008. A Prestressed Concrete Bridge Registered for the First Time as a Cultural Asset—The Daiiti Daidogawa Bridge on the Shigaraki Railway Line. *Journal of Prestressed Concrete, Japan* Vol. 50. No. 5, pp. 42–46 (in Japanese).

16. Suzuki, S., M. Ishimaru, F. Nemoto, et al. 1976. Construction of Hamana Bridge. *Journal of Japan Prestressed Concrete Engineering Association* Vol. 18, No. 6, pp. 1–12 (in Japanese).

17. Ikeda, S., H. Ikeda, K. Mizuguchi, et al. 2002. Design and Construction of Furukawa Viaduct. *Proceedings of the First International Fib Congress 2002* Session 2, pp. 21–28.

18. Higashida, N., T. Yasuzato, M. Horiguchi, et al. 2006. Launching Nose with Corrugated Steel Webs and Chords of Ultra-High-Performance Fiber-Reinfoced Concrete—Torisaki River Bridge. *Proceedings of the Second International Fib Congress 2006* Session 5, ID5-14.

19. Ashizuka, K., A. Takahashi, M. Tohma, et al. 2007. Design and Construction of Sugitanigawa Bridge—Application of New Erection Methods. *Journal of Prestressed Concrete, Japan* Vol. 49. No. 3, pp. 26–33 (in Japanese).

20. Minami, H., M. Ymamura, Y. Taira, et al. 2002. Design of Kinokawa Viaduct Composite Truss Bridge. *Proceedings of the First International Fib Congress 2002* Session 5, pp. 371–380.

21. Kurihara, H., I. Tokuyama, K. Tanabe, et al. 2000. Construction of Superstructure of the Ujina Bridge. *Bridge and Foundation Engineering* Vol. 34, No. 2, pp. 8–18 (in Japanese).

22. Ishiguro, K., and K. Takabayashi. 2003. Introduction of Construction of New Kitakyusyu Airport Access Bridge. *Topy Tekko Technical Report* No. 19, pp. 66–67 (in Japanese).

23. Miyazaki, Y. and T. Igarashi. 1947. Cantilever Method for Hokawazu Bridge. *Journal of Japan Prestressed Concrete Engineering Association* Vol. 16, No. 5, pp. 24–31 (in Japanese).

24. Ando, H., H. Mochizuki, K. Suzuki, et al. 2002. Aesthetic Design of Long Span Bridge—The Ikeda Hesokko Bridge. *Proceedings of the First International Fib Congress 2002* Session 14, pp. 119–124.

25. Ando, H., S. Kitakuni, K. Utsugi, et al. 2002. Long-Span Deck-Stiffened Concrete Arch Bridge—Ikeda Hesokko Ohashi. *National Report: Recent Works of Prestressed Concrete Structures* by Japan Prestressed Concrete Engineering Association. The First Fib Congress 2002, pp. 97–100.

26. Fukunaga, Y., K. Osada, M. Sadamitsu, et al. 2002. Planning and Design of the New Tomei Expressway Fujikawa Bridge. *Proceedings of the First International Fib Congress 2002* Session 1, pp. 141–150.

27. Mochizuki, H., K. Hanada, T. Nakagawa, et al. 2000. Design and Construction of a Cable-Trussed Girder Bridge. *Proceedings of International Bridge Engineering Conference* No. 5, pp. 293–298.

28. Irie, T. 2005. A Bridge Made of Wood. http://www.jsce.or.jp/kokusai/civil_engineering/2005/1-2.pdf

29. JICE. 1998. *Wooden Bridge Design and Construction Handbook*. Japan Institute of Construction Engineering. Tokyo, Japan (in Japanese).

30. Ohashi, H. 1999. Design Safety Check of the Tatara Bridge. *Honshi Technical Report* Vol. 23, No. 90, pp. 11–16 (in Japanese).

31. Takahashi, N. and M. Ikuma. 1970. Design and Construction of PC Cable-Stayed Bridge—Omotogawa Bridge on Kuji Line. *Bridge and Foundation Engineering* Vol. 80-3, pp. 1–8 (in Japanese).

32. Morimitsu, T., K. Ide, H. Noborita, et al. 2002. A Cable-Stayed Bridge with a Slender Segmental Concrete Superstructure—Oshiba Bridge. *National Report: Recent Works of Prestressed Concrete Structures* by Japan Prestressed Concrete Engineering Association. The First Fib Congress 2002, pp. 121–124.

33. Oguchi, H., M. Yokomine, A. Arikado, et al. 2006. Design of Yabegawa Bridge. *Journal of Prestressed Concrete, Japan* Vol. 48. No. 3, pp. 15–22 (in Japanese).

34. Terada, N., Y. Kamihigashi, T. Tsujimura, et al. Load-Carrying Capacity of the Hibrid Stay-Cable Anchor Structure—Experimental Investigations of Yahagigawa Bridge of the New Tomei Expressway. *Proceedings of the Second International Fib Congress 2006* Session 5, ID5-15.

35. Kadotani, T. 2002. The Innovated Technology on Prestressing System Developed by Japan Highway Public Corporation. *Proceedings of the First International Fib Congress 2002* Prenary, pp. 35–46.

36. Shirono, Y., I. Takuma, A. Kasuga, et al. 1993. The Design of an Extradosed Prestressed Concrete Bridge—The Odawara Port Bridge. *Proceedings of FIP Symposium 1993*, Kyoto, Japan, pp. 959–966.

37. Kuroiwa, T., K. Nishikawa, A. Kasuga, et al. 2006. Extradosed Bridge with Corrugated Steel Web—Himi Yume Bridge. *National Report: Recent Works of Prestressed Concrete Structures* by Japan Prestressed Concrete Engineering Association. The Second Fib Congress 2006, pp. 93–96.

38. Tamai, S., T. Suzuki, M. Kato, et al. 2008. Construction of Tohoku Shin-kansen Sannaimaruyama Bridge. *Journal of Prestressed Concrete, Japan* Vol. 50. No. 3, pp. 22–29 (in Japanese).

39. Ikeda, H., K. Nakamura, M. Nakasu, et al. 2002. Construction of the Superstructures of Kiso and Ibi River Bridges. *Proceedings of the First International Fib Congress 2002* Session 1, pp. 51–60.

40. Momoshima, H., T. Naito and Y. Tomita. 1969. Design and Execution of No. 9 Footway Bridge at EXPO'70. *Journal of Japan Prestressed Concrete Engineering Association* Vol. 11, No. 4, pp. 10–17 (in Japanese).

41. Kasuga, A., T. Noritsune, K. Yamazaki, et al. 2005. Design and Construction of Composite Truss Bridge Using Suspension Structure. *Proceedings of Fib Symposium "Keep Concrete Attractive,"* Budapest, pp. 168–173.

42. Kinoshita, K. 2001. Application of Thin-Titanium-Clad Steel Sheets to Yumemai Ohashi in Port of Osaka. *Titanium Japan* Vol. 49, No. 3, pp. 101–106 (in Japanese).

43. JSCE. 2006. Committee on Investigative Research for Reopening of Katidoki Bridge, Japan Society of Civil Engineers (JSCE). 2006. *Report on Investigative Research for Reopening of Katidoki Bridge.* Committee on Investigative Research for Reopening of Katidoki Bridge, Japan Society of Civil Engineers, Tokyo, Japan (in Japanese).

44. Hara, K., Y. Matsuo and M. Yamashita. 1961. Design of Miyagawa Bridge as Base Isolated Structure. *Proceedings of Annual Conference of the Japan Society of Civil Engineers* Vol. 46, pp. 1368–1369 (in Japanese).

45. Nagumo, H., T. Ichinomiya, Y. Ataka, et al. 2007. Design and Construction of Riverside Senshu Footbridge. *Bridge and Foundation Engineering* Vol. 41, No. 12, pp. 5–12 (in Japanese).

46. Fukunaga, Y., J. Ishizuka, T. Yamato, et al. 2009. Renewal Construction of RC Slab Bridge at Okubi River Bridge in the Okinawa Expressway. *Bridge and Foundation Engineering* Vol. 43-2, pp. 20–26 and Vol. 43-3, pp. 13–19 (in Japanese).

47. Fujishiro, M., K. Suda and Y. Nagata. 2008. Jointless Prestressed Concrete Viaduct using ECC. J. Walravan and D. Stoelhorst (eds)., *Tailor Made Concrete Structures.* Taylor & Francis Group. London, UK.

48. Long-Span Bridges Engineering Center Honshu-Shikoku Bridge Authority. 1999. Corrosion Protection of Inside of Box Girder with Dehumidified Air. *Newsletter on Long-Span Bridges*, No. 3.

49. Akanuma, M., N. Katayama, H. Tanaka, T. Saito, S. Kuroda and K. Ishii. 2007. Anticorrosion Technique for Steel Bridges Using Thermal Spraying, *Technical Report of Hokkaido Industrial Research Institute*, No. 306, pp. 165–169 (in Japanese).

50. Uchida, D., S. Inokuchi, A. Kawabata, M. Ishio and T. Tamakoshi. 2008. Field Investigations and Measurements of Orthotropic Steel Decks to Draft Efficient Method of Stock Management. Proceedings of International *Orthotropic Bridge Conference* August 25–29, Sacramento, USA.

51. Takasuka, T., T. Kumano, K. Kanda, A. Uemura and M. Nagai. 2006. Load-Carrying Characteristics of Steel–Concrete Composite Deck Using Deformed Flange T-shapes at Middle Support of Continuous Slab. *Journal of Constructional Steel* Vol. 14, pp. 1–8 (in Japanese).

52. Kimizu, T., K. Arai, T. Kasugai, J. Nagata and M. Nagai. 2002. Experimental Study on Crack Behavior of Composite Slab in Continuous Composite Girder Bridges. *Journal of Structural Engineeering* Vol. 55A, pp. 1417–1428 (in Japanese).

53. Kaneuji, M., N. Yamamoto, E. Watanabe, H. Furuta and K. Kobayashi. 2006. Bridge Management System Developed for the Local Government in Japan. *Proceedings of IABMAS Conference*, Porto, Portugal.

54. Matsumura, E., Y. Senoh, M. Sato, et al. 2006. Condition Evaluation Standards and Deterioration Prediction for BMS. *Proceedings of IABMAS Conference*, Porto, Portugal.

55. Miyashita T., H. Ishii, K. Kubota and Y. Fujino. 2007. Advanced Vibration Measurement System using Laser Doppler Vibrometers for Structural Monitoring. *Proc. of Int. Conf. on Experimental Vibration Analysis for Civil Engineering Structures*, pp. 133–142.

56. Kogure, Y., Y. Nakamura, and H. Okamoto 2007. Esthetic Design of Akiba Bridge: Pedestrian Bridge of Ultra-High-Strength Concrete. *Journal of Prestressed Concrete, Japan* Vol. 49, No. 6, pp. 57–63 (in Japanese).

57. Okamoto, H., I. Oda and T. Ichinomiya. 2006. Construction of the Akihabara Public Deck Using Low-Autogenous Shirinkage Ultra-High-Strength Concrete. *Proceedings of the Second International Fib Congress 2006* Session 2, ID2-8.

58. Maekawa. T., T. Ichoki and T. Niki. 2006. Development of Ultra-High-Strength Prestressing Strands. *Proceedings of the Second International Fib Congress 2006* Session 14, ID14-8.

59. Okiji, M., and M. Nishida. 2007. Ushioshinmachi Route Bridge Using FRP Composite Slabs. *Technical Report of Miyaji Iron Works* Vol. 22, pp. 111–114 (in Japanese).

60. Okura, I., N. Hagisawa, S. Iwata and K. Kitamura. 2004. Technological developments for realizing aluminum bridges. *Journal of Japan Institute of Light Metals* Vol. 54, No. 9, pp. 380–387 (in Japanese).

24

Bridge Engineering in Chinese Taipei

Yeong-Bin Yang
National Taiwan University

Dyi-Wei Chang
*CECI Engineering
Consultants, Inc.*

**Dzong-Chwang
Dzeng**
*CECI Engineering
Consultants, Inc.*

Ping-Hsun Huang
*CECI Engineering
Consultants, Inc.*

24.1 Introduction

The way bridges are constructed is closely related to the economic development of a country. In general, the materials, structural types, and techniques used in the construction of each bridge are reflective of the state-of-the-art technologies available at the time when the bridge was built. Historically, Taiwan* has been regarded as a remote island off the central mainland of China. As such, the development history of Taiwan is relatively short, that is, less than 200 years, compared with other parts of China that have a history of thousands of years. A brief retrospect of the construction history of bridges in Taiwan is helpful for us to comprehend the economic development curve of the residents in Taiwan since the last quarter of the nineteenth century, when ancestors started to cross the Taiwan Strait and cultivate the island.

This chapter is written based primarily on the review book of Taiwan bridges edited by Chang (2007). Most of the figures presented in this chapter were also adapted or modified from the same book via permission of the author, who has also acquired permission from various sources for the figures used. In this chapter, the source that was quoted in Chang (2007) for each figure will be directly quoted. However, due to the lack of time available, no effort will be made to verify the accuracy of such quotations. To avoid any distortion in translation, all translated Chinese terms, mainly the names of bridges, persons, places, highways, and railways, will be supplemented by their original Chinese characters in parentheses at their first appearance throughout the chapter. Except for those Chinese terms of which their traditional translations are widely accepted, all Chinese–English translations will be made based on the Hanyu Pinyin System (漢語拼音系統).

24.1.1 Historical Bridge Development

According to historical records, construction of bridges began at the time when Shen Baozhen (沈葆禎) was assigned by the Qing Dynasty (清朝) in 1874 as the governor of Taiwan. For the purpose of opening up barren lands for farming, he built the northern, central, and southern roadways in Taiwan. However, most of the bridges constructed at that time were made of either wood or stone and could only be used by animal-powered wagons or pedestrians. As of now, only one stone girder bridge and one stone arch bridge constructed in this period remain in some obsolete ancient roads in the rural areas.

The construction of railways in Taiwan started at the time when Liu Mingchuan (劉銘傳) was appointed as the governor in 1884. The railways from Taipei (臺北) to Keelung (基隆), 28.6 km long, were completed in 1887 and then extended to Hsinchu (新竹) in 1893. Among the 106.7 km of railways, there were more than 70 bridges, most of which were made of steel or wood. Since a more advanced technique level is required in the construction of railways compared with roadways, the completion of these railways symbolized that bridge engineering had grown quite well in Taiwan during that period.

As an important link of the first railways built between Keelung and Hsinchu, the Taipei Bridge (台北橋) crossing the Danshui River (淡水河) was completed in 1889. The upper part of the bridge was made of wood and iron, and the lower part was made of wood only. To meet the needs of the growing regional economy, the Taipei Bridge has been expanded or reconstructed several times. From the historical records of reconstruction of the Taipei Bridge, one can observe how foreign technologies were transferred to Taiwan for bridge construction. Starting with their occupation of the Taiwan island in 1895, the Japanese government introduced technologies from the Occident to build highway and railway networks. Over 2000 bridges were constructed at the time, until the sovereignty of Taiwan was returned to the Chinese Nationalist government in 1945. Most of the bridges constructed during the Japanese-controlled period were reinforced concrete T-girders, arch bridges, steel plate girder and truss bridges, along with some suspension bridges.

---

* Commonly designated as Chinese Taipei in some international organizations, such an Olympics and APEC.

To offer a global aesthetic landscape for the city of Taipei, four major bridges of different structural configurations were constructed for crossing Xindian Creek (新店溪), Danshui River, and Keelung River. For instance, steel plate girder bridges and suspension bridges were used for the Kawabata Bridge (川端橋), later renamed Zhongzheng Bridge (中正橋) after 1945, and the Syouwa Bridge (昭和橋), later renamed Guangfu Bridge (光復橋), across the upstream and downstream, respectively, of the Xindian River; a steel trussed bridge was used for the Taipei Bridge across the Danshui River; and a reinforced concrete arch bridge was used for the Meiji Bridge (明治橋), later renamed Zhongshan Bridge (中山橋). The above four bridges are the main gates for residents from other parts of Taiwan to enter the city of Taipei. Their special geometrical shapes gives visitors a vivid view of the growing strength of the capital city. Aside from the aforementioned bridges constructed during the Japanese-controlled period, there are two other bridges that deserve special mentioning. One is the Pingdong Bridge (屏東橋), completed in 1913, which had 24 arches, each of 63.4 m, and a total length of 1526 m. This bridge was the longest in the Far East at the time. The other bridge is the Taidong Bridge (台東橋), completed in 1936, which had a total length of 490 m and a central span of 330 m. This bridge was the longest suspension bridge in the Far East at the time.

Due to the lack of construction materials and difficulty in transportation, glutinous rice was grounded to yield a paste for use as the adhesion material for some brick and stone bridges in the old days. One typical example is the Yutengping Railway Bridge (魚藤坪鐵路橋), which was constructed as a red brick arch in Sanyi County (三義鄉) in 1907. This bridge was damaged by the Hsinchu Earthquake (新竹地震) in 1935 and later by the 921 Jiji Earthquake (集集地震) in 1999. The remaining parts of this bridge are now known as the Longteng Broken Bridge (龍騰斷橋), which remains a scenic and cultural spot of the Sanyi area.

The first reinforced concrete bridge ever constructed in Taiwan was the Liugong Bridge (瑠公橋), built over the Jingmei River (景美溪) in 1909. By adopting a box-type girder as the superstructure, this bridge was used not only for carrying vehicular traffic, but also for transporting water through the box. The other two reinforced concrete bridges that are still in use are the Pinglinwei Bridge (坪林尾橋), completed in 1912, and the Shanxia Bridge (三峽橋), completed in 1933.

The sovereignty of Taiwan was returned to the Nationalist government in 1945. The Xiluo Bridge (西螺大橋), completed in 1953 as a truss girder bridge with a total length of 2 km, was the first bridge built by the Nationalist government in Taiwan. Later, due to the rapid economic growth and the transfer of advanced technologies from Western countries, a new wave of bridge construction arrived in Taiwan. Before the 1980s, prestressed concrete bridges appeared to be most popular. However, the traditional methods of assembling the prestressed girders on site or cast-in-place were also used.

In the past two decades, a number of elevated multispan continuous concrete and steel girder bridges have been built island-wide for both highways and railways. To meet the needs of less construction time and cost, new methods originating from Europe, America, and Japan aimed at the automation of construction procedures have been employed. One feature in this regard is the use of standardized cross sections for bridge girders, while the construction facilities were significantly mechanized, allowing repeated, cyclic operations to be performed.

Prior to 1977, the steel to be used in bridge construction was imported mainly from foreign countries, and therefore was generally costly. As such, steel was restricted to use in suspension bridges in high mountains, truss bridges in plains, or steel girders for railways. Not until 1977, when the China Steel Corporation began its production lines, did steel plates, and later steel sections, gain wide use in all kinds of bridge construction, including freeway interchanges, urban elevated bridges, cross-valley bridges, large-span bridges, and some landmark bridges. The past three decades can be regarded as a booming period for prestressed concrete bridges and steel bridges, enhanced by national economic growth.

Only a brief historical review will be presented for each of the broad categories of prestressed concrete bridges and steel bridges. For bridges with specific geometric configurations, such as girder bridges, arch bridges, truss bridges, cable-stayed bridges, suspension bridges, bridges of other types, pedestrian footbridges, and so on, only some of the most magnificent bridges that are of historical and technical value will be introduced in a chronological, itemized manner in this chapter.

24.1.2 Historical Development of Prestressed Concrete Bridges

The first prestressed concrete (PC) bridge in Taiwan was the 18.4 m railway bridge built by the Taiwan Sugar Company in 1955 in their deployment yard of the Pingtong Sugar Factory (屏東糖廠), which uses concrete with a strength of 500 kg/cm². In 1956, a PC bridge with a span length of 30 m was built for the first time for the highway in Taoyuan (桃園), Provincial Route No. 3, and is still in use today. The successful completion of these two bridges led to wider application of PC bridges in Taiwan. Noteworthy is the completion of the Zhongxing Bridge (中興大橋), a PC bridge with a total length of 1055 m and a maximum span length of 40 m, over the Danshui River in 1958. This was the second-longest bridge ever completed in Taiwan at the time. The longest was the Xiluo Bridge. The Zhongxing Bridge was also the longest PC bridge ever built in the Far East at the time. The Changhong Bridge (長虹橋), completed in 1969 in Hualian (花蓮), has the largest span length of 120 m, which was also the first PC bridge ever built by the cantilever method at the time.

By 1978, more than 300 PC bridges had been constructed for the Sun Yatsen or First National Freeways (中山或第一高速公路). By 2004, the total length of PC bridges built for the freeways, highways, and railways reached a record of 1000 km. It is noteworthy that among the 345 km high-speed railways completed in 2006, bridges constitute a great portion of 251 km, of which 248 km are PC bridges. PC arch bridges have been widely adopted in various transportation links in Taiwan. The Bitan Bridge (碧潭橋), with a 160 m main span, completed in 1996 as part of the Northern Second Freeway, is an aesthetically pleasing, curved arch bridge. The Liyutan Bridge (鯉魚潭橋), with a 134 m main span and completed in 1998, is regarded as a landmark of the Taiwan railways. The longest span length of 187 m for PC arch bridges belongs to the Donshan River Bridge (冬山河橋), completed in 2005 for the Beiyi Expressway (北宜高速公路). In comparison, relatively few PC cable-stayed bridges have been constructed in Taiwan. The two examples are the Guangfu Bridge (光復橋), completed in 1977 in Taipei, which has a main span of 134 m, and the Jilu Bridge (集鹿大橋), completed in Nantou (南投), with a main span of 120 m, which was damaged by the Jiji Earthquake in 1999 and repaired in 2004.

The technologies for the design and construction of prestressed concrete bridges have reached a state of maturity, following the completion of a number of large-scale transportation engineering projects. In general, the previously popular PC I-girder bridges have been replaced by the aesthetically more attractive multispan continuous box girder bridges. Moreover, to reduce the construction cost while shortening the working period and enhancing the quality of the project, automated construction methods have been adopted for PC bridges in various projects, including the Second National Freeways, Beiyi Expressway, East–West Express Links, West Coast Expressways, and high-speed railways. The methods that have been adopted include the supported advancing method, segmental advancing method, equilibrating cantilever method, precast segmental method, and full-span assembling method, and so on.

24.1.3 Historical Development of Steel Bridges

After Taiwan was handed over to the Nationalist government in 1945, there was a general lack of steel for construction, as most steel was imported from foreign countries. Thus, steel was not considered a proper material for bridge construction at that time. In fact, it was used only in special cases, such as in high mountains as suspension bridges or in railways as truss bridges. Among those few steel bridges that were built in this period, the most famous two are the Loufu Suspension Bridge (羅浮吊橋), included as a part of the Northern Cross-Island Highway, and the Xiluo Bridge, with a total length of roughly 2 km. The foundation of the Xiluo Bridge was completed under Japanese governance, but the superstructure was constructed using steel donated by the American government. This bridge is a truss bridge with 31 spans, each with a length of roughly 60 m. This bridge has been an important scenic spot in central Taiwan for a long time, attracting visitors from many places.

In 1977, the China Steel Corporation (中國鋼鐵公司) began its operations, particularly producing steel for structural use in Taiwan. However, not until 1983, after the completion of Guandu Bridge (關渡大橋),

a steel arch bridge, was steel widely accepted in bridge construction. In the past two decades, due to rapid progress in analysis software and computing devices, enhanced by advances in construction technologies, the construction of steel bridges has entered a new era. This is especially true in the metropolitan areas, where elevated bridges, underground structures, or structures combined with other engineering works, such as riverbanks, are often adopted as approaches for dissolving congested traffic due to the lack of land available for construction. Today, steel bridges with double or multiple decks, or combined with other structures, are becoming popular in the city of Taipei and other metropolitan areas. A typical example in this regard is the completion of the Taipei Shimindadao Bridge (市民大道) in 1996, in which some underground spaces are used for parking and shopping.

To meet scenic demands while enhancing local cultures, various long-span steel arch bridges and cable-stayed bridges with appealing shapes have been constructed in various locations in the high mountains and valleys of Taiwan, standing as landmarks of some townships and country villages. Among these, the Gaoping Creek cable-stayed bridge (高屏溪斜張橋), with a span length of 330 m and completed in 1999 as part of the Second National Freeways project, deserves special mention. This bridge is the longest asymmetric single-pylon cable-stayed bridge ever built in Asia. It also represents advances in the design and construction technologies for steel bridges in Taiwan.

24.2 Design Practice

24.2.1 Historical Development

The first version of the Bridge Design Specifications was prepared in 1954 by the Taipei Branch of the Chinese Institute of Engineers (CIE), as part of the pocket-sized engineering handbook, second edition. The formal specifications were translated in 1956 from the Standard Specifications for Highway Bridges by the American Association of State Highway Officials (AASHO), and was officially reviewed and approved by the Ministry of Transportation and Communications (MOTC) in 1960 as the specifications for the design of highways. Later, the specifications published by the American Association of State Highway and Transportation Officials (AASHTO), with the enhancement of the 1983 edition, were adopted as the model for developing the specifications for Taiwan highway bridges. This version of the specifications, in which earthquake forces were specified based on Japanese codes, was issued by the MOTC in 1987. In 2001, the Design Specifications for Highway Bridges were modified again, as a reflection of the 1996 edition of the AASHTO specifications.

Concerning seismic forces for the design of bridges, some preliminary guidelines were given in the 1954 pocket-sized engineering handbook by the CIE. The seismic zones in Taiwan were classified into two regions, one with a seismic force coefficient of $K_h = 0.1$ and the other with $K_h = 0.15$. More detailed guidelines were given in the 1987 Design Specifications for Highway Bridges, which were based primarily on the 1980 specifications prepared by the Japanese Road Association.

When the First National Freeway system was being built in the 1960s and 1970s, three design seismic force coefficients were adopted, $K_h = 0.1$, 0.15, and 0.2, as recommended by the technical consultants of the project. Since then, there has been significant progress in earthquake engineering, especially for seismic-resistant design of structures, as well as accumulation of experiences for design against the damage and collapse of bridges. To reflect such a trend and the local seismic characteristics in Taiwan, the first version of the Seismic Resistant Design Specifications for Highway Bridges was formally issued by the MOTC in 1995. The Jiji Earthquake occurring on September 21, 1999, in central Taiwan provided a real test of the adequacy for the division of seismic zones, as well as a reason to draw a more reliable specification of the seismic forces for use in bridge design. Consequently, the Seismic Resistant Design Specifications for Highway Bridges were significantly revised in 2000, particularly with regard to ground acceleration coefficients, seismic zones, and response spectra. Further efforts were carried out to collect relevant earthquake data and used as the base for refinement of the codes published in 2008.

In comparison, relatively few revisions have been made for railway bridges, because the design of railway bridges had been based primarily on the codes issued by the Japanese government during their governance of Taiwan. Not until 1999 was the first version of the Seismic Design Specifications for Railway Bridges issued by the MOTC, which was further modified in 2006. For the first time in history, the Design Specifications for Railway Bridges were issued by the MOTC in 2004. In Section 24.2.2, a summary will be given of the design specifications for bridges that are currently in use in Taiwan.

24.2.2 Design Considerations for Bridges

An overall brief review will be given for the design principles, specifications and materials, highway loads, railway loads, and seismic design loads that are currently adopted by the bridge design offices in Taiwan.

24.2.2.1 Design Principles

In this section, the basic design principles adopted for the highway and railway bridges in Taiwan will be outlined. These include the design requirements, structural analysis and design check, and considerations for the various design aspects of a bridge.

24.2.2.1.1 *Design Requirements*

The requirements for the design of bridges are as follows:

- *Safety*: To ensure the safety of a bridge structure, various load combinations should be taken into account in the design of the bridge.
- *Aesthetics*: The selection of the geometrical configuration for a bridge should be consistent or in harmony with the natural surroundings.
- *Economy*: In selecting the bridge layout, materials and methods of construction, as well as the economic factor or cost of construction, should be taken into account.
- *Ease of construction*: In selecting the bridge types and materials, the constructability of the structure in all stages should be taken into account.
- *Ease of maintenance*: In the design stage of a bridge, considerations should be given to the ease of maintenance and replacement of components of the structure in service.
- *Site conditions*: In designing a bridge layout, factors such as the topographical, geological, and hydrological conditions, as well local transportation and environment, should all be taken into account.
- *Related facilities*: Requirements for auxiliary devices should also be considered in the design of bridges, to ensure the safety and comfort of the passing vehicles.

24.2.2.1.2 *Structural Analysis and Design Check*

There are three concerns here. First, the structural model used in the analysis of a bridge should at best reflect the true behavior of the structure, and should be capable of producing accurate stresses for the critical cross sections. The boundary conditions for the bridge should be modeled by equivalent soil springs or by other more accurate means. Second, the member stresses under various load combinations should be checked to ensure that they are below the allowable values given by the specifications. Third, to avoid an uneconomical design, the member stresses generated by a structural analysis should not be excessively or unduly lower than the allowable values.

24.2.2.1.3 *Various Design Aspects of a Bridge*

The following aspects should be considered in the design of a bridge according to the specifications:

- *Locations of piers*: The selection of the starting and ending points, including the locations of the piers, of a bridge should be based on the local topographical plots and surveys on site, considering the height, accessibility, and ease of construction of the bridge.

- *Layouts of horizontal and vertical planes*: In designing the horizontal and vertical layouts of a bridge, considerations should be given to the present and future possible usage of the space crossed by the bridge, for instance, the present river width and its future plan, road width, urban planning, and other related issues. The other consideration is the effect of geological conditions on the construction cost of the foundations.
- *Structural system*: In selecting the most proper bridge structural system, considerations should be given to the total length, span lengths, and height of the bridge with respect to the overall construction method and seismic design requirements.
- *Superstructure*: Two issues are of concern here. First, the structural type should be consistent with the aesthetic view of the adjacent roads in connection. Second, the road surface width of the bridge should meet the need for the predicted future traffic volume, while the depth of the girder(s) should be selected based on the span length and clear height required.
- *Supporting structure*: Two issues are of concern here. First, in designing the bridge columns, consideration should be given to the overall bottom width of the girder(s) and the provision of enough space for installing the bearings and antisliding devices. Second, in determining the geometry and dimensions of the bridge columns, it is recommended that consistency in the external shapes be maintained. The dimensions should not vary too much to guarantee the ease in construction. Special care should be taken of the pipes installed inside the columns for discharging rainwater.
- *Auxiliary devices*: All of the following auxiliary devices should be designed to meet the practical requirements: expansion joints, bearings, seismic isolation devices, antifalling devices, and water-discharging holes on the pavement surface.

24.2.2.2 Specifications and Materials

In general, the design of bridges in Taiwan are based on the following specifications published by the Ministry of Transportation and Communications (MOTC):

- *Design Specifications for Highway Bridges* (2001)
- *Seismic Design Specifications for Highway Bridges* (2008)
- *Design Specifications for Railway Bridges* (2004)
- *Seismic Design Specifications for Railway Bridges* (2006)

and the following guideline issued by the Ministry of Economy:

- *Review Considerations for Crossing-River Structures* (2006)

In case of controversy, the MOTC specifications are given a higher priority for compliance. The specifications of materials should follow those given in the China National Standards (CNS). In case of lack of relevant information, both the standards provided by the American Society for Testing and Materials (ASTM) and the Japanese Standards Association (JIS) may be adopted.

24.2.2.3 Design Loads for Highway Bridges

For the design of highway bridges, the following should be taken into account:

- *Loading considerations*: In designing bridge structures for highways, the following loads should be taken into account: static loads, vehicle live loads with impact effects, temperature variations and gradients, shrinkage and creep, settlements, earthquakes, wind forces, construction machinery, and so on. For the design of pedestrian or pedestrian–bicycle bridges, a uniform live load of 400 kgf/m² should be used.
- *Loading combinations:* Various load combinations are listed in the specifications for the design of segmental concrete bridges after completion based either on the allowable stress design or ultimate stress design. Provisions are given for load combinations for seismic design of bridges in the construction stage. For other types of bridge structures with other construction methods, relevant specifications are also given.

24.2.2.4 Design Loads for Railway Bridges

For the design of railway bridges, the following should be taken into account:

- *Loading considerations*: In designing bridge structures for railways, the following loads should be taken into account: static loads, train live loads with impact effects, lateral forces induced by moving carriages and wheels, braking forces and starting-up forces, axial forces on rails, other loads (such as shrinkage, creep, temperature, and deformations in bearings), earthquakes, wind forces, construction machinery, and so on.
- *Loading combinations*: Each component of the bridge structure, as well as the foundations, should be able to resist all the loading combinations provided for railway bridges in the specifications. Those load combinations related to earthquake forces are available in the Seismic Design Specifications for Railway Bridges (2006). Different load combinations are specified for the design of railway bridges for both the allowable stress design and ultimate strength design.

As for the slope and deflection checks of railway bridges, the live load is taken as that produced by a single line of train cars. For high-speed railways, it is necessary to consider the effect of impact loads caused by the moving vehicles. For railways with three lines or more, the live loads used should be able to reflect the conditions encountered in practice. In examining the slopes of railway bridges caused by live loads, only the live loads associated with the trains need to be considered.

24.2.2.5 Seismic Design Considerations

The objectives of the seismic design of bridges in Taiwan as stated in the related specifications are as follows:

- *Medium earthquake with a return period of roughly 30 years*: The structure should be able to remain elastic with no damages during a medium earthquake.
- *Design earthquake with a return period of roughly 475 years*: Plastic hinges may occur on the structure during an earthquake to exhibit its ductility. Damages on the structure should remain on the level of being repairable.
- *Maximum conceivable earthquake with a return period of roughly 2500 years*: The safety of the structure under such an earthquake should rely fully on its displacement ductility capacity. The bridge should not fall or collapse under such an earthquake.

24.3 Girder Bridges

24.3.1 Reconstruction of Taipei Bridge (台北橋)

The Taipei Bridge was rebuilt several times due to the continuous, increasing traffic volume crossing the Danshui River. The third time (or for the third generation), the bridge was rebuilt as a simply supported steel girder bridge with the original spans of 62.4 m, but with an expanded width of 28.5 m. The reconstruction was completed in 1969 (Jia 2004) (see Figure 24.1). It is a pity that the clear height of the third-generation Taipei Bridge did not meet the requirements for the mitigation of flood of the 200-year return period in the Great Taipei Basin. Thus, right after completion of the bridge in the 1960s, there was already planning for rebuilding the bridge for a fourth time to resolve not only the flooding concern, but to include this bridge as an efficient link for the global expressways of Taipei that were to be constructed along the riverside. The fourth-generation Taipei Bridge was constructed under the condition that ongoing traffic was not terminated, as it has been one of the busiest bridges crossing the Danshui River. It was finally completed in 1996 (Lin 1996) (see Figure 24.2).

FIGURE 24.1   Taipei Bridge (third generation, 1969). (Adapted from Chang 2007.)

FIGURE 24.2   Taipei Bridge (fourth generation, 1996). (Adapted from Lin 1996.)

24.3.2 Sea-Crossing Bridge in Penghu (澎湖跨海大橋)

There is a total of 64 islands in Penghu (澎湖) off the western coast of the main island of Taiwan, which is also known as the Pescadores by the Western people, among which Magong (馬公), Baisha (白沙), and Xiyu (西嶼) are the three biggest islands and therefore are the places with most concentrated residents. Transportation inside the group of islands was mainly by boat, which becomes quite difficult and dangerous during the windy seasons. To improve the transportation between the three islands, the central government began the construction of the Sea-Crossing Bridge in 1965. This bridge was made of prestressed concrete and had a total length of 2160 m and a standard width of 4.6 m. It also had seven transition zones with a width of 7.5 m that allowed vehicles from both directions to pass each other (see Figures 24.3 and 24.4). The construction of this bridge was generally difficult, as it was carried out in the windy sea (Taiwan Highway Bureau 1971).

Since its completion in 1970, this concrete girder bridge has deteriorated drastically in different parts of the structure, due to erosion caused by the surrounding sea. To replace the damaged parts and to meet the growing traffic demand, the bridge was rebuilt in 1984. This new version of the bridge, completed in

FIGURE 24.3   Sea-Crossing Bridge in Penghu (1970). (Adapted from Taiwan Highway Bureau 1971.)

FIGURE 24.4   Sea-Crossing Bridge in Penghu under construction (1970). (Adapted from Taiwan Highway Bureau 1971.)

1996, consists of 2 steel spans, 26 prestressed concrete spans, and embankment sections on the two ends. The standard width of the bridge was increased to 14 m, allowing traffic from both directions to move without interrupting each other. The total length of the bridge is 975 m.

24.3.3 Yuanshan Bridge (圓山橋)

With a total length of 1385 m, the Yuanshan Bridge (圓山橋) was built over the Keelung River (基隆河) near the Taipei Yuanshan region, where traffic has always been busy (see Figure 24.5). To accommodate the function of flood mitigation for the Keelung River, while taking into account the overall aesthetic concerns of the Zhongshan North Road (中山北路), the superstructure of the Yuanshan Bridge was constructed as a three-cell prestressed concrete girder of variable depths. This bridge was constructed using the cantilever method. Owing to variation in the depth of the bedrock, three different types of foundations were adopted, including the 40 cm × 40 cm square PC concrete piles, well foundations with

FIGURE 24.5 Yuanshan Bridge (1978). (Adapted from Hu 1984.)

diameters of 1.5 m and 2.5 m and depths of 11.6–16.7 m, and caissons with diameters of 6 m and depths of 17–25 m. This project was the most difficult part of the construction of the First National Freeways. It was completed in 1978 with the technical assistance of American and Japanese consultants (Hu 1984).

24.3.4 Liwu Creek Bridge (立霧溪橋)

For the purpose of improving railway traffic on the north and east coasts of Taiwan, the central government began construction of the Northern Link Railways (北迴鐵路) in 1973. This project has a total of over 80 km connecting Suao New Station (蘇澳新站) to Hualian New Station (花蓮新站). This project consists of 22 bridges, totaling a length of 5330 m, most of which are prestressed concrete I-girder bridges. Among these bridges, there are two bridges with a length of over 1000 m, including the Liwu Creek Bridge (with 58 spans each at 20 m) shown in Figure 24.6 (RSEA Engineering Corporation 1982).

24.3.5 Jianguo Elevated Bridge Crossing the Minzu Road (民族路口建國高架橋)

The Jianguo Elevated Bridge (建國高架橋) in Taipei was constructed along the Jianguo North–South Road and connected to the First National Freeways in the Yuanshan region. At the crossing of the Mingzu Road, the bridge was constructed as a three-span steel bridge; the bridge width is 20.5 m, the total length is 130 m, and the three span lengths are 37.94 m, 54.12 m, and 37.94 m. This bridge is located under the landing and take-off route of the Songshan Airport (松山機場). Thus, there is a restriction on the overall bridge height. To meet such a requirement, plus the clear height for undergoing traffic, while accommodating geographical variations, the bridge was designed as a single-column three-span curved steel bridge of the framed type, as shown in Figure 24.7 (Chang 1983). Completed in 1982, this bridge represents an advance in design technologies for bridges of the framed type.

24.3.6 Shuiyuan Expressway (水源快速道路)

The Shuiyuan Expressway runs mainly along the bank of the Xindian River. Its main bridge appears as an S-shaped five-span (80 m + 3 × 125 m + 80 m) continuous steel box girder, whose total length is 535 m (see Figure 24.8). For the part along the embankment of the river, the expressway was constructed as a single-column double-deck steel framed bridge with the height varying from 30 m to 50 m, mixed with the embankment (see Figure 24.9). The expressway was completed in 1992. The double-deck

FIGURE 24.6   Liwu Creek Bridge (1978). (Adapted from RSEA Engineering Corporation 1982.)

FIGURE 24.7   Jianguo Elevated Bridge crossing Minzu Road (1982). (Adapted from Chang 1983.)

FIGURE 24.8   Shuiyuan Expressway (1992). (Adapted from Tzeng et al. 1993.)

bridge, constructed mixed with the embankment, was the first of its kind in Taiwan. In addition, the main bridge had the longest span of steel bridges at the time. During the construction of this expressway, various analysis and design programs were developed to enhance the accuracy, to save computation time, and to reduce the complexity involved in the design works. This project also signified progress in the design practices of steel bridges in Taiwan (Tzeng et al. 1993).

FIGURE 24.9   Shuiyuan Expressway with embankment (1992). (Adapted from Tzeng et al. 1993.)

24.3.7 Shimindadao Bridge in Taipei (台北市市民大道)

The Shimindadao Bridge is an expressway running through the busiest district of Taipei in the East–West direction. Most of the expressway appears as an elevated bridge (with a length of 6.8 km) in combination with an underground structure for railways or shopping or parking areas. This is the only structure that combines underground or railways or buildings with an elevated bridge for an expressway. When completed in 1996, this bridge was the largest steel construction project ever carried out in Taiwan, consuming a total amount of steel of over 80,000 tons. Both the superstructure and columns are made of steel.

There are three distinct parts of the bridge. For the part integrated with the underground railways, the elevated bridge was just built over the tunnel used by the railways (see Figure 24.10). They work as a single structure. For the part of the bridge with no railways, the superstructure is combined with the underground structure to offer the parking space. For the part of the bridge adjacent to the Taipei Train Station, the underground structure is used mainly for shopping and parking purposes, and for connection to other transportation systems (see Figure 24.11). Because of its elevated nature, the Shimindadao Bridge offers a vivid, colorful global view of the city of Taipei. This is the first project in Taiwan that combines urban scenic design with civil engineering construction (Lin et al. 1994).

24.3.8 Main Bridge of Zhoumei Expressway (洲美快速道主橋)

The Zhoumei Expressway in Taipei has a total length of 4.1 km. The main bridge crossing the Keelung River was composed of a three-span continuous steel box girder (126 m + 168 m + 126 m). The approaches of the bridge were integrated with the embankments. In these parts, three- to five-span continuous steel box girders with spans ranging from 50 m to 77 m were adopted, all supported by steel columns. This bridge was completed in 2002. The total amount of steel used in this project was 62,000 tons. With a construction period of 1 year and 9 months, this bridge ranked first in Taiwan for its speed of construction.

The main span length of 168 m is the longest so far for steel bridges in Taiwan (see Figure 24.12). The piles for the main bridge have a diameter of 2 m and a length of 104 m, also record-breaking. The main bridge was erected using a barge, instead of the traditional temporary framework. To save construction time, the girders were constructed using scaffolds while the piles were also constructed. Facilities such as electric lines, water pipes, and bike ways were also provided on the elevated bridges (Chang et al. 2003).

FIGURE 24.10   Shimindadao Bridge in Taipei. (Adapted from Chang 2007.)

FIGURE 24.11   Typical cross section of the Shimindadao Bridge in Taipei (1996). (Adapted from Chang 2007.)

FIGURE 24.12   Main bridge of Zhoumei Expressway (2002). (Adapted from Chang et al. 2003.)

FIGURE 24.13 Xihu River Bridge (2002). (Adapted from Taiwan Area National Expressway Engineering Bureau 2003.)

24.3.9 Xihu River Bridge (西湖溪橋)

This bridge crosses the Xihu River in Miaoli County (苗栗縣) and has a total length of 3000 m and a height varying from 30 m to 40 m (see Figure 24.13). For its rather large scale, the bridge was constructed as prestressed concrete box girders by the segmental cantilever method. As part of the Second National Freeways project, this bridge ranked first for its span length (the main span length is 60 m) and girder width among the bridges constructed by the segmental cantilever method (Taiwan Area National Expressway Engineering Bureau 2003).

24.4 Arch Bridges

24.4.1 Sanxia Arch Bridge (三峽拱橋)

Completed in 1933, this bridge is located in the town of Sanxia (三峽鎮) in Taipei County, for which it was named the Sanxia Arch Bridge. This reinforced concrete bridge has three arches, resulting in a total length of 93.3 m (see Figure 24.14). This kind of through-type RC bridge was considered quite special at the time of construction. This bridge is still in use today and is regarded as a landmark of the village (Sanxia Township 2008).

24.4.2 Guandu Bridge (關渡大橋)

Guandu Bridge is located about 10 km in the upstream from the mouth of the DanShui River near Taipei (see Figure 24.15). This five-span through-type steel arch bridge was completed in 1983. The five span lengths are 44 m, 143 m, 165 m, 143 m, and 44 m. The clear surface width of the bridge is 19 m, with 14.5 m for traffic lanes and 2.25 m for a pedestrian walk on each side. The three central spans are made of steel arches and the two side spans are cantilevers. High-strength bolts were used to connect the arches with the main girders. Welding connections were used for the other joints. Since the bridge site is quite near the river mouth, great variations exist in the water levels due to the tidal sea. The kind of daily water level variation was utilized in handling a working barge for erecting the bridge during the construction stage (Sun et al. 1984).

FIGURE 24.14   Sanxia Arch Bridge (1933). (Adapted from Chang 2007.)

FIGURE 24.15   Guanduda Bridge (1983). (Adapted from Chang 2007.)

24.4.3 Liyutan Bridge (鯉魚潭橋)

The Liyutan Bridge was constructed in an effort to provide an additional railway track such that the single-track bottleneck between Sanyi (三義) and Houli (后里) was resolved and the northbound and southbound trains could each have their own tracks from the north end to the south end. This bridge is 790 m long and 40 m high. It is a continuous concrete arch bridge with four spans, each 134 m long (see Figure 24.16) (Wang et al. 1999).

24.4.4 Lizejian Bridge (利澤簡橋)

As part of the global project for renovation of the Dongshan River (冬山河) as a water recreational park in Yilan County (宜蘭縣), the Lizejian Bridge (利澤簡橋) was constructed in 1992 as a landmark structure. This region of the river segment has often been used for the International Dragon Boat Competition. This single-span through-type steel bridge has a total length of 148 m and a width of 21 m (see Figure 24.17) (Directorate General of Highways 1998).

24.4.5 Toushe Creek Arch Bridge (頭社溪橋)

Mingtan Lake (明潭) is a reservoir used to store the water pumped up in off-peak hours from the end pool discharged by the hydraulic generators in peak hours via two parallel tunnels. Inside each tunnel is a 6.8 m diameter steel pipe, which offers a rate of 246 m³/sec for pumping water. The Toushe Creek Arch

FIGURE 24.16 Liyutan Bridge (1988). (Adapted from Wang et al. 1999.)

FIGURE 24.17 Lizejian Bridge (1992). (Adapted from Directorate General of Highways 1998.)

Bridge was constructed to carry the two pipes over a creek as part of the transit for water transportation. This bridge is of the spandrel wall type, with a three-dimensional parabolic arch (see Figure 24.18). The span length of the arch is 70 m and the total length of the bridge is 130 m. The design loads carried by this arch is about 25 times that of arches used in traditional highways (Sinotech Engineering Consultants, Inc. 2008).

24.4.6 Bitan Bridge (碧潭橋)

Completed in 1996, the Bitan Bridge (Figure 24.19) is located in the vicinity of Taipei, 250 m downstream of the famous Bitan Suspension Bridge (碧潭吊橋), crossing the Xindian River, forming an important link of the Second National Freeways (Taiwan Area National Expressway Engineering Bureau 2003). This bridge is a three-dimensional prestressed concrete curved arch bridge. The northbound side has a length of 813.7 m and the southbound side 781.5 m. This bridge was designed as part of the effort to promote tourism in the Bitan scenic area of the Xindian City (新店市). The main span is a 160 m curved arch with a prestressed box girder. Above the main pier is a reverse triangular shape. It was constructed by both the cast-in-place and cantilever methods. No expansion joints were built on the pavement surface to enhance the comfort of high-speed moving vehicles. This bridge is now regarded as a landmark of the Bitan scenic area.

FIGURE 24.18 Toushe Creek Arch Bridge (1993). (Adapted from Sinotech Engineering Consultants, Inc. 2008.)

FIGURE 24.19 Bitan Bridge (1996). (Adapted from Taiwan Area National Expressway Engineering Bureau 2003.)

24.4.7 Taipei Second McArthur Bridge (台北市麥帥二橋)

To comply with the flood mitigation project of the Keelung River, a large-span Nielsen-type steel arch bridge was used for the Taipei Second McArthur Bridge, which has a length of 210 m and an arch height of 35 m (see Figure 24.20) (Peng 1995). Inclined crossing tie rods were used. The main girder was prestressed by tendons. Completed in 1996, this bridge is the first steel bridge constructed of the Nielsen type with the longest span in Taiwan.

24.4.8 Huandong Bridge (環東大橋)

Crossing the Keeling River in Taipei, the Huandong Bridge is a double-deck stiffened steel arch bridge of the Lohse type. With a span length of 166 m and an arch height of 30 m, this bridge was completed in 1998. A total amount of 3000 tons of steel was used in its construction. To add to the aesthetic view of the neighborhood, the bridge was installed with some lighting facilities (see Figures 24.21 and 24.22). This bridge was the first double-deck steel bridge constructed at the time. It was also the first bridge installed with lighting facilities of varying colors and intensities, controlled by computer programs to reflect season changes or the needs of festivals (Tzeng et al. 1996).

FIGURE 24.20 Taipei Second McArthur Bridge (1996). (Adapted from Chang 2007.)

FIGURE 24.21 Day view of Huandong Bridge (1998). (Adapted from Tzeng et al. 1996.)

FIGURE 24.22 Night view of Huandong Bridge (1998). (Adapted from Tzeng et al. 1996.)

24.4.9 Taipei First McArthur Bridge (台北市麥帥一橋)

With a span length of 170 m and an arch height of 30 m, the Taipei First McArthur Bridge was constructed in 2001 as a double-deck single-arch bridge. As can be seen from Figures 24.23 and 24.24, this double-deck steel arch bridge is composed of a single arch of the Lohse type and a Virrendeel truss. Due to its innovative design, this bridge is unique in Taiwan, and is also likely to be so in the world. Various sensors have been installed on the bridge to monitor the settlement, relative displacement, inclination, natural frequency of vibration, and earthquake-induced vibrations. All the data have been collected in real time for the purpose of monitoring the structural safety of the bridge and serving as a reference for long-term maintenance (Chang et al. 1998).

24.4.10 Maling Bridge (瑪陵橋)

In order to cross the Malingken Creek (瑪陵坑溪) and some adjacent valleys in Keelung City, while enhancing the aesthetic view of the rural environment, the main bridge was designed as two steel reinforced concrete arch bridges of span length 138 m, one for the northbound, which has a total length of 445 m, and the other for the southbound, which has a total length of 345 m (see Figure 24.25) (Taiwan Area National Expressway Engineering Bureau 1999, 2003). The central line of the arch is actually a cosine curve. The following are the data for the arch: thickness = 2.0–3.2 m, width = 12 m, and height = 33 m (roughly 60 m about the bottom of the creek). To overcome the difficulties encountered in the construction of the bridge in rough hilly areas, the main bridge was built by the concrete-lapping method with a pre-erected composite arch. In other words, the bottom chord made of the steel reinforced concrete member was used as the supporting frame for erecting the arch body of the upper part. The supported advancing method was also used to construct the plate girders of the superstructure. This bridge is regarded as one of the most beautiful bridges of the Second National Freeways for its harmonious blend with the neighboring environment.

24.4.11 Wanban Bridge (萬板大橋)

The Wanban Bridge was originally known as the Xizang Bridge (西藏大橋), as it was extended from the Xizang Street of Taipei (Figure 24.26). This bridge is a key link for connecting Taipei with Banqiao (板橋) in Taipei County. One purpose of this bridge is to alleviate the traffic carried by the Huajiang Bridge (華江橋) and Guangfu Bridge (光復橋), connected with the Taipei Outer Expressway System and the Taipei County

FIGURE 24.23   Taipei First McArthur Bridge (2001). (Adapted from Chang 2007.)

FIGURE 24.24 Night view of Taipei First McArthur Bridge (2001). (Adapted from Chang 2007.)

FIGURE 24.25 Maling Bridge (2000). (Adapted from Taiwan Area National Expressway Engineering Bureau 2003.)

FIGURE 24.26 Wanban Bridge (2000). (Adapted from Taipei County Government 2002.)

Along-River Expressway System, respectively, so as to elevate the overall efficiency of transportation of the entire area. Completed in 2000, the Wanban Bridge was built over Xindian Creek. This bridge is a five-span continuous arch bridge with the following span lengths: 55 m + 3 × 110 m + 55 m = 440 m. The piers have the shape of a diamond, as can be seen from Figure 24.26 (Taipei County Government 2002).

24.4.12 Donshan River Bridge (冬山河橋)

The Donshan River (冬山河) is one of the best-managed rivers for the ecological environment in Taiwan. For this reason, the Water Entertainment Park associated with this river has been a very attractive area for tourists. The bridge, constructed in 2005 over the river near the Water Entertainment Park, is called the Donshan River Bridge (see Figures 24.27 and 24.28). It consists of three-span prestressed stiffened

FIGURE 24.27 Day view of Donshan Bridge (2005). (Adapted from Taiwan Area National Expressway Engineering Bureau 2006; Lin et al. 2006.)

FIGURE 24.28 Night view of Donshan Bridge (2005). (Adapted from Taiwan Area National Expressway Engineering Bureau 2006; Lin et al. 2006.)

concrete arches. The central arch has clear height of 20 m and a span length of 187 m, the longest at the time in Taiwan for arch bridges. The pier is made of a reverse triangular shape, and the arches and girders are connected by several vertical columns. The bridge was constructed by a combination of cast-in-place and cantilever methods (Taiwan Area National Expressway Engineering Bureau 2006; Lin et al. 2006).

24.5 Truss Bridges

24.5.1 Daan Creek Bridge (大安溪橋) and Dajia Creek Bridge (大甲溪橋)

Completed in 1906, both the Daan Creek Bridge and Dajia Creek Bridge are steel truss bridges of the through type with a length of 62.4 m for each span. The Daan Bridge has a total length of 630 m divided into 10 spans (see Figures 24.29 and 24.30), and the Dajia Creek Bridge has a total length of 380 m divided into 6 spans (see Figure 24.31) (Ministry of Railways 1911). The cross sections of the bridge columns are elliptical in shape, with the exterior covered by bricks and carved stones and the interior made of graded aggregates and limes. The foundations are made of wooden piles or rail-section steel piles. The super trussed girders of both bridges were replaced in the 1960s by the Taiwan Railway Bureau with the foundations kept intact. Both bridges were used until 1998, when the new mountain-side railways were completed. The Dajia Creek Bridge is now used as a bike road by the local residents.

24.5.2 Second-Generation Taipei Bridge (台北橋)

The Taipei Bridge of the first generation for highway use was built in 1901 by the Qing Dynasty (清朝) by transferring the one for railway use. The second-generation Taipei Bridge was built in 1925 as a seven-span through-type steel truss bridge. Each span has a length of 62.4 m. The design vehicle load was the

大安溪橋樑

FIGURE 24.29 General layout of the Daan Creek Bridge (1906). (Adapted from Ministry of Railways 1911.)

FIGURE 24.30 Daan Creek Bridge (1906). (Adapted from Ministry of Railways 1911.)

FIGURE 24.31 Dajia Creek Bridge (1906). (Adapted from Ministry of Railways 1911.)

FIGURE 24.32 Second-generation Taipei Bridge (1925). (Adapted from He 2007.)

standard 8 ton load based on the old Japanese specifications. The supporting structure was constructed as gravity-type reinforced concrete columns supported by elliptic caissons with long diameter of 12 m and short diameter of 5.5 m. The foundation depth is 25.5 m (Taiwan Provincial Government Transportation Bureau 1987). This bridge was considered one of the eight scenic spots of Taipei for its poetic colorful view under the setting sun (see Figure 24.32) (He 2007). The bridge was rebuilt for the third time in 1969 due to serious corrosion on the steel truss and the lack of strength to sustain increasing traffic loads.

24.5.3 Xiluo Bridge (西螺大橋)

Completed in 1953, the Xiluo Bridge (西螺大橋) has a total length of roughly 2 km. This was the first long bridge completed in Taiwan after the handover to the Nationalist government. It was also the first long bridge ever constructed in the Far East after World War II. The foundation of the Xiluo Bridge was completed by the Japanese government, but the superstructure was constructed using steel supplied by the American government as part of its foreign aid. This truss bridge consists of 31 spans, each with a length of roughly 62.4 m (see Figure 24.33) (Directorate General of Highways 1998). For a long period, this bridge has been a scenic spot in central Taiwan.

24.5.4 Truss Bridges on High-Speed Railways

The high-speed railway began its commercial operation in 2006. Of the high-speed railway bridges, with a total length of 345 km, 72.8% were made of elevated bridges, which accounts for 251.3 km in length. Of these bridge segments, most were made of simply supported prestressed concrete girder bridges of length 25–35 m. To cross rivers or other major traffic routes, large-span prestressed concrete box girders or stiffened continuous truss girders were used. As can be seen from Figure 24.34, a stiffened truss girder is used to pass a river near the south end of the Taichung Station of the high-speed railway. This bridge has three spans of lengths 150 m, 120 m, and 100 m (Bridges 2006).

FIGURE 24.33 Xiluo Bridge (1953). (Adapted from Directorate General of Highways 1998.)

FIGURE 24.34 A truss bridge on a high-speed railway (2006). (Adapted from Chang 2007.)

24.6 Cable-Stayed Bridges

24.6.1 Guangfu Bridge (光復橋)

The Guangfu Bridge is actually a replacement of the Syouwa Bridge completed in 1933 for connecting Banqiao with Taipei. The latter was abandoned in 1968 for its lack of sufficient strength in carrying the continuously increasing traffic loads. The Guangfu Bridge was constructed as a three-span cable-stayed bridge with prestressed concrete girders in 1977 (see Figure 24.35), the first cable-stayed bridge ever built in Taiwan. The total length of the bridge is 402 m, divided into three equal spans. The height of the pylons is 18 m and the width of the bridge deck is 20 m. The approaches on the two ends are made of prestressed concrete I-girders of 310 m. Including the length of the two approaches, the total length of the bridge is 712 m (Sun 1999). Since its completion, this bridge has been a landmark of the region. However, there was a drawback with the Guangfu Bridge, as the approach on the Banqiao side was not high enough to accommodate a flood with a return period of 200 years, which was the case with the adjacent embankment. For this reason, the approach on the Banqiao side was later rebuilt.

24.6.2 Chongyang Bridge (重陽橋)

The Chongyang Bridge is one of the bridges crossing the Danshui River (see Figure 24.36). It is a three-span cable-stayed bridge with a total length of 385 m, divided into three spans of lengths 92.5 m, 200 m,

FIGURE 24.35 Guangfu Bridge (1977). (Adapted from Chang 2007.)

FIGURE 24.36 Chongyang Bridge (1989). (Adapted from Chang 2007.)

and 92.5 m. The girder is made of steel, the two H-shaped pylons (height 385 m) are made of reinforced concrete, and the stay cables are of the fan type. This bridge was completed in 1989. It was the first steel cable-stayed bridge ever built and also the one with the longest span at the time in Taiwan (Lin et al. 1989).

24.6.3 Gaoping Creek Bridge (高屏溪橋)

Completed in 1999, this cable-stayed bridge built over Gaoping Creek is a landmark of the Second National Freeway in southern Taiwan (see Figures 24.37 and 24.38). It offers a very pleasing aesthetic view for drivers entering or leaving Pingtung County (屏東縣), and is thus regarded as the main gate for Pingtung County. This unsymmetric single-pylon, single-plane, cable-stayed bridge has a total length of 510 m and a main span length of 330 m. The bridge girder is made of welded orthotropic steel box sections, with the side span of 180 m made of prestressed concrete box girder. The pylon has a total height of 183.5 m and is made of hollow reinforced concrete sections. The reverse "Y" shape of the pylon gives the bridge a sense of stability from the external appearance. This bridge is the longest-span cable-stayed

FIGURE 24.37 Day view of the Gaoping Creek Bridge (1999). (Adapted from Taiwan Area National Expressway Engineering Bureau 2003.)

FIGURE 24.38 **(See color insert)** Night view of the Gaoping Creek Bridge (1999). (Adapted from Taiwan Area National Expressway Engineering Bureau 2003.)

bridge with the highest pylon in Taiwan. When ranked among single-pylon cable-stayed bridges, it is the second-longest bridge in the world (Taiwan Area National Expressway Engineering Bureau 2003).

24.6.4 Maoluo Creek Cable-Stayed Bridge (貓羅溪斜張橋)

This bridge was designed to cross the Maoluo Creek River (貓羅溪) and connect to the adjacent tunnel right after crossing the river. Because of geographical constraints, the arch type was selected for the pylon of the bridge and located in a direction perpendicular to the direction of the bridge, while the bridge girder is horizontally curved. Such a geometry for the bridge, that is, a cable-stayed bridge with a single pylon of the arch type, is generally rare in the world. The bridge width varies from 17.2 m to 21.8 m and is divided into two spans of lengths 51 and 119 m (see Figure 24.39). The arch has a height of 60 m, and a width of 85 m. A link beam was added at the bottoms of the arch to reduce the forces induced. The superstructure of the bridge is composed of two nearly parallel steel box girders (Chang et al. 2001a, b). There is a total of 36 sets of stay cables with radial shapes. The bridge was completed in 2002.

24.6.5 Dazhi Bridge (大直橋)

The Dazhi Bridge is a cable-stayed bridge with a total length of 245 m, divided into three spans of lengths 50 m, 172 m, and 23 m (Figure 24.40). The pylon that looks like a fishing pole is made of box sections with thicknesses of 100 mm and a total height of 67 m. The stay cables in front of the pylon were arranged in a single-plane unsymmetric pattern, and those behind the pylon were anchored on the abutments located on both sides of the bridge. The main girder has a length of 172 m, as hung by the pylon through the stay cables. The bridge deck, varying from 28 m to 40 m in width, was made of plate girders welded around the entire cross sections. This bridge opened to traffic in July 2002. Its unique aesthetic appearance has entitled it as a new landmark of the city of Taipei (Chen 2001).

24.6.6 Dapeng Bridge (大鵬橋)

The Dapeng Bridge is located at the exit of the channel of the Dapeng Gulf (大鵬灣). It is a single pylon, asymmetric cable-stayed bridge. The two spans of the bridge are 55 m and 100 m long and the pylon is 72.9 m high. The pylon is a reinforced concrete structure with a side view with a yacht shape and a positive view of an "A" shape (see Figure 24.41). A rotatable bridge deck of span length 39.5 m is installed

FIGURE 24.39   Maoluo Creek Cable-Stayed Bridge (2002). (Adapted from Chang et al. 2001a, b.)

FIGURE 24.40 Dazhi Bridge (2002). (Adapted from Chang 2007.)

FIGURE 24.41 Dapeng Bridge (2011). (Adapted from Huang, P.H. 2011).

on the southern span of the bridge. This rotatable bridge can offer a waterway of net height of 17 m for navigation when closed, and a waterway of 20 m width with no limit on height when opened up. This bridge was completed in 2011. (Taiwan Area National Expressway Engineering Bureau 2007).

24.6.7 Shezi Bridge (社子大橋)

This bridge is used to connect Taipei with Shezi Island (社子島), crossing the Keelung River. It is designed as an asymmetric, single pylon, self-anchored cable-stayed bridge. The two spans have lengths of 70 m and 180 m. As part of the symbolic and cultural considerations, this bridge is designed to mimic the shape of an egret, which is a kind of bird popular in the adjacent villages and farms (see Figure 24.42). The pylon, with a height of 105 m, is integrated with the exterior girder of the short-span side. It is in this sense that the bridge is self-anchored. This bridge was completed in 2012. (Taipei City Government 2008).

FIGURE 24.42   Shezi Bridge (2012). (Adapted from Teo, E.H. 2012).

FIGURE 24.43   Syouwa Bridge (1933). (Adapted from He 2007.)

24.7 Suspension Bridges

24.7.1 Syouwa Bridge (昭和橋)

The Syouwa Bridge, crossing the downstream of the Xindian River (新店溪), was originally completed in 1933 by the Japanese government. With a total length of 367 m, this bridge was constructed as a three-span suspension bridge with stiffened truss girder (see Figure 24.43) (He 2007). It was the first bridge connecting Taipei with Banqiao (板橋市). To meet the increasing traffic volume and vehicle loads, while enhancing structural safety, the bridge was rebuilt in 1977 as the present Guangfu Bridge.

24.8 Bridges of Other Structural Types

Other types of bridge structures used in Taiwan include the extradosed bridge, corrugated steel web bridge, and the steel–prestressed concrete mixed bridge structure. Figure 24.44 shows one typical example of the extradosed bridge called the Nangang Creek Bridge (南港溪橋) on National Freeway No. 6, the first one of the type built in Taiwan. Completed in 2009, the span arrangement of the bridge is

FIGURE 24.44   Nangang Creek Bridge (2008). (Adapted from Lin 1996.)

FIGURE 24.45   Dali Creek Bridge (2010). (Adapted from Dzeng, D. C. 2010.)

80 m + 140 m + 80 m = 300 m (Fang et al. 2008). As for corrugated steel web bridges, one example is the Dali Creek Bridge (大里溪橋), constructed as part of the No. 4 Highway in the Taichung living circle (Figure 24.45). The span length of 145 m for the Dali Creek Bridge is the longest in the world for this type of bridge. This bridge was completed in 2010. (Chang et al. 2008).

24.9 Pedestrian Footbridges

24.9.1 Bitan Suspension Bridge (碧潭吊橋)

This bridge, used exclusively for pedestrians, is one of the most famous suspension bridges in Taiwan (see Figure 24.46). It was initially built in 1937 during Japanese governance for improving transportation between Xindian and its neighboring areas. This suspension bridge has a height of 15 m, width of 3.5 m, and length of 200 m. It is a two-plane single-span suspension bridge. The towers are made of reinforced concrete. At the time of completion, the bridge was the longest of its type. In the early years, small trucks for carrying coal and agricultural products were allowed to pass the bridge, but later were

FIGURE 24.46   Bitan Suspension Bridge (1937). (Adapted from Chang 2007.)

prohibited due to safety considerations. This bridge has been rehabilitated three times. Subsequently, it was damaged by several typhoons, particularly by the one in 1994, and is now regarded as an unsafe bridge. There have been arguments about how to maintain this bridge. The latest decision is to rebuild the bridge to its original shape. What we see today is a bridge that was rebuilt in 1999 with the enhancement of lighting to create different views for day and night time. This bridge has been an attractive place for residents and tourists. Boating facilities are also available at the embankments under the bridge (Yin 1994).

24.9.2 Pedestrian Bridge at Xinsheng South Road (新生南路人行橋)

Located at the intersection of the Xinsheng South Road and Heping East Road (和平東路), this pedestrian bridge, forming a square, was built in 1981 (see Figure 24.47) (Luo et al. 1980). It was the first Vierendeel steel truss ever built in Taiwan. On each of the four sides of the bridge, covers are provided to protect the pedestrians from rain and sun. There are a total of eight spiral staircases, arranged along the two directions of the corner of the square. The bridge was significantly rehabilitated in 2006 by the city government.

24.9.3 Sanxiantai Sea Bridge (三仙台跨海人行橋)

The Sanxiantai is a group of tiny coral islands located in the northeastern direction at 3 km off the coast of the town of Chenggong (成功鎮) in Taidong County (台東縣). It was originally a cape, but deteriorated into some tiny islands after long years of scouring by sea tides. The total area of the islands is 22 ha with the highest point at an elevation of 77 m above sea level. There are several geologically important spots on the islands, including the sea trench, pot-shaped holes, remaining columns, and indented walls, in addition to scarce near-coast plants. As such, it was classified as a natural preservation area. The three massive coral rocks on the islands were traditionally interpreted as three ancient fairy figures in Chinese history, and it is in this sense that the islands were named as Sanxiantai, which means "rocks for three fairy figures." In 1987, an eight-span concrete arch bridge was constructed to connect these tiny islands to the mainland for pedestrian use. The arch bridge has a total length of 320 m and 311 stairs, and looks like a series of waves (see Figure 24.48). This was the first time that

FIGURE 24.47 Pedestrian bridge at Xinsheng South Road (1981). (Adapted from Chang 2007.)

FIGURE 24.48 Sanxiantai Sea Bridge (1987). (Adapted from Chang 2007.)

epoxy-protected reinforced steel bars were used in a civil construction project in Taiwan for the purpose of guaranteeing the life span of a bridge (Yang et al. 1985).

24.9.4 Xiying Rainbow Bridge (西瀛虹橋)

This bridge is located in the Guanyin Pavilion Coast Resort (觀音亭濱海公園), a famous scenic spot to the west of the city of Magong (馬公市) in the Pescadores islands. It was built for pedestrians to cross from the southern to the western sea banks. The bridge is made of a steel arch raised above the ground with three spans of lengths 40 m, 100 m, and 40 m, plus approaches of 20 m on the two sides. It has a total length of 200 m and a width of 4.6 m. Due to its external appearance as a rainbow shape, it was named the Xiying Rainbow Bridge (see Figure 24.49). The bridge is especially pleasing during the night due to the colorful lighting. Completed in 2004, this bridge was the first arch ever built in the Pescadores islands and has become a landmark of the Xiying area (Penghu County 2008).

FIGURE 24.49   Xiying Rainbow Bridge (2004). (Adapted from Penghu County 2008.)

FIGURE 24.50   Lovers' Bridge (2003). (Adapted from Danshui Township 2008.)

24.9.5 Lovers' Bridge (情人橋)

The Lovers' Bridge is located at the Second Fisher's Wharf in Danshui. This is one of the most attractive places for tourists in the area. It was designed as a single pylon unsymmetric cable-stayed bridge with three span lengths of 42 m, 100 m, and 22 m. The white color elegant steel design of the bridge, along with the slanting single pylon, makes the bridge look like a sailing boat in the sea. Sitting inside the wharf, it also serves as a landmark for guiding the fishers' boats to enter the harbor. This bridge was the largest pedestrian bridge constructed in Taiwan at that time (see Figure 24.50) (Danshui Township 2008).

24.9.6 Ecologically Scenic Bridge (生態景觀橋)

Located in the Wanggong Fishers' Port (王功漁港) in Zhanghua County (彰化縣), the Ecologically Scenic Bridge spans Hougang Creek (後港溪) with a length of 90 m. This bridge was constructed as a steel trussed girder bridge, with reinforced concrete pavement for pedestrain walks and lightweight

FIGURE 24.51 Ecologically Scenic Bridge (2004). (Adapted from Zhanghua County 2008.)

coverage for the truss girder. The smooth, elegant appearance of the bridge has enabled it to attact numerous tourists from the neighborhood (Figure 24.51). From the bridge, visitors can examine at a close distance the ecology of the diverse sea species in the tidal regions, offering a good educational environment for visitors. This is where the bridge gets its name (Zhanghua County 2008).

24.10 Monitoring and Management Systems for Bridges

Most of the time, bridge design codes or specifications are strictly valid for bridges of regular shapes or simple types. For those that are not regular in shape, have very long spans, or are located in regions with variations in geographical configurations, special care should be taken to ensure that the bridge is safely designed and can function well during its service period. One typical example for this are the Xizhi-Wugu Expanded Elevated Bridges (汐止五股高架拓寬工程) constructed along the two sides of the First National Freeway near the Yuanshan regions, which became fully operational in 1997, offering additional lanes for both the southbound and northbound directions to relieve the frequently encountered traffic jams in the region (Yang et al. 1994).

The Xizhi-Wugu Expanded Elevated Bridges were constructed in a region where various design requirements had to be met for different purposes. For instance, to meet the aviation requirements from the adjacent Songshan Airport (松山機場), while accommodating the flood mitigation requirements of the Keelung River below, the centerlines of the bridges were designed to be curved in the three-dimensional space. Further, the geological conditions are quite irregular in this region. In some places the bedrock shows as outcrop, but in the other places, the bedrock was buried at a depth of 10 m or more. To reflect all of these considerations, the Expanded Elevated Bridges were designed as a series of continuous prestressed concrete girder beams of three to five spans, each with a length ranging from 70 m to 175 m. Due to the complexity in the structural configuration of the Expanded Elevated Bridges, a monitoring system was conceived in 1993 and installed on the northbound section from PU7J to PU10J and on the southbound section from PD9K to PD1L aimed at examining the static and dynamic behaviors of this part of the bridges for an initial period of 5 years. Such a monitoring system was believed to be the largest one ever installed in a single bridge in Taiwan at the time of installation (Yang et al. 1994).

The main purpose of this project is to monitor both the dynamic and static behaviors of the bridges for long-term evaluation. As for the dynamic part, the key concerns include: (1) the ambient vibration

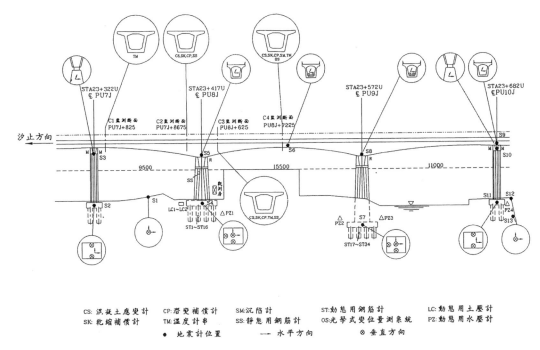

FIGURE 24.52   Locations of the sensors mounted on the northbound section (in Chinese).

properties of the bridges; (2) the dynamic response of the bridges under earthquakes; (3) the behaviors of the bridges under moving vehicle loads; and (4) the interaction of the bridges with the underlying soil. As for the static part, focus is placed on: (1) observing the mechanical behavior of the bridges under varying temperatures; (2) evaluating the effect of prestressed tendons by monitoring the strain gages installed on the tendons and inside the concrete; (3) evaluating the effect of creep and shrinkage on the concrete behaviors; and (4) understanding the combined effect of prestressing and reinforcing bars on the global behavior of the concrete girders. To carry out these investigations, a great number of measuring gages were installed permanently on the bridge, including seismometers (a total of 43 channels, some of the surface type and others of the embedded type), dynamic soil pressure gages (9 channels), dynamic water pressure gages (5 channels), dynamic rebar strain gages (34 channels), concrete thermocouples (66 channels), concrete strain gages (27 channels), shrinkage compensators (6 channels), settlement gages (1 channel), static rebar gages (36 channels), and creep compensators (6 channels).

All the sensors were installed on the northbound section (PU7J to PU10J) and southbound section (PD9K to PD1L) and connected to a central data acquisition house located under the bridge by wires. Figure 24.52 shows the locations of the sensors mounted on the northbound section, in which the following symbols are used: CS = concrete strain gage; SK = shrinkage compensation gage; CP = creep compensation gage; TM = thermometer; SM = settlement meter; SS = static strain gage for rebars; ST = dynamic strain gage for rebars; OS = optical displacement measurement gage; LC = dynamic soil pressure gage; PZ = dynamic water pressure gage; · location for seismometer; → horizontal direction; and ⊗ vertical direction.

References

Bureau of High Speed Rails, Ministry of Transportation and Communications. 2006. *Bridges and Introduction of Construction Methods*, Taipei, Taiwan (in Chinese).

Chang, D. W. (張荻薇). 1983. Design of the Jianguo Elevated Bridge Crossing the Minzu Road—The First Steel Framed Bridge in Taiwan, *Modern Construction*, 37 pp. 30–48 (in Chinese).

Chang, D. W. (張荻薇) (ed.). 2007. *Bridge Engineering.* Part of the series *Taiwan Civil Engineering History,* Chinese Institute of Civil and Hydraulic Engineering, Taipei, Taiwan (in Chinese).

Chang, D. W. (張荻薇), Sung, Y. C., (宋裕祺), and Teo, E. H. (張英發). 1998. Planning and Design of the First McArthur Bridge, *Civil Technologies,* 8 pp. 123–130 (in Chinese).

Chang, D. W. (張荻薇), Sung, Y. C., (宋裕祺), and Teo, E. H. (張英發). 2001a. Planning and Design of Cable-Stayed Bridge with Pylons Made of Steel Arches. *Conference on Advanced Bridge Engineering Technologies,* Taiwan Construction Research Institute, Taipei, Taiwan (in Chinese).

Chang, D. W. (張荻薇), Tang, H. X. (湯輝雄), Tomimoto, M. (富本信), Terada N. (寺田直樹), Matsumoto, A. (松本淳司), and Fukui, Y. (福井康夫). 2001b. Design and Construction of the Maoluo Creek Cable-Stayed Bridge. *Bridges and Foundations,* 35(10) pp. 2–8 (in Japanese).

Chang, D. W. (張荻薇), Wang, J. L. (王炤烈), Lin, C. W. (林正偉), and Wu, H. H. (吳宣欣). 2008. Comparison of the Design of Corrugated Steel Web Bridges with Purely Prestressed Concrete Bridges. *9th National Conference on Structural Engineering* August 22–24, (in Chinese).

Chang, D. W. (張荻薇), Wang, J. L. (王炤烈), Yang, X. L. (楊顯梁), and Chen, H. (陳輝). 2003. Design and Construction of the Steel Bridges for the Zhoumei Expressway in Taipei. *Proc. of 3rd Cross-Strait Conference on Steel and Metal Structures,* November 6–7, Hong Kong pp. 109–126 (in Chinese).

Chen, Z. J. (陳周駿). 2001. Design and Construction of the Dazhi Bridge in Zhongshan Area. *Conference on Advanced Bridge Engineering Technologies,* Taiwan Construction Research Institute, Taipei, Taiwan (in Chinese).

Danshui Township (淡水鎮公所). 2008. http://www.tamsui.gov.tw/ January 10 (in Chinese).

Directorate General of Highways. 1998. *Highways—A Rainbow over the Sun,* Ministry of Transportation and Communications, Taipei, Taiwan (in Chinese).

Fang, W. Z. (方文志), Chen, G. L. (陳國隆), Luo, C. Y. (羅財怡), and Liu, M. J. (劉懋基). 2008. Design and Construction of an Extradosed Bridge—Nan Creek Bridge in Ailan Exchanges of National Freeways No. 6. *9th National Conference on Structural Engineering* August 22–24, (in Chinese).

He, P. Q. (何培齊) 2007 *Taipei City under the Japanese Colony.* Part 1 of the Image Series of Taiwan, National Central Library, Taipei, Taiwan (in Chinese).

Hu, M. H. (胡美璜). 1984. *My Career of Forty Years in Highways—Construction of National Freeways* (in Chinese).

Jia, J. X. (賈駿祥). 2004. Development of Bridges in Taiwan Area. *Taiwan Highway Engineering,* 30(12) pp. 2–24 (in Chinese).

Lin, C. C. (林正喬), Chi, S. J. (戚樹人), Lin, C. F. (林勤福), and Cheng, M. Y. (鄭明源). 2006. Design of the Donshan River Bridge in Taipei-Yilan Freeways. *CECI Technologies,* CECI, Taipei, Taiwan (in Chinese).

Lin, S. J. (林樹柱), Tzeng, C. C. (曾清銓), and Chen, G. X. (陳冠雄). 1989. Wind Resistant Capability of the Chongyang Bridge. *CECI Technologies,* CECI, Taipei, Taiwan (in Chinese).

Lin, S. J. (林樹柱), Tzeng, C. C. (曾清銓), and Chang, D. W. (張荻薇). 1994. Planning and Design of the Elevated Bridges for the East–West Expressway in Taipei. *International Conference on Structural and Geotechnical Engineering,* Zhejiang University, Hangzhou, China October, pp. 681–686, (in Chinese).

Lin, Y. T. (林耀滄). 1996. Rebuilding Planning, Design and Construction of Taipei Bridge. *Civil Engineering Technology,* Civil Engineers Society of Taiwan Province, 6 (in Chinese).

Luo, R. G. (羅瑞剛) et al. 1980. *Pedestrian Bridge at the Intersection of the Xinsheng South Road and Heping East Road—Design Charts.* CECI, Taipei, Taiwan (in Chinese).

Ministry of Economy. 2006. *Review Considerations for Crossing-River Structures.* Taipei, Taiwan (in Chinese).

Ministry of Railways. 1911. *Taiwan Railways History,* Japan's Colonial Government in Taiwan (臺灣總督府).

Ministry of Transportation and Communications. 2001. *Design Specifications for Highway Bridges,* Taipei, Taiwan (in Chinese).

Ministry of Transportation and Communications. 2004. *Design Specifications for Railway Bridges,* Taipei, Taiwan (in Chinese).

Ministry of Transportation and Communications. 2006. *Seismic Design Specifications for Railway Bridges*, Taipei, Taiwan (in Chinese).

Ministry of Transportation and Communications. 2008. *Seismic Design Specifications for Highway Bridges*, Taipei, Taiwan (in Chinese).

Peng, K. Y. (彭康瑜). 1995. Study of Lateral Stability of the Arch Component of the Taipei Second McArthur Bridge. *Civil Engineering Technology*, Civil Engineers Society of Taiwan Province, 2 pp. 17–26 (in Chinese).

Penghu County Government (澎湖縣政府). 2008. http://www.penghu.gov.tw January 10 (in Chinese).

RSEA Engineering Corporation. 1982. *Finishing Report of Northern Link Railways*, Taipei, Taiwan (in Chinese).

Sanxia Township (三峽鎮公所). 2008. http://www.sanshia.tpc.gov.tw January 10 (in Chinese).

Sinotech Engineering Consultants, Inc. 2008. http://www.sinotech.com.tw January 10 (in Chinese).

Sun, G. X. (孫恭先) et al. 1984. *Special Issue for Construction of Guanduda Bridge*, Taiwan Highway Bureau, Taipei, Taiwan (in Chinese).

Sun, G. X. (孫恭先) 1999. Historical Development of Taiwan Bridges. *Taiwan Highway Engineering*, 26(1) pp. 2–20 (in Chinese).

Taipei City Government. 2008. Shezi Bridge: Document of the Bridge as a New Construction for Urban Design Review. New Construction Bureau (in Chinese).

Taipei County Government (台北縣政府). http://www.tpc.gov.tw (in Chinese) January 10, 2008.

Taiwan Area National Expressway Engineering Bureau. 1999. *Special Collection for Bridges in Second Freeways*, Ministry of Transportation and Communications, Taipei, Taiwan (in Chinese).

Taiwan Area National Expressway Engineering Bureau. 2003. *Special Collection in Celebration of Completion of Second National Freeways*, Ministry of Transportation and Communications, Taipei, Taiwan (in Chinese).

Taiwan Area National Expressway Engineering Bureau. 2006. *Special Collection for Taipei-Yilan Freeways*, Ministry of Transportation and Communications, Taipei, Taiwan (in Chinese) January 10, 2008.

Taiwan Area National Expressway Engineering Bureau. 2007. *Dapeng Bridge as Part of the National Scenic Area Roadway Engineering in Dapeng Gulf—A Design Report*, Ministry of Transportation and Communications, Taipei, Taiwan (in Chinese).

Taiwan Highway Bureau. 1971. *Penghu Sea-Crossing Bridge, Construction Report* (in Chinese).

Taiwan Provincial Government Transportation Bureau. 1987. *Transportation Engineering in Taiwan* 1987. Transportation Bureau, Taiwan Provincial Government (in Chinese).

Tzeng, C. C. (曾清銓), Chang, D. W. (張荻薇), and Feng, Y. Y. (馮怡園). 1993. Introduction of the Southern Extension Project of the Shuiyuan Expressway in Taipei. *Structural Engineering*, 8(2) pp. 3–11 (in Chinese).

Tzeng, Q. Q. (曾清銓), Chang, D. W. (張荻薇), and Dzeng, D. C. (曾榮川). 1996. Planning and Design of the Main Bridges for the Huandong Keelung River Expressway, *CECI Technologies*, CECI, 31 pp. 2–12 (in Chinese).

Wang, J. L. (王焔烈), Huang, M. J. (黃民仁), Chang, H. T. (張歡堂), and Chen, H. L. (陳鴻麟). 1999. Introduction of the Construction of the Liyutan Bridge for Railways. *Civil Engineering Technology*, 21 (in Chinese).

Yang, H. S. (楊漢生) et al. 1985. *Sanxiantai Bridge's Connection with the Main Island—Design*. Charts CECI, Taipei, Taiwan (in Chinese).

Yang, Y. B., Tsai, I. C., Loh, C. H., Chang, K. C., Chen, C. H., et al. 1994. Monitoring and Analysis of the Dynamic and Static Characteristics of the Yuan-Shan Bridge System, *1st Midterm Report*, Taiwan Area National Freeway Bureau (partly in Chinese).

Yin, Z. Y. (尹章義). 1994. *History of Xindian* (新店市誌), Xindian Office, Taipei, Taiwan (in Chinese).

Zhanghua County (彰化縣政府). 2008. http://www.chcg.gov.tw January 10 (in Chinese).

25

Bridge Engineering in Thailand

Ekasit Limsuwan
Chulalongkorn University

Amorn Pimanmas
Thammasat University

25.1 Introduction

25.1.1 Geographical Characteristics

Thailand is separated into six regions based on geographical characteristics. It consists of northern, southern, western, eastern, northeastern, and central parts (see Figure 25.1a). The characteristics of each part are unique. The central parts are mostly lowlands or basins, which are suitable for growing rice and for fisheries. There are many small canals, river branches, and main rivers. Therefore, the need for short- and long-span bridges is relatively high. The main river is the Chao Phraya River, the biggest river in Thailand. Chao Phraya River serves not only the main water transportation but also provides the water supply for agricultural activities. The rice ships, sand ships, and logs from the northern part of Thailand also need this river for distribution of goods to Bangkok. As a result, bridges in the central part, especially those crossing the Chao Phraya River, require high clearance and longer span lengths. In addition, ship impact analysis also needs to be taken into account in the bridge design. Thus, higher lateral stiffness and strength of the piers is needed to counteract the lateral forces from ship collision or floating logs along the river. In terms of soil condition, it is acknowledged that the central areas of Thailand, especially Bangkok, have low-quality soil as foundation materials. In particular, Bangkok clay has been used to emphasize the unique property of soil in the central parts of Thailand. As a result, long pile foundations or drilled shafts are usually required.

The geographical configurations of the northern and western parts of Thailand are characterized by raised areas and mountains. The soils are mostly rock and dense sand, which are good foundation conditions. Therefore, spread footing is typically recommended. However, in case of long-span and high-clearance bridges, pile foundations can be used. Sometimes batter piles are used if lateral stability is required. Agricultural products and construction materials (i.e., sand, rock, and log), especially from the northern part of the country, are transported to Bangkok by ships along the branch rivers. These rivers are later combined and become the Chao Phraya River. For small canals, short- to medium-span bridges are needed. The main function of these bridges is to facilitate highway traffic. There are no special design criteria. For long-span bridges over branch rivers, the design criteria for the substructure and clearance of the bridge are similar to those required for bridges in the central part of the country.

The geographical setting of the northeastern part, the largest region of Thailand, consists of flat terrain on the east side and mountainous areas on the west side (see Figure 25.1b). Most of the areas are quite dry, thus medium to large rivers are rarely found. Only short- to medium-span bridges are therefore usually required. In this area, the soil is covered by stiff sand. Hence, spread foundations were

North

Central

East

North-East

West

South

(a)

FIGURE 25.1   (a) Map of Thailand; (b) Thailand's geographical characteristics.

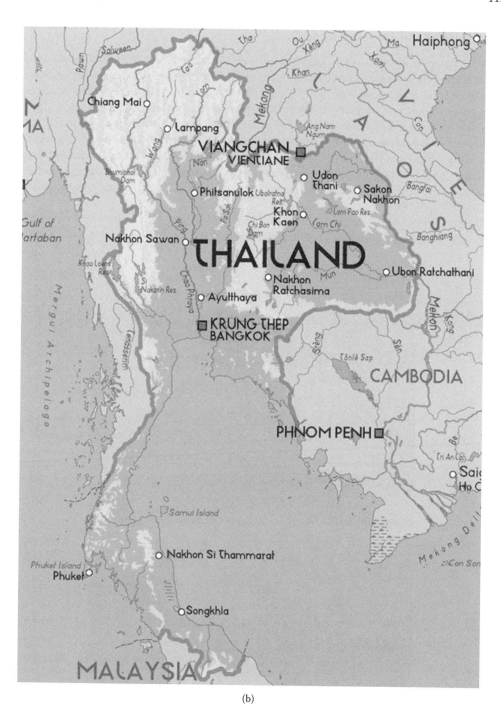

(b)

FIGURE 25.1   (*Continued*)

generally preferred. The eastern and southern parts are characterized by high terrain areas. However, there are some parts, especially in the lower regions, which are covered by long mountains. These regions consist of small and medium rivers, and therefore both small to medium bridges are recommended. The foundation type may be pile foundation and spread footing, depending on the location of bridge site construction and soil conditions.

25.1.2 Historical Development

The development of bridges in Thailand started over 750 years ago with the Sukhothai Kingdom. Typical bridges are short spans made from timber or masonry. The construction of bridges was influenced by religion, location, culture, politic, and transportation lines. Bridges influenced by religion can be seen from eminent temples and royal places, for example, Naga Bridge (Figure 25.2). The major function of this bridge is to fly over the pool around the main temple while the minor function, based on religious belief, is to separate the human area and temple area (area of gods). Therefore, Naga Bridge were created to represent the linkage between human earth and heaven. In the past, Thailand's people preferred to build houses and residences close to the river banks (Figure 25.3a). The main reason was that the river was the main transportation line. Hence, bridges were mostly elevated to facilitate boat movement.

These examples of bridge construction show the harmony between Thailand's lifestyle and civil engineering structures. Bridge construction technology in the past was not complicated because only

FIGURE 25.2   Naga Bridge.

FIGURE 25.3 Bridges influenced by old Thai styles: (a) the old Thai house style; (b–c) elevated bridge over the canal.

short-span bridges were required. However, the development of modern bridge engineering in Thailand has started solidly since the beginning of the Rattanakosin period.

25.1.2.1 Rattanakosin Bridges

When King Rama I established the Chakri Dynasty (1782–present), he also created the new capital city on the east side of the Chao Phraya River. This city was called Rattanakosin and later known as Bangkok. For reasons of security and delay of enemy invasion, the old Rattanakosin area (currently Bangkok's downtown) was surrounded by many small canals. Therefore, it was sometimes known as the Rattanakosin islands. Bridges were built to facilitate commerce and linkage between outbound communities and the Rattanakosin islands. Here, the technology of bridge engineering was progressing continuously. The main function of the bridge in the early Rattanakosin period was mainly to support the transportation of people and animals (i.e., horses and elephants). At this time, there were still no long-span bridges.

Bridges were typically constructed from thick timbers assembled to form a structure, so that they could be demolished or removed in case of war. By this technique, rivers and canals became natural barriers for the protection of the main islands. However, in case of the Chao Phraya River, even though there was some demand for long-span bridges, it was considered not economical to construct long-span bridges at that time. Thus, the solution was to adopt cross-river boats instead of bridges.

25.1.2.2 Early Bridges of Bangkok

When the technologies from European countries came to Thailand, the technology to construct small permanent bridges, movable bridges, and medium truss bridges became popular. Drawbridges built around the islands are examples (see Figure 25.4). At that time, the demand for medium-span bridges

(a)

(b)

FIGURE 25.4 Drawbridges of King Rama IV's period (1801–1868): (a) original drawbridge (b) recent drawbridge (not movable).

increased. Therefore, the need for stronger construction materials was evident. During this period, steel bridges began to be constructed around Bangkok. Recently, many original steel bridges were removed and changed to reinforced concrete (RC) bridges due to deterioration of the steel.

25.1.2.3 King Rama V's Bridges

Bridge construction increased rapidly in the period of King Rama V (1868–1910). The king visited Europe and brought many modern technologies to Thailand. Bridge construction technology was also imported and money for constructing permanent bridges over the small canals around the Rattanakosin islands was invested. This period was called The Golden Age of Bridge Building in Thailand. The great improvement of bridge construction in Thailand can be seen from series of bridges with prefix "Charlerm" in front of the bridge name. *Charlerm* in Thai means "to celebrate."

Bridges were constructed annually to celebrate the King Rama V's birthday. The budget was mainly supported by his majesty. The bridges in this series have a number at the end of the name which indicated the age of King Rama V. There are 17 total bridges in this series starting from Charlerm Sri 42 to Charlerm Sawan 58. Examples of bridges in the Charlerm series are shown in Figure 25.5. Other

(a)

(b)

FIGURE 25.5 Bridges of the Charlerm series: (a) Charlerm Sakdi 43, (b) Charlerm Wang 47, (c) Charlerm Loke 55, (d) Charlerm Yos 45, (e) Charlerm Phoep 50, and (f) Charlerm Dej 57.

(c)

(d)

(e)

FIGURE 25.5 (*Continued*)

(f)

FIGURE 25.5 *(Continued)*

important bridges were also made in this period. More than 100 bridges were built for Bangkok's people. The typical spans of the bridges are around 5–10 m. The bridges were constructed from both steel and concrete. Concrete arch bridges were used widely. In this series, Charlerm Loke 55 was recorded as the first reinforced concrete bridge in Thailand. Today, all steel bridges have been replaced by concrete.

25.1.2.4 King Rama VI's Bridges

King Rama VI (Vajiravudh, 1910–1925) followed the footsteps of his father (King Rama V) in donating funds to build bridges on the anniversary of his birthday. At that time, the bridges had "Charoen" as the prefix in the name, which means "prosperity" in Thai. There are six bridges (Charoen Ruj 31 to Charoen Sawasdi 36). Some bridges in this series are shown here (see Figure 25.6). Besides the bridges in the Charoen series, King Rama VI also built bridges around the Rattanakosin islands. In addition to short-span bridges, long-span bridges across the Chao Phraya River were also constructed in the period of King Rama VI.

25.1.2.5 Bridges before World War II

Before World War II (1939–1945), Thailand was ruled by an absolute monarchy. Civil infrastructure projects were decided under the judgment of the king. Many projects, such as railway, highway, electricity, and water supply, were authorized by the king (Rama V–Rama VII). The first long-span railway truss bridge across the Chao Phraya River was Rama VI Bridge (1927). This bridge linked the railway from the northern and southern line at Bangkok. The original bridge was destroyed in the war and the new bridge was rebuilt in the same place (1953). The new Rama VI Bridge consists of five spans and has a total length of 442.08 m. The width of the superstructure is 10 m (see Figure 25.7).

In the age of King Rama V–VI, many long-span bridges were constructed around Thailand. However, most of them were railway bridges. Another large highway bridge over the Chao Phraya River constructed before World War II is the Memorial Bridge (1932). This was the first highway bridge over the Chao Phraya River (see Section 25.4.1). Other important long-span bridges built in this period are Ratchadapisek Bridge in Lam Pang province (1893) and Ta Chom Pu Bridge at Lam Poon province (1920); see Section 25.6.2.1 and 25.6.2.2.

(a)

(b)

FIGURE 25.6   Bridges of the Charoen series: (a) Charoen Sri 34 and (b) Charoen Ruj 31.

25.1.2.6 Bridges after World War II

After World War II, Thailand's politics dramatically changed. Democracy replaced absolute monarchy. A lot of financial help from the United States was supplied to Thailand. Bridges were widely analyzed and designed by Thai engineers. However, supervision from experienced foreign engineers was still needed. Some of the bridges illustrated in Sections 25.3 through 25.6 were constructed in this period.

25.1.2.7 Recent and Future Bridges in Thailand

Recently, bridges in Thailand have been mostly designed and constructed by Thai engineers. Concrete slab-type bridges are widely used for short-span bridges (5–10 m). For medium-span bridges (20–40 m), steel composite I and box girders are also selected due to time limitations. However, prestressed concrete (PC) I-girder and box girder are generally preferred among Thailand's engineers and contractors. Many modern bridges across the Chao Phraya River were constructed by the cast-in-place balanced cantilever method. Cable-stayed bridges are also used for long spans. Bridge design and construction in Thailand is in continuous progression.

(a)

(b)

FIGURE 25.7   Rama VI Bridge: (a) original Rama VI Bridge and (b) present Rama VI Bridge.

25.2 Design Practice

The design criteria of bridges in Thailand are still not constant because higher load demands are requested from the private transportation sector. The brief development of bridge designs in Thailand is described next.

25.2.1 Analysis and Design Method

In the past, bridge designs in Thailand were conducted without reference to codes or design standards. Behaviors of bridges were analyzed by a rigorous theory of structural mechanics. The simple slab type was originally designed by assuming that the bridge behaved as simply supported beam. The axle loads were placed at critical locations in the span so as to produce maximum bending and shear forces. From

1960–1975, this method was changed to a more refined method by assuming that the bridge slab behaves as an RC plate with added strength and stiffness provided by the edge curbs. The method was known as the "stiffening curb" method. This method was originally developed at the University of Illinois. Thailand's officers who graduated from the University of Illinois adopted it and used as Thailand's design practice.

The slab thickness was less than those designed by conventional methods. Based on the new design method, the curbs of slab-type bridge were reinforced and cast monolithically with the slab to increase the strength and stiffness. The typical design of roadway width by this method was 6 m. The slab bridges designed by this approach were quite satisfactory. However, the drawback of this method was that the curbs cannot be removed. Therefore, when expansion of the bridge was required, the entire bridge had to be removed and replaced by a new larger bridge. This was quite inflexible and was later replaced by the American practice (AASHTO) after World War II. At present, curbs were considered in design to merely function as edge beams. In this approach, the extension of roadway from 6 m to 14 m became possible.

After AASHTO standards spread to Thailand, the American practice was rapidly adopted by Thai engineers. The design of conventional I-girder bridges was carried out by this approach. However, for special structures, such as Donmuang Tollway (see Figure 25.8), the grillage method with random placement of truck wheels with the help of microcomputers was adopted in Thailand's practice. Recently, the design of highway bridges in Thailand has been officially based on AASHTO standards and will gradually move to AASHTO LRFD standards.

25.2.2 Design Codes and Standards

As mentioned before, Thailand did not have an official national design standard. The analysis and design of bridges was based on structural mechanics theory to analyze the bridge under the actual loads that existed at the time. For example, the design of the first long-span steel truss bridge over the Chao Phraya River (Memorial Bridge, 1932) specified that (1) the bridge must be designed to carry convoy of 16 ton Thai truck placed on all lanes (four lanes) and a uniform load of 4 kN/m², and (2) 3 kN/m² of pedestrian loads on sidewalks. For typical bridges, a 12 ton Thai Truck was used to analyze the bridge. After World War II, there was a big change in Thai bridge design. The United States provided assistance to Thailand to construct a highway from the central to northeastern part. The project was named the Saraburi-Korat Highway Project and was later called the Friendship Highway (1936). In this project, the United States invested money and dispatched designers, contractors, and constructing equipment to Thailand. Therefore, the American standard AASHO (later AASHTO) was widely acknowledged by Thailand's engineers.

FIGURE 25.8 Donmuang tollway.

The bridges were designed based on H20-44 and H20-S16-44 (currently HS20-44). These criteria were equivalent to 18 tons Thai Truck (10 wheels Thai Truck) and became the Thai standard for vehicle weight on highway structures. Until 1976, transportation companies requested that the Department of Highways (DOH) increase the gross weight of 10 wheel trucks to 21 tons (maximum load on an axle shall be less than 8.2 tons). An 8.2 ton axle load is also equivalent to 18,000 lbs of equivalent single axle (ESA) load provided by AASHTO. The responses of the bridge (i.e., shear and moment) under a new load type with 10 wheels Thai Truck is comparable to 130% of HS20-44 provided by AASHTO. Therefore, the DOH and the Department of Public Works (DPW) recommended a factor of 1.3 to the ASSHTO load. However, the factor 1.3 is only satisfactory for short- and medium-span bridges. In case of long-span or special bridges, a higher factor such as 1.5 may be more appropriate.

25.3 Reinforced Concrete Girder Bridges

25.3.1 Slab-Type/Prestressed Plank Girder Bridges

Slab-type bridges are common in many areas in the countryside. These bridges are used when the obstruction is a small river or canal. The span length varies from 8 m to 10 m, and the angle of skew varies from 0° to 25°. When one single span (see Figure 25.9a) is not suitable or river width is greater than 10 m, multiple of simple span is required (see Figure 25.9b and c). In a simple span, the bridges are supported by abutments. When multiple spans are needed, pile bents (0.40 m × 0.40 m), sometimes masonry infilled, are designed. The function of these infilled walls is to increase the stiffness of the foundation against log attack if flooding occurs. The foundations are both spread footing and pile foundations depended on the supporting materials (i.e., soft soil, stiff soil, or rock). The short-span bridges on the riverbank are connected with the main span. The deck slab consists of solid precast beams with pretensioned strands 3/8 inch in diameter and an in situ reinforced concrete top slab (see Figure 25.9c). A proportion of the pretension strands are debonded at the ends of the beam with plastic sleeves.

25.3.2 I- and T-Girder Prestressed Bridges

These bridges are simple form and usually used for short to medium span lengths (20–30 m). Post-tensioned and pretensioned girders are both available. The selection is based on the construction criteria. Normally, when a longer span is required (30–40 m) and transportation of girders is limited, the post-tensioned girder is selected. However, in urban areas where traffic congestion should be avoided, pretensioned (15–30 m) is preferred. The eminent projects are listed next.

25.3.2.1 Uttraphimuk Expressway (Donmuang Tollway)

Uttraphimuk elevated tollway is an expressway in the vicinity of Bangkok (see Figure 25.8). The objective of the project is to relieve the traffic problem on Viphavadee-Rangsit and Paholyothin roads. In the first stage, the DOH wanted to enlarge the lanes of Viphavadee-Rangsit Road, which is congested during rush hour for travel to Donmuang Airport. However, it could not enlarge the lanes due to an obstacle along the perimeter route. The DOH then decided to build the viaduct above the median roadway, and gave its concession to a private company to manage. Donmuang Tollway Company won the bidding and administrates the route. The bridge was designed by a consortium of consulting engineers comprising Thai Engineering Consultants Co., Ltd. and Asian Engineering Consultants Corp., Ltd. There are three lanes in each traffic direction and 28 km total distance. The project opened to traffic on December 14, 1994. The typical span lengths are 33–35 m and superstructure width is 25.35 m. Giant Y-shaped piers were selected. These piers are connected to a pile cap which is supported by ⌀ 0.80 m spun piles. To minimize the deck formworks, which are time-consuming, PC T-girders were designed. This bridge is the first bridge fully designed by Thai engineers.

(a)

(b)

(c)

FIGURE 25.9   (a) Slab-type bridge with simple span; (b) slab type bridge with multiple of simple span; (c) PC planks of bridge deck at approach span.

25.3.2.2 Boromrachachonnanee Elevated Roadway

The Boromrachachonnanee is an elevated roadway (see Figure 25.10). In 1995, the king gave the idea to build expressway above the existing road, for relieving traffic at the Arun Ummarin and Boromrachachonnanee intersection. The bridge was designed by the DOH. There are two lanes of traffic in each direction and 14 km total distance. Construction started in April 1996 and the roadway opened to traffic on April 21, 1998. The bridge cost was approximately 4.4 Billion baht. The typical span length is 32.5 m and superstructure width is 16.5 m. In this project, I-girders with 2 m depth were designed. These piers are connected to pile caps which are supported by $\varnothing$ 0.80 m spun piles.

25.3.3 Cast-in-Situ Concrete Box Girders

When the transportation of box segments and resting yards are restricted, the cast-in-situ full girder (i.e., span-by-span construction) method or the incremental cast-in-situ based on balanced cantilever method is selected. Well-known projects constructed from these methods are described next.

25.3.3.1 Rama III (New Krungthep) Bridge

To alleviate traffic congestion on the Krungthep Bridge, a new bridge was constructed adjacent to the existing one and named New Krungthep Bridge, also known as the Rama III Bridge (see Figure 25.11). The project also involved the construction of ramps up to the new bridge and the rehabilitation of the existing bridge. The project started at Ratchadapisek Road on the Thonburi side and ended at the Rama III Road on the Bangkok side with approximately 3.6 km of roadway. The contract of the project was first awarded by the Bangkok Public Works Department in August 1996 and work started in October 1996. The entire project was completed by October 1999, with a total construction time of 1125 days. This was the tenth bridge crossing the Chao Phraya River in the city of Bangkok, and has been recorded as the fifth-highest bridge in the world.

The bridge is a cast-in-situ concrete box girder constructed by the free cantilever method. The length of the main span between the two towers is 226 m and the two spans on either side of the central one are

FIGURE 25.10   Boromrachachonnanee elevated roadway.

FIGURE 25.11   (See color insert.) Rama III Bridge.

125 m. The depth of the box girder varies from 12.5 m at the piers to a minimum of 2.5 m at midspan. The bridge width is 23 m, to accommodate three traffic lanes in each direction. The piers were designed and constructed as hollow columns, each with 22 bored piles supporting them. The piles have a diameter of 1.5 m and go down 55 m in depth. The minimum clearance of the bridge in the navigation channel is 32 m. Since the bridge was constructed beside the existing Krungthep Bridge, whose the main span can be opened for navigation, a high clearance for the Rama III Bridge was also required. The minimum clearance of the bridge in the navigation channel is 32 m. The bridge alleviates traffic congestion on adjacent sections of the road system and was designed to carry 127,000 vehicles per day across the bridge by 2011; however, it was already carrying a traffic volume of 163,000 vehicles per day in 2001, just two years after opening.

25.3.3.2 New Phra-Nangklao Bridge

The New Phra-Nangklao Bridge is constructed over the Chao Phraya River to become a part of Highway Route No. 302 (Ratanathibet Road) in Nonthaburi Province (see Figure 25.12). The purpose of the bridge is to alleviate traffic demand in Nonthaburi and Pathumthani provinces that exceeds the capacity of the old Phra-Nangklao Bridge. In addition to solving traffic congestion, the new bridge also supports Ratanathibet Road as a main concession highway that links the Western and Eastern Ring Roads

FIGURE 25.12 New Phra-Nangklao Bridge.

together. The upstream side of the bridge is Nonthaburi Bridge, located 13 km away, and the downstream side is Phra-Ram 5 Bridge, located 5 km away. The DOH is the project owner. The project began in June 2006 and finish at the end of 2008. When completed in 2008, the New Phra-Nangklao Bridge crossing the Chao Phraya River will be recorded as the longest bridge built by the balanced cantilever method in Thailand.

25.3.4 Precast Segmental Box Girder Bridges

Precast segmental construction is a suitable solution when the box segment can be delivered or the construction site has the space to store a temporary box segment. Some of the important projects are listed in the following sections.

25.3.4.1 Sri Rat Expressway

Sri Rat Expressway, the second stage expressway, is an elevated highway network in the vicinity of urban and suburban areas (Figure 25.13). The bridges were constructed as dual two and three lane elevated highway with a typical span length of 45 m, and a total length of approximately 38.5 km. The superstructure consists of single span precast segmental girder with externally prestressed and precast U-beams, to facilitate construction and to minimize disruption of the traffic flow in the surrounding areas. Superstructure depth is 2.4 m and width is 10.2 m. The concrete columns are supported by $\varnothing$ 1 m bored piles. The construction method was overhead gantry. The project was opened to traffic on September 2, 1993.

25.3.4.2 Burapha Withi Expressway

The Bang Na-Bang Pli–Bang Pakong Expressway, Bang Na Expressway for short, or officially Burapha Withi Expressway, runs above national highway route no. 34, Bang Na-Bang Pakong highway (see Figure 25.14). The project was completed in January 2000. This gigantic project provides an important link in the transportation system around Bangkok and plays a major role in the commercial development of Southeast Thailand. The design solution was to use precast segmental, span-by-span construction. The box girders were match-cast and post-tensioned in place with dry joints and external longitudinal tendons. The structure was designed by Jean Muller International and is owned by the Expressway Authority of Thailand (EXAT).

FIGURE 25.13 Sri Rat Expressway.

FIGURE 25.14 Burapha Withi Expressway.

The main viaduct is a precast segmental box girder constructed by the span-by-span method using six erection gantries. It has about 8052 precast segments routing to 910 spans. Long spans over intersections and expressways were cast in situ by the free cantilevering method using a form traveler system. An external post-tensioned system was used for the precast segmental spans, where over 4800 tons of prestressing steel were consumed. The average span length is 42 m and total length of the bridge is 54 km. The 27 m superstructure width and 2.6 m girder depth were designed. In this project, an H-shaped slender light appearance supported by 30 m of pile foundation was constructed.

25.3.4.3 BTS Skytrain Project

The Bangkok Mass Transit system (see Figure 25.15), commonly known as BTS, is the first elevated rapid transit system in Thailand. The project was initiated in order to alleviate the chronic Bangkok traffic problem by reducing the number of private car commuters in the city area. Because each station is surrounded by buildings and located over the existing road, the construction of 25 stations along the project also required special techniques and erection schemes. Construction began in 1994 and the project officially opened on December 5, 1999, by Princess Maha Chakri Sirindhorn in commemoration of the King's Bhumibol's golden jubilee.

The total length of the project is 28.7 km, comprising two main lines laying across the city from north to south and east to west. The Sukhumvit Line starts from Sukhumvit 81 (On Nut) to National Stadium while the Silom Line starts from Jatujak Depot to Taksin Bridge. A new extension line from Taksin Bridge crossing the Chao Phraya River to Thonburi opened to the public in May 2009. The Sukhumvit extension from On Nut to Bearing was also built and has been in operation since 2011. The main viaduct is a precast segmental box girder constructed with span-by-span method using six erection gantries. It has about 8052 precast segments routing to 910 spans. Long spans over intersections and expressways were cast in situ by the free cantilevering method using a form traveler system. An external post-tensioned system was used for precast segmental spans where over 4800 tons of prestressing steel were consumed.

25.3.4.4 Rama V Bridge

Rama V Bridge (see Figure 25.16) is the fourteenth bridge built to cross the Chao Phraya River in Bangkok and its perimeters. It is located in Nonthaburi province and is a part of Nakorn-In Road. The bridge was constructed using prestressed concrete and forms a continuous structure. It comprises three spans—the main span is 130 m and the two side spans are 95 m. The overall span length of the bridge is

FIGURE 25.15 BTS project.

FIGURE 25.16 Rama V Bridge.

320 m. There are 29.1 m of total road width, separated into 14.55 m and 14.55 m for each traffic direction. The bridge is separated into two parts for supporting the traffic in each direction. In each part, three lanes were designed to accommodate road traffic. Construction began on November 1, 1999, and the bridge officially opened on June 21, 2002.

Originally, it was called Wat Nakorn-In Bridge because the bridge was constructed close to Nakorn-In Temple and is also a part of Nakorn-In Road. Later, it was renamed Rama V Bridge to honor King Rama V. The opening day of the bridge was the same as the celebration of Nonthaburi province for its 453 year of establishment.

25.3.4.5 Rangsit Interchange

The Rangsit Interchange (see Figure 25.17) project was initiated to relieve traffic congestion around Rangsit Junction, which is one of the heaviest traffic areas for transportation towards the north and northeastern part of Thailand. In this project, five new ramps were constructed to connect the traffic among Bangkok,

FIGURE 25.17   Rangsit interchange.

Saraburi, Nakorn-Nayok, and Pathum-Thani. The Rangsit Interchange project is phase 2 of construction. It was finished in 2008. Phase I, constructed 10 years ago, was unable to accommodate the rapid increase in traffic volume. The budget of phase 2 construction was US$32 million. The typical span lengths are 30–40 m. The construction method was span-by-span segmental precast concrete girders.

25.4 Steel Truss Bridges

Rama VI Bridge can be recorded as the first long-span steel truss bridge across Chao Phraya River. After this bridge was constructed, the other steel truss bridges were created continuously. However, one important steel truss bridge is Memorial Bridge.

25.4.1 Memorial Bridge

Before 1782, the capital city of Thailand, which was then called Siam, was located at the city of Thonburi by King Taksin the Great. After King Phra Buddha Yodfa (Rama I) established the new "Chakri" dynasty, the capital city was moved to the other side of Chao Phraya River—Bangkok, the city of angels. To commemorate the anniversary of the Chakri Dynasty 150 years later, King Rama VII ordered a bridge be built across the Chao Phraya River to accommodate people traveling between Thonburi and Bangkok. In order to memorialize the benevolence and kindness of the King Phra Buddha Yodfa, it was named Memorial Bridge (see Figure 25.18). It is usually called Buddha Yodfa Bridge by Thai people.

The bridge was the first highway bridge across the Chao Phraya River in Bangkok. In the past, to accommodate ship transportation underneath, the middle span of bridge could be opened; however, it is now fully closed. In addition to the main bridge structure, the bridge environment was also intentionally designed. A statue of King Rama I was placed at the bridge garden on the Bangkok side. In addition, the public toilets under bridge's approaches were the first public toilets in Thailand. Construction of the bridge began on December 3, 1929, and it opened officially on April 6, 1932. The bridge consists of three spans (through truss for the end spans and half through truss for the opening span). Both end spans are 75.25 m and the opening, Bascule-type span is 60 m. The total length is 229.76 m. The width of the superstructure is 10 m. The piers were constructed from double-cell hollow piers and supported by caisson.

The main reason for constructing this bridge was not only to provide a route for people on both sides of the Chao Phraya River, but to serve political purposes during that time. The bridge was constructed in the period of King Rama VII, the last king of the absolute monarchy system. Therefore, another purpose of bridge construction was to emphasize the goodness of the king's system to the people during that time. The statue of King Rama I, the first monarch of the Chakri dynasty, was situated close to the bridge.

FIGURE 25.18 Memorial Bridge.

25.5 Cable-Supported Bridges

Usually, the span length of obstruction in Thailand (i.e., width of river or valley) is mostly moderate in length and therefore the demand for long span bridges may not be a critical consideration. Hence, balanced cantilever construction is expedient. However, cable-supported bridges have been occasionally considered when a wide clear span under the bridge is specially required. Most cable-supported bridges in Thailand, therefore, have been constructed close to harbors or congested ship traffic.

25.5.1 Cable-Stayed Bridges

There are four cable-stayed bridges in Thailand. All of them not only facilitate traffic but are also attractive locations for tourists.

25.5.1.1 Rama IX Bridge

Rama IX Bridge (see Figure 25.19) was the first modern cable-stayed bridge in Thailand. Construction began on October 1, 1987. In order to celebrate the auspicious occasion of His Majesty the King's sixtieth birthday anniversary, the bridge was named after King Rama IX to honor King Bhumibol Adulyadej of the Chakri Dynasty. The opening of the bridge was selected to be on the king's birthday, December 5, 1987. Rama IX Bridge provides a road over the Chao Phraya River linking Yan Nawa district and Rat Burana district as a part of Dao Khanong. The white pylons and black cables were the original color scheme of the bridge, but were replaced with an all yellow scheme to represent the king on his sixtieth anniversary celebration to the throne in 2006.

The Rama IX Bridge carries six lanes of expressway traffic across the Chao Phraya River in Bangkok. It was the second-longest cable-stayed span in the world when it was completed in 1987. To this day, it remains the longest span single-plane cable-stayed bridge in the world. There are three main spans consisting of 166 m, 450 m, and 166 m. The overall length is 2716 m. The width of the superstructure is 33 m

FIGURE 25.19 Rama IX Bridge.

and the depth is 4 m. The deck was constructed from steel orthotropic. A single steel pylon, a single plane fan type 87 m high, was designed.

25.5.1.2 Rama VIII Bridge

Rama VIII Bridge is the second modern cable-stayed bridge in Thailand (see Figure 25.20). The bridge is named after the eighth monarch of the Chakri Dynasty, King Ananda Mahidol. It was constructed in 1999 and officially opened on September 20, 2002. The bridge crosses the Chao Phraya River with the

FIGURE 25.20   **(See color insert.)** Rama VIII Bridge.

tower located on the river bank. The bridge consists of a single pylon located approximately at one-third of the bridge on Thonburi side. Golden suspension cables extend from this pylon to the road surface. The total bridge span is 475 m, with 300 m for main span. The height of the pylon is 165 m. The pylon shape was designed as an inverse "Y." The bridge is 2.45 km long including the approach spans. The river side stays are arranged in double plane and the back side stays in a single plane. The Rama VIII Bridge is another sculptural addition to Bangkok's skyline. The pylon is topped by a sort of flame, also painted a golden color. The series of golden threads of cable fan stretch out over the river to hold up the road surface which arches gracefully over the river.

25.5.1.3 Bhumibol Bridge

The Bhumibol Bridge (see Figure 25.21), also known as Industrial Ring Road (IRR) Bridge, is a royal scheme initiated by King Bhumibol Adulyadej that aims to solve traffic problems within Bangkok and the surrounding areas, especially the industrial areas around Khlong Toei Port in Southern Bangkok and Samut Prakarn province. The bridges are named after the king: Bhumibol I for the northern bridge and Bhumibol II for the southern bridge. Construction began in March 2003. The project was officially opened on the king's birthday, December 5, 2006. The inauguration was held with the presence of the king and Princess Maha Chakri Sirindhorn on November 24, 2010.

Bhumibol I consists of a 326 m main span and 128 m side spans. Bhumibol II consists of a 398 m main span and 152 m side spans. The width of the superstructure is 35.9 m and depth is 2.8 m. The bridge is supported on two diamond-shaped pylons on each side of the riverbank as required by the Harbor Department. The height of Bhumibol I and II are 164 m and 173 m, respectively. Both bridges provide dual three lane traffic with a navigation clearance of 50 m at midspan. Each segment of the steel deck is 12 m long and weighs 470 tons.

The diamond-shaped pylon is built on a pile cap of 40 m wide × 22 m long × 5 m deep dimension. The concrete bored piles have a diameter of 1.5 m and are embedded 55–60 m into the ground with the pile tip at the third sand layer. Each pile was designed with a safe load of 1200 tons. Forty piles are provided for each pile cap of the North Bridge (Bhumibol I) and 45 piles are provided for each pile cap of the South Bridge (Bhumibol II). Mass concreting of pile cap, approximate volume 3600 m³, was done at 100–200 m³/h to avoid cold joint. All four pylons are crowned with 12.5 m golden pinnacles. The bridge pinnacles are designed in a uniquely Thai style, inspired by the topmost segment of a pagoda or a Thai theatrical crown, which are symbolic of purity and preciousness, like a diamond granted by the king to his people. Eight sculptures stand at the ends of each span on both sides of the road representing kindness and care of the king to the Thai people.

FIGURE 25.21 Bhumibol Bridge.

25.5.1.4 Kanchanapisek Bridge

The Kanchanapisek cable-stayed bridge is a part of the Southern Outer Bangkok Ring Road crossing the Chao Phraya River to accomplish the loop of Road No. 9, located in Samuthprakarn province south of Bangkok (see Figure 25.22). The bridge was designed by a consortium of consulting engineers comprised of PB Asia Ltd., Asian Engineering Consultants Corp., Ltd., and Thai

FIGURE 25.22 Kanchanapisek Bridge.

Engineering Consultants Co., Ltd. Bridge construction started in January 2005 and it opened to traffic in September 2007. The name Kanchanapisek was given to celebrate King Bhumibol's sixtieth anniversary of accession to the throne.

At present, the bridge has the longest span over the river, with a 500 m long main span and three side spans on each side. The total length is 951 m. The width of the superstructure is 36.7 and the clearance is 50.5 m. The A-shaped pylon reaches a height of 187.64 m to accommodate the installation of 168 stay cables. The deck is a composite steel–concrete section. The pylon pile caps were redesigned by the prestressing method to solve right of way and working area issues.

25.5.2 Extradosed Bridges

The concept of extradose is now popular in Thailand. It presents good performance and suitability for Thai construction. One of the new bridge projects with this type is the New Nonthaburi Bridge.

25.5.2.1 New Nonthaburi Bridge

The New Nonthaburi Bridge crosses over the Chao Phraya River at Nonthaburi I Road (see Figure 25.23). The construction project was under the Department of Rural Roads. The bridge crosses the Chao Phraya River approximately in the middle between two existing bridges, Phra Nang Klao Bridge and Rama V Bridge. The bridge completes the link of the main road network in the area on the west side of the river, Wat Nakorn-In Bridge and the Connecting Road Construction Project, and Pak-Kret Bridge and the Connecting Road Construction Project (Rama IV Bridge). This project was assigned to the Department of Rural Roads and is now under construction. It is expected to be completed in 2014.

The project is located in Muang district in Nonthaburi province and is composed of five main components an interchange at Nonthaburi I Road, a flyover at Muang Nonthaburi Bypass Road, a six-lane extradosed prestressed concrete bridge, a six-lane at-grade road, and an interchange at Ratcha Phruk Road. The river crossing of the project is an extradosed prestressed concrete box girder bridge with two single pylons. Twelve pairs of cables anchored on the deck from the side spans passing through their supporting pylon via saddles to the bridge's midspan help support six lanes of inbound and outbound roadways with two side ways. The deck of the bridge is cast-in-place prestressed concrete with transverse concrete ribs constructed by the balanced cantilever technique. The pylons of the bridge are supported on footings with 36 bored piles 1.5 m in diameter, one in the river and one on the land. The structure of the pylon is reinforced concrete and on top is an architectural decoration imitating a crown, a lotus, and a flame, representing and honoring the king, religious purity, and the ongoing advance of Thai architecture and engineering. The bridge span is 200 m for the main girder and two 130 m for the side girders. The width and depth of the superstructure are 32.4 m and 3.3–6.5 m, respectively. There are two single pylons 64.32 m high. The foundations are supported by bored piles.

FIGURE 25.23 New Nonthaburi Bridge.

25.6 Miscellaneous Bridges

Uncommon bridge types are also available in Thailand. These bridges are grouped as miscellaneous, and examples are presented here.

25.6.1 Cantilever Bridges

When the span length of a simple support girder cannot be extended, the extension of the pier deck to support the main girder is an alternative. RC cantilever bridges can be found in the rural areas of Thailand (Figure 25.24). The anchor spans will be cast first. Then, the suspended span, usually built from PC girder, at the middle of the stream or river will be rested on each tip of the anchor spans. In most cases, this type of bridge configuration is required when the intermediate piers cannot be placed in the stream. This type of construction has become unpopular. Recently, when a bridge needs to have a longer clear span, steel truss or continuous PC girder are usually preferred. However, this kind of bridge construction is economical and only simple calculations are required.

FIGURE 25.24   Cantilever Bridges.

Cantilever bridges were popular for long-span railway bridges in the late 1800s. For example, before bombing during World War II, the original Rama VI Bridge (1926–1945) was built as a cantilever steel truss bridge. The advantage of cantilever construction is a substantial reduction of moment or forces in the suspended span. Therefore, the size of the main span over the river, the difficult part of construction, can be reduced. Moreover, the suspended span can be constructed without formworks, so it does not impede the river flow.

25.6.2 Vierendeel Bridges

There are three existing RC Vierendeel bridges in Thailand. All of them were constructed in periods of war when steel was hard to import from aboard. The designer needed to design and construct the bridge based on local materials (i.e., concrete).

25.6.2.1 Ratchadapisek Bridge

Ratchadapisek Bridge is one of the earliest long-span bridges in Thailand (Figure 25.25). The bridge crossed the Wang River at Lam Pang province. Originally, Ratchadapisek Bridge was constructed from timber. The bridge was constructed by the last Lam Pang's lord, Boonrawd Wongmanit (1857–1922), to

FIGURE 25.25 Ratchadapisek Bridge.

celebrate the twenty-fifth anniversary of Rama V's accession to the throne. It officially opened in 1893 and used the name of the occasion (in Thai, "ratchadapisek" means 25 years of throne accession, or silver jubilee). The bridge was replaced by RC in 1917. Almost 100 years has passed, and the bridge still serves traffic and is one of the most attractive places in Lampang.

The original Ratchadapisek Bridge was constructed from timber in 1893. At that time, the total span length of the bridge was 120 m, which was the longest bridge in Siam. In 1901, the first Ratchadapisek Bridge collapsed due to logs that flowed along the river. Therefore, the lord of Lam Pang, Boonrawd Wongmanit, wrote to the Ministry of Interior for the funds to construct a new bridge. Lam Pang citizens also donated to the fund. On January 17, 1905, the second Ratchadapisek Bridge opened officially. The new bridge was constructed from timber and steel. Similar to the first bridge, the new one also failed, in 1915. To eliminate the repeated problem, King Rama VI asked Prince Burachat Chaiyakorn, the commander of the State Railway of Siam, to construct the third bridge from reinforced concrete.

Construction started in 1916 and was completed in 1917. In this time, the architecture became eminent. The firm columns at the gate of the bridge represented the kindness of King Rama V while the red garuda on each column refers to the royal symbol. The white chicken beside the columns stands for Lord Boonrawd Wongmanit and is the symbol of Lam Pang province. During World Wars I and II, the bridge was painted to conceal it from bombing by the coalition. It was a surprise that even though bullets from the planes were spread around the bridge body, it stayed safe from bombs. The overall length of the bridge is 120 m. The bridge consists of four main spans, each 30 m in span length. The width of the superstructure is 6.5 m.

25.6.2.2 Ta Chom Pu Bridge

In the northern part of Thailand close to Chiang Mai province, there is a classic RC Vierendeel bridge, Ta Chom Pu Bridge, sometimes called the White Bridge by locals (see Figure 25.26). The bridge is located in the Mae Ta district in Lam Poon province. The bridge originally crossed the Mae Ta canal, which is now closed. The construction of this bridge reflects the development of Thailand's engineering. Ta Chom Pu Bridge was constructed in 1918 and opened officially in 1920.

Based on the development plan of King Rama V, transportation by train from Bangkok to important provinces was established. The main line was Bangkok-Chiang Mai, the biggest province in the northern part of Thailand. In this project, Ta Chom Pu Bridge was one of the main links. The first problem was the location of the bridge. The bridge was located on dry land and therefore heavy construction machines could not be supplied. The second problem was the effect of World War II, which

FIGURE 25.26   Ta Chom Pu Bridge.

caused difficulty in ordering steel from aboard. At that time, steel was very popular because it could form the truss bridge at the construction site without concrete casting and formworks. These problems forced Thailand's engineers to construct a long-span bridge by themselves. To solve the problems, Prince Burachat Chaiyakorn, the commander of Stage Railway of Siam, decided to construct a long-span bridge by reinforced concrete and used construction materials found in Thailand. However, without advanced technology and supervision from foreign engineers, Ta Chom Pu Bridge was a challenging task for Thailand's engineers. Today, the existence of the bridge shows the success of Thailand's engineers and the history of Thailand's engineering. Full information on the bridge is not available. Only the total length of 87.3 m (two spans) is known.

25.6.3 Bowstring Arch Bridges

Before 1960, this type of bridge was popular in Thailand. Because prestressed concrete technology was not admired by Thailand's engineers, the bowstring arch bridge was an alternative choice compared to steel truss, which needed to be ordered from overseas. There are three existing bowstring arch bridges in Thailand.

25.6.3.1 Pridi-Thamrong Bridge

Pridi-Thamrong Bridge (Figure 25.27) is one of the three classical bowstring arch bridges (Pridi-Thamrong, Dechatiwong, and Wuttikul bridges) in Thailand. Among them, this bridge was the first

FIGURE 25.27 Pridi-Thamrong Bridge.

constructed. It provided road transportation over the Chao Phraya River in Ayutthaya province. The bridge was constructed in 1940 and opened officially on July 14, 1943. This was the same day as the anniversary birthday of General Plaek Pibulsongkram, the Thai prime minister at that time. The name of the bridge was obtained from the name of the former Thai prime ministers who made contributions to Ayutthaya province. The first one is Professor Pridi Banomyong (1900–1983) and the second is Rear Admiral Thawal Thamrong Navaswadhi (1901–1988).

In order to accommodate river transportation such as rice, logs, sands, and agricultural and industrial products from northern cities to Bangkok, a 60 m clear span was required. This span length can be constructed easily using steel truss or prestressed concrete. However, back in the age where PC was new and steel was rare due to the effects of war, the technology to construct RC bridges with spans of 60 m was fantastic. It is interesting that all construction material and methods were developed in Thailand by Thai engineers under the consultation of a so-called Dr. Kruck, a German engineer employed by Department of Highways (DOH). A technique to reduce the thrust force transferred from the arch portion that produced a lot of tension in the deck was introduced by Dr. Kruck. The bridge was precambered during construction in order to counteract future deformation. In addition, the main reinforcements in the deck were pretensioned and released after the concrete had been cast. The technique, later known "partial prestressed concrete," was quite new at that time. The overall bridge length is 168.6 m. Only one main span was constructed and it has a 60 m length. The roadway width is 6.5 m.

25.6.3.2 Dechatiwong Bridge

This is the second masterpiece of work by DOH under the consultation of Dr. Kruck from Germany. Dechatiwong Bridge (see Figure 25.28) spans over the Chao Phraya River in Nakorn-Sawan province, at a junction of subrivers that merge into the Chao Phraya River. The bridge was named after Major Greee Dechatiwong (1941–1943), the fourth commander of the DOH, who supervised the construction of the bridge. Bridge construction began in 1942 and was completed on September 1, 1950. The configuration, span length, and construction techniques were similar to Pridi-Thamrong Bridge in Ayutthaya province. The overall bridge length is 404.5 m. There are four main spans, and each one has approximately 60 m span length. The roadway width is 6.5 m.

25.6.3.3 Wuttikul Bridge

Wuttikul Bridge (Figure 25.29) is the last in the series of bowstring arch bridges in Thailand. Originally, the configuration of the bridge was replicated from Pridi-Thamrong Bridge in Ayutthaya province and Dechatiwong Bridge in Nakorn-Sawan province. The bridge was fully supervised and constructed by Thai engineers because Dr. Kruck, the German engineer who designed the bridge, left Thailand due to World War II. Wuttikul Bridge crosses the Ping River in Tak province. The construction of Wuttikul Bridge began in early 1937 and opened officially on June 24, 1952. Wuttikul was named after Mr. Uthai Wuttikul, the tenth commander of the DOH, who supervised bridge construction until completion.

Chronologically, the Wuttikul Bridge was constructed before Dechatiwong Bridge in Nakorn-Sawan province, however due to many obstructions, the opening was delayed and took almost 15 years. Originally, the bridge was designed to be constructed by a simple technique, that is, the entire 50 m main span was sat on hollow piers which were placed over single huge caissons. The pile driving was difficult to perform, probably due to the bottom of the river, which is comprised purely of sand supported by bed rock. This situation was different from the Pridi-Thamrong and Dechatiwong bridges. The first construction failed. All caissons were blown away because the flow of the water in the river was so strong.

After that, a new construction method was proposed by the Italian contractor. They recommended that the two main spans at the center of the river be combined to reduce the supports in the river. Hence, the main spans were revised to 101 m. To comply with the construction method proposed by contractor, the configuration was automatically changed to a three-hinged arch. An enormous formwork was

FIGURE 25.28   Dechatiwong Bridge.

established. Due to the strong water flow and inadequacy of the formwork supports, the main formwork collapsed and 14 workers were killed. Reconstruction was conducted in 1950. The bridge construction was based on the original drawing (i.e., four main spans with 50 m of span length). The construction was finished in two years and the bridge is still in operation today.

25.6.4   Reinforced Concrete Arch Bridges

The RC arch bridge is popular in Thailand. However, most of them were constructed only for medium simple span lengths (8–15 m). An example of a long-span continuous RC arch bridge is the Nakorn Ping Bridge.

25.6.4.1   Nakorn Ping Bridge

Nakorn Ping Bridge (Figure 25.30) spans the Ping River in Chiang Mai province. It links the west side (Tye Wang district) and east side (Wat Gate district) of the Ping River. However, the history of the bridge was not fully recorded. It was constructed around 1956–1957. Nakorn Ping is the first

FIGURE 25.29   Wuttikul Bridge.

FIGURE 25.30   Nakorn Ping Bridge.

reinforced concrete bridge in Chiang Mai. Constructed before the New Naowarat Bridge, it supports the traffic between the west and the east side of the Ping River. Bridge piers are supported by pile foundations. The three-span continuous bridge is bent in an arch shape. Girders at the pier segments were cast as box-shaped while the intermediate spans were cast as parallel panel girders, thus leaving the bottom slab of the box girder open. Reinforced concrete diaphragms were cast equally along the intermediate spans. However, to strengthen the bridge, steel braces were added parallel to the existing RC diaphragm later. The total span length of the bridge is around 100 m and the roadway width is around 6.5 m.

26

Bridge Engineering
in Egypt

Mourad M. Bakhoum
Cairo University

26.1 Introduction

This chapter briefly presents the historical development of bridge engineering and summarizes the major bridges projects in Egypt. This section describes geographical features, climate, the river Nile, the Suez Canal, and bridge classifications.

26.1.1 Geography and Land

Egypt is located in the northeastern corner of the African continent. Egypt is a square-shaped country, and has a north-to-south distance of 1,024 km and east-to-west distance of approximately 1,240 km. The Suez Canal divides the total land into two parts: the mainland and the Sinai Peninsula. The mainland faces the Mediterranean Sea to the north and the Red Sea to the east (Figure 26.1).

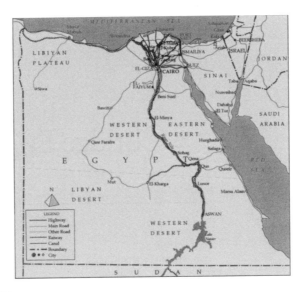

FIGURE 26.1 Map of Egypt.

Egypt occupies about one million square km, out of which the areas suitable for agriculture are limited only to the Nile delta and a narrow valley along the river Nile. The remaining parts are desert land comprising about 95% of the total land area. Sinai is located to the east of the Suez Canal.

26.1.2 Climate

Most of Egypt has a dry desert climate with little rainfall. There is a big temperature variation over the year and over a single day. The Egyptian climate features extremely high solar radiation. The coastal zone along the Mediterranean Sea and the land within 10 km of the Mediterranean Sea features a Mediterranean climate with about 500 mm of rainfall annually. The Nile delta is affected by both the Mediterranean climate and the desert climate. Summer weather prevails from May to September and winter weather prevails from November to March. The highest temperature around the capital, Cairo, reaches 40°C and the lowest temperature is around 7°C.

26.1.3 River Nile

In Egypt, about 95% of the population lives along the Nile River and the delta area. The river Nile extends mainly in the north—south direction, from its border with Sudan up to the Mediterranean Sea. Over 60 bridges were constructed over the river Nile. There are considered among the largest construction bridges.

26.1.4 Suez Canal

The Suez Canal, as shown in Figure 26.2, links the Mediterranean Sea in the north to the Red Sea (Gulf of Suez) in the south with a length of approximately 161 km. It is one of the most important routes for international marine transportation. Ships transit the canal in three convoys daily: two convoys from Port Said to Suez, and one convoy from Suez to Port Said.

26.1.5 Bridge Classification

Bridges in Egypt could be tentatively classified as shown in Table 26.1, based on different criteria.

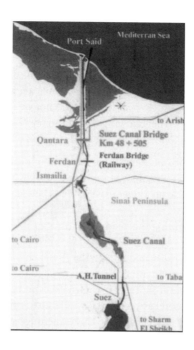

FIGURE 26.2   Suez Canal.

TABLE 26.1   Bridge Classification based on Different Criteria

	Bridges in Egypt are classified according to:	
Crossing Over	Suez Canal – River Nile – Branch water canals – Small canals – Roadways – Highways – Railways – Metro Line – Valleys – Urban Streets (Flyovers) – Urban Intersections	
Type	Overheads Bridges* – Movable Bridges – At Grade Bridges	
Material	Reinforced concrete – Prestressed concrete – Steel trusses – Steel beams – Orthotropic deck – Composite steel – Masonry	
Span	Short span – Medium span – Long span	
Purpose/Use	Roadway bridges – Railway bridges – Metro & LRT – Pedestrian bridges – Utility (Water – cables – …) – Combined bridges	
Structural Systems in longitudinal direction	*Most common:*	Slabs, beams, frames, steel trusses, cable-stayed
	Not very common:	Arches, suspected (only one bridge in Cario Zoo)

*With sufficient clearance for passing ships in the river Nile (typically 13 m in southern Egypt)—over most highways and railways line (typically 5.5 m)—over the Suez Canal (70 m for the cable-stayed bridge).

26.2 Historical Development

A brief review of the history of bridge construction in Egypt is presented in this section. Recent large bridge projects are also briefly discussed. It is interesting to note that despite the fact that Egypt encompasses some of the largest (in number and size), oldest, and most durable monuments and structures in the world, which were built throughout all the stages of Egypt's long history, it seems that very few major bridges were built until recently. Crossing the river Nile and the branch canals was definitely a required activity. It seems that it was done mainly using boats. The section briefly reviews old bridges that were known to have existed in Egypt. Related constructions such as aqueducts, barrages, and boats are also discussed. This section only provides introductory comments and some observations about the history of old bridges and related major constructions in Egypt, and does not claim to be a comprehensive review. A brief discussion of recent large bridges in Egypt is also mentioned.

26.2.1 Ancient Bridges and Related Structures

26.2.1.1 Ancient Bridges

It is quite strange that the ancient Egyptians with all their huge and advanced construction did not build bridges or have impressive ones. According to what is available about ancient Egyptian structures one can deduce that the idea of building bridges over the Nile was not an appealing idea to the Egyptians. But the Egyptians still had some bridges. The following is a summary of an article written by Rostem (1948) about bridges in ancient Egypt. The following are the bridges from which some remains were found:

- Five bridges that date back to the fourth dynasty are found in the Pyramid Complex in Giza. According to a map drawn by Perring, two of these bridges were contained in the causeway of Khufu. It is assumed that these no longer exist or that they may be located somewhere in the village of Nazlet El-Samman.
- Two other small bridges spanning the channel in front of the valley of Khefren can still be seen almost free from damage.
- As for the fifth bridge, it was excavated during the 1940s on behalf of the Antiquity Department. Figures 26.3 shows the Ancient Egyptian Bridge excavated in 1940.
- Remains of a bridge were found in Tell El-Amarna dating back to the eighteenth dynasty. The main street in Akhetaten passed between the Royal House and the palace, which were connected by a bridge spanning the road. The bridge had three openings and the lower parts of the piers were cleared by excavation of the site.
- Other remains of a small bridge that were found in Dahshur date back to the twelfth dynasty. They are a part of the brickwork of the causeway of the pyramid of Amenemhat III.
- The Karnak Bridge: In Karnak Temple there is a simple representation of a bridge spanning a canal near the eastern frontiers of the country (Figure 26.4).
- Another Egyptian representation of two bridges dates from the same period. In the scenes of Ramesses' great achievements in the famous battle, the city of Kadesh is shown surrounded by tow moats crossed by two bridges.

26.2.1.2 Boat Bridges

A bridge consisting of a series of wooden boats tied together (a pontoon bridge) was built at different periods on the small branch of the river Nile in Cairo, connecting Cairo to Roda Island. This bridge was also extended across the whole width of the Nile to reach Giza, on the west side of the river, by El-Zaher Beibars for the crossing of soldiers (Mosselhy 1988).

FIGURE 26.3   Stone Bridge from the 4th Dynasty near the Giza Pyramids (Rostem 1948).

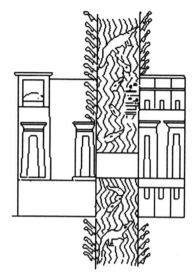

FIGURE 26.4 Representation of a bridge from the walls of Karnak Temple.

26.2.1.3 Bridges in Cairo over El-Khalig El-Masry from 970 to 1860

Fourteen bridges (*qantara*) were built over El-Khalig El-Masry (one of the main canals from the river Nile in Cairo) from 970 to 1810. Five others were built over El-Khalig El-Masry in the nineeenth century (Mosselhy 1988).

26.2.1.4 Movable Bridges

Movable bridges existed in Egypt, as reviewed by Abrahams (2000). Quoting from Abrahams' keynote paper at this conference:

> The earliest movable bridges can be found in Egypt. For example, the book *Egypt* mentions a Nubian fortress built along Egypt's southern border in the twelfth dynasty that utilized a dry moat in front of mudbrick walls. While not illustrated, one can only surmise that this type of construction would incorporate a draw bridge at the entrance. The same reference reports that draw bridges were used at fortified Coptic monasteries in Egypt, "Qasrs," although their date of construction is not known. But certainly the use of draw bridges would be consistent with the use of moats in the twelfth dynasty. Hovey's *Movable Bridges*, referring in turn to Knight's *American Mechanical Dictionary*, Boston, 1884, reports a draw bridge over a moat on an Egyptian monument's depiction of a 1355 BC victory over fortified cities by Ramses II and to "pontoon draw bridges" across the Nile. So we can conclude that the earliest movable or draw bridges can be associated with Egypt.

26.2.1.5 Related Structures

26.2.1.5.1 Pylons

In ancient Egypt, very impressive constructions were built, such as the Pyramids and pylons in the temples. In his article about bridges in ancient Egypt, Rostem (1948) concluded that: "From an architectural point of view, the passage that usually connects the two towers of a pylon over the doorway can be considered as a kind of bridge." In the *British Museum Dictionary,* the following definition for pylons states the conclusion of Rostem: "Massive ceremonial gateway consisting of two tapering towers lined by a bridge of masonry and surmounted by a cornice." The pylon is also discussed by Schlaich (2000). Figure 26.5 shows a photo of a pylon.

FIGURE 26.5   Pylons at Edfu Temple (Stierlin 1995).

26.2.1.5.2 Boats

Boat construction in ancient Egypt is one of the most interesting fields when discussing bridges. The reasons have a paradox nature. On one hand the advanced methods the ancient Egyptians used in building boats and the high level of technology they reached may be one of the strongest reasons why they did not build bridges. On the other hand one finds that the same advanced methods and techniques as those used in constructing some elements in their boats are the same those used now for building cable (hanging) bridges and cable structures. It is quite interesting to see how they reached this level with intuition and through trial and error, without the knowledge of the theories concerning stresses, forces, etc.

Jones (1995) presents a very interesting discussion of boats in ancient Egypt. Figure 26.6 shows a mast and rigging on a model sailing boat found in Tutankhamun's tomb. The sail boom that is carrying the sail is made out of wood and is carried with ropes (cables) and can therefore be of a light material and have a relatively small resistance.

26.2.1.5.3 Aqueducts

Concentrating on remarkable constructions that are similar to viaducts but are not, are the aqueducts. The idea of transporting water from the Nile to the Citadel was aroused during the time of Salah Al-Din in 1176. In Fum El-Khalig, the "octagonal tower was the intake for an aqueduct built originally in the era of Sultan Al-Nasir Ibn Qalaun in about 1311 and subsequently extended to its present extent by Sultan Al-Ghuri in 1505" (Parker 1975, p. 85). The main aim of this aqueduct (Figure 26.7) was to transport water from the Nile to the Citadel. "The intake tower housed huge waterwheels, of a type still used in Syria, which lifted water from the Nile up to the top of the tower, from whence it flowed to the base of the Citadel. It served as Cairo's principal water supply until 1872" (Parker 1975, p. 85). The construction is a composite construction built of stones and bricks. It was built on semicircular arches. The number of remaining arches of the tower are 271. The length of the aqueduct is 3100 m, which makes this construction a very long structure for this time.

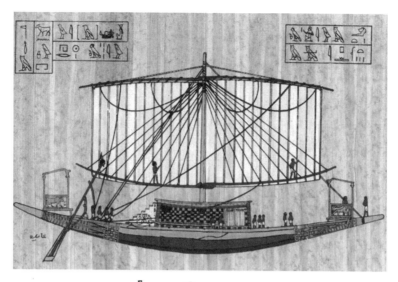

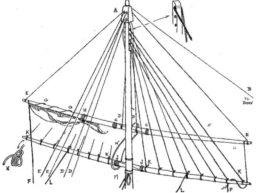

FIGURE 26.6 Pharaoh boats showing the cable-supported sail system (Jones 1995).

FIGURE 26.7 Renderings of columns (Stierlin 1995).

26.2.2 Old Bridges Over the River Nile in Egypt Recorded from 1798 to 1801

Figures 26.8 to 26.10 show old bridges over the river Nile in Egypt recorded from 1798 to 1801.

26.2.3 Recent Bridges in Egypt

With the increasing use of modern transportation techniques since the late 1800s and early 1900s, and the wish to connect the two sides of the Nile, the existence of bridges crossing the Nile became one of the priorities in governmental projects. The following are the first bridges built in Cairo over the Nile (most of them were movable steel bridges):

- Kasr El-Nil Bridge, 1871 (replaced in the 1930s by the existing bridge as shown in Figure 26.11).
- El-Gala'a Bridge, 1891 (in some other references 1877).
- El-Malek El-Saleh Bridge, 1901.
- Abbas Bridge at Giza, 1901 (replaced in 1970 by the existing Giza Bridge).
- Beau Laque Bridge (Abou El-Ela Bridge), 1908 (removed in 1998, currently being replaced).

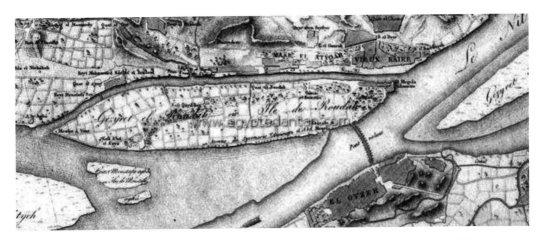

FIGURE 26.8 General Map Roudah Island, E. M. Vol. I, Pl. 15. (From *Description of Egypt (1821). Antiquities, Volume 2, (In French).)*

FIGURE 26.9 View of the short arm (Bridge) of the Nile Vis a Vis the island of Roudah, E. M. Vol. I, Pl. 17. From "Description of Egypt (1822), Modern State, Volume 2, Memory Meqyâs on the island of Roudah and entries contained in this monument. J. J. Marcel, Part Two, Chapter 1. The island of Roudah, Page 120, (In French)".

FIGURE 26.10 Bridge at the entrance of Syout Village, E. M. Vol. I, Pl. 3. View of surroundings of the city during a flood and a view of a bridge. From "Description of Egypt (1822), Antiquities, Volume 2, Chapter XX. Antiques Description The De La Ville De La Province and from Cairo By Mr. Jomard, Page 18-19, (In French)".

FIGURE 26.11 The 6th October Bridge and the Kasr El Nile Bridge over the main branch of the river Nile in Cairo.

- Old Zamalek Bridge, 1912 (removed in 1980, and replaced by 15th of May Bridge as shown in Figure 26.12). The piers of the old bridge in the middle of the river Nile, which were previously used as a pivot for the movable bridge, were used to support the scaffoldings for the construction of a new reinforced concrete bridge which has a span of 90 m. This is one of the longest reinforced concrete bridges (without prestressing) of the beam system in the world.

Figure 26.13 shows suggested a classification of bridges in Egypt in 1997. It is clear that the following changes have occurred with time in the systems of the bridges:

- The material used in the earlier bridges up to almost the year 1955 was steel, and the bridges were movable.
- After that, using steel in bridge construction remained, but the bridges were not movable.
- Since the early 1960s, most bridges over the river Nile are prestressed concrete bridges.
- Steel movable bridges had a low clearance (Table 26.2), while concrete bridges have a clearance of 13 m, to allow passing of boats in the Nile.

FIGURE 26.12   The 15th of May Bridge over the Nile.

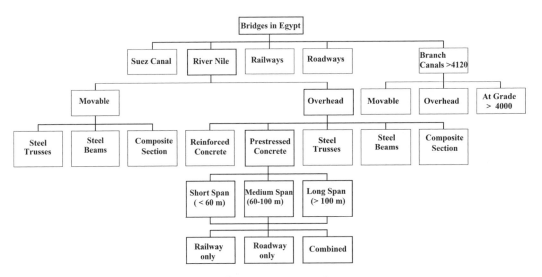

FIGURE 26.13   Suggested classification of bridges in Egypt prepared in 1997.

TABLE 26.2   New Kasr El Nil Bridge and the 6th of October Bridge

	New Kasr El-Nil Bridge	6th of October Bridge
Year of Construction	1936	1976
Span	20 m navigational span (movable), 40 m maximum span	110 m
Material	Steel	Concrete
Nature	Movable	Fixed
Clearance	Low	High (13 m)

26.3 Cable-stayed Bridge Over Suez Canal at Qantara

This section presents the main aspects of design and construction of the cable-stayed bridge that crosses over the Suez Canal almost 45 km south of the Mediterranean Sea, with a navigational clearance of 70 m (Sharaf, Ishitate, and Ishii. 2000). It is considered a vital crossing to Sinai (particularly the northern part), which helps in the implementation of the Egyptian National Project of the Development of Sinai (NDPS), and an important link in the international Northern Coastal Highway, which connects the countries of three continents, Africa, Asia, and Europe, on the Mediterranean Sea. The total length of the Suez Canal Bridge Project is 9 km, comprising about 4.1 km of bridges. The part over the Suez Canal is a cable-stayed bridge, with a main span of 404 m. The bridge is built in cooperation with Japan, which contributed 60% of the construction cost, in addition to the costs for the feasibility study and detailed design. Additional information may be found by Labib, and Bakhoum (2000); Fouad, et al. (2001); Sharaf, and Bakhoum (1997); Fouad and Bakhoum (2001); Mahdi, El-Kadi, and El-Kadi (2001).

26.3.1 Introduction

The Sinai Peninsula has a unique strategic location. It is the crossing link between Africa and Asia overlooking two main seas: the Mediterranean Sea to the north and Red Sea to the south. To the west, there is the Suez Canal, which is one of the most important sea routes in the world. NPDS aims to attract industrial, touristic, and agricultural activities in Sinai. The population of Sinai is estimated to be 3.2 million in 2017. One of the main obstacles that has hindered the development of Sinai's potential resources is the water barrier of the Suez Canal (with a length of approximately 162 km), which separates the peninsula from the rest of the country.

The crossing facilities available in 1996 included Ahmed Hamdy Tunnel in the south, and 7 ferry crossings at several locations along the canal that were not be able to meet the recent and future expected traffic demand in 2017, which is expected to reach more than 50,000 vehicles/day. The need for a permanent crossing structure is evident. According to the agreements between the governments of Japan and Egypt, feasibility studies and detailed design of the Suez Canal Bridge project were carried out by a Japanese study team (JICA, 1996 & 1997). The study was conducted by JICA (Japan International Cooperation Agency). GARBLT (General Authority for Roads, Bridges, and Land Transport) organized the project from the Egyptian side. The funding for the Suez Canal road bridge crossing was confirmed: Japan contributed 60%, and Egypt, the remaining 40%. The work allotment is as follows: The Japanese side work includes the main bridge, which is a cable-stayed bridge with a main span of 404 m; the approach bridges on both the east and west sides above the level of 49.5 m, called the central portion. In addition to the and construction supervision and review of the drawings for the whole project. The Egyptian side work allotment includes the Approach Bridges and roads on both the East side and the West side from the at-grade roads to level 49.5 m. The construction of Suez Canal Bridge started in June 1997 and was completed in October 2001.

To decide the best alternative regarding the location, type of crossing structure, and its main parameters, a feasibility study was carried out by the Japan International Cooperation Agency (JICA) based on a request from the Egyptian Government. The feasibility study, which was completed in October 1996, concluded that a bridge is more economical than a tunnel and its location should be at Qantara, to the north of Ismailia.

To implement the project, the Japanese Government provided a grant to cover approximately 60% of the construction cost. A work allotment for both sides was agreed upon that assigned the main bridge and the portions of the approach viaducts exceeding a height of 49.5 m on both sides of the canal to be implemented by the Japanese grant.

The bridge is considered one of the main elements of the Japanese contribution to peace and prosperity in the Middle East. The bridge will connect the two continents of Asia and Africa by a fixed first-class road link and thus enable smooth and efficient transportation of passengers and goods between the countries of the Old World. The bridge will be part of the North Africa Road extending from Morocco to Egypt and connecting to Europe through the East Mediterranean countries.

26.3.2 Description of the Project

26.3.2.1 Main Features

The road is classified as a primary desert road and has 2 lanes in each direction. The maximum vertical grade is 3.3% for smooth traffic flow. A clearance of 70 m above the high water level will assure free

FIGURE 26.14   Cable-stayed bridge over the Suez Canal Large Ship (Photo, Chodai Co. Ltd.)

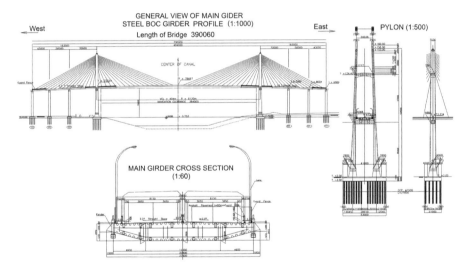

FIGURE 26.15   Main bridge.

navigation on the canal. The main bridge is a steel cable-stayed bridge with girder length of 730 m and central span of 404 m (Figures 26.14 and 26.15). The concrete pylons with a height of 154 m are designed to reflect an image of obelisk towers and to represent a gateway to the Nile valley. The approach viaducts consist of 7 span continuous prestressed concrete rigid frame structures with girder lengths of 280 m and spans of 40 m each (Figure 26.16). All foundations are supported by 1.5 m diameter reinforced concrete piles.

26.3.2 Project Implementation

The project was executed by the following organizations:

1. Executing and Sponsor Organizations
   a. General Authority for Roads, Bridges and Land Transport (GARBLT), Ministry of Transport, Egypt
   b. Japan International Cooperation Agency (JICA)

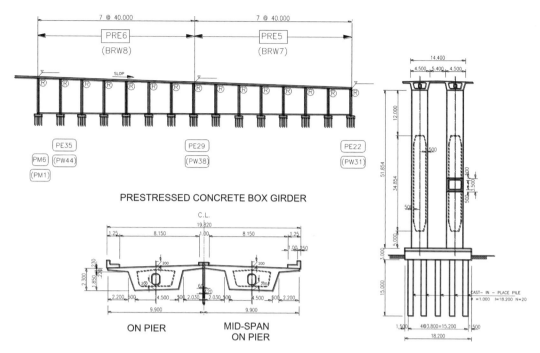

FIGURE 26.16 Approach viaducts.

2. Supporting Organizations
    a. Suez Canal Authority (SCA)
    b. Ismailia Governorate
3. Consultants (The Engineer) for construction supervision phase, Chodai Company, Ltd., PCI, Japan, in association with ACE, Egypt
4. Contractors
    a. West Portion: The General Nile Company for Roads and Bridges (NC), Egypt
    b. Central Portion: The Consortium of Kajima—NKK/Nippon Steel (KNN), Japan
    c. East Portion: The Arab Contractors (AC), Egypt

26.3.3 Design Aspects

The following summarizes the bridge structure system:

- Main bridge: Taking into consideration the length of the main span and the need for quick construction over the Canal, the following bridge types were compared: steel arch, steel truss, suspension with steel girders, and cable-stayed with prestressed concrete girders or steel girders. As a result of the comparison, the last option was found to be the most economic and viable alternative for the main bridge.
- Superstructure: Single-cell and two-cell steel box girder options were considered. The single box girder was selected due to its rigidity, reliability, streamlined appearance, and easy maintenance as compared to the other types of girders that may have lower exposed surface areas, which would be difficult for repainting and maintenance. Wind tunnel tests were conducted to confirm the aerodynamic stability of the proposed girder and deck system.
- Span arrangement: The main span of 404 m and the continuous side spans of 163 m (70 + 50 + 43 m) were selected as the span arrangement of the cable-stayed bridge. Two auxiliary piers were

introduced on the land side of each pylon to improve the load distribution characteristics of the bridge. In fact these auxiliary piers serve as anchors for the cable-stayed bridge

- Approach bridges: steel box girder, steel plate girder, prestressed concrete (PC) box girder, and PC I-section girder bridges were compared. The prestressed box girder was selected due to the following considerations: availability of materials, experience of the contractors, successful previous performance in Egypt, high rigidity, and aesthetic appearance. The approach viaducts consist of 7 span continuous prestressed concrete rigid frame structures with girder lengths of 280 m and spans of 40 m each.

26.3.4 Construction of the Main Bridge

26.3.4.1 Pylon

The general features of the construction sequence for the bridge are shown in Figure 26.17. The construction sequence of the pylon is shown in photos from construction in Figure 26.18. The slip-form method was adopted for the construction of the pylon legs from elevation +12.0 m to elevation +153.0 m. The

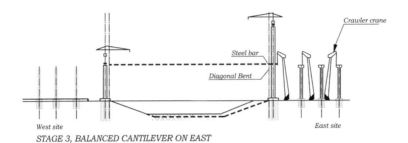

STAGE 3, BALANCED CANTILEVER ON EAST

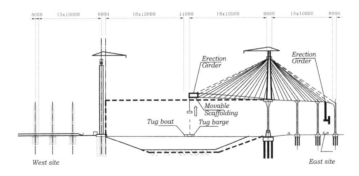

STAGE 4, BALANCED CANTILEVER ON WEST

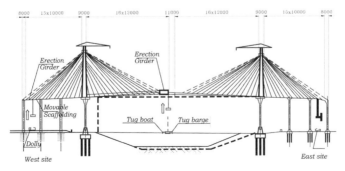

FIGURE 26.17 General features of the construction sequence.

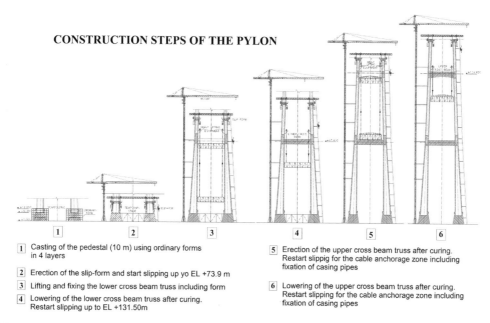

CONSTRUCTION STEPS OF THE PYLON

1. Casting of the pedestal (10 m) using ordinary forms in 4 layers

2. Erection of the slip-form and start slipping up yo EL +73.9 m

3. Lifting and fixing the lower cross beam truss including form

4. Lowering of the lower cross beam truss after curing. Restart slipping up to EL +131.50m

5. Erection of the upper cross beam truss after curing. Restart slippig for the cable anchorage zone including fixation of casing pipes

6. Lowering of the upper cross beam truss after curing. Restart slipping for the cable anchorage zone including fixation of casing pipes

FIGURE 26.18 Construction steps of the pylon.

slip-form method has several advantages. One of its major advantages is the rapid construction rate, which can reach 4 meters of vertical height every 24 hours. Another advantage of this system is its safety. Construction using the slip-form method is very safe, as the system is basically composed of one unit with the necessary scaffolding for the two legs (Figure 26.18).

The slip-form system is designed, fabricated, and supervised by Greitbau Gmbh, Austria. It has 3 levels of working platforms. The upper platform is used for the installation of the vertical reinforcement and the pouring of concrete into the distribution hoppers. The middle platform is used for fixing the reinforcement and casting and compacting the concrete. The lower platform is hanging scaffolding used for surface finishing and curing of the concrete. The lattice girder serves as a storage area and derrick crane operations as well as for stabilizing the slip-form system. The slip-form system is leveled and adjusted by semi-automatic leveling devices attached to each hydraulic jack, and the alignment of the slip-form is continuously monitored by laser systems. The geometry is checked by total stations every morning to avoid the effects of temperature. The tolerance of variations from the plumbs was specified to be less than 0.1%. The lower and upper cross beams are cast in situ on truss supports, which are lifted by jacks.

26.3.4.2 Steel Girder and Cables

Steel girders were fabricated in the workshop and assembled at the project site. The cross section of the steel girder is single box with a streamlined configuration to achieve high torsion rigidity and aerodynamic stability. The total weight of the steel girders is 7,400 tons. The boxes were divided into small panels for easy transportation and fabrication and were fabricated at NKK's Tsu Works in Japan and a workshop in Thailand. In order to ensure a high quality of fabrication the computer integrated fabrication system and the panel fabrication line, which equips welding robots, were used. This was proved successful following the assembly works on site. All fabricated panels were transported to the site and assembled there into 67 block units using a 150-ton crawler crane. Typical lengths and weights of each block are 10 m for side spans, 12 m for the center span, and 100 and 120 tons, respectively. Figure 26.19 shows the assembly sequence. The assembled girder blocks were then moved to the painting yard (Figure 26.20).

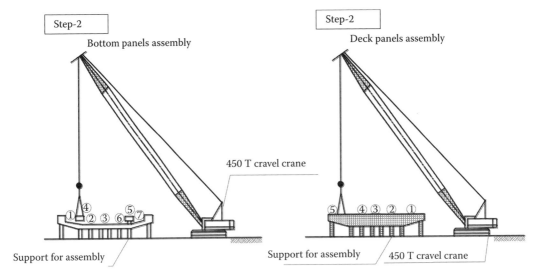

FIGURE 26.19 Assembly sequence.

FIGURE 26.20 Moving a steel box girder to the painting yard.

26.3.4.2.1 *Construction on Site*

- Girder erection commenced from the east pylon using a 450-ton crawler crane. The block on the pylon was divided into three sections (the weight of each section is approximately 40 tons) to meet the capacity of the 450-ton crawler crane.
- To preserve the stability of the bridge, the balancing cantilever erection method is adopted. Blocks for side spans and center span are lifted into position alternately. Two pairs of high-speed winches are used for lifting to minimize the erection time in the canal. The lifting speed is 6m/minute. Lifting up to a height of 75 m takes about 15 minutes.

- For cantilever erection, the assembled blocks were transported by dolly to a position just below the final position for the side spans and by a floating barge for the center span. The floating barge was operated in flea time zone between southward and northward convoys navigating the Suez Canal. To preserve the stability of the bridge, the balancing cantilever erection method was adopted. Blocks for the side spans and the enter span were lifted to position alternately. Figure 26.17 shows erection cyclic work through Cantilever erection. Figure 26.21 shows several erection stages.
- Stay cables are installed and tensioned as soon as a block is in place and connected by welding and high strength friction grip bolts. Figure 26.22 shows the installation procedure for stay cables. The fabricated cables were transported on reels and lifted onto the girder deck using the 450-ton

(a)

(b)

FIGURE 26.21 Girder erection photos from construction of the approach viaduct and the east part of the cable-stayed bridge. (a) Erection of Block 49, East Bank. (b) Erecting the girder block from the barge.

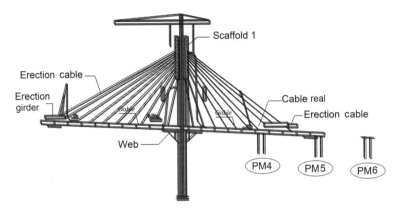

FIGURE 26.22   Cable installation.

crawler crane. A tower crane adjacent to the pylon and a mobile crane on the deck were used to develop the cables and to lead the sockets to a mouth of anchorage on the pylon and the girder. At first the dead end in the pylon was fixed, and finally the live end in the girder was fixed by bearing plate. A 400-ton hydraulic center-hole jack was used to provide the designated tension to the cables. All cable tensions are precisely adjusted at each erection step by the variable thickness of shim plates.

• Duration of each cyclic work was estimated as one week, and out of cyclic work was also estimated as one week. If two works were carried out in sequence, duration of cantilever erection was doubled. This procedure has contributed to minimize the construction schedule.

26.3.4.2.2 *Welding Quality Control on Site*

In order to ensure that the required weld quality is maintained during the assembly and erection works, the welding and inspection procedures were carried out as follows:

• Welding procedures qualified by conducting procedure tests in accordance with BS EN288 Part 3 "Specification and approval of welding procedure for metallic materials,"
• Automatic submerged arc welding and CO_2 semi-automatic welding were used to improve welding productivity as well as the quality of the weld.
• The welding consumables used are high-quality Japanese products.
• Welders engaged are those who have previously obtained welder qualification certificates in accordance with international standards such as AWS, ASME, and so on and have passed additional welder qualification tests conducted at the site in accordance with BS EN287 Part 1 "Approval testing of welders for fusion welding."
• Weld joints were properly prepared within the qualified range.
• Non-destructive testing (NDT) such as ultrasonic testing (UT), radiographic testing (RT), or magnetic particle testing (MT) were used depending on the joint type to test the welds.

26.4 Movable Swing Bridge Over the Suez Canal at Ferdan

The current bridge at Ferdan is the fifth movable bridge constructed at Ferdan. It is a double cantilever truss girder with a main span of 340 m (more than double that of the fourth bridge) and an overall length of 640 m. It provides a clear opening for shipping of 320 m and is driven at each pivot pier by an electromechanical slewing system. Figure 26.23 summarizes the project phases: planning, design, construction, and final bridge during operation (Tomlison et al. 2000; Mizon et al. 2000; Taha and Buckby 2000). More detailed information is provided by Schlecht, et al. (2000); Adrian, Krüger, and Hess (2000); Ramadan (2002); and Fuchs, Tomlinson and Buckby (2003).

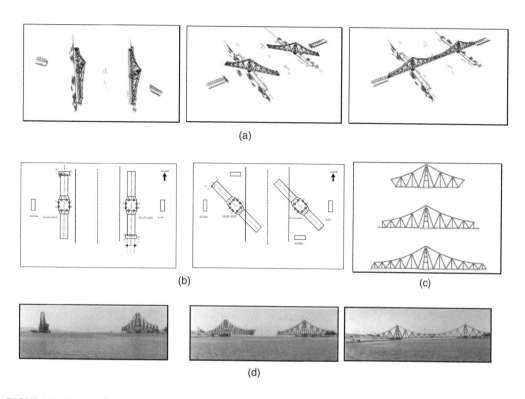

FIGURE 26.23 Ferdan Railway Movable Swing Bridge: (a) planning, (b) design, (c) construction, and (d) final bridge during operation.

26.4.1 Introduction

26.4.1.1 Suez Canal

The Suez Canal was completed in 1869, reducing the sailing distance from Europe to the Far East and to east Africa by thousands of kilometers. At that time, ships with draughts of up to 8 m could pass through the canal. Today, the canal remains one of the most important manmade shipping channels in the world, and revenue from ships passing through the canal continues to be an important part of the Egyptian national economy.

Over the years, the canal has been widened and deepened on a number of occasions and today, ships with water draughts of up to 18 m can pass through the canal. At the site of the new El Ferdan Bridge, 14 km north of Ismailia, the canal has recently been dredged to a depth of 27 m and its width increased to 320 m. Vessels pass though the canal in convoys. The average time taken to pass through the canal is 11 hours for southbound convoys and 16 hours for northbound convoys. This difference is because the northbound convoy has to wait in the Great Bitter Lake to allow the nonstop southbound convoy right of passage.

26.4.1.2 History of Railway Bridges Over the Suez Canal

The current bridge at El-Ferdan is the fifth rail bridge constructed at Ferdan. In common with earlier bridges it also provides a vehicular link on the Suez Canal.

- The first bridge, which was completed in 1920, was a five span truss girder bridge with a total length of 146 m. The bridge had two opening spans of 48 m and 28 m, which provided a maximum clear width for shipping of 42 m. The bridge was designed and constructed by Egyptian National Railways (ENR).

- The second bridge at El-Ferdan was completed in 1943. This bridge was a 151 m long truss-girder bridge with two opening spans, providing a maximum clear opening for shipping of 67 m. It had a manual slewing system and was constructed by ENR and a French construction company.
- The third bridge at El Ferdan was completed in 1952. It had a length of 210 m with two opening spans operated by an electro-mechanical slewing system. The maximum clear width for shipping was 96 m. It was designed and constructed for ENR by a Belgian company.
- The fourth bridge at El-Ferdan was completed in 1963. It was a double-cantilever truss girder bridge with an electro-mechanical operated opening swing span of 167.5 m. The bridge had an overall length of 318 m and provided a clear width for shipping of 148 m. The fourth bridge was designed and built by Krupp. It was severely damaged in 1967. At that time, the El-Ferdan Bridge was the largest opening span bridge in the world.

26.4.2 Description of the Project

26.4.2.1 General

The bridge provides a single-track rail crossing of the Suez Canal with a 3-meter wide roadway on each side of the rail track, which is fixed on wooden sleepers supported directly by the steel deck.

As a moveable bridge the scale is unprecedented. The cantilever trusses are 60 m deep at the pile caps and the main span is 340 m, which would be exceptional even for a static truss structure. All nodes in the truss superstructures have fully site-welded joints, necessitating a large force of qualified welders. Steel members were brought to the site in pieces by road from the works at Sixth of October City or from Germany through the ports of Alexandria and Port Said. Sufficiently well in advance of erection, these pieces were welded together to form the truss and deck members. Pre-assembly was a significant element of the steel group's site operation, requiring large working areas for layout, alignment, welding, and storage. Also, a workshop was established on the West Bank for fabricating temporary works, access ladders, walkways, and erection equipment.

Building a bridge of this type and scale involves a wide spectrum of activities, divided into three principal groups: civil, mechanical/electrical, and steel erection works. It was vital in the planning of the works to ensure that the interfaces between these disciplines were adequately managed. Much of the civil work involved casting steel components into concrete substructures for the attachment of precision mechanical elements, requiring design and detailing that took account of practically acceptable tolerances for concrete works whilst respecting the often precise mechanical engineering requirements.

Advantage is taken of the swing bridge form by building the bridge on dry land parallel to the Suez Canal, thus avoiding conflict with canal traffic. Precise surveying and geometric control is imperative to guarantee proper alignment when the two halves are rotated and joined in the center. Erection of the steel superstructure is carried out using conventional crawler crane travelling on compacted gravel roads and use was made of the protection jetty on each canal bank to provide temporary support to the steelwork at each erection stage.

In both the design and construction, measures were taken to provide adequate protection against the harmful effects of chloride attack on reinforced concrete and steelwork. A site laboratory provided under the contract, enabled chloride contents on rebar and steel plate to be measured and monitored, in addition to providing facilities for the many other quality control tests required by the contract.

26.4.2.2 Egyptian National Railways' Requirements

ENR specified certain basic requirements in the tender invitation documents. These included:

- Navigational clearance agreements
- Special requirements of the Canal Authority
- Minimum operational requirements for the bridge

- Provisions for future maintenance and inspection
- Design loading standards
- Material specification standards

The special requirements of the SCA were in relation to safeguarding the passage of ships on the canal during construction of the bridge.

ENR's operational requirements placed a limit on the maximum time allowed for slewing the bridge and included specifications for emergency operation in the event of a failure occurring under normal operating conditions. These were broadly based upon the specifications used for the fourth bridge. They were subsequently modified during the detail design period to give an improved specification which incorporated up-to-date control and operating systems and which reduced the overall time to open and close the bridge.

The contractor was also required to produce maintenance and operating manuals as part of his contract. ENR's requirements in relation to facilities for future maintenance and inspection of the bridge included a requirement for the contractor to provide a range of replacement parts for the operating system and detailed specifications for all structural, mechanical, and electrical components. This information will be stored on a computer-based bridge maintenance management system that ENR will maintain.

26.4.2.3 Project Implementation

The project was initialed by ENR. The design and construction was carried out by the design and build contractor, Consortium El-Ferdan Bridge, CEFB, with Halcrow as consultants to ENR, in the role of engineer-in-charge supervising the construction. Within the Consortium was a joint venture for the civil works between Besix of Belgium and Orascom of Egypt. Krupp Stahlbau Hannover of Germany and Orascom fabricated and erected the steelwork. The large bearing assemblies and other mechanical and electrical items were made and assembled by Krupp Fördertechnik of Germany. The tender, offer, and final design was worked out in close co-operation between the engineers of Krupp Stahlbau Hannover and the Design Office of Professor Weyer, Dortmund/Germany.

The final design phase was worked out in intensive coordination and cooperation with the bngineers of ENR and its consultant, Halcrow (Consulting Engineers & Architects) Ltd., Swindon/UK.

Of the 10,500 tons of structural steelwork required, Krupp fabricated approximately 4,000 tons in Hannover, Germany, with the remainder fabricated in Egypt by National Steel Fabricators in 6th October City. All of the St52 grade (now S355 J2 G3) plate required was supplied from Preussag Stahl's Ilsenburg facility in Germany.

26.4.3 Steel Structures

26.4.3.1 General

As is typical for a swing bridge, the steel structure works for El-Ferdan are a close amalgamation of structural steel and electromechanical works. Here, 1,200 tons of mechanical items are built into the 10,500 tons of the net steel superstructure. Moreover, highly sophisticated electronic equipment is needed for safe and fully automatic bridge operation.

Besides this PLC-controlled bridge motion, which has a net turning duration of 10 minutes, each electromechanical step can be carried out semi-automatically or even manually. The overall time for bridge opening or closing is 30 minutes.

26.4.3.2 Steel Superstructure

The superstructure of the El-Ferdan bridge (Figures 26.24 and 26.25) is made of high-grade steel S355 J2 G3 (St52) with corrosion protection, with the exception of the wind bracing system, which is made of rolled sections. The elements are fabricated from a welded plate.

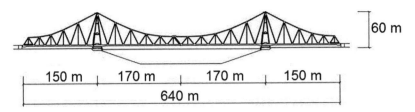

FIGURE 26.24 Elevation view of the superstructure.

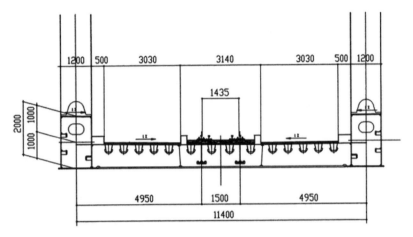

FIGURE 26.25 Cross-section of the road deck.

The bridge's net weight is approximately 1,150 tons of each superstructure, half of which is situated in the bridge rotation center. Here the 500 tons of pylon with a height of 60 m and a rigid 400 ton grillage system is needed to transmit the dead load of the balanced cantilevers into a 250 ton ring girder system, which has a diameter of 17 m, and is placed directly on the roller bearing on the concrete pier.

The truss system net weight is 2,800 tons. Each of the 320 m long road deck areas of the 150 m and 170 m long cantilever arms have a weight of 1,000 tons. The remaining 100 tons is needed for wind bracing and minor construction components.

The pylon and most of the remaining truss walls consist of fully welded box sections. Only the short posts and the tension diagonals at the bridge ends are designed as open sections (Figure 26.26).

The heaviest components are located below the bridge pylon. Here the 3.5 m deep lower chords have a bottom flange thickness of 190 mm and a final transport weight of 76 tons. Together with extremely rigid cross girders they form a grillage system to transmit the bridge dead load uniformly into the slewing system below. In addition the road deck is widened in this location to reduce the pressure on the roller bearing system underneath the ring girder.

The upper and lower chords at the bridge center span as well as the lower chords at the bridge ends are deepened to accommodate the electromechanical locking system.

To simplify transportation, the road deck member is divided into six deck panels with dimensions of 3.50 m × 18.00 m for each lower chord. The outer two deck panels form the road deck, whereas the center panel carries the railway track.

The longitudinal stiffening of the road decks is achieved by cold-formed trapezoidal stiffeners. Underneath the rail tracks built-up sections are used in addition. The welded cross girders transmit the road and railway loads into the main truss system.

All members were prefabricated in the workshops and brought to site with the maximum allowable transport dimensions. About 60% of the truss members have intermediate site weld joints to enable their marine and road transport.

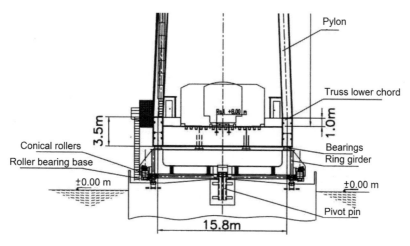

FIGURE 26.26   Cross-section at pylon.

26.4.3.3 Corrosion Protection

In the final tender stage, the client stated a preference for a coating system for corrosion protection instead of building the bridge in weathering steel. As a result the contractor was required to ensure a 30-year design life for the coating system. A special system is chosen, which consists of one prime coat and an additional three workshop topcoats. The final fourth topcoat is applied onsite.

26.4.3.4 Quality Control During Fabrication

The fabrication process was covered by an overall quality assurance system. For the structural steel that is completely of German origin, all material tests and chemical analyses were made in the mill. Besides this all plates were automatically ultrasonically checked against laminations. The test procedure is according to the requirements of the German railways.

Before the release of the components to the site all welds were checked by ultrasonic or X-ray according to the requirements of the German railway code DS804. Furthermore, dimensional checks were carried out.

Regarding the protective treatment surface roughness, checks are made before coating. Each coat thickness is measured and recorded before delivery to the site.

Two slewing drives, each having a rating of 45 kW, are used per span to rotate the bridge. They are mounted on the ring girder and transmit the necessary rotation force via gearwheels and rack-rails into the bearing base connected to the foundation (Figure 26.27).

At the bridge center two pairs of locking bolts situated in the lower chords of the west span and the upper chords of the east span transmit positive bending moments and shear forces. In the locked position the bolts are supported vertically by rollers built into the opposite bridge structure. Horizontal locking forces are transmitted by sliding supports of hardened steel. The 50 × 90 cm locking bolts in the bridge center are engaged and disengaged a 45 kW drive unit.

At the bridge end locking systems are incorporated in the lower chords. They engage with the abutments to fix the bridge ends against vertical and horizontal deflections. The 55 × 55 cm end locking bolts are moved by drive units having a rating of 37 kW each.

In the abutment for the closed position and in the locking heads for the parked position, so called locking boxes are cast into the concrete structures to transmit the support reactions into the foundation.

Automatic roadway flaps and rail moving systems in the center locking and end locking areas of the bridge ensure a smooth passage for rail and road traffic.

Two independent diesel generators per span power all bridge movements. They are placed in the widened deck areas at the pylons. Alternatively, if necessary, the bridge may be operated by the external power supply. All mechanical actions are controlled from the bridge control cabins situated in each pylon.

(a)

(b)

FIGURE 26.27 Pre-assembled slewing system. (a) Roller bearing base. (b) Ring girder and roller bearing.

26.4.4 Construction of Steel Structure

The erection of the El-Ferdan bridge was carried out by cantilever erection parallel to the shoreline on each canal bank, without any influence on the canal traffic passing by. The overall steelwork erection started by re-assembling, adjusting, and connecting the roller bearing base to the reinforcement in the concrete pile cap. In parallel the center pivot pin was placed and adjusted at the bearing base center. Both components were then cast in.

After placing and assembling the conical roller bearing sections, the ring girder assembly started.

On top of each ring girder eight bridge pot bearings were bolted. They were connected to the ring girder top flange by high strength friction bolts. The grillage components were then placed on the pot bearings, adjusted, and then welded together. For later adjustment the bearing top was left unconnected to the grillage lower flange during the whole erection phase.

The pylon erection is divided into four major steps according to the pylon rail levels and the assembly of the pylon head and top cross rails (Figure 26.28). For each rail level the pylon leg and the longitudinal rails are preassembled into frames. These two frames were lifted and located in the existing structures below by steel pins which are placed in lugs welded to the inside of the pylon leg. Then the transverse cross rails were lifted and connected to the two frames. Finally the whole assembly is adjusted, welded, and the temporary pin connection lugs are cut off.

(a)   (b)

FIGURE 26.28   Pylon erection. (a) First pylon rail level. (b) Fourth pylon rail level.

The erection sequence described above allows for continuous erection progress since the upper pylon levels can be placed before the levels below are completely welded.

Cantilever erection took place north and south of the pylon parallel to the shoreline. Here support frames were placed between the banks of the canal and the protection jetty constructed by the civil works Joint Venture. On top of these support frames the next lower chords to be erected were placed and the related road deck panels fitted in and connected (Figure 26.29).

The first 70 m long diagonals inclined to the pylon were lifted and temporarily hinged to the lower chord ends. At their top they are supported in the vertical and horizontal direction in the joint area at the pylon head.

Using jacks the lower chord elements together with the internal road deck were brought into their final alignment according to a predetermined bridge profile. The members are then completely welded.

Finally the jacks on top of the support frames were lowered and the structure made free for vertical movements. To ensure equilibrium of the dead load about the pylon all erection stages are made symmetrically.

For the erection of the upper chords and the diagonals inclined to the bridge ends a steel-rope tie-back system is used. Each pair of diagonals is interconnected by a temporary wind bracing system.

The two diagonals of each cantilever side were lifted with their bottom end hinged to the already built in lower chord. The diagonals are tied back to the pylon by use of steel ropes and tension jacks and adjusted to the correct inclination (Figure 26.30).

In the next stage the upper chords were lifted and connected to the diagonal ends and the already built up structure. After welding of the joints the tie-back system at the top and the hinge connection at the bottom were removed.

Before removing the temporary bracing system between the diagonals, the permanent bracing elements between the upper chords were erected with their final fitted bolt connections.

The sequence described above is adopted for all lower chords and diagonals inclined to the pylon as well as all upper chords and diagonals inclined to the bridge ends.

The end lower chords and the center members were lifted with the complete pre-assembled electro-mechanical items.

FIGURE 26.29 Supported lower chords during the erection of diagonals inclined to the pylon.

FIGURE 26.30 Tied-back system of diagonals during and before lifting of upper chords.

26.4.5 Global Structural Stability During Erection

During the erection phase both cantilevers of each bridge span remained free in the vertical direction. Strong winds could cause an overturning of the structure in this condition. Furthermore, the member weight of the components related to the 150 m span is less then the one of the 170 m cantilever.

To ensure global stability during erection a method of placing and rearranging the ballast on the bridge spans was used. In the first erection phases the ballast taken from the final counterweight is placed on the grillage in the pylon area. With the increase of the cantilever arms the ballast was moved progressively in three defined steps towards the bridge end. In the end area of the smaller cantilever the ballast is needed on the completed bridge as counterweight material to maintain a dead load equilibrium of both cantilever arms (Figure 26.31).

FIGURE 26.31 Bridge erection.

26.4.6 Bridge Movement

The contract specification was written on the assumption that the operating and signal interlocking systems would be similar to those used on previous swing bridges at Ferdan. This would provide control of the opening and closing operations by means of physical keys and interlocks to ensure that all joints were properly engaged and aligned before traffic could be allowed to cross the bridge.

When the details of such a system were worked out, it became clear that, although the time taken to swing the bridge through 90° would be about 12 minutes, the total time for a closing operation would be about 55 minutes (Figure 26.32). Following discussions between reviewer and Contractor, the Contractor proposed an alternative system, which would take advantage of modern electronic systems and would reduce this time to about 30 minutes. This proposal was accepted and the design has proceeded on this basis. It involves the use of intelligent proximity switches at all the critical joint closing positions. These send signals directly to the control rooms and to an electronic system, which interlocks the railway signaling system with the bridge locking systems. This ensures that the bridge can only be used when it is safe to do so.

FIGURE 26.32 Bridge positions during swing to 90° (taken after 12 minutes) and closing (taken after 55 minutes).

26.5 Aswan Cable-Stayed Bridge Over the River Nile

26.5.1 Introduction

The Aswan single-plane, cable-stayed bridge over the river Nile is located 11 km north of Aswan in southern Egypt (Labib 2000; Asmar and Huynh. 2000). The bridge has a total length of 977 m and consists of two approach viaducts (respectively 128 m and 349 m long from west to east) connected to the main cable-stayed bridge of 500 m long. The span arrangement of the cable stay part is 49 -76 -250 -76 -49 m, width 24.3 m. Figure 26.33 shows a view of the Aswan cable-stayed bridge. The project was completed in 2002. More detailed information are provided by Mahdi, El-Kadi and El-Kadi (2000); Farooq, Shohayeb and Jones (2001); Abou-Rayan (2004); and Kamal, et al. (2006).

The vertical profile is symmetrical with respect to the center span. The minimum clearance between the high water level and bridge soffit is equal to 13 m to allow ship navigation in the river Nile.

26.5.2 Design Aspects and General Description

26.5.2.1 Longitudinal Structure of the Cable-Stayed Bridge

The deck is suspended to the pylon by means of 14 pairs of stays per pylon disposed on one central plane (Figure 26.34). The central plane stay cables solution has been retained for the following reasons:

- Most economical solution.
- Construction of the pylon is easier than that of the double plane solution and necessitates less operation for tensioning of the cable stays.
- Aesthetics of the bridge. This solution avoids the unpleasant crossings that would have resulted from the presence of two lateral planes of cable stays.

FIGURE 26.33 Aswan cable-stayed bridge over the river Nile.

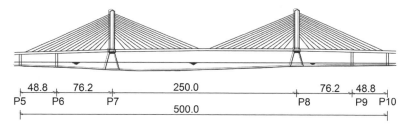

FIGURE 26.34 Elevation view of the central portion of the Aswan Bridge.

The deck is rigidly connected to the pylon and is supported by two lines of large elastomeric bearings on the top of the pier shaft. The intermediate piers in the side spans have are able to resist of the uplift reactions and to limit the vertical deflections when the main span is loaded by traffic loads. The deck is related to the pier by prestressed cables to resist to the uplift displacement. This restriction allows a free longitudinal movement.

The two Teflon bearings placed on each intermediary pier are selected to permit longitudinal movements and to limit horizontal deflections. One of the two Teflon bearings on each pier prevents transverse displacements. Thus, the intermediate piers resist the transversal seismic and wind effects.

26.5.2.2 Deck Cross Section

The box girder has a total height of 3.30 m. The deck width is 24.30 m, which allows 2 × 2 lanes of 3.50 m each and 2 sidewalks of 2 m wide (Figure 26.35).

This structure looks like the Brotonne Bridge (France) or the Sunshine Skyway Bridge (Florida, USA). This is the same family in terms of structural behavior and aesthetics although the Brotonne segments were cast in place with precast webs and the Sunshine segments were precast and lifted from a barge.

The general shape of the deck results from the axial suspension. The deck cross-section is designed to

- Distribute in the box girder the forces introduced by the stay cables anchored in the central nodes of the box-girder
- Support a large top slab by two inclined webs
- Give high torsional rigidity to withstand the torsion stresses resulting from asymmetrical loading
- Reduce wind forces on the bridge

A wind tunnel test was conducted in CSTB (Nantes, France) to determine the aeroelastic stability of the bridge deck. The study revealed that no risk for aeroelastic instability.

The thickness of the top slab is 22 cm. The top slab is stiffened in the center part of the box girder by a double longitudinal girder. The shape of the last one is designed to allow the anchorage of the stays on the deck level. The top slab is prestressed transversely by 4F15S tendons. The bottom slab is 20 cm thick and is stiffened transversely by 30 cm deep cross-beams.

The web thickness is 42 cm for the standard segments. The webs are thickened from 42 cm to 57 cm for the two segments adjacent, on each side, to the pylon. Blisters are placed on the angles formed by the web and the slabs (top and bottom) to allow anchorage of prestressing tendons.

26.5.2.3 Pylon and Piers

The pylon is a prestressed concrete structure. Its cross section is a box with outer sizes of 3 m × 6 m. The inner sizes (1 m × 3 m) allow future inspection of the anchorage heads (Figure 26.36).

The stays are anchored to the pylon tower on a vertical range of about 30 m. The stay anchorage is carried out inside the pylons to dissimulate the anchorage heads. They do not cross the pylon. The transfer of the stay forces to the pylon is performed by looped post-tensioning tendons. The prestressed solution was adopted to avoid any cracks on the concrete due to the horizontal forces of the cable stay.

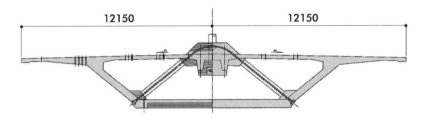

FIGURE 26.35   Typical deck Cross-section.

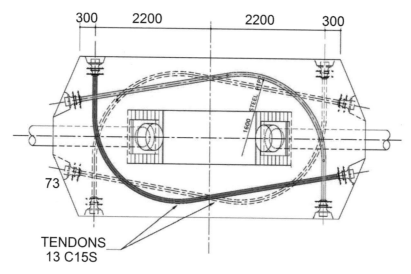

FIGURE 26.36 Pylon cross section.

The stay cables are constituted of 66 to 109 H15 strands according to their position in the bridge. The 109H15 cable stay is one of the most important stays for this type of bridges.

The anchorage spacing of stays on the deck is 7.81 m (i.e. every two segments). The stay arrangement is symmetrical towards the pylons.

The pylon shaft, of pyramidal shape, is a massive reinforced concrete structure (24 × 25 m at bottom). It is designed to withstand the ship impact. Each pylon shaft is founded on 88 piles diameter 1.10 m.

The intermediate piers and exterior piers are reinforced concrete box sections. These piers have an architectural shape with two convex faces (facing the stream flow) and two concave faces (perpendicular to the stream flow).

26.5.2.4 Method of Construction and Casting Curve

The Aswan Bridge was built by the cast-in-place balanced cantilever method using traveling forms, which is the method normally adopted in the construction of long span concrete bridges. The towers and the cable stays are constructed simultaneously with the deck. For the construction stability, a cable stay has to be tensioned every two new segments to support the new segments and to reduce the stress generated during construction.

The construction of the Aswan Bridge contains 16 cycles corresponding to the evolution of the construction. The first cycle, called cycle zero, corresponds to the erection of the first no-stayed balanced cantilever segments. The typical cycles, called cycles 1 to 14, correspond to the erection of the 14 pairs of successive stayed and no stayed segments. The last cycle, cycle 15, corresponds to the erection of the main span closure segment and the final stay tensioning adjustment.

During construction, the theoretical profile and alignment was compared carefully. Any deviation from the anticipated deflections were analyzed and interpreted as an unsatisfactory response of the structure.

The analysis of the deflections deviation is complex because the structure is highly indeterminate and the deflections of the deck and the pylon have a direct effect on the stay forces, bending moment, and shear force distribution in the structure.

Special care was taken to check the dead load weight of each segment (e.g., measurement of member dimensions, concrete specific weight).

26.5.2.5 Calculation Procedure

The basic concept of the cable-stayed bridge is simple. The deck and the stays behave as a triangular truss of which the bottom chord in compression is the deck and the member in tension consists of the stays, and the third member is the pylon (Figure 26.37).

The stays are stressed during construction and basically balance the weight of a deck segment with a length corresponding to the horizontal stay spacing.

In fact, the analysis of the triangular truss is much more complicated because of the stay vertical flexibility. The cable stays are often compared to a beam on elastic supports consisting of springs with vertical flexibility equal to the vertical flexibility of the stays.

It is normal to size the stays in balancing the weight of the deck segments (including a fraction of the live load). However, an adjustment is always required because of the addition of creep redistribution and live load bending moments, which cannot be fully taken by the longitudinal post-tensioning.

The longitudinal analysis of the Aswan Bridge was carried out by the computer program BC (BRIDGE CONSTRUCTION). This program is a specialized design tool giving necessary information at every phase of construction.

The input data are the geometric characteristics (spans, longitudinal and transverse dimensions), permanent loads, tendon and stay geometry, and material characteristics. The bridge is divided into as many members as there are segments. A smaller division is used in order to represent temporary or permanent supports.

The calculation procedure of the construction is subdivided into phases. In each phase the time is fixed and new segments can be erected or cast, supports can be added or removed, loads can be applied or removed, and one or more prestressing tendons or stays can be tensioned.

The program allows the possibility to close joints between two parts of the structure, to change the internal static system by releasing temporarily fixed hinges, and to add or to release temporary launching girder loads. Precast or cast in place segments can be intermixed.

The effects of concrete creep, shrinkage, and steel relaxation are taken into account for both the segments and the tendons. The anchorage set loss at tendon anchorage is incorporated together with all geometric and deferred losses.

The seismic analysis was carried out by applying the earthquake in each of three directions. The resultant elastic response forces and displacements were combined into three seismic directional load combinations.

The elastic seismic forces were determined according to the multimode response spectrum method with complete quadratic combination procedure. The studies were performed using a computer program SYSTUS.

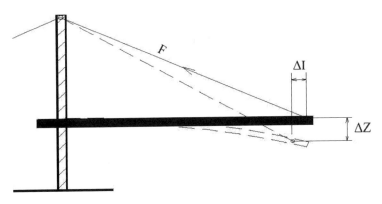

FIGURE 26.37 Cable-stayed equilibrium.

26.5.3 Particular Features of the Project

26.5.3.1 Stay Force Distribution on the Deck

For wide decks, as with the Aswan Bridge, the distribution of the stay forces from the anchorage to the cross section is an important factor in the deck strength. Assuming a distribution in the horizontal plan at 45°, the horizontal component of the stay force becomes regularly distributed to the deck section after a distribution length approximately equal to half of the deck width (Figure 26.38).

The bending analysis of the deck must take this problem into account in order to make sure that all the stresses are within the allowable range specified by the design criteria, and also that the nonuniform stress distribution does not affect the section.

The width of the Aswan Bridge is one of the widest widths for a one-plane cable-stayed bridge. The distribution of the cable stay force on the deck was treated with particular attention especially that the stays forces equivalent to 109H15 cable stay is very high and can lead to a local buckling. For the above reasons, the top slab was stiffened in the center part of the box girder by a double longitudinal girder and a transversal prestressing was added to increase the lateral continuity of the bridge. Quite clearly, the connection between this cable stay and the deck necessitate a high density of reinforcement due to the concentration of the force.

26.5.3.2 Loads and Stresses

The following curves show the maximum and minimum stresses (envelope stresses) on the upper and bottom fibers due to the permanent loads and live load (Figure 26.39).

The particular aspect of these curves is the stress range. The stress range on the bottom side attains 14 MPa due to the live load. This value has necessitated additional prestressed tendons to maintain the section in compression under the different case of loading.

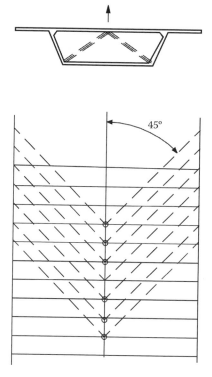

FIGURE 26.38 Distribution of stay forces.

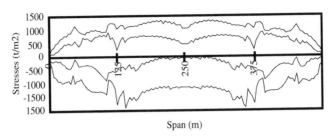

FIGURE 26.39   Maximum and minimum stresses (5000 days) on the upper and bottom fibers due to permanent loads and live load.

26.5.4 Construction of the Pier Segment and Pylon

26.5.4.1 Pier Segment

The external formworks of the form travelers are used to cast the pylon pier segment. They are supported by traditional props to provide an easy handling at the top of the column.

Traditional wooden formworks will be used to cast the bottom slab and to build the internal bulkhead.

Then, casting operations proceed in four stages: (1) the bottom slab, (2) the webs, (3) the diaphragms, and (4) the top slab. During this operation, the pier segment is resting on its permanent bearings and sand boxes. However during the construction of the cantilever the pier segment is resting on its permanent bearings only; as a consequence, the bearing conditions of the deck onto the column during construction are close to service conditions. Temporary stability systems are requested:

- Some vertical prestressing bars were tensioned at the pylon pier segment in order to clamp the deck to the columns during construction
- A horizontal rotational restrain was provided at the pylon pier segment to prevent instability during conditions under wind conditions.

26.5.4.2 Pylon

Once the pier segment concrete has reached the required strength, the pylon self-climbing formwork is installed and the construction of the pylon can start. One lift is about 3 m high.

The installation of the first stay requires the segment number 4 to be cast but also the pylon lift number 7 to have reached an appropriate strength to limit the permanent deformation of the pylon under the loading of the stay.

The steel rebars are placed in cages lifted by the tower crane: in the lower part of the mast (without stay anchorages) one cage corresponds to two lifts with 50% couplers. In the upper part (with stay anchorages) one cage corresponds to one lift with 100% couplers.

The stay cables formwork tubes and bearing plates, the prestressing tromplates are pre-assembled in the rebar panels.

The self-climbing formwork is about 3 lifts high and thus becomes fully assembled once in position to cast the third lift. The pylon is constructed in 20 lifts and one special casting stage required to cast the upper element.

26.5.5 Construction of the Cable-Stayed Bridge Deck

26.5.5.1 Structural Features

The prestressed segmental concrete deck consists of a single-cell trapezoidal box girder with interior stiffening struts and two inclined webs. A single plane of stays is connected with the box at the intersection of the two internal struts.

26.5.5.2 Construction Principles

Two pairs of travelers will be used to cast the typical segment and the closure joints and one set of specially designed formwork to cast the pier segments. The construction proceeds in cantilever from the pylons (Figure 26.40). Permanent prestressing cables are installed at each segment pour and stay cables every two segments.

The construction proceeds symmetrically for the first four segments. The side span will then progress with a 2 day cycle advance over the main span. When we reach the intermediate pier, the form traveler is used to cast the pier segment. Pier brackets are necessary to support the forms. The construction will continue until the side span closure and then mid span closure (Figure 26.41).

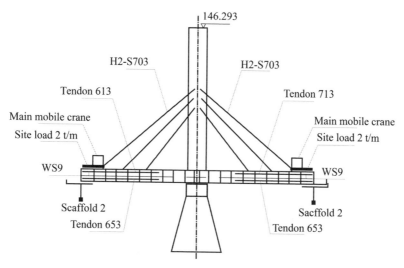

FIGURE 26.40 Typical segment.

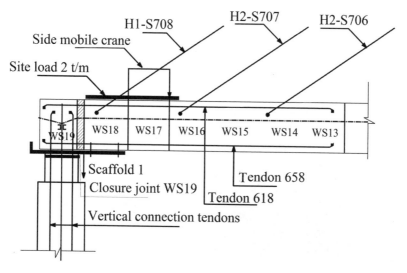

FIGURE 26.41 Side spans closure.

26.5.6 Project Implementation

The project was initiated by GARBLT (General Authority for Roads & Bridges and Land Transport). The client consulting engineers are RPT (UK) and ACE (Egypt). The cable-stayed part was designed by ARCADIS (Europe Etudes GECTI) in a joint venture with SYSTRA. The bridge was constructed by General Nile Company for Roads and Bridges with the assistance from Freyssinet for the cable stays part. The foundation of the bridge and the approaches were designed by the NECB and Dr. Sabry Samaan respectively.

26.6 Elevated Roads in Cairo—The 6th October Bridge

26.6.1 Introduction

The 6th October Bridge is an elevated highway in central Cairo, Egypt. With total length ramps it is approximately 20 km. The bridge crosses the Nile River twice. It connects west bank suburbs, east through Gezira Island to Downtown Cairo, and on to connect the city to the Cairo International Airport to the east. The flyover was constructed in 10 main phases between 1969 and 2002 (Figure 26.42).

26.6.2 Construction Phases

Table 26.3 summarizes the different phases. Over 30 years of construction, several materials and construction systems were used, such aseinforced concrete, prestressed concrte, and steel orthotropic. Construction methods included cantilever carriage, incremental launching of precast girder, and movable scaffolding (flying shuttering), in addition to traditional construction on scaffolding from the ground. Figures 26.43 to 26.53 show 10 phases of bridge projects.

FIGURE 26.42　October bridge crossing from the west bank to Gezira Island.

TABLE 26.3 Construction phases of the 6th October Bridge started at phase (1) in 1969 and ended at phase (10) in 2002

Phases	Description	Bridge Length and Ramps Length	Construction Started and Ended Date
1	Bridge spanned the smaller west branch of the Nile from Gezira to Agouza, width 34 m and ramps width 8 m	130 m 880 m	May 1969 to August 1972
2	Bridge over El Gezira island, from street El Gabalayan and Saray street, width 34 m and ramps width 8 m	725 m 250 m	July 1970 to January 1973
3	Bridge over El Agouza, from El Nile street, over Agriculture Museum width 14 m and ramps width 8 m	710 m 1000 m	January 1973 to October 1976
4	Bridge crossing the Nile, width 34 m, Ramps on Saraya El Gezira, Ramps width 8 m	480 m 750 m	January 1973 to October 1976
5	4 Ramps: one beside TV Buildting on Nile Corniche Street, one beside Hiliton Hotel and one to Abdel Manem Riad Sqare, Ramps width 8 m	740 m	January 1976 to July 1978
6	Bridges started from the end of the original bridge to Ramsis Square, over Galaa Street, width 18 m with Ramps, width 8 m	3000 m 850 m	February 1977 to February 1979
7	Surface Road with 3 Ramps over Adbel Manem Riad Square, width 8 m	700 m	February 1979 to January 1981
8	Bridge Started from Ramis Squarter to Ghamra Bridge, width 18 m, The bridge end with ramps in Ramsis Square, this phase includes 3 Ramps, width 8 m	1928 m 1850 m	January 1981 to March 1989
9	This is the longest phase, started from Ghamra Bridge to El Masr Road, it includes 5 Ramps	5300 m 2100 m	March 1989 to August 1999
10	Ramps beside El Saka El Hadid Club, width 8 m	700 m	July 2001 to March 2002

FIGURE 26.43 Phase (1)—Bridge spanned the west branch of the Nile from Gezira to Agouza.

FIGURE 26.44 Phase (2)—Bridge over El Gezira island, from street El Gabalaya and Saray street.

FIGURE 26.45 Phase (3)—Bridge over El Agouza, from El Nile Street, over Agricultural Museum.

FIGURE 26.46 Phase (4)—Bridge crossing the Nile, width 34 m. Ramps on Saraya El Gezira; ramps width 8 m.

FIGURE 26.47 Phase (5)—Ramps: one beside TV Building on Nile Corniche Street, one beside Hilton Hotel, and one to Abdel Manem Riad Square.

FIGURE 26.48   Phase (6)—Bridge started from the end of the original bridge to Ramsis Square, over Galaa Street.

FIGURE 26.49   Phase (7)—Surface road with 3 ramps over Abdel Manem Riad Square.

FIGURE 26.50 Phase (8)—bridge started from Ramsis Squater to Ghamra Bridge.

FIGURE 26.51 Phase (8)—Cable-stayed bridge at Ghamra.

FIGURE 26.52   Phase (9)—the longest phase – started from Ghamra Bridge to El Nasr Road.

FIGURE 26.53   Phase (10)—ramps beside El Saka El Hadid Club.

26.7 Structural System for Bridges

26.7.1 Bridges Over the River Nile

Figure 26.54 illustrates the development of bridges over the river Nile since the 1850s.

26.7.2 Short and Medium Span Bridges

Bridge systems in longitudinal directions for short and medium spans less than 40 m are shown in Figures 26.55 through 25.57.

26.7.3 Cable-Stayed Bridges

Table 26.4 summarizes cable-stayed bridges used in Egypt.

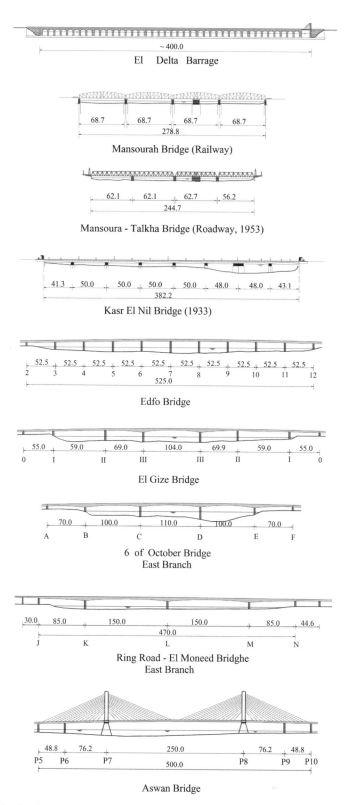

FIGURE 26.54 The development of bridges over the river Nile since the 1850s.

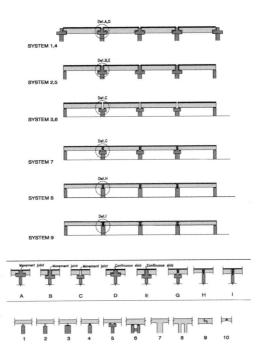

FIGURE 26.55 Simply supported bridges.

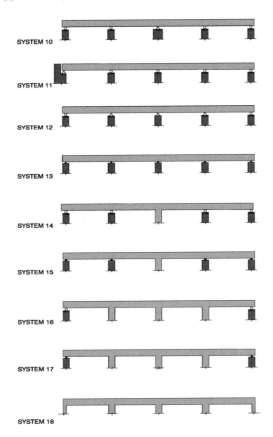

FIGURE 26.56 Continuous beams and frames.

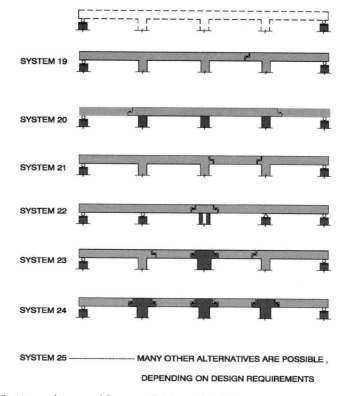

SYSTEM 19

SYSTEM 20

SYSTEM 21

SYSTEM 22

SYSTEM 23

SYSTEM 24

SYSTEM 25 ----------------- MANY OTHER ALTERNATIVES ARE POSSIBLE ,

DEPENDING ON DESIGN REQUIREMENTS

FIGURE 26.57 Continuous beams and frames with intermediate hinge.

TABLE 26.4 Cable-Stayed Bridge Systems (Bakhoum 1995)

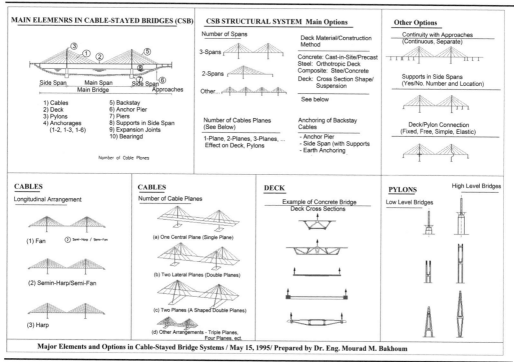

References

Abrahams, M. J. 2000. "Movable Bridges, Past, Present and Future", *The Proceedings of Bridge Engineering Conference*, March 26–30, Sharm El Sheikh, Egypt, pp2–31.

Abou-Rayan, A. M. 2004. "Static and Dynamic Characteristics of Aswan Cable-Stayed Bridge", *Mansoura University, Engineering Journal*, Vol. 29, Mansoura, Egypt.

Adrian, E., Krüger, H., and Hess, J. 2000. Mechanical Engineering, Drive and Control Technology of the El-Ferdan Railway Swing Bridge. *The Proceedings of Bridge Engineering Conference*, March 26–30, Sharm El Sheikh, Egypt, pp. 525–536.

Asmar, R. and A. Huynh, A. 2000. "The Aswan Cable-Stayed Bridge: Construction Methods and Specific Equipment", *The Proceedings of Bridge Engineering Conference*, March 26–30, Sharm El Sheikh, Egypt, pp. 609–616.

Bakhoum. M. M. 1995. "Some Recent Developments in Cable-Stayed Bridges," *Arab Roads Magazine*, Vol. 43, No.2. Cairo, Egypt.

Farooq, A., Shohayeb, M. and Jones, V. 2001. "Aswan Cable-Stayed Bridge", *IABSE Conference, Cable-Supported Bridges—Challenging Technical Limits*, June 12–14, Seoul, Korea, pp. 8–18.

Fouad, A. et al. 2001. "Design and Construction Aspects of the Suez Canal Cable Stayed Bridge", *IABSE Conference, Cable-Supported Bridges—Challenging Technical Limits*, June 12–14, Seoul, Korea, pp. 47–54.

Fouad, A., and Bakhoum. M. M. 2001. "Planning, design and construction aspects of the Suez Canal Cable Stayed", *Strait Crossing Conference—Fourth Symposium*, Sept. 2–5, Bergen, Norway.

Fuchs, N., Tomlinson, K. and Buckby, R. 2003. "El Ferdan Bridge, Egypt: The world's longest swing bridge", *Proceedings of the Institution of Civil Engineers, Bridge Engineering*, March, 21–30, ICE, London, UK, pp. 21–30.

Jones, D. 1995. *BOATS London*: British Museum Press, Egyptian Book Shelf. London, UK.

Kamal, A., et al. 2006. "Ambient Vibration Test of Aswan Cable Stayed Bridge", *Journal of Applied Mechanics, JSCE* (August) Vol. 9, pp. 85–93.

Labib, A. 2000. "Aswan Bridge Over the Nile", *The Proceedings of Bridge Engineering Conference*, March 26–30, Sharm El Sheikh, Egypt, pp. 601–608.

Labib, S. and Bakhoum, M. M. 2000. "General Aspects of Suez Canal Bridge: A Bridge for Peace and Prosperity in the Middle East" *The Proceedings of Bridge Engineering Conference*, March 26–30, Sharm El Sheikh, Egypt, pp. 651–668.

Mahdi, H. A., El-Kadi, A., and El-Kadi, F. I. 2000. "Innovative Cofferdam for the Pylons' Construction of Aswan Cable Stayed Bridge Over the Nile", *The Proceedings of Bridge Engineering Conference*, March 26–30, Sharm El Sheikh, Egypt, pp. 617–628.

Mahdi, H. A., El-Kadi, A. F., and El-Kadi, F. I. 2001. "Precise Structural Analysis and Construction of Prestressed Concrete Box-Girder of Suez Canal Bridge", *Fourth Alexandria International Conference on Structural And Geotechnical Engineering*, Alexandria, Egypt.

Mizon, D. H., Mohaammed, D., Binder, B. 2000. El-Ferdan Bridge—Construction. *The Proceedings of Bridge Engineering Conference*, March 26–30, Sharm El Sheikh, Egypt, pp. 577–590.

Mosselhy, F. M. 1988. *Evolution of the Egyptian Capital and the great Cairo*, Published by the Author, Cairo, Egypt.

Parker, R. B. 1975. A Practical Guide to Islamic Monument in Cairo, American University in Cairo Press, Cairo, Egypt.

Ramadan, O. M. O. 2002. "Seismic Analysis of Laterally Loaded Piles Near Ground Slopes: Case Study of El Ferdan Swing Bridge Over Suez Canal, Ground and Water: Theory to Practice", *55th Canadian Geotechnical and 3rd Joint IAH-CNC and CGS Groundwater Specially Conference*, October 20–23, Niagara Falls, Ontario, Canada.

Rostom, O. R. 1948. *Bridges in Ancient Egypt*, with a report on a newly excavated bridge from the old Kingdom, Giza Annales du Service Des Antiquites de L'Egypte.

Schlaich, J. 2000. "Conceptual Design of Bridges—More Variety!", *The Proceedings of Bridge Engineering Conference*, March 26–30, Sharm El Sheikh, Egypt, pp. 1–25.

Schlecht, B., et al. 2000. "Multibody-System-Simulation of the El-Ferdan Swing Steel Railway Bridge", *The Proceedings of Bridge Engineering Conference*, March 26–30, Sharm El Sheikh, Egypt, pp. 537–548.

Sharaf, M., and Bakhoum, M. M. 1997. "Seismic Design Considerations for the Suez Canal Cable-Stayed Road Bridge Project, Egyquake 2, *The Second Egyptian Conference on Earthquake Engineering*, Aswan, Egypt, pp. 103–116.

Sharaf, M, Ishitate, N. and Ishii, T. 2000. "Construction of Suez Canal Bridge (Central Portion)", *The Proceedings of Bridge Engineering Conference*, March 26–30, Sharm El Sheikh, Egypt, pp. 629–640.

Stierlin, H. 1995. *The PHARAOHS master-builders*, Italy, Terrail, 1st Edition in English, 224 pp.

Taha, N. and Buckby, R. 2000. "The El-Ferdan Bridge Over the Suez Canal Rail Link to the Sinai", *The Proceedings of Bridge Engineering Conference*, Sharm El Sheikh, Egypt, 26–30 March, pp. 549–558.

Tomlinson, G. K., et al. 2000. "El-Ferdan Bridge—Design", *The Proceedings of Bridge Engineering Conference*, March 26–30, Sharm El Sheikh, Egypt, pp. 559–576.

27

Benchmark Designs of Highway Composite Girder Bridges

Shouji Toma
Kokkai Gakuen University

27.1 Introduction

At the IABSE 2007 Symposium in Weimar, Germany, a comparative design project for a highway girder bridge by benchmark was proposed (Toma and Duan 2007). A simple type of bridge, that is, a simply supported steel composite girder bridge, was adopted for the benchmark in order to make the comparative study easier. Resultantly, 10 benchmark designs from different countries were collected for the project. A comparison of these 10 bridges, designed basically under the same criteria for the proposed benchmark bridge, their characteristics, and thoughts on the steel composite girder designs of the world will be presented in this chapter, from which a rational bridge design can be derived. The design standards in those countries are also compared. The countries (and the organizations) of the 10 participating engineers are: United States of America (government), Belgium (university), China (university), Czech Republic (university), Egypt (university), Ireland or United Kingdom (consultant), Italy (consultant), Japan (university), Korea (university), and Russia (government). Professor of university is the occupation most of the participants, and the others are engineers in governments and consulting companies. Name of the nations is shown only by the symbol A to J among which Japan is denoted by H.

The study of the benchmark design found that girder height (web depth) took various values. Girder height is the most important factor in girder design because it will affect the web thickness and flange sizes, and thus the entire sectional area and the stiffness of the girder. The optimum girder height, which gives the minimum sectional area of the girder, is studied in Section 27.10 for the span lengths $L = 20$ m, 30 m, and 40 m. The details of the design results will be described in Section 27.10, from which designers can get ideas for deciding the girder height at initial design.

27.2 Design Criteria for the Benchmark Bridge

The design criteria proposed for the benchmark project are as follows (Toma and Duan 2007):

1. Span length = 30 m
2. Road width = 8.5 m
3. Support conditions: simply supported
4. Number of girders = 4
5. Thickness of concrete slab = 0.24 m
6. Thickness of asphalt pavement = 0.08 m
7. Weight of curb (base for hand rail) = 4.85 kN/m
8. Weight of hand rail (or barrier) = 0.5 kN/m
9. Weight of steel girder (exterior and interior) = 3.3 kN/m
10. Weight of haunch (exterior and interior girders) = 1.5 kN/m
11. Weight of form work = 1.0 kN/m^2 (per unit road area, to be removed for composite section after the concrete slab hardens)
12. Yield strength of steel = 315 N/mm^2 (SM490) or equivalent

SM490 is a designation by the Japanese Industrial Standards (JIS). Two kinds of design conditions are proposed in the project: a free design (Design A) that specifies only first three fundamental conditions 1 to 3 of the above criteria, and a conditional design (Design B) that specifies all the conditions from 1 to 12. There is no pedestrian walkway. In the free design, except the first three fundamental conditions, engineers can take any criteria for rest of the conditions. Therefore, the basic design concept, such as number of girders, can be compared in the free design (Design A). In the conditional design (Design B), other criteria such as dead loads and material strength are also specified in addition to the fundamental conditions of the free design. The conditional design intends to compare the effect of live load. Figure 27.1 shows a general plan of the conditional design, which stipulates four main girders.

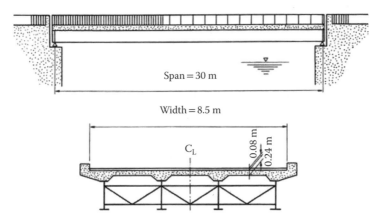

FIGURE 27.1   General plan of benchmark bridge.

When the same dead loads are used, the results of each design may be understood to reflect the policy of safety. Naturally, the design procedures and live loads used in each country are different. However, the traffic conditions in the 10 countries that participated in this benchmark project can be regarded as similar in the present modern society, where globalization is advanced. If traffic situations in the participating countries can be considered similar, even though the design live loads in the design standards are different, the live load condition can be substantially considered as nearly equivalent. Namely, design results such as girder sections may be regarded as a kind of indicator to reflect the policy of safety.

27.3 Live Loads Used in Benchmark Designs

27.3.1 Belgium

The applied design code in the benchmark design of Belgium is Eurocode 1 Part 3. The live load by the Eurocode is summarized in Figure 27.2. The road width is divided into theoretical lanes based on the

Width of the road	Number of theoretical lanes	Width of the lanes	Remaining area
$w < 5.4$ m	$n_1 = 1$	3 m	$w - 3$ m
5.4 m $\leq .w < 6$ m	$n_1 = 2$	$w/3$	0
6 m $\leq w$	$n_1 = \text{Int}(w/3)$	3 m	$w - n_1 \times 3$ m

Number and width of the theoretical lanes

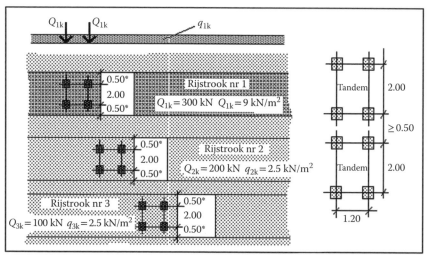

Live loads according to load model 1 (Eurocode)

Lane	Concentrated load Q_{1k}	Uniform load q_{1k}
1	300 kN	9 kN/m^2
2	200 kN	2.5 kN/m^2
3	100 kN	2.5 kN/m^2
Remaining lanes	0 kN	2.5 kN/m^2
Remaining area	0 kN	2.5 kN/m^2

Live loads according to load model 1 (Eurocode)

FIGURE 27.2 Live load in Belgium.

table in Figure 27.2. The number of lanes for an 8.5 m road width of the benchmark bridge is three. The live loads consist of the concentrated load Q_k and the distributed load q_k. They are loaded to each lane in such a way to maximize the sectional forces of the girder under the design.

27.3.2 China

The benchmark bridge is designed according to the Chinese General Code for Design of Highway Bridges and Culverts (JTJ021-89) and the Chinese Design Specifications of Steel and Wood Structures for Highway Bridges (JTJ025-86). The live loads in JTJ021-89 consist of a vehicle load and a trailer load. Vehicle load is used for designing and the trailer load is for checking. According to the grade of traffic, the vehicle load is divided into QC-C20, Q-20, and Q-10, and the trailer load is into G-120, G-100, G-80, and G-50. The width of the lane is 3.75 m. The loads of QC-C20 and G-120 used in the benchmark design are shown in Figure 27.3a–c, in which the units are m and kN.

In a 2004 edition of the highway bridge design code (JTG D60-2004), the live loads are changed into the form of lane loads and standard truck loads. The lane load is used for whole bridge calculation in the design, which is classified as highway-I lane load and highway-II lane load. In the General Code for Design of Highway Bridges and Culverts (JTG D60-2004), the highway-I lane load is composed of an even load and a concentrated load, as shown Figure 27.4a. The concentrated load in the highway-I lane load is 180 kN when the calculated span is no more than 5 m and is 360 kN when the calculated span is no less than 50 m. When the calculated span length is between 5 m and 50 m, the concentrated load is calculated by linear interpolating between 180 kN and 360 kN. Multiplying 0.75 to the load of highway-I lane load will obtain highway-II lane load.

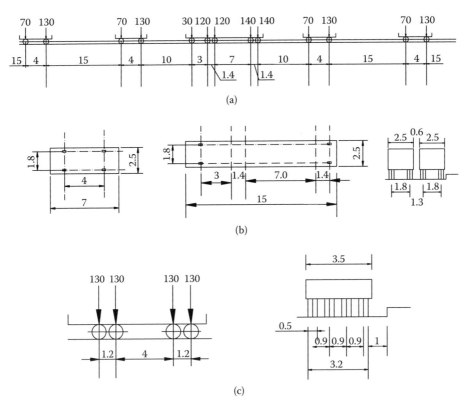

FIGURE 27.3 Live load in China (1989). (a) Distribution of QC-C20 in the longitudinal direction; (b) dimension of the car and distribution in the transverse direction in QC-C20; (c) load distribution of G-120 in the longitudinal and transverse directions.

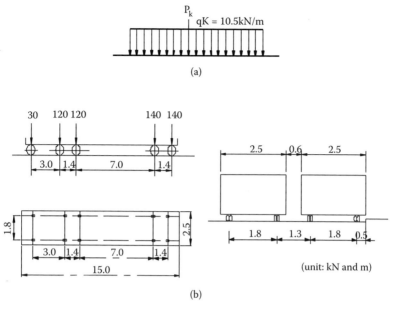

FIGURE 27.4 Live load in China (2004). (a) Highway-I lane load; (b) standard truck load.

The standard truck loading is used for local loading, transversal bridge deck loading, and soil pressure behind the abutments. The standard truck load for highway class I and II has the same total weight of 550 kN, as shown in Figure 27.4a. When calculating the load transverse distribution factors, the truck load is arranged as Figure 27.4b. The design can be updated based on the design code of JTG D60-2004, but the result is not expected to be greatly different since the lane loads in JTG D60-2004 are equivalent to the live load adopted in JTJ021-89.

27.3.3 Czech Republic

The Czech Republic is now in the transition stage of transforming the Czech national codes into the Eurocodes. Because of that, it is now possible to use three sets of codes for the design of steel composite bridges: Czech codes, Eurocodes, and "Technical Conditions" for the design of composite steel–concrete bridges. Nominal values of traffic loads are considered in the benchmark design according to the Czech codes, and they are shown in Figure 27.5.

27.3.4 Egypt

The benchmark design is based on the Egyptian Code for Design of Steel Bridges ECP205 (2001). The roadway is divided into traffic lanes of 3 m width; the most critical lane for designing the structural member is called the main lane. Two types of loads are specified in the codes:

1. Truck loads: This load is intended to represent the extreme effects of heavy vehicles. It consists of a 60 ton (588 kN) truck in the main lane and a 30 ton (294 kN) truck in the secondary lane, next to the main lane. The arrangement of wheel loads is shown in Figure 27.6. The locations of the main and secondary lanes are chosen so as to produce maximum effect on the member under consideration.
2. Uniform distributed load: This load simulates the effects of normal permitted vehicles. It is applied in the traffic lanes and over the lengths that give the maximum values of the internal force being considered, which may be continuous or discontinuous. It consists of a 500 kg/m² (4.9 kN/m²) uniform load in the main lane in front and back of the main truck and 300 kg/m² (2.94 kN/m²) in the remaining bridge floor area excluding the truck area, as shown in Figure 27.6.

Load group I: (dimensions in meters)

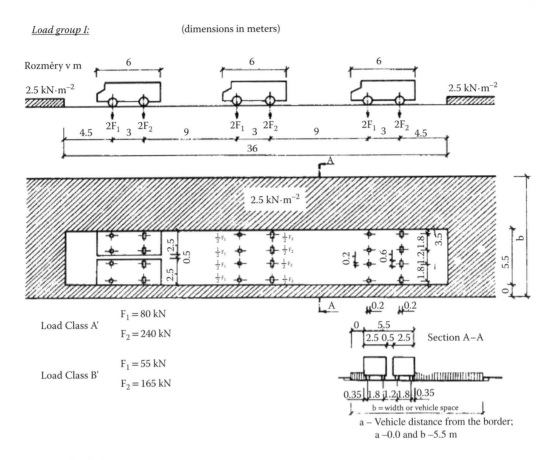

F₁ = 80 kN

Load Class A'

$F_1 = 80$ kN

$F_2 = 240$ kN

$F_1 = 55$ kN

Load Class B'

$F_2 = 165$ kN

a – Vehicle distance from the border;
a –0.0 and b –5.5 m

Four-axle vehicle:

Dimensions in m

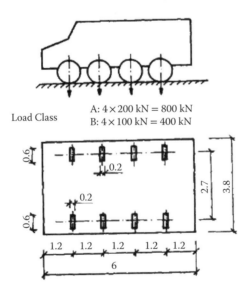

Load Class

A: 4 × 200 kN = 800 kN
B: 4 × 100 kN = 400 kN

FIGURE 27.5   Live load in Czech Republic.

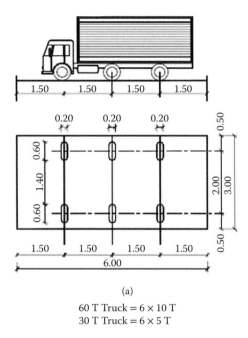

(a)

60 T Truck = 6 × 10 T
30 T Truck = 6 × 5 T

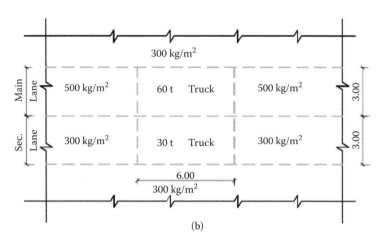

(b)

FIGURE 27.6 Live load in Egypt: (a) wheel arrangement, (b) loading plan.

27.3.5 Italy

The benchmark design is based on the design codes Ministerial Decree 16.01.1996, "Technical regulations pertaining safety verification of structures and loads" and Ministerial Decree 4.05.1990, "Traffic loads on bridges." Following the criteria of the Italian standard, the carriageway is to be subdivided into notional lanes 3.5 m wide, to be positioned transversely in order to obtain the most adverse effects on the structure. On each lane the following loads have to be considered:

- Load Q1a: "Heavy" vehicle load, having nominal intensity of 600 kN, formed by three two-wheel axles (100kN/wheel), longitudinally spaced 1.5 m apart and with a total length of 15 m as shown in Figure 27.7.
- Load Q1b: Uniformly distributed load having nominal intensity of 8.57 kN/m² (total 30 kN/m/ lane), to be positioned out of the clearance of the vehicle Q1a.

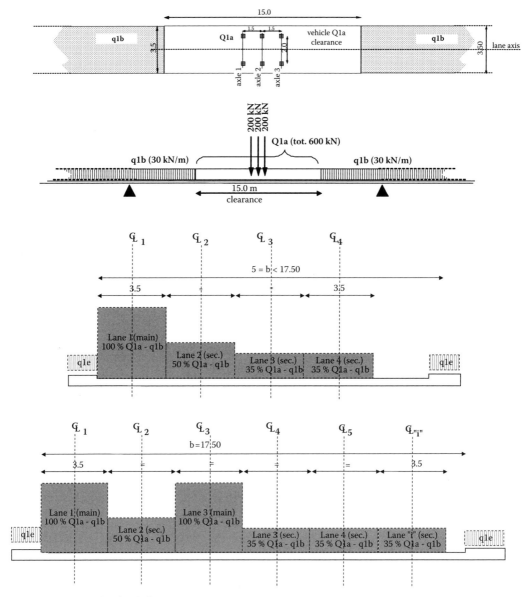

FIGURE 27.7 Live Load in Italy.

Figure 27.8a,b is a live load distribution used in this particular benchmark design by applying the above standards.

27.3.6 Korea

The benchmark design is based on the Korean Highway Bridge Specifications. The live load consists of the truck load (DB-24), which has three axles, and the lane load (DL-24), as shown in Figure 27.9. The rear wheel spacing varies from 4.2 m to 9 m. Different values of the lane load P_m and P_s are used for calculating moment and shear.

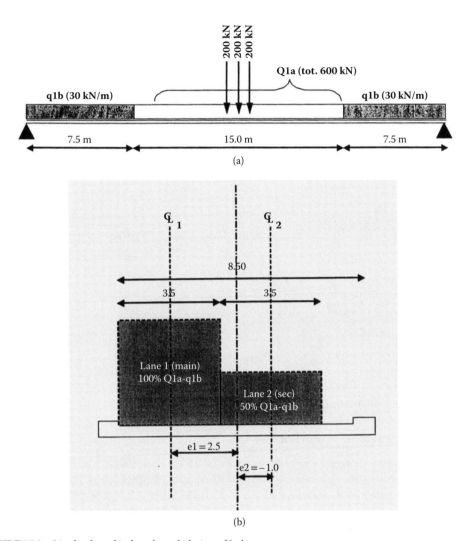

FIGURE 27.8   Live load used in benchmark design of Italy.

27.3.7 Russia

The Russian new code GOST R 52748-2007 "Automobile roads of the general use: Standard loads, loading systems and clearance approaches" for live loads was published in 2008. This code determines live loads for newly constructed roads or for reconstructed roads. These loads for bridges are A14 and NK-103 (except timber bridges). Today such bridges are very few (in Perm Territory, West Urals there are no such bridges). Thus, the live loads (except timber bridges) in Russia today are as follows:

- For newly constructed and for reconstructed bridges: A14 and NK-103 under GOST R 52748-2007
- For other bridges: A11 and NK-80 under SNiP 2.05.03-84 "Bridges and Culverts"

For roads in Moscow only, there is code MGSN 5.02-99 "Design of municipal road bridges." In this code there is live load NK-176 (except above loads).

Due to the economic depression road building is mostly stopped now except for large infrastructure projects (Sochi, Golden Horn Bay, etc.). Replacing existing bridges is currently too expensive. Therefore,

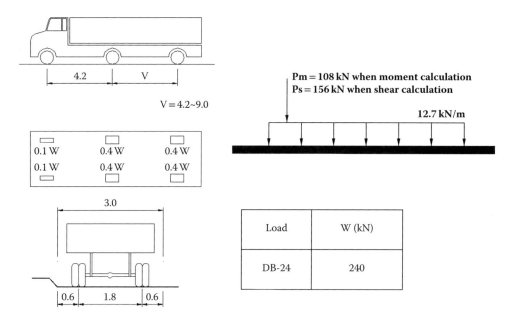

FIGURE 27.9 Live load in Korea.

loads A11 and NK-80 will be used for the next 10 years. The Russian codes for bridges, SNiP 2.05.03-84 "Bridges and culverts," are applied for live loads in the benchmark project. The live loads A11 and NK-80 are shown in Figure 27.10. In the benchmark design, live load A11 controls the strength requirements.

27.3.8 United States of America

The benchmark bridge is designed according to AASHTO LRFD Bridge Design Specifications, Fourth Edition (AASHTO 2007). The vehicular live load HL93 consists of either one of the truck loads with three axles or the tandem load, and the uniform lane load as shown in Figure 27.11. The number of design lanes should be determined by taking the integer part of the ratio $w/3600$, where w is the clear roadway width in mm. It is unlikely that three adjacent lanes will be loaded simultaneously with heavy trucks. To account for this effect, an adjustment factor called the multiple presence factor is considered according to the number of loaded lanes. In the United States the Departments of Transportation of some states like Pennsylvania or California provides a permit vehicle load which models the traffic characteristic of their particular state. This permit load generally gives larger force effects to bridges than AASHTO loads.

27.3.9 United Kingdom

As per BD 37/01, two types of loads are applicable on the highway bridges: HA + KEL and HB loading. BD 37 is a composite version of BS 5400 part 2. HA loading intensity varies per span length of the bridge, and is less for shorter spans. HB loading is applicable for major highways and based on the number of units of HB loading. For highway bridges 45 units of HB loading are applicable. The position of the load for maximum bending moment in the outer girder is shown in Figure 27.12

HA loading
The load intensity calculation is given below:

$$\text{Notional lane width} = \frac{8.5}{3} = 2.833\,\text{m}$$

A11

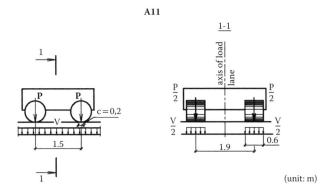

(unit: m)

for strength calculations:
L = 30m:

· V = k × 15,81 kN/m (1st lane – k = 1, other lanes – k = 0.6)
· P = 158.24 kN

· L = 15m

· V = k × 16,63 kN/m (1st lane – k = 1, other lanes – k = 0.6)
· P = 187.34 kN

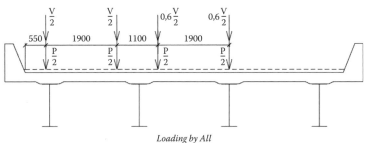

Loading by All

HK-80

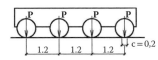 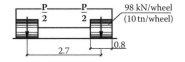

(unit: m)

P = 785/4 = 196.25 kN (20 tn)
Dynamic coefficients: 1 + μ = 1.1
Reliability coefficients: γ$_f$ = 1.0
Thus for strength calculations: P = 215.87 kN
According codes an equivalent load for NK-80 and L = 30 m is 48.15 kN/m. Thus for strength calculations: p = 52.96 kN/m, for L = 15 m – p = 87.87 × 1.1 = 96.66 kN/m.

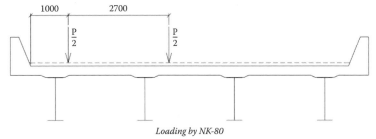

Loading by NK-80

FIGURE 27.10 Live load in Russia (A11 and HK-80).

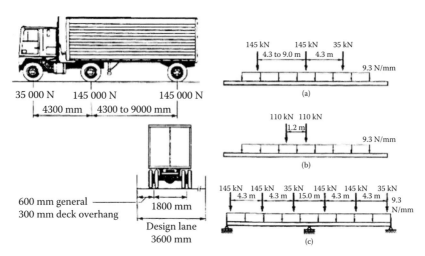

FIGURE 27.11   Live load in the United States (HL93). (a) Design truck and lane load; (b) Design tanden and lane load; (c) Live load application for continuous beam.

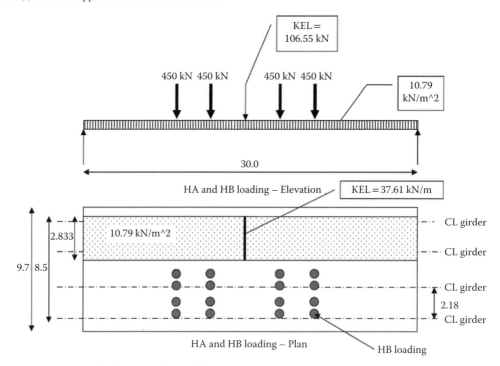

FIGURE 27.12   Live load in United Kingdom.

$$\text{Intensity of HA loading} = 336\left(\frac{1}{30}\right)^{0.67} = 34.4 \text{ kN/m}$$

Applicable on a lane width of 2.833 m

$$\text{Load per sqm} = \frac{34.4}{2.833} = 12.142 \text{ kN/m}^2$$

$$\text{HA lane factor } \alpha^2 = 0.0137(2.833(40-30)+3.65(30-20)) = 0.888$$

$$\text{Net intensity} = 0.888 \times 12.142 = 10.79 \text{ kN/m}^2$$

KEL (Knife-edge load) is applicable along with HA loading on full lane width

$$\text{Intensity} = \frac{120}{2.833} \times 0.888 = 37.61 \text{ kN/m}$$

HB Loading

No. of HB loading units = 45
Lad per axle = 10 kN
Total load per axle = $10 \times 45 = 450$ kN
The detail of application of load is given in Figure 27.13 of BD 37.

27.3.10 Japan

Highway bridges are designed according to the Japanese Specifications for Highway Bridges (JSHB) (JRA 2002). The live load to the design main girders consists of the partially distributed load p_1 and the overall distributed load p_2, as shown in Figure 27.13. The partially distributed load p_1 is intended to represent the extreme effects of heavy vehicles. The location is chosen to produce the maximum effect on the member considered. The overall distributed load p_2 simulates the effects of normal vehicles. It is applied in the traffic lanes and over the lengths that give the maximum values of the internal force for

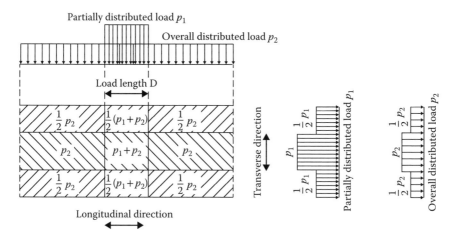

Loads	Primary live load						Secondary live load
	Partial distributed load p_1			Allover distributed load p_2			
	Load length D (m)	Load (kN/m²)		Load (KN/m²)			
		For bending moment	Four shear force	L ≦ 80	80 < L ≦ 130	130 < L	
A-Load	6	10	12	3.5	4.3 – 0.01L	3	50% of primary load
B-Load	10						

L : Span length (m)
Live loads of JSHB

FIGURE 27.13 Live load in Japan.

the member being considered. Both p_1 and p_2 are divided into two types according to traffic density: primary live load and secondary live load. The primary live load is applied to 5.5 m width for two traffic lanes and the secondary live load, which is half of the primary live load, is applied to the remaining lanes. Further, the live load is categorized into A load and B load according to the traffic conditions: B load is used for the design of ordinary bridges and A load is for bridges with less traffic. B load is used in this study.

27.4 Dead Loads Used in Benchmark Designs (Conditional Design)

The dead loads for exterior and interior girders can be calculated according to the design criteria in Section 27.2. The resultant dead loads used in the benchmark design (conditional design) of each country are shown in Table 27.1. There is no design data for Nations A or G. Although the dead loads tried to be set the same in the benchmark project, unfortunately the customary design procedures of calculating the dead load intensity in each country bring somewhat different results. As generally expected, the slab weight of the exterior and interior girders is not taken the same except Nation F. In most of the designs, the weights of the girders and haunch are loaded equally to the exterior and interior girders. It should be pointed out that the dead load of Nation E is significantly smaller compared to the other countries.

The dead loads for the composite girder consist of the pavement and the barrier. The weight of the pavement is simply considered to distribute over the road width and to be supported evenly by all girders in around half of the designs. It can be seen in Table 27.1 that the barrier weight is taken quite differently in each design even though it is given in the design criteria. Further, since the barrier is located at the side ends of the road, it is a common practice to be supported by the exterior girders. However, some designs consider that the barrier is also supported by all girders evenly.

In the design criteria, the weight of the formwork used for pouring the concrete slab, which should be removed after the concrete has hardened to complete the composite girder, is provided. Therefore, first the form weight has to be considered for noncomposite girder, and then subtracted by applying negative weight (uplift) for composite girder as shown in Nation H (Japan) in Table 27.1. However, many of the designs do not consider this.

27.5 Comparison of Free Designs

The design results of the 10 countries (Nations A to J) are shown in Table 27.2 for the free and conditional designs. The upper part of the table is for the free design (Design A) and the lower is for the conditional design (Design B). There are no free designs for Nations B and F. Although there is a slight difference in steel yield strength, this will not significantly influence the comparisons of this study because the design results scatter widely beyond the slight difference in steel strength. The sizes of the girders are at midspan of the bridge. There is no distinction for exterior or interior girders since both have similar sizes.

The number of girders is not specified for the free design and is to be determined by the designers. Table 27.2 shows that fewer-girder designs, such as two- or three-girder designs, are predominant in the world. The girder spacing ranges from 5 m to 6 m for a two-girder design. The reason for having very wide girder spacing in Nation I (7.5 m) is that the bridge has two stringers between the main girders. From this fact, it might be presumable that the composite slab or the prestressed slab, which is commonly used for two-girder bridges in order to reduce the dead weight of the slab, is not used in Nation I. There are a few nations, such as A and J, which adopt four or five girders in the free design. An engineer of Nation J explains that the erection during construction is easier than a design with fewer girders.

TABLE 27.1 Dead Loads for Exterior and Interior Girders (kN/m)

Girder	Portion	Nation B		Nation C		Nation D		Nation E	
		Exterior	Interior	Exterior	Interior	Exterior	Interior	Exterior	Interior
Noncomposite	Slab	15.77	15.84	12.9	15.6	12.244	17.545	8.9	13.0
	Haunch	1.65	1.65	1.5	1.5			1.0	1.0
	Girder	3.63	3.63	3.27	3.27	5.769	5.769	2.4	2.7
	Formwork	2.63	2.64	2.38	2.38	–	–	–	–
	Total	23.68	23.76	20.05	22.75	18.013	23.314	12.3	16.7
	Grand Total	47.44		42.8		41.327		29	
Composite	Pavement	4.42	5.74	4.68	4.68	12.433	2.924	3.83	3.83
	Barrier	5.88	0	1.69	1.69	12.433		1.34	1.34
	Total	10.3	5.74	6.37	6.37	12.433	2.924	5.17	5.17
	Grand Total	16.04		12.74		15.357		10.34	

Girder	Portion	Nation F		Nation H		Nation I		Nation J	
		Exterior	Interior	Exterior	Interior	Exterior	Interior	Exterior	Interior
Noncomposite	Slab	13.965	13.965	14.247	15.288	13.444	15.056	19.64	14.19
	Haunch	2.205	2.205	1.5	1.5	1.5	1.5		
	Girder	3.762	3.762	3.3	3.3	3.3	3.3	4.71	4.71
	Formwork	–	–	2.423	2.6	2.241	2.509	–	–
	Total	19.932	19.932	21.47	22.688	20.485	22.365	24.35	18.9
	Grand Total	39.864		44.158		42.85		43.25	
Composite	Pavement	3.832	3.832	3.013	4.68	3.585	4.015	3.74	3.74
	Barrier	2.675	2.675	7.699	0	7.21	0	4.038	1.312
	Total	6.507	6.507	8.289*	3.08*	10.795	4.015	7.778	5.052
	Grand Total	13.014		11.369		14.81		12.83	

*including form removed

TABLE 27.2 Design Results (Free Design: Design A, and Conditional Design: Design B)

Design A	A	B	C	D	E	F	G	H	I	J
Top flange	250*20	—	500*30	800*26	356*25.4	—	800*20	400*22	400*20	425*20
(Sectional area cm²)	(50.0)	—	(150.0)	(208.0)	(90.4)	—	(160.0)	(88.0)	(80.0)	(85.0)
Web	1945*20	—	1580*14	1250*26	1168*15.9	—	1850*14	1700*9	2250*14	1000*14
(Sectional area cm²)	(389.0)	—	(221.2)	(325.0)	(185.7)	—	(259.0)	(153.0)	(315.0)	(140.0)
Bottom flange	250*35	—	750*40	800*26	406*32	—	800*30	550*47	600*38	750*40
(Sectional area cm²)	(87.5)	—	(300.0)	(208.0)	(129.9)	—	(240.0)	(258.5)	(228.0)	(300.0)
Total sectional area (cm²)	526.5	—	671.2	741.0	406.1	—	659.0	499.5	623.0	525.0
Number of girders	5	—	2	3	3	—	2	2	2	4
Total sectional area of all girders (cm²)	2632.5	—	1342.4	2223.0	1218.2	—	1318.0	999.0	1246.0	2100.0
Girder spacing	1.7 m	—	6.0 m	3.0 m	3.5 m	—	5.0 m	5.5 m	7.5 m (string 3 × 2.5 m)	2.18 m
Yield strength (N/mm²)	315	—	—	315	315	—	—	315	315	355
Depth of web/span length	1/15.4	—	1/19	1/24	1/25.7	—	1/16.2	1/17.6	1/15	1/30
Design B										
Top flange	350*20	420*16	400*20	800*26	305*19	600*20	600*20	320*18	300*12	425*25
(sectional area cm²)	(70.0)	(67.2)	(80.0)	(208.0)	(58.0)	(120.0)	(120.0)	(57.6)	(36.0)	(106.3)
Web	1925*20	1268*12	1495*12	1000*26	1067*15.9	1350*14	1650*10	1600*9	2000*12	1150*16
(sectional area cm²)	(385.0)	(152.2)	(179.4)	(260.0)	(169.7)	(189.0)	(165.0)	(144.0)	(240.0)	(184.0)
Bottom-flange	350*55	480*32	450*35	800*26	356*32	600*30	600*30	520*32	500*32	760*40
(sectional area cm²)	(192.5)	(153.6)	(157.5)	(208.0)	(113.9)	(180.0)	(180.0)	(166.4)	(160.0)	(304.0)
Total sectional area (cm²)	647.5	373.0	416.9	676.0	341.5	489.0	465.0	368.0	436.0	594.3
Number of girders	4	4	4	4	4	4	4	4	4	4
Total sectional area of all girders (cm²)	2590.0	1491.8	1667.6	2704.0	1366.1	1956.0	1860.0	1472.0	1744.0	2377.0
Girder spacing	1.97 m	2.4 m	2.6 m	2.4 m	2.29 m	2.5 m	2.5 m	2.6 m	2.5 m	2.18 m
Yield strength (Nmm²)	315	295	315	315	315	345	315	315	315	315
Depth of web/span length	1/15.6	1/23.7	1/20.1	1/30	1/28.1	1/22.2	1/18.2	1/18.8	1/15	1/30

In Table 27.2, it can be seen that sizes of the flanges in each design vary significantly. Since the upper flange is subjected to compressive force, the width-to-thickness ratio should be limited to avoid local buckling. However, the cantilever ratio of the plate (the ratio of half of the flange width to the thickness) in the free design of Nation G is 20, which exceeds the limit of 16 given by the Japanese Specifications for Highway Bridges (JSHB) (JRA 2002). The widths of the upper and lower flanges is the same in Nations A, D, and G, whose designs try to simplify the structural features. The design of Nation D has identical upper and lower flange sizes, which seem to ignore the composite action with a concrete slab for simplicity. In the design of Nation J, there is a difference only in the steel strength between the free and conditional designs. The web depth of the free design, which uses slightly stronger steel, is a little lower than the conditional design.

27.6 Comparison of Conditional Designs

27.6.1 Girder Spacing

In the conditional design (Design B), whose results are given in Table 27.2, the number of girders is stipulated to four and the dead loads are basically the same values. The section plans of each design for the conditional design are shown in Figure 27.14. Although the road width is set as 8.5 m in the conditional design, Nations A and D take it for the overall width of bridge. It is interesting to see that the girder spacing scatters from 1.97 m to 2.6 even though the bridges have basically the same road width and the same number of girders. Excluding Nation A's 1.97 m, which is the narrowest road width, the average girder spacing of the other nations is 2.44 m. The girder spacing can be determined from the balance of slab strength for the continuous portion between the girders and the cantilever portion outside the exterior girder. According to JSHB (JRA 2002) the optimum girder spacing is calculated as 2.52 m (Nakai et al. 2005). Though the curb (base of hand rail) weight is given in the design criteria, the sizes are different in each country as can be seen in Figure 27.14.

27.6.2 Sectional Area

The total cross-sectional areas of all main girders are plotted in Figure 27.15 for the free design (Design A) and the conditional design (Design B). There are no data for the free design (Design A) for Nations B and F, which only show the conditional design (Design B) in Figure 27.15. If the use of stiffeners is disregarded, the cross-sectional area of the main girder can be considered one of the basic indices that indicate economic efficiency, rationality of design, or factor of safety. From the comparison in Figure 27.15, it can be found that the total cross-sectional areas of each design vary widely and the difference between the maximum and the minimum is over two times. When the conditional design (Design B) is compared with the free design (Design A), naturally the former has a larger total cross-sectional area than the latter because the conditional design has more restrictions in the design. Nation A has almost same total cross-sectional area for both designs because, unlike other countries, the number of girders for the free design is five, which is more than the four of the conditional design.

Figure 27.15 shows that the total cross-sectional areas of Nations A and D are large, and on the contrary those of Nations E and H are small. Especially in the free design (Design A) of Nation H, the total cross-sectional area is the smallest in comparison with other countries. The conditional design of Nation H also has a small sectional area but is not an exceptional design and rather can be regarded as an ordinary design in Japan when taking into consideration the sample designs in the text books (Nakai and Kitada 2003; Hayashikawa 2002; Miyamoto 2004).

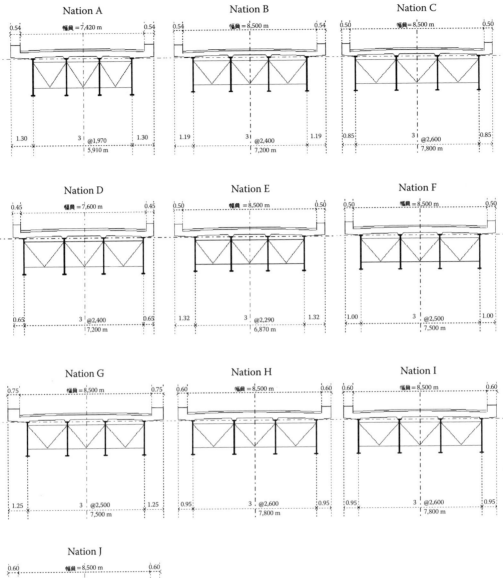

FIGURE 27.14 Section plans (conditional design).

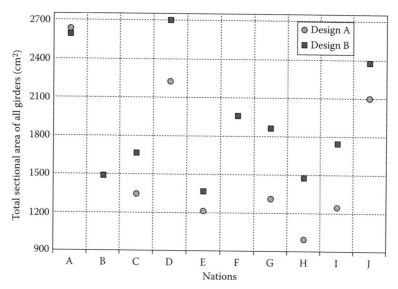

FIGURE 27.15 Comparisons of total sectional areas.

It should be noted that the dead loads of Nation E are significantly small compared to other countries, as pointed out in Section 27.4. This would be a reason for the rather small sectional area of Nation E. As for the designs of Nations A and D, the cross-sectional areas are large because of the large thickness of the web (see Table 27.2). However, the design of Nation D has a shallow web depth in order to neglect the stiffeners, and thus may intend to reduce the fabrication cost as will see in the next section.

27.6.3 Web Design

Various data regarding the web designs of the conditional design (Design B), such as the ratios of web depth to span length (h/L), web thickness to depth (t/h), and the minimum limiting values to neglect vertical (transverse) and longitudinal (horizontal) stiffeners, which are prescribed in JSHB (JRA 2002), are shown in Table 27.3. The web depth is the most important key factor in girder design. As Table 27.3 shows, the web depth ranges widely from 1000 mm to 2000 mm. Although the basic design conditions, that is, span length, road width, number of girders, and dead loads, are taken as the same, the web depths differ extensively, which indicates that there is a conceptual difference in bridge design in each country. In Japan it is common practice to take the ratio of web depth to span length (h/L) as between 1/18 to 1/20 for economical reason (Nakai and Kitada 2003). Applying this to the benchmark design for the span length $L = 30$ m gives 1.5 m to 1.67 m as the optimum web depth. It is found in Table 27.3 that the web depths in about half of the designs are lower than this range. Namely, Japanese design practice tends to have deeper webs than other countries.

The web depths in Table 27.3 are compared in Figure 27.16, in which the limits to neglect the stiffeners when JSHB is applied for the steel grade with yield strength 315 N/mm² (JRA 2002) are also shown. If the web depth is located below the marks of the limit depth, stiffeners are not required according to the Japanese design standards. The design of Nation D would need neither the vertical nor the horizontal stiffeners, and Nations E and J are very close to the limit mark.

On the other hand, the designs of Nations G, H, and I obviously need both stiffeners. The Japanese design pursues to reduce the steel weight with thin web by using the stiffeners. This tendency can be seen by the smallest sectional area in Figure 27.13. The designs in other countries tend to have thicker webs and neglect the horizontal stiffener. However, recently engineers in Japan have also been trying to reduce the number of member components such as stiffeners because it is more effective to cut the

TABLE 27.3 Sizes of Web (Conditional Design) (mm)

Nations	h (Web Depth)	t (Web Thickness)	h/L Ratio	t/h Ratio	h w/o Vertical Stiffener	t w/o Vertical Stiffener	h w/o Horizontal Stiffener	t w/o Horizontal Stiffener
A	1925	20	1/15.6	1/96	1200	32.1	2600	14.8
B	1268	12	1/23.7	1/106	720	21.1	1560	9.8
C	1495	12	1/20.1	1/125	720	24.9	1560	11.5
D	1000	26	1/30	1/39	1560	16.7	3380	7.7
E	1067	15.9	1/28.1	1/67	954	17.8	2067	8.2
F	1350	14	1/22.2	1/96	840	22.5	1820	10.4
G	1650	10	1/18.2	1/165	600	27.5	1300	12.7
H	1600	9	1/18.8	1/178	540	26.7	1170	12.3
I	2000	12	1/15	1/167	720	33.3	1560	15.4
J	1150	16	1/26.1	1/72	960	19.2	2080	8.8

L: Span.

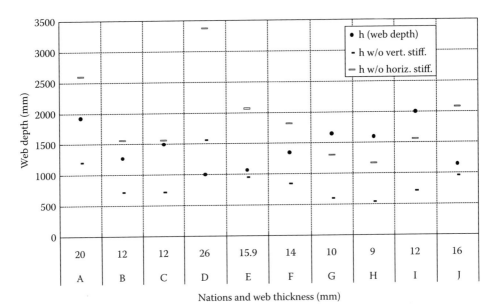

FIGURE 27.16 Web depth (conditional design).

fabrication cost. Other designs of Nations A, B, C, and F would not need the horizontal stiffener but would require the vertical stiffener.

The ratio of web thickness to depth (t/h) is plotted in the vertical axis in Figure 27.17. The limits to neglect the horizontal and vertical stiffeners when JSHB is applied for the steel of yield strength 315 N/mm² (JRA 2002), which are given in Table 27.3, are also shown in Figure 27.17. If the ratio is located above the limit lines, the corresponding stiffener is not required. As previously seen in Figure 27.16, the designs of Nations G, H, and I have small (t/h) ratios and thus need both stiffeners. On the other hand, the design of Nation D has a large (t/h) ratio and does not need both stiffeners. Nations E and J may not require both stiffeners either. The other designs of Nations A, B, C, and F are located in between, that is, they require the vertical stiffener but not the horizontal stiffener.

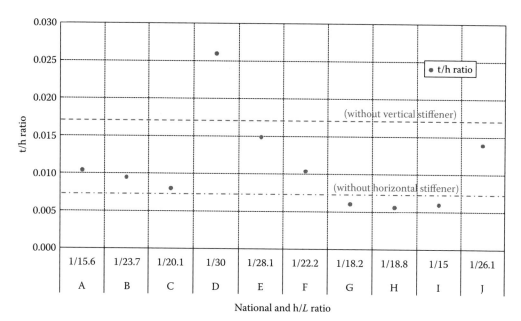

FIGURE 27.17 Ratio of web thickness to depth (conditional design).

27.7 Deflections

The stiffness of the girders is significantly dependent on the girder depth and directly related to the girder deflection. The calculated deflections and the deflection limits in each design are shown in Table 27.4. It can be found that the deflections of all designs have a sufficient margin to the limit, except Nation G, whose design seems to have a different policy than the others in this regard. Accordingly, there will be no problem if the lower girder depth is taken, which results in smaller girder stiffness. Another important point is whether the stiffeners are required or not in conjunction with the web thickness. When the web depth is smaller, the possibility of neglecting the stiffeners increases. Namely, it would be better to design a smaller girder depth and have no stiffeners as much as possible.

The equations for the deflection limit generally take a simple linear form to the span length (L), except Nations D and H, as can be seen in Table 27.4. The deflection limits widely vary from $L/300$ to $L/800$. However, they are not simply compared because the load conditions to calculate the deflection would not be the same. The deflection limits in the countries are shown in Figure 27.18. As stated above, the deflections generally have sufficient margins to the limits, which leads to the conclusion that the stiffness of the girder generally does not control the girder design. It is interesting to note that Nation J does not have a deflection limit.

27.8 Impact Allowances

The live load impact effect is generally expressed by the allowance ratio to the live load. The equations of the impact allowance around the world are summarized in Table 27.5 and plotted in Figure 27.19 (JSCE 2008). A variety of forms from simple constant to complex hyperbolic equations are adopted. Some are based more on the results of theoretical or experimental research, and some take more advantage of simplicity. AASHTO LRFD in the United States (AASHTO 2007) has a simple constant allowance of 0.33.

TABLE 27.4 Comparison of Deflections (Free Design: Design A, and Conditional Design: Design B) (mm)

Nations	A	B	C	D	E	F	G	H	I	J
Design A: Deflection	10.18	–	26.63	23.00	18.50	–	99.70	18.04	32.00	79.00
Design B: Deflection	7.93	28.00	26.72	29.28	19.00	14.00	99.70	20.20	24.00	62.40
Deflection Limit	43	75	60	45	37.5	50	100	45	50	–
	$L/700$	$L/400$	$L/500$	$\dfrac{L}{20000/L}$	$L/800$	$L/600$	$L/300$	$\dfrac{L}{20000/L}$	$L/600$	None

L: Span

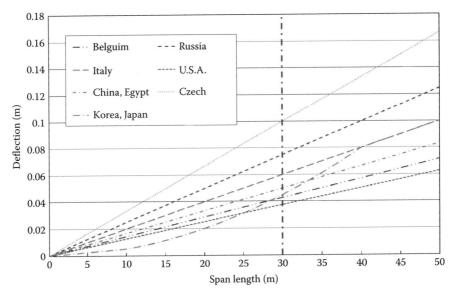

FIGURE 27.18 Deflection limits for live load.

TABLE 27.5 Impact Allowances [L = span length (m)]

Nations	United States		China	Russia	Italy	Korea
Codes	AASHTO		JTJ021-89	SNiP 2.05.03-84	Eurocode	KHBS
	Standard Spec.	LRFD				
Allowance	$\dfrac{15.24}{(L+38)} \le 0.3$	0.33 const.	$\dfrac{15}{(37.5+L)}$		$0.4 - \dfrac{L-10}{150} \le 0.4$	$\dfrac{15}{(40+L)} \le 0.3$
Nations	Czech	France	German	Egypt	Japan	
Codes	ČSN	Fascicule 61 Titre II	DIN 1072	ECP 205	JHBS	
Allowance	$\dfrac{1}{0.95 - (1.4L)^{-0.6}}$	$\dfrac{0.4}{(1+0.2L)} + \dfrac{0.4}{(1+4G/P)}$	$0.4 - 0.008L$		$\dfrac{20}{(50+L)}$	

On the contrary, the equation of France is a complex form in which G/P is a ratio of dead load to live load and is assumed as 3.0 in Figure 27.19. The impact of Russia, shown in Figure 27.19 and Table 27.5, is for the live load A11 (see Figure 27.10) but for NK-80 it should take a constant value 0.1. In the benchmark design, the span length $L = 30$ m gives the lowest impact around 0.1 to the highest 0.33, in which the difference is quite large.

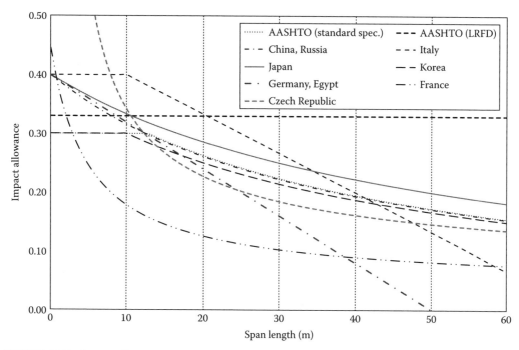

FIGURE 27.19 Impact allowances.

27.9 Limit State Design and Allowable Stress Design

The limit state design is more common in the world today than the allowable stress design. The limit state design intends to ensure the plastic moment capacity by using the compact section for girders. Apparently, it is easier to have a small web depth and large thickness for the section to obtain the fully plastic moment. Figure 27.17 shows that the designs of Nations D, E, and J have large ratios of web thickness to depth, and thus have the compact section. It should be noted that Nation E, in spite of the large ratio of web thickness to depth, has a small sectional area, as seen in Figure 27.15, which means it is an economical design with a compact section. On the other hand, as discussed in Section 27.6.3 (Figure 27.17), the designs of Nations G, H, and I use the noncompact section equipped with vertical and/or longitudinal stiffeners in which the yield stress can be reached only at the extreme fiber of the section. In these designs the allowable stress design may possibly be adopted. The slender section is not usually allowed in bridge designs.

27.10 Optimum Girder Height

27.10.1 General

The benchmark project has been studied by comparing 10 participating designs from around the world. As a result, it has found that many countries tend to use lower girder heights (web depth) without stiffeners. This section studies the optimum girder height by comparing the girder sections for various web depths so as to minimize the sectional area. The Specifications for Highway Bridges in Japan (JSHB) (JRA 2002) is applied in this study. Therefore, the live load of JSHB shown in Figure 27.13 is used. The computer program used in this section is Preliminary Design of Composite Girder Bridge developed by JIP Techno Science Inc. in Tokyo, Japan.

TABLE 27.6   Girder Height and Stiffeners for Study Cases

Stiffeners	Horizontal	Vertical	Horizontal	Vertical	Horizontal	Vertical
Span	×	×	×	○	○	○
20m	800~1400 mm		800~1300 mm			
30m	900~1600 mm		1200~1700 mm		1400~2000 mm	
40m			1500~2100 mm		1700~2300 mm	

× no stiffener, ○ with stiffener

Table 27.6 shows case studies for various span lengths and girder heights with and without stiffeners. The span length is taken for three cases: 20 m, 30 m, and 40 m. The appropriate strength of steel is used in the designs: SM400 (the yield strength 235 N/mm²) and SM490 (the yield strength 315 N/mm²) for $L = 20$ m, SM490 for $L = 30$ m, and SM490Y (the yield strength 355 N/mm²) for $L = 40$ m (SM400, SM490, and SM490Y are designations for structural steel of the Japanese Industrial Standards). The girder spacing is 2.6 m, which is the same as the benchmark design of Nation H. The other criteria also follow the benchmark design given in Section 27.2. The same slab sizes are used in all design cases. The designs in this section may be considered another benchmark provided to study the optimum design from a different point of view.

27.10.2 Web Thickness and Stiffener Requirements

The results of the web designs and the stiffener requirements in the benchmark project are shown in Figure 27.17. The horizontal (longitudinal) stiffener would not be required except for Nations G, H, and I. The vertical (transverse) stiffener would be needed for most of the designs. As explained in Section 27.7, there is a sufficient margin to the deflection limit, which means the stiffness of the girder does not control the design, and the Japanese design (denoted by H in Figure 27.17) might be improved by designing a lower girder height with no stiffeners. It is worthwhile to study the optimum girder height in conjunction with the stiffener requirements.

27.10.3 Optimum Girder Height for Span Length *L* = 20 m

27.10.3.1 Optimum Girder Height

A design comparison is made when thicker web is used by neglecting vertical and horizontal stiffeners that are used in the previous benchmark design. Table 27.7 shows the design results of exterior and interior girder sections for various web heights (depths) when the steel material SM400 (yield strength 235 N/mm²) and no stiffeners are used. The girder height (web depth) varies from 800 mm to 1300 mm. Sectional areas of the exterior and interior girders and the total area of the four girders are plotted in Figure 27.20. The total sectional areas can be approximated by a second-order polynomial equation with respect to the girder height. The equation is expressed as follows:

$$A = 0.001648 \times h^2 - 3.535 \times h + 3300 \tag{27.1}$$

where A = total area of four girders (cm²) and h = web height (mm). The optimum girder height is obtained by differentiating Equation 27.1.

$$\frac{\partial A}{\partial h} = 0.003296 \times h - 3.535 = 0 \tag{27.2}$$

$$h = 1070 \text{ mm}$$

The derivative of the approximation equation gives the optimum girder height $h_{min} = 1070$ mm and the least sectional area $A_{min} = 1400$ cm², as can be seen in Figure 27.20. The ratio of the optimum height to the span length is $h_{min}/L = 18.7$.

TABLE 27.7 Design results for $L = 20$ m (No Stiffeners, SM400)

Web Height (mm)		800	900	1000	1100	1200	1300
Interior Girder	Top flange	400*18	360*16	320*13	280*11	260*10	240*10
	(sectional area cm²)	(72)	(58)	(42)	(31)	(26)	(24)
	Web	800*12	900*13	1000*15	1100*16	1200*18	1300*19
	(sectional area cm²)	(96)	(117)	(150)	(176)	(216)	(247)
	Bottom flange	620*34	580*33	560*29	530*27	500*24	440*23
	(sectional area cm²)	(211)	(191)	(162)	(143)	(120)	(101)
	Sectional area (cm²)	379	366	354	350	362	372
	Is (cm⁴)	448,200	508,400	551,400	611,700	711,900	823,200
	Iv (cm⁴)	1,962,000	2,259,000	2,493,000	2,782,000	3,069,000	3,368,000
	Deflection δ_L (mm)	12.4	11.1	9.9	8.9	8.0	7.3
	Deflection δ_d (mm)	44.3	39.7	36.4	32.9	28.5	24.7
Exterior Girder	Top flange	420*19	370*17	330*14	290*11	260*10	240*10
	(sectional area cm²)	(80)	(63)	(46)	(32)	(26)	(24)
	Web	800*12	900*13	1000*15	1100*16	1200*18	1300*19
	(sectional area cm²)	(96)	(117)	(150)	(176)	(216)	(247)
	Bottom flange	620*34	570*31	570*28	540*26	490*24	430*23
	(sectional area cm²)	(211)	(177)	(160)	(140)	(118)	(99)
	Sectional area (cm²)	387	357	356	348	360	370
	Is (cm⁴)	472,900	514,200	569,000	613,000	706,900	816,900
	Iv (cm⁴)	1,893,000	2,072,000	2,384,000	2,663,000	2,937,000	3,223,000
	Deflection δ_L (mm)	10.5	9.4	8.4	7.6	6.9	6.3
	Deflection δ_d (mm)	47.0	42.0	38.7	35.3	30.9	26.8
Total sectional area (cm²)		1531	1445	1420	1396	1443	1484

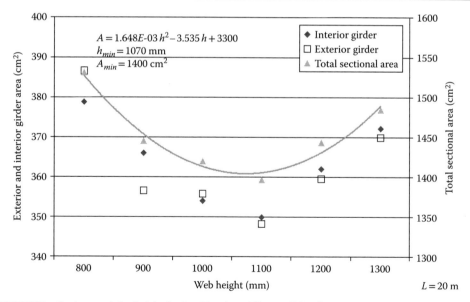

FIGURE 27.20 Optimum girder height for $L = 20$ m (no stiffeners, SM400).

Table 27.8 and Figure 27.21 show the design results when the steel material SM490 (yield strength 315 N/mm²) and no stiffeners are used. Similar to Figure 27.20, the optimum girder height and the corresponding least sectional area are obtained and shown in Figure 27.21. As a higher strength of material is used than the previous case, the optimum depth and the corresponding sectional area are smaller. The web thickness for the higher-strength steel must be larger than the lower-strength steel

TABLE 27.8   Design Results for $L = 20$ m (No Stiffeners, SM490)

Web Height (mm)		700	800	900	1000	1100	1200
Interior Girder	Top flange	350*16	330*13	290*13	250*14	240*13	240*11
	(sectional area cm²)	(58)	(43)	(38)	(35)	(31)	(26)
	Web	700*12	800*14	900*15	1000*17	1100*19	1200*20
	(sectional area cm²)	(84)	(112)	(135)	(170)	(209)	(240)
	Bottom flange	540*33	480*32	480*27	460*23	360*24	320*22
	(sectional area cm²)	(178)	(154)	(130)	(106)	(86)	(70)
	Sectional area (cm²)	320	309	302	311	327	337
	Is (cm⁴)	281,200	326,600	387,400	465,929	550,500	619,700
	Iv (cm⁴)	1,414,000	1,625,000	1,807,000	2,007,000	2,257,000	2,485,000
	Deflection δ_L (mm)	17.6	15.4	13.7	12.1	11.0	10.0
	Deflection δ_d (mm)	71.0	61.6	52.2	44.5	36.9	33.0
Exterior Girder	Top flange	360*17	350*13	290*13	250*14	230*14	240*11
	(sectional area cm²)	(61)	(46)	(38)	(35)	(32)	(26)
	Web	700*12	800*14	900*15	1000*17	1100*19	1200*20
	(sectional area cm²)	(84)	(112)	(135)	(170)	(209)	(240)
	Bottom flange	550*32	480*31	470*27	460*23	360*23	310*21
	(sectional area cm²)	(176)	(149)	(127)	(106)	(83)	(65)
	Sectional area (cm²)	321	306	300	311	324	332
	Is (cm⁴)	292,700	330,600	384,600	465,929	546,600	610,400
	Iv (cm⁴)	1,354,000	1,541,000	1,729,000	1,943,000	2,146,000	2,366,000
	Deflection δ_L (mm)	14.8	12.9	11.6	10.4	9.4	8.5
	Deflection δ_d (mm)	74.9	65.5	56.4	47.4	40.3	35.9
Total sectional area (cm²)		1282	1230	1204	1243	1301	1337

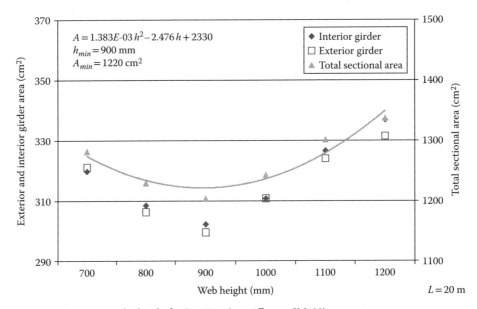

$A = 1.383E\text{-}03\,h^2 - 2.476\,h + 2330$
$h_{min} = 900$ mm
$A_{min} = 1220$ cm²

- ◆ Interior girder
- □ Exterior girder
- ▲ Total sectional area

FIGURE 27.21   Optimum girder height for $L = 20$ m (no stiffeners, SM490).

because the web with higher strength will be subjected to higher stress which may increase the possibility of shear buckling.

Table 27.9 and Figure 27.22 show the design results when the steel material SM490 (yield strength 315N/mm²) and vertical (transverse) stiffeners are used. When the vertical stiffener is used to prevent shear buckling, the web thickness can be thinner, which results in more economical girder sections. The optimum girder height (web depth) is larger than the previous cases; thinner web will bear a larger depth in order not to have a larger sectional area but to maintain the stiffness of the girders. The horizontal (longitudinal) stiffener is not effective in this case because of the short span length. The horizontal stiffener is normally useful to reduce the sectional area of web when large stiffness and large girder height are required for long spans.

TABLE 27.9   Design Results for $L = 20$ m (with vertical stiffener, SM490)

	Web Height (mm)	800	900	1000	1100	1200	1300	1400
Interior Girder	Top flange	320*16	320*14	310*12	290*11	260*11	230*13	220*13
	(sectional area cm²)	(51)	(45)	(37)	(32)	(29)	(30)	(29)
	Web	800*9	900*9	1000*9	1100*10	1200*10	1300*10	1400*11
	(sectional area cm²)	(72)	(81)	(90)	(99)	(120)	(130)	(154)
	Bottom flange	520*30	510*27	470*26	470*23	470*20	400*22	360*19
	(sectional area cm²)	(156)	(138)	(122)	(108)	(94)	(88)	(68)
	Sectional area (cm²)	279	264	249	239	243	248	251
	Is (cm⁴)	323,600	372,800	414,800	461,800	532,600	636,600	706,700
	Iv (cm⁴)	1,552,000	1,730,000	1,916,000	2,096,000	2,311,000	2,597,000	2,715,000
	Deflection δ_L (mm)	15.8	14.1	12.8	11.7	10.6	10.4	8.8
	Deflection δ_d (mm)	59.3	51.9	45.9	40.4	35.7	31.4	28.8
Exterior Girder	Top flange	320*17	310*14	310*12	260*14	260*11	230*13	230*11
	(sectional area cm²)	(54)	(43)	(37)	(36)	(29)	(30)	(25)
	Web	800*9	900*9	1000*9	1100*10	1200*10	1300*10	1400*11
	(sectional area cm²)	(72)	(81)	(90)	(99)	(120)	(130)	(154)
	Bottom flange	540*29	510*27	470*26	470*23	470*20	370*21	350*20
	(sectional area cm²)	(157)	(138)	(122)	(108)	(94)	(78)	(70)
	Sectional area (cm²)	283	262	249	244	243	238	249
	Is (cm⁴)	334,000	367,400	414,800	486,500	532,600	608,700	689,700
	Iv (cm⁴)	1,506,000	1,680,000	1,863,000	2,041,000	2,251,000	2,365,000	2,676,000
	Deflection δ_L (mm)	12.8	11.5	10.3	9.5	8.7	7.2	7.4
	Deflection δ_d (mm)	61.7	55.3	48.7	42.4	38.2	35.5	30.9
Total sectional area (cm²)		1124	1051	998	965	970	971	1001

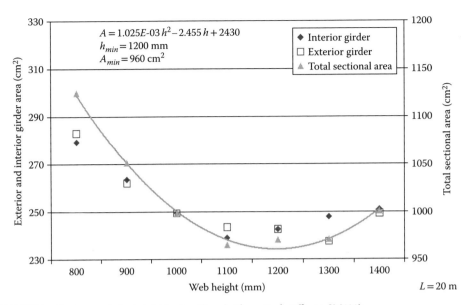

FIGURE 27.22 Optimum girder height for $L = 20$ m (with vertical stiffener, SM490).

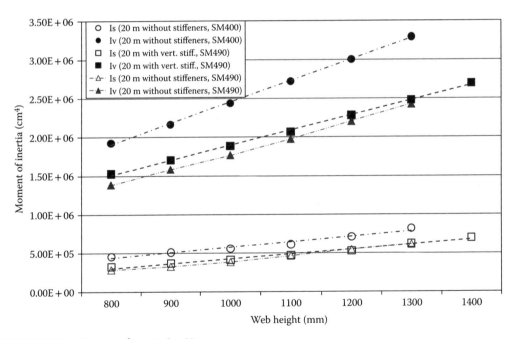

FIGURE 27.23 Moment of inertia, $L = 20$ m.

27.10.3.2 Stiffness and Deflections

Figure 27.23 shows a comparison of the moment of inertia of the steel section I_s and the composite section I_v. It can be seen that the lower strength requires a larger moment of inertia. The ratio of I_v to I_s lies between 4.0 to 4.5. Deflections due to the live load are plotted in Figure 27.24 for the exterior and interior girders. Since the flexural stiffness of the girder depends on the moment of inertia of the cross section, it should be inversely proportional to the deflection: the larger moment of inertia, that is, bending stiffness, in Figure 27.23 results in smaller deflection in Figure 27.24. According to JSHB (JRA 2002),

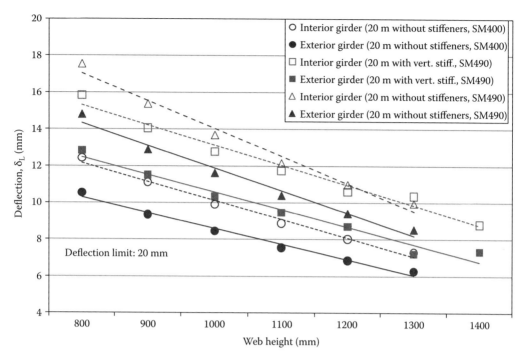

FIGURE 27.24 Deflections due to live load, *L* = 20 m.

as seen previously in Table 27.4, the deflection limit for live load is calculated as 20 mm for the span length *L* = 20 m, which is well above the calculated deflections in the entire range of the web height in Figure 27.24.

27.10.4 Optimum Girder Height for Span Length *L* = 30 m

27.10.4.1 Optimum Girder Height

In case of the span length *L* = 30 m, the design results are given in Tables 27.10 through 27.12 for different conditions of stiffeners: no stiffeners, with vertical stiffener only, and with both vertical and horizontal stiffeners, respectively. The sectional areas of girders are plotted in Figures 27.25 through 27.27 for each stiffener condition. The steel material SM490 (yield stress 315 N/mm²) is used in this case. It should be noted that Table 27.12 is comparable to the Japanese design (Nation H) of Design B of the benchmark design in Table 27.2.

The optimum girder height which gives the minimum sectional area is shown in Figures 27.25 through 27.27. As the stiffeners are used, the optimum girder height becomes larger. For example, if no stiffeners are used (Table 27.10 and Figure 27.25) the optimum girder height is 1285 mm, which is small compared to other stiffener conditions. However, the web thickness is unrealistically large, as can be seen in Table 27.10. Namely, the minimum thickness for web height 1600 mm is 27 mm, which is about two times the thickness of 13 mm for a case with a vertical stiffener. This means that in the case of large web height a design neglecting the vertical stiffener is not practical. Therefore, the vertical stiffener should at least be used to have a practical design as shown in Table 27.11 and Figure 27.26. The optimum height *h* = 1285 mm gives the ratio to the span length h_{min}/L = 23.3. When both vertical and horizontal stiffeners are used as shown in Table 27.12 and Figure 27.27, the sectional area becomes small compared to other cases but the number of members increases, which will cause greater fabrication cost. The minimum thickness of the steel plate is 9 mm in this case.

TABLE 27.10 Design Results for $L = 30$ m (with No Stiffeners, SM490)

Web Height (mm)		900	1000	1100	1200	1300	1400	1500	1600
Interior Girder	Top flange	520*23	480*20	390*19	340*18	340*14	300*15	320*11	330*10
	(sectional area cm²)	(120)	(96)	(74)	(61)	(48)	(45)	(35)	(33)
	Web	900*16	1000*17	1100*19	1200*20	1300*22	1400*24	1500*25	1600*27
	(sectional area cm²)	(144)	(170)	(209)	(240)	(286)	(336)	(375)	(432)
	Bottom flange	520*60	520*54	510*48	520*39	460*39	450*33	470*27	400*24
	(sectional area cm²)	(312)	(281)	(245)	(203)	(179)	(149)	(127)	(96)
	Sectional area (cm²)	576	547	528	504	513	530	537	561
	Is (cm⁴)	916,000	993,000	1,064,000	1,139,000	1,259,000	1,432,000	1,552,000	1,721,000
	Iv (cm⁴)	3,222,000	3,607,000	3,960,000	4,182,000	4,644,000	5,004,000	5,438,000	5,797,000
	Deflection δ_L (mm)	28.8	25.8	23.4	22.3	19.3	18.4	18.2	15.9
	Deflection δ_d (mm)	120.9	109.8	102.6	96.9	85.9	79.2	76.2	63.7
Exterior Girder	Top flange	520*25	480*23	390*22	340*19	340*13	300*14	300*12	300*11
	(sectional area cm²)	(130)	(110)	(86)	(65)	(44)	(42)	(36)	(33)
	Web	900*16	1000*17	1100*19	1200*20	1300*22	1400*24	1500*25	1600*27
	(sectional area cm²)	(144)	(170)	(209)	(240)	(286)	(336)	(375)	(432)
	Bottom flange	520*64	520*57	520*51	520*45	480*43	480*35	460*30	420*27
	(sectional area cm²)	(333)	(296)	(265)	(234)	(206)	(161)	(138)	(113)
	Sectional area (cm²)	607	577	560	539	537	539	549	578
	Is (cm⁴)	984,000	1,070,000	1,165,000	1,225,000	1,301,000	1,451,000	1,605,000	1,812,000
	Iv (cm⁴)	3,223,000	3,480,000	3,988,000	4,354,000	4,788,000	4,981,000	5,402,000	5,868,000
	Deflection δ_L (mm)	28.8	26.1	23.3	22.1	19.2	18.3	18.3	15.9
	Deflection δ_d (mm)	125.5	114.1	106.2	101.0	89.8	82.6	79.6	66.0
Total sectional area (cm²)		2365	2247	2176	2085	2099	2137	2172	2279

TABLE 27.11 Design Results for $L = 30$ m (with Vertical Stiffener, SM490)

Web Height (mm)		900	1000	1100	1200	1300	1400	1500	1600	1700
Interior Girder	Top flange	470*29	460*25	460*22	400*22	400*19	330*19	320*16	300*16	290*14
	(sectional area cm²)	(136)	(115)	(101)	(88)	(76)	(63)	(51)	(48)	(41)
	Web	900*9	1000*9	1100*9	1200*10	1300*10	1400*11	1500*12	1600*13	1700*14
	(sectional area cm²)	(81)	(90)	(99)	(120)	(130)	(154)	(180)	(208)	(238)
	Bottom flange	550*58	550*52	550*47	520*45	520*38	500*36	510*35	500*30	460*28
	(sectional area cm²)	(319)	(286)	(259)	(234)	(198)	(180)	(179)	(150)	(129)
	Sectional area (cm²)	536	491	459	442	404	397	410	406	407
	Is (cm⁴)	932,000	999,000	1,087,000	1,189,000	1,231,000	1,314,000	1,448,000	1,582,000	1,689,000
	Iv (cm⁴)	3,151,000	3,469,000	3,799,000	4,170,000	4,296,000	4,697,000	5,397,000	5,612,000	5,947,000
	Deflection δ_L (mm)	29.0	26.5	24.3	22.2	21.4	19.6	18.3	17.0	15.8
	Deflection δ_d (mm)	119.8	110.0	102.3	94.7	90.2	84.3	76.2	67.2	67.1
Exterior Girder	Top flange	520*29	500*25	500*22	430*22	430*19	370*19	320*17	300*15	290*14
	(sectional area cm²)	(151)	(125)	(110)	(95)	(82)	(70)	(54)	(45)	(41)
	Web	900*9	1000*9	1100*9	1200*10	1300*10	1400*11	1500*12	1600*13	1700*14
	(sectional area cm²)	(81)	(90)	(99)	(120)	(130)	(154)	(180)	(208)	(238)
	Bottom flange	550*63	550*55	550*50	520*49	520*44	510*38	510*34	500*32	500*29
	(sectional area cm²)	(347)	(303)	(275)	(255)	(229)	(194)	(173)	(160)	(145)
	Sectional area (cm²)	578	518	484	469	441	418	408	413	424
	Is (cm⁴)	1,023,000	1,070,000	1,164,000	1,273,000	1,346,000	1,416,000	1,464,000	1,588,000	1,761,000
	Iv (cm⁴)	3,220,000	3,480,000	3,829,000	4,267,000	4,577,000	4,763,000	5,122,000	5,616,000	6,097,000
	Deflection δ_L (mm)	29.4	26.5	24.4	22.3	21.5	19.6	18.2	16.7	15.3
	Deflection δ_d (mm)	124.2	114.2	106.0	98.1	93.4	87.2	79.7	70.3	69.7
Total sectional area (cm²)		2229	2017	1885	1823	1688	1630	1635	1638	1662

TABLE 27.12 Design Results for $L = 30$ m (with Vertical and Horizontal Stiffeners, SM490)

Web Height (mm)		1400	1500	1600	1700	1800	1900	2000
Interior Girder	Top flange	330*17	320*19	310*18	310*18	310*18	300*18	300*18
	(sectional area cm²)	(56)	(61)	(56)	(56)	(56)	(54)	(54)
	Web	1400*9	1500*9	1600*9	1700*9	1800*9	1900*9	2000*10
	(sectional area cm²)	(126)	(135)	(144)	(153)	(162)	(171)	(200)
	Bottom flange	530*31	530*32	510*30	480*30	450*29	450*27	440*24
	(sectional area cm²)	(164)	(170)	(153)	(144)	(131)	(122)	(106)
	Sectional area (cm²)	346	365	353	353	348	347	360
	Is (cm⁴)	1,154,000	1,407,000	1,539,000	1,691,000	1,802,000	1,916,000	1,916,000
	Iv (cm⁴)	4,284,000	4,989,000	5,351,000	5,731,000	6,120,000	6,595,000	6,595,000
	Deflection δ_L (mm)	20.7	18.6	17.5	16.7	15.3	14.2	14.2
	Deflection δ_d (mm)	86.2	78.2	73.2	70.6	64.2	60.5	60.5
Exterior Girder	Top flange	360*25	330*20	310*19	290*18	310*15	290*17	290*17
	(sectional area cm²)	(90)	(66)	(59)	(52)	(47)	(49)	(49)
	Web	1400*9	1500*9	1600*9	1700*9	1800*9	1900*9	2000*10
	(sectional area cm²)	(126)	(135)	(144)	(153)	(162)	(171)	(200)
	Bottom flange	550*39	530*34	520*33	510*30	500*29	470*28	450*26
	(sectional area cm²)	(215)	(180)	(172)	(153)	(145)	(132)	(117)
	Sectional area (cm²)	431	381	375	358	354	352	366
	Is (cm⁴)	1,584,000	1,491,000	1,593,000	1,684,000	1,782,000	1,898,000	1,898,000
	Iv (cm⁴)	4,980,000	5,032,000	5,504,000	5,754,000	6,231,000	6,631,000	6,631,000
	Deflection δ_L (mm)	19.5	18.4	16.9	15.5	14.8	14.0	14.0
	Deflection δ_d (mm)	85.7	81.4	75.8	71.1	67.2	63.2	63.2
Total sectional area (cm²)		1554	1493	1455	1422	1404	1397	1452

In conclusion, the use of a vertical stiffener but no horizontal stiffener as shown in Table 27.11 and Figure 27.26 is recommendable for the span length $L = 30$ m. The optimum height $h = 1525$ mm gives the ratio to the span length $h_{min}/L = 19.7$. When the web height is lower than 1525 mm, the increase of flange areas is more than the decrease of the web area and consequently the total area gradually increases, as can be seen in Table 27.11 and Figure 27.26.

27.10.4.2 Stiffness and Deflections

The moment of inertia of the external and internal girders for the steel section, I_s, and the composite section, I_v, are plotted in Figure 27.28 when the girder height varies. It can be seen that the moment of inertia is proportional with respect to the girder height. From Figure 27.28, it is found that the ratio of the moment of inertia of the composite girder to the steel girder lies in a narrow range from 3.3 to 3.6.

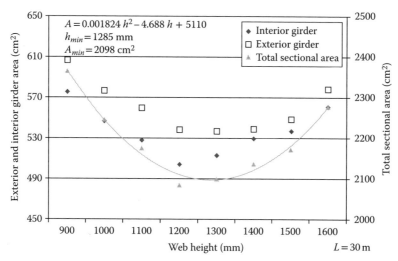

FIGURE 27.25   Optimum girder height for $L = 30$ m (no stiffeners, SM490).

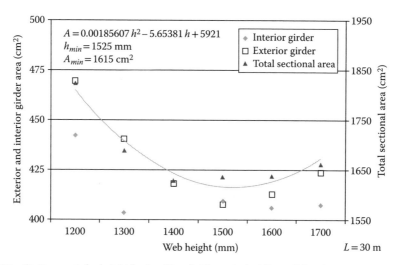

FIGURE 27.26   Optimum girder height for $L = 30$ m (with vertical stiffener, SM490).

This ratio is smaller than the span length $L = 20$ m, which is given as 4.0 to 4.5 in Section 27.10.3.2. This is because the same slab is used in both designs. The deflections due to the live and dead loads are plotted in Figure 27.29, which shows a clear inverse proportion to the web height. Similar to the span length $L = 20$ m, the deflection limit 45 mm for live load by JSHB (JRA 2002) for the span length 30 m is above the actual deflections. This implies that the deflection (stiffness) does not control the design.

27.10.5 Optimum Girder Height for Span Length $L = 40$ m

27.10.5.1 Optimum Girder Height

In case of the span length $L = 40$ m, the design results are given in Tables 27.13 and 27.14 for two different conditions of stiffener: with vertical stiffener only and with both vertical and horizontal stiffeners, respectively. The sectional areas of girders are plotted in Figures 27.30 and 27.31 for different stiffener conditions, correspondingly. The steel material SM490Y (yield stress 355 N/mm²) is used in this case. The optimum girder height that gives the minimum sectional area is shown in Figures 27.30 and 27.31.

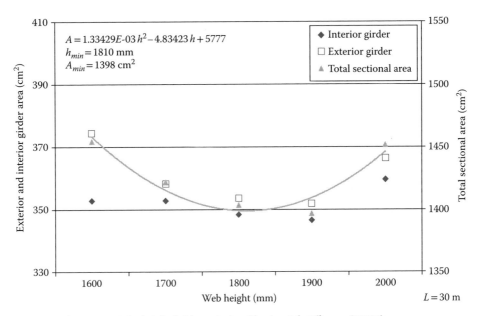

FIGURE 27.27 Optimum girder height (with vertical and horizontal stiffeners, SM490).

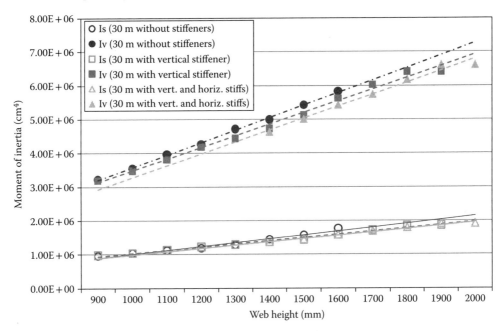

FIGURE 27.28 Moment of inertia, $L = 30$ m.

As more stiffeners are used, the optimum girder height becomes large. For example, if only a vertical stiffener is used (Table 27.13 and Figure 27.30), the optimum girder height is 1830 mm, which gives the ratio to the span length $h_{min}/L = 21.9$ for $L = 40$ m.

The web thickness becomes unrealistically large for large-span bridges if no stiffeners are used. Therefore, at least the vertical stiffener should be used to have a practical design. When both vertical and horizontal stiffeners are used, as shown in Table 27.14 and Figure 27.31, the sectional area becomes small compared to the design with a vertical stiffener only, but the number of members increases, which will cause greater fabrication cost.

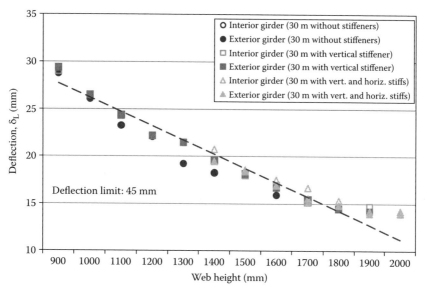

FIGURE 27.29 Deflections due to live load, $L = 30$ m.

TABLE 27.13 Design Results for $L = 40$ m (with Vertical Stiffener, SM490Y)

	Web Height (mm)	1500	1600	1700	1800	1900	2000	2100
Interior Girder	Top flange	410*26	400*23	380*21	360*18	350*16	320*17	320*15
	(sectional area cm²)	(107)	(92)	(80)	(65)	(56)	(54)	(48)
	Web	1500*13	1600*14	1700*14	1800*15	1900*16	2000*17	2100*18
	(sectional area cm²)	(195)	(224)	(238)	(270)	(304)	(340)	(378)
	Bottom flange	600*43	640*37	610*36	630*32	620*30	580*28	560*26
	(sectional area cm²)	(284)	(237)	(220)	(202)	(186)	(162)	(146)
	Sectional area (cm²)	585	553	537	536	546	557	572
	Is (cm⁴)	2,352,000	2,414,000	2,542,000	2,661,000	2,871,000	3,140,000	3,382,000
	Iv (cm⁴)	7,276,000	7,412,000	7,960,000	8,576,000	9,273,000	9,786,000	10,470,000
	Deflection δ_L (mm)	33.3	32.8	30.5	28.3	26.5	24.8	23.3
	Deflection δ_d (mm)	156.2	153.0	145.0	138.0	130.0	121.4	111.5
Exterior Girder	Top flange	420*9	420*24	380*23	380*19	340*17	320*17	310*16
	(sectional area cm²)	(122)	(101)	(87)	(72)	(58)	(54)	(50)
	Web	1500*13	1600*14	1700*14	1800*15	1900*16	2000*17	2100*18
	(sectional area cm²)	(195)	(224)	(238)	(270)	(304)	(340)	(378)
	Bottom flange	680*45	640*40	660*36	660*33	600*33	580*31	570*28
	(sectional area cm²)	(306)	(256)	(238)	(218)	(198)	(180)	(160)

(Continued)

TABLE 27.13 (*Continued*)   Design Results for $L = 40$ m (with Vertical Stiffener, SM490Y)

Web Height (mm)	1500	1600	1700	1800	1900	2000	2100
Sectional area (cm²)	623	581	563	560	560	574	587
Is (cm⁴)	2,517,000	2,582,000	2,708,000	2,836,000	2,965,000	3,257,000	3,517,000
Iv (cm⁴)	7,320,000	7,479,000	8,023,000	8,608,000	9,206,000	9,851,000	10,462,000
Deflection δ_L (mm)	37.1	36.5	33.9	31.5	29.4	27.5	25.9
Deflection δ_d (mm)	159.6	156.3	148.2	141.0	133.1	124.1	114.2
Total sectional area (cm²)	2416	2267	2201	2193	2212	2262	2318

TABLE 27.14   Design Results for $L = 40$ m (with Vertical and Horizontal Stiffeners, SM490Y)

	Web Height (mm)	1700	1800	1900	2000	2100	2200	2300
Interior Girder	Top flange (sectional area cm²)	410*24 (98)	390*23 (90)	360*21 (76)	340*20 (68)	330*18 (59)	340*16 (54)	340*14 (48)
	Web (sectional area cm²)	1800*9 (162)	1800*9 (162)	1900*10 (190)	2000*10 (200)	2100*11 (231)	2200*11 (242)	2300*12 (276)
	Bottom flange (sectional area cm²)	660*35 (231)	620*35 (217)	630*32 (202)	590*32 (189)	550*31 (171)	520*33 (172)	480*31 (149)
	Sectional area (cm²)	491	469	467	457	461	468	472
	Is (cm⁴)	2,561,000	2,716,000	2,831,000	2,977,000	3,145,000	3,415,000	3,579,000
	Iv (cm⁴)	7,729,000	8,274,000	8,900,000	9,450,000	10,000,000	11,028,000	11,430,000
	Deflection δ_L (mm)	31.3	29.4	27.4	25.8	24.3	22.9	21.5
	Deflection δ_d (mm)	144.0	137.4	130.6	124.3	119.3	112.7	106.1
Exterior Girder	Top flange (sectional area cm²)	410*26 (107)	400*24 (96)	360*23 (83)	360*21 (76)	360*17 (61)	350*15 (53)	330*14 (46)
	Web (sectional area cm²)	1800*9 (162)	1800*9 (162)	1900*10 (190)	2000*10 (200)	2100*11 (231)	2200*11 (242)	2300*12 (276)
	Bottom flange (sectional area cm²)	690*36 (248)	630*37 (233)	620*35 (217)	600*34 (204)	590*32 (189)	560*31 (174)	510*31 (158)
	Sectional area (cm²)	517	491	490	480	481	468	480
	Is (cm⁴)	2,733,000	2,877,000	3,021,000	3,191,000	3,293,000	3,388,000	3,627,000
	Iv (cm⁴)	7,815,000	8,369,000	8,974,000	9,566,000	10,212,000	10,693,000	11,362,000
	Deflection δ_L (mm)	34.9	32.8	30.6	28.7	26.5	25.0	23.6
	Deflection δ_d (mm)	147.1	140.4	133.5	127.0	121.9	115.7	109.0
	Total sectional area (cm²)	2017	1920	1914	1873	1884	1872	1905

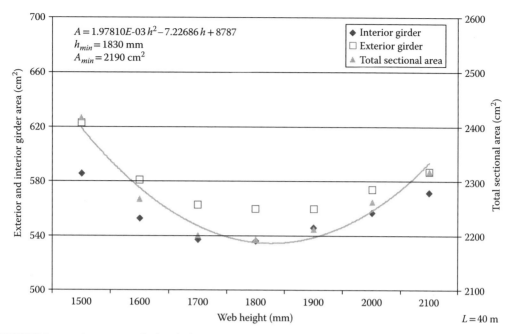

FIGURE 27.30 Optimum girder height for $L = 40$ m (with vertical stiffener, SM490Y).

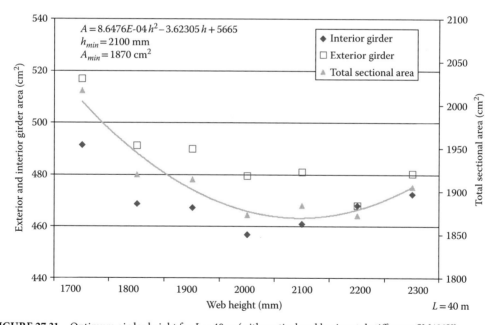

FIGURE 27.31 Optimum girder height for $L = 40$ m (with vertical and horizontal stiffeners, SM490Y).

27.10.5.2 Stiffness and Deflections

The moment of inertia of the external and internal girders for the steel section, I_s, and the composite section, I_v, are plotted in Figure 27.32 when the girder height varies. As seen in the previous cases, the moment of inertia is proportional with respect to the girder height. From Figure 27.32, it is found that the ratio of the moment of inertia of the composite girder to the steel girder lies in a narrow range from 2.9 to 3.2. This ratio is smaller than the span lengths $L = 20$ m and 30 m because the same slab is

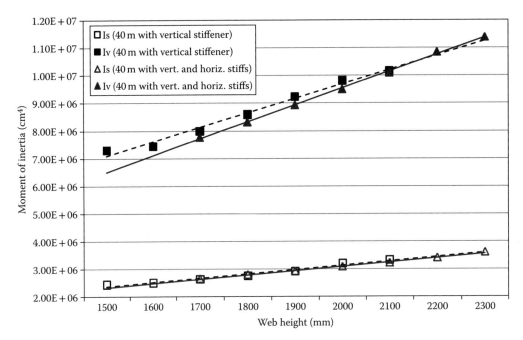

FIGURE 27.32   Moment of inertia, $L = 40$ m.

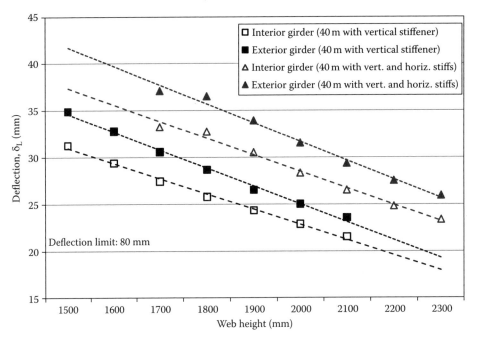

FIGURE 27.33   Deflections due to live load, $L = 40$ m.

used in those designs. The deflections due to the live and dead loads are plotted in Figure 27.33, which shows a clear inverse proportion to the web height. Similar to the other span lengths, the deflection limit 80 mm for live load by JSHB (JRA 2002) for the span length 40 m is above the actual deflections. This implies that the deflection (stiffness) does not control the design.

27.10.6 Summary

Based on the studies to investigate the optimum girder height for the benchmark bridges, the results are summarized in Figure 27.34 for the span lengths L = 20 m, 30 m, and 40 m. Note that different strengths of material are used for each span length. It can be seen that the optimum girder height differs depending on the span length and the stiffener conditions: (1) the optimum girder height if using both vertical and horizontal stiffeners is given by the ratio of girder height to span length h_{min}/L = 1/16 to 1/20 depending on the span length; (2) for the case of a web with a vertical stiffener but without a horizontal stiffener, the optimum girder height is h_{min}/L = 1/16 to 1/22; and (3) for the case of the girder without both vertical and horizontal stiffeners, the optimum web height is h_{min}/L = 1/18 to 1/24.

It is found that the shorter the span length, the larger the ratio of the optimum girder height to the span length. This is because neglecting the stiffeners makes the web thickness large, which leads to a lower girder height to reduce the sectional area of girders. Lowering the stiffness of the girders would not be a problem to meet the deflection limit because a sufficient margin is included. In principle, neglecting the horizontal stiffener is recommended to reduce the number of steel member components, thus reducing the fabrication cost, but the vertical stiffener should be used for the span lengths L = 30 m and 40 m. If the vertical stiffener is neglected for large-span bridges, the web thickness will become unrealistically large.

The total sectional areas of the four girders at the optimum girder heights are plotted in Figure 27.35. It can be seen that when more stiffeners are used the sectional area becomes smaller for the same span length. It should be noted that only the section at midspan is considered in this study. The optimum web depth would be slightly lower if the entire span length was considered to minimize the total weight instead of considering the sectional area at midspan only.

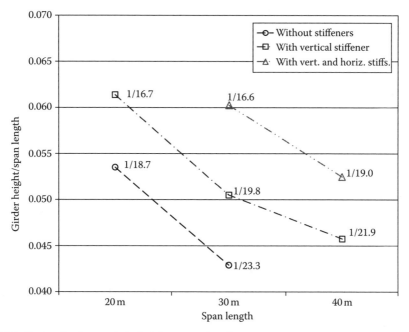

FIGURE 27.34 Ratio of optimum girder height to span length.

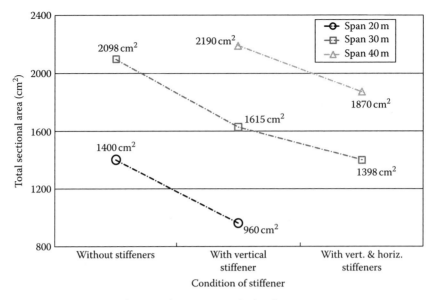

FIGURE 27.35   Minimum sectional areas with optimum girder height.

27.11 Conclusions

The benchmark design project was proposed at the IABSE 2007 Symposium in Weimar, Germany, for a simply supported composite girder highway bridge, in which the design criteria were provided (Toma and Duan 2007). There are two types of designs in the proposal: the free design and the conditional design. The benchmark bridge is to be designed under the same design criteria. Ten designs from around the world participated in the project and were compared in this chapter. The conclusions are as follows:

1. A small number of girders, such as two or three, is a more common design with a road width of 8.5 m.
2. Fundamental design results such as girder sections and spacing vary significantly, although the design criteria are substantially the same.
3. The web design with a vertical (transverse) stiffener but without a longitudinal (horizontal) stiffener is more common for span lengths of 30 m.
4. It would be more economical to neglect the stiffeners by using as low web depth and large thickness as possible within the code limitations.
5. Deflection would not be a problem even if the girder stiffness is decreased due to lower girder depth.

It should be pointed out that the designs presented in this chapter do not necessarily represent official design of their respective countries, but rather are designs by those individual engineers themselves based on their own bridge design codes. Therefore, the designs would possibly differ if other engineers designed the same bridge in that country. Nevertheless, from the consideration of the project's intent, it is believed that comparable standard designs have been provided for the project from the engineers of the world.

As another benchmark design, the optimum girder heights (web depth) for the span lengths $L = 20$ m, 30 m, and 40 m were studied in Section 27.10. Design details for each span length were provided for various girder heights. It was found that the ratio of the optimum girder height to the span length differs depending on the stiffener conditions and the span length: the shorter the span length, the larger the ratio. This study will give a good indication to decide the girder height at the initial design when the strength of steel and the span length are specified. It is expected that the results of the benchmark design comparisons presented in this chapter will give perspective views from different angles and precious data to engineers when they examine their design and the bridge codes in their countries.

Acknowledgments

Needless to say, this chapter could not be completed without the active participation of bridge engineers from around the world. The author would like to express his deepest gratitude to those engineers who participated to the project with great efforts as listed below.

United States: Dr. Lian Duan and Mr. Yusuf Saleh, California Department of Transportation
Belgium: Prof. Van Bogaert and Amelie Outtier, Gent University
China: Prof. Chunsheng Wang and Mr. Lan Duan, Chang'an University
Czech Republic: Roman Safar, Ph D., Czech Technical University
Italy: Guido Furlanetto and Lucio Ferretti Torricelli, SPEA Ingegneria Europea S.p.A.
Korea: Prof. Dong-Ho Choi, Hanyang University
Egypt: Prof. Metwally Abu-Hamd, Cairo University
Russia: Mr. Michael Timofeev, NIL Transmost
Ireland: Mr. Harminder Singh, consultant engineer

Also, the author would like to thank his former graduate student, Mr. Jun-ya Maeda, and undergraduate student, Mr. Yu Miyagawa, who worked on the comparative studies in Section 27.10 in their dissertations.

References

AASHTO. 2007. *LRFD Bridge Design Specifications*, 4th edition, American Association of State Highway and Transportation Officials.

Hayashikawa, Toshiro. 2000. *Bridge Engineering*, Asakura Pub. Co., Tokyo, pp. 222–251 (in Japanese).

JRA. 2002. *Specifications for Design of Highway Bridges (JSHB)*, Japan Road Association, Tokyo, March (in Japanese).

JSCE. 2008. *Guideline of Load Actions for Civil Engineering Structures in Performance Based Design*, Japan Society of Civil Engineers (in Japanese).

Nakai, Hiroshi, and Kitada, Toshiyuki. 2003. *Bridge Engineering*, Kyoritsu Pub. Co., Tokyo, pp. 126–127, pp. 258–277 (in Japanese).

Nakai, Hiroshi, Toma, Shouji, and Niwa, Kazuhisa. 2005. *Bridge Design Simulation by Dialog with Computer*, Kyoritsu Pub. Co., Tokyo, pp. 9–12 (in Japanese).

Miyamoto, Hiroshi. 2004. *Bridge Engineering*, Gihodo Pub. Co., Tokyo, pp. 289–312 (in Japanese).

Toma, Shouji, and Duan, Lian. 2007. "Comparative Study of Highway Steel Girder Bridge between Japan and USA." *International Association for Bridge and Structural Engineering* (IABSE), Weimar, Germany, Paper ID 118, September 19–21.

Highest Bridges

Eric Sakowski
Technicolor

28.1 Introduction

Throughout the history of bridges, the discussion of size has almost always focused on length—usually of the main span. This is natural given that the design and construction of a long span is usually the greatest challenge to an engineer who is attempting to cross over a deep body of water with a limited number of supports and the least amount of material. Of less interest to bridge engineers but just as effective aesthetically is the height of a bridge. A vast, vertical void below the structure seems to make the mass of the bridge more gravity defying, more surreal, and more magical.

The winner of a recent poll asking the world's bridge experts to name the greatest twentieth century bridge engineering achievement chose the Salginatobel Bridge in Switzerland, built in 1930. Designed by the earliest and arguably greatest master of concrete arch bridges, Robert Maillart, the sleek span was celebrated for its light, graceful lines resulting from a slender, three-hinge design. Was Maillart's Tavanasa Bridge, built 25 years earlier, not an equal feat of engineering? To quote *Bridges: Three Thousand Years of Defying Nature* book author David J. Brown (Brown 2005), "Though by no means the climax of his career, Maillart's Salginatobel bridge has gained prominence on account of the spectacular grandeur of its site spanning a precipitous gorge." Germany's great, modern day bridge engineer Fritz Leonhardt (1984) added "These Maillart-type arch bridges look good in special situations as here over a gorge and against a mountainous background." Had the design for one of Maillart's least-known spans, such as Rossgraben, been used for the Salginatobel site, you can be sure its fame would have been as great as the bridge that was built. Like a fancy picture frame, the vertical cliffs and tree-filled alpine foothills of the Swiss valley gave Salginatobel a breathtaking elegance that became a worldwide inspiration to both engineers and the public alike.

A *tall* bridge measurement is from the highest point of the structure, such as the top of a suspension bridge tower, to the ground or water below it. Only a bird or maintenance worker can visit the lofty heights of these towering structures. A *high* bridge measurement is from the road or rail deck elevation to the ground below it—the maximum level at which you and I are at while traveling across the span. There are many variables that should be considered when attempting to find the true height of a bridge. The most obvious of these is the level of the water below the bridge. Elevation drawings drafted by bridge engineers usually refer to the water level surface below a bridge as NW for normal or mean water level. The heights of the bridges within this chapter are often between 5 ft and 10 ft (3 m) higher or lower than the mean water level figures that are listed.

While bridges that cross harbors, large rivers, and other navigable waterways generally have accurate and carefully measured height figures, bridges that span creeks, gorges, and other uneven terrain often have measurements that are inaccurate or ignored altogether. Reaching a conclusive height for such bridges can also be difficult since the ground can have two or three measurement points, referred to as left ground line, center ground line, and right ground line. For this chapter, the downstream side or lowest elevation ground line figure available is the one that is used.

Table 28.1 is a list of the 100 highest international bridges. Most of the bridges discussed in this chapter are from China and Mexico and were recently constructed, so readers will see bridges that are new to them. For detailed coverage of the world's highest bridges, please visit http://www.highestbridges.com.

TABLE 28.1 Top 100 Highest International Bridges

	Rank	Name	Height ft. (m)	Main Span Length ft. (m)	Total Length ft. (m)	Completed	Location	Country
	1	Beipanjiang Bridge 北盘江特大桥	1,850 (564)	2,362 (720)	4,400 (1,341)	2016	Duge Bouyei, Guizhou	China
	2	Siduhe Bridge 四渡河特大桥	1,627 (496)	2,952 (900)	4,009 (1,222)	2009	Yesanguanzhen, Hubei	China
	3	Hegigio Gorge Pipeline Bridge	1,289 (393)	1,542 (470)	1,772 (540)	2005	Otoma, Southern Highlands	Papua New Guinea
	4	Baluarte Bridge Puente Baluarte	1,280 (390)	1,706 (520)	3,688 (1,124)	2012	El Palmito, Sinaloa	Mexico
	5	Balinghe Bridge 坝陵河特大桥	1,214 (370)	3,570 (1,088)	7,339 (2,237)	2009	Guanling, Guizhou	China
	6	Beipanjiang Bridge 北盘江特大桥	1,200 (366)	1,273 (388)	1,594 (486)	2003	Xingbeizhen, Guizhou	China
	7	Dimuhe Bridge 抵母河特大桥	1,181 (360)	1,765 (538)	2,388 (728)	2015	Liupanshui, Guizhou	China
	8	Puli Bridge 普立特大桥	1,115 (340)	2,060 (628)	3,176 (968)	2015	Pulixiang, Yunnan	China

TABLE 28.1 (*Continued*)   Top 100 Highest International Bridges

	Rank	Name	Height ft. (m)	Main Span Length ft. (m)	Total Length ft. (m)	Completed	Location	Country
	9	Aizhai Bridge 矮寨特大桥	1,102 (336)	3,858 (1,176)	5,033 (1,534)	2012	Jishou, Hunan	China
	10	Liuchonghe Bridge 六冲河特大桥	1,102 (336)	1,437 (438)	4,948 (1,508)	2013	Zhilin, Guizhou	China
	11	Lishuihe Bridge 澧水大桥	1,083 (330)	2,808 (856)	4,633 (1,412)	2013	Zhangjiajie, Hunan	China
	12	Chenab Bridge	1,053 (321)	1,509 (460)	4,314 (1,315)	2017	Katra, Jammu-Kashmir	India
	13	Najiehe High Speed Railway Bridge 纳界河特大桥	1,050 (320)	1,155 (352)	2,657 (810)	2016	Ziqiangxiang, Guizhou	China
	14	Beipanjiang Bridge 北盘江特大桥	1,043 (318)	2,087 (636)	3,425 (1,044)	2009	Qinglong, Guizhou	China
	15	Liuguanghe Bridge 六广河大桥	975 (297)	787 (240)	1,850 (564)	2001	Liu Guangzhen, Guizhou	China
	16	Zhijinghe Bridge 支井河大桥	960 (293)	1,411 (430)	1,790 (545)	2009	Dazhipingzhen, Hubei	China
	17	Longjiang Bridge 龙江大桥	958 (292)	3,924 (1,196)	8,077 (2,462)	2015	Wuhexiang, Yunnan	China
	18	Royal Gorge Bridge	955 (291)	938 (286)	1,258 (383)	1929	Cañon City, Colorado	United States
	19	Beipanjiang High Speed Railway Bridge 北盘江铁路桥	928 (283)	1,460 (445)	2,386 (727)	2015	Qinglong, Guizhou	China
	20	Millau Viaduct Viaduc Millau	909 (277)	1,122 (342)	8,071 (2,460)	2004	Millau, Midi-Pyrénées	France
	21	Beipanjiang Railway Bridge 北盘江铁路桥	902 (275)	771 (235)	1,535 (468)	2001	Fa'er Bouyei, Guizhou	China
	22	Lancanjiang Railway Bridge 与澜沧江大桥	890 (271)	1,122 (342)	1,733 (528)	2014	Shanyangxiang, Yunnan	China

(Continued)

TABLE 28.1 (*Continued*) Top 100 Highest International Bridges

	Rank	Name	Height ft. (m)	Main Span Length ft. (m)	Total Length ft. (m)	Completed	Location	Country
	23	Mike O'Callaghan-Pat Tillman Memorial Bridge	890 (271)	1,060 (323)	1,896 (578)	2010	Boulder City, Nevada	United States
	24	New River Gorge Bridge	876 (267)	1,700 (518)	3,030 (924)	1977	Fayetteville, West Virginia	United States
	25	Zongqihe Bridge 总溪河特大桥	866 (264)	1,181 (360)	3,035 (925)	2015	Weixinzhen, Guizhou	China
	26	Wulingshan Bridge 武陵山大桥	863 (263)	1,181 (360)	2,756 (840)	2009	Pengshui, Chongqing	China
	27	Nanpanjiang High Speed Railway Bridge 南盘江大桥	860 (262)	1,365 (416)	2,746 (837)	2015	Qiubei, Yunnan	China
	28	Long Bridge 龙桥特大桥	853 (260)	879 (268)	1,654 (504)	2014	Xuan'en, Hubei	China
	29	Zhongjianhe Bridge 忠建河大桥	853 (260)	1,312 (400)	3,488 (1,063)	2014	Xuan'en, Hubei	China
	30	Italia Viaduct Viadotto Italia	850 (259)	574 (175)	3,809 (1,161)	1974	Laino Borgo, Calabria	Italy
	31	Jiangjiehe Bridge 江界河大桥	840 (256)	1,083 (330)	1,512 (461)	1995	Weng'an, Guizhou	China
	32	Sfalassa Bridge Viadotto Sfalassà	820 (250)	1,181 (360)	2,536 (773)	1974	Bagnara Calabra, Calabria	Italy
	33	Beipanjiang Bridge 北盘江特大桥	820 (250)	951 (290)	4,137 (1,261)	2013	Fa'er Bouyei, Guizhou	China
	34	Azhihe Bridge 阿志河大桥	810 (247)	928 (283)	1,706 (520)	2003	Changliuxiang, Guizhou	China
	35	Erlanghe Bridge 二郎河特大桥	804 (245)	656 (200)	2,300 (701)	2013	Erlangxiang, Guizhou	China
	36	Malinghe Bridge 马岭河大桥	787 (240)	1,181 (360)	4,547 (1,386)	2011	Xingyi, Guizhou	China
	37	Furongjiang Bridge 芙蓉江特大桥为	787 (240)	755 (230)	1,768 (539)	2009	Haokouxiang, Chongqing	China

TABLE 28.1 (*Continued*) Top 100 Highest International Bridges

	Rank	Name	Height ft. (m)	Main Span Length ft. (m)	Total Length ft. (m)	Completed	Location	Country
	38	Karun 3 Dam Bridge سد کارون 3	771 (235)	827 (252)	1,102 (336)	2005	Bajool, Khuzestan	Iran
	39	Mengdong Bridge 猛洞河特大桥	761 (232)	837 (255)	1,962 (598)	2013	Yongshun, Hunan	China
	40	Nanjiang Railway Bridge 南江特大桥	755 (230)	577 (176)	2,133 (650)	2014	Kaiyang, Guizhou	China
	41	Labajin Bridge 腊八斤特大桥	750 (229)	656 (200)	3,740 (1,140)	2012	Yingjing, Sichuan	China
	42	Tongzihe Bridge 桐梓特大桥	741 (226)	656 (200)	3,714 (1,132)	2013	Erlangxiang, Guizhou	China
	43	Shennongxi Bridge 神农溪大桥	738 (225)	1,050 (320)	3,773 (1,150)	2013	Yanduhezhen, Hubei	China
	44	Zhuchanghe Bridge 朱昌河特大桥	735 (224)	656 (200)	2,205 (672)	2009	Sanbanqiao, Guizhou	China
	45	Auburn-Foresthill Bridge	730 (223)	862 (263)	2,428 (740)	1973	Auburn, California	United States
	46	San Marcos Bridge Puente San Marcos	722 (220)	591 (180)	2,789 (850)	2013	Xicotepec de Juárez, Puebla	Mexico
	47	Corgo Viaduct Viaduto Corgo	722 (220)	984 (300)	9,137 (2,785)	2013	Vila Real	Portugal
	48	Platano Viaduct Viadotto Platano	722 (220)	955 (291)	2,041 (622)	1978	Romagnano al Monte, Campania	Italy
	49	Mangjiedu Bridge 漭街渡大桥	722 (220)	722 (220)	2,654 (809)	2009	Yongxin Yizu Xiang, Yunnan	China
	50	Weijiazhou Bridge 魏家州特大桥	722 (220)	656 (200)	1,818 (554)	2009	Gaojiayanzhen, Hubei	China
	51	Mashuihe Viaduct 马水河特大桥	720 (219)	656 (200)	3,261 (994)	2008	Baiyangpingxiang, Hubei	China
	52	Grand Canyon Skywalk	720 (219)	70 (21)	180 (55)	2007	Grand Canyon West, Arizona	United States

(Continued)

TABLE 28.1 (*Continued*)   Top 100 Highest International Bridges

	Rank	Name	Height ft. (m)	Main Span Length ft. (m)	Total Length ft. (m)	Completed	Location	Country
	53	Daninghe Bridge 大宁河桥	719 (219)	1,312 (400)	2,238 (682)	2010	Wushan, Chongqing	China
	54	Houzihe Bridge 猴子河特大桥	705 (215)	722 (220)	2,890 (881)	2010	Sandu Suizu, Guizhou	China
	55	Xixi Bridge 西溪大桥	715 (218)	1,109 (338)	2,297 (700)	2001	Linquanzhen, Guizhou	China
	56	Xisha Bridge 细沙河特大桥	710 (216)	623 (190)	1,250 (381)	2010	Lianghezhen, Chongqing	China
	57	Bloukrans Bridge	708 (216)	892 (272)	1,480 (451)	1984	Nature's Valley, Western Cape	South Africa
	58	Phil G McDonald Bridge	700 (213)	784 (239)	2,190 (668)	1988	Beckley, West Virginia	United States
	59	Glen Canyon Dam Bridge	700 (213)	1,028 (313)	1,228 (374)	1959	Page, Arizona	United States
	60	Hezhang Bridge 赫章特大桥	690 (210)	591 (180)	3,522 (1,074)	2013	Hezhang, Guizhou	China
	61	Yesanhe Bridge 野三河特大桥	689 (210)	656 (200)	3,268 (996)	2009	Gaopingzhen, Hubei	China
	62	Tieluoping Bridge 铁罗特大桥	685 (209)	1,056 (322)	2,897 (883)	2009	Langpingzhen, Hubei	China
	63	Hutiaohe Viaduct 虎跳河大桥	685 (209)	738 (225)	6,424 (1,958)	2009	Sanbanqiao, Guizhou	China
	64	Falanggou Bridge 法朗沟特大桥	679 (207)	738 (225)	2,254 (687)	2015	Linkouzhen, Guizhou	China
	65	Mi River Bridge 洣水河大桥	679 (207)	591 (180)	1,778 (542)	2013	Yanling, Hunan	China
	66	Xiaohe Bridge 小河特大桥	675 (206)	1,109 (338)	1,652 (504)	2010	Enshi, Hubei	China
	67	Shuanghekou Bridge 双河口特大桥	666 (203)	558 (170)	2,825 (861)	2009	Langpingzhen, Hubei	China

TABLE 28.1 (*Continued*)   Top 100 Highest International Bridges

	Rank	Name	Height ft. (m)	Main Span Length ft. (m)	Total Length ft. (m)	Completed	Location	Country
	68	Shintabisoko Bridge 新旅足橋	656 (200)	722 (220)	1,516 (462)	2010	Yaotsu, Gifu	Japan
	69	San Cristóbal Bridge Puente San Cristóbal	656 (200)	584 (178)	1,063 (324)	2006	Ajtectic, Chiapas	Mexico
	70	El Carrizo Bridge Puente El Carrizo	650 (198)	712 (217)	1,424 (434)	2013	El Palmito, Sinaloa	Mexico
	71	Mala Rijeka Viaduct	650 (198)	495 (151)	1,637 (499)	1973	Podgorica, Montenegro	Montenegro
	72	Longtanhe River Viaduct 龙潭河大桥	645 (197)	656 (200)	3,878 (1,182)	2009	Langpingzhen, Hubei	China
	73	Sanshuihe Viaduct 三水河特大桥	640 (195)	607 (185)	5,538 (1,688)	2016	Xunyi, Shaanxi	China
	74	Wujiang Bridge 乌江河大桥	640 (195)	722 (220)	3,855 (1,175)	2012	Sinan, Guizhou	China
	75	Hiroshima Airport Bridge 広島空港大橋	640 (195)	1,247 (380)	2,625 (800)	2010	Hiroshima	Japan
	76	Bidwell Bar Bridge	627 (191)	1,108 (338)	1,793 (546)	1965	Oroville, California	United States
	77	Wushan Bridge 湖山大桥	625 (191)	558 (170)	2,561 (781)	2009	Pengshui, Chongqing	China
	78	Akaishi Viaduct 赤石特大桥	623 (190)	1,247 (380)	7,457 (2,273)	2013	Chenzhou, Hunan	China
	79	Zhegao Bridge 者告河特大桥	623 (190)	705 (215)	2,907 (886)	2012	Ceheng, Guizhou	China
	80	Europa Bridge Europabrücke	623 (190)	650 (198)	2,690 (820)	1964	Innsbruck, Tirol	Austria
	81	Niouc Bridge Pont Niouc	623 (190)	623 (190)	623 (190)	1922	Niouc, Valais	Switzerland
	82	Buliuhe Bridge 布柳河大桥	623 (190)	771 (235)	1,761 (537)	2006	Xiangyangzhen, Guanxi	China

(Continued)

TABLE 28.1 (*Continued*)　Top 100 Highest International Bridges

	Rank	Name	Height ft. (m)	Main Span Length ft. (m)	Total Length ft. (m)	Completed	Location	Country
	83	Nianziping Bridge 碾子坪特大桥	620 (189)	623 (190)	2,372 (723)	2015	Linkouzhen, Guizhou	China
	84	Trient Bridge Pont Trient	620 (189)	323 (98)	552 (168)	1933	Gueuroz, Valais	Switzerland
	85	New Trient Bridge Pont Neuf Trient	620 (189)	356 (109)	559 (170)	1994	Gueuroz, Valais	Switzerland
	86	Yanjinhe 1995 Bridge 盐津河拱桥	620 (189)	571 (174)	1,037 (316)	1995	Renhuai, Guizhou	China
	87	Lancanjiang Pipeline Bridge 与澜沧江大桥	614 (187)	919 (280)		2012	Shanyangxiang, Yunnan	China
	88	Chishuihe Bridge 赤水河特大桥	610 (186)	814 (248)	3,196 (974)	2011	Chishui, Chongqing	China
	89	Anjikhad Railway Bridge	610 (186)	869 (265)	2,156 (657)	2016	Katra, Jammu-Kashmir	India
	90	Kochertal Viaduct Kochertal brücke	607 (185)	453 (138)	3,684 (1,123)	1979	Braunsbach, Baden-Württemberg	Germany
	91	Beipanjiang Bridge 北盘江特大桥	607 (185)	722 (220)	3,661 (1,116)	2013	Zhenfeng, Guizhou	China
	92	Sente Bridge Viadotto Sente	607 (185)	656 (200)		1977	Belmonte del Sannio, Molise	Italy
	93	Piave Viaduct Viadotto Piave	604 (184)	836 (255)	1,770 (539)	1985	Caralte, Veneto	Italy
	94	Wuxi Bridge 乌溪大桥	600 (183)	525 (160)	1,161 (354)	2001	Huangnitangzhen, Guizhou	China
	95	Luojiaohe Bridge 落脚河大桥	600 (183)	879 (268)	1,667 (508)	2001	Dafang, Guizhou	China
	96	Longhe Bridge 龙河特大桥	590 (180)	787 (240)	3,875 (1,181)	2013	Fengdu, Chongqing	China
	97	Qingshuihe River Railway Bridge 清水河大桥	590 (180)	420 (128)	1,183 (360)	2000	Qingshuihezhe, Guizhou	China

TABLE 28.1 (*Continued*) Top 100 Highest International Bridges

	Rank	Name	Height ft. (m)	Main Span Length ft. (m)	Total Length ft. (m)	Completed	Location	Country
	98	Maguohe Bridge 马过河特大桥	591 (180)	623 (190)	2,480 (756)	2012	Kunming, Guizhou	China
	99	Wushan Yangtze Bridge 巫山长江大桥	590 (180)	1,509 (460)	2,009 (612)	2005	Wushan, Chongqing	China
	100	Guozigou Bridge 果子沟大桥	590 (180)	1,181 (360)		2011	Ili Kazak, Xinjiang	China

28.2 Historical Development

Up until the new millennium, there were few bridges on earth that even came close to challenging the height record of the Royal Gorge suspension bridge in the state of Colorado. Built in 1929, the wood-planked suspension span nearly became the dictionary definition of how spectacular a high bridge could be. That all changed in 2001 when an astounding succession of 10 higher bridges began opening in the decade that followed. Most of this activity occurred in China, a country that as recently as 1994 had just one bridge over 400 ft (122 m). The Chinese have embarked on what can only be called an explosion of bridge construction. Their goal is nothing less than the complete connection of every major and minor city in the country with a full web of high-speed expressways and rail lines. Even more incredible is the desire to have much of it done in just 20 years—half the time it would have taken in any other industrialized nation. This breathtaking pace has resulted in an average of 10,000 new bridge openings per year. This is no small task when you consider that the entire western half of China is mountainous. China is already home to two-thirds of the world's 100 highest bridges!

The highest of China's many high bridges are located primarily in the four western provinces of Guizhou, Hubei, Chongqing, and Yunnan. This region contains an astounding 10 of the worlds 12 highest bridges. All 10 of these exceed 900 ft (275 m) in height, while 4 have held the world record for highest road or rail bridge, including the latest Guinness book champ, the stunning 1627 ft high (496 m) Siduhe (Sidu River) Bridge. More than 50 others in the region exceed 500 ft (152 m) in height. The big cities of Yichang, Chongqing, and Guiyang are ground zero for those looking to travel to these bridges.

Siduhe Bridge (Figures 28.1 and 28.2) is the latest of just 11 bridges in history to hold the title of world's highest. The first was the famous Pont Du Gard Stone Aqueduct (Figure 28.3) located near Nimes, France. Completed around 90 BC, the Roman structure has three levels of stacked arches that reach a height of 160 ft (49 m). More than 200 years passed before the Puente Alcántara became the new record holder with five semicircular arches that cross 164 ft (50 m) above the Tajo River west of Caceres, Spain. Italy's massive Ponte Delle Torri in Spoleto surpassed the Spanish bridge with a height of 269 ft (82 m); the exact completion date of the aqueduct is unknown but probably dates to the thirteenth or fourteenth century. Spain would take the title back several hundred years later when the 322 ft (98 m) high Puente Nuevo or New Ronda Bridge opened across a deep chasm in Ronda, Spain. It would be the last time a stone or masonry arch bridge would hold the height record.

By the early 1800s, suspension bridges were becoming the bridge of choice for spans of more than 200 ft (60 m). So in 1839, French engineer E. Belin had no choice but to cross the 482 ft (147 m) deep chasm of the Usses River with a suspension bridge in the mountainous Rhône-Alpes region of France. Named after Charles-Albert, the king of Piedmont, the 656 ft (200 m) long bridge is also referred to as the Pont Caille. The French bridge kept the height record in Europe for 73 years before a higher

FIGURE 28.1 **(See color insert.)** Siduhe Bridge.

FIGURE 28.2 Siduhe Bridge tower.

suspension bridge opened across the Rhumel River gorge in the North African country of Algeria in the city of Constantine. Designed by renowned French bridge engineer Ferdinand Arnodin, the Sidi M'Cid suspension/cable stayed bridge (Figure 28.4) hybrid connects two cliffs more than 574 ft (175 m) above the river. In 1929 the height record nearly doubled when the Royal Gorge suspension bridge opened in the state of Colorado with a deck 955 ft (291 m) above the Arkansas River. The popular tourist attraction

FIGURE 28.3 **(See color insert.)** Pont Du Gard stone aqueduct.

FIGURE 28.4 Sidi M'Cid Bridge.

became the first bridge known around the world for its great height and was listed in the Guinness Book of Records for more than 40 years.

After China, the country with the greatest number of high bridges is Italy. For a region that is only the size of the state of California, it is simply amazing how many towering viaducts or *viadottos* are spread out among Italy's expansive array of *Autopistas*. Built mostly in the 1970s and 1980s, this web of bridge-filled motorways includes the A3, A10, A12, A15, and A26. The most recent addition to this family is the breathtaking A20 motorway that opened in 2004 on the island of Sicily between Palermo and Messina.

Perched hundreds of feet above the Mediterranean coast, this four-lane marvel crosses dozens of colossal beam bridges that are nearly always wedged between two large tunnels. Italy is also home to 6 of the 10 tallest bridge piers in Europe.

Japan ranks third among the world's high bridge countries, even though the island nation has only three entries on the list of the world's 100 highest. This is possible because Japan is rich with bridges in the lower height range of 300 ft (90 m) to 500 ft (150 m). After Japan, the United States would rank fourth followed by France, Spain, Mexico, Germany, Austria, and Switzerland. Beyond these 10 countries, few others have more than two or three bridges above 300 ft (90 m) except for Canada, Algeria, and South Africa.

28.3 China's New Highway in the Sky

With a roadway 1627 ft (496 m) above the water, the Siduhe Bridge is the latest Chinese champ to take the record as the highest bridge in the world. Opened on November 15, 2009, it is the third Chinese bridge in less than a decade to claim the title of world's highest bridge and is a symbol of just how fast and how far China's highway infrastructure has come in such a short period of time.

Located about 50 mi (80 km) south of the famous Three Gorges region of the Yangtze River in China's mountainous Hubei Province, it would only be appropriate that the highest bridge in the world is also on the greatest bridge highway in the world, the G50. Stretching 300 mi (483 km) from Yichang and the Three Gorges dam in the east to the city of Zhongxian in the west, the most difficult section of this four-lane engineering marvel has more than half a dozen spectacular bridges that exceed 500 ft (150 m) in height, including Zhijinghe (Zhijing River), the highest roadway arch bridge in the world. This more direct route bypasses one of the toughest and most mountainous stretches of the Yangtze River. What once took more than a day of travel on dangerous mountain roads or a Yangtze River boat can now be safely traversed in 5 hours.

28.4 The Soaring Siduhe Suspension Bridge

With a span of 2952 ft (900 m), the Siduhe Bridge is a fairly typical long-span Chinese suspension bridge with H-frame concrete towers, a truss stiffened road deck, and unsuspended side spans. One aspect of its construction that was not so ordinary relates to its extreme height. Due to the remote location where the deep gorge offered no easy access, the engineers decided to use their expertise in fireworks and tied a 4265 ft (1300 m) long tether made of chinlon yarn to the back end of a 4 ft (1.5 m) long rocket and blasted the first dragging line cable across the vast river canyon (Figures 28.5 through 28.8). Since the eastern

FIGURE 28.5 Siduhe Bridge truss view.

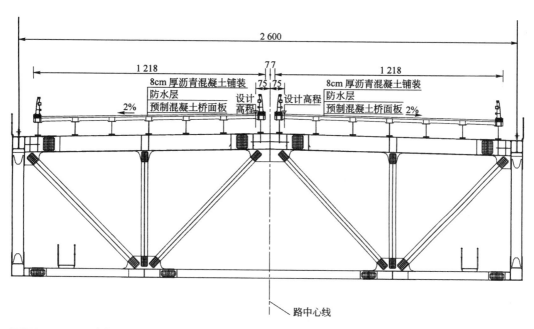

FIGURE 28.6 Siduhe Bridge truss drawing.

FIGURE 28.7 Siduhe Bridge—rocket blasting across gorge.

end of the bridge meets the ground on a steep mountain slope, the engineers had no choice but to terminate the two massive suspension cables within a deep circular tunnel foundation that extends more than 130 ft (40 m) underground (Figure 28.9). Each of the main cables (Figure 28.10) is composed of 127 hexagon shaped bundles that are each made up of 127 wires 5.1 mm thick for a total of 16,129 wires per suspension cable. The two main cables stretch over three spans that measure 114 m, 900 m, and 208 m.

Siduhe is so high that its closest rival in Mexico is nearly 328 ft lower (100 m). It is 673 ft (205 m) higher than Colorado's Royal Gorge Bridge. The structure also has the unique distinction of being the only bridge in the world where a person falling from the deck can reach terminal velocity—the speed at which a falling object will no longer accelerate. With so many unprecedented superlatives, the Siduhe

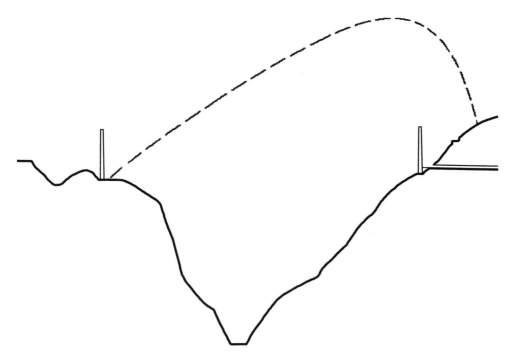

FIGURE 28.8   Siduhe rocket launch.

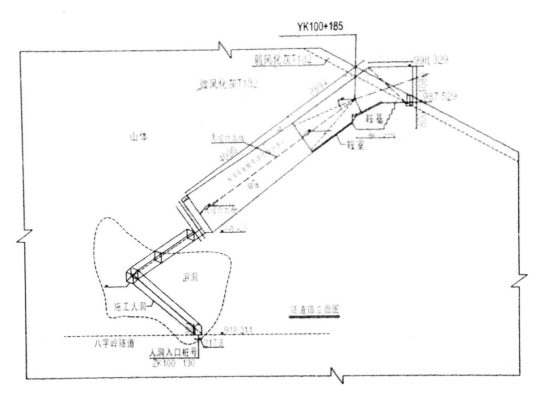

FIGURE 28.9   Siduhe tunnel anchorage.

FIGURE 28.10 Siduhe cable bundle.

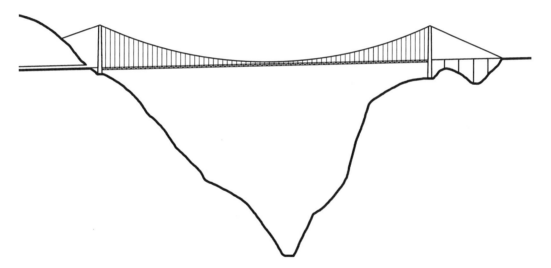

FIGURE 28.11 Siduhe Bridge elevation drawing.

Bridge is unlikely to lose its lofty suspension bridge height record for decades to come. "Siduhe" translates to Four Crossings River. "Si" means four, while "du" means crossings and "he" means river. The name probably came from the Red Army's famous Four Crossings of the Chishui River in Guizhou Province during Chairman Mao Tse Tung's Long March in 1935.

A trip along the highway to Siduhe Bridge (Figures 28.11 and 28.12) is a truly spectacular experience as you will be treated to an assortment of huge bridges of every size and variety. West of Siduhe there is the Zhijinghe (Zhijing River) Bridge, the world's highest roadway arch that soars 965 ft (294 m) above the river gorge. East of the bridge there is the incredible parade of 200 m high monsters including the Tieluoping cable-stayed bridge, the Shuanghekou beam bridge, and finally the Longtanhe (Longtan River) viaduct, a multispan concrete beam bridge that exceeded Germany's Kochertal as the world's largest beam bridge viaduct (Frances Millau is still the world's highest viaduct overall).

FIGURE 28.12   Siduhe Bridge long view.

28.5 Zhijinge Bridge: The World's Highest Arch Bridge

Located about 20 mi (32 km) west of the Siduhe suspension bridge is Zhijinghe Bridge (Xie 2009), the highest arch bridge on earth. Opened on November 28, 2009 with a deck 965 ft (294 m) in height, it surpassed the Beipanjiang (Beipan River) railway arch bridge by 40 ft (12 m) and West Virginia's New River Gorge arch bridge by 90 ft (27 m). India's Chenab River Bridge will be higher when it opens sometime after 2015, but Zhijinghe Bridge will still retain the honor of highest roadway arch, since Chenab is a railway bridge. Wedged between two cliffs, Zhijinghe Bridge is also one of the world's highest tunnel-to-tunnel bridges. The soaring red span is also impressive for its 1411 ft (430 m) span length, ranking it among the world's ten longest arches. Figure 28.13 shows Zhijinghe Bridge under construction.

The type of steel truss arch used for the Zhijinghe Bridge may look common, but the Chinese have taken it one step further, developing it into a unique type of structure all its own. Known as a concrete-filled steel tubular (CFST) bridge, the eight large steel tubes that run along the underside of the Zhijinghe arch were initially hollow (Figure 28.14). Once the arch was closed, concrete was pumped into these tubes from the bottom up. First used by the Chinese in 1990, they have refined and improved the technique and now use it on the majority of their steel arch bridges (Chen 2008). Depending on the length of the span and the width of the bridge, different styles of tubing are used. For Zhijinghe Bridge, they adopted an array of single tubes spaced apart from each other. Other configurations include dumbbells with two tubes closely connected, a mix of a dumbbell and single tubes such as was done on the Beipanjiang railway bridge, or a tight cluster of tubes known as multiple contiguous. Once hardened, the concrete solidifies and stiffens the arch, improving the compressive strength of the entire structure.

The erection process for CFST bridges is also easier since the hollow steel tubes can be built thinner and lighter than a steel-only arch bridge, which is a great advantage within a deep gorge where every piece must be moved into position with a highline. With a concrete–steel arch hybrid, there is a cost reduction as there is less steel within the structure. To minimize weight on such an unusually long span as Zhijinghe Bridge (Figures 28.15 through 28.17), the spandrel columns were all built of steel.

FIGURE 28.13 Zhijinghe Bridge under construction.

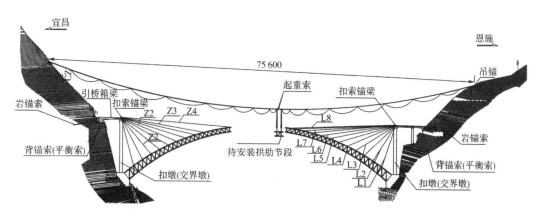

图 1　钢管拱吊装系统总体布置

FIGURE 28.14 Zhijinghe Bridge arch details.

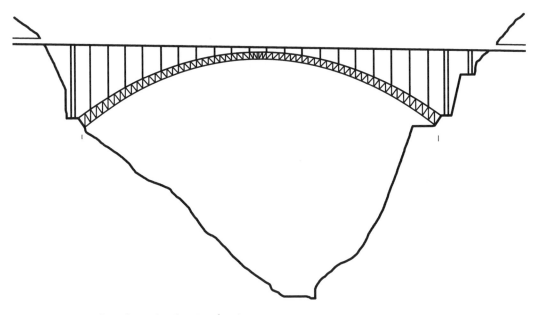

FIGURE 28.15   Zhijinghe Bridge elevation drawing.

FIGURE 28.16   Zhijinghe Bridge arch footing.

FIGURE 28.17 Zhijinghe Bridge arch view.

28.6 Longtanhe Bridge and Mashuihe Bridge Reach for the Sky

Two more record-breaking structures opened in 2009 on the Chinese G50 Hurong highway in the form of Longtanhe (Longtan River) Bridge and Masuhuihe (Masuhui River) Bridge, the two highest concrete beam bridge viaducts in the world. The larger of the two is the gargantuan Longtanhe Viaduct (Figure 28.18) located just 6 mi (10 km) east of Siduhe, the highest suspension bridge in the world. The Longtanhe Bridge is tied with Germany's Kochertal as the world's largest beam bridge viaduct. The cable-stayed Millau Viaduct in France is the world's largest viaduct overall.

The definition of what constitutes a viaduct may be open to debate, but Longtanhe Bridge follows the form with three repetitive and equal-sized spans of 656 ft (200 m) rather than just one central span flanked by much smaller approaches. The 584 ft (178 m) tall piers (Figure 28.19) are the fourth highest in China and tied with Germany's Kochertal as the seventh highest in the world.

Like Longtanhe, the equally spectacular Mashuihe Viaduct (Figures 28.20 and 28.21) consists of three main spans of 656 ft (200 m) supported on piers as high as 466 ft (142 m). The main spans were constructed using the balanced cantilever method (Yan 2007). The eastbound viaduct is the lower and shorter of the two structures with a total length of 878 m. The westbound viaduct is 994 m long.

28.7 Connecting Mountain Ranges: The Balinghe and Aizhai Bridges

High bridge history was made on December 23, 2009 when the Balinghe (Baling River) Bridge opened near Guanling in Guizhou province. Never before had a bridge crossed a gorge with a span so long. Measuring 3570 ft (1,088 m) between towers, Balinghe Bridge (Figure 28.22) has the second-longest span length of the world's 400 known bridges that exceed 100 m in height. The Golden Gate–sized bridge gap is so great the span appears to float among the clouds as it connects two mountain ranges. Balinghe Bridge is one of four high Chinese suspension bridges with extremely long spans. The other three are the Siduhe, Aizhai, and Beipanjiang highway bridges, built in 2009. Prior to these crossings, the longest span of any notable height was the New River Gorge Bridge in West Virginia with a soaring arch of 1700 ft (518 m).

The design of Balinghe Bridge (Figure 28.23) is similar to some of China's other long-span suspension bridges that cross the lower end of the Yangtze River, but with a few subtle changes. The most notable of these is the sleek, trapezoid-like shape of the tower openings and the lighter, less bulky triangular-shaped

FIGURE 28.18   Longtanhe Viaduct overview.

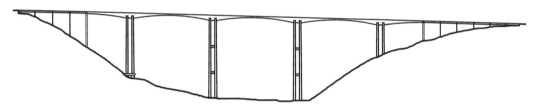

FIGURE 28.19   Longtanhe Viaduct elevation.

FIGURE 28.20   Mashuihe Viaduct overview.

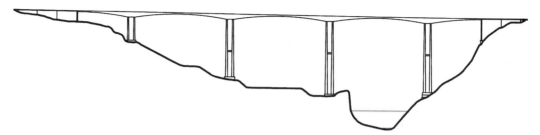

FIGURE 28.21 Mashuihe Viaduct elevation.

FIGURE 28.22 Balinghe Bridge under construction.

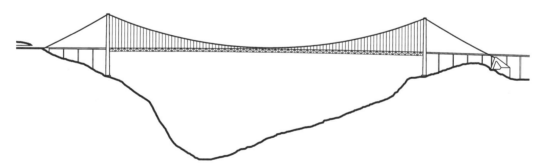

FIGURE 28.23 Balinghe Bridge elevation drawing.

east anchorage. The saddle covers are also modern looking and somewhat reminiscent of the Tsing Ma Bridge in Hong Kong. The four-lane deck rests on an extremely deep 33 ft (10 m) thick stiffening truss which harks back to the classic, post-Tacoma U.S. suspension bridges of the 1950s, like the Mackinac Bridge in Michigan. The truss (Figure 28.24) was built in 8 months by two traveling cranes that worked from either end before meeting in the middle 1215 ft (370 m) above the Baling River. The final price tag was 1.48 billion Chinese Yuan.

While a suspension bridge might seem to be a no-brainer for such a long, high crossing, Chinese engineers did design a concrete beam viaduct and several multispan cable-stayed bridges. With a central

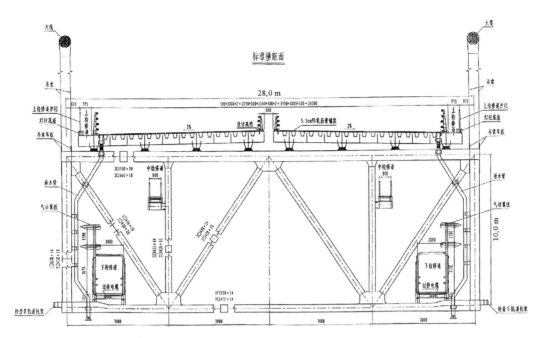

FIGURE 28.24   Balinghe Bridge truss cross section.

tower rising approximately 1476 ft (450 m), any one of the cable-stayed schemes (Figures 28.25 and 28.26) would have surpassed France's Millau bridge as the worlds tallest. With so much concrete required for so many tall piers, there simply was not enough natural sand found at or near the bridge site, as man-made sand is not suitable for concrete mixing. If natural sand had been transported from a far city, the cost of the project would have sky rocketed.

Aizhai Bridge (Figure 28.27), a suspension bridge in Hunan province that has a main span of 3858 ft (1176 m), was completed in 2012 and is 88 m longer then Balinghe Bridge, even though the deck is not be as high at 336 m. Aizhai is also the world's highest tunnel-to-tunnel bridge and the fourth suspension bridge in China to cross a valley so wide it seems to be connecting two mountain ranges. The first three were the Siduhe, Balinghe, and Beipanjiang bridges, built in 2009. Located deep in the heart of China's Hunan province near the city of Jishou, the suspension bridge is the largest structure on the Jishou to Chadong expressway crossing high above the DeHang Canyon.

With such an unusually long span locked between two tunnels that go right to the edge of a cliff, the engineers had to devise a new method to build the massive roadway that hangs from the cables. They devised the world's first "suspended truss roll." By pulling wheel bogies over horizontal cables at the bottom of each suspender cable, the massive truss sections were rolled out towards the center of the bridge and then lifted into place by a traveling crane (Figure 28.28). Eight horizontal cables were run across the canyon with four beneath each of the main cables on each side for the bogies to roll along. Once the segments met their destination, a jack under a movable crane lifted the truss girder to the correct height and the procedure was repeated with the next segment (Figures 28.29 through 28.33).

The two tunnels on either side of the Aizhai Bridge allowed the engineers to use the mountain top for the location of one of the towers, reducing its height to just 165 ft (50 m), unusually short for a bridge with a span nearly a mile long. In addition to cost savings, the stubby support also allows the bridge to blend more naturally into its surroundings. The taller bridge tower is no less unique, with side-span cables that soar down the backside of a mountain, making first-time visitors quizzical as to what exactly lies ahead. With most of the structure hidden from view, the bridge comes as a jaw-dropping surprise whether you enter the canyon from either tunnel. Due to a gap of approximately 328 ft (100 m) between

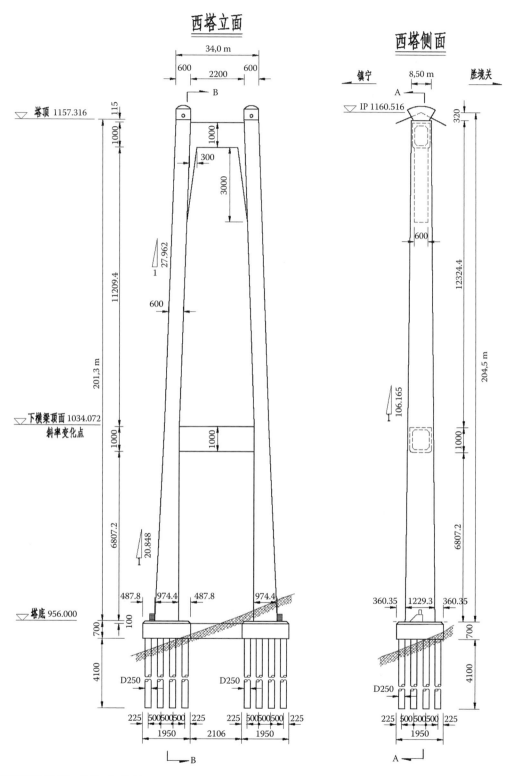

FIGURE 28.25 Balinghe Bridge tower.

FIGURE 28.26 **(See color insert.)** Balinghe cable-stayed bridge proposal.

FIGURE 28.27 Aizhai Bridge wide view.

the last truss suspenders and the tops of the bridge towers, the engineers added some additional ground-anchored suspenders to stabilize the two massive suspension cables and reduce any oscillations that could damage other components of the bridge. An overlook and visitors' center offer additional views of the broad valley (Figures 28.34 and 28.35).

One Aizhai Bridge proposal would have required an unusually long main span of 1300 m and a lengthy side span, but both towers would have been short (Figure 28.36). Another Aizhai Bridge proposal had a main span of 1228 m but required a tower almost 200 m high as well as an anchorage resting on a steep mountain slope (Figure 28.37).

FIGURE 28.28 Aizhai Bridge—men and bogies.

FIGURE 28.29 Aizhai Bridge—bogie view.

FIGURE 28.30 Aizhai Bridge—another bogie view.

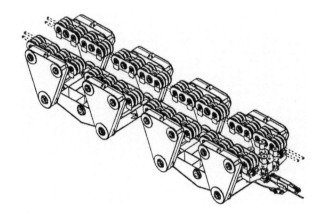

FIGURE 28.31 Aizhai Bridge—roller diagram.

FIGURE 28.32 Aizhai Bridge—first stiffening truss under construction.

FIGURE 28.33 Aizhai Bridge—stiffening truss under construction.

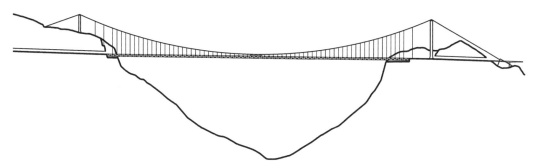

FIGURE 28.34 Aizhai Bridge elevation.

FIGURE 28.35 Aizhai Bridge short tower side.

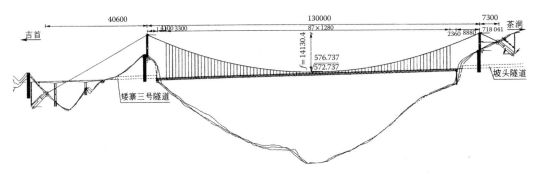

FIGURE 28.36 Aizhai Bridge proposal with long side span.

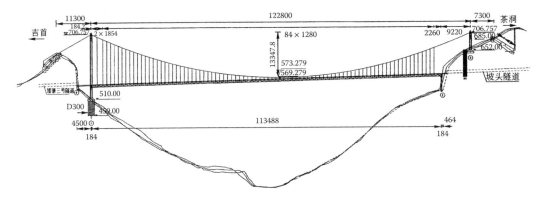

FIGURE 28.37 Aizhai Bridge proposal with 200 m tower.

28.8 Liuguanghe Bridge: The World's Highest Beam Bridge

Liuguanghe Bridge (Figure 28.38) entered the record books in 2001 when it became the world's highest bridge at 975 ft (297 m), toppling the 72-year-old record long held by Colorado's 955 ft (291 m) high Royal Gorge Bridge. Even though Liuguanghe's record would stand for just 2 years, it will always have the distinction of being the only beam bridge in history that held the top spot among high bridges. The decision not to use an arch or a suspension bridge was probably motivated by the deep height of the gorge, where the tall piers of a prestressed concrete beam bridge would be easier to construct since the two sides of the bridge could be cantilevered outward without any temporary cable stays or highlines.

The Liuguanghe Bridge is the crown jewel in a highway that is best described as a museum of high Chinese bridges. Located near the city of Guiyang in China's western province of Guizhou, this 100 mi (161 km) stretch of two-lane highway contains two suspension bridges, one 650 ft (198 m) and one 550 ft (168 m) high, as well as another concrete beam bridge 600 ft (183 m) high and two arches, 380 ft (116 m) and 360 ft (110 m) high. Although Liuguanghe Bridge is named after a nearby town, the bridge actually crosses the upper end of the Wujiang (Wu River), a large tributary of the Yangtze River.

The scale of the Liuguanghe Bridge (Figures 28.39 and 28.40) is not always evident from photographs until you realize that the main span of the bridge is 787 ft (240 m) between piers—longer than any beam bridge span that has ever been built in the United States. The pier on the west side of the bridge is the tallest point of the structure, standing 295 ft (90 m) in height. Resting on top of the two piers is a single-cell box girder with a height of 44 ft (13.4 m) over the piers and 13.5 ft (4.1 m) at midspan. On the southeast end of the bridge there is a temporary pullout along the shoulder to park. From there you can walk across the bridge and peer into the void over the 4 ft (1.5 m) high concrete barrier.

FIGURE 28.38 Liuguanghe Bridge overview 1.

FIGURE 28.39 Liuguanghe Bridge overview 2.

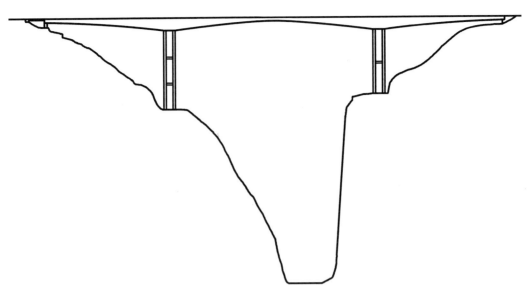

FIGURE 28.40 Liuguanghe Bridge elevation drawing.

28.9 Rotating a Railway Bridge over the Beipan River

When it opened in 2001, the 902 ft (275 m) high Beipanjiang (Beipan River) Railway Bridge (Yong 2002) became the highest arch bridge on earth, unseating the 23-year-old record long held by West Virginia's New River Gorge bridge. It also became the world's highest railway bridge, toppling the quarter-century record of the Mala Rijeka viaduct in Podgorica, Montenegro. In 2009, the Beipanjiang Railway Bridge lost its highest arch title to another Chinese span, the 965 ft (294 m) high Zhijinghe road bridge, while sometime after 2015, the opening of India's 1053 ft (321 m) high Chenab Bridge and a second high-speed railway bridge over the Beipanjiang will push it into third place overall among the world's highest railway bridges.

Located just west of the city of Liupanshui near the north end of China's Guizhou province on the Shuibai Railway, the Beipanjiang Railway Bridge (Figure 28.41) crosses an exceptionally deep gorge with vertical cliffs rising hundreds of feet from either side of the Beipan River. The bridge was the first of four Beipanjiang crossings to have been among the world's 12 highest. The word Beipanjiang (pronounced bay-pan-gee-ang) translates into North Winding River, with the word "bei" meaning north and "pan" meaning winding. Cutting a huge swath from the northwest end of Guizhou Province to the southwest, where it becomes the Hongshuihe River at the border of Guangxi Province, the Beipan River traverses through some of China's most spectacular mountain gorges. The river remains the biggest obstacle between the city of Guiyang and the city of Kunming.

The main 771 ft (235 m) arch was built using a method developed by the Chinese that had never been used before on a bridge so large. Instead of building two temporary towers to support a highline to assist in building the two sides of the arch outward until they met in the middle, the two halves of the bridge were built separately on falsework constructed just above ground on the hillsides at the edge of the canyon. Once completed, each side was then rotated horizontally outward over the river in one delicate maneuver and then connected at the crown. The central bearing located on top of each of the foundations consisted of a pair of closely fitted 11.5 ft (3.5 m) diameter concave spherical sections with a radius of 26 ft (8 m). On top of the lower bearing, between the two sections, 610 pieces of 60 mm × 18 mm teflon flakes were used to keep friction to a minimum. A massive water tank was installed on the back of the counter weighting pier to assist in finding an accurate center of gravity as well as preventing the system from overturning. Once the two halves of the arch were closed at the crown (Figure 28.42), the rotatable

FIGURE 28.41   Beipanjiang Bridge rotation.

FIGURE 28.42   Beipanjiang Bridge arch closure.

foundations were entombed in tons of concrete. This unique method of rotating the arch halves during construction has been used on other large arch bridges in China, including those that are built on flat terrain just above the level of the river.

The two steel arch ribs of the Beipanjiang Bridge, one of China's largest CFST bridges, were initially hollow during construction. Concrete was then pumped inside of them from the foundations upward to the crown. First used by the Chinese in 1990, they have refined and improved this technique and now use it on the majority of their steel arch bridges. Depending on the length of the span and the width of the bridge, different styles of tubing are used. For the Beipanjiang, a mix of a horizontal dumbbell and single tubes was adopted. Once hardened, the concrete solidifies and stiffens the arch, improving the compressive strength of the entire structure.

28.10 Canyon Crossings of Concrete: The Colorado and Wu River Bridges

Although they are located half a world apart from each other, the Mike O'Callaghan–Pat Tillman Memorial Bridge in the United States and the Jiangjiehe Bridge in China share one thing in common: they are two of the highest concrete arch bridges ever built. Nearly identical in height and length, the two spans are an interesting contrast in design.

The tight switchbacks that descend into the Colorado River canyon and across the Hoover Dam had long been a major bottleneck on busy Route 93, which links Las Vegas, Nevada with Kingman, Arizona and Interstate 10. Since the completion of the dam in 1935, large trucks had to negotiate several tight switchbacks along the two-lane road before dodging hordes of tourists visiting the Hoover Dam and visitor center. After 2001, trucks were diverted south to Laughlin on Route 95 but the congestion had reached a point where it became obvious a high-level bypass would be the only real solution. The resulting bridge opened in late 2010 as America's second-highest ever at 890 ft (271 m). Officially named the Mike O'Callaghan–Pat Tillman Memorial Bridge (Figure 28.43), the main span of 1060 ft (323 m) ranks seventh among all concrete arch bridges (Goodyear, Klamerus, and Turton 2005). Several steel and concrete arch designs were considered before it was decided to go with a composite concrete deck arch. They included a solid steel rib, a steel Vierendeel arch with inclined spandrels that would have seemed to radiate out of the arch in a sunburst pattern, and a steel trussed rib proposal similar to West Virginia's New River Gorge Bridge.

To construct the arch over the deep gorge of the Colorado River without falsework, the arch rib sections were cast in 24 ft (7 m) increments and were suspended over the canyon with cable stays until the two sides of the arch could be closed at the crown (Figure 28.44). The pier sections and pylons for the bridge were all built off site and trucked in. The project had its biggest setback on September 15, 2006, when all four of the 50-ton highline cable towers collapsed from the wind. The mishap delayed erection of the arch by nearly 2 years before a newly designed highline was built. Figure 28.45 shows a Hoover Dam Bridge strut and Figure 28.46 shows the Hoover Dam Bridge and Jiangjiehe Bridge elevation drawings. The best design trait of all might well have been the simple decision to include a walkway on the north side of the bridge. From here, the majesty of the Hoover Dam can finally be seen straight on from a vantage point normally reserved for planes or helicopters. Since the bridge crosses a state line, the construction (Figure 28.47) was undertaken by the Arizona and Nevada Departments of Transportation along

FIGURE 28.43 Hoover Dam Bridge construction.

FIGURE 28.44 Hoover Dam arch traveler.

FIGURE 28.45 Hoover Dam Bridge strut.

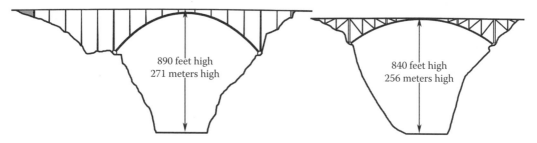

FIGURE 28.46 Hoover Dam Bridge and Jiangjiehe Bridge elevation drawings.

FIGURE 28.47   Hoover Dam Bridge completion view in July 2010.

FIGURE 28.48   Jiangjiehe Bridge view 1.

with several private contractors including T.Y. Lin International, Sverdrup Civil, Edward Kraemer & Sons, R.E. Monks Construction, Vastco, Obayashi Corp, and PSM Construction USA.

Completed in 1995, the Jiangjiehe Bridge (Figures 28.48 and 28.49) is a concrete truss-arch span that crosses 840 ft (256 m) above China's Wujiang (Wu River) near Weng'an in the western province of Guizhou. With its signature N-shaped spandrel openings, the concrete truss arch was a style of bridge that had become very popular in the 1990s, but has not been used much in China since 2000. As of 2008, China had more than 200 concrete arch bridges with spans in excess of 328 ft (100 m). Jiangjiehe Bridge

FIGURE 28.49 Jiangjiehe Bridge view 2.

was built by cantilevering both halves of the bridge out from the sides of the gorge. Temperature-related expansion and contraction of the massive 1083 ft (330 m) main span is allowed through a gap on either side of the center third of the bridge.

28.11 Upgrading Sfalassa: Italy's Mammoth Strut Frame Bridge

Crisscrossing the Lucanian and Calabrian Apennines mountain range in southern Italy is the Autostrada Napoli-Reggio Calabria or A3 motorway. Built in several sections throughout the 1970s, it stretches 300 mi (483 km) from Napoli to the tip of Italy's boot near the island of Sicily. With dozens of tunnels and towering bridges, it is a showcase of Italian engineering. The last few miles of the motorway traverse some of the most spectacular coastal terrain in all of Italy. Crossing the deepest of these ravines is the Sfalassa Gorge Bridge (Figure 28.50), the highest and longest-span frame bridge in the world. Rising 820 ft (250 m) above the canyon floor, the bridge was the third highest of any kind upon its opening in 1974. The main span measures 1181 ft (360 m) between the pins of the two 500 ft (152 m) long angled box beam struts.

A frame bridge combines elements of an arch bridge and a beam bridge. In an arch bridge, the support follows a continuous curve from one foundation to another. In a frame bridge, the road deck is supported by two inclined piers that are straight. These two struts usually support three horizontal beam spans that carry the roadway on top. To construct Sfalassa Bridge (Romeo 2009), the two struts were built vertically, like a 50-story skyscraper, and then lowered out over either side of the gorge to an angle of approximately 50 degrees. Held back by a large temporary truss and several cable stays, the struts finally supported the roadway after the central span was completed. The design is credited to Silvano Zorzi, Lucio Lonardo, and Sabatino Procaccia.

In recent years, parts of the A3 were becoming congested as well as unsafe with no shoulders and curves that are too tight. Now underway is a huge reconstruction of the most troublesome stretches. New tunnels, wider curves, and wider lanes and shoulders in both directions will make the original Autostrada Napoli-Reggio Calabria a thing of the past. As part of the reconstruction, Sfalassa Bridge was widened (Figure 28.51) in 2009 and 2010. A customized traveler moved across the older 63 ft (19 m) wide deck to add wider wings to the cantilevered roadway, making the new Sfalassa 82 ft (25 m) wide. For each of the approaches, a new three-span configuration with wider viaducts for the north and south lanes will replace the older spans (Figures 28.52 and 28.53).

FIGURE 28.50   Sfalassa Bridge.

FIGURE 28.51   Sfalassa Bridge widening.

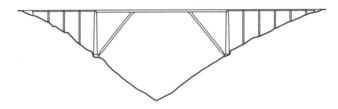

FIGURE 28.52   Sfalassa old elevation drawing.

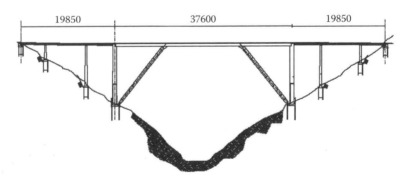

FIGURE 28.53   Sfalassa new elevation drawing.

28.12 Guizhou Family of Four Suspension Spans

No other region on earth has as many high bridges as China's western province of Guizhou. While the majority of these spans are on major four-lane highways, several high suspension bridges were constructed on secondary two-lane highways between 2001 and 2003. Since all four of these Guizhou suspension bridges were of a similar length, it was economical and efficient to have them designed at the same time by the China Zhongtie Major Bridge Reconnaissance & Design Institute Co., of Wuhan, Hubei. The bridges all share common traits including an extremely thin 2 ft (0.6 m) thick prestressed concrete slab deck and cable suspenders just 16 ft (5 m) apart. All four bridges are in similar gorge settings and all four rank among the world's 60 highest bridges.

The highest and longest of the four spans is the Beipanjiang (Beipan River) Suspension Bridge (Figure 28.54) that opened in late 2003 as the highest bridge in the world, with a deck 1200 ft (366 m) above the river. The Beipanjiang Bridge (Figure 28.55) also became the first bridge to break the 1000 ft and 300 m height thresholds as well as the first suspension bridge in the world to surpass the height of Colorado's Royal Gorge Bridge after a 74-year reign. The Beipanjiang Suspension Bridge would finally be surpassed in 2009 by the Siduhe Bridge in Hubei and the nearby Balinghe Bridge in Guizhou. The underside of the Beipanjiang Suspension Bridge shows the thin concrete deck (Figure 28.56). Despite its fall from the top spot among China's highest spans, the Beipanjiang Bridge of 2003 is still the most vertigo-inducing of all, with a drop-off that plummets into a void that seems to have no bottom, with the west side cliff nearly vertical for 800 ft (244 m).

Also completed in 2003 is the Azhihe (Azhi River) Suspension Bridge (Figure 28.57) near Changliuxiang, Guizhou. Like the Beipanjiang Suspension Bridge, the extremely thin 2 ft (0.6 m) thick prestressed concrete slab deck is supported by cable suspenders just 16 ft (5 m) apart. With an elevated side span on the north side of the bridge, the cable anchorages are about 100 ft (30 m) below the level of the road deck. The south side span of the bridge is less interesting but earns points for being at the entrance to a tunnel. Of course the real magic of Azhihe is the location of the bridge above a precipitously deep gorge with an unusually narrow throat. The Azhihe Bridge (Figures 28.58 and 28.59) spans an 810 ft (247 m) deep gorge on the highway between Anshun and Liupanshui in a remote part of Guizhou.

The Xixi Suspension Bridge (Figure 28.60) is one of many high bridges along the Guiyang-Bijie highway in China's Guizhou Province. This two-lane route is home to Liuguanghe, the highest beam bridge on earth. Also on this former toll road is the Luojiaohe Suspension Bridge. Both were completed in 2001 with the signature thin slab deck design of the China Zhongtie Major Bridge Reconnaissance and

FIGURE 28.54 Beipanjiang Suspension Bridge.

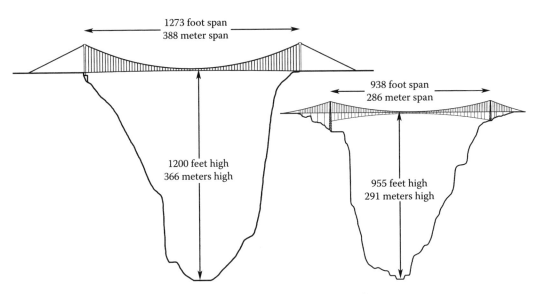

FIGURE 28.55 Beipanjiang Bridge elevation drawing with Royal Gorge Bridge.

FIGURE 28.56 Beipanjiang Bridge deck underside.

Design Institute. The Xixi Bridge (Figure 28.61) is most similar to Azhihe Bridge, with an elevated side span on the east side of the bridge that leaves the cable anchorages about 100 ft (30 m) below the level of the road deck. The gorge itself is most spectacular on the west side where it drops vertically to the water for nearly its entire 540 ft (165 m) height. Interestingly, a recently built dam just downstream of the bridge has created a reservoir below the bridge. Prior to the dam's construction, the bridge was over 700 ft (213 m) high! The main span measures 1109 ft (338 m) between towers.

FIGURE 28.57 Azhihe Bridge.

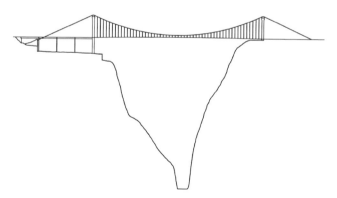

FIGURE 28.58 Azhihe Bridge elevation.

Located about 20 mi (32 km) west of the Xixi Bridge and on the Gui-Bi highway near Fafang, Guizhou, the slightly smaller Luojiaohe (Luojia River) Bridge (Figures 28.62 and 28.63) crosses a similar-sized gorge with a main span of 879 ft (268 m). A dam was also built downstream of Luojiaohe Bridge, reducing its height to 480 ft (146 m) at full pool level. The gorge itself is most spectacular on the west side, where it drops vertically to the water for nearly its entire height.

28.13 Crossing the Devil's Backbone: Mexico's Baluarte Bridge

When it was completed in 2012, the Baluarte River Bridge became the highest bridge in North America but also the highest cable-stayed bridge in the world. It is the crown jewel of the greatest bridge and tunnel highway project ever undertaken in North America. Known as the Durango–Mazatlán highway,

FIGURE 28.59 Azhihe Bridge deck view.

FIGURE 28.60 Xixi Bridge.

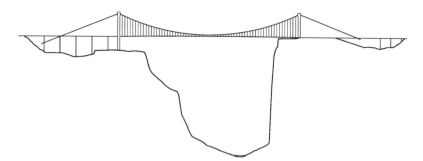

FIGURE 28.61   Xixi Bridge elevation drawing.

FIGURE 28.62   Luojiaohe Bridge.

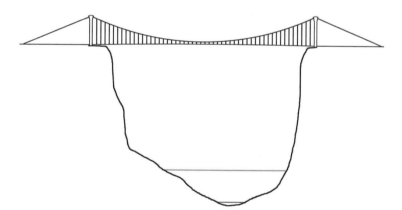

FIGURE 28.63   Luojiaohe Bridge elevation.

it is now the only crossing for more than 500 mi (800 km) between the Pacific coast and the interior of Mexico. The path of this new highway roughly parallels the famous Devil's Backbone, a narrow road that earned its nickname from the way it follows the precarious ridge crest of the jagged peaks of the Sierra Madre Occidental Mountains. The dangerous road is a seemingly endless onslaught of twisting, terrifying turns that are so tight there are times the road nearly spirals back into itself.

(a)

(b)

(c)

FIGURE 28.64   (a) Baluarte Bridge aerial view; (b) Baluarte Bridge tower and deck; (c) Baluarte Bridge traveler.

Figure 28.64 shows an aerial view of the Baluarte Bridge in the mountains near El Palmito, Sinaloa, Mexico. By cutting a safer, more direct route through the mountains, the highway department of Mexico hopes to improve trade and increase tourism between the city of Durango and the coastal city of Mazatlán (Figure 28.65). To achieve this connection, Mexican engineers were forced to design an *autopista* with no less than 63 tunnels—nearly 10 times more than have ever been built on any road in North America. For big bridge fans, the highway is no less amazing, with a parade of towering concrete beam bridges. Including Baluarte, there will be eight bridges that exceed 300 ft (90 m) in height, including Santa Lucia, Neverías, La Pinta, Chico, Botijas, Pueblo Nuevo, and El Carrizo. Only a few highways in China and on Italy's A3 have a greater collection of high bridges.

Forming the border between Sinaloa and Durango states, the Baluarte River is the most formidable obstacle on the route, with a gorge more than a quarter mile in height. To cross it, Mexican engineers decided to go with a cable-stayed bridge (Figure 28.66). It would allow construction to proceed outward from a single tower on either side of the canyon, avoiding the difficult and expensive construction of temporary falsework. The central span is composed of 36 steel roadway segments that are 12 m long with one set of stays per segment. Once completed, the final height of 1280 ft (390 m) will make it the second-highest

Mexico's 4 tallest bridges

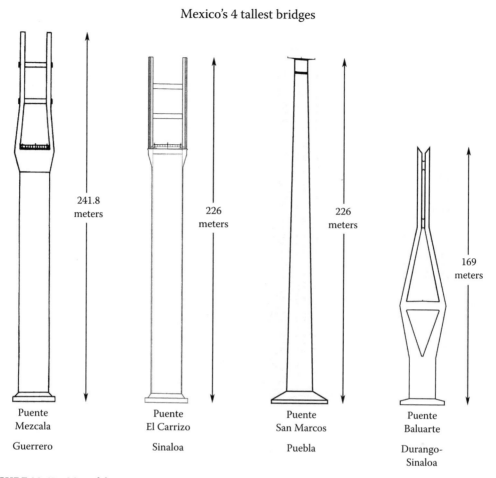

| Puente Mezcala | Puente El Carrizo | Puente San Marcos | Puente Baluarte |
| Guerrero | Sinaloa | Puebla | Durango-Sinaloa |

241.8 meters · 226 meters · 226 meters · 169 meters

FIGURE 28.65 Map of the Durango–Mazatlán Highway.

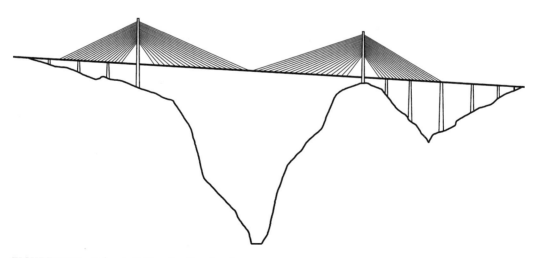

FIGURE 28.66 Baluarte Bridge elevation drawing.

roadway bridge on earth. It will also have the longest span of any cable-stayed bridge in North America at 1706 ft (520 m), exceeding the John James Audubon bridge in St. Francisville, Louisiana by 123 ft (37 m).

28.14 Mexico's Towering Titans: The San Marcos Bridge

Opened to traffic in 2013, the soaring San Marcos Bridge is the largest and tallest bridge on the final Nuevo Necaxa–Tihuatlán section of the México–Tuxpan highway. Extending from Mexico City to the Gulf of Mexico, the first and last thirds of the 182 mi (293 km) highway were finished in 2005. The difficulty was in completing the central Nuevo Necaxa–Ávila Camacho section that runs for 23 miles (37 km) along the mountainous San Marcos River gorge. To push this modern *carretera* through the steep terrain, engineers constructed six tunnels and several high bridges. Three exceed 100 meters in height, including Xicotepec, Texcapa I, and San Marcos.

Curving 722 feet (220 meters) above the San Marcos River, the prestressed concrete beam bridge (Figures 28.68 and 28.69) will have the second-highest bridge pier in the world after the Millau Viaduct in France. Rising 682 ft (208 meters) from the top of the foundation to the underside of the beam, pier number 4 supports 2 spans of 180 meters. Figure 28.67 shows a comparison of the world's 10 highest bridge piers.

Interestingly, the original design for the San Marcos River crossing was for a fin back bridge (Figure 28.70). A prestressed beam bridge with a highly variable depth of prestressing, the fin back is unique for having the internal cables at their highest as they pass over the piers, enclosed in a wall or fin of concrete. The hump-like profile may look similar to a cable-stayed or extradosed bridge, but the engineering has more in common with a beam bridge. Many consider the lower profile to be more attractive than a conventional prestressed beam bridge. Mexico has two other large fin back bridges: the Texcapa Bridge, also located on the México–Tuxpan highway, and the Papagayo Bridge located on the México City–Acapulco highway. Opened in 2005, the Texcapa Bridge (Figures 28.71 and 28.72) is the largest of Mexico's fin back bridges with a main span of 561 ft (171 meters) between the center of the piers. The first fin back bridge was built in Nuremberg, Germany for a commuter railway in 1969. In the United States, the only one is the Barton Creek Bridge near Austin, Texas, built in 1987 with a main span of 340 ft (103.5 meters).

28.15 Jungle Pipeline: The Hegigio Gorge Bridge

Located deep in the Southern Highlands Province of Papua New Guinea is the Hegigio Gorge Suspension Bridge (Figure 28.73), the highest pipeline bridge in the world. Upon its opening in November 2005, the 1289 ft (393 m) high bridge also became the highest bridge of any kind in the world, taking the record away from China's two-year-old, 1200 ft high (366 m) Beipanjiang Suspension Bridge, built in 2003. China took the title back again in 2009 when the 1627 ft (496 m) high Siduhe suspension bridge opened. Some may argue that the Hegigio span is not a true bridge since it was not built for people, but this third of a mile long web of wire and steel is a substantial structure capable of supporting several hundred tons of weight. The precarious look of the light and airy crossing is accentuated by the steep canyon walls that vary between 70 and 90 degrees. The bridge is a major component of the Southeast Mananda field, a remote expanse of oil that was discovered in 1991. The small size of the field and the extremely remote jungle terrain made it a difficult petroleum project to undertake, with many technical obstacles to overcome. The challenge of economically extracting the oil was undertaken by a partnership of four oil companies: Oil Search Ltd., AGL Gas Developments, Merlin Petroleum, and Petroleum Resources Kutubu.

Well fluids are transported from the Southeast Mananda field to the Agogo production facility via a 10 mi pipeline (16 km). The biggest physical barrier along the route is the deep limestone gorge of the Hegigio River. Once the fluids cross over the river, they flow into the Agogo facility where the oil is then fed into the main Moran pipeline where it travels more than 160 mi (258 km) further to the Kumul terminal in the Gulf of Papua. The bridge also supports a gas lift line that goes from the Agogo facility to the Southeast Mananda field. The bridge is designed to support two future flow lines when more

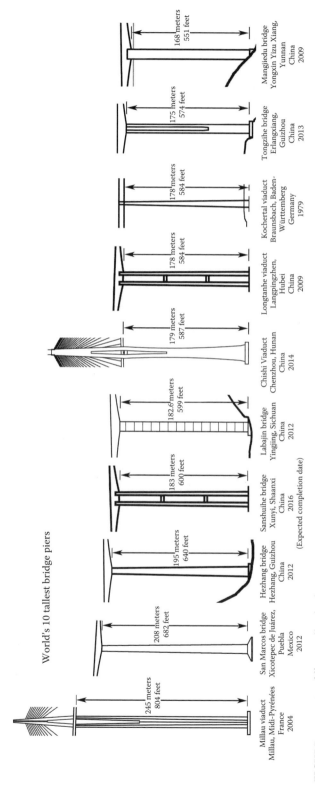

FIGURE 28.67 World's ten tallest bridge piers drawing.

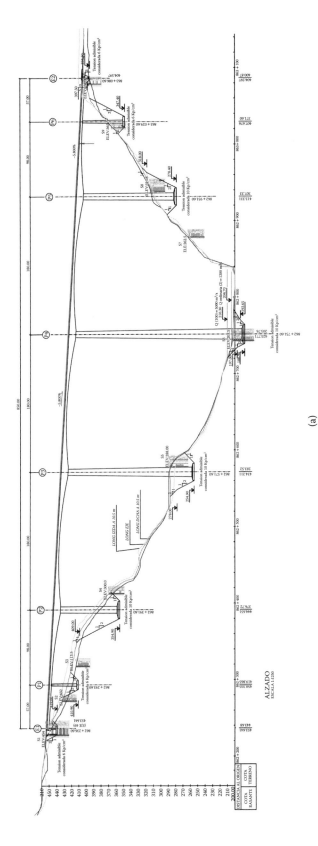

FIGURE 28.68 San Marcos Bridge elevation drawing with foundations. (a) Elevation. (b) Typical section at midspan. (c) Typical section at pier.

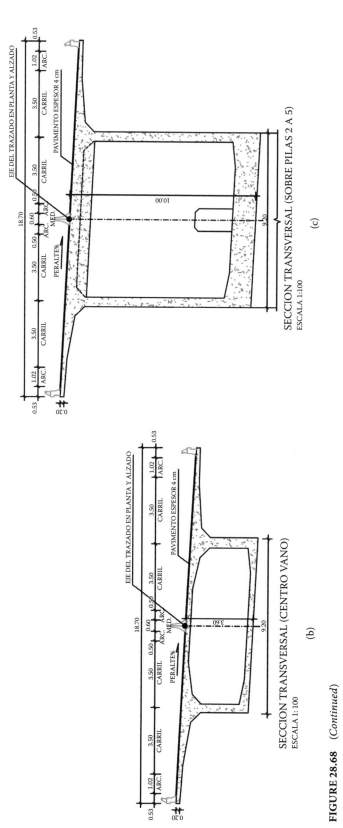

FIGURE 28.68 *(Continued)*

FIGURE 28.69   San Marcos Bridge aerial view.

fields are developed along the Mananda ridge. Oil first went across the bridge in March of 2006, with full production capacity achieved a few months later.

The assignment of building a bridge across the river went to Kellogg Brown & Root, a company owned by Halliburton that has constructed many pipeline-related engineering projects over the years. Designed by engineer Ken Ross and built by Clough construction, the main span consists of three catenary cables, two of them horizontal and one vertical. The vertical cable consists of 87 individual wire strands that are 15.2 mm thick, while each of the wind cables consists of 35 strands. The south side of the bridge is comprised of a 108 ft (33 m) high tower while the north side cable terminates just above the ground at the top of a rocky outcropping. The steep terrain on the north side required a 100 ft (30 m) high terrace to be carved into the rock in 16 ft (5 m) wide steps so the wind anchors could be installed. The north side of the bridge was also difficult to reach since there was no road access in place like there was on the south side. The first two cables were strung across by a helicopter while successively thicker cables were then winched across. Another construction hurdle came from the harsh weather, which even today can be problematic to the operation of the pipeline. There is no pedestrian walkway on the bridge. If necessary, personnel access for maintenance is from a trolley that rides along two rails. Figures 28.74 through 28.78 show the bridge under construction.

28.16 A Grande Crossing on Reunion Island

Réunion Island is a French territory located east of Madagascar Island in the Indian Ocean. Just 35 mi (56 km) across, the region has become a popular tourist getaway as well as a permanent home to more than 800,000 residents. Figure 28.79 shows a construction view of the Grande Ravine Bridge on Réunion Island in the Indian Ocean. Travel between the many coastal communities on Réunion Island has always been through a network of older, two-lane roads that traverse in and around many deep rifts cut from rain runoff tumbling down the tall volcanic peaks that created the island. To relieve congestion on the west side, it was decided to build a 30 mi (50 km) north–south highway called the Route des Tamarins. Crossing dozens of huge ravines, the new route required the construction of three tunnels, nine interchanges, and four major bridges including the Saint Paul, the Trois Bassins, and the Ravine Fontaine viaducts. The longest and highest crossing of all is the one over Grande Ravine (Figure 28.80).

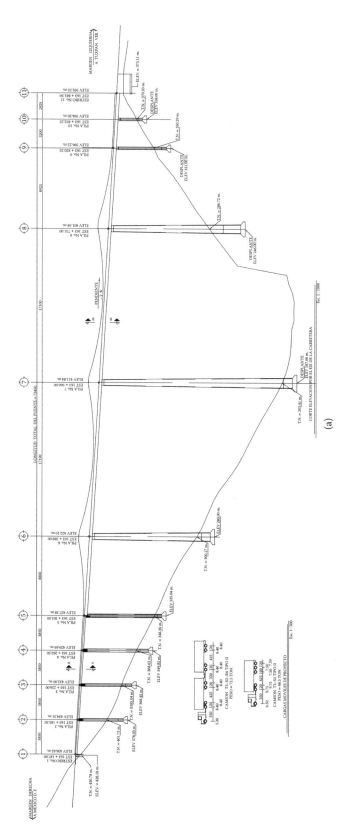

FIGURE 28.70 San Marcos Bridge. (a) Fin back elevation. (b) Section B-B at main spans. (c) Section A-A at side spans.

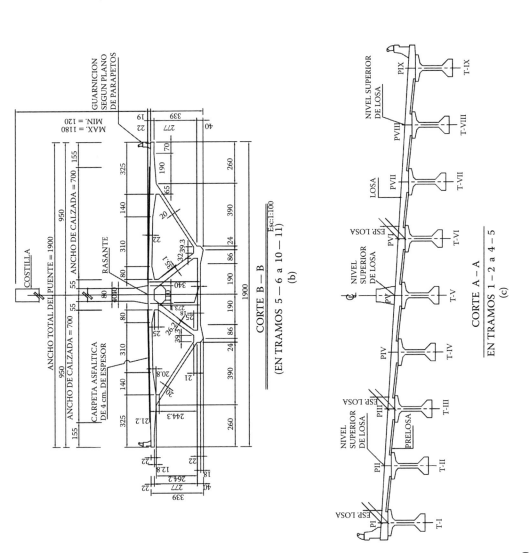

FIGURE 28.70 *(Continued)*

FIGURE 28.71 Texcapa Bridge.

FIGURE 28.72 Texcapa Bridge by Grupo Triada.

Dropping almost vertical for much of its 532 ft (162 m) depth, the Grande Ravine site is perfect for a frame bridge. The design chosen by the engineers is a sleek, stealthy-looking span with a 940 ft (286.5 m) long span box girder resting on two angled struts of just 20 degrees—an incline so shallow they almost look as level as the road deck! The concrete struts are part of a cantilever that is counterweighted on the back sides by two large abutments filled with tons of soil. Like many strut frame bridges that are hard to categorize among bridge types, Grande Ravine Bridge (Figure 28.81) is especially complex, with an arch effect only occurring under service loads. To construct the main span without using wind-prone towers and a highline, the famous French bridge company Freyssinet

FIGURE 28.73   Hegigio Gorge pipeline bridge aerial view.

FIGURE 28.74   Hegigio Gorge Bridge tower construction.

FIGURE 28.75 Hegigio Gorge Bridge cable connection.

FIGURE 28.76 Hegigio Gorge Bridge wind cables.

FIGURE 28.77 Hegigio Gorge Bridge anchorage.

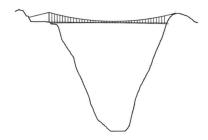

FIGURE 28.78   Hegigio Gorge Bridge elevation.

FIGURE 28.79   Grande Ravine Bridge construction.

FIGURE 28.80   Grande Ravine Bridge.

designed a unique stay cable system to support the two half decks as they were launched out from either side of the canyon (Russell 2008).

Less obtrusive than an arch, but just as elegant, the Grande Ravine is a rare type of bridge we can only hope to see more of in the future. The talent behind the Grande Ravine Bridge includes designer Setec, architect Alain Spielmann, and builders Dodin, VINCI Construction, Eiffel, and Freyssinet.

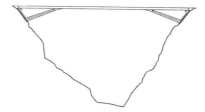

FIGURE 28.81   Grande Ravine elevation.

References

Brown, D. J. (2005). *Bridges: Three Thousand Years of Defying Nature*, MBI Publishing Company, St. Paul, MN.

Chen, B. (2008). *New Development of Long-Span CFST Arch Bridges in China*, Fuzhou University, Fujian, China.

Goodyear, D., Klamerus, B., Turton, R. (2005). New Colorado River Arch Bridge at the Hoover Dam (PDF). *International Bridge Conference 2005* (Pittsburgh: Engineers' Society of Western Pennsylvania): 16–18.

Leonhardt, F. (1984). *Bridges: Aesthetics and Design*, The MIT Press, Cambridge, MA, USA, P217.

Romeo, F. (2009). *The Giant Of Costa Viola*, Strade & Autostrade, Italy.

Russell, H. (2008). Ravine Rendez-vous. *Bridge Design & Engineering Magazine*, November, England.

Xie, G.-Y. (2009). *Key Techniques for Construction of Zhijing River Concrete-Filled Steel Tube Arch Bridge*, Chang'an University, Xi'an, China.

Yan, T. (2007). *Analysis of Bridgehead Slope Stability for the Mashuihe Bridge*, Transportation Science & Technology, China.

Yong, X. (2002). *Steel Tube with Concrete Filling Arch Design of Beipanjiang Large Bridge*, The 2nd Railway Survey & Design Institute, Chengdu, Sichuan, China.

Relevant Websites

http://www.highestbridges.com

Siduhe Bridge http://www.ilib.cn/A-ISSN~0451-0712%282006%2903-0184-04.html

Balinghe Bridge http://www.ilib.cn/A-QCode~zgkjxx200620014.html

http://www.ilib.cn/A-ISSN~1009-6825%282009%2911-0098-03.html

Longtanhe Bridge http://www.ilib.cn/A-ISSN~1005-6033%282005%2912-0238-03.html

29

Longest Bridges and Bridge Spans

Lian Duan
*California Department
of Transportation*

29.1 Introduction

The total length of a bridge and the longest span (main span) length are two major features of a bridge project. However, there is not a standard method to measure the total length of a bridge. "Span" usually refers to a portion of bridge length without any direct supports and its length is measured from the centerlines of its supports. The total length often reflects a project size, while the span length commonly correlates with the engineering complexity involved in designing and constructing of the bridge. Based on the best of information available, Section 29.2 lists the top 100 longest bridges in the world. Sections 29.3 through 29.11 list the top 20 longest bridge spans for various bridge types including suspension bridges, cable-stayed bridges, extradosed bridges, truss bridges, arch bridges (steel, concrete filled in steel tube, concrete, and masonry), girder bridges (steel and concrete), movable bridges (vertical lift, swing and bascule), stress ribbon bridges, and timber bridges. Section 29.12 lists the top 15 longest floating bridges. It should be noted that these tables are just for information only, not ranking bridges. Also, the lengths are reported in both SI units (meters) and U.S. customary units (feet) and may not be precise due to conversion round off. All Internet references were accessed on May 10, 2013. Abbreviations such as concrete filled in steel tube (CFST), cast-in-place (CIP), prestressed (PS), precast (PC), high speed rail (HSR), railway (Rwy), highway (Hwy), and pedestrian (Ped) are used in following tables.

29.2 Top 100 Longest Bridges

Rank	Bridge Name	Total Length m (ft)	Structure Features	Longest Span m (ft)	Year Opened	Usage	Country/ Area	References
1	Dankun Bridge	164,800 (540,682)	PS box girder	80 (263)	2011	HSR	China	Dankun Bridge (2013)
2	Bagua Mt.–Renwu Viaduct	157,315 (516,125)	PS box girder	30.35 (99.6)	2007	HSR	Taiwan	Taiwan HSR (2013)
3	Alf Priolo Bridge	131,966 (432,960)			2014	Hwy	UK	Alf Priolo Bridge (2013)
4	Tianjin Bridge	113,690 (373,000)	PS box girder		2011	HSR	China	Tianjin Bridge (2013)
5	Cangde Bridge	105,900 (347,441)	PS box girder	128 (420)	2011	HSR	China	Beijing–Shanghai HSR (2013)
6	Huai River Bridge	85,400 (280,184)	PS box girder		2011	HSR	China	Beijing–Shanghai HSR (2013)
7	Weinan We River Bridge	79,732 (261,588)	PS box girder	80 (263)	2010	HSR	China	Zhengzhou–Xi'an HSR (2013)
8	Jiaxing-Shaoxing River-Sea-Crossing Bridge	69,462 (227,894)	PS box girder/ Cable-stayed	428 (1,404)	2013	Hwy	China	Qin, et al. (2013) Chapter 20
9	Sui River Bridge	65,600 (215,223)	PS box girder	80 (263)	2011	HSR	China	Beijing–Shanghai HSR (2013)
10	Bang Na Expressway	54,000 (177,165)	PS box girder	42 (138)	2000	Hwy	Thailand	Limsuwan (2013) Chapter 5
11	Beijing Bridge	48,153 (157,982)	PS box girder, arch	108 (354)	2011	HSR	China	Beijing Bridge (2013)
12	Qingdao Bay Bridge	41,580 (136,417)	Box girder, suspension	260 (853)	2011	Hwy	China	Qingdao Bay Bridge (2013)
13	Qatar Bahrain Causeway	40,000 (131,200)	Girder, cable-stayed	400 (1312)	2015	Hwy/Rwy	Qatar/ Bahrain	Qatar Bahrain Causeway (2013)
14	Sri Rat Expressway	38,500 (126,312)	PS girder	45 (148)	1993	Hwy	Thailand	Limsuwan (2013) Chapter 5
15	Lake Pontchartrain Causeway	38,442 (126,122)	Trestle, bascule two parallel	46 (151)	1956 (SB) 1969 (NB)	Hwy	USA	Lake Pontchartrain Causeway (2013)
16	Manchac Swamp Bridge	36,690 (120,374)	Concrete trestle, bascule		1979	Hwy	USA	Manchac Swamp Bridge (2013)
17	Yangcun Bridge	35,812 (117,493)	PS box girder	32 (105)	2007	HSR	China	Beijing–Tianjin Intercity Railway (2013)
18	Hangzhou Bay Bridge	35,673 (117,037)	Box girder, cable-stayed	448 (1470)	2007	Hwy	China	Hangzhou Bay Bridge (2013)
19	Runyang Bridge	35,660 (116,995)	PS box girder, cable-stayed, suspension	1490 (4889)	2005	Hwy	China	Runyang Bridge (2013)
20	Donghai Bridge	32,500 (106,626)	Box girder, suspension	420 (1378)	2005	Hwy	China	Donghai Bridge (2013)
21	Emsland Test Facility	31,500 (103,347)			1985	HSR (Maglev)	Germany	Emsland Test Facility (2013)
22	Shanghai Maglev Line	30,500 (985,892)			2004	HSR (Maglev)	China	Shanghai Maglev Line (2013)
23	Louisiana Highway 1 Expressway	30,578 (100,320)	Elevated highway		2011	Hwy	USA	LA 1 Project (2013)
24	Atchafalaya Basin Bridge	29,290 (96,095)	Two parallel bridges		1973	Hwy	USA	Atchafalaya Basin Bridge (2013)
25	Fouth Nanjing Yangtze River Bridge	28,996 (95,131.2)	Three spans	1418 (4652.2)	2013	Hwy	China	Fourth Nanjing Yangtze River Bridge (2013)

(Continued)

Rank	Bridge Name	Total Length m (ft)	Structure Features	Longest Span m (ft)	Year Opened	Usage	Country/ Area	References
26	Louisiana Highway 1 Bridge	28,968 (95,040)			2011	Hwy	USA	LA 1 Project (2013)
27	Bangkok Mass Transit System	28,700 (94,160)	PS box girder	37 (121)	1999	Rwy	Thailand	Limsuwan (2013) Chapter 5
28	Suvarnabhumi Airport Link	28,600 (93,832)			2011	Rwy	Thailand	Suvarnabhumi Airport Link (2013)
29	Yanshi Bridge	28,543 (93,645)	PS box girders	32 (105)	2009	HSR	China	Yanshi Bridge (2013)
30	Chesapeake Bay Bridge Tunnel	28,324 (92,928)	Girder, trestle		1964 (northbound) 1999 (southbound)	Hwy	USA	CBBT (2013)
31	Uttraphimuk Tollway	28,000 (91,864)	Precast T-girder	35 (115)	1994	Hwy	Thailand	Limsuwan (2013) Chapter 5
32	Qingcang Bridge	27,900 (91,535)	PS box girder		2011	HSR	China	Beijing–Shanghai HSR (2013)
33	Jingtang Bridge	26,540	Cable-stayed	620 (2034)	2009	Hwy	China	Jingtang Bridge (2013)
34	Jinbin Light Rail No. 1 Bridge	25,313 (83,048)	PS box girder, composite girder	62 (203)	2004	Metro	China	Sun and Li (2003)
35	King Fahd Causeway	25,000 (82,020)			1986	Hwy	Saudi Arabia and Bahrain	King Fahd Causeway (2013)
36	Penang Second Bridge	23,370 (76,673)	Cable-stayed	250 (820)	2012	Hwy	Malaysia	Penang Second Bridge (2013)
37	Yunzaobang Bridge	23,300 (76,444)	PS box girder		2011	HSR	China	Beijing–Shanghai HSR (2013)
38	Hong Kong– Zhuhai–Macau Bridge	22,800 (74,803)	Girder, cable-stayed	460 (1509)	2016	Hwy	China	Hong Kong– Zhuhai–Macau Bridge (2013)
39	Deyu Bridge	22,100 (72,507)	PS box girder		2011	HSR	China	Beijing–Shanghai HSR (2013)
40	Liangshui River Bridge	21,563 (70,745)	PS box girder		2008	HSR	China	Liangshui River Bridge (2013)
41	Incheon Bridge	21,384 (70,157)	Cable-stayed	800 (2625)	2009	Hwy	Korea	Incheon Bridge (2013)
42	Second Nanjing Yangtze River Bridge	21,337 (70,003)	Cable-stayed	628 (2060)	2001	Hwy	China	Yangtze River Bridges (2013)
43	Dawen River Bridge	21,100 (69,226)	PS box girder	80 (263)	2011	HSR	China	Beijing–Shanghai HSR (2013)
44	Sixth October Bridge	20,500 (67,257)		141 (463)	1996	Hwy	Egypt	Bakhoum (2013)
45	Baltic Sea Bridge	20,000 (66,617)			2018	Hwy/Rwy	Germany/ Denmark	Baltic Sea Bridge (2013)
46	C215 Viaduct	20,000 (66,617)	PS box girder		2005	HSR	Taiwan	C215 Viaduct (2013)
47	Youngding New River Bridge	19,725 (64,715)	PS box girder	32 (105)	2010	HSR	China	Youngding New River Bridge (2013)
48	Hami Bridge	19,287 (63,276)	PS box girder	32 (105)	2013	Rwy	China	Hami Bridge (2013)
49	Chennai Port–Maduravoyal Expressway	19,000 (62,336)			2013	Hwy	India	Chennai Port–Maduravoyal Expressway (2013)

(Continued)

(Continued)

Rank	Bridge Name	Total Length m (ft)	Structure Features	Longest Span m (ft)	Year Opened	Usage	Country/ Area	References
50	Bonnet Carré Spillway Bridge of I-10	17,702 (58,077)			1960s	Hwy	USA	Longest Bridges (2013)
51	Mumbai Trans-Harbor Link	17,500 (57,415)			2014	Hwy	India	MSRDC (2013)
52	Vascoda Gama Bridge	17,200 (56,430)	Cable-stayed	420 (1,378)	1998	Hwy	Portugal	Vascoda Gama Bridge (2013)
53	Si River Bridge	17,185 (56,381)		56 (184)	2011	HSR	China	Beijing–Shanghai HSR (2013)
54	Saigon–Trung Luong Skyway	16,000 (52,493)			2010	Hwy	Vietnam	Longest Bridges (2013)
55	Lianggang Bridge	15,672 (51,417)	PS box girder		2011	HSR	China	Lianggang Bridge (2013)
56	Third Nanjing Yangtze River Bridge	15,600 (51,181)	Cable-stayed	648 (2126)	2005	Hwy	China	Yangtze River Bridges (2013)
57	Beijing Ring Road Overcrossing Bridge	15,595 (51,165)	PS box girder	128 (420)	2007	HSR	China	Beijing–Tianjin Intercity Railway (2013).
58	Dashengguan Bridge	14,789 (48,520)	Girder, truss-arch	336 (1102)	2010	HSR	China	Dashengguan Bridge (2013)
59	Jinpu Railway Overcrossing Bridge	14,121 (46,329)			2011	HSR	China	Jinpu Railway Overcrossing Bridge (2013)
60	Boromratcha- chonnanee	14,000 (45,932)	Precast I-girder	32.5 (107)	1998	Hwy	Thailand	Limsuwan (2013) Chapter 25
61	Kama Bridge	13,967 (45,824)			2002	Hwy	Russia	Kama Bridge (2013)
62	San Francisco– Oakland Bay Bridge	13,519 (44,352)	Double decks, suspension, truss, girder	704 (2310)	1936	Hwy	USA	SFOBB (2013b)
63	Penang Bridge	13,500 (44,291)	Cable-stayed	225 (738)	1985	Hwy	Malaysia	Penang Bridge (2013)
64	Kam Sheung Road–Tuen Mun Viaduct	13,400 (43,963)			2003	Rwy	China	Longest Bridges (2013)
65	Wuppertal Schwebebahn	13,300 (43,635)		33 (108)	1903	Suspended monorail track	Germany	Wuppertal Schwebebahn (2013)
66	Rio–Niterói Bridge	13,290 (43,602)	PS box girder	300 (984)	1974	Hwy	Brazil	Vasconcelos, et al. (2013).
67	Seto–Ohashi Bridge	13,100 (42,979)	Double decks, cable-stayed, suspension, girders	1118 (3668)	1988	Hwy/HSR	Japan	Seto–Ohashi Bridge (2013)
68	Bhumibol Bridge	13,000 (42,651)	Cable-stayed, girders	398 (1306)	2006	Hwy	Thailand	Bhumibol Bridge (2013)
69	Okinawa Monorail Bridge	13,000 (42,651)	Steel box		2003	Rwy	Japan	JASBC (2004)
70	New Ulyanovsk Bridge	12,980 (42,585)	Truss	220 (722)	2009	Hwy and light metro	Russia	New Ulyanovsk Bridge (2013)
71	Confederation Bridge	12,900 (42,323)	PS girder	250 (820)	1997	Hwy	Canada	Confederation Bridge (2013)
72	Novyi SaratovskiyBridge	12,800 (41,995)			2000	Hwy	Russia	Novyi Saratovskiy Bridge (2013)

(Continued)

Rank	Bridge Name	Total Length m (ft)	Structure Features	Longest Span m (ft)	Year Opened	Usage	Country/ Area	References
73	Qinhuai River Bridge	12,626 (41,424)		80 (263)	2011	HSR	China	Beijing–Shanghai HSR (2013)
74	Yellow Sea Bridge	12,600 (41,339)	Girder		2008	Hwy	China	Yellow Sea Bridge (2013)
75	Hanzhuang Canal Bridge	12,480 (40,945)	PS girder	100 (328)	2011	HSR	China	Hanzhuang Canal Bridge (2013)
76	Quanzhou Bay Bridge	12,451 (40,850)	PS girder, cable-stayed	400 (1312)	2015	Hwy	China	Quanzhou Bay Bridge (2013)
77	Jubilee Parkway	12,070 (39,600)	Concrete girder		1978	Hwy	USA	Jubilee Parkway (2013)
78	Jinghang Canal Bridge	11,885 (38,993)	PS concrete girder, CFST arch	180 (591)	2011	HSR	China	Jinghang Canal Bridge (2013)
79	Third Mainland Bridge	11,800 (38,714)			1990	Hwy	Nigeria	Third Mainland Bridge (2013)
80	Qingshui River Bridge	11,700 (38,386)	PS girder		2006	Rwy	China	Qingshui River Bridge (2013)
81	Leziria Bridge	11,670 (38,287)		133 (436)	2007		Portugal	Leziria Bridge (2013)
82	P.V. Narasimha Rao Elevated Expressway	11,600 (38,058)			2009	Hwy	India	Elevated Expressways in Hyderabad (2013)
83	San Mateo–Hayward Bridge	11,265 (36,959)	Steel girder, concrete trestle	229 (750)	1967	Hwy	USA	San Mateo–Hayward Bridge (2013)
84	Mercosur Bridge	11,125 (36,499)	Cable-stayed, truss	360 (1181)	2011	Hwy/Rwy	Venezuela	Mercosur Bridge (2013)
85	Seven Mile Bridge	10,931 (35,862)	PS box girder	41 (135)	1982	Hwy	USA	Seven Mile Bridge (2013)
86	Changdong Yellow River Bridge	10,282 (33,734)	Truss, girder		1986	Rwy	China	Changdong Yellow River Bridge (2013)
87	Wuhu Yangtze River Bridge	10,521 (34,518)	Truss, cable-stayed	312 (1024)	2000	Hwy/Rwy	China	Wuhu Yangtze River Bridge (2013)
88	Wuhan Yangluo Yangtze River Bridge	10,000 (32,808)	Suspension	1280 (4200)	2007	Hwy/Rwy	China	Wuhan Yangluo Yangtze River Bridge (2013)
89	Bangalore Elevated Tollway	9,985 (32,759)	PS box girder		2010	Hwy	India	Hosur Road (2013)
90	Shanghai Yangtze River Bridge	9,970 (32,710)	Cable-stayed, PS box girder	730 (2395)	2009	Hwy/future Rwy	China	Shanghai Yangtze River Bridge (2013)
91	General W.K. Wilson Jr. Bridge	9,786 (32,106)	Girder, arch		1978	Hwy	USA	Top 10 Longest Bridges in USA (2013)
92	Norfolk Southern Lake Pontchartrain Bridge	9,334 (30,624)			1995	Rwy	USA	Norfolk Southern Lake Pontchartrain Bridge (2013)
93	Chacahoula Swamp Bridge	9,005 (29,544)			1995	Hwy	USA	Longest Bridges (2013)
94	Richmond–San Rafael Bridge	8,851 (29,040)	Truss	317 (1040)	1956	Hwy	USA	Richmond–San Rafael Bridge (2013)
95	El Mansoura Bridge	8,750 (28,707)					Egypt	*Bakhoum (2013)* Chapter 26

(Continued)

(Continued)

Rank	Bridge Name	Total Length m (ft)	Structure Features	Longest Span m (ft)	Year Opened	Usage	Country/ Area	References
96	I-10 Twin Span bridge	8,690 (28,512)	Girder, bascule		1962 (original); 2009 (new westbound); 2011 (new eastbound)	Hwy	USA	I-10 Twin Span bridge (2013)
97	General Rafael Urdaneta Bridge	8,678 (28,471)	PS concrete, cable-stayed	235 (771)	1962	Hwy	Venezuela	General Rafael Urdaneta Bridge (2013)
98	Virginia Dare Memorial Bridge	8,369 (27,456)			2002	Hwy	USA	Virginia Dare Memorial Bridge (2013)
99	Yangpu Bridge	8,354 (27,408)	Cable-stayed	602 (1975)	1993	Hwy	China	Yangpu Bridge (2013)
100	Xiasha Bridge	8,230 (27,001)	PS concrete	232 (761)	2002	Hwy	China	Xiasha Bridge (2013)

29.3 Top 20 Longest Suspension Bridge Spans

Rank	Bridge Name	Main Span Length m (ft)	Structure Features	Total Length m (ft)	Year Opened	Usage	Country	References
Earth-Anchored Suspension Bridges								
1	Akashi Kaikyō Bridge	1991 (6532.2)	Stiffening truss	3,911 (12,831.4)	1998	Hwy	Japan	Nagai, et al. (2013) Chapter 23
2	Xihoumen Bridge	1650 (5413.4)	Steel box girder	5,300 (17,388.5)	2009	Hwy	China	Qin et al. (2013) Chapter 20
3	Great Belt Bridge (East Bridge)	1624 (5328.1)	Steel box girder	6,790 (22,276.9)	1998	Hwy	Denmark	Great Belt Fixed Link (2013)
4	Yi Sun-sin Bridge	1545 (5068.9)	Steel box girder	6,600 (21,653.5)	2012	Hwy	Korea	Yi Sun-sin Bridge (2013)
5	Runyang Bridge	1490 (4888.5)	Steel box girder	35,660 (116,995)	2005	Hwy	China	Qin et al. (2013) Chapter 20
6	Fourth Nanjing Yangtze River Bridge	1418 (4652.2)	Three spans	28,996 (95,131.2)	2013	Hwy	China	Fourth Nanjing Yangtze River Bridge (2013)
7	Humber Bridge	1410 (4626)		2219.9 (7283)	1981	Hwy	UK	Humber Bridge (2013)
8	Jiangyin Bridge	1385 (4453)	Steel box girder	3,071 (10,075.5)	1999	Hwy	China	Jiangyin Bridge (2013)
9	Tsing Ma Bridge	1377 (4517.7)	Stiffening truss	2200 (7217.8)	1997	Hwy/Rwy	China	Tsing Ma Bridge (2013)
10	Hardanger Bridge	1310 (4297.9)	Steel box girder	1380 (4527.6)	2013	Hwy	Norway	Hardanger Bridge (2013)
11	Verrazano Narrows Bridge	1298.5 (4260)	Double decks	2039.1 (6690)	1964	Hwy	USA	Verrazano Narrows Bridge (2013)
12	Golden Gate Bridge	1280.2 (4200)	Stiffening truss	2788.9 (9150)	1937	Hwy	USA	Lwin and Kulicki (2013) Chapter 2

(Continued)

Rank	Bridge Name	Main Span Length m (ft)	Structure Features	Total Length m (ft)	Year Opened	Usage	Country	References
13	Yangluo Bridge	1280 (4199.5)	Steel box girder	10,000 (32,808.4)	2007	Hwy	China	Yangluo Bridge (2013)
14	High Coast Bridge	1210 (3969.8)	Steel box girder	1867 (6125.3)	1997	Hwy	Sweden	High Coast Bridge (2013)
15	Aizhai Bridge	1176 (3858.3)	Stiffening truss	1534 (5032.8)	2012	Hwy	China	Aizhai Bridge (2013)
16	Mackinac Bridge	1158.2 (3800)	Stiffening truss	8,038.2 (26,372)	1957	Hwy	USA	Mackinac Bridge (2013)
17	Huangpu Bridge	1108 (3635.2)		7,016 (23,018.4)	2008	Hwy	China	Huangpu Bridge (2013)
18	Minami Bisan–Seto Bridge	1100 (3608.9)	Stiffening truss	1648 (5406.8)	1988		Japan	Minami Bisan–Seto Bridge (2013)
19	Fatih Sultan Mehmet Bridge (Second Bosporus Bridge)	1090 (3576.1)		1510 (4954.1)	1988	Hwy	Turkey	Yilmaz et al. (2013) Chapter 18
20	Baling River Bridge	1,088 (3,569.6)	Stiffening truss	2237 (7667.3)	2009	Hwy	China	Baling River Bridge (2013)

Self-Anchored Suspension Bridges

Rank	Bridge Name	Main Span Length m (ft)	Structure Features	Total Length m (ft)	Year Opened	Usage	Country	References
1	San Francisco–Oakland Bay Bridge Self-Anchored Suspension Span	385 (1263.1)	Single tower steel box girder	624.4 (2048.6)	2013	Hwy	USA	Lwin and Kulicki (2013) Chapter 2
2	Pingsheng Bridge	350 (1148.3)	Four cables	3,200 (10,498.7)	2010	Hwy	China	China Bridges (2013)
3	Sachaji Bridge	328 (1076.1)		1577 (5173.9)	2006	Hwy	China	Sachaji Bridge (2013)
4	Konohana Bridge	300 (984.3)	Single cable	540 (1771.6)	1990	Hwy	Japan	Konohana Bridge (2013)
5	Yeongjong Grand Bridge	300 (984.3)		4,420 (14,501.3)	2001	Hwy	Korea	Yeongjong Bridge (2013)
6	Jiangdong Bridge	260 (853)	Steel box girder	4,330 (14,206)	2008	Hwy	China	Jiangdong Bridge (2013)
7	Sorok Bridge	250 (820.2)	Single cable	470 (1542)	2007	Hwy	Korea	Koh and Choo (2011)
8	Gushan Bridge	235 (771)		1520 (4986.9)	2010	Hwy	China	Tian (2008)
9	Liede Bridge	219 (718.5)		4,300 (14,107.6)	2008	Hwy	China	Liede Bridge (2013)
10	Great Loire River Bridge	200 (656.2)		270 (885.8)	2008	Hwy	France	Great Loire River Bridge (2013)
11	Zhuang River Bridge	200 (656.2)		340 (1115.5)	2011	Hwy	China	Zhuang River Bridge (2013)
12	Qilin Bridge	180 (590.6)	Concrete box girder	508.32 (1667.7)	2009	Hwy	China	Qilin Bridge (2013)
13	Luozhou Bridge	168 (551.2)		4,950 (16,240.2)	2012	Hwy	China	Luozhou Bridge (2013)
14	Seventh Street Bridge	134.7 (442)		323.4 (1061)	1926	Hwy	USA	Three Sisters (2013)

(Continued)

(Continued)

Rank	Bridge Name	Total Length m (ft)	Structure Features	Longest Span m (ft)	Year Opened	Usage	Country/ Area	References
15	Sixth Street Bridge	131.1 (430)		303.3 (995)	1928	Hwy	USA	Three Sisters (2013)
16	Ninth Street Bridge	125 (410)		303.3 (995)	1925	Hwy	USA	Three Sisters (2013)
17	Ziya River Bridge	115 (377.3)	Steel box girder	215 (705.4)	2004	Hwy	China	Ziya River Bridge (2013)
18	Chelsea Bridge	101.2 (322)		212.8 (698)	1937	Hwy	UK	Chelsea Bridge (2013)
19	Peace Bridge	101 (331.4)	S-shape	315 (1033.5)	2011	Ped	UK	Peace Bridge (2013b)
20	Kiyosu Bridge	91.4 (299.9)		183 (600.4)	1928	Hwy	Japan	Nagai, et al. (2013) Chapter 23

29.4 Top 20 Longest Cable-Stayed Bridge Spans

Rank	Bridge Name	Main Span Length m (ft)	Structure Features	Total Length m (ft)	Year Opened	Usage	Country	References
			Multi Tower Cable-Stayed Bridges					
1	Russky Bridge	1104 (3622)	Steel box girder	3,100 (10,170.6)	2012	Hwy	Russia	Blank and Seliverstov (2013) Chapter 15
2	Sutong Bridge	1088 (3569.6)	Steel box girder	8,206 (26,922.6)	2008	Hwy	China	Sutong Bridge (2011b)
3	Stonecutters Bridge	1018 (3339.9)	Steel box girder	1596 (5236.2)	2009	Hwy	China	Stonecutters Bridge (2013)
4	Edong Yangtze River Bridge	926 (3038.1)	Steel box girder	6,300 (20,669.3)	2010	Hwy	China	Edong Yangtze River Bridge (2013)
5	Tatara Bridge	890 (2920)	Steel box girder	1480 (4855.6)	1999	Hwy	Japan	Nagai, et al. (2013) Chapter 23
6	Pont de Normandie Bridge	856 (2808.4)	Steel box girder	2143.21 (7031.5)	1995	Hwy	France	Calgaro (2013) Chapter 11
7	Jiujiang Yangtze River Highway Bridge	818 (2683.7)		25,145 (82,496.7)	2012	Hwy	China	Jiujiang Yangtze River Highway Bridge (2013)
8	Jingyue Yangtze River Bridge	816 (2677.2)		5,400 (17,716.5)	2010	Hwy	China	Yangtze River Bridge (2013)
9	Incheon Bridge	800 (2624.7)		18,384 (60,315)	2009	Hwy	Korea	Incheon Bridge (2013)
10	Zolotoy Rog Bridge	737 (2418)		2100 (6890)	2011	Hwy	Russia	Zolotoy Rog Bridge (2013)
11	Shanghai Yangtze River Bridge	730 (2395)		9,970 (32,710)	2009	Hwy	China	Yangtze River Bridge (2013)
12	Minpu Bridge	708 (2322.8)	Double decks	3,610 (11,843.8)	2010	Hwy	China	Minpu Bridge (2013)

(*Continued*)

Rank	Bridge Name	Main Span Length m (ft)	Structure Features	Total Length m (ft)	Year Opened	Usage	Country	References
13	Third Nanjing Yangtze Bridge	648 (2126)	Steel box girder	4,700 (15,419.9)	2005	Hwy	China	Yangtze River Bridge (2013)
14	Second Nanjing Yangtze Bridge	628 (2060.4)		21,227 (69,642.4)	2001	Hwy	China	Yangtze River Bridge (2013)
15	Jintang Bridge	620 (2034.1)		26,540 (87,073.5)	2009	Hwy	China	Jintang Bridge (2013)
16	Baishazhou Yangtze River Bridge	618 (2027.6)	Steel box girder	3,589 (11,774.9)	2000	Hwy	China	Yangtze River Bridge (2013)
17	Qingzhou Bridge	605 (1984.9)	Composite girder	3,587 (11,768.4)	2001	Hwy	China	Qingzhou Bridge (2013)
18	Yangpu Bridge	602 (1975.1)	Composite girder	8,354 (27,048.1)	1993	Hwy	China	Yangpu Bridge (2011)
19	Xupu Bridge	590 (1935.7)	Composite girder	6,017 (19,740.8)	1997	Hwy	China	Xupu Bridge (2011)
20	Meiko–Chuo Bridge	590 (1935.7)		1170 (3838.6)	1998	Hwy	Japan	Nagai, et al. (2013)

Single Tower Cable–Stayed Bridge Spans

Rank	Bridge Name	Main Span Length m (ft)	Structure Features*	Total Length m (ft)	Year Opened	Usage	Country	References
1	Serebryanyi Bor (Zhivopisny) Bridge	410 (1,345.1)	Arch pylon with two cable planes	1,025.6 (3,364.8)	2007	Hwy	Russia	Blank and Seliverstov (2013) Chapter 15
2	Surgut Ob River Bridge	408 (1,338.6)	Two cable planes	2,110 (6,922.6)	2000	Hwy	Russia	Blank and Seliverstov (2013) Chapter 15
3	New Taipei Bridge	400 (1,312.3)	Two cable planes	1,100 (3,608.9)	2010	Hwy	Chinese Taipei	New Taipei Bridge (2013)
4	Jinan Huanghe Bridge	386 (1,226.4)	Two cable planes	766 (2,513.1)	2008	Hwy	China	Jinan Huanghe Bridge (2013)
5	South Span– Huangpu Bridge	383 (1,256.6)	Two cable planes	7,016 (23,018.4)	2008	Hwy	China	South Span –Huang puBridge (2013)
6	Sava (Ada) Bridge	375.75 (1,232.8)	Two cable planes	929 (3,047.9)	2012	Hwy/ Rwy/Ped	Serbia	Folić (2013) Chapter 16 Sava Bridge (2013)
7	Flehe Bridge	368 (1,207.3)	Single cable plane	1,166 (3,825.5)	1979	Hwy	Germany	Flehe Bridge (2013)
8	Kao-Ping Hsi Bridge	330 (1,082.7)	Single cable plane	2,617 (8,586.0)	1999	Hwy	Chinese Taipei	Kao-Ping Hsi Bridge (2013)
9	Karnali River Bridge	325 (1,066.3)	Two cable planes	500 (1,640.4)	1993	Hwy	Nepal	Karnali River Bridge (2013)
10	Rheinknie bridge in Dusseldorf	319 (1,046.6)	Two cable planes	1,519 (4,983.6)	1969	Hwy	Germany	Rheinknie Bridge (2013)
11	South Channel Bridge– Hangzhou Bay Bridge	318 (1,043.3)	Two cable planes	578 (1,896.3)	2008	Hwy	China	Wang and Meng (2009)

(*Continued*)

(*Continued*)

Rank	Bridge Name	Main Span Length m (ft)	Structure Features	Total Length m (ft)	Year Opened	Usage	Country	References
12	Vansu Bridge	312 (1,023.6)	Single cable plane	595 (1,952.1)	1981	Hwy	Latvia	Vansu Bridge (2013) Baláž and Ding (2006)
13	Taijin Haihe Bridge	310 (1,017.1)	Two cable planes./ Twin bridge	2,650 (8,694.2)	2002/2011	Hwy	China	Tianjin Haihe Bridge (2013)
14	Grenland Bridge	305 (1000.7)	Two cable planes	608 (1,994.8)	1996	Hwy	Norway	Grenland Bridge (2013)
15	Dubrovnik Bridge	304.5 (999.0)	Two cable planes	518 (1,699.5)	2002	Hwy	Croatia	Radić and Puž (2013) Chapter 7
16	SNP (New) Bridge	303 (994.1)	Single cable plane	431.8 (1,416.7)	1972	Hwy	Slovak Republic	Baláž (2013b) Chapter 17
17	Severin Bridge	302 (990.8)	Two cable planes	691 (2,267.1)	1961	Hwy	Germany	Vejrum, T. et al. (2013) Severin Bridge (2013)
18	Moskovsky Bridge	300 (984.3)	Two cable planes	448 (1,469.8)	1976	Hwy	Ukraine	Mykhailo K. (2013)
18	Rama VIII Bridge	300 (984.3)	Three cable planes	440 (1,443.6)	1975	Hwy	Germany	Deggendorf-Deggenau Bridge (2013)
19	Deggendorf-Deggenau bridge	290 (951.4)	Three cable planes	440 (1,443.6)	1975	Hwy	Germany	Deggendorf-Deggenau Bridge (2013)
20	Kurt Schumacher Bridge (Mannheim)	287 (941.6)	Two cable planes	412 (1,351.7)	1971	Hwy	Germany	Kurt Schumacher Bridge (2013)

Note: *Number of cable planes is the number of lines or curves in which cables are connected to the bridge deck. Number of cable planes shown in table is for the main span only.

29.5 Top 20 Longest Extradosed Bridge Spans

Rank	Bridge Name	Main Span Length m (ft)	Structure Features	Total Length m (ft)	Year Opened	Usage	Country	References
1	Wuhu Yangtze River Bridge	312 (1,023.6)	Double Decks	10,521 (34,517.7)	2000	Hwy/ Rwy	China	Qin, et al. (2013) Chapter 20
2	Kiso River Bridge	275 (902.2)		1,145 (3,756.6)	2001	Hwy	Japan	Kiso River Bridge (2013)
3	Ibi River Bridge	271.5 (890.7)		1,397 (4,583.3.3)	2001	Hwy	Japan	Ibi River Bridge (2013)
4	Keong-An Bridge	270 (885.8)			2009	Hwy	Korea	Keong-An Bridge (2013)
5	Jiayue Bridge	250 (820.2)		778 (2,552.5)	2010	Hwy	China	Tang (2011)
6	Qiancao Bridge	248 (813.7)		1,190 (3,904.2)	2011	Hwy	China	Qiancao Bridge (2013)
7	Korror Babeldoap Bridge	247 (810.4)		413 (1,35.05)	2002	Hwy	Palau	Korror Babeldoap Bridge (2013)
8	Golden Ears Bridge	242 (794.0)		2410 (7,906.9)	2009	Hwy	Canada	Golden Ears Bridge (2013)

(Continued)

Rank	Bridge Name	Main Span Length m (ft)	Structure Features	Total Length m (ft)	Year Opened	Usage	Country	References
9	Tokunoyama Hattoku Bridge	220 (721.8)	146 m high	526.6 (1,727.8)	2006	Hwy	Japan	Tokunoyama Hattoku Bridge (2013)
10	Shah Amanat Bridge	200 (656.2)		950 (3,116.8)	2010	Hwy	Bangladesh	Shah Amanat Bridge (2013)
11	Sannohe-Boukyo Bridge	200 (656.2)		400 (1,312.3)	2005	Hwy	Japan	Sannohe-Boukyo Bridge (2013)
12	Second Mactan-Mandaue Bridge	185 (607.0)		1010 (3,313.6)	1999	Hwy	Philippines	Second Mactan-Mandaue Bridge (2013)
13	Gumgang No,1 Bridge	185 (607.0)		735 (2,411.4)	2012	Hwy	Korea	BD&E (2008)
14	Tsukuhara Bridge	180 (590.6)	Single-cell PS box girder	323 (1,059.7)	1997	Hwy	Japan	Ogawa et al. (1998)
15	Syoyo (Kanisawa) Bridge	180 (590.7)		380.1 (1,247)	1998	Hwy	Japan	Kasuga (2013)
16	Himi Yume Bridge	180 (590.6)	Corrugated steel plate webs	365 (1,197.5)	2004	Hwy	Japan	Hino (2005)
17	North Arm Bridge	180 (590.6)		562 (1,843.8)	2009	Rwy	Canada	North Arm Bridge (2013)
18	Calafat-Vidin Bridge	180 (590.6)		3598 (11,804.5)	2013	Hwy/Rwy	Romania and Bulgaria	Calafat-Vidin Bridge (2013)
19	Ohmi Ohtori (Rittoh) Bridge	170 (557.7)	Corrugated steel plate webs	490.2 (1,608.3)	2007	Hwy	Japan	Kasuga (2013)
20	Sannai-Maruyama Bridge	150 (492.1)	Concrete box girder	450 (1,476.4)	2008	HSR	Japan	Sannai-Maruyama Bridge (2013)
20	Ningjiang Songhua River Bridge	150 (492.1)		2236 (7,336.0)	2013	Hwy	China	Ningjiang Songhua River Bridge (2013)

29.6 Top 20 Longest Truss Bridge Spans

Rank	Bridge Name	Main Span Length m (ft)	Structure Features	Total Length m (ft)	Year Opened	Usage	Country	References
1	Québec Bridge	549 (1801.2)		987 (3238.2)	1919	Hwy/Rwy	Canada	Québec Bridge (2013)
2	Firth of Forth Bridge	521.3 (1710.3)	Two spans	2528.7 (8296.2)	1890	Rwy	UK	Firth of Forth Bridge (2013)
3	Minato Bridge	510 (1673.2)	Double decks	983 (3225.1)	1974	Hwy	Japan	Minato Bridge (2013)
4	Commodore Barry Bridge	501.1 (1644)		4,240.4 (13,912)	1964	Hwy	USA	Commodore Barry Bridge (2013)
5	Crescent City Connection	480.1 (1575)	Twin bridges	4,092.9 (13,428)	1958 (east) 1988 (west)	Hwy	USA	Crescent City Connection (2013)

(Continued)

(Continued)

Rank	Bridge Name	Main Span Length m (ft)	Structure Features	Total Length m (ft)	Year Opened	Usage	Country	References
6	Howrah Bridge	457.5 (1501)		755.8 (2479.7)	1943	Hwy	India	Howrah Bridge (2013)
7	Tokyo Gate Bridge	440 (1443.6)	Welded box	2933 (9622.7)	2011	Hwy	Japan	Tokyo Gate Bridge (2013)
8	Veterans Memorial Bridge	445 (1460)		945.2 (3101)	1995	Hwy	USA	Veterans Memorial Bridge (2013)
9	San Francisco–Oakland Bay Bridge East Span	426.7 (1400)	Double decks	3,101.6 (10,176)	1936	Hwy	USA	SFOBB (2013a)
10	Ikitsuki Bridge	400 (1312.3)		960 (3149.6)	1991	Hwy	Japan	Ikitsuki Bridge (2013)
11	Astoria–Megler Bridge	375.5 (1232)		6,545.3 (21,474)	1966	Hwy	USA	Astoria–Megler Bridge (2013)
12	Horace Wilkinson Bridge	376.4 (1235)		1386.8 (4550)	1968	Hwy	USA	Horace Wilkinson Bridge (2013)
13	Tappan Zee Bridge	369.4 (1212)		4,880.8 (16,013)	1955	Hwy	USA	Tappan Zee Bridge (2013)
14	Lewis and Clark Bridge	365.8 (1200)		829.7 (2722)	1930	Hwy	USA	Lewis and Clark Bridge (2013)
15	Francis Scott Key Bridge	365.8 (1200)		2632.3 (8636)	1997	Hwy	USA	Francis Scott Key Bridge (2013)
16	Queensboro Bridge	360.3 (1182)	Double decks	1135.2 (3724.5)	1909	Hwy	USA	Queensboro Bridge (2013)
17	Carquinez Bridge	335.3 (1100)		1005.8 (3300)	1958	Hwy	USA	Carquinez Bridge (2013)
18	Hart Bridge	331.6 (1088)		1171.7 (3844)	1967	Hwy	USA	Hart Bridge (2013)
19	Richmond–San Rafael Bridge	326.1 (1070)	Double decks	8,851.4 (29,040)	1956	Hwy	USA	Richmond–San Rafael Bridge (2013)
20	Newburgh–Beacon Bridge	304.8 (1000)		2374.1 (7789) 2403.3 (7885)	1963 (westbound) 1980 (eastbound)	Hwy	USA	Newburgh–Beacon Bridge (2013)

29.7 Top 20 Longest Arch Bridge Spans

Rank	Bridge Name	Main Span Length m (ft)	Structure Features	Total Length m (ft)	Year Opened	Usage	Country	References
				Steel Arch Bridges				
1	Sheikh Rashid bin Saeed Crossing	667 (2188.3)		1600 (5249.3)		Hwy/Rwy	United Arab Emirates	Sheikh Rashid bin Saeed Crossing (2013)
2	Chaotianmen Bridge	552 (1811)	Truss arch	1741 (5711.9)	2009	Hwy	China	Chaotianmen Bridge (2013)

(Continued)

Rank	Bridge Name	Main Span Length m (ft)	Structure Features	Total Length m (ft)	Year Opened	Usage	Country	References
3	Lupu Bridge	550 (1804.5)	Box tied arch	3,900 (12,795.3)	2003	Hwy	China	Qin et al. (2013) Chapter 20
4	New River Gorge Bridge	518.2 (1700)	Truss arch	923.5 (3030)	1977	Hwy	USA	New River Gorge Bridge (2013)
5	Bayonne Bridge	510.5 (1675)	Truss arch	1761.7 (5780)	1931	Hwy	USA	Bayonne Bridge (2013)
6	Sydney Harbor Bridge	503 (1650.3)	Truss arch	1149 (3769.7)	1932	Hwy	Australia	Sydney Harbor Bridge (2013)
7	Chenab Bridge	465 (1525.6)	Truss arch	1315 (4314.3)	2015	Rwy	India	Chenab Bridge (2013)
8	Mingzhou Bridge	450 (1476.4)	Boxes	1250 (4101.1)	2011	Hwy	China	Mingzhou Bridge (2013)
9	Xingguang Bridge	428 (1404.2)	Three arches	1082 (3549.9)	2007	Hwy	China	Xingguang Bridge (2013)
10	Caiyuanba Yangtze River Bridge	420 (1378)	Double decks, box	1866 (6122.1)	2007	Hwy/Rwy	China	Caiyuanba Yangtze River Bridge (2013)
11	Daning River Bridge	400 (1312.3)	Truss arch	682 (2237.5)	2010	Hwy	China	Daning River Bridge (2013)
12	Fremont Bridge	382.5 (1255)	Tied arch	656.5 (2154)	1973	Hwy	USA	Fremont Bridge (2013)
13	Hiroshima Airport Bridge	380 (1246.7)	Truss arch	500 (1640.4)	2010	Hwy	Japan	Hiroshima Airport Bridge (2013)
14	Sloboda Bridge	373 (1223.8)	Deck arch	806 (2644.4)	1959	Hwy	Croatia	Pilot (2007)
15	Port Mann Bridge	366 (1200.8)	Tied arch	2093 (6866.8)	1964	Hwy	Canada	Port Mann Bridge (2013)
16	Wanzhou Yangtze River Railway Bridge	360 (1181.1)	Truss arch	1106.3 (3629.6)	2010	Rwy	China	Wanzhou Yangtze River Railway Bridge (2013)
17	Bridge of the Americas	344 (1128.6)	Tied arch	1654 (5426.5)	1962	Hwy	Panama	Bridge of the Americas (2013)
18	Dashengguan Railway Bridge	336 (1102)	Truss tied arches	14,789 (48,520)	2010	HSR	China	Dashengguan Bridge (2013)
19	Ironworkers Memorial Bridge	335 (1099)	Deck truss arch	1292 (4239)	1960	Hwy	Canada	Ironworkers Memorial Bridge (2013)
20	Laviolette Bridge	335 (1099.1)	Truss arch	2707 (8881.2)	1967	Hwy	Canada	Laviolette Bridge (2013)

Concrete-Filled Steel Tube (CFST) Arch Bridges								
1	First He Jiang Yangtze River Bridge	530 (1,738.9)		841 (2,759.2)	2012	Hwy	China	First He Jiang Yangtze River Bridge (2013)
2	Wushan Yangtze River Bridge	492 (1614.2)	Half through arch	612.2 (2008.5)	2005	Hwy	China	Wushan Yangtze River Bridge (2013)
3	Zhijing River Bridge	430 (1410.8)	Deck arch	545.54 (1789.8)	2009	Hwy	China	Zhijing River Bridge (2013)

(Continued)

(Continued)

Rank	Bridge Name	Main Span Length m (ft)	Structure Features	Total Length m (ft)	Year Opened	Usage	Country	References
4	Lianxiang Bridge	400 (1312.3)	Cable-stayed	640 (2099.7)	2007	Hwy	China	Lianxiang Bridge (2013)
5	Maocao Street Bridge	368 (1207.3)		7,959.7 (26,114.5)	2007	Hwy	China	Maocao Street Bridge (2013)
6	Zhaohua Jialing River Bridge	364 (1194.2)	Deck arch	864 (2834.6)	2012	Hwy	China	Zhaohua Jialing River Bridge (2013)
7	Yajisha Bridge	360 (1181.1)		1084 (3556.4)	2000	Hwy	China	Yajisha Bridge (2013)
8	Taiping Lake Bridge	336 (1102.4)	Half through	504 (1653.5)	2007	Hwy	China	Chen (2008)
9	Nanning Yonghe Bridge	335.4 (1100.4)	Half through	398.7 (1308.1)	2004	Hwy	China	Nanning Yonghe Bridge (2013)
10	Guangxi Yongning Bridge	312 (1203.6)	Half through	485.5 (1592.8)	1996	Hwy	China	Guangxi Yongning Bridge (2013)
11	Chunan Napu Bridge	308 (1010.5)	Half through	330 (1082.7)	2003	Hwy	China	Chunan Napu Bridge (2013)
12	Meixi River Bridge	288 (944.9)	Deck arch	491 (1610.9)	2001	Hwy	China	Meixi River Bridge (2013)
13	Dongguan Shuidao Bridge	280 (918.7)		400 (1312.3)	2005	Hwy	China	Dongguan Shuidao Bridge (2013)
14	Wuhan Qingchuan Bridge	280 (918.7)	Through arch	2148 (7047.2)	2001	Hwy	China	Wuhan Qingchuan Bridge (2013)
15	Sanmenkou Bridge	270 (885.8)	Two arches	1472 (4829.4)	2008	Hwy	China	Sanmenkou Bridge (2013)
16	Nanning Sanan Yi River Bridge	270 (885.8)	Half through	352 (1154.9)	1998	Hwy	China	Nanning Sanan Yi River Bridge (2013)
17	Yichang Yangtze River Railway Bridge	264 (866.1)	Through deck	2526.7 (8289.7)	2011	Rwy	China	Yichang Yangtze River Railway Bridge (2013)
18	Rongzhou Jinsha River Bridge	260 (853.0)	Half through	506 (1660.1)	2004	Hwy	China	Rongzhou Jinsha River Bridge (2013)
19	First Qiandao Lake Bridge	252 (826.8)	Deck-arch	1,343 (4,406.2)	2005	Hwy	China	First Qiandao Lake Bridge (2013)
20	Zigui Qinggan River Bridge	248 (813.6)	Half through	3,120 (10,236.2)	2002	Hwy	China	Zigui Qinggan River Bridge (2013)
Concrete Arch Bridges								
1	Wanzhou Yangtze River Highway Bridge	420 (1378.0)	Deck arch	864.12 (2835.0)	1997	Hwy	China	Wanzhou Yangtze River Highway Bridge (2013)
2	Krk Bridge	390 (1279.5)	Two arch spans	1430 (4691.6)	1980	Hwy	Croatia	Folić (2013) Chapter 7

(Continued)

Rank	Bridge Name	Main Span Length m (ft)	Structure Features	Total Length m (ft)	Year Opened	Usage	Country	References
3	Jiangjie River Bridge	330 (1082.7)	Deck truss arch	461 (1512.5)	1995	Hwy	China	Jiangjie River Bridge (2013)
4	Mike O'Callaghan–Pat Tillman Memorial Bridge	323.1 (1060)	Deck arch	579.1 (1900)	2010	Hwy	USA	Lwin and Kulicki (2013) Chapter 2
5	Gladesville Bridge	300 (984.3)	Deck arch	488 (1601.1)	1964	Hwy	Australia	Gladesville Bridge (2013)
6	Friendship Bridge (Paraguay–Brazil)	290 (951.4)	Deck arch	552.4 (1812.3)	1965	Hwy	Paraguay/ Brazil	Friendship Bridge (2013)
7	Infante D. Henrique Bridge	280 (918.6)	Deck arch	371 (1217.2)	2002	Hwy	Portugal	Infante D. Henrique Bridge (2013)
8	Bloukrans Bridge	272 (892.4)	Deck arch	451 (1479.7)	1984	Hwy	South Africa	Bloukrans Bridge (2013)
9	Arrabida Bridge	270 (885.8)	Deck arch	493 (1617.5)	1963	Hwy	Portugal	Arrabida Bridge (2013)
10	Viaduct Froschgrundsee	270 (885.8)	Deck arch	798 (2618.1)	2010	Rwy	Germany	Viaduct Froschgrundsee (2013)
11	Grümpental Bridge	270 (885.8)	Deck arch	1104 (3622.1)	2011	Rwy	Germany	Grümpental Bridge (2013)
12	Fujikawa Bridge	265 (869.4)	Deck arch, two arches	381 (1250.0)	2005	Hwy	Japan	Imai (2013)
13	Sandö Bridge	264 (866.1)	Deck arch	810 (2657.5)	1943	Hwy	Sweden	Sandö Bridge (2013)
14	Chateaubriand Bridge	261 (856.3)	Deck arch	424 (1391.1)	1991	Hwy	France	Chateaubriand Bridge (2013)
15	Tensho (Takamatu) Bridge	260 (853.0)	Deck arch	463.2 (1519.7)	2000	Hwy	Japan	Tensho (Takamatu) Bridge (2013)
16	Kishiwada Bridge	255 (836.6)	Deck arch	445 (1460.0)	1993	Hwy	Japan	Kishiwada Bridge (2013)
17	Los Tilos Arch	255 (836.6)	Deck arch	319 (1046.6)	2004	Hwy	Spain	Los Tilos Arch (2013)
18	Wild Gera Viaduct	252 (826.8)	Deck arch	552 (1811.0)	2000	Hwy	Germany	Wild Gera Viaduct (2013)
19	Svinesund Bridge	247 (810.4)	Half through	704 (2309.7)	2005	Hwy	Sweden/ Norway	Svinesund Bridge (2013)
20	Šibenik Bridge	246.4 (808.4)		390 (1279.5)	1966	Hwy	Croatia	Šibenik Bridge (2013)
	Masonry Arch Bridges							
1	New Danhe (DanRiver) Bridge	146 (479.0)	Stone	413.2 (1,355.6)	2000	Hwy	China	Qin, et al. (2013) Chapter 20
2	Wuchaohe Bridge	120 (393.7)	Stone	241 (790.7)	1990	Hwy	China	Wuchaoher Bridge (2013)

(Continued)

(*Continued*)

Rank	Bridge Name	Main Span Length m (ft)	Structure Features	Total Length m (ft)	Year Opened	Usage	Country	References
3	Jiuxigou Bridge	116 (380.6)	Stone	140 (459.3)	1972	Hwy	China	Jiuxigou Bridge (2013)
4	Changhong Bridge	112.5 (369.1)	Stone	171.2 (561.7)	1961	Hwy	China	Changhong Bridge (2013)
5	Fushun Tuojiang Bridge	111 (364.2)	Stone	277.2 (909.4)	1968	Hwy	China	Fushun Tuojiang Bridge (2013)
6	Shengli Bridge	108 (354.3)	Stone		1989	Hwy	China	Chen (2009)
7	No. 1 Sizhuang Bridge	108 (354.3)	Stone		1996	Hwy	China	Zhou (1997)
8	Danhe Bridge	105 (344.5)	Stone		1983	Hwy	China	Chen (2009)
9	Jiangpinghe Bridghe	105 (344.5)	Stone		1990	Hwy	China	Chen. (2009)
10	Huwan Bridge	105 (344.5)	Stone		1972	Hwy	China	Chen (2009)
11	New Tongshan Bridge	105 (344.5)	Stone		1977	Hwy	China	Chen (2009)
12	Yugong Bridge	102 (334.6)	Stone		1970	Hwy	China	Chen (2009)
13	Gongtan Bridge	100 (328.1)	Stone		1954	Hwy	China	Chen (2009)
14	Youduhe Bridge	100 (328.1)	Stone		1973	Hwy	China	Chen (2009)
15	Hongdu Bridge	100 (328.1)	Stone		1977	Hwy	China	Chen. (2009)
16	Longwu Bridge	100 (328.1)	Stone		1979	Hwy	China	Chen (2009)
17	Fujin Bridge	100 (328.1)	Stone	153 (502.0)	2003	Hwy	China	Chang (2004)
18	Jin-shan Bridge	99 (324.8)	Stone	161 (528.2)	1972	Hwy	China	Jin-shan Bridge (2013)
19	Pont de la Libération Bridge	96.25 (315.8)	Plan Concrete		1919	Hwy	France	Pont de la Libération Bridge (2013)
20	Peace Bridge	90 (295.3)	Stone		1905	Hwy	Germany	Peace Bridge (2013a)

29.8 Top 20 Longest Girder Bridge Spans

Rank	Bridge Name	Main Span Length m (ft)	Structure Features	Total Length m (ft)	Year Opened	Usage	Country	References
			Steel Girder Bridges (*Orthotropic* Deck)					
1	Ponte Costa e Silva Bridge (Rio–Niteroi Bridge)	300 (984.3)	Box	13,290 (43,602.4)	1974	Hwy	Brazil	Vasconcelos, et al. (2013). Chapter 4
2	Neckar Valley Viaduct	263.2 (863.5)	Box	900 (2952.8)	1978	Hwy	Germany	Neckar Valley Viaduct (2013)
3	Sava River Bridge	261 (856.3)	Plate girder	411 (1348.4)	1956	Hwy	Serbia	Folić (2013) Chapter 16
4	Ponte de Vitoria III Bridge	260 (853)	Box	3,300 (10,826.8)	1989	Hwy	Brazil	Third Bridge (2013)
5	Zoo Bridge	259 (849.7)	Box	597 (1958.7)	1962	Hwy	Germany	Zoo Bridge (2013)
6	Kaita Bridge	250 (820.2)	Box	550 (1804.5)	1990	Hwy	Japan	Kaita Bridge (2013)
7	Namihaya Bridge	250 (820.2)	Box	1572 (5157.5)	1994	Hwy	Japan	Namihaya Bridge (2013)
8	Gazelle Bridge	249.9 (819.9)	Box	475.5 (1560)	1971	Hwy	Serbia	Gazelle Bridge (2013)
9	Auckland Harbor Bridge	243.8 (799.9)	Box truss	1020 (3346.5)	1959	Hwy	New Zealand	Auckland Harbor Bridge
10	Trans-Tokyo Bay Bridge	240 (787.4)	Box	11,000 (36,089.2)	1997	Hwy	Japan	Nagai, et al. (2013) Chapter 23
11	Shorenji–Gawa Bridge	235 (771)	Box	534.5 (1753.6)	1989	Hwy	Japan	Shorenji–Gawa Bridge (2013)
12	Grand Duchess Charlotte	234.1 (768)	Box	355 (1164.7)	1966	Hwy	France	Grand Duchess Charlotte (2013)
13	Konrad Adenauer Bridge	230 (754.6)	Box	480 (1574.8)	1972	Hwy	Germany	Konrad Adenauer Bridge (2013)
14	San Mateo-Hayward	228.6 (750)	Box	11,265 (36,959)	1967	Hwy	USA	Lwin and Kulicki (2013) Chapter 2

(Continued)

(Continued)

Rank	Bridge Name	Main Span Length m (ft)	Structure Features	Total Length m (ft)	Year Opened	Usage	Country	References
15	Rader Island Bridge	221.54 (726.8)		1497.5 (4913.1)	1972	Hwy	Germany	Rader Island Bridge (2013)
16	Winningen Bridge	228.2 (748.7)	Box	935 (3067.6)	1972	Hwy	Germany	Winningen Bridge (2013)
17	Umemachi Bridge	215 (705.4)		390 (1279.5)	1989	Hwy	Japan	Umemachi Bridge (2013)
18	Second Maya Bridge	210 (689)	Box	360 (1181.1)	1975	Hwy	Japan	Second Maya Bridge (2013)
19	Düsseldorf–Neuss Bridge	206 (675.9)	Box		1951	Hwy	Germany	Düsseldorf–Neuss Bridge (2013)
20	Schierstein Bridge	205 (672.6)	Plate girder	1282 (4206)	1962	Hwy	Germany	Schierstein Bridge (2013)
Steel Girder Bridges (Concrete Composite Deck)								
1	Ulla Viaduct	240 (787.4)	Box	1620 (5255.9)	2005	HSR	Spain	Millanes (2008)
2	Angosturita Bridge	213.75 (701.3)	Box	468.75 (1537.8)	1992	Hwy	Venezuela	Saul and Humpf (2007)
3	Sidney Sherman Bridge	183 (600.4)	Box	1921 (6302.5)	1973	Hwy	USA	Houston Freeways (2013)
4	Lower Buffalo Bridge	160 (525)	Box	563.9 (1850)	1998	Hwy	USA	Lower Buffalo Bridge (2013)
5	Volgograd Volga River Bridge	155 (508.5)		1200 (3937)	2009	Hwy	Russia	Blank and Seliverstov (2013) Chapter 15
6	Chusovaya River Bridges	147 (482.3)	Plate girder	1504 (4934.8)	1997	Hwy	Russia	Blank and Seliverstov (2013) Chapter 15
7	Wintergreen Gorge Bridge	142 (466)	Plate girder	339.5 (1114)	2003	Hwy	USA	Macioce, et al. (2013)
8	Antioch Bridge	140.2 (460)	Plate girder	2896.8 (9504)	1978	Hwy	USA	Antioch Bridge (2013)
9	Verrières Viaduct	140 (459.3)	Box	720 (2362.2)	2002	Hwy	France	Verrières Viaduct (2013)

(Continued)

Rank	Bridge Name	Main Span Length m (ft)	Structure Features	Total Length m (ft)	Year Opened	Usage	Country	References
10	White Bird Canyon Bridge	137.5 (451)	Box	246.9 (810)	1976	Hwy	USA	White Bird Canyon Bridge (2013)
11	Waverly Bridge (Missouri)	137.2 (450)	Plate girder	419.1 (1375)	2004	Hwy	USA	NSBA (2007)
12	Kanawha River Bridge	134.1 (440)	Plate girder		1974	Hwy	USA	Rodriguez (2009)
13	Vehmersalmi Bridge	130 (426.5)	Plate girder	370 (1213.9)	2001	Hwy	Finland	WSP (2013)
14	Morris Bridge	125 (410)	Plate girder	527.3 (1730)	2002	Hwy	USA	Morris Bridge (2013)
15	Star City Bridge	(412)	Plate girder	(1004)	2002	Hwy	USA	NSBA (2007)
16	Oka River Bridge	126	Twin box	998	1993	Hwy	Russia	Blank and Seliverstov (2013) Chapter 15
18	Morris Illinois River Bridge	(403.5)		(1730)		Hwy	USA	NSBA (2007)
19	Yazoo River Bridge	121.9 (400)	Plate girder	301.8 (990)	2007	Hwy	USA	NSBA (2007)
20	White's Hill Bridge	115.8 (380)	Plate girder	115.8 (380)	2003	Hwy	USA	Caltrans (2013)
19	Shippingsport Bridge	112.8 (370)	Plate girder	541 (1775)	2003	Hwy	USA	NSBA (2007)
20	Sacramento River (Bryte Bend) Bridge	112.8 (370)	Box	1234.4 (4050)	1974	Hwy	USA	Caltrans (2013)
Concrete Girder Bridges								
1	Shibanpo Bridge	330 (1082.7)	Box	1104 (3622.1)	2011	Hwy	China	Shibanpo Bridge (2013)
2	Stolma Bridge	301 (987.5)	Box	467 (1532.2)	1998	Hwy	Norway	Stolma Bridge (2013)
3	Raftsundet Bridge	298 (977.7)	Box	711 (2332.7)	1998	Hwy	Norway	Raftsundet Bridge (2013)
4	Sundoy Bridge	298 (977.7)	Box	538 (1765.1)	2003	Hwy	Norway	Sundoy Bridge (2013)
5	Humen Bridge Approach Span	270 (885.8)	Box	15,760 (51,706)	1997	Hwy	China	Humen Bridge (2013)

(Continued)

(*Continued*)

				Concrete Girder Bridges				
Rank	Bridge Name	Main Span Length m (ft)	Structure Features	Total Length m (ft)	Year Opened	Usage	Country	References
6	Sutong Bridge Approach Span	268 (879.3)	Box	8,206 (26,922.6)	2009	Hwy	China	Sutong Bridge (2013a)
7	Hong River Bridge	265 (869.4)	Box	801 (2628)	2003	Hwy	China	Hong River Bridge (2013)
8	First Sir Leo Hielscher Bridge	260 (853)	Box	1627 (5337.9)	1986	Hwy	Australia	Sir Leo Hielscher Bridges (2013)
9	Varodd Bridge	260 (853)	Box	660 (2165.4)	1994	Hwy	Norway	Varodd Bridge (2013)
10	Xiabaishi Bridge	260 (853)	Box	999.6 (3279.5)	2003	Hwy	China	Xiabaishi Bridge (2013)
11	Yudong Yangtze River Bridge	260 (853)	Box	1514.6 (4969.2)	2008	Hwy/Rwy	China	Yudong Yangtze River Bridge (2013)
12	Second Sir Leo Hielscher Bridges	260 (853)	Box	1627 (5337.9)	2010	Hwy	Australia	Sir Leo Hielscher Bridges (2013)
13	Second Luzhou Yangtze River Bridge	252 (826.8)	Box	1408 (4619.4)	2000	Hwy	China	Second Luzhou Yangtze River Bridge (2013)
14	Schottwien Bridge	250 (820.2)	Box	632.5 (2075.1)	1991	Hwy	Austria	Schottwien Bridge (2013)
15	John Bridge (Ponte de São João)	250 (820.2)	Box	1140 (3740.2)	1991	Rwy	Portugal	John Bridge (2013)
16	Skye Bridge	250 (820.2)	Box	570 (1870.1)	1995	Hwy	UK	Skye Bridge (2013)
17	Confederation-Bridge	250 (820.2)	Box	12,900 (42,323.8)	1997	Hwy	Canada	Confederation Bridge (2013)
18	Huanghua Yuan Bridge	250 (820.2)	Box	4,400 (14,435.7)	1999	Hwy	China	Huanghua Yuan Bridge (2013)
19	Hamana Bridge	240 (787.4)	Box	630 (2066.9)	1976	Hwy	Japan	Hamana Bridge (2013)
20	Hikoshima Bridge	236 (774.3)	Box	710 (2329.4)	1972	Hwy	Japan	Hikoshima Bridge (2013)

29.9 Top 20 Longest Movable Bridge Spans

Vertical Lift Bridges							
Rank	Bridge Name	Lift Span Length m (ft)	Structure Features	Total Length m (ft)	Year Opened	Usage	Country	References
1	Arthur Kill Vertical Lift Bridge	170.1 (558)	Truss		1959	Rwy	USA	Arthur Kill Vertical Lift Bridge (2013)
2	Cape Cod Canal Railroad Bridge	165.8 (544)	Truss		1935	Rwy	USA	Cape Cod Canal Railroad Bridge (2013)
3	Marine Parkway–Gil Hodges Memorial Bridge	164.6 (540)	Truss	1225.9 (4022)	1937	Hwy	USA	Marine Parkway–Gil Hodges Memorial Bridge (2013)
4	Burlington–Bristol Bridge	164.6 (540)	Truss	701.3 (2301)	1931	Hwy	USA	Burlington–Bristol Bridge (2013)
5	Illinois River TP&W Railroad Bridge	140.2 (460)	Truss	329.2 (1080)	1909	Rwy	USA	Illinois River TP&W Railroad Bridge (2013)
6	Harry S. Truman Bridge	130.1 (427)	Truss			Rwy	USA	Harry S. Truman Bridge (2013)
7	Roosevelt Island Bridge	127.4 (418)	Truss	285.5 (936.7)	1955	Hwy	USA	Roosevelt Island Bridge (2013)
8	James River Bridge (Isle of Wight County)	126.5 (415)		7,071.8 (23,201.3)	1980	Hwy	USA	James River Bridge (Isle of Wight County) (2013)
9	Cape Fear Memorial Bridge	124.4 (408.2)	Truss	924.5 (3033.3)	1969	Hwy	USA	Cape Fear Memorial Bridge (2013)
10	Wabash Railroad Bridge	123.7 (406)	Truss		1993	Rwy	USA	Wabash Railroad Bridge (2013)
11	Rochefort/ Martrou Bridge	139 (456)	Truss	175 (574.1)	1900	Ped	France	Pilot (2007)

(Continued)

(Continued)

Vertical Lift Bridges

Rank	Bridge Name	Lift Span Length m (ft)	Structure Features	Total Length m (ft)	Year Opened	Usage	Country	References
12	Duluth Aerial Lift Bridge	117.5 (385.8)	Truss	152.9 (501.7)	1930	Hwy	USA	Duluth Aerial Lift Bridge (2013)
13	International Railway Bridge	110.9 (363.9)	Truss			Rwy	USA/Canada	International Railway Bridge (2013)
14	James River Bridge (Prince George County)	110.9 (363.9)	Truss	1360.4 (4463.1)	1967	Hwy	USA	James River Bridge (Prince George County) (2013)
15	Main Street Bridge	111.3 (365)	Truss	512.1 (1680)	1941	Hwy	USA	Main Street Bridge (2013)
16	Claiborne Avenue Bridge	109.7 (359.8)	Truss	736.8 (2417.3)	1957	Hwy	USA	Claiborne Avenue Bridge (2013)
17	Passaic River Bridge	101.5 (333)	Truss	611.1 (2005)	1939	Hwy	USA	Passaic River Bridge (2013)
18	Danziger Bridge	100.6 (329.9)	Truss	996.4 (3269.1)	1989	Hwy	USA	Danziger Bridge (2013)
19	Kattwyk Liftbridge at Hamburg	100 (328.1)	Truss	285 (935)	1973	Hwy	Germany	Humpf and Saul (2008)
20	Pont Gustave–Flaubert	100 (328.1)	Box girder	670 (2198.2)	2008	Hwy	France	Pont Gustave–Flaubert (2011)

Swing Bridges

Rank	Bridge Name	Swing Span Length m (ft)	Structure Features	Total Length m (ft)	Year Opened	Usage	Country	References
1	El Ferdan Railway Bridge	340 (1115.5)	Truss double swing	640 (2099.7)	2001	Rwy	Egypt	El Ferdan Railway Bridge (2013)
2	Ford Island Bridge	283.5 (930)	Floating swing	1609.3 (5280)	1998	Hwy	USA	Abrahams (1996)
3	Yumemai Bridge	280 (918.6)	Floating arch	410 (1345.1)	2001	Hwy	Japan	Maruyama and Kawamura (1999)
4	Han River Bridge	201 (659.4)	Cable-stayed swing		2000	Hwy	Vietnam	Han River Bridge (2013)
5	Hood Canal Bridge	182.9 (600)	Floating swing	1990.3 (6530)	1962	Hwy	USA	Lwin and Kulicki (2013). Chapter 2

(*Continued*)

		Swing Span Length m (ft)	Structure Features	Total Length m (ft)	Year Opened	Usage	Country	References
			Swing Bridges					
Rank	Bridge Name	Swing Span Length m (ft)	Structure Features	Total Length m (ft)	Year Opened	Usage	Country	References
6	Port of Barcelona Bridge	180 (590.6)	Cable-stayed swing	255 (836.6)	1997	Hwy	Spain	Humpf and Saul (2008)
7	Kaiser Wilhelm Bridge	159 (521.7)	Double swing		1907	Hwy	Germany	Kaiser Wilhelm Bridge (2013)
8	Spokane Street Bridge	146.3 (480)	Concrete double swing	254.8 (836)	1991	Hwy	USA	Spokane Street Bridge (2013)
9	Kedzie Avenue CN Railroad Bridge	146.1 (479.4)	Truss	146.1 (479.4)	1899	Rwy	USA	Kedzie Avenue CN Railroad Bridge (2013)
10	East Haddam Bridge	139 (456)	Truss	269 (881)	1999	Hwy	USA	East Haddam Bridge (2013)
11	George P. Coleman Memorial Bridge	137.1 (450)	Truss double swing	1143 (3750)	1995	Hwy	USA	George P. Coleman Memorial Bridge (2013)
12	Macombs Dam Bridge	124.4 (408)		774.2 (2540)	1895	Hwy	USA	Macombs Dam Bridge (2013)
13	Fort Pike Bridge	121.9 (399.8)	Truss	1182 (3877.9)	2008	Hwy	USA	Fort Pike Bridge (2013)
14	Samuel Beckett Bridge	120 (393.7)	Cable-stayed swing			Ped	Ireland	Samuel Beckett Bridge (2013)
15	Lemont Railroad Bridge	114.3 (375)	Truss	114.3 (375)	1898	Rwy	USA	Lemont Railroad Bridge (2013)
16	Harlem Avenue Railroad Bridge	113.4 (372.5)	Truss	113.5 (372.5)	1899	Rwy	USA	Harlem Avenue Railroad Bridge (2013)
17	Third Avenue Bridge	110.9 (364)		853.4 (2800)	1898/2004	Hwy	USA	Third Avenue Bridge (2013)

(*Continued*)

(Continued)

		Swing Bridges						
Rank	Bridge Name	Swing Span Length m (ft)	Structure Features	Total Length m (ft)	Year Opened	Usage	Country	References
18	Willis Avenue Bridge	106.7 (350)	Truss	979 (3212)	1987/2010	Hwy	USA	Willis Avenue Bridge (2013)
19	Cicero Avenue Railroad Bridge	102 (334.5)	Truss	102 (334.5)	1900	Rwy	USA	Cicero Avenue Railroad Bridge (2013)
20	Ouse Swing Bridge	100 (328.1)	Cable-stayed swing		2005	Hwy	UK	Ouse Swing Bridge (2013)

		Bascule Bridges						
Rank	Bridge Name	Bascule Span Length m (ft)	Structure Features	Total Length m (ft)	Year Opened	Usage	Country	References
1	Duwamish River West Bridge	115.5 (379)		88.25 (2895.2)	1996	Hwy	USA	Duwamish River West Bridge (2013)
2	Knippel Bridge	115 (377.3)			1937	Hwy	Denmark	Knippel Bridge (2013)
3	10th Street Bridge	114.3 (375)			1958	Hwy	USA	10th Street Bridge (2013)
4	John Ross Bridge	109.4 (358.8)	Arch truss	577.4 (1894.5)	1917	Hwy	USA	John Ross Bridge (2013)
5	Morrison Street Bridge	103.6 (339.9)	Truss	231.6 (759.9)	1958	Hwy	USA	Morrison Street Bridge (2013)
6	Charles Berry Memorial Bridge	101.5 (333)	Truss	320.6 (1052)	1940	Hwy	USA	Charles Berry Memorial Bridge (2013)
7	Curtis Creek Bridge	97.5 (319.9)		973.4 (3075.3)	1978	Hwy	USA	Curtis Creek Bridge (2013)
8	Pennington Avenue Bridge	96.6 (316.9)		584 (1916.1)	1976	Hwy	USA	Pennington Avenue Bridge (2013)
9	SW Second Avenue Bascule Bridge	92.5 (303.5)	Double leaf		2003	Hwy	USA	AISC (2002)
11	Broadway Bridge (Portland)	92.7 (304)	Double leaf	531 (1742)	1913	Hwy	USA	Broadway Bridge (Portland) (2013)

(*Continued*)

				Bascule Bridges				
Rank	Bridge Name	Bascule Span Length m (ft)	Structure Features	Total Length m (ft)	Year Opened	Usage	Country	References
12	Matosinhos–Leça-Porto Movable Bridge	92 (301.8)	Double leaf		2007	Hwy	Portugal	Matosinhos–Leça-Porto Movable Bridge (2013)
13	José Leon de Carranza Bridge	90 (295.3)		1400 (4593.2)	1969	Hwy	Spain	José Leon de Carranza Bridge (2013)
14	Fore River Bridge	87.5 (287.1)		1447.3 (4748.3)	1998	Hwy	USA	Fore River Bridge (2013)
15	Casco Bay Bridge	86.9 (285)		1447.2 (4748)	1996	Hwy	USA	Casco Bay Bridge (2013)
16	High Rise Bridge	85.3 (279.9)		1470.8 (4825.4)	1969	Hwy	USA	High Rise Bridge (2013)
17	Intracoastal Canal Bridge	84.1 (275.9)		1680.5 (5513.4)	2005	Hwy	USA	Intracoastal Canal Bridge (2013)
18	7th Street Bridge	82.9 (272)	Single leaf		1933	Hwy	USA	7th Street Bridge (2013)
19	Woodrow Wilson Memorial Bridge	82.3 (270)		1855.4 (6087.2)	2008	Hwy	USA	Woodrow Wilson Memorial Bridge (2013)
20	Columbus Drive Bridge	82 (269)		118.6 (389)	1982	Hwy	USA	Columbus Drive Bridge (2013)

29.10 Top 20 Longest Stress Ribbon Bridge Spans

				Stress Ribbon Bridges				
Rank	Bridge Name	Main Span Length m (ft)	Structure Features	Total Length m (ft)	Year Opened	Usage	Country	References
1	Rowing Canal Bridge	150 (492.1)	PS concrete	246 (807.1)	1989	Ped	Bulgaria	Partov and Dinev (2013) Chapter 6
2	Yumetsuri Bridge	147.6 (484.3)	PS concrete	172.6 (566.3)	1996	Ped	Japan	Strasky (2005)

(*Continued*)

(Continued)

Stress Ribbon Bridges								
Rank	Bridge Name	Main Span Length m (ft)	Structure Features	Total Length m (ft)	Year Opened	Usage	Country	References
3	Lignon–Loex Bridge	136 (446.2)	PS concrete	136 (446.2)	1971	Ped	Switzerland	Strasky (2005)
4	Mori-no-wakuwaku Hashi Bridge	128.5 (421.6)	External PS	165.5 (543)	2001	Ped	Japan	Strasky (2005)
5	Redding Bridge	127.4 (418)	PS concrete	127.4 (418)	1990	Ped	USA	Strasky (2005)
6	Rio Colorado Bridge	124 (406.8)	PS concrete	216 (708.7)	1974	Hwy	Costa Rica	Rio Colorado Bridge (2013)
7	Umenoki–Todoro Park Bridge	105 (344.5)	PS concrete	116 (354.3)		Ped	Japan	Strasky (2005)
8	Lake Hodges Bridge	100.6 (330)	PS concrete	302 (990)	2009	Ped	USA	Lwin and Kulicki (2013) Chapter 2
9	Nymbuk Bridge	102 (334.7)	PS concrete	219 (718.5)	1985	Ped	Czech	Strasky (2005)
10	Prague–Troja Bridge	96 (315)	PS concrete	249 (816.9)	1984	Ped	Czech	Strasky (2005)
11	Puente de la Barrade Maldonado	90 (295.3)	PS	150 (492.1)		Hwy	Uruguay	Puente de la Barrade Maldonado (2013)
12	Grants Pass Bridge	84.7 (278)	PS	200.6 (658)	2000	Ped	USA	Strasky (2005)
13	Tonbo No Hashi Bridge	80 (262.5)	CIP	150 (492.1)	1996	Ped	Japan	Strasky (2005)
14	Pasarella de Sant Pere	80 (262.5)	Prestressed concrete/ PS	80 (262.5)	2004		Spain	Pasarella de Sant Pere (2013)
15	Brno–Komin Bridge	78 (255.9)	PS concrete	97 (318.2)	1985	Ped	Czech	Strasky (2005)
16	Blue Valley Ranch Bridge	76.81	CIP/PS	76.81	2001	Ped	USA	Strasky (2005)
17	Glacis Bridge	76 (249.3)		174 (570.9)	1999	Ped	Germany	Glacis Bridge (2013)
18	Essing Bridges	73 (239.5)	Timber	192 (629.9)	1986	Ped	Germany	Culling (2009)

(Continued)

				Stress Ribbon Bridges				
Rank	Bridge Name	Main Span Length m (ft)	Structure Features	Total Length m (ft)	Year Opened	Usage	Country	References
19	Shin Mominoki Suspension Bridge	72 (236.2)	PS	72 (236.2)	1988	Ped	Japan	Shin Mominoki Suspension Bridge (2013)
20	Hureai Bridge	70 (229.7)					Japan	Hureai Bridge (2013)

29.11 Top 20 Longest Timber Bridge Spans

Rank	Bridge Name	Main Span Length m (ft)	Structure Features	Total Length m (ft)	Year Opened	Usage	Country	References
1	Remseck Neckar Bridge	80 (262.5)	Three Trusses	85 (278.9)	1990	Ped	Germany	Remseck Neckar Bridge (2013)
2	Essing Bridge	73 (239.5)	Stress ribbon	192 (629.9)	1992	Ped	Germany	Culling (2009)
3	Stuttgart–Bad Cannstatt Bridge	72 (239.5)	Stress ribbon	158 (518.4)	1997	Ped	Germany	Baláž (2013a) Stuttgart–Bad Cannstatt Bridge (2013)
4	Flisa Bridge	71 (232.9)	Truss, arch	181.5 (595.6)	2003	Hwy	Norway	Flisa Bridge (2013)
5	Tynset Bridge	70 (229.7)	Arch-truss	124 (406.8)	2001	Hwy	Norway	Nordic (2013)
6	Maribyrnong River Footbridge	68 (223.1)	Two-hinged arch	68 (223.1)	1995	Ped	Australia	BSC (2013)
7	Sioux Narrows Bridge	64 (210.0)	Howe Truss	64 (210.0)	1936		Canada	Sioux Narrows Bridge (2013)
8	Bridgeport Covered Bridge	63.4 (208)	Arch	71 (233)	1862	Hwy	USA	Bridgeport Covered Bridge (2013)
9	Jackson Covered Bridge	58.4 (191.5)	Covered	61 (200)	2000		USA	Jackson Covered Bridge (2013)

(Continued)

(*Continued*)

Rank	Bridge Name	Main Span Length m (ft)	Structure Features	Total Length m (ft)	Year Opened	Usage	Country	References
10	Hartland Bridge	55.86 (181.3)	Howe truss	391 (1282.8)	1901	Hwy	Canada	Hartland Bridge (2013)
11	Dragon's Tail Bridge	55 (180.4)	Stress ribbon	225 (738.2)	2007	Ped	Germany	Dragon's Tail Bridge (2013)
12	Office Bridge	54.9 (180)	Howe truss		1944	Hwy	USA	Oregon (2013)
13	Erdberger Steg Bridge	52.5 (172.2)	Rigid Frame	85.2 (279.5)	2003	Ped	Austria	Baláž (2013a) Erdberger Steg Bridge (2013)
14	Greensborough Footbridge	50 (164)	Three-hinged arch	50 (164)	1975	Ped	Australia	BSC (2013)
15	Great Karikobozu Bridge	50 (164)	King-post truss	140 (459.3)	2002	Hwy	Japan	Nagai, et al. (2013)
16	Måsør Bridge	50		83	2005	Hwy	Norway	Aasheim (2011)
17	Vaires-sur-Marne Footbridge	49 (160.8)	Covered arch	75 (246.1)	2004	Ped	France	Baláž (2013a) Vaires-sur-Marne Footbridge (2013)
18	Dell'Accad:mia Bridge	48 (157.5)	Arch	48 (157.5)	1933	Ped	Italy	Pilot (2007)
19	Kicking Horse Bridge	46 (150)	Burr arch	46 (150)	2001	Ped	Canada	Golden (2011)
20	Smolen–Gulf Bridge	45.7 (150)	Truss	186.8 (613)	2008	Hwy	USA	Smolen–Gulf Bridge (2011)

29.12 Top 15 Longest Floating Bridge Spans

Rank	Bridge Name	Total Length m (ft)	Structure Features	Draw Span m (ft)	Year Opened	Usage	Country	References
1	Albert D. Rosselini Bridge	2309.8 (7578)		61 (200)	1963	Hwy	USA	Lwin and Kulicki (2013) Chapter 2
2	First Lake Washington Bridge	1999.8 (6561)		61 (200)	1940	Hwy	USA	Lwin and Kulicki (2013)
3	New Lacey V. Murrow Bridge	1999.8 (6561)		None	1993	Hwy	USA	Lwin and Kulicki (2013)
4	Hood Canal Bridge	1990.3 (6530)		182.9 (600)	1962	Hwy	USA	Lwin and Kulicki (2013)

(Continued)

Rank	Bridge Name	Total Length m (ft)	Structure Features	Draw Span m (ft)	Year Opened	Usage	Country	References
5	Demerara Harbor Bridge	1851 (6072.8)			1978	Hwy	Guyana	Demerara Harbor Bridge (2013)
6	Homer M. Hadley Bridge	1748.3 (5736)		None	1989	Hwy	USA	Lwin and Kulicki (2013) Chapter 2
7	Nordhordland Bridge	1246 (4087.9)		None	1994	Hwy	Norway	Nordhordland Bridge (2013)
8	Bergøysund Floating Bridge	933 (3061)		None	1992	Hwy	Norway	Bergøysund Floating Bridge (2013)
9	William R. Bennett Bridge	690 (2263.8)		None	2008	Hwy	Canada	William R. Bennett Bridge (2013)
10	Yumemai Bridge	410 (1345.1)	Arch	280 (918.6)	2001	Hwy	Japan	Maruyama and Kawamura (1999)
11	Dongjin Bridge	400 (1312.3)	Wood, boat	None	1173	Ped	China	Dongjin Bridge (2013)
12	Eastbank Esplanade	365.8 (1200)		None	2001	Ped	USA	Eastbank Esplanade (2013)
13	Dubai Floating Bridge	365 (1197.5)		None	2007	Hwy	United Arab Emirates	Dubai Floating Bridge (2013)
14	Ford Island Bridge	283.5 (930)		283.5 (930)	1998	Hwy	USA	Abrahams, M. J. (1996)
15	Queen Emma Bridge	167 (547.9)		None	1939	Hwy	Curaçao	Queen Emma Bridge (2013)

Acknowledgements

I would like to thank suggestions and recommendations during the development of the lists of single tower cable-stayed bridges, extradosed bridges, masonry arch bridges and timber bridges by Professor Ivan Baláž, Slovak University of Technology, Slovak Republic; and extradosed bridges by Mr. Akio Kasuga, Sumitomo Mitsui Construction, and Japan. Advice and information provided from Professor Baláž is greatly appreciated.

References

7th Street Bridge. (2013). http://www.historicbridges.org/other/7th.

10th Street Bridge. (2013). http://www.historicbridges.org/other/10th.

Abrahams, M. J. (1996). "Ford Island Bridge, Pearl Harbor, Hawaii," *Sixth Biennial Symposium, Heavy Movable Structures, Inc.*, October 30–November 1, Clearwater Beach, FL. http://heavymovablestructures.org/hms_papers/Symposium_6/36.pdf.

Adolphe Bridge. (2013). http://en.wikipedia.org/wiki/Adolphe_Bridge.

AISC. (2002). *Modern Steel Construction*, American Institute of Steel Constriction, Chicago, IL.

Aizhai Bridge. (2013). http://www.highestbridges.com/wiki/index.php?title=Aizhai_Bridge.

Alf Priolo Bridge. (2013). http://www.liverpoolecho.co.uk/liverpool-news/local-news/2008/04/01/we-ll-build-longest-bridge-in-the-world-100252-20699209.

Antioch Bridge. (2013). http://bata.mtc.ca.gov/bridges/antioch.htm.

Arrabida Bridge. (2013). http://www.britannica.com/EBchecked/topic/35988/Arrabida-Highway-Bridge.

Arthur Kill Vertical Lift Bridge. (2013). http://en.wikipedia.org/wiki/Arthur_Kill_Vertical_Lift_Bridge.

Aasheim, E. (2011). "Timber in Construction: Move Forward", Keynote Address, Nov. 17, Kuala Lumpur, Malaysia.

Astoria–Megler Bridge. (2013). http://en.wikipedia.org/wiki/Astoria%E2%80%93Megler_Bridge.

Atchafalaya Basin Bridge. (2013). http://en.wikipedia.org/wiki/Atchafalaya_Basin_Bridge.

Auckland Harbor Bridge. (2013). http://en.wikipedia.org/wiki/Auckland_Harbour_Bridge.

Bakhoum, M. M. (2013). "Chapter 26, Bridge Engineering in Egypt", *Handbook of International Bridge Engineering*, Ed, Chen, W.F. and Duan, L. CRC Press, Boca Raton, FL."

Baláž, I. (2013a). "Timber Bridges with Record Spans", Eurostav, No. 5, pp. 62–63. (in Slovak)

Baláž, I. (2013b). "Chapter 17 Bridge Engineering in Slovak Republic", *Handbook of International Bridge Engineering*, Ed, Chen, W.F. and Duan, L. CRC Press, Boca Raton, FL.

Baláž, I. and Ding D. (2006). "Cable-Stayed Bridges", *Proceedings of International Colloquium on Stability and Ductility of Steel Structures*, Ed., Camotim, D. et al., Sept. 6-8, Lisboa, Portugal, pp. 1083–1090.

Baling River Bridge. (2013). http://en.wikipedia.org/wiki/Balinghe_Bridge.

Baltic Sea Bridge. (2013). http://www.bloomberg.com/apps/news?pid=newsarchive&sid=amTE6UqwgoYE.

Bayonne Bridge. (2013). http://www.panynj.gov/bridges-tunnels/bayonne-bridge-facts-info.html.

BD&E. (2008). " Korea's Newest City to Have Two Dramatic Bridges," *Bridge Design & Engineering*, 2nd Quarter, 14(51):35.

Beijing Bridge. (2013). http://news.sohu.com/20090730/n265587893.shtml (in Chinese).

Beijing–Shanghai HSR. (2013). http://jc8858.blog.163.com/blog/static/111837469200921293649154 (in Chinese).

Beijing–Tianjin Intercity Railway. (2013). http://www.crecc.com.cn/zhuanti/lunwen/jingjin.htm (in Chinese).

Bergøysund Floating Bridge. (2013). http://en.structurae.de/structures/data/index.cfm?id=s0005140.

Bhumibol Bridge. (2013). http://en.wikipedia.org/wiki/Bhumibol_Bridge.

Blank S. A. and Seliverstov, A. A. (2013). "Chapter 15 Bridge Engineering in Russia", *Handbook of International Bridge Engineering*, Ed, Chen, W.F. and Duan, L. CRC Press, Boca Raton, FL.

Blank, S. A. and Seliverstov, V. A. (2013). "Chapter 15, Bridge Engineering in Russia", *Handbook of International Bridge Engineering*, Ed, Chen, W.F. and Duan, L. CRC Press, Boca Raton, FL.

Bloukrans Bridge. (2013). http://en.wikipedia.org/wiki/Bloukrans_Bridge.

Bridge of the Americas. (2013). http://en.wikipedia.org/wiki/Bridge_of_the_Americas.

Bridgeport Covered Bridge. (2013). http://en.wikipedia.org/wiki/Bridgeport_Covered_Bridge.

Broadway Bridge (Portland). (2013). http://en.wikipedia.org/wiki/Broadway_Bridge_(Portland).

BSC. (2013). http://www.bscconsulting.com.au/municipal_engineering.php.

Burlington–Bristol Bridge. (2013). http://en.wikipedia.org/wiki/Burlington-Bristol_Bridge.

C215 Viaduct. (2013). http://en.wikipedia.org/wiki/C215_Viaduct.

Caiyuanba Yangtze River Bridge. (2013). http://news.163.com/07/1029/04/3RUOQEF10001124J.html (in Chinese).

Calafat-Vidin Bridge. (2013.) http://en.structurae.de/structures/data/index.cfm?id=s0004900

Calgaro, J. A. (2013). "Chapter 11, Bridge Engineering in France", *Handbook of International Bridge Engineering*, Ed, Chen, W.F. and Duan, L. CRC Press, Boca Raton, FL.

Caltrans. (2013). http://dot.ca.gov/hq/structur/strmaint/brlog/logpdf/logd03.pdf.

Cape Cod Canal Railroad Bridge. (2013). http://en.wikipedia.org/wiki/Cape_Cod_Canal_Railroad_ Bridge.

Cape Fear Memorial Bridge. (2013). http://bridgehunter.com/nc/new-hanover/1290013.

Carquinez Bridge. (2013). http://en.wikipedia.org/wiki/Carquinez_Bridge.

Casco Bay Bridge. (2013). http://en.wikipedia.org/wiki/Casco_Bay_Bridge.

Carquinez Bridge. (2013). http://en.wikipedia.org/wiki/Carquinez_Bridge.

Casco Bay Bridge. (2013). http://en.wikipedia.org/wiki/Casco_Bay_Bridge.

CBBT. (2013). http://www.cbbt.com/facts.html.

Chang, W.W. (2004). "Construction of 100 m Stone Arch Bridge",. *Shanxi Transportation Technology*, No. 4. (In Chinese)

Changhong Bridge. (2013). http://zhidao.baidu.com/question/129626690.html (in Chinese).

Changdong Yellow River Bridge. (2013). http://baike.baidu.com/view/1582277.htm (in Chinese).

Chaotianmen Bridge. (2013). http://en.wikipedia.org/wiki/Chaotianmen_Bridge.

Charles Berry Memorial Bridge. (2013). http://www.dot.state.oh.us/districts/D03/Documents/Fact%20Sheet%20-%20Bascule%20Bridge-102108.pdf.

Chateaubriand Bridge. (2013). http://en.structurae.de/structures/data/index.cfm?id=s0002926.

Chelsea Bridge. (2013). http://en.wikipedia.org/wiki/Chelsea_Bridge.

Chen, B. C. (2009). "Construction Methods of Arch Bridges in China," *Chinese-Croatian JointColloquium*, Oct. 05-09, Fuzhou, China.

Chen, B. C. (2013). "Long-Span Arch Bridges in China," *Chinese-Croatian Joint Colloquium,* July 10–14, Brijuni Islands, Criatia.

Chenab Bridge. (2013). http://www.kotu.oulu.fi/find2008/esitykset/29.5/pulkkinen.pdf.

Chennai Port–Maduravoyal Expressway. (2013). http://en.wikipedia.org/wiki/Chennai_Port_%E2%80%93_Maduravoyal_Expressway.

China Bridges. (2013). http://www.chinakzw.com/BBS/forum.php?mod=viewthread&tid=25789#lastpost (in Chinese).

Chunan Napu Bridge. (2013). http://www.sneb6.cn/rongyu/show.asp?id=101.

Cicero Avenue Railroad Bridge. (2013). http://www.historicbridges.org/bridges/browser/?bridgebrowser=illinois/cicerorr.

Claiborne Avenue Bridge. (2013). http://bridgehunter.com/la/orleans/23600463102221.

Columbus Drive Bridge. (2013). http://bridgehunter.com/il/cook/columbus-drive.

Commodore Barry Bridge. (2013). http://en.wikipedia.org/wiki/Commodore_Barry_Bridge.

Confederation Bridge. (2013). http://www.confederationbridge.com/en/design_construction.php.

Crescent City Connection. (2013). http://en.wikipedia.org/wiki/Crescent_City_Connection.

Culling D. (2009). "Critical Analysis of the Essing Timber Bridge, Germany," *Proceedings of Bridge Engineering 2 Conference*, April 2009, University of Bath, Bath, UK. http://www.bath.ac.uk/ace/uploads/StudentProjects/Bridgeconference2009/Papers/CULLING.pdf

Culling, D. (2009). "Critical Analysis of the Essing Timber Bridge, Germany," *Proceedings of Bridge Engineering 2 Conference*, April, University of Bath, UK. http://www.bath.ac.uk/ace/uploads/StudentProjects/Bridgeconference2009/Papers/CULLING.pdf.

Curtis Creek Bridge. (2013). http://bridgehunter.com/md/baltimore-city/300000BCZ001015.

Daning River Bridge. (2013). http://www.sxcoal.com/coal/1190049/articlenew.html (in Chinese).

Dankun Bridge. (2013). http://news.xinhuanet.com/politics/2008-04/22/content_8025633.htm (in Chinese).

Danziger Bridge. (2013). http://bridgehunter.com/la/orleans/danziger.

Dashengguan Bridge. (2013). http://baike.baidu.com/view/1415725.html (in Chinese).

Deggendorf-Deggenau Bridge. (2013). http://hexahedron.hu/personal/m_ivanyi/danube-bridges/d/09/comp.htm

Demerara Harbor Bridge. (2013). http://en.wikipedia.org/wiki/Demerara_Harbour_Bridge.

Dongguan Shuidao Bridge. (2013). http://wenku.baidu.com/view/e2439deb172ded630b1cb681.html (in Chinese).

Donghai Bridge. (2013). http://baike.baidu.com/view/122000.htm (in Chinese).

Dongjin Bridge. (2013). http://baike.baidu.com/view/173518.htm (in Chinese).

Dragon's Tail Bridge. (2013). http://www.koeppl-ingenieure.de/cms/index.php?option=com_content&view=article&id=151&Itemid=174 (in German).

Dubai Floating Bridge. (2013). http://en.wikipedia.org/wiki/Floating_Bridge_Dubai.

Duluth Aerial Lift Bridge. (2013). http://bridgehunter.com/mn/st-louis/aerial.

Düsseldorf–Neuss Bridge. (2013). http://en.structurae.de/structures/data/index.cfm?id=s0000329.

Duwamish River West Bridge. (2013). http://bridgehunter.com/wa/king/14459A0000000.

East Haddam Bridge. (2013). http://www.kurumi.com/roads/ct/ct82.htm. Eastbank Esplanade. (2013). http://en.wikipedia.org/wiki/Eastbank_Esplanade.

Edong Yangtze River Bridge. (2013). http://baike.baidu.com/view/3963034.htm.

El Ferdan Railway Bridge. (2013). http://en.structurae.de/structures/data/index.cfm?ID=s0002510.

Elevated Expressways in Hyderabad. (2013). http://en.wikipedia.org/wiki/Elevated_Expressways_in_Hyderabad.

Emsland Test Facility. (2013). http://googlesightseeing.com/2008/01/maglev-test-track.

Erdberger Steg Bridge. (2013). http://en.structurae.de/structures/data/index.cfm?id=s0019561

First He Jiang Yangtze River Bridge. (2013). http://baike.baidu.com/view/5075940.html (in Chinese).

Firth of Forth Bridge. (2013). http://en.wikipedia.org/wiki/Forth_Bridge.

Flehe Bridge. (2013). http://en.wikipedia.org/wiki/Flehe_Bridge

Flisa Bridge. (2013). http://en.structurae.de/structures/data/index.cfm?id=s0009160.

Folić, R. J. (2013). "Chapter 16, Bridge Engineering in Serbia", *Handbook of International Bridge Engineering*, Ed, Chen, W.F. and Duan, L. CRC Press, Boca Raton, FL.

Fore River Bridge. (2013). http://bridgehunter.com/me/cumberland/5900.

Fort Pike Bridge. (2013). http://en.wikipedia.org/wiki/Fort_Pike_Bridge.

Fourth Nanjing Yangtze River Bridge. (2013). http://baike.baidu.com/view/2432170.htm (in Chinese).

Francis Scott Key Bridge. (2013). http://www.dcroads.net/crossings/key-MD.

Fremont Bridge. (2013). http://en.wikipedia.org/wiki/Fremont_Bridge_(Portland).

Friendship Bridge. (2013). http://en.wikipedia.org/wiki/Friendship_Bridge_(Paraguay%E2%80%93Brazil).

Fushun Tuojiang Bridge. (2013) http://hi.baidu.com/chengchong2010/item/fb2b76d956f77eefb2f777f1

Gazelle Bridge. (2013). http://en.structurae.de/structures/data/index.cfm?ID=s0003925.

General Rafael Urdaneta Bridge. (2013). http://en.wikipedia.org/wiki/General_Rafael_Urdaneta_Bridge.

George P. Coleman Memorial Bridge. (2013). http://en.wikipedia.org/wiki/George_P._Coleman_Memorial_Bridge.

Glacis Bridge. (2011). http://en.structurae.de/structures/data/index.cfm?id=s0001027

Gladesville Bridge. (2013). Construction Projects in Japan. http://www.bridgepros.com/projects/Gladesville/Gladsesville.htm.

Golden Ears Bridge. (2013). http://en.wikipedia.org/wiki/Golden_Ears_Bridge

Golden. (2013). http://en.wikipedia.org/wiki/Golden_British_Columbia.

Grand Duchess Charlotte. (2013). http://en.structurae.de/structures/data/index.cfm?ID=s0000520.

Great Belt Fixed Link. (2013). http://en.wikipedia.org/wiki/Great_Belt_Fixed_Link.

Great Loire River Bridge. (2013). http://en.structurae.net/structures/data/index.cfm?id=s0026038.

Grenland Bridge. (2013). http://en.wikipedia.org/wiki/Grenland_Bridge

Grümpental Bridge. (2013). http://de.wikipedia.org/wiki/Gr%C3%BCmpentalbr%C3%BCcke (in German).

Guangxi Yongning Bridge. (2013). http://oldweb.cqvip.com/qk/90739X/199709/2836622.html (in Chinese).

Hamana Bridge. (2013). http://en.structurae.de/structures/data/index.cfm?id=s0006637.

Hami Bridge. (2013). http://www.cnr.cn/allnews/201011/t20101108_507287855.html (in Chinese).

Han River Bridge. (2013). http://en.wikipedia.org/wiki/Han_River_Bridge. Hangzhou Bay Bridge. (2013). http://baike.baidu.com/view/50119.htm (in Chinese).

Hanzhuang Canal Bridge. (2013). http://www.cityphotos.cn/picture/albumpic.aspx?aid=18942 (in Chinese).

Hardanger Bridge. (2013). http://www.roadtraffic-technology.com/projects/hardangerbridge.

Hartland Bridge. (2011). http://en.wikipedia.org/wiki/Hartland_Bridge

Harlem Avenue Railroad Bridge. (2013). http://www.historicbridges.org/bridges/browser/?bridgebrowser=illinois/harlemrr.

Harry S. Truman Bridge. (2013). http://en.wikipedia.org/wiki/Harry_S._Truman_Bridge.

Hart Bridge. (2013). http://en.wikipedia.org/wiki/Hart_Bridge.

Hartland Bridge. (2013). http://en.wikipedia.org/wiki/Hartland_Bridge.

High Coast Bridge. (2013). http://en.wikipedia.org/wiki/H%C3%B6ga_Kusten_Bridge.

High Rise Bridge. (2013). http://bridgehunter.com/va/chesapeake-city/high-rise.

Hikoshima Bridge. (2013). http://www.hikoshima.com/sea/hikoshimaohashi.htm (in Japanese).

Hino, S. 2005. "The Great Himiyume Bridge, Highway Entrance to Nagasaki, The World First Extradosed Bridge Using Corrugated Steel Plate Webs", www.jsce.or.jp/kokusai/civil_engineering/2005/4-2.pdf

Hiroshima Airport Bridge. (2013). http:// www.highestbridges.com/wiki/index.php?title=Hiroshima_Airport_Bridge.

Hong Kong–Zhuhai–Macau Bridge. (2013). http://en.wikipedia.org/wiki/Hong_Kong-Zhuhai-Macau_Bridge.

Hong River Bridge. (2013). http://baike.baidu.com/view/2568168.htm (in Chinese).

Horace Wilkinson Bridge. (2013). http://en.wikipedia.org/wiki/Horace_Wilkinson_Bridge.

Hosur Road. (2013). http://en.wikipedia.org/wiki/Hosur_Road.

Houston Freeways. (2013). http://oscarmail.net/houstonfreeways/ebook/CH7_bridges_and_tunnels_pp340-375_72.pdf.

Howrah Bridge. (2013). http://en.wikipedia.org/wiki/Howrah_Bridge.

Huanghua Yuan Bridge. (2013). http://baike.baidu.com/view/1488824.htm (in Chinese).

Huangpu Bridge. (2013). http://en.wikipedia.org/wiki/Huangpu_Bridge.

Humber Bridge. (2013). http://en.wikipedia.org/wiki/Humber_Bridge.

Humen Bridge. (2013). http://baike.baidu.com/view/195368.htm (in Chinese).

Humpf, K. and Saul, R. (2008). "Innovative Design of Movable Bridges," *12th Biennial Symposium, Heavy Movable Structures, Inc.*, November 3–6, Orlando, FL. http://www.lap-consult.com/pdf-files/deutsch/sonderdrucke/sdr550.pdf.

Hureai Bridge. (2013). http://en.structurae.de/structures/data/index.cfm?id=s0011828

I-10 Twin Span bridge. (2013). http://en.wikipedia.org/wiki/I-10_Twin_Span_Bridge.

Ikitsuki Bridge. (2013). http://ja.wikipedia.org/wiki/%E7%94%9F%E6%9C%88%E5%A4%A7%E6%A9%8B (in Japanese).

Iller Bridges Kempten. (2013). http://en.structurae.net/structures/data/index.cfm?id=s0012036.

Illinois River TP&W Railroad Bridge. (2013). http://bridgehunter.com/il/peoria/bh38089.

Imai, Y. (2013). http://www.jsce.or.jp/committee/concrete/newsletter/newsletter02/newsletter02f/ 1-Mongolia%20(Imai).pdf.

Incheon Bridge. (2013). http://en.wikipedia.org/wiki/Incheon_Bridge

Infante D. Henrique Bridge. (2013). http://en.structurae.net/structures/data/index.cfm?id=s0004697.

International Railway Bridge. (2013). http://www.historicbridges.org/truss/ibrailvertical/index.htm.

Intracoastal Canal Bridge. (2013). http://bridgehunter.com/la/st-mary/intracoastal-canal.

Ironworkers Memorial Bridge. (2013). http://en.wikipedia.org/wiki/Ironworkers_Memorial_Second_Narrows_Crossing.

Jackson Covered Bridge. (2013). http://bridgehunter.com/il/cumberland/jackson.

James River Bridge (Isle of Wight County). (2013). http://bridgehunter.com/va/isle-of-wight/10364.

James River Bridge (Prince George County). (2013). http://bridgehunter.com/va/prince-george/14069.

JASBC. (2004). *Bridges in Japan (History of Iron and Steel Bridges)*, Japan Association of Steel Bridge Construction, Tokyo, Japan.

Jinan Huanghe Bridge. (2013). http://www.whfasten.com/gsyjqygs.html (in Chinese)

Jiangdong Bridge. (2013). http://baike.baidu.com/view/2112515.htm#1 (in Chinese).

Jiangjie River Bridge. (2013). http://baike.baidu.com/view/2318095.htm (in Chinese).

Jiangyin Bridge. (2013). http://en.wikipedia.org/wiki/Jiangyin_Suspension_Bridge.

Jinghang Canal Bridge. (2013). http://www.ztsj.com.cn/gc/gcl.html (in Chinese).

Jingtang Bridge. (2013). http://en.wikipedia.org/wiki/Jintang_Bridge (in Chinese).

Jinpu Railway Overcrossing Bridge. (2013). http://wenku.baidu.com/view/9f412cf7f61fb7360b4c658d .html (in Chinese).

Jintang Bridge. (2013). http://en.wikipedia.org/wiki/Jintang_Bridge.

Jin-shan Bridge. (2013). http://www.gxcic.net/tksp/PictureDetails.aspx?pid=449

Jiujiang Yangtze River Highway Bridge. (2013). http://baike.baidu.com/view/2606371.htm (in Chinese).

Jiuxigou Bridge. (2013). http://baike.baidu.com/view/326283.htm (in Chinese).

John Bridge. (2013). http://pt.wikipedia.org/wiki/Ponte_de_S%C3%A3o_Jo%C3%A3o (in Portuguese).

John Ross Bridge. (2013). http://en.wikipedia.org/wiki/Market_Street_Bridge_(Chattanooga).

José Leon de Carranza Bridge. (2013). http://en.wikipedia.org/wiki/Jos%C3%A9_Leon_de_Carranza_Bridge.

Jubilee Parkway. (2013). http://en.wikipedia.org/wiki/Jubilee_Parkway.

Kaiser Wilhelm Bridge. (2013). http://en.wikipedia.org/wiki/Kaiser-Wilhelm-Br%C3%BCcke.

Kaita Bridge. (2013). http://www.civil.eng.osaka-u.ac.jp/str/rec/kaita/kaita_eng.html.

Kama Bridge. (2013). http://en.wikipedia.org/wiki/Kama_Bridge.

Kao-Ping Hsi Bridge. (2013.) http://www.flickr.com/photos/little-wang/6052573309/

Karnali River Bridge. (2013). http://en.structurae.de/structures/data/index.cfm?id=s0004660

Kasuga, A. (2013). "Chapter 11. Extradosed Bridges", Bridge Engineering *Handbook, 2nd Edition, Volume 2-Superstructure Design*, Ed, Chen, W.F. and Duan, L. CRC Press, Boca Raton, FL

Kedzie Avenue CN Railroad Bridge. (2013). http://www.historicbridges.org/bridges/browser/? bridgebrowser=illinois/kedziecn.

Keong-An Bridge. (2013). http://en.structurae.de/structures/data/index.cfm?id=s0002145

King Fahd Causeway. (2013). http://en.wikipedia.org/wiki/King_Fahd_Causeway.

Kishiwada Bridge. (2013). http://en.structurae.de/structures/data/index.cfm?id=s0004438.

Kiso River Bridge. (2013). http://en.structurae.de/structures/data/index.cfm?id=s0000844

Knippel Bridge. (2013). http://en.wikipedia.org/wiki/Knippelsbro.

Koh, H. M. and Choo, J. F. (2011). *Recent Bridges in Korea*, http://bscw-app1.ethz.ch/pub/bscw.cgi/ d376861/06-H.M.%20Koh.pdf.

Konohana Bridge. (2013). http://en.wikipedia.org/wiki/Konohana_Bridge.

Konrad Adenauer Bridge. (2013). http://en.structurae.de/structures/data/index.cfm?id=s0001726.

Korror Babeldoap Bridge. (2013). http://en.wikipedia.org/wiki/Koror%E2%80%93Babeldaob_Bridge

Kurt Schumacher Bridge. (2013). http://en.structurae.de/structures/data/index.cfm?id=s0001096

Lake Pontchartrain Causeway. (2013). http://en.wikipedia.org/wiki/Lake_Pontchartrain_Causeway.

Laviolette Bridge. (2013). http://en.wikipedia.org/wiki/Laviolette_Bridge.

LemontRailroadBridge.(2013).http://www.historicbridges.org/bridges/browser/?bridgebrowser=illinois/ lemont.

Lewis and Clark Bridge. (2013). http://en.wikipedia.org/wiki/Lewis_and_Clark_Bridge_(Columbia_ River).

Leziria Bridge. (2013). http://www.wolframalpha.com/entities/bridges/zp/jx/bj.

Lianggang Bridge. (2013). http://baike.baidu.com/view/1762278.htm (in Chinese).

Liangshui River Bridge. (2013). http://news.china.com/zh_cn/news100/11038989/20060807/13521040 .html (in Chinese).

Lianxiang Bridge. (2013). http://www.arch-bridges.cn/Show.asp?PaperID=178 (in Chinese).

Liede Bridge. (2013). http://wenku.baidu.com/view/5696946e58fafab069dc028d.html (in Chinese).

Limsuwan, E. (2013). "Chapter 25, Bridge Engineering in Thailand", *Handbook of International Bridge Engineering*, Ed, Chen, W.F. and Duan, L. CRC Press, Boca Raton, FL.

Longest Bridges. (2013). http://en.wikipedia.org/wiki/List_of_longest_bridges_in_the_world.

Los Tilos Arch. (2013). http://en.structurae.net/structures/data/index.cfm?id=s0014425.

LA 1 Project. (2013). http://en.wikipedia.org/wiki/Louisiana_Highway_1.

Lower Buffalo Bridge. (2013). http://www.transportation.wv.gov/communications/bridge_facts/Modern-Bridges/Pages/LowerBuffalo.aspx.

Luozhou Bridge. (2013). http://www.tfol.com/10026/12705/12706/2009/10/7/10815163.shtml (in Chinese).

Lwin, M. and Kulicki, J. M. (2013). "Chapter 2, Bridge Engineering in he United States", *Handbook of International Bridge Engineering*, Ed, Chen, W.F. and Duan, L. CRC Press, Boca Raton, FL.

Macioce, T. P., Thompson, B. G., and Zielinski, B. J. (2001). *Pennsylvania's Experience and Future Plans for High-Performance Steel Bridges*, Pennsylvania Department of Transportation, Harrisburg, PA.

Mackinac Bridge. (2013). http://www.mackinacbridge.org/facts-figures-16.

Macombs Dam Bridge. (2013). http://en.wikipedia.org/wiki/Macombs_Dam_Bridge.

Main Street Bridge. (2013). http://en.wikipedia.org/wiki/Main_Street_Bridge_(Jacksonville).

Manchac Swamp Bridge. (2013). http://en.wikipedia.org/wiki/Manchac_Swamp_bridge.

Maocao Street Bridge. (2013). http://baike.baidu.com/view/1893999.htm (in Chinese).

Marine Parkway-Gil Hodges Memorial Bridge. (2013). http://en.wikipedia.org/wiki/Marine_Parkway-Gil_Hodges_Memorial_Bridge.

Maruyama, T. and Kawamura, Y. (1999). Construction of a Floating Swing Bridge-Yumemai Bridge, http://bridgeworld.net/wordpress/archives/docs/yumemai%20bridge.pdf.

Matosinhos–Leça-Porto Movable Bridge. (2013). http://www.waymarking.com/waymarks/WMAKYQ_Ponte_Mvel_de_Matosinhos_Lea_Porto.

Meixi River Bridge. (2013). http://www.cnbridge.cn/2010/0708/5770.html (in Chinese).

Mercosur Bridge. (2013). http://trid.trb.org/view.aspx?id=982123.

Millanes, F. (2008). "Outstanding Composite Steel-Concrete Bridges in the Spanish HSRL," *7th International Conference on Steel Bridges*, June 4–6, Guimaiaes, Portugal.

Minami Bisan–Seto Bridge. (2013). http://en.wikipedia.org/wiki/Minami_Bisan-Seto_Bridge.

Minato Bridge. (2013). http://en.wikipedia.org/wiki/Minato_Bridge.

Mingzhou Bridge. (2013). http://www.chinahighway.com/zt/zt_info.php?id=265 (in Chinese).

Minpu Bridge. (2013). http://www.china.org.cn/china/photos/2009-07/24/content_18196705.htm.

Morris Bridge. (2013). http://www.johnweeks.com/river_illinois/pages/illC11.html.

Morrison Street Bridge. (2013). http://bridgehunter.com/or/multnomah/morrison-street.

MSRDC. (2013). http://www.msrdc.org/Projects/Mumbai_trans_harbour.aspx.

Mykhailo K. (2013). "Chapter19 Bridge Engineering in Ukraine", *Handbook of International Bridge Engineering*, Ed, Chen, W.F. and Duan, L. CRC Press, Boca Raton, FL.

Nagai, M. et al. (2013). "Chapter 22 Bridge Engineering in Japan", *Handbook of International Bridge Engineering*, Ed. Chen, W.F. and Duan, L., CRC Press, Roca Baton, FL.

Nagai, M. et al. (2013). "Chapter 23, Bridge Engineering in Japan", *Handbook of International Bridge Engineering*, Ed, Chen, W.F. and Duan, L. CRC Press, Boca Raton, FL.

Namihaya Bridge. (2013). http://en.structurae.de/structures/data/index.cfm?id=s0005087.

Nanning Sanan Yi River Bridge. (2013). http://baike.baidu.com/view/2244444.htm (in Chinese).

Nanning Yonghe Bridge. (2013). http://baike.baidu.com/view/2247020.htm (in Chinese).

Neckar Valley Viaduct. (2013). http://nisee.berkeley.edu/elibrary/getpkg?id=GoddenG43-50.

New River Gorge Bridge. (2013). http://en.wikipedia.org/wiki/New_River_Gorge_Bridge.

New Ulyanovsk Bridge. (2013). http://en.wikipedia.org/wiki/New_Ulyanovsk_Bridge.

Newburgh–New Beacon Bridge. (2013). http://en.wikipedia.org/wiki/Newburgh%E2%80%93Beacon_Bridge.

New Taipei Bridge. (2013). http://zh.wikipedia.org/wiki/%E6%96%B0%E5%8C%97%E5%A4%A7%E6%A9%8B (in Chinese)

Ningjiang Songhua River Bridge. (2013). http://www.jljt.gov.cn/bdzx/yb_1/wz/201108/t20110819_35602.html (In Chinese)

North Arm Bridge. (2013). http://en.wikipedia.org/wiki/North_Arm_Bridge

Nordhordland Bridge. (2013). http://en.wikipedia.org/wiki/Nordhordland_Bridge.

Nordic. (2013). http://www.nordicroads.com/website/index.asp?pageID=64.

Norfolk Southern Lake Pontchartrain Bridge. (2013). http://en.wikipedia.org/wiki/Norfolk_Southern_Lake_Pontchartrain_Bridge.

Novyi Saratovskiy Bridge. (2013). http://www.wolframalpha.com/entities/bridges/9j/we/q1.

NSBA. (2007). *HPS Scoreboard*, National Steel Bridge Alliance, Chicago, IL.

Oregon. (2013). http://www.oregon.gov/ODOT/HIGHWAY/GEOENVIRONMENTAL/historic_bridges_covered1.shtml.

Ogawa A., Matsuda T. and Kasuga, A. (1998). "The Tsukuhara Extradosed Bridge near Kobe", *Structural Engineering International*, August, 3(8): 172–173.

Ouse Swing Bridge. (2013). http://skanska-sustainability-case-studies.com/pdfs/50/50_Selby_v001.pdf.

Partov, D. N. and Dinev, D. (2013), "Chapter 6, Bridge Engineering in Bulgaria", *Handbook of International Bridge Engineering*, Ed, Chen, W.F. and Duan, L. CRC Press, Boca Raton, FL.

Passaic River Bridge. (2013). http://bridgehunter.com/nj/essex/705151.

Pasarella de Sant Pere. (2011). http://en.structurae.de/structures/data/index.cfm?id=s0012011

Peace Bridge. (2013). http://en.structurae.eu/structures/data/index.cfm?id=s0001703

Peace Bridge. (2013a). http://en.structurae.eu/structures/data/index.cfm?id=s0001703.

Peace Bridge. (2013b). http://tigger1.host.ulster.ac.uk/ceni/e107_images/Case%20Study%20-Final.pdf.

Penang Bridge. (2013). http://en.wikipedia.org/wiki/Penang_Bridge.

Penang Second Bridge. (2013). http://en.wikipedia.org/wiki/Penang_Second_Bridge.

Pennington Avenue Bridge. (2013). http://bridgehunter.com/md/baltimore-city/200000BC5217010.

Pilot, G. (2007). *Bridges in Europe*, presented at ECCE Athens meeting, October, Athens, Greece.

Pont de la Libération Bridge. (2013). http://en.structurae.de/structures/data/index.cfm?id=s0001221.

Pont Gustave-Flaubert. (2013). http://en.wikipedia.org/wiki/Pont_Gustave-Flaubert.

Port Mann Bridge. (2013). http://en.wikipedia.org/wiki/Port_Mann_Bridge.

Puente de la Barra de Maldonado. (2011). http://en.structurae.de/structures/data/index.cfm?id=s0002210

Qatar Bahrain Causeway. (2013). http://en.wikipedia.org/wiki/Qatar_Bahrain_Causeway.

Qiancao Bridge. (2013). http://baike.baidu.com/view/1933073.htm (in Chinese), also http://highest-bridges.com/wiki/index.php?title=2012_High_Bridge_Trip_Photo_Album

Qilin Bridge. (2013). http://news.qq.com/a/20090911/003133.htm (in Chinese).

Qin, Q., Mei, G. and Xu G. (2013). "Chapter 20, Bridge Engineering in China", *Handbook of International Bridge Engineering*, Ed, Chen, W.F. and Duan, L. CRC Press, Boca Raton, FL.

Qingdao Bay Bridge. (2013). http://baike.baidu.com/view/754373.htm (in Chinese).

Qingshui River Bridge. (2013). http://baike.baidu.com/view/1027276.htm (in Chinese).

Qingzhou Bridge. (2013). http://baike.baidu.com/view/4256597.htm (in Chinese).

Quanzhou Bay Bridge. (2013). http://baike.baidu.com/view/3132887.htm (in Chinese).

Québec Bridge. (2013). http://en.wikipedia.org/wiki/Quebec_Bridge.

Queen Emma Bridge. (2013). http://en.wikipedia.org/wiki/Queen_Emma_Bridge.

Queensboro Bridge. (2013). http://www.nycroads.com/crossings/queensboro.

Rader Island Bridge. (2013). http://de.wikipedia.org/wiki/Europabr%C3%BCcke_(Rendsburg) (in German).

Radić, J. and Puž, G. (2013). "Chapter 7 Bridge Engineering in Croatia", *Handbook of International Bridge Engineering*, Ed, Chen, W.F. and Duan, L. CRC Press, Boca Raton, FL.

Raftsundet Bridge. (2013). http://en.structurae.de/structures/data/index.cfm?id=s0001759.

Remseck Neckar Bridge (2013). http://www.karl-gotsch.de/Album/Neckar4.htm (in Germany)

Rheinknie Bridge. (2013). http://de.wikipedia.org/wiki/Rheinkniebr%C3%BCcke (in German)

Richmond–San Rafael Bridge. (2013). http://bata.mtc.ca.gov/bridges/richmond-sr.htm.

Rio Colorado Bridge. (2011). http://highestbridges.com/wiki/index.php?title=Rio_Colorado_Bridge

Rodriguez, S. (2013). "A Record Span Kanawha River Bridge," *ASPIRE*, Winter, http://www.aspirebridge.org/pdfs/magazine/issue_09/kanawha_win09.pdf.

Rongzhou Jinsha River Bridge. (2013). http://www.hb.xinhuanet.com/zhuanti/2007-10/15/content_11401476_8.htm (in Chinese).

Roosevelt Island Bridge. (2013). http://bridgehunter.com/ny/new-york/2240640.

Runyang Bridge. (2013). http://en.wikipedia.org/wiki/Runyang_Bridge.

Sachaji Bridge. (2013). http://wenku.baidu.com/view/4970aa573c1ec5da50e27079.html (in Chinese).

Samuel Beckett Bridge. (2013). http://en.wikipedia.org/wiki/Samuel_Beckett_Bridge.

San Mateo–Hayward Bridge. (2013). http://www.dot.ca.gov/hq/esc/tollbridge/SM-Hay/SMfacts.html.

Sandö Bridge. (2013). http://en.wikipedia.org/wiki/Sand%C3%B6_Bridge.

Sannai-Maruyama Bridge. (2013). http://www.jsce.or.jp/committee/concrete/e/newsletter/newsletter19/award.html

Sannohe-Boukyo Bridge. (2013). http://en.structurae.de/structures/data/index.cfm?id=s0034430

Sanmenkou Bridge. (2013). http://www.nbxs.gov.cn/gb/nbxs/node1212/node1226/node1607.

Saul, R. and Humpf, K. (2007). "The Contribution of Latin America to the Development of Long Span Bridges," IABSE Symposium, September 19–21, Weimar, Germany. http://www.lap-consult.com/pdf-files/deutsch/sonderdrucke/sdr537.pdf.

Sava Bridge. (2013). www.savabridge.com

Sava (Ada) Bridge. (2013). www.savabridge.com

Schierstein Bridge. (2013). http://en.wikipedia.org/wiki/Schierstein_Bridge.

Schottwien Bridge. (2013). http://en.structurae.de/structures/data/index.cfm?id=s0007931.

Second Luzhou Yangtze River Bridge. (2013). http://www.cnbridge.cn/2010/0719/5989.html (in Chinese).

Second Maya Bridge. (2013). http://en.structurae.net/structures/data/index.cfm?id=s0005409.

Second Mactan-Mandaue Bridge. (2013). http://en.structurae.de/structures/data/index.cfm?id=s0005543

Seto–Ohashi Bridge. (2013). http://en.wikipedia.org/wiki/Great_Seto_Bridge.

Seven Mile Bridge. (2013). http://en.wikipedia.org/wiki/Seven_Mile_Bridge.

Severin Bridge. (2013). http://de.wikipedia.org/wiki/Severinsbr%C3%BCcke (in German)

SFOBB. (2013a). http://en.wikipedia.org/wiki/San_Francisco_%E2%80%93_Oakland_Bay_Bridge.

SFOBB. (2013b). http://www.dot.ca.gov/hq/esc/tollbridge/SFOBB/Sfobbfacts.html.

Shanghai Maglev Line. (2013). http://en.wikipedia.org/wiki/Shanghai_Maglev_Train.

Shanghai Yangtze River Bridge. (2013). http://en.wikipedia.org/wiki/Shanghai_Yangtze_River_Tunnel_and_Bridge.

Shah Amanat Bridge. (2013). http://en.wikipedia.org/wiki/Shah_Amanat_Bridge

Sheikh Rashid bin Saeed Crossing. (2013). http://en.wikipedia.org/wiki/Sheikh_Rashid_bin_Saeed_Crossing.

Shibanpo Bridge. (2013). http://www.dormanlongtechnology.com/en/projects/Shibanpo.htm.

Shin Mominoki Suspension Bridge. (2011). http://en.structurae.de/structures/data/index.cfm?id=s0006618

Shorenji-Gawa Bridge. (2013). http://www.jasbc.or.jp/kyoryodb/detail.cgi?id=1689 (in Japanese).

Šibenik Bridge. (2013). http://en.structurae.de/structures/data/index.cfm?id=s0002270.

Sioux Narrows Bridge (2013) http://en.wikipedia.org/wiki/Sioux_Narrows_Bridge

Sir Leo Hielscher Bridges. (2013). http://en.wikipedia.org/wiki/Sir_Leo_Hielscher_Bridges.

Skye Bridge. (2013). http://www.scottish-places.info/features/featurefirst7918.html.

Smolen–Gulf Bridge. (2013). http://en.wikipedia.org/wiki/Smolen-Gulf_Bridge.

South Span – Huangpu Bridge. (2013). http://en.wikipedia.org/wiki/Huangpu_Bridge

Spokane Street Bridge. (2013). http://bridgehunter.com/wa/king/85944000000000.

Stolma Bridge. (2013). http://en.wikipedia.org/wiki/Stolma_Bridge.

Strasky J. (2005). *Stress Ribbon and Cable-supported Pedestrian Bridges*, Thomas Telford, London, UK

Stonecutters Bridge. (2013). http://en.wikipedia.org/wiki/Stonecutters_Bridge.

Stuttgart–Bad Cannstatt Bridge. (2013.) http://www.waymarking.com/waymarks/WM9AGY_Bad_Cannstatt_Holzsteg

Sun, Y. F. and Li, X. J. (2003). "Overall Design of Viaduct for Jinbin Light Rail," *Railway Standard Design* 2003(8): 19–21. http://wenku.baidu.com/view/07a524ec102de2bd96058895.html (in Chinese).

Sundoy Bridge. (2013). http://en.wikipedia.org/wiki/Sund%C3%B8y_Bridge.

Sutong Bridge. (2013a). http://baike.baidu.com/view/532716.htm (in Chinese).

Sutong Bridge. (2013b). http://en.wikipedia.org/wiki/Sutong_Bridge.

Suvarnabhumi Airport Link. (2013). http://en.wikipedia.org/wiki/Suvarnabhumi_Airport_Link.

Svinesund Bridge. (2013). http://en.wikipedia.org/wiki/Svinesund_Bridge.

Sydney Harbor Bridge. (2013). http://en.wikipedia.org/wiki/Sydney_Harbour_Bridge.

Taiwan HSR. (2013). http://tw.myblog.yahoo.com/jw!8LEg9RSUBRV43yV_RQsDsA-/article?mid=-2& prev=500&l=f&fid=13 (in Chinese).

Tang M.C. (2011). "Poised For Growth", Civil Engineering, ASCE, September, 81(9): 64–67,73.

Tappan Zee Bridge. (2013). http://en.wikipedia.org/wiki/Tappan_Zee_Bridge.

Tensho (Takamatu) Bridge. (2013). http://en.structurae.de/structures/data/index.cfm?id=s0004197.

Third Avenue Bridge. (2013). http://www.nycroads.com/crossings/third-avenue.

Third Bridge. (2013). http://en.wikipedia.org/wiki/Third_Bridge.

Third Mainland Bridge. (2013). http://en.wikipedia.org/wiki/Third_Mainland_Bridge.

Three Sisters. (2013). http://en.wikipedia.org/wiki/Three_Sisters_(Pittsburgh).

Tianjin Haihe Bridge. (2013). http://baike.baidu.com/view/1599645.htm (in Chinese)

Tian, Y. R. (2013). "General Design of Fuzhou Gushan Bridge, City Road," *Bridge and Flooding Control* 2, February (in Chinese). http://wenku.baidu.com/view/1d41d518227916888486d79f.html.

Tianjin Bridge. (2013). http://home.51.com/seogtw984/diary/item/10054098.html (in Chinese).

Tokyo Gate Bridge. (2008). http://en.wikipedia.org/wiki/Tokyo_Gate_Bridge.

Tokunoyama Hattoku Bridge. (2013). http://www.highestbridges.com/wiki/index.php?title= Tokunoyamahattoku_Bridge

Top 10 Longest Bridges in USA. (2013). http://listosaur.com/miscellaneous/top-10-longest-bridges-in-the-united-states.html.

Tsing Ma Bridge. (2013). http://en.wikipedia.org/wiki/Tsing_Ma_Bridge.

Umemachi Bridge. (2013). http://www.daido-it.ac.jp/~doboku/miki/japbrg3.html.

Vansu Bridge. (2013). http://en.wikipedia.org/wiki/Van%C5%A1u_Bridge

Vaires-sur-Marne Footbridge. (2013). http://en.structurae.de/structures/data/index.cfm?id=s0016522

Varodd Bridge. (2013). http://en.structurae.de/structures/data/index.cfm?ID=s0008130.

Vascoda Gama Bridge. (2013). http://en.wikipedia.org/wiki/Vasco_da_Gama_Bridge.

Vasconcelos, A., et al. (2013). "Chapter 4, Bridge Engineering in Brazil", *Handbook of International Bridge Engineering*, Ed, Chen, W.F. and Duan, L. CRC Press, Boca Raton, FL.

Vejrum, T. et al. (2013). "Chapter 10 – Cable-stayed Bridges", B*ridge Engineering Handbook, Volume 2-Superstructure Design, 2nd Edition*, Ed, Chen, W.F. and Duan, L. CRC Press, Boca Raton, FL.

Verrazano Narrows Bridge. (2013). http://en.wikipedia.org/wiki/Verrazano-Narrows_Bridge.

Verrières Viaduct. (2013). http://en.structurae.de/structures/data/index.cfm?id=s0000687.

Veterans Memorial Bridge. (2013). http://en.wikipedia.org/wiki/Gramercy_Bridge.

Viaduct Froschgrundsee. (2013). http://de.wikipedia.org/wiki/Talbr%C3%BCcke_Froschgrundsee (in German).

Virginia Dare Memorial Bridge. (2013). http://en.wikipedia.org/wiki/Virginia_Dare_Memorial_Bridge.

Wabash Railroad Bridge. (2013). http://bridgehunter.com/mo/marion/hannibal-rr.

Wang, R. and Meng, F. (2009). "Hangzhou Bay Bridge– a 36 km shortcut between Shanghai and Ningbo, Recent Major Bridges", *IABSE Workshop - Recent Major Bridge*, May 11–20, 2009, Shanghai, China.

Wanzhou Yangtze River Highway Bridge. (2013). http://en.wikipedia.org/wiki/Wanxian_Bridge.

Wanzhou Yangtze River Railway Bridge. (2013). http://news.sina.com.cn/c/2005-06-13/08456155010s.shtml (in Chinese).

White Bird Canyon Bridge. (2013). http://users.bentonrea.com/~tinear/wb-bridge.htm.

Wild Gera Viaduct. (2013). http://en.structurae.de/structures/data/index.cfm?id=s0000331.

William R. Bennett Bridge. (2013). http://www.b-t.com/menu/project/designbuild/Pages/William-R-Bennett-Bridge.aspx.

Willis Avenue Bridge. (2013). http://www.nycroads.com/crossings/willis-avenue.

Winningen Bridge. (2013). http://en.structurae.de/structures/data/index.cfm?id=s0000686.

Woodrow Wilson Memorial Bridge. (2013). http://bridgehunter.com/md/prince-georges/bh48616.

WSP. (2013). http://www.wspgroup.com/upload/documents/PDF/capabilitystatements/bridgedesign.pdf.

Wuchao River Bridge. (2013). http://en.structurae.de/structures/data/index.cfm?id=s0001904.

Wuchaoher Bridge. (2013). http://en.structurae.de/structures/data/index.cfm?id=s0001904

Wuhan Qingchuan Bridge. (2013). http://baike.baidu.com/view/2202121.htm (in Chinese).

Wuhan Yangluo Yangtze River Bridge. (2013). http://baike.baidu.com/view/1407881.htm (in Chinese).

Wuhu Yangtze River Bridge. (2013). http://en.wikipedia.org/wiki/Wuhu_Yangtze_River_Bridge.

Wuppertal Schwebebahn. (2013). http://www.flickr.com/photos/dandelion-and-burdock/4890986167.

Wushan Yangtze River Bridge. (2013). http://baike.baidu.com/view/212678.htm (in Chinese).

Xiabaishi Bridge. (2013). http://www.cnbridge.cn/2010/1013/8058.html (in Chinese).

Xiasha Bridge. (2013). http://en.structurae.de/structures/data/index.cfm?ID=s0006002.

Xingguang Bridge. (2013). http://www.cces.net.cn/guild/sites/tmxh/detail.asp?i=bzjlxm&id=19853 (in Chinese).

Xupu Bridge. (2013). http://baike.baidu.com/view/73836.htm (in Chinese).

Yajisha Bridge. (2013). http://baike.baidu.com/view/1488130.htm (in Chinese).

Yangluo Bridge. (2013). http://en.wikipedia.org/wiki/Yangluo_Bridge.

Yangpu Bridge. (2013). http://en.wikipedia.org/wiki/Yangpu_Bridge.

Yangtze River Bridges. (2013). http://en.wikipedia.org/wiki/Yangtze_River_bridges_and_tunnels.

Yanshi Bridge. (2013). http://www.lyd.com.cn/xianqu/content.asp?id=7761 (in Chinese).

Yellow Sea Bridge. (2013). http://www.ntykg.com/site/jsjz/dltd/4906669251.shtml (in Chinese).

Yeongjong Bridge. (2013). http://en.wikipedia.org/wiki/Yeongjong_Bridge.

Yi Sun-sin Bridge. (2013). http://world.kbs.co.kr/english/news/news_zoom_detail.htm?No=3939.

Yichang Yangtze River Railway Bridge. (2013). http://baike.baidu.com/view/1060385.htm.

Yilmaz, C. et al. (2013). "Chapter 18, Bridge Engineering in Turkey", *Handbook of International Bridge Engineering*, Ed, Chen, W.F. and Duan, L. CRC Press, Boca Raton, FL.

Youngding New River Bridge. (2013). http://www.peoplerail.com/jianshe/2010418/n41358211.html (in Chinese).

Yudong Yangtze River Bridge. (2013). http://baike.baidu.com/view/211659.htm%202010-5-28 (in Chinese).

Zhaohua Jialing River Bridge. (2013). http://www.sc.gov.cn/zwgk/zwdt/szdt/201009/t20100917_1028923.shtml (in Chinese).

Zhengzhou–Xi'an HSR. (2013). http://www.cqvip.com/Main/Detail.aspx?id=32021061 (in Chinese).

Zhijing River Bridge. (2013). http://www.cnbridge.cn/2010/0713/5864.html (in Chinese).

Zhou, W.Q. (1997). "Design of 108 meter Stone Arch Bridge", *Shanxi Transportation Technology*, No. 7. (In Chinese)

Zhuang River Bridge. (2013). http://www.ln.gov.cn/zfxx/qsgd/dls/zhs/200910/t20091009_429323.html (in Chinese).

Ziya River Bridge. (2013). http://www.chinahighway.com/news/2003/49243.php (in Chinese).

Zolotoy Rog Bridge. (2013). http://ru.wikipedia.org/wiki (in Russian).

Zoo Bridge. (2013). http://de.wikipedia.org/wiki/Zoobr%C3%BCcke (in German).

Index

A

M